Elementary
Linear
Algebra

10TH EDITION

Elementary Linear Algebra

HOWARD ANTON

Professor Emeritus, Drexel University

WILEY

VICE PRESIDENT AND PUBLISHER	Laurie Rosatone
SENIOR ACQUISITIONS EDITOR	David Dietz
ASSISTANT EDITOR	Jeff Benson
PRODUCTION SERVICES MANAGER	Dorothy Sinclair
PRODUCTION EDITOR	Janet Foxman
SENIOR MARKETING MANAGER	Sarah Davis
CREATIVE DIRECTOR	Harry Nolan
TEXT AND COVER DESIGNER	Madelyn Lesure
TECHNICAL ILLUSTRATION	Brian Haughwout/Techsetters, Inc.
PHOTO EDITOR	Sheena Goldstein
MEDIA EDITOR	Melissa Edwards
PRODUCTION SERVICES	Carol Sawyer/The Perfect Proof
COVER ART	Norm Christiansen

This book was set in Times New Roman PS by Techsetters, Inc., and printed and bound by World Color USA/Versailles. The cover was printed by World Color USA/Versailles.

This book is printed on acid-free paper. ∞

Library of Congress Cataloging-in-Publication Data

Anton, Howard.
  Elementary linear algebra / Howard Anton. -- 10th ed.
    p. cm.
  ISBN 978-0-470-45821-1 (cloth)
1. Algebras, Linear--Textbooks. I. Title.
  QA184.2.A58 2010
  512'.5--dc22

                            2010000340

ISBN 978-0-470-45821-1 (Main Book)
ISBN 978-0-470-55991-8 (Binder-Ready Version)

Printed in the United States of America

10 9 8 7 6 5 4 3 2 1

To

My wife, Pat

My children, Brian, David, and Lauren

My parents, Shirley and Benjamin

My benefactor, Stephen Girard (1750–1831),
   whose philanthropy changed my life

Howard Anton

Howard Anton obtained his B.A. from Lehigh University, his M.A. from the University of Illinois, and his Ph.D. from the Polytechnic University of Brooklyn, all in mathematics. In the early 1960s he worked for Burroughs Corporation and Avco Corporation at Cape Canaveral, Florida, where he was involved with the manned space program. In 1968 he joined the Mathematics Department at Drexel University, where he taught full time until 1983. Since then he has devoted the majority of his time to textbook writing and activities for mathematical associations. Dr. Anton was president of the EPADEL Section of the Mathematical Association of America (MAA), served on the Board of Governors of that organization, and guided the creation of the Student Chapters of the MAA. In addition to various pedagogical articles, he has published numerous research papers in functional analysis, approximation theory, and topology. He is best known for his textbooks in mathematics, which are among the most widely used in the world. There are currently more than 150 versions of his books, including translations into Spanish, Arabic, Portuguese, Italian, Indonesian, French, Japanese, Chinese, Hebrew, and German. For relaxation, Dr. Anton enjoys travel and photography.

This edition of *Elementary Linear Algebra* gives an introductory treatment of linear algebra that is suitable for a first undergraduate course. Its aim is to present the fundamentals of linear algebra in the clearest possible way—sound pedagogy is the main consideration. Although calculus is not a prerequisite, there is some optional material that is clearly marked for students with a calculus background. If desired, that material can be omitted without loss of continuity.

Technology is not required to use this text, but for instructors who would like to use MATLAB, *Mathematica*, Maple, or calculators with linear algebra capabilities, we have posted some supporting material that can be accessed at either of the following Web sites:

www.howardanton.com
www.wiley.com/college/anton

Summary of Changes in this Edition

This edition is a major revision of its predecessor. In addition to including some new material, some of the old material has been streamlined to ensure that the major topics can all be covered in a standard course. These are the most significant changes:

- **Vectors in 2-space, 3-space, and *n*-space**   Chapters 3 and 4 of the previous edition have been combined into a single chapter. This has enabled us to eliminate some duplicate exposition and to juxtapose concepts in *n*-space with those in 2-space and 3-space, thereby conveying more clearly how *n*-space ideas generalize those already familiar to the student.

- **New Pedagogical Elements**   Each section now ends with a *Concept Review* and a *Skills* mastery that provide the student a convenient reference to the main ideas in that section.

- **New Exercises**   Many new exercises have been added, including a set of True/False exercises at the end of most sections.

- **Earlier Coverage of Eigenvalues and Eigenvectors**   The chapter on eigenvalues and eigenvectors, which was Chapter 7 in the previous edition, is Chapter 5 in this edition.

- **Complex Vector Spaces**   The chapter entitled *Complex Vector Spaces* in the previous edition has been completely revised. The most important ideas are now covered in Sections 5.3 and 7.5 in the context of matrix diagonalization. A brief review of complex numbers is included in the Appendix.

- **Quadratic Forms**   This material has been extensively rewritten to focus more precisely on the most important ideas.

- **New Chapter on Numerical Methods**   In the previous edition an assortment of topics appeared in the last chapter. That chapter has been replaced by a new chapter that focuses *exclusively* on numerical methods of linear algebra. We achieved this by moving those topics not concerned with numerical methods elsewhere in the text.

- **Singular-Value Decomposition**   In recognition of its growing importance, a new section on *Singular-Value Decomposition* has been added to the chapter on numerical methods.

- **Internet Search and the Power Method**   A new section on the *Power Method* and its application to Internet search engines has been added to the chapter on numerical methods.

- **Applications**   There is an expanded version of this text by Howard Anton and Chris Rorres entitled *Elementary Linear Algebra: Applications Version*, 10th edition

(ISBN 9780470432051), whose purpose is to supplement this version with an extensive body of applications. However, to accommodate instructors who asked us to include some applications in this version of the text, we have done so. These are generally less detailed than those appearing in the Anton/Rorres text and can be omitted without loss of continuity.

Hallmark Features

- **Relationships Among Concepts**   One of our main pedagogical goals is to convey to the student that linear algebra is a cohesive subject and not simply a collection of isolated definitions and techniques. One way in which we do this is by using a crescendo of *Equivalent Statements* theorems that continually revisit relationships among systems of equations, matrices, determinants, vectors, linear transformations, and eigenvalues. To get a general sense of how we use this technique see Theorems 1.5.3, 1.6.4, 2.2.8, 4.8.10, 4.10.4 and then Theorem 5.1.6, for example.

- **Smooth Transition to Abstraction**   Because the transition from R^n to general vector spaces is difficult for many students, considerable effort is devoted to explaining the purpose of abstraction and helping the student to "visualize" abstract ideas by drawing analogies to familiar geometric ideas.

- **Mathematical Precision**   When reasonable, we try to be mathematically precise. In keeping with the level of student audience, proofs are presented in a patient style that is tailored for beginners. There is a brief section in the Appendix on how to read proof statements, and there are various exercises in which students are guided through the steps of a proof and asked for justification.

- **Suitability for a Diverse Audience**   This text is designed to serve the needs of students in engineering, computer science, biology, physics, business, and economics as well as those majoring in mathematics.

- **Historical Notes**   To give the students a sense of mathematical history and to convey that real people created the mathematical theorems and equations they are studying, we have included numerous *Historical Notes* that put the topic being studied in historical perspective.

About the Exercises

- **Graded Exercise Sets**   Each exercise set begins with routine drill problems and progresses to problems with more substance.

- **True/False Exercises**   Most exercise sets end with a set of True/False exercises that are designed to check conceptual understanding and logical reasoning. To avoid pure guessing, the students are required to justify their responses in some way.

- **Supplementary Exercise Sets**   Most chapters end with a set of supplementary exercises that tend to be more challenging and force the student to draw on ideas from the *entire* chapter rather than a specific section.

Supplementary Materials for Students

- **Student Solutions Manual**   This supplement provides detailed solutions to most theoretical exercises and to at least one nonroutine exercise of every type (ISBN 9780470458228).

- **Technology Exercises and Data Files**   The technology exercises that appeared in the previous edition have been moved to the Web site that accompanies this text. Those exercises are designed to be solved using MATLAB, *Mathematica*, or Maple and are accompanied by data files in all three formats. The exercises and data can be downloaded from either of the following Web sites:

www.howardanton.com
www.wiley.com/college/anton

Supplementary Materials for Instructors

- **Instructor's Solutions Manual**   This supplement provides worked-out solutions to most exercises in the text (ISBN 9780470458235).
- **WileyPLUS™**   This is Wiley's proprietary online teaching and learning environment that integrates a digital version of this textbook with instructor and student resources to fit a variety of teaching and learning styles. WileyPLUS will help your students master concepts in a rich and structured environment that is available to them 24/7. It will also help you to personalize and manage your course more effectively with student assessments, assignments, grade tracking, and other useful tools.
  - Your students will receive timely access to resources that address their individual needs and will receive immediate feedback and remediation resources when needed.
  - There are also self-assessment tools that are linked to the relevant portions of the text that will enable your students to take control of their own learning and practice.
  - WileyPLUS will help you to identify those students who are falling behind and to intervene in a timely manner without waiting for scheduled office hours.

  More information about WileyPLUS can be obtained from your Wiley representative.

A Guide for the Instructor

Although linear algebra courses vary widely in content and philosophy, most courses fall into two categories—those with about 35–40 lectures and those with about 25–30 lectures. Accordingly, we have created long and short templates as possible starting points for constructing a course outline. Of course, these are just guides, and you will certainly want to customize them to fit your local interests and requirements. Neither of these sample templates includes applications. Those can be added, if desired, as time permits.

	Long Template	Short Template
Chapter 1: Systems of Linear Equations and Matrices	7 lectures	6 lectures
Chapter 2: Determinants	3 lectures	2 lectures
Chapter 3: Euclidean Vector Spaces	4 lectures	3 lectures
Chapter 4: General Vector Spaces	10 lectures	10 lectures
Chapter 5: Eigenvalues and Eigenvectors	3 lectures	3 lectures
Chapter 6: Inner Product Spaces	3 lectures	1 lecture
Chapter 7: Diagonalization and Quadratic Forms	4 lectures	3 lectures
Chapter 8: Linear Transformations	3 lectures	2 lectures
Total:	**37 lectures**	**30 lectures**

Acknowledgements

I would like to express my appreciation to the following people whose helpful guidance has greatly improved the text:

Reviewers and Contributors

Don Allen, *Texas A&M University*
John Alongi, *Northwestern University*
John Beachy, *Northern Illinois University*
Przemslaw Bogacki, *Old Dominion University*

Robert Buchanan, *Millersville University of Pennsylvania*
Ralph Byers, *University of Kansas*
Evangelos A. Coutsias, *University of New Mexico*
Joshua Du, *Kennesaw State University*
Fatemeh Emdad, *Michigan Technological University*
Vincent Ervin, *Clemson University*
Anda Gadidov, *Kennesaw State University*
Guillermo Goldsztein, *Georgia Institute of Technology*
Tracy Hamilton, *California State University, Sacramento*
Amanda Hattway, *Wentworth Institute of Technology*
Heather Hulett, *University of Wisconsin–La Crosse*
David Hyeon, *Northern Illinois University*
Matt Insall, *Missouri University of Science and Technology*
Mic Jackson, *Earlham College*
Anton Kaul, *California Polytechnic Institute, San Luis Obispo*
Harihar Khanal, *Embry-Riddle University*
Hendrik Kuiper, *Arizona State University*
Kouok Law, *Georgia Perimeter College*
James McKinney, *California State University, Pomona*
Eric Schmutz, *Drexel University*
Qin Sheng, *Baylor University*
Adam Sikora, *State University of New York at Buffalo*
Allan Silberger, *Cleveland State University*
Dana Williams, *Dartmouth College*

Mathematical Advisors

Special thanks are due to a number of talented teachers and mathematicians who provided pedagogical guidance, provided help with answers and exercises, or provided detailed checking or proofreading:

John Alongi, *Northwestern University*
Scott Annin, *California State University, Fullerton*
Anton Kaul, *California Polytechnic State University*
Sarah Streett
Cindy Trimble, *C Trimble and Associates*
Brad Davis, *C Trimble and Associates*

The Wiley Support Team

David Dietz, Senior Acquisitions Editor
Jeff Benson, Assistant Editor
Pamela Lashbrook, Senior Editorial Assistant
Janet Foxman, Production Editor
Maddy Lesure, Senior Designer
Laurie Rosatone, Vice President and Publisher
Sarah Davis, Senior Marketing Manager
Diana Smith, Marketing Assistant
Melissa Edwards, Media Editor
Lisa Sabatini, Media Project Manager
Sheena Goldstein, Photo Editor
Carol Sawyer, Production Manager
Lilian Brady, Copyeditor

Special Contributions   The talents and dedication of many individuals are required to produce a book such as this, and I am fortunate to have benefited from the expertise of the following people:

David Dietz – my editor, for his attention to detail, his sound judgment, and his unwavering faith in me.

Jeff Benson – my assistant editor, who did an unbelievable job in organizing and coordinating the many threads required to make this edition a reality.

Carol Sawyer – of *The Perfect Proof*, who coordinated the myriad of details in the production process. It will be a pleasure to finally delete from my computer the hundreds of emails we exchanged in the course of working together on this book.

Scott Annin – *California State University at Fullerton*, who critiqued the previous edition and provided valuable ideas on how to improve the text. I feel fortunate to have had the benefit of Prof. Annin's teaching expertise and insights.

Dan Kirschenbaum – of *The Art of Arlene and Dan Kirschenbaum*, whose artistic and technical expertise resolved some difficult and critical illustration issues.

Bill Tuohy – who read parts of the manuscript and whose critical eye for detail had an important influence on the evolution of the text.

Pat Anton – who proofread manuscript, when needed, and shouldered the burden of household chores to free up time for me to work on this edition.

Maddy Lesure – our text and cover designer whose unerring sense of elegant design is apparent in the pages of this book.

Rena Lam – of *Techsetters, Inc.*, who did an absolutely amazing job of wading through a nightmare of author edits, scribbles, and last-minute changes to produce a beautiful book.

John Rogosich – of *Techsetters, Inc.*, who skillfully programmed the design elements of the book and resolved numerous thorny typesetting issues.

Lilian Brady – my copyeditor of many years, whose eye for typography and whose knowledge of language never ceases to amaze me.

The Wiley Team – There are many other people at Wiley who worked behind the scenes and to whom I owe a debt of gratitude: Laurie Rosatone, Ann Berlin, Dorothy Sinclair, Janet Foxman, Sarah Davis, Harry Nolan, Sheena Goldstein, Melissa Edwards, and Norm Christiansen. Thanks to you all.

CONTENTS

CHAPTER 1   Systems of Linear Equations and Matrices   1

1.1 Introduction to Systems of Linear Equations   2
1.2 Gaussian Elimination   11
1.3 Matrices and Matrix Operations   25
1.4 Inverses; Algebraic Properties of Matrices   38
1.5 Elementary Matrices and a Method for Finding A^{-1} 51
1.6 More on Linear Systems and Invertible Matrices   60
1.7 Diagonal, Triangular, and Symmetric Matrices   66
1.8 **Application:** Applications of Linear Systems   73
1.9 **Application:** Leontief Input-Output Models   85

CHAPTER 2   Determinants   93

2.1 Determinants by Cofactor Expansion   93
2.2 Evaluating Determinants by Row Reduction   100
2.3 Properties of Determinants; Cramer's Rule   106

CHAPTER 3   Euclidean Vector Spaces   119

3.1 Vectors in 2-Space, 3-Space, and n-Space   119
3.2 Norm, Dot Product, and Distance in R^n 130
3.3 Orthogonality   143
3.4 The Geometry of Linear Systems   152
3.5 Cross Product   161

CHAPTER 4   General Vector Spaces   171

4.1 Real Vector Spaces   171
4.2 Subspaces   179
4.3 Linear Independence   190
4.4 Coordinates and Basis   200
4.5 Dimension   209
4.6 Change of Basis   217
4.7 Row Space, Column Space, and Null Space   225
4.8 Rank, Nullity, and the Fundamental Matrix Spaces   237
4.9 Matrix Transformations from R^n to R^m 247
4.10 Properties of Matrix Transformations   263
4.11 **Application:** Geometry of Matrix Operators on R^2 273
4.12 **Application:** Dynamical Systems and Markov Chains   282

CHAPTER 5   Eigenvalues and Eigenvectors   295

5.1 Eigenvalues and Eigenvectors   295
5.2 Diagonalization   305

5.3 Complex Vector Spaces   315
5.4 **Application:** Differential Equations   327

C H A P T E R 6   Inner Product Spaces   335

6.1 Inner Products   335
6.2 Angle and Orthogonality in Inner Product Spaces   345
6.3 Gram–Schmidt Process; QR-Decomposition   352
6.4 Best Approximation; Least Squares   366
6.5 **Application:** Least Squares Fitting to Data   376
6.6 **Application:** Function Approximation; Fourier Series   382

C H A P T E R 7   Diagonalization and Quadratic Forms   389

7.1 Orthogonal Matrices   389
7.2 Orthogonal Diagonalization   397
7.3 Quadratic Forms   405
7.4 Optimization Using Quadratic Forms   417
7.5 Hermitian, Unitary, and Normal Matrices   424

C H A P T E R 8   Linear Transformations   433

8.1 General Linear Transformations   433
8.2 Isomorphism   445
8.3 Compositions and Inverse Transformations   452
8.4 Matrices for General Linear Transformations   458
8.5 Similarity   468

C H A P T E R 9   Numerical Methods   477

9.1 LU-Decompositions   477
9.2 The Power Method   487
9.3 **Application:** Internet Search Engines   496
9.4 Comparison of Procedures for Solving Linear Systems   501
9.5 Singular Value Decomposition   506
9.6 **Application:** Data Compression Using Singular Value Decomposition   514

A P P E N D I X A   How to Read Theorems   519

A P P E N D I X B   Complex Numbers   521

Answers to Exercises   529
Index   559

CHAPTER 1

Systems of Linear Equations and Matrices

CHAPTER CONTENTS   1.1 Introduction to Systems of Linear Equations 2
                    1.2 Gaussian Elimination 11
                    1.3 Matrices and Matrix Operations 25
                    1.4 Inverses; Algebraic Properties of Matrices 38
                    1.5 Elementary Matrices and a Method for Finding A^{-1} 51
                    1.6 More on Linear Systems and Invertible Matrices 60
                    1.7 Diagonal, Triangular, and Symmetric Matrices 66
                    1.8 Applications of Linear Systems 73
                         • Network Analysis (Traffic Flow) 73
                         • Electrical Circuits 76
                         • Balancing Chemical Equations 78
                         • Polynomial Interpolation 80
                    1.9 Leontief Input-Output Models 85

INTRODUCTION   Information in science, business, and mathematics is often organized into rows and
                columns to form rectangular arrays called "matrices" (plural of "matrix"). Matrices
                often appear as tables of numerical data that arise from physical observations, but they
                occur in various mathematical contexts as well. For example, we will see in this
                chapter that all of the information required to solve a system of equations such as

$$5x + y = 3$$
$$2x - y = 4$$

is embodied in the matrix

$$\begin{bmatrix} 5 & 1 & 3 \\ 2 & -1 & 4 \end{bmatrix}$$

and that the solution of the system can be obtained by performing appropriate
operations on this matrix. This is particularly important in developing computer
programs for solving systems of equations because computers are well suited for
manipulating arrays of numerical information. However, matrices are not simply a
notational tool for solving systems of equations; they can be viewed as mathematical
objects in their own right, and there is a rich and important theory associated with them
that has a multitude of practical applications. It is the study of matrices and related
topics that forms the mathematical field that we call "linear algebra." In this chapter we
will begin our study of matrices.

1.1 Introduction to Systems of Linear Equations

Systems of linear equations and their solutions constitute one of the major topics that we will study in this course. In this first section we will introduce some basic terminology and discuss a method for solving such systems.

Linear Equations Recall that in two dimensions a line in a rectangular xy-coordinate system can be represented by an equation of the form

$$ax + by = c \quad (a, b \text{ not both } 0)$$

and in three dimensions a plane in a rectangular xyz-coordinate system can be represented by an equation of the form

$$ax + by + cz = d \quad (a, b, c \text{ not all } 0)$$

These are examples of "linear equations," the first being a linear equation in the variables x and y and the second a linear equation in the variables x, y, and z. More generally, we define a ***linear equation*** in the n variables $x_1, x_2, \ldots, x_n$ to be one that can be expressed in the form

$$a_1 x_1 + a_2 x_2 + \cdots + a_n x_n = b \tag{1}$$

where $a_1, a_2, \ldots, a_n$ and b are constants, and the a's are not all zero. In the special cases where $n = 2$ or $n = 3$, we will often use variables without subscripts and write linear equations as

$$a_1 x + a_2 y = b \quad (a_1, a_2 \text{ not both } 0) \tag{2}$$

$$a_1 x + a_2 y + a_3 z = b \quad (a_1, a_2, a_3 \text{ not all } 0) \tag{3}$$

In the special case where $b = 0$, Equation (1) has the form

$$a_1 x_1 + a_2 x_2 + \cdots + a_n x_n = 0 \tag{4}$$

which is called a ***homogeneous linear equation*** in the variables $x_1, x_2, \ldots, x_n$.

▶ **EXAMPLE 1 Linear Equations**

Observe that a linear equation does not involve any products or roots of variables. All variables occur only to the first power and do not appear, for example, as arguments of trigonometric, logarithmic, or exponential functions. The following are linear equations:

$$x + 3y = 7 \qquad\qquad x_1 - 2x_2 - 3x_3 + x_4 = 0$$
$$\tfrac{1}{2}x - y + 3z = -1 \qquad\qquad x_1 + x_2 + \cdots + x_n = 1$$

The following are not linear equations:

$$x + 3y^2 = 4 \qquad\qquad 3x + 2y - xy = 5$$
$$\sin x + y = 0 \qquad\qquad \sqrt{x_1} + 2x_2 + x_3 = 1 \;\blacktriangleleft$$

A finite set of linear equations is called a ***system of linear equations*** or, more briefly, a ***linear system***. The variables are called ***unknowns***. For example, system (5) that follows has unknowns x and y, and system (6) has unknowns x_1, x_2, and x_3.

$$\begin{array}{ll} 5x + y = 3 & 4x_1 - x_2 + 3x_3 = -1 \\ 2x - y = 4 & 3x_1 + x_2 + 9x_3 = -4 \end{array} \tag{5--6}$$

The double subscripting on the coefficients a_{ij} of the unknowns gives their location in the system—the first subscript indicates the equation in which the coefficient occurs, and the second indicates which unknown it multiplies. Thus, a_{12} is in the first equation and multiplies x_2.

A general linear system of m equations in the n unknowns $x_1, x_2, \ldots, x_n$ can be written as

$$
\begin{aligned}
a_{11}x_1 + a_{12}x_2 + \cdots + a_{1n}x_n &= b_1 \\
a_{21}x_1 + a_{22}x_2 + \cdots + a_{2n}x_n &= b_2 \\
\vdots \qquad \vdots \qquad\quad \vdots \qquad \vdots \\
a_{m1}x_1 + a_{m2}x_2 + \cdots + a_{mn}x_n &= b_m
\end{aligned}
\tag{7}
$$

A *solution* of a linear system in n unknowns $x_1, x_2, \ldots, x_n$ is a sequence of n numbers $s_1, s_2, \ldots, s_n$ for which the substitution

$$x_1 = s_1, \quad x_2 = s_2, \ldots, \quad x_n = s_n$$

makes each equation a true statement. For example, the system in (5) has the solution

$$x = 1, \quad y = -2$$

and the system in (6) has the solution

$$x_1 = 1, \quad x_2 = 2, \quad x_3 = -1$$

These solutions can be written more succinctly as

$$(1, -2) \quad \text{and} \quad (1, 2, -1)$$

in which the names of the variables are omitted. This notation allows us to interpret these solutions geometrically as points in two-dimensional and three-dimensional space. More generally, a solution

$$x_1 = s_1, \quad x_2 = s_2, \ldots, \quad x_n = s_n$$

of a linear system in n unknowns can be written as

$$(s_1, s_2, \ldots, s_n)$$

which is called an *ordered n-tuple*. With this notation it is understood that all variables appear in the same order in each equation. If $n = 2$, then the n-tuple is called an *ordered pair*, and if $n = 3$, then it is called an *ordered triple*.

Linear Systems with Two and Three Unknowns

Linear systems in two unknowns arise in connection with intersections of lines. For example, consider the linear system

$$
\begin{aligned}
a_1x + b_1y &= c_1 \\
a_2x + b_2y &= c_2
\end{aligned}
$$

in which the graphs of the equations are lines in the xy-plane. Each solution (x, y) of this system corresponds to a point of intersection of the lines, so there are three possibilities (Figure 1.1.1):

1. The lines may be parallel and distinct, in which case there is no intersection and consequently no solution.

2. The lines may intersect at only one point, in which case the system has exactly one solution.

3. The lines may coincide, in which case there are infinitely many points of intersection (the points on the common line) and consequently infinitely many solutions.

In general, we say that a linear system is *consistent* if it has at least one solution and *inconsistent* if it has no solutions. Thus, a *consistent* linear system of two equations in

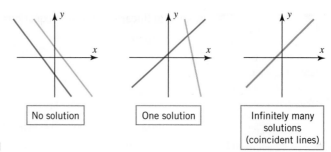

▶ Figure 1.1.1

two unknowns has either one solution or infinitely many solutions—there are no other possibilities. The same is true for a linear system of three equations in three unknowns

$$a_1x + b_1y + c_1z = d_1$$
$$a_2x + b_2y + c_2z = d_2$$
$$a_3x + b_3y + c_3z = d_3$$

in which the graphs of the equations are planes. The solutions of the system, if any, correspond to points where all three planes intersect, so again we see that there are only three possibilities—no solutions, one solution, or infinitely many solutions (Figure 1.1.2).

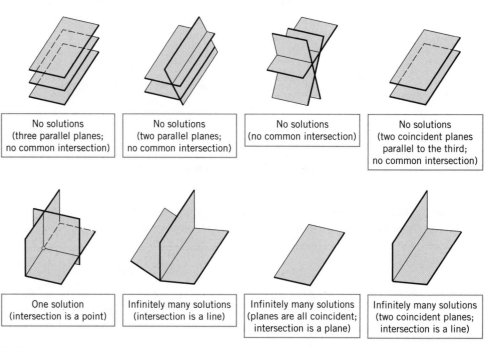

▲ Figure 1.1.2

We will prove later that our observations about the number of solutions of linear systems of two equations in two unknowns and linear systems of three equations in three unknowns actually hold for *all* linear systems. That is:

Every system of linear equations has zero, one, or infinitely many solutions. There are no other possibilities.

▶ **EXAMPLE 2 A Linear System with One Solution**

Solve the linear system

$$x - y = 1$$
$$2x + y = 6$$

Solution We can eliminate x from the second equation by adding -2 times the first equation to the second. This yields the simplified system

$$x - y = 1$$
$$3y = 4$$

From the second equation we obtain $y = \frac{4}{3}$, and on substituting this value in the first equation we obtain $x = 1 + y = \frac{7}{3}$. Thus, the system has the unique solution

$$x = \tfrac{7}{3}, \quad y = \tfrac{4}{3}$$

Geometrically, this means that the lines represented by the equations in the system intersect at the single point $\left(\frac{7}{3}, \frac{4}{3}\right)$. We leave it for you to check this by graphing the lines.

▶ **EXAMPLE 3 A Linear System with No Solutions**

Solve the linear system

$$x + y = 4$$
$$3x + 3y = 6$$

Solution We can eliminate x from the second equation by adding -3 times the first equation to the second equation. This yields the simplified system

$$x + y = 4$$
$$0 = -6$$

The second equation is contradictory, so the given system has no solution. Geometrically, this means that the lines corresponding to the equations in the original system are parallel and distinct. We leave it for you to check this by graphing the lines or by showing that they have the same slope but different y-intercepts.

▶ **EXAMPLE 4 A Linear System with Infinitely Many Solutions**

Solve the linear system

$$4x - 2y = 1$$
$$16x - 8y = 4$$

Solution We can eliminate x from the second equation by adding -4 times the first equation to the second. This yields the simplified system

$$4x - 2y = 1$$
$$0 = 0$$

The second equation does not impose any restrictions on x and y and hence can be omitted. Thus, the solutions of the system are those values of x and y that satisfy the single equation

$$4x - 2y = 1 \tag{8}$$

Geometrically, this means the lines corresponding to the two equations in the original system coincide. One way to describe the solution set is to solve this equation for x in terms of y to obtain $x = \frac{1}{4} + \frac{1}{2}y$ and then assign an arbitrary value t (called a ***parameter***)

In Example 4 we could have also obtained parametric equations for the solutions by solving (8) for y in terms of x, and letting $x = t$ be the parameter. The resulting parametric equations would look different but would define the same solution set.

to y. This allows us to express the solution by the pair of equations (called *parametric equations*)

$$x = \tfrac{1}{4} + \tfrac{1}{2}t, \quad y = t$$

We can obtain specific numerical solutions from these equations by substituting numerical values for the parameter. For example, $t = 0$ yields the solution $\left(\tfrac{1}{4}, 0\right)$, $t = 1$ yields the solution $\left(\tfrac{3}{4}, 1\right)$, and $t = -1$ yields the solution $\left(-\tfrac{1}{4}, -1\right)$. You can confirm that these are solutions by substituting the coordinates into the given equations.

▶ **EXAMPLE 5 A Linear System with Infinitely Many Solutions**

Solve the linear system

$$\begin{aligned} x - y + 2z &= 5 \\ 2x - 2y + 4z &= 10 \\ 3x - 3y + 6z &= 15 \end{aligned}$$

Solution This system can be solved by inspection, since the second and third equations are multiples of the first. Geometrically, this means that the three planes coincide and that those values of x, y, and z that satisfy the equation

$$x - y + 2z = 5 \tag{9}$$

automatically satisfy all three equations. Thus, it suffices to find the solutions of (9). We can do this by first solving (9) for x in terms of y and z, then assigning arbitrary values r and s (parameters) to these two variables, and then expressing the solution by the three parametric equations

$$x = 5 + r - 2s, \quad y = r, \quad z = s$$

Specific solutions can be obtained by choosing numerical values for the parameters r and s. For example, taking $r = 1$ and $s = 0$ yields the solution $(6, 1, 0)$. ◀

Augmented Matrices and Elementary Row Operations

As the number of equations and unknowns in a linear system increases, so does the complexity of the algebra involved in finding solutions. The required computations can be made more manageable by simplifying notation and standardizing procedures. For example, by mentally keeping track of the location of the $+$'s, the x's, and the $=$'s in the linear system

$$\begin{aligned} a_{11}x_1 + a_{12}x_2 + \cdots + a_{1n}x_n &= b_1 \\ a_{21}x_1 + a_{22}x_2 + \cdots + a_{2n}x_n &= b_2 \\ \vdots \qquad \vdots \qquad\qquad \vdots \qquad \vdots \\ a_{m1}x_1 + a_{m2}x_2 + \cdots + a_{mn}x_n &= b_m \end{aligned}$$

we can abbreviate the system by writing only the rectangular array of numbers

$$\begin{bmatrix} a_{11} & a_{12} & \cdots & a_{1n} & b_1 \\ a_{21} & a_{22} & \cdots & a_{2n} & b_2 \\ \vdots & \vdots & & \vdots & \vdots \\ a_{m1} & a_{m2} & \cdots & a_{mn} & b_m \end{bmatrix}$$

As noted in the introduction to this chapter, the term "matrix" is used in mathematics to denote a rectangular array of numbers. In a later section we will study matrices in detail, but for now we will only be concerned with augmented matrices for linear systems.

This is called the *augmented matrix* for the system. For example, the augmented matrix for the system of equations

$$\begin{aligned} x_1 + x_2 + 2x_3 &= 9 \\ 2x_1 + 4x_2 - 3x_3 &= 1 \\ 3x_1 + 6x_2 - 5x_3 &= 0 \end{aligned} \quad \text{is} \quad \begin{bmatrix} 1 & 1 & 2 & 9 \\ 2 & 4 & -3 & 1 \\ 3 & 6 & -5 & 0 \end{bmatrix}$$

The basic method for solving a linear system is to perform algebraic operations on the system that do not alter the solution set and that produce a succession of increasingly simpler systems, until a point is reached where it can be ascertained whether the system is consistent, and if so, what its solutions are. Typically, the algebraic operations are as follows:

1. Multiply an equation through by a nonzero constant.
2. Interchange two equations.
3. Add a constant times one equation to another.

Since the rows (horizontal lines) of an augmented matrix correspond to the equations in the associated system, these three operations correspond to the following operations on the rows of the augmented matrix:

1. Multiply a row through by a nonzero constant.
2. Interchange two rows.
3. Add a constant times one row to another.

These are called ***elementary row operations*** on a matrix.

In the following example we will illustrate how to use elementary row operations and an augmented matrix to solve a linear system in three unknowns. Since a systematic procedure for solving linear systems will be developed in the next section, do not worry about how the steps in the example were chosen. Your objective here should be simply to understand the computations.

▶ **EXAMPLE 6 Using Elementary Row Operations**

In the left column we solve a system of linear equations by operating on the equations in the system, and in the right column we solve the same system by operating on the rows of the augmented matrix.

$$
\begin{aligned}
x + y + 2z &= 9 \\
2x + 4y - 3z &= 1 \\
3x + 6y - 5z &= 0
\end{aligned}
\qquad
\begin{bmatrix}
1 & 1 & 2 & 9 \\
2 & 4 & -3 & 1 \\
3 & 6 & -5 & 0
\end{bmatrix}
$$

Add -2 times the first equation to the second to obtain

Add -2 times the first row to the second to obtain

$$
\begin{aligned}
x + y + 2z &= 9 \\
2y - 7z &= -17 \\
3x + 6y - 5z &= 0
\end{aligned}
\qquad
\begin{bmatrix}
1 & 1 & 2 & 9 \\
0 & 2 & -7 & -17 \\
3 & 6 & -5 & 0
\end{bmatrix}
$$

Maxime Bôcher
(1867–1918)

Historical Note The first known use of augmented matrices appeared between 200 B.C. and 100 B.C. in a Chinese manuscript entitled *Nine Chapters of Mathematical Art*. The coefficients were arranged in columns rather than in rows, as today, but remarkably the system was solved by performing a succession of operations on the columns. The actual use of the term *augmented matrix* appears to have been introduced by the American mathematician Maxime Bôcher in his book *Introduction to Higher Algebra*, published in 1907. In addition to being an outstanding research mathematician and an expert in Latin, chemistry, philosophy, zoology, geography, meteorology, art, and music, Bôcher was an outstanding expositor of mathematics whose elementary textbooks were greatly appreciated by students and are still in demand today.

[*Image: Courtesy of the American Mathematical Society*]

Add -3 times the first equation to the third to obtain

$$\begin{aligned}
x + y + 2z &= 9 \\
2y - 7z &= -17 \\
3y - 11z &= -27
\end{aligned}$$

Add -3 times the first row to the third to obtain

$$\begin{bmatrix} 1 & 1 & 2 & 9 \\ 0 & 2 & -7 & -17 \\ 0 & 3 & -11 & -27 \end{bmatrix}$$

Multiply the second equation by $\frac{1}{2}$ to obtain

$$\begin{aligned}
x + y + 2z &= 9 \\
y - \tfrac{7}{2}z &= -\tfrac{17}{2} \\
3y - 11z &= -27
\end{aligned}$$

Multiply the second row by $\frac{1}{2}$ to obtain

$$\begin{bmatrix} 1 & 1 & 2 & 9 \\ 0 & 1 & -\tfrac{7}{2} & -\tfrac{17}{2} \\ 0 & 3 & -11 & -27 \end{bmatrix}$$

Add -3 times the second equation to the third to obtain

$$\begin{aligned}
x + y + 2z &= 9 \\
y - \tfrac{7}{2}z &= -\tfrac{17}{2} \\
-\tfrac{1}{2}z &= -\tfrac{3}{2}
\end{aligned}$$

Add -3 times the second row to the third to obtain

$$\begin{bmatrix} 1 & 1 & 2 & 9 \\ 0 & 1 & -\tfrac{7}{2} & -\tfrac{17}{2} \\ 0 & 0 & -\tfrac{1}{2} & -\tfrac{3}{2} \end{bmatrix}$$

Multiply the third equation by -2 to obtain

$$\begin{aligned}
x + y + 2z &= 9 \\
y - \tfrac{7}{2}z &= -\tfrac{17}{2} \\
z &= 3
\end{aligned}$$

Multiply the third row by -2 to obtain

$$\begin{bmatrix} 1 & 1 & 2 & 9 \\ 0 & 1 & -\tfrac{7}{2} & -\tfrac{17}{2} \\ 0 & 0 & 1 & 3 \end{bmatrix}$$

Add -1 times the second equation to the first to obtain

$$\begin{aligned}
x + \tfrac{11}{2}z &= \tfrac{35}{2} \\
y - \tfrac{7}{2}z &= -\tfrac{17}{2} \\
z &= 3
\end{aligned}$$

Add -1 times the second row to the first to obtain

$$\begin{bmatrix} 1 & 0 & \tfrac{11}{2} & \tfrac{35}{2} \\ 0 & 1 & -\tfrac{7}{2} & -\tfrac{17}{2} \\ 0 & 0 & 1 & 3 \end{bmatrix}$$

Add $-\tfrac{11}{2}$ times the third equation to the first and $\tfrac{7}{2}$ times the third equation to the second to obtain

$$\begin{aligned}
x &= 1 \\
y &= 2 \\
z &= 3
\end{aligned}$$

Add $-\tfrac{11}{2}$ times the third row to the first and $\tfrac{7}{2}$ times the third row to the second to obtain

$$\begin{bmatrix} 1 & 0 & 0 & 1 \\ 0 & 1 & 0 & 2 \\ 0 & 0 & 1 & 3 \end{bmatrix}$$

The solution $x = 1$, $y = 2$, $z = 3$ is now evident. ◀

Concept Review

- Linear equation
- Homogeneous linear equation
- System of linear equations
- Solution of a linear system

- Ordered n-tuple
- Consistent linear system
- Inconsistent linear system
- Parameter

- Parametric equations
- Augmented matrix
- Elementary row operations

Skills

- Determine whether a given equation is linear.
- Determine whether a given n-tuple is a solution of a linear system.
- Find the augmented matrix of a linear system.
- Find the linear system corresponding to a given augmented matrix.

- Perform elementary row operations on a linear system and on its corresponding augmented matrix.
- Determine whether a linear system is consistent or inconsistent.
- Find the set of solutions to a consistent linear system.

Exercise Set 1.1

1. In each part, determine whether the equation is linear in x_1, x_2, and x_3.

   (a) $x_1 + 5x_2 - \sqrt{2}x_3 = 1$ (b) $x_1 + 3x_2 + x_1x_3 = 2$

   (c) $x_1 = -7x_2 + 3x_3$ (d) $x_1^{-2} + x_2 + 8x_3 = 5$

   (e) $x_1^{3/5} - 2x_2 + x_3 = 4$

   (f) $\pi x_1 - \sqrt{2}x_2 + \frac{1}{3}x_3 = 7^{1/3}$

2. In each part, determine whether the equations form a linear system.

   (a) $-2x + 4y + z = 2$
   $$3x - \frac{2}{y} = 0$$

   (b) $x = 4$
   $$2x = 8$$

   (c) $4x - y + 2z = -1$
   $$-x + (\ln 2)y - 3z = 0$$

   (d) $3z + x = -4$
   $$y + 5z = 1$$
   $$6x + 2z = 3$$
   $$-x - y - z = 4$$

3. In each part, determine whether the equations form a linear system.

   (a) $2x_1 - x_4 = 5$
   $$-x_1 + 5x_2 + 3x_3 - 2x_4 = -1$$

   (b) $\sin(2x_1 + x_3) = \sqrt{5}$
   $$e^{-2x_2 - 2x_4} = \frac{1}{x_2}$$
   $$4x_4 = 4$$

   (c) $7x_1 - x_2 + 2x_3 = 0$ (d) $x_1 + x_2 = x_3 + x_4$
   $$2x_1 + x_2 - x_3x_4 = 3$$
   $$-x_1 + 5x_2 - x_4 = -1$$

4. For each system in Exercise 2 that is linear, determine whether it is consistent.

5. For each system in Exercise 3 that is linear, determine whether it is consistent.

6. Write a system of linear equations consisting of three equations in three unknowns with

   (a) no solutions.

   (b) exactly one solution.

   (c) infinitely many solutions.

7. In each part, determine whether the given vector is a solution of the linear system

   $$2x_1 - 4x_2 - x_3 = 1$$
   $$x_1 - 3x_2 + x_3 = 1$$
   $$3x_1 - 5x_2 - 3x_3 = 1$$

   (a) $(3, 1, 1)$ (b) $(3, -1, 1)$ (c) $(13, 5, 2)$

   (d) $\left(\frac{13}{2}, \frac{5}{2}, 2\right)$ (e) $(17, 7, 5)$

8. In each part, determine whether the given vector is a solution of the linear system

   $$x_1 + 2x_2 - 2x_3 = 3$$
   $$3x_1 - x_2 + x_3 = 1$$
   $$-x_1 + 5x_2 - 5x_3 = 5$$

   (a) $\left(\frac{5}{7}, \frac{8}{7}, 1\right)$ (b) $\left(\frac{5}{7}, \frac{8}{7}, 0\right)$ (c) $(5, 8, 1)$

   (d) $\left(\frac{5}{7}, \frac{10}{7}, \frac{2}{7}\right)$ (e) $\left(\frac{5}{7}, \frac{22}{7}, 2\right)$

9. In each part, find the solution set of the linear equation by using parameters as necessary.

   (a) $7x - 5y = 3$

   (b) $-8x_1 + 2x_2 - 5x_3 + 6x_4 = 1$

10. In each part, find the solution set of the linear equation by using parameters as necessary.

    (a) $3x_1 - 5x_2 + 4x_3 = 7$

    (b) $3v - 8w + 2x - y + 4z = 0$

11. In each part, find a system of linear equations corresponding to the given augmented matrix.

(a) $\begin{bmatrix} 2 & 0 & 0 \\ 3 & -4 & 0 \\ 0 & 1 & 1 \end{bmatrix}$
(b) $\begin{bmatrix} 3 & 0 & -2 & 5 \\ 7 & 1 & 4 & -3 \\ 0 & -2 & 1 & 7 \end{bmatrix}$

(c) $\begin{bmatrix} 7 & 2 & 1 & -3 & 5 \\ 1 & 2 & 4 & 0 & 1 \end{bmatrix}$

(d) $\begin{bmatrix} 1 & 0 & 0 & 0 & 7 \\ 0 & 1 & 0 & 0 & -2 \\ 0 & 0 & 1 & 0 & 3 \\ 0 & 0 & 0 & 1 & 4 \end{bmatrix}$

12. In each part, find a system of linear equations corresponding to the given augmented matrix.

(a) $\begin{bmatrix} 2 & -1 \\ -4 & -6 \\ 1 & -1 \\ 3 & 0 \end{bmatrix}$
(b) $\begin{bmatrix} 0 & 3 & -1 & -1 & -1 \\ 5 & 2 & 0 & -3 & -6 \end{bmatrix}$

(c) $\begin{bmatrix} 1 & 2 & 3 & 4 \\ -4 & -3 & -2 & -1 \\ 5 & -6 & 1 & 1 \\ -8 & 0 & 0 & 3 \end{bmatrix}$

(d) $\begin{bmatrix} 3 & 0 & 1 & -4 & 3 \\ -4 & 0 & 4 & 1 & -3 \\ -1 & 3 & 0 & -2 & -9 \\ 0 & 0 & 0 & -1 & -2 \end{bmatrix}$

13. In each part, find the augmented matrix for the given system of linear equations.

(a) $-2x_1 = 6$
$3x_1 = 8$
$9x_1 = -3$

(b) $6x_1 - x_2 + 3x_3 = 4$
$5x_2 - x_3 = 1$

(c) $2x_2 - 3x_4 + x_5 = 0$
$-3x_1 - x_2 + x_3 = -1$
$6x_1 + 2x_2 - x_3 + 2x_4 - 3x_5 = 6$

(d) $x_1 - x_5 = 7$

14. In each part, find the augmented matrix for the given system of linear equations.

(a) $3x_1 - 2x_2 = -1$
$4x_1 + 5x_2 = 3$
$7x_1 + 3x_2 = 2$

(b) $2x_1 + 2x_3 = 1$
$3x_1 - x_2 + 4x_3 = 7$
$6x_1 + x_2 - x_3 = 0$

(c) $x_1 + 2x_2 - x_4 + x_5 = 1$
$3x_2 + x_3 - x_5 = 2$
$x_3 + 7x_4 = 1$

(d) $x_1 = 1$
$x_2 = 2$
$x_3 = 3$

15. The curve $y = ax^2 + bx + c$ shown in the accompanying figure passes through the points (x_1, y_1), (x_2, y_2), and (x_3, y_3).

Show that the coefficients a, b, and c are a solution of the system of linear equations whose augmented matrix is

$$\begin{bmatrix} x_1^2 & x_1 & 1 & y_1 \\ x_2^2 & x_2 & 1 & y_2 \\ x_3^2 & x_3 & 1 & y_3 \end{bmatrix}$$

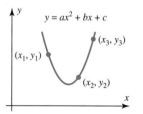

$y = ax^2 + bx + c$

(x_3, y_3)

(x_1, y_1)

(x_2, y_2)

◀ Figure Ex-15

16. Explain why each of the three elementary row operations does not affect the solution set of a linear system.

17. Show that if the linear equations

$$x_1 + kx_2 = c \quad \text{and} \quad x_1 + lx_2 = d$$

have the same solution set, then the two equations are identical (i.e., $k = l$ and $c = d$).

True-False Exercises

In parts (a)–(h) determine whether the statement is true or false, and justify your answer.

(a) A linear system whose equations are all homogeneous must be consistent.

(b) Multiplying a linear equation through by zero is an acceptable elementary row operation.

(c) The linear system

$$x - y = 3$$
$$2x - 2y = k$$

cannot have a unique solution, regardless of the value of k.

(d) A single linear equation with two or more unknowns must always have infinitely many solutions.

(e) If the number of equations in a linear system exceeds the number of unknowns, then the system must be inconsistent.

(f) If each equation in a consistent linear system is multiplied through by a constant c, then all solutions to the new system can be obtained by multiplying solutions from the original system by c.

(g) Elementary row operations permit one equation in a linear system to be subtracted from another.

(h) The linear system with corresponding augmented matrix

$$\begin{bmatrix} 2 & -1 & 4 \\ 0 & 0 & -1 \end{bmatrix}$$

is consistent.

1.2 Gaussian Elimination

In this section we will develop a systematic procedure for solving systems of linear equations. The procedure is based on the idea of performing certain operations on the rows of the augmented matrix that simplifies it to a form from which the solution of the system can be ascertained by inspection.

Considerations in Solving Linear Systems

When considering methods for solving systems of linear equations, it is important to distinguish between large systems that must be solved by computer and small systems that can be solved by hand. For example, there are many applications that lead to linear systems in thousands or even millions of unknowns. Large systems require special techniques to deal with issues of memory size, roundoff errors, solution time, and so forth. Such techniques are studied in the field of ***numerical analysis*** and will only be touched on in this text. However, almost all of the methods that are used for large systems are based on the ideas that we will develop in this section.

Echelon Forms

In Example 6 of the last section, we solved a linear system in the unknowns x, y, and z by reducing the augmented matrix to the form

$$\begin{bmatrix} 1 & 0 & 0 & 1 \\ 0 & 1 & 0 & 2 \\ 0 & 0 & 1 & 3 \end{bmatrix}$$

from which the solution $x = 1$, $y = 2$, $z = 3$ became evident. This is an example of a matrix that is in ***reduced row echelon form***. To be of this form, a matrix must have the following properties:

1. If a row does not consist entirely of zeros, then the first nonzero number in the row is a 1. We call this a ***leading 1***.

2. If there are any rows that consist entirely of zeros, then they are grouped together at the bottom of the matrix.

3. In any two successive rows that do not consist entirely of zeros, the leading 1 in the lower row occurs farther to the right than the leading 1 in the higher row.

4. Each column that contains a leading 1 has zeros everywhere else in that column.

A matrix that has the first three properties is said to be in ***row echelon form***. (Thus, a matrix in reduced row echelon form is of necessity in row echelon form, but not conversely.)

▶ **EXAMPLE 1 Row Echelon and Reduced Row Echelon Form**

The following matrices are in reduced row echelon form.

$$\begin{bmatrix} 1 & 0 & 0 & 4 \\ 0 & 1 & 0 & 7 \\ 0 & 0 & 1 & -1 \end{bmatrix}, \quad \begin{bmatrix} 1 & 0 & 0 \\ 0 & 1 & 0 \\ 0 & 0 & 1 \end{bmatrix}, \quad \begin{bmatrix} 0 & 1 & -2 & 0 & 1 \\ 0 & 0 & 0 & 1 & 3 \\ 0 & 0 & 0 & 0 & 0 \\ 0 & 0 & 0 & 0 & 0 \end{bmatrix}, \quad \begin{bmatrix} 0 & 0 \\ 0 & 0 \end{bmatrix}$$

The following matrices are in row echelon form but not reduced row echelon form.

$$\begin{bmatrix} 1 & 4 & -3 & 7 \\ 0 & 1 & 6 & 2 \\ 0 & 0 & 1 & 5 \end{bmatrix}, \quad \begin{bmatrix} 1 & 1 & 0 \\ 0 & 1 & 0 \\ 0 & 0 & 0 \end{bmatrix}, \quad \begin{bmatrix} 0 & 1 & 2 & 6 & 0 \\ 0 & 0 & 1 & -1 & 0 \\ 0 & 0 & 0 & 0 & 1 \end{bmatrix}$$

▶ **EXAMPLE 2 More on Row Echelon and Reduced Row Echelon Form**

As Example 1 illustrates, a matrix in row echelon form has zeros below each leading 1, whereas a matrix in reduced row echelon form has zeros below *and above* each leading 1. Thus, with any real numbers substituted for the ∗'s, all matrices of the following types are in row echelon form:

$$
\begin{bmatrix} 1 & * & * & * \\ 0 & 1 & * & * \\ 0 & 0 & 1 & * \\ 0 & 0 & 0 & 1 \end{bmatrix},
\begin{bmatrix} 1 & * & * & * \\ 0 & 1 & * & * \\ 0 & 0 & 1 & * \\ 0 & 0 & 0 & 0 \end{bmatrix},
\begin{bmatrix} 1 & * & * & * \\ 0 & 1 & * & * \\ 0 & 0 & 0 & 0 \\ 0 & 0 & 0 & 0 \end{bmatrix},
\begin{bmatrix} 0 & 1 & * & * & * & * & * & * & * \\ 0 & 0 & 0 & 1 & * & * & * & * & * \\ 0 & 0 & 0 & 0 & 1 & * & * & * & * \\ 0 & 0 & 0 & 0 & 0 & 1 & * & * & * \\ 0 & 0 & 0 & 0 & 0 & 0 & 0 & 1 & * \end{bmatrix}
$$

All matrices of the following types are in reduced row echelon form:

$$
\begin{bmatrix} 1 & 0 & 0 & 0 \\ 0 & 1 & 0 & 0 \\ 0 & 0 & 1 & 0 \\ 0 & 0 & 0 & 1 \end{bmatrix},
\begin{bmatrix} 1 & 0 & 0 & * \\ 0 & 1 & 0 & * \\ 0 & 0 & 1 & * \\ 0 & 0 & 0 & 0 \end{bmatrix},
\begin{bmatrix} 1 & 0 & * & * \\ 0 & 1 & * & * \\ 0 & 0 & 0 & 0 \\ 0 & 0 & 0 & 0 \end{bmatrix},
\begin{bmatrix} 0 & 1 & * & 0 & 0 & 0 & * & * & 0 & * \\ 0 & 0 & 0 & 1 & 0 & 0 & * & * & 0 & * \\ 0 & 0 & 0 & 0 & 1 & 0 & * & * & 0 & * \\ 0 & 0 & 0 & 0 & 0 & 1 & * & * & 0 & * \\ 0 & 0 & 0 & 0 & 0 & 0 & 0 & 0 & 1 & * \end{bmatrix} \blacktriangleleft
$$

If, by a sequence of elementary row operations, the augmented matrix for a system of linear equations is put in *reduced* row echelon form, then the solution set can be obtained either by inspection or by converting certain linear equations to parametric form. Here are some examples.

▶ **EXAMPLE 3 Unique Solution**

Suppose that the augmented matrix for a linear system in the unknowns x_1, x_2, x_3, and x_4 has been reduced by elementary row operations to

$$
\begin{bmatrix} 1 & 0 & 0 & 0 & 3 \\ 0 & 1 & 0 & 0 & -1 \\ 0 & 0 & 1 & 0 & 0 \\ 0 & 0 & 0 & 1 & 5 \end{bmatrix}
$$

This matrix is in reduced row echelon form and corresponds to the equations

$$
\begin{aligned}
x_1 &= 3 \\
x_2 &= -1 \\
x_3 &= 0 \\
x_4 &= 5
\end{aligned}
$$

In Example 3 we could, if desired, express the solution more succinctly as the 4-tuple $(3, -1, 0, 5)$.

Thus, the system has a unique solution, namely, $x_1 = 3$, $x_2 = -1$, $x_3 = 0$, $x_4 = 5$.

▶ **EXAMPLE 4 Linear Systems in Three Unknowns**

In each part, suppose that the augmented matrix for a linear system in the unknowns x, y, and z has been reduced by elementary row operations to the given reduced row echelon form. Solve the system.

$$
\text{(a)} \begin{bmatrix} 1 & 0 & 0 & 0 \\ 0 & 1 & 2 & 0 \\ 0 & 0 & 0 & 1 \end{bmatrix} \quad
\text{(b)} \begin{bmatrix} 1 & 0 & 3 & -1 \\ 0 & 1 & -4 & 2 \\ 0 & 0 & 0 & 0 \end{bmatrix} \quad
\text{(c)} \begin{bmatrix} 1 & -5 & 1 & 4 \\ 0 & 0 & 0 & 0 \\ 0 & 0 & 0 & 0 \end{bmatrix}
$$

Solution (a) The equation that corresponds to the last row of the augmented matrix is

$$0x + 0y + 0z = 1$$

Since this equation is not satisfied by any values of x, y, and z, the system is inconsistent.

Solution (b) The equation that corresponds to the last row of the augmented matrix is

$$0x + 0y + 0z = 0$$

This equation can be omitted since it imposes no restrictions on x, y, and z; hence, the linear system corresponding to the augmented matrix is

$$x \quad\;\; + 3z = -1$$
$$y - 4z = \quad 2$$

Since x and y correspond to the leading 1's in the augmented matrix, we call these the ***leading variables***. The remaining variables (in this case z) are called ***free variables***. Solving for the leading variables in terms of the free variables gives

$$x = -1 - 3z$$
$$y = 2 + 4z$$

From these equations we see that the free variable z can be treated as a parameter and assigned an arbitrary value t, which then determines values for x and y. Thus, the solution set can be represented by the parametric equations

$$x = -1 - 3t, \quad y = 2 + 4t, \quad z = t$$

By substituting various values for t in these equations we can obtain various solutions of the system. For example, setting $t = 0$ yields the solution

$$x = -1, \quad y = 2, \quad z = 0$$

and setting $t = 1$ yields the solution

$$x = -4, \quad y = 6, \quad z = 1$$

Solution (c) As explained in part (b), we can omit the equations corresponding to the zero rows, in which case the linear system associated with the augmented matrix consists of the single equation

$$x - 5y + z = 4 \tag{1}$$

from which we see that the solution set is a plane in three-dimensional space. Although (1) is a valid form of the solution set, there are many applications in which it is preferable to express the solution set in parametric form. We can convert (1) to parametric form by solving for the leading variable x in terms of the free variables y and z to obtain

$$x = 4 + 5y - z$$

From this equation we see that the free variables can be assigned arbitrary values, say $y = s$ and $z = t$, which then determine the value of x. Thus, the solution set can be expressed parametrically as

$$x = 4 + 5s - t, \quad y = s, \quad z = t \;\blacktriangleleft \tag{2}$$

We will usually denote parameters in a general solution by the letters $r, s, t, \ldots$, but any letters that do not conflict with the names of the unknowns can be used. For systems with more than three unknowns, subscripted letters such as $t_1, t_2, t_3, \ldots$ are convenient.

Formulas, such as (2), that express the solution set of a linear system parametrically have some associated terminology.

DEFINITION 1 If a linear system has infinitely many solutions, then a set of parametric equations from which all solutions can be obtained by assigning numerical values to the parameters is called a ***general solution*** of the system.

Elimination Methods   We have just seen how easy it is to solve a system of linear equations once its augmented matrix is in reduced row echelon form. Now we will give a step-by-step *elimination procedure* that can be used to reduce any matrix to reduced row echelon form. As we state each step in the procedure, we illustrate the idea by reducing the following matrix to reduced row echelon form.

$$\begin{bmatrix} 0 & 0 & -2 & 0 & 7 & 12 \\ 2 & 4 & -10 & 6 & 12 & 28 \\ 2 & 4 & -5 & 6 & -5 & -1 \end{bmatrix}$$

Step 1. Locate the leftmost column that does not consist entirely of zeros.

$$\begin{bmatrix} 0 & 0 & -2 & 0 & 7 & 12 \\ 2 & 4 & -10 & 6 & 12 & 28 \\ 2 & 4 & -5 & 6 & -5 & -1 \end{bmatrix}$$

   ↑
   └── **Leftmost nonzero column**

Step 2. Interchange the top row with another row, if necessary, to bring a nonzero entry to the top of the column found in Step 1.

$$\begin{bmatrix} 2 & 4 & -10 & 6 & 12 & 28 \\ 0 & 0 & -2 & 0 & 7 & 12 \\ 2 & 4 & -5 & 6 & -5 & -1 \end{bmatrix}$$
⟵ ——— The first and second rows in the preceding matrix were interchanged.

Step 3. If the entry that is now at the top of the column found in Step 1 is a, multiply the first row by $1/a$ in order to introduce a leading 1.

$$\begin{bmatrix} 1 & 2 & -5 & 3 & 6 & 14 \\ 0 & 0 & -2 & 0 & 7 & 12 \\ 2 & 4 & -5 & 6 & -5 & -1 \end{bmatrix}$$
⟵ ——— The first row of the preceding matrix was multiplied by $\frac{1}{2}$.

Step 4. Add suitable multiples of the top row to the rows below so that all entries below the leading 1 become zeros.

$$\begin{bmatrix} 1 & 2 & -5 & 3 & 6 & 14 \\ 0 & 0 & -2 & 0 & 7 & 12 \\ 0 & 0 & 5 & 0 & -17 & -29 \end{bmatrix}$$
⟵ ——— -2 times the first row of the preceding matrix was added to the third row.

Step 5. Now cover the top row in the matrix and begin again with Step 1 applied to the submatrix that remains. Continue in this way until the *entire* matrix is in row echelon form.

$$\begin{bmatrix} 1 & 2 & -5 & 3 & 6 & 14 \\ 0 & 0 & -2 & 0 & 7 & 12 \\ 0 & 0 & 5 & 0 & -17 & -29 \end{bmatrix}$$

    ↑
    └── **Leftmost nonzero column**
      in the submatrix

$$\begin{bmatrix} 1 & 2 & -5 & 3 & 6 & 14 \\ 0 & 0 & 1 & 0 & -\frac{7}{2} & -6 \\ 0 & 0 & 5 & 0 & -17 & -29 \end{bmatrix}$$
⟵ ——— The first row in the submatrix was multiplied by $-\frac{1}{2}$ to introduce a leading 1.

$$\begin{bmatrix} 1 & 2 & -5 & 3 & 6 & 14 \\ 0 & 0 & 1 & 0 & -\frac{7}{2} & -6 \\ 0 & 0 & 0 & 0 & \frac{1}{2} & 1 \end{bmatrix}$$

← -5 times the first row of the submatrix was added to the second row of the submatrix to introduce a zero below the leading 1.

$$\begin{bmatrix} 1 & 2 & -5 & 3 & 6 & 14 \\ 0 & 0 & 1 & 0 & -\frac{7}{2} & -6 \\ 0 & 0 & 0 & 0 & \frac{1}{2} & 1 \end{bmatrix}$$

← The top row in the submatrix was covered, and we returned again to Step 1.

Leftmost nonzero column in the new submatrix

$$\begin{bmatrix} 1 & 2 & -5 & 3 & 6 & 14 \\ 0 & 0 & 1 & 0 & -\frac{7}{2} & -6 \\ 0 & 0 & 0 & 0 & 1 & 2 \end{bmatrix}$$

← The first (and only) row in the new submatrix was multiplied by 2 to introduce a leading 1.

The *entire* matrix is now in row echelon form. To find the reduced row echelon form we need the following additional step.

Step 6. Beginning with the last nonzero row and working upward, add suitable multiples of each row to the rows above to introduce zeros above the leading 1's.

$$\begin{bmatrix} 1 & 2 & -5 & 3 & 6 & 14 \\ 0 & 0 & 1 & 0 & 0 & 1 \\ 0 & 0 & 0 & 0 & 1 & 2 \end{bmatrix}$$

← $\frac{7}{2}$ times the third row of the preceding matrix was added to the second row.

$$\begin{bmatrix} 1 & 2 & -5 & 3 & 0 & 2 \\ 0 & 0 & 1 & 0 & 0 & 1 \\ 0 & 0 & 0 & 0 & 1 & 2 \end{bmatrix}$$

← -6 times the third row was added to the first row.

$$\begin{bmatrix} 1 & 2 & 0 & 3 & 0 & 7 \\ 0 & 0 & 1 & 0 & 0 & 1 \\ 0 & 0 & 0 & 0 & 1 & 2 \end{bmatrix}$$

← 5 times the second row was added to the first row.

The last matrix is in reduced row echelon form.

The procedure (or algorithm) we have just described for reducing a matrix to reduced row echelon form is called ***Gauss–Jordan elimination***. This algorithm consists of two parts, a ***forward phase*** in which zeros are introduced below the leading 1's and a ***backward phase*** in which zeros are introduced above the leading 1's. If only the forward phase

Carl Friedrich Gauss (1777–1855)

Wilhelm Jordan (1842–1899)

Historical Note Although versions of Gaussian elimination were known much earlier, the power of the method was not recognized until the great German mathematician Carl Friedrich Gauss used it to compute the orbit of the asteroid Ceres from limited data. What happened was this: On January 1, 1801 the Sicilian astronomer Giuseppe Piazzi (1746–1826) noticed a dim celestial object that he believed might be a "missing planet." He named the object Ceres and made a limited number of positional observations but then lost the object as it neared the Sun. Gauss undertook the problem of computing the orbit from the limited data and the procedure that we now call Gaussian elimination. The work of Gauss caused a sensation when Ceres reappeared a year later in the constellation Virgo at almost the precise position that Gauss predicted! The method was further popularized by the German engineer Wilhelm Jordan in his handbook on geodesy (the science of measuring Earth shapes) entitled *Handbuch der Vermessungskunde* and published in 1888.

[Images: Granger Collection (Gauss); wikipedia (Jordan)]

is used, then the procedure produces a row echelon form only and is called *Gaussian elimination*. For example, in the preceding computations a row echelon form was obtained at the end of Step 5.

▶ **EXAMPLE 5 Gauss–Jordan Elimination**

Solve by Gauss–Jordan elimination.

$$
\begin{aligned}
x_1 + 3x_2 - 2x_3 \quad\quad + 2x_5 \quad\quad &= \quad 0 \\
2x_1 + 6x_2 - 5x_3 - \; 2x_4 + 4x_5 - \; 3x_6 &= -1 \\
5x_3 + 10x_4 \quad\quad + 15x_6 &= \quad 5 \\
2x_1 + 6x_2 \quad\quad + \; 8x_4 + 4x_5 + 18x_6 &= \quad 6
\end{aligned}
$$

Solution The augmented matrix for the system is

$$
\begin{bmatrix}
1 & 3 & -2 & 0 & 2 & 0 & 0 \\
2 & 6 & -5 & -2 & 4 & -3 & -1 \\
0 & 0 & 5 & 10 & 0 & 15 & 5 \\
2 & 6 & 0 & 8 & 4 & 18 & 6
\end{bmatrix}
$$

Adding -2 times the first row to the second and fourth rows gives

$$
\begin{bmatrix}
1 & 3 & -2 & 0 & 2 & 0 & 0 \\
0 & 0 & -1 & -2 & 0 & -3 & -1 \\
0 & 0 & 5 & 10 & 0 & 15 & 5 \\
0 & 0 & 4 & 8 & 0 & 18 & 6
\end{bmatrix}
$$

Multiplying the second row by -1 and then adding -5 times the new second row to the third row and -4 times the new second row to the fourth row gives

$$
\begin{bmatrix}
1 & 3 & -2 & 0 & 2 & 0 & 0 \\
0 & 0 & 1 & 2 & 0 & 3 & 1 \\
0 & 0 & 0 & 0 & 0 & 0 & 0 \\
0 & 0 & 0 & 0 & 0 & 6 & 2
\end{bmatrix}
$$

Interchanging the third and fourth rows and then multiplying the third row of the resulting matrix by $\frac{1}{6}$ gives the row echelon form

$$
\begin{bmatrix}
1 & 3 & -2 & 0 & 2 & 0 & 0 \\
0 & 0 & 1 & 2 & 0 & 3 & 1 \\
0 & 0 & 0 & 0 & 0 & 1 & \frac{1}{3} \\
0 & 0 & 0 & 0 & 0 & 0 & 0
\end{bmatrix}
$$

This completes the forward phase since there are zeros below the leading 1's.

Adding -3 times the third row to the second row and then adding 2 times the second row of the resulting matrix to the first row yields the reduced row echelon form

$$
\begin{bmatrix}
1 & 3 & 0 & 4 & 2 & 0 & 0 \\
0 & 0 & 1 & 2 & 0 & 0 & 0 \\
0 & 0 & 0 & 0 & 0 & 1 & \frac{1}{3} \\
0 & 0 & 0 & 0 & 0 & 0 & 0
\end{bmatrix}
$$

This completes the backward phase since there are zeros above the leading 1's.

The corresponding system of equations is

$$\begin{aligned} x_1 + 3x_2 \qquad\quad + 4x_4 + 2x_5 \qquad\; &= 0 \\ x_3 + 2x_4 \qquad\qquad\; &= 0 \\ x_6 &= \tfrac{1}{3} \end{aligned} \qquad (3)$$

Note that in constructing the linear system in (3) we ignored the row of zeros in the corresponding augmented matrix. Why is this justified?

Solving for the leading variables we obtain

$$\begin{aligned} x_1 &= -3x_2 - 4x_4 - 2x_5 \\ x_3 &= -2x_4 \\ x_6 &= \tfrac{1}{3} \end{aligned}$$

Finally, we express the general solution of the system parametrically by assigning the free variables x_2, x_4, and x_5 arbitrary values r, s, and t, respectively. This yields

$$x_1 = -3r - 4s - 2t, \quad x_2 = r, \quad x_3 = -2s, \quad x_4 = s, \quad x_5 = t, \quad x_6 = \tfrac{1}{3} \; \blacktriangleleft$$

Homogeneous Linear Systems

A system of linear equations is said to be ***homogeneous*** if the constant terms are all zero; that is, the system has the form

$$\begin{aligned} a_{11}x_1 + a_{12}x_2 + \cdots + a_{1n}x_n &= 0 \\ a_{21}x_1 + a_{22}x_2 + \cdots + a_{2n}x_n &= 0 \\ \vdots \qquad \vdots \qquad\quad \vdots \quad\; \vdots \\ a_{m1}x_1 + a_{m2}x_2 + \cdots + a_{mn}x_n &= 0 \end{aligned}$$

Every homogeneous system of linear equations is consistent because all such systems have $x_1 = 0$, $x_2 = 0$, ..., $x_n = 0$ as a solution. This solution is called the ***trivial solution***; if there are other solutions, they are called ***nontrivial solutions***.

Because a homogeneous linear system always has the trivial solution, there are only two possibilities for its solutions:

- The system has only the trivial solution.

- The system has infinitely many solutions in addition to the trivial solution.

In the special case of a homogeneous linear system of two equations in two unknowns, say

$$\begin{aligned} a_1 x + b_1 y = 0 \quad &(a_1, b_1 \text{ not both zero}) \\ a_2 x + b_2 y = 0 \quad &(a_2, b_2 \text{ not both zero}) \end{aligned}$$

the graphs of the equations are lines through the origin, and the trivial solution corresponds to the point of intersection at the origin (Figure 1.2.1).

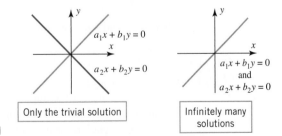

Only the trivial solution   Infinitely many solutions

▶ **Figure 1.2.1**

There is one case in which a homogeneous system is assured of having nontrivial solutions—namely, whenever the system involves more unknowns than equations. To see why, consider the following example of four equations in six unknowns.

▶ **EXAMPLE 6 A Homogeneous System**

Use Gauss–Jordan elimination to solve the homogeneous linear system

$$\begin{aligned}
x_1 + 3x_2 - 2x_3 \qquad\quad + 2x_5 \qquad\quad &= 0 \\
2x_1 + 6x_2 - 5x_3 - 2x_4 + 4x_5 - 3x_6 &= 0 \\
5x_3 + 10x_4 \qquad\quad + 15x_6 &= 0 \\
2x_1 + 6x_2 \qquad\quad + 8x_4 + 4x_5 + 18x_6 &= 0
\end{aligned} \tag{4}$$

Solution Observe first that the coefficients of the unknowns in this system are the same as those in Example 5; that is, the two systems differ only in the constants on the right side. The augmented matrix for the given homogeneous system is

$$\begin{bmatrix}
1 & 3 & -2 & 0 & 2 & 0 & 0 \\
2 & 6 & -5 & -2 & 4 & -3 & 0 \\
0 & 0 & 5 & 10 & 0 & 15 & 0 \\
2 & 6 & 0 & 8 & 4 & 18 & 0
\end{bmatrix} \tag{5}$$

which is the same as the augmented matrix for the system in Example 5, except for zeros in the last column. Thus, the reduced row echelon form of this matrix will be the same as that of the augmented matrix in Example 5, except for the last column. However, a moment's reflection will make it evident that a column of zeros is not changed by an elementary row operation, so the reduced row echelon form of (5) is

$$\begin{bmatrix}
1 & 3 & 0 & 4 & 2 & 0 & 0 \\
0 & 0 & 1 & 2 & 0 & 0 & 0 \\
0 & 0 & 0 & 0 & 0 & 1 & 0 \\
0 & 0 & 0 & 0 & 0 & 0 & 0
\end{bmatrix} \tag{6}$$

The corresponding system of equations is

$$\begin{aligned}
x_1 + 3x_2 \qquad\quad + 4x_4 + 2x_5 \qquad &= 0 \\
x_3 + 2x_4 \qquad\quad &= 0 \\
x_6 &= 0
\end{aligned}$$

Solving for the leading variables we obtain

$$\begin{aligned}
x_1 &= -3x_2 - 4x_4 - 2x_5 \\
x_3 &= -2x_4 \\
x_6 &= 0
\end{aligned} \tag{7}$$

If we now assign the free variables x_2, x_4, and x_5 arbitrary values r, s, and t, respectively, then we can express the solution set parametrically as

$$x_1 = -3r - 4s - 2t, \quad x_2 = r, \quad x_3 = -2s, \quad x_4 = s, \quad x_5 = t, \quad x_6 = 0$$

Note that the trivial solution results when $r = s = t = 0$. ◀

Free Variables in Homogeneous Linear Systems

Example 6 illustrates two important points about solving homogeneous linear systems:

1. Elementary row operations do not alter columns of zeros in a matrix, so the reduced row echelon form of the augmented matrix for a homogeneous linear system has a final column of zeros. This implies that the linear system corresponding to the reduced row echelon form is homogeneous, just like the original system.

2. When we constructed the homogeneous linear system corresponding to augmented matrix (6), we ignored the row of zeros because the corresponding equation

$$0x_1 + 0x_2 + 0x_3 + 0x_4 + 0x_5 + 0x_6 = 0$$

does not impose any conditions on the unknowns. Thus, depending on whether or not the reduced row echelon form of the augmented matrix for a homogeneous linear system has any rows of zero, the linear system corresponding to that reduced row echelon form will either have the same number of equations as the original system or it will have fewer.

Now consider a general homogeneous linear system with n unknowns, and suppose that the reduced row echelon form of the augmented matrix has r nonzero rows. Since each nonzero row has a leading 1, and since each leading 1 corresponds to a leading variable, the homogeneous system corresponding to the reduced row echelon form of the augmented matrix must have r leading variables and $n - r$ free variables. Thus, this system is of the form

$$
\begin{aligned}
x_{k_1} \quad & \quad + \sum(\ ) = 0 \\
x_{k_2} \quad & \quad + \sum(\ ) = 0 \\
\ddots \quad & \quad \vdots \\
x_{k_r} & + \sum(\ ) = 0
\end{aligned}
\tag{8}
$$

where in each equation the expression $\sum(\ )$ denotes a sum that involves the free variables, if any [see (7), for example]. In summary, we have the following result.

THEOREM 1.2.1 Free Variable Theorem for Homogeneous Systems

If a homogeneous linear system has n unknowns, and if the reduced row echelon form of its augmented matrix has r nonzero rows, then the system has n − r free variables.

Theorem 1.2.1 has an important implication for homogeneous linear systems with more unknowns than equations. Specifically, if a homogeneous linear system has m equations in n unknowns, and if $m < n$, then it must also be true that $r < n$ (why?). This being the case, the theorem implies that there is at least one free variable, and this implies that the system has infinitely many solutions. Thus, we have the following result.

> Note that Theorem 1.2.2 applies only to homogeneous systems—a *nonhomogeneous* system with more unknowns than equations need not be consistent. However, we will prove later that if a nonhomogeneous system with more unknowns then equations is consistent, then it has infinitely many solutions.

THEOREM 1.2.2 *A homogeneous linear system with more unknowns than equations has infinitely many solutions.*

In retrospect, we could have anticipated that the homogeneous system in Example 6 would have infinitely many solutions since it has four equations in six unknowns.

Gaussian Elimination and Back-Substitution

For small linear systems that are solved by hand (such as most of those in this text), Gauss–Jordan elimination (reduction to reduced row echelon form) is a good procedure to use. However, for large linear systems that require a computer solution, it is generally more efficient to use Gaussian elimination (reduction to row echelon form) followed by a technique known as ***back-substitution*** to complete the process of solving the system. The next example illustrates this technique.

▶ **EXAMPLE 7 Example 5 Solved by Back-Substitution**

From the computations in Example 5, a row echelon form of the augmented matrix is

$$
\begin{bmatrix}
1 & 3 & -2 & 0 & 2 & 0 & 0 \\
0 & 0 & 1 & 2 & 0 & 3 & 1 \\
0 & 0 & 0 & 0 & 0 & 1 & \frac{1}{3} \\
0 & 0 & 0 & 0 & 0 & 0 & 0
\end{bmatrix}
$$

To solve the corresponding system of equations

$$
\begin{aligned}
x_1 + 3x_2 - 2x_3 \quad\quad + 2x_5 \quad\quad &= 0\\
x_3 + 2x_4 \quad\quad + 3x_6 &= 1\\
x_6 &= \tfrac{1}{3}
\end{aligned}
$$

we proceed as follows:

Step 1. Solve the equations for the leading variables.

$$
\begin{aligned}
x_1 &= -3x_2 + 2x_3 - 2x_5\\
x_3 &= 1 - 2x_4 - 3x_6\\
x_6 &= \tfrac{1}{3}
\end{aligned}
$$

Step 2. Beginning with the bottom equation and working upward, successively substitute each equation into all the equations above it.
Substituting $x_6 = \tfrac{1}{3}$ into the second equation yield

$$
\begin{aligned}
x_1 &= -3x_2 + 2x_3 - 2x_5\\
x_3 &= -2x_4\\
x_6 &= \tfrac{1}{3}
\end{aligned}
$$

Substituting $x_3 = -2x_4$ into the first equation yields

$$
\begin{aligned}
x_1 &= -3x_2 - 4x_4 - 2x_5\\
x_3 &= -2x_4\\
x_6 &= \tfrac{1}{3}
\end{aligned}
$$

Step 3. Assign arbitrary values to the free variables, if any.

If we now assign x_2, x_4, and x_5 the arbitrary values r, s, and t, respectively, the general solution is given by the formulas

$$
x_1 = -3r - 4s - 2t, \quad x_2 = r, \quad x_3 = -2s, \quad x_4 = s, \quad x_5 = t, \quad x_6 = \tfrac{1}{3}
$$

This agrees with the solution obtained in Example 5.

▶ **EXAMPLE 8**

Suppose that the matrices below are augmented matrices for linear systems in the unknowns x_1, x_2, x_3, and x_4. These matrices are all in row echelon form but not reduced row echelon form. Discuss the existence and uniqueness of solutions to the corresponding linear systems

$$
(a)\begin{bmatrix} 1 & -3 & 7 & 2 & 5\\ 0 & 1 & 2 & -4 & 1\\ 0 & 0 & 1 & 6 & 9\\ 0 & 0 & 0 & 0 & 1 \end{bmatrix}
\quad
(b)\begin{bmatrix} 1 & -3 & 7 & 2 & 5\\ 0 & 1 & 2 & -4 & 1\\ 0 & 0 & 1 & 6 & 9\\ 0 & 0 & 0 & 0 & 0 \end{bmatrix}
\quad
(c)\begin{bmatrix} 1 & -3 & 7 & 2 & 5\\ 0 & 1 & 2 & -4 & 1\\ 0 & 0 & 1 & 6 & 9\\ 0 & 0 & 0 & 1 & 0 \end{bmatrix}
$$

Solution (a) The last row corresponds to the equation

$$ 0x_1 + 0x_2 + 0x_3 + 0x_4 = 1 $$

from which it is evident that the system is inconsistent.

Solution (b) The last row corresponds to the equation

$$ 0x_1 + 0x_2 + 0x_3 + 0x_4 = 0 $$

which has no effect on the solution set. In the remaining three equations the variables x_1, x_2, and x_3 correspond to leading 1's and hence are leading variables. The variable x_4

is a free variable. With a little algebra, the leading variables can be expressed in terms of the free variable, and the free variable can be assigned an arbitrary value. Thus, the system must have infinitely many solutions.

Solution (c) The last row corresponds to the equation

$$x_4 = 0$$

which gives us a numerical value for x_4. If we substitute this value into the third equation, namely,

$$x_3 + 6x_4 = 9$$

we obtain $x_3 = 9$. You should now be able to see that if we continue this process and substitute the known values of x_3 and x_4 into the equation corresponding to the second row, we will obtain a unique numerical value for x_2; and if, finally, we substitute the known values of x_4, x_3, and x_2 into the equation corresponding to the first row, we will produce a unique numerical value for x_1. Thus, the system has a unique solution. ◀

Some Facts About Echelon Forms

There are three facts about row echelon forms and reduced row echelon forms that are important to know but we will not prove:

1. Every matrix has a unique reduced row echelon form; that is, regardless of whether you use Gauss–Jordan elimination or some other sequence of elementary row operations, the same reduced row echelon form will result in the end.[*]

2. Row echelon forms are not unique; that is, different sequences of elementary row operations can result in different row echelon forms.

3. Although row echelon forms are not unique, all row echelon forms of a matrix A have the same number of zero rows, and the leading 1's always occur in the same positions in the row echelon forms of A. Those are called the ***pivot positions*** of A. A column that contains a pivot position is called a ***pivot column*** of A.

▶ **EXAMPLE 9 Pivot Positions and Columns**

Earlier in this section (immediately after Definition 1) we found a row echelon form of

$$A = \begin{bmatrix} 0 & 0 & -2 & 0 & 7 & 12 \\ 2 & 4 & -10 & 6 & 12 & 28 \\ 2 & 4 & -5 & 6 & -5 & -1 \end{bmatrix}$$

to be

$$\begin{bmatrix} 1 & 2 & -5 & 3 & 6 & 14 \\ 0 & 0 & 1 & 0 & -\frac{7}{2} & -6 \\ 0 & 0 & 0 & 0 & 1 & 2 \end{bmatrix}$$

The leading 1's occur in positions (row 1, column 1), (row 2, column 3), and (row 3, column 5). These are the pivot positions. The pivot columns are columns 1, 3, and 5. ◀

Roundoff Error and Instability

There is often a gap between mathematical theory and its practical implementation—Gauss–Jordan elimination and Gaussian elimination being good examples. The problem is that computers generally approximate numbers, thereby introducing ***roundoff*** errors,

---

[*]A proof of this result can be found in the article "The Reduced Row Echelon Form of a Matrix Is Unique: A Simple Proof," by Thomas Yuster, *Mathematics Magazine*, Vol. 57, No. 2, 1984, pp. 93–94.

so unless precautions are taken, successive calculations may degrade an answer to a degree that makes it useless. Algorithms (procedures) in which this happens are called **unstable**. There are various techniques for minimizing roundoff error and instability. For example, it can be shown that for large linear systems Gauss–Jordan elimination involves roughly 50% more operations than Gaussian elimination, so most computer algorithms are based on the latter method. Some of these matters will be considered in Chapter 9.

Concept Review

- Reduced row echelon form
- Row echelon form
- Leading 1
- Leading variables
- Free variables
- General solution to a linear system

- Gaussian elimination
- Gauss–Jordan elimination
- Forward phase
- Backward phase
- Homogeneous linear system
- Trivial solution

- Nontrivial solution
- Dimension Theorem for Homogeneous Systems
- Back-substitution

Skills

- Recognize whether a given matrix is in row echelon form, reduced row echelon form, or neither.
- Construct solutions to linear systems whose corresponding augmented matrices that are in row echelon form or reduced row echelon form.
- Use Gaussian elimination to find the general solution of a linear system.

- Use Gauss–Jordan elimination to find the general solution of a linear system.
- Analyze homogeneous linear systems using the Free Variable Theorem for Homogeneous Systems.

Exercise Set 1.2

1. In each part, determine whether the matrix is in row echelon form, reduced row echelon form, both, or neither.

(a) $\begin{bmatrix} 1 & 0 & 0 \\ 0 & 1 & 0 \\ 0 & 0 & 1 \end{bmatrix}$ (b) $\begin{bmatrix} 1 & 0 & 0 \\ 0 & 1 & 0 \\ 0 & 0 & 0 \end{bmatrix}$ (c) $\begin{bmatrix} 0 & 1 & 0 \\ 0 & 0 & 1 \\ 0 & 0 & 0 \end{bmatrix}$

(d) $\begin{bmatrix} 1 & 0 & 3 & 1 \\ 0 & 1 & 2 & 4 \end{bmatrix}$ (e) $\begin{bmatrix} 1 & 2 & 0 & 3 & 0 \\ 0 & 0 & 1 & 1 & 0 \\ 0 & 0 & 0 & 0 & 1 \\ 0 & 0 & 0 & 0 & 0 \end{bmatrix}$

(f) $\begin{bmatrix} 0 & 0 \\ 0 & 0 \\ 0 & 0 \end{bmatrix}$ (g) $\begin{bmatrix} 1 & -7 & 5 & 5 \\ 0 & 1 & 3 & 2 \end{bmatrix}$

2. In each part, determine whether the matrix is in row echelon form, reduced row echelon form, both, or neither.

(a) $\begin{bmatrix} 1 & 2 & 0 \\ 0 & 1 & 0 \\ 0 & 0 & 0 \end{bmatrix}$ (b) $\begin{bmatrix} 1 & 0 & 0 \\ 0 & 1 & 0 \\ 0 & 2 & 0 \end{bmatrix}$ (c) $\begin{bmatrix} 1 & 3 & 4 \\ 0 & 0 & 1 \\ 0 & 0 & 0 \end{bmatrix}$

(d) $\begin{bmatrix} 1 & 5 & -3 \\ 0 & 1 & 1 \\ 0 & 0 & 0 \end{bmatrix}$ (e) $\begin{bmatrix} 1 & 2 & 3 \\ 0 & 0 & 0 \\ 0 & 0 & 1 \end{bmatrix}$

(f) $\begin{bmatrix} 1 & 2 & 3 & 4 & 5 \\ 1 & 0 & 7 & 1 & 3 \\ 0 & 0 & 0 & 0 & 1 \\ 0 & 0 & 0 & 0 & 0 \end{bmatrix}$ (g) $\begin{bmatrix} 1 & -2 & 0 & 1 \\ 0 & 0 & 1 & -2 \end{bmatrix}$

3. In each part, suppose that the augmented matrix for a system of linear equations has been reduced by row operations to the given row echelon form. Solve the system.

(a) $\begin{bmatrix} 1 & -3 & 4 & 7 \\ 0 & 1 & 2 & 2 \\ 0 & 0 & 1 & 5 \end{bmatrix}$

(b) $\begin{bmatrix} 1 & 0 & 8 & -5 & 6 \\ 0 & 1 & 4 & -9 & 3 \\ 0 & 0 & 1 & 1 & 2 \end{bmatrix}$

(c) $\begin{bmatrix} 1 & 7 & -2 & 0 & -8 & -3 \\ 0 & 0 & 1 & 1 & 6 & 5 \\ 0 & 0 & 0 & 1 & 3 & 9 \\ 0 & 0 & 0 & 0 & 0 & 0 \end{bmatrix}$

(d) $\begin{bmatrix} 1 & -3 & 7 & 1 \\ 0 & 1 & 4 & 0 \\ 0 & 0 & 0 & 1 \end{bmatrix}$

4. In each part, suppose that the augmented matrix for a system of linear equations has been reduced by row operations to the given row echelon form. Solve the system.

(a) $\begin{bmatrix} 1 & 0 & 0 & -3 \\ 0 & 1 & 0 & 0 \\ 0 & 0 & 1 & 7 \end{bmatrix}$

(b) $\begin{bmatrix} 1 & 0 & 0 & -7 & 8 \\ 0 & 1 & 0 & 3 & 2 \\ 0 & 0 & 1 & 1 & -5 \end{bmatrix}$

(c) $\begin{bmatrix} 1 & -6 & 0 & 0 & 3 & -2 \\ 0 & 0 & 1 & 0 & 4 & 7 \\ 0 & 0 & 0 & 1 & 5 & 8 \\ 0 & 0 & 0 & 0 & 0 & 0 \end{bmatrix}$

(d) $\begin{bmatrix} 1 & -3 & 0 & 0 \\ 0 & 0 & 1 & 0 \\ 0 & 0 & 0 & 1 \end{bmatrix}$

▶ In Exercises **5–8**, solve the linear system by Gauss–Jordan elimination. ◀

5. $\begin{aligned} x_1 + x_2 + 2x_3 &= 8 \\ -x_1 - 2x_2 + 3x_3 &= 1 \\ 3x_1 - 7x_2 + 4x_3 &= 10 \end{aligned}$

6. $\begin{aligned} 2x_1 + 2x_2 + 2x_3 &= 0 \\ -2x_1 + 5x_2 + 2x_3 &= 1 \\ 8x_1 + x_2 + 4x_3 &= -1 \end{aligned}$

7. $\begin{aligned} x - y + 2z - w &= -1 \\ 2x + y - 2z - 2w &= -2 \\ -x + 2y - 4z + w &= 1 \\ 3x - 3w &= -3 \end{aligned}$

8. $\begin{aligned} -2b + 3c &= 1 \\ 3a + 6b - 3c &= -2 \\ 6a + 6b + 3c &= 5 \end{aligned}$

▶ In Exercises **9–12**, solve the linear system by Gaussian elimination. ◀

9. Exercise 5

10. Exercise 6

11. Exercise 7

12. Exercise 8

▶ In Exercises **13–16**, determine whether the homogeneous system has nontrivial solutions by inspection (without pencil and paper). ◀

13. $\begin{aligned} 2x_1 - 3x_2 + 4x_3 - x_4 &= 0 \\ 7x_1 + x_2 - 8x_3 + 9x_4 &= 0 \\ 2x_1 + 8x_2 + x_3 - x_4 &= 0 \end{aligned}$

14. $\begin{aligned} x_1 + 3x_2 - x_3 &= 0 \\ x_2 - 8x_3 &= 0 \\ 4x_3 &= 0 \end{aligned}$

15. $\begin{aligned} a_{11}x_1 + a_{12}x_2 + a_{13}x_3 &= 0 \\ a_{21}x_1 + a_{22}x_2 + a_{23}x_3 &= 0 \end{aligned}$

16. $\begin{aligned} 3x_1 - 2x_2 &= 0 \\ 6x_1 - 4x_2 &= 0 \end{aligned}$

▶ In Exercises **17–24**, solve the given linear system by any method. ◀

17. $\begin{aligned} 2x_1 + x_2 + 3x_3 &= 0 \\ x_1 + 2x_2 &= 0 \\ x_2 + x_3 &= 0 \end{aligned}$

18. $\begin{aligned} 2x - y - 3z &= 0 \\ -x + 2y - 3z &= 0 \\ x + y + 4z &= 0 \end{aligned}$

19. $\begin{aligned} 3x_1 + x_2 + x_3 + x_4 &= 0 \\ 5x_1 - x_2 + x_3 - x_4 &= 0 \end{aligned}$

20. $\begin{aligned} v + 3w - 2x &= 0 \\ 2u + v - 4w + 3x &= 0 \\ 2u + 3v + 2w - x &= 0 \\ -4u - 3v + 5w - 4x &= 0 \end{aligned}$

21. $\begin{aligned} 2x + 2y + 4z &= 0 \\ w - y - 3z &= 0 \\ 2w + 3x + y + z &= 0 \\ -2w + x + 3y - 2z &= 0 \end{aligned}$

22. $\begin{aligned} x_1 + 3x_2 + x_4 &= 0 \\ x_1 + 4x_2 + 2x_3 &= 0 \\ -2x_2 - 2x_3 - x_4 &= 0 \\ 2x_1 - 4x_2 + x_3 + x_4 &= 0 \\ x_1 - 2x_2 - x_3 + x_4 &= 0 \end{aligned}$

23. $\begin{aligned} 2I_1 - I_2 + 3I_3 + 4I_4 &= 9 \\ I_1 - 2I_3 + 7I_4 &= 11 \\ 3I_1 - 3I_2 + I_3 + 5I_4 &= 8 \\ 2I_1 + I_2 + 4I_3 + 4I_4 &= 10 \end{aligned}$

24. $\begin{aligned} Z_3 + Z_4 + Z_5 &= 0 \\ -Z_1 - Z_2 + 2Z_3 - 3Z_4 + Z_5 &= 0 \\ Z_1 + Z_2 - 2Z_3 - Z_5 &= 0 \\ 2Z_1 + 2Z_2 - Z_3 + Z_5 &= 0 \end{aligned}$

▶ In Exercises **25–28**, determine the values of a for which the system has no solutions, exactly one solution, or infinitely many solutions. ◀

25. $\begin{aligned} x + 2y - 3z &= 4 \\ 3x - y + 5z &= 2 \\ 4x + y + (a^2 - 14)z &= a + 2 \end{aligned}$

26. $\begin{aligned} x + 2y + z &= 2 \\ 2x - 2y + 3z &= 1 \\ x + 2y - (a^2 - 3)z &= a \end{aligned}$

27. $\begin{aligned} x + 2y &= 1 \\ 2x + (a^2 - 5)y &= a - 1 \end{aligned}$

28. $\begin{aligned} x + y + 7z &= -7 \\ 2x + 3y + 17z &= -16 \\ x + 2y + (a^2 + 1)z &= 3a \end{aligned}$

▶ In Exercises 29–30, solve the following systems, where a, b, and c are constants. ◀

29. $2x + y = a$
    $3x + 6y = b$

30. $x_1 + x_2 + x_3 = a$
    $2x_1 \quad\quad + 2x_3 = b$
    $\quad\quad 3x_2 + 3x_3 = c$

31. Find two different row echelon forms of

$$\begin{bmatrix} 1 & 3 \\ 2 & 7 \end{bmatrix}$$

This exercise shows that a matrix can have multiple row echelon forms.

32. Reduce

$$\begin{bmatrix} 2 & 1 & 3 \\ 0 & -2 & -29 \\ 3 & 4 & 5 \end{bmatrix}$$

to reduced row echelon form without introducing fractions at any intermediate stage.

33. Show that the following nonlinear system has 18 solutions if $0 \le \alpha \le 2\pi, 0 \le \beta \le 2\pi$, and $0 \le \gamma < 2\pi$.

$$\sin \alpha + 2 \cos \beta + 3 \tan \gamma = 0$$
$$2 \sin \alpha + 5 \cos \beta + 3 \tan \gamma = 0$$
$$- \sin \alpha - 5 \cos \beta + 5 \tan \gamma = 0$$

[*Hint:* Begin by making the substitutions $x = \sin \alpha$, $y = \cos \beta$, and $z = \tan \gamma$.]

34. Solve the following system of nonlinear equations for the unknown angles α, β, and γ, where $0 \le \alpha \le 2\pi, 0 \le \beta \le 2\pi$, and $0 \le \gamma < \pi$.

$$2 \sin \alpha - \cos \beta + 3 \tan \gamma = 3$$
$$4 \sin \alpha + 2 \cos \beta - 2 \tan \gamma = 2$$
$$6 \sin \alpha - 3 \cos \beta + \tan \gamma = 9$$

35. Solve the following system of nonlinear equations for x, y, and z.

$$x^2 + y^2 + z^2 = 6$$
$$x^2 - y^2 + 2z^2 = 2$$
$$2x^2 + y^2 - z^2 = 3$$

[*Hint:* Begin by making the substitutions $X = x^2$, $Y = y^2$, $Z = z^2$.]

36. Solve the following system for x, y, and z.

$$\frac{1}{x} + \frac{2}{y} - \frac{4}{z} = 1$$

$$\frac{2}{x} + \frac{3}{y} + \frac{8}{z} = 0$$

$$-\frac{1}{x} + \frac{9}{y} + \frac{10}{z} = 5$$

37. Find the coefficients a, b, c, and d so that the curve shown in the accompanying figure is the graph of the equation $y = ax^3 + bx^2 + cx + d$.

◀ Figure Ex-37

38. Find the coefficients a, b, c, and d so that the curve shown in the accompanying figure is given by the equation $ax^2 + ay^2 + bx + cy + d = 0$.

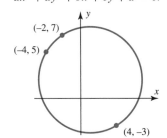

◀ Figure Ex-38

39. If the linear system

$$a_1x + b_1y + c_1z = 0$$
$$a_2x - b_2y + c_2z = 0$$
$$a_3x + b_3y - c_3z = 0$$

has only the trivial solution, what can be said about the solutions of the following system?

$$a_1x + b_1y + c_1z = 3$$
$$a_2x - b_2y + c_2z = 7$$
$$a_3x + b_3y - c_3z = 11$$

40. (a) If A is a 3×5 matrix, then what is the maximum possible number of leading 1's in its reduced row echelon form?

(b) If B is a 3×6 matrix whose last column has all zeros, then what is the maximum possible number of parameters in the general solution of the linear system with augmented matrix B?

(c) If C is a 5×3 matrix, then what is the minimum possible number of rows of zeros in any row echelon form of C?

41. (a) Prove that if $ad - bc \ne 0$, then the reduced row echelon form of

$$\begin{bmatrix} a & b \\ c & d \end{bmatrix} \quad \text{is} \quad \begin{bmatrix} 1 & 0 \\ 0 & 1 \end{bmatrix}$$

(b) Use the result in part (a) to prove that if $ad - bc \ne 0$, then the linear system

$$ax + by = k$$
$$cx + dy = l$$

has exactly one solution.

42. Consider the system of equations

$$ax + by = 0$$
$$cx + dy = 0$$
$$ex + fy = 0$$

Discuss the relative positions of the lines $ax + by = 0$, $cx + dy = 0$, and $ex + fy = 0$ when (a) the system has only the trivial solution, and (b) the system has nontrivial solutions.

43. Describe all possible reduced row echelon forms of

(a) $\begin{bmatrix} a & b & c \\ d & e & f \\ g & h & i \end{bmatrix}$
(b) $\begin{bmatrix} a & b & c & d \\ e & f & g & h \\ i & j & k & l \\ m & n & p & q \end{bmatrix}$

True-False Exercises

In parts (a)–(i) determine whether the statement is true or false, and justify your answer.

(a) If a matrix is in reduced row echelon form, then it is also in row echelon form.

(b) If an elementary row operation is applied to a matrix that is in row echelon form, the resulting matrix will still be in row echelon form.

(c) Every matrix has a unique row echelon form.

(d) A homogeneous linear system in n unknowns whose corresponding augmented matrix has a reduced row echelon form with r leading 1's has $n - r$ free variables.

(e) All leading 1's in a matrix in row echelon form must occur in different columns.

(f) If every column of a matrix in row echelon form has a leading 1 then all entries that are not leading 1's are zero.

(g) If a homogeneous linear system of n equations in n unknowns has a corresponding augmented matrix with a reduced row echelon form containing n leading 1's, then the linear system has only the trivial solution.

(h) If the reduced row echelon form of the augmented matrix for a linear system has a row of zeros, then the system must have infinitely many solutions.

(i) If a linear system has more unknowns than equations, then it must have infinitely many solutions.

1.3 Matrices and Matrix Operations

Rectangular arrays of real numbers arise in contexts other than as augmented matrices for linear systems. In this section we will begin to study matrices as objects in their own right by defining operations of addition, subtraction, and multiplication on them.

Matrix Notation and Terminology

In Section 1.2 we used rectangular arrays of numbers, called *augmented matrices*, to abbreviate systems of linear equations. However, rectangular arrays of numbers occur in other contexts as well. For example, the following rectangular array with three rows and seven columns might describe the number of hours that a student spent studying three subjects during a certain week:

	Mon.	Tues.	Wed.	Thurs.	Fri.	Sat.	Sun.
Math	2	3	2	4	1	4	2
History	0	3	1	4	3	2	2
Language	4	1	3	1	0	0	2

If we suppress the headings, then we are left with the following rectangular array of numbers with three rows and seven columns, called a "matrix":

$$\begin{bmatrix} 2 & 3 & 2 & 4 & 1 & 4 & 2 \\ 0 & 3 & 1 & 4 & 3 & 2 & 2 \\ 4 & 1 & 3 & 1 & 0 & 0 & 2 \end{bmatrix}$$

More generally, we make the following definition.

> **DEFINITION 1** A *matrix* is a rectangular array of numbers. The numbers in the array are called the *entries* in the matrix.

▶ EXAMPLE 1 Examples of Matrices

Some examples of matrices are

$$\begin{bmatrix} 1 & 2 \\ 3 & 0 \\ -1 & 4 \end{bmatrix}, \quad [2 \quad 1 \quad 0 \quad -3], \quad \begin{bmatrix} e & \pi & -\sqrt{2} \\ 0 & \tfrac{1}{2} & 1 \\ 0 & 0 & 0 \end{bmatrix}, \quad \begin{bmatrix} 1 \\ 3 \end{bmatrix}, \quad [4] \quad ◄$$

A matrix with only one column is called a *column vector* or a *column matrix*, and a matrix with only one row is called a *row vector* or a *row matrix*. In Example 1, the 2×1 matrix is a column vector, the 1×4 matrix is a row vector, and the 1×1 matrix is both a row vector and a column vector.

The *size* of a matrix is described in terms of the number of rows (horizontal lines) and columns (vertical lines) it contains. For example, the first matrix in Example 1 has three rows and two columns, so its size is 3 by 2 (written 3×2). In a size description, the first number always denotes the number of rows, and the second denotes the number of columns. The remaining matrices in Example 1 have sizes 1×4, 3×3, 2×1, and 1×1, respectively.

We will use capital letters to denote matrices and lowercase letters to denote numerical quantities; thus we might write

$$A = \begin{bmatrix} 2 & 1 & 7 \\ 3 & 4 & 2 \end{bmatrix} \quad \text{or} \quad C = \begin{bmatrix} a & b & c \\ d & e & f \end{bmatrix}$$

When discussing matrices, it is common to refer to numerical quantities as *scalars*. Unless stated otherwise, *scalars will be real numbers*; complex scalars will be considered later in the text.

The entry that occurs in row i and column j of a matrix A will be denoted by a_{ij}. Thus a general 3×4 matrix might be written as

$$A = \begin{bmatrix} a_{11} & a_{12} & a_{13} & a_{14} \\ a_{21} & a_{22} & a_{23} & a_{24} \\ a_{31} & a_{32} & a_{33} & a_{34} \end{bmatrix}$$

Matrix brackets are often omitted from 1×1 matrices, making it impossible to tell, for example, whether the symbol 4 denotes the number "four" or the matrix [4]. This rarely causes problems because it is usually possible to tell which is meant from the context.

and a general $m \times n$ matrix as

$$A = \begin{bmatrix} a_{11} & a_{12} & \cdots & a_{1n} \\ a_{21} & a_{22} & \cdots & a_{2n} \\ \vdots & \vdots & & \vdots \\ a_{m1} & a_{m2} & \cdots & a_{mn} \end{bmatrix} \tag{1}$$

When a compact notation is desired, the preceding matrix can be written as

$$[a_{ij}]_{m \times n} \quad \text{or} \quad [a_{ij}]$$

the first notation being used when it is important in the discussion to know the size, and the second when the size need not be emphasized. Usually, we will match the letter denoting a matrix with the letter denoting its entries; thus, for a matrix B we would generally use b_{ij} for the entry in row i and column j, and for a matrix C we would use the notation c_{ij}.

The entry in row i and column j of a matrix A is also commonly denoted by the symbol $(A)_{ij}$. Thus, for matrix (1) above, we have

$$(A)_{ij} = a_{ij}$$

and for the matrix

$$A = \begin{bmatrix} 2 & -3 \\ 7 & 0 \end{bmatrix}$$

we have $(A)_{11} = 2$, $(A)_{12} = -3$, $(A)_{21} = 7$, and $(A)_{22} = 0$.

Row and column vectors are of special importance, and it is common practice to denote them by boldface lowercase letters rather than capital letters. For such matrices, double subscripting of the entries is unnecessary. Thus a general $1 \times n$ row vector $\mathbf{a}$ and a general $m \times 1$ column vector $\mathbf{b}$ would be written as

$$\mathbf{a} = \begin{bmatrix} a_1 & a_2 & \cdots & a_n \end{bmatrix} \quad \text{and} \quad \mathbf{b} = \begin{bmatrix} b_1 \\ b_2 \\ \vdots \\ b_m \end{bmatrix}$$

A matrix A with n rows and n columns is called a ***square matrix of order n***, and the shaded entries $a_{11}, a_{22}, \ldots, a_{nn}$ in (2) are said to be on the ***main diagonal*** of A.

$$\begin{bmatrix} a_{11} & a_{12} & \cdots & a_{1n} \\ a_{21} & a_{22} & \cdots & a_{2n} \\ \vdots & \vdots & & \vdots \\ a_{n1} & a_{n2} & \cdots & a_{nn} \end{bmatrix} \tag{2}$$

Operations on Matrices

So far, we have used matrices to abbreviate the work in solving systems of linear equations. For other applications, however, it is desirable to develop an "arithmetic of matrices" in which matrices can be added, subtracted, and multiplied in a useful way. The remainder of this section will be devoted to developing this arithmetic.

DEFINITION 2 Two matrices are defined to be ***equal*** if they have the same size and their corresponding entries are equal.

The equality of two matrices

$$A = [a_{ij}] \quad \text{and} \quad B = [b_{ij}]$$

of the same size can be expressed either by writing

$$(A)_{ij} = (B)_{ij}$$

or by writing

$$a_{ij} = b_{ij}$$

where it is understood that the equalities hold for all values of i and j.

▶ **EXAMPLE 2 Equality of Matrices**

Consider the matrices

$$A = \begin{bmatrix} 2 & 1 \\ 3 & x \end{bmatrix}, \quad B = \begin{bmatrix} 2 & 1 \\ 3 & 5 \end{bmatrix}, \quad C = \begin{bmatrix} 2 & 1 & 0 \\ 3 & 4 & 0 \end{bmatrix}$$

If $x = 5$, then $A = B$, but for all other values of x the matrices A and B are not equal, since not all of their corresponding entries are equal. There is no value of x for which $A = C$ since A and C have different sizes. ◀

DEFINITION 3 If A and B are matrices of the same size, then the ***sum $A + B$*** is the matrix obtained by adding the entries of B to the corresponding entries of A, and the ***difference $A - B$*** is the matrix obtained by subtracting the entries of B from the corresponding entries of A. Matrices of different sizes cannot be added or subtracted.

In matrix notation, if $A = [a_{ij}]$ and $B = [b_{ij}]$ have the same size, then

$$(A + B)_{ij} = (A)_{ij} + (B)_{ij} = a_{ij} + b_{ij} \quad \text{and} \quad (A - B)_{ij} = (A)_{ij} - (B)_{ij} = a_{ij} - b_{ij}$$

▶ **EXAMPLE 3** **Addition and Subtraction**

Consider the matrices

$$
A = \begin{bmatrix} 2 & 1 & 0 & 3 \\ -1 & 0 & 2 & 4 \\ 4 & -2 & 7 & 0 \end{bmatrix}, \quad B = \begin{bmatrix} -4 & 3 & 5 & 1 \\ 2 & 2 & 0 & -1 \\ 3 & 2 & -4 & 5 \end{bmatrix}, \quad C = \begin{bmatrix} 1 & 1 \\ 2 & 2 \end{bmatrix}
$$

Then

$$
A + B = \begin{bmatrix} -2 & 4 & 5 & 4 \\ 1 & 2 & 2 & 3 \\ 7 & 0 & 3 & 5 \end{bmatrix} \quad \text{and} \quad A - B = \begin{bmatrix} 6 & -2 & -5 & 2 \\ -3 & -2 & 2 & 5 \\ 1 & -4 & 11 & -5 \end{bmatrix}
$$

The expressions $A + C$, $B + C$, $A - C$, and $B - C$ are undefined. ◀

DEFINITION 4 If A is any matrix and c is any scalar, then the **product** cA is the matrix obtained by multiplying each entry of the matrix A by c. The matrix cA is said to be a **scalar multiple** of A.

In matrix notation, if $A = [a_{ij}]$, then

$$
(cA)_{ij} = c(A)_{ij} = ca_{ij}
$$

▶ **EXAMPLE 4** **Scalar Multiples**

For the matrices

$$
A = \begin{bmatrix} 2 & 3 & 4 \\ 1 & 3 & 1 \end{bmatrix}, \quad B = \begin{bmatrix} 0 & 2 & 7 \\ -1 & 3 & -5 \end{bmatrix}, \quad C = \begin{bmatrix} 9 & -6 & 3 \\ 3 & 0 & 12 \end{bmatrix}
$$

we have

$$
2A = \begin{bmatrix} 4 & 6 & 8 \\ 2 & 6 & 2 \end{bmatrix}, \quad (-1)B = \begin{bmatrix} 0 & -2 & -7 \\ 1 & -3 & 5 \end{bmatrix}, \quad \tfrac{1}{3}C = \begin{bmatrix} 3 & -2 & 1 \\ 1 & 0 & 4 \end{bmatrix}
$$

It is common practice to denote $(-1)B$ by $-B$. ◀

Thus far we have defined multiplication of a matrix by a scalar but not the multiplication of two matrices. Since matrices are added by adding corresponding entries and subtracted by subtracting corresponding entries, it would seem natural to define multiplication of matrices by multiplying corresponding entries. However, it turns out that such a definition would not be very useful for most problems. Experience has led mathematicians to the following more useful definition of matrix multiplication.

DEFINITION 5 If A is an $m \times r$ matrix and B is an $r \times n$ matrix, then the **product** AB is the $m \times n$ matrix whose entries are determined as follows: To find the entry in row i and column j of AB, single out row i from the matrix A and column j from the matrix B. Multiply the corresponding entries from the row and column together, and then add up the resulting products.

▶ **EXAMPLE 5** **Multiplying Matrices**

Consider the matrices

$$
A = \begin{bmatrix} 1 & 2 & 4 \\ 2 & 6 & 0 \end{bmatrix}, \quad B = \begin{bmatrix} 4 & 1 & 4 & 3 \\ 0 & -1 & 3 & 1 \\ 2 & 7 & 5 & 2 \end{bmatrix}
$$

Since A is a 2×3 matrix and B is a 3×4 matrix, the product AB is a 2×4 matrix. To determine, for example, the entry in row 2 and column 3 of AB, we single out row 2 from A and column 3 from B. Then, as illustrated below, we multiply corresponding entries together and add up these products.

$$\begin{bmatrix} 1 & 2 & 4 \\ 2 & 6 & 0 \end{bmatrix} \begin{bmatrix} 4 & 1 & 4 & 3 \\ 0 & -1 & 3 & 1 \\ 2 & 7 & 5 & 2 \end{bmatrix} = \begin{bmatrix} \square & \square & \square & \square \\ \square & \square & 26 & \square \end{bmatrix}$$

$$(2 \cdot 4) + (6 \cdot 3) + (0 \cdot 5) = 26$$

The entry in row 1 and column 4 of AB is computed as follows:

$$\begin{bmatrix} 1 & 2 & 4 \\ 2 & 6 & 0 \end{bmatrix} \begin{bmatrix} 4 & 1 & 4 & 3 \\ 0 & -1 & 3 & 1 \\ 2 & 7 & 5 & 2 \end{bmatrix} = \begin{bmatrix} \square & \square & \square & 13 \\ \square & \square & \square & \square \end{bmatrix}$$

$$(1 \cdot 3) + (2 \cdot 1) + (4 \cdot 2) = 13$$

The computations for the remaining entries are

$$
\begin{aligned}
(1 \cdot 4) + (2 \cdot 0) + (4 \cdot 2) &= 12 \\
(1 \cdot 1) - (2 \cdot 1) + (4 \cdot 7) &= 27 \\
(1 \cdot 4) + (2 \cdot 3) + (4 \cdot 5) &= 30 \\
(2 \cdot 4) + (6 \cdot 0) + (0 \cdot 2) &= 8 \\
(2 \cdot 1) - (6 \cdot 1) + (0 \cdot 7) &= -4 \\
(2 \cdot 3) + (6 \cdot 1) + (0 \cdot 2) &= 12
\end{aligned}
\qquad
AB = \begin{bmatrix} 12 & 27 & 30 & 13 \\ 8 & -4 & 26 & 12 \end{bmatrix} \blacktriangleleft
$$

The definition of matrix multiplication requires that the number of columns of the first factor A be the same as the number of rows of the second factor B in order to form the product AB. If this condition is not satisfied, the product is undefined. A convenient way to determine whether a product of two matrices is defined is to write down the size of the first factor and, to the right of it, write down the size of the second factor. If, as in (3), the inside numbers are the same, then the product is defined. The outside numbers then give the size of the product.

$$
\begin{array}{ccc}
A & B & AB \\
m \times r & r \times n = & m \times n
\end{array}
$$

Inside

Outside

(3)

**Gotthold Eisenstein
(1823–1852)**

Historical Note The concept of matrix multiplication is due to the German mathematician Gotthold Eisenstein, who introduced the idea around 1844 to simplify the process of making substitutions in linear systems. The idea was then expanded on and formalized by Cayley in his *Memoir on the Theory of Matrices* that was published in 1858. Eisenstein was a pupil of Gauss, who ranked him as the equal of Isaac Newton and Archimedes. However, Eisenstein, suffering from bad health his entire life, died at age 30, so his potential was never realized.

[*Image: wikipedia*]

▶ **EXAMPLE 6 Determining Whether a Product Is Defined**

Suppose that A, B, and C are matrices with the following sizes:

$$\begin{array}{ccc} A & B & C \\ 3 \times 4 & 4 \times 7 & 7 \times 3 \end{array}$$

Then by (3), AB is defined and is a 3×7 matrix; BC is defined and is a 4×3 matrix; and CA is defined and is a 7×4 matrix. The products AC, CB, and BA are all undefined. ◀

In general, if $A = [a_{ij}]$ is an $m \times r$ matrix and $B = [b_{ij}]$ is an $r \times n$ matrix, then, as illustrated by the shading in (4),

$$AB = \begin{bmatrix} a_{11} & a_{12} & \cdots & a_{1r} \\ a_{21} & a_{22} & \cdots & a_{2r} \\ \vdots & \vdots & & \vdots \\ a_{i1} & a_{i2} & \cdots & a_{ir} \\ \vdots & \vdots & & \vdots \\ a_{m1} & a_{m2} & \cdots & a_{mr} \end{bmatrix} \begin{bmatrix} b_{11} & b_{12} & \cdots & b_{1j} & \cdots & b_{1n} \\ b_{21} & b_{22} & \cdots & b_{2j} & \cdots & b_{2n} \\ \vdots & \vdots & & \vdots & & \vdots \\ b_{r1} & b_{r2} & \cdots & b_{rj} & \cdots & b_{rn} \end{bmatrix} \tag{4}$$

the entry $(AB)_{ij}$ in row i and column j of AB is given by

$$(AB)_{ij} = a_{i1}b_{1j} + a_{i2}b_{2j} + a_{i3}b_{3j} + \cdots + a_{ir}b_{rj} \tag{5}$$

Partitioned Matrices A matrix can be subdivided or ***partitioned*** into smaller matrices by inserting horizontal and vertical rules between selected rows and columns. For example, the following are three possible partitions of a general 3×4 matrix A—the first is a partition of A into four ***submatrices*** A_{11}, A_{12}, A_{21}, and A_{22}; the second is a partition of A into its row vectors $\mathbf{r}_1$, $\mathbf{r}_2$, and $\mathbf{r}_3$; and the third is a partition of A into its column vectors $\mathbf{c}_1$, $\mathbf{c}_2$, $\mathbf{c}_3$, and $\mathbf{c}_4$:

$$A = \left[\begin{array}{ccc|c} a_{11} & a_{12} & a_{13} & a_{14} \\ a_{21} & a_{22} & a_{23} & a_{24} \\ \hline a_{31} & a_{32} & a_{33} & a_{34} \end{array}\right] = \begin{bmatrix} A_{11} & A_{12} \\ A_{21} & A_{22} \end{bmatrix}$$

$$A = \left[\begin{array}{cccc} a_{11} & a_{12} & a_{13} & a_{14} \\ \hline a_{21} & a_{22} & a_{23} & a_{24} \\ \hline a_{31} & a_{32} & a_{33} & a_{34} \end{array}\right] = \begin{bmatrix} \mathbf{r}_1 \\ \mathbf{r}_2 \\ \mathbf{r}_3 \end{bmatrix}$$

$$A = \left[\begin{array}{c|c|c|c} a_{11} & a_{12} & a_{13} & a_{14} \\ a_{21} & a_{22} & a_{23} & a_{24} \\ a_{31} & a_{32} & a_{33} & a_{34} \end{array}\right] = [\mathbf{c}_1 \quad \mathbf{c}_2 \quad \mathbf{c}_3 \quad \mathbf{c}_4]$$

Matrix Multiplication by Columns and by Rows Partitioning has many uses, one of which is for finding particular rows or columns of a matrix product AB without computing the entire product. Specifically, the following formulas, whose proofs are left as exercises, show how individual column vectors of AB can be obtained by partitioning B into column vectors and how individual row vectors of AB can be obtained by partitioning A into row vectors.

$$AB = A[\mathbf{b}_1 \quad \mathbf{b}_2 \quad \cdots \quad \mathbf{b}_n] = [A\mathbf{b}_1 \quad A\mathbf{b}_2 \quad \cdots \quad A\mathbf{b}_n] \tag{6}$$

(AB computed column by column)

$$AB = \begin{bmatrix} \mathbf{a}_1 \\ \mathbf{a}_2 \\ \vdots \\ \mathbf{a}_m \end{bmatrix} B = \begin{bmatrix} \mathbf{a}_1 B \\ \mathbf{a}_2 B \\ \vdots \\ \mathbf{a}_m B \end{bmatrix} \tag{7}$$

(AB computed row by row)

In words, these formulas state that

$$j\text{th column vector of } AB = A[\,j\text{th column vector of } B] \tag{8}$$

$$i\text{th row vector of } AB = [i\text{th row vector of } A]B \tag{9}$$

▶ **EXAMPLE 7 Example 5 Revisited**

If A and B are the matrices in Example 5, then from (8) the second column vector of AB can be obtained by the computation

$$\begin{bmatrix} 1 & 2 & 4 \\ 2 & 6 & 0 \end{bmatrix} \begin{bmatrix} 1 \\ -1 \\ 7 \end{bmatrix} = \begin{bmatrix} 27 \\ -4 \end{bmatrix}$$

Second column of B Second column of AB

and from (9) the first row vector of AB can be obtained by the computation

$$[1 \quad 2 \quad 4] \begin{bmatrix} 4 & 1 & 4 & 3 \\ 0 & -1 & 3 & 1 \\ 2 & 7 & 5 & 2 \end{bmatrix} = [12 \quad 27 \quad 30 \quad 13] \quad ◀$$

First row of A First row of AB

Matrix Products as Linear Combinations   We have discussed three methods for computing a matrix product AB—entry by entry, column by column, and row by row. The following definition provides yet another way of thinking about matrix multiplication.

DEFINITION 6 If $A_1, A_2, \ldots, A_r$ are matrices of the same size, and if $c_1, c_2, \ldots, c_r$ are scalars, then an expression of the form

$$c_1 A_1 + c_2 A_2 + \cdots + c_r A_r$$

is called a ***linear combination*** of $A_1, A_2, \ldots, A_r$ with ***coefficients*** $c_1, c_2, \ldots, c_r$.

To see how matrix products can be viewed as linear combinations, let A be an $m \times n$ matrix and $\mathbf{x}$ an $n \times 1$ column vector, say

$$A = \begin{bmatrix} a_{11} & a_{12} & \cdots & a_{1n} \\ a_{21} & a_{22} & \cdots & a_{2n} \\ \vdots & \vdots & & \vdots \\ a_{m1} & a_{m2} & \cdots & a_{mn} \end{bmatrix} \quad \text{and} \quad \mathbf{x} = \begin{bmatrix} x_1 \\ x_2 \\ \vdots \\ x_n \end{bmatrix}$$

Then

$$A\mathbf{x} = \begin{bmatrix} a_{11}x_1 + a_{12}x_2 + \cdots + a_{1n}x_n \\ a_{21}x_1 + a_{22}x_2 + \cdots + a_{2n}x_n \\ \vdots & \vdots & & \vdots \\ a_{m1}x_1 + a_{m2}x_2 + \cdots + a_{mn}x_n \end{bmatrix} = x_1 \begin{bmatrix} a_{11} \\ a_{21} \\ \vdots \\ a_{m1} \end{bmatrix} + x_2 \begin{bmatrix} a_{12} \\ a_{22} \\ \vdots \\ a_{m2} \end{bmatrix} + \cdots + x_n \begin{bmatrix} a_{1n} \\ a_{2n} \\ \vdots \\ a_{mn} \end{bmatrix}$$

$$(10)$$

This proves the following theorem.

THEOREM 1.3.1 *If A is an $m \times n$ matrix, and if $\mathbf{x}$ is an $n \times 1$ column vector, then the product $A\mathbf{x}$ can be expressed as a linear combination of the column vectors of A in which the coefficients are the entries of $\mathbf{x}$.*

▶ **EXAMPLE 8** Matrix Products as Linear Combinations

The matrix product

$$\begin{bmatrix} -1 & 3 & 2 \\ 1 & 2 & -3 \\ 2 & 1 & -2 \end{bmatrix} \begin{bmatrix} 2 \\ -1 \\ 3 \end{bmatrix} = \begin{bmatrix} 1 \\ -9 \\ -3 \end{bmatrix}$$

can be written as the following linear combination of column vectors

$$2 \begin{bmatrix} -1 \\ 1 \\ 2 \end{bmatrix} - 1 \begin{bmatrix} 3 \\ 2 \\ 1 \end{bmatrix} + 3 \begin{bmatrix} 2 \\ -3 \\ -2 \end{bmatrix} = \begin{bmatrix} 1 \\ -9 \\ -3 \end{bmatrix}$$

▶ **EXAMPLE 9** Columns of a Product *AB* as Linear Combinations

We showed in Example 5 that

$$AB = \begin{bmatrix} 1 & 2 & 4 \\ 2 & 6 & 0 \end{bmatrix} \begin{bmatrix} 4 & 1 & 4 & 3 \\ 0 & -1 & 3 & 1 \\ 2 & 7 & 5 & 2 \end{bmatrix} = \begin{bmatrix} 12 & 27 & 30 & 13 \\ 8 & -4 & 26 & 12 \end{bmatrix}$$

It follows from Formula (6) and Theorem 1.3.1 that the jth column vector of AB can be expressed as a linear combination of the column vectors of A in which the coefficients in the linear combination are the entries from the jth column of B. The computations are as follows:

$$\begin{bmatrix} 12 \\ 8 \end{bmatrix} = 4\begin{bmatrix} 1 \\ 2 \end{bmatrix} + 0\begin{bmatrix} 2 \\ 6 \end{bmatrix} + 2\begin{bmatrix} 4 \\ 0 \end{bmatrix}$$

$$\begin{bmatrix} 27 \\ -4 \end{bmatrix} = \begin{bmatrix} 1 \\ 2 \end{bmatrix} - \begin{bmatrix} 2 \\ 6 \end{bmatrix} + 7\begin{bmatrix} 4 \\ 0 \end{bmatrix}$$

$$\begin{bmatrix} 30 \\ 26 \end{bmatrix} = 4\begin{bmatrix} 1 \\ 2 \end{bmatrix} + 3\begin{bmatrix} 2 \\ 6 \end{bmatrix} + 5\begin{bmatrix} 4 \\ 0 \end{bmatrix}$$

$$\begin{bmatrix} 13 \\ 12 \end{bmatrix} = 3\begin{bmatrix} 1 \\ 2 \end{bmatrix} + \begin{bmatrix} 2 \\ 6 \end{bmatrix} + 2\begin{bmatrix} 4 \\ 0 \end{bmatrix} \quad \blacktriangleleft$$

Matrix Form of a Linear System

Matrix multiplication has an important application to systems of linear equations. Consider a system of m linear equations in n unknowns:

$$\begin{aligned}
a_{11}x_1 + a_{12}x_2 + \cdots + a_{1n}x_n &= b_1 \\
a_{21}x_1 + a_{22}x_2 + \cdots + a_{2n}x_n &= b_2 \\
\vdots \qquad \vdots \qquad \qquad \vdots \quad &\;\; \vdots \\
a_{m1}x_1 + a_{m2}x_2 + \cdots + a_{mn}x_n &= b_m
\end{aligned}$$

Since two matrices are equal if and only if their corresponding entries are equal, we can replace the m equations in this system by the single matrix equation

$$\begin{bmatrix} a_{11}x_1 + a_{12}x_2 + \cdots + a_{1n}x_n \\ a_{21}x_1 + a_{22}x_2 + \cdots + a_{2n}x_n \\ \vdots \qquad \vdots \qquad \qquad \vdots \\ a_{m1}x_1 + a_{m2}x_2 + \cdots + a_{mn}x_n \end{bmatrix} = \begin{bmatrix} b_1 \\ b_2 \\ \vdots \\ b_m \end{bmatrix}$$

The $m \times 1$ matrix on the left side of this equation can be written as a product to give

$$\begin{bmatrix} a_{11} & a_{12} & \cdots & a_{1n} \\ a_{21} & a_{22} & \cdots & a_{2n} \\ \vdots & \vdots & & \vdots \\ a_{m1} & a_{m2} & \cdots & a_{mn} \end{bmatrix} \begin{bmatrix} x_1 \\ x_2 \\ \vdots \\ x_n \end{bmatrix} = \begin{bmatrix} b_1 \\ b_2 \\ \vdots \\ b_m \end{bmatrix}$$

If we designate these matrices by A, $\mathbf{x}$, and $\mathbf{b}$, respectively, then we can replace the original system of m equations in n unknowns by the single matrix equation

$$A\mathbf{x} = \mathbf{b}$$

The matrix A in this equation is called the ***coefficient matrix*** of the system. The ***augmented matrix*** for the system is obtained by adjoining $\mathbf{b}$ to A as the last column; thus the augmented matrix is

The vertical bar in $[A \mid \mathbf{b}]$ is a convenient way to separate A from $\mathbf{b}$ visually; it has no mathematical significance.

$$[A \mid \mathbf{b}] = \left[\begin{array}{cccc|c} a_{11} & a_{12} & \cdots & a_{1n} & b_1 \\ a_{21} & a_{22} & \cdots & a_{2n} & b_2 \\ \vdots & \vdots & & \vdots & \vdots \\ a_{m1} & a_{m2} & \cdots & a_{mn} & b_m \end{array}\right]$$

Transpose of a Matrix

We conclude this section by defining two matrix operations that have no analogs in the arithmetic of real numbers.

> **DEFINITION 7** If A is any $m \times n$ matrix, then the ***transpose of A***, denoted by A^T, is defined to be the $n \times m$ matrix that results by interchanging the rows and columns of A; that is, the first column of A^T is the first row of A, the second column of A^T is the second row of A, and so forth.

▶ **EXAMPLE 10 Some Transposes**

The following are some examples of matrices and their transposes.

$$A = \begin{bmatrix} a_{11} & a_{12} & a_{13} & a_{14} \\ a_{21} & a_{22} & a_{23} & a_{24} \\ a_{31} & a_{32} & a_{33} & a_{34} \end{bmatrix}, \quad B = \begin{bmatrix} 2 & 3 \\ 1 & 4 \\ 5 & 6 \end{bmatrix}, \quad C = [1 \quad 3 \quad 5], \quad D = [4]$$

$$A^T = \begin{bmatrix} a_{11} & a_{21} & a_{31} \\ a_{12} & a_{22} & a_{32} \\ a_{13} & a_{23} & a_{33} \\ a_{14} & a_{24} & a_{34} \end{bmatrix}, \quad B^T = \begin{bmatrix} 2 & 1 & 5 \\ 3 & 4 & 6 \end{bmatrix}, \quad C^T = \begin{bmatrix} 1 \\ 3 \\ 5 \end{bmatrix}, \quad D^T = [4] \quad ◀$$

Observe that not only are the columns of A^T the rows of A, but the rows of A^T are the columns of A. Thus the entry in row i and column j of A^T is the entry in row j and column i of A; that is,

$$(A^T)_{ij} = (A)_{ji} \tag{11}$$

Note the reversal of the subscripts.

In the special case where A is a square matrix, the transpose of A can be obtained by interchanging entries that are symmetrically positioned about the main diagonal. In (12) we see that A^T can also be obtained by "reflecting" A about its main diagonal.

$$A = \begin{bmatrix} 1 & -2 & 4 \\ 3 & 7 & 0 \\ -5 & 8 & 6 \end{bmatrix} \rightarrow \begin{bmatrix} 1 & -2 & 4 \\ 3 & 7 & 0 \\ -5 & 8 & 6 \end{bmatrix} \rightarrow A^T = \begin{bmatrix} 1 & 3 & -5 \\ -2 & 7 & 8 \\ 4 & 0 & 6 \end{bmatrix} \tag{12}$$

Interchange entries that are symmetrically positioned about the main diagonal.

James Sylvester
(1814–1897)

Arthur Cayley
(1821–1895)

Historical Note The term *matrix* was first used by the English mathematician James Sylvester, who defined the term in 1850 to be an "oblong arrangement of terms." Sylvester communicated his work on matrices to a fellow English mathematician and lawyer named Arthur Cayley, who then introduced some of the basic operations on matrices in a book entitled *Memoir on the Theory of Matrices* that was published in 1858. As a matter of interest, Sylvester, who was Jewish, did not get his college degree because he refused to sign a required oath to the Church of England. He was appointed to a chair at the University of Virginia in the United States but resigned after swatting a student with a stick because he was reading a newspaper in class. Sylvester, thinking he had killed the student, fled back to England on the first available ship. Fortunately, the student was not dead, just in shock!

[*Images: The Granger Collection, New York*]

> **DEFINITION 8** If A is a square matrix, then the ***trace of A***, denoted by tr(A), is defined to be the sum of the entries on the main diagonal of A. The trace of A is undefined if A is not a square matrix.

▶ **EXAMPLE 11 Trace of a Matrix**

The following are examples of matrices and their traces.

$$A = \begin{bmatrix} a_{11} & a_{12} & a_{13} \\ a_{21} & a_{22} & a_{23} \\ a_{31} & a_{32} & a_{33} \end{bmatrix}, \quad B = \begin{bmatrix} -1 & 2 & 7 & 0 \\ 3 & 5 & -8 & 4 \\ 1 & 2 & 7 & -3 \\ 4 & -2 & 1 & 0 \end{bmatrix}$$

$$\text{tr}(A) = a_{11} + a_{22} + a_{33} \qquad \text{tr}(B) = -1 + 5 + 7 + 0 = 11 \quad ◀$$

In the exercises you will have some practice working with the transpose and trace operations.

Concept Review

- Matrix
- Entries
- Column vector (or column matrix)
- Row vector (or row matrix)
- Square matrix
- Main diagonal
- Equal matrices

- Matrix operations: sum, difference, scalar multiplication
- Linear combination of matrices
- Product of matrices (matrix multiplication)
- Partitioned matrices
- Submatrices

- Row-column method
- Column method
- Row method
- Coefficient matrix of a linear system
- Transpose
- Trace

Skills

- Determine the size of a given matrix.
- Identify the row vectors and column vectors of a given matrix.
- Perform the arithmetic operations of matrix addition, subtraction, scalar multiplication, and multiplication.
- Determine whether the product of two given matrices is defined.
- Compute matrix products using the row-column method, the column method, and the row method.

- Express the product of a matrix and a column vector as a linear combination of the columns of the matrix.
- Express a linear system as a matrix equation, and identify the coefficient matrix.
- Compute the transpose of a matrix.
- Compute the trace of a square matrix.

Exercise Set 1.3

1. Suppose that A, B, C, D, and E are matrices with the following sizes:

$$\begin{array}{ccccc} A & B & C & D & E \\ (4 \times 5) & (4 \times 5) & (5 \times 2) & (4 \times 2) & (5 \times 4) \end{array}$$

In each part, determine whether the given matrix expression is defined. For those that are defined, give the size of the resulting matrix.

(a) BA

(b) $AC + D$

(c) $AE + B$

(d) $AB + B$

(e) $E(A + B)$

(f) $E(AC)$

(g) $E^T A$

(h) $(A^T + E)D$

2. Suppose that A, B, C, D, and E are matrices with the following sizes:

$$
\begin{array}{ccccc}
A & B & C & D & E \\
(3 \times 1) & (3 \times 6) & (6 \times 2) & (2 \times 6) & (1 \times 3)
\end{array}
$$

In each part, determine whether the given matrix expression is defined. For those that are defined, give the size of the resulting matrix.

(a) EA (b) AB^T (c) $B^T(A + E^T)$

(d) $2A + C$ (e) $(C^T + D)B^T$ (f) $CD + B^T E^T$

(g) $(BD^T)C^T$ (h) $DC + EA$

3. Consider the matrices

$$
A = \begin{bmatrix} 3 & 0 \\ -1 & 2 \\ 1 & 1 \end{bmatrix}, \quad
B = \begin{bmatrix} 4 & -1 \\ 0 & 2 \end{bmatrix}, \quad
C = \begin{bmatrix} 1 & 4 & 2 \\ 3 & 1 & 5 \end{bmatrix},
$$

$$
D = \begin{bmatrix} 1 & 5 & 2 \\ -1 & 0 & 1 \\ 3 & 2 & 4 \end{bmatrix}, \quad
E = \begin{bmatrix} 6 & 1 & 3 \\ -1 & 1 & 2 \\ 4 & 1 & 3 \end{bmatrix}
$$

In each part, compute the given expression (where possible).

(a) $D + E$ (b) $D - E$ (c) $5A$

(d) $-7C$ (e) $2B - C$ (f) $4E - 2D$

(g) $-3(D + 2E)$ (h) $A - A$ (i) $\operatorname{tr}(D)$

(j) $\operatorname{tr}(D - 3E)$ (k) $4\operatorname{tr}(7B)$ (l) $\operatorname{tr}(A)$

4. Using the matrices in Exercise 3, in each part compute the given expression (where possible).

(a) $2A^T + C$ (b) $D^T - E^T$ (c) $(D - E)^T$

(d) $B^T + 5C^T$ (e) $\frac{1}{2}C^T - \frac{1}{4}A$ (f) $B - B^T$

(g) $2E^T - 3D^T$ (h) $(2E^T - 3D^T)^T$ (i) $(CD)E$

(j) $C(BA)$ (k) $\operatorname{tr}(DE^T)$ (l) $\operatorname{tr}(BC)$

5. Using the matrices in Exercise 3, in each part compute the given expression (where possible).

(a) AB (b) BA (c) $(3E)D$

(d) $(AB)C$ (e) $A(BC)$ (f) CC^T

(g) $(DA)^T$ (h) $(C^TB)A^T$ (i) $\operatorname{tr}(DD^T)$

(j) $\operatorname{tr}(4E^T - D)$ (k) $\operatorname{tr}(C^TA^T + 2E^T)$ (l) $\operatorname{tr}((EC^T)^TA)$

6. Using the matrices in Exercise 3, in each part compute the given expression (where possible).

(a) $(2D^T - E)A$ (b) $(4B)C + 2B$

(c) $(-AC)^T + 5D^T$ (d) $(BA^T - 2C)^T$

(e) $B^T(CC^T - A^TA)$ (f) $D^TE^T - (ED)^T$

7. Let

$$
A = \begin{bmatrix} 3 & -2 & 7 \\ 6 & 5 & 4 \\ 0 & 4 & 9 \end{bmatrix} \quad \text{and} \quad
B = \begin{bmatrix} 6 & -2 & 4 \\ 0 & 1 & 3 \\ 7 & 7 & 5 \end{bmatrix}
$$

Use the row method or column method (as appropriate) to find

(a) the first row of AB. (b) the third row of AB.

(c) the second column of AB. (d) the first column of BA.

(e) the third row of AA. (f) the third column of AA.

8. Referring to the matrices in Exercise 7, use the row method or column method (as appropriate) to find

(a) the first column of AB. (b) the third column of BB.

(c) the second row of BB. (d) the first column of AA.

(e) the third column of AB. (f) the first row of BA.

9. Referring to the matrices in Exercise 7 and Example 9,

(a) express each column vector of AA as a linear combination of the column vectors of A.

(b) express each column vector of BB as a linear combination of the column vectors of B.

10. Referring to the matrices in Exercise 7 and Example 9,

(a) express each column vector of AB as a linear combination of the column vectors of A.

(b) express each column vector of BA as a linear combination of the column vectors of B.

11. In each part, find matrices A, $\mathbf{x}$, and $\mathbf{b}$ that express the given system of linear equations as a single matrix equation $A\mathbf{x} = \mathbf{b}$, and write out this matrix equation.

(a) $\begin{aligned} 2x_1 - 3x_2 + 5x_3 &= 7 \\ 9x_1 - x_2 + x_3 &= -1 \\ x_1 + 5x_2 + 4x_3 &= 0 \end{aligned}$

(b) $\begin{aligned} 4x_1 \quad\quad - 3x_3 + x_4 &= 1 \\ 5x_1 + x_2 \quad\quad - 8x_4 &= 3 \\ 2x_1 - 5x_2 + 9x_3 - x_4 &= 0 \\ 3x_2 - x_3 + 7x_4 &= 2 \end{aligned}$

12. In each part, find matrices A, $\mathbf{x}$, and $\mathbf{b}$ that express the given system of linear equations as a single matrix equation $A\mathbf{x} = \mathbf{b}$, and write out this matrix equation.

(a) $\begin{aligned} x_1 - 2x_2 + 3x_3 &= -3 \\ 2x_1 + x_2 \quad\quad &= 0 \\ - 3x_2 + 4x_3 &= 1 \\ x_1 \quad\quad + x_3 &= 5 \end{aligned}$

(b) $\begin{aligned} 3x_1 + 3x_2 + 3x_3 &= -3 \\ -x_1 - 5x_2 - 2x_3 &= 3 \\ - 4x_2 + x_3 &= 0 \end{aligned}$

13. In each part, express the matrix equation as a system of linear equations.

(a) $\begin{bmatrix} 5 & 6 & -7 \\ -1 & -2 & 3 \\ 0 & 4 & -1 \end{bmatrix} \begin{bmatrix} x_1 \\ x_2 \\ x_3 \end{bmatrix} = \begin{bmatrix} 2 \\ 0 \\ 3 \end{bmatrix}$

(b) $\begin{bmatrix} 1 & 1 & 1 \\ 2 & 3 & 0 \\ 5 & -3 & -6 \end{bmatrix} \begin{bmatrix} x_1 \\ x_2 \\ x_3 \end{bmatrix} = \begin{bmatrix} 2 \\ 2 \\ -9 \end{bmatrix}$

14. In each part, express the matrix equation as a system of linear equations.

(a) $\begin{bmatrix} 3 & -1 & 2 \\ 4 & 3 & 7 \\ -2 & 1 & 5 \end{bmatrix} \begin{bmatrix} x_1 \\ x_2 \\ x_3 \end{bmatrix} = \begin{bmatrix} 2 \\ -1 \\ 4 \end{bmatrix}$

(b) $\begin{bmatrix} 3 & -2 & 0 & 1 \\ 5 & 0 & 2 & -2 \\ 3 & 1 & 4 & 7 \\ -2 & 5 & 1 & 6 \end{bmatrix} \begin{bmatrix} w \\ x \\ y \\ z \end{bmatrix} = \begin{bmatrix} 0 \\ 0 \\ 0 \\ 0 \end{bmatrix}$

▶ In Exercises **15–16**, find all values of k, if any, that satisfy the equation. ◀

15. $\begin{bmatrix} k & 1 & 1 \end{bmatrix} \begin{bmatrix} 1 & 1 & 0 \\ 1 & 0 & 2 \\ 0 & 2 & -3 \end{bmatrix} \begin{bmatrix} k \\ 1 \\ 1 \end{bmatrix} = 0$

16. $\begin{bmatrix} 2 & 2 & k \end{bmatrix} \begin{bmatrix} 1 & 2 & 0 \\ 2 & 0 & 3 \\ 0 & 3 & 1 \end{bmatrix} \begin{bmatrix} 2 \\ 2 \\ k \end{bmatrix} = 0$

▶ In Exercises **17–18**, solve the matrix equation for a, b, c, and d.

17. $\begin{bmatrix} a & 3 \\ -1 & a+b \end{bmatrix} = \begin{bmatrix} 4 & d-2c \\ d+2c & -2 \end{bmatrix}$

18. $\begin{bmatrix} a-b & b+a \\ 3d+c & 2d-c \end{bmatrix} = \begin{bmatrix} 8 & 1 \\ 7 & 6 \end{bmatrix}$

19. Let A be any $m \times n$ matrix and let 0 be the $m \times n$ matrix each of whose entries is zero. Show that if $kA = 0$, then $k = 0$ or $A = 0$.

20. (a) Show that if AB and BA are both defined, then AB and BA are square matrices.

(b) Show that if A is an $m \times n$ matrix and $A(BA)$ is defined, then B is an $n \times m$ matrix.

21. Prove: If A and B are $n \times n$ matrices, then

$$\operatorname{tr}(A + B) = \operatorname{tr}(A) + \operatorname{tr}(B)$$

22. (a) Show that if A has a row of zeros and B is any matrix for which AB is defined, then AB also has a row of zeros.

(b) Find a similar result involving a column of zeros.

23. In each part, find a 6×6 matrix $[a_{ij}]$ that satisfies the stated condition. Make your answers as general as possible by using letters rather than specific numbers for the nonzero entries.

(a) $a_{ij} = 0$ if $i \neq j$ (b) $a_{ij} = 0$ if $i > j$

(c) $a_{ij} = 0$ if $i < j$

(d) $a_{ij} = 0$ if $|i - j| > 1$

24. Find the 4×4 matrix $A = [a_{ij}]$ whose entries satisfy the stated condition.

(a) $a_{ij} = i + j$ (b) $a_{ij} = i^{j-1}$

(c) $a_{ij} = \begin{cases} 1 & \text{if} \quad |i - j| > 1 \\ -1 & \text{if} \quad |i - j| \leq 1 \end{cases}$

25. Consider the function $y = f(x)$ defined for 2×1 matrices x by $y = Ax$, where

$$A = \begin{bmatrix} 1 & 1 \\ 0 & 1 \end{bmatrix}$$

Plot $f(x)$ together with x in each case below. How would you describe the action of f?

(a) $x = \begin{pmatrix} 1 \\ 1 \end{pmatrix}$ (b) $x = \begin{pmatrix} 2 \\ 0 \end{pmatrix}$

(c) $x = \begin{pmatrix} 4 \\ 3 \end{pmatrix}$ (d) $x = \begin{pmatrix} 2 \\ -2 \end{pmatrix}$

26. Let I be the $n \times n$ matrix whose entry in row i and column j is

$$\begin{cases} 1 & \text{if} \quad i = j \\ 0 & \text{if} \quad i \neq j \end{cases}$$

Show that $AI = IA = A$ for every $n \times n$ matrix A.

27. How many 3×3 matrices A can you find such that

$$A \begin{bmatrix} x \\ y \\ z \end{bmatrix} = \begin{bmatrix} x+y \\ x-y \\ 0 \end{bmatrix}$$

for all choices of x, y, and z?

28. How many 3×3 matrices A can you find such that

$$A \begin{bmatrix} x \\ y \\ z \end{bmatrix} = \begin{bmatrix} xy \\ 0 \\ 0 \end{bmatrix}$$

for all choices of x, y, and z?

29. A matrix B is said to be a *square root* of a matrix A if $BB = A$.

(a) Find two square roots of $A = \begin{bmatrix} 2 & 2 \\ 2 & 2 \end{bmatrix}$.

(b) How many different square roots can you find of
$$A = \begin{bmatrix} 5 & 0 \\ 0 & 9 \end{bmatrix}?$$

(c) Do you think that every 2×2 matrix has at least one square root? Explain your reasoning.

30. Let 0 denote a 2×2 matrix, each of whose entries is zero.

(a) Is there a 2×2 matrix A such that $A \neq 0$ and $AA = 0$? Justify your answer.

(b) Is there a 2×2 matrix A such that $A \neq 0$ and $AA = A$? Justify your answer.

True-False Exercises

In parts (a)–(o) determine whether the statement is true or false, and justify your answer.

(a) The matrix $\begin{bmatrix} 1 & 2 & 3 \\ 4 & 5 & 6 \end{bmatrix}$ has no main diagonal.

(b) An $m \times n$ matrix has m column vectors and n row vectors.

(c) If A and B are 2×2 matrices, then $AB = BA$.

(d) The ith row vector of a matrix product AB can be computed by multiplying A by the ith row vector of B.

(e) For every matrix A, it is true that $(A^T)^T = A$.

(f) If A and B are square matrices of the same order, then $\text{tr}(AB) = \text{tr}(A)\text{tr}(B)$.

(g) If A and B are square matrices of the same order, then $(AB)^T = A^T B^T$.

(h) For every square matrix A, it is true that $\text{tr}(A^T) = \text{tr}(A)$.

(i) If A is a 6×4 matrix and B is an $m \times n$ matrix such that $B^T A^T$ is a 2×6 matrix, then $m = 4$ and $n = 2$.

(j) If A is an $n \times n$ matrix and c is a scalar, then $\text{tr}(cA) = c\,\text{tr}(A)$.

(k) If A, B, and C are matrices of the same size such that $A - C = B - C$, then $A = B$.

(l) If A, B, and C are square matrices of the same order such that $AC = BC$, then $A = B$.

(m) If $AB + BA$ is defined, then A and B are square matrices of the same size.

(n) If B has a column of zeros, then so does AB if this product is defined.

(o) If B has a column of zeros, then so does BA if this product is defined.

1.4 Inverses; Algebraic Properties of Matrices

In this section we will discuss some of the algebraic properties of matrix operations. We will see that many of the basic rules of arithmetic for real numbers hold for matrices, but we will also see that some do not.

Properties of Matrix Addition and Scalar Multiplication

The following theorem lists the basic algebraic properties of the matrix operations.

THEOREM 1.4.1 Properties of Matrix Arithmetic

Assuming that the sizes of the matrices are such that the indicated operations can be performed, the following rules of matrix arithmetic are valid.

(a) $A + B = B + A$ **(Commutative law for addition)**

(b) $A + (B + C) = (A + B) + C$ **(Associative law for addition)**

(c) $A(BC) = (AB)C$ **(Associative law for multiplication)**

(d) $A(B + C) = AB + AC$ **(Left distributive law)**

(e) $(B + C)A = BA + CA$ **(Right distributive law)**

(f) $A(B - C) = AB - AC$

(g) $(B - C)A = BA - CA$

(h) $a(B + C) = aB + aC$

(i) $a(B - C) = aB - aC$

(j) $(a + b)C = aC + bC$

(k) $(a - b)C = aC - bC$

(l) $a(bC) = (ab)C$

(m) $a(BC) = (aB)C = B(aC)$

To prove any of the equalities in this theorem we must show that the matrix on the left side has the same size as that on the right and that the corresponding entries on the two sides are the same. Most of the proofs follow the same pattern, so we will prove part (*d*) as a sample. The proof of the associative law for multiplication is more complicated than the rest and is outlined in the exercises.

Proof (d) We must show that $A(B + C)$ and $AB + AC$ have the same size and that corresponding entries are equal. To form $A(B + C)$, the matrices B and C must have the same size, say $m \times n$, and the matrix A must then have m columns, so its size must be of the form $r \times m$. This makes $A(B + C)$ an $r \times n$ matrix. It follows that $AB + AC$ is also an $r \times n$ matrix and, consequently, $A(B + C)$ and $AB + AC$ have the same size.

Suppose that $A = [a_{ij}]$, $B = [b_{ij}]$, and $C = [c_{ij}]$. We want to show that corresponding entries of $A(B + C)$ and $AB + AC$ are equal; that is,

$$[A(B + C)]_{ij} = [AB + AC]_{ij}$$

for all values of i and j. But from the definitions of matrix addition and matrix multiplication, we have

$$[A(B + C)]_{ij} = a_{i1}(b_{1j} + c_{1j}) + a_{i2}(b_{2j} + c_{2j}) + \cdots + a_{im}(b_{mj} + c_{mj})$$
$$= (a_{i1}b_{1j} + a_{i2}b_{2j} + \cdots + a_{im}b_{mj}) + (a_{i1}c_{1j} + a_{i2}c_{2j} + \cdots + a_{im}c_{mj})$$
$$= [AB]_{ij} + [AC]_{ij} = [AB + AC]_{ij} \quad \blacktriangleleft$$

> There are three basic ways to prove that two matrices of the same size are equal—prove that corresponding entries are the same, prove that corresponding row vectors are the same, or prove that corresponding column vectors are the same.

Remark Although the operations of matrix addition and matrix multiplication were defined for pairs of matrices, associative laws (*b*) and (*c*) enable us to denote sums and products of three matrices as $A + B + C$ and ABC without inserting any parentheses. This is justified by the fact that no matter how parentheses are inserted, the associative laws guarantee that the same end result will be obtained. In general, *given any sum or any product of matrices, pairs of parentheses can be inserted or deleted anywhere within the expression without affecting the end result.*

▶ **EXAMPLE 1 Associativity of Matrix Multiplication**

As an illustration of the associative law for matrix multiplication, consider

$$A = \begin{bmatrix} 1 & 2 \\ 3 & 4 \\ 0 & 1 \end{bmatrix}, \quad B = \begin{bmatrix} 4 & 3 \\ 2 & 1 \end{bmatrix}, \quad C = \begin{bmatrix} 1 & 0 \\ 2 & 3 \end{bmatrix}$$

Then

$$AB = \begin{bmatrix} 1 & 2 \\ 3 & 4 \\ 0 & 1 \end{bmatrix}\begin{bmatrix} 4 & 3 \\ 2 & 1 \end{bmatrix} = \begin{bmatrix} 8 & 5 \\ 20 & 13 \\ 2 & 1 \end{bmatrix} \quad \text{and} \quad BC = \begin{bmatrix} 4 & 3 \\ 2 & 1 \end{bmatrix}\begin{bmatrix} 1 & 0 \\ 2 & 3 \end{bmatrix} = \begin{bmatrix} 10 & 9 \\ 4 & 3 \end{bmatrix}$$

Thus

$$(AB)C = \begin{bmatrix} 8 & 5 \\ 20 & 13 \\ 2 & 1 \end{bmatrix}\begin{bmatrix} 1 & 0 \\ 2 & 3 \end{bmatrix} = \begin{bmatrix} 18 & 15 \\ 46 & 39 \\ 4 & 3 \end{bmatrix}$$

and

$$A(BC) = \begin{bmatrix} 1 & 2 \\ 3 & 4 \\ 0 & 1 \end{bmatrix}\begin{bmatrix} 10 & 9 \\ 4 & 3 \end{bmatrix} = \begin{bmatrix} 18 & 15 \\ 46 & 39 \\ 4 & 3 \end{bmatrix}$$

so $(AB)C = A(BC)$, as guaranteed by Theorem 1.4.1(*c*). ◀

Properties of Matrix Multiplication

Do not let Theorem 1.4.1 lull you into believing that *all* laws of real arithmetic carry over to matrix arithmetic. For example, you know that in real arithmetic it is always true that $ab = ba$, which is called the *commutative law for multiplication*. In matrix arithmetic, however, the equality of AB and BA can fail for three possible reasons:

1. AB may be defined and BA may not (for example, if A is 2×3 and B is 3×4).

2. AB and BA may both be defined, but they may have different sizes (for example, if A is 2×3 and B is 3×2).

3. AB and BA may both be defined and have the same size, but the two matrices may be different (as illustrated in the next example).

▶ **EXAMPLE 2 Order Matters in Matrix Multiplication**

Consider the matrices

$$A = \begin{bmatrix} -1 & 0 \\ 2 & 3 \end{bmatrix} \quad \text{and} \quad B = \begin{bmatrix} 1 & 2 \\ 3 & 0 \end{bmatrix}$$

Do not read too much into Example 2—it does not rule out the possibility that AB and BA may be equal in *certain* cases, just that they are not equal in *all* cases. If it so happens that $AB = BA$, then we say that AB and BA ***commute***.

Multiplying gives

$$AB = \begin{bmatrix} -1 & -2 \\ 11 & 4 \end{bmatrix} \quad \text{and} \quad BA = \begin{bmatrix} 3 & 6 \\ -3 & 0 \end{bmatrix}$$

Thus, $AB \neq BA$. ◀

Zero Matrices

A matrix whose entries are all zero is called a ***zero matrix***. Some examples are

$$\begin{bmatrix} 0 & 0 \\ 0 & 0 \end{bmatrix}, \quad \begin{bmatrix} 0 & 0 & 0 \\ 0 & 0 & 0 \\ 0 & 0 & 0 \end{bmatrix}, \quad \begin{bmatrix} 0 & 0 & 0 & 0 \\ 0 & 0 & 0 & 0 \end{bmatrix}, \quad \begin{bmatrix} 0 \\ 0 \\ 0 \\ 0 \end{bmatrix}, \quad [0]$$

We will denote a zero matrix by 0 unless it is important to specify its size, in which case we will denote the $m \times n$ zero matrix by $0_{m \times n}$.

It should be evident that if A and 0 are matrices with the same size, then

$$A + 0 = 0 + A = A$$

Thus, 0 plays the same role in this matrix equation that the number 0 plays in the numerical equation $a + 0 = 0 + a = a$.

The following theorem lists the basic properties of zero matrices. Since the results should be self-evident, we will omit the formal proofs.

THEOREM 1.4.2 Properties of Zero Matrices

If c is a scalar, and if the sizes of the matrices are such that the operations can be perfomed, then:

(a) $A + 0 = 0 + A = A$

(b) $A - 0 = A$

(c) $A - A = A + (-A) = 0$

(d) $0A = 0$

(e) *If $cA = 0$, then $c = 0$ or $A = 0$.*

Since we know that the commutative law of real arithmetic is not valid in matrix arithmetic, it should not be surprising that there are other rules that fail as well. For example, consider the following two laws of real arithmetic:

- If $ab = bc$ and $a \neq 0$, then $b = c$. **[The cancellation law]**
- If $ab = 0$, then at least one of the factors on the left is 0.

The next two examples show that these laws are not universally true in matrix arithmetic.

▶ **EXAMPLE 3 Failure of the Cancellation Law**

Consider the matrices

$$A = \begin{bmatrix} 0 & 1 \\ 0 & 2 \end{bmatrix}, \quad B = \begin{bmatrix} 1 & 1 \\ 3 & 4 \end{bmatrix}, \quad C = \begin{bmatrix} 2 & 5 \\ 3 & 4 \end{bmatrix}$$

We leave it for you to confirm that

$$AB = AC = \begin{bmatrix} 3 & 4 \\ 6 & 8 \end{bmatrix}$$

Although $A \neq 0$, canceling A from both sides of the equation $AB = AC$ would lead to the incorrect conclusion that $B = C$. Thus, the cancellation law does not hold, in general, for matrix multiplication.

▶ **EXAMPLE 4 A Zero Product with Nonzero Factors**

Here are two matrices for which $AB = 0$, but $A \neq 0$ and $B \neq 0$:

$$A = \begin{bmatrix} 0 & 1 \\ 0 & 2 \end{bmatrix}, \quad B = \begin{bmatrix} 3 & 7 \\ 0 & 0 \end{bmatrix} \blacktriangleleft$$

Identity Matrices   A square matrix with 1's on the main diagonal and zeros elsewhere is called an ***identity matrix***. Some examples are

$$[1], \quad \begin{bmatrix} 1 & 0 \\ 0 & 1 \end{bmatrix}, \quad \begin{bmatrix} 1 & 0 & 0 \\ 0 & 1 & 0 \\ 0 & 0 & 1 \end{bmatrix}, \quad \begin{bmatrix} 1 & 0 & 0 & 0 \\ 0 & 1 & 0 & 0 \\ 0 & 0 & 1 & 0 \\ 0 & 0 & 0 & 1 \end{bmatrix}$$

An identity matrix is denoted by the letter I. If it is important to emphasize the size, we will write I_n for the $n \times n$ identity matrix.

To explain the role of identity matrices in matrix arithmetic, let us consider the effect of multiplying a general 2×3 matrix A on each side by an identity matrix. Multiplying on the right by the 3×3 identity matrix yields

$$AI_3 = \begin{bmatrix} a_{11} & a_{12} & a_{13} \\ a_{21} & a_{22} & a_{23} \end{bmatrix} \begin{bmatrix} 1 & 0 & 0 \\ 0 & 1 & 0 \\ 0 & 0 & 1 \end{bmatrix} = \begin{bmatrix} a_{11} & a_{12} & a_{13} \\ a_{21} & a_{22} & a_{23} \end{bmatrix} = A$$

and multiplying on the left by the 2×2 identity matrix yields

$$I_2 A = \begin{bmatrix} 1 & 0 \\ 0 & 1 \end{bmatrix} \begin{bmatrix} a_{11} & a_{12} & a_{13} \\ a_{21} & a_{22} & a_{23} \end{bmatrix} = \begin{bmatrix} a_{11} & a_{12} & a_{13} \\ a_{21} & a_{22} & a_{23} \end{bmatrix} = A$$

The same result holds in general; that is, if A is any $m \times n$ matrix, then

$$AI_n = A \quad \text{and} \quad I_m A = A$$

Thus, the identity matrices play the same role in these matrix equations that the number 1 plays in the numerical equation $a \cdot 1 = 1 \cdot a = a$.

As the next theorem shows, identity matrices arise naturally in studying reduced row echelon forms of *square* matrices.

THEOREM 1.4.3 *If R is the reduced row echelon form of an $n \times n$ matrix A, then either R has a row of zeros or R is the identity matrix I_n.*

Proof Suppose that the reduced row echelon form of A is

$$R = \begin{bmatrix} r_{11} & r_{12} & \cdots & r_{1n} \\ r_{21} & r_{22} & \cdots & r_{2n} \\ \vdots & \vdots & & \vdots \\ r_{n1} & r_{n2} & \cdots & r_{nn} \end{bmatrix}$$

Either the last row in this matrix consists entirely of zeros or it does not. If not, the matrix contains no zero rows, and consequently each of the n rows has a leading entry of 1. Since these leading 1's occur progressively farther to the right as we move down the matrix, each of these 1's must occur on the main diagonal. Since the other entries in the same column as one of these 1's are zero, R must be I_n. Thus, either R has a row of zeros or $R = I_n$. ◄

Inverse of a Matrix   In real arithmetic every nonzero number a has a reciprocal $a^{-1}(= 1/a)$ with the property

$$a \cdot a^{-1} = a^{-1} \cdot a = 1$$

The number a^{-1} is sometimes called the *multiplicative inverse* of a. Our next objective is to develop an analog of this result for matrix arithmetic. For this purpose we make the following definition.

DEFINITION 1   If A is a square matrix, and if a matrix B of the same size can be found such that $AB = BA = I$, then A is said to be *invertible* (or *nonsingular*) and B is called an *inverse* of A. If no such matrix B can be found, then A is said to be *singular*.

Remark   The relationship $AB = BA = I$ is not changed by interchanging A and B, so if A is invertible and B is an inverse of A, then it is also true that B is invertible, and A is an inverse of B. Thus, when

$$AB = BA = I$$

we say that A and B are *inverses of one another.*

► **EXAMPLE 5 An Invertible Matrix**

Let

$$A = \begin{bmatrix} 2 & -5 \\ -1 & 3 \end{bmatrix} \quad \text{and} \quad B = \begin{bmatrix} 3 & 5 \\ 1 & 2 \end{bmatrix}$$

Then

$$AB = \begin{bmatrix} 2 & -5 \\ -1 & 3 \end{bmatrix}\begin{bmatrix} 3 & 5 \\ 1 & 2 \end{bmatrix} = \begin{bmatrix} 1 & 0 \\ 0 & 1 \end{bmatrix} = I$$

$$BA = \begin{bmatrix} 3 & 5 \\ 1 & 2 \end{bmatrix}\begin{bmatrix} 2 & -5 \\ -1 & 3 \end{bmatrix} = \begin{bmatrix} 1 & 0 \\ 0 & 1 \end{bmatrix} = I$$

Thus, A and B are invertible and each is an inverse of the other. ◄

▶ **EXAMPLE 6 A Class of Singular Matrices**

In general, a square matrix with a row or column of zeros is singular. To help understand why this is so, consider the matrix

$$A = \begin{bmatrix} 1 & 4 & 0 \\ 2 & 5 & 0 \\ 3 & 6 & 0 \end{bmatrix}$$

To prove that A is singular we must show that there is no 3×3 matrix B such that $AB = BA = I$. For this purpose let $\mathbf{c}_1, \mathbf{c}_2, \mathbf{0}$ be the column vectors of A. Thus, for any 3×3 matrix B we can express the product BA as

$$BA = B[\mathbf{c}_1 \quad \mathbf{c}_2 \quad \mathbf{0}] = [B\mathbf{c}_1 \quad B\mathbf{c}_2 \quad \mathbf{0}] \quad \text{[Formula (6) of Section 1.3]}$$

The column of zeros shows that $BA \neq I$ and hence that A is singular. ◀

Properties of Inverses It is reasonable to ask whether an invertible matrix can have more than one inverse. The next theorem shows that the answer is no—*an invertible matrix has exactly one inverse.*

THEOREM 1.4.4 *If B and C are both inverses of the matrix A, then $B = C$.*

Proof Since B is an inverse of A, we have $BA = I$. Multiplying both sides on the right by C gives $(BA)C = IC = C$. But it is also true that $(BA)C = B(AC) = BI = B$, so $C = B$. ◀

As a consequence of this important result, we can now speak of "the" inverse of an invertible matrix. If A is invertible, then its inverse will be denoted by the symbol A^{-1}. Thus,

$$AA^{-1} = I \quad \text{and} \quad A^{-1}A = I \tag{1}$$

The inverse of A plays much the same role in matrix arithmetic that the reciprocal a^{-1} plays in the numerical relationships $aa^{-1} = 1$ and $a^{-1}a = 1$.

In the next section we will develop a method for computing the inverse of an invertible matrix of any size. For now we give the following theorem that specifies conditions under which a 2×2 matrix is invertible and provides a simple formula for its inverse.

The quantity $ad - bc$ in Theorem 1.4.5 is called the **determinant** of the 2×2 matrix A and is denoted by

$$\det(A) = ad - bc$$

or alternatively by

$$\begin{vmatrix} a & b \\ c & d \end{vmatrix} = ad - bc$$

THEOREM 1.4.5 *The matrix*

$$A = \begin{bmatrix} a & b \\ c & d \end{bmatrix}$$

is invertible if and only if $ad - bc \neq 0$, in which case the inverse is given by the formula

$$A^{-1} = \frac{1}{ad - bc}\begin{bmatrix} d & -b \\ -c & a \end{bmatrix} \tag{2}$$

We will omit the proof, because we will study a more general version of this theorem later. For now, you should at least confirm the validity of Formula (2) by showing that $AA^{-1} = A^{-1} = I$.

Historical Note The formula for A^{-1} given in Theorem 1.4.5 first appeared (in a more general form) in Arthur Cayley's 1858 *Memoir on the Theory of Matrices*. The more general result that Cayley discovered will be studied later.

$$\det(A) = \begin{vmatrix} a & b \\ c & d \end{vmatrix} = ad - bc$$

▲ Figure 1.4.1

Remark Figure 1.4.1 illustrates that the determinant of a 2×2 matrix A is the product of the entries on its main diagonal minus the product of the entries *off* its main diagonal. In words, Theorem 1.4.5 states that a 2×2 matrix A is invertible if and only if its determinant is nonzero, and if invertible, then its inverse can be obtained by interchanging its diagonal entries, reversing the signs of its off-diagonal entries, and multiplying the entries by the reciprocal of the determinant of A.

▶ **EXAMPLE 7** Calculating the Inverse of a 2 × 2 Matrix

In each part, determine whether the matrix is invertible. If so, find its inverse.

$$\text{(a) } A = \begin{bmatrix} 6 & 1 \\ 5 & 2 \end{bmatrix} \qquad \text{(b) } A = \begin{bmatrix} -1 & 2 \\ 3 & -6 \end{bmatrix}$$

Solution (a) The determinant of A is $\det(A) = (6)(2) - (1)(5) = 7$, which is nonzero. Thus, A is invertible, and its inverse is

$$A^{-1} = \frac{1}{7}\begin{bmatrix} 2 & -1 \\ -5 & 6 \end{bmatrix} = \begin{bmatrix} \frac{2}{7} & -\frac{1}{7} \\ -\frac{5}{7} & \frac{6}{7} \end{bmatrix}$$

We leave it for you to confirm that $AA^{-1} = A^{-1}A = I$.

Solution (b) The matrix is not invertible since $\det(A) = (-1)(-6) - (2)(3) = 0$.

▶ **EXAMPLE 8** Solution of a Linear System by Matrix Inversion

A problem that arises in many applications is to solve a pair of equations of the form

$$u = ax + by$$
$$v = cx + dy$$

for x and y in terms of u and v. One approach is to treat this as a linear system of two equations in the unknowns x and y and use Gauss–Jordan elimination to solve for x and y. However, because the coefficients of the unknowns are *literal* rather than *numerical*, this procedure is a little clumsy. As an alternative approach, let us replace the two equations by the single matrix equation

$$\begin{bmatrix} u \\ v \end{bmatrix} = \begin{bmatrix} ax + by \\ cx + dy \end{bmatrix}$$

which we can rewrite as

$$\begin{bmatrix} u \\ v \end{bmatrix} = \begin{bmatrix} a & b \\ c & d \end{bmatrix}\begin{bmatrix} x \\ y \end{bmatrix}$$

If we assume that the 2×2 matrix is invertible (i.e., $ad - bc \neq 0$), then we can multiply through on the left by the inverse and rewrite the equation as

$$\begin{bmatrix} a & b \\ c & d \end{bmatrix}^{-1}\begin{bmatrix} u \\ v \end{bmatrix} = \begin{bmatrix} a & b \\ c & d \end{bmatrix}^{-1}\begin{bmatrix} a & b \\ c & d \end{bmatrix}\begin{bmatrix} x \\ y \end{bmatrix}$$

which simplifies to

$$\begin{bmatrix} a & b \\ c & d \end{bmatrix}^{-1}\begin{bmatrix} u \\ v \end{bmatrix} = \begin{bmatrix} x \\ y \end{bmatrix}$$

Using Theorem 1.4.5, we can rewrite this equation as

$$\frac{1}{ad - bc}\begin{bmatrix} d & -b \\ -c & a \end{bmatrix}\begin{bmatrix} u \\ v \end{bmatrix} = \begin{bmatrix} x \\ y \end{bmatrix}$$

from which we obtain

$$x = \frac{du - bv}{ad - bc}, \quad y = \frac{av - cu}{ad - bc} \quad \blacktriangleleft$$

The next theorem is concerned with inverses of matrix products.

THEOREM 1.4.6 *If A and B are invertible matrices with the same size, then AB is invertible and*

$$(AB)^{-1} = B^{-1}A^{-1}$$

Proof We can establish the invertibility and obtain the stated formula at the same time by showing that

$$(AB)(B^{-1}A^{-1}) = (B^{-1}A^{-1})(AB) = I$$

But

$$(AB)(B^{-1}A^{-1}) = A(BB^{-1})A^{-1} = AIA^{-1} = AA^{-1} = I$$

and similarly, $(B^{-1}A^{-1})(AB) = I$. $\blacktriangleleft$

Although we will not prove it, this result can be extended to three or more factors:

A product of any number of invertible matrices is invertible, and the inverse of the product is the product of the inverses in the reverse order.

▶ **EXAMPLE 9 The Inverse of a Product**

Consider the matrices

$$A = \begin{bmatrix} 1 & 2 \\ 1 & 3 \end{bmatrix}, \quad B = \begin{bmatrix} 3 & 2 \\ 2 & 2 \end{bmatrix}$$

We leave it for you to show that

$$AB = \begin{bmatrix} 7 & 6 \\ 9 & 8 \end{bmatrix}, \quad (AB)^{-1} = \begin{bmatrix} 4 & -3 \\ -\frac{9}{2} & \frac{7}{2} \end{bmatrix}$$

and also that

$$A^{-1} = \begin{bmatrix} 3 & -2 \\ -1 & 1 \end{bmatrix}, \quad B^{-1} = \begin{bmatrix} 1 & -1 \\ -1 & \frac{3}{2} \end{bmatrix}, \quad B^{-1}A^{-1} = \begin{bmatrix} 1 & -1 \\ -1 & \frac{3}{2} \end{bmatrix}\begin{bmatrix} 3 & -2 \\ -1 & 1 \end{bmatrix} = \begin{bmatrix} 4 & -3 \\ -\frac{9}{2} & \frac{7}{2} \end{bmatrix}$$

Thus, $(AB)^{-1} = B^{-1}A^{-1}$ as guaranteed by Theorem 1.4.6. $\blacktriangleleft$

If a product of matrices is singular, then at least one of the factors must be singular. Why?

Powers of a Matrix If A is a *square* matrix, then we define the nonnegative integer powers of A to be

$$A^0 = I \quad \text{and} \quad A^n = AA \cdots A \quad \text{[n factors]}$$

and if A is invertible, then we define the negative integer powers of A to be

$$A^{-n} = (A^{-1})^n = A^{-1}A^{-1} \cdots A^{-1} \quad \text{[n factors]}$$

Because these definitions parallel those for real numbers, the usual laws of nonnegative exponents hold; for example,

$$A^r A^s = A^{r+s} \quad \text{and} \quad (A^r)^s = A^{rs}$$

In addition, we have the following properties of negative exponents.

> **THEOREM 1.4.7** *If A is invertible and n is a nonnegative integer, then:*
> (a) A^{-1} *is invertible and* $(A^{-1})^{-1} = A$.
> (b) A^n *is invertible and* $(A^n)^{-1} = A^{-n} = (A^{-1})^n$.
> (c) kA *is invertible for any nonzero scalar* k, *and* $(kA)^{-1} = k^{-1}A^{-1}$.

We will prove part (c) and leave the proofs of parts (a) and (b) as exercises.

Proof (c)   Property (c) in Theorem 1.4.1 and property (f) in Theorem 1.4.2 imply that

$$(kA)(k^{-1}A^{-1}) = k^{-1}(kA)A^{-1} = (k^{-1}k)AA^{-1} = (1)I = I$$

and similarly, $(k^{-1}A^{-1})(kA) = I$. Thus, kA is invertible and $(kA)^{-1} = k^{-1}A^{-1}$. ◄

▶ **EXAMPLE 10 Properties of Exponents**

Let A and A^{-1} be the matrices in Example 9; that is,

$$A = \begin{bmatrix} 1 & 2 \\ 1 & 3 \end{bmatrix} \quad \text{and} \quad A^{-1} = \begin{bmatrix} 3 & -2 \\ -1 & 1 \end{bmatrix}$$

Then

$$A^{-3} = (A^{-1})^3 = \begin{bmatrix} 3 & -2 \\ -1 & 1 \end{bmatrix}\begin{bmatrix} 3 & -2 \\ -1 & 1 \end{bmatrix}\begin{bmatrix} 3 & -2 \\ -1 & 1 \end{bmatrix} = \begin{bmatrix} 41 & -30 \\ -15 & 11 \end{bmatrix}$$

Also,

$$A^3 = \begin{bmatrix} 1 & 2 \\ 1 & 3 \end{bmatrix}\begin{bmatrix} 1 & 2 \\ 1 & 3 \end{bmatrix}\begin{bmatrix} 1 & 2 \\ 1 & 3 \end{bmatrix} = \begin{bmatrix} 11 & 30 \\ 15 & 41 \end{bmatrix}$$

so, as expected from Theorem 1.4.7(b),

$$(A^3)^{-1} = \frac{1}{(11)(41) - (30)(15)}\begin{bmatrix} 41 & -30 \\ -15 & 11 \end{bmatrix} = \begin{bmatrix} 41 & -30 \\ -15 & 11 \end{bmatrix} = (A^{-1})^3$$

▶ **EXAMPLE 11 The Square of a Matrix Sum**

In real arithmetic, where we have a commutative law for multiplication, we can write

$$(a + b)^2 = a^2 + ab + ba + b^2 = a^2 + ab + ab + b^2 = a^2 + 2ab + b^2$$

However, in matrix arithmetic, where we have no commutative law for multiplication, the best we can do is to write

$$(A + B)^2 = A^2 + AB + BA + B^2$$

It is only in the special case where A and B *commute* (i.e., $AB = BA$) that we can go a step further and write

$$(A + B)^2 = A^2 + 2AB + B^2 \quad ◄$$

Matrix Polynomials   If A is a square matrix, say $n \times n$, and if

$$p(x) = a_0 + a_1x + a_2x^2 + \cdots + a_mx^m$$

is any polynomial, then we define the $n \times n$ matrix $p(A)$ to be

$$p(A) = a_0I + a_1A + a_2A^2 + \cdots + a_mA^m \tag{3}$$

where I is the $n \times n$ identity matrix; that is, $p(A)$ is obtained by substituting A for x and replacing the constant term a_0 by the matrix $a_0 I$. An expression of form (3) is called a **matrix polynomial in A**.

▶ **EXAMPLE 12 A Matrix Polynomial**

Find $p(A)$ for

$$p(x) = x^2 - 2x - 3 \quad \text{and} \quad A = \begin{bmatrix} -1 & 2 \\ 0 & 3 \end{bmatrix}$$

Solution

$$p(A) = A^2 - 2A - 3I$$

$$= \begin{bmatrix} -1 & 2 \\ 0 & 3 \end{bmatrix}^2 - 2\begin{bmatrix} -1 & 2 \\ 0 & 3 \end{bmatrix} - 3\begin{bmatrix} 1 & 0 \\ 0 & 1 \end{bmatrix}$$

$$= \begin{bmatrix} 1 & 4 \\ 0 & 9 \end{bmatrix} - \begin{bmatrix} -2 & 4 \\ 0 & 6 \end{bmatrix} - \begin{bmatrix} 3 & 0 \\ 0 & 3 \end{bmatrix} = \begin{bmatrix} 0 & 0 \\ 0 & 0 \end{bmatrix}$$

or more briefly, $p(A) = 0$. ◀

Remark It follows from the fact that $A^r A^s = A^{r+s} = A^{s+r} = A^s A^r$ that powers of a square matrix commute, and since a matrix polynomial in A is built up from powers of A, any two matrix polynomials in A also commute; that is, for any polynomials p_1 and p_2 we have

$$p_1(A)p_2(A) = p_2(A)p_1(A) \tag{4}$$

Properties of the Transpose   The following theorem lists the main properties of the transpose.

THEOREM 1.4.8 *If the sizes of the matrices are such that the stated operations can be performed, then*:

(a) $(A^T)^T = A$

(b) $(A + B)^T = A^T + B^T$

(c) $(A - B)^T = A^T - B^T$

(d) $(kA)^T = kA^T$

(e) $(AB)^T = B^T A^T$

If you keep in mind that transposing a matrix interchanges its rows and columns, then you should have little trouble visualizing the results in parts (a)–(d). For example, part (a) states the obvious fact that interchanging rows and columns twice leaves a matrix unchanged; and part (b) states that adding two matrices and then interchanging the rows and columns produces the same result as interchanging the rows and columns before adding. We will omit the formal proofs. Part (e) is a less obvious, but for brevity we will omit its proof as well. The result in that part can be extended to three or more factors and restated as:

The transpose of a product of any number of matrices is the product of the transposes in the reverse order.

The following theorem establishes a relationship between the inverse of a matrix and the inverse of its transpose.

THEOREM 1.4.9 *If A is an invertible matrix, then A^T is also invertible and*

$$(A^T)^{-1} = (A^{-1})^T$$

Proof We can establish the invertibility and obtain the formula at the same time by showing that

$$A^T(A^{-1})^T = (A^{-1})^T A^T = I$$

But from part (*e*) of Theorem 1.4.8 and the fact that $I^T = I$, we have

$$A^T(A^{-1})^T = (A^{-1}A)^T = I^T = I$$
$$(A^{-1})^T A^T = (AA^{-1})^T = I^T = I$$

which completes the proof. ◄

▶ **EXAMPLE 13 Inverse of a Transpose**

Consider a general 2×2 invertible matrix and its transpose:

$$A = \begin{bmatrix} a & b \\ c & d \end{bmatrix} \quad \text{and} \quad A^T = \begin{bmatrix} a & c \\ b & d \end{bmatrix}$$

Since A is invertible, its determinant $ad - bc$ is nonzero. But the determinant of A^T is also $ad - bc$ (verify), so A^T is also invertible. It follows from Theorem 1.4.5 that

$$(A^T)^{-1} = \begin{bmatrix} \dfrac{d}{ad - bc} & -\dfrac{c}{ad - bc} \\ -\dfrac{b}{ad - bc} & \dfrac{a}{ad - bc} \end{bmatrix}$$

which is the same matrix that results if A^{-1} is transposed (verify). Thus,

$$(A^T)^{-1} = (A^{-1})^T$$

as guaranteed by Theorem 1.4.9. ◄

Concept Review

- Commutative law for matrix addition
- Associative law for matrix addition
- Associative law for matrix multiplication
- Left and right distributive laws
- Zero matrix
- Identity matrix
- Inverse of a matrix
- Invertible matrix
- Nonsingular matrix
- Singular matrix
- Determinant
- Power of a matrix
- Matrix polynomial

Skills

- Know the arithmetic properties of matrix operations.
- Be able to prove arithmetic properties of matrices.
- Know the properties of zero matrices.
- Know the properties of identity matrices.
- Be able to recognize when two square matrices are inverses of each other.
- Be able to determine whether a 2×2 matrix is invertible.
- Be able to solve a linear system of two equations in two unknowns whose coefficient matrix is invertible.
- Be able to prove basic properties involving invertible matrices.
- Know the properties of the matrix transpose and its relationship with invertible matrices.

Exercise Set 1.4

1. Let

$$A = \begin{bmatrix} 2 & -1 & 3 \\ 0 & 4 & 5 \\ -2 & 1 & 4 \end{bmatrix}, \quad B = \begin{bmatrix} 8 & -3 & -5 \\ 0 & 1 & 2 \\ 4 & -7 & 6 \end{bmatrix},$$

$$C = \begin{bmatrix} 0 & -2 & 3 \\ 1 & 7 & 4 \\ 3 & 5 & 9 \end{bmatrix}, \quad a = 4, \quad b = -7$$

Show that

(a) $A + (B + C) = (A + B) + C$

(b) $(AB)C = A(BC)$ (c) $(a + b)C = aC + bC$

(d) $a(B - C) = aB - aC$

2. Using the matrices and scalars in Exercise 1, verify that

(a) $a(BC) = (aB)C = B(aC)$

(b) $A(B - C) = AB - AC$ (c) $(B + C)A = BA + CA$

(d) $a(bC) = (ab)C$

3. Using the matrices and scalars in Exercise 1, verify that

(a) $(A^T)^T = A$ (b) $(A + B)^T = A^T + B^T$

(c) $(aC)^T = aC^T$ (d) $(AB)^T = B^T A^T$

▶ In Exercises 4–7, use Theorem 1.4.5 to compute the inverses of the following matrices. ◀

4. $A = \begin{bmatrix} 3 & 1 \\ 5 & 2 \end{bmatrix}$ **5.** $B = \begin{bmatrix} 2 & -3 \\ 4 & 4 \end{bmatrix}$

6. $C = \begin{bmatrix} 6 & 4 \\ -2 & -1 \end{bmatrix}$ **7.** $D = \begin{bmatrix} 2 & 0 \\ 0 & 3 \end{bmatrix}$

8. Find the inverse of

$$\begin{bmatrix} \cos\theta & \sin\theta \\ -\sin\theta & \cos\theta \end{bmatrix}$$

9. Find the inverse of

$$\begin{bmatrix} \frac{1}{2}(e^x + e^{-x}) & \frac{1}{2}(e^x - e^{-x}) \\ \frac{1}{2}(e^x - e^{-x}) & \frac{1}{2}(e^x + e^{-x}) \end{bmatrix}$$

10. Use the matrix A in Exercise 4 to verify that $(A^T)^{-1} = (A^{-1})^T$.

11. Use the matrix B in Exercise 5 to verify that $(B^T)^{-1} = (B^{-1})^T$.

12. Use the matrices A and B in Exercises 4 and 5 to verify that $(AB)^{-1} = B^{-1}A^{-1}$.

13. Use the matrices A, B, and C in Exercises 4–6 to verify that $(ABC)^{-1} = C^{-1}B^{-1}A^{-1}$.

▶ In Exercises 14–17, use the given information to find A. ◀

14. $A^{-1} = \begin{bmatrix} 2 & -1 \\ 3 & 5 \end{bmatrix}$ **15.** $(7A)^{-1} = \begin{bmatrix} -3 & 7 \\ 1 & -2 \end{bmatrix}$

16. $(5A^T)^{-1} = \begin{bmatrix} -3 & -1 \\ 5 & 2 \end{bmatrix}$ **17.** $(I + 2A)^{-1} = \begin{bmatrix} -1 & 2 \\ 4 & 5 \end{bmatrix}$

18. Let A be the matrix

$$\begin{bmatrix} 2 & 0 \\ 4 & 1 \end{bmatrix}$$

In each part, compute the given quantity.

(a) A^3 (b) A^{-3} (c) $A^2 - 2A + I$

(d) $p(A)$, where $p(x) = x - 2$

(e) $p(A)$, where $p(x) = 2x^2 - x + 1$

(f) $p(A)$, where $p(x) = x^3 - 2x + 4$

19. Repeat Exercise 18 for the matrix

$$A = \begin{bmatrix} 3 & 1 \\ 2 & 1 \end{bmatrix}$$

20. Repeat Exercise 18 for the matrix

$$A = \begin{bmatrix} 3 & 0 & -1 \\ 0 & -2 & 0 \\ 5 & 0 & 2 \end{bmatrix}$$

21. Repeat Exercise 18 for the matrix

$$A = \begin{bmatrix} 3 & 0 & 0 \\ 0 & -1 & 3 \\ 0 & -3 & -1 \end{bmatrix}$$

▶ In Exercises 22–24, let $p_1(x) = x^2 - 9$, $p_2(x) = x + 3$, and $p_3(x) = x - 3$. Show that $p_1(A) = p_2(A)p_3(A)$ for the given matrix. ◀

22. The matrix A in Exercise 18.

23. The matrix A in Exercise 21.

24. An arbitrary square matrix A.

25. Show that if $p(x) = x^2 - (a + d)x + (ad - bc)$ and

$$A = \begin{bmatrix} a & b \\ c & d \end{bmatrix}$$

then $p(A) = 0$.

26. Show that if $p(x) = x^2 - (a + b + e)x + a(be - cd)$ and

$$A = \begin{bmatrix} a & 0 & 0 \\ 0 & b & c \\ 0 & d & e \end{bmatrix}$$

then $p(A) = 0$.

27. Consider the matrix

$$A = \begin{bmatrix} a_{11} & 0 & \cdots & 0 \\ 0 & a_{22} & \cdots & 0 \\ \vdots & \vdots & & \vdots \\ 0 & 0 & \cdots & a_{nn} \end{bmatrix}$$

where $a_{11}a_{22}\cdots a_{nn} \neq 0$. Show that A is invertible and find its inverse.

28. Show that if a square matrix A satisfies $A^2 - 3A + I = 0$, then $A^{-1} = 3I - A$.

29. (a) Show that a matrix with a row of zeros cannot have an inverse.

(b) Show that a matrix with a column of zeros cannot have an inverse.

30. Assuming that all matrices are $n \times n$ and invertible, solve for D.
$$ABC^T DBA^T C = AB^T$$

31. Assuming that all matrices are $n \times n$ and invertible, solve for D.
$$C^T B^{-1} A^2 BAC^{-1} DA^{-2} B^T C^{-2} = C^T$$

32. If A is a square matrix and n is a positive integer, is it true that $(A^n)^T = (A^T)^n$? Justify your answer.

33. Simplify:
$$(AB)^{-1}(AC^{-1})(D^{-1}C^{-1})^{-1}D^{-1}$$

34. Simplify:
$$(AC^{-1})^{-1}(AC^{-1})(AC^{-1})^{-1}AD^{-1}$$

▶ In Exercises 35–37, determine whether A is invertible, and if so, find the inverse. [*Hint:* Solve $AX = I$ for X by equating corresponding entries on the two sides.] ◀

35. $A = \begin{bmatrix} 1 & 0 & 1 \\ 1 & 1 & 0 \\ 0 & 1 & 1 \end{bmatrix}$

36. $A = \begin{bmatrix} 1 & 1 & 1 \\ 1 & 0 & 0 \\ 0 & 1 & 1 \end{bmatrix}$ **37.** $A = \begin{bmatrix} 0 & 0 & 1 \\ 1 & 1 & 0 \\ -1 & 1 & 1 \end{bmatrix}$

38. Prove Theorem 1.4.2.

▶ In Exercises 39–42, use the method of Example 8 to find the unique solution of the given linear system. ◀

39. $3x_1 - 2x_2 = -1$
      $4x_1 + 5x_2 = 3$

40. $-x_1 + 5x_2 = 4$
      $-x_1 - 3x_2 = 1$

41. $6x_1 + x_2 = 0$
      $4x_1 - 3x_2 = -2$

42. $2x_1 - 2x_2 = 4$
      $x_1 + 4x_2 = 4$

43. Prove part (a) of Theorem 1.4.1.

44. Prove part (c) of Theorem 1.4.1.

45. Prove part (f) of Theorem 1.4.1.

46. Prove part (b) of Theorem 1.4.2.

47. Prove part (c) of Theorem 1.4.2.

48. Verify Formula (4) in the text by a direct calculation.

49. Prove part (d) of Theorem 1.4.8.

50. Prove part (e) of Theorem 1.4.8.

51. (a) Show that if A is invertible and $AB = AC$, then $B = C$.

(b) Explain why part (a) and Example 3 do not contradict one another.

52. Show that if A is invertible and k is any nonzero scalar, then $(kA)^n = k^n A^n$ for all integer values of n.

53. (a) Show that if A, B, and $A + B$ are invertible matrices with the same size, then
$$A(A^{-1} + B^{-1})B(A + B)^{-1} = I$$

(b) What does the result in part (a) tell you about the matrix $A^{-1} + B^{-1}$?

54. A square matrix A is said to be ***idempotent*** if $A^2 = A$.

(a) Show that if A is idempotent, then so is $I - A$.

(b) Show that if A is idempotent, then $2A - I$ is invertible and is its own inverse.

55. Show that if A is a square matrix such that $A^k = 0$ for some positive integer k, then the matrix A is invertible and
$$(I - A)^{-1} = I + A + A^2 + \cdots + A^{k-1}$$

True-False Exercises

In parts (a)–(k) determine whether the statement is true or false, and justify your answer.

(a) Two $n \times n$ matrices, A and B, are inverses of one another if and only if $AB = BA = 0$.

(b) For all square matrices A and B of the same size, it is true that $(A + B)^2 = A^2 + 2AB + B^2$.

(c) For all square matrices A and B of the same size, it is true that $A^2 - B^2 = (A - B)(A + B)$.

(d) If A and B are invertible matrices of the same size, then AB is invertible and $(AB)^{-1} = A^{-1}B^{-1}$.

(e) If A and B are matrices such that AB is defined, then it is true that $(AB)^T = A^T B^T$.

(f) The matrix
$$A = \begin{bmatrix} a & b \\ c & d \end{bmatrix}$$
is invertible if and only if $ad - bc \neq 0$.

(g) If A and B are matrices of the same size and k is a constant, then $(kA + B)^T = kA^T + B^T$.

(h) If A is an invertible matrix, then so is A^T.

(i) If $p(x) = a_0 + a_1x + a_2x^2 + \cdots + a_mx^m$ and I is an identity matrix, then $p(I) = a_0 + a_1 + a_2 + \cdots + a_m$.

(j) A square matrix containing a row or column of zeros cannot be invertible.

(k) The sum of two invertible matrices of the same size must be invertible.

1.5 Elementary Matrices and a Method for Finding A^{-1}

In this section we will develop an algorithm for finding the inverse of a matrix, and we will discuss some of the basic properties of invertible matrices.

In Section 1.1 we defined three elementary row operations on a matrix A:

1. Multiply a row by a nonzero constant c.
2. Interchange two rows.
3. Add a constant c times one row to another.

It should be evident that if we let B be the matrix that results from A by performing one of the operations in this list, then the matrix A can be recovered from B by performing the corresponding operation in the following list:

1. Multiply the same row by $1/c$.
2. Interchange the same two rows.
3. If B resulted by adding c times row r_1 of A to row r_2, then add $-c$ times r_2 to row r_1.

It follows that if B is obtained from A by performing a sequence of elementary row operations, then there is a second sequence of elementary row operations, which when applied to B recovers A (Exercise 43). Accordingly, we make the following definition.

> **DEFINITION 1** Matrices A and B are said to be **row equivalent** if either (hence each) can be obtained from the other by a sequence of elementary row operations.

Our next goal is to show how matrix multiplication can be used to carry out an elementary row operation.

> **DEFINITION 2** An $n \times n$ matrix is called an **elementary matrix** if it can be obtained from the $n \times n$ identity matrix I_n by performing a *single* elementary row operation.

▶ **EXAMPLE 1 Elementary Matrices and Row Operations**

Listed below are four elementary matrices and the operations that produce them.

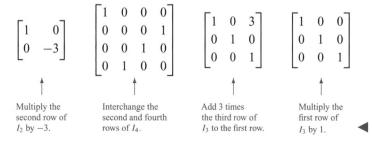

$$\begin{bmatrix} 1 & 0 \\ 0 & -3 \end{bmatrix} \qquad \begin{bmatrix} 1 & 0 & 0 & 0 \\ 0 & 0 & 0 & 1 \\ 0 & 0 & 1 & 0 \\ 0 & 1 & 0 & 0 \end{bmatrix} \qquad \begin{bmatrix} 1 & 0 & 3 \\ 0 & 1 & 0 \\ 0 & 0 & 1 \end{bmatrix} \qquad \begin{bmatrix} 1 & 0 & 0 \\ 0 & 1 & 0 \\ 0 & 0 & 1 \end{bmatrix}$$

Multiply the second row of I_2 by -3.

Interchange the second and fourth rows of I_4.

Add 3 times the third row of I_3 to the first row.

Multiply the first row of I_3 by 1. ◀

The following theorem, whose proof is left as an exercise, shows that when a matrix A is multiplied on the *left* by an elementary matrix E, the effect is to perform an elementary row operation on A.

> **THEOREM 1.5.1 Row Operations by Matrix Multiplication**
>
> *If the elementary matrix E results from performing a certain row operation on I_m and if A is an $m \times n$ matrix, then the product EA is the matrix that results when this same row operation is performed on A.*

▶ **EXAMPLE 2 Using Elementary Matrices**

Consider the matrix

$$A = \begin{bmatrix} 1 & 0 & 2 & 3 \\ 2 & -1 & 3 & 6 \\ 1 & 4 & 4 & 0 \end{bmatrix}$$

and consider the elementary matrix

$$E = \begin{bmatrix} 1 & 0 & 0 \\ 0 & 1 & 0 \\ 3 & 0 & 1 \end{bmatrix}$$

which results from adding 3 times the first row of I_3 to the third row. The product EA is

$$EA = \begin{bmatrix} 1 & 0 & 2 & 3 \\ 2 & -1 & 3 & 6 \\ 4 & 4 & 10 & 9 \end{bmatrix}$$

which is precisely the matrix that results when we add 3 times the first row of A to the third row. ◀

> Theorem 1.5.1 will be a useful tool for developing new results about matrices, but as a practical matter it is usually preferable to perform row operations directly.

We know from the discussion at the beginning of this section that if E is an elementary matrix that results from performing an elementary row operation on an identity matrix I, then there is a second elementary row operation, which when applied to E, produces I back again. Table 1 lists these operations. The operations on the right side of the table are called the ***inverse operations*** of the corresponding operations on the left.

Table 1

Row Operation on I That Produces E	Row Operation on E That Reproduces I
Multiply row i by $c \neq 0$	Multiply row i by $1/c$
Interchange rows i and j	Interchange rows i and j
Add c times row i to row j	Add $-c$ times row i to row j

▶ **EXAMPLE 3 Row Operations and Inverse Row Operations**

In each of the following, an elementary row operation is applied to the 2×2 identity matrix to obtain an elementary matrix E, then E is restored to the identity matrix by applying the inverse row operation.

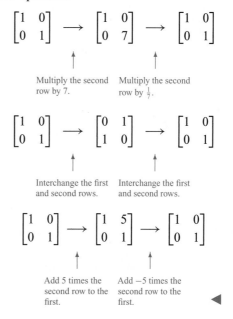

Multiply the second Multiply the second
row by 7. row by $\frac{1}{7}$.

Interchange the first Interchange the first
and second rows. and second rows.

Add 5 times the Add -5 times the
second row to the second row to the
first. first.

The next theorem is a key result about invertibility of elementary matrices. It will be a building block for many results that follow.

THEOREM 1.5.2 *Every elementary matrix is invertible, and the inverse is also an elementary matrix.*

Proof If E is an elementary matrix, then E results by performing some row operation on I. Let E_0 be the matrix that results when the inverse of this operation is performed on I. Applying Theorem 1.5.1 and using the fact that inverse row operations cancel the effect of each other, it follows that

$$E_0 E = I \quad \text{and} \quad E E_0 = I$$

Thus, the elementary matrix E_0 is the inverse of E. ◄

Equivalence Theorem One of our objectives as we progress through this text is to show how seemingly diverse ideas in linear algebra are related. The following theorem, which relates results we have obtained about invertibility of matrices, homogeneous linear systems, reduced row echelon forms, and elementary matrices, is our first step in that direction. As we study new topics, more statements will be added to this theorem.

THEOREM 1.5.3 Equivalent Statements

If A is an $n \times n$ matrix, then the following statements are equivalent, that is, all true or all false.

(*a*) *A is invertible.*

(*b*) *$A\mathbf{x} = \mathbf{0}$ has only the trivial solution.*

(*c*) *The reduced row echelon form of A is I_n.*

(*d*) *A is expressible as a product of elementary matrices.*

It may make the logic of our proof of Theorem 1.5.3 more apparent by writing the implications

$$(a) \Rightarrow (b) \Rightarrow (c) \Rightarrow (d) \Rightarrow (a)$$

This makes it evident visually that the validity of any one statement implies the validity of all the others, and hence that the falsity of any one implies the falsity of the others.

Proof   We will prove the equivalence by establishing the chain of implications: $(a) \Rightarrow (b) \Rightarrow (c) \Rightarrow (d) \Rightarrow (a)$.

(a) $\Rightarrow$ (b)   Assume A is invertible and let $\mathbf{x}_0$ be any solution of $A\mathbf{x} = \mathbf{0}$. Multiplying both sides of this equation by the matrix A^{-1} gives $A^{-1}(A\mathbf{x}_0) = A^{-1}\mathbf{0}$, or $(A^{-1}A)\mathbf{x}_0 = \mathbf{0}$, or $I\mathbf{x}_0 = \mathbf{0}$, or $\mathbf{x}_0 = \mathbf{0}$. Thus, $A\mathbf{x} = \mathbf{0}$ has only the trivial solution.

(b) $\Rightarrow$ (c)   Let $A\mathbf{x} = \mathbf{0}$ be the matrix form of the system

$$
\begin{aligned}
a_{11}x_1 + a_{12}x_2 + \cdots + a_{1n}x_n &= 0 \\
a_{21}x_1 + a_{22}x_2 + \cdots + a_{2n}x_n &= 0 \\
\vdots \qquad \vdots \qquad\qquad \vdots \qquad \vdots \\
a_{n1}x_1 + a_{n2}x_2 + \cdots + a_{nn}x_n &= 0
\end{aligned}
\tag{1}
$$

and assume that the system has only the trivial solution. If we solve by Gauss–Jordan elimination, then the system of equations corresponding to the reduced row echelon form of the augmented matrix will be

$$
\begin{aligned}
x_1 \qquad\qquad\qquad\quad &= 0 \\
x_2 \qquad\qquad\quad &= 0 \\
\ddots \qquad\qquad \\
x_n &= 0
\end{aligned}
\tag{2}
$$

Thus the augmented matrix

$$
\begin{bmatrix}
a_{11} & a_{12} & \cdots & a_{1n} & 0 \\
a_{21} & a_{22} & \cdots & a_{2n} & 0 \\
\vdots & \vdots & & \vdots & \vdots \\
a_{n1} & a_{n2} & \cdots & a_{nn} & 0
\end{bmatrix}
$$

for (1) can be reduced to the augmented matrix

$$
\begin{bmatrix}
1 & 0 & 0 & \cdots & 0 & 0 \\
0 & 1 & 0 & \cdots & 0 & 0 \\
0 & 0 & 1 & \cdots & 0 & 0 \\
\vdots & \vdots & \vdots & & \vdots & \vdots \\
0 & 0 & 0 & \cdots & 1 & 0
\end{bmatrix}
$$

for (2) by a sequence of elementary row operations. If we disregard the last column (all zeros) in each of these matrices, we can conclude that the reduced row echelon form of A is I_n.

(c) $\Rightarrow$ (d)   Assume that the reduced row echelon form of A is I_n, so that A can be reduced to I_n by a finite sequence of elementary row operations. By Theorem 1.5.1, each of these operations can be accomplished by multiplying on the left by an appropriate elementary matrix. Thus we can find elementary matrices $E_1, E_2, \ldots, E_k$ such that

$$
E_k \cdots E_2 E_1 A = I_n
\tag{3}
$$

By Theorem 1.5.2, $E_1, E_2, \ldots, E_k$ are invertible. Multiplying both sides of Equation (3) on the left successively by $E_k^{-1}, \ldots, E_2^{-1}, E_1^{-1}$ we obtain

$$A = E_1^{-1}E_2^{-1} \cdots E_k^{-1}I_n = E_1^{-1}E_2^{-1} \cdots E_k^{-1} \tag{4}$$

By Theorem 1.5.2, this equation expresses A as a product of elementary matrices.

(d) $\Rightarrow$ **(a)** If A is a product of elementary matrices, then from Theorems 1.4.7 and 1.5.2, the matrix A is a product of invertible matrices and hence is invertible. ◀

A Method for Inverting Matrices As a first application of Theorem 1.5.3, we will develop a procedure (or algorithm) that can be used to tell whether a given matrix is invertible, and if so, produce its inverse. To derive this algorithm, assume for the moment, that A is an invertible $n \times n$ matrix. In Equation (3), the elementary matrices execute a sequence of row operations that reduce A to I_n. If we multiply both sides of this equation on the right by A^{-1} and simplify, we obtain

$$A^{-1} = E_k \cdots E_2 E_1 I_n$$

But this equation tells us that *the same sequence of row operations that reduces A to I_n will transform I_n to A^{-1}.* Thus, we have established the following result.

Inversion Algorithm To find the inverse of an invertible matrix A, find a sequence of elementary row operations that reduces A to the identity and then perform that same sequence of operations on I_n to obtain A^{-1}.

A simple method for carrying out this procedure is given in the following example.

▶ **EXAMPLE 4 Using Row Operations to Find A^{-1}**

Find the inverse of

$$A = \begin{bmatrix} 1 & 2 & 3 \\ 2 & 5 & 3 \\ 1 & 0 & 8 \end{bmatrix}$$

Solution We want to reduce A to the identity matrix by row operations and simultaneously apply these operations to I to produce A^{-1}. To accomplish this we will adjoin the identity matrix to the right side of A, thereby producing a partitioned matrix of the form

$$[A \mid I]$$

Then we will apply row operations to this matrix until the left side is reduced to I; these operations will convert the right side to A^{-1}, so the final matrix will have the form

$$[I \mid A^{-1}]$$

The computations are as follows:

$$\begin{bmatrix} 1 & 2 & 3 & | & 1 & 0 & 0 \\ 2 & 5 & 3 & | & 0 & 1 & 0 \\ 1 & 0 & 8 & | & 0 & 0 & 1 \end{bmatrix}$$

$$\begin{bmatrix} 1 & 2 & 3 & | & 1 & 0 & 0 \\ 0 & 1 & -3 & | & -2 & 1 & 0 \\ 0 & -2 & 5 & | & -1 & 0 & 1 \end{bmatrix} \longleftarrow \quad \text{We added } -2 \text{ times the first row to the second and } -1 \text{ times the first row to the third.}$$

$$\begin{bmatrix} 1 & 2 & 3 & | & 1 & 0 & 0 \\ 0 & 1 & -3 & | & -2 & 1 & 0 \\ 0 & 0 & -1 & | & -5 & 2 & 1 \end{bmatrix} \longleftarrow \quad \text{We added 2 times the second row to the third.}$$

$$\begin{bmatrix} 1 & 2 & 3 & | & 1 & 0 & 0 \\ 0 & 1 & -3 & | & -2 & 1 & 0 \\ 0 & 0 & 1 & | & 5 & -2 & -1 \end{bmatrix} \longleftarrow \quad \text{We multiplied the third row by } -1.$$

$$\begin{bmatrix} 1 & 2 & 0 & | & -14 & 6 & 3 \\ 0 & 1 & 0 & | & 13 & -5 & -3 \\ 0 & 0 & 1 & | & 5 & -2 & -1 \end{bmatrix} \longleftarrow \quad \text{We added 3 times the third row to the second and } -3 \text{ times the third row to the first.}$$

$$\begin{bmatrix} 1 & 0 & 0 & | & -40 & 16 & 9 \\ 0 & 1 & 0 & | & 13 & -5 & -3 \\ 0 & 0 & 1 & | & 5 & -2 & -1 \end{bmatrix} \longleftarrow \quad \text{We added } -2 \text{ times the second row to the first.}$$

Thus,

$$A^{-1} = \begin{bmatrix} -40 & 16 & 9 \\ 13 & -5 & -3 \\ 5 & -2 & -1 \end{bmatrix} \quad \blacktriangleleft$$

Often it will not be known in advance if a given $n \times n$ matrix A is invertible. However, if it is not, then by parts (*a*) and (*c*) of Theorem 1.5.3 it will be impossible to reduce A to I_n by elementary row operations. This will be signaled by a row of zeros appearing on the *left side* of the partition at some stage of the inversion algorithm. If this occurs, then you can stop the computations and conclude that A is not invertible.

▶ **EXAMPLE 5 Showing That a Matrix Is Not Invertible**

Consider the matrix

$$A = \begin{bmatrix} 1 & 6 & 4 \\ 2 & 4 & -1 \\ -1 & 2 & 5 \end{bmatrix}$$

Applying the procedure of Example 4 yields

$$\left[\begin{array}{ccc|ccc} 1 & 6 & 4 & 1 & 0 & 0 \\ 2 & 4 & -1 & 0 & 1 & 0 \\ -1 & 2 & 5 & 0 & 0 & 1 \end{array}\right]$$

$$\left[\begin{array}{ccc|ccc} 1 & 6 & 4 & 1 & 0 & 0 \\ 0 & -8 & -9 & -2 & 1 & 0 \\ 0 & 8 & 9 & 1 & 0 & 1 \end{array}\right]$$ ◄———— We added -2 times the first row to the second and added the first row to the third.

$$\left[\begin{array}{ccc|ccc} 1 & 6 & 4 & 1 & 0 & 0 \\ 0 & -8 & -9 & -2 & 1 & 0 \\ 0 & 0 & 0 & -1 & 1 & 1 \end{array}\right]$$ ◄———— We added the second row to the third.

Since we have obtained a row of zeros on the left side, A is not invertible.

▶ **EXAMPLE 6 Analyzing Homogeneous Systems**

Use Theorem 1.5.3 to determine whether the given homogeneous system has nontrivial solutions.

$$\begin{array}{ll} x_1 + 2x_2 + 3x_3 = 0 & \qquad x_1 + 6x_2 + 4x_3 = 0 \\ \text{(a)} \ \ 2x_1 + 5x_2 + 3x_3 = 0 & \text{(b)} \ \ 2x_1 + 4x_2 - x_3 = 0 \\ \quad\ x_1 \qquad\quad + 8x_3 = 0 & \quad -x_1 + 2x_2 + 5x_3 = 0 \end{array}$$

Solution From parts (*a*) and (*b*) of Theorem 1.5.3 a homogeneous linear system has only the trivial solution if and only if its coefficient matrix is invertible. From Examples 4 and 5 the coefficient matrix of system (a) is invertible and that of system (b) is not. Thus, system (a) has only the trivial solution whereas system (b) has nontrivial solutions.   ◄

Concept Review

- Row equivalent matrices
- Elementary matrix
- Inverse operations
- Inversion algorithm

Skills

- Determine whether a given square matrix is an elementary.
- Determine whether two square matrices are row equivalent.
- Apply the inverse of a given elementary row operation to a matrix.
- Apply elementary row operations to reduce a given square matrix to the identity matrix.
- Understand the relationships between statements that are equivalent to the invertibility of a square matrix (Theorem 1.5.3).
- Use the inversion algorithm to find the inverse of an invertible matrix.
- Express an invertible matrix as a product of elementary matrices.

Exercise Set 1.5

1. Decide whether each matrix below is an elementary matrix.

(a) $\begin{bmatrix} 1 & 0 \\ -5 & 1 \end{bmatrix}$

(b) $\begin{bmatrix} -5 & 1 \\ 1 & 0 \end{bmatrix}$

(c) $\begin{bmatrix} 1 & 1 & 0 \\ 0 & 0 & 1 \\ 0 & 0 & 0 \end{bmatrix}$

(d) $\begin{bmatrix} 2 & 0 & 0 & 2 \\ 0 & 1 & 0 & 0 \\ 0 & 0 & 1 & 0 \\ 0 & 0 & 0 & 1 \end{bmatrix}$

2. Decide whether each matrix below is an elementary matrix.

(a) $\begin{bmatrix} 1 & 0 \\ 0 & \sqrt{3} \end{bmatrix}$

(b) $\begin{bmatrix} 0 & 0 & 1 \\ 0 & 1 & 0 \\ 1 & 0 & 0 \end{bmatrix}$

(c) $\begin{bmatrix} 1 & 0 & 0 \\ 0 & 1 & 9 \\ 0 & 0 & 1 \end{bmatrix}$

(d) $\begin{bmatrix} -1 & 0 & 0 \\ 0 & 0 & 1 \\ 0 & 1 & 0 \end{bmatrix}$

3. Find a row operation and the corresponding elementary matrix that will restore the given elementary matrix to the identity matrix.

(a) $\begin{bmatrix} 1 & -3 \\ 0 & 1 \end{bmatrix}$

(b) $\begin{bmatrix} -7 & 0 & 0 \\ 0 & 1 & 0 \\ 0 & 0 & 1 \end{bmatrix}$

(c) $\begin{bmatrix} 1 & 0 & 0 \\ 0 & 1 & 0 \\ -5 & 0 & 1 \end{bmatrix}$

(d) $\begin{bmatrix} 0 & 0 & 1 & 0 \\ 0 & 1 & 0 & 0 \\ 1 & 0 & 0 & 0 \\ 0 & 0 & 0 & 1 \end{bmatrix}$

4. Find a row operation and the corresponding elementary matrix that will restore the given elementary matrix to the identity matrix.

(a) $\begin{bmatrix} 1 & 0 \\ -3 & 1 \end{bmatrix}$

(b) $\begin{bmatrix} 1 & 0 & 0 \\ 0 & 1 & 0 \\ 0 & 0 & 3 \end{bmatrix}$

(c) $\begin{bmatrix} 0 & 0 & 0 & 1 \\ 0 & 1 & 0 & 0 \\ 0 & 0 & 1 & 0 \\ 1 & 0 & 0 & 0 \end{bmatrix}$

(d) $\begin{bmatrix} 1 & 0 & -\frac{1}{7} & 0 \\ 0 & 1 & 0 & 0 \\ 0 & 0 & 1 & 0 \\ 0 & 0 & 0 & 1 \end{bmatrix}$

5. In each part, an elementary matrix E and a matrix A are given. Write down the row operation corresponding to E and show that the product EA results from applying the row operation to A.

(a) $E = \begin{bmatrix} 0 & 1 \\ 1 & 0 \end{bmatrix}$, $A = \begin{bmatrix} -1 & -2 & 5 & -1 \\ 3 & -6 & -6 & -6 \end{bmatrix}$

(b) $E = \begin{bmatrix} 1 & 0 & 0 \\ 0 & 1 & 0 \\ 0 & -3 & 1 \end{bmatrix}$, $A = \begin{bmatrix} 2 & -1 & 0 & -4 & -4 \\ 1 & -3 & -1 & 5 & 3 \\ 2 & 0 & 1 & 3 & -1 \end{bmatrix}$

(c) $E = \begin{bmatrix} 1 & 0 & 4 \\ 0 & 1 & 0 \\ 0 & 0 & 1 \end{bmatrix}$, $A = \begin{bmatrix} 1 & 4 \\ 2 & 5 \\ 3 & 6 \end{bmatrix}$

6. In each part, an elementary matrix E and a matrix A are given. Write down the row operation corresponding to E and show that the product EA results from applying the row operation to A.

(a) $E = \begin{bmatrix} -6 & 0 \\ 0 & 1 \end{bmatrix}$, $A = \begin{bmatrix} -1 & -2 & 5 & -1 \\ 3 & -6 & -6 & -6 \end{bmatrix}$

(b) $E = \begin{bmatrix} 1 & 0 & 0 \\ -4 & 1 & 0 \\ 0 & 0 & 1 \end{bmatrix}$, $A = \begin{bmatrix} 2 & -1 & 0 & -4 & -4 \\ 1 & -3 & -1 & 5 & 3 \\ 2 & 0 & 1 & 3 & -1 \end{bmatrix}$

(c) $E = \begin{bmatrix} 1 & 0 & 0 \\ 0 & 5 & 0 \\ 0 & 0 & 1 \end{bmatrix}$, $A = \begin{bmatrix} 1 & 4 \\ 2 & 5 \\ 3 & 6 \end{bmatrix}$

▶ In Exercises 7–8, use the following matrices.

$$A = \begin{bmatrix} 3 & 4 & 1 \\ 2 & -7 & -1 \\ 8 & 1 & 5 \end{bmatrix},\quad B = \begin{bmatrix} 8 & 1 & 5 \\ 2 & -7 & -1 \\ 3 & 4 & 1 \end{bmatrix}$$

$$C = \begin{bmatrix} 3 & 4 & 1 \\ 2 & -7 & -1 \\ 2 & -7 & 3 \end{bmatrix},\quad D = \begin{bmatrix} 8 & 1 & 5 \\ -6 & 21 & 3 \\ 3 & 4 & 1 \end{bmatrix}$$

$$F = \begin{bmatrix} 8 & 1 & 5 \\ 8 & 1 & 1 \\ 3 & 4 & 1 \end{bmatrix} \blacktriangleleft$$

7. Find an elementary matrix E that satisfies the equation.

(a) $EA = B$

(b) $EB = A$

(c) $EA = C$

(d) $EC = A$

8. Find an elementary matrix E that satisfies the equation.

(a) $EB = D$

(b) $ED = B$

(c) $EB = F$

(d) $EF = B$

▶ In Exercises 9–24, use the inversion algorithm to find the inverse of the given matrix, if the inverse exists. ◀

9. $\begin{bmatrix} 1 & 4 \\ 2 & 7 \end{bmatrix}$

10. $\begin{bmatrix} -3 & 6 \\ 4 & 5 \end{bmatrix}$

11. $\begin{bmatrix} -1 & 3 \\ 3 & -2 \end{bmatrix}$

12. $\begin{bmatrix} 6 & -4 \\ -3 & 2 \end{bmatrix}$

13. $\begin{bmatrix} 3 & 4 & -1 \\ 1 & 0 & 3 \\ 2 & 5 & -4 \end{bmatrix}$

14. $\begin{bmatrix} 1 & 2 & 0 \\ 2 & 1 & 2 \\ 0 & 2 & 1 \end{bmatrix}$

15. $\begin{bmatrix} -1 & 3 & -4 \\ 2 & 4 & 1 \\ -4 & 2 & -9 \end{bmatrix}$

16. $\begin{bmatrix} \frac{1}{5} & \frac{1}{5} & -\frac{2}{5} \\ \frac{1}{5} & \frac{1}{5} & \frac{1}{10} \\ \frac{1}{5} & -\frac{4}{5} & \frac{1}{10} \end{bmatrix}$

17. $\begin{bmatrix} 1 & 0 & 1 \\ 0 & 1 & 1 \\ 1 & 1 & 0 \end{bmatrix}$

18. $\begin{bmatrix} \sqrt{2} & 3\sqrt{2} & 0 \\ -4\sqrt{2} & \sqrt{2} & 0 \\ 0 & 0 & 1 \end{bmatrix}$

19. $\begin{bmatrix} 2 & 6 & 6 \\ 2 & 7 & 6 \\ 2 & 7 & 7 \end{bmatrix}$

20. $\begin{bmatrix} 1 & 0 & 0 & 0 \\ 1 & 3 & 0 & 0 \\ 1 & 3 & 5 & 0 \\ 1 & 3 & 5 & 7 \end{bmatrix}$

21. $\begin{bmatrix} 2 & -4 & 0 & 0 \\ 1 & 2 & 12 & 0 \\ 0 & 0 & 2 & 0 \\ 0 & -1 & -4 & -5 \end{bmatrix}$

22. $\begin{bmatrix} -8 & 17 & 2 & \frac{1}{3} \\ 4 & 0 & \frac{2}{5} & -9 \\ 0 & 0 & 0 & 0 \\ -1 & 13 & 4 & 2 \end{bmatrix}$

23. $\begin{bmatrix} -1 & 0 & 1 & 0 \\ 2 & 3 & -2 & 6 \\ 0 & -1 & 2 & 0 \\ 0 & 0 & 1 & 5 \end{bmatrix}$

24. $\begin{bmatrix} 0 & 0 & 2 & 0 \\ 1 & 0 & 0 & 1 \\ 0 & -1 & 3 & 0 \\ 2 & 1 & 5 & -3 \end{bmatrix}$

▷ In Exercises **25–26**, find the inverse of each of the following 4×4 matrices, where k_1, k_2, k_3, k_4, and k are all nonzero. ◁

25. (a) $\begin{bmatrix} k_1 & 0 & 0 & 0 \\ 0 & k_2 & 0 & 0 \\ 0 & 0 & k_3 & 0 \\ 0 & 0 & 0 & k_4 \end{bmatrix}$ **(b)** $\begin{bmatrix} k & 1 & 0 & 0 \\ 0 & 1 & 0 & 0 \\ 0 & 0 & k & 1 \\ 0 & 0 & 0 & 1 \end{bmatrix}$

26. (a) $\begin{bmatrix} 0 & 0 & 0 & k_1 \\ 0 & 0 & k_2 & 0 \\ 0 & k_3 & 0 & 0 \\ k_4 & 0 & 0 & 0 \end{bmatrix}$ **(b)** $\begin{bmatrix} k & 0 & 0 & 0 \\ 1 & k & 0 & 0 \\ 0 & 1 & k & 0 \\ 0 & 0 & 1 & k \end{bmatrix}$

▷ In Exercises **27–28**, find all values of c, if any, for which the given matrix is invertible. ◁

27. $\begin{bmatrix} c & c & c \\ 1 & c & c \\ 1 & 1 & c \end{bmatrix}$

28. $\begin{bmatrix} c & 1 & 0 \\ 1 & c & 1 \\ 0 & 1 & c \end{bmatrix}$

▷ In Exercises **29–32**, write the given matrix as a product of elementary matrices. ◁

29. $\begin{bmatrix} -3 & 1 \\ 2 & 2 \end{bmatrix}$

30. $\begin{bmatrix} 1 & 0 \\ -5 & 2 \end{bmatrix}$

31. $\begin{bmatrix} 1 & 0 & -2 \\ 0 & 4 & 3 \\ 0 & 0 & 1 \end{bmatrix}$

32. $\begin{bmatrix} 1 & 1 & 0 \\ 1 & 1 & 1 \\ 0 & 1 & 1 \end{bmatrix}$

▷ In Exercises **33–36**, write the *inverse* of the given matrix as a product of elementary matrices. ◁

33. The matrix in Exercise 29.

34. The matrix in Exercise 30.

35. The matrix in Exercise 31.

36. The matrix in Exercise 32.

▷ In Exercises **37–38**, show that the given matrices A and B are row equivalent, and find a sequence of elementary row operations that produces B from A. ◁

37. $A = \begin{bmatrix} 1 & 2 & 3 \\ 1 & 4 & 1 \\ 2 & 1 & 9 \end{bmatrix}$, $B = \begin{bmatrix} 1 & 0 & 5 \\ 0 & 2 & -2 \\ 1 & 1 & 4 \end{bmatrix}$

38. $A = \begin{bmatrix} 2 & 1 & 0 \\ -1 & 1 & 0 \\ 3 & 0 & -1 \end{bmatrix}$, $B = \begin{bmatrix} 6 & 9 & 4 \\ -5 & -1 & 0 \\ -1 & -2 & -1 \end{bmatrix}$

39. Show that if

$$A = \begin{bmatrix} 1 & 0 & 0 \\ 0 & 1 & 0 \\ a & b & c \end{bmatrix}$$

is an elementary matrix, then at least one entry in the third row must be zero.

40. Show that

$$A = \begin{bmatrix} 0 & a & 0 & 0 & 0 \\ b & 0 & c & 0 & 0 \\ 0 & d & 0 & e & 0 \\ 0 & 0 & f & 0 & g \\ 0 & 0 & 0 & h & 0 \end{bmatrix}$$

is not invertible for any values of the entries.

41. Prove that if A and B are $m \times n$ matrices, then A and B are row equivalent if and only if A and B have the same reduced row echelon form.

42. Prove that if A is an invertible matrix and B is row equivalent to A, then B is also invertible.

43. Show that if B is obtained from A by performing a sequence of elementary row operations, then there is a second sequence of elementary row operations, which when applied to B recovers A.

True-False Exercises

In parts (a)–(g) determine whether the statement is true or false, and justify your answer.

(a) The product of two elementary matrices of the same size must be an elementary matrix.

(b) Every elementary matrix is invertible.

(c) If A and B are row equivalent, and if B and C are row equivalent, then A and C are row equivalent.

(d) If A is an $n \times n$ matrix that is not invertible, then the linear system $A\mathbf{x} = 0$ has infinitely many solutions.

(e) If A is an $n \times n$ matrix that is not invertible, then the matrix obtained by interchanging two rows of A cannot be invertible.

(f) If A is invertible and a multiple of the first row of A is added to the second row, then the resulting matrix is invertible.

(g) An expression of the invertible matrix A as a product of elementary matrices is unique.

1.6 More on Linear Systems and Invertible Matrics

In this section we will show how the inverse of a matrix can be used to solve a linear system and we will develop some more results about invertible matrices.

Number of Solutions of a Linear System

In Section 1.1 we made the statement (based on Figures 1.1.1 and 1.1.2) that every linear system has either no solutions, has exactly one solution, or has infinitely many solutions. We are now in a position to prove this fundamental result.

> **THEOREM 1.6.1** *A system of linear equations has zero, one, or infinitely many solutions. There are no other possibilities.*

Proof If $A\mathbf{x} = \mathbf{b}$ is a system of linear equations, exactly one of the following is true: (a) the system has no solutions, (b) the system has exactly one solution, or (c) the system has more than one solution. The proof will be complete if we can show that the system has infinitely many solutions in case (c).

Assume that $A\mathbf{x} = \mathbf{b}$ has more than one solution, and let $\mathbf{x}_0 = \mathbf{x}_1 - \mathbf{x}_2$, where $\mathbf{x}_1$ and $\mathbf{x}_2$ are any two distinct solutions. Because $\mathbf{x}_1$ and $\mathbf{x}_2$ are distinct, the matrix $\mathbf{x}_0$ is nonzero; moreover,

$$A\mathbf{x}_0 = A(\mathbf{x}_1 - \mathbf{x}_2) = A\mathbf{x}_1 - A\mathbf{x}_2 = \mathbf{b} - \mathbf{b} = 0$$

If we now let k be any scalar, then

$$A(\mathbf{x}_1 + k\mathbf{x}_0) = A\mathbf{x}_1 + A(k\mathbf{x}_0) = A\mathbf{x}_1 + k(A\mathbf{x}_0)$$
$$= \mathbf{b} + k0 = \mathbf{b} + 0 = \mathbf{b}$$

But this says that $\mathbf{x}_1 + k\mathbf{x}_0$ is a solution of $A\mathbf{x} = \mathbf{b}$. Since $\mathbf{x}_0$ is nonzero and there are infinitely many choices for k, the system $A\mathbf{x} = \mathbf{b}$ has infinitely many solutions. ◀

Solving Linear Systems by Matrix Inversion

Thus far we have studied two *procedures* for solving linear systems—Gauss–Jordan elimination and Gaussian elimination. The following theorem provides an actual *formula* for the solution of a linear system of n equations in n unknowns in the case where the coefficient matrix is invertible.

> **THEOREM 1.6.2** *If A is an invertible $n \times n$ matrix, then for each $n \times 1$ matrix $\mathbf{b}$, the system of equations $A\mathbf{x} = \mathbf{b}$ has exactly one solution, namely, $\mathbf{x} = A^{-1}\mathbf{b}$.*

Proof Since $A(A^{-1}\mathbf{b}) = \mathbf{b}$, it follows that $\mathbf{x} = A^{-1}\mathbf{b}$ is a solution of $A\mathbf{x} = \mathbf{b}$. To show that this is the only solution, we will assume that $\mathbf{x}_0$ is an arbitrary solution and then show that $\mathbf{x}_0$ must be the solution $A^{-1}\mathbf{b}$.

If $\mathbf{x}_0$ is any solution of $A\mathbf{x} = \mathbf{b}$, then $A\mathbf{x}_0 = \mathbf{b}$. Multiplying both sides of this equation by A^{-1}, we obtain $\mathbf{x}_0 = A^{-1}\mathbf{b}$. ◄

▶ **EXAMPLE 1 Solution of a Linear System Using A^{-1}**

Consider the system of linear equations

$$
\begin{aligned}
x_1 + 2x_2 + 3x_3 &= 5 \\
2x_1 + 5x_2 + 3x_3 &= 3 \\
x_1 \qquad\quad + 8x_3 &= 17
\end{aligned}
$$

In matrix form this system can be written as $A\mathbf{x} = \mathbf{b}$, where

$$
A = \begin{bmatrix} 1 & 2 & 3 \\ 2 & 5 & 3 \\ 1 & 0 & 8 \end{bmatrix}, \quad \mathbf{x} = \begin{bmatrix} x_1 \\ x_2 \\ x_3 \end{bmatrix}, \quad \mathbf{b} = \begin{bmatrix} 5 \\ 3 \\ 17 \end{bmatrix}
$$

In Example 4 of the preceding section, we showed that A is invertible and

$$
A^{-1} = \begin{bmatrix} -40 & 16 & 9 \\ 13 & -5 & -3 \\ 5 & -2 & -1 \end{bmatrix}
$$

By Theorem 1.6.2, the solution of the system is

$$
\mathbf{x} = A^{-1}\mathbf{b} = \begin{bmatrix} -40 & 16 & 9 \\ 13 & -5 & -3 \\ 5 & -2 & -1 \end{bmatrix} \begin{bmatrix} 5 \\ 3 \\ 17 \end{bmatrix} = \begin{bmatrix} 1 \\ -1 \\ 2 \end{bmatrix}
$$

Keep in mind that the method of Example 1 only applies when the system has as many equations as unknowns and the coefficient matrix is invertible.

or $x_1 = 1, x_2 = -1, x_3 = 2$. ◄

Linear Systems with a Common Coefficient Matrix

Frequently, one is concerned with solving a sequence of systems

$$
A\mathbf{x} = \mathbf{b}_1, \quad A\mathbf{x} = \mathbf{b}_2, \quad A\mathbf{x} = \mathbf{b}_3, \dots, \quad A\mathbf{x} = \mathbf{b}_k
$$

each of which has the same square coefficient matrix A. If A is invertible, then the solutions

$$
\mathbf{x}_1 = A^{-1}\mathbf{b}_1, \quad \mathbf{x}_2 = A^{-1}\mathbf{b}_2, \quad \mathbf{x}_3 = A^{-1}\mathbf{b}_3, \dots, \quad \mathbf{x}_k = A^{-1}\mathbf{b}_k
$$

can be obtained with one matrix inversion and k matrix multiplications. An efficient way to do this is to form the partitioned matrix

$$
[A \mid \mathbf{b}_1 \mid \mathbf{b}_2 \mid \cdots \mid \mathbf{b}_k] \tag{1}
$$

in which the coefficient matrix A is "augmented" by all k of the matrices $\mathbf{b}_1, \mathbf{b}_2, \dots, \mathbf{b}_k$, and then reduce (1) to reduced row echelon form by Gauss–Jordan elimination. In this way we can solve all k systems at once. This method has the added advantage that it applies even when A is not invertible.

▶ **EXAMPLE 2 Solving Two Linear Systems at Once**

Solve the systems

(a) $x_1 + 2x_2 + 3x_3 = 4$ (b) $x_1 + 2x_2 + 3x_3 = 1$
$\ 2x_1 + 5x_2 + 3x_3 = 5$ $\ 2x_1 + 5x_2 + 3x_3 = 6$
$\ x_1 + 8x_3 = 9$ $\ x_1 + 8x_3 = -6$

Solution The two systems have the same coefficient matrix. If we augment this co-efficient matrix with the columns of constants on the right sides of these systems, we obtain

$$\left[\begin{array}{ccc|c|c} 1 & 2 & 3 & 4 & 1 \\ 2 & 5 & 3 & 5 & 6 \\ 1 & 0 & 8 & 9 & -6 \end{array}\right]$$

Reducing this matrix to reduced row echelon form yields (verify)

$$\left[\begin{array}{ccc|c|c} 1 & 0 & 0 & 1 & 2 \\ 0 & 1 & 0 & 0 & 1 \\ 0 & 0 & 1 & 1 & -1 \end{array}\right]$$

It follows from the last two columns that the solution of system (a) is $x_1 = 1$, $x_2 = 0$, $x_3 = 1$ and the solution of system (b) is $x_1 = 2$, $x_2 = 1$, $x_3 = -1$. ◀

Properties of Invertible Matrices Up to now, to show that an $n \times n$ matrix A is invertible, it has been necessary to find an $n \times n$ matrix B such that

$$AB = I \quad \text{and} \quad BA = I$$

The next theorem shows that if we produce an $n \times n$ matrix B satisfying *either* condition, then the other condition holds automatically.

> **THEOREM 1.6.3** *Let A be a square matrix.*
>
> (a) *If B is a square matrix satisfying $BA = I$, then $B = A^{-1}$.*
> (b) *If B is a square matrix satisfying $AB = I$, then $B = A^{-1}$.*

We will prove part (*a*) and leave part (*b*) as an exercise.

Proof (a) Assume that $BA = I$. If we can show that A is invertible, the proof can be completed by multiplying $BA = I$ on both sides by A^{-1} to obtain

$$BAA^{-1} = IA^{-1} \quad \text{or} \quad BI = IA^{-1} \quad \text{or} \quad B = A^{-1}$$

To show that A is invertible, it suffices to show that the system $A\mathbf{x} = \mathbf{0}$ has only the trivial solution (see Theorem 1.5.3). Let $\mathbf{x}_0$ be any solution of this system. If we multiply both sides of $A\mathbf{x}_0 = \mathbf{0}$ on the left by B, we obtain $BA\mathbf{x}_0 = B\mathbf{0}$ or $I\mathbf{x}_0 = \mathbf{0}$ or $\mathbf{x}_0 = \mathbf{0}$. Thus, the system of equations $A\mathbf{x} = \mathbf{0}$ has only the trivial solution. ◀

Equivalence Theorem We are now in a position to add two more statements to the four given in Theorem 1.5.3.

> **THEOREM 1.6.4 Equivalent Statements**
>
> *If A is an n × n matrix, then the following are equivalent.*
>
> (a) *A is invertible.*
>
> (b) *$A\mathbf{x} = \mathbf{0}$ has only the trivial solution.*
>
> (c) *The reduced row echelon form of A is I_n.*
>
> (d) *A is expressible as a product of elementary matrices.*
>
> (e) *$A\mathbf{x} = \mathbf{b}$ is consistent for every n × 1 matrix **b**.*
>
> (f) *$A\mathbf{x} = \mathbf{b}$ has exactly one solution for every n × 1 matrix **b**.*

Proof Since we proved in Theorem 1.5.3 that (*a*), (*b*), (*c*), and (*d*) are equivalent, it will be sufficient to prove that $(a) \Rightarrow (f) \Rightarrow (e) \Rightarrow (a)$.

(a) ⇒ (f) This was already proved in Theorem 1.6.2.

(f) ⇒ (e) This is almost self-evident, for if $A\mathbf{x} = \mathbf{b}$ has exactly one solution for every $n \times 1$ matrix **b**, then $A\mathbf{x} = \mathbf{b}$ is consistent for every $n \times 1$ matrix **b**.

(e) ⇒ (a) If the system $A\mathbf{x} = \mathbf{b}$ is consistent for every $n \times 1$ matrix **b**, then, in particular, this is so for the systems

$$A\mathbf{x} = \begin{bmatrix} 1 \\ 0 \\ 0 \\ \vdots \\ 0 \end{bmatrix}, \quad A\mathbf{x} = \begin{bmatrix} 0 \\ 1 \\ 0 \\ \vdots \\ 0 \end{bmatrix}, \dots, \quad A\mathbf{x} = \begin{bmatrix} 0 \\ 0 \\ 0 \\ \vdots \\ 1 \end{bmatrix}$$

Let $\mathbf{x}_1, \mathbf{x}_2, \dots, \mathbf{x}_n$ be solutions of the respective systems, and let us form an $n \times n$ matrix C having these solutions as columns. Thus C has the form

$$C = [\mathbf{x}_1 \mid \mathbf{x}_2 \mid \cdots \mid \mathbf{x}_n]$$

As discussed in Section 1.3, the successive columns of the product AC will be

$$A\mathbf{x}_1, A\mathbf{x}_2, \dots, A\mathbf{x}_n$$

[see Formula (8) of Section 1.3]. Thus,

It follows from the equivalency of parts (*e*) and (*f*) that if you can show that $A\mathbf{x} = \mathbf{b}$ has at *least one* solution for every $n \times 1$ matrix **b**, then you can conclude that it has *exactly one* solution for every $n \times 1$ matrix **b**.

$$AC = [A\mathbf{x}_1 \mid A\mathbf{x}_2 \mid \cdots \mid A\mathbf{x}_n] = \begin{bmatrix} 1 & 0 & \cdots & 0 \\ 0 & 1 & \cdots & 0 \\ 0 & 0 & \cdots & 0 \\ \vdots & \vdots & & \vdots \\ 0 & 0 & \cdots & 1 \end{bmatrix} = I$$

By part (*b*) of Theorem 1.6.3, it follows that $C = A^{-1}$. Thus, A is invertible. ◀

We know from earlier work that invertible matrix factors produce an invertible product. Conversely, the following theorem shows that if the product of square matrices is invertible, then the factors themselves must be invertible.

> **THEOREM 1.6.5** *Let A and B be square matrices of the same size. If AB is invertible, then A and B must also be invertible.*

In our later work the following fundamental problem will occur frequently in various contexts.

A Fundamental Problem   Let A be a fixed $m \times n$ matrix. Find all $m \times 1$ matrices $\mathbf{b}$ such that the system of equations $A\mathbf{x} = \mathbf{b}$ is consistent.

If A is an invertible matrix, Theorem 1.6.2 completely solves this problem by asserting that for *every* $m \times 1$ matrix $\mathbf{b}$, the linear system $A\mathbf{x} = \mathbf{b}$ has the unique solution $\mathbf{x} = A^{-1}\mathbf{b}$. If A is not square, or if A is square but not invertible, then Theorem 1.6.2 does not apply. In these cases the matrix $\mathbf{b}$ must usually satisfy certain conditions in order for $A\mathbf{x} = \mathbf{b}$ to be consistent. The following example illustrates how the methods of Section 1.2 can be used to determine such conditions.

▶ **EXAMPLE 3   Determining Consistency by Elimination**

What conditions must b_1, b_2, and b_3 satisfy in order for the system of equations

$$
\begin{aligned}
x_1 + x_2 + 2x_3 &= b_1 \\
x_1 \quad\quad\; + x_3 &= b_2 \\
2x_1 + x_2 + 3x_3 &= b_3
\end{aligned}
$$

to be consistent?

Solution   The augmented matrix is

$$
\begin{bmatrix}
1 & 1 & 2 & b_1 \\
1 & 0 & 1 & b_2 \\
2 & 1 & 3 & b_3
\end{bmatrix}
$$

which can be reduced to row echelon form as follows:

$$
\begin{bmatrix}
1 & 1 & 2 & b_1 \\
0 & -1 & -1 & b_2 - b_1 \\
0 & -1 & -1 & b_3 - 2b_1
\end{bmatrix}
$$
    ◀——— -1 times the first row was added to the second and -2 times the first row was added to the third.

$$
\begin{bmatrix}
1 & 1 & 2 & b_1 \\
0 & 1 & 1 & b_1 - b_2 \\
0 & -1 & -1 & b_3 - 2b_1
\end{bmatrix}
$$
    ◀——— The second row was multiplied by -1.

$$
\begin{bmatrix}
1 & 1 & 2 & b_1 \\
0 & 1 & 1 & b_1 - b_2 \\
0 & 0 & 0 & b_3 - b_2 - b_1
\end{bmatrix}
$$
    ◀——— The second row was added to the third.

It is now evident from the third row in the matrix that the system has a solution if and only if b_1, b_2, and b_3 satisfy the condition

$$
b_3 - b_2 - b_1 = 0 \quad \text{or} \quad b_3 = b_1 + b_2
$$

To express this condition another way, $A\mathbf{x} = \mathbf{b}$ is consistent if and only if $\mathbf{b}$ is a matrix of the form

$$
\mathbf{b} = \begin{bmatrix} b_1 \\ b_2 \\ b_1 + b_2 \end{bmatrix}
$$

where b_1 and b_2 are arbitrary.

▶ **EXAMPLE 4** Determining Consistency by Elimination

What conditions must b_1, b_2, and b_3 satisfy in order for the system of equations

$$x_1 + 2x_2 + 3x_3 = b_1$$
$$2x_1 + 5x_2 + 3x_3 = b_2$$
$$x_1 \qquad + 8x_3 = b_3$$

to be consistent?

Solution The augmented matrix is

$$\begin{bmatrix} 1 & 2 & 3 & b_1 \\ 2 & 5 & 3 & b_2 \\ 1 & 0 & 8 & b_3 \end{bmatrix}$$

Reducing this to reduced row echelon form yields (verify)

$$\begin{bmatrix} 1 & 0 & 0 & -40b_1 + 16b_2 + 9b_3 \\ 0 & 1 & 0 & 13b_1 - 5b_2 - 3b_3 \\ 0 & 0 & 1 & 5b_1 - 2b_2 - b_3 \end{bmatrix} \qquad (2)$$

In this case there are no restrictions on b_1, b_2, and b_3, so the system has the unique solution

$$x_1 = -40b_1 + 16b_2 + 9b_3, \quad x_2 = 13b_1 - 5b_2 - 3b_3, \quad x_3 = 5b_1 - 2b_2 - b_3 \quad (3)$$

for all values of b_1, b_2, and b_3. ◀

What does the result in Example 4 tell you about the coefficient matrix of the system?

Skills

- Determine whether a linear system of equations has no solutions, exactly one solution, or infinitely many solutions.

- Solve linear systems by inverting its coefficient matrix.

- Solve multiple linear systems with the same coefficient matrix simultaneously.

- Be familiar with the additional conditions of invertibility stated in the Equivalence Theorem.

Exercise Set 1.6

▶ In Exercises **1–8**, solve the system by inverting the coefficient matrix and using Theorem 1.6.2. ◀

1. $x_1 + x_2 = 2$
$5x_1 + 6x_2 = 9$

2. $4x_1 - 3x_2 = -3$
$2x_1 - 5x_2 = 9$

3. $x_1 + 3x_2 + x_3 = 4$
$2x_1 + 2x_2 + x_3 = -1$
$2x_1 + 3x_2 + x_3 = 3$

4. $5x_1 + 3x_2 + 2x_3 = 4$
$3x_1 + 3x_2 + 2x_3 = 2$
$x_2 + x_3 = 5$

5. $x + y + z = 5$
$x + y - 4z = 10$
$-4x + y + z = 0$

6. $- x - 2y - 3z = 0$
$w + x + 4y + 4z = 7$
$w + 3x + 7y + 9z = 4$
$-w - 2x - 4y - 6z = 6$

7. $3x_1 + 5x_2 = b_1$
$x_1 + 2x_2 = b_2$

8. $x_1 + 2x_2 + 3x_3 = b_1$
$2x_1 + 5x_2 + 5x_3 = b_2$
$3x_1 + 5x_2 + 8x_3 = b_3$

▶ In Exercises **9–12**, solve the linear systems together by reducing the appropriate augmented matrix. ◀

9. $x_1 - 5x_2 = b_1$
$3x_1 + 2x_2 = b_2$
(i) $b_1 = 1$, $b_2 = 4$ (ii) $b_1 = -2$, $b_2 = 5$

10. $-x_1 + 4x_2 + x_3 = b_1$
$x_1 + 9x_2 - 2x_3 = b_2$
$6x_1 + 4x_2 - 8x_3 = b_3$
(i) $b_1 = 0$, $b_2 = 1$, $b_3 = 0$
(ii) $b_1 = -3$, $b_2 = 4$, $b_3 = -5$

11. $4x_1 - 7x_2 = b_1$
$x_1 + 2x_2 = b_2$
(i) $b_1 = 0$, $b_2 = 1$ (ii) $b_1 = -4$, $b_2 = 6$
(iii) $b_1 = -1$, $b_2 = 3$ (iv) $b_1 = -5$, $b_2 = 1$

12.
$$x_1 + 3x_2 + 5x_3 = b_1$$
$$-x_1 - 2x_2 \quad\quad = b_2$$
$$2x_1 + 5x_2 + 4x_3 = b_3$$

(i) $b_1 = 1, \quad b_2 = 0, \quad b_3 = -1$
(ii) $b_1 = 0, \quad b_2 = 1, \quad b_3 = 1$
(iii) $b_1 = -1, \quad b_2 = -1, \quad b_3 = 0$

▶ In Exercises **13–17**, determine conditions on the b_i's, if any, in order to guarantee that the linear system is consistent. ◀

13.
$$x_1 + 3x_2 = b_1$$
$$-2x_1 + x_2 = b_2$$

14.
$$6x_1 - 4x_2 = b_1$$
$$3x_1 - 2x_2 = b_2$$

15.
$$x_1 - 2x_2 + 5x_3 = b_1$$
$$4x_1 - 5x_2 + 8x_3 = b_2$$
$$-3x_1 + 3x_2 - 3x_3 = b_3$$

16.
$$x_1 - 2x_2 - x_3 = b_1$$
$$-4x_1 + 5x_2 + 2x_3 = b_2$$
$$-4x_1 + 7x_2 + 4x_3 = b_3$$

17.
$$x_1 - x_2 + 3x_3 + 2x_4 = b_1$$
$$-2x_1 + x_2 + 5x_3 + x_4 = b_2$$
$$-3x_1 + 2x_2 + 2x_3 - x_4 = b_3$$
$$4x_1 - 3x_2 + x_3 + 3x_4 = b_4$$

18. Consider the matrices

$$A = \begin{bmatrix} 2 & 1 & 2 \\ 2 & 2 & -2 \\ 3 & 1 & 1 \end{bmatrix} \quad \text{and} \quad \mathbf{x} = \begin{bmatrix} x_1 \\ x_2 \\ x_3 \end{bmatrix}$$

(a) Show that the equation $A\mathbf{x} = \mathbf{x}$ can be rewritten as $(A - I)\mathbf{x} = \mathbf{0}$ and use this result to solve $A\mathbf{x} = \mathbf{x}$ for $\mathbf{x}$.

(b) Solve $A\mathbf{x} = 4\mathbf{x}$.

▶ In Exercises **19–20**, solve the given matrix equation for X. ◀

19.
$$\begin{bmatrix} 1 & -1 & 1 \\ 2 & 3 & 0 \\ 0 & 2 & -1 \end{bmatrix} X = \begin{bmatrix} 2 & -1 & 5 & 7 & 8 \\ 4 & 0 & -3 & 0 & 1 \\ 3 & 5 & -7 & 2 & 1 \end{bmatrix}$$

20.
$$\begin{bmatrix} -2 & 0 & 1 \\ 0 & -1 & -1 \\ 1 & 1 & -4 \end{bmatrix} X = \begin{bmatrix} 4 & 3 & 2 & 1 \\ 6 & 7 & 8 & 9 \\ 1 & 3 & 7 & 9 \end{bmatrix}$$

21. Let $A\mathbf{x} = \mathbf{0}$ be a homogeneous system of n linear equations in n unknowns that has only the trivial solution. Show that if k is any positive integer, then the system $A^k\mathbf{x} = \mathbf{0}$ also has only the trivial solution.

22. Let $A\mathbf{x} = \mathbf{0}$ be a homogeneous system of n linear equations in n unknowns, and let Q be an invertible $n \times n$ matrix. Show that $A\mathbf{x} = \mathbf{0}$ has just the trivial solution if and only if $(QA)\mathbf{x} = \mathbf{0}$ has just the trivial solution.

23. Let $A\mathbf{x} = \mathbf{b}$ be any consistent system of linear equations, and let $\mathbf{x}_1$ be a fixed solution. Show that every solution to the system can be written in the form $\mathbf{x} = \mathbf{x}_1 + \mathbf{x}_0$, where $\mathbf{x}_0$ is a solution to $A\mathbf{x} = \mathbf{0}$. Show also that every matrix of this form is a solution.

24. Use part (*a*) of Theorem 1.6.3 to prove part (*b*).

True-False Exercises

In parts (a)–(g) determine whether the statement is true or false, and justify your answer.

(a) It is impossible for a system of linear equations to have exactly two solutions.

(b) If the linear system $A\mathbf{x} = \mathbf{b}$ has a unique solution, then the linear system $A\mathbf{x} = \mathbf{c}$ also must have a unique solution.

(c) If A and B are $n \times n$ matrices such that $AB = I_n$, then $BA = I_n$.

(d) If A and B are row equivalent matrices, then the linear systems $A\mathbf{x} = \mathbf{0}$ and $B\mathbf{x} = \mathbf{0}$ have the same solution set.

(e) If A is an $n \times n$ matrix and S is an $n \times n$ invertible matrix, then if $\mathbf{x}$ is a solution to the linear system $(S^{-1}AS)\mathbf{x} = \mathbf{b}$, then $S\mathbf{x}$ is a solution to the linear system $A\mathbf{y} = S\mathbf{b}$.

(f) Let A be an $n \times n$ matrix. The linear system $A\mathbf{x} = 4\mathbf{x}$ has a unique solution if and only if $A - 4I$ is an invertible matrix.

(g) Let A and B be $n \times n$ matrices. If A or B (or both) are not invertible, then neither is AB.

1.7 Diagonal, Triangular, and Symmetric Matrices

In this section we will discuss matrices that have various special forms. These matrices arise in a wide variety of applications and will play an important role in our subsequent work.

Diagonal Matrices   A square matrix in which all the entries off the main diagonal are zero is called a *diagonal matrix*. Here are some examples:

$$\begin{bmatrix} 0 & 0 \\ 0 & 0 \end{bmatrix}, \quad \begin{bmatrix} 2 & 0 \\ 0 & -5 \end{bmatrix}, \quad \begin{bmatrix} 1 & 0 & 0 \\ 0 & 1 & 0 \\ 0 & 0 & 1 \end{bmatrix}, \quad \begin{bmatrix} 6 & 0 & 0 & 0 \\ 0 & -4 & 0 & 0 \\ 0 & 0 & 0 & 0 \\ 0 & 0 & 0 & 8 \end{bmatrix}$$

A general $n \times n$ diagonal matrix D can be written as

$$D = \begin{bmatrix} d_1 & 0 & \cdots & 0 \\ 0 & d_2 & \cdots & 0 \\ \vdots & \vdots & & \vdots \\ 0 & 0 & \cdots & d_n \end{bmatrix} \tag{1}$$

A diagonal matrix is invertible if and only if all of its diagonal entries are nonzero; in this case the inverse of (1) is

Confirm Formula (2) by showing that

$$DD^{-1} = D^{-1}D = I$$

$$D^{-1} = \begin{bmatrix} 1/d_1 & 0 & \cdots & 0 \\ 0 & 1/d_2 & \cdots & 0 \\ \vdots & \vdots & & \vdots \\ 0 & 0 & \cdots & 1/d_n \end{bmatrix} \tag{2}$$

Powers of diagonal matrices are easy to compute; we leave it for you to verify that if D is the diagonal matrix (1) and k is a positive integer, then

$$D^k = \begin{bmatrix} d_1^k & 0 & \cdots & 0 \\ 0 & d_2^k & \cdots & 0 \\ \vdots & \vdots & & \vdots \\ 0 & 0 & \cdots & d_n^k \end{bmatrix} \tag{3}$$

▶ **EXAMPLE 1 Inverses and Powers of Diagonal Matrices**

If

$$A = \begin{bmatrix} 1 & 0 & 0 \\ 0 & -3 & 0 \\ 0 & 0 & 2 \end{bmatrix}$$

then

$$A^{-1} = \begin{bmatrix} 1 & 0 & 0 \\ 0 & -\frac{1}{3} & 0 \\ 0 & 0 & \frac{1}{2} \end{bmatrix}, \quad A^5 = \begin{bmatrix} 1 & 0 & 0 \\ 0 & -243 & 0 \\ 0 & 0 & 32 \end{bmatrix}, \quad A^{-5} = \begin{bmatrix} 1 & 0 & 0 \\ 0 & -\frac{1}{243} & 0 \\ 0 & 0 & \frac{1}{32} \end{bmatrix}$$

◀

Matrix products that involve diagonal factors are especially easy to compute. For example,

$$\begin{bmatrix} d_1 & 0 & 0 \\ 0 & d_2 & 0 \\ 0 & 0 & d_3 \end{bmatrix} \begin{bmatrix} a_{11} & a_{12} & a_{13} & a_{14} \\ a_{21} & a_{22} & a_{23} & a_{24} \\ a_{31} & a_{32} & a_{33} & a_{34} \end{bmatrix} = \begin{bmatrix} d_1 a_{11} & d_1 a_{12} & d_1 a_{13} & d_1 a_{14} \\ d_2 a_{21} & d_2 a_{22} & d_2 a_{23} & d_2 a_{24} \\ d_3 a_{31} & d_3 a_{32} & d_3 a_{33} & d_3 a_{34} \end{bmatrix}$$

$$\begin{bmatrix} a_{11} & a_{12} & a_{13} \\ a_{21} & a_{22} & a_{23} \\ a_{31} & a_{32} & a_{33} \\ a_{41} & a_{42} & a_{43} \end{bmatrix} \begin{bmatrix} d_1 & 0 & 0 \\ 0 & d_2 & 0 \\ 0 & 0 & d_3 \end{bmatrix} = \begin{bmatrix} d_1 a_{11} & d_2 a_{12} & d_3 a_{13} \\ d_1 a_{21} & d_2 a_{22} & d_3 a_{23} \\ d_1 a_{31} & d_2 a_{32} & d_3 a_{33} \\ d_1 a_{41} & d_2 a_{42} & d_3 a_{43} \end{bmatrix}$$

In words, *to multiply a matrix A on the left by a diagonal matrix D, one can multiply successive rows of A by the successive diagonal entries of D, and to multiply A on the right by D, one can multiply successive columns of A by the successive diagonal entries of D.*

Triangular Matrices   A square matrix in which all the entries above the main diagonal are zero is called ***lower triangular***, and a square matrix in which all the entries below the main diagonal are zero is called ***upper triangular***. A matrix that is either upper triangular or lower triangular is called ***triangular***.

▶ **EXAMPLE 2**   **Upper and Lower Triangular Matrices**

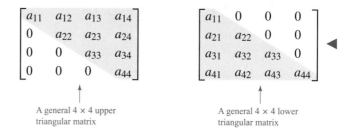

A general 4 × 4 upper triangular matrix

A general 4 × 4 lower triangular matrix

Remark   Observe that diagonal matrices are both upper triangular and lower triangular since they have zeros below and above the main diagonal. Observe also that a *square* matrix in row echelon form is upper triangular since it has zeros below the main diagonal.

Properties of Triangular Matrices   Example 2 illustrates the following four facts about triangular matrices that we will state without formal proof:

- A square matrix $A = [a_{ij}]$ is upper triangular if and only if all entries to the left of the main diagonal are zero; that is, $a_{ij} = 0$ if $i > j$ (Figure 1.7.1).

- A square matrix $A = [a_{ij}]$ is lower triangular if and only if all entries to the right of the main diagonal are zero; that is, $a_{ij} = 0$ if $i < j$ (Figure 1.7.1).

- A square matrix $A = [a_{ij}]$ is upper triangular if and only if the ith row starts with at least $i - 1$ zeros for every i.

- A square matrix $A = [a_{ij}]$ is lower triangular if and only if the jth column starts with at least $j - 1$ zeros for every j.

▲ Figure 1.7.1

The following theorem lists some of the basic properties of triangular matrices.

THEOREM 1.7.1

(*a*)   *The transpose of a lower triangular matrix is upper triangular, and the transpose of an upper triangular matrix is lower triangular.*

(*b*)   *The product of lower triangular matrices is lower triangular, and the product of upper triangular matrices is upper triangular.*

(*c*)   *A triangular matrix is invertible if and only if its diagonal entries are all nonzero.*

(*d*)   *The inverse of an invertible lower triangular matrix is lower triangular, and the inverse of an invertible upper triangular matrix is upper triangular.*

Part (*a*) is evident from the fact that transposing a square matrix can be accomplished by reflecting the entries about the main diagonal; we omit the formal proof. We will prove (*b*), but we will defer the proofs of (*c*) and (*d*) to the next chapter, where we will have the tools to prove those results more efficiently.

Proof (b) We will prove the result for lower triangular matrices; the proof for upper triangular matrices is similar. Let $A = [a_{ij}]$ and $B = [b_{ij}]$ be lower triangular $n \times n$ matrices, and let $C = [c_{ij}]$ be the product $C = AB$. We can prove that C is lower triangular by showing that $c_{ij} = 0$ for $i < j$. But from the definition of matrix multiplication,

$$c_{ij} = a_{i1}b_{1j} + a_{i2}b_{2j} + \cdots + a_{in}b_{nj}$$

If we assume that $i < j$, then the terms in this expression can be grouped as follows:

$$c_{ij} = \underbrace{a_{i1}b_{1j} + a_{i2}b_{2j} + \cdots + a_{i(j-1)}b_{(j-1)j}}_{\substack{\text{Terms in which the row} \\ \text{number of } b \text{ is less than} \\ \text{the column number of } b}} + \underbrace{a_{ij}b_{j} + \cdots + a_{in}b_{nj}}_{\substack{\text{Terms in which the row} \\ \text{number of } a \text{ is less than} \\ \text{the column number of } a}}$$

In the first grouping all of the b factors are zero since B is lower triangular, and in the second grouping all of the a factors are zero since A is lower triangular. Thus, $c_{ij} = 0$, which is what we wanted to prove. ◄

▶ **EXAMPLE 3 Computations with Triangular Matrices**

Consider the upper triangular matrices

$$A = \begin{bmatrix} 1 & 3 & -1 \\ 0 & 2 & 4 \\ 0 & 0 & 5 \end{bmatrix}, \quad B = \begin{bmatrix} 3 & -2 & 2 \\ 0 & 0 & -1 \\ 0 & 0 & 1 \end{bmatrix}$$

It follows from part (c) of Theorem 1.7.1 that the matrix A is invertible but the matrix B is not. Moreover, the theorem also tells us that A^{-1}, AB, and BA must be upper triangular. We leave it for you to confirm these three statements by showing that

$$A^{-1} = \begin{bmatrix} 1 & -\frac{3}{2} & \frac{7}{5} \\ 0 & \frac{1}{2} & -\frac{2}{5} \\ 0 & 0 & \frac{1}{5} \end{bmatrix}, \quad AB = \begin{bmatrix} 3 & -2 & -2 \\ 0 & 0 & 2 \\ 0 & 0 & 5 \end{bmatrix}, \quad BA = \begin{bmatrix} 3 & 5 & -1 \\ 0 & 0 & -5 \\ 0 & 0 & 5 \end{bmatrix} \quad ◄$$

Symmetric Matrices

DEFINITION 1 A square matrix A is said to be ***symmetric*** if $A = A^T$.

It is easy to recognize a symmetric matrix by inspection: The entries on the main diagonal have no restrictions, but mirror images of entries *across* the main diagonal must be equal. Here is a picture using the second matrix in Example 4:

$$\begin{bmatrix} 1 & 4 & 5 \\ 4 & -3 & 0 \\ 5 & 0 & 7 \end{bmatrix}$$

All diagonal matrices, such as the third matrix in Example 4, have this property.

▶ **EXAMPLE 4 Symmetric Matrices**

The following matrices are symmetric, since each is equal to its own transpose (verify).

$$\begin{bmatrix} 7 & -3 \\ -3 & 5 \end{bmatrix}, \quad \begin{bmatrix} 1 & 4 & 5 \\ 4 & -3 & 0 \\ 5 & 0 & 7 \end{bmatrix}, \quad \begin{bmatrix} d_1 & 0 & 0 & 0 \\ 0 & d_2 & 0 & 0 \\ 0 & 0 & d_3 & 0 \\ 0 & 0 & 0 & d_4 \end{bmatrix} \quad ◄$$

Remark It follows from Formula (11) of Section 1.3 that a square matrix $A = [a_{ij}]$ is symmetric if and only if

$$(A)_{ij} = (A)_{ji} \tag{4}$$

for all values of i and j.

The following theorem lists the main algebraic properties of symmetric matrices. The proofs are direct consequences of Theorem 1.4.8 and are omitted.

THEOREM 1.7.2 *If A and B are symmetric matrices with the same size, and if k is any scalar, then:*

(a) A^T *is symmetric.*

(b) $A + B$ *and* $A - B$ *are symmetric.*

(c) kA *is symmetric.*

It is not true, in general, that the product of symmetric matrices is symmetric. To see why this is so, let A and B be symmetric matrices with the same size. Then it follows from part (*e*) of Theorem 1.4.8 and the symmetry of A and B that

$$(AB)^T = B^T A^T = BA$$

Thus, $(AB)^T = AB$ if and only if $AB = BA$, that is, if and only if A and B commute. In summary, we have the following result.

THEOREM 1.7.3 *The product of two symmetric matrices is symmetric if and only if the matrices commute.*

Use the result in Theorem 1.4.5 to confirm Theorem 1.7.3 in the case where the symmetric matrix

$$\begin{bmatrix} a & b \\ b & d \end{bmatrix}$$

is invertible.

▶ **EXAMPLE 5 Products of Symmetric Matrices**

The first of the following equations shows a product of symmetric matrices that *is not* symmetric, and the second shows a product of symmetric matrices that *is* symmetric. We conclude that the factors in the first equation do not commute, but those in the second equation do. We leave it for you to verify that this is so.

$$\begin{bmatrix} 1 & 2 \\ 2 & 3 \end{bmatrix} \begin{bmatrix} -4 & 1 \\ 1 & 0 \end{bmatrix} = \begin{bmatrix} -2 & 1 \\ -5 & 2 \end{bmatrix}$$

$$\begin{bmatrix} 1 & 2 \\ 2 & 3 \end{bmatrix} \begin{bmatrix} -4 & 3 \\ 3 & -1 \end{bmatrix} = \begin{bmatrix} 2 & 1 \\ 1 & 3 \end{bmatrix} ◀$$

Invertibility of Symmetric Matrices

In general, a symmetric matrix need not be invertible. For example, a diagonal matrix with a zero on the main diagonal is symmetric but not invertible. However, the following theorem shows that if a symmetric matrix happens to be invertible, then its inverse must also be symmetric.

THEOREM 1.7.4 *If A is an invertible symmetric matrix, then A^{-1} is symmetric.*

Proof Assume that A is symmetric and invertible. From Theorem 1.4.9 and the fact that $A = A^T$, we have

$$(A^{-1})^T = (A^T)^{-1} = A^{-1}$$

which proves that A^{-1} is symmetric. ◀

Products AA^T and A^TA

Matrix products of the form AA^T and A^TA arise in a variety of applications. If A is an $m \times n$ matrix, then A^T is an $n \times m$ matrix, so the products AA^T and A^TA are both square matrices—the matrix AA^T has size $m \times m$, and the matrix A^TA has size $n \times n$. Such products are always symmetric since

$$(AA^T)^T = (A^T)^T A^T = AA^T \quad \text{and} \quad (A^TA)^T = A^T(A^T)^T = A^TA$$

▶ **EXAMPLE 6 The Product of a Matrix and Its Transpose Is Symmetric**

Let A be the 2×3 matrix

$$A = \begin{bmatrix} 1 & -2 & 4 \\ 3 & 0 & -5 \end{bmatrix}$$

Then

$$A^T A = \begin{bmatrix} 1 & 3 \\ -2 & 0 \\ 4 & -5 \end{bmatrix} \begin{bmatrix} 1 & -2 & 4 \\ 3 & 0 & -5 \end{bmatrix} = \begin{bmatrix} 10 & -2 & -11 \\ -2 & 4 & -8 \\ -11 & -8 & 41 \end{bmatrix}$$

$$A A^T = \begin{bmatrix} 1 & -2 & 4 \\ 3 & 0 & -5 \end{bmatrix} \begin{bmatrix} 1 & 3 \\ -2 & 0 \\ 4 & -5 \end{bmatrix} = \begin{bmatrix} 21 & -17 \\ -17 & 34 \end{bmatrix}$$

Observe that $A^T A$ and $A A^T$ are symmetric as expected. ◀

Later in this text, we will obtain general conditions on A under which $A A^T$ and $A^T A$ are invertible. However, in the special case where A is *square*, we have the following result.

THEOREM 1.7.5 *If A is an invertible matrix, then $A A^T$ and $A^T A$ are also invertible.*

Proof Since A is invertible, so is A^T by Theorem 1.4.9. Thus $A A^T$ and $A^T A$ are invertible, since they are the products of invertible matrices. ◀

Concept Review

- Diagonal matrix
- Lower triangular matrix
- Upper triangular matrix
- Triangular matrix
- Symmetric matrix

Skills

- Determine whether a diagonal matrix is invertible with no computations.
- Compute matrix products involving diagonal matrices by inspection.
- Determine whether a matrix is triangular.
- Understand how the transpose operation affects diagonal and triangular matrices.
- Understand how inversion affects diagonal and triangular matrices.
- Determine whether a matrix is a symmetric matrix.

Exercise Set 1.7

▶ In Exercises **1–4**, determine whether the given matrix is invertible. ◀

1. $\begin{bmatrix} 2 & 0 \\ 0 & -5 \end{bmatrix}$

2. $\begin{bmatrix} 4 & 0 & 0 \\ 0 & 0 & 0 \\ 0 & 0 & 5 \end{bmatrix}$

3. $\begin{bmatrix} -1 & 0 & 0 \\ 0 & 2 & 0 \\ 0 & 0 & \frac{1}{3} \end{bmatrix}$

4. $\begin{bmatrix} -1 & 0 & 0 & 0 \\ 0 & 3 & 0 & 0 \\ 0 & 0 & -3 & 0 \\ 0 & 0 & 0 & -2 \end{bmatrix}$

▶ In Exercises **5–8**, determine the product by inspection. ◀

5. $\begin{bmatrix} 3 & 0 & 0 \\ 0 & -1 & 0 \\ 0 & 0 & 2 \end{bmatrix} \begin{bmatrix} 2 & 1 \\ -4 & 1 \\ 2 & 5 \end{bmatrix}$

6. $\begin{bmatrix} 1 & 2 & -5 \\ -3 & -1 & 0 \end{bmatrix} \begin{bmatrix} -4 & 0 & 0 \\ 0 & 3 & 0 \\ 0 & 0 & 2 \end{bmatrix}$

7. $\begin{bmatrix} 5 & 0 & 0 \\ 0 & 2 & 0 \\ 0 & 0 & -3 \end{bmatrix} \begin{bmatrix} -3 & 2 & 0 & 4 & -4 \\ 1 & -5 & 3 & 0 & 3 \\ -6 & 2 & 2 & 2 & 2 \end{bmatrix}$

8. $\begin{bmatrix} 2 & 0 & 0 \\ 0 & -1 & 0 \\ 0 & 0 & 4 \end{bmatrix} \begin{bmatrix} 4 & -1 & 3 \\ 1 & 2 & 0 \\ -5 & 1 & -2 \end{bmatrix} \begin{bmatrix} -3 & 0 & 0 \\ 0 & 5 & 0 \\ 0 & 0 & 2 \end{bmatrix}$

▶ In Exercises 9–12, find A^2, A^{-2}, and A^{-k} (where k is any integer) by inspection. ◀

9. $A = \begin{bmatrix} 1 & 0 \\ 0 & -2 \end{bmatrix}$ **10.** $A = \begin{bmatrix} -6 & 0 & 0 \\ 0 & 3 & 0 \\ 0 & 0 & 5 \end{bmatrix}$

11. $A = \begin{bmatrix} \frac{1}{2} & 0 & 0 \\ 0 & \frac{1}{3} & 0 \\ 0 & 0 & \frac{1}{4} \end{bmatrix}$ **12.** $A = \begin{bmatrix} -2 & 0 & 0 & 0 \\ 0 & -4 & 0 & 0 \\ 0 & 0 & -3 & 0 \\ 0 & 0 & 0 & 2 \end{bmatrix}$

▶ In Exercises 13–19, decide whether the given matrix is symmetric. ◀

13. $\begin{bmatrix} -8 & -8 \\ 0 & 0 \end{bmatrix}$ **14.** $\begin{bmatrix} 2 & -1 \\ 1 & 2 \end{bmatrix}$ **15.** $\begin{bmatrix} 0 & -7 \\ -7 & 7 \end{bmatrix}$

16. $\begin{bmatrix} 3 & 4 \\ 4 & 0 \end{bmatrix}$ **17.** $\begin{bmatrix} 0 & 1 & 2 \\ 1 & 5 & -6 \\ 2 & 6 & 6 \end{bmatrix}$

18. $\begin{bmatrix} 2 & -1 & 3 \\ -1 & 5 & 1 \\ 3 & 1 & 7 \end{bmatrix}$ **19.** $\begin{bmatrix} 0 & 0 & 1 \\ 0 & 2 & 0 \\ 3 & 0 & 0 \end{bmatrix}$

▶ In Exercises 20–22, decide by inspection whether the given matrix is invertible. ◀

20. $\begin{bmatrix} -1 & 2 & 4 \\ 0 & 3 & 0 \\ 0 & 0 & 5 \end{bmatrix}$ **21.** $\begin{bmatrix} 0 & 1 & -2 & 5 \\ 0 & 1 & 5 & 6 \\ 0 & 0 & -3 & 1 \\ 0 & 0 & 0 & 5 \end{bmatrix}$

22. $\begin{bmatrix} 2 & 0 & 0 & 0 \\ -3 & -1 & 0 & 0 \\ -4 & -6 & 0 & 0 \\ 0 & 3 & 8 & -5 \end{bmatrix}$

▶ In Exercises 23–24, find all values of the unknown constant(s) in order for A to be symmetric. ◀

23. $A = \begin{bmatrix} 4 & -3 \\ a+5 & -1 \end{bmatrix}$

24. $A = \begin{bmatrix} 2 & a-2b+2c & 2a+b+c \\ 3 & 5 & a+c \\ 0 & -2 & 7 \end{bmatrix}$

▶ In Exercises 25–26, find all values of x in order for A to be invertible. ◀

25. $A = \begin{bmatrix} x-1 & x^2 & x^4 \\ 0 & x+2 & x^3 \\ 0 & 0 & x-4 \end{bmatrix}$

26. $A = \begin{bmatrix} x-\frac{1}{2} & 0 & 0 \\ x & x-\frac{1}{3} & 0 \\ x^2 & x^3 & x+\frac{1}{4} \end{bmatrix}$

▶ In Exercises 27–28, find a diagonal matrix A that satisfies the given condition. ◀

27. $A^5 = \begin{bmatrix} 1 & 0 & 0 \\ 0 & -1 & 0 \\ 0 & 0 & -1 \end{bmatrix}$ **28.** $A^{-2} = \begin{bmatrix} 9 & 0 & 0 \\ 0 & 4 & 0 \\ 0 & 0 & 1 \end{bmatrix}$

29. Verify Theorem 1.7.1(b) for the product AB, where

$$A = \begin{bmatrix} -1 & 2 & 5 \\ 0 & 1 & 3 \\ 0 & 0 & -4 \end{bmatrix}, \quad B = \begin{bmatrix} 2 & -8 & 0 \\ 0 & 2 & 1 \\ 0 & 0 & 3 \end{bmatrix}$$

30. Verify Theorem 1.7.1(d) for the matrices A and B in Exercise 29.

31. Verify Theorem 1.7.4 for the given matrix A.

(a) $A = \begin{bmatrix} 2 & -1 \\ -1 & 3 \end{bmatrix}$ (b) $A = \begin{bmatrix} 1 & -2 & 3 \\ -2 & 1 & -7 \\ 3 & -7 & 4 \end{bmatrix}$

32. Let A be an $n \times n$ symmetric matrix.

(a) Show that A^2 is symmetric.

(b) Show that $2A^2 - 3A + I$ is symmetric.

33. Prove: If $A^T A = A$, then A is symmetric and $A = A^2$.

34. Find all 3×3 diagonal matrices A that satisfy $A^2 - 3A - 4I = 0$.

35. Let $A = [a_{ij}]$ be an $n \times n$ matrix. Determine whether A is symmetric.

(a) $a_{ij} = i^2 + j^2$ (b) $a_{ij} = i^2 - j^2$

(c) $a_{ij} = 2i + 2j$ (d) $a_{ij} = 2i^2 + 2j^3$

36. On the basis of your experience with Exercise 35, devise a general test that can be applied to a formula for a_{ij} to determine whether $A = [a_{ij}]$ is symmetric.

37. A square matrix A is called **skew-symmetric** if $A^T = -A$. Prove:

(a) If A is an invertible skew-symmetric matrix, then A^{-1} is skew-symmetric.

(b) If A and B are skew-symmetric matrices, then so are A^T, $A + B$, $A - B$, and kA for any scalar k.

(c) Every square matrix A can be expressed as the sum of a symmetric matrix and a skew-symmetric matrix. [*Hint:* Note the identity $A = \frac{1}{2}(A + A^T) + \frac{1}{2}(A - A^T)$.]

▶ In Exercises 38–39, fill in the missing entries (marked with ×) to produce a skew-symmetric matrix. ◀

38. $A = \begin{bmatrix} \times & \times & 4 \\ 0 & \times & \times \\ \times & -1 & \times \end{bmatrix}$ **39.** $A = \begin{bmatrix} \times & 0 & \times \\ \times & \times & -4 \\ 8 & \times & \times \end{bmatrix}$

40. Find all values of a, b, c, and d for which A is skew-symmetric.

$$A = \begin{bmatrix} 0 & 2a - 3b + c & 3a - 5b + 5c \\ -2 & 0 & 5a - 8b + 6c \\ -3 & -5 & d \end{bmatrix}$$

41. We showed in the text that the product of symmetric matrices is symmetric if and only if the matrices commute. Is the product of commuting skew-symmetric matrices skew-symmetric? Explain. [*Note:* See Exercise 37 for the definition of *skew-symmetric*.]

42. If the $n \times n$ matrix A can be expressed as $A = LU$, where L is a lower triangular matrix and U is an upper triangular matrix, then the linear system $A\mathbf{x} = \mathbf{b}$ can be expressed as $LU\mathbf{x} = \mathbf{b}$ and can be solved in two steps:

Step 1. Let $U\mathbf{x} = \mathbf{y}$, so that $LU\mathbf{x} = \mathbf{b}$ can be expressed as $L\mathbf{y} = \mathbf{b}$. Solve this system.

Step 2. Solve the system $U\mathbf{x} = \mathbf{y}$ for $\mathbf{x}$.

In each part, use this two-step method to solve the given system.

(a) $\begin{bmatrix} 1 & 0 & 0 \\ -2 & 3 & 0 \\ 2 & 4 & 1 \end{bmatrix} \begin{bmatrix} 2 & -1 & 3 \\ 0 & 1 & 2 \\ 0 & 0 & 4 \end{bmatrix} \begin{bmatrix} x_1 \\ x_2 \\ x_3 \end{bmatrix} = \begin{bmatrix} 1 \\ -2 \\ 0 \end{bmatrix}$

(b) $\begin{bmatrix} 2 & 0 & 0 \\ 4 & 1 & 0 \\ -3 & -2 & 3 \end{bmatrix} \begin{bmatrix} 3 & -5 & 2 \\ 0 & 4 & 1 \\ 0 & 0 & 2 \end{bmatrix} \begin{bmatrix} x_1 \\ x_2 \\ x_3 \end{bmatrix} = \begin{bmatrix} 4 \\ -5 \\ 2 \end{bmatrix}$

43. Find an upper triangular matrix that satisfies

$$A^3 = \begin{bmatrix} 1 & 30 \\ 0 & -8 \end{bmatrix}$$

True-False Exercises

In parts (a)–(m) determine whether the statement is true or false, and justify your answer.

(a) The transpose of a diagonal matrix is a diagonal matrix.

(b) The transpose of an upper triangular matrix is an upper triangular matrix.

(c) The sum of an upper triangular matrix and a lower triangular matrix is a diagonal matrix.

(d) All entries of a symmetric matrix are determined by the entries occurring on and above the main diagonal.

(e) All entries of an upper triangular matrix are determined by the entries occurring on and above the main diagonal.

(f) The inverse of an invertible lower triangular matrix is an upper triangular matrix.

(g) A diagonal matrix is invertible if and only if all of its diagonal entries are positive.

(h) The sum of a diagonal matrix and a lower triangular matrix is a lower triangular matrix.

(i) A matrix that is both symmetric and upper triangular must be a diagonal matrix.

(j) If A and B are $n \times n$ matrices such that $A + B$ is symmetric, then A and B are symmetric.

(k) If A and B are $n \times n$ matrices such that $A + B$ is upper triangular, then A and B are upper triangular.

(l) If A^2 is a symmetric matrix, then A is a symmetric matrix.

(m) If kA is a symmetric matrix for some $k \neq 0$, then A is a symmetric matrix.

1.8 Applications of Linear Systems

In this section we will discuss some brief applications of linear systems. These are but a small sample of the wide variety of real-world problems to which our study of linear systems is applicable.

Network Analysis   The concept of a *network* appears in a variety of applications. Loosely stated, a **network** is a set of **branches** through which something "flows." For example, the branches might be electrical wires through which electricity flows, pipes through which water or oil flows, traffic lanes through which vehicular traffic flows, or economic linkages through which money flows, to name a few possibilities.

In most networks, the branches meet at points, called **nodes** or **junctions**, where the flow divides. For example, in an electrical network, nodes occur where three or more wires join, in a traffic network they occur at street intersections, and in a financial network they occur at banking centers where incoming money is distributed to individuals or other institutions.

In the study of networks, there is generally some numerical measure of the rate at which the medium flows through a branch. For example, the flow rate of electricity is often measured in amperes, the flow rate of water or oil in gallons per minute, the flow rate of traffic in vehicles per hour, and the flow rate of European currency in millions of Euros per day. We will restrict our attention to networks in which there is *flow conservation* at each node, by which we mean that *the rate of flow into any node is equal to the rate of flow out of that node.* This ensures that the flow medium does not build up at the nodes and block the free movement of the medium through the network.

A common problem in network analysis is to use known flow rates in certain branches to find the flow rates in all of the branches. Here is an example.

▶ **EXAMPLE 1   Network Analysis Using Linear Systems**

Figure 1.8.1 shows a network with four nodes in which the flow rate and direction of flow in certain branches are known. Find the flow rates and directions of flow in the remaining branches.

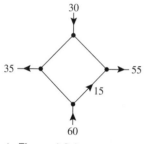

▲ **Figure 1.8.1**

Solution   As illustrated in Figure 1.8.2, we have assigned arbitrary directions to the unknown flow rates x_1, x_2, and x_3. We need not be concerned if some of the directions are incorrect, since an incorrect direction will be signaled by a negative value for the flow rate when we solve for the unknowns.

It follows from the conservation of flow at node A that

$$x_1 + x_2 = 30$$

Similarly, at the other nodes we have

$$x_2 + x_3 = 35 \quad \text{(node } B\text{)}$$
$$x_3 + 15 = 60 \quad \text{(node } C\text{)}$$
$$x_1 + 15 = 55 \quad \text{(node } D\text{)}$$

These four conditions produce the linear system

$$
\begin{aligned}
x_1 + x_2 \quad\quad &= 30 \\
x_2 + x_3 &= 35 \\
x_3 &= 45 \\
x_1 \quad\quad\quad &= 40
\end{aligned}
$$

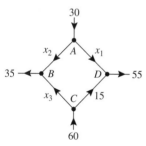

▲ **Figure 1.8.2**

which we can now try to solve for the unknown flow rates. In this particular case the system is sufficiently simple that it can be solved by inspection (work from the bottom up). We leave it for you to confirm that the solution is

$$x_1 = 40, \quad x_2 = -10, \quad x_3 = 45$$

The fact that x_2 is negative tells us that the direction assigned to that flow in Figure 1.8.2 is incorrect; that is, the flow in that branch is *into* node A.

▶ **EXAMPLE 2   Design of Traffic Patterns**

The network in Figure 1.8.3 shows a proposed plan for the traffic flow around a new park that will house the Liberty Bell in Philadelphia, Pennsylvania. The plan calls for a computerized traffic light at the north exit on Fifth Street, and the diagram indicates the average number of vehicles per hour that are expected to flow in and out of the streets that border the complex. All streets are one-way.

(a) How many vehicles per hour should the traffic light let through to ensure that the average number of vehicles per hour flowing into the complex is the same as the average number of vehicles flowing out?

(b) Assuming that the traffic light has been set to balance the total flow in and out of the complex, what can you say about the average number of vehicles per hour that will flow along the streets that border the complex?

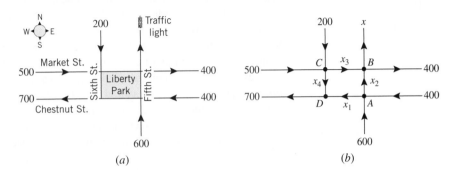

(a)   (b)

▶ **Figure 1.8.3**

Solution (a) If, as indicated in Figure 1.8.3b, we let x denote the number of vehicles per hour that the traffic light must let through, then the total number of vehicles per hour that flow in and out of the complex will be

Flowing in: $500 + 400 + 600 + 200 = 1700$

Flowing out: $x + 700 + 400$

Equating the flows in and out shows that the traffic light should let $x = 600$ vehicles per hour pass through.

Solution (b) To avoid traffic congestion, the flow in must equal the flow out at each intersection. For this to happen, the following conditions must be satisfied:

Intersection	Flow In		Flow Out
A	$400 + 600$	$=$	$x_1 + x_2$
B	$x_2 + x_3$	$=$	$400 + x$
C	$500 + 200$	$=$	$x_3 + x_4$
D	$x_1 + x_4$	$=$	700

Thus, with $x = 600$, as computed in part (a), we obtain the following linear system:

$$
\begin{aligned}
x_1 + x_2 \quad\quad\quad\quad &= 1000 \\
x_2 + x_3 \quad\quad &= 1000 \\
x_3 + x_4 &= 700 \\
x_1 \quad\quad\quad\quad + x_4 &= 700
\end{aligned}
$$

We leave it for you to show that the system has infinitely many solutions and that these are given by the parametric equations

$$x_1 = 700 - t, \quad x_2 = 300 + t, \quad x_3 = 700 - t, \quad x_4 = t \tag{1}$$

However, the parameter t is not completely arbitrary here, since there are physical constraints to be considered. For example, the average flow rates must be nonnegative since we have assumed the streets to be one-way, and a negative flow rate would indicate a flow in the wrong direction. This being the case, we see from (1) that t can be any real number that satisfies $0 \le t \le 700$, which implies that the average flow rates along the streets will fall in the ranges

$$0 \le x_1 \le 700, \quad 300 \le x_2 \le 1000, \quad 0 \le x_3 \le 700, \quad 0 \le x_4 \le 700 \quad \blacktriangleleft$$

Electrical Circuits

▲ Figure 1.8.4

Next, we will show how network analysis can be used to analyze electrical circuits consisting of batteries and resistors. A **battery** is a source of electric energy, and a **resistor**, such as a lightbulb, is an element that dissipates electric energy. Figure 1.8.4 shows a schematic diagram of a circuit with one battery (represented by the symbol ⊣⊢), one resistor (represented by the symbol ⌐w⌐), and a switch. The battery has a **positive pole** (+) and a **negative pole** (−). When the switch is closed, electrical current is considered to flow from the positive pole of the battery, through the resistor, and back to the negative pole (indicated by the arrowhead in the figure).

Electrical current, which is a flow of electrons through wires, behaves much like the flow of water through pipes. A battery acts like a pump that creates "electrical pressure" to increase the flow rate of electrons, and a resistor acts like a restriction in a pipe that reduces the flow rate of electrons. The technical term for electrical pressure is **electrical potential**; it is commonly measured in **volts** (V). The degree to which a resistor reduces the electrical potential is called its **resistance** and is commonly measured in **ohms** (Ω). The rate of flow of electrons in a wire is called **current** and is commonly measured in **amperes** (also called **amps**) (A). The precise effect of a resistor is given by the following law:

> **Ohm's Law** If a current of I amperes passes through a resistor with a resistance of R ohms, then there is a resulting drop of E volts in electrical potential that is the product of the current and resistance; that is,
>
> $$E = IR$$

▲ Figure 1.8.5

A typical electrical network will have multiple batteries and resistors joined by some configuration of wires. A point at which three or more wires in a network are joined is called a **node** (or **junction point**). A **branch** is a wire connecting two nodes, and a **closed loop** is a succession of connected branches that begin and end at the same node. For example, the electrical network in Figure 1.8.5 has two nodes and three closed loops—two inner loops and one outer loop. As current flows through an electrical network, it undergoes increases and decreases in electrical potential, called **voltage rises** and **voltage drops**, respectively. The behavior of the current at the nodes and around closed loops is governed by two fundamental laws:

> **Kirchhoff's Current Law** The sum of the currents flowing into any node is equal to the sum of the currents flowing out.

▲ Figure 1.8.6

> **Kirchhoff's Voltage Law** In one traversal of any closed loop, the sum of the voltage rises equals the sum of the voltage drops.

Kirchhoff's current law is a restatement of the principle of flow conservation at a node that was stated for general networks. Thus, for example, the currents at the top node in Figure 1.8.6 satisfy the equation $I_1 = I_2 + I_3$.

In circuits with multiple loops and batteries there is usually no way to tell in advance which way the currents are flowing, so the usual procedure in circuit analysis is to assign *arbitrary* directions to the current flows in the branches and let the mathematical computations determine whether the assignments are correct. In addition to assigning directions to the current flows, Kirchhoff's voltage law requires a direction of travel for each closed loop. The choice is arbitrary, but for consistency we will always take this direction to be *clockwise* (Figure 1.8.7). We also make the following conventions:

Clockwise closed-loop convention with arbitrary direction assignments to currents in the branches

▲ Figure 1.8.7

- A voltage drop occurs at a resistor if the direction assigned to the current through the resistor is the same as the direction assigned to the loop, and a voltage rise occurs at a resistor if the direction assigned to the current through the resistor is the opposite to that assigned to the loop.
- A voltage rise occurs at a battery if the direction assigned to the loop is from − to + through the battery, and a voltage drop occurs at a battery if the direction assigned to the loop is from + to − through the battery.

If you follow these conventions when calculating currents, then those currents whose directions were assigned correctly will have positive values and those whose directions were assigned incorrectly will have negative values.

▶ **EXAMPLE 3** **A Circuit with One Closed Loop**

Determine the current I in the circuit shown in Figure 1.8.8.

Solution Since the direction assigned to the current through the resistor is the same as the direction of the loop, there is a voltage drop at the resistor. By Ohm's law this voltage drop is $E = IR = 3I$. Also, since the direction assigned to the loop is from − to + through the battery, there is a voltage rise of 6 volts at the battery. Thus, it follows from Kirchhoff's voltage law that

$$3I = 6$$

from which we conclude that the current is $I = 2$ A. Since I is positive, the direction assigned to the current flow is correct.

▲ Figure 1.8.8

▶ **EXAMPLE 4** **A Circuit with Three Closed Loops**

Determine the currents I_1, I_2, and I_3 in the circuit shown in Figure 1.8.9.

Solution Using the assigned directions for the currents, Kirchhoff's current law provides one equation for each node:

Node	Current In	Current Out
A	$I_1 + I_2$ =	I_3
B	I_3 =	$I_1 + I_2$

However, these equations are really the same, since both can be expressed as

$$I_1 + I_2 - I_3 = 0 \tag{2}$$

▲ Figure 1.8.9

Gustav Kirchhoff
(1824–1887)

Historical Note The German physicist Gustav Kirchhoff was a student of Gauss. His work on Kirchhoff's laws, announced in 1854, was a major advance in the calculation of currents, voltages, and resistances of electrical circuits. Kirchhoff was severely disabled and spent most of his life on crutches or in a wheelchair.
[*Image: ©SSPL/The Image Works*]

To find unique values for the currents we will need two more equations, which we will obtain from Kirchhoff's voltage law. We can see from the network diagram that there are three closed loops, a left inner loop containing the 50 V battery, a right inner loop containing the 30 V battery, and an outer loop that contains both batteries. Thus, Kirchhoff's voltage law will actually produce three equations. With a clockwise traversal of the loops, the voltage rises and drops in these loops are as follows:

	Voltage Rises	**Voltage Drops**
Left Inside Loop	50	$5I_1 + 20I_3$
Right Inside Loop	$30 + 10I_2 + 20I_3$	0
Outside Loop	$30 + 50 + 10I_2$	$5I_1$

These conditions can be rewritten as

$$\begin{aligned} 5I_1 \qquad\quad + 20I_3 &= 50 \\ 10I_2 + 20I_3 &= -30 \\ 5I_1 - 10I_2 \qquad\quad &= 80 \end{aligned} \qquad (3)$$

However, the last equation is superfluous, since it is the difference of the first two. Thus, if we combine (2) and the first two equations in (3), we obtain the following linear system of three equations in the three unknown currents:

$$\begin{aligned} I_1 + I_2 - I_3 &= 0 \\ 5I_1 \qquad\quad + 20I_3 &= 50 \\ 10I_2 + 20I_3 &= -30 \end{aligned}$$

We leave it for you to solve this system and show that $I_1 = 6$ A, $I_2 = -5$ A, and $I_3 = 1$ A. The fact that I_2 is negative tells us that the direction of this current is opposite to that indicated in Figure 1.8.9. ◀

Balancing Chemical Equations

Chemical compounds are represented by ***chemical formulas*** that describe the atomic makeup of their molecules. For example, water is composed of two hydrogen atoms and one oxygen atom, so its chemical formula is H_2O; and stable oxygen is composed of two oxygen atoms, so its chemical formula is O_2.

When chemical compounds are combined under the right conditions, the atoms in their molecules rearrange to form new compounds. For example, when methane burns, the methane (CH_4) and stable oxygen (O_2) react to form carbon dioxide (CO_2) and water (H_2O). This is indicated by the ***chemical equation***

$$CH_4 + O_2 \longrightarrow CO_2 + H_2O \qquad (4)$$

The molecules to the left of the arrow are called the ***reactants*** and those to the right the ***products***. In this equation the plus signs serve to separate the molecules and are not intended as algebraic operations. However, this equation does not tell the whole story, since it fails to account for the proportions of molecules required for a ***complete reaction*** (no reactants left over). For example, we can see from the right side of (4) that to produce one molecule of carbon dioxide and one molecule of water, one needs *three* oxygen atoms for each carbon atom. However, from the left side of (4) we see that one molecule of methane and one molecule of stable oxygen have only *two* oxygen atoms for each carbon atom. Thus, on the reactant side the ratio of methane to stable oxygen cannot be one-to-one in a complete reaction.

A chemical equation is said to be **balanced** if for each type of atom in the reaction, the same number of atoms appears on each side of the arrow. For example, the balanced version of Equation (4) is

$$CH_4 + 2O_2 \longrightarrow CO_2 + 2H_2O \tag{5}$$

by which we mean that one methane molecule combines with two stable oxygen molecules to produce one carbon dioxide molecule and two water molecules. In theory, one could multiply this equation through by any positive integer. For example, multiplying through by 2 yields the balanced chemical equation

$$2CH_4 + 4O_2 \longrightarrow 2CO_2 + 4H_2O$$

However, the standard convention is to use the smallest positive integers that will balance the equation.

Equation (4) is sufficiently simple that it could have been balanced by trial and error, but for more complicated chemical equations we will need a systematic method. There are various methods that can be used, but we will give one that uses systems of linear equations. To illustrate the method let us reexamine Equation (4). To balance this equation we must find positive integers, $x_1, x_2, x_3,$ and x_4 such that

$$x_1\,(CH_4) + x_2\,(O_2) \longrightarrow x_3\,(CO_2) + x_4\,(H_2O) \tag{6}$$

For each of the atoms in the equation, the number of atoms on the left must be equal to the number of atoms on the right. Expressing this in tabular form we have

	Left Side		**Right Side**
Carbon	x_1	$=$	x_3
Hydrogen	$4x_1$	$=$	$2x_4$
Oxygen	$2x_2$	$=$	$2x_3 + x_4$

from which we obtain the homogeneous linear system

$$\begin{aligned} x_1 \quad - \quad x_3 \qquad\quad &= 0 \\ 4x_1 \qquad\qquad - 2x_4 &= 0 \\ 2x_2 - 2x_3 - \quad x_4 &= 0 \end{aligned}$$

The augmented matrix for this system is

$$\begin{bmatrix} 1 & 0 & -1 & 0 & 0 \\ 4 & 0 & 0 & -2 & 0 \\ 0 & 2 & -2 & -1 & 0 \end{bmatrix}$$

We leave it for you to show that the reduced row echelon form of this matrix is

$$\begin{bmatrix} 1 & 0 & 0 & -\frac{1}{2} & 0 \\ 0 & 1 & 0 & -1 & 0 \\ 0 & 0 & 1 & -\frac{1}{2} & 0 \end{bmatrix}$$

from which we conclude that the general solution of the system is

$$x_1 = t/2, \quad x_2 = t, \quad x_3 = t/2, \quad x_4 = t$$

where t is arbitrary. The smallest positive integer values for the unknowns occur when we let $t = 2$, so the equation can be balanced by letting $x_1 = 1, x_2 = 2, x_3 = 1, x_4 = 2$. This agrees with our earlier conclusions, since substituting these values into Equation (6) yields Equation (5).

▶ **EXAMPLE 5** Balancing Chemical Equations Using Linear Systems

Balance the chemical equation

$$HCl \quad + \quad Na_3PO_4 \quad \longrightarrow \quad H_3PO_4 \quad + \quad NaCl$$

[hydrochloric acid] + [sodium phosphate] $\longrightarrow$ [phosphoric acid] + [sodium chloride]

Solution Let x_1, x_2, x_3, and x_4 be positive integers that balance the equation

$$x_1 \,(HCl) + x_2 \,(Na_3PO_4) \longrightarrow x_3 \,(H_3PO_4) + x_4 \,(NaCl) \tag{7}$$

Equating the number of atoms of each type on the two sides yields

$$1x_1 = 3x_3 \quad \textbf{Hydrogen (H)}$$
$$1x_1 = 1x_4 \quad \textbf{Chlorine (Cl)}$$
$$3x_2 = 1x_4 \quad \textbf{Sodium (Na)}$$
$$1x_2 = 1x_3 \quad \textbf{Phosphorous (P)}$$
$$4x_2 = 4x_3 \quad \textbf{Oxygen (O)}$$

from which we obtain the homogeneous linear system

$$
\begin{aligned}
x_1 \quad\quad\quad - 3x_3 \quad\quad\quad &= 0 \\
x_1 \quad\quad\quad\quad\quad - x_4 &= 0 \\
3x_2 \quad\quad\quad - x_4 &= 0 \\
x_2 - \quad x_3 \quad\quad\quad &= 0 \\
4x_2 - 4x_3 \quad\quad\quad &= 0
\end{aligned}
$$

We leave it for you to show that the reduced row echelon form of the augmented matrix for this system is

$$
\begin{bmatrix}
1 & 0 & 0 & -1 & 0 \\
0 & 1 & 0 & -\frac{1}{3} & 0 \\
0 & 0 & 1 & -\frac{1}{3} & 0 \\
0 & 0 & 0 & 0 & 0 \\
0 & 0 & 0 & 0 & 0
\end{bmatrix}
$$

from which we conclude that the general solution of the system is

$$x_1 = t, \quad x_2 = t/3, \quad x_3 = t/3, \quad x_4 = t$$

where t is arbitrary. To obtain the smallest positive integers that balance the equation, we let $t = 3$, in which case we obtain $x_1 = 3, x_2 = 1, x_3 = 1$, and $x_4 = 3$. Substituting these values in (7) produces the balanced equation

$$3HCl + Na_3PO_4 \longrightarrow H_3PO_4 + 3NaCl \quad \blacktriangleleft$$

Polynomial Interpolation

▲ Figure 1.8.10

An important problem in various applications is to find a polynomial whose graph passes through a specified set of points in the plane; this is called an ***interpolating polynomial*** for the points. The simplest example of such a problem is to find a linear polynomial

$$p(x) = ax + b \tag{8}$$

whose graph passes through two known distinct points, (x_1, y_1) and (x_2, y_2), in the xy-plane (Figure 1.8.10). You have probably encountered various methods in analytic geometry for finding the equation of a line through two points, but here we will give a method based on linear systems that can be adapted to general polynomial interpolation.

The graph of (8) is the line $y = ax + b$, and for this line to pass through the points (x_1, y_1) and (x_2, y_2), we must have

$$y_1 = ax_1 + b \quad \text{and} \quad y_2 = ax_2 + b$$

Therefore, the unknown coefficients a and b can be obtained by solving the linear system

$$ax_1 + b = y_1$$
$$ax_2 + b = y_2$$

We don't need any fancy methods to solve this system—the value of a can be obtained by subtracting the equations to eliminate b, and then the value of a can be substituted into either equation to find b. We leave it as an exercise for you to find a and b and then show that they can be expressed in the form

$$a = \frac{y_2 - y_1}{x_2 - x_1} \quad \text{and} \quad b = \frac{y_1 x_2 - y_2 x_1}{x_2 - x_1} \tag{9}$$

provided $x_1 \neq x_2$. Thus, for example, the line $y = ax + b$ that passes through the points

$$(2, 1) \quad \text{and} \quad (5, 4)$$

can be obtained by taking $(x_1, y_1) = (2, 1)$ and $(x_2, y_2) = (5, 4)$, in which case (9) yields

$$a = \frac{4 - 1}{5 - 2} = 1 \quad \text{and} \quad b = \frac{(1)(5) - (4)(2)}{5 - 2} = -1$$

Therefore, the equation of the line is

$$y = x - 1$$

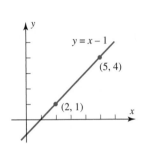

▲ Figure 1.8.11

(Figure 1.8.11).

Now let us consider the more general problem of finding a polynomial whose graph passes through n points with distinct x-coordinates

$$(x_1, y_1), \quad (x_2, y_2), \quad (x_3, y_3), \dots, \quad (x_n, y_n) \tag{10}$$

Since there are n conditions to be satisfied, intuition suggests that we should begin by looking for a polynomial of the form

$$p(x) = a_0 + a_1 x + a_2 x^2 + \cdots + a_{n-1} x^{n-1} \tag{11}$$

since a polynomial of this form has n coefficients that are at our disposal to satisfy the n conditions. However, we want to allow for cases where the points may lie on a line or have some other configuration that would make it possible to use a polynomial whose degree is less than $n - 1$; thus, we allow for the possibility that a_{n-1} and other coefficients in (11) may be zero.

The following theorem, which we will prove later in the text, is the basic result on polynomial interpolation.

THEOREM 1.8.1 Polynomial Interpolation

Given any n points in the xy-plane that have distinct x-coordinates, there is a unique polynomial of degree n − 1 or less whose graph passes through those points.

Let us now consider how we might go about finding the interpolating polynomial (11) whose graph passes through the points in (10). Since the graph of this polynomial is the graph of the equation

$$y = a_0 + a_1 x + a_2 x^2 + \cdots + a_{n-1} x^{n-1} \tag{12}$$

it follows that the coordinates of the points must satisfy

$$a_0 + a_1 x_1 + a_2 x_1^2 + \cdots + a_{n-1} x_1^{n-1} = y_1$$
$$a_0 + a_1 x_2 + a_2 x_2^2 + \cdots + a_{n-1} x_2^{n-1} = y_2 \tag{13}$$
$$\vdots \quad \vdots \quad \vdots \qquad \vdots \qquad \vdots$$
$$a_0 + a_1 x_n + a_2 x_n^2 + \cdots + a_{n-1} x_n^{n-1} = y_n$$

In these equations the values of x's and y's are assumed to be known, so we can view this as a linear system in the unknowns $a_0, a_1, \ldots, a_{n-1}$. From this point of view the augmented matrix for the system is

$$\begin{bmatrix} 1 & x_1 & x_1^2 & \cdots & x_1^{n-1} & y_1 \\ 1 & x_2 & x_2^2 & \cdots & x_2^{n-1} & y_2 \\ \vdots & \vdots & \vdots & & \vdots & \vdots \\ 1 & x_n & x_n^2 & \cdots & x_n^{n-1} & y_n \end{bmatrix} \tag{14}$$

and hence the interpolating polynomial can be found by reducing this matrix to reduced row echelon form (Gauss–Jordan elimination).

▶ **EXAMPLE 6 Polynomial Interpolation by Gauss–Jordan Elimination**

Find a cubic polynomial whose graph passes through the points

$$(1, 3), \quad (2, -2), \quad (3, -5), \quad (4, 0)$$

Solution Since there are four points, we will use an interpolating polynomial of degree $n = 3$. Denote this polynomial by

$$p(x) = a_0 + a_1 x + a_2 x^2 + a_3 x^3$$

and denote the x- and y-coordinates of the given points by

$$x_1 = 1, \quad x_2 = 2, \quad x_3 = 3, \quad x_4 = 4 \quad \text{and} \quad y_1 = 3, \quad y_2 = -2, \quad y_3 = -5, \quad y_4 = 0$$

Thus, it follows from (14) that the augmented matrix for the linear system in the unknowns a_0, a_1, a_2, and a_3 is

$$\begin{bmatrix} 1 & x_1 & x_1^2 & x_1^3 & y_1 \\ 1 & x_2 & x_2^2 & x_2^3 & y_2 \\ 1 & x_3 & x_3^2 & x_3^3 & y_3 \\ 1 & x_4 & x_4^2 & x_4^3 & y_4 \end{bmatrix} = \begin{bmatrix} 1 & 1 & 1 & 1 & 3 \\ 1 & 2 & 4 & 8 & -2 \\ 1 & 3 & 9 & 27 & -5 \\ 1 & 4 & 16 & 64 & 0 \end{bmatrix}$$

We leave it for you to confirm that the reduced row echelon form of this matrix is

$$\begin{bmatrix} 1 & 0 & 0 & 0 & 4 \\ 0 & 1 & 0 & 0 & 3 \\ 0 & 0 & 1 & 0 & -5 \\ 0 & 0 & 0 & 1 & 1 \end{bmatrix}$$

from which it follows that $a_0 = 4, a_1 = 3, a_2 = -5, a_3 = 1$. Thus, the interpolating polynomial is

$$p(x) = 4 + 3x - 5x^2 + x^3$$

The graph of this polynomial and the given points are shown in Figure 1.8.12. ◀

▲ Figure 1.8.12

Remark Later we will give a more efficient method for finding interpolating polynomials that is better suited for problems in which the number of data points is large.

▶ **EXAMPLE 7 Approximate Integration**

There is no way to evaluate the integral

$$\int_0^1 \sin\left(\frac{\pi x^2}{2}\right) dx$$

directly since there is no way to express an antiderivative of the integrand in terms of elementary functions. This integral could be approximated by Simpson's rule or some comparable method, but an alternative approach is to approximate the integrand by an interpolating polynomial and integrate the approximating polynomial. For example, let us consider the five points

$$x_0 = 0, \quad x_1 = 0.25, \quad x_2 = 0.5, \quad x_3 = 0.75, \quad x_4 = 1$$

that divide the interval $[0, 1]$ into four equally spaced subintervals. The values of

$$f(x) = \sin\left(\frac{\pi x^2}{2}\right)$$

at these points are approximately

$$f(0) = 0, \quad f(0.25) = 0.098017, \quad f(0.5) = 0.382683,$$
$$f(0.75) = 0.77301, \quad f(1) = 1$$

The interpolating polynomial is (verify)

$$p(x) = 0.098796x + 0.762356x^2 + 2.14429x^3 - 2.00544x^4 \qquad (15)$$

and

$$\int_0^1 p(x)\, dx \approx 0.438501 \qquad (16)$$

As shown in Figure 1.8.13, the graphs of f and p match very closely over the interval $[0, 1]$, so the approximation is quite good. ◀

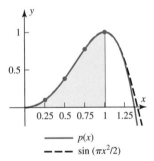

—— $p(x)$
--- $\sin(\pi x^2/2)$

▲ Figure 1.8.13

Concept Review

- Network
- Branches
- Nodes
- Flow conservation

- Electrical circuits: battery, resistor, poles (positive and negative), electrical potential, Ohm's law, Kirchhoff's current law, Kirchhoff's voltage law

- Chemical equations: reactants, products, balanced equation
- Interpolating polynomial

Skills

- Find the flow rates and directions of flow in branches of a network.
- Find the amount of current flowing through parts of an electrical circuit.

- Write a balanced chemical equation for a given chemical reaction.
- Find an interpolating polynomial for a graph passing through a given collection of points.

Exercise Set 1.8

1. The accompanying figure shows a network in which the flow rate and direction of flow in certain branches are known. Find the flow rates and directions of flow in the remaining branches.

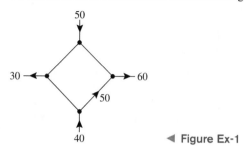

◀ **Figure Ex-1**

2. The accompanying figure shows known flow rates of hydrocarbons into and out of a network of pipes at an oil refinery.

(a) Set up a linear system whose solution provides the unknown flow rates.

(b) Solve the system for the unknown flow rates.

(c) Find the flow rates and directions of flow if $x_4 = 50$ and $x_6 = 0$.

◀ **Figure Ex-2**

3. The accompanying figure shows a network of one-way streets with traffic flowing in the directions indicated. The flow rates along the streets are measured as the average number of vehicles per hour.

(a) Set up a linear system whose solution provides the unknown flow rates.

(b) Solve the system for the unknown flow rates.

(c) If the flow along the road from A to B must be reduced for construction, what is the minimum flow that is required to keep traffic flowing on all roads?

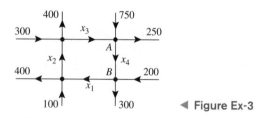

◀ **Figure Ex-3**

4. The accompanying figure shows a network of one-way streets with traffic flowing in the directions indicated. The flow rates along the streets are measured as the average number of vehicles per hour.

(a) Set up a linear system whose solution provides the unknown flow rates.

(b) Solve the system for the unknown flow rates.

(c) Is it possible to close the road from A to B for construction and keep traffic flowing on the other streets? Explain.

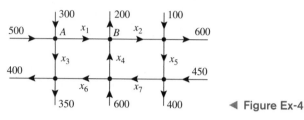

◀ **Figure Ex-4**

▶ In Exercises **5–8**, analyze the given electrical circuits by finding the unknown currents. ◀

5.

6.

7.

8.

▶ In Exercises **9–12**, write a balanced equation for the given chemical reaction. ◀

9. $C_3H_8 + O_2 \rightarrow CO_2 + H_2O$ (propane combustion)

10. $C_6H_{12}O_6 \rightarrow CO_2 + C_2H_5OH$ (fermentation of sugar)

11. $CH_3COF + H_2O \rightarrow CH_3COOH + HF$

12. $CO_2 + H_2O \rightarrow C_6H_{12}O_6 + O_2$ (photosynthesis)

13. Find the quadratic polynomial whose graph passes through the points $(1, 1)$, $(2, 2)$, and $(3, 5)$.

14. Find the quadratic polynomial whose graph passes through the points $(0, 0)$, $(-1, 1)$, and $(1, 1)$.

15. Find the cubic polynomial whose graph passes through the points $(-1, -1)$, $(0, 1)$, $(1, 3)$, $(4, -1)$.

16. The accompanying figure shows the graph of a cubic polynomial. Find the polynomial.

◄ **Figure Ex-16**

17. (a) Find an equation that represents the family of all second-degree polynomials that pass through the points $(0, 1)$ and $(1, 2)$. [*Hint:* The equation will involve one arbitrary parameter that produces the members of the family when varied.]

(b) By hand, or with the help of a graphing utility, sketch four curves in the family.

18. In this section we have selected only a few applications of linear systems. Using the Internet as a search tool, try to find some more real-world applications of such systems. Select one that is of interest to you, and write a paragraph about it.

True-False Exercises

In parts (a)–(e) determine whether the statement is true or false, and justify your answer.

(a) In any network, the sum of the flows out of a node must equal the sum of the flows into a node.

(b) When a current passes through a resistor, there is an increase in the electrical potential in a circuit.

(c) Kirchhoff's current law states that the sum of the currents flowing into a node equals the sum of the currents flowing out of the node.

(d) A chemical equation is called balanced if the total number of atoms on each side of the equation is the same.

(e) Given any n points in the xy-plane, there is a unique polynomial of degree $n - 1$ or less whose graph passes through those points.

1.9 Leontief Input-Output Models

In 1973 the economist Wassily Leontief was awarded the Nobel prize for his work on economic modeling in which he used matrix methods to study the relationships between different sectors in an economy. In this section we will discuss some of the ideas developed by Leontief.

Inputs and Outputs in an Economy

One way to analyze an economy is to divide it into **sectors** and study how the sectors interact with one another. For example, a simple economy might be divided into three sectors—manufacturing, agriculture, and utilities. Typically, a sector will produce certain **outputs** but will require **inputs** from the other sectors and itself. For example, the agricultural sector may produce wheat as an output but will require inputs of farm machinery from the manufacturing sector, electrical power from the utilities sector, and food from its own sector to feed its workers. Thus, we can imagine an economy to be a network in which inputs and outputs flow in and out of the sectors; the study of such flows is called **input-output analysis**. Inputs and outputs are commonly measured in monetary units (dollars or millions of dollars, for example) but other units of measurement are also possible.

The flows between sectors of a real economy are not always obvious. For example, in World War II the United States had a demand for 50,000 new airplanes that required the construction of many new aluminum manufacturing plants. This produced an unexpectedly large demand for certain copper electrical components, which in turn produced a copper shortage. The problem was eventually resolved by using silver borrowed from Fort Knox as a copper substitute. In all likelihood modern input-output analysis would have anticipated the copper shortage.

Most sectors of an economy will produce outputs, but there may exist sectors that consume outputs without producing anything themselves (the consumer market, for example). Those sectors that do not produce outputs are called *open sectors*. Economies with no open sectors are called *closed economies*, and economies with one or more open sectors are called *open economies* (Figure 1.9.1). In this section we will be concerned with economies with one open sector, and our primary goal will be to determine the output levels that are required for the productive sectors to sustain themselves and satisfy the demand of the open sector.

Leontief Model of an Open Economy

Let us consider a simple open economy with one open sector and three product-producing sectors: manufacturing, agriculture, and utilities. Assume that inputs and outputs are measured in dollars and that the inputs required by the productive sectors to produce one dollar's worth of output are in accordance with Table 1.

Manufacturing Agriculture

Utilities

▲ Figure 1.9.1

Table 1

		Input Required per Dollar Output		
		Manufacturing	**Agriculture**	**Utilities**
	Manufacturing	$ 0.50	$ 0.10	$ 0.10
Provider	**Agriculture**	$ 0.20	$ 0.50	$ 0.30
	Utilities	$ 0.10	$ 0.30	$ 0.40

Usually, one would suppress the labeling and express this matrix as

$$C = \begin{bmatrix} 0.5 & 0.1 & 0.1 \\ 0.2 & 0.5 & 0.3 \\ 0.1 & 0.3 & 0.4 \end{bmatrix} \tag{1}$$

This is called the *consumption matrix* (or sometimes the *technology matrix*) for the economy. The column vectors

$$c_1 = \begin{bmatrix} 0.5 \\ 0.2 \\ 0.1 \end{bmatrix}, \quad c_2 = \begin{bmatrix} 0.1 \\ 0.5 \\ 0.3 \end{bmatrix}, \quad c_3 = \begin{bmatrix} 0.1 \\ 0.3 \\ 0.4 \end{bmatrix}$$

Historical Note It is somewhat ironic that it was the Russian-born Wassily Leontief who won the Nobel prize in 1973 for pioneering the modern methods for analyzing free-market economies. Leontief was a precocious student who entered the University of Leningrad at age 15. Bothered by the intellectual restrictions of the Soviet system, he was put in jail for anti-Communist activities, after which he headed for the University of Berlin, receiving his Ph.D. there in 1928. He came to the United States in 1931, where he held professorships at Harvard and then New York University.

[*Image: ©Bettmann/©Corbis*]

Wassily Leontief
(1906–1999)

in C list the inputs required by the manufacturing, agricultural, and utilities sectors, respectively, to produce $1.00 worth of output. These are called the ***consumption vectors*** of the sectors. For example, $\mathbf{c}_1$ tells us that to produce $1.00 worth of output the manufacturing sector needs $0.50 worth of manufacturing output, $0.20 worth of agricultural output, and $0.10 worth of utilities output.

What is the economic significance of the row sums of the consumption matrix?

Continuing with the above example, suppose that the open sector wants the economy to supply it manufactured goods, agricultural products, and utilities with dollar values:

d_1 dollars of manufactured goods

d_2 dollars of agricultural products

d_3 dollars of utilities

The column vector $\mathbf{d}$ that has these numbers as successive components is called the ***outside demand vector***. Since the product-producing sectors consume some of their own output, the dollar value of their output must cover their own needs plus the outside demand. Suppose that the dollar values required to do this are

x_1 dollars of manufactured goods

x_2 dollars of agricultural products

x_3 dollars of utilities

The column vector $\mathbf{x}$ that has these numbers as successive components is called the ***production vector*** for the economy. For the economy with consumption matrix (1), that portion of the production vector $\mathbf{x}$ that will be consumed by the three productive sectors is

$$x_1 \begin{bmatrix} 0.5 \\ 0.2 \\ 0.1 \end{bmatrix} + x_2 \begin{bmatrix} 0.1 \\ 0.5 \\ 0.3 \end{bmatrix} + x_3 \begin{bmatrix} 0.1 \\ 0.3 \\ 0.4 \end{bmatrix} = \begin{bmatrix} 0.5 & 0.1 & 0.1 \\ 0.2 & 0.5 & 0.3 \\ 0.1 & 0.3 & 0.4 \end{bmatrix} \begin{bmatrix} x_1 \\ x_2 \\ x_3 \end{bmatrix} = C\mathbf{x}$$

| Fractions consumed by manufacturing | Fractions consumed by agriculture | Fractions consumed by utilities |

The vector $C\mathbf{x}$ is called the ***intermediate demand vector*** for the economy. Once the intermediate demand is met, the portion of the production that is left to satisfy the outside demand is $\mathbf{x} - C\mathbf{x}$. Thus, if the outside demand vector is $\mathbf{d}$, then $\mathbf{x}$ must satisfy the equation

$$\mathbf{x} \quad - \quad C\mathbf{x} \quad = \quad \mathbf{d}$$

| Amount produced | Intermediate demand | Outside demand |

which we will find convenient to rewrite as

$$(I - C)\mathbf{x} = \mathbf{d} \tag{2}$$

The matrix $I - C$ is called the ***Leontief matrix*** and (2) is called the ***Leontief equation***.

▶ EXAMPLE 1 Satisfying Outside Demand

Consider the economy described in Table 1. Suppose that the open sector has a demand for $7900 worth of manufacturing products, $3950 worth of agricultural products, and $1975 worth of utilities.

(a)   Can the economy meet this demand?

(b)   If so, find a production vector $\mathbf{x}$ that will meet it exactly.

Solution The consumption matrix, production vector, and outside demand vector are

$$C = \begin{bmatrix} 0.5 & 0.1 & 0.1 \\ 0.2 & 0.5 & 0.3 \\ 0.1 & 0.3 & 0.4 \end{bmatrix}, \quad \mathbf{x} = \begin{bmatrix} x_1 \\ x_2 \\ x_3 \end{bmatrix}, \quad \mathbf{d} = \begin{bmatrix} 7900 \\ 3950 \\ 1975 \end{bmatrix} \quad (3)$$

To meet the outside demand, the vector $\mathbf{x}$ must satisfy the Leontief equation (2), so the problem reduces to solving the linear system

$$\underbrace{\begin{bmatrix} 0.5 & -0.1 & -0.1 \\ -0.2 & 0.5 & -0.3 \\ -0.1 & -0.3 & 0.6 \end{bmatrix}}_{I-C} \underbrace{\begin{bmatrix} x_1 \\ x_2 \\ x_3 \end{bmatrix}}_{\mathbf{x}} = \underbrace{\begin{bmatrix} 7900 \\ 3950 \\ 1975 \end{bmatrix}}_{\mathbf{d}} \quad (4)$$

(if consistent). We leave it for you to show that the reduced row echelon form of the augmented matrix for this system is

$$\begin{bmatrix} 1 & 0 & 0 & 27{,}500 \\ 0 & 1 & 0 & 33{,}750 \\ 0 & 0 & 1 & 24{,}750 \end{bmatrix}$$

This tells us that (4) is consistent, and the economy can satisfy the demand of the open sector exactly by producing \$27,500 worth of manufacturing output, \$33,750 worth of agricultural output, and \$24,750 worth of utilities output. ◀

Productive Open Economies

In the preceding discussion we considered an open economy with three product-producing sectors; the same ideas apply to an open economy with n product-producing sectors. In this case, the consumption matrix, production vector, and outside demand vector have the form

$$C = \begin{bmatrix} c_{11} & c_{12} & \cdots & c_{1n} \\ c_{21} & c_{22} & \cdots & c_{2n} \\ \vdots & \vdots & & \vdots \\ c_{n1} & c_{n2} & \cdots & c_{nn} \end{bmatrix}, \quad \mathbf{x} = \begin{bmatrix} x_1 \\ x_2 \\ \vdots \\ x_n \end{bmatrix}, \quad \mathbf{d} = \begin{bmatrix} d_1 \\ d_2 \\ \vdots \\ d_n \end{bmatrix}$$

where all entries are nonnegative and

c_{ij} = the monetary value of the output of the ith sector that is needed by the jth sector to produce one unit of output

x_i = the monetary value of the output of the ith sector

d_i = the monetary value of the output of the ith sector that is required to meet the demand of the open sector

Remark Note that the jth column vector of C contains the monetary values that the jth sector requires of the other sectors to produce one monetary unit of output, and the ith row vector of C contains the monetary values required of the ith sector by the other sectors for each of them to produce one monetary unit of output.

As discussed in our example above, a production vector $\mathbf{x}$ that meets the demand $\mathbf{d}$ of the outside sector must satisfy the Leontief equation

$$(I - C)\mathbf{x} = \mathbf{d}$$

If the matrix $I - C$ is invertible, then this equation has the unique solution

$$\mathbf{x} = (I - C)^{-1}\mathbf{d} \quad (5)$$

for every demand vector **d**. However, for **x** to be a valid production vector it must have nonnegative entries, so the problem of importance in economics is to determine conditions under which the Leontief equation has a solution with nonnegative entries.

It is evident from the form of (5) that if $I - C$ is invertible, and if $(I - C)^{-1}$ has nonnegative entries, then for every demand vector **d** the corresponding **x** will also have nonnegative entries, and hence will be a valid production vector for the economy. Economies for which $(I - C)^{-1}$ has nonnegative entries are said to be **productive**. Such economies are desirable because demand can always be met by some level of production. The following theorem, whose proof can be found in many books on economics, gives conditions under which open economies are productive.

THEOREM 1.9.1 *If C is the consumption matrix for an open economy, and if all of the column sums are less than* 1, *then the matrix* $I - C$ *is invertible, the entries of* $(I - C)^{-1}$ *are nonnegative, and the economy is productive.*

Remark   The jth column sum of C represents the total dollar value of input that the jth sector requires to produce \$1 of output, so if the jth column sum is less than 1, then the jth sector requires less than \$1 of input to produce \$1 of output; in this case we say that the jth sector is *profitable*. Thus, Theorem 1.9.1 states that if all product-producing sectors of an open economy are profitable, then the economy is productive. In the exercises we will ask you to show that an open economy is productive if all of the row sums of C are less than 1 (Exercise 11). Thus, an open economy is productive if *either* all of the column sums or all of the row sums of C are less than 1.

▶ **EXAMPLE 2 An Open Economy Whose Sectors Are All Profitable**

The column sums of the consumption matrix C in (1) are less than 1, so $(I - C)^{-1}$ exists and has nonnegative entries. Use a calculating utility to confirm this, and use this inverse to solve Equation (4) in Example 1.

Solution   We leave it for you to show that

$$(I - C)^{-1} \approx \begin{bmatrix} 2.65823 & 1.13924 & 1.01266 \\ 1.89873 & 3.67089 & 2.15190 \\ 1.39241 & 2.02532 & 2.91139 \end{bmatrix}$$

This matrix has nonnegative entries, and

$$\mathbf{x} = (I - C)^{-1}\mathbf{d} \approx \begin{bmatrix} 2.65823 & 1.13924 & 1.01266 \\ 1.89873 & 3.67089 & 2.15190 \\ 1.39241 & 2.02532 & 2.91139 \end{bmatrix} \begin{bmatrix} 7900 \\ 3950 \\ 1975 \end{bmatrix} \approx \begin{bmatrix} 27,500 \\ 33,750 \\ 24,750 \end{bmatrix}$$

which is consistent with the solution in Example 1. ◀

Concept Review

- Sectors
- Inputs
- Outputs
- Input-output analysis
- Open sector

- Economies: open, closed
- Consumption (technology) matrix
- Consumption vector
- Outside demand vector
- Production vector

- Intermediate demand vector
- Leontief matrix
- Leontief equation

Skills

- Construct a consumption matrix for an economy.

- Understand the relationships among the vectors of a sector of an economy: consumption, outside demand, production, and intermediate demand.

Exercise Set 1.9

1. An automobile mechanic (M) and a body shop (B) use each other's services. For each $1.00 of business that M does, it uses $0.50 of its own services and $0.25 of B's services, and for each $1.00 of business that B does it uses $0.10 of its own services and $0.25 of M's services.

(a) Construct a consumption matrix for this economy.

(b) How much must M and B each produce to provide customers with $7000 worth of mechanical work and $14,000 worth of body work?

2. A simple economy produces food (F) and housing (H). The production of $1.00 worth of food requires $0.30 worth of food and $0.10 worth of housing, and the production of $1.00 worth of housing requires $0.20 worth of food and $0.60 worth of housing.

(a) Construct a consumption matrix for this economy.

(b) What dollar value of food and housing must be produced for the economy to provide consumers $130,000 worth of food and $130,000 worth of housing?

3. Consider the open economy described by the accompanying table, where the input is in dollars needed for $1.00 of output.

(a) Find the consumption matrix for the economy.

(b) Suppose that the open sector has a demand for $1930 worth of housing, $3860 worth of food, and $5790 worth of utilities. Use row reduction to find a production vector that will meet this demand exactly.

Table Ex-3

Input Required per Dollar Output

	Housing	Food	Utilities
Housing	$ 0.10	$ 0.60	$ 0.40
Food	$ 0.30	$ 0.20	$ 0.30
Utilities	$ 0.40	$ 0.10	$ 0.20

(Provider label at left spanning rows)

4. A company produces Web design, software, and networking services. View the company as an open economy described by the accompanying table, where input is in dollars needed for $1.00 of output.

(a) Find the consumption matrix for the company.

(b) Suppose that the customers (the open sector) have a demand for $5400 worth of Web design, $2700 worth of software, and $900 worth of networking. Use row reduction to find a production vector that will meet this demand exactly.

Table Ex-4

Input Required per Dollar Output

	Web Design	Software	Networking
Web Design	$ 0.40	$ 0.20	$ 0.45
Software	$ 0.30	$ 0.35	$ 0.30
Networking	$ 0.15	$ 0.10	$ 0.20

(Provider label at left spanning rows)

▶ In Exercises 5–6, use matrix inversion to find the production vector **x** that meets the demand **d** for the consumption matrix C.

5. $C = \begin{bmatrix} 0.1 & 0.3 \\ 0.5 & 0.4 \end{bmatrix}$; $\mathbf{d} = \begin{bmatrix} 50 \\ 60 \end{bmatrix}$

6. $C = \begin{bmatrix} 0.3 & 0.1 \\ 0.3 & 0.7 \end{bmatrix}$; $\mathbf{d} = \begin{bmatrix} 22 \\ 14 \end{bmatrix}$

7. Consider an open economy with consumption matrix

$$C = \begin{bmatrix} \frac{1}{2} & 0 \\ 0 & 1 \end{bmatrix}$$

(a) Show that the economy can meet a demand of $d_1 = 2$ units from the first sector and $d_2 = 0$ units from the second sector, but it cannot meet a demand of $d_1 = 2$ units from the first sector and $d_2 = 1$ unit from the second sector.

(b) Give both a mathematical and an economic explanation of the result in part (a).

8. Consider an open economy with consumption matrix

$$C = \begin{bmatrix} \frac{1}{2} & \frac{1}{4} & \frac{1}{4} \\ \frac{1}{2} & \frac{1}{8} & \frac{1}{4} \\ \frac{1}{2} & \frac{1}{4} & \frac{1}{8} \end{bmatrix}$$

If the open sector demands the same dollar value from each product-producing sector, which such sector must produce the greatest dollar value to meet the demand?

9. Consider an open economy with consumption matrix

$$C = \begin{bmatrix} c_{11} & c_{12} \\ c_{21} & 0 \end{bmatrix}$$

Show that the Leontief equation $\mathbf{x} - C\mathbf{x} = \mathbf{d}$ has a unique solution for every demand vector $\mathbf{d}$ if $c_{21}c_{12} < 1 - c_{11}$.

10. (a) Consider an open economy with a consumption matrix C whose column sums are less than 1, and let $\mathbf{x}$ be the production vector that satisfies an outside demand $\mathbf{d}$; that is, $(I - C)^{-1}\mathbf{d} = \mathbf{x}$. Let $\mathbf{d}_j$ be the demand vector that is obtained by increasing the jth entry of $\mathbf{d}$ by 1 and leaving the other entries fixed. Prove that the production vector

$\mathbf{x}_j$ that meets this demand is

$$\mathbf{x}_j = \mathbf{x} + j\text{th column vector of } (I - C)^{-1}$$

(b) In words, what is the economic significance of the jth column vector of $(I - C)^{-1}$? [*Hint:* Look at $\mathbf{x}_j - \mathbf{x}$.]

11. Prove: If C is an $n \times n$ matrix whose entries are nonnegative and whose row sums are less than 1, then $I - C$ is invertible and has nonnegative entries. [*Hint:* $(A^T)^{-1} = (A^{-1})^T$ for any invertible matrix A.]

True-False Exercises

In parts (a)–(e) determine whether the statement is true or false, and justify your answer.

(a) Sectors of an economy that produce outputs are called open sectors.

(b) A closed economy is an economy that has no open sectors.

(c) The rows of a consumption matrix represent the outputs in a sector of an economy.

(d) If the column sums of the consumption matrix are all less than 1, then the Leontif matrix is invertible.

(e) The Leontif equation relates the production vector for an economy to the outside demand vector.

Chapter 1   Supplementary Exercises

▶ In Exercises 1–4 the given matrix represents an augmented matrix for a linear system. Write the corresponding set of linear equations for the system, and use Gaussian elimination to solve the linear system. Introduce free parameters as necessary. ◀

1. $\begin{bmatrix} 3 & -1 & 0 & 4 & 1 \\ 2 & 0 & 3 & 3 & -1 \end{bmatrix}$

2. $\begin{bmatrix} 1 & 4 & -1 \\ -2 & -8 & 2 \\ 3 & 12 & -3 \\ 0 & 0 & 0 \end{bmatrix}$

3. $\begin{bmatrix} 2 & -4 & 1 & 6 \\ -4 & 0 & 3 & -1 \\ 0 & 1 & -1 & 3 \end{bmatrix}$

4. $\begin{bmatrix} 3 & 1 & -2 \\ -9 & -3 & 6 \\ 6 & 2 & 1 \end{bmatrix}$

5. Use Gauss–Jordan elimination to solve for x' and y' in terms of x and y.

$$x = \tfrac{3}{5}x' - \tfrac{4}{5}y'$$
$$y = \tfrac{4}{5}x' + \tfrac{3}{5}y'$$

6. Use Gauss–Jordan elimination to solve for x' and y' in terms of x and y.

$$x = x' \cos\theta - y' \sin\theta$$
$$y = x' \sin\theta + y' \cos\theta$$

7. Find positive integers that satisfy

$$x + y + z = 9$$
$$x + 5y + 10z = 44$$

8. A box containing pennies, nickels, and dimes has 13 coins with a total value of 83 cents. How many coins of each type are in the box?

9. Let

$$\begin{bmatrix} a & 0 & b & 2 \\ a & a & 4 & 4 \\ 0 & a & 2 & b \end{bmatrix}$$

be the augmented matrix for a linear system. Find for what values of a and b the system has

(a) a unique solution.

(b) a one-parameter solution.

(c) a two-parameter solution.   (d) no solution.

10. For which value(s) of a does the following system have zero solutions? One solution? Infinitely many solutions?

$$x_1 + x_2 + x_3 = 4$$
$$x_3 = 2$$
$$(a^2 - 4)x_3 = a - 2$$

11. Find a matrix K such that $AKB = C$ given that

$$A = \begin{bmatrix} 1 & 4 \\ -2 & 3 \\ 1 & -2 \end{bmatrix}, \quad B = \begin{bmatrix} 2 & 0 & 0 \\ 0 & 1 & -1 \end{bmatrix},$$

$$C = \begin{bmatrix} 8 & 6 & -6 \\ 6 & -1 & 1 \\ -4 & 0 & 0 \end{bmatrix},$$

12. How should the coefficients a, b, and c be chosen so that the system

$$ax + by - 3z = -3$$
$$-2x - by + cz = -1$$
$$ax + 3y - cz = -3$$

has the solution $x = 1$, $y = -1$, and $z = 2$?

13. In each part, solve the matrix equation for X.

(a) $X \begin{bmatrix} -1 & 0 & 1 \\ 1 & 1 & 0 \\ 3 & 1 & -1 \end{bmatrix} = \begin{bmatrix} 1 & 2 & 0 \\ -3 & 1 & 5 \end{bmatrix}$

(b) $X \begin{bmatrix} 1 & -1 & 2 \\ 3 & 0 & 1 \end{bmatrix} = \begin{bmatrix} -5 & -1 & 0 \\ 6 & -3 & 7 \end{bmatrix}$

(c) $\begin{bmatrix} 3 & 1 \\ -1 & 2 \end{bmatrix} X - X \begin{bmatrix} 1 & 4 \\ 2 & 0 \end{bmatrix} = \begin{bmatrix} 2 & -2 \\ 5 & 4 \end{bmatrix}$

14. Let A be a square matrix.

(a) Show that $(I - A)^{-1} = I + A + A^2 + A^3$ if $A^4 = 0$.

(b) Show that

$$(I - A)^{-1} = I + A + A^2 + \cdots + A^n$$

if $A^{n+1} = 0$.

15. Find values of a, b, and c such that the graph of the polynomial $p(x) = ax^2 + bx + c$ passes through the points $(1, 2)$, $(-1, 6)$, and $(2, 3)$.

16. (*Calculus required*) Find values of a, b, and c such that the graph of $p(x) = ax^2 + bx + c$ passes through the point $(-1, 0)$ and has a horizontal tangent at $(2, -9)$.

17. Let J_n be the $n \times n$ matrix each of whose entries is 1. Show that if $n > 1$, then

$$(I - J_n)^{-1} = I - \frac{1}{n-1} J_n$$

18. Show that if a square matrix A satisfies

$$A^3 + 4A^2 - 2A + 7I = 0$$

then so does A^T.

19. Prove: If B is invertible, then $AB^{-1} = B^{-1}A$ if and only if $AB = BA$.

20. Prove: If A is invertible, then $A + B$ and $I + BA^{-1}$ are both invertible or both not invertible.

21. Prove: If A is an $m \times n$ matrix and B is the $n \times 1$ matrix each of whose entries is $1/n$, then

$$AB = \begin{bmatrix} \bar{r}_1 \\ \bar{r}_2 \\ \vdots \\ \bar{r}_m \end{bmatrix}$$

where $\bar{r}_i$ is the average of the entries in the ith row of A.

22. (*Calculus required*) If the entries of the matrix

$$C = \begin{bmatrix} c_{11}(x) & c_{12}(x) & \cdots & c_{1n}(x) \\ c_{21}(x) & c_{22}(x) & \cdots & c_{2n}(x) \\ \vdots & \vdots & & \vdots \\ c_{m1}(x) & c_{m2}(x) & \cdots & c_{mn}(x) \end{bmatrix}$$

are differentiable functions of x, then we define

$$\frac{dC}{dx} = \begin{bmatrix} c'_{11}(x) & c'_{12}(x) & \cdots & c'_{1n}(x) \\ c'_{21}(x) & c'_{22}(x) & \cdots & c'_{2n}(x) \\ \vdots & \vdots & & \vdots \\ c'_{m1}(x) & c'_{m2}(x) & \cdots & c'_{mn}(x) \end{bmatrix}$$

Show that if the entries in A and B are differentiable functions of x and the sizes of the matrices are such that the stated operations can be performed, then

(a) $\dfrac{d}{dx}(kA) = k\dfrac{dA}{dx}$

(b) $\dfrac{d}{dx}(A + B) = \dfrac{dA}{dx} + \dfrac{dB}{dx}$

(c) $\dfrac{d}{dx}(AB) = \dfrac{dA}{dx}B + A\dfrac{dB}{dx}$

23. (*Calculus required*) Use part (c) of Exercise 22 to show that

$$\frac{dA^{-1}}{dx} = -A^{-1}\frac{dA}{dx}A^{-1}$$

State all the assumptions you make in obtaining this formula.

24. Assuming that the stated inverses exist, prove the following equalities.

(a) $(C^{-1} + D^{-1})^{-1} = C(C + D)^{-1}D$

(b) $(I + CD)^{-1}C = C(I + DC)^{-1}$

(c) $(C + DD^T)^{-1}D = C^{-1}D(I + D^TC^{-1}D)^{-1}$

CHAPTER 2

Determinants

CHAPTER CONTENTS
2.1 Determinants by Cofactor Expansion 93
2.2 Evaluating Determinants by Row Reduction 100
2.3 Properties of Determinants; Cramer's Rule 106

INTRODUCTION In this chapter we will study "determinants" or, more precisely, "determinant functions." Unlike real-valued functions, such as $f(x) = x^2$, that assign a real number to a real variable x, determinant functions assign a real number $f(A)$ to a matrix variable A. Although determinants first arose in the context of solving systems of linear equations, they are no longer used for that purpose in real-world applications. Although they can be useful for solving very small linear systems (say two or three unknowns), our main interest in them stems from the fact that they link together various concepts in linear algebra and provide a useful formula for the inverse of a matrix.

2.1 Determinants by Cofactor Expansion

In this section we will define the notion of a "determinant." This will enable us to give a specific formula for the inverse of an invertible matrix, whereas up to now we have had only a computational procedure for finding it. This, in turn, will eventually provide us with a formula for solutions of certain kinds of linear systems.

Recall from Theorem 1.4.5 that the 2×2 matrix

$$A = \begin{bmatrix} a & b \\ c & d \end{bmatrix}$$

is invertible if and only if $ad - bc \neq 0$ and that the expression $ad - bc$ is called the **determinant** of the matrix A. Recall also that this determinant is denoted by writing

$$\det(A) = ad - bc \quad \text{or} \quad \begin{vmatrix} a & b \\ c & d \end{vmatrix} = ad - bc \tag{1}$$

> **WARNING** It is important to keep in mind that $\det(A)$ is a *number*, whereas A is a *matrix*.

and that the inverse of A can be expressed in terms of the determinant as

$$A^{-1} = \frac{1}{\det(A)} \begin{bmatrix} d & -b \\ -c & a \end{bmatrix} \tag{2}$$

Minors and Cofactors One of our main goals in this chapter is to obtain an analog of Formula (2) that is applicable to square matrices of *all orders*. For this purpose we will find it convenient to use subscripted entries when writing matrices or determinants. Thus, if we denote a 2×2 matrix as

$$A = \begin{bmatrix} a_{11} & a_{12} \\ a_{21} & a_{22} \end{bmatrix}$$

We define the determinant of a 1×1 matrix $A = [a_{11}]$ as

$$\det[A] = \det[a_{11}] = a_{11}$$

then the two equations in (1) take the form

$$\det(A) = \begin{vmatrix} a_{11} & a_{12} \\ a_{21} & a_{22} \end{vmatrix} = a_{11}a_{22} - a_{12}a_{21} \tag{3}$$

The following definition will be key to our goal of extending the definition of a determinant to higher order matrices.

DEFINITION 1 If A is a square matrix, then the ***minor of entry*** a_{ij} is denoted by M_{ij} and is defined to be the determinant of the submatrix that remains after the ith row and jth column are deleted from A. The number $(-1)^{i+j} M_{ij}$ is denoted by C_{ij} and is called the ***cofactor of entry*** a_{ij}.

▶ **EXAMPLE 1** Finding Minors and Cofactors

Let

$$A = \begin{bmatrix} 3 & 1 & -4 \\ 2 & 5 & 6 \\ 1 & 4 & 8 \end{bmatrix}$$

The minor of entry a_{11} is

WARNING We have followed the standard convention of using capital letters to denote minors and cofactors even though they are numbers, not matrices.

$$M_{11} = \begin{vmatrix} 3 & 1 & -4 \\ 2 & 5 & 6 \\ 1 & 4 & 8 \end{vmatrix} = \begin{vmatrix} 5 & 6 \\ 4 & 8 \end{vmatrix} = 16$$

The cofactor of a_{11} is

$$C_{11} = (-1)^{1+1} M_{11} = M_{11} = 16$$

Similarly, the minor of entry a_{32} is

$$M_{32} = \begin{vmatrix} 3 & 1 & -4 \\ 2 & 5 & 6 \\ 1 & 4 & 8 \end{vmatrix} = \begin{vmatrix} 3 & -4 \\ 2 & 6 \end{vmatrix} = 26$$

The cofactor of a_{32} is

$$C_{32} = (-1)^{3+2} M_{32} = -M_{32} = -26 \quad ◀$$

Historical Note The term *determinant* was first introduced by the German mathematician Carl Friedrich Gauss in 1801 (see p. 15), who used them to "determine" properties of certain kinds of functions. Interestingly, the term *matrix* is derived from a Latin word for "womb" because it was viewed as a container of determinants.

Historical Note The term *minor* is apparently due to the English mathematician James Sylvester (see p. 34), who wrote the following in a paper published in 1850: "Now conceive any one line and any one column be struck out, we get...a square, one term less in breadth and depth than the original square; and by varying in every possible selection of the line and column excluded, we obtain, supposing the original square to consist of n lines and n columns, n^2 such minor squares, each of which will represent what I term a "First Minor Determinant" relative to the principal or complete determinant."

Remark   Note that a minor M_{ij} and its corresponding cofactor C_{ij} are either the same or negatives of each other and that the relating sign $(-1)^{i+j}$ is either $+1$ or -1 in accordance with the pattern in the "checkerboard" array

$$\begin{bmatrix} + & - & + & - & + & \cdots \\ - & + & - & + & - & \cdots \\ + & - & + & - & + & \cdots \\ - & + & - & + & - & \cdots \\ \vdots & \vdots & \vdots & \vdots & \vdots & \end{bmatrix}$$

For example,

$$C_{11} = M_{11}, \quad C_{21} = -M_{21}, \quad C_{22} = M_{22}$$

and so forth. Thus, it is never really necessary to calculate $(-1)^{i+j}$ to calculate C_{ij}—you can simply compute the minor M_{ij} and then adjust the sign in accordance with the checkerboard pattern. Try this in Example 1.

▶ **EXAMPLE 2   Cofactor Expansions of a 2 × 2 Matrix**

The checkerboard pattern for a 2×2 matrix $A = [a_{ij}]$ is

$$\begin{bmatrix} + & - \\ - & + \end{bmatrix}$$

so that

$$C_{11} = M_{11} = a_{22} \qquad C_{12} = -M_{12} = -a_{21}$$
$$C_{21} = -M_{21} = -a_{12} \qquad C_{22} = M_{22} = a_{11}$$

We leave it for you to use Formula (3) to verify that $\det(A)$ can be expressed in terms of cofactors in the following four ways:

$$\det(A) = \begin{vmatrix} a_{11} & a_{12} \\ a_{21} & a_{22} \end{vmatrix}$$
$$= a_{11}C_{11} + a_{12}C_{12}$$
$$= a_{21}C_{21} + a_{22}C_{22} \qquad (4)$$
$$= a_{11}C_{11} + a_{21}C_{21}$$
$$= a_{12}C_{12} + a_{22}C_{22}$$

Each of last four equations is called a *cofactor expansion* of $\det[A]$. In each cofactor expansion the entries and cofactors all come from the same row or same column of A. For example, in the first equation the entries and cofactors all come from the first row of A, in the second they all come from the second row of A, in the third they all come from the first column of A, and in the fourth they all come from the second column of A. ◀

Definition of a General   Formula (4) is a special case of the following general result, which we will state without
Determinant   proof.

> **THEOREM 2.1.1**   *If A is an n × n matrix, then regardless of which row or column of A is chosen, the number obtained by multiplying the entries in that row or column by the corresponding cofactors and adding the resulting products is always the same.*

This result allows us to make the following definition.

DEFINITION 2 If A is an $n \times n$ matrix, then the number obtained by multiplying the entries in any row or column of A by the corresponding cofactors and adding the resulting products is called the ***determinant of*** A, and the sums themselves are called ***cofactor expansions of*** A. That is,

$$\det(A) = a_{1j}C_{1j} + a_{2j}C_{2j} + \cdots + a_{nj}C_{nj} \qquad (5)$$

[cofactor expansion along the jth column]

and

$$\det(A) = a_{i1}C_{i1} + a_{i2}C_{i2} + \cdots + a_{in}C_{in} \qquad (6)$$

[cofactor expansion along the ith row]

▶ **EXAMPLE 3** Cofactor Expansion Along the First Row

Find the determinant of the matrix

$$A = \begin{bmatrix} 3 & 1 & 0 \\ -2 & -4 & 3 \\ 5 & 4 & -2 \end{bmatrix}$$

by cofactor expansion along the first row.

Solution

$$\det(A) = \begin{vmatrix} 3 & 1 & 0 \\ -2 & -4 & 3 \\ 5 & 4 & -2 \end{vmatrix} = 3 \begin{vmatrix} -4 & 3 \\ 4 & -2 \end{vmatrix} - 1 \begin{vmatrix} -2 & 3 \\ 5 & -2 \end{vmatrix} + 0 \begin{vmatrix} -2 & -4 \\ 5 & 4 \end{vmatrix}$$

$$= 3(-4) - (1)(-11) + 0 = -1$$

▶ **EXAMPLE 4** Cofactor Expansion Along the First Column

Let A be the matrix in Example 3, and evaluate $\det(A)$ by cofactor expansion along the first column of A.

Solution

$$\det(A) = \begin{vmatrix} 3 & 1 & 0 \\ -2 & -4 & 3 \\ 5 & 4 & -2 \end{vmatrix} = 3 \begin{vmatrix} -4 & 3 \\ 4 & -2 \end{vmatrix} - (-2) \begin{vmatrix} 1 & 0 \\ 4 & -2 \end{vmatrix} + 5 \begin{vmatrix} 1 & 0 \\ -4 & 3 \end{vmatrix}$$

$$= 3(-4) - (-2)(-2) + 5(3) = -1$$

This agrees with the result obtained in Example 3.

Note that in Example 4 we had to compute three cofactors, whereas in Example 3 only two were needed because the third was multiplied by zero. As a rule, the best strategy for cofactor expansion is to expand along a row or column with the most zeros.

Charles Lutwidge Dodgson (Lewis Carroll) (1832–1898)

Historical Note Cofactor expansion is not the only method for expressing the determinant of a matrix in terms of determinants of lower order. For example, although it is not well known, the English mathematician Charles Dodgson, who was the author of *Alice's Adventures in Wonderland* and *Through the Looking Glass* under the pen name of Lewis Carroll, invented such a method, called "*condensation*." That method has recently been resurrected from obscurity because of its suitability for parallel processing on computers.

[*Image: Time & Life Pictures/Getty Images, Inc.*]

▶ **EXAMPLE 5 Smart Choice of Row or Column**

If A is the 4×4 matrix

$$A = \begin{bmatrix} 1 & 0 & 0 & -1 \\ 3 & 1 & 2 & 2 \\ 1 & 0 & -2 & 1 \\ 2 & 0 & 0 & 1 \end{bmatrix}$$

then to find $\det(A)$ it will be easiest to use cofactor expansion along the second column, since it has the most zeros:

$$\det(A) = 1 \cdot \begin{vmatrix} 1 & 0 & -1 \\ 1 & -2 & 1 \\ 2 & 0 & 1 \end{vmatrix}$$

For the 3×3 determinant, it will be easiest to use cofactor expansion along its second column, since it has the most zeros:

$$\det(A) = 1 \cdot -2 \cdot \begin{vmatrix} 1 & -1 \\ 2 & 1 \end{vmatrix}$$
$$= -2(1 + 2)$$
$$= -6$$

▶ **EXAMPLE 6 Determinant of an Upper Triangular Matrix**

The following computation shows that the determinant of a 4×4 upper triangular matrix is the product of its diagonal entries. Each part of the computation uses a cofactor expansion along the first row.

$$\begin{vmatrix} a_{11} & 0 & 0 & 0 \\ a_{21} & a_{22} & 0 & 0 \\ a_{31} & a_{32} & a_{33} & 0 \\ a_{41} & a_{42} & a_{43} & a_{44} \end{vmatrix} = a_{11} \begin{vmatrix} a_{22} & 0 & 0 \\ a_{32} & a_{33} & 0 \\ a_{42} & a_{43} & a_{44} \end{vmatrix}$$

$$= a_{11} a_{22} \begin{vmatrix} a_{33} & 0 \\ a_{43} & a_{44} \end{vmatrix}$$

$$= a_{11} a_{22} a_{33} |a_{44}| = a_{11} a_{22} a_{33} a_{44} \blacktriangleleft$$

The method illustrated in Example 6 can be easily adapted to prove the following general result.

THEOREM 2.1.2 *If A is an $n \times n$ triangular matrix (upper triangular, lower triangular, or diagonal), then $\det(A)$ is the product of the entries on the main diagonal of the matrix; that is, $\det(A) = a_{11} a_{22} \cdots a_{nn}$.*

A Useful Technique for Evaluating 2×2 and 3×3 Determinants

Determinants of 2×2 and 3×3 matrices can be evaluated very efficiently using the pattern suggested in Figure 2.1.1.

▶ Figure 2.1.1

In the 2×2 case, the determinant can be computed by forming the product of the entries on the rightward arrow and subtracting the product of the entries on the leftward arrow. In the 3×3 case we first recopy the first and second columns as shown in the figure, after which we can compute the determinant by summing the products of the entries on the rightward arrows and subtracting the products on the leftward arrows. These procedures execute the computations

$$\begin{vmatrix} a_{11} & a_{12} \\ a_{21} & a_{22} \end{vmatrix} = a_{11}a_{22} - a_{12}a_{21}$$

$$\begin{vmatrix} a_{11} & a_{12} & a_{13} \\ a_{21} & a_{22} & a_{23} \\ a_{31} & a_{32} & a_{33} \end{vmatrix} = a_{11}\begin{vmatrix} a_{22} & a_{23} \\ a_{32} & a_{33} \end{vmatrix} - a_{12}\begin{vmatrix} a_{21} & a_{23} \\ a_{31} & a_{33} \end{vmatrix} + a_{13}\begin{vmatrix} a_{21} & a_{22} \\ a_{31} & a_{32} \end{vmatrix}$$

$$= a_{11}(a_{22}a_{33} - a_{23}a_{32}) - a_{12}(a_{21}a_{33} - a_{23}a_{31}) + a_{13}(a_{21}a_{32} - a_{22}a_{31})$$

$$= a_{11}a_{22}a_{33} + a_{12}a_{23}a_{31} + a_{13}a_{21}a_{32} - a_{13}a_{22}a_{31} - a_{12}a_{21}a_{33} - a_{11}a_{23}a_{32}$$

which agrees with the cofactor expansions along the first row.

▶ **EXAMPLE 7 A Technique for Evaluating 2 × 2 and 3 × 3 Determinants**

$$\begin{vmatrix} 3 & 1 \\ 4 & -2 \end{vmatrix} = (3)(-2) - (1)(4) = -10$$

$$\begin{vmatrix} 1 & 2 & 3 \\ -4 & 5 & 6 \\ 7 & -8 & 9 \end{vmatrix} =$$

$$= [45 + 84 + 96] - [105 - 48 - 72] = 240 \quad ◀$$

Concept Review

- Determinant
- Minor
- Cofactor
- Cofactor expansion

Skills

- Find the minors and cofactors of a square matrix.
- Use cofactor expansion to evaluate the determinant of a square matrix.
- Use the arrow technique to evaluate the determinant of a 2 × 2 or 3 × 3 matrix.
- Use the determinant of a 2 × 2 invertible matrix to find the inverse of that matrix.
- Find the determinant of an upper triangular, lower triangular, or diagonal matrix by inspection.

Exercise Set 2.1

▶ In Exercises 1–2, find all the minors and cofactors of the matrix A. ◀

1. $A = \begin{bmatrix} 1 & -2 & 3 \\ 6 & 7 & -1 \\ -3 & 1 & 4 \end{bmatrix}$

2. $A = \begin{bmatrix} 1 & 1 & 2 \\ 3 & 3 & 6 \\ 0 & 1 & 4 \end{bmatrix}$

3. Let

$$A = \begin{bmatrix} 4 & -1 & 1 & 6 \\ 0 & 0 & -3 & 3 \\ 4 & 1 & 0 & 14 \\ 4 & 1 & 3 & 2 \end{bmatrix}$$

Find

(a) M_{13} and C_{13}.

(b) M_{23} and C_{23}.

(c) M_{22} and C_{22}.

(d) M_{21} and C_{21}.

4. Let

$$A = \begin{bmatrix} 2 & 3 & -1 & 1 \\ -3 & 2 & 0 & 3 \\ 3 & -2 & 1 & 0 \\ 3 & -2 & 1 & 4 \end{bmatrix}$$

Find

(a) M_{32} and C_{32}.　　　　(b) M_{44} and C_{44}.

(c) M_{41} and C_{41}.　　　　(d) M_{24} and C_{24}.

▶ In Exercises **5–8**, evaluate the determinant of the given matrix. If the matrix is invertible, use Equation (3) to find its inverse. ◀

5. $\begin{bmatrix} 3 & 5 \\ -2 & 4 \end{bmatrix}$　　　　**6.** $\begin{bmatrix} 4 & 1 \\ 8 & 2 \end{bmatrix}$

7. $\begin{bmatrix} -5 & 7 \\ -7 & -2 \end{bmatrix}$　　　　**8.** $\begin{bmatrix} \sqrt{2} & \sqrt{6} \\ 4 & \sqrt{3} \end{bmatrix}$

▶ In Exercises **9–14**, use the arrow technique to evaluate the determinant of the given matrix. ◀

9. $\begin{bmatrix} a-3 & 5 \\ -3 & a-2 \end{bmatrix}$　　　　**10.** $\begin{bmatrix} -2 & 7 & 6 \\ 5 & 1 & -2 \\ 3 & 8 & 4 \end{bmatrix}$

11. $\begin{bmatrix} -2 & 1 & 4 \\ 3 & 5 & -7 \\ 1 & 6 & 2 \end{bmatrix}$　　　　**12.** $\begin{bmatrix} -1 & 1 & 2 \\ 3 & 0 & -5 \\ 1 & 7 & 2 \end{bmatrix}$

13. $\begin{bmatrix} 3 & 0 & 0 \\ 2 & -1 & 5 \\ 1 & 9 & -4 \end{bmatrix}$　　　　**14.** $\begin{bmatrix} c & -4 & 3 \\ 2 & 1 & c^2 \\ 4 & c-1 & 2 \end{bmatrix}$

▶ In Exercises **15–18**, find all values of λ for which $\det(A) = 0$.

15. $A = \begin{bmatrix} \lambda - 2 & 1 \\ -5 & \lambda + 4 \end{bmatrix}$　　　　**16.** $A = \begin{bmatrix} \lambda - 4 & 0 & 0 \\ 0 & \lambda & 2 \\ 0 & 3 & \lambda - 1 \end{bmatrix}$

17. $A = \begin{bmatrix} \lambda - 1 & 0 \\ 2 & \lambda + 1 \end{bmatrix}$　　　　**18.** $A = \begin{bmatrix} \lambda - 4 & 4 & 0 \\ -1 & \lambda & 0 \\ 0 & 0 & \lambda - 5 \end{bmatrix}$

19. Evaluate the determinant of the matrix in Exercise 13 by a cofactor expansion along

(a) the first row.　　　　(b) the first column.

(c) the second row.　　　　(d) the second column.

(e) the third row.　　　　(f) the third column.

20. Evaluate the determinant of the matrix in Exercise 12 by a cofactor expansion along

(a) the first row.　　　　(b) the first column.

(c) the second row.　　　　(d) the second column.

(e) the third row.　　　　(f) the third column.

▶ In Exercises **21–26**, evaluate $\det(A)$ by a cofactor expansion along a row or column of your choice. ◀

21. $A = \begin{bmatrix} -3 & 0 & 7 \\ 2 & 5 & 1 \\ -1 & 0 & 5 \end{bmatrix}$　　**22.** $A = \begin{bmatrix} 3 & 3 & 1 \\ 1 & 0 & -4 \\ 1 & -3 & 5 \end{bmatrix}$

23. $A = \begin{bmatrix} 1 & k & k^2 \\ 1 & k & k^2 \\ 1 & k & k^2 \end{bmatrix}$　　**24.** $A = \begin{bmatrix} k+1 & k-1 & 7 \\ 2 & k-3 & 4 \\ 5 & k+1 & k \end{bmatrix}$

25. $A = \begin{bmatrix} 3 & 3 & 0 & 5 \\ 2 & 2 & 0 & -2 \\ 4 & 1 & -3 & 0 \\ 2 & 10 & 3 & 2 \end{bmatrix}$

26. $A = \begin{bmatrix} 4 & 0 & 0 & 1 & 0 \\ 3 & 3 & 3 & -1 & 0 \\ 1 & 2 & 4 & 2 & 3 \\ 9 & 4 & 6 & 2 & 3 \\ 2 & 2 & 4 & 2 & 3 \end{bmatrix}$

▶ In Exercises **27–32**, evaluate the determinant of the given matrix by inspection. ◀

27. $\begin{bmatrix} 1 & 0 & 0 \\ 0 & -1 & 0 \\ 0 & 0 & 1 \end{bmatrix}$　　**28.** $\begin{bmatrix} 2 & 0 & 0 \\ 0 & 2 & 0 \\ 0 & 0 & 2 \end{bmatrix}$

29. $\begin{bmatrix} 0 & 0 & 0 & 0 \\ 1 & 2 & 0 & 0 \\ 0 & 4 & 3 & 0 \\ 1 & 2 & 3 & 8 \end{bmatrix}$　　**30.** $\begin{bmatrix} 1 & 1 & 1 & 1 \\ 0 & 2 & 2 & 2 \\ 0 & 0 & 3 & 3 \\ 0 & 0 & 0 & 4 \end{bmatrix}$

31. $\begin{bmatrix} 1 & 2 & 7 & -3 \\ 0 & 1 & -4 & 1 \\ 0 & 0 & 2 & 7 \\ 0 & 0 & 0 & 3 \end{bmatrix}$　　**32.** $\begin{bmatrix} -3 & 0 & 0 & 0 \\ 1 & 2 & 0 & 0 \\ 40 & 10 & -1 & 0 \\ 100 & 200 & -23 & 3 \end{bmatrix}$

33. Show that the value of the following determinant is independent of θ.

$$\begin{vmatrix} \sin(\theta) & \cos(\theta) & 0 \\ -\cos(\theta) & \sin(\theta) & 0 \\ \sin(\theta) - \cos(\theta) & \sin(\theta) + \cos(\theta) & 1 \end{vmatrix}$$

34. Show that the matrices

$$A = \begin{bmatrix} a & b \\ 0 & c \end{bmatrix} \quad \text{and} \quad B = \begin{bmatrix} d & e \\ 0 & f \end{bmatrix}$$

commute if and only if

$$\begin{vmatrix} b & a-c \\ e & d-f \end{vmatrix} = 0$$

35. By inspection, what is the relationship between the following determinants?

$$d_1 = \begin{vmatrix} a & b & c \\ d & 1 & f \\ g & 0 & 1 \end{vmatrix} \quad \text{and} \quad d_2 = \begin{vmatrix} a + \lambda & b & c \\ d & 1 & f \\ g & 0 & 1 \end{vmatrix}$$

36. Show that

$$\det(A) = \frac{1}{2} \begin{vmatrix} \text{tr}(A) & 1 \\ \text{tr}(A^2) & \text{tr}(A) \end{vmatrix}$$

for every 2×2 matrix A.

37. What can you say about an nth-order determinant all of whose entries are 1? Explain your reasoning.

38. What is the maximum number of zeros that a 3×3 matrix can have without having a zero determinant? Explain your reasoning.

39. What is the maximum number of zeros that a 4×4 matrix can have without having a zero determinant? Explain your reasoning.

40. Prove that (x_1, y_1), (x_2, y_2), and (x_3, y_3) are collinear points if and only if

$$\begin{vmatrix} x_1 & y_1 & 1 \\ x_2 & y_2 & 1 \\ x_3 & y_3 & 1 \end{vmatrix} = 0$$

41. Prove: If the equation of the line through the distinct points (a_1, b_1) and (a_2, b_2) can be written as

$$\begin{vmatrix} x & y & 1 \\ a_1 & b_1 & 1 \\ a_2 & b_2 & 1 \end{vmatrix} = 0$$

42. Prove that if A is upper triangular and B_{ij} is the matrix that results when the ith row and jth column of A are deleted, then B_{ij} is upper triangular if $i < j$.

True-False Exercises

In parts (a)–(j) determine whether the statement is true or false, and justify your answer.

(a) The determinant of the 2×2 matrix $\begin{bmatrix} a & b \\ c & d \end{bmatrix}$ is $ad + bc$.

(b) Two square matrices A and B can have the same determinant only if they are the same size.

(c) The minor M_{ij} is the same as the cofactor C_{ij} if and only if $i + j$ is even.

(d) If A is a 3×3 symmetric matrix, then $C_{ij} = C_{ji}$ for all i and j.

(e) The value of a cofactor expansion of a matrix A is independent of the row or column chosen for the expansion.

(f) If A is a matrix whose minors all have the same value, then $\det(A) = 0$.

(g) The determinant of a lower triangular matrix is the sum of the entries along its main diagonal.

(h) For every square matrix A and every scalar c, we have $\det(cA) = c \det(A)$.

(i) For all square matrices A and B, we have $\det(A + B) = \det(A) + \det(B)$.

(j) For every 2×2 matrix A, we have $\det(A^2) = (\det(A))^2$.

2.2 Evaluating Determinants by Row Reduction

In this section we will show how to evaluate a determinant by reducing the associated matrix to row echelon form. In general, this method requires less computation than cofactor expansion and hence is the method of choice for large matrices.

A Basic Theorem   We begin with a fundamental theorem that will lead us to an efficient procedure for evaluating the determinant of a square matrix of any size.

> **THEOREM 2.2.1** *Let A be a square matrix. If A has a row of zeros or a column of zeros, then $\det(A) = 0$.*

Proof   Since the determinant of A can be found by a cofactor expansion along any row or column, we can use the row or column of zeros. Thus, if we let $C_1, C_2, \ldots, C_n$ denote the cofactors of A along that row or column, then it follows from Formula (5) or (6) in Section 2.1 that

$$\det(A) = 0 \cdot C_1 + 0 \cdot C_2 + \cdots + 0 \cdot C_n = 0 \quad \blacktriangleleft$$

The following useful theorem relates the determinant of a matrix and the determinant of its transpose.

Because transposing a matrix changes its columns to rows and its rows to columns, almost every theorem about the rows of a determinant has a companion version about columns, and vice versa.

THEOREM 2.2.2 *Let A be a square matrix. Then* $\det(A) = \det(A^T)$.

Proof   Since transposing a matrix changes its columns to rows and its rows to columns, the cofactor expansion of A along any row is the same as the cofactor expansion of A^T along the corresponding column. Thus, both have the same determinant. ◀

Elementary Row Operations

The next theorem shows how an elementary row operation on a square matrix affects the value of its determinant. In place of a formal proof we have provided a table to illustrate the ideas in the 3×3 case (see Table 1).

THEOREM 2.2.3 *Let A be an n × n matrix.*

(a)   *If B is the matrix that results when a single row or single column of A is multiplied by a scalar k, then* $\det(B) = k \det(A)$.

(b)   *If B is the matrix that results when two rows or two columns of A are interchanged, then* $\det(B) = -\det(A)$.

(c)   *If B is the matrix that results when a multiple of one row of A is added to another row or when a multiple of one column is added to another column, then* $\det(B) = \det(A)$.

Table 1

The first panel of Table 1 shows that you can bring a common factor from any row (column) of a determinant through the determinant sign. This is a slightly different way of thinking about part (a) of Theorem 2.2.3.

Relationship	Operation
$\begin{vmatrix} ka_{11} & ka_{12} & ka_{13} \\ a_{21} & a_{22} & a_{23} \\ a_{31} & a_{32} & a_{33} \end{vmatrix} = k \begin{vmatrix} a_{11} & a_{12} & a_{13} \\ a_{21} & a_{22} & a_{23} \\ a_{31} & a_{32} & a_{33} \end{vmatrix}$ $\det(B) = k\det(A)$	The first row of A is multiplied by k.
$\begin{vmatrix} a_{21} & a_{22} & a_{23} \\ a_{11} & a_{12} & a_{13} \\ a_{31} & a_{32} & a_{33} \end{vmatrix} = - \begin{vmatrix} a_{11} & a_{12} & a_{13} \\ a_{21} & a_{22} & a_{23} \\ a_{31} & a_{32} & a_{33} \end{vmatrix}$ $\det(B) = -\det(A)$	The first and second rows of A are interchanged.
$\begin{vmatrix} a_{11}+ka_{21} & a_{12}+ka_{22} & a_{13}+ka_{23} \\ a_{21} & a_{22} & a_{23} \\ a_{31} & a_{32} & a_{33} \end{vmatrix} = \begin{vmatrix} a_{11} & a_{12} & a_{13} \\ a_{21} & a_{22} & a_{23} \\ a_{31} & a_{32} & a_{33} \end{vmatrix}$ $\det(B) = \det(A)$	A multiple of the second row of A is added to the first row.

We will verify the first equation in Table 1 and leave the other two for you. To start, note that the determinants on the two sides of the equation differ only in the first row, so these determinants have the same cofactors, C_{11}, C_{12}, C_{13}, along that row (since those cofactors depend only on the entries in the *second* two rows). Thus, expanding the left side by cofactors along the first row yields

$$
\begin{vmatrix} ka_{11} & ka_{12} & ka_{13} \\ a_{21} & a_{22} & a_{23} \\ a_{31} & a_{32} & a_{33} \end{vmatrix} = ka_{11}C_{11} + ka_{12}C_{12} + ka_{33}C_{13}
$$

$$
= k(a_{11}C_{11} + a_{12}C_{12} + a_{33}C_{13})
$$

$$
= k \begin{vmatrix} a_{11} & a_{12} & a_{13} \\ a_{21} & a_{22} & a_{23} \\ a_{31} & a_{32} & a_{33} \end{vmatrix}
$$

Elementary Matrices   It will be useful to consider the special case of Theorem 2.2.3 in which $A = I_n$ is the $n \times n$ identity matrix and E (rather than B) denotes the elementary matrix that results when the row operation is performed on I_n. In this special case Theorem 2.2.3 implies the following result.

THEOREM 2.2.4 *Let E be an $n \times n$ elementary matrix.*

(a) *If E results from multiplying a row of I_n by a nonzero number k, then $\det(E) = k$.*

(b) *If E results from interchanging two rows of I_n, then $\det(E) = -1$.*

(c) *If E results from adding a multiple of one row of I_n to another, then $\det(E) = 1$.*

▶ **EXAMPLE 1 Determinants of Elementary Matrices**

The following determinants of elementary matrices, which are evaluated by inspection, illustrate Theorem 2.2.4.

Observe that the determinant of an elementary matrix cannot be zero.

$$
\begin{vmatrix} 1 & 0 & 0 & 0 \\ 0 & 3 & 0 & 0 \\ 0 & 0 & 1 & 0 \\ 0 & 0 & 0 & 1 \end{vmatrix} = 3,
\qquad
\begin{vmatrix} 0 & 0 & 0 & 1 \\ 0 & 1 & 0 & 0 \\ 0 & 0 & 1 & 0 \\ 1 & 0 & 0 & 0 \end{vmatrix} = -1,
\qquad
\begin{vmatrix} 1 & 0 & 0 & 7 \\ 0 & 1 & 0 & 0 \\ 0 & 0 & 1 & 0 \\ 0 & 0 & 0 & 1 \end{vmatrix} = 1 ◀
$$

The second row of I_4 was multiplied by 3.	The first and last rows of I_4 were interchanged.	7 times the last row of I_4 was added to the first row.

Matrices with Proportional Rows or Columns   If a square matrix A has two proportional rows, then a row of zeros can be introduced by adding a suitable multiple of one of the rows to the other. Similarly for columns. But adding a multiple of one row or column to another does not change the determinant, so from Theorem 2.2.1, we must have $\det(A) = 0$. This proves the following theorem.

THEOREM 2.2.5 *If A is a square matrix with two proportional rows or two proportional columns, then $\det(A) = 0$.*

▶ **EXAMPLE 2 Introducing Zero Rows**

The following computation shows how to introduce a row of zeros when there are two proportional rows.

$$
\begin{vmatrix} 1 & 3 & -2 & 4 \\ 2 & 6 & -4 & 8 \\ 3 & 9 & 1 & 5 \\ 1 & 1 & 4 & 8 \end{vmatrix} = \begin{vmatrix} 1 & 3 & -2 & 4 \\ 0 & 0 & 0 & 0 \\ 3 & 9 & 1 & 5 \\ 1 & 1 & 4 & 8 \end{vmatrix} = 0
$$

The second row is 2 times the first, so we added -2 times the first row to the second to introduce a row of zeros.

Each of the following matrices has two proportional rows or columns; thus, each has a determinant of zero.

$$\begin{bmatrix} -1 & 4 \\ -2 & 8 \end{bmatrix}, \qquad \begin{bmatrix} 1 & -2 & 7 \\ -4 & 8 & 5 \\ 2 & -4 & 3 \end{bmatrix}, \qquad \begin{bmatrix} 3 & -1 & 4 & -5 \\ 6 & -2 & 5 & 2 \\ 5 & 8 & 1 & 4 \\ -9 & 3 & -12 & 15 \end{bmatrix} \blacktriangleleft$$

Evaluating Determinants by Row Reduction

We will now give a method for evaluating determinants that involves substantially less computation than cofactor expansion. The idea of the method is to reduce the given matrix to upper triangular form by elementary row operations, then compute the determinant of the upper triangular matrix (an easy computation), and then relate that determinant to that of the original matrix. Here is an example.

▶ **EXAMPLE 3** Using Row Reduction to Evaluate a Determinant

Evaluate det(A) where

$$A = \begin{bmatrix} 0 & 1 & 5 \\ 3 & -6 & 9 \\ 2 & 6 & 1 \end{bmatrix}$$

Solution We will reduce A to row echelon form (which is upper triangular) and then apply Theorem 2.1.2.

$$\det(A) = \begin{vmatrix} 0 & 1 & 5 \\ 3 & -6 & 9 \\ 2 & 6 & 1 \end{vmatrix} = - \begin{vmatrix} 3 & -6 & 9 \\ 0 & 1 & 5 \\ 2 & 6 & 1 \end{vmatrix}$$

The first and second rows of A were interchanged.

Even with today's fastest computers it would take millions of years to calculate a 25 × 25 determinant by cofactor expansion, so methods based on row reduction are often used for large determinants. For determinants of small size (such as those in this text), cofactor expansion is often a reasonable choice.

$$= -3 \begin{vmatrix} 1 & -2 & 3 \\ 0 & 1 & 5 \\ 2 & 6 & 1 \end{vmatrix}$$

A common factor of 3 from the first row was taken through the determinant sign.

$$= -3 \begin{vmatrix} 1 & -2 & 3 \\ 0 & 1 & 5 \\ 0 & 10 & -5 \end{vmatrix}$$

−2 times the first row was added to the third row.

$$= -3 \begin{vmatrix} 1 & -2 & 3 \\ 0 & 1 & 5 \\ 0 & 0 & -55 \end{vmatrix}$$

−10 times the second row was added to the third row.

$$= (-3)(-55) \begin{vmatrix} 1 & -2 & 3 \\ 0 & 1 & 5 \\ 0 & 0 & 1 \end{vmatrix}$$

A common factor of −55 from the last row was taken through the determinant sign.

$$= (-3)(-55)(1) = 165$$

▶ **EXAMPLE 4** Using Column Operations to Evaluate a Determinant

Compute the determinant of

$$A = \begin{bmatrix} 1 & 0 & 0 & 3 \\ 2 & 7 & 0 & 6 \\ 0 & 6 & 3 & 0 \\ 7 & 3 & 1 & -5 \end{bmatrix}$$

Solution This determinant could be computed as above by using elementary row operations to reduce A to row echelon form, but we can put A in lower triangular form in one step by adding -3 times the first column to the fourth to obtain

$$\det(A) = \det \begin{bmatrix} 1 & 0 & 0 & 0 \\ 2 & 7 & 0 & 0 \\ 0 & 6 & 3 & 0 \\ 7 & 3 & 1 & -26 \end{bmatrix} = (1)(7)(3)(-26) = -546 \quad \blacktriangleleft$$

Example 4 points out that it is always wise to keep an eye open for column operations that can shorten computations.

Cofactor expansion and row or column operations can sometimes be used in combination to provide an effective method for evaluating determinants. The following example illustrates this idea.

▶ **EXAMPLE 5 Row Operations and Cofactor Expansion**

Evaluate $\det(A)$ where

$$A = \begin{bmatrix} 3 & 5 & -2 & 6 \\ 1 & 2 & -1 & 1 \\ 2 & 4 & 1 & 5 \\ 3 & 7 & 5 & 3 \end{bmatrix}$$

Solution By adding suitable multiples of the second row to the remaining rows, we obtain

$$\det(A) = \begin{vmatrix} 0 & -1 & 1 & 3 \\ 1 & 2 & -1 & 1 \\ 0 & 0 & 3 & 3 \\ 0 & 1 & 8 & 0 \end{vmatrix}$$

$$= - \begin{vmatrix} -1 & 1 & 3 \\ 0 & 3 & 3 \\ 1 & 8 & 0 \end{vmatrix} \qquad \longleftarrow \quad \text{Cofactor expansion along the first column}$$

$$= - \begin{vmatrix} -1 & 1 & 3 \\ 0 & 3 & 3 \\ 0 & 9 & 3 \end{vmatrix} \qquad \longleftarrow \quad \text{We added the first row to the third row.}$$

$$= -(-1) \begin{vmatrix} 3 & 3 \\ 9 & 3 \end{vmatrix} \qquad \longleftarrow \quad \text{Cofactor expansion along the first column}$$

$$= -18 \quad \blacktriangleleft$$

Skills

- Know the effect of elementary row operations on the value of a determinant.
- Know the determinants of the three types of elementary matrices.
- Know how to introduce zeros into the rows or columns of a matrix to facilitate the evaluation of its determinant.

- Use row reduction to evaluate the determinant of a matrix.
- Use column operations to evaluate the determinant of a matrix.
- Combine the use of row reduction and cofactor expansion to evaluate the determinant of a matrix.

Exercise Set 2.2

▶ In Exercises 1–4, verify that $\det(A) = \det(A^T)$. ◀

1. $A = \begin{bmatrix} -2 & 3 \\ 1 & 4 \end{bmatrix}$ **2.** $A = \begin{bmatrix} -6 & 1 \\ 2 & -2 \end{bmatrix}$

3. $A = \begin{bmatrix} 2 & -1 & 3 \\ 1 & 2 & 4 \\ 5 & -3 & 6 \end{bmatrix}$ **4.** $A = \begin{bmatrix} 4 & 2 & -1 \\ 0 & 2 & -3 \\ -1 & 1 & 5 \end{bmatrix}$

▶ In Exercises 5–9, find the determinant of the given elementary matrix by inspection. ◀

5. $\begin{bmatrix} 1 & 0 & 0 & 0 \\ 0 & 1 & 0 & 0 \\ 0 & 0 & -5 & 0 \\ 0 & 0 & 0 & 1 \end{bmatrix}$ **6.** $\begin{bmatrix} 1 & 0 & 0 \\ 0 & 1 & 0 \\ -5 & 0 & 1 \end{bmatrix}$

7. $\begin{bmatrix} 1 & 0 & 0 & 0 \\ 0 & 0 & 1 & 0 \\ 0 & 1 & 0 & 0 \\ 0 & 0 & 0 & 1 \end{bmatrix}$ **8.** $\begin{bmatrix} 1 & 0 & 0 & 0 \\ 0 & -\frac{1}{3} & 0 & 0 \\ 0 & 0 & 1 & 0 \\ 0 & 0 & 0 & 1 \end{bmatrix}$

9. $\begin{bmatrix} 1 & 0 & 0 & 0 \\ 0 & 1 & 0 & -9 \\ 0 & 0 & 1 & 0 \\ 0 & 0 & 0 & 1 \end{bmatrix}$

▶ In Exercises 10–17, evaluate the determinant of the given matrix by reducing the matrix to row echelon form. ◀

10. $\begin{bmatrix} 3 & 6 & -9 \\ 0 & 0 & -2 \\ -2 & 1 & 5 \end{bmatrix}$ **11.** $\begin{bmatrix} 0 & 3 & 1 \\ 1 & 1 & 2 \\ 3 & 2 & 4 \end{bmatrix}$

12. $\begin{bmatrix} 1 & -3 & 0 \\ -2 & 4 & 1 \\ 5 & -2 & 2 \end{bmatrix}$ **13.** $\begin{bmatrix} 3 & -6 & 9 \\ -2 & 7 & -2 \\ 0 & 1 & 5 \end{bmatrix}$

14. $\begin{bmatrix} 1 & -2 & 3 & 1 \\ 5 & -9 & 6 & 3 \\ -1 & 2 & -6 & -2 \\ 2 & 8 & 6 & 1 \end{bmatrix}$ **15.** $\begin{bmatrix} 2 & 1 & 3 & 1 \\ 1 & 0 & 1 & 1 \\ 0 & 2 & 1 & 0 \\ 0 & 1 & 2 & 3 \end{bmatrix}$

16. $\begin{bmatrix} 0 & 1 & 1 & 1 \\ \frac{1}{2} & \frac{1}{2} & 1 & \frac{1}{2} \\ \frac{2}{3} & \frac{1}{3} & \frac{1}{3} & 0 \\ -\frac{1}{3} & \frac{2}{3} & 0 & 0 \end{bmatrix}$

17. $\begin{bmatrix} 1 & 3 & 1 & 5 & 3 \\ -2 & -7 & 0 & -4 & 2 \\ 0 & 0 & 1 & 0 & 1 \\ 0 & 0 & 2 & 1 & 1 \\ 0 & 0 & 0 & 1 & 1 \end{bmatrix}$

18. Repeat Exercises 10–13 by using a combination of row reduction and cofactor expansion.

19. Repeat Exercises 14–17 by using a combination of row reduction and cofactor expansion.

▶ In Exercises 20–27, evaluate the determinant, given that

$$\begin{vmatrix} a & b & c \\ d & e & f \\ g & h & i \end{vmatrix} = -6 \quad ◀$$

20. $\begin{vmatrix} g & h & i \\ d & e & f \\ a & b & c \end{vmatrix}$ **21.** $\begin{vmatrix} d & e & f \\ g & h & i \\ a & b & c \end{vmatrix}$ **22.** $\begin{vmatrix} a & b & c \\ d & e & f \\ 2a & 2b & 2c \end{vmatrix}$

23. $\begin{vmatrix} 3a & 3b & 3c \\ -d & -e & -f \\ 4g & 4h & 4i \end{vmatrix}$ **24.** $\begin{vmatrix} a+d & b+e & c+f \\ -d & -e & -f \\ g & h & i \end{vmatrix}$

25. $\begin{vmatrix} a+g & b+h & c+i \\ d & e & f \\ g & h & i \end{vmatrix}$ **26.** $\begin{vmatrix} a & b & c \\ 2d & 2e & 2f \\ g+3a & h+3b & i+3c \end{vmatrix}$

27. $\begin{vmatrix} -3a & -3b & -3c \\ d & e & f \\ g-4d & h-4e & i-4f \end{vmatrix}$

28. Show that

(a) $\det \begin{bmatrix} 0 & 0 & a_{13} \\ 0 & a_{22} & a_{23} \\ a_{31} & a_{32} & a_{33} \end{bmatrix} = -a_{13}a_{22}a_{31}$

(b) $\det \begin{bmatrix} 0 & 0 & 0 & a_{14} \\ 0 & 0 & a_{23} & a_{24} \\ 0 & a_{32} & a_{33} & a_{34} \\ a_{41} & a_{42} & a_{43} & a_{44} \end{bmatrix} = a_{14}a_{23}a_{32}a_{41}$

29. Use row reduction to show that

$$\begin{vmatrix} 1 & 1 & 1 \\ a & b & c \\ a^2 & b^2 & c^2 \end{vmatrix} = (b-a)(c-a)(c-b)$$

▶ In Exercises 30–33, confirm the identities without evaluating the determinants directly. ◀

30. $\begin{vmatrix} a_1+b_1t & a_2+b_2t & a_3+b_3t \\ a_1t+b_1 & a_2t+b_2 & a_3t+b_3 \\ c_1 & c_2 & c_3 \end{vmatrix} = (1-t^2) \begin{vmatrix} a_1 & a_2 & a_3 \\ b_1 & b_2 & b_3 \\ c_1 & c_2 & c_3 \end{vmatrix}$

31. $\begin{vmatrix} a_1 & b_1 & a_1+b_1+c_1 \\ a_2 & b_2 & a_2+b_2+c_2 \\ a_3 & b_3 & a_3+b_3+c_3 \end{vmatrix} = \begin{vmatrix} a_1 & b_1 & c_1 \\ a_2 & b_2 & c_2 \\ a_3 & b_3 & c_3 \end{vmatrix}$

32. $\begin{vmatrix} a_1 & b_1 + ta_1 & c_1 + rb_1 + sa_1 \\ a_2 & b_2 + ta_2 & c_2 + rb_2 + sa_2 \\ a_3 & b_3 + ta_3 & c_3 + rb_3 + sa_3 \end{vmatrix} = \begin{vmatrix} a_1 & a_2 & a_3 \\ b_1 & b_2 & b_3 \\ c_1 & c_2 & c_3 \end{vmatrix}$

33. $\begin{vmatrix} a_1 + b_1 & a_1 - b_1 & c_1 \\ a_2 + b_2 & a_2 - b_2 & c_2 \\ a_3 + b_3 & a_3 - b_3 & c_3 \end{vmatrix} = -2 \begin{vmatrix} a_1 & b_1 & c_1 \\ a_2 & b_2 & c_2 \\ a_3 & b_3 & c_3 \end{vmatrix}$

34. Find the determinant of the following matrix.

$$\begin{bmatrix} a & b & b & b \\ b & a & b & b \\ b & b & a & b \\ b & b & b & a \end{bmatrix}$$

▶ In Exercises 35–36, show that $\det(A) = 0$ without directly evaluating the determinant. ◀

35. $A = \begin{bmatrix} -2 & 8 & 1 & 4 \\ 3 & 2 & 5 & 1 \\ 1 & 10 & 6 & 5 \\ 4 & -6 & 4 & -3 \end{bmatrix}$

36. $A = \begin{bmatrix} -4 & 1 & 1 & 1 & 1 \\ 1 & -4 & 1 & 1 & 1 \\ 1 & 1 & -4 & 1 & 1 \\ 1 & 1 & 1 & -4 & 1 \\ 1 & 1 & 1 & 1 & -4 \end{bmatrix}$

True-False Exercises

In parts (a)–(f) determine whether the statement is true or false, and justify your answer.

(a) If A is a 4×4 matrix and B is obtained from A by interchanging the first two rows and then interchanging the last two rows, then $\det(B) = \det(A)$.

(b) If A is a 3×3 matrix and B is obtained from A by multiplying the first column by 4 and multiplying the third column by $\frac{3}{4}$, then $\det(B) = 3 \det(A)$.

(c) If A is a 3×3 matrix and B is obtained from A by adding 5 times the first row to each of the second and third rows, then $\det(B) = 25 \det(A)$.

(d) If A is an $n \times n$ matrix and B is obtained from A by multiplying each row of A by its row number, then

$$\det(B) = \frac{n(n+1)}{2} \det(A)$$

(e) If A is a square matrix with two identical columns, then $\det(A) = 0$.

(f) If the sum of the second and fourth row vectors of a 6×6 matrix A is equal to the last row vector, then $\det(A) = 0$.

2.3 Properties of Determinants; Cramer's Rule

In this section we will develop some fundamental properties of matrices, and we will use these results to derive a formula for the inverse of an invertible matrix and formulas for the solutions of certain kinds of linear systems.

Basic Properties of Determinants

Suppose that A and B are $n \times n$ matrices and k is any scalar. We begin by considering possible relationships between $\det(A)$, $\det(B)$, and

$$\det(kA), \quad \det(A + B), \quad \text{and} \quad \det(AB)$$

Since a common factor of any row of a matrix can be moved through the determinant sign, and since each of the n rows in kA has a common factor of k, it follows that

$$\det(kA) = k^n \det(A) \tag{1}$$

For example,

$$\begin{vmatrix} ka_{11} & ka_{12} & ka_{13} \\ ka_{21} & ka_{22} & ka_{23} \\ ka_{31} & ka_{32} & ka_{33} \end{vmatrix} = k^3 \begin{vmatrix} a_{11} & a_{12} & a_{13} \\ a_{21} & a_{22} & a_{23} \\ a_{31} & a_{32} & a_{33} \end{vmatrix}$$

Unfortunately, no simple relationship exists among $\det(A)$, $\det(B)$, and $\det(A + B)$. In particular, we emphasize that $\det(A + B)$ will usually *not* be equal to $\det(A) + \det(B)$. The following example illustrates this fact.

▶ **EXAMPLE 1** det(A + B) ≠ det(A) + det(B)

Consider

$$A = \begin{bmatrix} 1 & 2 \\ 2 & 5 \end{bmatrix}, \quad B = \begin{bmatrix} 3 & 1 \\ 1 & 3 \end{bmatrix}, \quad A + B = \begin{bmatrix} 4 & 3 \\ 3 & 8 \end{bmatrix}$$

We have det(A) = 1, det(B) = 8, and det(A + B) = 23; thus

$$\det(A + B) \neq \det(A) + \det(B) \quad ◀$$

In spite of the previous example, there is a useful relationship concerning sums of determinants that is applicable when the matrices involved are the same except for *one* row (column). For example, consider the following two matrices that differ only in the second row:

$$A = \begin{bmatrix} a_{11} & a_{12} \\ a_{21} & a_{22} \end{bmatrix} \quad \text{and} \quad B = \begin{bmatrix} a_{11} & a_{12} \\ b_{21} & b_{22} \end{bmatrix}$$

Calculating the determinants of A and B we obtain

$$\det(A) + \det(B) = (a_{11}a_{22} - a_{12}a_{21}) + (a_{11}b_{22} - a_{12}b_{21})$$
$$= a_{11}(a_{22} + b_{22}) - a_{12}(a_{21} + b_{21})$$
$$= \det \begin{bmatrix} a_{11} & a_{12} \\ a_{21} + b_{21} & a_{22} + b_{22} \end{bmatrix}$$

Thus

$$\det \begin{bmatrix} a_{11} & a_{12} \\ a_{21} & a_{22} \end{bmatrix} + \det \begin{bmatrix} a_{11} & a_{12} \\ b_{21} & b_{22} \end{bmatrix} = \det \begin{bmatrix} a_{11} & a_{12} \\ a_{21} + b_{21} & a_{22} + b_{22} \end{bmatrix}$$

This is a special case of the following general result.

THEOREM 2.3.1 *Let A, B, and C be $n \times n$ matrices that differ only in a single row, say the rth, and assume that the rth row of C can be obtained by adding corresponding entries in the rth rows of A and B. Then*

$$\det(C) = \det(A) + \det(B)$$

The same result holds for columns.

▶ **EXAMPLE 2** Sums of Determinants

We leave it to you to confirm the following equality by evaluating the determinants.

$$\det \begin{bmatrix} 1 & 7 & 5 \\ 2 & 0 & 3 \\ 1+0 & 4+1 & 7+(-1) \end{bmatrix} = \det \begin{bmatrix} 1 & 7 & 5 \\ 2 & 0 & 3 \\ 1 & 4 & 7 \end{bmatrix} + \det \begin{bmatrix} 1 & 7 & 5 \\ 2 & 0 & 3 \\ 0 & 1 & -1 \end{bmatrix} \quad ◀$$

Determinant of a Matrix Product

Considering the complexity of the formulas for determinants and matrix multiplication, it would seem unlikely that a simple relationship should exist between them. This is what makes the simplicity of our next result so surprising. We will show that if A and B are square matrices of the same size, then

$$\det(AB) = \det(A) \det(B) \tag{2}$$

The proof of this theorem is fairly intricate, so we will have to develop some preliminary results first. We begin with the special case of (2) in which A is an elementary matrix. Because this special case is only a prelude to (2), we call it a lemma.

> **LEMMA 2.3.2** *If B is an $n \times n$ matrix and E is an $n \times n$ elementary matrix, then*
>
> $$\det(EB) = \det(E)\det(B)$$

Proof   We will consider three cases, each in accordance with the row operation that produces the matrix E.

Case 1   If E results from multiplying a row of I_n by k, then by Theorem 1.5.1, EB results from B by multiplying the corresponding row by k; so from Theorem 2.2.3(a) we have

$$\det(EB) = k\det(B)$$

But from Theorem 2.2.4(a) we have $\det(E) = k$, so

$$\det(EB) = \det(E)\det(B)$$

Cases 2 and 3   The proofs of the cases where E results from interchanging two rows of I_n or from adding a multiple of one row to another follow the same pattern as Case 1 and are left as exercises.   ◄

Remark   It follows by repeated applications of Lemma 2.3.2 that if B is an $n \times n$ matrix and $E_1, E_2, \ldots, E_r$ are $n \times n$ elementary matrices, then

$$\det(E_1 E_2 \cdots E_r B) = \det(E_1)\det(E_2)\cdots\det(E_r)\det(B) \tag{3}$$

Determinant Test for Invertibility

Our next theorem provides an important criterion for determining whether a matrix is invertible. It also takes us a step closer to establishing Formula (2).

> **THEOREM 2.3.3**   *A square matrix A is invertible if and only if $\det(A) \neq 0$.*

Proof   Let R be the reduced row echelon form of A. As a preliminary step, we will show that $\det(A)$ and $\det(R)$ are both zero or both nonzero: Let $E_1, E_2, \ldots, E_r$ be the elementary matrices that correspond to the elementary row operations that produce R from A. Thus

$$R = E_r \cdots E_2 E_1 A$$

and from (3),

$$\det(R) = \det(E_r)\cdots\det(E_2)\det(E_1)\det(A) \tag{4}$$

We pointed out in the margin note that accompanies Theorem 2.2.4 that the determinant of an elementary matrix is nonzero. Thus, it follows from Formula (4) that $\det(A)$ and $\det(R)$ are either both zero or both nonzero, which sets the stage for the main part of the proof. If we assume first that A is invertible, then it follows from Theorem 1.6.4 that $R = I$ and hence that $\det(R) = 1\,(\neq 0)$. This, in turn, implies that $\det(A) \neq 0$, which is what we wanted to show.

Conversely, assume that $\det(A) \neq 0$. It follows from this that $\det(R) \neq 0$, which tells us that R cannot have a row of zeros. Thus, it follows from Theorem 1.4.3 that $R = I$ and hence that A is invertible by Theorem 1.6.4.   ◄

It follows from Theorems 2.3.3 and 2.2.5 that a square matrix with two proportional rows or two proportional columns is not invertible.

▶ **EXAMPLE 3** **Determinant Test for Invertibility**

Since the first and third rows of

$$A = \begin{bmatrix} 1 & 2 & 3 \\ 1 & 0 & 1 \\ 2 & 4 & 6 \end{bmatrix}$$

are proportional, $\det(A) = 0$. Thus A is not invertible. ◀

We are now ready for the main result concerning products of matrices.

> **THEOREM 2.3.4** *If A and B are square matrices of the same size, then*
>
> $$\det(AB) = \det(A)\det(B)$$

Proof We divide the proof into two cases that depend on whether or not A is invertible. If the matrix A is not invertible, then by Theorem 1.6.5 neither is the product AB. Thus, from Theorem 2.3.3, we have $\det(AB) = 0$ and $\det(A) = 0$, so it follows that $\det(AB) = \det(A)\det(B)$.

Now

assume that A is invertible. By Theorem 1.6.4, the matrix A is expressible as a product of elementary matrices, say

$$A = E_1 E_2 \cdots E_r \tag{5}$$

so

$$AB = E_1 E_2 \cdots E_r B$$

Applying (3) to this equation yields

$$\det(AB) = \det(E_1)\det(E_2)\cdots\det(E_r)\det(B)$$

and applying (3) again yields

$$\det(AB) = \det(E_1 E_2 \cdots E_r)\det(B)$$

which, from (5), can be written as $\det(AB) = \det(A)\det(B)$. ◀

**Augustin Louis Cauchy
(1789–1857)**

▶ **EXAMPLE 4** **Verifying That det(*AB*) = det(*A*) det(*B*)**

Consider the matrices

$$A = \begin{bmatrix} 3 & 1 \\ 2 & 1 \end{bmatrix}, \quad B = \begin{bmatrix} -1 & 3 \\ 5 & 8 \end{bmatrix}, \quad AB = \begin{bmatrix} 2 & 17 \\ 3 & 14 \end{bmatrix}$$

We leave it for you to verify that

$$\det(A) = 1, \quad \det(B) = -23, \quad \text{and} \quad \det(AB) = -23$$

Thus $\det(AB) = \det(A)\det(B)$, as guaranteed by Theorem 2.3.4. ◀

The following theorem gives a useful relationship between the determinant of an invertible matrix and the determinant of its inverse.

> **THEOREM 2.3.5** *If A is invertible, then*
> $$\det(A^{-1}) = \frac{1}{\det(A)}$$

Proof Since $A^{-1}A = I$, it follows that $\det(A^{-1}A) = \det(I)$. Therefore, we must have $\det(A^{-1})\det(A) = 1$. Since $\det(A) \neq 0$, the proof can be completed by dividing through by $\det(A)$. ◄

Adjoint of a Matrix

It follows from Theorems 2.3.5 and 2.1.2 that

$$\det(A^{-1}) = \frac{1}{a_{11}}\frac{1}{a_{22}}\cdots\frac{1}{a_{nn}}$$

Moreover, by using the adjoint formula it is possible to show that

$$\frac{1}{a_{11}},\ \frac{1}{a_{22}},\ \ldots,\ \frac{1}{a_{nn}}$$

are actually the successive diagonal entries of A^{-1} (compare A and A^{-1} in Example 3 of Section 1.7).

In a cofactor expansion we compute $\det(A)$ by multiplying the entries in a row or column by their cofactors and adding the resulting products. It turns out that if one multiplies the entries in any row by the corresponding cofactors from a *different* row, the sum of these products is always zero. (This result also holds for columns.) Although we omit the general proof, the next example illustrates the idea of the proof in a special case.

▶ **EXAMPLE 5** **Entries and Cofactors from Different Rows**

Let

$$A = \begin{bmatrix} a_{11} & a_{12} & a_{13} \\ a_{21} & a_{22} & a_{23} \\ a_{31} & a_{32} & a_{33} \end{bmatrix}$$

Consider the quantity

$$a_{11}C_{31} + a_{12}C_{32} + a_{13}C_{33}$$

that is formed by multiplying the entries in the first row by the cofactors of the corresponding entries in the third row and adding the resulting products. We can show that this quantity is equal to zero by the following trick: Construct a new matrix A' by replacing the third row of A with another copy of the first row. That is,

$$A' = \begin{bmatrix} a_{11} & a_{12} & a_{13} \\ a_{21} & a_{22} & a_{23} \\ a_{11} & a_{12} & a_{13} \end{bmatrix}$$

Let $C'_{31}, C'_{32}, C'_{33}$ be the cofactors of the entries in the third row of A'. Since the first two rows of A and A' are the same, and since the computations of $C_{31}, C_{32}, C_{33}, C'_{31}, C'_{32}$, and C'_{33} involve only entries from the first two rows of A and A', it follows that

$$C_{31} = C'_{31}, \quad C_{32} = C'_{32}, \quad C_{33} = C'_{33}$$

Since A' has two identical rows, it follows from (3) that

$$\det(A') = 0 \tag{6}$$

On the other hand, evaluating $\det(A')$ by cofactor expansion along the third row gives

$$\det(A') = a_{11}C'_{31} + a_{12}C'_{32} + a_{13}C'_{33} = a_{11}C_{31} + a_{12}C_{32} + a_{13}C_{33} \tag{7}$$

From (6) and (7) we obtain

$$a_{11}C_{31} + a_{12}C_{32} + a_{13}C_{33} = 0 \quad ◄$$

DEFINITION 1   If A is any $n \times n$ matrix and C_{ij} is the cofactor of a_{ij}, then the matrix

$$\begin{bmatrix} C_{11} & C_{12} & \cdots & C_{1n} \\ C_{21} & C_{22} & \cdots & C_{2n} \\ \vdots & \vdots & & \vdots \\ C_{n1} & C_{n2} & \cdots & C_{nn} \end{bmatrix}$$

is called the ***matrix of cofactors from A***. The transpose of this matrix is called the ***adjoint of A*** and is denoted by adj(A).

Leonard Eugene Dickson (1874–1954)

Historical Note   The use of the term *adjoint* for the transpose of the matrix of cofactors appears to have been introduced by the American mathematician L. E. Dickson in a research paper that he published in 1902.
[*Image: Courtesy of the American Mathematical Society*]

▶ **EXAMPLE 6   Adjoint of a 3 × 3 Matrix**

Let

$$A = \begin{bmatrix} 3 & 2 & -1 \\ 1 & 6 & 3 \\ 2 & -4 & 0 \end{bmatrix}$$

The cofactors of A are

$$\begin{array}{lll} C_{11} = 12 & C_{12} = 6 & C_{13} = -16 \\ C_{21} = 4 & C_{22} = 2 & C_{23} = 16 \\ C_{31} = 12 & C_{32} = -10 & C_{33} = 16 \end{array}$$

so the matrix of cofactors is

$$\begin{bmatrix} 12 & 6 & -16 \\ 4 & 2 & 16 \\ 12 & -10 & 16 \end{bmatrix}$$

and the adjoint of A is

$$\text{adj}(A) = \begin{bmatrix} 12 & 4 & 12 \\ 6 & 2 & -10 \\ -16 & 16 & 16 \end{bmatrix} \blacktriangleleft$$

In Theorem 1.4.5 we gave a formula for the inverse of a 2×2 invertible matrix. Our next theorem extends that result to $n \times n$ invertible matrices.

THEOREM 2.3.6   Inverse of a Matrix Using Its Adjoint

If A is an invertible matrix, then

$$A^{-1} = \frac{1}{\det(A)} \text{adj}(A) \qquad (8)$$

Proof   We show first that

$$A\,\text{adj}(A) = \det(A)I$$

Consider the product

$$A\,\text{adj}(A) = \begin{bmatrix} a_{11} & a_{12} & \cdots & a_{1n} \\ a_{21} & a_{22} & \cdots & a_{2n} \\ \vdots & \vdots & & \vdots \\ a_{i1} & a_{i2} & \cdots & a_{in} \\ \vdots & \vdots & & \vdots \\ a_{n1} & a_{n2} & \cdots & a_{nn} \end{bmatrix} \begin{bmatrix} C_{11} & C_{21} & \cdots & C_{j1} & \cdots & C_{n1} \\ C_{12} & C_{22} & \cdots & C_{j2} & \cdots & C_{n2} \\ \vdots & \vdots & & \vdots & & \vdots \\ C_{1n} & C_{2n} & \cdots & C_{jn} & \cdots & C_{nn} \end{bmatrix}$$

The entry in the ith row and jth column of the product $A \, \mathrm{adj}(A)$ is

$$a_{i1}C_{j1} + a_{i2}C_{j2} + \cdots + a_{in}C_{jn} \tag{9}$$

(see the shaded lines above).

If $i = j$, then (9) is the cofactor expansion of $\det(A)$ along the ith row of A (Theorem 2.1.1), and if $i \neq j$, then the a's and the cofactors come from different rows of A, so the value of (9) is zero. Therefore,

$$A \, \mathrm{adj}(A) = \begin{bmatrix} \det(A) & 0 & \cdots & 0 \\ 0 & \det(A) & \cdots & 0 \\ \vdots & \vdots & & \vdots \\ 0 & 0 & \cdots & \det(A) \end{bmatrix} = \det(A)I \tag{10}$$

Since A is invertible, $\det(A) \neq 0$. Therefore, Equation (10) can be rewritten as

$$\frac{1}{\det(A)}[A \, \mathrm{adj}(A)] = I \quad \text{or} \quad A\left[\frac{1}{\det(A)}\mathrm{adj}(A)\right] = I$$

Multiplying both sides on the left by A^{-1} yields

$$A^{-1} = \frac{1}{\det(A)}\mathrm{adj}(A) \quad \blacktriangleleft$$

▶ **EXAMPLE 7** **Using the Adjoint to Find an Inverse Matrix**

Use (8) to find the inverse of the matrix A in Example 6.

Solution We leave it for you to check that $\det(A) = 64$. Thus

$$A^{-1} = \frac{1}{\det(A)}\mathrm{adj}(A) = \frac{1}{64}\begin{bmatrix} 12 & 4 & 12 \\ 6 & 2 & -10 \\ -16 & 16 & 16 \end{bmatrix} = \begin{bmatrix} \frac{12}{64} & \frac{4}{64} & \frac{12}{64} \\ \frac{6}{64} & \frac{2}{64} & -\frac{10}{64} \\ -\frac{16}{64} & \frac{16}{64} & \frac{16}{64} \end{bmatrix} \quad \blacktriangleleft$$

Cramer's Rule   Our next theorem uses the formula for the inverse of an invertible matrix to produce a formula, called **Cramer's rule**, for the solution of a linear system $A\mathbf{x} = \mathbf{b}$ of n equations in n unknowns in the case where the coefficient matrix A is invertible (or, equivalently, when $\det(A) \neq 0$).

THEOREM 2.3.7 **Cramer's Rule**

If $A\mathbf{x} = \mathbf{b}$ is a system of n linear equations in n unknowns such that $\det(A) \neq 0$, then the system has a unique solution. This solution is

$$x_1 = \frac{\det(A_1)}{\det(A)}, \quad x_2 = \frac{\det(A_2)}{\det(A)}, \ldots, \quad x_n = \frac{\det(A_n)}{\det(A)}$$

where A_j is the matrix obtained by replacing the entries in the jth column of A by the entries in the matrix

$$\mathbf{b} = \begin{bmatrix} b_1 \\ b_2 \\ \vdots \\ b_n \end{bmatrix}$$

Proof If $\det(A) \neq 0$, then A is invertible, and by Theorem 1.6.2, $\mathbf{x} = A^{-1}\mathbf{b}$ is the unique solution of $A\mathbf{x} = \mathbf{b}$. Therefore, by Theorem 2.3.6 we have

$$\mathbf{x} = A^{-1}\mathbf{b} = \frac{1}{\det(A)}\text{adj}(A)\mathbf{b} = \frac{1}{\det(A)}\begin{bmatrix} C_{11} & C_{21} & \cdots & C_{n1} \\ C_{12} & C_{22} & \cdots & C_{n2} \\ \vdots & \vdots & & \vdots \\ C_{1n} & C_{2n} & \cdots & C_{nn} \end{bmatrix}\begin{bmatrix} b_1 \\ b_2 \\ \vdots \\ b_n \end{bmatrix}$$

Multiplying the matrices out gives

$$\mathbf{x} = \frac{1}{\det(A)}\begin{bmatrix} b_1 C_{11} + b_2 C_{21} + \cdots + b_n C_{n1} \\ b_1 C_{12} + b_2 C_{22} + \cdots + b_n C_{n2} \\ \vdots & \vdots & \vdots \\ b_1 C_{1n} + b_2 C_{2n} + \cdots + b_n C_{nn} \end{bmatrix}$$

The entry in the jth row of $\mathbf{x}$ is therefore

$$x_j = \frac{b_1 C_{1j} + b_2 C_{2j} + \cdots + b_n C_{nj}}{\det(A)} \tag{11}$$

Now let

$$A_j = \begin{bmatrix} a_{11} & a_{12} & \cdots & a_{1j-1} & b_1 & a_{1j+1} & \cdots & a_{1n} \\ a_{21} & a_{22} & \cdots & a_{2j-1} & b_2 & a_{2j+1} & \cdots & a_{2n} \\ \vdots & \vdots & & \vdots & \vdots & \vdots & & \vdots \\ a_{n1} & a_{n2} & \cdots & a_{nj-1} & b_n & a_{nj+1} & \cdots & a_{nn} \end{bmatrix}$$

Since A_j differs from A only in the jth column, it follows that the cofactors of entries $b_1, b_2, \ldots, b_n$ in A_j are the same as the cofactors of the corresponding entries in the jth column of A. The cofactor expansion of $\det(A_j)$ along the jth column is therefore

$$\det(A_j) = b_1 C_{1j} + b_2 C_{2j} + \cdots + b_n C_{nj}$$

Substituting this result in (11) gives

$$x_j = \frac{\det(A_j)}{\det(A)} \quad \blacktriangleleft$$

Gabriel Cramer
(1704–1752)

▶ **EXAMPLE 8** Using Cramer's Rule to Solve a Linear System

Use Cramer's rule to solve

$$\begin{aligned} x_1 + \quad\ + 2x_3 &= 6 \\ -3x_1 + 4x_2 + 6x_3 &= 30 \\ -x_1 - 2x_2 + 3x_3 &= 8 \end{aligned}$$

Solution

$$A = \begin{bmatrix} 1 & 0 & 2 \\ -3 & 4 & 6 \\ -1 & -2 & 3 \end{bmatrix}, \quad A_1 = \begin{bmatrix} 6 & 0 & 2 \\ 30 & 4 & 6 \\ 8 & -2 & 3 \end{bmatrix},$$

$$A_2 = \begin{bmatrix} 1 & 6 & 2 \\ -3 & 30 & 6 \\ -1 & 8 & 3 \end{bmatrix}, \quad A_3 = \begin{bmatrix} 1 & 0 & 6 \\ -3 & 4 & 30 \\ -1 & -2 & 8 \end{bmatrix}$$

For $n > 3$, it is usually more efficient to solve a linear system with n equations in n unknowns by Gauss–Jordan elimination than by Cramer's rule. Its main use is for obtaining properties of solutions of a linear system without actually solving the system.

Therefore,

$$x_1 = \frac{\det(A_1)}{\det(A)} = \frac{-40}{44} = \frac{-10}{11}, \quad x_2 = \frac{\det(A_2)}{\det(A)} = \frac{72}{44} = \frac{18}{11},$$

$$x_3 = \frac{\det(A_3)}{\det(A)} = \frac{152}{44} = \frac{38}{11} \quad \blacktriangleleft$$

Equivalence Theorem   In Theorem 1.6.4 we listed five results that are equivalent to the invertibility of a matrix A. We conclude this section by merging Theorem 2.3.3 with that list to produce the following theorem that relates all of the major topics we have studied thus far.

THEOREM 2.3.8 Equivalent Statements

If A is an $n \times n$ matrix, then the following statements are equivalent.

(*a*)   *A is invertible.*

(*b*)   *$A\mathbf{x} = \mathbf{0}$ has only the trivial solution.*

(*c*)   *The reduced row echelon form of A is I_n.*

(*d*)   *A can be expressed as a product of elementary matrices.*

(*e*)   *$A\mathbf{x} = \mathbf{b}$ is consistent for every $n \times 1$ matrix $\mathbf{b}$.*

(*f*)   *$A\mathbf{x} = \mathbf{b}$ has exactly one solution for every $n \times 1$ matrix $\mathbf{b}$.*

(*g*)   *$\det(A) \neq 0$.*

OPTIONAL

We now have all of the machinery necessary to prove the following two results, which we stated without proof in Theorem 1.7.1:

- **Theorem 1.7.1(*c*)** A triangular matrix is invertible if and only if its diagonal entries are all nonzero.

- **Theorem 1.7.1(*d*)** The inverse of an invertible lower triangular matrix is lower triangular, and the inverse of an invertible upper triangular matrix is upper triangular.

Proof of Theorem 1.7.1(c)   Let $A = [a_{ij}]$ be a triangular matrix, so that its diagonal entries are

$$a_{11}, a_{22}, \ldots, a_{nn}$$

From Theorem 2.1.2, the matrix A is invertible if and only if

$$\det(A) = a_{11}a_{22} \cdots a_{nn}$$

is nonzero, which is true if and only if the diagonal entries are all nonzero.

Proof of Theorem 1.7.1(d)   We will prove the result for upper triangular matrices and leave the lower triangular case for you. Assume that A is upper triangular and invertible. Since

$$A^{-1} = \frac{1}{\det(A)} \operatorname{adj}(A)$$

we can prove that A^{-1} is upper triangular by showing that $\operatorname{adj}(A)$ is upper triangular or, equivalently, that the matrix of cofactors is lower triangular. We can do this by showing that every cofactor C_{ij} with $i < j$ (i.e., above the main diagonal) is zero. Since

$$C_{ij} = (-1)^{i+j} M_{ij}$$

it suffices to show that each minor M_{ij} with $i < j$ is zero. For this purpose, let B_{ij} be the matrix that results when the ith row and jth column of A are deleted, so

$$M_{ij} = \det(B_{ij}) \tag{12}$$

From the assumption that $i < j$, it follows that B_{ij} is upper triangular (see Figure 1.7.1). Since A is upper triangular, its $(i + 1)$-st row begins with at least i zeros. But the ith row of B_{ij} is the $(i + 1)$-st row of A with the entry in the jth column removed. Since $i < j$, none of the first i zeros is removed by deleting the jth column; thus the ith row of B_{ij} starts with at least i zeros, which implies that this row has a zero on the main diagonal. It now follows from Theorem 2.1.2 that $\det(B_{ij}) = 0$ and from (12) that $M_{ij} = 0$. ◀

Concept Review

- Determinant test for invertibility
- Matrix of cofactors
- Adjoint of a matrix
- Cramer's rule
- Equivalent statements about an invertible matrix

Skills

- Know how determinants behave with respect to basic arithmetic operations, as given in Equation (1), Theorem 2.3.1, Lemma 2.3.2, and Theorem 2.3.4.
- Use the determinant to test a matrix for invertibility.
- Know how $\det(A)$ and $\det(A^{-1})$ are related.
- Compute the matrix of cofactors for a square matrix A.
- Compute $\text{adj}(A)$ for a square matrix A.
- Use the adjoint of an invertible matrix to find its inverse.
- Use Cramer's rule to solve linear systems of equations.
- Know the equivalent characterizations of an invertible matrix given in Theorem 2.3.8.

Exercise Set 2.3

▶ In Exercises 1–4, verify that $\det(kA) = k^n \det(A)$. ◀

1. $A = \begin{bmatrix} -1 & 2 \\ 3 & 4 \end{bmatrix}$; $k = 2$

2. $A = \begin{bmatrix} 2 & 2 \\ 5 & -2 \end{bmatrix}$; $k = -4$

3. $A = \begin{bmatrix} 2 & -1 & 3 \\ 3 & 2 & 1 \\ 1 & 4 & 5 \end{bmatrix}$; $k = -2$

4. $A = \begin{bmatrix} 1 & 1 & 1 \\ 0 & 2 & 3 \\ 0 & 1 & -2 \end{bmatrix}$; $k = 3$

▶ In Exercises 5–6, verify that $\det(AB) = \det(BA)$ and determine whether the equality $\det(A + B) = \det(A) + \det(B)$ holds.

5. $A = \begin{bmatrix} 2 & 1 & 0 \\ 3 & 4 & 0 \\ 0 & 0 & 2 \end{bmatrix}$ and $B = \begin{bmatrix} 1 & -1 & 3 \\ 7 & 1 & 2 \\ 5 & 0 & 1 \end{bmatrix}$

6. $A = \begin{bmatrix} -1 & 8 & 2 \\ 1 & 0 & -1 \\ -2 & 2 & 2 \end{bmatrix}$ and $B = \begin{bmatrix} 2 & -1 & -4 \\ 1 & 1 & 3 \\ 0 & 3 & -1 \end{bmatrix}$

▶ In Exercises 7–14, use determinants to decide whether the given matrix is invertible. ◀

7. $A = \begin{bmatrix} 2 & 5 & 5 \\ -1 & -1 & 0 \\ 2 & 4 & 3 \end{bmatrix}$

8. $A = \begin{bmatrix} 2 & 0 & 3 \\ 0 & 3 & 2 \\ -2 & 0 & -4 \end{bmatrix}$

9. $A = \begin{bmatrix} 2 & -3 & 5 \\ 0 & 1 & -3 \\ 0 & 0 & 2 \end{bmatrix}$

10. $A = \begin{bmatrix} -3 & 0 & 1 \\ 5 & 0 & 6 \\ 8 & 0 & 3 \end{bmatrix}$

11. $A = \begin{bmatrix} 4 & 2 & 8 \\ -2 & 1 & -4 \\ 3 & 1 & 6 \end{bmatrix}$

12. $A = \begin{bmatrix} 1 & 0 & -1 \\ 9 & -1 & 4 \\ 8 & 9 & -1 \end{bmatrix}$

13. $A = \begin{bmatrix} 2 & 0 & 0 \\ 8 & 1 & 0 \\ -5 & 3 & 6 \end{bmatrix}$

14. $A = \begin{bmatrix} \sqrt{2} & -\sqrt{7} & 0 \\ 3\sqrt{2} & -3\sqrt{7} & 0 \\ 5 & -9 & 0 \end{bmatrix}$

▶ In Exercises 15–18, find the values of k for which A is invertible. ◀

15. $A = \begin{bmatrix} k-3 & -2 \\ -2 & k-2 \end{bmatrix}$

16. $A = \begin{bmatrix} k & 2 \\ 2 & k \end{bmatrix}$

17. $A = \begin{bmatrix} 1 & 2 & 4 \\ 3 & 1 & 6 \\ k & 3 & 2 \end{bmatrix}$

18. $A = \begin{bmatrix} 1 & 2 & 0 \\ k & 1 & k \\ 0 & 2 & 1 \end{bmatrix}$

▶ In Exercises 19–23, decide whether the given matrix is invertible, and if so, use the adjoint method to find its inverse. ◀

19. $A = \begin{bmatrix} 2 & 5 & 5 \\ -1 & -1 & 0 \\ 2 & 4 & 3 \end{bmatrix}$

20. $A = \begin{bmatrix} 2 & 0 & 3 \\ 0 & 3 & 2 \\ -2 & 0 & -4 \end{bmatrix}$

21. $A = \begin{bmatrix} 2 & -3 & 5 \\ 0 & 1 & -3 \\ 0 & 0 & 2 \end{bmatrix}$

22. $A = \begin{bmatrix} 2 & 0 & 0 \\ 8 & 1 & 0 \\ -5 & 3 & 6 \end{bmatrix}$

23. $A = \begin{bmatrix} 1 & 3 & 1 & 1 \\ 2 & 5 & 2 & 2 \\ 1 & 3 & 8 & 9 \\ 1 & 3 & 2 & 2 \end{bmatrix}$

▶ In Exercises 24–29, solve by Cramer's rule, where it applies.

24. $7x_1 - 2x_2 = 3$
$\quad\ \ 3x_1 + x_2 = 5$

25. $4x + 5y \qquad\ \ = 2$
$\quad\ \ 11x + y + 2z = 3$
$\quad\ \ x + 5y + 2z = 1$

26. $x - 4y + z = 6$
$\quad\ \ 4x - y + 2z = -1$
$\quad\ \ 2x + 2y - 3z = -20$

27. $x_1 - 3x_2 + x_3 = 4$
$\quad\ \ 2x_1 - x_2 \qquad = -2$
$\quad\ \ 4x_1 \qquad - 3x_3 = 0$

28. $-x_1 - 4x_2 + 2x_3 + x_4 = -32$
$\quad\ \ 2x_1 - x_2 + 7x_3 + 9x_4 = \ \ 14$
$\quad -x_1 + x_2 + 3x_3 + x_4 = \ \ 11$
$\quad\ \ \ x_1 - 2x_2 + x_3 - 4x_4 = \ -4$

29. $3x_1 - x_2 + x_3 = 4$
$\quad -x_1 + 7x_2 - 2x_3 = 1$
$\quad\ \ 2x_1 + 6x_2 - x_3 = 5$

30. Show that the matrix

$$A = \begin{bmatrix} \cos\theta & \sin\theta & 0 \\ -\sin\theta & \cos\theta & 0 \\ 0 & 0 & 1 \end{bmatrix}$$

is invertible for all values of θ; then find A^{-1} using Theorem 2.3.6.

31. Use Cramer's rule to solve for y without solving for the unknowns x, z, and w.

$$4x + y + z + w = \ \ 6$$
$$3x + 7y - z + w = \ \ 1$$
$$7x + 3y - 5z + 8w = -3$$
$$x + y + z + 2w = \ \ 3$$

32. Let $A\mathbf{x} = \mathbf{b}$ be the system in Exercise 31.

(a) Solve by Cramer's rule.

(b) Solve by Gauss–Jordan elimination.

(c) Which method involves fewer computations?

33. Prove that if $\det(A) = 1$ and all the entries in A are integers, then all the entries in A^{-1} are integers.

34. Let $A\mathbf{x} = \mathbf{b}$ be a system of n linear equations in n unknowns with integer coefficients and integer constants. Prove that if $\det(A) = 1$, the solution $\mathbf{x}$ has integer entries.

35. Let

$$A = \begin{bmatrix} a & b & c \\ d & e & f \\ g & h & i \end{bmatrix}$$

Assuming that $\det(A) = -7$, find

(a) $\det(3A)$ (b) $\det(A^{-1})$ (c) $\det(2A^{-1})$

(d) $\det((2A)^{-1})$ (e) $\det\begin{bmatrix} a & g & d \\ b & h & e \\ c & i & f \end{bmatrix}$

36. In each part, find the determinant given that A is a 4×4 matrix for which $\det(A) = -2$.

(a) $\det(-A)$ (b) $\det(A^{-1})$ (c) $\det(2A^T)$ (d) $\det(A^3)$

37. In each part, find the determinant given that A is a 3×3 matrix for which $\det(A) = 7$.

(a) $\det(3A)$ (b) $\det(A^{-1})$

(c) $\det(2A^{-1})$ (d) $\det((2A)^{-1})$

38. Prove that a square matrix A is invertible if and only if $A^T A$ is invertible.

39. Show that if A is a square matrix, then $\det(A^T A) = \det(A A^T)$.

True-False Exercises

In parts (a)–(l) determine whether the statement is true or false, and justify your answer.

(a) If A is a 3×3 matrix, then $\det(2A) = 2\det(A)$.

(b) If A and B are square matrices of the same size such that $\det(A) = \det(B)$, then $\det(A + B) = 2\det(A)$.

(c) If A and B are square matrices of the same size and A is invertible, then

$$\det(A^{-1}BA) = \det(B)$$

(d) A square matrix A is invertible if and only if $\det(A) = 0$.

(e) The matrix of cofactors of A is precisely $[\mathrm{adj}(A)]^T$.

(f) For every $n \times n$ matrix A, we have

$$A \cdot \mathrm{adj}(A) = (\det(A))I_n$$

(g) If A is a square matrix and the linear system $A\mathbf{x} = \mathbf{0}$ has multiple solutions for $\mathbf{x}$, then $\det(A) = 0$.

(h) If A is an $n \times n$ matrix and there exists an $n \times 1$ matrix $\mathbf{b}$ such that the linear system $A\mathbf{x} = \mathbf{b}$ has no solutions, then the reduced row echelon form of A cannot be I_n.

(i) If E is an elementary matrix, then $E\mathbf{x} = \mathbf{0}$ has only the trivial solution.

(j) If A is an invertible matrix, then the linear system $A\mathbf{x} = \mathbf{0}$ has only the trivial solution if and only if the linear system $A^{-1}\mathbf{x} = \mathbf{0}$ has only the trivial solution.

(k) If A is invertible, then $\mathrm{adj}(A)$ must also be invertible.

(l) If A has a row of zeros, then so does $\mathrm{adj}(A)$.

Chapter 2   Supplementary Exercises

▶ In Exercises 1–8, evaluate the determinant of the given matrix by (a) cofactor expansion and (b) using elementary row operations to introduce zeros into the matrix. ◀

1. $\begin{bmatrix} -4 & 2 \\ 3 & 3 \end{bmatrix}$

2. $\begin{bmatrix} 7 & -1 \\ -2 & -6 \end{bmatrix}$

3. $\begin{bmatrix} -1 & 5 & 2 \\ 0 & 2 & -1 \\ -3 & 1 & 1 \end{bmatrix}$

4. $\begin{bmatrix} -1 & -2 & -3 \\ -4 & -5 & -6 \\ -7 & -8 & -9 \end{bmatrix}$

5. $\begin{bmatrix} 3 & 0 & -1 \\ 1 & 1 & 1 \\ 0 & 4 & 2 \end{bmatrix}$

6. $\begin{bmatrix} -5 & 1 & 4 \\ 3 & 0 & 2 \\ 1 & -2 & 2 \end{bmatrix}$

7. $\begin{bmatrix} 3 & 6 & 0 & 1 \\ -2 & 3 & 1 & 4 \\ 1 & 0 & -1 & 1 \\ -9 & 2 & -2 & 2 \end{bmatrix}$

8. $\begin{bmatrix} -1 & -2 & -3 & -4 \\ 4 & 3 & 2 & 1 \\ 1 & 2 & 3 & 4 \\ -4 & -3 & -2 & -1 \end{bmatrix}$

9. Evaluate the determinants in Exercises 3–6 by using the arrow technique (see Example 7 in Section 2.1).

10. (a) Construct a 4×4 matrix whose determinant is easy to compute using cofactor expansion but hard to evaluate using elementary row operations.

(b) Construct a 4×4 matrix whose determinant is easy to compute using elementary row operations but hard to evaluate using cofactor expansion.

11. Use the determinant to decide whether the matrices in Exercises 1–4 are invertible.

12. Use the determinant to decide whether the matrices in Exercises 5–8 are invertible.

▶ In Exercises 13–15, find the determinant of the given matrix by any method. ◀

13. $\begin{vmatrix} 5 & b-3 \\ b-2 & -3 \end{vmatrix}$

14. $\begin{vmatrix} 3 & -4 & a \\ a^2 & 1 & 2 \\ 2 & a-1 & 4 \end{vmatrix}$

15. $\begin{vmatrix} 0 & 0 & 0 & 0 & -3 \\ 0 & 0 & 0 & -4 & 0 \\ 0 & 0 & -1 & 0 & 0 \\ 0 & 2 & 0 & 0 & 0 \\ 5 & 0 & 0 & 0 & 0 \end{vmatrix}$

16. Solve for x.

$$\begin{vmatrix} x & -1 \\ 3 & 1-x \end{vmatrix} = \begin{vmatrix} 1 & 0 & -3 \\ 2 & x & -6 \\ 1 & 3 & x-5 \end{vmatrix}$$

▶ In Exercises 17–24, use the adjoint method (Theorem 2.3.6) to find the inverse of the given matrix, if it exists. ◀

17. The matrix in Exercise 1.   **18.** The matrix in Exercise 2.

19. The matrix in Exercise 3.   **20.** The matrix in Exercise 4.

21. The matrix in Exercise 5.   **22.** The matrix in Exercise 6.

23. The matrix in Exercise 7.   **24.** The matrix in Exercise 8.

25. Use Cramer's rule to solve for x' and y' in terms of x and y.

$$x = \tfrac{3}{5}x' - \tfrac{4}{5}y'$$
$$y = \tfrac{4}{5}x' + \tfrac{3}{5}y'$$

26. Use Cramer's rule to solve for x' and y' in terms of x and y.

$$x = x'\cos\theta - y'\sin\theta$$
$$y = x'\sin\theta + y'\cos\theta$$

27. By examining the determinant of the coefficient matrix, show that the following system has a nontrivial solution if and only if $\alpha = \beta$.

$$x + y + \alpha z = 0$$
$$x + y + \beta z = 0$$
$$\alpha x + \beta y + z = 0$$

28. Let A be a 3×3 matrix, each of whose entries is 1 or 0. What is the largest possible value for $\det(A)$?

29. (a) For the triangle in the accompanying figure, use trigonometry to show that

$$b \cos\gamma + c \cos\beta = a$$
$$c \cos\alpha + a \cos\gamma = b$$
$$a \cos\beta + b \cos\alpha = c$$

and then apply Cramer's rule to show that

$$\cos\alpha = \frac{b^2 + c^2 - a^2}{2bc}$$

(b) Use Cramer's rule to obtain similar formulas for $\cos\beta$ and $\cos\gamma$.

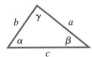
◀ Figure Ex-29

30. Use determinants to show that for all real values of λ, the only solution of

$$x - 2y = \lambda x$$
$$x - y = \lambda y$$

is $x = 0$, $y = 0$.

31. Prove: If A is invertible, then $\text{adj}(A)$ is invertible and

$$[\text{adj}(A)]^{-1} = \frac{1}{\det(A)} A = \text{adj}(A^{-1})$$

32. Prove: If A is an $n \times n$ matrix, then

$$\det[\mathrm{adj}(A)] = [\det(A)]^{n-1}$$

33. Prove: If the entries in each row of an $n \times n$ matrix A add up to zero, then the determinant of A is zero. [*Hint:* Consider the product AX, where X is the $n \times 1$ matrix, each of whose entries is one.]

34. (a) In the accompanying figure, the area of the triangle ABC can be expressed as

area ABC = area $ADEC$ + area $CEFB$ − area $ADFB$

Use this and the fact that the area of a trapezoid equals $\frac{1}{2}$ the altitude times the sum of the parallel sides to show that

$$\text{area } ABC = \frac{1}{2} \begin{vmatrix} x_1 & y_1 & 1 \\ x_2 & y_2 & 1 \\ x_3 & y_3 & 1 \end{vmatrix}$$

[*Note:* In the derivation of this formula, the vertices are labeled such that the triangle is traced counterclockwise proceeding from (x_1, y_1) to (x_2, y_2) to (x_3, y_3). For a clockwise orientation, the determinant above yields the *negative* of the area.]

   (b) Use the result in (a) to find the area of the triangle with vertices $(3, 3)$, $(4, 0)$, $(-2, -1)$.

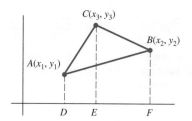

◀ **Figure Ex-34**

35. Use the fact that $21{,}375$, $38{,}798$, $34{,}162$, $40{,}223$, and $79{,}154$ are all divisible by 19 to show that

$$\begin{vmatrix} 2 & 1 & 3 & 7 & 5 \\ 3 & 8 & 7 & 9 & 8 \\ 3 & 4 & 1 & 6 & 2 \\ 4 & 0 & 2 & 2 & 3 \\ 7 & 9 & 1 & 5 & 4 \end{vmatrix}$$

is divisible by 19 without directly evaluating the determinant.

36. Without directly evaluating the determinant, show that

$$\begin{vmatrix} \sin\alpha & \cos\alpha & \sin(\alpha+\delta) \\ \sin\beta & \cos\beta & \sin(\beta+\delta) \\ \sin\gamma & \cos\gamma & \sin(\gamma+\delta) \end{vmatrix} = 0$$

Euclidean Vector Spaces

CHAPTER CONTENTS

3.1 Vectors in 2-Space, 3-Space, and n-Space 119
3.2 Norm, Dot Product, and Distance in R^n 130
3.3 Orthogonality 143
3.4 The Geometry of Linear Systems 152
3.5 Cross Product 161

INTRODUCTION   Engineers and physicists distinguish between two types of physical quantities— *scalars*, which are quantities that can be described by a numerical value alone, and *vectors*, which are quantities that require both a number and a direction for their complete physical description. For example, temperature, length, and speed are scalars because they can be fully described by a number that tells "how much"—a temperature of 20°C, a length of 5 cm, or a speed of 75 km/h. In contrast, velocity and force are vectors because they require a number that tells "how much" and a direction that tells "which way"—say, a boat moving at 10 knots in a direction 45° northeast, or a force of 100 lb acting vertically. Although the notions of vectors and scalars that we will study in this text have their origins in physics and engineering, we will be more concerned with using them to build mathematical structures and then applying those structures to such diverse fields as genetics, computer science, economics, telecommunications, and environmental science.

3.1 Vectors in 2-Space, 3-Space, and n-Space

Linear algebra is concerned with two kinds of mathematical objects, "matrices" and "vectors." We are already familiar with the basic ideas about matrices, so in this section we will introduce some of the basic ideas about vectors. As we progress through this text we will see that vectors and matrices are closely related and that much of linear algebra is concerned with that relationship.

Geometric Vectors   Engineers and physicists represent vectors in two dimensions (also called **2-*space***) or in three dimensions (also called **3-*space***) by arrows. The direction of the arrowhead specifies the ***direction*** of the vector and the ***length*** of the arrow specifies the magnitude. Mathematicians call these ***geometric vectors***. The tail of the arrow is called the ***initial point*** of the vector and the tip the ***terminal point*** (Figure 3.1.1).

In this text we will denote vectors in boldface type such as **a**, **b**, **v**, **w**, and **x**, and we will denote scalars in lowercase italic type such as a, k, v, w, and x. When we want to indicate that a vector **v** has initial point A and terminal point B, then, as shown in Figure 3.1.2, we will write

$$\mathbf{v} = \overrightarrow{AB}$$

▲ Figure 3.1.1

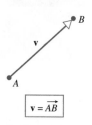

$$v = \vec{AB}$$

▲ Figure 3.1.2

Vectors with the same length and direction, such as those in Figure 3.1.3, are said to be *equivalent*. Since we want a vector to be determined solely by its length and direction, equivalent vectors are regarded to be the same vector even though they may be in different positions. Equivalent vectors are also said to be *equal*, which we indicate by writing

$$\mathbf{v} = \mathbf{w}$$

The vector whose initial and terminal points coincide has length zero, so we call this the *zero vector* and denote it by **0**. The zero vector has no natural direction, so we will agree that it can be assigned any direction that is convenient for the problem at hand.

Vector Addition

Equivalent vectors

▲ Figure 3.1.3

There are a number of important algebraic operations on vectors, all of which have their origin in laws of physics.

> **Parallelogram Rule for Vector Addition** If **v** and **w** are vectors in 2-space or 3-space that are positioned so their initial points coincide, then the two vectors form adjacent sides of a parallelogram, and the *sum* **v** + **w** is the vector represented by the arrow from the common initial point of **v** and **w** to the opposite vertex of the parallelogram (Figure 3.1.4a).

Here is another way to form the sum of two vectors.

> **Triangle Rule for Vector Addition** If **v** and **w** are vectors in 2-space or 3-space that are positioned so the initial point of **w** is at the terminal point of **v**, then the *sum* **v** + **w** is represented by the arrow from the initial point of **v** to the terminal point of **w** (Figure 3.1.4b).

In Figure 3.1.4c we have constructed the sums **v** + **w** and **w** + **v** by the triangle rule. This construction makes it evident that

$$\mathbf{v} + \mathbf{w} = \mathbf{w} + \mathbf{v} \tag{1}$$

and that the sum obtained by the triangle rule is the same as the sum obtained by the parallelogram rule.

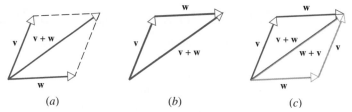

▶ Figure 3.1.4   (a)   (b)   (c)

Vector addition can also be viewed as a process of translating points.

> **Vector Addition Viewed as Translation** If **v**, **w**, and **v** + **w** are positioned so their initial points coincide, then the terminal point of **v** + **w** can be viewed in two ways:
>
> 1. The terminal point of **v** + **w** is the point that results when the terminal point of **v** is translated in the direction of **w** by a distance equal to the length of **w** (Figure 3.1.5a).
>
> 2. The terminal point of **v** + **w** is the point that results when the terminal point of **w** is translated in the direction of **v** by a distance equal to the length of **v** (Figure 3.1.5b).
>
> Accordingly, we say that **v** + **w** is the *translation of* **v** *by* **w** or, alternatively, the *translation of* **w** *by* **v**.

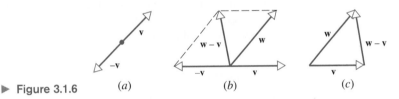

▶ **Figure 3.1.5**   *(a)*   *(b)*

Vector Subtraction   In ordinary arithmetic we can write $a - b = a + (-b)$, which expresses subtraction in terms of addition. There is an analogous idea in vector arithmetic.

> **Vector Subtraction** The ***negative*** of a vector **v**, denoted by $-\mathbf{v}$, is the vector that has the same length as **v** but is oppositely directed (Figure 3.1.6*a*), and the ***difference*** of **v** from **w**, denoted by $\mathbf{w} - \mathbf{v}$, is taken to be the sum
>
> $$\mathbf{w} - \mathbf{v} = \mathbf{w} + (-\mathbf{v}) \tag{2}$$

The difference of **v** from **w** can be obtained geometrically by the parallelogram method shown in Figure 3.1.6*b*, or more directly by positioning **w** and **v** so their initial points coincide and drawing the vector from the terminal point of **v** to the terminal point of **w** (Figure 3.1.6*c*).

▶ **Figure 3.1.6**   *(a)*   *(b)*   *(c)*

Scalar Multiplication   Sometimes there is a need to change the length of a vector or change its length and reverse its direction. This is accomplished by a type of multiplication in which vectors are multiplied by scalars. As an example, the product $2\mathbf{v}$ denotes the vector that has the same direction as **v** but twice the length, and the product $-2\mathbf{v}$ denotes the vector that is oppositely directed to **v** and has twice the length. Here is the general result.

> **Scalar Multiplication** If **v** is a nonzero vector in 2-space or 3-space, and if k is a nonzero scalar, then we define the ***scalar product of*** **v** ***by*** k to be the vector whose length is $|k|$ times the length of **v** and whose direction is the same as that of **v** if k is positive and opposite to that of **v** if k is negative. If $k = 0$ or $\mathbf{v} = \mathbf{0}$, then we define $k\mathbf{v}$ to be **0**.

▲ **Figure 3.1.7**

Figure 3.1.7 shows the geometric relationship between a vector **v** and some of its scalar multiples. In particular, observe that $(-1)\mathbf{v}$ has the same length as **v** but is oppositely directed; therefore,

$$(-1)\mathbf{v} = -\mathbf{v} \tag{3}$$

Parallel and Collinear Vectors   Suppose that **v** and **w** are vectors in 2-space or 3-space with a common initial point. If one of the vectors is a scalar multiple of the other, then the vectors lie on a common line, so it is reasonable to say that they are *collinear* (Figure 3.1.8*a*). However, if we translate one of the vectors, as indicated in Figure 3.1.8*b*, then the vectors are *parallel* but no longer collinear. This creates a linguistic problem because translating a vector does not change it. The only way to resolve this problem is to agree that the terms *parallel* and

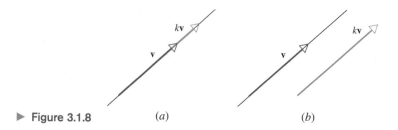

► Figure 3.1.8   (a)   (b)

collinear mean the same thing when applied to vectors. Although the vector **0** has no clearly defined direction, we will regard it to be parallel to all vectors when convenient.

Sums of Three or More Vectors

Vector addition satisfies the ***associative law for addition***, meaning that when we add three vectors, say **u**, **v**, and **w**, it does not matter which two we add first; that is,

$$\mathbf{u} + (\mathbf{v} + \mathbf{w}) = (\mathbf{u} + \mathbf{v}) + \mathbf{w}$$

It follows from this that there is no ambiguity in the expression **u** + **v** + **w** because the same result is obtained no matter how the vectors are grouped.

A simple way to construct **u** + **v** + **w** is to place the vectors "tip to tail" in succession and then draw the vector from the initial point of **u** to the terminal point of **w** (Figure 3.1.9*a*). The tip-to-tail method also works for four or more vectors (Figure 3.1.9*b*). The tip-to-tail method also makes it evident that if **u**, **v**, and **w** are vectors in 3-space with a *common initial point*, then **u** + **v** + **w** is the diagonal of the parallelepiped that has the three vectors as adjacent sides (Figure 3.1.9*c*).

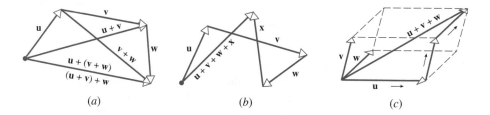

► Figure 3.1.9   (a)   (b)   (c)

Vectors in Coordinate Systems

Up until now we have discussed vectors without reference to a coordinate system. However, as we will soon see, computations with vectors are much simpler to perform if a coordinate system is present to work with.

If a vector **v** in 2-space or 3-space is positioned with its initial point at the origin of a rectangular coordinate system, then the vector is completely determined by the coordinates of its terminal point (Figure 3.1.10). We call these coordinates the ***components*** of **v** relative to the coordinate system. We will write $\mathbf{v} = (v_1, v_2)$ to denote a vector **v** in 2-space with components (v_1, v_2), and $\mathbf{v} = (v_1, v_2, v_3)$ to denote a vector **v** in 3-space with components (v_1, v_2, v_3).

The component forms of the zero vector are $\mathbf{0} = (0, 0)$ in 2-space and $\mathbf{0} = (0, 0, 0)$ in 3-space.

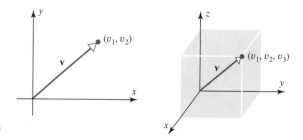

► Figure 3.1.10

It should be evident geometrically that two vectors in 2-space or 3-space are equivalent if and only if they have the same terminal point when their initial points are at the origin. Algebraically, this means that two vectors are equivalent if and only if their corresponding components are equal. Thus, for example, the vectors

$$\mathbf{v} = (v_1, v_2, v_3) \quad \text{and} \quad \mathbf{w} = (w_1, w_2, w_3)$$

in 3-space are equivalent if and only if

$$v_1 = w_1, \quad v_2 = w_2, \quad v_3 = w_3$$

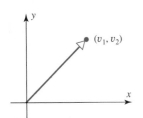

▲ **Figure 3.1.11** The ordered pair (v_1, v_2) can represent a point or a vector.

Remark   It may have occurred to you that an ordered pair (v_1, v_2) can represent either a vector with *components* v_1 and v_2 or a point with *components* v_1 and v_2 (and similarly for ordered triples). Both are valid geometric interpretations, so the appropriate choice will depend on the geometric viewpoint that we want to emphasize (Figure 3.1.11).

Vectors Whose Initial Point Is Not at the Origin

It is sometimes necessary to consider vectors whose initial points are not at the origin. If $\overrightarrow{P_1 P_2}$ denotes the vector with initial point $P_1(x_1, y_1)$ and terminal point $P_2(x_2, y_2)$, then the components of this vector are given by the formula

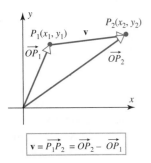

$$\overrightarrow{P_1 P_2} = (x_2 - x_1, y_2 - y_1) \tag{4}$$

▲ **Figure 3.1.12**

That is, the components of $\overrightarrow{P_1 P_2}$ are obtained by subtracting the coordinates of the initial point from the coordinates of the terminal point. For example, in Figure 3.1.12 the vector $\overrightarrow{P_1 P_2}$ is the difference of vectors $\overrightarrow{OP_2}$ and $\overrightarrow{OP_1}$, so

$$\overrightarrow{P_1 P_2} = \overrightarrow{OP_2} - \overrightarrow{OP_1} = (x_2, y_2) - (x_1, y_1) = (x_2 - x_1, y_2 - y_1)$$

As you might expect, the components of a vector in 3-space that has initial point $P_1(x_1, y_1, z_1)$ and terminal point $P_2(x_2, y_2, z_2)$ are given by

$$\overrightarrow{P_1 P_2} = (x_2 - x_1, y_2 - y_1, z_2 - z_1) \tag{5}$$

▶ **EXAMPLE 1 Finding the Components of a Vector**

The components of the vector $\mathbf{v} = \overrightarrow{P_1 P_2}$ with initial point $P_1(2, -1, 4)$ and terminal point $P_2(7, 5, -8)$ are

$$\mathbf{v} = (7 - 2, 5 - (-1), (-8) - 4) = (5, 6, -12) \blacktriangleleft$$

n-Space

The idea of using ordered pairs and triples of real numbers to represent points in two-dimensional space and three-dimensional space was well known in the eighteenth and nineteenth centuries. By the dawn of the twentieth century, mathematicians and physicists were exploring the use of "higher-dimensional" spaces in mathematics and physics. Today, even the layman is familiar with the notion of time as a fourth dimension, an idea used by Albert Einstein in developing the general theory of relativity. Today, physicists working in the field of "string theory" commonly use 11-dimensional space in their quest for a unified theory that will explain how the fundamental forces of nature work. Much of the remaining work in this section is concerned with extending the notion of space to *n*-dimensions.

To explore these ideas further, we start with some terminology and notation. The set of all real numbers can be viewed geometrically as a line. It is called the ***real line*** and is denoted by R or R^1. The superscript reinforces the intuitive idea that a line is one-dimensional. The set of all ordered pairs of real numbers (called **2-*tuples***) and the set of all ordered triples of real numbers (called **3-*tuples***) are denoted by R^2 and

R^3, respectively. The superscript reinforces the idea that the ordered pairs correspond to points in the plane (two-dimensional) and ordered triples to points in space (three-dimensional). The following definition extends this idea.

> **DEFINITION 1** If n is a positive integer, then an ***ordered n-tuple*** is a sequence of n real numbers $(v_1, v_2, \ldots, v_n)$. The set of all ordered n-tuples is called ***n-space*** and is denoted by R^n.

Remark You can think of the numbers in an n-tuple $(v_1, v_2, \ldots, v_n)$ as either the coordinates of a *generalized point* or the components of a *generalized vector*, depending on the geometric image you want to bring to mind—the choice makes no difference mathematically, since it is the algebraic properties of n-tuples that are of concern.

Here are some typical applications that lead to n-tuples.

- **Experimental Data**—A scientist performs an experiment and makes n numerical measurements each time the experiment is performed. The result of each experiment can be regarded as a vector $\mathbf{y} = (y_1, y_2, \ldots, y_n)$ in R^n in which $y_1, y_2, \ldots, y_n$ are the measured values.

- **Storage and Warehousing**—A national trucking company has 15 depots for storing and servicing its trucks. At each point in time the distribution of trucks in the service depots can be described by a 15-tuple $\mathbf{x} = (x_1, x_2, \ldots, x_{15})$ in which x_1 is the number of trucks in the first depot, x_2 is the number in the second depot, and so forth.

- **Electrical Circuits**—A certain kind of processing chip is designed to receive four input voltages and produces three output voltages in response. The input voltages can be regarded as vectors in R^4 and the output voltages as vectors in R^3. Thus, the chip can be viewed as a device that transforms an input vector $\mathbf{v} = (v_1, v_2, v_3, v_4)$ in R^4 into an output vector $\mathbf{w} = (w_1, w_2, w_3)$ in R^3.

- **Graphical Images**—One way in which color images are created on computer screens is by assigning each pixel (an addressable point on the screen) three numbers that describe the ***hue***, ***saturation***, and ***brightness*** of the pixel. Thus, a complete color image can be viewed as a set of 5-tuples of the form $\mathbf{v} = (x, y, h, s, b)$ in which x and y are the screen coordinates of a pixel and h, s, and b are its hue, saturation, and brightness.

- **Economics**—One approach to economic analysis is to divide an economy into sectors (manufacturing, services, utilities, and so forth) and measure the output of each sector by a dollar value. Thus, in an economy with 10 sectors the economic output of the entire economy can be represented by a 10-tuple $\mathbf{s} = (s_1, s_2, \ldots, s_{10})$ in which the numbers $s_1, s_2, \ldots, s_{10}$ are the outputs of the individual sectors.

**Albert Einstein
(1879–1955)**

- **Mechanical Systems**—Suppose that six particles move along the same coordinate line so that at time t their coordinates are $x_1, x_2, \ldots, x_6$ and their velocities are $v_1, v_2, \ldots, v_6$, respectively. This information can be represented by the vector

$$\mathbf{v} = (x_1, x_2, x_3, x_4, x_5, x_6, v_1, v_2, v_3, v_4, v_5, v_6, t)$$

in R^{13}. This vector is called the *state* of the particle system at time t.

Operations on Vectors in R^n
Our next goal is to define useful operations on vectors in R^n. These operations will all be natural extensions of the familiar operations on vectors in R^2 and R^3. We will denote a vector $\mathbf{v}$ in R^n using the notation

$$\mathbf{v} = (v_1, v_2, \ldots, v_n)$$

and we will call $\mathbf{0} = (0, 0, \ldots, 0)$ the *zero vector*.

We noted earlier that in R^2 and R^3 two vectors are equivalent (equal) if and only if their corresponding components are the same. Thus, we make the following definition.

DEFINITION 2 Vectors $\mathbf{v} = (v_1, v_2, \ldots, v_n)$ and $\mathbf{w} = (w_1, w_2, \ldots, w_n)$ in R^n are said to be *equivalent* (also called *equal*) if

$$v_1 = w_1, \quad v_2 = w_2, \ldots, \quad v_n = w_n$$

We indicate this by writing $\mathbf{v} = \mathbf{w}$.

▶ **EXAMPLE 2 Equality of Vectors**

$$(a, b, c, d) = (1, -4, 2, 7)$$

if and only if $a = 1, b = -4, c = 2$, and $d = 7$. ◀

Our next objective is to define the operations of addition, subtraction, and scalar multiplication for vectors in R^n. To motivate these ideas, we will consider how these operations can be performed on vectors in R^2 using components. By studying Figure 3.1.13 you should be able to deduce that if $\mathbf{v} = (v_1, v_2)$ and $\mathbf{w} = (w_1, w_2)$, then

$$\mathbf{v} + \mathbf{w} = (v_1 + w_1, v_2 + w_2) \qquad (6)$$
$$k\mathbf{v} = (kv_1, kv_2) \qquad (7)$$

In particular, it follows from (7) that

$$-\mathbf{v} = (-1)\mathbf{v} = (-v_1, -v_2) \qquad (8)$$

and hence that

$$\mathbf{w} - \mathbf{v} = \mathbf{w} + (-\mathbf{v}) = (w_1 - v_1, w_2 - v_2) \qquad (9)$$

▶ **Figure 3.1.13**

Motivated by Formulas (6)–(9), we make the following definition.

DEFINITION 3 If $\mathbf{v} = (v_1, v_2, \ldots, v_n)$ and $\mathbf{w} = (w_1, w_2, \ldots, w_n)$ are vectors in R^n, and if k is any scalar, then we define

$$\mathbf{v} + \mathbf{w} = (v_1 + w_1, v_2 + w_2, \ldots, v_n + w_n) \tag{10}$$

$$k\mathbf{v} = (kv_1, kv_2, \ldots, kv_n) \tag{11}$$

$$-\mathbf{v} = (-v_1, -v_2, \ldots, -v_n) \tag{12}$$

$$\mathbf{w} - \mathbf{v} = \mathbf{w} + (-\mathbf{v}) = (w_1 - v_1, w_2 - v_2, \ldots, w_n - v_n) \tag{13}$$

In words, vectors are added (or subtracted) by adding (or subtracting) their corresponding components, and a vector is multiplied by a scalar by multiplying each component by that scalar.

▶ **EXAMPLE 3 Algebraic Operations Using Components**

If $\mathbf{v} = (1, -3, 2)$ and $\mathbf{w} = (4, 2, 1)$, then

$$\mathbf{v} + \mathbf{w} = (5, -1, 3), \qquad 2\mathbf{v} = (2, -6, 4)$$
$$-\mathbf{w} = (-4, -2, -1), \qquad \mathbf{v} - \mathbf{w} = \mathbf{v} + (-\mathbf{w}) = (-3, -5, 1) \ ◀$$

The following theorem summarizes the most important properties of vector operations.

THEOREM 3.1.1 *If $\mathbf{u}$, $\mathbf{v}$, and $\mathbf{w}$ are vectors in R^n, and if k and m are scalars, then*:

(a) $\mathbf{u} + \mathbf{v} = \mathbf{v} + \mathbf{u}$

(b) $(\mathbf{u} + \mathbf{v}) + \mathbf{w} = \mathbf{u} + (\mathbf{v} + \mathbf{w})$

(c) $\mathbf{u} + \mathbf{0} = \mathbf{0} + \mathbf{u} = \mathbf{u}$

(d) $\mathbf{u} + (-\mathbf{u}) = \mathbf{0}$

(e) $k(\mathbf{u} + \mathbf{v}) = k\mathbf{u} + k\mathbf{v}$

(f) $(k + m)\mathbf{u} = k\mathbf{u} + m\mathbf{u}$

(g) $k(m\mathbf{u}) = (km)\mathbf{u}$

(h) $1\mathbf{u} = \mathbf{u}$

We will prove part (b) and leave some of the other proofs as exercises.

Proof (b) Let $\mathbf{u} = (u_1, u_2, \ldots, u_n)$, $\mathbf{v} = (v_1, v_2, \ldots, v_n)$, and $\mathbf{w} = (w_1, w_2, \ldots, w_n)$. Then

$$\begin{aligned}
(\mathbf{u} + \mathbf{v}) + \mathbf{w} &= \big((u_1, u_2, \ldots, u_n) + (v_1, v_2, \ldots, v_n)\big) + (w_1, w_2, \ldots, w_n) \\
&= (u_1 + v_1, u_2 + v_2, \ldots, u_n + v_n) + (w_1, w_2, \ldots, w_n) && \text{[Vector addition]} \\
&= \big((u_1 + v_1) + w_1, (u_2 + v_2) + w_2, \ldots, (u_n + v_n) + w_n\big) && \text{[Vector addition]} \\
&= \big(u_1 + (v_1 + w_1), u_2 + (v_2 + w_2), \ldots, u_n + (v_n + w_n)\big) && \text{[Regroup]} \\
&= (u_1, u_2, \ldots, u_n) + (v_1 + w_1, v_2 + w_2, \ldots, v_n + w_n) && \text{[Vector addition]} \\
&= \mathbf{u} + (\mathbf{v} + \mathbf{w}) \ ◀
\end{aligned}$$

The following additional properties of vectors in R^n can be deduced easily by expressing the vectors in terms of components (verify).

THEOREM 3.1.2 *If $\mathbf{v}$ is a vector in R^n and k is a scalar, then*:

(a) $0\mathbf{v} = \mathbf{0}$

(b) $k\mathbf{0} = \mathbf{0}$

(c) $(-1)\mathbf{v} = -\mathbf{v}$

Calculating Without Components

One of the powerful consequences of Theorems 3.1.1 and 3.1.2 is that they allow calculations to be performed without expressing the vectors in terms of components. For example, suppose that **x**, **a**, and **b** are vectors in R^n, and we want to solve the vector equation **x** + **a** = **b** for the vector **x** without using components. We could proceed as follows:

$$\mathbf{x} + \mathbf{a} = \mathbf{b} \qquad \text{[Given]}$$
$$(\mathbf{x} + \mathbf{a}) + (-\mathbf{a}) = \mathbf{b} + (-\mathbf{a}) \qquad \text{[Add the negative of a to both sides]}$$
$$\mathbf{x} + (\mathbf{a} + (-\mathbf{a})) = \mathbf{b} - \mathbf{a} \qquad \text{[Part (\textit{b}) of Theorem 3.1.1]}$$
$$\mathbf{x} + \mathbf{0} = \mathbf{b} - \mathbf{a} \qquad \text{[Part (\textit{d}) of Theorem 3.1.1]}$$
$$\mathbf{x} = \mathbf{b} - \mathbf{a} \qquad \text{[Part (\textit{c}) of Theorem 3.1.1]}$$

While this method is obviously more cumbersome than computing with components in R^n, it will become important later in the text where we will encounter more general kinds of vectors.

Linear Combinations

Addition, subtraction, and scalar multiplication are frequently used in combination to form new vectors. For example, if $\mathbf{v}_1$, $\mathbf{v}_2$, and $\mathbf{v}_3$ are vectors in R^n, then the vectors

$$\mathbf{u} = 2\mathbf{v}_1 + 3\mathbf{v}_2 + \mathbf{v}_3 \quad \text{and} \quad \mathbf{w} = 7\mathbf{v}_1 - 6\mathbf{v}_2 + 8\mathbf{v}_3$$

are formed in this way. In general, we make the following definition.

> **DEFINITION 4** If **w** is a vector in R^n, then **w** is said to be a *linear combination* of the vectors $\mathbf{v}_1, \mathbf{v}_2, \ldots, \mathbf{v}_r$ in R^n if it can be expressed in the form
>
> $$\mathbf{w} = k_1\mathbf{v}_1 + k_2\mathbf{v}_2 + \cdots + k_r\mathbf{v}_r \tag{14}$$
>
> where $k_1, k_2, \ldots, k_r$ are scalars. These scalars are called the *coefficients* of the linear combination. In the case where $r = 1$, Formula (14) becomes $\mathbf{w} = k_1\mathbf{v}_1$, so that a linear combination of a single vector is just a scalar muliple of that vector.

Note that this definition of a linear combination is consistent with that given in the context of matrices (see Definition 6 in Section 1.3).

Application of Linear Combinations to Color Models

Colors on computer monitors are commonly based on what is called the **RGB** *color model*. Colors in this system are created by adding together percentages of the primary colors red (R), green (G), and blue (B). One way to do this is to identify the primary colors with the vectors

$$\mathbf{r} = (1, 0, 0) \quad \text{(pure red)},$$
$$\mathbf{g} = (0, 1, 0) \quad \text{(pure green)},$$
$$\mathbf{b} = (0, 0, 1) \quad \text{(pure blue)}$$

in R^3 and to create all other colors by forming linear combinations of **r**, **g**, and **b** using coefficients between 0 and 1, inclusive; these coefficients represent the percentage of each pure color in the mix.

The set of all such color vectors is called **RGB** *space* or the **RGB** *color cube* (Figure 3.1.14). Thus, each color vector **c** in this cube is expressible as a linear combination of the form

$$\begin{aligned} \mathbf{c} &= k_1\mathbf{r} + k_2\mathbf{g} + k_3\mathbf{b} \\ &= k_1(1, 0, 0) + k_2(0, 1, 0) + k_3(0, 0, 1) \\ &= (k_1, k_2, k_3) \end{aligned}$$

where $0 \le k_i \le 1$. As indicated in the figure, the corners of the cube represent the pure primary colors together with the colors black, white, magenta, cyan, and yellow. The vectors along the diagonal running from black to white correspond to shades of gray.

▶ Figure 3.1.14

Alternative Notations for Vectors

Up to now we have been writing vectors in R^n using the notation

$$\mathbf{v} = (v_1, v_2, \ldots, v_n) \tag{15}$$

We call this the **comma-delimited** form. However, since a vector in R^n is just a list of its n components in a specific order, any notation that displays those components in the correct order is a valid way of representing the vector. For example, the vector in (15) can be written as

$$\mathbf{v} = [v_1 \quad v_2 \quad \cdots \quad v_n] \tag{16}$$

which is called **row-matrix** form, or as

$$\mathbf{v} = \begin{bmatrix} v_1 \\ v_2 \\ \vdots \\ v_n \end{bmatrix} \tag{17}$$

which is called **column-matrix** form. The choice of notation is often a matter of taste or convenience, but sometimes the nature of a problem will suggest a preferred notation. Notations (15), (16), and (17) will all be used at various places in this text.

Concept Review

- Geometric vector
- Direction
- Length
- Initial point
- Terminal point
- Equivalent vectors
- Zero vector

- Vector addition: parallelogram rule and triangle rule
- Vector subtraction
- Negative of a vector
- Scalar multiplication
- Collinear (i.e., parallel) vectors
- Components of a vector

- Coordinates of a point
- n-tuple
- n-space
- Vector operations in n-space: addition, subtraction, scalar multiplication
- Linear combination of vectors

Skills

- Perform geometric operations on vectors: addition, subtraction, and scalar multiplication.
- Perform algebraic operations on vectors: addition, subtraction, and scalar multiplication.
- Determine whether two vectors are equivalent.
- Determine whether two vectors are collinear.

- Sketch vectors whose initial and terminal points are given.
- Find components of a vector whose initial and terminal points are given.
- Prove basic algebraic properties of vectors (Theorems 3.1.1 and 3.1.2).

Exercise Set 3.1

▶ In Exercises **1–2**, draw a coordinate system (as in Figure 3.1.10) and locate the points whose coordinates are given. ◀

1. (a) $(3, 4, 5)$ (b) $(-3, 4, 5)$ (c) $(3, -4, 5)$
    (d) $(3, 4, -5)$ (e) $(-3, -4, 5)$ (f) $(-3, 4, -5)$

2. (a) $(0, 3, -3)$ (b) $(3, -3, 0)$ (c) $(-3, 0, 0)$
    (d) $(3, 0, 3)$ (e) $(0, 0, -3)$ (f) $(0, 3, 0)$

▶ In Exercises **3–4**, sketch the following vectors with the initial points located at the origin. ◀

3. (a) $\mathbf{v}_1 = (3, 6)$ (b) $\mathbf{v}_2 = (-4, -8)$
    (c) $\mathbf{v}_3 = (-4, -3)$ (d) $\mathbf{v}_4 = (3, 4, 5)$
    (e) $\mathbf{v}_5 = (3, 3, 0)$ (f) $\mathbf{v}_6 = (-1, 0, 2)$

4. (a) $\mathbf{v}_1 = (5, -4)$ (b) $\mathbf{v}_2 = (3, 0)$
    (c) $\mathbf{v}_3 = (0, -7)$ (d) $\mathbf{v}_4 = (0, 0, -3)$
    (e) $\mathbf{v}_5 = (0, 4, -1)$ (f) $\mathbf{v}_6 = (2, 2, 2)$

▷ In Exercises **5–6**, sketch the following vectors with the initial points located at the origin. ◁

5. (a) $P_1(4, 8)$, $P_2(3, 7)$

(b) $P_1(3, -5)$, $P_2(-4, -7)$

(c) $P_1(3, -7, 2)$, $P_2(-2, 5, -4)$

6. (a) $P_1(-5, 0)$, $P_2(-3, 1)$

(b) $P_1(0, 0)$, $P_2(3, 4)$

(c) $P_1(-1, 0, 2)$, $P_2(0, -1, 0)$

(d) $P_1(2, 2, 2)$, $P_2(0, 0, 0)$

▷ In Exercises **7–8**, find the components of the vector $\overrightarrow{P_1 P_2}$. ◁

7. (a) $P_1(3, 5)$, $P_2(2, 8)$

(b) $P_1(5, -2, 1)$, $P_2(2, 4, 2)$

8. (a) $P_1(-6, 2)$, $P_2(-4, -1)$

(b) $P_1(0, 0, 0)$, $P_2(-1, 6, 1)$

9. (a) Find the terminal point of the vector that is equivalent to $\mathbf{u} = (1, 2)$ and whose initial point is $A(1, 1)$.

(b) Find the initial point of the vector that is equivalent to $\mathbf{u} = (1, 1, 3)$ and whose terminal point is $B(-1, -1, 2)$.

10. (a) Find the initial point of the vector that is equivalent to $\mathbf{u} = (1, 2)$ and whose terminal point is $B(2, 0)$.

(b) Find the terminal point of the vector that is equivalent to $\mathbf{u} = (1, 1, 3)$ and whose initial point is $A(0, 2, 0)$.

11. Find a nonzero vector $\mathbf{u}$ with terminal point $Q(3, 0, -5)$ such that

(a) $\mathbf{u}$ has the same direction as $\mathbf{v} = (4, -2, -1)$.

(b) $\mathbf{u}$ is oppositely directed to $\mathbf{v} = (4, -2, -1)$.

12. Find a nonzero vector $\mathbf{u}$ with initial point $P(-1, 3, -5)$ such that

(a) $\mathbf{u}$ has the same direction as $\mathbf{v} = (6, 7, -3)$.

(b) $\mathbf{u}$ is oppositely directed to $\mathbf{v} = (6, 7, -3)$.

13. Let $\mathbf{u} = (4, -1)$, $\mathbf{v} = (0, 5)$, and $\mathbf{w} = (-3, -3)$. Find the components of

(a) $\mathbf{u} + \mathbf{w}$ (b) $\mathbf{v} - 3\mathbf{u}$

(c) $2(\mathbf{u} - 5\mathbf{w})$ (d) $3\mathbf{v} - 2(\mathbf{u} + 2\mathbf{w})$

(e) $-3(\mathbf{w} - 2\mathbf{u} + \mathbf{v})$ (f) $(-2\mathbf{u} - \mathbf{v}) - 5(\mathbf{v} + 3\mathbf{w})$

14. Let $\mathbf{u} = (-3, 1, 2)$, $\mathbf{v} = (4, 0, -8)$, and $\mathbf{w} = (6, -1, -4)$. Find the components of

(a) $\mathbf{v} - \mathbf{w}$ (b) $6\mathbf{u} + 2\mathbf{v}$

(c) $-\mathbf{v} + \mathbf{u}$ (d) $5(\mathbf{v} - 4\mathbf{u})$

(e) $-3(\mathbf{v} - 8\mathbf{w})$ (f) $(2\mathbf{u} - 7\mathbf{w}) - (8\mathbf{v} + \mathbf{u})$

15. Let $\mathbf{u} = (-3, 2, 1, 0)$, $\mathbf{v} = (4, 7, -3, 2)$, and $\mathbf{w} = (5, -2, 8, 1)$. Find the components of

(a) $\mathbf{v} - \mathbf{w}$ (b) $2\mathbf{u} + 7\mathbf{v}$

(c) $-\mathbf{u} + (\mathbf{v} - 4\mathbf{w})$ (d) $6(\mathbf{u} - 3\mathbf{v})$

(e) $-\mathbf{v} - \mathbf{w}$ (f) $(6\mathbf{v} - \mathbf{w}) - (4\mathbf{u} + \mathbf{v})$

16. Let $\mathbf{u}$, $\mathbf{v}$, and $\mathbf{w}$ be the vectors in Exercise 15. Find the vector $\mathbf{x}$ that satisfies $5\mathbf{x} - 2\mathbf{v} = 2(\mathbf{w} - 5\mathbf{x})$.

17. Let $\mathbf{u} = (5, -1, 0, 3, -3)$, $\mathbf{v} = (-1, -1, 7, 2, 0)$, and $\mathbf{w} = (-4, 2, -3, -5, 2)$. Find the components of

(a) $\mathbf{w} - \mathbf{u}$ (b) $2\mathbf{v} + 3\mathbf{u}$

(c) $-\mathbf{w} + 3(\mathbf{v} - \mathbf{u})$ (d) $5(-\mathbf{v} + 4\mathbf{u} - \mathbf{w})$

(e) $-2(3\mathbf{w} + \mathbf{v}) + (2\mathbf{u} + \mathbf{w})$ (f) $\frac{1}{2}(\mathbf{w} - 5\mathbf{v} + 2\mathbf{u}) + \mathbf{v}$

18. Let $\mathbf{u} = (1, 2, -3, 5, 0)$, $\mathbf{v} = (0, 4, -1, 1, 2)$, and $\mathbf{w} = (7, 1, -4, -2, 3)$. Find the components of

(a) $\mathbf{v} + \mathbf{w}$ (b) $3(2\mathbf{u} - \mathbf{v})$

(c) $(3\mathbf{u} - \mathbf{v}) - (2\mathbf{u} + 4\mathbf{w})$

19. Let $\mathbf{u} = (-3, 1, 2, 4, 4)$, $\mathbf{v} = (4, 0, -8, 1, 2)$, and $\mathbf{w} = (6, -1, -4, 3, -5)$. Find the components of

(a) $\mathbf{v} - \mathbf{w}$ (b) $6\mathbf{u} + 2\mathbf{v}$

(c) $(2\mathbf{u} - 7\mathbf{w}) - (8\mathbf{v} + \mathbf{u})$

20. Let $\mathbf{u}$, $\mathbf{v}$, and $\mathbf{w}$ be the vectors in Exercise 18. Find the components of the vector $\mathbf{x}$ that satisfies the equation $3\mathbf{u} + \mathbf{v} - 2\mathbf{w} = 3\mathbf{x} + 2\mathbf{w}$.

21. Let $\mathbf{u}$, $\mathbf{v}$, and $\mathbf{w}$ be the vectors in Exercise 19. Find the components of the vector $\mathbf{x}$ that satisfies the equation $2\mathbf{u} - \mathbf{v} + \mathbf{x} = 7\mathbf{x} + \mathbf{w}$.

22. For what value(s) of t, if any, is the given vector parallel to $\mathbf{u} = (4, -1)$?

(a) $(8t, -2)$ (b) $(8t, 2t)$ (c) $(1, t^2)$

23. Which of the following vectors in R^6 are parallel to $\mathbf{u} = (-2, 1, 0, 3, 5, 1)$?

(a) $(4, 2, 0, 6, 10, 2)$

(b) $(4, -2, 0, -6, -10, -2)$

(c) $(0, 0, 0, 0, 0, 0)$

24. Let $\mathbf{u} = (2, 1, 0, 1, -1)$ and $\mathbf{v} = (-2, 3, 1, 0, 2)$. Find scalars a and b so that $a\mathbf{u} + b\mathbf{v} = (-8, 8, 3, -1, 7)$.

25. Let $\mathbf{u} = (1, -1, 3, 5)$ and $\mathbf{v} = (2, 1, 0, -3)$. Find scalars a and b so that $a\mathbf{u} + b\mathbf{v} = (1, -4, 9, 18)$.

26. Find all scalars c_1, c_2, and c_3 such that

$$c_1(1, 2, 0) + c_2(2, 1, 1) + c_3(0, 3, 1) = (0, 0, 0)$$

27. Find all scalars c_1, c_2, and c_3 such that

$$c_1(1, -1, 0) + c_2(3, 2, 1) + c_3(0, 1, 4) = (-1, 1, 19)$$

28. Find all scalars c_1, c_2, and c_3 such that

$$c_1(-1, 0, 2) + c_2(2, 2, -2) + c_3(1, -2, 1) = (-6, 12, 4)$$

29. Let $\mathbf{u}_1 = (-1, 3, 2, 0)$, $\mathbf{u}_2 = (2, 0, 4, -1)$, $\mathbf{u}_3 = (7, 1, 1, 4)$, and $\mathbf{u}_4 = (6, 3, 1, 2)$. Find scalars c_1, c_2, c_3, and c_4 such that $c_1\mathbf{u}_1 + c_2\mathbf{u}_2 + c_3\mathbf{u}_3 + c_4\mathbf{u}_4 = (0, 5, 6, -3)$.

30. Show that there do not exist scalars c_1, c_2, and c_3 such that
$$c_1(1, 0, 1, 0) + c_2(1, 0, -2, 1) + c_3(2, 0, 1, 2) = (1, -2, 2, 3)$$

31. Show that there do not exist scalars c_1, c_2, and c_3 such that
$$c_1(-2, 9, 6) + c_2(-3, 2, 1) + c_3(1, 7, 5) = (0, 5, 4)$$

32. Consider Figure 3.1.12. Discuss a geometric interpretation of the vector
$$\mathbf{u} = \overrightarrow{OP_1} + \tfrac{1}{2}(\overrightarrow{OP_2} - \overrightarrow{OP_1})$$

33. Let P be the point $(2, 3, -2)$ and Q the point $(7, -4, 1)$.

(a) Find the midpoint of the line segment connecting P and Q.

(b) Find the point on the line segment connecting P and Q that is $\tfrac{3}{4}$ of the way from P to Q.

34. Let P be the point $(1, 3, 7)$. If the point $(4, 0, -6)$ is the midpoint of the line segment connecting P and Q, what is Q?

35. Prove parts (a), (c), and (d) of Theorem 3.1.1.

36. Prove parts (e)–(h) of Theorem 3.1.1.

37. Prove parts (a)–(c) of Theorem 3.1.2.

True-False Exercises

In parts (a)–(k) determine whether the statement is true or false, and justify your answer.

(a) Two equivalent vectors must have the same initial point.

(b) The vectors (a, b) and $(a, b, 0)$ are equivalent.

(c) If k is a scalar and $\mathbf{v}$ is a vector, then $\mathbf{v}$ and $k\mathbf{v}$ are parallel if and only if $k \geq 0$.

(d) The vectors $\mathbf{v} + (\mathbf{u} + \mathbf{w})$ and $(\mathbf{w} + \mathbf{v}) + \mathbf{u}$ are the same.

(e) If $\mathbf{u} + \mathbf{v} = \mathbf{u} + \mathbf{w}$, then $\mathbf{v} = \mathbf{w}$.

(f) If a and b are scalars such that $a\mathbf{u} + b\mathbf{v} = \mathbf{0}$, then $\mathbf{u}$ and $\mathbf{v}$ are parallel vectors.

(g) Collinear vectors with the same length are equal.

(h) If $(a, b, c) + (x, y, z) = (x, y, z)$, then (a, b, c) must be the zero vector.

(i) If k and m are scalars and $\mathbf{u}$ and $\mathbf{v}$ are vectors, then
$$(k + m)(\mathbf{u} + \mathbf{v}) = k\mathbf{u} + m\mathbf{v}$$

(j) If the vectors $\mathbf{v}$ and $\mathbf{w}$ are given, then the vector equation
$$3(2\mathbf{v} - \mathbf{x}) = 5\mathbf{x} - 4\mathbf{w} + \mathbf{v}$$
can be solved for $\mathbf{x}$.

(k) The linear combinations $a_1\mathbf{v}_1 + a_2\mathbf{v}_2$ and $b_1\mathbf{v}_1 + b_2\mathbf{v}_2$ can only be equal if $a_1 = b_1$ and $a_2 = b_2$.

3.2 Norm, Dot Product, and Distance in R^n

In this section we will be concerned with the notions of length and distance as they relate to vectors. We will first discuss these ideas in R^2 and R^3 and then extend them algebraically to R^n.

Norm of a Vector In this text we will denote the length of a vector $\mathbf{v}$ by the symbol $\|\mathbf{v}\|$, which is read as the *norm* of $\mathbf{v}$, the *length* of $\mathbf{v}$, or the *magnitude* of $\mathbf{v}$ (the term "norm" being a common mathematical synonym for length). As suggested in Figure 3.2.1a, it follows from the Theorem of Pythagoras that the norm of a vector (v_1, v_2) in R^2 is

$$\|\mathbf{v}\| = \sqrt{v_1^2 + v_2^2} \tag{1}$$

Similarly, for a vector (v_1, v_2, v_3) in R^3, it follows from Figure 3.2.1b and two applications of the Theorem of Pythagoras that

$$\|\mathbf{v}\|^2 = (OR)^2 + (RP)^2 = (OQ)^2 + (QR)^2 + (RP)^2 = v_1^2 + v_2^2 + v_3^2$$

and hence that

$$\|\mathbf{v}\| = \sqrt{v_1^2 + v_2^2 + v_3^2} \tag{2}$$

Motivated by the pattern of Formulas (1) and (2) we make the following definition.

▲ Figure 3.2.1

> **DEFINITION 1** If $\mathbf{v} = (v_1, v_2, \ldots, v_n)$ is a vector in R^n, then the **norm** of $\mathbf{v}$ (also called the **length** of $\mathbf{v}$ or the **magnitude** of $\mathbf{v}$) is denoted by $\|\mathbf{v}\|$, and is defined by the formula
>
> $$\|\mathbf{v}\| = \sqrt{v_1^2 + v_2^2 + v_3^2 + \cdots + v_n^2} \qquad (3)$$

▶ **EXAMPLE 1 Calculating Norms**

It follows from Formula (2) that the norm of the vector $\mathbf{v} = (-3, 2, 1)$ in R^3 is

$$\|\mathbf{v}\| = \sqrt{(-3)^2 + 2^2 + 1^2} = \sqrt{14}$$

and it follows from Formula (3) that the norm of the vector $\mathbf{v} = (2, -1, 3, -5)$ in R^4 is

$$\|\mathbf{v}\| = \sqrt{2^2 + (-1)^2 + 3^2 + (-5)^2} = \sqrt{39} \quad ◀$$

Our first theorem in this section will generalize to R^n the following three familiar facts about vectors in R^2 and R^3:

- Distances are nonnegative.
- The zero vector is the only vector of length zero.
- Multiplying a vector by a scalar multiplies its length by the absolute value of that scalar.

It is important to recognize that just because these results hold in R^2 and R^3 does not guarantee that they hold in R^n—their validity in R^n must be *proved* using algebraic properties of n-tuples.

> **THEOREM 3.2.1** *If $\mathbf{v}$ is a vector in R^n, and if k is any scalar, then:*
> (a) $\|\mathbf{v}\| \geq 0$
> (b) $\|\mathbf{v}\| = 0$ *if and only if* $\mathbf{v} = \mathbf{0}$
> (c) $\|k\mathbf{v}\| = |k|\|\mathbf{v}\|$

We will prove part (c) and leave (a) and (b) as exercises.

Proof (c)   If $\mathbf{v} = (v_1, v_2, \ldots, v_n)$, then $k\mathbf{v} = (kv_1, kv_2, \ldots, kv_n)$, so

$$\|k\mathbf{v}\| = \sqrt{(kv_1)^2 + (kv_2)^2 + \cdots + (kv_n)^2}$$
$$= \sqrt{(k^2)(v_1^2 + v_2^2 + \cdots + v_n^2)}$$
$$= |k|\sqrt{v_1^2 + v_2^2 + \cdots + v_n^2}$$
$$= |k|\|\mathbf{v}\| \quad ◀$$

Unit Vectors   A vector of norm 1 is called a **unit vector**. Such vectors are useful for specifying a direction when length is not relevant to the problem at hand. You can obtain a unit vector in a desired direction by choosing any *nonzero* vector $\mathbf{v}$ in that direction and multiplying $\mathbf{v}$ by the reciprocal of its length. For example, if $\mathbf{v}$ is a vector of length 2 in R^2 or R^3, then $\frac{1}{2}\mathbf{v}$ is a unit vector in the same direction as $\mathbf{v}$. More generally, if $\mathbf{v}$ is any nonzero vector in R^n, then

$$\mathbf{u} = \frac{1}{\|\mathbf{v}\|}\mathbf{v} \qquad (4)$$

defines a unit vector that is in the same direction as **v**. We can confirm that (4) is a unit vector by applying part (*c*) of Theorem 3.2.1 with $k = 1/\|\mathbf{v}\|$ to obtain

$$\|\mathbf{u}\| = \|k\mathbf{v}\| = |k|\|\mathbf{v}\| = k\|\mathbf{v}\| = \frac{1}{\|\mathbf{v}\|}\|\mathbf{v}\| = 1$$

The process of multiplying a nonzero vector by the reciprocal of its length to obtain a unit vector is called **normalizing v**.

▶ **EXAMPLE 2 Normalizing a Vector**

Find the unit vector **u** that has the same direction as $\mathbf{v} = (2, 2, -1)$.

Solution The vector **v** has length

$$\|\mathbf{v}\| = \sqrt{2^2 + 2^2 + (-1)^2} = 3$$

Thus, from (4)

$$\mathbf{u} = \tfrac{1}{3}(2, 2, -1) = \left(\tfrac{2}{3}, \tfrac{2}{3}, -\tfrac{1}{3}\right)$$

As a check, you may want to confirm that $\|\mathbf{u}\| = 1$. ◀

The Standard Unit Vectors

When a rectangular coordinate system is introduced in R^2 or R^3, the unit vectors in the positive directions of the coordinate axes are called the **standard unit vectors**. In R^2 these vectors are denoted by

$$\mathbf{i} = (1, 0) \quad \text{and} \quad \mathbf{j} = (0, 1)$$

and in R^3 by

$$\mathbf{i} = (1, 0, 0), \quad \mathbf{j} = (0, 1, 0), \quad \text{and} \quad \mathbf{k} = (0, 0, 1)$$

(Figure 3.2.2). Every vector $\mathbf{v} = (v_1, v_2)$ in R^2 and every vector $\mathbf{v} = (v_1, v_2, v_3)$ in R^3 can be expressed as a linear combination of standard unit vectors by writing

$$\mathbf{v} = (v_1, v_2) = v_1(1, 0) + v_2(0, 1) = v_1\mathbf{i} + v_2\mathbf{j} \tag{5}$$

$$\mathbf{v} = (v_1, v_2, v_3) = v_1(1, 0, 0) + v_2(0, 1, 0) + v_3(0, 0, 1) = v_1\mathbf{i} + v_2\mathbf{j} + v_3\mathbf{k} \tag{6}$$

Moreover, we can generalize these formulas to R^n by defining the **standard unit vectors in R^n** to be

$$\mathbf{e}_1 = (1, 0, 0, \ldots, 0), \quad \mathbf{e}_2 = (0, 1, 0, \ldots, 0), \ldots, \quad \mathbf{e}_n = (0, 0, 0, \ldots, 1) \tag{7}$$

in which case every vector $\mathbf{v} = (v_1, v_2, \ldots, v_n)$ in R^n can be expressed as

$$\mathbf{v} = (v_1, v_2, \ldots, v_n) = v_1\mathbf{e}_1 + v_2\mathbf{e}_2 + \cdots + v_n\mathbf{e}_n \tag{8}$$

(a)

(b)

▲ Figure 3.2.2

▶ **EXAMPLE 3 Linear Combinations of Standard Unit Vectors**

$$(2, -3, 4) = 2\mathbf{i} - 3\mathbf{j} + 4\mathbf{k}$$
$$(7, 3, -4, 5) = 7\mathbf{e}_1 + 3\mathbf{e}_2 - 4\mathbf{e}_3 + 5\mathbf{e}_4 \quad ◀$$

Distance in R^n

If P_1 and P_2 are points in R^2 or R^3, then the length of the vector $\overrightarrow{P_1 P_2}$ is equal to the distance d between the two points (Figure 3.2.3). Specifically, if $P_1(x_1, y_1)$ and $P_2(x_2, y_2)$ are points in R^2, then Formula (4) of Section 3.1 implies that

$$d = \|\overrightarrow{P_1 P_2}\| = \sqrt{(x_2 - x_1)^2 + (y_2 - y_1)^2} \tag{9}$$

$d = \|\overrightarrow{P_1 P_2}\|$

▲ Figure 3.2.3

We noted in the previous section that n-tuples can be viewed either as vectors or points in R^n. In Definition 2 we chose to describe them as points, as that seemed the more natural interpretation.

This is the familiar distance formula from analytic geometry. Similarly, the distance between the points $P_1(x_1, y_1, z_1)$ and $P_2(x_2, y_2, z_2)$ in 3-space is

$$d(\mathbf{u}, \mathbf{v}) = \|\overrightarrow{P_1 P_2}\| = \sqrt{(x_2 - x_1)^2 + (y_2 - y_1)^2 + (z_2 - z_1)^2} \qquad (10)$$

Motivated by Formulas (9) and (10), we make the following definition.

DEFINITION 2 If $\mathbf{u} = (u_1, u_2, \ldots, u_n)$ and $\mathbf{v} = (v_1, v_2, \ldots, v_n)$ are points in R^n, then we denote the ***distance*** between $\mathbf{u}$ and $\mathbf{v}$ by $d(\mathbf{u}, \mathbf{v})$ and define it to be

$$d(\mathbf{u}, \mathbf{v}) = \|\mathbf{u} - \mathbf{v}\| = \sqrt{(u_1 - v_1)^2 + (u_2 - v_2)^2 + \cdots + (u_n - v_n)^2} \qquad (11)$$

▶ **EXAMPLE 4 Calculating Distance in R^n**

If
$$\mathbf{u} = (1, 3, -2, 7) \quad \text{and} \quad \mathbf{v} = (0, 7, 2, 2)$$
then the distance between $\mathbf{u}$ and $\mathbf{v}$ is
$$d(\mathbf{u}, \mathbf{v}) = \sqrt{(1 - 0)^2 + (3 - 7)^2 + (-2 - 2)^2 + (7 - 2)^2} = \sqrt{58} \ \blacktriangleleft$$

Dot Product Our next objective is to define a useful multiplication operation on vectors in R^2 and R^3 and then extend that operation to R^n. To do this we will first need to define exactly what we mean by the "angle" between two vectors in R^2 or R^3. For this purpose, let $\mathbf{u}$ and $\mathbf{v}$ be nonzero vectors in R^2 or R^3 that have been positioned so that their initial points coincide. We define the ***angle between $\mathbf{u}$ and $\mathbf{v}$*** to be the angle θ determined by $\mathbf{u}$ and $\mathbf{v}$ that satisfies the inequalities $0 \le \theta \le \pi$ (Figure 3.2.4).

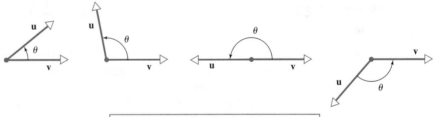

The angle θ between $\mathbf{u}$ and $\mathbf{v}$ satisfies $0 \le \theta \le \pi$.

▶ Figure 3.2.4

DEFINITION 3 If $\mathbf{u}$ and $\mathbf{v}$ are nonzero vectors in R^2 or R^3, and if θ is the angle between $\mathbf{u}$ and $\mathbf{v}$, then the ***dot product*** (also called the ***Euclidean inner product***) of $\mathbf{u}$ and $\mathbf{v}$ is denoted by $\mathbf{u} \cdot \mathbf{v}$ and is defined as

$$\mathbf{u} \cdot \mathbf{v} = \|\mathbf{u}\| \|\mathbf{v}\| \cos \theta \qquad (12)$$

If $\mathbf{u} = \mathbf{0}$ or $\mathbf{v} = \mathbf{0}$, then we define $\mathbf{u} \cdot \mathbf{v}$ to be 0.

The sign of the dot product reveals information about the angle θ that we can obtain by rewriting Formula (12) as

$$\cos \theta = \frac{\mathbf{u} \cdot \mathbf{v}}{\|\mathbf{u}\| \|\mathbf{v}\|} \qquad (13)$$

Since $0 \le \theta \le \pi$, it follows from Formula (13) and properties of the cosine function studied in trigonometry that

- θ is acute if $\mathbf{u} \cdot \mathbf{v} > 0$.
- θ is obtuse if $\mathbf{u} \cdot \mathbf{v} < 0$.
- $\theta = \pi/2$ if $\mathbf{u} \cdot \mathbf{v} = 0$.

▲ Figure 3.2.5

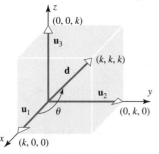

▲ Figure 3.2.6

Note that the angle θ obtained in Example 6 does not involve k. Why was this to be expected?

Component Form of the Dot Product

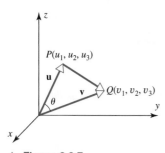

▲ Figure 3.2.7

▶ **EXAMPLE 5** **Dot Product**

Find the dot product of the vectors shown in Figure 3.2.5.

Solution The lengths of the vectors are

$$\|\mathbf{u}\| = 1 \quad \text{and} \quad \|\mathbf{v}\| = \sqrt{8} = 2\sqrt{2}$$

and the cosine of the angle θ between them is

$$\cos(45°) = 1/\sqrt{2}$$

Thus, it follows from Formula (12) that

$$\mathbf{u} \cdot \mathbf{v} = \|\mathbf{u}\|\|\mathbf{v}\| \cos\theta = (1)(2\sqrt{2})(1/\sqrt{2}) = 2$$

▶ **EXAMPLE 6** **A Geometry Problem Solved Using Dot Product**

Find the angle between a diagonal of a cube and one of its edges.

Solution Let k be the length of an edge and introduce a coordinate system as shown in Figure 3.2.6. If we let $\mathbf{u}_1 = (k, 0, 0)$, $\mathbf{u}_2 = (0, k, 0)$, and $\mathbf{u}_3 = (0, 0, k)$, then the vector

$$\mathbf{d} = (k, k, k) = \mathbf{u}_1 + \mathbf{u}_2 + \mathbf{u}_3$$

is a diagonal of the cube. It follows from Formula (13) that the angle θ between $\mathbf{d}$ and the edge $\mathbf{u}_1$ satisfies

$$\cos\theta = \frac{\mathbf{u}_1 \cdot \mathbf{d}}{\|\mathbf{u}_1\|\|\mathbf{d}\|} = \frac{k^2}{(k)(\sqrt{3k^2})} = \frac{1}{\sqrt{3}}$$

With the help of a calculator we obtain

$$\theta = \cos^{-1}\left(\frac{1}{\sqrt{3}}\right) \approx 54.74° \quad \blacktriangleleft$$

For computational purposes it is desirable to have a formula that expresses the dot product of two vectors in terms of components. We will derive such a formula for vectors in 3-space; the derivation for vectors in 2-space is similar.

Let $\mathbf{u} = (u_1, u_2, u_3)$ and $\mathbf{v} = (v_1, v_2, v_3)$ be two nonzero vectors. If, as shown in Figure 3.2.7, θ is the angle between $\mathbf{u}$ and $\mathbf{v}$, then the law of cosines yields

$$\|\overrightarrow{PQ}\|^2 = \|\mathbf{u}\|^2 + \|\mathbf{v}\|^2 - 2\|\mathbf{u}\|\|\mathbf{v}\| \cos\theta \tag{14}$$

Josiah Willard Gibbs
(1839–1903)

Historical Note The dot product notation was first introduced by the American physicist and mathematician J. Willard Gibbs in a pamphlet distributed to his students at Yale University in the 1880s. The product was originally written on the baseline, rather than centered as today, and was referred to as the *direct product*. Gibbs's pamphlet was eventually incorporated into a book entitled *Vector Analysis* that was published in 1901 and coauthored with one of his students. Gibbs made major contributions to the fields of thermodynamics and electromagnetic theory and is generally regarded as the greatest American physicist of the nineteenth century.
[*Image: The Granger Collection, New York*]

Since $\overrightarrow{PQ} = \mathbf{v} - \mathbf{u}$, we can rewrite (14) as

$$\|\mathbf{u}\|\,\|\mathbf{v}\|\cos\theta = \tfrac{1}{2}(\|\mathbf{u}\|^2 + \|\mathbf{v}\|^2 - \|\mathbf{v} - \mathbf{u}\|^2)$$

or

$$\mathbf{u} \cdot \mathbf{v} = \tfrac{1}{2}(\|\mathbf{u}\|^2 + \|\mathbf{v}\|^2 - \|\mathbf{v} - \mathbf{u}\|^2)$$

Substituting

$$\|\mathbf{u}\|^2 = u_1^2 + u_2^2 + u_3^2, \qquad \|\mathbf{v}\|^2 = v_1^2 + v_2^2 + v_3^2$$

and

$$\|\mathbf{v} - \mathbf{u}\|^2 = (v_1 - u_1)^2 + (v_2 - u_2)^2 + (v_3 - u_3)^2$$

we obtain, after simplifying,

$$\mathbf{u} \cdot \mathbf{v} = u_1 v_1 + u_2 v_2 + u_3 v_3 \tag{15}$$

| Although we derived Formula (15) and its 2-space companion under the assumption that $\mathbf{u}$ and $\mathbf{v}$ are nonzero, it turned out that these formulas are also applicable if $\mathbf{u} = \mathbf{0}$ or $\mathbf{v} = \mathbf{0}$ (verify).

The companion formula for vectors in 2-space is

$$\mathbf{u} \cdot \mathbf{v} = u_1 v_1 + u_2 v_2 \tag{16}$$

Motivated by the pattern in Formulas (15) and (16), we make the following definition.

| In words, *to calculate the dot product (Euclidean inner product) multiply corresponding components and add the resulting products.*

DEFINITION 4 If $\mathbf{u} = (u_1, u_2, \ldots, u_n)$ and $\mathbf{v} = (v_1, v_2, \ldots, v_n)$ are vectors in R^n, then the **dot product** (also called the **Euclidean inner product**) of $\mathbf{u}$ and $\mathbf{v}$ is denoted by $\mathbf{u} \cdot \mathbf{v}$ and is defined by

$$\mathbf{u} \cdot \mathbf{v} = u_1 v_1 + u_2 v_2 + \cdots + u_n v_n \tag{17}$$

▶ **EXAMPLE 7 Calculating Dot Products Using Components**

(a)   Use Formula (15) to compute the dot product of the vectors $\mathbf{u}$ and $\mathbf{v}$ in Example 5.

(b)   Calculate $\mathbf{u} \cdot \mathbf{v}$ for the following vectors in R^4:

$$\mathbf{u} = (-1, 3, 5, 7), \quad \mathbf{v} = (-3, -4, 1, 0)$$

Solution (a) The component forms of the vectors are $\mathbf{u} = (0, 0, 1)$ and $\mathbf{v} = (0, 2, 2)$. Thus,

$$\mathbf{u} \cdot \mathbf{v} = (0)(0) + (0)(2) + (1)(2) = 2$$

which agrees with the result obtained geometrically in Example 5.

Solution (b)

$$\mathbf{u} \cdot \mathbf{v} = (-1)(-3) + (3)(-4) + (5)(1) + (7)(0) = -4 \;\blacktriangleleft$$

Algebraic Properties of the Dot Product

In the special case where $\mathbf{u} = \mathbf{v}$ in Definition 4, we obtain the relationship

$$\mathbf{v} \cdot \mathbf{v} = v_1^2 + v_2^2 + \cdots + v_n^2 = \|\mathbf{v}\|^2 \tag{18}$$

This yields the following formula for expressing the length of a vector in terms of a dot product:

$$\|\mathbf{v}\| = \sqrt{\mathbf{v} \cdot \mathbf{v}} \tag{19}$$

Dot products have many of the same algebraic properties as products of real numbers.

THEOREM 3.2.2 *If* $\mathbf{u}$, $\mathbf{v}$, *and* $\mathbf{w}$ *are vectors in* R^n, *and if* k *is a scalar, then*:

(*a*) $\mathbf{u} \cdot \mathbf{v} = \mathbf{v} \cdot \mathbf{u}$ [Symmetry property]

(*b*) $\mathbf{u} \cdot (\mathbf{v} + \mathbf{w}) = \mathbf{u} \cdot \mathbf{v} + \mathbf{u} \cdot \mathbf{w}$ [Distributive property]

(*c*) $k(\mathbf{u} \cdot \mathbf{v}) = (k\mathbf{u}) \cdot \mathbf{v}$ [Homogeneity property]

(*d*) $\mathbf{v} \cdot \mathbf{v} \geq 0$ *and* $\mathbf{v} \cdot \mathbf{v} = 0$ *if and only if* $\mathbf{v} = \mathbf{0}$ [Positivity property]

We will prove parts (*c*) and (*d*) and leave the other proofs as exercises.

Proof (c)   Let $\mathbf{u} = (u_1, u_2, \ldots, u_n)$ and $\mathbf{v} = (v_1, v_2, \ldots, v_n)$. Then

$$k(\mathbf{u} \cdot \mathbf{v}) = k(u_1 v_1 + u_2 v_2 + \cdots + u_n v_n)$$
$$= (k u_1) v_1 + (k u_2) v_2 + \cdots + (k u_n) v_n = (k\mathbf{u}) \cdot \mathbf{v}$$

Proof (d)   The result follows from parts (*a*) and (*b*) of Theorem 3.2.1 and the fact that

$$\mathbf{v} \cdot \mathbf{v} = v_1 v_1 + v_2 v_2 + \cdots + v_n v_n = v_1^2 + v_2^2 + \cdots + v_n^2 = \|\mathbf{v}\|^2 \quad \blacktriangleleft$$

The next theorem gives additional properties of dot products. The proofs can be obtained either by expressing the vectors in terms of components or by using the algebraic properties established in Theorem 3.2.2.

THEOREM 3.2.3 *If* $\mathbf{u}$, $\mathbf{v}$, *and* $\mathbf{w}$ *are vectors in* R^n, *and if* k *is a scalar, then*:

(*a*) $\mathbf{0} \cdot \mathbf{v} = \mathbf{v} \cdot \mathbf{0} = 0$

(*b*) $(\mathbf{u} + \mathbf{v}) \cdot \mathbf{w} = \mathbf{u} \cdot \mathbf{w} + \mathbf{v} \cdot \mathbf{w}$

(*c*) $\mathbf{u} \cdot (\mathbf{v} - \mathbf{w}) = \mathbf{u} \cdot \mathbf{v} - \mathbf{u} \cdot \mathbf{w}$

(*d*) $(\mathbf{u} - \mathbf{v}) \cdot \mathbf{w} = \mathbf{u} \cdot \mathbf{w} - \mathbf{v} \cdot \mathbf{w}$

(*e*) $k(\mathbf{u} \cdot \mathbf{v}) = \mathbf{u} \cdot (k\mathbf{v})$

We will show how Theorem 3.2.2 can be used to prove part (*b*) without breaking the vectors into components. The other proofs are left as exercises.

Proof (b)

$$(\mathbf{u} + \mathbf{v}) \cdot \mathbf{w} = \mathbf{w} \cdot (\mathbf{u} + \mathbf{v}) \quad \text{[By symmetry]}$$
$$= \mathbf{w} \cdot \mathbf{u} + \mathbf{w} \cdot \mathbf{v} \quad \text{[By distributivity]}$$
$$= \mathbf{u} \cdot \mathbf{w} + \mathbf{v} \cdot \mathbf{w} \quad \text{[By symmetry]} \quad \blacktriangleleft$$

Formulas (18) and (19) together with Theorems 3.2.2 and 3.2.3 make it possible to manipulate expressions involving dot products using familiar algebraic techniques.

▶ **EXAMPLE 8 Calculating with Dot Products**

$$(\mathbf{u} - 2\mathbf{v}) \cdot (3\mathbf{u} + 4\mathbf{v}) = \mathbf{u} \cdot (3\mathbf{u} + 4\mathbf{v}) - 2\mathbf{v} \cdot (3\mathbf{u} + 4\mathbf{v})$$
$$= 3(\mathbf{u} \cdot \mathbf{u}) + 4(\mathbf{u} \cdot \mathbf{v}) - 6(\mathbf{v} \cdot \mathbf{u}) - 8(\mathbf{v} \cdot \mathbf{v})$$
$$= 3\|\mathbf{u}\|^2 - 2(\mathbf{u} \cdot \mathbf{v}) - 8\|\mathbf{v}\|^2 \quad \blacktriangleleft$$

Cauchy–Schwarz Inequality and Angles in R^n

Our next objective is to extend to R^n the notion of "angle" between nonzero vectors $\mathbf{u}$ and $\mathbf{v}$. We will do this by starting with the formula

$$\theta = \cos^{-1}\left(\frac{\mathbf{u} \cdot \mathbf{v}}{\|\mathbf{u}\|\|\mathbf{v}\|}\right) \tag{20}$$

which we previously derived for nonzero vectors in R^2 and R^3. Since dot products and norms have been defined for vectors in R^n, it would seem that this formula has all the ingredients to serve as a *definition* of the angle θ between two vectors, $\mathbf{u}$ and $\mathbf{v}$, in R^n. However, there is a fly in the ointment, the problem being that the inverse cosine in Formula (20) is not defined unless its argument satisfies the inequalities

$$-1 \le \frac{\mathbf{u} \cdot \mathbf{v}}{\|\mathbf{u}\|\|\mathbf{v}\|} \le 1 \tag{21}$$

Fortunately, these inequalities *do* hold for all nonzero vectors in R^n as a result of the following fundamental result known as the ***Cauchy–Schwarz inequality***.

THEOREM 3.2.4 Cauchy–Schwarz Inequality

If $\mathbf{u} = (u_1, u_2, \ldots, u_n)$ *and* $\mathbf{v} = (v_1, v_2, \ldots, v_n)$ *are vectors in* R^n, *then*

$$|\mathbf{u} \cdot \mathbf{v}| \le \|\mathbf{u}\|\|\mathbf{v}\| \tag{22}$$

or in terms of components

$$|u_1v_1 + u_2v_2 + \cdots + u_nv_n| \le (u_1^2 + u_2^2 + \cdots + u_n^2)^{1/2}(v_1^2 + v_2^2 + \cdots + v_n^2)^{1/2} \tag{23}$$

We will omit the proof of this theorem because later in the text we will prove a more general version of which this will be a special case. Our goal for now will be to use this theorem to prove that the inequalities in (21) hold for all nonzero vectors in R^n. Once that is done we will have established all the results required to use Formula (20) as our *definition* of the angle between nonzero vectors $\mathbf{u}$ and $\mathbf{v}$ in R^n.

To prove that the inequalities in (21) hold for all nonzero vectors in R^n, divide both sides of Formula (22) by the product $\|\mathbf{u}\|\|\mathbf{v}\|$ to obtain

$$\frac{|\mathbf{u} \cdot \mathbf{v}|}{\|\mathbf{u}\|\|\mathbf{v}\|} \le 1 \quad \text{or equivalently} \quad \left|\frac{\mathbf{u} \cdot \mathbf{v}}{\|\mathbf{u}\|\|\mathbf{v}\|}\right| \le 1$$

from which (21) follows.

Hermann Amandus Schwarz (1843–1921)

Viktor Yakovlevich Bunyakovsky (1804–1889)

Geometry in R^n

$$\|\mathbf{u} + \mathbf{v}\| \leq \|\mathbf{u}\| + \|\mathbf{v}\|$$

▲ Figure 3.2.8

Earlier in this section we extended various concepts to R^n with the idea that familiar results that we can visualize in R^2 and R^3 might be valid in R^n as well. Here are two fundamental theorems from plane geometry whose validity extends to R^n:

- The sum of the lengths of two side of a triangle is at least as large as the third (Figure 3.2.8).

- The shortest distance between two points is a straight line (Figure 3.2.9).

The following theorem generalizes these theorems to R^n.

THEOREM 3.2.5 *If* $\mathbf{u}$, $\mathbf{v}$, *and* $\mathbf{w}$ *are vectors in* R^n, *and if* k *is any scalar, then*:

(*a*) $\|\mathbf{u} + \mathbf{v}\| \leq \|\mathbf{u}\| + \|\mathbf{v}\|$ **[Triangle inequality for vectors]**

(*b*) $d(\mathbf{u}, \mathbf{v}) \leq d(\mathbf{u}, \mathbf{w}) + d(\mathbf{w}, \mathbf{v})$ **[Triangle inequality for distances]**

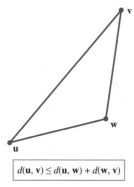

$$d(\mathbf{u}, \mathbf{v}) \leq d(\mathbf{u}, \mathbf{w}) + d(\mathbf{w}, \mathbf{v})$$

▲ Figure 3.2.9

Proof (a)

$$
\begin{aligned}
\|\mathbf{u} + \mathbf{v}\|^2 &= (\mathbf{u} + \mathbf{v}) \cdot (\mathbf{u} + \mathbf{v}) = (\mathbf{u} \cdot \mathbf{u}) + 2(\mathbf{u} \cdot \mathbf{v}) + (\mathbf{v} \cdot \mathbf{v}) \\
&= \|\mathbf{u}\|^2 + 2(\mathbf{u} \cdot \mathbf{v}) + \|\mathbf{v}\|^2 \\
&\leq \|\mathbf{u}\|^2 + 2|\mathbf{u} \cdot \mathbf{v}| + \|\mathbf{v}\|^2 \qquad \longleftarrow \text{Property of absolute value} \\
&\leq \|\mathbf{u}\|^2 + 2\|\mathbf{u}\|\|\mathbf{v}\| + \|\mathbf{v}\|^2 \qquad \longleftarrow \text{Cauchy–Schwarz inequality} \\
&= (\|\mathbf{u}\| + \|\mathbf{v}\|)^2
\end{aligned}
$$

Proof (b)   It follows from part (*a*) and Formula (11) that

$$
\begin{aligned}
d(\mathbf{u}, \mathbf{v}) = \|\mathbf{u} - \mathbf{v}\| &= \|(\mathbf{u} - \mathbf{w}) + (\mathbf{w} - \mathbf{v})\| \\
&\leq \|\mathbf{u} - \mathbf{w}\| + \|\mathbf{w} - \mathbf{v}\| = d(\mathbf{u}, \mathbf{w}) + d(\mathbf{w}, \mathbf{v}) \quad \blacktriangleleft
\end{aligned}
$$

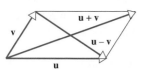

▲ Figure 3.2.10

It is proved in plane geometry that for any parallelogram the sum of the squares of the diagonals is equal to the sum of the squares of the four sides (Figure 3.2.10). The following theorem generalizes that result to R^n.

THEOREM 3.2.6 Parallelogram Equation for Vectors

If $\mathbf{u}$ *and* $\mathbf{v}$ *are vectors in* R^n, *then*

$$\|\mathbf{u} + \mathbf{v}\|^2 + \|\mathbf{u} - \mathbf{v}\|^2 = 2\left(\|\mathbf{u}\|^2 + \|\mathbf{v}\|^2\right) \tag{24}$$

Proof

$$
\begin{aligned}
\|\mathbf{u} + \mathbf{v}\|^2 + \|\mathbf{u} - \mathbf{v}\|^2 &= (\mathbf{u} + \mathbf{v}) \cdot (\mathbf{u} + \mathbf{v}) + (\mathbf{u} - \mathbf{v}) \cdot (\mathbf{u} - \mathbf{v}) \\
&= 2(\mathbf{u} \cdot \mathbf{u}) + 2(\mathbf{v} \cdot \mathbf{v}) \\
&= 2\left(\|\mathbf{u}\|^2 + \|\mathbf{v}\|^2\right) \quad \blacktriangleleft
\end{aligned}
$$

We could state and prove many more theorems from plane geometry that generalize to R^n, but the ones already given should suffice to convince you that R^n is not so different from R^2 and R^3 even though we cannot visualize it directly. The next theorem establishes a fundamental relationship between the dot product and norm in R^n.

> **THEOREM 3.2.7** *If* **u** *and* **v** *are vectors in* R^n *with the Euclidean inner product, then*
> $$\mathbf{u} \cdot \mathbf{v} = \tfrac{1}{4}\|\mathbf{u} + \mathbf{v}\|^2 - \tfrac{1}{4}\|\mathbf{u} - \mathbf{v}\|^2 \tag{25}$$

Note that Formula (25) expresses the dot product in terms of norms.

Proof
$$\|\mathbf{u} + \mathbf{v}\|^2 = (\mathbf{u} + \mathbf{v}) \cdot (\mathbf{u} + \mathbf{v}) = \|\mathbf{u}\|^2 + 2(\mathbf{u} \cdot \mathbf{v}) + \|\mathbf{v}\|^2$$
$$\|\mathbf{u} - \mathbf{v}\|^2 = (\mathbf{u} - \mathbf{v}) \cdot (\mathbf{u} - \mathbf{v}) = \|\mathbf{u}\|^2 - 2(\mathbf{u} \cdot \mathbf{v}) + \|\mathbf{v}\|^2$$

from which (25) follows by simple algebra. ◀

Dot Products as Matrix Multiplication

There are various ways to express the dot product of vectors using matrix notation. The formulas depend on whether the vectors are expressed as row matrices or column matrices. Here are the possibilities.

Table 1

Form	Dot Product	Example
u a column matrix and **v** a column matrix	$\mathbf{u} \cdot \mathbf{v} = \mathbf{u}^T\mathbf{v} = \mathbf{v}^T\mathbf{u}$	$\mathbf{u} = \begin{bmatrix} 1 \\ -3 \\ 5 \end{bmatrix}$ $\quad$ $\mathbf{v} = \begin{bmatrix} 5 \\ 4 \\ 0 \end{bmatrix}$	$\mathbf{u}^T\mathbf{v} = \begin{bmatrix} 1 & -3 & 5 \end{bmatrix}\begin{bmatrix} 5 \\ 4 \\ 0 \end{bmatrix} = -7$ $\quad$ $\mathbf{v}^T\mathbf{u} = \begin{bmatrix} 5 & 4 & 0 \end{bmatrix}\begin{bmatrix} 1 \\ -3 \\ 5 \end{bmatrix} = -7$
u a row matrix and **v** a column matrix	$\mathbf{u} \cdot \mathbf{v} = \mathbf{u}\mathbf{v} = \mathbf{v}^T\mathbf{u}^T$	$\mathbf{u} = \begin{bmatrix} 1 & -3 & 5 \end{bmatrix}$ $\quad$ $\mathbf{v} = \begin{bmatrix} 5 \\ 4 \\ 0 \end{bmatrix}$	$\mathbf{u}\mathbf{v} = \begin{bmatrix} 1 & -3 & 5 \end{bmatrix}\begin{bmatrix} 5 \\ 4 \\ 0 \end{bmatrix} = -7$ $\quad$ $\mathbf{v}^T\mathbf{u}^T = \begin{bmatrix} 5 & 4 & 0 \end{bmatrix}\begin{bmatrix} 1 \\ -3 \\ 5 \end{bmatrix} = -7$
u a column matrix and **v** a row matrix	$\mathbf{u} \cdot \mathbf{v} = \mathbf{v}\mathbf{u} = \mathbf{u}^T\mathbf{v}^T$	$\mathbf{u} = \begin{bmatrix} 1 \\ -3 \\ 5 \end{bmatrix}$ $\quad$ $\mathbf{v} = \begin{bmatrix} 5 & 4 & 0 \end{bmatrix}$	$\mathbf{v}\mathbf{u} = \begin{bmatrix} 5 & 4 & 0 \end{bmatrix}\begin{bmatrix} 1 \\ -3 \\ 5 \end{bmatrix} = -7$ $\quad$ $\mathbf{u}^T\mathbf{v}^T = \begin{bmatrix} 1 & -3 & 5 \end{bmatrix}\begin{bmatrix} 5 \\ 4 \\ 0 \end{bmatrix} = -7$
u a row matrix and **v** a row matrix	$\mathbf{u} \cdot \mathbf{v} = \mathbf{u}\mathbf{v}^T = \mathbf{v}\mathbf{u}^T$	$\mathbf{u} = \begin{bmatrix} 1 & -3 & 5 \end{bmatrix}$ $\quad$ $\mathbf{v} = \begin{bmatrix} 5 & 4 & 0 \end{bmatrix}$	$\mathbf{u}\mathbf{v}^T = \begin{bmatrix} 1 & -3 & 5 \end{bmatrix}\begin{bmatrix} 5 \\ 4 \\ 0 \end{bmatrix} = -7$ $\quad$ $\mathbf{v}\mathbf{u}^T = \begin{bmatrix} 5 & 4 & 0 \end{bmatrix}\begin{bmatrix} 1 \\ -3 \\ 5 \end{bmatrix} = -7$

If A is an $n \times n$ matrix and **u** and **v** are $n \times 1$ matrices, then it follows from the first row in Table 1 and properties of the transpose that
$$A\mathbf{u} \cdot \mathbf{v} = \mathbf{v}^T(A\mathbf{u}) = (\mathbf{v}^T A)\mathbf{u} = (A^T\mathbf{v})^T\mathbf{u} = \mathbf{u} \cdot A^T\mathbf{v}$$
$$\mathbf{u} \cdot A\mathbf{v} = (A\mathbf{v})^T\mathbf{u} = (\mathbf{v}^T A^T)\mathbf{u} = \mathbf{v}^T(A^T\mathbf{u}) = A^T\mathbf{u} \cdot \mathbf{v}$$

The resulting formulas

$$Au \cdot v = u \cdot A^T v \tag{26}$$

$$u \cdot Av = A^T u \cdot v \tag{27}$$

provide an important link between multiplication by an $n \times n$ matrix A and multiplication by A^T.

▶ **EXAMPLE 9 Verifying That $Au \cdot v = u \cdot A^T v$**

Suppose that

$$A = \begin{bmatrix} 1 & -2 & 3 \\ 2 & 4 & 1 \\ -1 & 0 & 1 \end{bmatrix}, \quad u = \begin{bmatrix} -1 \\ 2 \\ 4 \end{bmatrix}, \quad v = \begin{bmatrix} -2 \\ 0 \\ 5 \end{bmatrix}$$

Then

$$Au = \begin{bmatrix} 1 & -2 & 3 \\ 2 & 4 & 1 \\ -1 & 0 & 1 \end{bmatrix} \begin{bmatrix} -1 \\ 2 \\ 4 \end{bmatrix} = \begin{bmatrix} 7 \\ 10 \\ 5 \end{bmatrix}$$

$$A^T v = \begin{bmatrix} 1 & 2 & -1 \\ -2 & 4 & 0 \\ 3 & 1 & 1 \end{bmatrix} \begin{bmatrix} -2 \\ 0 \\ 5 \end{bmatrix} = \begin{bmatrix} -7 \\ 4 \\ -1 \end{bmatrix}$$

from which we obtain

$$Au \cdot v = 7(-2) + 10(0) + 5(5) = 11$$
$$u \cdot A^T v = (-1)(-7) + 2(4) + 4(-1) = 11$$

Thus, $Au \cdot v = u \cdot A^T v$ as guaranteed by Formula (8). We leave it for you to verify that Formula (27) also holds. ◀

A Dot Product View of Matrix Multiplication

Dot products provide another way of thinking about matrix multiplication. Recall that if $A = [a_{ij}]$ is an $m \times r$ matrix and $B = [b_{ij}]$ is an $r \times n$ matrix, then the ijth entry of AB is

$$a_{i1}b_{1j} + a_{i2}b_{2j} + \cdots + a_{ir}b_{rj}$$

which is the dot product of the ith row vector of A

$$[a_{i1} \quad a_{i2} \quad \cdots \quad a_{ir}]$$

and the jth column vector of B

$$\begin{bmatrix} b_{1j} \\ b_{2j} \\ \vdots \\ b_{rj} \end{bmatrix}$$

Thus, if the row vectors of A are $r_1, r_2, \ldots, r_m$ and the column vectors of B are $c_1, c_2, \ldots, c_n$, then the matrix product AB can be expressed as

$$AB = \begin{bmatrix} r_1 \cdot c_1 & r_1 \cdot c_2 & \cdots & r_1 \cdot c_n \\ r_2 \cdot c_1 & r_2 \cdot c_2 & \cdots & r_2 \cdot c_n \\ \vdots & \vdots & & \vdots \\ r_m \cdot c_1 & r_m \cdot c_2 & \cdots & r_m \cdot c_n \end{bmatrix} \tag{28}$$

Application of Dot Products to ISBN Numbers

Although the system has recently changed, most books published in the last 25 years have been assigned a unique 10-digit number called an *International Standard Book Number* or ISBN. The first nine digits of this number are split into three groups—the first group representing the country or group of countries in which the book originates, the second identifying the publisher, and the third assigned to the book title itself. The tenth and final digit, called a *check digit*, is computed from the first nine digits and is used to ensure that an electronic transmission of the ISBN, say over the Internet, occurs without error.

To explain how this is done, regard the first nine digits of the ISBN as a vector **b** in R^9, and let **a** be the vector

$$\mathbf{a} = (1, 2, 3, 4, 5, 6, 7, 8, 9)$$

Then the check digit c is computed using the following procedure:

1. Form the dot product $\mathbf{a} \cdot \mathbf{b}$.

2. Divide $\mathbf{a} \cdot \mathbf{b}$ by 11, thereby producing a remainder c that is an integer between 0 and 10, inclusive. The check digit is taken to be c, with the proviso that $c = 10$ is written as X to avoid double digits.

For example, the ISBN of the brief edition of *Calculus*, sixth edition, by Howard Anton is

$$0\text{-}471\text{-}15307\text{-}9$$

which has a check digit of 9. This is consistent with the first nine digits of the ISBN, since

$$\mathbf{a} \cdot \mathbf{b} = (1, 2, 3, 4, 5, 6, 7, 8, 9) \cdot (0, 4, 7, 1, 1, 5, 3, 0, 7) = 152$$

Dividing 152 by 11 produces a quotient of 13 and a remainder of 9, so the check digit is $c = 9$. If an electronic order is placed for a book with a certain ISBN, then the warehouse can use the above procedure to verify that the check digit is consistent with the first nine digits, thereby reducing the possibility of a costly shipping error.

Concept Review

- Norm (or length or magnitude) of a vector
- Unit vector
- Normalized vector
- Standard unit vectors
- Distance between points in R^n
- Angle between two vectors in R^n
- Dot product (or Euclidean inner product) of two vectors in R^n
- Cauchy–Schwarz inequality
- Triangle inequality
- Parallelogram equation for vectors

Skills

- Compute the norm of a vector in R^n.
- Determine whether a given vector in R^n is a unit vector.
- Normalize a nonzero vector in R^n.
- Determine the distance between two vectors in R^n.
- Compute the dot product of two vectors in R^n.
- Compute the angle between two nonzero vectors in R^n.
- Prove basic properties pertaining to norms and dot products (Theorems 3.2.1–3.2.3 and 3.2.5–3.2.7).

Exercise Set 3.2

▷ In Exercises 1–2, find the norm of **v**, a unit vector that has the same direction as **v**, and a unit vector that is oppositely directed to **v**. ◁

1. (a) $\mathbf{v} = (4, -3)$ (b) $\mathbf{v} = (2, 2, 2)$
   (c) $\mathbf{v} = (1, 0, 2, 1, 3)$

2. (a) $\mathbf{v} = (-5, 12)$ (b) $\mathbf{v} = (1, -1, 2)$
   (c) $\mathbf{v} = (-2, 3, 3, -1)$

▷ In Exercises 3–4, evaluate the given expression with $\mathbf{u} = (2, -2, 3)$, $\mathbf{v} = (1, -3, 4)$, and $\mathbf{w} = (3, 6, -4)$. ◁

3. (a) $\|\mathbf{u} + \mathbf{v}\|$ (b) $\|\mathbf{u}\| + \|\mathbf{v}\|$
   (c) $\|-2\mathbf{u} + 2\mathbf{v}\|$ (d) $\|3\mathbf{u} - 5\mathbf{v} + \mathbf{w}\|$

4. (a) $\|\mathbf{u} + \mathbf{v} + \mathbf{w}\|$ (b) $\|\mathbf{u} - \mathbf{v}\|$
   (c) $\|3\mathbf{v}\| - 3\|\mathbf{v}\|$ (d) $\|\mathbf{u}\| - \|\mathbf{v}\|$

▷ In Exercises 5–6, evaluate the given expression with $\mathbf{u} = (-2, -1, 4, 5)$, $\mathbf{v} = (3, 1, -5, 7)$, and $\mathbf{w} = (-6, 2, 1, 1)$. ◁

5. (a) $\|3\mathbf{u} - 5\mathbf{v} + \mathbf{w}\|$ (b) $\|3\mathbf{u}\| - 5\|\mathbf{v}\| + \|\mathbf{w}\|$
   (c) $\|-\|\mathbf{u}\|\mathbf{v}\|$

6. (a) $\|\mathbf{u}\| - 2\|\mathbf{v}\| - 3\|\mathbf{w}\|$ (b) $\|\mathbf{u}\| + \|-2\mathbf{v}\| + \|-3\mathbf{w}\|$
   (c) $\|\|\mathbf{u} - \mathbf{v}\|\mathbf{w}\|$

7. Let $\mathbf{v} = (-2, 3, 0, 6)$. Find all scalars k such that $\|k\mathbf{v}\| = 5$.

8. Let $\mathbf{v} = (1, 1, 2, -3, 1)$. Find all scalars k such that $\|k\mathbf{v}\| = 4$.

▶ In Exercises **9–10**, find $\mathbf{u} \cdot \mathbf{v}$, $\mathbf{u} \cdot \mathbf{u}$, and $\mathbf{v} \cdot \mathbf{v}$. ◀

9. (a) $\mathbf{u} = (3, 1, 4)$, $\mathbf{v} = (2, 2, -4)$

  (b) $\mathbf{u} = (1, 1, 4, 6)$, $\mathbf{v} = (2, -2, 3, -2)$

10. (a) $\mathbf{u} = (1, 1, -2, 3)$, $\mathbf{v} = (-1, 0, 5, 1)$

  (b) $\mathbf{u} = (2, -1, 1, 0, -2)$, $\mathbf{v} = (1, 2, 2, 2, 1)$

▶ In Exercises **11–12**, find the Euclidean distance between $\mathbf{u}$ and $\mathbf{v}$. ◀

11. (a) $\mathbf{u} = (3, 3, 3)$, $\mathbf{v} = (1, 0, 4)$

  (b) $\mathbf{u} = (0, -2, -1, 1)$, $\mathbf{v} = (-3, 2, 4, 4)$

  (c) $\mathbf{u} = (3, -3, -2, 0, -3, 13, 5)$,
    $\mathbf{v} = (-4, 1, -1, 5, 0, -11, 4)$

12. (a) $\mathbf{u} = (1, 2, -3, 0)$, $\mathbf{v} = (5, 1, 2, -2)$

  (b) $\mathbf{u} = (2, -1, -4, 1, 0, 6, -3, 1)$,
    $\mathbf{v} = (-2, -1, 0, 3, 7, 2, -5, 1)$

  (c) $\mathbf{u} = (0, 1, 1, 1, 2)$, $\mathbf{v} = (2, 1, 0, -1, 3)$

13. Find the cosine of the angle between the vectors in each part of Exercise 11, and then state whether the angle is acute, obtuse, or 90°.

14. Find the cosine of the angle between the vectors in each part of Exercise 12, and then state whether the angle is acute, obtuse, or 90°.

15. Suppose that a vector $\mathbf{a}$ in the xy-plane has a length of 9 units and points in a direction that is 120° counterclockwise from the positive x-axis, and a vector $\mathbf{b}$ in that plane has a length of 5 units and points in the positive y-direction. Find $\mathbf{a} \cdot \mathbf{b}$.

16. Suppose that a vector $\mathbf{a}$ in the xy-plane points in a direction that is 47° counterclockwise from the positive x-axis, and a vector $\mathbf{b}$ in that plane points in a direction that is 43° clockwise from the positive x-axis. What can you say about the value of $\mathbf{a} \cdot \mathbf{b}$?

▶ In Exercises **17–18**, determine whether the expression makes sense mathematically. If not, explain why. ◀

17. (a) $\mathbf{u} \cdot (\mathbf{v} \cdot \mathbf{w})$ (b) $\mathbf{u} \cdot (\mathbf{v} + \mathbf{w})$

  (c) $\|\mathbf{u} \cdot \mathbf{v}\|$ (d) $(\mathbf{u} \cdot \mathbf{v}) - \|\mathbf{u}\|$

18. (a) $\|\mathbf{u}\| \cdot \|\mathbf{v}\|$ (b) $(\mathbf{u} \cdot \mathbf{v}) - \mathbf{w}$

  (c) $(\mathbf{u} \cdot \mathbf{v}) - k$ (d) $k \cdot \mathbf{u}$

19. Find a unit vector that has the same direction as the given vector.

  (a) $(-4, -3)$ (b) $(1, 7)$

  (c) $(-3, 2, \sqrt{3})$ (d) $(1, 2, 3, 4, 5)$

20. Find a unit vector that is oppositely directed to the given vector.

  (a) $(-12, -5)$ (b) $(3, -3, -3)$

  (c) $(-6, 8)$ (d) $(-3, 1, \sqrt{6}, 3)$

21. State a procedure for finding a vector of a specified length m that points in the same direction as a given vector $\mathbf{v}$.

22. If $\|\mathbf{v}\| = 2$ and $\|\mathbf{w}\| = 3$, what are the largest and smallest values possible for $\|\mathbf{v} - \mathbf{w}\|$? Give a geometric explanation of your results.

23. Find the cosine of the angle θ between $\mathbf{u}$ and $\mathbf{v}$.

  (a) $\mathbf{u} = (2, 3)$, $\mathbf{v} = (5, -7)$

  (b) $\mathbf{u} = (-6, -2)$, $\mathbf{v} = (4, 0)$

  (c) $\mathbf{u} = (1, -5, 4)$, $\mathbf{v} = (3, 3, 3)$

  (d) $\mathbf{u} = (-2, 2, 3)$, $\mathbf{v} = (1, 7, -4)$

24. Find the radian measure of the angle θ (with $0 \le \theta \le \pi$) between $\mathbf{u}$ and $\mathbf{v}$.

  (a) $(1, -7)$ and $(21, 3)$ (b) $(0, 2)$ and $(3, -3)$

  (c) $(-1, 1, 0)$ and $(0, -1, 1)$ (d) $(1, -1, 0)$ and $(1, 0, 0)$

▶ In Exercises **25–26**, verify that the Cauchy–Schwarz inequality holds. ◀

25. (a) $\mathbf{u} = (3, 2)$, $\mathbf{v} = (4, -1)$

  (b) $\mathbf{u} = (-3, 1, 0)$, $\mathbf{v} = (2, -1, 3)$

  (c) $\mathbf{u} = (0, 2, 2, 1)$, $\mathbf{v} = (1, 1, 1, 1)$

26. (a) $\mathbf{u} = (4, 1, 1)$, $\mathbf{v} = (1, 2, 3)$

  (b) $\mathbf{u} = (1, 2, 1, 2, 3)$, $\mathbf{v} = (0, 1, 1, 5, -2)$

  (c) $\mathbf{u} = (1, 3, 5, 2, 0, 1)$, $\mathbf{v} = (0, 2, 4, 1, 3, 5)$

27. Let $\mathbf{p}_0 = (x_0, y_0, z_0)$ and $\mathbf{p} = (x, y, z)$. Describe the set of all points (x, y, z) for which $\|\mathbf{p} - \mathbf{p}_0\| = 1$.

28. (a) Show that the components of the vector $\mathbf{v} = (v_1, v_2)$ in Figure Ex-28a are $v_1 = \|\mathbf{v}\| \cos \theta$ and $v_2 = \|\mathbf{v}\| \sin \theta$.

  (b) Let $\mathbf{u}$ and $\mathbf{v}$ be the vectors in Figure Ex-28b. Use the result in part (a) to find the components of $4\mathbf{u} - 5\mathbf{v}$.

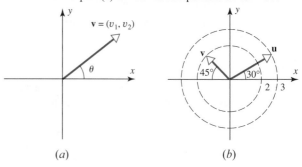

  (a)   (b)

▲ **Figure Ex-28**

29. Prove parts (*a*) and (*b*) of Theorem 3.2.1.

30. Prove parts (*a*) and (*c*) of Theorem 3.2.3.

31. Prove parts (*d*) and (*e*) of Theorem 3.2.3.

32. Under what conditions will the triangle inequality (Theorem 3.2.5*a*) be an equality? Explain your answer geometrically.

33. What can you say about two nonzero vectors, $\mathbf{u}$ and $\mathbf{v}$, that satisfy the equation $\|\mathbf{u} + \mathbf{v}\| = \|\mathbf{u}\| + \|\mathbf{v}\|$?

34. (a) What relationship must hold for the point $\mathbf{p} = (a, b, c)$ to be equidistant from the origin and the xz-plane? Make sure that the relationship you state is valid for positive and negative values of a, b, and c.

   (b) What relationship must hold for the point $\mathbf{p} = (a, b, c)$ to be farther from the origin than from the xz-plane? Make sure that the relationship you state is valid for positive and negative values of a, b, and c

True-False Exercises

In parts (a)–(j) determine whether the statement is true or false, and justify your answer.

(a) If each component of a vector in R^3 is doubled, the norm of that vector is doubled.

(b) In R^2, the vectors of norm 5 whose initial points are at the origin have terminal points lying on a circle of radius 5 centered at the origin.

(c) Every vector in R^n has a positive norm.

(d) If $\mathbf{v}$ is a nonzero vector in R^n, there are exactly two unit vectors that are parallel to $\mathbf{v}$.

(e) If $\|\mathbf{u}\| = 2$, $\|\mathbf{v}\| = 1$, and $\mathbf{u} \cdot \mathbf{v} = 1$, then the angle between $\mathbf{u}$ and $\mathbf{v}$ is $\pi/3$ radians.

(f) The expressions $(\mathbf{u} \cdot \mathbf{v}) + \mathbf{w}$ and $\mathbf{u} \cdot (\mathbf{v} + \mathbf{w})$ are both meaningful and equal to each other.

(g) If $\mathbf{u} \cdot \mathbf{v} = \mathbf{u} \cdot \mathbf{w}$, then $\mathbf{v} = \mathbf{w}$.

(h) If $\mathbf{u} \cdot \mathbf{v} = 0$, then either $\mathbf{u} = \mathbf{0}$ or $\mathbf{v} = \mathbf{0}$.

(i) In R^2, if $\mathbf{u}$ lies in the first quadrant and $\mathbf{v}$ lies in the third quadrant, then $\mathbf{u} \cdot \mathbf{v}$ cannot be positive.

(j) For all vectors $\mathbf{u}$, $\mathbf{v}$, and $\mathbf{w}$ in R^n, we have

$$\|\mathbf{u} + \mathbf{v} + \mathbf{w}\| \leq \|\mathbf{u}\| + \|\mathbf{v}\| + \|\mathbf{w}\|$$

3.3 Orthogonality

In the last section we defined the notion of "angle" between vectors in R^n. In this section we will focus on the notion of "perpendicularity." Perpendicular vectors in R^n play an important role in a wide variety of applications.

Orthogonal Vectors   Recall from Formula (20) in the previous section that the angle θ between two *nonzero* vectors $\mathbf{u}$ and $\mathbf{v}$ in R^n is defned by the formula

$$\theta = \cos^{-1}\left(\frac{\mathbf{u} \cdot \mathbf{v}}{\|\mathbf{u}\|\|\mathbf{v}\|}\right)$$

It follows from this that $\theta = \pi/2$ if and only if $\mathbf{u} \cdot \mathbf{v} = 0$. Thus, we make the following definition.

DEFINITION 1 Two nonzero vectors $\mathbf{u}$ and $\mathbf{v}$ in R^n are said to be **orthogonal** (or **perpendicular**) if $\mathbf{u} \cdot \mathbf{v} = 0$. We will also agree that the zero vector in R^n is orthogonal to *every* vector in R^n. A nonempty set of vectors in R^n is called an **orthogonal set** if all pairs of distinct vectors in the set are orthogonal. An orthogonal set of unit vectors is called an **orthonormal set**.

▶ **EXAMPLE 1 Orthogonal Vectors**

(a) Show that $\mathbf{u} = (-2, 3, 1, 4)$ and $\mathbf{v} = (1, 2, 0, -1)$ are orthogonal vectors in R^4.

(b) Show that the set $S = \{\mathbf{i}, \mathbf{j}, \mathbf{k}\}$ of standard unit vectors is an orthogonal set in R^3.

Solution (a) The vectors are orthogonal since

$$\mathbf{u} \cdot \mathbf{v} = (-2)(1) + (3)(2) + (1)(0) + (4)(-1) = 0$$

In Example 1 there is no need to check that

$$\mathbf{j} \cdot \mathbf{i} = \mathbf{k} \cdot \mathbf{i} = \mathbf{k} \cdot \mathbf{j} = 0$$

since this follows from computations in the example and the symmetry property of the dot product.

Solution (b) We must show that all pairs of distinct vectors are orthogonal, that is,

$$\mathbf{i} \cdot \mathbf{j} = \mathbf{i} \cdot \mathbf{k} = \mathbf{j} \cdot \mathbf{k} = 0$$

This is evident geometrically (Figure 3.2.2), but it can be seen as well from the computations

$$\mathbf{i} \cdot \mathbf{j} = (1, 0, 0) \cdot (0, 1, 0) = 0$$
$$\mathbf{i} \cdot \mathbf{k} = (1, 0, 0) \cdot (0, 0, 1) = 0$$
$$\mathbf{j} \cdot \mathbf{k} = (0, 1, 0) \cdot (0, 0, 1) = 0 \blacktriangleleft$$

Lines and Planes Determined by Points and Normals

One learns in analytic geometry that a line in R^2 is determined uniquely by its slope and one of its points, and that a plane in R^3 is determined uniquely by its "inclination" and one of its points. One way of specifying slope and inclination is to use a *nonzero* vector **n**, called a ***normal***, that is orthogonal to the line or plane in question. For example, Figure 3.3.1 shows the line through the point $P_0(x_0, y_0)$ that has normal $\mathbf{n} = (a, b)$ and the plane through the point $P_0(x_0, y_0, z_0)$ that has normal $\mathbf{n} = (a, b, c)$. Both the line and the plane are represented by the vector equation

$$\mathbf{n} \cdot \overrightarrow{P_0 P} = 0 \qquad (1)$$

where P is either an arbitrary point (x, y) on the line or an arbitrary point (x, y, z) in the plane. The vector $\overrightarrow{P_0 P}$ can be expressed in terms of components as

$$\overrightarrow{P_0 P} = (x - x_0, y - y_0) \qquad \text{[line]}$$
$$\overrightarrow{P_0 P} = (x - x_0, y - y_0, z - z_0) \qquad \text{[plane]}$$

Thus, Equation (1) can be written as

$$a(x - x_0) + b(y - y_0) = 0 \qquad \text{[line]} \qquad (2)$$

$$a(x - x_0) + b(y - y_0) + c(z - z_0) = 0 \qquad \text{[plane]} \qquad (3)$$

These are called the ***point-normal*** equations of the line and plane.

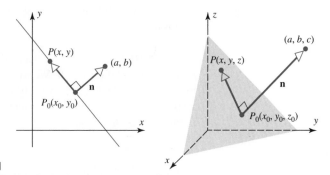

▶ Figure 3.3.1

▶ **EXAMPLE 2 Point-Normal Equations**

It follows from (2) that in R^2 the equation

$$6(x - 3) + (y + 7) = 0$$

represents the line through the point $(3, -7)$ with normal $\mathbf{n} = (6, 1)$; and it follows from (3) that in R^3 the equation

$$4(x - 3) + 2y - 5(z - 7) = 0$$

represents the plane through the point $(3, 0, 7)$ with normal $\mathbf{n} = (4, 2, -5)$. ◄

When convenient, the terms in Equations (2) and (3) can be multiplied out and the constants combined. This leads to the following theorem.

THEOREM 3.3.1

(a) *If a and b are constants that are not both zero, then an equation of the form*

$$ax + by + c = 0 \tag{4}$$

represents a line in R^2 with normal $\mathbf{n} = (a, b)$.

(b) *If a, b, and c are constants that are not all zero, then an equation of the form*

$$ax + by + cz + d = 0 \tag{5}$$

represents a plane in R^3 with normal $\mathbf{n} = (a, b, c)$.

▶ **EXAMPLE 3 Vectors Orthogonal to Lines and Planes Through the Origin**

(a) The equation $ax + by = 0$ represents a line through the origin in R^2. Show that the vector $\mathbf{n}_1 = (a, b)$ formed from the coefficients of the equation is orthogonal to the line, that is, orthogonal to every vector along the line.

(b) The equation $ax + by + cz = 0$ represents a plane through the origin in R^3. Show that the vector $\mathbf{n}_2 = (a, b, c)$ formed from the coefficients of the equation is orthogonal to the plane, that is, orthogonal to every vector that lies in the plane.

Solution We will solve both problems together. The two equations can be written as

$$(a, b) \cdot (x, y) = 0 \quad \text{and} \quad (a, b, c) \cdot (x, y, z) = 0$$

or, alternatively, as

$$\mathbf{n}_1 \cdot (x, y) = 0 \quad \text{and} \quad \mathbf{n}_2 \cdot (x, y, z) = 0$$

These equations show that $\mathbf{n}_1$ is orthogonal to every vector (x, y) on the line and that $\mathbf{n}_2$ is orthogonal to every vector (x, y, z) in the plane (Figure 3.3.1). ◄

Recall that

$$ax + by = 0 \quad \text{and} \quad ax + by + cz = 0$$

Referring to Table 1 of Section 3.2, in what other ways can you write (6) if $\mathbf{n}$ and $\mathbf{x}$ are expressed in matrix form?

are called *homogeneous equations*. Example 3 illustrates that homogeneous equations in two or three unknowns can be written in the vector form

$$\mathbf{n} \cdot \mathbf{x} = 0 \tag{6}$$

where $\mathbf{n}$ is the vector of coefficients and $\mathbf{x}$ is the vector of unknowns. In R^2 this is called the *vector form of a line* through the origin, and in R^3 it is called the *vector form of a plane* through the origin.

Orthogonal Projections   In many applications it is necessary to "decompose" a vector $\mathbf{u}$ into a sum of two terms, one term being a scalar multiple of a specified nonzero vector $\mathbf{a}$ and the other term being orthogonal to $\mathbf{a}$. For example, if $\mathbf{u}$ and $\mathbf{a}$ are vectors in R^2 that are positioned so their initial points coincide at a point Q, then we can create such a decomposition as follows (Figure 3.3.2):

- Drop a perpendicular from the tip of $\mathbf{u}$ to the line through $\mathbf{a}$.
- Construct the vector $\mathbf{w}_1$ from Q to the foot of the perpendicular.
- Construct the vector $\mathbf{w}_2 = \mathbf{u} - \mathbf{w}_1$.

(a)   (b)   (c)   (d)

▲ **Figure 3.3.2** In parts (b) through (d), $\mathbf{u} = \mathbf{w}_1 + \mathbf{w}_2$, where $\mathbf{w}_1$ is parallel to $\mathbf{a}$ and $\mathbf{w}_2$ is orthogonal to $\mathbf{a}$.

Since

$$\mathbf{w}_1 + \mathbf{w}_2 = \mathbf{w}_1 + (\mathbf{u} - \mathbf{w}_1) = \mathbf{u}$$

we have decomposed $\mathbf{u}$ into a sum of two orthogonal vectors, the first term being a scalar multiple of $\mathbf{a}$ and the second being orthogonal to $\mathbf{a}$.

The following theorem shows that the foregoing results, which we illustrated using vectors in R^2, apply as well in R^n.

THEOREM 3.3.2 Projection Theorem

If $\mathbf{u}$ and $\mathbf{a}$ are vectors in R^n, and if $\mathbf{a} \neq 0$, then $\mathbf{u}$ can be expressed in exactly one way in the form $\mathbf{u} = \mathbf{w}_1 + \mathbf{w}_2$, where $\mathbf{w}_1$ is a scalar multiple of $\mathbf{a}$ and $\mathbf{w}_2$ is orthogonal to $\mathbf{a}$.

Proof   Since the vector $\mathbf{w}_1$ is to be a scalar multiple of $\mathbf{a}$, it must have the form

$$\mathbf{w}_1 = k\mathbf{a} \qquad (7)$$

Our goal is to find a value of the scalar k and a vector $\mathbf{w}_2$ that is orthogonal to $\mathbf{a}$ such that

$$\mathbf{u} = \mathbf{w}_1 + \mathbf{w}_2 \qquad (8)$$

We can determine k by using (7) to rewrite (8) as

$$\mathbf{u} = \mathbf{w}_1 + \mathbf{w}_2 = k\mathbf{a} + \mathbf{w}_2$$

and then applying Theorems 3.2.2 and 3.2.3 to obtain

$$\mathbf{u} \cdot \mathbf{a} = (k\mathbf{a} + \mathbf{w}_2) \cdot \mathbf{a} = k\|\mathbf{a}\|^2 + (\mathbf{w}_2 \cdot \mathbf{a}) \qquad (9)$$

Since $\mathbf{w}_2$ is to be orthogonal to $\mathbf{a}$, the last term in (9) must be 0, and hence k must satisfy the equation

$$\mathbf{u} \cdot \mathbf{a} = k\|\mathbf{a}\|^2$$

from which we obtain

$$k = \frac{\mathbf{u} \cdot \mathbf{a}}{\|\mathbf{a}\|^2}$$

as the only possible value for k. The proof can be completed by rewriting (8) as

$$\mathbf{w}_2 = \mathbf{u} - \mathbf{w}_1 = \mathbf{u} - k\mathbf{a} = \mathbf{u} - \frac{\mathbf{u} \cdot \mathbf{a}}{\|\mathbf{a}\|^2}\mathbf{a}$$

and then confirming that $\mathbf{w}_2$ is orthogonal to $\mathbf{a}$ by showing that $\mathbf{w}_2 \cdot \mathbf{a} = 0$ (we leave the details for you). ◄

The vectors $\mathbf{w}_1$ and $\mathbf{w}_2$ in the Projection Theorem have associated names—the vector $\mathbf{w}_1$ is called the ***orthogonal projection of*** $\mathbf{u}$ ***on*** $\mathbf{a}$ or sometimes ***the vector component of*** $\mathbf{u}$ ***along*** $\mathbf{a}$, and the vector $\mathbf{w}_2$ is called the vector ***component of*** $\mathbf{u}$ ***orthogonal to*** $\mathbf{a}$. The vector $\mathbf{w}_1$ is commonly denoted by the symbol $\text{proj}_\mathbf{a}\mathbf{u}$, in which case it follows from (8) that $\mathbf{w}_2 = \mathbf{u} - \text{proj}_\mathbf{a}\mathbf{u}$. In summary,

$$\text{proj}_\mathbf{a}\mathbf{u} = \frac{\mathbf{u} \cdot \mathbf{a}}{\|\mathbf{a}\|^2}\mathbf{a} \quad \textit{(vector component of } \mathbf{u} \textit{ along } \mathbf{a}) \tag{10}$$

$$\mathbf{u} - \text{proj}_\mathbf{a}\mathbf{u} = \mathbf{u} - \frac{\mathbf{u} \cdot \mathbf{a}}{\|\mathbf{a}\|^2}\mathbf{a} \quad \textit{(vector component of } \mathbf{u} \textit{ orthogonal to } \mathbf{a}) \tag{11}$$

▶ **EXAMPLE 4** Orthogonal Projection on a Line

Find the orthogonal projections of the vectors $\mathbf{e}_1 = (1, 0)$ and $\mathbf{e}_2 = (0, 1)$ on the line L that makes an angle θ with the positive x-axis in R^2.

Solution As illustrated in Figure 3.3.3, $\mathbf{a} = (\cos\theta, \sin\theta)$ is a unit vector along the line L, so our first problem is to find the orthogonal projection of $\mathbf{e}_1$ along $\mathbf{a}$. Since

$$\|\mathbf{a}\| = \sqrt{\sin^2\theta + \cos^2\theta} = 1 \quad \text{and} \quad \mathbf{e}_1 \cdot \mathbf{a} = (1, 0) \cdot (\cos\theta, \sin\theta) = \cos\theta$$

it follows from Formula (10) that this projection is

$$\text{proj}_\mathbf{a}\mathbf{e}_1 = \frac{\mathbf{e}_1 \cdot \mathbf{a}}{\|\mathbf{a}\|^2}\mathbf{a} = (\cos\theta)(\cos\theta, \sin\theta) = (\cos^2\theta, \sin\theta\cos\theta)$$

Similarly, since $\mathbf{e}_2 \cdot \mathbf{a} = (0, 1) \cdot (\cos\theta, \sin\theta) = \sin\theta$, it follows from Formula (10) that

$$\text{proj}_\mathbf{a}\mathbf{e}_2 = \frac{\mathbf{e}_2 \cdot \mathbf{a}}{\|\mathbf{a}\|^2}\mathbf{a} = (\sin\theta)(\cos\theta, \sin\theta) = (\sin\theta\cos\theta, \sin^2\theta)$$

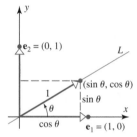

▲ Figure 3.3.3

▶ **EXAMPLE 5** Vector Component of u Along a

Let $\mathbf{u} = (2, -1, 3)$ and $\mathbf{a} = (4, -1, 2)$. Find the vector component of $\mathbf{u}$ along $\mathbf{a}$ and the vector component of $\mathbf{u}$ orthogonal to $\mathbf{a}$.

Solution
$$\mathbf{u} \cdot \mathbf{a} = (2)(4) + (-1)(-1) + (3)(2) = 15$$
$$\|\mathbf{a}\|^2 = 4^2 + (-1)^2 + 2^2 = 21$$

Thus the vector component of $\mathbf{u}$ along $\mathbf{a}$ is

$$\text{proj}_\mathbf{a}\mathbf{u} = \frac{\mathbf{u} \cdot \mathbf{a}}{\|\mathbf{a}\|^2}\mathbf{a} = \tfrac{15}{21}(4, -1, 2) = \left(\tfrac{20}{7}, -\tfrac{5}{7}, \tfrac{10}{7}\right)$$

and the vector component of $\mathbf{u}$ orthogonal to $\mathbf{a}$ is

$$\mathbf{u} - \text{proj}_\mathbf{a}\mathbf{u} = (2, -1, 3) - \left(\tfrac{20}{7}, -\tfrac{5}{7}, \tfrac{10}{7}\right) = \left(-\tfrac{6}{7}, -\tfrac{2}{7}, \tfrac{11}{7}\right)$$

As a check, you may wish to verify that the vectors $\mathbf{u} - \text{proj}_\mathbf{a}\mathbf{u}$ and $\mathbf{a}$ are perpendicular by showing that their dot product is zero. ◄

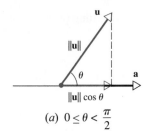

(a) $0 \le \theta < \dfrac{\pi}{2}$

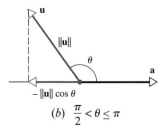

(b) $\dfrac{\pi}{2} < \theta \le \pi$

▲ Figure 3.3.4

Sometimes we will be more interested in the *norm* of the vector component of **u** along **a** than in the vector component itself. A formula for this norm can be derived as follows:

$$\|\mathrm{proj}_\mathbf{a}\mathbf{u}\| = \left\| \frac{\mathbf{u} \cdot \mathbf{a}}{\|\mathbf{a}\|^2} \mathbf{a} \right\| = \left| \frac{\mathbf{u} \cdot \mathbf{a}}{\|\mathbf{a}\|^2} \right| \|\mathbf{a}\| = \frac{|\mathbf{u} \cdot \mathbf{a}|}{\|\mathbf{a}\|^2} \|\mathbf{a}\|$$

where the second equality follows from part (*c*) of Theorem 3.2.1 and the third from the fact that $\|\mathbf{a}\|^2 > 0$. Thus,

$$\|\mathrm{proj}_\mathbf{a}\mathbf{u}\| = \frac{|\mathbf{u} \cdot \mathbf{a}|}{\|\mathbf{a}\|} \tag{12}$$

If θ denotes the angle between **u** and **a**, then $\mathbf{u} \cdot \mathbf{a} = \|\mathbf{u}\|\|\mathbf{a}\|\cos\theta$, so (12) can also be written as

$$\|\mathrm{proj}_\mathbf{a}\mathbf{u}\| = \|\mathbf{u}\||\cos\theta| \tag{13}$$

(Verify.) A geometric interpretation of this result is given in Figure 3.3.4.

The Theorem of Pythagoras

In Section 3.2 we found that many theorems about vectors in R^2 and R^3 also hold in R^n. Another example of this is the following generalization of the Theorem of Pythagoras (Figure 3.3.5).

▲ Figure 3.3.5

> **THEOREM 3.3.3** Theorem of Pythagoras in R^n
>
> *If* **u** *and* **v** *are orthogonal vectors in* R^n *with the Euclidean inner product, then*
>
> $$\|\mathbf{u} + \mathbf{v}\|^2 = \|\mathbf{u}\|^2 + \|\mathbf{v}\|^2 \tag{14}$$

Proof Since **u** and **v** are orthogonal, we have $\mathbf{u} \cdot \mathbf{v} = 0$, from which it follows that
$$\|\mathbf{u} + \mathbf{v}\|^2 = (\mathbf{u} + \mathbf{v}) \cdot (\mathbf{u} + \mathbf{v}) = \|\mathbf{u}\|^2 + 2(\mathbf{u} \cdot \mathbf{v}) + \|\mathbf{v}\|^2 = \|\mathbf{u}\|^2 + \|\mathbf{v}\|^2 \ \blacktriangleleft$$

▶ **EXAMPLE 6** Theorem of Pythagoras in R^4

We showed in Example 1 that the vectors

$$\mathbf{u} = (-2, 3, 1, 4) \quad \text{and} \quad \mathbf{v} = (1, 2, 0, -1)$$

are orthogonal. Verify the Theorem of Pythagoras for these vectors.

Solution We leave it for you to confirm that

$$\mathbf{u} + \mathbf{v} = (-1, 5, 1, 3)$$
$$\|\mathbf{u} + \mathbf{v}\|^2 = 36$$
$$\|\mathbf{u}\|^2 + \|\mathbf{v}\|^2 = 30 + 6$$

Thus, $\|\mathbf{u} + \mathbf{v}\|^2 = \|\mathbf{u}\|^2 + \|\mathbf{v}\|^2 \ \blacktriangleleft$

OPTIONAL

Distance Problems

We will now show how orthogonal projections can be used to solve the following three distance problems:

Problem 1. Find the distance between a point and a line in R^2.

Problem 2. Find the distance between a point and a plane in R^3.

Problem 3. Find the distance between two parallel planes in R^3.

A method for solving the first two problems is provided by the next theorem. Since the proofs of the two parts are similar, we will prove part (*b*) and leave part (*a*) as an exercise.

THEOREM 3.3.4

(*a*) *In R^2 the distance D between the point $P_0(x_0, y_0)$ and the line $ax + by + c = 0$ is*

$$D = \frac{|ax_0 + by_0 + c|}{\sqrt{a^2 + b^2}} \tag{15}$$

(*b*) *In R^3 the distance D between the point $P_0(x_0, y_0, z_0)$ and the plane $ax + by + cz + d = 0$ is*

$$D = \frac{|ax_0 + by_0 + cz_0 + d|}{\sqrt{a^2 + b^2 + c^2}} \tag{16}$$

Distance from P_0 to plane.

▲ **Figure 3.3.6**

Proof (b) Let $Q(x_1, y_1, z_1)$ be any point in the plane. Position the normal $\mathbf{n} = (a, b, c)$ so that its initial point is at Q. As illustrated in Figure 3.3.6, the distance D is equal to the length of the orthogonal projection of $\overrightarrow{QP_0}$ on $\mathbf{n}$. Thus, it follows from Formula (16) that

$$D = \|\text{proj}_{\mathbf{n}} \overrightarrow{QP_0}\| = \frac{|\overrightarrow{QP_0} \cdot \mathbf{n}|}{\|\mathbf{n}\|}$$

But

$$\overrightarrow{QP_0} = (x_0 - x_1, y_0 - y_1, z_0 - z_1)$$

$$\overrightarrow{QP_0} \cdot \mathbf{n} = a(x_0 - x_1) + b(y_0 - y_1) + c(z_0 - z_1)$$

$$\|\mathbf{n}\| = \sqrt{a^2 + b^2 + c^2}$$

Thus

$$D = \frac{|a(x_0 - x_1) + b(y_0 - y_1) + c(z_0 - z_1)|}{\sqrt{a^2 + b^2 + c^2}} \tag{17}$$

Since the point $Q(x_1, y_1, z_1)$ lies in the given plane, its coordinates satisfy the equation of that plane; thus

$$ax_1 + by_1 + cz_1 + d = 0$$

or

$$d = -ax_1 - by_1 - cz_1$$

Substituting this expression in (17) yields (16). ◀

▶ **EXAMPLE 7 Distance Between a Point and a Plane**

Find the distance D between the point $(1, -4, -3)$ and the plane $2x - 3y + 6z = -1$.

Solution Since the distance formulas in Theorem 3.3.4 require that the equations of the line and plane be written with zero on the right side, we first need to rewrite the equation of the plane as

$$2x - 3y + 6z + 1 = 0$$

from which we obtain

$$D = \frac{|2(1) + (-3)(-4) + 6(-3) + 1|}{\sqrt{2^2 + (-3)^2 + 6^2}} = \frac{|-3|}{7} = \frac{3}{7} \quad ◀$$

▲ Figure 3.3.7 The distance between the parallel planes V and W is equal to the distance between P_0 and W.

The third distance problem posed above is to find the distance between two parallel planes in R^3. As suggested in Figure 3.3.7, the distance between a plane V and a plane W can be obtained by finding any point P_0 in one of the planes, and computing the distance between that point and the other plane. Here is an example.

▶ EXAMPLE 8 Distance Between Parallel Planes

The planes

$$x + 2y - 2z = 3 \quad \text{and} \quad 2x + 4y - 4z = 7$$

are parallel since their normals, $(1, 2, -2)$ and $(2, 4, -4)$, are parallel vectors. Find the distance between these planes.

Solution To find the distance D between the planes, we can select an arbitrary point in one of the planes and compute its distance to the other plane. By setting $y = z = 0$ in the equation $x + 2y - 2z = 3$, we obtain the point $P_0(3, 0, 0)$ in this plane. From (16), the distance between P_0 and the plane $2x + 4y - 4z = 7$ is

$$D = \frac{|2(3) + 4(0) + (-4)(0) - 7|}{\sqrt{2^2 + 4^2 + (-4)^2}} = \frac{1}{6} \quad \blacktriangleleft$$

Concept Review

- Orthogonal (perpendicular) vectors
- Orthogonal set of vectors
- Normal to a line
- Normal to a plane

- Point-normal equations
- Vector form of a line
- Vector form of a plane
- Orthogonal projection of $\mathbf{u}$ on $\mathbf{a}$
- Vector component of $\mathbf{u}$ along $\mathbf{a}$

- Vector component of $\mathbf{u}$ orthogonal to $\mathbf{a}$
- Theorem of Pythagoras

Skills

- Determine whether two vectors are orthogonal.
- Determine whether a given set of vectors forms an orthogonal set.
- Find equations for lines (or planes) by using a normal vector and a point on the line (or plane).
- Find the vector form of a line or plane through the origin.

- Compute the vector component of $\mathbf{u}$ along $\mathbf{a}$ and orthogonal to $\mathbf{a}$.
- Find the distance between a point and a line in R^2 or R^3.
- Find the distance between two parallel planes in R^3.
- Find the distance between a point and a plane.

Exercise Set 3.3

▶ In Exercises 1–2, determine whether $\mathbf{u}$ and $\mathbf{v}$ are orthogonal vectors. ◀

1. (a) $\mathbf{u} = (6, 1, 4)$, $\mathbf{v} = (2, 0, -3)$

    (b) $\mathbf{u} = (0, 0, -1)$, $\mathbf{v} = (1, 1, 1)$

    (c) $\mathbf{u} = (-6, 0, 4)$, $\mathbf{v} = (3, 1, 6)$

    (d) $\mathbf{u} = (2, 4, -8)$, $\mathbf{v} = (5, 3, 7)$

2. (a) $\mathbf{u} = (2, 3)$, $\mathbf{v} = (5, -7)$

    (b) $\mathbf{u} = (-6, -2)$, $\mathbf{v} = (4, 0)$

    (c) $\mathbf{u} = (1, -5, 4)$, $\mathbf{v} = (3, 3, 3)$

    (d) $\mathbf{u} = (-2, 2, 3)$, $\mathbf{v} = (1, 7, -4)$

▶ In Exercises 3–4, determine whether the vectors form an orthogonal set. ◀

3. (a) $\mathbf{v}_1 = (2, 3)$, $\mathbf{v}_2 = (3, 2)$

    (b) $\mathbf{v}_1 = (-1, 1)$, $\mathbf{v}_2 = (1, 1)$

    (c) $\mathbf{v}_1 = (-2, 1, 1)$, $\mathbf{v}_2 = (1, 0, 2)$, $\mathbf{v}_3 = (-2, -5, 1)$

    (d) $\mathbf{v}_1 = (-3, 4, -1)$, $\mathbf{v}_2 = (1, 2, 5)$, $\mathbf{v}_3 = (4, -3, 0)$

4. (a) $\mathbf{v}_1 = (2, 3)$, $\mathbf{v}_2 = (-3, 2)$

    (b) $\mathbf{v}_1 = (1, -2)$, $\mathbf{v}_2 = (-2, 1)$

    (c) $\mathbf{v}_1 = (1, 0, 1)$, $\mathbf{v}_2 = (1, 1, 1)$, $\mathbf{v}_3 = (-1, 0, 1)$

    (d) $\mathbf{v}_1 = (2, -2, 1)$, $\mathbf{v}_2 = (2, 1, -2)$, $\mathbf{v}_3 = (1, 2, 2)$

5. Find a unit vector that is orthogonal to both $\mathbf{u} = (1, 0, 1)$ and $\mathbf{v} = (0, 1, 1)$.

6. (a) Show that $\mathbf{v} = (a, b)$ and $\mathbf{w} = (-b, a)$ are orthogonal vectors.

(b) Use the result in part (a) to find two vectors that are orthogonal to $\mathbf{v} = (2, -3)$.

(c) Find two unit vectors that are orthogonal to $(-3, 4)$.

7. Do the points $A(1, 1, 1)$, $B(-2, 0, 3)$, and $C(-3, -1, 1)$ form the vertices of a right triangle? Explain your answer.

8. Repeat Exercise 7 for the points $A(3, 0, 2)$, $B(4, 3, 0)$, and $C(8, 1, -1)$.

▷ In Exercises 9–12, find a point-normal form of the equation of the plane passing through P and having $\mathbf{n}$ as a normal. ◁

9. $P(-1, 3, -2)$; $\mathbf{n} = (-2, 1, -1)$

10. $P(1, 1, 4)$; $\mathbf{n} = (1, 9, 8)$ **11.** $P(2, 0, 0)$; $\mathbf{n} = (0, 0, 2)$

12. $P(0, 0, 0)$; $\mathbf{n} = (1, 2, 3)$

▷ In Exercises 13–16, determine whether the given planes are parallel. ◁

13. $4x - y + 2z = 5$ and $7x - 3y + 4z = 8$

14. $x - 4y - 3z - 2 = 0$ and $3x - 12y - 9z - 7 = 0$

15. $2y = 8x - 4z + 5$ and $x = \frac{1}{2}z + \frac{1}{4}y$

16. $(-4, 1, 2) \cdot (x, y, z) = 0$ and $(8, -2, -4) \cdot (x, y, z) = 0$

▷ In Exercises 17–18, determine whether the given planes are perpendicular. ◁

17. $3x - y + z - 4 = 0$, $x + 2z = -1$

18. $x - 2y + 3z = 4$, $-2x + 5y + 4z = -1$

▷ In Exercises 19–20, find $\|\text{proj}_{\mathbf{a}}\mathbf{u}\|$. ◁

19. (a) $\mathbf{u} = (1, -2)$, $\mathbf{a} = (-4, -3)$

(b) $\mathbf{u} = (3, 0, 4)$, $\mathbf{a} = (2, 3, 3)$

20. (a) $\mathbf{u} = (5, 6)$, $\mathbf{a} = (2, -1)$

(b) $\mathbf{u} = (3, -2, 6)$, $\mathbf{a} = (1, 2, -7)$

▷ In Exercises 21–28, find the vector component of $\mathbf{u}$ along $\mathbf{a}$ and the vector component of $\mathbf{u}$ orthogonal to $\mathbf{a}$. ◁

21. $\mathbf{u} = (6, 2)$, $\mathbf{a} = (3, -9)$ **22.** $\mathbf{u} = (-1, -2)$, $\mathbf{a} = (-2, 3)$

23. $\mathbf{u} = (3, 1, -7)$, $\mathbf{a} = (1, 0, 5)$

24. $\mathbf{u} = (1, 0, 0)$, $\mathbf{a} = (4, 3, 8)$

25. $\mathbf{u} = (1, 1, 1)$, $\mathbf{a} = (0, 2, -1)$

26. $\mathbf{u} = (2, 0, 1)$, $\mathbf{a} = (1, 2, 3)$

27. $\mathbf{u} = (2, 1, 1, 2)$, $\mathbf{a} = (4, -4, 2, -2)$

28. $\mathbf{u} = (5, 0, -3, 7)$, $\mathbf{a} = (2, 1, -1, -1)$

▷ In Exercises 29–32, find the distance between the point and the line. ◁

29. $4x + 3y + 4 = 0$; $(-3, 1)$

30. $x - 3y + 2 = 0$; $(-1, 4)$

31. $y = -4x + 2$; $(2, -5)$

32. $3x + y = 5$; $(1, 8)$

▷ In Exercises 33–36, find the distance between the point and the plane. ◁

33. $(3, 1, -2)$; $x + 2y - 2z = 4$

34. $(-1, -1, 2)$; $2x + 5y - 6z = 4$

35. $(-1, 2, 1)$; $2x + 3y - 4z = 1$

36. $(0, 3, -2)$; $x - y - z = 3$

▷ In Exercises 37–40, find the distance between the given parallel planes. ◁

37. $2x - y - z = 5$ and $-4x + 2y + 2z = 12$

38. $3x - 4y + z = 1$ and $6x - 8y + 2z = 3$

39. $-4x + y - 3z = 0$ and $8x - 2y + 6z = 0$

40. $2x - y + z = 1$ and $2x - y + z = -1$

41. Let $\mathbf{i}$, $\mathbf{j}$, and $\mathbf{k}$ be unit vectors along the positive x, y, and z axes of a rectangular coordinate system in 3-space. If $\mathbf{v} = (a, b, c)$ is a nonzero vector, then the angles α, β, and γ between $\mathbf{v}$ and the vectors $\mathbf{i}$, $\mathbf{j}$, and $\mathbf{k}$, respectively, are called the *direction angles* of $\mathbf{v}$ (Figure Ex-41), and the numbers $\cos \alpha$, $\cos \beta$, and $\cos \gamma$ are called the *direction cosines* of $\mathbf{v}$.

(a) Show that $\cos \alpha = a/\|\mathbf{v}\|$.

(b) Find $\cos \beta$ and $\cos \gamma$.

(c) Show that $\mathbf{v}/\|\mathbf{v}\| = (\cos \alpha, \cos \beta, \cos \gamma)$.

(d) Show that $\cos^2 \alpha + \cos^2 \beta + \cos^2 \gamma = 1$.

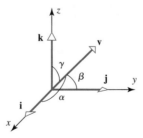

◁ Figure Ex-41

42. Use the result in Exercise 41 to estimate, to the nearest degree, the angles that a diagonal of a box with dimensions 10 cm × 15 cm × 25 cm makes with the edges of the box.

43. Show that if $\mathbf{v}$ is orthogonal to both $\mathbf{w}_1$ and $\mathbf{w}_2$, then $\mathbf{v}$ is orthogonal to $k_1\mathbf{w}_1 + k_2\mathbf{w}_2$ for all scalars k_1 and k_2.

44. Let $\mathbf{u}$ and $\mathbf{v}$ be nonzero vectors in 2- or 3-space, and let $k = \|\mathbf{u}\|$ and $l = \|\mathbf{v}\|$. Show that the vector $\mathbf{w} = l\mathbf{u} + k\mathbf{v}$ bisects the angle between $\mathbf{u}$ and $\mathbf{v}$.

45. Prove part (*a*) of Theorem 3.3.4.

46. Is it possible to have

$$\text{proj}_a \mathbf{u} = \text{proj}_u \mathbf{a}?$$

Explain your reasoning.

True-False Exercises

In parts (a)–(g) determine whether the statement is true or false, and justify your answer.

(a) The vectors $(3, -1, 2)$ and $(0, 0, 0)$ are orthogonal.

(b) If $\mathbf{u}$ and $\mathbf{v}$ are orthogonal vectors, then for all nonzero scalars k and m, $k\mathbf{u}$ and $m\mathbf{v}$ are orthogonal vectors.

(c) The orthogonal projection of $\mathbf{u}$ along $\mathbf{a}$ is perpendicular to the vector component of $\mathbf{u}$ orthogonal to $\mathbf{a}$.

(d) If $\mathbf{a}$ and $\mathbf{b}$ are orthogonal vectors, then for every nonzero vector $\mathbf{u}$, we have

$$\text{proj}_a (\text{proj}_b (\mathbf{u})) = \mathbf{0}$$

(e) If $\mathbf{a}$ and $\mathbf{u}$ are nonzero vectors, then

$$\text{proj}_a (\text{proj}_a (\mathbf{u})) = \text{proj}_a (\mathbf{u})$$

(f) If the relationship

$$\text{proj}_a \mathbf{u} = \text{proj}_a \mathbf{v}$$

holds for some nonzero vector $\mathbf{a}$, then $\mathbf{u} = \mathbf{v}$.

(g) For all vectors $\mathbf{u}$ and $\mathbf{v}$, it is true that

$$\|\mathbf{u} + \mathbf{v}\| = \|\mathbf{u}\| + \|\mathbf{v}\|$$

3.4 The Geometry of Linear Systems

In this section we will use parametric and vector methods to study general systems of linear equations. This work will enable us to interpret solution sets of linear systems with n unknowns as geometric objects in R^n just as we interpreted solution sets of linear systems with two and three unknowns as points, lines, and planes in R^2 and R^3.

Vector and Parametric Equations of Lines in R^2 and R^3

In the last section we derived equations of lines and planes that are determined by a point and a normal vector. However, there are other useful ways of specifying lines and planes. For example, a unique line in R^2 or R^3 is determined by a point $\mathbf{x}_0$ on the line and a nonzero vector $\mathbf{v}$ parallel to the line, and a unique plane in R^3 is determined by a point $\mathbf{x}_0$ in the plane and two noncollinear vectors $\mathbf{v}_1$ and $\mathbf{v}_2$ parallel to the plane. The best way to visualize this is to translate the vectors so their initial points are at $\mathbf{x}_0$ (Figure 3.4.1).

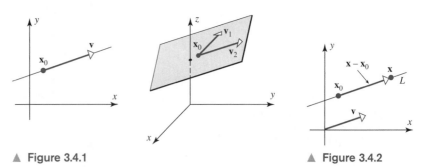

▲ Figure 3.4.1　　　　　▲ Figure 3.4.2

Let us begin by deriving an equation for the line L that contains the point $\mathbf{x}_0$ and is parallel to $\mathbf{v}$. If $\mathbf{x}$ is a general point on such a line, then, as illustrated in Figure 3.4.2, the vector $\mathbf{x} - \mathbf{x}_0$ will be some scalar multiple of $\mathbf{v}$, say

$$\mathbf{x} - \mathbf{x}_0 = t\mathbf{v} \quad \text{or equivalently} \quad \mathbf{x} = \mathbf{x}_0 + t\mathbf{v}$$

As the variable t (called a ***parameter***) varies from $-\infty$ to ∞, the point $\mathbf{x}$ traces out the line L. Accordingly, we have the following result.

Although it is not stated explicitly, it is understood in Formulas (1) and (2) that the parameter t varies from $-\infty$ to ∞. This applies to all vector and parametric equations in this text except where stated otherwise.

THEOREM 3.4.1 *Let L be the line in R^2 or R^3 that contains the point $\mathbf{x}_0$ and is parallel to the nonzero vector $\mathbf{v}$. Then the equation of the line through $\mathbf{x}_0$ that is parallel to $\mathbf{v}$ is*

$$\mathbf{x} = \mathbf{x}_0 + t\mathbf{v} \tag{1}$$

If $\mathbf{x}_0 = \mathbf{0}$, then the line passes through the origin and the equation has the form

$$\mathbf{x} = t\mathbf{v} \tag{2}$$

Vector and Parametric Equations of Planes in R^3

▲ Figure 3.4.3

Next we will derive an equation for the plane W that contains the point $\mathbf{x}_0$ and is parallel to the noncollinear vectors $\mathbf{v}_1$ and $\mathbf{v}_2$. As shown in Figure 3.4.3, if $\mathbf{x}$ is any point in the plane, then by forming suitable scalar multiples of $\mathbf{v}_1$ and $\mathbf{v}_2$, say $t_1\mathbf{v}_1$ and $t_2\mathbf{v}_2$, we can create a parallelogram with diagonal $\mathbf{x} - \mathbf{x}_0$ and adjacent sides $t_1\mathbf{v}_1$ and $t_2\mathbf{v}_2$. Thus, we have

$$\mathbf{x} - \mathbf{x}_0 = t_1\mathbf{v}_1 + t_2\mathbf{v}_2 \quad \text{or equivalently} \quad \mathbf{x} = \mathbf{x}_0 + t_1\mathbf{v}_1 + t_2\mathbf{v}_2$$

As the variables t_1 and t_2 (called ***parameters***) vary independently from $-\infty$ to ∞, the point $\mathbf{x}$ varies over the entire plane W. Accordingly, we make the following definition.

THEOREM 3.4.2 *Let W be the plane in R^3 that contains the point $\mathbf{x}_0$ and is parallel to the noncollinear vectors $\mathbf{v}_1$ and $\mathbf{v}_2$. Then an equation of the plane through $\mathbf{x}_0$ that is parallel to $\mathbf{v}_1$ and $\mathbf{v}_2$ is given by*

$$\mathbf{x} = \mathbf{x}_0 + t_1\mathbf{v}_1 + t_2\mathbf{v}_2 \tag{3}$$

If $\mathbf{x}_0 = \mathbf{0}$, then the plane passes through the origin and the equation has the form

$$\mathbf{x} = t_1\mathbf{v}_1 + t_2\mathbf{v}_2 \tag{4}$$

Remark Observe that the line through $\mathbf{x}_0$ represented by Equation (1) is the translation by $\mathbf{x}_0$ of the line through the origin represented by Equation (2) and that the plane through $\mathbf{x}_0$ represented by Equation (3) is the translation by $\mathbf{x}_0$ of the plane through the origin represented by Equation (4) (Figure 3.4.4).

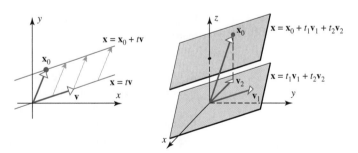

▶ Figure 3.4.4

Motivated by the forms of Formulas (1) to (4), we can extend the notions of line and plane to R^n by making the following definitions.

DEFINITION 1 If $\mathbf{x}_0$ and $\mathbf{v}$ are vectors in R^n, and if $\mathbf{v}$ is nonzero, then the equation

$$\mathbf{x} = \mathbf{x}_0 + t\mathbf{v} \tag{5}$$

defines the ***line through $\mathbf{x}_0$ that is parallel to $\mathbf{v}$***. In the special case where $\mathbf{x}_0 = \mathbf{0}$, the line is said to ***pass through the origin***.

> **DEFINITION 2** If x_0, v_1, and v_2 are vectors in R^n, and if v_1 and v_2 are not collinear, then the equation
>
> $$\mathbf{x} = \mathbf{x}_0 + t_1\mathbf{v}_1 + t_2\mathbf{v}_2 \qquad (6)$$
>
> defines the **plane through x_0 that is parallel to v_1 and v_2**. In the special case where $\mathbf{x}_0 = \mathbf{0}$, the plane is said to **pass through the origin**.

Equations (5) and (6) are called **vector forms** of a line and plane in R^n. If the vectors in these equations are expressed in terms of their components and the corresponding components on each side are equated, then the resulting equations are called **parametric equations** of the line and plane. Here are some examples.

▶ **EXAMPLE 1 Vector and Parametric Equations of Lines in R^2 and R^3**

(a) Find a vector equation and parametric equations of the line in R^2 that passes through the origin and is parallel to the vector $\mathbf{v} = (-2, 3)$.

(b) Find a vector equation and parametric equations of the line in R^3 that passes through the point $P_0(1, 2, -3)$ and is parallel to the vector $\mathbf{v} = (4, -5, 1)$.

(c) Use the vector equation obtained in part (b) to find two points on the line that are different from P_0.

Solution (a) It follows from (5) with $\mathbf{x}_0 = \mathbf{0}$ that a vector equation of the line is $\mathbf{x} = t\mathbf{v}$. If we let $\mathbf{x} = (x, y)$, then this equation can be expressed in vector form as

$$(x, y) = t(-2, 3)$$

Equating corresponding components on the two sides of this equation yields the parametric equations

$$x = -2t, \quad y = 3t$$

Solution (b) It follows from (5) that a vector equation of the line is $\mathbf{x} = \mathbf{x}_0 + t\mathbf{v}$. If we let $\mathbf{x} = (x, y, z)$, and if we take $\mathbf{x}_0 = (1, 2, -3)$, then this equation can be expressed in vector form as

$$(x, y, z) = (1, 2, -3) + t(4, -5, 1) \qquad (7)$$

Equating corresponding components on the two sides of this equation yields the parametric equations

$$x = 1 + 4t, \quad y = 2 - 5t, \quad z = -3 + t$$

Solution (c) A point on the line represented by Equation (7) can be obtained by substituting a specific numerical value for the parameter t. However, since $t = 0$ produces $(x, y, z) = (1, 2, -3)$, which is the point P_0, this value of t does not serve our purpose. Taking $t = 1$ produces the point $(5, -3, -2)$ and taking $t = -1$ produces the point $(-3, 7, -4)$. Any other distinct values for t (except $t = 0$) would work just as well. ◀

▶ **EXAMPLE 2 Vector and Parametric Equations of a Plane in R^3**

Find vector and parametric equations of the plane $x - y + 2z = 5$.

Solution We will find the parametric equations first. We can do this by solving the equation for any one of the variables in terms of the other two and then using those two variables as parameters. For example, solving for x in terms of y and z yields

$$x = 5 + y - 2z \qquad (8)$$

We would have obtained different parametric and vector equations in Example 2 had we solved (8) for y or z rather than x. However, one can show the same plane results in all three cases as the parameters vary from $-\infty$ to ∞.

and then using y and z as parameters t_1 and t_2, respectively, yields the parametric equations

$$x = 5 + t_1 - 2t_2, \quad y = t_1, \quad z = t_2$$

To obtain a vector equation of the plane we rewrite these parametric equations as

$$(x, y, z) = (5 + t_1 - 2t_2, t_1, t_2)$$

or, equivalently, as

$$(x, y, z) = (5, 0, 0) + t_1(1, 1, 0) + t_2(-2, 0, 1)$$

▶ **EXAMPLE 3 Vector and Parametric Equations of Lines and Planes in R^4**

(a) Find vector and parametric equations of the line through the origin of R^4 that is parallel to the vector $\mathbf{v} = (5, -3, 6, 1)$.

(b) Find vector and parametric equations of the plane in R^4 that passes through the point $\mathbf{x}_0 = (2, -1, 0, 3)$ and is parallel to both $\mathbf{v}_1 = (1, 5, 2, -4)$ and $\mathbf{v}_2 = (0, 7, -8, 6)$.

Solution (a) If we let $\mathbf{x} = (x_1, x_2, x_3, x_4)$, then the vector equation $\mathbf{x} = t\mathbf{v}$ can be expressed as

$$(x_1, x_2, x_3, x_4) = t(5, -3, 6, 1)$$

Equating corresponding components yields the parametric equations

$$x_1 = 5t, \quad x_2 = -3t, \quad x_3 = 6t, \quad x_4 = t$$

Solution (b) The vector equation $\mathbf{x} = \mathbf{x}_0 + t_1\mathbf{v}_1 + t_2\mathbf{v}_2$ can be expressed as

$$(x_1, x_2, x_3, x_4) = (2, -1, 0, 3) + t_1(1, 5, 2, -4) + t_2(0, 7, -8, 6)$$

which yields the parametric equations

$$x_1 = 2 + t_1$$
$$x_2 = -1 + 5t_1 + 7t_2$$
$$x_3 = 2t_1 - 8t_2$$
$$x_4 = 3 - 4t_1 + 6t_2 \quad ◀$$

Lines Through Two Points in R^n

If $\mathbf{x}_0$ and $\mathbf{x}_1$ are distinct points in R^n, then the line determined by these points is parallel to the vector $\mathbf{v} = \mathbf{x}_1 - \mathbf{x}_0$ (Figure 3.4.5), so it follows from (5) that the line can be expressed in vector form as

$$\mathbf{x} = \mathbf{x}_0 + t(\mathbf{x}_1 - \mathbf{x}_0) \tag{9}$$

or, equivalently, as

$$\mathbf{x} = (1 - t)\mathbf{x}_0 + t\mathbf{x}_1 \tag{10}$$

▲ Figure 3.4.5

These are called the *two-point vector equations* of a line in R^n.

▶ **EXAMPLE 4 A Line Through Two Points in R^2**

Find vector and parametric equations for the line in R^2 that passes through the points $P(0, 7)$ and $Q(5, 0)$.

Solution We will see below that it does not matter which point we take to be $\mathbf{x}_0$ and which we take to be $\mathbf{x}_1$, so let us choose $\mathbf{x}_0 = (0, 7)$ and $\mathbf{x}_1 = (5, 0)$. It follows that $\mathbf{x}_1 - \mathbf{x}_0 = (5, -7)$ and hence that

$$(x, y) = (0, 7) + t(5, -7) \tag{11}$$

which we can rewrite in parametric form as

$$x = 5t, \quad y = 7 - 7t$$

Had we reversed our choices and taken $\mathbf{x}_0 = (5, 0)$ and $\mathbf{x}_1 = (0, 7)$, then the resulting vector equation would have been

$$(x, y) = (5, 0) + t(-5, 7) \tag{12}$$

and the parametric equations would have been

$$x = 5 - 5t, \quad y = 7t$$

(verify). Although (11) and (12) look different, they both represent the line whose equation in rectangular coordinates is

$$7x + 5y = 35$$

(Figure 3.4.6). This can be seen by eliminating the parameter t from the parametric equations (verify). ◀

▲ Figure 3.4.6

The point $\mathbf{x} = (x, y)$ in Equations (10) and (11) traces an entire line in R^2 as the parameter t varies over the interval $(-\infty, \infty)$. If, however, we restrict the parameter to vary from $t = 0$ to $t = 1$, then $\mathbf{x}$ will not trace the entire line but rather just the *line segment* joining the points $\mathbf{x}_0$ and $\mathbf{x}_1$. The point $\mathbf{x}$ will start at $\mathbf{x}_0$ when $t = 0$ and end at $\mathbf{x}_1$ when $t = 1$. Accordingly, we make the following definition.

DEFINITION 3 If $\mathbf{x}_0$ and $\mathbf{x}_1$ are vectors in R^n, then the equation

$$\mathbf{x} = \mathbf{x}_0 + t(\mathbf{x}_1 - \mathbf{x}_0) \quad (0 \le t \le 1) \tag{13}$$

defines the **line segment from $\mathbf{x}_0$ to $\mathbf{x}_1$**. When convenient, Equation (13) can be written as

$$\mathbf{x} = (1 - t)\mathbf{x}_0 + t\mathbf{x}_1 \quad (0 \le t \le 1) \tag{14}$$

▶ **EXAMPLE 5** A Line Segment from One Point to Another in R^2

It follows from (13) and (14) that the line segment in R^2 from $\mathbf{x}_0 = (1, -3)$ to $\mathbf{x}_1 = (5, 6)$ can be represented either by the equation

$$\mathbf{x} = (1, -3) + t(4, 9) \quad (0 \le t \le 1)$$

or by

$$\mathbf{x} = (1 - t)(1, -3) + t(5, 6) \quad (0 \le t \le 1) \quad ◀$$

Dot Product Form of a Linear System   Our next objective is to show how to express linear equations and linear systems in dot product notation. This will lead us to some important results about orthogonality and linear systems.

Recall that a *linear equation* in the variables $x_1, x_2, \ldots, x_n$ has the form

$$a_1 x_1 + a_2 x_2 + \cdots + a_n x_n = b \quad (a_1, a_2, \ldots, a_n \text{ not all zero}) \tag{15}$$

and that the corresponding *homogeneous* equation is

$$a_1 x_1 + a_2 x_2 + \cdots + a_n x_n = 0 \quad (a_1, a_2, \ldots, a_n \text{ not all zero}) \tag{16}$$

These equations can be rewritten in vector form by letting

$$\mathbf{a} = (a_1, a_2, \ldots, a_n) \quad \text{and} \quad \mathbf{x} = (x_1, x_2, \ldots, x_n)$$

in which case Formula (15) can be written as

$$\mathbf{a} \cdot \mathbf{x} = b \tag{17}$$

and Formula (16) as

$$\mathbf{a} \cdot \mathbf{x} = 0 \tag{18}$$

Except for a notational change from $\mathbf{n}$ to $\mathbf{a}$, Formula (18) is the extension to R^n of Formula (6) in Section 3.3. This equation reveals that *each solution vector* $\mathbf{x}$ *of a homogeneous equation is orthogonal to the coefficient vector* $\mathbf{a}$. To take this geometric observation a step further, consider the homogeneous system

$$
\begin{aligned}
a_{11}x_1 + a_{12}x_2 + \cdots + a_{1n}x_n &= 0 \\
a_{21}x_1 + a_{22}x_2 + \cdots + a_{2n}x_n &= 0 \\
\vdots \qquad \vdots \qquad\qquad \vdots \quad\ \ \vdots \\
a_{m1}x_1 + a_{m2}x_2 + \cdots + a_{mn}x_n &= 0
\end{aligned}
$$

If we denote the successive row vectors of the coefficient matrix by $\mathbf{r}_1, \mathbf{r}_2, \ldots, \mathbf{r}_m$, then we can rewrite this system in dot product form as

$$
\begin{aligned}
\mathbf{r}_1 \cdot \mathbf{x} &= \mathbf{0} \\
\mathbf{r}_2 \cdot \mathbf{x} &= \mathbf{0} \\
\vdots \qquad \vdots \\
\mathbf{r}_m \cdot \mathbf{x} &= \mathbf{0}
\end{aligned}
\tag{19}
$$

from which we see that every solution vector $\mathbf{x}$ is orthogonal to every row vector of the coefficient matrix. In summary, we have the following result.

> **THEOREM 3.4.3** *If A is an $m \times n$ matrix, then the solution set of the homogeneous linear system $A\mathbf{x} = \mathbf{0}$ consists of all vectors in R^n that are orthogonal to every row vector of A.*

▶ **EXAMPLE 6** Orthogonality of Row Vectors and Solution Vectors

We showed in Example 6 of Section 1.2 that the general solution of the homogeneous linear system

$$
\begin{bmatrix}
1 & 3 & -2 & 0 & 2 & 0 \\
2 & 6 & -5 & -2 & 4 & -3 \\
0 & 0 & 5 & 10 & 0 & 15 \\
2 & 6 & 0 & 8 & 4 & 18
\end{bmatrix}
\begin{bmatrix}
x_1 \\ x_2 \\ x_3 \\ x_4 \\ x_5 \\ x_6
\end{bmatrix}
=
\begin{bmatrix}
0 \\ 0 \\ 0 \\ 0
\end{bmatrix}
$$

is

$$x_1 = -3r - 4s - 2t, \quad x_2 = r, \quad x_3 = -2s, \quad x_4 = s, \quad x_5 = t, \quad x_6 = 0$$

which we can rewrite in vector form as

$$\mathbf{x} = (-3r - 4s - 2t, r, -2s, s, t, 0)$$

According to Theorem 3.4.3, the vector $\mathbf{x}$ must be orthogonal to each of the row vectors

$$
\begin{aligned}
\mathbf{r}_1 &= (1, 3, -2, 0, 2, 0) \\
\mathbf{r}_2 &= (2, 6, -5, -2, 4, -3) \\
\mathbf{r}_3 &= (0, 0, 5, 10, 0, 15) \\
\mathbf{r}_4 &= (2, 6, 0, 8, 4, 18)
\end{aligned}
$$

We will confirm that $\mathbf{x}$ is orthogonal to $\mathbf{r}_1$, and leave it for you to verify that $\mathbf{x}$ is orthogonal to the other three row vectors as well. The dot product of $\mathbf{r}_1$ and $\mathbf{x}$ is

$$\mathbf{r}_1 \cdot \mathbf{x} = 1(-3r - 4s - 2t) + 3(r) + (-2)(-2s) + 0(s) + 2(t) + 0(0) = 0$$

which establishes the orthogonality. ◀

The Relationship Between
$A\mathbf{x} = \mathbf{0}$ and $A\mathbf{x} = \mathbf{b}$

We will conclude this section by exploring the relationship between the solutions of a homogeneous linear system $A\mathbf{x} = \mathbf{0}$ and the solutions (if any) of a nonhomogeneous linear system $A\mathbf{x} = \mathbf{b}$ that has the same coefficient matrix. These are called ***corresponding linear systems***.

To motivate the result we are seeking, let us compare the solutions of the corresponding linear systems

$$\begin{bmatrix} 1 & 3 & -2 & 0 & 2 & 0 \\ 2 & 6 & -5 & -2 & 4 & -3 \\ 0 & 0 & 5 & 10 & 0 & 15 \\ 2 & 6 & 0 & 8 & 4 & 18 \end{bmatrix} \begin{bmatrix} x_1 \\ x_2 \\ x_3 \\ x_4 \\ x_5 \\ x_6 \end{bmatrix} = \begin{bmatrix} 0 \\ 0 \\ 0 \\ 0 \end{bmatrix} \quad \text{and} \quad \begin{bmatrix} 1 & 3 & -2 & 0 & 2 & 0 \\ 2 & 6 & -5 & -2 & 4 & -3 \\ 0 & 0 & 5 & 10 & 0 & 15 \\ 2 & 6 & 0 & 8 & 4 & 18 \end{bmatrix} \begin{bmatrix} x_1 \\ x_2 \\ x_3 \\ x_4 \\ x_5 \\ x_6 \end{bmatrix} = \begin{bmatrix} 0 \\ -1 \\ 5 \\ 6 \end{bmatrix}$$

We showed in Examples 5 and 6 of Section 1.2 that the general solutions of these linear systems can be written in parametric form as

homogeneous $\longrightarrow$ $x_1 = -3r - 4s - 2t, \quad x_2 = r, \quad x_3 = -2s, \quad x_4 = s, \quad x_5 = t, \quad x_6 = 0$

nonhomogeneous $\longrightarrow$ $x_1 = -3r - 4s - 2t, \quad x_2 = r, \quad x_3 = -2s, \quad x_4 = s, \quad x_5 = t, \quad x_6 = \frac{1}{3}$

which we can then rewrite in vector form as

homogeneous $\longrightarrow$ $(x_1, x_2, x_3, x_4, x_5) = (-3r - 4s - 2t, r, -2s, s, t, 0)$

nonhomogeneous $\longrightarrow$ $(x_1, x_2, x_3, x_4, x_5) = \left(-3r - 4s - 2t, r, -2s, s, t, \frac{1}{3}\right)$

By splitting the vectors on the right apart and collecting terms with like parameters, we can rewrite these equations as

homogeneous $\longrightarrow$ $(x_1, x_2, x_3, x_4, x_5) = r(-3, 1, 0, 0, 0) + s(-4, 0, -2, 1, 0, 0) + t(-2, 0, 0, 0, 1, 0)$ (20)

nonhomogeneous $\longrightarrow$ $(x_1, x_2, x_3, x_4, x_5) = r(-3, 1, 0, 0, 0) + s(-4, 0, -2, 1, 0, 0)$
$$+ t(-2, 0, 0, 0, 1, 0) + \left(0, 0, 0, 0, 0, \frac{1}{3}\right) \quad (21)$$

Formulas (20) and (21) reveal that each solution of the nonhomogeneous system can be obtained by adding the fixed vector $\left(0, 0, 0, 0, 0, \frac{1}{3}\right)$ to the corresponding solution of the homogeneous system. This is a special case of the following general result.

> **THEOREM 3.4.4** *The general solution of a consistent linear system $A\mathbf{x} = \mathbf{b}$ can be obtained by adding any specific solution of $A\mathbf{x} = \mathbf{b}$ to the general solution of $A\mathbf{x} = \mathbf{0}$.*

Proof Let $\mathbf{x}_0$ be any specific solution of $A\mathbf{x} = \mathbf{b}$, let W denote the solution set of $A\mathbf{x} = \mathbf{0}$, and let $\mathbf{x}_0 + W$ denote the set of all vectors that result by adding $\mathbf{x}_0$ to each vector in W. We must show that if $\mathbf{x}$ is a vector in $\mathbf{x}_0 + W$, then $\mathbf{x}$ is a solution of $A\mathbf{x} = \mathbf{b}$, and conversely, that every solution of $A\mathbf{x} = \mathbf{b}$ is in the set $\mathbf{x}_0 + W$.

Assume first that $\mathbf{x}$ is a vector in $\mathbf{x}_0 + W$. This implies that $\mathbf{x}$ is expressible in the form $\mathbf{x} = \mathbf{x}_0 + \mathbf{w}$, where $A\mathbf{x}_0 = \mathbf{b}$ and $A\mathbf{w} = \mathbf{0}$. Thus,

$$A\mathbf{x} = A(\mathbf{x}_0 + \mathbf{w}) = A\mathbf{x}_0 + A\mathbf{w} = \mathbf{b} + \mathbf{0} = \mathbf{b}$$

which shows that $\mathbf{x}$ is a solution of $A\mathbf{x} = \mathbf{b}$.

Conversely, let **x** be any solution of $A\mathbf{x} = \mathbf{b}$. To show that **x** is in the set $\mathbf{x}_0 + W$ we must show that **x** is expressible in the form

$$\mathbf{x} = \mathbf{x}_0 + \mathbf{w} \tag{22}$$

where **w** is in W (i.e., $A\mathbf{w} = \mathbf{0}$). We can do this by taking $\mathbf{w} = \mathbf{x} - \mathbf{x}_0$. This vector obviously satisfies (22), and it is in W since

$$A\mathbf{w} = A(\mathbf{x} - \mathbf{x}_0) = A\mathbf{x} - A\mathbf{x}_0 = \mathbf{b} - \mathbf{b} = \mathbf{0} \quad \blacktriangleleft$$

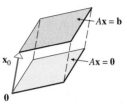

▲ **Figure 3.4.7** The solution set of $A\mathbf{x} = \mathbf{b}$ is a translation of the solution space of $A\mathbf{x} = \mathbf{0}$.

Remark Theorem 3.4.4 has a useful geometric interpretation that is illustrated in Figure 3.4.7. If, as discussed in Section 3.1, we interpret vector addition as translation, then the theorem states that if $\mathbf{x}_0$ is *any* specific solution of $A\mathbf{x} = \mathbf{b}$, then the *entire* solution set of $A\mathbf{x} = \mathbf{b}$ can be obtained by translating the solution set of $A\mathbf{x} = \mathbf{0}$ by the vector $\mathbf{x}_0$.

Concept Review

- Parameters
- Parametric equations of lines
- Parametric equations of planes
- Two-point vector equations of a line
- Vector equation of a line
- Vector equation of a plane

Skills

- Express the equations of lines in R^2 and R^3 using either vector or parametric equations.
- Express the equations of planes in R^n using either vector or parametric equations.
- Express the equation of a line containing two given points in R^2 or R^3 using either vector or parametric equations.
- Find equations of a line and a line segment.

- Verify the orthogonality of the row vectors of a linear system of equations and a solution vector.
- Use a specific solution to the nonhomogeneous linear system $A\mathbf{x} = \mathbf{b}$ and the general solution of the corresponding linear system $A\mathbf{x} = \mathbf{0}$ to obtain the general solution to $A\mathbf{x} = \mathbf{b}$.

Exercise Set 3.4

▷ In Exercises 1–4, find vector and parametric equations of the line containing the point and parallel to the vector. ◁

1. Point: $(-4, 1)$; vector: $\mathbf{v} = (0, -8)$

2. Point: $(2, -1)$; vector: $\mathbf{v} = (-4, -2)$

3. Point: $(0, 0, 0)$; vector: $\mathbf{v} = (-3, 0, 1)$

4. Point: $(-9, 3, 4)$; vector: $\mathbf{v} = (-1, 6, 0)$

▷ In Exercises 5–8, use the given equation of a line to find a point on the line and a vector parallel to the line. ◁

5. $\mathbf{x} = (3 - 5t, -6 - t)$ **6.** $(x, y, z) = (4t, 7, 4 + 3t)$

7. $\mathbf{x} = (1 - t)(4, 6) + t(-2, 0)$

8. $\mathbf{x} = (1 - t)(0, -5, 1)$

▷ In Exercises 9–12, find vector and parametric equations of the plane containing the given point and parallel vectors. ◁

9. Point: $(-3, 1, 0)$; vectors: $\mathbf{v}_1 = (0, -3, 6)$ and $\mathbf{v}_2 = (-5, 1, 2)$

10. Point: $(0, 6, -2)$; vectors: $\mathbf{v}_1 = (0, 9, -1)$ and $\mathbf{v}_2 = (0, -3, 0)$

11. Point: $(-1, 1, 4)$; vectors: $\mathbf{v}_1 = (6, -1, 0)$ and $\mathbf{v}_2 = (-1, 3, 1)$

12. Point: $(0, 5, -4)$; vectors: $\mathbf{v}_1 = (0, 0, -5)$ and $\mathbf{v}_2 = (1, -3, -2)$

▷ In Exercises 13–14, find vector and parametric equations of the line in R^2 that passes through the origin and is orthogonal to **v**.

13. $\mathbf{v} = (-2, 3)$ **14.** $\mathbf{v} = (1, -4)$

▷ In Exercises 15–16, find vector and parametric equations of the plane in R^2 that passes through the origin and is orthogonal to **v**. ◁

15. $\mathbf{v} = (4, 0, -5)$ [*Hint:* Construct two nonparallel vectors to **v** in R^3].

16. $\mathbf{v} = (3, 1, -6)$

▷ In Exercises 17–20, find the general solution to the linear system and confirm that the row vectors of the coefficient matrix are orthogonal to the solution vectors. ◁

17.
$$x_1 + x_2 + x_3 = 0$$
$$2x_1 + 2x_2 + 2x_3 = 0$$
$$3x_1 + 3x_2 + 3x_3 = 0$$

18.
$$x_1 + 3x_2 - 4x_3 = 0$$
$$2x_1 + 6x_2 - 8x_3 = 0$$

19.
$$x_1 + 5x_2 + x_3 + 2x_4 - x_5 = 0$$
$$x_1 - 2x_2 - x_3 + 3x_4 + 2x_5 = 0$$

20.
$$x_1 + 3x_2 - 4x_3 = 0$$
$$x_1 + 2x_2 + 3x_3 = 0$$

21. (a) The equation $x + y + z = 1$ can be viewed as a linear system of one equation in three unknowns. Express a general solution of this equation as a particular solution plus a general solution of the associated homogeneous system.

(b) Give a geometric interpretation of the result in part (a).

22. (a) The equation $x + y = 1$ can be viewed as a linear system of one equation in two unknowns. Express a general solution of this equation as a particular solution plus a general solution of the associated homogeneous system.

(b) Give a geometric interpretation of the result in part (a).

23. (a) Find a homogeneous linear system of two equations in three unknowns whose solution space consists of those vectors in R^3 that are orthogonal to $\mathbf{a} = (1, 1, 1)$ and $\mathbf{b} = (-2, 3, 0)$.

(b) What kind of geometric object is the solution space?

(c) Find a general solution of the system obtained in part (a), and confirm that Theorem 3.4.3 holds.

24. (a) Find a homogeneous linear system of two equations in three unknowns whose solution space consists of those vectors in R^3 that are orthogonal to $\mathbf{a} = (-3, 2, -1)$ and $\mathbf{b} = (0, -2, -2)$.

(b) What kind of geometric object is the solution space?

(c) Find a general solution of the system obtained in part (a), and confirm that Theorem 3.4.3 holds.

25. Consider the linear systems

$$\begin{bmatrix} 3 & 2 & -1 \\ 6 & 4 & -2 \\ -3 & -2 & 1 \end{bmatrix} \begin{bmatrix} x_1 \\ x_2 \\ x_3 \end{bmatrix} = \begin{bmatrix} 0 \\ 0 \\ 0 \end{bmatrix}$$

and

$$\begin{bmatrix} 3 & 2 & -1 \\ 6 & 4 & -2 \\ -3 & -2 & 1 \end{bmatrix} \begin{bmatrix} x_1 \\ x_2 \\ x_3 \end{bmatrix} = \begin{bmatrix} 2 \\ 4 \\ -2 \end{bmatrix}$$

(a) Find a general solution of the homogeneous system.

(b) Confirm that $x_1 = 1, x_2 = 0, x_3 = 1$ is a solution of the nonhomogeneous system.

(c) Use the results in parts (a) and (b) to find a general solution of the nonhomogeneous system.

(d) Check your result in part (c) by solving the nonhomogeneous system directly.

26. Consider the linear systems

$$\begin{bmatrix} 1 & -2 & 3 \\ 2 & 1 & 4 \\ 1 & -7 & 5 \end{bmatrix} \begin{bmatrix} x_1 \\ x_2 \\ x_3 \end{bmatrix} = \begin{bmatrix} 0 \\ 0 \\ 0 \end{bmatrix}$$

and

$$\begin{bmatrix} 1 & -2 & 3 \\ 2 & 1 & 4 \\ 1 & -7 & 5 \end{bmatrix} \begin{bmatrix} x_1 \\ x_2 \\ x_3 \end{bmatrix} = \begin{bmatrix} 2 \\ 7 \\ -1 \end{bmatrix}$$

(a) Find a general solution of the homogeneous system.

(b) Confirm that $x_1 = 1, x_2 = 1, x_3 = 1$ is a solution of the nonhomogeneous system.

(c) Use the results in parts (a) and (b) to find a general solution of the nonhomogeneous system.

(d) Check your result in part (c) by solving the nonhomogeneous system directly.

▶ In Exercises 27–28, find a general solution of the system, and use that solution to find a general solution of the associated homogeneous system and a particular solution of the given system.

27.
$$\begin{bmatrix} 3 & 4 & 1 & 2 \\ 6 & 8 & 2 & 5 \\ 9 & 12 & 3 & 10 \end{bmatrix} \begin{bmatrix} x_1 \\ x_2 \\ x_3 \\ x_4 \end{bmatrix} = \begin{bmatrix} 3 \\ 7 \\ 13 \end{bmatrix}$$

28.
$$\begin{bmatrix} 9 & -3 & 5 & 6 \\ 6 & -2 & 3 & 1 \\ 3 & -1 & 3 & 14 \end{bmatrix} \begin{bmatrix} x_1 \\ x_2 \\ x_3 \\ x_4 \end{bmatrix} = \begin{bmatrix} 4 \\ 5 \\ -8 \end{bmatrix}$$

True-False Exercises

In parts (a)–(f) determine whether the statement is true or false, and justify your answer.

(a) The vector equation of a line can be determined from any point lying on the line and a nonzero vector parallel to the line.

(b) The vector equation of a plane can be determined from any point lying in the plane and a nonzero vector parallel to the plane.

(c) The points lying on a line through the origin in R^2 or R^3 are all scalar multiples of any nonzero vector on the line.

(d) All solution vectors of the linear system $A\mathbf{x} = \mathbf{b}$ are orthogonal to the row vectors of the matrix A if and only if $\mathbf{b} = \mathbf{0}$.

(e) The general solution of the nonhomogeneous linear system $A\mathbf{x} = \mathbf{b}$ can be obtained by adding $\mathbf{b}$ to the general solution of the homogeneous linear system $A\mathbf{x} = \mathbf{0}$.

(f) If $\mathbf{x}_1$ and $\mathbf{x}_2$ are two solutions of the nonhomogeneous linear system $A\mathbf{x} = \mathbf{b}$, then $\mathbf{x}_1 - \mathbf{x}_2$ is a solution of the corresponding homogeneous linear system.

3.5 Cross Product

This optional section is concerned with properties of vectors in 3-space that are important to physicists and engineers. It can be omitted, if desired, since subsequent sections do not depend on its content. Among other things, we define an operation that provides a way of constructing a vector in 3-space that is perpendicular to two given vectors, and we give a geometric interpretation of 3×3 determinants.

Cross Product of Vectors   In Section 3.2 we defined the dot product of two vectors **u** and **v** in n-space. That operation produced a *scalar* as its result. We will now define a type of vector multiplication that produces a *vector* as the result but which is applicable only to vectors in 3-space.

DEFINITION 1   If $\mathbf{u} = (u_1, u_2, u_3)$ and $\mathbf{v} = (v_1, v_2, v_3)$ are vectors in 3-space, then the **cross product** $\mathbf{u} \times \mathbf{v}$ is the vector defined by

$$\mathbf{u} \times \mathbf{v} = (u_2 v_3 - u_3 v_2, \, u_3 v_1 - u_1 v_3, \, u_1 v_2 - u_2 v_1)$$

or, in determinant notation,

$$\mathbf{u} \times \mathbf{v} = \left(\begin{vmatrix} u_2 & u_3 \\ v_2 & v_3 \end{vmatrix}, \, -\begin{vmatrix} u_1 & u_3 \\ v_1 & v_3 \end{vmatrix}, \, \begin{vmatrix} u_1 & u_2 \\ v_1 & v_2 \end{vmatrix} \right) \tag{1}$$

Remark   Instead of memorizing (1), you can obtain the components of $\mathbf{u} \times \mathbf{v}$ as follows:

- Form the 2×3 matrix $\begin{bmatrix} u_1 & u_2 & u_3 \\ v_1 & v_2 & v_3 \end{bmatrix}$ whose first row contains the components of **u** and whose second row contains the components of **v**.

- To find the first component of $\mathbf{u} \times \mathbf{v}$, delete the first column and take the determinant; to find the second component, delete the second column and take the negative of the determinant; and to find the third component, delete the third column and take the determinant.

▶ **EXAMPLE 1   Calculating a Cross Product**

Find $\mathbf{u} \times \mathbf{v}$, where $\mathbf{u} = (1, 2, -2)$ and $\mathbf{v} = (3, 0, 1)$.

Solution   From either (1) or the mnemonic in the preceding remark, we have

$$\mathbf{u} \times \mathbf{v} = \left(\begin{vmatrix} 2 & -2 \\ 0 & 1 \end{vmatrix}, \, -\begin{vmatrix} 1 & -2 \\ 3 & 1 \end{vmatrix}, \, \begin{vmatrix} 1 & 2 \\ 3 & 0 \end{vmatrix} \right)$$
$$= (2, -7, -6) \blacktriangleleft$$

The following theorem gives some important relationships between the dot product and cross product and also shows that $\mathbf{u} \times \mathbf{v}$ is orthogonal to both **u** and **v**.

Historical Note   The cross product notation $A \times B$ was introduced by the American physicist and mathematician J. Willard Gibbs, (see p. 134) in a series of unpublished lecture notes for his students at Yale University. It appeared in a published work for the first time in the second edition of the book *Vector Analysis*, by Edwin Wilson (1879–1964), a student of Gibbs. Gibbs originally referred to $A \times B$ as the "skew product."

THEOREM 3.5.1 **Relationships Involving Cross Product and Dot Product**

If **u**, **v**, *and* **w** *are vectors in 3-space, then*

(a) $\mathbf{u} \cdot (\mathbf{u} \times \mathbf{v}) = 0$ (**u** × **v** *is orthogonal to* **u**)

(b) $\mathbf{v} \cdot (\mathbf{u} \times \mathbf{v}) = 0$ (**u** × **v** *is orthogonal to* **v**)

(c) $\|\mathbf{u} \times \mathbf{v}\|^2 = \|\mathbf{u}\|^2 \|\mathbf{v}\|^2 - (\mathbf{u} \cdot \mathbf{v})^2$ (*Lagrange's identity*)

(d) $\mathbf{u} \times (\mathbf{v} \times \mathbf{w}) = (\mathbf{u} \cdot \mathbf{w})\mathbf{v} - (\mathbf{u} \cdot \mathbf{v})\mathbf{w}$ (*relationship between cross and dot products*)

(e) $(\mathbf{u} \times \mathbf{v}) \times \mathbf{w} = (\mathbf{u} \cdot \mathbf{w})\mathbf{v} - (\mathbf{v} \cdot \mathbf{w})\mathbf{u}$ (*relationship between cross and dot products*)

Proof (a)   Let $\mathbf{u} = (u_1, u_2, u_3)$ and $\mathbf{v} = (v_1, v_2, v_3)$. Then

$$\mathbf{u} \cdot (\mathbf{u} \times \mathbf{v}) = (u_1, u_2, u_3) \cdot (u_2 v_3 - u_3 v_2, u_3 v_1 - u_1 v_3, u_1 v_2 - u_2 v_1)$$
$$= u_1(u_2 v_3 - u_3 v_2) + u_2(u_3 v_1 - u_1 v_3) + u_3(u_1 v_2 - u_2 v_1) = 0$$

Proof (b)   Similar to (a).

Proof (c)   Since

$$\|\mathbf{u} \times \mathbf{v}\|^2 = (u_2 v_3 - u_3 v_2)^2 + (u_3 v_1 - u_1 v_3)^2 + (u_1 v_2 - u_2 v_1)^2 \tag{2}$$

and

$$\|\mathbf{u}\|^2 \|\mathbf{v}\|^2 - (\mathbf{u} \cdot \mathbf{v})^2 = (u_1^2 + u_2^2 + u_3^2)(v_1^2 + v_2^2 + v_3^2) - (u_1 v_1 + u_2 v_2 + u_3 v_3)^2 \tag{3}$$

the proof can be completed by "multiplying out" the right sides of (2) and (3) and verifying their equality.

Proof (d) and (e)   See Exercises 38 and 39. ◀

▶ **EXAMPLE 2** **u × v Is Perpendicular to u and to v**

Consider the vectors

$$\mathbf{u} = (1, 2, -2) \quad \text{and} \quad \mathbf{v} = (3, 0, 1)$$

In Example 1 we showed that

$$\mathbf{u} \times \mathbf{v} = (2, -7, -6)$$

Since

$$\mathbf{u} \cdot (\mathbf{u} \times \mathbf{v}) = (1)(2) + (2)(-7) + (-2)(-6) = 0$$

and

$$\mathbf{v} \cdot (\mathbf{u} \times \mathbf{v}) = (3)(2) + (0)(-7) + (1)(-6) = 0$$

u × **v** is orthogonal to both **u** and **v**, as guaranteed by Theorem 3.5.1. ◀

**Joseph Louis Lagrange
(1736–1813)**

The main arithmetic properties of the cross product are listed in the next theorem.

THEOREM 3.5.2 **Properties of Cross Product**

If **u**, **v**, *and* **w** *are any vectors in 3-space and k is any scalar, then*:

(*a*) $\mathbf{u} \times \mathbf{v} = -(\mathbf{v} \times \mathbf{u})$

(*b*) $\mathbf{u} \times (\mathbf{v} + \mathbf{w}) = (\mathbf{u} \times \mathbf{v}) + (\mathbf{u} \times \mathbf{w})$

(*c*) $(\mathbf{u} + \mathbf{v}) \times \mathbf{w} = (\mathbf{u} \times \mathbf{w}) + (\mathbf{v} \times \mathbf{w})$

(*d*) $k(\mathbf{u} \times \mathbf{v}) = (k\mathbf{u}) \times \mathbf{v} = \mathbf{u} \times (k\mathbf{v})$

(*e*) $\mathbf{u} \times \mathbf{0} = \mathbf{0} \times \mathbf{u} = \mathbf{0}$

(*f*) $\mathbf{u} \times \mathbf{u} = \mathbf{0}$

The proofs follow immediately from Formula (1) and properties of determinants; for example, part (*a*) can be proved as follows.

Proof (*a*) Interchanging **u** and **v** in (1) interchanges the rows of the three determinants on the right side of (1) and hence changes the sign of each component in the cross product. Thus $\mathbf{u} \times \mathbf{v} = -(\mathbf{v} \times \mathbf{u})$. ◀

The proofs of the remaining parts are left as exercises.

▶ **EXAMPLE 3** **Standard Unit Vectors**

Consider the vectors

$$\mathbf{i} = (1, 0, 0), \quad \mathbf{j} = (0, 1, 0), \quad \mathbf{k} = (0, 0, 1)$$

These vectors each have length 1 and lie along the coordinate axes (Figure 3.5.1). They are called the ***standard unit vectors*** in 3-space. Every vector $\mathbf{v} = (v_1, v_2, v_3)$ in 3-space is expressible in terms of **i**, **j**, and **k** since we can write

$$\mathbf{v} = (v_1, v_2, v_3) = v_1(1, 0, 0) + v_2(0, 1, 0) + v_3(0, 0, 1) = v_1\mathbf{i} + v_2\mathbf{j} + v_3\mathbf{k}$$

For example,

$$(2, -3, 4) = 2\mathbf{i} - 3\mathbf{j} + 4\mathbf{k}$$

From (1) we obtain

$$\mathbf{i} \times \mathbf{j} = \left(\begin{vmatrix} 0 & 0 \\ 1 & 0 \end{vmatrix}, -\begin{vmatrix} 1 & 0 \\ 0 & 0 \end{vmatrix}, \begin{vmatrix} 1 & 0 \\ 0 & 1 \end{vmatrix} \right) = (0, 0, 1) = \mathbf{k} \;\blacktriangleleft$$

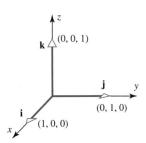

▲ **Figure 3.5.1** The standard unit vectors.

You should have no trouble obtaining the following results:

$$\begin{array}{lll}
\mathbf{i} \times \mathbf{i} = \mathbf{0} & \mathbf{j} \times \mathbf{j} = \mathbf{0} & \mathbf{k} \times \mathbf{k} = \mathbf{0} \\
\mathbf{i} \times \mathbf{j} = \mathbf{k} & \mathbf{j} \times \mathbf{k} = \mathbf{i} & \mathbf{k} \times \mathbf{i} = \mathbf{j} \\
\mathbf{j} \times \mathbf{i} = -\mathbf{k} & \mathbf{k} \times \mathbf{j} = -\mathbf{i} & \mathbf{i} \times \mathbf{k} = -\mathbf{j}
\end{array}$$

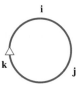

▲ **Figure 3.5.2**

Figure 3.5.2 is helpful for remembering these results. Referring to this diagram, the cross product of two consecutive vectors going clockwise is the next vector around, and the cross product of two consecutive vectors going counterclockwise is the negative of the next vector around.

Determinant Form of Cross Product

It is also worth noting that a cross product can be represented symbolically in the form

$$\mathbf{u} \times \mathbf{v} = \begin{vmatrix} \mathbf{i} & \mathbf{j} & \mathbf{k} \\ u_1 & u_2 & u_3 \\ v_1 & v_2 & v_3 \end{vmatrix} = \begin{vmatrix} u_2 & u_3 \\ v_2 & v_3 \end{vmatrix} \mathbf{i} - \begin{vmatrix} u_1 & u_3 \\ v_1 & v_3 \end{vmatrix} \mathbf{j} + \begin{vmatrix} u_1 & u_2 \\ v_1 & v_2 \end{vmatrix} \mathbf{k} \tag{4}$$

For example, if $\mathbf{u} = (1, 2, -2)$ and $\mathbf{v} = (3, 0, 1)$, then

$$\mathbf{u} \times \mathbf{v} = \begin{vmatrix} \mathbf{i} & \mathbf{j} & \mathbf{k} \\ 1 & 2 & -2 \\ 3 & 0 & 1 \end{vmatrix} = 2\mathbf{i} - 7\mathbf{j} - 6\mathbf{k}$$

which agrees with the result obtained in Example 1.

WARNING It is not true in general that $\mathbf{u} \times (\mathbf{v} \times \mathbf{w}) = (\mathbf{u} \times \mathbf{v}) \times \mathbf{w}$. For example,

$$\mathbf{i} \times (\mathbf{j} \times \mathbf{j}) = \mathbf{i} \times \mathbf{0} = \mathbf{0}$$

and

$$(\mathbf{i} \times \mathbf{j}) \times \mathbf{j} = \mathbf{k} \times \mathbf{j} = -\mathbf{i}$$

so

$$\mathbf{i} \times (\mathbf{j} \times \mathbf{j}) \neq (\mathbf{i} \times \mathbf{j}) \times \mathbf{j}$$

▲ Figure 3.5.3

We know from Theorem 3.5.1 that $\mathbf{u} \times \mathbf{v}$ is orthogonal to both $\mathbf{u}$ and $\mathbf{v}$. If $\mathbf{u}$ and $\mathbf{v}$ are nonzero vectors, it can be shown that the direction of $\mathbf{u} \times \mathbf{v}$ can be determined using the following "right-hand rule" (Figure 3.5.3): Let θ be the angle between $\mathbf{u}$ and $\mathbf{v}$, and suppose $\mathbf{u}$ is rotated through the angle θ until it coincides with $\mathbf{v}$. If the fingers of the right hand are cupped so that they point in the direction of rotation, then the thumb indicates (roughly) the direction of $\mathbf{u} \times \mathbf{v}$.

You may find it instructive to practice this rule with the products

$$\mathbf{i} \times \mathbf{j} = \mathbf{k}, \quad \mathbf{j} \times \mathbf{k} = \mathbf{i}, \quad \mathbf{k} \times \mathbf{i} = \mathbf{j}$$

Geometric Interpretation of Cross Product

If $\mathbf{u}$ and $\mathbf{v}$ are vectors in 3-space, then the norm of $\mathbf{u} \times \mathbf{v}$ has a useful geometric interpretation. Lagrange's identity, given in Theorem 3.5.1, states that

$$\|\mathbf{u} \times \mathbf{v}\|^2 = \|\mathbf{u}\|^2 \|\mathbf{v}\|^2 - (\mathbf{u} \cdot \mathbf{v})^2 \tag{5}$$

If θ denotes the angle between $\mathbf{u}$ and $\mathbf{v}$, then $\mathbf{u} \cdot \mathbf{v} = \|\mathbf{u}\| \|\mathbf{v}\| \cos\theta$, so (5) can be rewritten as

$$\begin{aligned} \|\mathbf{u} \times \mathbf{v}\|^2 &= \|\mathbf{u}\|^2 \|\mathbf{v}\|^2 - \|\mathbf{u}\|^2 \|\mathbf{v}\|^2 \cos^2\theta \\ &= \|\mathbf{u}\|^2 \|\mathbf{v}\|^2 (1 - \cos^2\theta) \\ &= \|\mathbf{u}\|^2 \|\mathbf{v}\|^2 \sin^2\theta \end{aligned}$$

Since $0 \leq \theta \leq \pi$, it follows that $\sin\theta \geq 0$, so this can be rewritten as

$$\|\mathbf{u} \times \mathbf{v}\| = \|\mathbf{u}\| \|\mathbf{v}\| \sin\theta \tag{6}$$

▲ Figure 3.5.4

But $\|\mathbf{v}\| \sin\theta$ is the altitude of the parallelogram determined by $\mathbf{u}$ and $\mathbf{v}$ (Figure 3.5.4). Thus, from (6), the area A of this parallelogram is given by

$$A = \text{(base)(altitude)} = \|\mathbf{u}\| \|\mathbf{v}\| \sin\theta = \|\mathbf{u} \times \mathbf{v}\|$$

This result is even correct if **u** and **v** are collinear, since the parallelogram determined by **u** and **v** has zero area and from (6) we have $\mathbf{u} \times \mathbf{v} = \mathbf{0}$ because $\theta = 0$ in this case. Thus we have the following theorem.

> **THEOREM 3.5.3 Area of a Parallelogram**
>
> *If **u** and **v** are vectors in 3-space, then $\|\mathbf{u} \times \mathbf{v}\|$ is equal to the area of the parallelogram determined by **u** and **v**.*

▶ **EXAMPLE 4 Area of a Triangle**

Find the area of the triangle determined by the points $P_1(2, 2, 0)$, $P_2(-1, 0, 2)$, and $P_3(0, 4, 3)$.

Solution The area A of the triangle is $\frac{1}{2}$ the area of the parallelogram determined by the vectors $\overrightarrow{P_1P_2}$ and $\overrightarrow{P_1P_3}$ (Figure 3.5.5). Using the method discussed in Example 1 of Section 3.1, $\overrightarrow{P_1P_2} = (-3, -2, 2)$ and $\overrightarrow{P_1P_3} = (-2, 2, 3)$. It follows that

$$\overrightarrow{P_1P_2} \times \overrightarrow{P_1P_3} = (-10, 5, -10)$$

(verify) and consequently that

$$A = \tfrac{1}{2}\|\overrightarrow{P_1P_2} \times \overrightarrow{P_1P_3}\| = \tfrac{1}{2}(15) = \tfrac{15}{2} \quad \blacktriangleleft$$

▲ Figure 3.5.5

> **DEFINITION 2** If **u**, **v**, and **w** are vectors in 3-space, then
>
> $$\mathbf{u} \cdot (\mathbf{v} \times \mathbf{w})$$
>
> is called the *scalar triple product* of **u**, **v**, and **w**.

The scalar triple product of $\mathbf{u} = (u_1, u_2, u_3)$, $\mathbf{v} = (v_1, v_2, v_3)$, and $\mathbf{w} = (w_1, w_2, w_3)$ can be calculated from the formula

$$\mathbf{u} \cdot (\mathbf{v} \times \mathbf{w}) = \begin{vmatrix} u_1 & u_2 & u_3 \\ v_1 & v_2 & v_3 \\ w_1 & w_2 & w_3 \end{vmatrix} \quad (7)$$

This follows from Formula (4) since

$$\mathbf{u} \cdot (\mathbf{v} \times \mathbf{w}) = \mathbf{u} \cdot \left(\begin{vmatrix} v_2 & v_3 \\ w_2 & w_3 \end{vmatrix} \mathbf{i} - \begin{vmatrix} v_1 & v_3 \\ w_1 & w_3 \end{vmatrix} \mathbf{j} + \begin{vmatrix} v_1 & v_2 \\ w_1 & w_2 \end{vmatrix} \mathbf{k} \right)$$

$$= \begin{vmatrix} v_2 & v_3 \\ w_2 & w_3 \end{vmatrix} u_1 - \begin{vmatrix} v_1 & v_3 \\ w_1 & w_3 \end{vmatrix} u_2 + \begin{vmatrix} v_1 & v_2 \\ w_1 & w_2 \end{vmatrix} u_3$$

$$= \begin{vmatrix} u_1 & u_2 & u_3 \\ v_1 & v_2 & v_3 \\ w_1 & w_2 & w_3 \end{vmatrix}$$

▶ **EXAMPLE 5 Calculating a Scalar Triple Product**

Calculate the scalar triple product $\mathbf{u} \cdot (\mathbf{v} \times \mathbf{w})$ of the vectors

$$\mathbf{u} = 3\mathbf{i} - 2\mathbf{j} - 5\mathbf{k}, \quad \mathbf{v} = \mathbf{i} + 4\mathbf{j} - 4\mathbf{k}, \quad \mathbf{w} = 3\mathbf{j} + 2\mathbf{k}$$

Solution   From (7),

$$\mathbf{u} \cdot (\mathbf{v} \times \mathbf{w}) = \begin{vmatrix} 3 & -2 & -5 \\ 1 & 4 & -4 \\ 0 & 3 & 2 \end{vmatrix}$$

$$= 3 \begin{vmatrix} 4 & -4 \\ 3 & 2 \end{vmatrix} - (-2) \begin{vmatrix} 1 & -4 \\ 0 & 2 \end{vmatrix} + (-5) \begin{vmatrix} 1 & 4 \\ 0 & 3 \end{vmatrix}$$

$$= 60 + 4 - 15 = 49 \blacktriangleleft$$

Remark   The symbol $(\mathbf{u} \cdot \mathbf{v}) \times \mathbf{w}$ makes no sense because we cannot form the cross product of a scalar and a vector. Thus, no ambiguity arises if we write $\mathbf{u} \cdot \mathbf{v} \times \mathbf{w}$ rather than $\mathbf{u} \cdot (\mathbf{v} \times \mathbf{w})$. However, for clarity we will usually keep the parentheses.

▲ Figure 3.5.6

It follows from (7) that

$$\mathbf{u} \cdot (\mathbf{v} \times \mathbf{w}) = \mathbf{w} \cdot (\mathbf{u} \times \mathbf{v}) = \mathbf{v} \cdot (\mathbf{w} \times \mathbf{u})$$

since the 3×3 determinants that represent these products can be obtained from one another by *two* row interchanges. (Verify.) These relationships can be remembered by moving the vectors $\mathbf{u}$, $\mathbf{v}$, and $\mathbf{w}$ clockwise around the vertices of the triangle in Figure 3.5.6.

Geometric Interpretation of Determinants   The next theorem provides a useful geometric interpretation of 2×2 and 3×3 determinants.

THEOREM 3.5.4

(*a*) *The absolute value of the determinant*

$$\det \begin{bmatrix} u_1 & u_2 \\ v_1 & v_2 \end{bmatrix}$$

is equal to the area of the parallelogram in 2-space determined by the vectors $\mathbf{u} = (u_1, u_2)$ *and* $\mathbf{v} = (v_1, v_2)$. *(See Figure 3.5.7a.)*

(*b*) *The absolute value of the determinant*

$$\det \begin{bmatrix} u_1 & u_2 & u_3 \\ v_1 & v_2 & v_3 \\ w_1 & w_2 & w_3 \end{bmatrix}$$

is equal to the volume of the parallelepiped in 3-space determined by the vectors $\mathbf{u} = (u_1, u_2, u_3)$, $\mathbf{v} = (v_1, v_2, v_3)$, *and* $\mathbf{w} = (w_1, w_2, w_3)$. *(See Figure 3.5.7b.)*

Proof (a)   The key to the proof is to use Theorem 3.5.3. However, that theorem applies to vectors in 3-space, whereas $\mathbf{u} = (u_1, u_2)$ and $\mathbf{v} = (v_1, v_2)$ are vectors in 2-space. To circumvent this "dimension problem," we will view $\mathbf{u}$ and $\mathbf{v}$ as vectors in the xy-plane of an xyz-coordinate system (Figure 3.5.7c), in which case these vectors are expressed as $\mathbf{u} = (u_1, u_2, 0)$ and $\mathbf{v} = (v_1, v_2, 0)$. Thus

$$\mathbf{u} \times \mathbf{v} = \begin{vmatrix} \mathbf{i} & \mathbf{j} & \mathbf{k} \\ u_1 & u_2 & 0 \\ v_1 & v_2 & 0 \end{vmatrix} = \begin{vmatrix} u_1 & u_2 \\ v_1 & v_2 \end{vmatrix} \mathbf{k} = \det \begin{bmatrix} u_1 & u_2 \\ v_1 & v_2 \end{bmatrix} \mathbf{k}$$

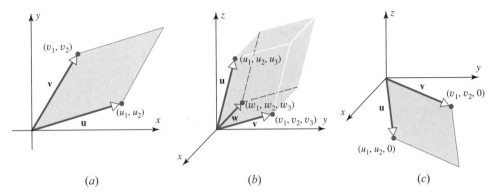

▲ Figure 3.5.7

It now follows from Theorem 3.5.3 and the fact that $\|\mathbf{k}\| = 1$ that the area A of the parallelogram determined by $\mathbf{u}$ and $\mathbf{v}$ is

$$A = \|\mathbf{u} \times \mathbf{v}\| = \left\| \det \begin{bmatrix} u_1 & u_2 \\ v_1 & v_2 \end{bmatrix} \mathbf{k} \right\| = \left| \det \begin{bmatrix} u_1 & u_2 \\ v_1 & v_2 \end{bmatrix} \right| \|\mathbf{k}\| = \left| \det \begin{bmatrix} u_1 & u_2 \\ v_1 & v_2 \end{bmatrix} \right|$$

which completes the proof.

▲ Figure 3.5.8

Proof (b)   As shown in Figure 3.5.8, take the base of the parallelepiped determined by $\mathbf{u}$, $\mathbf{v}$, and $\mathbf{w}$ to be the parallelogram determined by $\mathbf{v}$ and $\mathbf{w}$. It follows from Theorem 3.5.3 that the area of the base is $\|\mathbf{v} \times \mathbf{w}\|$ and, as illustrated in Figure 3.5.8, the height h of the parallelepiped is the length of the orthogonal projection of $\mathbf{u}$ on $\mathbf{v} \times \mathbf{w}$. Therefore, by Formula (12) of Section 3.3,

$$h = \|\mathrm{proj}_{\mathbf{v} \times \mathbf{w}} \mathbf{u}\| = \frac{|\mathbf{u} \cdot (\mathbf{v} \times \mathbf{w})|}{\|\mathbf{v} \times \mathbf{w}\|}$$

It follows that the volume V of the parallelepiped is

$$V = (\text{area of base}) \cdot \text{height} = \|\mathbf{v} \times \mathbf{w}\| \frac{|\mathbf{u} \cdot (\mathbf{v} \times \mathbf{w})|}{\|\mathbf{v} \times \mathbf{w}\|} = |\mathbf{u} \cdot (\mathbf{v} \times \mathbf{w})|$$

so from (7),

$$V = \left| \det \begin{bmatrix} u_1 & u_2 & u_3 \\ v_1 & v_2 & v_3 \\ w_1 & w_2 & w_3 \end{bmatrix} \right| \tag{8}$$

which completes the proof.   ◀

Remark   If V denotes the volume of the parallelepiped determined by vectors $\mathbf{u}$, $\mathbf{v}$, and $\mathbf{w}$, then it follows from Formulas (7) and (8) that

$$V = \begin{bmatrix} \text{volume of parallelepiped} \\ \text{determined by } \mathbf{u}, \mathbf{v}, \text{ and } \mathbf{w} \end{bmatrix} = |\mathbf{u} \cdot (\mathbf{v} \times \mathbf{w})| \tag{9}$$

From this result and the discussion immediately following Definition 3 of Section 3.2, we can conclude that

$$\mathbf{u} \cdot (\mathbf{v} \times \mathbf{w}) = \pm V$$

where the $+$ or $-$ results depending on whether $\mathbf{u}$ makes an acute or an obtuse angle with $\mathbf{v} \times \mathbf{w}$.

Formula (9) leads to a useful test for ascertaining whether three given vectors lie in the same plane. Since three vectors not in the same plane determine a parallelepiped of positive volume, it follows from (9) that $|\mathbf{u} \cdot (\mathbf{v} \times \mathbf{w})| = 0$ if and only if the vectors $\mathbf{u}$, $\mathbf{v}$, and $\mathbf{w}$ lie in the same plane. Thus we have the following result.

THEOREM 3.5.5 *If the vectors* $\mathbf{u} = (u_1, u_2, u_3)$, $\mathbf{v} = (v_1, v_2, v_3)$, *and* $\mathbf{w} = (w_1, w_2, w_3)$ *have the same initial point, then they lie in the same plane if and only if*

$$\mathbf{u} \cdot (\mathbf{v} \times \mathbf{w}) = \begin{vmatrix} u_1 & u_2 & u_3 \\ v_1 & v_2 & v_3 \\ w_1 & w_2 & w_3 \end{vmatrix} = 0$$

Concept Review

- Cross product of two vectors
- Determinant form of cross product
- Scalar triple product

Skills

- Compute the cross product of two vectors $\mathbf{u}$ and $\mathbf{v}$ in R^3.
- Know the geometric relationship between $\mathbf{u} \times \mathbf{v}$ to $\mathbf{u}$ and $\mathbf{v}$.
- Know the properties of the cross product (listed in Theorem 3.5.2).
- Compute the scalar triple product of three vectors in 3-space.
- Know the geometric interpretation of the scalar triple product.
- Compute the areas of triangles and parallelograms determined by two vectors or three points in 2-space or 3-space.
- Use the scalar triple product to determine whether three given vectors in 3-space are collinear.

Exercise Set 3.5

▶ In Exercises 1–2, let $\mathbf{u} = (3, 2, -1)$, $\mathbf{v} = (0, 2, -3)$, and $\mathbf{w} = (2, 6, 7)$. Compute the indicated vectors. ◀

1. (a) $\mathbf{v} \times \mathbf{w}$ (b) $\mathbf{u} \times (\mathbf{v} \times \mathbf{w})$ (c) $(\mathbf{u} \times \mathbf{v}) \times \mathbf{w}$

2. (a) $(\mathbf{u} \times \mathbf{v}) \times (\mathbf{v} \times \mathbf{w})$ (b) $\mathbf{u} \times (\mathbf{v} - 2\mathbf{w})$

   (c) $(\mathbf{u} \times \mathbf{v}) - 2\mathbf{w}$

▶ In Exercises 3–6, use the cross product to find a vector that is orthogonal to both $\mathbf{u}$ and $\mathbf{v}$. ◀

3. $\mathbf{u} = (-6, 4, 2)$, $\mathbf{v} = (3, 1, 5)$

4. $\mathbf{u} = (1, 1, -2)$, $\mathbf{v} = (2, -1, 2)$

5. $\mathbf{u} = (-2, 1, 5)$, $\mathbf{v} = (3, 0, -3)$

6. $\mathbf{u} = (3, 3, 1)$, $\mathbf{v} = (0, 4, 2)$

▶ In Exercises 7–10, find the area of the parallelogram determined by the given vectors $\mathbf{u}$ and $\mathbf{v}$. ◀

7. $\mathbf{u} = (1, -1, 2)$, $\mathbf{v} = (0, 3, 1)$

8. $\mathbf{u} = (3, -1, 4)$, $\mathbf{v} = (6, -2, 8)$

9. $\mathbf{u} = (2, 3, 0)$, $\mathbf{v} = (-1, 2, -2)$

10. $\mathbf{u} = (1, 1, 1)$, $\mathbf{v} = (3, 2, -5)$

▶ In Exercises 11–12, find the area of the parallelogram with the given vertices. ◀

11. $P_1(1, 2)$, $P_2(4, 4)$, $P_3(7, 5)$, $P_4(4, 3)$

12. $P_1(3, 2)$, $P_2(5, 4)$, $P_3(9, 4)$, $P_4(7, 2)$

▶ In Exercises 13–14, find the area of the triangle with the given vertices. ◀

13. $A(2, 0)$, $B(3, 4)$, $C(-1, 2)$

14. $A(1, 1)$, $B(2, 2)$, $C(3, -3)$

▶ In Exercises 15–16, find the area of the triangle in 3-space that has the given vertices. ◀

15. $P_1(2, 6, -1)$, $P_2(1, 1, 1)$, $P_3(4, 6, 2)$

16. $P(1, -1, 2)$, $Q(0, 3, 4)$, $R(6, 1, 8)$

▶ In Exercises 17–18, find the volume of the parallelepiped with sides $\mathbf{u}$, $\mathbf{v}$, and $\mathbf{w}$. ◀

17. $\mathbf{u} = (2, -6, 2)$, $\mathbf{v} = (0, 4, -2)$, $\mathbf{w} = (2, 2, -4)$

18. $\mathbf{u} = (3, 1, 2)$, $\mathbf{v} = (4, 5, 1)$, $\mathbf{w} = (1, 2, 4)$

▶ In Exercises 19–20, determine whether $\mathbf{u}$, $\mathbf{v}$, and $\mathbf{w}$ lie in the same plane when positioned so that their initial points coincide.

19. $\mathbf{u} = (-1, -2, 1)$, $\mathbf{v} = (3, 0, -2)$, $\mathbf{w} = (5, -4, 0)$

20. $\mathbf{u} = (5, -2, 1)$, $\mathbf{v} = (4, -1, 1)$, $\mathbf{w} = (1, -1, 0)$

▷ In Exercises **21–24**, compute the scalar triple product $\mathbf{u} \cdot (\mathbf{v} \times \mathbf{w})$. ◁

21. $\mathbf{u} = (-2, 0, 6)$, $\mathbf{v} = (1, -3, 1)$, $\mathbf{w} = (-5, -1, 1)$

22. $\mathbf{u} = (-1, 2, 4)$, $\mathbf{v} = (3, 4, -2)$, $\mathbf{w} = (-1, 2, 5)$

23. $\mathbf{u} = (a, 0, 0)$, $\mathbf{v} = (0, b, 0)$, $\mathbf{w} = (0, 0, c)$

24. $\mathbf{u} = (3, -1, 6)$, $\mathbf{v} = (2, 4, 3)$, $\mathbf{w} = (5, -1, 2)$

▷ In Exercises **25–26**, suppose that $\mathbf{u} \cdot (\mathbf{v} \times \mathbf{w}) = 3$. Find

25. (a) $\mathbf{u} \cdot (\mathbf{w} \times \mathbf{v})$ (b) $(\mathbf{v} \times \mathbf{w}) \cdot \mathbf{u}$ (c) $\mathbf{w} \cdot (\mathbf{u} \times \mathbf{v})$

26. (a) $\mathbf{v} \cdot (\mathbf{u} \times \mathbf{w})$ (b) $(\mathbf{u} \times \mathbf{w}) \cdot \mathbf{v}$ (c) $\mathbf{v} \cdot (\mathbf{w} \times \mathbf{w})$

27. (a) Find the area of the triangle having vertices $A(1, 0, 1)$, $B(0, 2, 3)$, and $C(2, 1, 0)$.

    (b) Use the result of part (a) to find the length of the altitude from vertex C to side AB.

28. Use the cross product to find the sine of the angle between the vectors $\mathbf{u} = (2, 3, -6)$ and $\mathbf{v} = (2, 3, 6)$.

29. Simplify $(\mathbf{u} + \mathbf{v}) \times (\mathbf{u} - \mathbf{v})$.

30. Let $\mathbf{a} = (a_1, a_2, a_3)$, $\mathbf{b} = (b_1, b_2, b_3)$, $\mathbf{c} = (c_1, c_2, c_3)$, and $\mathbf{d} = (d_1, d_2, d_3)$. Show that

$$(\mathbf{a} + \mathbf{d}) \cdot (\mathbf{b} \times \mathbf{c}) = \mathbf{a} \cdot (\mathbf{b} \times \mathbf{c}) + \mathbf{d} \cdot (\mathbf{b} \times \mathbf{c})$$

31. Let $\mathbf{u}$, $\mathbf{v}$, and $\mathbf{w}$ be nonzero vectors in 3-space with the same initial point, but such that no two of them are collinear. Show that

    (a) $\mathbf{u} \times (\mathbf{v} \times \mathbf{w})$ lies in the plane determined by $\mathbf{v}$ and $\mathbf{w}$.

    (b) $(\mathbf{u} \times \mathbf{v}) \times \mathbf{w}$ lies in the plane determined by $\mathbf{u}$ and $\mathbf{v}$.

32. Prove the following identities.

    (a) $(\mathbf{u} + k\mathbf{v}) \times \mathbf{v} = \mathbf{u} \times \mathbf{v}$

    (b) $\mathbf{u} \cdot (\mathbf{v} \times \mathbf{z}) = -(\mathbf{u} \times \mathbf{z}) \cdot \mathbf{v}$

33. Prove: If $\mathbf{a}$, $\mathbf{b}$, $\mathbf{c}$, and $\mathbf{d}$ lie in the same plane, then $(\mathbf{a} \times \mathbf{b}) \times (\mathbf{c} \times \mathbf{d}) = \mathbf{0}$.

34. Prove: If θ is the angle between $\mathbf{u}$ and $\mathbf{v}$ and $\mathbf{u} \cdot \mathbf{v} \neq 0$, then $\tan \theta = \|\mathbf{u} \times \mathbf{v}\| / (\mathbf{u} \cdot \mathbf{v})$.

35. Show that if $\mathbf{u}$, $\mathbf{v}$, and $\mathbf{w}$ are vectors in R^3, no two of which are collinear, then $\mathbf{u} \times (\mathbf{v} \times \mathbf{w})$ lies in the plane determined by $\mathbf{v}$ and $\mathbf{w}$.

36. It is a theorem of solid geometry that the volume of a tetrahedron is $\frac{1}{3}$(area of base) · (height). Use this result to prove

that the volume of a tetrahedron whose sides are the vectors $\mathbf{a}$, $\mathbf{b}$, and $\mathbf{c}$ is $\frac{1}{6}|\mathbf{a} \cdot (\mathbf{b} \times \mathbf{c})|$ (see the accompanying figure).

◁ **Figure Ex-36**

37. Use the result of Exercise 36 to find the volume of the tetrahedron with vertices P, Q, R, S.

    (a) $P(-1, 2, 0)$, $Q(2, 1, -3)$, $R(1, 1, 1)$, $S(3, -2, 3)$

    (b) $P(0, 0, 0)$, $Q(1, 2, -1)$, $R(3, 4, 0)$, $S(-1, -3, 4)$

38. Prove part (d) of Theorem 3.5.1. [*Hint:* First prove the result in the case where $\mathbf{w} = \mathbf{i} = (1, 0, 0)$, then when $\mathbf{w} = \mathbf{j} = (0, 1, 0)$, and then when $\mathbf{w} = \mathbf{k} = (0, 0, 1)$. Finally, prove it for an arbitrary vector $\mathbf{w} = (w_1, w_2, w_3)$ by writing $\mathbf{w} = w_1\mathbf{i} + w_2\mathbf{j} + w_3\mathbf{k}$.]

39. Prove part (e) of Theorem 3.5.1. [*Hint:* Apply part (a) of Theorem 3.5.2 to the result in part (d) of Theorem 3.5.1.]

40. Prove:

    (a) Prove (b) of Theorem 3.5.2.

    (b) Prove (c) of Theorem 3.5.2.

    (c) Prove (d) of Theorem 3.5.2.

    (d) Prove (e) of Theorem 3.5.2.

    (e) Prove (f) of Theorem 3.5.2.

True-False Exercises

In parts (a)–(f) determine whether the statement is true or false, and justify your answer.

(a) The cross product of two nonzero vectors $\mathbf{u}$ and $\mathbf{v}$ is a nonzero vector if and only if $\mathbf{u}$ and $\mathbf{v}$ are not parallel.

(b) A normal vector to a plane can be obtained by taking the cross product of two nonzero and noncollinear vectors lying in the plane.

(c) The scalar triple product of $\mathbf{u}$, $\mathbf{v}$, and $\mathbf{w}$ determines a vector whose length is equal to the volume of the parallelepiped determined by $\mathbf{u}$, $\mathbf{v}$, and $\mathbf{w}$.

(d) If $\mathbf{u}$ and $\mathbf{v}$ are vectors in 3-space, then $\|\mathbf{v} \times \mathbf{u}\|$ is equal to the area of the parallelogram determined by $\mathbf{u}$ and $\mathbf{v}$.

(e) For all vectors $\mathbf{u}$, $\mathbf{v}$, and $\mathbf{w}$ in 3-space, the vectors $(\mathbf{u} \times \mathbf{v}) \times \mathbf{w}$ and $\mathbf{u} \times (\mathbf{v} \times \mathbf{w})$ are the same.

(f) If $\mathbf{u}$, $\mathbf{v}$, and $\mathbf{w}$ are vectors in R^3, where $\mathbf{u}$ is nonzero and $\mathbf{u} \times \mathbf{v} = \mathbf{u} \times \mathbf{w}$, then $\mathbf{v} = \mathbf{w}$.

Chapter 3 Supplementary Exercises

1. Let $\mathbf{u} = (-2, 0, 4)$, $\mathbf{v} = (3, -1, 6)$, and $\mathbf{w} = (2, -5, -5)$. Compute

(a) $3\mathbf{v} - 2\mathbf{u}$ 　　　　 (b) $\|\mathbf{u} + \mathbf{v} + \mathbf{w}\|$

(c) the distance between $-3\mathbf{u}$ and $\mathbf{v} + 5\mathbf{w}$

(d) $\text{proj}_\mathbf{w}\mathbf{u}$ 　　　　 (e) $\mathbf{u} \cdot (\mathbf{v} \times \mathbf{w})$

(f) $(-5\mathbf{v} + \mathbf{w}) \times ((\mathbf{u} \cdot \mathbf{v})\mathbf{w})$

2. Repeat Exercise 1 for the vectors $\mathbf{u} = 3\mathbf{i} - 5\mathbf{j} + \mathbf{k}$, $\mathbf{v} = -2\mathbf{i} + 2\mathbf{k}$, and $\mathbf{w} = -\mathbf{j} + 4\mathbf{k}$.

3. Repeat parts (a)–(d) of Exercise 1 for the vectors $\mathbf{u} = (-2, 6, 2, 1)$, $\mathbf{v} = (-3, 0, 8, 0)$, and $\mathbf{w} = (9, 1, -6, -6)$.

4. Repeat parts (a)–(d) of Exercise 1 for the vectors $\mathbf{u} = (0, 5, 0, -1, -2)$, $\mathbf{v} = (1, -1, 6, -2, 0)$, and $\mathbf{w} = (-4, -1, 4, 0, 2)$.

▶ In Exercises 5–6, determine whether the given set of vectors forms an orthogonal set. If so, normalize each vector to form an orthonormal set. ◀

5. $(-32, -1, 19)$, $(3, -1, 5)$, $(1, 6, 2)$

6. $(-2, 0, 1)$, $(1, 1, 2)$, $(1, -5, 2)$

7. (a) The set of all vectors in R^2 that are orthogonal to a nonzero vector is what kind of geometric object?

(b) The set of all vectors in R^3 that are orthogonal to a nonzero vector is what kind of geometric object?

(c) The set of all vectors in R^2 that are orthogonal to two noncollinear vectors is what kind of geometric object?

(d) The set of all vectors in R^3 that are orthogonal to two noncollinear vectors is what kind of geometric object?

8. Show that $\mathbf{v}_1 = \left(\frac{2}{3}, \frac{1}{3}, \frac{2}{3}\right)$ and $\mathbf{v}_2 = \left(\frac{1}{3}, \frac{2}{3}, -\frac{2}{3}\right)$ are orthonormal vectors, and find a third vector $\mathbf{v}_3$ for which $\{\mathbf{v}_1, \mathbf{v}_2, \mathbf{v}_3\}$ is an orthonormal set.

9. *True or False:* If $\mathbf{u}$ and $\mathbf{v}$ are nonzero vectors such that $\|\mathbf{u} + \mathbf{v}\|^2 = \|\mathbf{u}\|^2 + \|\mathbf{v}\|^2$, then $\mathbf{u}$ and $\mathbf{v}$ are orthogonal.

10. *True or False:* If $\mathbf{u}$ is orthogonal to $\mathbf{v} + \mathbf{w}$, then $\mathbf{u}$ is orthogonal to $\mathbf{v}$ and $\mathbf{w}$.

11. Consider the points $P(3, -1, 4)$, $Q(6, 0, 2)$, and $R(5, 1, 1)$. Find the point S in R^3 whose first component is -1 and such that $\overrightarrow{PQ}$ is parallel to $\overrightarrow{RS}$.

12. Consider the points $P(-3, 1, 0, 6)$, $Q(0, 5, 1, -2)$, and $R(-4, 1, 4, 0)$. Find the point S in R^4 whose third component is 6 and such that $\overrightarrow{PQ}$ is parallel to $\overrightarrow{RS}$.

13. Using the points in Exercise 11, find the cosine of the angle between the vectors $\overrightarrow{PQ}$ and $\overrightarrow{PR}$.

14. Using the points in Exercise 12, find the cosine of the angle between the vectors $\overrightarrow{PQ}$ and $\overrightarrow{PR}$.

15. Find the distance between the point $P(-3, 1, 3)$ and the plane $5x + z = 3y - 4$.

16. Show that the planes $3x - y + 6z = 7$ and $-6x + 2y - 12z = 1$ are parallel, and find the distance between the planes.

▶ In Exercises 17–22, find vector and parametric equations for the line or plane in question. ◀

17. The plane in R^3 that contains the points $P(-2, 1, 3)$, $Q(-1, -1, 1)$, and $R(3, 0, -2)$.

18. The line in R^3 that contains the point $P(-1, 6, 0)$ and is orthogonal to the plane $4x - z = 5$.

19. The line in R^2 that is parallel to the vector $\mathbf{v} = (8, -1)$ and contains the point $P(0, -3)$.

20. The plane in R^3 that contains the point $P(-2, 1, 0)$ and parallel to the plane $-8x + 6y - z = 4$.

21. The line in R^2 with equation $y = 3x - 5$.

22. The plane in R^3 with equation $2x - 6y + 3z = 5$.

▶ In Exercises 23–25, find a point-normal equation for the given plane. ◀

23. The plane that is represented by the vector equation $(x, y, z) = (-1, 5, 6) + t_1(0, -1, 3) + t_2(2, -1, 0)$.

24. The plane that contains the point $P(-5, 1, 0)$ and is orthogonal to the line with parametric equations $x = 3 - 5t$, $y = 2t$, and $z = 7$.

25. The plane that passes through the points $P(9, 0, 4)$, $Q(-1, 4, 3)$, and $R(0, 6, -2)$.

26. Suppose that $\{\mathbf{v}_1, \mathbf{v}_2, \mathbf{v}_3\}$ and $\{\mathbf{w}_1, \mathbf{w}_2\}$ are two sets of vectors such that $\mathbf{v}_i$ and $\mathbf{w}_j$ are orthogonal for all i and j. Prove that if a_1, a_2, a_3, b_1, b_2 are any scalars, then the vectors $\mathbf{v} = a_1\mathbf{v}_1 + a_2\mathbf{v}_2 + a_3\mathbf{v}_3$ and $\mathbf{w} = b_1\mathbf{w}_1 + b_2\mathbf{w}_2$ are orthogonal.

27. Prove that if two vectors $\mathbf{u}$ and $\mathbf{v}$ in R^2 are orthogonal to a nonzero vector $\mathbf{w}$ in R^2, then $\mathbf{u}$ and $\mathbf{v}$ are scalar multiples of each other.

28. Prove that $\|\mathbf{u} + \mathbf{v}\| = \|\mathbf{u}\| + \|\mathbf{v}\|$ if and only if $\mathbf{u}$ and $\mathbf{v}$ are parallel vectors.

29. The equation $Ax + By = 0$ represents a line through the origin in R^2 if A and B are not both zero. What does this equation represent in R^3 if you think of it as $Ax + By + 0z = 0$? Explain.

CHAPTER 4

General Vector Spaces

CHAPTER CONTENTS

4.1   Real Vector Spaces 171
4.2   Subspaces 179
4.3   Linear Independence 190
4.4   Coordinates and Basis 200
4.5   Dimension 209
4.6   Change of Basis 217
4.7   Row Space, Column Space, and Null Space 225
4.8   Rank, Nullity, and the Fundamental Matrix Spaces 237
4.9   Matrix Transformations from R^n to R^m 247
4.10 Properties of Matrix Transformations 263
4.11 Geometry of Matrix Operators on R^2 273
4.12 Dynamical Systems and Markov Chains 282

INTRODUCTION   Recall that we began our study of vectors by viewing them as directed line segments (arrows). We then extended this idea by introducing rectangular coordinate systems, which enabled us to view vectors as ordered pairs and ordered triples of real numbers. As we developed properties of these vectors we noticed patterns in various formulas that enabled us to extend the notion of a vector to an n-tuple of real numbers. Although n-tuples took us outside the realm of our "visual experience," it gave us a valuable tool for understanding and studying systems of linear equations. In this chapter we will extend the concept of a vector yet again by using the most important algebraic properties of vectors in R^n as axioms. These axioms, if satisfied by a set of objects, will enable us to think of those objects as vectors.

4.1 Real Vector Spaces

In this section we will extend the concept of a vector by using the basic properties of vectors in R^n as axioms, which if satisfied by a set of objects, guarantee that those objects behave like familiar vectors.

Vector Space Axioms   The following definition consists of ten axioms, eight of which are properties of vectors in R^n that were stated in Theorem 3.1.1. It is important to keep in mind that one does not *prove* axioms; rather, they are assumptions that serve as the starting point for proving theorems.

171

> **DEFINITION 1** Let V be an arbitrary nonempty set of objects on which two operations are defined: addition, and multiplication by scalars. By **addition** we mean a rule for associating with each pair of objects $\mathbf{u}$ and $\mathbf{v}$ in V an object $\mathbf{u} + \mathbf{v}$, called the **sum** of $\mathbf{u}$ and $\mathbf{v}$; by **scalar multiplication** we mean a rule for associating with each scalar k and each object $\mathbf{u}$ in V an object $k\mathbf{u}$, called the **scalar multiple** of $\mathbf{u}$ by k. If the following axioms are satisfied by all objects $\mathbf{u}$, $\mathbf{v}$, $\mathbf{w}$ in V and all scalars k and m, then we call V a **vector space** and we call the objects in V **vectors**.
>
> 1. If $\mathbf{u}$ and $\mathbf{v}$ are objects in V, then $\mathbf{u} + \mathbf{v}$ is in V.
> 2. $\mathbf{u} + \mathbf{v} = \mathbf{v} + \mathbf{u}$
> 3. $\mathbf{u} + (\mathbf{v} + \mathbf{w}) = (\mathbf{u} + \mathbf{v}) + \mathbf{w}$
> 4. There is an object $\mathbf{0}$ in V, called a **zero vector** for V, such that $\mathbf{0} + \mathbf{u} = \mathbf{u} + \mathbf{0} = \mathbf{u}$ for all $\mathbf{u}$ in V.
> 5. For each $\mathbf{u}$ in V, there is an object $-\mathbf{u}$ in V, called a **negative** of $\mathbf{u}$, such that $\mathbf{u} + (-\mathbf{u}) = (-\mathbf{u}) + \mathbf{u} = \mathbf{0}$.
> 6. If k is any scalar and $\mathbf{u}$ is any object in V, then $k\mathbf{u}$ is in V.
> 7. $k(\mathbf{u} + \mathbf{v}) = k\mathbf{u} + k\mathbf{v}$
> 8. $(k + m)\mathbf{u} = k\mathbf{u} + m\mathbf{u}$
> 9. $k(m\mathbf{u}) = (km)(\mathbf{u})$
> 10. $1\mathbf{u} = \mathbf{u}$

Vector space scalars can be real numbers or complex numbers. Vector spaces with real scalars are called **real vector spaces** and those with complex scalars are called **complex vector spaces**. For now we will be concerned exclusively with real vector spaces. We will consider complex vector spaces later.

Observe that the definition of a vector space does not specify the nature of the vectors or the operations. Any kind of object can be a vector, and the operations of addition and scalar multiplication need not have any relationship to those on R^n. The only requirement is that the ten vector space axioms be satisfied. In the examples that follow we will use four basic steps to show that a set with two operations is a vector space.

> **To Show that a Set with Two Operations is a Vector Space**
>
> **Step 1.** Identify the set V of objects that will become vectors.
>
> **Step 2.** Identify the addition and scalar multiplication operations on V.
>
> **Step 3.** Verify Axioms 1 and 6; that is, adding two vectors in V produces a vector in V, and multiplying a vector in V by a scalar also produces a vector in V. Axiom 1 is called **closure under addition**, and Axiom 6 is called **closure under scalar multiplication**.
>
> **Step 4.** Confirm that Axioms 2, 3, 4, 5, 7, 8, 9, and 10 hold.

Historical Note The notion of an "abstract vector space" evolved over many years and had many contributors. The idea crystallized with the work of the German mathematician H. G. Grassmann, who published a paper in 1862 in which he considered abstract systems of unspecified elements on which he defined formal operations of addition and scalar multiplication. Grassmann's work was controversial, and others, including Augustin Cauchy (p. 137), laid reasonable claim to the idea.

[*Image:* ©Sueddeutsche Zeitung Photo/The Image Works]

Hermann Günther Grassmann (1809–1877)

Our first example is the simplest of all vector spaces in that it contains only one object. Since Axiom 4 requires that every vector space contain a zero vector, the object will have to be that vector.

▶ **EXAMPLE 1 The Zero Vector Space**

Let V consist of a single object, which we denote by $\mathbf{0}$, and define

$$\mathbf{0} + \mathbf{0} = \mathbf{0} \quad \text{and} \quad k\mathbf{0} = \mathbf{0}$$

for all scalars k. It is easy to check that all the vector space axioms are satisfied. We call this the ***zero vector space***. ◀

Our second example is one of the most important of all vector spaces—the familiar space R^n. It should not be surprising that the operations on R^n satisfy the vector space axioms because those axioms were based on known properties of operations on R^n.

▶ **EXAMPLE 2** R^n **Is a Vector Space**

Let $V = R^n$, and define the vector space operations on V to be the usual operations of addition and scalar multiplication of n-tuples; that is,

$$\mathbf{u} + \mathbf{v} = (u_1, u_2, \ldots, u_n) + (v_1, v_2, \ldots, v_n) = (u_1 + v_1, u_2 + v_2, \ldots, u_n + v_n)$$
$$k\mathbf{u} = (ku_1, ku_2, \ldots, ku_n)$$

The set $V = R^n$ is closed under addition and scalar multiplication because the foregoing operations produce n-tuples as their end result, and these operations satisfy Axioms 2, 3, 4, 5, 7, 8, 9, and 10 by virtue of Theorem 3.1.1. ◀

Our next example is a generalization of R^n in which we allow vectors to have infinitely many components.

▶ **EXAMPLE 3 The Vector Space of Infinite Sequences of Real Numbers**

Let V consist of objects of the form

$$\mathbf{u} = (u_1, u_2, \ldots, u_n, \ldots)$$

in which $u_1, u_2, \ldots, u_n, \ldots$ is an infinite sequence of real numbers. We define two infinite sequences to be *equal* if their corresponding components are equal, and we define addition and scalar multiplication componentwise by

$$\begin{aligned}
\mathbf{u} + \mathbf{v} &= (u_1, u_2, \ldots, u_n, \ldots) + (v_1, v_2, \ldots, v_n, \ldots) \\
&= (u_1 + v_1, u_2 + v_2, \ldots, u_n + v_n, \ldots) \\
k\mathbf{u} &= (ku_1, ku_2, \ldots, ku_n, \ldots)
\end{aligned}$$

We leave it as an exercise to confirm that V with these operations is a vector space. We will denote this vector space by the symbol R^∞. ◀

In the next example our vectors will be matrices. This may be a little confusing at first because matrices are composed of rows and columns, which are themselves vectors (row vectors and column vectors). However, here we will not be concerned with the individual rows and columns but rather with the properties of the matrix operations as they relate to the matrix as a whole.

▶ **EXAMPLE 4 A Vector Space of 2 × 2 Matrices**

Let V be the set of 2×2 matrices with real entries, and take the vector space operations on V to be the usual operations of matrix addition and scalar multiplication; that is,

$$\mathbf{u} + \mathbf{v} = \begin{bmatrix} u_{11} & u_{12} \\ u_{21} & u_{22} \end{bmatrix} + \begin{bmatrix} v_{11} & v_{12} \\ v_{21} & v_{22} \end{bmatrix} = \begin{bmatrix} u_{11} + v_{11} & u_{12} + v_{12} \\ u_{21} + v_{21} & u_{22} + v_{22} \end{bmatrix} \tag{1}$$

$$k\mathbf{u} = k \begin{bmatrix} u_{11} & u_{12} \\ u_{21} & u_{22} \end{bmatrix} = \begin{bmatrix} ku_{11} & ku_{12} \\ ku_{21} & ku_{22} \end{bmatrix}$$

Note that Equation (1) involves *three* different addition operations: the addition operation on vectors, the addition operation on matrices, and the addition operation on real numbers.

The set V is closed under addition and scalar multiplication because the foregoing operations produce 2×2 matrices as the end result. Thus, it remains to confirm that Axioms 2, 3, 4, 5, 7, 8, 9, and 10 hold. Some of these are standard properties of matrix operations. For example, Axiom 2 follows from Theorem 1.4.1*a* since

$$\mathbf{u} + \mathbf{v} = \begin{bmatrix} u_{11} & u_{12} \\ u_{21} & u_{22} \end{bmatrix} + \begin{bmatrix} v_{11} & v_{12} \\ v_{21} & v_{22} \end{bmatrix} = \begin{bmatrix} v_{11} & v_{12} \\ v_{21} & v_{22} \end{bmatrix} + \begin{bmatrix} u_{11} & u_{12} \\ u_{21} & u_{22} \end{bmatrix} = \mathbf{v} + \mathbf{u}$$

Similarly, Axioms 3, 7, 8, and 9 follow from parts (*b*), (*h*), (*j*), and (*e*), respectively, of that theorem (verify). This leaves Axioms 4, 5, and 10 that remain to be verified.

To confirm that Axiom 4 is satisfied, we must find a 2×2 matrix $\mathbf{0}$ in V for which $\mathbf{u} + \mathbf{0} = \mathbf{0} + \mathbf{u}$ for all 2×2 matrices in V. We can do this by taking

$$\mathbf{0} = \begin{bmatrix} 0 & 0 \\ 0 & 0 \end{bmatrix}$$

With this definition,

$$\mathbf{0} + \mathbf{u} = \begin{bmatrix} 0 & 0 \\ 0 & 0 \end{bmatrix} + \begin{bmatrix} u_{11} & u_{12} \\ u_{21} & u_{22} \end{bmatrix} = \begin{bmatrix} u_{11} & u_{12} \\ u_{21} & u_{22} \end{bmatrix} = \mathbf{u}$$

and similarly $\mathbf{u} + \mathbf{0} = \mathbf{u}$. To verify that Axiom 5 holds we must show that each object $\mathbf{u}$ in V has a negative $-\mathbf{u}$ in V such that $\mathbf{u} + (-\mathbf{u}) = \mathbf{0}$ and $(-\mathbf{u}) + \mathbf{u} = \mathbf{0}$. This can be done by defining the negative of $\mathbf{u}$ to be

$$-\mathbf{u} = \begin{bmatrix} -u_{11} & -u_{12} \\ -u_{21} & -u_{22} \end{bmatrix}$$

With this definition,

$$\mathbf{u} + (-\mathbf{u}) = \begin{bmatrix} u_{11} & u_{12} \\ u_{21} & u_{22} \end{bmatrix} + \begin{bmatrix} -u_{11} & -u_{12} \\ -u_{21} & -u_{22} \end{bmatrix} = \begin{bmatrix} 0 & 0 \\ 0 & 0 \end{bmatrix} = \mathbf{0}$$

and similarly $(-\mathbf{u}) + \mathbf{u} = \mathbf{0}$. Finally, Axiom 10 holds because

$$1\mathbf{u} = 1 \begin{bmatrix} u_{11} & u_{12} \\ u_{21} & u_{22} \end{bmatrix} = \begin{bmatrix} u_{11} & u_{12} \\ u_{21} & u_{22} \end{bmatrix} = \mathbf{u}$$

▶ **EXAMPLE 5 The Vector Space of $m \times n$ Matrices**

Example 4 is a special case of a more general class of vector spaces. You should have no trouble adapting the argument used in that example to show that the set V of all $m \times n$ matrices with the usual matrix operations of addition and scalar multiplication is a vector space. We will denote this vector space by the symbol M_{mn}. Thus, for example, the vector space in Example 4 is denoted as M_{22}.

▶ **EXAMPLE 6 The Vector Space of Real-Valued Functions**

Let V be the set of real-valued functions that are defined at each x in the interval $(-\infty, \infty)$. If $\mathbf{f} = f(x)$ and $\mathbf{g} = g(x)$ are two functions in V and if k is any scalar, then define the operations of addition and scalar multiplication by

$$(\mathbf{f} + \mathbf{g})(x) = f(x) + g(x) \tag{2}$$

$$(k\mathbf{f})(x) = kf(x) \tag{3}$$

One way to think about these operations is to view the numbers $f(x)$ and $g(x)$ as "components" of $\mathbf{f}$ and $\mathbf{g}$ at the point x, in which case Equations (2) and (3) state that two functions are added by adding corresponding components, and a function is multiplied by a scalar by multiplying each component by that scalar—exactly as in R^n and R^∞. This idea is illustrated in parts (a) and (b) of Figure 4.1.1. The set V with these operations is denoted by the symbol $F(-\infty, \infty)$. We can prove that this is a vector space as follows:

Axioms 1 and 6: These closure axioms require that if we add two functions that are defined at each x in the interval $(-\infty, \infty)$, then sums and scalar multiples of those functions are also defined at each x in the interval $(-\infty, \infty)$. This follows from Formulas (2) and (3).

Axiom 4: This axiom requires that there exists a function $\mathbf{0}$ in $F(-\infty, \infty)$, which when added to any other function $\mathbf{f}$ in $F(-\infty, \infty)$ produces $\mathbf{f}$ back again as the result. The function, whose value at every point x in the interval $(-\infty, \infty)$ is zero, has this property. Geometrically, the graph of the function $\mathbf{0}$ is the line that coincides with the x-axis.

In Example 6 the functions were defined on the entire interval $(-\infty, \infty)$. However, the arguments used in that example apply as well on all subintervals of $(-\infty, \infty)$, such as a closed interval $[a, b]$ or an open interval (a, b). We will denote the vector spaces of functions on these intervals by $F[a, b]$ and $F(a, b)$, respectively.

Axiom 5: This axiom requires that for each function $\mathbf{f}$ in $F(-\infty, \infty)$ there exists a function $-\mathbf{f}$ in $F(-\infty, \infty)$, which when added to $\mathbf{f}$ produces the function $\mathbf{0}$. The function defined by $-\mathbf{f}(x) = -f(x)$ has this property. The graph of $-\mathbf{f}$ can be obtained by reflecting the graph of $\mathbf{f}$ about the x-axis (Figure 4.1.1c).

Axioms 2, 3, 7, 8, 9, 10: The validity of each of these axioms follows from properties of real numbers. For example, if $\mathbf{f}$ and $\mathbf{g}$ are functions in $F(-\infty, \infty)$, then Axiom 2 requires that $\mathbf{f} + \mathbf{g} = \mathbf{g} + \mathbf{f}$. This follows from the computation

$$(\mathbf{f} + \mathbf{g})(x) = \mathbf{f}(x) + \mathbf{g}(x) = \mathbf{g}(x) + \mathbf{f}(x) = (\mathbf{g} + \mathbf{f})(x)$$

in which the first and last equalities follow from (2), and the middle equality is a property of real numbers. We will leave the proofs of the remaining parts as exercises. ◀

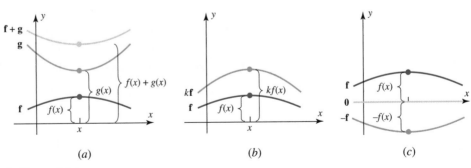

(a) (b) (c)

▲ Figure 4.1.1

It is important to recognize that you cannot impose any two operations on any set V and expect the vector space axioms to hold. For example, if V is the set of n-tuples with *positive* components, and if the standard operations from R^n are used, then V is not closed under scalar multiplication, because if $\mathbf{u}$ is a nonzero n-tuple in V, then $(-1)\mathbf{u}$ has at least one negative component and hence is not in V. The following is a less obvious example in which only one of the ten vector space axioms fails to hold.

▶ **EXAMPLE 7 A Set That Is Not a Vector Space**

Let $V = R^2$ and define addition and scalar multiplication operations as follows: If $\mathbf{u} = (u_1, u_2)$ and $\mathbf{v} = (v_1, v_2)$, then define

$$\mathbf{u} + \mathbf{v} = (u_1 + v_1, u_2 + v_2)$$

and if k is any real number, then define

$$k\mathbf{u} = (ku_1, 0)$$

For example, if $\mathbf{u} = (2, 4)$, $\mathbf{v} = (-3, 5)$, and $k = 7$, then

$$\mathbf{u} + \mathbf{v} = (2 + (-3), 4 + 5) = (-1, 9)$$
$$k\mathbf{u} = 7\mathbf{u} = (7 \cdot 2, 0) = (14, 0)$$

The addition operation is the standard one from R^2, but the scalar multiplication is not. In the exercises we will ask you to show that the first nine vector space axioms are satisfied. However, Axiom 10 fails to hold for certain vectors. For example, if $\mathbf{u} = (u_1, u_2)$ is such that $u_2 \neq 0$, then

$$1\mathbf{u} = 1(u_1, u_2) = (1 \cdot u_1, 0) = (u_1, 0) \neq \mathbf{u}$$

Thus, V is not a vector space with the stated operations. ◀

Our final example will be an unusual vector space that we have included to illustrate how varied vector spaces can be. Since the objects in this space will be real numbers, it will be important for you to keep track of which operations are intended as vector operations and which ones as ordinary operations on real numbers.

▶ **EXAMPLE 8 An Unusual Vector Space**

Let V be the set of positive real numbers, and define the operations on V to be

$$u + v = uv \qquad \text{[Vector addition is numerical multiplication.]}$$
$$ku = u^k \qquad \text{[Scalar multiplication is numerical exponentiation.]}$$

Thus, for example, $1 + 1 = 1$ and $(2)(1) = 1^2 = 1$—strange indeed, but nevertheless the set V with these operations satisfies the 10 vector space axioms and hence is a vector space. We will confirm Axioms 4, 5, and 7, and leave the others as exercises.

- Axiom 4—The zero vector in this space is the number 1 (i.e., $\mathbf{0} = 1$) since

$$u + 1 = u \cdot 1 = u$$

- Axiom 5—The negative of a vector u is its reciprocal (i.e., $-u = 1/u$) since

$$u + \frac{1}{u} = u\left(\frac{1}{u}\right) = 1 (= \mathbf{0})$$

- Axiom 7—$k(u + v) = (uv)^k = u^k v^k = (ku) + (kv)$. ◀

Some Properties of Vectors   The following is our first theorem about general vector spaces. As you will see, its proof is very formal with each step being justified by a vector space axiom or a known property of real numbers. There will not be many rigidly formal proofs of this type in the text, but we have included these to reinforce the idea that the familiar properties of vectors can all be derived from the vector space axioms.

THEOREM 4.1.1 *Let V be a vector space, **u** a vector in V, and k a scalar; then:*

(a) $0\mathbf{u} = \mathbf{0}$

(b) $k\mathbf{0} = \mathbf{0}$

(c) $(-1)\mathbf{u} = -\mathbf{u}$

(d) *If* $k\mathbf{u} = \mathbf{0}$, *then* $k = 0$ *or* $\mathbf{u} = \mathbf{0}$.

We will prove parts (a) and (c) and leave proofs of the remaining parts as exercises.

Proof (a)   We can write

$$0\mathbf{u} + 0\mathbf{u} = (0 + 0)\mathbf{u} \quad \text{[Axiom 8]}$$
$$= 0\mathbf{u} \qquad\qquad \text{[Property of the number 0]}$$

By Axiom 5 the vector $0\mathbf{u}$ has a negative, $-0\mathbf{u}$. Adding this negative to both sides above yields

$$[0\mathbf{u} + 0\mathbf{u}] + (-0\mathbf{u}) = 0\mathbf{u} + (-0\mathbf{u})$$

or

$$0\mathbf{u} + [0\mathbf{u} + (-0\mathbf{u})] = 0\mathbf{u} + (-0\mathbf{u}) \quad \text{[Axiom 3]}$$
$$0\mathbf{u} + \mathbf{0} = \mathbf{0} \qquad\qquad\qquad \text{[Axiom 5]}$$
$$0\mathbf{u} = \mathbf{0} \qquad\qquad\qquad\quad \text{[Axiom 4]}$$

Proof (c)   To prove that $(-1)\mathbf{u} = -\mathbf{u}$, we must show that $\mathbf{u} + (-1)\mathbf{u} = \mathbf{0}$. The proof is as follows:

$$\mathbf{u} + (-1)\mathbf{u} = 1\mathbf{u} + (-1)\mathbf{u} \quad \text{[Axiom 10]}$$
$$= (1 + (-1))\mathbf{u} \quad \text{[Axiom 8]}$$
$$= 0\mathbf{u} \qquad\qquad \text{[Property of numbers]}$$
$$= \mathbf{0} \qquad\qquad\; \text{[Part (a) of this theorem]} \quad \blacktriangleleft$$

A Closing Observation   This section of the text is very important to the overall plan of linear algebra in that it establishes a common thread between such diverse mathematical objects as geometric vectors, vectors in R^n, infinite sequences, matrices, and real-valued functions, to name a few. As a result, whenever we discover a new theorem about general vector spaces, we will at the same time be discovering a theorem about geometric vectors, vectors in R^n, sequences, matrices, real-valued functions, and about any new kinds of vectors that we might discover.

To illustrate this idea, consider what the rather innocent-looking result in part (a) of Theorem 4.1.1 says about the vector space in Example 8. Keeping in mind that the vectors in that space are positive real numbers, that scalar multiplication means numerical exponentiation, and that the zero vector is the number 1, the equation

$$0\mathbf{u} = \mathbf{0}$$

is a statement of the fact that if u is a positive real number, then

$$u^0 = 1$$

Concept Review

- Vector space
- Closure under addition
- Closure under scalar multiplication
- Examples of vector spaces

Skills

- Determine whether a given set with two operations is a vector space.
- Show that a set with two operations is not a vector space by demonstrating that at least one of the vector space axioms fails.

Exercise Set 4.1

1. Let V be the set of all ordered pairs of real numbers, and consider the following addition and scalar multiplication operations on $\mathbf{u} = (u_1, u_2)$ and $\mathbf{v} = (v_1, v_2)$:

$$\mathbf{u} + \mathbf{v} = (u_1 + v_1, u_2 + v_2), \quad k\mathbf{u} = (0, ku_2)$$

   (a) Compute $\mathbf{u} + \mathbf{v}$ and $k\mathbf{u}$ for $\mathbf{u} = (-1, 2)$, $\mathbf{v} = (3, 4)$, and $k = 3$.

   (b) In words, explain why V is closed under addition and scalar multiplication.

   (c) Since addition on V is the standard addition operation on R^2, certain vector space axioms hold for V because they are known to hold for R^2. Which axioms are they?

   (d) Show that Axioms 7, 8, and 9 hold.

   (e) Show that Axiom 10 fails and hence that V is not a vector space under the given operations.

2. Let V be the set of all ordered pairs of real numbers, and consider the following addition and scalar multiplication operations on $\mathbf{u} = (u_1, u_2)$ and $\mathbf{v} = (v_1, v_2)$:

$$\mathbf{u} + \mathbf{v} = (u_1 + v_1 + 1, u_2 + v_2 + 1), \quad k\mathbf{u} = (ku_1, ku_2)$$

   (a) Compute $\mathbf{u} + \mathbf{v}$ and $k\mathbf{u}$ for $\mathbf{u} = (0, 4)$, $\mathbf{v} = (1, -3)$, and $k = 2$.

   (b) Show that $(0, 0) \neq \mathbf{0}$.

   (c) Show that $(-1, -1) = \mathbf{0}$.

   (d) Show that Axiom 5 holds by producing an ordered pair $-\mathbf{u}$ such that $\mathbf{u} + (-\mathbf{u}) = \mathbf{0}$ for $\mathbf{u} = (u_1, u_2)$.

   (e) Find two vector space axioms that fail to hold.

▶ In Exercises 3–12, determine whether each set equipped with the given operations is a vector space. For those that are not vector spaces identify the vector space axioms that fail. ◀

3. The set of all real numbers with the standard operations of addition and multiplication.

4. The set of all pairs of real numbers of the form $(x, 0)$ with the standard operations on R^2.

5. The set of all pairs of real numbers of the form (x, y), where $x \geq 0$, with the standard operations on R^2.

6. The set of all n-tuples of real numbers that have the form $(x, x, \ldots, x)$ with the standard operations on R^n.

7. The set of all triples of real numbers with the standard vector addition but with scalar multiplication defined by

$$k(x, y, z) = (k^2 x, k^2 y, k^2 z)$$

8. The set of all 2×2 invertible matrices with the standard matrix addition and scalar multiplication.

9. The set of all 2×2 matrices of the form

$$\begin{bmatrix} a & 0 \\ 0 & b \end{bmatrix}$$

   with the standard matrix addition and scalar multiplication.

10. The set of all real-valued functions f defined everywhere on the real line and such that $f(1) = 0$ with the operations used in Example 6.

11. The set of all pairs of real numbers of the form $(1, x)$ with the operations

$$(1, y) + (1, y') = (1, y + y') \quad \text{and} \quad k(1, y) = (1, ky)$$

12. The set of polynomials of the form $a + bx$ with the operations

$$(a_0 + a_1 x) + (b_0 + b_1 x) = (a_0 + b_0) + (a_1 + b_1)x$$

   and

$$k(a_0 + a_1 x) = (ka_0) + (ka_1)x$$

13. Verify Axioms 3, 7, 8, and 9 for the vector space given in Example 4.

14. Verify Axioms 1, 2, 3, 7, 8, 9, and 10 for the vector space given in Example 6.

15. With the addition and scalar multiplication operations defined in Example 7, show that $V = R^2$ satisfies Axioms 1–9.

16. Verify Axioms 1, 2, 3, 6, 8, 9, and 10 for the vector space given in Example 8.

17. Show that the set of all points in R^2 lying on a line is a vector space with respect to the standard operations of vector addition and scalar multiplication if and only if the line passes through the origin.

18. Show that the set of all points in R^3 lying in a plane is a vector space with respect to the standard operations of vector addition and scalar multiplication if and only if the plane passes through the origin.

▶ In Exercises **19–21**, prove that the given set with the stated operations is a vector space. ◀

19. The set $V = \{0\}$ with the operations of addition and scalar multiplication given in Example 1.

20. The set R^∞ of all infinite sequences of real numbers with the operations of addition and scalar multiplication given in Example 3.

21. The set M_{mn} of all $m \times n$ matrices with the usual operations of addition and scalar multiplication.

22. Prove part (d) of Theorem 4.1.1.

23. The argument that follows proves that if $\mathbf{u}$, $\mathbf{v}$, and $\mathbf{w}$ are vectors in a vector space V such that $\mathbf{u} + \mathbf{w} = \mathbf{v} + \mathbf{w}$, then $\mathbf{u} = \mathbf{v}$ (the *cancellation law* for vector addition). As illustrated, justify the steps by filling in the blanks.

$\mathbf{u} + \mathbf{w} = \mathbf{v} + \mathbf{w}$	Hypothesis
$(\mathbf{u} + \mathbf{w}) + (-\mathbf{w}) = (\mathbf{v} + \mathbf{w}) + (-\mathbf{w})$	Add $-\mathbf{w}$ to both sides.
$\mathbf{u} + [\mathbf{w} + (-\mathbf{w})] = \mathbf{v} + [\mathbf{w} + (-\mathbf{w})]$	
$\mathbf{u} + \mathbf{0} = \mathbf{v} + \mathbf{0}$	
$\mathbf{u} = \mathbf{v}$	

24. Let $\mathbf{v}$ be any vector in a vector space V. Prove that $0\mathbf{v} = \mathbf{0}$.

25. Below is a seven-step proof of part (b) of Theorem 4.1.1. Justify each step either by stating that it is true by *hypothesis* or by specifying which of the ten vector space axioms applies.

Hypothesis: Let $\mathbf{u}$ be any vector in a vector space V, let $\mathbf{0}$ be the zero vector in V, and let k be a scalar.

Conclusion: Then $k\mathbf{0} = \mathbf{0}$.

Proof: (1) $k\mathbf{0} + k\mathbf{u} = k(\mathbf{0} + \mathbf{u})$
(2) $\qquad\qquad = k\mathbf{u}$
(3) Since $k\mathbf{u}$ is in V, $-k\mathbf{u}$ is in V.
(4) Therefore, $(k\mathbf{0} + k\mathbf{u}) + (-k\mathbf{u}) = k\mathbf{u} + (-k\mathbf{u})$.
(5) $\qquad\quad k\mathbf{0} + (k\mathbf{u} + (-k\mathbf{u})) = k\mathbf{u} + (-k\mathbf{u})$
(6) $\qquad\qquad\qquad\qquad k\mathbf{0} + \mathbf{0} = \mathbf{0}$
(7) $\qquad\qquad\qquad\qquad\qquad k\mathbf{0} = \mathbf{0}$

26. Let $\mathbf{v}$ be any vector in a vector space V. Prove that $-\mathbf{v} = (-1)\mathbf{v}$.

27. Prove: If $\mathbf{u}$ is a vector in a vector space V and k a scalar such that $k\mathbf{u} = \mathbf{0}$, then either $k = 0$ or $\mathbf{u} = \mathbf{0}$. [*Suggestion:* Show that if $k\mathbf{u} = \mathbf{0}$ and $k \neq 0$, then $\mathbf{u} = \mathbf{0}$. The result then follows as a logical consequence of this.]

True-False Exercises

In parts (a)–(e) determine whether the statement is true or false, and justify your answer.

(a) A vector is a directed line segment (an arrow).

(b) A vector is an n-tuple of real numbers.

(c) A vector is any element of a vector space.

(d) There is a vector space consisting of exactly two distinct vectors.

(e) The set of polynomials with degree exactly 1 is a vector space under the operations defined in Exercise 12.

4.2 Subspaces

It is possible for one vector space to be contained within another. We will explore this idea in this section, we will discuss how to recognize such vector spaces, and we will give a variety of examples that will be used in our later work.

We will begin with some terminology.

> **DEFINITION 1** A subset W of a vector space V is called a *subspace* of V if W is itself a vector space under the addition and scalar multiplication defined on V.

In general, to show that a nonempty set W with two operations is a vector space one must verify the ten vector space axioms. However, if W is a subspace of a known vector space V, then certain axioms need not be verified because they are "inherited" from V. For example, it is *not* necessary to verify that $\mathbf{u} + \mathbf{v} = \mathbf{v} + \mathbf{u}$ holds in W because it holds for all vectors in V including those in W. On the other hand, it *is* necessary to verify that W is closed under addition and scalar multiplication since it is possible that adding two vectors in W or multiplying a vector in W by a scalar produces a vector in V that is outside of W (Figure 4.2.1).

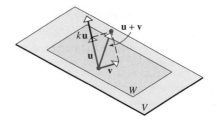

▶ **Figure 4.2.1** The vectors **u** and **v** are in W, but the vectors **u** + **v** and k**u** are not.

Those axioms that are *not* inherited by W are

Axiom 1—Closure of W under addition

Axiom 4—Existence of a zero vector in W

Axiom 5—Existence of a negative in W for every vector in W

Axiom 6—Closure of W under scalar multiplication

so these must be verified to prove that it is a subspace of V. However, the following theorem shows that if Axiom 1 and Axiom 6 hold in W, then Axioms 4 and 5 hold in W as a consequence and hence need not be verified.

THEOREM 4.2.1 *If W is a set of one or more vectors in a vector space V, then W is a subspace of V if and only if the following conditions hold.*

*(a) If **u** and **v** are vectors in W, then **u** + **v** is in W.*

*(b) If k is any scalar and **u** is any vector in W, then k**u** is in W.*

Proof If W is a subspace of V, then all the vector space axioms hold in W, including Axioms 1 and 6, which are precisely conditions (*a*) and (*b*).

Conversely, assume that conditions (*a*) and (*b*) hold. Since these are Axioms 1 and 6, and since Axioms 2, 3, 7, 8, 9, and 10 are inherited from V, we only need to show that Axioms 4 and 5 hold in W. For this purpose, let **u** be any vector in W. It follows from condition (*b*) that k**u** is a vector in W for every scalar k. In particular, 0**u** = **0** and (-1)**u** = −**u** are in W, which shows that Axioms 4 and 5 hold in W. ◀

In words, Theorem 4.2.1 states that W is a subspace of V if and only if it is closed under addition and scalar multiplication.

▶ **EXAMPLE 1 The Zero Subspace**

If V is any vector space, and if $W = \{\mathbf{0}\}$ is the subset of V that consists of the zero vector only, then W is closed under addition and scalar multiplication since

$$\mathbf{0} + \mathbf{0} = \mathbf{0} \quad \text{and} \quad k\mathbf{0} = \mathbf{0}$$

for any scalar k. We call W the ***zero subspace*** of V.

Note that every vector space has at least two subspaces, itself and its zero subspace.

▶ **EXAMPLE 2 Lines Through the Origin Are Subspaces of R^2 and of R^3**

If W is a line through the origin of either R^2 or R^3, then adding two vectors on the line W or multiplying a vector on the line W by a scalar produces another vector on the line W, so W is closed under addition and scalar multiplication (see Figure 4.2.2 for an illustration in R^3).

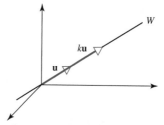

(a) W is closed under addition.   (b) W is closed under scalar multiplication.

◄ Figure 4.2.2

► **EXAMPLE 3** **Planes Through the Origin Are Subspaces of R^3**

If $\mathbf{u}$ and $\mathbf{v}$ are vectors in a plane W through the origin of R^3, then it is evident geometrically that $\mathbf{u} + \mathbf{v}$ and $k\mathbf{u}$ lie in the same plane W for any scalar k (Figure 4.2.3). Thus W is closed under addition and scalar multiplication. ◄

Table 1 that follows gives a list of subspaces of R^2 and of R^3 that we have encountered thus far. We will see later that these are the only subspaces of R^2 and of R^3.

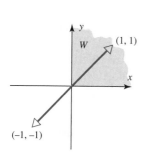

▲ Figure 4.2.3 The vectors $\mathbf{u} + \mathbf{v}$ and $k\mathbf{u}$ both lie in the same plane as $\mathbf{u}$ and $\mathbf{v}$.

Table 1

Subspaces of R^2	Subspaces of R^3
• {0}	• {0}
• Lines through the origin	• Lines through the origin
• R^2	• Planes through the origin
	• R^3

► **EXAMPLE 4** **A Subset of R^2 That Is Not a Subspace**

Let W be the set of all points (x, y) in R^2 for which $x \geq 0$ and $y \geq 0$ (the shaded region in Figure 4.2.4). This set is not a subspace of R^2 because it is not closed under scalar multiplication. For example, $\mathbf{v} = (1, 1)$ is a vector in W, but $(-1)\mathbf{v} = (-1, -1)$ is not.

▲ Figure 4.2.4 W is not closed under scalar multiplication.

► **EXAMPLE 5** **Subspaces of M_{nn}**

We know from Theorem 1.7.2 that the sum of two symmetric $n \times n$ matrices is symmetric and that a scalar multiple of a symmetric $n \times n$ matrix is symmetric. Thus, the set of symmetric $n \times n$ matrices is closed under addition and scalar multiplication and hence is a subspace of M_{nn}. Similarly, the sets of upper triangular matrices, lower triangular matrices, and diagonal matrices are subspaces of M_{nn}.

► **EXAMPLE 6** **A Subset of M_{nn} That Is Not a Subspace**

The set W of invertible $n \times n$ matrices is not a subspace of M_{nn}, failing on two counts—it is not closed under addition and not closed under scalar multiplication. We will illustrate this with an example in M_{22} that you can readily adapt to M_{nn}. Consider the matrices

$$U = \begin{bmatrix} 1 & 2 \\ 2 & 5 \end{bmatrix} \quad \text{and} \quad V = \begin{bmatrix} -1 & 2 \\ -2 & 5 \end{bmatrix}$$

The matrix $0U$ is the 2×2 zero matrix and hence is not invertible, and the matrix $U + V$ has a column of zeros, so it also is not invertible.

▶ **EXAMPLE 7 The Subspace $C(-\infty, \infty)$**

There is a theorem in calculus which states that a sum of continuous functions is continuous and that a constant times a continuous function is continuous. Rephrased in vector language, the set of continuous functions on $(-\infty, \infty)$ is a subspace of $F(-\infty, \infty)$. We will denote this subspace by $C(-\infty, \infty)$.

▶ **EXAMPLE 8 Functions with Continuous Derivatives**

A function with a continuous derivative is said to be *continuously differentiable*. There is a theorem in calculus which states that the sum of two continuously differentiable functions is continuously differentiable and that a constant times a continuously differentiable function is continuously differentiable. Thus, the functions that are continuously differentiable on $(-\infty, \infty)$ form a subspace of $F(-\infty, \infty)$. We will denote this subspace by $C^1(-\infty, \infty)$, where the superscript emphasizes that the *first* derivative is continuous. To take this a step further, the set of functions with m continuous derivatives on $(-\infty, \infty)$ is a subspace of $F(-\infty, \infty)$ as is the set of functions with derivatives of all orders on $(-\infty, \infty)$. We will denote these subspaces by $C^m(-\infty, \infty)$ and $C^\infty(-\infty, \infty)$, respectively.

▶ **EXAMPLE 9 The Subspace of All Polynomials**

Recall that a ***polynomial*** is a function that can be expressed in the form

$$p(x) = a_0 + a_1 x + \cdots + a_n x^n \tag{1}$$

where $a_0, a_1, \ldots, a_n$ are constants. It is evident that the sum of two polynomials is a polynomial and that a constant times a polynomial is a polynomial. Thus, the set W of all polynomials is closed under addition and scalar multiplication and hence is a subspace of $F(-\infty, \infty)$. We will denote this space by P_∞.

In this text we regard all constants to be polynomials of degree zero. Be aware, however, that some authors do not assign a degree to the constant 0.

▶ **EXAMPLE 10 The Subspace of Polynomials of Degree $\leq n$**

Recall that the ***degree*** of a polynomial is the highest power of the variable that occurs with a nonzero coefficient. Thus, for example, if $a_n \neq 0$ in Formula (1), then that polynomial has degree n. It is *not* true that the set W of polynomials with positive degree n is a subspace of $F(-\infty, \infty)$ because that set is not closed under addition. For example, the polynomials

$$1 + 2x + 3x^2 \quad \text{and} \quad 5 + 7x - 3x^2$$

both have degree 2, but their sum has degree 1. What *is* true, however, is that for each nonnegative integer n the polynomials of degree n *or less* form a subspace of $F(-\infty, \infty)$. We will denote this space by P_n. ◀

The Hierarchy of Function Spaces

It is proved in calculus that polynomials are continuous functions and have continuous derivatives of all orders on $(-\infty, \infty)$. Thus, it follows that P_∞ is not only a subspace of $F(-\infty, \infty)$, as previously observed, but is also a subspace of $C^\infty(-\infty, \infty)$. We leave it for you to convince yourself that the vector spaces discussed in Examples 7 to 10 are "nested" one inside the other as illustrated in Figure 4.2.5.

Remark In our previous examples, and as illustrated in Figure 4.2.5, we have only considered functions that are defined at all points of the interval $(-\infty, \infty)$. Sometimes we will want to consider functions that are only defined on some subinterval of $(-\infty, \infty)$, say the closed interval $[a, b]$ or the open interval (a, b). In such cases we will make an appropriate notation change. For example, $C[a, b]$ is the space of continuous functions on $[a, b]$ and $C(a, b)$ is the space of continuous functions on (a, b).

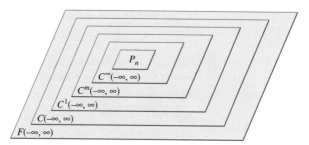

▶ Figure 4.2.5

Building Subspaces   The following theorem provides a useful way of creating a new subspace from known subspaces.

THEOREM 4.2.2 *If $W_1, W_2, \ldots, W_r$ are subspaces of a vector space V, then the intersection of these subspaces is also a subspace of V.*

Proof   Let W be the intersection of the subspaces $W_1, W_2, \ldots, W_r$. This set is not empty because each of these subspaces contains the zero vector of V, and hence so does their intersection. Thus, it remains to show that W is closed under addition and scalar multiplication.

To prove closure under addition, let $\mathbf{u}$ and $\mathbf{v}$ be vectors in W. Since W is the intersection of $W_1, W_2, \ldots, W_r$, it follows that $\mathbf{u}$ and $\mathbf{v}$ also lie in each of these subspaces. Since these subspaces are all closed under addition, they all contain the vector $\mathbf{u} + \mathbf{v}$ and hence so does their intersection W. This proves that W is closed under addition. We leave the proof that W is closed under scalar multiplication to you. ◀

> Note that the first step in proving Theorem 4.2.2 was to establish that W contained at least one vector. This is important, for otherwise the subsequent argument might be logically correct but meaningless.

Sometimes we will want to find the "smallest" subspace of a vector space V that contains all of the vectors in some set of interest. The following definition, which generalizes Definition 4 of Section 3.1, will help us to do that.

DEFINITION 2   If $\mathbf{w}$ is a vector in a vector space V, then $\mathbf{w}$ is said to be a ***linear combination*** of the vectors $\mathbf{v}_1, \mathbf{v}_2, \ldots, \mathbf{v}_r$ in V if $\mathbf{w}$ can be expressed in the form

$$\mathbf{w} = k_1\mathbf{v}_1 + k_2\mathbf{v}_2 + \cdots + k_r\mathbf{v}_r \qquad (2)$$

where $k_1, k_2, \ldots, k_r$ are scalars. These scalars are called the ***coefficients*** of the linear combination.

> If $k = 1$, then Equation (2) has the form $\mathbf{w} = k_1\mathbf{v}_1$, in which case the linear combination is just a scalar multiple of $\mathbf{v}_1$.

THEOREM 4.2.3 *If $S = \{\mathbf{w}_1, \mathbf{w}_2, \ldots, \mathbf{w}_r\}$ is a nonempty set of vectors in a vector space V, then:*

(a)   The set W of all possible linear combinations of the vectors in S is a subspace of V.

(b)   The set W in part (a) is the "smallest" subspace of V that contains all of the vectors in S in the sense that any other subspace that contains those vectors contains W.

Proof (a)   Let W be the set of all possible linear combinations of the vectors in S. We must show that S is closed under addition and scalar multiplication. To prove closure under addition, let

$$\mathbf{u} = c_1\mathbf{w}_1 + c_2\mathbf{w}_2 + \cdots + c_r\mathbf{w}_r \quad \text{and} \quad \mathbf{v} = k_1\mathbf{w}_1 + k_2\mathbf{w}_2 + \cdots + k_r\mathbf{w}_r$$

be two vectors in S. It follows that their sum can be written as

$$\mathbf{u} + \mathbf{v} = (c_1 + k_1)\mathbf{w}_1 + (c_2 + k_2)\mathbf{w}_2 + \cdots + (c_r + k_r)\mathbf{w}_r$$

which is a linear combination of the vectors in S. Thus, W is closed under addition. We leave it for you to prove that W is also closed under scalar multiplication and hence is a subspace of V.

Proof (b)   Let W' be any subspace of V that contains all of the vectors in S. Since W' is closed under addition and scalar multiplication, it contains all linear combinations of the vectors in S and hence contains W. ◄

The following definition gives some important notation and terminology related to Theorem 4.2.3.

DEFINITION 3 The subspace of a vector space V that is formed from all possible linear combinations of the vectors in a nonempty set S is called the **span of S**, and we say that the vectors in S **span** that subspace. If $S = \{\mathbf{w}_1, \mathbf{w}_2, \ldots, \mathbf{w}_r\}$, then we denote the span of S by

$$\text{span}\{\mathbf{w}_1, \mathbf{w}_2, \ldots, \mathbf{w}_r\} \quad \text{or} \quad \text{span}(S)$$

▶ **EXAMPLE 11 The Standard Unit Vectors Span R^n**

Recall that the standard unit vectors in R^n are

$$\mathbf{e}_1 = (1, 0, 0, \ldots, 0), \quad \mathbf{e}_2 = (0, 1, 0, \ldots, 0), \ldots, \quad \mathbf{e}_n = (0, 0, 0, \ldots, 1)$$

These vectors span R^n since every vector $\mathbf{v} = (v_1, v_2, \ldots, v_n)$ in R^n can be expressed as

$$\mathbf{v} = v_1\mathbf{e}_1 + v_2\mathbf{e}_2 + \cdots + v_n\mathbf{e}_n$$

which is a linear combination of $\mathbf{e}_1, \mathbf{e}_2, \ldots, \mathbf{e}_n$. Thus, for example, the vectors

$$\mathbf{i} = (1, 0, 0), \quad \mathbf{j} = (0, 1, 0), \quad \mathbf{k} = (0, 0, 1)$$

span R^3 since every vector $\mathbf{v} = (a, b, c)$ in this space can be expressed as

$$\mathbf{v} = (a, b, c) = a(1, 0, 0) + b(0, 1, 0) + c(0, 0, 1) = a\mathbf{i} + b\mathbf{j} + c\mathbf{k}.$$

▶ **EXAMPLE 12 A Geometric View of Spanning in R^2 and R^3**

(a)   If $\mathbf{v}$ is a nonzero vector in R^2 or R^3 that has its initial point at the origin, then span$\{\mathbf{v}\}$, which is the set of all scalar multiples of $\mathbf{v}$, is the line through the origin determined by $\mathbf{v}$. You should be able to visualize this from Figure 4.2.6a by observing that the

George William Hill
(1838–1914)

Historical Note   The terms *linearly independent* and *linearly dependent* were introduced by Maxime Bôcher (see p. 7) in his book *Introduction to Higher Algebra*, published in 1907. The term *linear combination* is due to the American mathematician G. W. Hill, who introduced it in a research paper on planetary motion published in 1900. Hill was a "loner" who preferred to work out of his home in West Nyack, New York, rather than in academia, though he did try lecturing at Columbia University for a few years. Interestingly, he apparently returned the teaching salary, indicating that he did not need the money and did not want to be bothered looking after it. Although technically a mathematician, Hill had little interest in modern developments of mathematics and worked almost entirely on the theory of planetary orbits.

[*Image: Courtesy of the American Mathematical Society*]

tip of the vector $k\mathbf{v}$ can be made to fall at any point on the line by choosing the value of k appropriately.

(b) If $\mathbf{v}_1$ and $\mathbf{v}_2$ are nonzero vectors in R^3 that have their initial points at the origin, then span$\{\mathbf{v}_1, \mathbf{v}_2\}$, which consists of all linear combinations of $\mathbf{v}_1$ and $\mathbf{v}_2$, is the plane through the origin determined by these two vectors. You should be able to visualize this from Figure 4.2.6b by observing that the tip of the vector $k_1\mathbf{v}_1 + k_2\mathbf{v}_2$ can be made to fall at any point in the plane by adjusting the scalars k_1 and k_2 to lengthen, shorten, or reverse the directions of the vectors $k_1\mathbf{v}_1$ and $k_2\mathbf{v}_2$ appropriately.

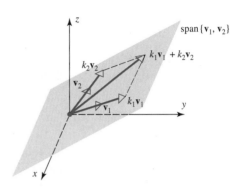

(a) Span$\{\mathbf{v}\}$ is the line through the origin determined by $\mathbf{v}$.

(b) Span$\{\mathbf{v}_1, \mathbf{v}_2\}$ is the plane through the origin determined by $\mathbf{v}_1$ and $\mathbf{v}_2$.

▶ Figure 4.2.6

▶ EXAMPLE 13 A Spanning Set for P_n

The polynomials $1, x, x^2, \ldots, x^n$ span the vector space P_n defined in Example 10 since each polynomial $\mathbf{p}$ in P_n can be written as

$$\mathbf{p} = a_0 + a_1 x + \cdots + a_n x^n$$

which is a linear combination of $1, x, x^2, \ldots, x^n$. We can denote this by writing

$$P_n = \text{span}\{1, x, x^2, \ldots, x^n\} \quad \blacktriangleleft$$

The next two examples are concerned with two important types of problems:

• Given a set S of vectors in R^n and a vector $\mathbf{v}$ in R^n, determine whether $\mathbf{v}$ is a linear combination of the vectors in S.

• Given a set S of vectors in R^n, determine whether the vectors span R^n.

▶ EXAMPLE 14 Linear Combinations

Consider the vectors $\mathbf{u} = (1, 2, -1)$ and $\mathbf{v} = (6, 4, 2)$ in R^3. Show that $\mathbf{w} = (9, 2, 7)$ is a linear combination of $\mathbf{u}$ and $\mathbf{v}$ and that $\mathbf{w}' = (4, -1, 8)$ is *not* a linear combination of $\mathbf{u}$ and $\mathbf{v}$.

Solution In order for $\mathbf{w}$ to be a linear combination of $\mathbf{u}$ and $\mathbf{v}$, there must be scalars k_1 and k_2 such that $\mathbf{w} = k_1\mathbf{u} + k_2\mathbf{v}$; that is,

$$(9, 2, 7) = k_1(1, 2, -1) + k_2(6, 4, 2)$$

or

$$(9, 2, 7) = (k_1 + 6k_2, 2k_1 + 4k_2, -k_1 + 2k_2)$$

Equating corresponding components gives

$$k_1 + 6k_2 = 9$$
$$2k_1 + 4k_2 = 2$$
$$-k_1 + 2k_2 = 7$$

Solving this system using Gaussian elimination yields $k_1 = -3$, $k_2 = 2$, so

$$\mathbf{w} = -3\mathbf{u} + 2\mathbf{v}$$

Similarly, for $\mathbf{w}'$ to be a linear combination of $\mathbf{u}$ and $\mathbf{v}$, there must be scalars k_1 and k_2 such that $\mathbf{w}' = k_1\mathbf{u} + k_2\mathbf{v}$; that is,

$$(4, -1, 8) = k_1(1, 2, -1) + k_2(6, 4, 2)$$

or

$$(4, -1, 8) = (k_1 + 6k_2, 2k_1 + 4k_2, -k_1 + 2k_2)$$

Equating corresponding components gives

$$k_1 + 6k_2 = 4$$
$$2k_1 + 4k_2 = -1$$
$$-k_1 + 2k_2 = 8$$

This system of equations is inconsistent (verify), so no such scalars k_1 and k_2 exist. Consequently, $\mathbf{w}'$ is not a linear combination of $\mathbf{u}$ and $\mathbf{v}$.

▶ EXAMPLE 15 Testing for Spanning

Determine whether $\mathbf{v}_1 = (1, 1, 2)$, $\mathbf{v}_2 = (1, 0, 1)$, and $\mathbf{v}_3 = (2, 1, 3)$ span the vector space R^3.

Solution We must determine whether an arbitrary vector $\mathbf{b} = (b_1, b_2, b_3)$ in R^3 can be expressed as a linear combination

$$\mathbf{b} = k_1\mathbf{v}_1 + k_2\mathbf{v}_2 + k_3\mathbf{v}_3$$

of the vectors $\mathbf{v}_1$, $\mathbf{v}_2$, and $\mathbf{v}_3$. Expressing this equation in terms of components gives

$$(b_1, b_2, b_3) = k_1(1, 1, 2) + k_2(1, 0, 1) + k_3(2, 1, 3)$$

or

$$(b_1, b_2, b_3) = (k_1 + k_2 + 2k_3, k_1 + k_3, 2k_1 + k_2 + 3k_3)$$

or

$$k_1 + k_2 + 2k_3 = b_1$$
$$k_1 \qquad + k_3 = b_2$$
$$2k_1 + k_2 + 3k_3 = b_3$$

Thus, our problem reduces to ascertaining whether this system is consistent for all values of b_1, b_2, and b_3. One way of doing this is to use parts (e) and (g) of Theorem 2.3.8, which state that the system is consistent if and only if its coefficient matrix

$$A = \begin{bmatrix} 1 & 1 & 2 \\ 1 & 0 & 1 \\ 2 & 1 & 3 \end{bmatrix}$$

has a nonzero determinant. But this is *not* the case here; we leave it for you to confirm that $\det(A) = 0$, so $\mathbf{v}_1$, $\mathbf{v}_2$, and $\mathbf{v}_3$ do not span R^3. ◀

Solution Spaces of Homogeneous Systems   The solutions of a homogeneous linear system $A\mathbf{x} = \mathbf{0}$ of m equations in n unknowns can be viewed as vectors in R^n. The following theorem provides a useful insight into the geometric structure of the solution set.

> **THEOREM 4.2.4** *The solution set of a homogeneous linear system $A\mathbf{x} = \mathbf{0}$ in n unknowns is a subspace of R^n.*

Proof   Let W be the solution set for the system. The set W is not empty because it contains at least the trivial solution $\mathbf{x} = \mathbf{0}$.

To show that W is a subspace of R^n, we must show that it is closed under addition and scalar multiplication. To do this, let $\mathbf{x}_1$ and $\mathbf{x}_2$ be vectors in W. Since these vectors are solutions of $A\mathbf{x} = \mathbf{0}$, we have

$$A\mathbf{x}_1 = \mathbf{0} \quad \text{and} \quad A\mathbf{x}_2 = \mathbf{0}$$

It follows from these equations and the distributive property of matrix multiplication that

$$A(\mathbf{x}_1 + \mathbf{x}_2) = A\mathbf{x}_1 + A\mathbf{x}_2 = \mathbf{0} + \mathbf{0} = \mathbf{0}$$

so W is closed under addition. Similarly, if k is any scalar then

$$A(k\mathbf{x}_1) = kA\mathbf{x}_1 = k\mathbf{0} = \mathbf{0}$$

so W is also closed under scalar multiplication.   ◀

> Because the solution set of a homogeneous system in n unknowns is actually a subspace of R^n, we will generally refer to it as the **solution space** of the system.

▶ **EXAMPLE 16   Solution Spaces of Homogeneous Systems**

Consider the linear systems

(a) $\begin{bmatrix} 1 & -2 & 3 \\ 2 & -4 & 6 \\ 3 & -6 & 9 \end{bmatrix} \begin{bmatrix} x \\ y \\ z \end{bmatrix} = \begin{bmatrix} 0 \\ 0 \\ 0 \end{bmatrix}$ (b) $\begin{bmatrix} 1 & -2 & 3 \\ -3 & 7 & -8 \\ -2 & 4 & -6 \end{bmatrix} \begin{bmatrix} x \\ y \\ z \end{bmatrix} = \begin{bmatrix} 0 \\ 0 \\ 0 \end{bmatrix}$

(c) $\begin{bmatrix} 1 & -2 & 3 \\ -3 & 7 & -8 \\ 4 & 1 & 2 \end{bmatrix} \begin{bmatrix} x \\ y \\ z \end{bmatrix} = \begin{bmatrix} 0 \\ 0 \\ 0 \end{bmatrix}$ (d) $\begin{bmatrix} 0 & 0 & 0 \\ 0 & 0 & 0 \\ 0 & 0 & 0 \end{bmatrix} \begin{bmatrix} x \\ y \\ z \end{bmatrix} = \begin{bmatrix} 0 \\ 0 \\ 0 \end{bmatrix}$

Solution

(a)   We leave it for you to verify that the solutions are

$$x = 2s - 3t, \quad y = s, \quad z = t$$

from which it follows that

$$x = 2y - 3z \quad \text{or} \quad x - 2y + 3z = 0$$

This is the equation of a plane through the origin that has $\mathbf{n} = (1, -2, 3)$ as a normal.

(b)   We leave it for you to verify that the solutions are

$$x = -5t, \quad y = -t, \quad z = t$$

which are parametric equations for the line through the origin that is parallel to the vector $\mathbf{v} = (-5, -1, 1)$.

(c)   We leave it for you to verify that the only solution is $x = 0$, $y = 0$, $z = 0$, so the solution space is $\{\mathbf{0}\}$.

(d)   This linear system is satisfied by all real values of x, y, and z, so the solution space is all of R^3.   ◀

Remark Whereas the solution set of every *homogeneous* system of m equations in n unknowns is a subspace of R^n, it is *never* true that the solution set of a *nonhomogeneous* system of m equations in n unknowns is a subspace of R^n. There are two possible scenarios: first, the system may not have any solutions at all, and second, if there are solutions, then the solution set will not be closed under either addition or under scalar multiplication (Exercise 18).

A Concluding Observation   It is important to recognize that spanning sets are not unique. For example, any nonzero vector on the line in Figure 4.2.6*a* will span that line, and any two noncollinear vectors in the plane in Figure 4.2.6*b* will span that plane. The following theorem, whose proof we leave as an exercise, states conditions under which two sets of vectors will span the same space.

THEOREM 4.2.5 *If $S = \{v_1, v_2, \ldots, v_r\}$ and $S' = \{w_1, w_2, \ldots, w_k\}$ are nonempty sets of vectors in a vector space V, then*

$$span\{v_1, v_2, \ldots, v_r\} = span\{w_1, w_2, \ldots, w_k\}$$

if and only if each vector in S is a linear combination of those in S', and each vector in S' is a linear combination of those in S.

Concept Review

- Subspace
- Zero subspace
- Examples of subspaces
- Linear combination
- Span
- Solution space

Skills

- Determine whether a subset of a vector space is a subspace.
- Show that a subset of a vector space is a subspace.
- Show that a nonempty subset of a vector space is not a subspace by demonstrating that the set is either not closed under addition or not closed under scalar multiplication.

- Given a set S of vectors in R^n and a vector v in R^n, determine whether v is a linear combination of the vectors in S.
- Given a set S of vectors in R^n, determine whether the vectors in S span R^n.
- Determine whether two nonempty sets of vectors in a vector space V span the same subspace of V.

Exercise Set 4.2

1. Use Theorem 4.2.1 to determine which of the following are subspaces of R^3.

(a) All vectors of the form $(a, 0, 0)$.

(b) All vectors of the form $(a, 1, 1)$.

(c) All vectors of the form (a, b, c), where $b = a + c$.

(d) All vectors of the form (a, b, c), where $b = a + c + 1$.

(e) All vectors of the form $(a, b, 0)$.

2. Use Theorem 4.2.1 to determine which of the following are subspaces of M_{nn}.

(a) The set of all diagonal $n \times n$ matrices.

(b) The set of all $n \times n$ matrices A such that $\det(A) = 0$.

(c) The set of all $n \times n$ matrices A such that $\operatorname{tr}(A) = 0$.

(d) The set of all symmetric $n \times n$ matrices.

(e) The set of all $n \times n$ matrices A such that $A^T = -A$.

(f) The set of all $n \times n$ matrices A for which $Ax = 0$ has only the trivial solution.

(g) The set of all $n \times n$ matrices A such that $AB = BA$ for some fixed $n \times n$ matrix B.

3. Use Theorem 4.2.1 to determine which of the following are subspaces of P_3.

(a) All polynomials $a_0 + a_1 x + a_2 x^2 + a_3 x^3$ for which $a_0 = 0$.

(b) All polynomials $a_0 + a_1 x + a_2 x^2 + a_3 x^3$ for which $a_0 + a_1 + a_2 + a_3 = 0$.

(c) All polynomials of the form $a_0 + a_1 x + a_2 x^2 + a_3 x^3$ in which $a_0, a_1, a_2,$ and a_3 are integers.

(d) All polynomials of the form $a_0 + a_1 x$, where a_0 and a_1 are real numbers.

4. Which of the following are subspaces of $F(-\infty, \infty)$?

(a) All functions f in $F(-\infty, \infty)$ for which $f(0) = 0$.

(b) All functions f in $F(-\infty, \infty)$ for which $f(0) = 1$.

(c) All functions f in $F(-\infty, \infty)$ for which $f(-x) = f(x)$.

(d) All polynomials of degree 2.

5. Which of the following are subspaces of R^∞?

(a) All sequences $\mathbf{v}$ in R^∞ of the form
$\mathbf{v} = (v, 0, v, 0, v, 0, \ldots)$.

(b) All sequences $\mathbf{v}$ in R^∞ of the form
$\mathbf{v} = (v, 1, v, 1, v, 1, \ldots)$.

(c) All sequences $\mathbf{v}$ in R^∞ of the form
$\mathbf{v} = (v, 2v, 4v, 8v, 16v, \ldots)$.

(d) All sequences in R^∞ whose components are 0 from some point on.

6. A line L through the origin in R^3 can be represented by parametric equations of the form $x = at$, $y = bt$, and $z = ct$. Use these equations to show that L is a subspace of R^3 by showing that if $\mathbf{v}_1 = (x_1, y_1, z_1)$ and $\mathbf{v}_2 = (x_2, y_2, z_2)$ are points on L and k is any real number, then $k\mathbf{v}_1$ and $\mathbf{v}_1 + \mathbf{v}_2$ are also points on L.

7. Which of the following are linear combinations of $\mathbf{u} = (0, -2, 2)$ and $\mathbf{v} = (1, 3, -1)$?

(a) $(2, 2, 2)$ (b) $(3, 1, 5)$ (c) $(0, 4, 5)$ (d) $(0, 0, 0)$

8. Express the following as linear combinations of $\mathbf{u} = (2, 1, 4)$, $\mathbf{v} = (1, -1, 3)$, and $\mathbf{w} = (3, 2, 5)$.

(a) $(-9, -7, -15)$ (b) $(6, 11, 6)$

(c) $(0, 0, 0)$ (d) $(7, 8, 9)$

9. Which of the following are linear combinations of

$$A = \begin{bmatrix} 4 & 0 \\ -2 & -2 \end{bmatrix}, \quad B = \begin{bmatrix} 1 & -1 \\ 2 & 3 \end{bmatrix}, \quad C = \begin{bmatrix} 0 & 2 \\ 1 & 4 \end{bmatrix}?$$

(a) $\begin{bmatrix} 6 & -8 \\ -1 & -8 \end{bmatrix}$ (b) $\begin{bmatrix} 0 & 0 \\ 0 & 0 \end{bmatrix}$

(c) $\begin{bmatrix} 6 & 0 \\ 3 & 8 \end{bmatrix}$ (d) $\begin{bmatrix} -1 & 5 \\ 7 & 1 \end{bmatrix}$

10. In each part express the vector as a linear combination of $\mathbf{p}_1 = 2 + x + 4x^2$, $\mathbf{p}_2 = 1 - x + 3x^2$, and $\mathbf{p}_3 = 3 + 2x + 5x^2$.

(a) $-9 - 7x - 15x^2$ (b) $6 + 11x + 6x^2$

(c) 0 (d) $7 + 8x + 9x^2$

11. In each part, determine whether the given vectors span R^3.

(a) $\mathbf{v}_1 = (2, 2, 2)$, $\mathbf{v}_2 = (0, 0, 3)$, $\mathbf{v}_3 = (0, 1, 1)$

(b) $\mathbf{v}_1 = (2, -1, 3)$, $\mathbf{v}_2 = (4, 1, 2)$, $\mathbf{v}_3 = (8, -1, 8)$

(c) $\mathbf{v}_1 = (3, 1, 4)$, $\mathbf{v}_2 = (2, -3, 5)$, $\mathbf{v}_3 = (5, -2, 9)$, $\mathbf{v}_4 = (1, 4, -1)$

(d) $\mathbf{v}_1 = (1, 2, 6)$, $\mathbf{v}_2 = (3, 4, 1)$, $\mathbf{v}_3 = (4, 3, 1)$, $\mathbf{v}_4 = (3, 3, 1)$

12. Suppose that $\mathbf{v}_1 = (2, 1, 0, 3)$, $\mathbf{v}_2 = (3, -1, 5, 2)$, and $\mathbf{v}_3 = (-1, 0, 2, 1)$. Which of the following vectors are in span$\{\mathbf{v}_1, \mathbf{v}_2, \mathbf{v}_3\}$?

(a) $(2, 3, -7, 3)$ (b) $(0, 0, 0, 0)$

(c) $(1, 1, 1, 1)$ (d) $(-4, 6, -13, 4)$

13. Determine whether the following polynomials span P_2.

$$\mathbf{p}_1 = 1 - x + 2x^2, \quad \mathbf{p}_2 = 3 + x,$$
$$\mathbf{p}_3 = 5 - x + 4x^2, \quad \mathbf{p}_4 = -2 - 2x + 2x^2$$

14. Let $\mathbf{f} = \cos^2 x$ and $\mathbf{g} = \sin^2 x$. Which of the following lie in the space spanned by $\mathbf{f}$ and $\mathbf{g}$?

(a) $\cos 2x$ (b) $3 + x^2$ (c) 1 (d) $\sin x$ (e) 0

15. Determine whether the solution space of the system $A\mathbf{x} = \mathbf{0}$ is a line through the origin, a plane through the origin, or the origin only. If it is a plane, find an equation for it. If it is a line, find parametric equations for it.

(a) $A = \begin{bmatrix} -1 & 1 & 1 \\ 3 & -1 & 0 \\ 2 & -4 & -5 \end{bmatrix}$ (b) $A = \begin{bmatrix} 1 & -2 & 3 \\ -3 & 6 & 9 \\ -2 & 4 & -6 \end{bmatrix}$

(c) $A = \begin{bmatrix} 1 & 2 & 3 \\ 2 & 5 & 3 \\ 1 & 0 & 8 \end{bmatrix}$ (d) $A = \begin{bmatrix} 1 & 2 & -6 \\ 1 & 4 & 4 \\ 3 & 10 & 6 \end{bmatrix}$

(e) $A = \begin{bmatrix} 1 & -1 & 1 \\ 2 & -1 & 4 \\ 3 & 1 & 11 \end{bmatrix}$ (f) $A = \begin{bmatrix} 1 & -3 & 1 \\ 2 & -6 & 2 \\ 3 & -9 & 3 \end{bmatrix}$

16. (*Calculus required*) Show that the following sets of functions are subspaces of $F(-\infty, \infty)$.

(a) All continuous functions on $(-\infty, \infty)$.

(b) All differentiable functions on $(-\infty, \infty)$.

(c) All differentiable functions on $(-\infty, \infty)$ that satisfy $\mathbf{f}' + 2\mathbf{f} = \mathbf{0}$.

17. (*Calculus required*) Show that the set of continuous functions $\mathbf{f} = f(x)$ on $[a, b]$ such that

$$\int_a^b f(x)\, dx = 0$$

is a subspace of $C[a, b]$.

18. Show that the solution vectors of a consistent nonhomogeneous system of m linear equations in n unknowns do not form a subspace of R^n.

19. Prove Theorem 4.2.5.

20. Use Theorem 4.2.5 to show that the vectors $\mathbf{v}_1 = (1, 6, 4)$, $\mathbf{v}_2 = (2, 4, -1)$, $\mathbf{v}_3 = (-1, 2, 5)$, and the vectors $\mathbf{w}_1 = (1, -2, -5)$, $\mathbf{w}_2 = (0, 8, 9)$ span the same subspace of R^3.

True-False Exercises

In parts (a)–(k) determine whether the statement is true or false, and justify your answer.

(a) Every subspace of a vector space is itself a vector space.

(b) Every vector space is a subspace of itself.

(c) Every subset of a vector space V that contains the zero vector in V is a subspace of V.

(d) The set R^2 is a subspace of R^3.

(e) The solution set of a consistent linear system $A\mathbf{x} = \mathbf{b}$ of m equations in n unknowns is a subspace of R^n.

(f) The span of any finite set of vectors in a vector space is closed under addition and scalar multiplication.

(g) The intersection of any two subspaces of a vector space V is a subspace of V.

(h) The union of any two subspaces of a vector space V is a subspace of V.

(i) Two subsets of a vector space V that span the same subspace of V must be equal.

(j) The set of upper triangular $n \times n$ matrices is a subspace of the vector space of all $n \times n$ matrices.

(k) The polynomials $x - 1$, $(x - 1)^2$, and $(x - 1)^3$ span P_3.

4.3 Linear Independence

In this section we will consider the question of whether the vectors in a given set are interrelated in the sense that one or more of them can be expressed as a linear combination of the others. This is important to know in applications because the existence of such relationships often signals that some kind of complication is likely to occur.

Extraneous Vectors

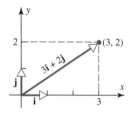

▲ Figure 4.3.1

In a rectangular xy-coordinate system every vector in the plane can be expressed in exactly one way as a linear combination of the standard unit vectors. For example, the only way to express the vector $(3, 2)$ as a linear combination of $\mathbf{i} = (1, 0)$ and $\mathbf{j} = (0, 1)$ is

$$(3, 2) = 3(1, 0) + 2(0, 1) = 3\mathbf{i} + 2\mathbf{j} \tag{1}$$

(Figure 4.3.1). Suppose, however, that we were to introduce a third coordinate axis that makes an angle of $45°$ with the x-axis. Call it the w-axis. As illustrated in Figure 4.3.2, the unit vector along the w-axis is

$$\mathbf{w} = \left(\frac{1}{\sqrt{2}}, \frac{1}{\sqrt{2}} \right)$$

▲ Figure 4.3.2

Whereas Formula (1) shows the only way to express the vector $(3, 2)$ as a linear combination of $\mathbf{i}$ and $\mathbf{j}$, there are infinitely many ways to express this vector as a linear combination of $\mathbf{i}$, $\mathbf{j}$, and $\mathbf{w}$. Three possibilities are

$$(3, 2) = 3(1, 0) + 2(0, 1) + 0 \left(\frac{1}{\sqrt{2}}, \frac{1}{\sqrt{2}} \right) = 3\mathbf{i} + 2\mathbf{j} + 0\mathbf{w}$$

$$(3, 2) = 2(1, 0) + (0, 1) + \sqrt{2} \left(\frac{1}{\sqrt{2}}, \frac{1}{\sqrt{2}} \right) = 3\mathbf{i} + \mathbf{j} + \sqrt{2}\mathbf{w}$$

$$(3, 2) = 4(1, 0) + 3(0, 1) - \sqrt{2} \left(\frac{1}{\sqrt{2}}, \frac{1}{\sqrt{2}} \right) = 4\mathbf{i} + 3\mathbf{j} - \sqrt{2}\mathbf{w}$$

In short, by introducing a superfluous axis we created the complication of having multiple ways of assigning coordinates to points in the plane. What makes the vector $\mathbf{w}$ superfluous is the fact that it can be expressed as a linear combination of the vectors $\mathbf{i}$ and $\mathbf{j}$, namely,

$$\mathbf{w} = \left(\frac{1}{\sqrt{2}}, \frac{1}{\sqrt{2}} \right) = \frac{1}{\sqrt{2}}\mathbf{i} + \frac{1}{\sqrt{2}}\mathbf{j}$$

Thus, one of our main tasks in this section will be to develop ways of ascertaining whether one vector in a set S is a linear combination of other vectors in S.

Linear Independence and Dependence

> **DEFINITION 1** If $S = \{\mathbf{v}_1, \mathbf{v}_2, \ldots, \mathbf{v}_r\}$ is a nonempty set of vectors in a vector space V, then the vector equation
>
> $$k_1\mathbf{v}_1 + k_2\mathbf{v}_2 + \cdots + k_r\mathbf{v}_r = \mathbf{0}$$
>
> has at least one solution, namely,
>
> $$k_1 = 0, \quad k_2 = 0, \ldots, \quad k_r = 0$$
>
> We call this the ***trivial solution***. If this is the only solution, then S is said to be a ***linearly independent set***. If there are solutions in addition to the trivial solution, then S is said to be a ***linearly dependent set***.

We will often apply the terms *linearly independent* and *linearly dependent* to the vectors themselves rather than to the set.

▶ **EXAMPLE 1 Linear Independence of the Standard Unit Vectors in R^n**

The most basic linearly independent set in R^n is the set of standard unit vectors

$$\mathbf{e}_1 = (1, 0, 0, \ldots, 0), \quad \mathbf{e}_2 = (0, 1, 0, \ldots, 0), \ldots, \quad \mathbf{e}_n = (0, 0, 0, \ldots, 1)$$

For notational simplicity, we will prove the linear independence in R^3 of

$$\mathbf{i} = (1, 0, 0), \quad \mathbf{j} = (0, 1, 0), \quad \mathbf{k} = (0, 0, 1)$$

The linear independence or linear dependence of these vectors is determined by whether there exist nontrivial solutions of the vector equation

$$k_1\mathbf{i} + k_2\mathbf{j} + k_3\mathbf{k} = \mathbf{0} \tag{2}$$

Since the component form of this equation is

$$(k_1, k_2, k_3) = (0, 0, 0)$$

it follows that $k_1 = k_2 = k_3 = 0$. This implies that (2) has only the trivial solution and hence that the vectors are linearly independent.

▶ **EXAMPLE 2 Linear Independence in R^3**

Determine whether the vectors

$$\mathbf{v}_1 = (1, -2, 3), \quad \mathbf{v}_2 = (5, 6, -1), \quad \mathbf{v}_3 = (3, 2, 1)$$

are linearly independent or linearly dependent in R^3.

Solution The linear independence or linear dependence of these vectors is determined by whether there exist nontrivial solutions of the vector equation

$$k_1\mathbf{v}_1 + k_2\mathbf{v}_2 + k_3\mathbf{v}_3 = \mathbf{0} \tag{3}$$

or, equivalently, of

$$k_1(1, -2, 3) + k_2(5, 6, -1) + k_3(3, 2, 1) = (0, 0, 0)$$

Equating corresponding components on the two sides yields the homogeneous linear system

$$\begin{aligned} k_1 + 5k_2 + 3k_3 &= 0 \\ -2k_1 + 6k_2 + 2k_3 &= 0 \\ 3k_1 - k_2 + k_3 &= 0 \end{aligned} \tag{4}$$

Thus, our problem reduces to determining whether this system has nontrivial solutions. There are various ways to do this; one possibility is to simply solve the system, which yields

$$k_1 = -\tfrac{1}{2}t, \quad k_2 = -\tfrac{1}{2}t, \quad k_3 = t$$

(we omit the details). This shows that the system has nontrivial solutions and hence that the vectors are linearly dependent. A second method for obtaining the same result is to compute the determinant of the coefficient matrix

$$A = \begin{bmatrix} 1 & 5 & 3 \\ -2 & 6 & 2 \\ 3 & -1 & 1 \end{bmatrix}$$

In Example 2, what relationship do you see between the components of $\mathbf{v}_1$, $\mathbf{v}_2$, and $\mathbf{v}_3$ and the columns of the coefficient matrix A?

and use parts (b) and (g) of Theorem 2.3.8. We leave it for you to verify that $\det(A) = 0$, from which it follows (3) has nontrivial solutions and the vectors are linearly dependent.

▶ **EXAMPLE 3 Linear Independence in R^4**

Determine whether the vectors

$$\mathbf{v}_1 = (1, 2, 2, -1), \quad \mathbf{v}_2 = (4, 9, 9, -4), \quad \mathbf{v}_3 = (5, 8, 9, -5)$$

in R^4 are linearly dependent or linearly independent.

Solution The linear independence or linear dependence of these vectors is determined by whether there exist nontrivial solutions of the vector equation

$$k_1\mathbf{v}_1 + k_2\mathbf{v}_2 + k_3\mathbf{v}_3 = \mathbf{0}$$

or, equivalently, of

$$k_1(1, 2, 2, -1) + k_2(4, 9, 9, -4) + k_3(5, 8, 9, -5) = (0, 0, 0, 0)$$

Equating corresponding components on the two sides yields the homogeneous linear system

$$k_1 + 4k_2 + 5k_3 = 0$$
$$2k_1 + 9k_2 + 8k_3 = 0$$
$$2k_1 + 9k_2 + 9k_3 = 0$$
$$-k_1 - 4k_2 - 5k_3 = 0$$

We leave it for you to show that this system has only the trivial solution

$$k_1 = 0, \quad k_2 = 0, \quad k_3 = 0$$

from which you can conclude that $\mathbf{v}_1$, $\mathbf{v}_2$, and $\mathbf{v}_3$ are linearly independent.

▶ **EXAMPLE 4 An Important Linearly Independent Set in P_n**

Show that the polynomials

$$1, \quad x, \quad x^2, \ldots, \quad x^n$$

form a linearly independent set in P_n.

Solution For convenience, let us denote the polynomials as

$$\mathbf{p}_0 = 1, \quad \mathbf{p}_1 = x, \quad \mathbf{p}_2 = x^2, \ldots, \quad \mathbf{p}_n = x^n$$

We must show that the vector equation

$$a_0\mathbf{p}_0 + a_1\mathbf{p}_1 + a_2\mathbf{p}_2 + \cdots + a_n\mathbf{p}_n = \mathbf{0} \tag{5}$$

has only the trivial solution

$$a_0 = a_1 = a_2 = \cdots = a_n = 0$$

But (5) is equivalent to the statement that

$$a_0 + a_1 x + a_2 x^2 + \cdots + a_n x^n = 0 \tag{6}$$

for all x in $(-\infty, \infty)$, so we must show that this holds if and only if each coefficient in (6) is zero. To see that this is so, recall from algebra that a nonzero polynomial of degree n has at most n distinct roots. That being the case, each coefficient in (6) must be zero, for otherwise the left side of the equation would be a nonzero polynomial with infinitely many roots. Thus, (5) has only the trivial solution. ◀

The following example shows that the problem of determining whether a given set of vectors in P_n is linearly independent or linearly dependent can be reduced to determining whether a certain set of vectors in R^n is linearly dependent or independent.

▶ **EXAMPLE 5 Linear Independence of Polynomials**

Determine whether the polynomials

$$\mathbf{p}_1 = 1 - x, \quad \mathbf{p}_2 = 5 + 3x - 2x^2, \quad \mathbf{p}_3 = 1 + 3x - x^2$$

are linearly dependent or linearly independent in P_2.

Solution The linear independence or linear dependence of these vectors is determined by whether there exist nontrivial solutions of the vector equation

$$k_1 \mathbf{p}_1 + k_2 \mathbf{p}_2 + k_3 \mathbf{p}_3 = \mathbf{0} \tag{7}$$

This equation can be written as

$$k_1(1 - x) + k_2(5 + 3x - 2x^2) + k_3(1 + 3x - x^2) = 0 \tag{8}$$

or, equivalently, as

$$(k_1 + 5k_2 + k_3) + (-k_1 + 3k_2 + 3k_3)x + (-2k_2 - k_3)x^2 = 0$$

Since this equation must be satisfied by all x in $(-\infty, \infty)$, each coefficient must be zero (as explained in the previous example). Thus, the linear dependence or independence of the given polynomials hinges on whether the following linear system has a nontrivial solution:

$$\begin{aligned} k_1 + 5k_2 + k_3 &= 0 \\ -k_1 + 3k_2 + 3k_3 &= 0 \\ -2k_2 - k_3 &= 0 \end{aligned} \tag{9}$$

In Example 5, what relationship do you see between the coefficients of the given polynomials and the column vectors of the coefficient matrix of system (9)?

We leave it for you to show that this linear system has a nontrivial solutions either by solving it directly or by showing that the coefficient matrix has determinant zero. Thus, the set $\{\mathbf{p}_1, \mathbf{p}_2, \mathbf{p}_3\}$ is linearly dependent. ◀

An Alternative Interpretation of Linear Independence

The terms *linearly dependent* and *linearly independent* are intended to indicate whether the vectors in a given set are interrelated in some way. The following theorem, whose proof is deferred to the end of this section, makes this idea more precise.

THEOREM 4.3.1 *A set S with two or more vectors is*

(a) *Linearly dependent if and only if at least one of the vectors in S is expressible as a linear combination of the other vectors in S.*

(b) *Linearly independent if and only if no vector in S is expressible as a linear combination of the other vectors in S.*

▶ **EXAMPLE 6** **Example 1 Revisited**

In Example 1 we showed that the standard unit vectors in R^n are linearly independent. Thus, it follows from Theorem 4.3.1 that none of these vectors is expressible as a linear combination of the other two. To illustrate this in R^3, suppose, for example, that

$$\mathbf{k} = k_1\mathbf{i} + k_2\mathbf{j}$$

or in terms of components that

$$(0, 0, 1) = (k_1, k_2, 0)$$

Since this equation cannot be satisfied by any values of k_1 and k_2, there is no way to express $\mathbf{k}$ as a linear combination of $\mathbf{i}$ and $\mathbf{j}$. Similarly, $\mathbf{i}$ is not expressible as a linear combination of $\mathbf{j}$ and $\mathbf{k}$, and $\mathbf{j}$ is not expressible as a linear combination of $\mathbf{i}$ and $\mathbf{k}$.

▶ **EXAMPLE 7** **Example 2 Revisited**

In Example 2 we saw that the vectors

$$\mathbf{v}_1 = (1, -2, 3), \quad \mathbf{v}_2 = (5, 6, -1), \quad \mathbf{v}_3 = (3, 2, 1)$$

are linearly dependent. Thus, it follows from Theorem 4.3.1 that at least one of these vectors is expressible as a linear combination of the other two. We leave it for you to confirm that these vectors satisfy the equation

$$\tfrac{1}{2}\mathbf{v}_1 + \tfrac{1}{2}\mathbf{v}_2 - \mathbf{v}_3 = \mathbf{0}$$

from which it follows, for example, that

$$\mathbf{v}_3 = \tfrac{1}{2}\mathbf{v}_1 + \tfrac{1}{2}\mathbf{v}_2 \quad ◀$$

Sets with One or Two Vectors

The following basic theorem is concerned with the linear independence and linear dependence of sets with one or two vectors and sets that contain the zero vector.

THEOREM 4.3.2

(a) *A finite set that contains $\mathbf{0}$ is linearly dependent.*

(b) *A set with exactly one vector is linearly independent if and only if that vector is not $\mathbf{0}$.*

(c) *A set with exactly two vectors is linearly independent if and only if neither vector is a scalar multiple of the other.*

Józef Hoëné de Wroński
(1778–1853)

Historical Note The Polish-French mathematician Józef Hoëné de Wroński was born Józef Hoëné and adopted the name Wroński after he married. Wroński's life was fraught with controversy and conflict, which some say was due to his psychopathic tendencies and his exaggeration of the importance of his own work. Although Wroński's work was dismissed as rubbish for many years, and much of it was indeed erroneous, some of his ideas contained hidden brilliance and have survived. Among other things, Wroński designed a caterpillar vehicle to compete with trains (though it was never manufactured) and did research on the famous problem of determining the longitude of a ship at sea. His final years were spent in poverty.

[*Image: wikipedia*]

We will prove part (a) and leave the rest as exercises.

Proof (a)   For any vectors $\mathbf{v}_1, \mathbf{v}_2, \ldots, \mathbf{v}_r$, the set $S = \{\mathbf{v}_1, \mathbf{v}_2, \ldots, \mathbf{v}_r, \mathbf{0}\}$ is linearly dependent since the equation

$$0\mathbf{v}_1 + 0\mathbf{v}_2 + \cdots + 0\mathbf{v}_r + 1(\mathbf{0}) = \mathbf{0}$$

expresses $\mathbf{0}$ as a linear combination of the vectors in S with coefficients that are not all zero. ◄

▶ **EXAMPLE 8 Linear Independence of Two Functions**

The functions $\mathbf{f}_1 = x$ and $\mathbf{f}_2 = \sin x$ are linearly independent vectors in $F(-\infty, \infty)$ since neither function is a scalar multiple of the other. On the other hand, the two functions $\mathbf{g}_1 = \sin 2x$ and $\mathbf{g}_2 = \sin x \cos x$ are linearly dependent because the trigonometric identity $\sin 2x = 2 \sin x \cos x$ reveals that $\mathbf{g}_1$ and $\mathbf{g}_2$ are scalar multiples of each other. ◄

A Geometric Interpretation of Linear Independence

Linear independence has the following useful geometric interpretations in R^2 and R^3:

- Two vectors in R^2 or R^3 are linearly independent if and only if they do not lie on the same line when they have their initial points at the origin. Otherwise one would be a scalar multiple of the other (Figure 4.3.3).

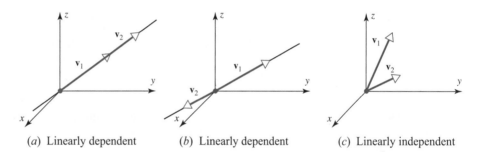

▶ **Figure 4.3.3**   (a) Linearly dependent   (b) Linearly dependent   (c) Linearly independent

- Three vectors in R^3 are linearly independent if and only if they do not lie in the same plane when they have their initial points at the origin. Otherwise at least one would be a linear combination of the other two (Figure 4.3.4).

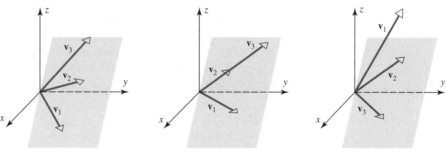

▶ **Figure 4.3.4**   (a) Linearly dependent   (b) Linearly dependent   (c) Linearly independent

At the beginning of this section we observed that a third coordinate axis in R^2 is superfluous by showing that a unit vector along such an axis would have to be expressible as a linear combination of unit vectors along the positive x- and y-axis. That result is a consequence of the next theorem, which shows that there can be at most n vectors in any linearly independent set R^n.

THEOREM 4.3.3 *Let $S = \{\mathbf{v}_1, \mathbf{v}_2, \ldots, \mathbf{v}_r\}$ be a set of vectors in R^n. If $r > n$, then S is linearly dependent.*

Proof Suppose that

$$\mathbf{v}_1 = (v_{11}, v_{12}, \ldots, v_{1n})$$
$$\mathbf{v}_2 = (v_{21}, v_{22}, \ldots, v_{2n})$$
$$\vdots \qquad\qquad \vdots$$
$$\mathbf{v}_r = (v_{r1}, v_{r2}, \ldots, v_{rn})$$

and consider the equation

$$k_1\mathbf{v}_1 + k_2\mathbf{v}_2 + \cdots + k_r\mathbf{v}_r = \mathbf{0}$$

It follows from Theorem 4.3.3, for example, that a set in R^2 with more than two vectors is linearly dependent and a set in R^3 with more than three vectors is linearly dependent.

If we express both sides of this equation in terms of components and then equate the corresponding components, we obtain the system

$$v_{11}k_1 + v_{21}k_2 + \cdots + v_{r1}k_r = 0$$
$$v_{12}k_1 + v_{22}k_2 + \cdots + v_{r2}k_r = 0$$
$$\vdots \qquad \vdots \qquad\qquad \vdots \qquad \vdots$$
$$v_{1n}k_1 + v_{2n}k_2 + \cdots + v_{rn}k_r = 0$$

This is a homogeneous system of n equations in the r unknowns $k_1, \ldots, k_r$. Since $r > n$, it follows from Theorem 1.2.2 that the system has nontrivial solutions. Therefore, $S = \{\mathbf{v}_1, \mathbf{v}_2, \ldots, \mathbf{v}_r\}$ is a linearly dependent set. ◄

CALCULUS REQUIRED

Linear Independence of Functions

Sometimes linear dependence of functions can be deduced from known identities. For example, the functions

$$\mathbf{f}_1 = \sin^2 x, \quad \mathbf{f}_2 = \cos^2 x, \quad \text{and} \quad \mathbf{f}_3 = 5$$

form a linearly dependent set in $F(-\infty, \infty)$, since the equation

$$5\mathbf{f}_1 + 5\mathbf{f}_2 - \mathbf{f}_3 = 5\sin^2 x + 5\cos^2 x - 5$$
$$= 5(\sin^2 x + \cos^2 x) - 5 = \mathbf{0}$$

expresses $\mathbf{0}$ as a linear combination of $\mathbf{f}_1$, $\mathbf{f}_2$, and $\mathbf{f}_3$ with coefficients that are not all zero.

Unfortunately, there is no *general* method that can be used to determine whether a set of functions is linearly independent or linearly dependent. However, there does exist a theorem that is useful for establishing linear independence in certain circumstances. The following definition will be useful for discussing that theorem.

DEFINITION 2 If $\mathbf{f}_1 = f_1(x), \mathbf{f}_2 = f_2(x), \ldots, \mathbf{f}_n = f_n(x)$ are functions that are $n - 1$ times differentiable on the interval $(-\infty, \infty)$, then the determinant

$$W(x) = \begin{vmatrix} f_1(x) & f_2(x) & \cdots & f_n(x) \\ f_1'(x) & f_2'(x) & \cdots & f_n'(x) \\ \vdots & \vdots & & \vdots \\ f_1^{(n-1)}(x) & f_2^{(n-1)}(x) & \cdots & f_n^{(n-1)}(x) \end{vmatrix}$$

is called the **Wronskian** of $f_1, f_2, \ldots, f_n$.

Suppose for the moment that $\mathbf{f}_1 = f_1(x), \mathbf{f}_2 = f_2(x), \ldots, \mathbf{f}_n = f_n(x)$ are *linearly dependent* vectors in $C^{(n-1)}(-\infty, \infty)$. This implies that for certain values of the coefficients the vector equation

$$k_1 \mathbf{f}_1 + k_2 \mathbf{f}_2 + \cdots + k_n \mathbf{f}_n = \mathbf{0}$$

has a nontrivial solution, or equivalently that the equation

$$k_1 f_1(x) + k_2 f_2(x) + \cdots + k_n f_n(x) = 0$$

is satisfied for all x in $(-\infty, \infty)$. Using this equation together with those that result by differentiating it $n - 1$ times yields the linear system

$$
\begin{array}{llll}
k_1 f_1(x) & + k_2 f_2(x) & + \cdots + k_n f_n(x) & = 0 \\
k_1 f_1'(x) & + k_2 f_2'(x) & + \cdots + k_n f_n'(x) & = 0 \\
\ \ \vdots & \ \ \ \vdots & \ \ \ \ \ \vdots & \vdots \\
k_1 f_1^{(n-1)}(x) & + k_2 f_2^{(n-1)}(x) & + \cdots + k_n f_n^{(n-1)}(x) & = 0
\end{array}
$$

Thus, the linear dependence of $\mathbf{f}_1, \mathbf{f}_2, \ldots, \mathbf{f}_n$ implies that the linear system

$$
\begin{bmatrix}
f_1(x) & f_2(x) & \cdots & f_n(x) \\
f_1'(x) & f_2'(x) & \cdots & f_n'(x) \\
\vdots & \vdots & & \vdots \\
f_1^{(n-1)}(x) & f_2^{(n-1)}(x) & \cdots & f_n^{(n-1)}(x)
\end{bmatrix}
\begin{bmatrix}
k_1 \\ k_2 \\ \vdots \\ k_n
\end{bmatrix}
=
\begin{bmatrix}
0 \\ 0 \\ \vdots \\ 0
\end{bmatrix}
\tag{10}
$$

has a nontrivial solution. But this implies that the determinant of the coefficient matrix of (10) is nonzero for every such x. Since this determinant is the Wronskian of $f_1, f_2, \ldots, f_n$, we have established the following result.

THEOREM 4.3.4 *If the functions $\mathbf{f}_1, \mathbf{f}_2, \ldots, \mathbf{f}_n$ have $n - 1$ continuous derivatives on the interval $(-\infty, \infty)$, and if the Wronskian of these functions is not identically zero on $(-\infty, \infty)$, then these functions form a linearly independent set of vectors in $C^{(n-1)}(-\infty, \infty)$.*

In Example 8 we showed that x and $\sin x$ are linearly independent functions by observing that neither is a scalar multiple of the other. The following example shows how to obtain the same result using the Wronskian (though it is a more complicated procedure in this particular case).

▶ **EXAMPLE 9 Linear Independence Using the Wronskian**

Use the Wronskian to show that $\mathbf{f}_1 = x$ and $\mathbf{f}_2 = \sin x$ are linearly independent.

Solution The Wronskian is

$$
W(x) = \begin{vmatrix} x & \sin x \\ 1 & \cos x \end{vmatrix} = x \cos x - \sin x
$$

This function is not identically zero on the interval $(-\infty, \infty)$ since, for example,

$$
W\left(\frac{\pi}{2}\right) = \frac{\pi}{2} \cos\left(\frac{\pi}{2}\right) - \sin\left(\frac{\pi}{2}\right) = \frac{\pi}{2}
$$

Thus, the functions are linearly independent.

WARNING The converse of Theorem 4.3.4 is false. If the Wronskian of $\mathbf{f}_1, \mathbf{f}_2, \ldots, \mathbf{f}_n$ is identically zero on $(-\infty, \infty)$, then no conclusion can be reached about the linear independence of $\{\mathbf{f}_1, \mathbf{f}_2, \ldots, \mathbf{f}_n\}$—this set of vectors may be linearly independent or linearly dependent.

▶ **EXAMPLE 10** **Linear Independence Using the Wronskian**

Use the Wronskian to show that $\mathbf{f}_1 = 1$, $\mathbf{f}_2 = e^x$, and $\mathbf{f}_3 = e^{2x}$ are linearly independent.

Solution The Wronskian is

$$W(x) = \begin{vmatrix} 1 & e^x & e^{2x} \\ 0 & e^x & 2e^{2x} \\ 0 & e^x & 4e^{2x} \end{vmatrix} = 2e^{3x}$$

This function is obviously not identically zero on $(-\infty, \infty)$, so $\mathbf{f}_1$, $\mathbf{f}_2$, and $\mathbf{f}_3$ form a linearly independent set. ◀

OPTIONAL

We will close this section by proving part (a) of Theorem 4.3.1. We will leave the proof of part (b) as an exercise.

Proof of Theorem 4.3.1(a) Let $S = \{\mathbf{v}_1, \mathbf{v}_2, \ldots, \mathbf{v}_r\}$ be a set with two or more vectors. If we assume that S is linearly dependent, then there are scalars $k_1, k_2, \ldots, k_r$, not all zero, such that

$$k_1\mathbf{v}_1 + k_2\mathbf{v}_2 + \cdots + k_r\mathbf{v}_r = \mathbf{0} \qquad (11)$$

To be specific, suppose that $k_1 \neq 0$. Then (11) can be rewritten as

$$\mathbf{v}_1 = \left(-\frac{k_2}{k_1}\right)\mathbf{v}_2 + \cdots + \left(-\frac{k_r}{k_1}\right)\mathbf{v}_r$$

which expresses $\mathbf{v}_1$ as a linear combination of the other vectors in S. Similarly, if $k_j \neq 0$ in (11) for some $j = 2, 3, \ldots, r$, then $\mathbf{v}_j$ is expressible as a linear combination of the other vectors in S.

Conversely, let us assume that at least one of the vectors in S is expressible as a linear combination of the other vectors. To be specific, suppose that

$$\mathbf{v}_1 = c_2\mathbf{v}_2 + c_3\mathbf{v}_3 + \cdots + c_r\mathbf{v}_r$$

so

$$\mathbf{v}_1 - c_2\mathbf{v}_2 - c_3\mathbf{v}_3 - \cdots - c_r\mathbf{v}_r = \mathbf{0}$$

It follows that S is linearly dependent since the equation

$$k_1\mathbf{v}_1 + k_2\mathbf{v}_2 + \cdots + k_r\mathbf{v}_r = \mathbf{0}$$

is satisfied by

$$k_1 = 1, \quad k_2 = -c_2, \ldots, \quad k_r = -c_r$$

which are not all zero. The proof in the case where some vector other than $\mathbf{v}_1$ is expressible as a linear combination of the other vectors in S is similar. ◀

Concept Review

- Trivial solution
- Linearly independent set
- Linearly dependent set
- Wronskian

Skills

- Determine whether a set of vectors is linearly independent or linearly dependent.
- Express one vector in a linearly dependent set as a linear combination of the other vectors in the set.
- Use the Wronskian to show that a set of functions is linearly independent.

Exercise Set 4.3

1. Explain why the following are linearly dependent sets of vectors. (Solve this problem by inspection.)

(a) $\mathbf{u}_1 = (-1, 2, 4)$ and $\mathbf{u}_2 = (5, -10, -20)$ in R^3

(b) $\mathbf{u}_1 = (3, -1)$, $\mathbf{u}_2 = (4, 5)$, $\mathbf{u}_3 = (-4, 7)$ in R^2

(c) $\mathbf{p}_1 = 3 - 2x + x^2$ and $\mathbf{p}_2 = 6 - 4x + 2x^2$ in P_2

(d) $A = \begin{bmatrix} -3 & 4 \\ 2 & 0 \end{bmatrix}$ and $B = \begin{bmatrix} 3 & -4 \\ -2 & 0 \end{bmatrix}$ in M_{22}

2. Which of the following sets of vectors in R^3 are linearly dependent?

(a) $(4, -1, 2), (-4, 10, 2)$

(b) $(-3, 0, 4), (5, -1, 2), (1, 1, 3)$

(c) $(8, -1, 3), (4, 0, 1)$

(d) $(-2, 0, 1), (3, 2, 5), (6, -1, 1), (7, 0, -2)$

3. Which of the following sets of vectors in R^4 are linearly dependent?

(a) $(3, 8, 7, -3), (1, 5, 3, -1), (2, -1, 2, 6), (1, 4, 0, 3)$

(b) $(0, 0, 2, 2), (3, 3, 0, 0), (1, 1, 0, -1)$

(c) $(0, 3, -3, -6), (-2, 0, 0, -6), (0, -4, -2, -2), (0, -8, 4, -4)$

(d) $(3, 0, -3, 6), (0, 2, 3, 1), (0, -2, -2, 0), (-2, 1, 2, 1)$

4. Which of the following sets of vectors in P_2 are linearly dependent?

(a) $2 - x + 4x^2, 3 + 6x + 2x^2, 2 + 10x - 4x^2$

(b) $3 + x + x^2, 2 - x + 5x^2, 4 - 3x^2$

(c) $6 - x^2, 1 + x + 4x^2$

(d) $1 + 3x + 3x^2, x + 4x^2, 5 + 6x + 3x^2, 7 + 2x - x^2$

5. Assume that $\mathbf{v}_1$, $\mathbf{v}_2$, and $\mathbf{v}_3$ are vectors in R^3 that have their initial points at the origin. In each part, determine whether the three vectors lie in a plane.

(a) $\mathbf{v}_1 = (2, -2, 0)$, $\mathbf{v}_2 = (6, 1, 4)$, $\mathbf{v}_3 = (2, 0, -4)$

(b) $\mathbf{v}_1 = (-6, 7, 2)$, $\mathbf{v}_2 = (3, 2, 4)$, $\mathbf{v}_3 = (4, -1, 2)$

6. Assume that $\mathbf{v}_1$, $\mathbf{v}_2$, and $\mathbf{v}_3$ are vectors in R^3 that have their initial points at the origin. In each part, determine whether the three vectors lie on the same line.

(a) $\mathbf{v}_1 = (-1, 2, 3)$, $\mathbf{v}_2 = (2, -4, -6)$, $\mathbf{v}_3 = (-3, 6, 0)$

(b) $\mathbf{v}_1 = (2, -1, 4)$, $\mathbf{v}_2 = (4, 2, 3)$, $\mathbf{v}_3 = (2, 7, -6)$

(c) $\mathbf{v}_1 = (4, 6, 8)$, $\mathbf{v}_2 = (2, 3, 4)$, $\mathbf{v}_3 = (-2, -3, -4)$

7. (a) Show that the three vectors $\mathbf{v}_1 = (0, 3, 1, -1)$, $\mathbf{v}_2 = (6, 0, 5, 1)$, and $\mathbf{v}_3 = (4, -7, 1, 3)$ form a linearly dependent set in R^4.

(b) Express each vector in part (a) as a linear combination of the other two.

8. (a) Show that the three vectors $\mathbf{v}_1 = (1, 2, 3, 4)$, $\mathbf{v}_2 = (0, 1, 0, -1)$, and $\mathbf{v}_3 = (1, 3, 3, 3)$ form a linearly dependent set in R^4.

(b) Express each vector in part (a) as a linear combination of the other two.

9. For which real values of λ do the following vectors form a linearly dependent set in R^3?

$$\mathbf{v}_1 = \left(\lambda, -\tfrac{1}{2}, -\tfrac{1}{2}\right), \quad \mathbf{v}_2 = \left(-\tfrac{1}{2}, \lambda, -\tfrac{1}{2}\right), \quad \mathbf{v}_3 = \left(-\tfrac{1}{2}, -\tfrac{1}{2}, \lambda\right)$$

10. Show that if $\{\mathbf{v}_1, \mathbf{v}_2, \mathbf{v}_3\}$ is a linearly independent set of vectors, then so are $\{\mathbf{v}_1, \mathbf{v}_2\}$, $\{\mathbf{v}_1, \mathbf{v}_3\}$, $\{\mathbf{v}_2, \mathbf{v}_3\}$, $\{\mathbf{v}_1\}$, $\{\mathbf{v}_2\}$, and $\{\mathbf{v}_3\}$.

11. Show that if $S = \{\mathbf{v}_1, \mathbf{v}_2, \ldots, \mathbf{v}_r\}$ is a linearly independent set of vectors, then so is every nonempty subset of S.

12. Show that if $S = \{\mathbf{v}_1, \mathbf{v}_2, \mathbf{v}_3\}$ is a linearly dependent set of vectors in a vector space V, and $\mathbf{v}_4$ is any vector in V that is not in S, then $\{\mathbf{v}_1, \mathbf{v}_2, \mathbf{v}_3, \mathbf{v}_4\}$ is also linearly dependent.

13. Show that if $S = \{\mathbf{v}_1, \mathbf{v}_2, \ldots, \mathbf{v}_r\}$ is a linearly dependent set of vectors in a vector space V, and if $\mathbf{v}_{r+1}, \ldots, \mathbf{v}_n$ are any vectors in V that are not in S, then $\{\mathbf{v}_1, \mathbf{v}_2, \ldots, \mathbf{v}_r, \mathbf{v}_{r+1}, \ldots, \mathbf{v}_n\}$ is also linearly dependent.

14. Show that in P_2 every set with more than three vectors is linearly dependent.

15. Show that if $\{\mathbf{v}_1, \mathbf{v}_2\}$ is linearly independent and $\mathbf{v}_3$ does not lie in span$\{\mathbf{v}_1, \mathbf{v}_2\}$, then $\{\mathbf{v}_1, \mathbf{v}_2, \mathbf{v}_3\}$ is linearly independent.

16. Prove: For any vectors $\mathbf{u}$, $\mathbf{v}$, and $\mathbf{w}$ in a vector space V, the vectors $\mathbf{u} - \mathbf{v}$, $\mathbf{v} - \mathbf{w}$, and $\mathbf{w} - \mathbf{u}$ form a linearly dependent set.

17. Prove: The space spanned by two vectors in R^3 is a line through the origin, a plane through the origin, or the origin itself.

18. Under what conditions is a set with one vector linearly independent?

19. Are the vectors $\mathbf{v}_1$, $\mathbf{v}_2$, and $\mathbf{v}_3$ in part (a) of the accompanying figure linearly independent? What about those in part (b)? Explain.

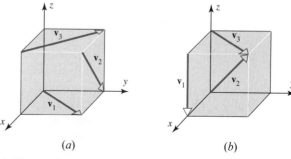

(a)   (b)

▲ Figure Ex-19

20. By using appropriate identities, where required, determine which of the following sets of vectors in $F(-\infty, \infty)$ are linearly dependent.

(a) $6,\ 3\sin^2 x,\ 2\cos^2 x$ (b) $x,\ \cos x$

(c) $1,\ \sin x,\ \sin 2x$ (d) $\cos 2x,\ \sin^2 x,\ \cos^2 x$

(e) $(3-x)^2,\ x^2 - 6x,\ 5$ (f) $0,\ \cos^3 \pi x,\ \sin^5 3\pi x$

21. The functions $f_1(x) = x$ and $f_2(x) = \cos x$ are linearly independent in $F(-\infty, \infty)$ because neither function is a scalar multiple of the other. Confirm the linear independence using Wroński's test.

22. The functions $f_1(x) = \sin x$ and $f_2(x) = \cos x$ are linearly independent in $F(-\infty, \infty)$ because neither function is a scalar multiple of the other. Confirm the linear independence using Wroński's test.

23. (*Calculus required*) Use the Wronskian to show that the following sets of vectors are linearly independent.

(a) $1,\ x,\ e^x$ (b) $1,\ x,\ x^2$

24. Use Wroński's test to show that the functions $f_1(x) = e^x$, $f_2(x) = xe^x$, and $f_3(x) = x^2 e^x$ span a three-dimensional subspace of $F(-\infty, \infty)$.

25. Use Wroński's test to show that the functions $f_1(x) = \sin x$, $f_2(x) = \cos x$, and $f_3(x) = x\cos x$ span a three-dimensional subspace of $F(-\infty, \infty)$.

26. Use part (*a*) of Theorem 4.3.1 to prove part (*b*).

27. Prove part (*b*) of Theorem 4.3.2.

28. (a) In Example 1 we showed that the mutually orthogonal vectors **i**, **j**, and **k** form a linearly independent set of vectors in R^3. Do you think that every set of three nonzero mutually orthogonal vectors in R^3 is linearly independent? Justify your conclusion with a geometric argument.

(b) Justify your conclusion with an algebraic argument. [*Hint:* Use dot products.]

True-False Exercises

In parts (a)–(h) determine whether the statement is true or false, and justify your answer.

(a) A set containing a single vector is linearly independent.

(b) The set of vectors $\{\mathbf{v}, k\mathbf{v}\}$ is linearly dependent for every scalar k.

(c) Every linearly dependent set contains the zero vector.

(d) If the set of vectors $\{\mathbf{v}_1, \mathbf{v}_2, \mathbf{v}_3\}$ is linearly independent, then $\{k\mathbf{v}_1, k\mathbf{v}_2, k\mathbf{v}_3\}$ is also linearly independent for every nonzero scalar k.

(e) If $\mathbf{v}_1, \ldots, \mathbf{v}_n$ are linearly dependent nonzero vectors, then at least one vector $\mathbf{v}_k$ is a unique linear combination of $\mathbf{v}_1, \ldots, \mathbf{v}_{k-1}$.

(f) The set of 2×2 matrices that contain exactly two 1's and two 0's is a linearly independent set in M_{22}.

(g) The three polynomials $(x-1)(x+2), x(x+2)$, and $x(x-1)$ are linearly independent.

(h) The functions f_1 and f_2 are linearly dependent if there is a real number x so that $k_1 f_1(x) + k_2 f_2(x) = 0$ for some scalars k_1 and k_2.

4.4 Coordinates and Basis

We usually think of a line as being one-dimensional, a plane as two-dimensional, and the space around us as three-dimensional. It is the primary goal of this section and the next to make this intuitive notion of dimension precise. In this section we will discuss coordinate systems in general vector spaces and lay the groundwork for a precise definition of dimension in the next section.

Coordinate Systems in Linear Algebra

In analytic geometry we learned to use *rectangular* coordinate systems to create a one-to-one correspondence between points in 2-space and ordered pairs of real numbers and between points in 3-space and ordered triples of real numbers (Figure 4.4.1). Although rectangular coordinate systems are common, they are not essential. For example, Figure 4.4.2 shows coordinate systems in 2-space and 3-space in which the coordinate axes are not mutually perpendicular.

In linear algebra coordinate systems are commonly specified using vectors rather than coordinate axes. For example, in Figure 4.4.3 we have recreated the coordinate systems in Figure 4.4.2 by using unit vectors to identify the positive directions and then attaching coordinates to a point P using the scalar coefficients in the equations

$$\overrightarrow{OP} = a\mathbf{u}_1 + b\mathbf{u}_2 \quad \text{and} \quad \overrightarrow{OP} = a\mathbf{u}_1 + b\mathbf{u}_2 + c\mathbf{u}_3$$

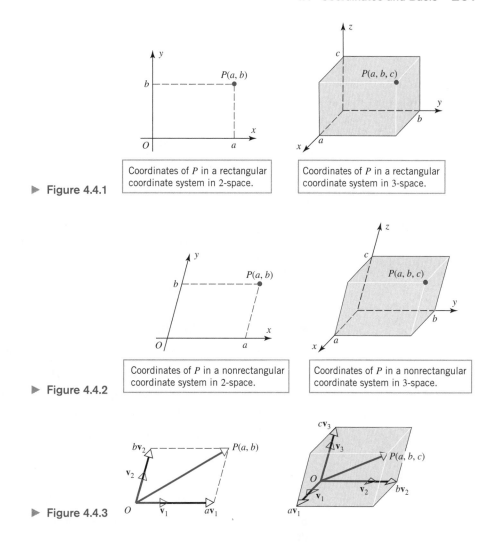

▶ **Figure 4.4.1**

Coordinates of P in a rectangular coordinate system in 2-space.

Coordinates of P in a rectangular coordinate system in 3-space.

▶ **Figure 4.4.2**

Coordinates of P in a nonrectangular coordinate system in 2-space.

Coordinates of P in a nonrectangular coordinate system in 3-space.

▶ **Figure 4.4.3**

Units of measurement are essential ingredients of any coordinate system. In geometry problems one tries to use the same unit of measurement on all axes to avoid distorting the shapes of figures. This is less important in applications where coordinates represent physical quantities with diverse units (for example, time in seconds on one axis and temperature in degrees Celsius on another axis). To allow for this level of generality, we will relax the requirement that *unit* vectors be used to identify the positive directions and require only that those vectors be linearly independent. We will refer to these as the "basis vectors" for the coordinate system. In summary, it is the directions of the basis vectors that establish the positive directions, and it is the lengths of the basis vectors that establish the spacing between the integer points on the axes (Figure 4.4.4).

Basis for a Vector Space

The following definition will make the preceding ideas more precise and will enable us to extend the concept of a coordinate system to general vector spaces.

Note that in Definition 1 we have required a basis to have finitely many vectors. Some authors call this a *finite basis*, but we will not use this terminology.

DEFINITION 1 If V is any vector space and $S = \{\mathbf{v}_1, \mathbf{v}_2, \ldots, \mathbf{v}_n\}$ is a finite set of vectors in V, then S is called a *basis* for V if the following two conditions hold:

(a) S is linearly independent.

(b) S spans V.

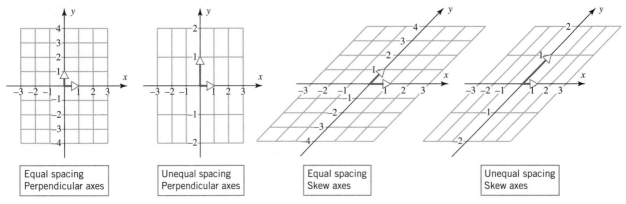

| Equal spacing Perpendicular axes | Unequal spacing Perpendicular axes | Equal spacing Skew axes | Unequal spacing Skew axes |

▲ Figure 4.4.4

If you think of a basis as describing a coordinate system for a vector space in V, then part (a) of this definition guarantees that there is no interrelationship between the basis vectors, and part (b) guarantees that there are enough basis vectors to provide coordinates for all vectors in V. Here are some examples.

▶ **EXAMPLE 1** **The Standard Basis for R^n**

Recall from Example 11 of Section 4.2 that the standard unit vectors

$$\mathbf{e}_1 = (1, 0, 0, \ldots, 0), \quad \mathbf{e}_2 = (0, 1, 0, \ldots, 0), \ldots, \quad \mathbf{e}_n = (0, 0, 0, \ldots, 1)$$

span R^n and from Example 1 of Section 4.3 that they are linearly independent. Thus, they form a basis for R^n that we call the ***standard basis for R^n***. In particular,

$$\mathbf{i} = (1, 0, 0), \quad \mathbf{j} = (0, 1, 0), \quad \mathbf{k} = (0, 0, 1)$$

is the standard basis for R^3.

▶ **EXAMPLE 2** **The Standard Basis for P_n**

Show that $S = \{1, x, x^2, \ldots, x^n\}$ is a basis for the vector space P_n of polynomials of degree n or less.

Solution We must show that the polynomials in S are linearly independent and span P_n. Let us denote these polynomials by

$$\mathbf{p}_0 = 1, \quad \mathbf{p}_1 = x, \quad \mathbf{p}_2 = x^2, \ldots, \quad \mathbf{p}_n = x^n$$

We showed in Example 13 of Section 4.2 that these vectors span P_n and in Example 4 of Section 4.3 that they are linearly independent. Thus, they form a basis for P_n that we call the ***standard basis for P_n***.

▶ **EXAMPLE 3** **Another Basis for R^3**

Show that the vectors $\mathbf{v}_1 = (1, 2, 1)$, $\mathbf{v}_2 = (2, 9, 0)$, and $\mathbf{v}_3 = (3, 3, 4)$ form a basis for R^3.

Solution We must show that these vectors are linearly independent and span R^3. To prove linear independence we must show that the vector equation

$$c_1\mathbf{v}_1 + c_2\mathbf{v}_2 + c_3\mathbf{v}_3 = \mathbf{0} \tag{1}$$

has only the trivial solution; and to prove that the vectors span R^3 we must show that every vector $\mathbf{b} = (b_1, b_2, b_3)$ in R^3 can be expressed as

$$c_1\mathbf{v}_1 + c_2\mathbf{v}_2 + c_3\mathbf{v}_3 = \mathbf{b} \tag{2}$$

By equating corresponding components on the two sides, these two equations can be expressed as the linear systems

$$
\begin{aligned}
c_1 + 2c_2 + 3c_3 &= 0 \\
2c_1 + 9c_2 + 3c_3 &= 0 \\
c_1 \quad\quad + 4c_3 &= 0
\end{aligned}
\quad \text{and} \quad
\begin{aligned}
c_1 + 2c_2 + 3c_3 &= b_1 \\
2c_1 + 9c_2 + 3c_3 &= b_2 \\
c_1 \quad\quad + 4c_3 &= b_3
\end{aligned}
\tag{3}
$$

(verify). Thus, we have reduced the problem to showing that in (3) the homogeneous system has only the trivial solution and that the nonhomogeneous system is consistent for all values of b_1, b_2, and b_3. But the two systems have the same coefficient matrix

$$
A = \begin{bmatrix} 1 & 2 & 3 \\ 2 & 9 & 3 \\ 1 & 0 & 4 \end{bmatrix}
$$

so it follows from parts (b), (e), and (g) of Theorem 2.3.8 that we can prove both results at the same time by showing that $\det(A) \neq 0$. We leave it for you to confirm that $\det(A) = -1$, which proves that the vectors v_1, v_2, and v_3 form a basis for R^3.

▶ **EXAMPLE 4 The Standard Basis for M_{mn}**

Show that the matrices

$$
M_1 = \begin{bmatrix} 1 & 0 \\ 0 & 0 \end{bmatrix}, \quad
M_2 = \begin{bmatrix} 0 & 1 \\ 0 & 0 \end{bmatrix}, \quad
M_3 = \begin{bmatrix} 0 & 0 \\ 1 & 0 \end{bmatrix}, \quad
M_4 = \begin{bmatrix} 0 & 0 \\ 0 & 1 \end{bmatrix}
$$

form a basis for the vector space M_{22} of 2×2 matrices.

Solution We must show that the matrices are linearly independent and span M_{22}. To prove linear independence we must show that the equation

$$
c_1 M_1 + c_2 M_2 + c_3 M_3 + c_4 M_4 = 0
\tag{4}
$$

has only the trivial solution, where 0 is the 2×2 zero matrix; and to prove that the matrices span M_{22} we must show that every 2×2 matrix

$$
B = \begin{bmatrix} a & b \\ c & d \end{bmatrix}
$$

can be expressed as

$$
c_1 M_1 + c_2 M_2 + c_3 M_3 + c_4 M_4 = B
\tag{5}
$$

The matrix forms of Equations (4) and (5) are

$$
c_1 \begin{bmatrix} 1 & 0 \\ 0 & 0 \end{bmatrix} + c_2 \begin{bmatrix} 0 & 1 \\ 0 & 0 \end{bmatrix} + c_3 \begin{bmatrix} 0 & 0 \\ 1 & 0 \end{bmatrix} + c_4 \begin{bmatrix} 0 & 0 \\ 0 & 1 \end{bmatrix} = \begin{bmatrix} 0 & 0 \\ 0 & 0 \end{bmatrix}
$$

and

$$
c_1 \begin{bmatrix} 1 & 0 \\ 0 & 0 \end{bmatrix} + c_2 \begin{bmatrix} 0 & 1 \\ 0 & 0 \end{bmatrix} + c_3 \begin{bmatrix} 0 & 0 \\ 1 & 0 \end{bmatrix} + c_4 \begin{bmatrix} 0 & 0 \\ 0 & 1 \end{bmatrix} = \begin{bmatrix} a & b \\ c & d \end{bmatrix}
$$

which can be rewritten as

$$
\begin{bmatrix} c_1 & c_2 \\ c_3 & c_4 \end{bmatrix} = \begin{bmatrix} 0 & 0 \\ 0 & 0 \end{bmatrix}
\quad \text{and} \quad
\begin{bmatrix} c_1 & c_2 \\ c_3 & c_4 \end{bmatrix} = \begin{bmatrix} a & b \\ c & d \end{bmatrix}
$$

Since the first equation has only the trivial solution

$$
c_1 = c_2 = c_3 = c_4 = 0
$$

the matrices are linearly independent, and since the second equation has the solution

$$
c_1 = a, \quad c_2 = b, \quad c_3 = c, \quad c_4 = d
$$

the matrices span M_{22}. This proves that the matrices M_1, M_2, M_3, M_4 form a basis for M_{22}. More generally, the mn different matrices whose entries are zero except for a single entry of 1 form a basis for M_{mn} called the ***standard basis for*** M_{mn}. ◄

Some writers define the empty set to be a basis for the zero vector space, but we will not do so.

It is not true that every vector space has a basis in the sense of Definition 1. The simplest example is the zero vector space, which contains no linearly independent sets and hence no basis. The following is an example of a nonzero vector space that has no basis in the sense of Definition 1 because it cannot be spanned by finitely many vectors.

▶ **EXAMPLE 5** **A Vector Space That Has No Finite Spanning Set**

Show that the vector space of P_∞ of all polynomials with real coefficients has no finite spanning set.

Solution If there were a finite spanning set, say $S = \{\mathbf{p}_1, \mathbf{p}_2, \ldots, \mathbf{p}_r\}$, then the degrees of the polynomials in S would have a maximum value, say n; and this in turn would imply that any linear combination of the polynomials in S would have degree at most n. Thus, there would be no way to express the polynomial x^{n+1} as a linear combination of the polynomials in S, contradicting the fact that the vectors in S span P_∞. ◄

For reasons that will become clear shortly, a vector space that cannot be spanned by finitely many vectors is said to be ***infinite-dimensional***, whereas those that can are said to be ***finite-dimensional***.

▶ **EXAMPLE 6** **Some Finite- and Infinite-Dimensional Spaces**

In Examples 1, 2, and 4 we found bases for R^n, P_n, and M_{mn}, so these vector spaces are finite-dimensional. We showed in Example 5 that the vector space P_∞ is not spanned by finitely many vectors and hence is infinite-dimensional. In the exercises of this section and the next we will ask you to show that the vector spaces R^∞, $F(-\infty, \infty)$, $C(-\infty, \infty)$, $C^m(-\infty, \infty)$, and $C^\infty(-\infty, \infty)$ are infinite-dimensional. ◄

Coordinates Relative to a Basis

Earlier in this section we drew an informal analogy between basis vectors and coordinate systems. Our next goal is to make this informal idea precise by defining the notion of a coordinate system in a general vector space. The following theorem will be our first step in that direction.

THEOREM 4.4.1 **Uniqueness of Basis Representation**

If $S = \{\mathbf{v}_1, \mathbf{v}_2, \ldots, \mathbf{v}_n\}$ is a basis for a vector space V, then every vector $\mathbf{v}$ in V can be expressed in the form $\mathbf{v} = c_1\mathbf{v}_1 + c_2\mathbf{v}_2 + \cdots + c_n\mathbf{v}_n$ in exactly one way.

Proof Since S spans V, it follows from the definition of a spanning set that every vector in V is expressible as a linear combination of the vectors in S. To see that there is only *one* way to express a vector as a linear combination of the vectors in S, suppose that some vector $\mathbf{v}$ can be written as

$$\mathbf{v} = c_1\mathbf{v}_1 + c_2\mathbf{v}_2 + \cdots + c_n\mathbf{v}_n$$

and also as

$$\mathbf{v} = k_1\mathbf{v}_1 + k_2\mathbf{v}_2 + \cdots + k_n\mathbf{v}_n$$

Subtracting the second equation from the first gives

$$\mathbf{0} = (c_1 - k_1)\mathbf{v}_1 + (c_2 - k_2)\mathbf{v}_2 + \cdots + (c_n - k_n)\mathbf{v}_n$$

Since the right side of this equation is a linear combination of vectors in S, the linear independence of S implies that

$$c_1 - k_1 = 0, \quad c_2 - k_2 = 0, \ldots, \quad c_n - k_n = 0$$

that is,

$$c_1 = k_1, \quad c_2 = k_2, \ldots, \quad c_n = k_n$$

Thus, the two expressions for $\mathbf{v}$ are the same. ◀

We now have all of the ingredients required to define the notion of "coordinates" in a general vector space V. For motivation, observe that in R^3, for example, the coordinates (a, b, c) of a vector $\mathbf{v}$ are precisely the coefficients in the formula

$$\mathbf{v} = a\mathbf{i} + b\mathbf{j} + c\mathbf{k}$$

▲ Figure 4.4.5

that expresses $\mathbf{v}$ as a linear combination of the standard basis vectors for R^3 (see Figure 4.4.5). The following definition generalizes this idea.

DEFINITION 2 If $S = \{\mathbf{v}_1, \mathbf{v}_2, \ldots, \mathbf{v}_n\}$ is a basis for a vector space V, and

$$\mathbf{v} = c_1\mathbf{v}_1 + c_2\mathbf{v}_2 + \cdots + c_n\mathbf{v}_n$$

is the expression for a vector $\mathbf{v}$ in terms of the basis S, then the scalars $c_1, c_2, \ldots, c_n$ are called the ***coordinates*** of $\mathbf{v}$ relative to the basis S. The vector $(c_1, c_2, \ldots, c_n)$ in R^n constructed from these coordinates is called the ***coordinate vector of $\mathbf{v}$ relative to S***; it is denoted by

$$(\mathbf{v})_S = (c_1, c_2, \ldots, c_n) \tag{6}$$

Sometimes it will be desirable to write a coordinate vector as a column matrix, in which case we will denote it using square brackets as

$$[\mathbf{v}]_S = \begin{bmatrix} c_1 \\ c_2 \\ \vdots \\ c_n \end{bmatrix}$$

We will refer to $[\mathbf{v}]_S$ as a ***coordinate matrix*** and reserve the terminology ***coordinate vector*** for the comma delimited form $(\mathbf{v})_S$.

Remark Recall that two *sets* are considered to be the same if they have the same members, even if those members are written in a different order. However, if $S = \{\mathbf{v}_1, \mathbf{v}_2, \ldots, \mathbf{v}_n\}$ is a set of *basis vectors*, then changing the order in which the vectors are written would change the order of the entries in $(\mathbf{v})_S$, possibly producing a different coordinate vector. To avoid this complication, we will make the convention that in any discussion involving a basis S the order of the vectors in S remains fixed. Some authors call a set of basis vectors with this restriction an ***ordered basis***. However, we will use this terminology only when emphasis on the order is required for clarity.

Observe that $(\mathbf{v})_S$ is a vector in R^n, so that once basis S is given for a vector space V, Theorem 4.4.1 establishes a one-to-one correspondence between vectors in V and vectors in R^n (Figure 4.4.6).

A one-to-one correspondence

▶ Figure 4.4.6

▶ **EXAMPLE 7 Coordinates Relative to the Standard Basis for R^n**

In the special case where $V = R^n$ and S is the *standard basis*, the coordinate vector $(\mathbf{v})_S$ and the vector $\mathbf{v}$ are the same; that is,

$$\mathbf{v} = (\mathbf{v})_S$$

For example, in R^3 the representation of a vector $\mathbf{v} = (a, b, c)$ as a linear combination of the vectors in the standard basis $S = \{\mathbf{i}, \mathbf{j}, \mathbf{k}\}$ is

$$\mathbf{v} = a\mathbf{i} + b\mathbf{j} + c\mathbf{k}$$

so the coordinate vector relative to this basis is $(\mathbf{v})_S = (a, b, c)$, which is the same as the vector $\mathbf{v}$.

▶ **EXAMPLE 8 Coordinate Vectors Relative to Standard Bases**

(a) Find the coordinate vector for the polynomial

$$\mathbf{p}(x) = c_0 + c_1 x + c_2 x^2 + \cdots + c_n x^n$$

relative to the standard basis for the vector space P_n.

(b) Find the coordinate vector of

$$B = \begin{bmatrix} a & b \\ c & d \end{bmatrix}$$

relative to the standard basis for M_{22}.

Solution (a) The given formula for $\mathbf{p}(x)$ expresses this polynomial as a linear combination of the standard basis vectors $S = \{1, x, x^2, \ldots, x^n\}$. Thus, the coordinate vector for $\mathbf{p}$ relative to S is

$$(\mathbf{p})_S = (c_0, c_1, c_2, \ldots, c_n)$$

Solution (b) We showed in Example 4 that the representation of a vector

$$B = \begin{bmatrix} a & b \\ c & d \end{bmatrix}$$

as a linear combination of the standard basis vectors is

$$B = \begin{bmatrix} a & b \\ c & d \end{bmatrix} = a \begin{bmatrix} 1 & 0 \\ 0 & 0 \end{bmatrix} + b \begin{bmatrix} 0 & 1 \\ 1 & 0 \end{bmatrix} + c \begin{bmatrix} 0 & 0 \\ 1 & 0 \end{bmatrix} + d \begin{bmatrix} 0 & 0 \\ 0 & 1 \end{bmatrix}$$

so the coordinate vector of B relative to S is

$$(B)_S = (a, b, c, d)$$

▶ **EXAMPLE 9 Coordinates in R^3**

(a) We showed in Example 3 that the vectors

$$\mathbf{v}_1 = (1, 2, 1), \quad \mathbf{v}_2 = (2, 9, 0), \quad \mathbf{v}_3 = (3, 3, 4)$$

form a basis for R^3. Find the coordinate vector of $\mathbf{v} = (5, -1, 9)$ relative to the basis $S = \{\mathbf{v}_1, \mathbf{v}_2, \mathbf{v}_3\}$.

(b) Find the vector $\mathbf{v}$ in R^3 whose coordinate vector relative to S is $(\mathbf{v})_S = (-1, 3, 2)$.

Solution (a) To find $(\mathbf{v})_S$ we must first express $\mathbf{v}$ as a linear combination of the vectors in S; that is, we must find values of c_1, c_2, and c_3 such that

$$\mathbf{v} = c_1 \mathbf{v}_1 + c_2 \mathbf{v}_2 + c_3 \mathbf{v}_3$$

or, in terms of components,

$$(5, -1, 9) = c_1(1, 2, 1) + c_2(2, 9, 0) + c_3(3, 3, 4)$$

Equating corresponding components gives

$$\begin{aligned} c_1 + 2c_2 + 3c_3 &= 5 \\ 2c_1 + 9c_2 + 3c_3 &= -1 \\ c_1 \qquad\quad + 4c_3 &= 9 \end{aligned}$$

Solving this system we obtain $c_1 = 1$, $c_2 = -1$, $c_3 = 2$ (verify). Therefore,

$$(\mathbf{v})_S = (1, -1, 2)$$

Solution (b) Using the definition of $(\mathbf{v})_S$, we obtain

$$\mathbf{v} = (-1)\mathbf{v}_1 + 3\mathbf{v}_2 + 2\mathbf{v}_3$$
$$= (-1)(1, 2, 1) + 3(2, 9, 0) + 2(3, 3, 4) = (11, 31, 7) \; \blacktriangleleft$$

Concept Review

- Basis
- Standard bases for R^n, P_n, M_{mn}
- Finite-dimensional
- Infinite-dimensional
- Coordinates
- Coordinate vector

Skills

- Show that a set of vectors is a basis for a vector space.
- Find the coordinates of a vector relative to a basis.
- Find the coordinate vector of a vector relative to a basis.

Exercise Set 4.4

1. In words, explain why the following sets of vectors are *not* bases for the indicated vector spaces.

(a) $\mathbf{u}_1 = (1, 2)$, $\mathbf{u}_2 = (0, 3)$, $\mathbf{u}_3 = (2, 7)$ for R^2

(b) $\mathbf{u}_1 = (-1, 3, 2)$, $\mathbf{u}_2 = (6, 1, 1)$ for R^3

(c) $\mathbf{p}_1 = 1 + x + x^2$, $\mathbf{p}_2 = x - 1$ for P_2

(d) $A = \begin{bmatrix} 1 & 1 \\ 2 & 3 \end{bmatrix}$, $B = \begin{bmatrix} 6 & 0 \\ -1 & 4 \end{bmatrix}$, $C = \begin{bmatrix} 3 & 0 \\ 1 & 7 \end{bmatrix}$,

$D = \begin{bmatrix} 5 & 1 \\ 4 & 2 \end{bmatrix}$, $E = \begin{bmatrix} 7 & 1 \\ 2 & 9 \end{bmatrix}$ for M_{22}

2. Which of the following sets of vectors are bases for R^2?

(a) $\{(2, 1), (3, 0)\}$

(b) $\{(4, 1), (-7, -8)\}$

(c) $\{(0, 0), (1, 3)\}$

(d) $\{(3, 9), (-4, -12)\}$

3. Which of the following sets of vectors are bases for R^3?

(a) $\{(1, 0, 0), (2, 2, 0), (3, 3, 3)\}$

(b) $\{(3, 1, -4), (2, 5, 6), (1, 4, 8)\}$

(c) $\{(2, -3, 1), (4, 1, 1), (0, -7, 1)\}$

(d) $\{(1, 6, 4), (2, 4, -1), (-1, 2, 5)\}$

4. Which of the following form bases for P_2?

(a) $1 - 3x + 2x^2$, $1 + x + 4x^2$, $1 - 7x$

(b) $4 + 6x + x^2$, $-1 + 4x + 2x^2$, $5 + 2x - x^2$

(c) $1 + x + x^2$, $x + x^2$, x^2

(d) $-4 + x + 3x^2$, $6 + 5x + 2x^2$, $8 + 4x + x^2$

5. Show that the following matrices form a basis for M_{22}.

$\begin{bmatrix} 3 & 6 \\ 3 & -6 \end{bmatrix}$, $\begin{bmatrix} 0 & -1 \\ -1 & 0 \end{bmatrix}$, $\begin{bmatrix} 0 & -8 \\ -12 & -4 \end{bmatrix}$, $\begin{bmatrix} 1 & 0 \\ -1 & 2 \end{bmatrix}$

6. Let V be the space spanned by $\mathbf{v}_1 = \cos^2 x$, $\mathbf{v}_2 = \sin^2 x$, $\mathbf{v}_3 = \cos 2x$.

(a) Show that $S = \{\mathbf{v}_1, \mathbf{v}_2, \mathbf{v}_3\}$ is not a basis for V.

(b) Find a basis for V.

7. Find the coordinate vector of $\mathbf{w}$ relative to the basis $S = \{\mathbf{u}_1, \mathbf{u}_2\}$ for R^2.

(a) $\mathbf{u}_1 = (1, 0)$, $\mathbf{u}_2 = (0, 1)$; $\mathbf{w} = (3, -7)$

(b) $\mathbf{u}_1 = (2, -4)$, $\mathbf{u}_2 = (3, 8)$; $\mathbf{w} = (1, 1)$

(c) $\mathbf{u}_1 = (1, 1)$, $\mathbf{u}_2 = (0, 2)$; $\mathbf{w} = (a, b)$

8. Find the coordinate vector of $\mathbf{w}$ relative to the basis $S = \{\mathbf{u}_1, \mathbf{u}_2\}$ of R^2.

(a) $\mathbf{u}_1 = (1, -1)$, $\mathbf{u}_2 = (1, 1)$; $\mathbf{w} = (1, 0)$

(b) $\mathbf{u}_1 = (1, -1)$, $\mathbf{u}_2 = (1, 1)$; $\mathbf{w} = (0, 1)$

(c) $\mathbf{u}_1 = (1, -1)$, $\mathbf{u}_2 = (1, 1)$; $\mathbf{w} = (1, 1)$

9. Find the coordinate vector of $\mathbf{v}$ relative to the basis $S = \{\mathbf{v}_1, \mathbf{v}_2, \mathbf{v}_3\}$.

(a) $\mathbf{v} = (2, -1, 3)$; $\mathbf{v}_1 = (1, 0, 0)$, $\mathbf{v}_2 = (2, 2, 0)$, $\mathbf{v}_3 = (3, 3, 3)$

(b) $\mathbf{v} = (5, -12, 3)$; $\mathbf{v}_1 = (1, 2, 3)$, $\mathbf{v}_2 = (-4, 5, 6)$, $\mathbf{v}_3 = (7, -8, 9)$

10. Find the coordinate vector of $\mathbf{p}$ relative to the basis $S = \{\mathbf{p}_1, \mathbf{p}_2, \mathbf{p}_3\}$.

(a) $\mathbf{p} = 4 - 3x + x^2$; $\mathbf{p}_1 = 1$, $\mathbf{p}_2 = x$, $\mathbf{p}_3 = x^2$

(b) $\mathbf{p} = 2 - x + x^2$; $\mathbf{p}_1 = 1 + x$, $\mathbf{p}_2 = 1 + x^2$, $\mathbf{p}_3 = x + x^2$

11. Find the coordinate vector of A relative to the basis $S = \{A_1, A_2, A_3, A_4\}$.

$$A = \begin{bmatrix} 2 & 0 \\ -1 & 3 \end{bmatrix}; \quad A_1 = \begin{bmatrix} -1 & 1 \\ 0 & 0 \end{bmatrix}, \quad A_2 = \begin{bmatrix} 1 & 1 \\ 0 & 0 \end{bmatrix},$$

$$A_3 = \begin{bmatrix} 0 & 0 \\ 1 & 0 \end{bmatrix}, \quad A_4 = \begin{bmatrix} 0 & 0 \\ 0 & 1 \end{bmatrix}$$

▶ In Exercises **12–13**, show that $\{A_1, A_2, A_3, A_4\}$ is a basis for M_{22}, and express A as a linear combination of the basis vectors.

12. $A_1 = \begin{bmatrix} 1 & 0 \\ 1 & 0 \end{bmatrix}, \quad A_2 = \begin{bmatrix} 1 & 1 \\ 0 & 0 \end{bmatrix}, \quad A_3 = \begin{bmatrix} 1 & 0 \\ 0 & 1 \end{bmatrix},$

$A_4 = \begin{bmatrix} 0 & 0 \\ 1 & 0 \end{bmatrix}; \quad A = \begin{bmatrix} 6 & 2 \\ 5 & 3 \end{bmatrix}$

13. $A_1 = \begin{bmatrix} 1 & 1 \\ 1 & 1 \end{bmatrix}, \quad A_2 = \begin{bmatrix} 0 & 1 \\ 1 & 1 \end{bmatrix}, \quad A_3 = \begin{bmatrix} 0 & 0 \\ 1 & 1 \end{bmatrix},$

$A_4 = \begin{bmatrix} 0 & 0 \\ 0 & 1 \end{bmatrix}; \quad A = \begin{bmatrix} 1 & 0 \\ 1 & 0 \end{bmatrix}$

▶ In Exercises **14–15**, show that $\{\mathbf{p}_1, \mathbf{p}_2, \mathbf{p}_3\}$ is a basis for P_2, and express $\mathbf{p}$ as a linear combination of the basis vectors. ◀

14. $\mathbf{p}_1 = 1 + 2x + x^2, \ \mathbf{p}_2 = 2 + 9x, \ \mathbf{p}_3 = 3 + 3x + 4x^2;$
$\mathbf{p} = 2 + 17x - 3x^2$

15. $\mathbf{p}_1 = 1 + x + x^2, \ \mathbf{p}_2 = x + x^2, \ \mathbf{p}_3 = x^2;$
$\mathbf{p} = 7 - x + 2x^2$

16. The accompanying figure shows a rectangular xy-coordinate system and an $x'y'$-coordinate system with skewed axes. Assuming that 1-unit scales are used on all the axes, find the $x'y'$-coordinates of the points whose xy-coordinates are given.

(a) $(1, 1)$ (b) $(1, 0)$ (c) $(0, 1)$ (d) (a, b)

◀ Figure Ex-16

17. The accompanying figure shows a rectangular xy-coordinate system determined by the unit basis vectors $\mathbf{i}$ and $\mathbf{j}$ and an $x'y'$-coordinate system determined by unit basis vectors $\mathbf{u}_1$ and $\mathbf{u}_2$. Find the $x'y'$-coordinates of the points whose xy-coordinates are given.

(a) $(\sqrt{3}, 1)$ (b) $(1, 0)$ (c) $(0, 1)$ (d) (a, b)

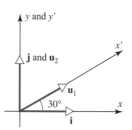

◀ Figure Ex-17

18. The basis that we gave for M_{22} in Example 4 consisted of noninvertible matrices. Do you think that there is a basis for M_{22} consisting of invertible matrices? Justify your answer.

19. Prove that R^∞ is infinite-dimensional.

True-False Exercises

In parts (a)–(e) determine whether the statement is true or false, and justify your answer.

(a) If $V = \text{span}\{\mathbf{v}_1, \ldots, \mathbf{v}_n\}$, then $\{\mathbf{v}_1, \ldots, \mathbf{v}_n\}$ is a basis for V.

(b) Every linearly independent subset of a vector space V is a basis for V.

(c) If $\{\mathbf{v}_1, \mathbf{v}_2, \ldots, \mathbf{v}_n\}$ is a basis for a vector space V, then every vector in V can be expressed as a linear combination of $\mathbf{v}_1, \mathbf{v}_2, \ldots, \mathbf{v}_n$.

(d) The coordinate vector of a vector $\mathbf{x}$ in R^n relative to the standard basis for R^n is $\mathbf{x}$.

(e) Every basis of P_4 contains at least one polynomial of degree 3 or less.

4.5 Dimension

We showed in the previous section that the standard basis R^n has n vectors and hence that the standard basis for R^3 has three vectors, the standard basis for R^2 has two vectors, and the standard basis for $R^1 (= R)$ has one vector. Since we think of space as three dimensional, a plane as two dimensional, and a line as one dimensional, there seems to be a link between the number of vectors in a basis and the dimension of a vector space. We will develop this idea in this section.

Number of Vectors in a Basis

Our first goal in this section is to establish the following fundamental theorem.

> **THEOREM 4.5.1** *All bases for a finite-dimensional vector space have the same number of vectors.*

To prove this theorem we will need the following preliminary result, whose proof is deferred to the end of the section.

> **THEOREM 4.5.2** *Let V be a finite-dimensional vector space, and let* $\{v_1, v_2, \ldots, v_n\}$ *be any basis.*
>
> (a) *If a set has more than n vectors, then it is linearly dependent.*
>
> (b) *If a set has fewer than n vectors, then it does not span V.*

We can now see rather easily why Theorem 4.5.1 is true; for if

$$S = \{v_1, v_2, \ldots, v_n\}$$

is an *arbitrary* basis for V, then the linear independence of S implies that any set in V with more than n vectors is linearly dependent and any set in V with fewer than n vectors does not span V. Thus, unless a set in V has exactly n vectors it cannot be a basis.

Some writers regard the empty set to be a basis for the zero vector space. This is consistent with our definition of dimension, since the empty set has no vectors and the zero vector space has dimension zero.

We noted in the introduction to this section that for certain familiar vector spaces the intuitive notion of dimension coincides with the number of vectors in a basis. The following definition makes this idea precise.

> **DEFINITION 1** The ***dimension*** of a finite-dimensional vector space V is denoted by $\dim(V)$ and is defined to be the number of vectors in a basis for V. In addition, the zero vector space is defined to have dimension zero.

▶ **EXAMPLE 1 Dimensions of Some Familiar Vector Spaces**

$$\dim(R^n) = n \qquad \text{The standard basis has } n \text{ vectors.}$$
$$\dim(P_n) = n + 1 \qquad \text{The standard basis has } n + 1 \text{ vectors.}$$
$$\dim(M_{mn}) = mn \qquad \text{The standard basis has } mn \text{ vectors.}$$

Engineers often use the term ***degrees of freedom*** as a synonym for dimension.

▶ **EXAMPLE 2 Dimension of Span(S)**

If $S = \{v_1, v_2, \ldots, v_r\}$ is a *linearly independent* set in a vector space V, then S is automatically a basis for span(S) (why?), and this implies that

$$\dim[\text{span}(S)] = r$$

In words, the dimension of the space spanned by a linearly independent set of vectors is equal to the number of vectors in that set.

▶ **EXAMPLE 3** Dimension of a Solution Space

Find a basis for and the dimension of the solution space of the homogeneous system

$$
\begin{aligned}
2x_1 + 2x_2 - x_3 + x_5 &= 0 \\
-x_1 - x_2 + 2x_3 - 3x_4 + x_5 &= 0 \\
x_1 + x_2 - 2x_3 - x_5 &= 0 \\
x_3 + x_4 + x_5 &= 0
\end{aligned}
$$

Solution We leave it for you to solve this system by Gauss–Jordan elimination and show that its general solution is

$$x_1 = -s - t, \quad x_2 = s, \quad x_3 = -t, \quad x_4 = 0, \quad x_5 = t$$

which can be written in vector form as

$$(x_1, x_2, x_3, x_4, x_5) = (-s - t, s, -t, 0, t)$$

or, alternatively, as

$$(x_1, x_2, x_3, x_4, x_5) = s(-1, 1, 0, 0, 0) + t(-1, 0, -1, 0, 1)$$

This shows that the vectors $\mathbf{v}_1 = (-1, 1, 0, 0, 0)$ and $\mathbf{v}_2 = (-1, 0, -1, 0, 1)$ span the solution space. Since neither vector is a scalar multiple of the other, they are linearly independent and hence form a basis for the solution space. Thus, the solution space has dimension 2.

▶ **EXAMPLE 4** Dimension of a Solution Space

Find a basis for and the dimension of the solution space of the homogeneous system

$$
\begin{aligned}
x_1 + 3x_2 - 2x_3 + 2x_5 &= 0 \\
2x_1 + 6x_2 - 5x_3 - 2x_4 + 4x_5 - 3x_6 &= 0 \\
5x_3 + 10x_4 + 15x_6 &= 0 \\
2x_1 + 6x_2 + 8x_4 + 4x_5 + 18x_6 &= 0
\end{aligned}
$$

Solution In Example 6 of Section 1.2 we found the solution of this system to be

$$x_1 = -3r - 4s - 2t, \quad x_2 = r, \quad x_3 = -2s, \quad x_4 = s, \quad x_5 = t, \quad x_6 = 0$$

which can be written in vector form as

$$(x_1, x_2, x_3, x_4, x_5, x_6) = (-3r - 4s - 2t, r, -2s, s, t, 0)$$

or, alternatively, as

$$(x_1, x_2, x_3, x_4, x_5) = r(-3, 1, 0, 0, 0, 0) + s(-4, 0, -2, 1, 0, 0) + t(-2, 0, 0, 0, 1, 0)$$

This shows that the vectors

$$\mathbf{v}_1 = (-3, 1, 0, 0, 0, 0), \quad \mathbf{v}_2 = (-4, 0, -2, 1, 0, 0), \quad \mathbf{v}_3 = (-2, 0, 0, 0, 1, 0)$$

span the solution space. We leave it for you to check that these vectors are linearly independent by showing that none of them is a linear combination of the other two (but see the remark that follows). Thus, the solution space has dimension 3. ◀

Remark It can be shown that for a homogeneous linear system, the method of the last example *always* produces a basis for the solution space of the system. We omit the formal proof.

Some Fundamental
Theorems

We will devote the remainder of this section to a series of theorems that reveal the subtle interrelationships among the concepts of linear independence, basis, and dimension. These theorems are not simply exercises in mathematical theory—they are essential to the understanding of vector spaces and the applications that build on them.

We will start with a theorem (proved at the end of this section) that is concerned with the effect on linear independence and spanning if a vector is added to or removed from a given nonempty set of vectors. Informally stated, if you start with a linearly independent set S and adjoin to it a vector that is not a linear combination of those in S, then the enlarged set will still be linearly independent. Also, if you start with a set S of two or more vectors in which one of the vectors is a linear combination of the others, then that vector can be removed from S without affecting span(S) (Figure 4.5.1).

| The vector outside the plane can be adjoined to the other two without affecting their linear independence. | Any of the vectors can be removed, and the remaining two will still span the plane. | Either of the collinear vectors can be removed, and the remaining two will still span the plane. |

▲ **Figure 4.5.1**

THEOREM 4.5.3 Plus/Minus Theorem

Let S be a nonempty set of vectors in a vector space V.

(a) If S is a linearly independent set, and if $\mathbf{v}$ is a vector in V that is outside of span(S), then the set $S \cup \{\mathbf{v}\}$ that results by inserting $\mathbf{v}$ into S is still linearly independent.

(b) If $\mathbf{v}$ is a vector in S that is expressible as a linear combination of other vectors in S, and if $S - \{\mathbf{v}\}$ denotes the set obtained by removing $\mathbf{v}$ from S, then S and $S - \{\mathbf{v}\}$ span the same space; that is,

$$\text{span}(S) = \text{span}(S - \{\mathbf{v}\})$$

▶ **EXAMPLE 5 Applying the Plus/Minus Theorem**

Show that $\mathbf{p}_1 = 1 - x^2$, $\mathbf{p}_2 = 2 - x^2$, and $\mathbf{p}_3 = x^3$ are linearly independent vectors.

Solution The set $S = \{\mathbf{p}_1, \mathbf{p}_2\}$ is linearly independent, since neither vector in S is a scalar multiple of the other. Since the vector $\mathbf{p}_3$ cannot be expressed as a linear combination of the vectors in S (why?), it can be adjoined to S to produce a linearly independent set $S' = \{\mathbf{p}_1, \mathbf{p}_2, \mathbf{p}_3\}$. ◀

In general, to show that a set of vectors $\{\mathbf{v}_1, \mathbf{v}_2, \ldots, \mathbf{v}_n\}$ is a basis for a vector space V, we must show that the vectors are linearly independent and span V. However, if we happen to know that V has dimension n (so that $\{\mathbf{v}_1, \mathbf{v}_2, \ldots, \mathbf{v}_n\}$ contains the right number of vectors for a basis), then it suffices to check *either* linear independence *or* spanning— the remaining condition will hold automatically. This is the content of the following theorem.

THEOREM 4.5.4 *Let V be an n-dimensional vector space, and let S be a set in V with exactly n vectors. Then S is a basis for V if and only if S spans V or S is linearly independent.*

Proof Assume that S has exactly n vectors and spans V. To prove that S is a basis, we must show that S is a linearly independent set. But if this is not so, then some vector $\mathbf{v}$ in S is a linear combination of the remaining vectors. If we remove this vector from S, then it follows from Theorem 4.5.3b that the remaining set of $n - 1$ vectors still spans V. But this is impossible, since it follows from Theorem 4.5.2b that no set with fewer than n vectors can span an n-dimensional vector space. Thus S is linearly independent.

Assume that S has exactly n vectors and is a linearly independent set. To prove that S is a basis, we must show that S spans V. But if this is not so, then there is some vector $\mathbf{v}$ in V that is not in span(S). If we insert this vector into S, then it follows from Theorem 4.5.3a that this set of $n + 1$ vectors is still linearly independent. But this is impossible, since Theorem 4.5.2a states that no set with more than n vectors in an n-dimensional vector space can be linearly independent. Thus S spans V. ◄

▶ **EXAMPLE 6** Bases by Inspection

(a) By inspection, explain why $\mathbf{v}_1 = (-3, 7)$ and $\mathbf{v}_2 = (5, 5)$ form a basis for R^2.

(b) By inspection, explain why $\mathbf{v}_1 = (2, 0, -1)$, $\mathbf{v}_2 = (4, 0, 7)$, and $\mathbf{v}_3 = (-1, 1, 4)$ form a basis for R^3.

Solution (a) Since neither vector is a scalar multiple of the other, the two vectors form a linearly independent set in the two-dimensional space R^2, and hence they form a basis by Theorem 4.5.4.

Solution (b) The vectors $\mathbf{v}_1$ and $\mathbf{v}_2$ form a linearly independent set in the xz-plane (why?). The vector $\mathbf{v}_3$ is outside of the xz-plane, so the set $\{\mathbf{v}_1, \mathbf{v}_2, \mathbf{v}_3\}$ is also linearly independent. Since R^3 is three-dimensional, Theorem 4.5.4 implies that $\{\mathbf{v}_1, \mathbf{v}_2, \mathbf{v}_3\}$ is a basis for R^3. ◄

The next theorem (whose proof is deferred to the end of this section) reveals two important facts about the vectors in a finite-dimensional vector space V:

1. Every spanning set for a subspace is either a basis for that subspace or has a basis as a subset.

2. Every linearly independent set in a subspace is either a basis for that subspace or can be extended to a basis for it.

THEOREM 4.5.5 *Let S be a finite set of vectors in a finite-dimensional vector space V.*

(a) *If S spans V but is not a basis for V, then S can be reduced to a basis for V by removing appropriate vectors from S.*

(b) *If S is a linearly independent set that is not already a basis for V, then S can be enlarged to a basis for V by inserting appropriate vectors into S.*

We conclude this section with a theorem that relates the dimension of a vector space to the dimensions of its subspaces.

> **THEOREM 4.5.6** *If W is a subspace of a finite-dimensional vector space V, then*:
> (a) *W is finite-dimensional.*
> (b) $\dim(W) \leq \dim(V)$.
> (c) *W = V if and only if* $\dim(W) = \dim(V)$.

Proof (a)   We will leave the proof of this part for the exercises.

Proof (b)   Part (*a*) shows that W is finite-dimensional, so it has a basis

$$S = \{\mathbf{w}_1, \mathbf{w}_2, \ldots, \mathbf{w}_m\}$$

Either S is also a basis for V or it is not. If so, then $\dim(V) = m$, which means that $\dim(V) = \dim(W)$. If not, then because S is a linearly independent set it can be enlarged to a basis for V by part (*b*) of Theorem 4.5.5. But this implies that $\dim(W) < \dim(V)$, so we have shown that $\dim(W) \leq \dim(V)$ in all cases.

Proof (c)   Assume that $\dim(W) = \dim(V)$ and that

$$S = \{\mathbf{w}_1, \mathbf{w}_2, \ldots, \mathbf{w}_m\}$$

is a basis for W. If S is not also a basis for V, then being linearly independent S can be extended to a basis for V by part (*b*) of Theorem 4.5.5. But this would mean that $\dim(V) > \dim(W)$, which contradicts our hypothesis. Thus S must also be a basis for V, which means that $\dim(W) = \dim(V)$. ◄

Figure 4.5.2 illustrates the geometric relationship between the subspaces of R^3 in order of increasing dimension.

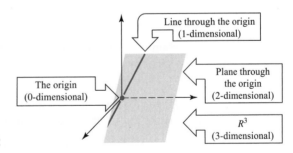

► Figure 4.5.2

We conclude this section with optional proofs of Theorems 4.5.2, 4.5.3, and 4.5.5.

Proof of Theorem 4.5.2(a)   Let $S' = \{\mathbf{w}_1, \mathbf{w}_2, \ldots, \mathbf{w}_m\}$ be any set of m vectors in V, where $m > n$. We want to show that S' is linearly dependent. Since $S = \{\mathbf{v}_1, \mathbf{v}_2, \ldots, \mathbf{v}_n\}$ is a basis, each $\mathbf{w}_i$ can be expressed as a linear combination of the vectors in S, say

$$\begin{aligned}
\mathbf{w}_1 &= a_{11}\mathbf{v}_1 + a_{21}\mathbf{v}_2 + \cdots + a_{n1}\mathbf{v}_n \\
\mathbf{w}_2 &= a_{12}\mathbf{v}_1 + a_{22}\mathbf{v}_2 + \cdots + a_{n2}\mathbf{v}_n \\
&\ \ \vdots \qquad\quad \vdots \qquad\quad \vdots \qquad\qquad \vdots \\
\mathbf{w}_m &= a_{1m}\mathbf{v}_1 + a_{2m}\mathbf{v}_2 + \cdots + a_{nm}\mathbf{v}_n
\end{aligned} \tag{1}$$

To show that S' is linearly dependent, we must find scalars $k_1, k_2, \ldots, k_m$, not all zero, such that

$$k_1\mathbf{w}_1 + k_2\mathbf{w}_2 + \cdots + k_m\mathbf{w}_m = \mathbf{0} \tag{2}$$

Using the equations in (1), we can rewrite (2) as

$$(k_1a_{11} + k_2a_{12} + \cdots + k_ma_{1m})\mathbf{v}_1$$
$$+ (k_1a_{21} + k_2a_{22} + \cdots + k_ma_{2m})\mathbf{v}_2$$
$$\ddots$$
$$+ (k_1a_{n1} + k_2a_{n2} + \cdots + k_ma_{nm})\mathbf{v}_n = \mathbf{0}$$

Thus, from the linear independence of S, the problem of proving that S' is a linearly dependent set reduces to showing there are scalars $k_1, k_2, \ldots, k_m$, not all zero, that satisfy

$$\begin{aligned}
a_{11}k_1 + a_{12}k_2 + \cdots + a_{1m}k_m &= 0 \\
a_{21}k_1 + a_{22}k_2 + \cdots + a_{2m}k_m &= 0 \\
\vdots \qquad \vdots \qquad\qquad \vdots \qquad \vdots \\
a_{n1}k_1 + a_{n2}k_2 + \cdots + a_{nm}k_m &= 0
\end{aligned} \tag{3}$$

But (3) has more unknowns than equations, so the proof is complete since Theorem 1.2.2 guarantees the existence of nontrivial solutions.

Proof of Theorem 4.5.2(b) Let $S' = \{\mathbf{w}_1, \mathbf{w}_2, \ldots, \mathbf{w}_m\}$ be any set of m vectors in V, where $m < n$. We want to show that S' does not span V. We will do this by showing that the assumption that S' spans V leads to a contradiction of the linear independence of $\{\mathbf{v}_1, \mathbf{v}_2, \ldots, \mathbf{v}_n\}$.

If S' spans V, then every vector in V is a linear combination of the vectors in S'. In particular, each basis vector $\mathbf{v}_i$ is a linear combination of the vectors in S', say

$$\begin{aligned}
\mathbf{v}_1 &= a_{11}\mathbf{w}_1 + a_{21}\mathbf{w}_2 + \cdots + a_{m1}\mathbf{w}_m \\
\mathbf{v}_2 &= a_{12}\mathbf{w}_1 + a_{22}\mathbf{w}_2 + \cdots + a_{m2}\mathbf{w}_m \\
\vdots \qquad \vdots \qquad \vdots \qquad\qquad \vdots \\
\mathbf{v}_n &= a_{1n}\mathbf{w}_1 + a_{2n}\mathbf{w}_2 + \cdots + a_{mn}\mathbf{w}_m
\end{aligned} \tag{4}$$

To obtain our contradiction, we will show that there are scalars $k_1, k_2, \ldots, k_n$, not all zero, such that

$$k_1\mathbf{v}_1 + k_2\mathbf{v}_2 + \cdots + k_n\mathbf{v}_n = \mathbf{0} \tag{5}$$

But (4) and (5) have the same form as (1) and (2) except that m and n are interchanged and the $\mathbf{w}$'s and $\mathbf{v}$'s are interchanged. Thus, the computations that led to (3) now yield

$$\begin{aligned}
a_{11}k_1 + a_{12}k_2 + \cdots + a_{1n}k_n &= 0 \\
a_{21}k_1 + a_{22}k_2 + \cdots + a_{2n}k_n &= 0 \\
\vdots \qquad \vdots \qquad\qquad \vdots \qquad \vdots \\
a_{m1}k_1 + a_{m2}k_2 + \cdots + a_{mn}k_n &= 0
\end{aligned}$$

This linear system has more unknowns than equations and hence has nontrivial solutions by Theorem 1.2.2.

Proof of Theorem 4.5.3(a) Assume that $S = \{\mathbf{v}_1, \mathbf{v}_2, \ldots, \mathbf{v}_r\}$ is a linearly independent set of vectors in V, and $\mathbf{v}$ is a vector in V outside of span(S). To show that $S' = \{\mathbf{v}_1, \mathbf{v}_2, \ldots, \mathbf{v}_r, \mathbf{v}\}$ is a linearly independent set, we must show that the only scalars that satisfy

$$k_1\mathbf{v}_1 + k_2\mathbf{v}_2 + \cdots + k_r\mathbf{v}_r + k_{r+1}\mathbf{v} = \mathbf{0} \tag{6}$$

are $k_1 = k_2 = \cdots = k_r = k_{r+1} = 0$. But it must be true that $k_{r+1} = 0$ for otherwise we could solve (6) for $\mathbf{v}$ as a linear combination of $\mathbf{v}_1, \mathbf{v}_2, \ldots, \mathbf{v}_r$, contradicting the assumption that $\mathbf{v}$ is outside of span(S). Thus, (6) simplifies to

$$k_1\mathbf{v}_1 + k_2\mathbf{v}_2 + \cdots + k_r\mathbf{v}_r = \mathbf{0} \qquad (7)$$

which, by the linear independence of $\{\mathbf{v}_1, \mathbf{v}_2, \ldots, \mathbf{v}_r\}$, implies that

$$k_1 = k_2 = \cdots = k_r = 0$$

Proof Theorem 4.5.3(b) Assume that $S = \{\mathbf{v}_1, \mathbf{v}_2, \ldots, \mathbf{v}_r\}$ is a set of vectors in V, and (to be specific) suppose that $\mathbf{v}_r$ is a linear combination of $\mathbf{v}_1, \mathbf{v}_2, \ldots, \mathbf{v}_{r-1}$, say

$$\mathbf{v}_r = c_1\mathbf{v}_1 + c_2\mathbf{v}_2 + \cdots + c_{r-1}\mathbf{v}_{r-1} \qquad (8)$$

We want to show that if $\mathbf{v}_r$ is removed from S, then the remaining set of vectors $\{\mathbf{v}_1, \mathbf{v}_2, \ldots, \mathbf{v}_{r-1}\}$ still spans S; that is, we must show that every vector $\mathbf{w}$ in span(S) is expressible as a linear combination of $\{\mathbf{v}_1, \mathbf{v}_2, \ldots, \mathbf{v}_{r-1}\}$. But if $\mathbf{w}$ is in span(S), then $\mathbf{w}$ is expressible in the form

$$\mathbf{w} = k_1\mathbf{v}_1 + k_2\mathbf{v}_2 + \cdots + k_{r-1}\mathbf{v}_{r-1} + k_r\mathbf{v}_r$$

or, on substituting (8),

$$\mathbf{w} = k_1\mathbf{v}_1 + k_2\mathbf{v}_2 + \cdots + k_{r-1}\mathbf{v}_{r-1} + k_r(c_1\mathbf{v}_1 + c_2\mathbf{v}_2 + \cdots + c_{r-1}\mathbf{v}_{r-1})$$

which expresses $\mathbf{w}$ as a linear combination of $\mathbf{v}_1, \mathbf{v}_2, \ldots, \mathbf{v}_{r-1}$.

Proof of Theorem 4.5.5(a) If S is a set of vectors that spans V but is not a basis for V, then S is a linearly dependent set. Thus some vector $\mathbf{v}$ in S is expressible as a linear combination of the other vectors in S. By the Plus/Minus Theorem (4.5.3b), we can remove $\mathbf{v}$ from S, and the resulting set S' will still span V. If S' is linearly independent, then S' is a basis for V, and we are done. If S' is linearly dependent, then we can remove some appropriate vector from S' to produce a set S'' that still spans V. We can continue removing vectors in this way until we finally arrive at a set of vectors in S that is linearly independent and spans V. This subset of S is a basis for V.

Proof of Theorem 4.5.5(b) Suppose that $\dim(V) = n$. If S is a linearly independent set that is not already a basis for V, then S fails to span V, so there is some vector $\mathbf{v}$ in V that is not in span(S). By the Plus/Minus Theorem (4.5.3a), we can insert $\mathbf{v}$ into S, and the resulting set S' will still be linearly independent. If S' spans V, then S' is a basis for V, and we are finished. If S' does not span V, then we can insert an appropriate vector into S' to produce a set S'' that is still linearly independent. We can continue inserting vectors in this way until we reach a set with n linearly independent vectors in V. This set will be a basis for V by Theorem 4.5.4. ◀

Concept Review

- Dimension

- Relationships among the concepts of linear independence, basis, and dimension

Skills

- Find a basis for and the dimension of the solution space of a homogeneous linear system.

- Use dimension to determine whether a set of vectors is a basis for a finite-dimensional vector space.

- Extend a linearly independent set to a basis.

Exercise Set 4.5

▷ In Exercises 1–6, find a basis for the solution space of the homogeneous linear system, and find the dimension of that space.

1.
$$x_1 + x_2 - x_3 = 0$$
$$-2x_1 - x_2 + 2x_3 = 0$$
$$-x_1 \quad\quad + x_3 = 0$$

2. $3x_1 + x_2 + x_3 + x_4 = 0$
$$5x_1 - x_2 + x_3 - x_4 = 0$$

3. $x_1 - 4x_2 + 3x_3 - x_4 = 0$
$$2x_1 - 8x_2 + 6x_3 - 2x_4 = 0$$

4.
$$x_1 - 3x_2 + x_3 = 0$$
$$2x_1 - 6x_2 + 2x_3 = 0$$
$$3x_1 - 9x_2 + 3x_3 = 0$$

5. $2x_1 + x_2 + 3x_3 = 0$
$$x_1 \quad\quad + 5x_3 = 0$$
$$x_2 + x_3 = 0$$

6. $x + y + z = 0$
$$3x + 2y - 2z = 0$$
$$4x + 3y - z = 0$$
$$6x + 5y + z = 0$$

7. Find bases for the following subspaces of R^3.

(a) The plane $3x - 2y + 5z = 0$.

(b) The plane $x - y = 0$.

(c) The line $x = 2t$, $y = -t$, $z = 4t$.

(d) All vectors of the form (a, b, c), where $b = a + c$.

8. Find the dimensions of the following subspaces of R^4.

(a) All vectors of the form $(a, b, c, 0)$.

(b) All vectors of the form (a, b, c, d), where $d = a + b$ and $c = a - b$.

(c) All vectors of the form (a, b, c, d), where $a = b = c = d$.

9. Find the dimension of each of the following vector spaces.

(a) The vector space of all diagonal $n \times n$ matrices.

(b) The vector space of all symmetric $n \times n$ matrices.

(c) The vector space of all upper triangular $n \times n$ matrices.

10. Find the dimension of the subspace of P_3 consisting of all polynomials $a_0 + a_1x + a_2x^2 + a_3x^3$ for which $a_0 = 0$.

11. (a) Show that the set W of all polynomials in P_2 such that $p(1) = 0$ is a subspace of P_2.

(b) Make a conjecture about the dimension of W.

(c) Confirm your conjecture by finding a basis for W.

12. Find a standard basis vector for R^3 that can be added to the set $\{v_1, v_2\}$ to produce a basis for R^3.

(a) $v_1 = (-1, 2, 3)$, $v_2 = (1, -2, -2)$

(b) $v_1 = (1, -1, 0)$, $v_2 = (3, 1, -2)$

13. Find standard basis vectors for R^4 that can be added to the set $\{v_1, v_2\}$ to produce a basis for R^4.

$$v_1 = (1, -4, 2, -3), \quad v_2 = (-3, 8, -4, 6)$$

14. Let $\{v_1, v_2, v_3\}$ be a basis for a vector space V. Show that $\{u_1, u_2, u_3\}$ is also a basis, where $u_1 = v_1$, $u_2 = v_1 + v_2$, and $u_3 = v_1 + v_2 + v_3$.

15. The vectors $v_1 = (1, -2, 3)$ and $v_2 = (0, 5, -3)$ are linearly independent. Enlarge $\{v_1, v_2\}$ to a basis for R^3.

16. The vectors $v_1 = (1, -2, 3, -5)$ and $v_2 = (0, -1, 2, -3)$ are linearly independent. Enlarge $\{v_1, v_2\}$ to a basis for R^4.

17. (a) Show that for every positive integer n, one can find $n + 1$ linearly independent vectors in $F(-\infty, \infty)$. [*Hint:* Look for polynomials.]

(b) Use the result in part (a) to prove that $F(-\infty, \infty)$ is infinite-dimensional.

(c) Prove that $C(-\infty, \infty)$, $C^m(-\infty, \infty)$, and $C^\infty(-\infty, \infty)$ are infinite-dimensional vector spaces.

18. Let S be a basis for an n-dimensional vector space V. Show that if $v_1, v_2, \ldots, v_r$ form a linearly independent set of vectors in V, then the coordinate vectors $(v_1)_S$, $(v_2)_S$, $\ldots$, $(v_r)_S$ form a linearly independent set in R^n, and conversely.

19. Using the notation from Exercise 18, show that if the vectors $v_1, v_2, \ldots, v_r$ span V, then the coordinate vectors $(v_1)_S$, $(v_2)_S$, $\ldots$, $(v_r)_S$ span R^n, and conversely.

20. Find a basis for the subspace of P_2 spanned by the given vectors.

(a) $-1 + x - 2x^2$, $3 + 3x + 6x^2$, 9

(b) $1 + x$, x^2, $-2 + 2x^2$, $-3x$

(c) $1 + x - 3x^2$, $2 + 2x - 6x^2$, $3 + 3x - 9x^2$

[*Hint:* Let S be the standard basis for P_2, and work with the coordinate vectors relative to S as in Exercises 18 and 19.]

21. Prove: A subspace of a finite-dimensional vector space is finite-dimensional.

22. State the two parts of Theorem 4.5.2 in contrapositive form.

True-False Exercises

In parts (a)–(j) determine whether the statement is true or false, and justify your answer.

(a) The zero vector space has dimension zero.

(b) There is a set of 17 linearly independent vectors in R^{17}.

(c) There is a set of 11 vectors that span R^{17}.

(d) Every linearly independent set of five vectors in R^5 is a basis for R^5.

(e) Every set of five vectors that spans R^5 is a basis for R^5.

(f) Every set of vectors that spans R^n contains a basis for R^n.

(g) Every linearly independent set of vectors in R^n is contained in some basis for R^n.

(h) There is a basis for M_{22} consisting of invertible matrices.

(i) If A has size $n \times n$ and $I_n, A, A^2, \ldots, A^{n^2}$ are distinct matrices, then $\{I_n, A, A^2, \ldots, A^{n^2}\}$ is linearly dependent.

(j) There are at least two distinct three-dimensional subspaces of P_2.

4.6 Change of Basis

A basis that is suitable for one problem may not be suitable for another, so it is a common process in the study of vector spaces to change from one basis to another. Because a basis is the vector space generalization of a coordinate system, changing bases is akin to changing coordinate axes in R^2 and R^3. In this section we will study problems related to change of basis.

Coordinate Maps   If $S = \{\mathbf{v}_1, \mathbf{v}_2, \ldots, \mathbf{v}_n\}$ is a basis for a finite-dimensional vector space V, and if

$$(\mathbf{v})_S = (c_1, c_2, \ldots, c_n)$$

is the coordinate vector of $\mathbf{v}$ relative to S, then, as observed in Section 4.4, the mapping

$$\mathbf{v} \rightarrow (\mathbf{v})_S \tag{1}$$

creates a connection (a one-to-one correspondence) between vectors in the *general* vector space V and vectors in the *familiar* vector space R^n. We call (1) the **coordinate map** from V to R^n. In this section we will find it convenient to express coordinate vectors in the matrix form

$$[\mathbf{v}]_S = \begin{bmatrix} c_1 \\ c_2 \\ \vdots \\ c_n \end{bmatrix} \tag{2}$$

where the square brackets emphasize the matrix notation (Figure 4.6.1).

Coordinate map

▲ Figure 4.6.1

Change of Basis   There are many applications in which it is necessary to work with more than one coordinate system. In such cases it becomes important to know how the coordinates of a fixed vector relative to each coordinate system are related. This leads to the following problem.

> **The Change-of-Basis Problem** If $\mathbf{v}$ is a vector in a finite-dimensional vector space V, and if we change the basis for V from a basis B to a basis B', how are the coordinate vectors $[\mathbf{v}]_B$ and $[\mathbf{v}]_{B'}$ related?

Remark To solve this problem, it will be convenient to refer to B as the "old basis" and B' as the "new basis." Thus, our objective is to find a relationship between the old and new coordinates of a fixed vector $\mathbf{v}$ in V.

For simplicity, we will solve this problem for two-dimensional spaces. The solution for n-dimensional spaces is similar. Let

$$B = \{\mathbf{u}_1, \mathbf{u}_2\} \quad \text{and} \quad B' = \{\mathbf{u}_1', \mathbf{u}_2'\}$$

be the old and new bases, respectively. We will need the coordinate vectors for the new basis vectors relative to the old basis. Suppose they are

$$[\mathbf{u}_1']_B = \begin{bmatrix} a \\ b \end{bmatrix} \quad \text{and} \quad [\mathbf{u}_2']_B = \begin{bmatrix} c \\ d \end{bmatrix} \tag{3}$$

That is,

$$\begin{aligned} \mathbf{u}_1' &= a\mathbf{u}_1 + b\mathbf{u}_2 \\ \mathbf{u}_2' &= c\mathbf{u}_1 + d\mathbf{u}_2 \end{aligned} \tag{4}$$

Now let $\mathbf{v}$ be any vector in V, and let

$$[\mathbf{v}]_{B'} = \begin{bmatrix} k_1 \\ k_2 \end{bmatrix} \tag{5}$$

be the new coordinate vector, so that

$$\mathbf{v} = k_1\mathbf{u}_1' + k_2\mathbf{u}_2' \tag{6}$$

In order to find the old coordinates of $\mathbf{v}$, we must express $\mathbf{v}$ in terms of the old basis B. To do this, we substitute (4) into (6). This yields

$$\mathbf{v} = k_1(a\mathbf{u}_1 + b\mathbf{u}_2) + k_2(c\mathbf{u}_1 + d\mathbf{u}_2)$$

or

$$\mathbf{v} = (k_1 a + k_2 c)\mathbf{u}_1 + (k_1 b + k_2 d)\mathbf{u}_2$$

Thus, the old coordinate vector for $\mathbf{v}$ is

$$[\mathbf{v}]_B = \begin{bmatrix} k_1 a + k_2 c \\ k_1 b + k_2 d \end{bmatrix}$$

which, by using (5), can be written as

$$[\mathbf{v}]_B = \begin{bmatrix} a & c \\ b & d \end{bmatrix}\begin{bmatrix} k_1 \\ k_2 \end{bmatrix} = \begin{bmatrix} a & c \\ b & d \end{bmatrix}[\mathbf{v}]_{B'}$$

This equation states that the old coordinate vector $[\mathbf{v}]_B$ results when we multiply the new coordinate vector $[\mathbf{v}]_{B'}$ on the left by the matrix

$$P = \begin{bmatrix} a & c \\ b & d \end{bmatrix}$$

Since the columns of this matrix are the coordinates of the new basis vectors relative to the old basis [see (3)] we have the following solution of the change-of-basis problem.

Solution of the Change-of-Basis Problem If we change the basis for a vector space V from an old basis $B = \{\mathbf{u}_1, \mathbf{u}_2, \ldots, \mathbf{u}_n\}$ to a new basis $B' = \{\mathbf{u}_1', \mathbf{u}_2', \ldots, \mathbf{u}_n'\}$, then for each vector $\mathbf{v}$ in V, the old coordinate vector $[\mathbf{v}]_B$ is related to the new coordinate vector $[\mathbf{v}]_{B'}$ by the equation

$$[\mathbf{v}]_B = P[\mathbf{v}]_{B'} \tag{7}$$

where the columns of P are the coordinate vectors of the new basis vectors relative to the old basis; that is, the column vectors of P are

$$[\mathbf{u}_1']_B, \quad [\mathbf{u}_2']_B, \ldots, \quad [\mathbf{u}_n']_B \tag{8}$$

Transition Matrices   The matrix P in Equation (7) is called the ***transition matrix*** from B' to B. For emphasis, we will often denote it by $P_{B' \to B}$. It follows from (8) that this matrix can be expressed in terms of its column vectors as

$$P_{B' \to B} = \left[[\mathbf{u}'_1]_B \mid [\mathbf{u}'_2]_B \mid \cdots \mid [\mathbf{u}'_n]_B \right] \tag{9}$$

Similarly, the transition matrix from B to B' can be expressed in terms of its column vectors as

$$P_{B \to B'} = \left[[\mathbf{u}_1]_{B'} \mid [\mathbf{u}_2]_{B'} \mid \cdots \mid [\mathbf{u}_n]_{B'} \right] \tag{10}$$

Remark   There is a simple way to remember both of these formulas using the terms "old basis" and "new basis" defined earlier in this section: In Formula (9) the old basis is B' and the new basis is B, whereas in Formula (10) the old basis is B and the new basis is B'. Thus, both formulas can be restated as follows:

> *The columns of the transition matrix from an old basis to a new basis are the coordinate vectors of the old basis relative to the new basis.*

▶ **EXAMPLE 1 Finding Transition Matrices**

Consider the bases $B = \{\mathbf{u}_1, \mathbf{u}_2\}$ and $B' = \{\mathbf{u}'_1, \mathbf{u}'_2\}$ for R^2, where

$$\mathbf{u}_1 = (1, 0), \quad \mathbf{u}_2 = (0, 1), \quad \mathbf{u}'_1 = (1, 1), \quad \mathbf{u}'_2 = (2, 1)$$

(a) Find the transition matrix $P_{B' \to B}$ from B' to B.

(b) Find the transition matrix $P_{B \to B'}$ from B to B'.

Solution (a)   Here the old basis vectors are $\mathbf{u}'_1$ and $\mathbf{u}'_2$ and the new basis vectors are $\mathbf{u}_1$ and $\mathbf{u}_2$. We want to find the coordinate matrices of the old basis vectors $\mathbf{u}'_1$ and $\mathbf{u}'_2$ relative to the new basis vectors $\mathbf{u}_1$ and $\mathbf{u}_2$. To do this, first we observe that

$$\mathbf{u}'_1 = \mathbf{u}_1 + \mathbf{u}_2$$
$$\mathbf{u}'_2 = 2\mathbf{u}_1 + \mathbf{u}_2$$

from which it follows that

$$[\mathbf{u}'_1]_B = \begin{bmatrix} 1 \\ 1 \end{bmatrix} \quad \text{and} \quad [\mathbf{u}'_2]_B = \begin{bmatrix} 2 \\ 1 \end{bmatrix}$$

and hence that

$$P_{B' \to B} = \begin{bmatrix} 1 & 2 \\ 1 & 1 \end{bmatrix}$$

Solution (b)   Here the old basis vectors are $\mathbf{u}_1$ and $\mathbf{u}_2$ and the new basis vectors are $\mathbf{u}'_1$ and $\mathbf{u}'_2$. As in part (a), we want to find the coordinate matrices of the old basis vectors $\mathbf{u}'_1$ and $\mathbf{u}'_2$ relative to the new basis vectors $\mathbf{u}_1$ and $\mathbf{u}_2$. To do this, observe that

$$\mathbf{u}_1 = -\mathbf{u}'_1 + \mathbf{u}'_2$$
$$\mathbf{u}_2 = 2\mathbf{u}'_1 - \mathbf{u}'_2$$

from which it follows that

$$[\mathbf{u}_1]_{B'} = \begin{bmatrix} -1 \\ 1 \end{bmatrix} \quad \text{and} \quad [\mathbf{u}_2]_{B'} = \begin{bmatrix} 2 \\ -1 \end{bmatrix}$$

and hence that

$$P_{B \to B'} = \begin{bmatrix} -1 & 2 \\ 1 & -1 \end{bmatrix} \quad ◀$$

Suppose now that B and B' are bases for a finite-dimensional vector space V. Since multiplication by $P_{B' \to B}$ maps coordinate vectors relative to the basis B' into coordinate vectors relative to a basis B, and $P_{B \to B'}$ maps coordinate vectors relative to B into coordinate vectors relative to B', it follows that for every vector $\mathbf{v}$ in V we have

$$[\mathbf{v}]_B = P_{B' \to B}[\mathbf{v}]_{B'} \tag{11}$$

$$[\mathbf{v}]_{B'} = P_{B \to B'}[\mathbf{v}]_B \tag{12}$$

▶ **EXAMPLE 2 Computing Coordinate Vectors**

Let B and B' be the bases in Example 1. Use an appropriate formula to find $[\mathbf{v}]_B$ given that

$$[\mathbf{v}]_{B'} = \begin{bmatrix} -3 \\ 5 \end{bmatrix}$$

Solution To find $[\mathbf{v}]_B$ we need to make the transition from B' to B. It follows from Formula (11) and part (a) of Example 1 that

$$[\mathbf{v}]_B = P_{B' \to B}[\mathbf{v}]_{B'} = \begin{bmatrix} 1 & 2 \\ 1 & 1 \end{bmatrix} \begin{bmatrix} -3 \\ 5 \end{bmatrix} = \begin{bmatrix} 7 \\ 2 \end{bmatrix} \quad \blacktriangleleft$$

Invertibility of Transition Matrices

If B and B' are bases for a finite-dimensional vector space V, then

$$(P_{B' \to B})(P_{B \to B'}) = P_{B \to B}$$

because multiplication by $(P_{B' \to B})(P_{B \to B'})$ first maps B-coordinates of a vector into B'-coordinates, and then maps those B'-coordinates back into the original B-coordinates. Since the net effect of the two operations is to leave each coordinate vector unchanged, we are led to conclude that $P_{B \to B}$ must be the identity matrix, that is,

$$(P_{B' \to B})(P_{B \to B'}) = I \tag{13}$$

(we omit the formal proof). For example, for the transition matrices obtained in Example 1 we have

$$(P_{B' \to B})(P_{B \to B'}) = \begin{bmatrix} 1 & 2 \\ 1 & 1 \end{bmatrix} \begin{bmatrix} -1 & 2 \\ 1 & -1 \end{bmatrix} = \begin{bmatrix} 1 & 0 \\ 0 & 1 \end{bmatrix} = I$$

It follows from (13) that $P_{B' \to B}$ is invertible and that its inverse is $P_{B \to B'}$. Thus, we have the following theorem.

THEOREM 4.6.1 *If P is the transition matrix from a basis B' to a basis B for a finite-dimensional vector space V, then P is invertible and P^{-1} is the transition matrix from B to B'.*

An Efficient Method for Computing Transition Matrices for R^n

Our next objective is to develop an efficient procedure for computing transition matrices *between bases for R^n*. As illustrated in Example 1, the first step in computing a transition matrix is to express each new basis vector as a linear combination of the old basis vectors. For R^n this involves solving n linear systems of n equations in n unknowns, each of which has the same coefficient matrix (why?). An efficient way to do this is by the method illustrated in Example 2 of Section 1.6, which is as follows:

A Procedure for Computing $P_{B \to B'}$

Step 1. Form the matrix $[B' \mid B]$.

Step 2. Use elementary row operations to reduce the matrix in Step 1 to reduced row echelon form.

Step 3. The resulting matrix will be $[I \mid P_{B \to B'}]$.

Step 4. Extract the matrix $P_{B \to B'}$ from the right side of the matrix in Step 3.

This procedure is captured in the following diagram.

$$[\text{new basis} \mid \text{old basis}] \xrightarrow{\text{row operations}} [I \mid \text{transition from old to new}] \qquad (14)$$

▶ **EXAMPLE 3 Example 1 Revisited**

In Example 1 we considered the bases $B = \{\mathbf{u}_1, \mathbf{u}_2\}$ and $B' = \{\mathbf{u}'_1, \mathbf{u}'_2\}$ for R^2, where

$$\mathbf{u}_1 = (1, 0), \quad \mathbf{u}_2 = (0, 1), \quad \mathbf{u}'_1 = (1, 1), \quad \mathbf{u}'_2 = (2, 1)$$

(a) Use Formula (14) to find the transition matrix from B' to B.

(b) Use Formula (14) to find the transition matrix from B to B'.

Solution (a) Here B' is the old basis and B is the new basis, so

$$[\text{new basis} \mid \text{old basis}] = \left[\begin{array}{cc|cc} 1 & 0 & 1 & 2 \\ 0 & 1 & 1 & 1 \end{array}\right]$$

Since the left side is already the identity matrix, no reduction is needed. We see by inspection that the transition matrix is

$$P_{B' \to B} = \begin{bmatrix} 1 & 2 \\ 1 & 1 \end{bmatrix}$$

which agrees with the result in Example 1.

Solution (b) Here B is the old basis and B' is the new basis, so

$$[\text{new basis} \mid \text{old basis}] = \left[\begin{array}{cc|cc} 1 & 2 & 1 & 0 \\ 1 & 1 & 0 & 1 \end{array}\right]$$

By reducing this matrix, so the left side becomes the identity we obtain (verify)

$$[I \mid \text{transition from old to new}] = \left[\begin{array}{cc|cc} 1 & 0 & -1 & 2 \\ 0 & 1 & 1 & -1 \end{array}\right]$$

so the transition matrix is

$$P_{B \to B'} = \begin{bmatrix} -1 & 2 \\ 1 & -1 \end{bmatrix}$$

which also agrees with the result in Example 1. ◀

Transition to the Standard Basis for R^n

Note that in part (a) of the last example the column vectors of the matrix that made the transition from the basis B' to the standard basis turned out to be the vectors in B' written in column form. This illustrates the following general result.

THEOREM 4.6.2 *Let* $B' = \{\mathbf{u}_1, \mathbf{u}_2, \ldots, \mathbf{u}_n\}$ *be any basis for the vector space* R^n *and let* $S = \{\mathbf{e}_1, \mathbf{e}_2, \ldots, \mathbf{e}_n\}$ *be the standard basis for* R^n. *If the vectors in these bases are written in column form, then*

$$P_{B' \to S} = [\mathbf{u}_1 \mid \mathbf{u}_2 \mid \cdots \mid \mathbf{u}_n] \tag{15}$$

It follows from this theorem that if

$$A = [\mathbf{u}_1 \mid \mathbf{u}_2 \mid \cdots \mid \mathbf{u}_n]$$

is *any* invertible $n \times n$ matrix, then A can be viewed as the transition matrix from the basis $\{\mathbf{u}_1, \mathbf{u}_2, \ldots, \mathbf{u}_n\}$ for R^n to the standard basis for R^n. Thus, for example, the matrix

$$A = \begin{bmatrix} 1 & 2 & 3 \\ 2 & 5 & 3 \\ 1 & 0 & 8 \end{bmatrix}$$

which was shown to be invertible in Example 4 of Section 1.5, is the transition matrix from the basis

$$\mathbf{u}_1 = (1, 2, 1), \quad \mathbf{u}_2 = (2, 5, 0), \quad \mathbf{u}_3 = (3, 3, 8)$$

to the basis

$$\mathbf{e}_1 = (1, 0, 0), \quad \mathbf{e}_2 = (0, 1, 0), \quad \mathbf{e}_3 = (0, 0, 1)$$

Concept Review
- Coordinate map
- Change-of-basis problem
- Transition matrix

Skills
- Find coordinate vectors relative to a given basis directly.
- Find the transition matrix from one basis to another.
- Use the transition matrix to compute coordinate vectors.

Exercise Set 4.6

1. Find the coordinate vector for $\mathbf{w}$ relative to the basis $S = \{\mathbf{u}_1, \mathbf{u}_2\}$ for R^2.
   (a) $\mathbf{u}_1 = (1, 0)$, $\mathbf{u}_2 = (0, 1)$; $\mathbf{w} = (3, -7)$
   (b) $\mathbf{u}_1 = (2, -4)$, $\mathbf{u}_2 = (3, 8)$; $\mathbf{w} = (1, 1)$
   (c) $\mathbf{u}_1 = (1, 1)$, $\mathbf{u}_2 = (0, 2)$; $\mathbf{w} = (a, b)$

2. Find the coordinate vector for $\mathbf{v}$ relative to the basis $S = \{\mathbf{v}_1, \mathbf{v}_2, \mathbf{v}_3\}$ for R^3.
   (a) $\mathbf{v} = (2, -1, 3)$; $\mathbf{v}_1 = (1, 0, 0)$, $\mathbf{v}_2 = (2, 2, 0)$, $\mathbf{v}_3 = (3, 3, 3)$
   (b) $\mathbf{v} = (5, -12, 3)$; $\mathbf{v}_1 = (1, 2, 3)$, $\mathbf{v}_2 = (-4, 5, 6)$, $\mathbf{v}_3 = (7, -8, 9)$

3. Find the coordinate vector for $\mathbf{p}$ relative to the basis $S = \{\mathbf{p}_1, \mathbf{p}_2, \mathbf{p}_3\}$ for P_2.
   (a) $\mathbf{p} = 4 - 3x + x^2$; $\mathbf{p}_1 = 1$, $\mathbf{p}_2 = x$, $\mathbf{p}_3 = x^2$
   (b) $\mathbf{p} = 2 - x + x^2$; $\mathbf{p}_1 = 1 + x$, $\mathbf{p}_2 = 1 + x^2$, $\mathbf{p}_3 = x + x^2$

4. Find the coordinate vector for A relative to the basis $S = \{A_1, A_2, A_3, A_4\}$ for M_{22}.

$$A = \begin{bmatrix} 2 & 0 \\ -1 & 3 \end{bmatrix}, \quad A_1 = \begin{bmatrix} -1 & 1 \\ 0 & 0 \end{bmatrix}, \quad A_2 = \begin{bmatrix} 1 & 1 \\ 0 & 0 \end{bmatrix},$$

$$A_3 = \begin{bmatrix} 0 & 0 \\ 1 & 0 \end{bmatrix}, \quad A_4 = \begin{bmatrix} 0 & 0 \\ 0 & 1 \end{bmatrix}$$

5. Consider the coordinate vectors

$$[\mathbf{w}]_S = \begin{bmatrix} 6 \\ -1 \\ 4 \end{bmatrix}, \quad [\mathbf{q}]_S = \begin{bmatrix} 3 \\ 0 \\ 4 \end{bmatrix}, \quad [B]_S = \begin{bmatrix} -8 \\ 7 \\ 6 \\ 3 \end{bmatrix}$$

   (a) Find $\mathbf{w}$ if S is the basis in Exercise 2(a).
   (b) Find $\mathbf{q}$ if S is the basis in Exercise 3(a).
   (c) Find B if S is the basis in Exercise 4.

6. Consider the bases $B = \{\mathbf{u}_1, \mathbf{u}_2\}$ and $B' = \{\mathbf{u}_1', \mathbf{u}_2'\}$ for R^2, where

$$\mathbf{u}_1 = \begin{bmatrix} 1 \\ 0 \end{bmatrix}, \quad \mathbf{u}_2 = \begin{bmatrix} 0 \\ 1 \end{bmatrix}, \quad \mathbf{u}_1' = \begin{bmatrix} 2 \\ 1 \end{bmatrix}, \quad \mathbf{u}_2' = \begin{bmatrix} -3 \\ 4 \end{bmatrix}$$

(a) Find the transition matrix from B' to B.

(b) Find the transition matrix from B to B'.

(c) Compute the coordinate vector $[\mathbf{w}]_B$, where

$$\mathbf{w} = \begin{bmatrix} 3 \\ -5 \end{bmatrix}$$

and use (10) to compute $[\mathbf{w}]_{B'}$.

(d) Check your work by computing $[\mathbf{w}]_{B'}$ directly.

7. Repeat the directions of Exercise 6 with the same vector $\mathbf{w}$ but with

$$\mathbf{u}_1 = \begin{bmatrix} 2 \\ 2 \end{bmatrix}, \quad \mathbf{u}_2 = \begin{bmatrix} 4 \\ -1 \end{bmatrix}, \quad \mathbf{u}_1' = \begin{bmatrix} 1 \\ 3 \end{bmatrix}, \quad \mathbf{u}_2' = \begin{bmatrix} -1 \\ -1 \end{bmatrix}$$

8. Consider the bases $B = \{\mathbf{u}_1, \mathbf{u}_2, \mathbf{u}_3\}$ and $B' = \{\mathbf{u}_1', \mathbf{u}_2', \mathbf{u}_3'\}$ for R^3, where

$$\mathbf{u}_1 = \begin{bmatrix} -3 \\ 0 \\ -3 \end{bmatrix}, \quad \mathbf{u}_2 = \begin{bmatrix} -3 \\ 2 \\ -1 \end{bmatrix}, \quad \mathbf{u}_3 = \begin{bmatrix} 1 \\ 6 \\ -1 \end{bmatrix}$$

$$\mathbf{u}_1' = \begin{bmatrix} -6 \\ -6 \\ 0 \end{bmatrix}, \quad \mathbf{u}_2' = \begin{bmatrix} -2 \\ -6 \\ 4 \end{bmatrix}, \quad \mathbf{u}_3' = \begin{bmatrix} -2 \\ -3 \\ 7 \end{bmatrix}$$

(a) Find the transition matrix from B to B'.

(b) Compute the coordinate vector $[\mathbf{w}]_B$, where

$$\mathbf{w} = \begin{bmatrix} -5 \\ 8 \\ -5 \end{bmatrix}$$

and use (10) to compute $[\mathbf{w}]_{B'}$.

(c) Check your work by computing $[\mathbf{w}]_{B'}$ directly.

9. Repeat the directions of Exercise 8 with the same vector $\mathbf{w}$, but with

$$\mathbf{u}_1 = \begin{bmatrix} 2 \\ 1 \\ 1 \end{bmatrix}, \quad \mathbf{u}_2 = \begin{bmatrix} 2 \\ -1 \\ 1 \end{bmatrix}, \quad \mathbf{u}_3 = \begin{bmatrix} 1 \\ 2 \\ 1 \end{bmatrix}$$

$$\mathbf{u}_1' = \begin{bmatrix} 3 \\ 1 \\ -5 \end{bmatrix}, \quad \mathbf{u}_2' = \begin{bmatrix} 1 \\ 1 \\ -3 \end{bmatrix}, \quad \mathbf{u}_3' = \begin{bmatrix} -1 \\ 0 \\ 2 \end{bmatrix}$$

10. Consider the bases $B = \{\mathbf{p}_1, \mathbf{p}_2\}$ and $B' = \{\mathbf{q}_1, \mathbf{q}_2\}$ for P_1, where

$$\mathbf{p}_1 = 6 + 3x, \quad \mathbf{p}_2 = 10 + 2x, \quad \mathbf{q}_1 = 2, \quad \mathbf{q}_2 = 3 + 2x$$

(a) Find the transition matrix from B' to B.

(b) Find the transition matrix from B to B'.

(c) Compute the coordinate vector $[\mathbf{p}]_B$, where $\mathbf{p} = -4 + x$, and use (10) to compute $[\mathbf{p}]_{B'}$.

(d) Check your work by computing $[\mathbf{p}]_{B'}$ directly.

11. Let V be the space spanned by $\mathbf{f}_1 = \sin x$ and $\mathbf{f}_2 = \cos x$.

(a) Show that $\mathbf{g}_1 = 2 \sin x + \cos x$ and $\mathbf{g}_2 = 3 \cos x$ form a basis for V.

(b) Find the transition matrix from $B' = \{\mathbf{g}_1, \mathbf{g}_2\}$ to $B = \{\mathbf{f}_1, \mathbf{f}_2\}$.

(c) Find the transition matrix from B to B'.

(d) Compute the coordinate vector $[\mathbf{h}]_B$, where $\mathbf{h} = 2 \sin x - 5 \cos x$, and use (10) to obtain $[\mathbf{h}]_{B'}$.

(e) Check your work by computing $[\mathbf{h}]_{B'}$ directly.

12. Let S be the standard basis for R^2, and let $B = \{\mathbf{v}_1, \mathbf{v}_2\}$ be the basis in which $\mathbf{v}_1 = (2, 1)$ and $\mathbf{v}_2 = (-3, 4)$.

(a) Find the transition matrix $P_{B \to S}$ by inspection.

(b) Use Formula (14) to find the transition matrix $P_{S \to B}$.

(c) Confirm that $P_{B \to S}$ and $P_{S \to B}$ are inverses of one another.

(d) Let $\mathbf{w} = (5, -3)$. Find $[\mathbf{w}]_B$ and then use Formula (11) to compute $[\mathbf{w}]_S$.

(e) Let $\mathbf{w} = (3, -5)$. Find $[\mathbf{w}]_S$ and then use Formula (12) to compute $[\mathbf{w}]_B$.

13. Let S be the standard basis for R^3, and let $B = \{\mathbf{v}_1, \mathbf{v}_2, \mathbf{v}_3\}$ be the basis in which $\mathbf{v}_1 = (1, 2, 1)$, $\mathbf{v}_2 = (2, 5, 0)$, and $\mathbf{v}_3 = (3, 3, 8)$.

(a) Find the transition matrix $P_{B \to S}$ by inspection.

(b) Use Formula (14) to find the transition matrix $P_{S \to B}$.

(c) Confirm that $P_{B \to S}$ and $P_{S \to B}$ are inverses of one another.

(d) Let $\mathbf{w} = (5, -3, 1)$. Find $[\mathbf{w}]_B$ and then use Formula (11) to compute $[\mathbf{w}]_S$.

(e) Let $\mathbf{w} = (3, -5, 0)$. Find $[\mathbf{w}]_S$ and then use Formula (12) to compute $[\mathbf{w}]_B$.

14. Let $B_1 = \{\mathbf{u}_1, \mathbf{u}_2\}$ and $B_2 = \{\mathbf{v}_1, \mathbf{v}_2\}$ be the bases for R^2 in which $\mathbf{u}_1 = (2, 2)$, $\mathbf{u}_2 = (4, -1)$, $\mathbf{v}_1 = (1, 3)$, and $\mathbf{v}_2 = (-1, -1)$.

(a) Use Formula (14) to find the transition matrix $P_{B_2 \to B_1}$.

(b) Use Formula (14) to find the transition matrix $P_{B_1 \to B_2}$.

(c) Confirm that $P_{B_2 \to B_1}$ and $P_{B_1 \to B_2}$ are inverses of one another.

(d) Let $\mathbf{w} = (5, -3)$. Find $[\mathbf{w}]_{B_1}$ and then use the matrix $P_{B_1 \to B_2}$ to compute $[\mathbf{w}]_{B_2}$ from $[\mathbf{w}]_{B_1}$.

(e) Let $\mathbf{w} = (3, -5)$. Find $[\mathbf{w}]_{B_2}$ and then use the matrix $P_{B_2 \to B_1}$ to compute $[\mathbf{w}]_{B_1}$ from $[\mathbf{w}]_{B_2}$.

15. Let $B_1 = \{\mathbf{u}_1, \mathbf{u}_2\}$ and $B_2 = \{\mathbf{v}_1, \mathbf{v}_2\}$ be the bases for R^2 in which $\mathbf{u}_1 = (1, 2)$, $\mathbf{u}_2 = (2, 3)$, $\mathbf{v}_1 = (1, 3)$, and $\mathbf{v}_2 = (1, 4)$.

(a) Use Formula (14) to find the transition matrix $P_{B_2 \to B_1}$.

(b) Use Formula (14) to find the transition matrix $P_{B_1 \to B_2}$.

(c) Confirm that $P_{B_2 \to B_1}$ and $P_{B_1 \to B_2}$ are inverses of one another.

(d) Let $\mathbf{w} = (0, 1)$. Find $[\mathbf{w}]_{B_1}$ and then use the matrix $P_{B_1 \to B_2}$ to compute $[\mathbf{w}]_{B_2}$ from $[\mathbf{w}]_{B_1}$.

(e) Let $\mathbf{w} = (2, 5)$. Find $[\mathbf{w}]_{B_2}$ and then use the matrix $P_{B_2 \to B_1}$ to compute $[\mathbf{w}]_{B_1}$ from $[\mathbf{w}]_{B_2}$.

16. Let $B_1 = \{\mathbf{u}_1, \mathbf{u}_2, \mathbf{u}_3\}$ and $B_2 = \{\mathbf{v}_1, \mathbf{v}_2, \mathbf{v}_3\}$ be the bases for R^3 in which $\mathbf{u}_1 = (-3, 0, -3)$, $\mathbf{u}_2 = (-3, 2, -1)$, $\mathbf{u}_3 = (1, 6, -1)$, $\mathbf{v}_1 = (-6, -6, 0)$, $\mathbf{v}_2 = (-2, -6, 4)$, and $\mathbf{v}_3 = (-2, -3, 7)$.

(a) Find the transition matrix $P_{B_1 \to B_2}$.

(b) Let $\mathbf{w} = (-5, 8, -5)$. Find $[\mathbf{w}]_{B_1}$ and then use the transition matrix obtained in part (a) to compute $[\mathbf{w}]_{B_2}$ by matrix multiplication.

(c) Check the result in part (b) by computing $[\mathbf{w}]_{B_2}$ directly.

17. Follow the directions of Exercise 16 with the same vector $\mathbf{w}$ but with $\mathbf{u}_1 = (2, 1, 1)$, $\mathbf{u}_2 = (2, -1, 1)$, $\mathbf{u}_3 = (1, 2, 1)$, $\mathbf{v}_1 = (3, 1, -5)$, $\mathbf{v}_2 = (1, 1, -3)$, and $\mathbf{v}_3 = (-1, 0, 2)$.

18. Let $S = \{\mathbf{e}_1, \mathbf{e}_2\}$ be the standard basis for R^2, and let $B = \{\mathbf{v}_1, \mathbf{v}_2\}$ be the basis that results when the vectors in S are reflected about the line $y = x$.

(a) Find the transition matrix $P_{B \to S}$.

(b) Let $P = P_{B \to S}$ and show that $P^T = P_{S \to B}$.

19. Let $S = \{\mathbf{e}_1, \mathbf{e}_2\}$ be the standard basis for R^2, and let $B = \{\mathbf{v}_1, \mathbf{v}_2\}$ be the basis that results when the vectors in S are reflected about the line that makes an angle θ with the positive x-axis.

(a) Find the transition matrix $P_{B \to S}$.

(b) Let $P = P_{B \to S}$ and show that $P^T = P_{S \to B}$.

20. If B_1, B_2, and B_3 are bases for R^2, and if

$$P_{B_1 \to B_2} = \begin{bmatrix} 3 & 1 \\ 5 & 2 \end{bmatrix} \quad \text{and} \quad P_{B_2 \to B_3} = \begin{bmatrix} 7 & 2 \\ 4 & -1 \end{bmatrix}$$

then $P_{B_3 \to B_1} = $ _____.

21. If P is the transition matrix from a basis B' to a basis B, and Q is the transition matrix from B to a basis C, what is the transition matrix from B' to C? What is the transition matrix from C to B'?

22. To write the coordinate vector for a vector, it is necessary to specify an order for the vectors in the basis. If P is the transition matrix from a basis B' to a basis B, what is the effect on P if we reverse the order of vectors in B from $\mathbf{v}_1, \ldots, \mathbf{v}_n$ to $\mathbf{v}_n, \ldots, \mathbf{v}_1$? What is the effect on P if we reverse the order of vectors in both B' and B?

23. Consider the matrix

$$P = \begin{bmatrix} 1 & 1 & 0 \\ 1 & 0 & 2 \\ 0 & 2 & 1 \end{bmatrix}$$

(a) P is the transition matrix from what basis B to the standard basis $S = \{\mathbf{e}_1, \mathbf{e}_2, \mathbf{e}_3\}$ for R^3?

(b) P is the transition matrix from the standard basis $S = \{\mathbf{e}_1, \mathbf{e}_2, \mathbf{e}_3\}$ to what basis B for R^3?

24. The matrix

$$P = \begin{bmatrix} 1 & 0 & 0 \\ 0 & 3 & 2 \\ 0 & 1 & 1 \end{bmatrix}$$

is the transition matrix from what basis B to the basis $\{(1, 1, 1), (1, 1, 0), (1, 0, 0)\}$ for R^3?

25. Let B be a basis for R^n. Prove that the vectors $\mathbf{v}_1, \mathbf{v}_2, \ldots, \mathbf{v}_k$ form a linearly independent set in R^n if and only if the vectors $[\mathbf{v}_1]_B, [\mathbf{v}_2]_B, \ldots, [\mathbf{v}_k]_B$ form a linearly independent set in R^n.

26. Let B be a basis for R^n. Prove that the vectors $\mathbf{v}_1, \mathbf{v}_2, \ldots, \mathbf{v}_k$ span R^n if and only if the vectors $[\mathbf{v}_1]_B, [\mathbf{v}_2]_B, \ldots, [\mathbf{v}_k]_B$ span R^n.

27. If $[\mathbf{w}]_B = \mathbf{w}$ holds for all vectors $\mathbf{w}$ in R^n, what can you say about the basis B?

True-False Exercises

In parts (a)–(f) determine whether the statement is true or false, and justify your answer.

(a) If B_1 and B_2 are bases for a vector space V, then there exists a transition matrix from B_1 to B_2.

(b) Transition matrices are invertible.

(c) If B is a basis for a vector space R^n, then $P_{B \to B}$ is the identity matrix.

(d) If $P_{B_1 \to B_2}$ is a diagonal matrix, then each vector in B_2 is a scalar multiple of some vector in B_1.

(e) If each vector in B_2 is a scalar multiple of some vector in B_1, then $P_{B_1 \to B_2}$ is a diagonal matrix.

(f) If A is a square matrix, then $A = P_{B_1 \to B_2}$ for some bases B_1 and B_2 for R^n.

4.7 Row Space, Column Space, and Null Space

In this section we will study some important vector spaces that are associated with matrices. Our work here will provide us with a deeper understanding of the relationships between the solutions of a linear system and properties of its coefficient matrix.

Row Space, Column Space, and Null Space

Recall that vectors can be written in comma-delimited form or in matrix form as either row vectors or column vectors. In this section we will use the latter two.

DEFINITION 1 For an $m \times n$ matrix

$$A = \begin{bmatrix} a_{11} & a_{12} & \cdots & a_{1n} \\ a_{21} & a_{22} & \cdots & a_{2n} \\ \vdots & \vdots & & \vdots \\ a_{m1} & a_{m2} & \cdots & a_{mn} \end{bmatrix}$$

the vectors

$$\mathbf{r}_1 = [a_{11} \quad a_{12} \quad \cdots \quad a_{1n}]$$
$$\mathbf{r}_2 = [a_{21} \quad a_{22} \quad \cdots \quad a_{2n}]$$
$$\vdots$$
$$\mathbf{r}_m = [a_{m1} \quad a_{m2} \quad \cdots \quad a_{mn}]$$

in R^n that are formed from the rows of A are called the *row vectors* of A, and the vectors

$$\mathbf{c}_1 = \begin{bmatrix} a_{11} \\ a_{21} \\ \vdots \\ a_{m1} \end{bmatrix}, \quad \mathbf{c}_2 = \begin{bmatrix} a_{12} \\ a_{22} \\ \vdots \\ a_{m2} \end{bmatrix}, \dots, \quad \mathbf{c}_n = \begin{bmatrix} a_{1n} \\ a_{2n} \\ \vdots \\ a_{mn} \end{bmatrix}$$

in R^m formed from the columns of A are called the *column vectors* of A.

▶ **EXAMPLE 1 Row and Column Vectors of a 2 × 3 Matrix**

Let

$$A = \begin{bmatrix} 2 & 1 & 0 \\ 3 & -1 & 4 \end{bmatrix}$$

The row vectors of A are

$$\mathbf{r}_1 = [2 \quad 1 \quad 0] \quad \text{and} \quad \mathbf{r}_2 = [3 \quad -1 \quad 4]$$

and the column vectors of A are

$$\mathbf{c}_1 = \begin{bmatrix} 2 \\ 3 \end{bmatrix}, \quad \mathbf{c}_2 = \begin{bmatrix} 1 \\ -1 \end{bmatrix}, \quad \text{and} \quad \mathbf{c}_3 = \begin{bmatrix} 0 \\ 4 \end{bmatrix} \quad ◄$$

The following definition defines three important vector spaces associated with a matrix.

DEFINITION 2 If A is an $m \times n$ matrix, then the subspace of R^n spanned by the row vectors of A is called the *row space* of A, and the subspace of R^m spanned by the column vectors of A is called the *column space* of A. The solution space of the homogeneous system of equations $A\mathbf{x} = \mathbf{0}$, which is a subspace of R^n, is called the *null space* of A.

In this section and the next we will be concerned with two general questions:

> **Question 1.** What relationships exist among the solutions of a linear system $A\mathbf{x} = \mathbf{b}$ and the row space, column space, and null space of the coefficient matrix A?
>
> **Question 2.** What relationships exist among the row space, column space, and null space of a matrix?

Starting with the first question, suppose that

$$A = \begin{bmatrix} a_{11} & a_{12} & \cdots & a_{1n} \\ a_{21} & a_{22} & \cdots & a_{2n} \\ \vdots & \vdots & & \vdots \\ a_{m1} & a_{m2} & \cdots & a_{mn} \end{bmatrix} \quad \text{and} \quad \mathbf{x} = \begin{bmatrix} x_1 \\ x_2 \\ \vdots \\ x_n \end{bmatrix}$$

It follows from Formula (10) of Section 1.3 that if $\mathbf{c}_1, \mathbf{c}_2, \ldots, \mathbf{c}_n$ denote the column vectors of A, then the product $A\mathbf{x}$ can be expressed as a linear combination of these vectors with coefficients from $\mathbf{x}$; that is,

$$A\mathbf{x} = x_1\mathbf{c}_1 + x_2\mathbf{c}_2 + \cdots + x_n\mathbf{c}_n \tag{1}$$

Thus, a linear system, $A\mathbf{x} = \mathbf{b}$, of m equations in n unknowns can be written as

$$x_1\mathbf{c}_1 + x_2\mathbf{c}_2 + \cdots + x_n\mathbf{c}_n = \mathbf{b} \tag{2}$$

from which we conclude that $A\mathbf{x} = \mathbf{b}$ is consistent if and only if $\mathbf{b}$ is expressible as a linear combination of the column vectors of A. This yields the following theorem.

> **THEOREM 4.7.1** *A system of linear equations $A\mathbf{x} = \mathbf{b}$ is consistent if and only if $\mathbf{b}$ is in the column space of A.*

▶ **EXAMPLE 2   A Vector b in the Column Space of A**

Let $A\mathbf{x} = \mathbf{b}$ be the linear system

$$\begin{bmatrix} -1 & 3 & 2 \\ 1 & 2 & -3 \\ 2 & 1 & -2 \end{bmatrix} \begin{bmatrix} x_1 \\ x_2 \\ x_3 \end{bmatrix} = \begin{bmatrix} 1 \\ -9 \\ -3 \end{bmatrix}$$

Show that $\mathbf{b}$ is in the column space of A by expressing it as a linear combination of the column vectors of A.

Solution Solving the system by Gaussian elimination yields (verify)

$$x_1 = 2, \quad x_2 = -1, \quad x_3 = 3$$

It follows from this and Formula (2) that

$$2\begin{bmatrix} -1 \\ 1 \\ 2 \end{bmatrix} - \begin{bmatrix} 3 \\ 2 \\ 1 \end{bmatrix} + 3\begin{bmatrix} 2 \\ -3 \\ -2 \end{bmatrix} = \begin{bmatrix} 1 \\ -9 \\ -3 \end{bmatrix} \quad ◀$$

Recall from Theorem 3.4.4 that the general solution of a consistent linear system $A\mathbf{x} = \mathbf{b}$ can be obtained by adding any specific solution of this system to the general solution of the corresponding homogeneous system $A\mathbf{x} = \mathbf{0}$. Keeping in mind that the null space of A is the same as the solution space of $A\mathbf{x} = \mathbf{0}$, we can rephrase that theorem in the following vector form.

THEOREM 4.7.2 *If $\mathbf{x}_0$ is any solution of a consistent linear system $A\mathbf{x} = \mathbf{b}$, and if $S = \{\mathbf{v}_1, \mathbf{v}_2, \ldots, \mathbf{v}_k\}$ is a basis for the null space of A, then every solution of $A\mathbf{x} = \mathbf{b}$ can be expressed in the form*

$$\mathbf{x} = \mathbf{x}_0 + c_1\mathbf{v}_1 + c_2\mathbf{v}_2 + \cdots + c_k\mathbf{v}_k \tag{3}$$

Conversely, for all choices of scalars $c_1, c_2, \ldots, c_k$, the vector $\mathbf{x}$ in this formula is a solution of $A\mathbf{x} = \mathbf{b}$.

Equation (3) gives a formula for the ***general solution of*** $A\mathbf{x} = \mathbf{b}$. The vector $\mathbf{x}_0$ in that formula is called a ***particular solution of*** $A\mathbf{x} = \mathbf{b}$, and the remaining part of the formula is called the ***general solution of*** $A\mathbf{x} = \mathbf{0}$. In words, this formula tells us that:

The general solution of a consistent linear system can be expressed as the sum of a particular solution of that system and the general solution of the corresponding homogeneous system.

Geometrically, the solution set of $A\mathbf{x} = \mathbf{b}$ can be viewed as the translation by $\mathbf{x}_0$ of the solution space of $A\mathbf{x} = \mathbf{0}$ (Figure 4.7.1).

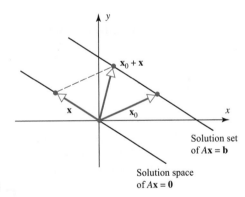

► Figure 4.7.1

► EXAMPLE 3 General Solution of a Linear System $A\mathbf{x} = \mathbf{b}$

In the concluding subsection of Section 3.4 we compared solutions of the linear systems

$$\begin{bmatrix} 1 & 3 & -2 & 0 & 2 & 0 \\ 2 & 6 & -5 & -2 & 4 & -3 \\ 0 & 0 & 5 & 10 & 0 & 15 \\ 2 & 6 & 0 & 8 & 4 & 18 \end{bmatrix} \begin{bmatrix} x_1 \\ x_2 \\ x_3 \\ x_4 \\ x_5 \\ x_6 \end{bmatrix} = \begin{bmatrix} 0 \\ 0 \\ 0 \\ 0 \end{bmatrix} \quad \text{and} \quad \begin{bmatrix} 1 & 3 & -2 & 0 & 2 & 0 \\ 2 & 6 & -5 & -2 & 4 & -3 \\ 0 & 0 & 5 & 10 & 0 & 15 \\ 2 & 6 & 0 & 8 & 4 & 18 \end{bmatrix} \begin{bmatrix} x_1 \\ x_2 \\ x_3 \\ x_4 \\ x_5 \\ x_6 \end{bmatrix} = \begin{bmatrix} 0 \\ -1 \\ 5 \\ 6 \end{bmatrix}$$

and deduced that the general solution $\mathbf{x}$ of the nonhomogeneous system and the general solution $\mathbf{x}_h$ of the corresponding homogeneous system (when written in column-vector form) are related by

$$
\underbrace{\begin{bmatrix} x_1 \\ x_2 \\ x_3 \\ x_4 \\ x_5 \\ x_6 \end{bmatrix}}_{\mathbf{x}} = \underbrace{\begin{bmatrix} -3r - 4s - 2t \\ r \\ -2s \\ s \\ t \\ \frac{1}{3} \end{bmatrix}}_{} = \underbrace{\begin{bmatrix} 0 \\ 0 \\ 0 \\ 0 \\ 0 \\ \frac{1}{3} \end{bmatrix}}_{\mathbf{x}_0} + \underbrace{r\begin{bmatrix} -3 \\ 1 \\ 0 \\ 0 \\ 0 \\ 0 \end{bmatrix} + s\begin{bmatrix} -4 \\ 0 \\ -2 \\ 1 \\ 0 \\ 0 \end{bmatrix} + t\begin{bmatrix} -2 \\ 0 \\ 0 \\ 0 \\ 1 \\ 0 \end{bmatrix}}_{\mathbf{x}_h}
$$

◀

Recall from the Remark following Example 4 of Section 4.5 that the vectors in $\mathbf{x}_h$ form a basis for the solution space of $A\mathbf{x} = \mathbf{0}$.

Bases for Row Spaces, Column Spaces, and Null Spaces

We first developed elementary row operations for the purpose of solving linear systems, and we know from that work that performing an elementary row operation on an augmented matrix does not change the solution set of the corresponding linear system. It follows that applying an elementary row operation to a matrix A does not change the solution set of the corresponding linear system $A\mathbf{x} = \mathbf{0}$, or, stated another way, it does not change the null space of A. Thus we have the following theorem.

THEOREM 4.7.3 *Elementary row operations do not change the null space of a matrix.*

The following theorem, whose proof is left as an exercise, is a companion to Theorem 4.7.3.

THEOREM 4.7.4 *Elementary row operations do not change the row space of a matrix.*

Theorems 4.7.3 and 4.7.4 might tempt you into *incorrectly* believing that elementary row operations do not change the column space of a matrix. To see why this is *not* true, compare the matrices

$$ A = \begin{bmatrix} 1 & 3 \\ 2 & 6 \end{bmatrix} \quad \text{and} \quad B = \begin{bmatrix} 1 & 3 \\ 0 & 0 \end{bmatrix} $$

The matrix B can be obtained from A by adding -2 times the first row to the second. However, this operation has changed the column space of A, since that column space consists of all scalar multiples of

$$ \begin{bmatrix} 1 \\ 2 \end{bmatrix} $$

whereas the column space of B consists of all scalar multiples of

$$ \begin{bmatrix} 1 \\ 0 \end{bmatrix} $$

and the two are different spaces.

▶ **EXAMPLE 4 Finding a Basis for the Null Space of a Matrix**

Find a basis for the null space of the matrix

$$A = \begin{bmatrix} 1 & 3 & -2 & 0 & 2 & 0 \\ 2 & 6 & -5 & -2 & 4 & -3 \\ 0 & 0 & 5 & 10 & 0 & 15 \\ 2 & 6 & 0 & 8 & 4 & 18 \end{bmatrix}$$

Solution The null space of A is the solution space of the homogeneous linear system $Ax = 0$, which, as shown in Example 3, has the basis

$$\mathbf{v}_1 = \begin{bmatrix} -3 \\ 1 \\ 0 \\ 0 \\ 0 \\ 0 \end{bmatrix}, \quad \mathbf{v}_2 = \begin{bmatrix} -4 \\ 0 \\ -2 \\ 1 \\ 0 \\ 0 \end{bmatrix}, \quad \mathbf{v}_3 = \begin{bmatrix} -2 \\ 0 \\ 0 \\ 0 \\ 1 \\ 0 \end{bmatrix} \quad ◀$$

Remark Observe that the basis vectors $\mathbf{v}_1$, $\mathbf{v}_2$, and $\mathbf{v}_3$ in the last example are the vectors that result by successively setting one of the parameters in the general solution equal to 1 and the others equal to 0.

The following theorem makes it possible to find bases for the row and column spaces of a matrix in row echelon form by inspection.

THEOREM 4.7.5 *If a matrix R is in row echelon form, then the row vectors with the leading 1's (the nonzero row vectors) form a basis for the row space of R, and the column vectors with the leading 1's of the row vectors form a basis for the column space of R.*

The proof involves little more than an analysis of the positions of the 0's and 1's of R. We omit the details.

▶ **EXAMPLE 5 Bases for Row and Column Spaces**

The matrix

$$R = \begin{bmatrix} 1 & -2 & 5 & 0 & 3 \\ 0 & 1 & 3 & 0 & 0 \\ 0 & 0 & 0 & 1 & 0 \\ 0 & 0 & 0 & 0 & 0 \end{bmatrix}$$

is in row echelon form. From Theorem 4.7.5, the vectors

$$\mathbf{r}_1 = \begin{bmatrix} 1 & -2 & 5 & 0 & 3 \end{bmatrix}$$
$$\mathbf{r}_2 = \begin{bmatrix} 0 & 1 & 3 & 0 & 0 \end{bmatrix}$$
$$\mathbf{r}_3 = \begin{bmatrix} 0 & 0 & 0 & 1 & 0 \end{bmatrix}$$

form a basis for the row space of R, and the vectors

$$\mathbf{c}_1 = \begin{bmatrix} 1 \\ 0 \\ 0 \\ 0 \end{bmatrix}, \quad \mathbf{c}_2 = \begin{bmatrix} -2 \\ 1 \\ 0 \\ 0 \end{bmatrix}, \quad \mathbf{c}_4 = \begin{bmatrix} 0 \\ 0 \\ 1 \\ 0 \end{bmatrix}$$

form a basis for the column space of R.

▶ **EXAMPLE 6 Basis for a Row Space by Row Reduction**

Find a basis for the row space of the matrix

$$A = \begin{bmatrix} 1 & -3 & 4 & -2 & 5 & 4 \\ 2 & -6 & 9 & -1 & 8 & 2 \\ 2 & -6 & 9 & -1 & 9 & 7 \\ -1 & 3 & -4 & 2 & -5 & -4 \end{bmatrix}$$

Solution Since elementary row operations do not change the row space of a matrix, we can find a basis for the row space of A by finding a basis for the row space of any row echelon form of A. Reducing A to row echelon form, we obtain (verify)

$$R = \begin{bmatrix} 1 & -3 & 4 & -2 & 5 & 4 \\ 0 & 0 & 1 & 3 & -2 & -6 \\ 0 & 0 & 0 & 0 & 1 & 5 \\ 0 & 0 & 0 & 0 & 0 & 0 \end{bmatrix}$$

By Theorem 4.7.5, the nonzero row vectors of R form a basis for the row space of R and hence form a basis for the row space of A. These basis vectors are

$$\mathbf{r}_1 = [1 \quad -3 \quad 4 \quad -2 \quad 5 \quad 4]$$
$$\mathbf{r}_2 = [0 \quad 0 \quad 1 \quad 3 \quad -2 \quad -6]$$
$$\mathbf{r}_3 = [0 \quad 0 \quad 0 \quad 0 \quad 1 \quad 5] \blacktriangleleft$$

The problem of finding a basis for the column space of a matrix A in Example 6 is complicated by the fact that an elementary row operation can alter its column space. However, the good news is that *elementary row operations do not alter dependence relationships among the column vectors*. To make this more precise, suppose that $\mathbf{w}_1, \mathbf{w}_2, \ldots, \mathbf{w}_k$ are linearly dependent column vectors of A, so there are scalars $c_1, c_2, \ldots, c_k$ that are not all zero and such that

$$c_1\mathbf{w}_1 + c_2\mathbf{w}_2 + \cdots + c_k\mathbf{w}_k = \mathbf{0} \tag{4}$$

If we perform an elementary row operation on A, then these vectors will be changed into new column vectors $\mathbf{w}_1', \mathbf{w}_2', \ldots, \mathbf{w}_k'$. At first glance it would seem possible that the transformed vectors might be linearly independent. However, this is not so, since it can be proved that these new column vectors will be linear dependent and, in fact, related by an equation

$$c_1\mathbf{w}_1' + c_2\mathbf{w}_2' + \cdots + c_k\mathbf{w}_k' = \mathbf{0}$$

that has exactly the same coefficients as (4). It follows from the fact that elementary row operations are reversible that they also preserve linear independence among column vectors (why?). The following theorem summarizes all of these results.

THEOREM 4.7.6 *If A and B are row equivalent matrices, then*:

(a) *A given set of column vectors of A is linearly independent if and only if the corresponding column vectors of B are linearly independent.*

(b) *A given set of column vectors of A forms a basis for the column space of A if and only if the corresponding column vectors of B form a basis for the column space of B.*

▶ **EXAMPLE 7** **Basis for a Column Space by Row Reduction**

Find a basis for the column space of the matrix

$$A = \begin{bmatrix} 1 & -3 & 4 & -2 & 5 & 4 \\ 2 & -6 & 9 & -1 & 8 & 2 \\ 2 & -6 & 9 & -1 & 9 & 7 \\ -1 & 3 & -4 & 2 & -5 & -4 \end{bmatrix}$$

Solution We observed in Example 6 that the matrix

$$R = \begin{bmatrix} 1 & -3 & 4 & -2 & 5 & 4 \\ 0 & 0 & 1 & 3 & -2 & -6 \\ 0 & 0 & 0 & 0 & 1 & 5 \\ 0 & 0 & 0 & 0 & 0 & 0 \end{bmatrix}$$

is a row echelon form of A. Keeping in mind that A and R can have different column spaces, we cannot find a basis for the column space of A directly from the column vectors of R. However, it follows from Theorem 4.7.6b that if we can find a set of column vectors of R that forms a basis for the column space of R, then the *corresponding* column vectors of A will form a basis for the column space of A.

Since the first, third, and fifth columns of R contain the leading 1's of the row vectors, the vectors

$$\mathbf{c}_1' = \begin{bmatrix} 1 \\ 0 \\ 0 \\ 0 \end{bmatrix}, \quad \mathbf{c}_3' = \begin{bmatrix} 4 \\ 1 \\ 0 \\ 0 \end{bmatrix}, \quad \mathbf{c}_5' = \begin{bmatrix} 5 \\ -2 \\ 1 \\ 0 \end{bmatrix}$$

form a basis for the column space of R. Thus, the corresponding column vectors of A, which are

$$\mathbf{c}_1 = \begin{bmatrix} 1 \\ 2 \\ 2 \\ -1 \end{bmatrix}, \quad \mathbf{c}_3 = \begin{bmatrix} 4 \\ 9 \\ 9 \\ -4 \end{bmatrix}, \quad \mathbf{c}_5 = \begin{bmatrix} 5 \\ 8 \\ 9 \\ -5 \end{bmatrix}$$

form a basis for the column space of A. ◀

Up to now we have focused on methods for finding bases associated with matrices. Those methods can readily be adapted to the more general problem of finding a basis for the space spanned by a set of vectors in R^n.

▶ **EXAMPLE 8 Basis for a Vector Space Using Row Operations**

Find a basis for the subspace of R^5 spanned by the vectors

$$\mathbf{v}_1 = (1, -2, 0, 0, 3), \quad \mathbf{v}_2 = (2, -5, -3, -2, 6),$$
$$\mathbf{v}_3 = (0, 5, 15, 10, 0), \quad \mathbf{v}_4 = (2, 6, 18, 8, 6)$$

Solution The space spanned by these vectors is the row space of the matrix

$$\begin{bmatrix} 1 & -2 & 0 & 0 & 3 \\ 2 & -5 & -3 & -2 & 6 \\ 0 & 5 & 15 & 10 & 0 \\ 2 & 6 & 18 & 8 & 6 \end{bmatrix}$$

Reducing this matrix to row echelon form, we obtain

$$\begin{bmatrix} 1 & -2 & 0 & 0 & 3 \\ 0 & 1 & 3 & 2 & 0 \\ 0 & 0 & 1 & 1 & 0 \\ 0 & 0 & 0 & 0 & 0 \end{bmatrix}$$

The nonzero row vectors in this matrix are

$$\mathbf{w}_1 = (1, -2, 0, 0, 3), \quad \mathbf{w}_2 = (0, 1, 3, 2, 0), \quad \mathbf{w}_3 = (0, 0, 1, 1, 0)$$

These vectors form a basis for the row space and consequently form a basis for the subspace of R^5 spanned by $\mathbf{v}_1$, $\mathbf{v}_2$, $\mathbf{v}_3$, and $\mathbf{v}_4$. ◀

Bases Formed from Row and Column Vectors of a Matrix

In all of the examples we have considered thus far we have looked for bases in which no restrictions were imposed on the individual vectors in the basis. We now want to focus on the problem of finding a basis for the row space of a matrix A consisting entirely of row vectors from A and a basis for the column space of A consisting entirely of column vectors of A.

Looking back on our earlier work, we see that the procedure followed in Example 7 did, in fact, produce a basis for the column space of A consisting of column vectors of A, whereas the procedure used in Example 6 produced a basis for the row space of A, but that basis did not consist of row vectors of A. The following example shows how to adapt the procedure from Example 7 to find a basis for the row space of a matrix that is formed from its row vectors.

▶ **EXAMPLE 9 Basis for the Row Space of a Matrix**

Find a basis for the row space of

$$A = \begin{bmatrix} 1 & -2 & 0 & 0 & 3 \\ 2 & -5 & -3 & -2 & 6 \\ 0 & 5 & 15 & 10 & 0 \\ 2 & 6 & 18 & 8 & 6 \end{bmatrix}$$

consisting entirely of row vectors from A.

Solution We will transpose A, thereby converting the row space of A into the column space of A^T; then we will use the method of Example 7 to find a basis for the column space of A^T; and then we will transpose again to convert column vectors back to row vectors. Transposing A yields

$$A^T = \begin{bmatrix} 1 & 2 & 0 & 2 \\ -2 & -5 & 5 & 6 \\ 0 & -3 & 15 & 18 \\ 0 & -2 & 10 & 8 \\ 3 & 6 & 0 & 6 \end{bmatrix}$$

Reducing this matrix to row echelon form yields

$$\begin{bmatrix} 1 & 2 & 0 & 2 \\ 0 & 1 & -5 & -10 \\ 0 & 0 & 0 & 1 \\ 0 & 0 & 0 & 0 \\ 0 & 0 & 0 & 0 \end{bmatrix}$$

The first, second, and fourth columns contain the leading 1's, so the corresponding column vectors in A^T form a basis for the column space of A^T; these are

$$c_1 = \begin{bmatrix} 1 \\ -2 \\ 0 \\ 0 \\ 3 \end{bmatrix}, \quad c_2 = \begin{bmatrix} 2 \\ -5 \\ -3 \\ -2 \\ 6 \end{bmatrix}, \quad \text{and} \quad c_4 = \begin{bmatrix} 2 \\ 6 \\ 18 \\ 8 \\ 6 \end{bmatrix}$$

Transposing again and adjusting the notation appropriately yields the basis vectors

$$r_1 = [1 \quad -2 \quad 0 \quad 0 \quad 3], \qquad r_2 = [2 \quad -5 \quad -3 \quad -2 \quad 6],$$

and

$$r_4 = [2 \quad 6 \quad 18 \quad 8 \quad 6]$$

for the row space of A. ◄

Next, we will give an example that adapts the methods we have developed above to solve the following general problem in R^n:

Problem Given a set of vectors $S = \{v_1, v_2, \ldots, v_k\}$ in R^n, find a subset of these vectors that forms a basis for span(S), and express those vectors that are not in that basis as a linear combination of the basis vectors.

▶ **EXAMPLE 10 Basis and Linear Combinations**

(a) Find a subset of the vectors

$$v_1 = (1, -2, 0, 3), \quad v_2 = (2, -5, -3, 6),$$
$$v_3 = (0, 1, 3, 0), \quad v_4 = (2, -1, 4, -7), \quad v_5 = (5, -8, 1, 2)$$

that forms a basis for the space spanned by these vectors.

(b) Express each vector not in the basis as a linear combination of the basis vectors.

Solution (a) We begin by constructing a matrix that has $\mathbf{v}_1, \mathbf{v}_2, \ldots, \mathbf{v}_5$ as its column vectors:

$$\begin{bmatrix} 1 & 2 & 0 & 2 & 5 \\ -2 & -5 & 1 & -1 & -8 \\ 0 & -3 & 3 & 4 & 1 \\ 3 & 6 & 0 & -7 & 2 \end{bmatrix} \tag{5}$$

$$\begin{array}{ccccc} \uparrow & \uparrow & \uparrow & \uparrow & \uparrow \\ \mathbf{v}_1 & \mathbf{v}_2 & \mathbf{v}_3 & \mathbf{v}_4 & \mathbf{v}_5 \end{array}$$

The first part of our problem can be solved by finding a basis for the column space of this matrix. Reducing the matrix to *reduced* row echelon form and denoting the column vectors of the resulting matrix by $\mathbf{w}_1, \mathbf{w}_2, \mathbf{w}_3, \mathbf{w}_4,$ and $\mathbf{w}_5$ yields

$$\begin{bmatrix} 1 & 0 & 2 & 0 & 1 \\ 0 & 1 & -1 & 0 & 1 \\ 0 & 0 & 0 & 1 & 1 \\ 0 & 0 & 0 & 0 & 0 \end{bmatrix} \tag{6}$$

$$\begin{array}{ccccc} \uparrow & \uparrow & \uparrow & \uparrow & \uparrow \\ \mathbf{w}_1 & \mathbf{w}_2 & \mathbf{w}_3 & \mathbf{w}_4 & \mathbf{w}_5 \end{array}$$

The leading 1's occur in columns 1, 2, and 4, so by Theorem 4.7.5,

$$\{\mathbf{w}_1, \mathbf{w}_2, \mathbf{w}_4\}$$

is a basis for the column space of (6), and consequently,

$$\{\mathbf{v}_1, \mathbf{v}_2, \mathbf{v}_4\}$$

is a basis for the column space of (5).

Solution (b) We will start by expressing $\mathbf{w}_3$ and $\mathbf{w}_5$ as linear combinations of the basis vectors $\mathbf{w}_1, \mathbf{w}_2, \mathbf{w}_4$. The simplest way of doing this is to express $\mathbf{w}_3$ and $\mathbf{w}_5$ in terms of basis vectors with smaller subscripts. Accordingly, we will express $\mathbf{w}_3$ as a linear combination of $\mathbf{w}_1$ and $\mathbf{w}_2$, and we will express $\mathbf{w}_5$ as a linear combination of $\mathbf{w}_1, \mathbf{w}_2,$ and $\mathbf{w}_4$. By inspection of (6), these linear combinations are

$$\mathbf{w}_3 = 2\mathbf{w}_1 - \mathbf{w}_2$$
$$\mathbf{w}_5 = \mathbf{w}_1 + \mathbf{w}_2 + \mathbf{w}_4$$

We call these the **dependency equations**. The corresponding relationships in (5) are

$$\mathbf{v}_3 = 2\mathbf{v}_1 - \mathbf{v}_2$$
$$\mathbf{v}_5 = \mathbf{v}_1 + \mathbf{v}_2 + \mathbf{v}_4 \quad \blacktriangleleft$$

The following is a summary of the steps that we followed in our last example to solve the problem posed above.

Basis for Span(S)

Step 1. Form the matrix A having vectors in $S = \{\mathbf{v}_1, \mathbf{v}_2, \ldots, \mathbf{v}_k\}$ as column vectors.

Step 2. Reduce the matrix A to reduced row echelon form R.

Step 3. Denote the column vectors of R by $\mathbf{w}_1, \mathbf{w}_2, \ldots, \mathbf{w}_k$.

Step 4. Identify the columns of R that contain the leading 1's. The corresponding column vectors of A form a basis for span(S).

This completes the first part of the problem.

Step 5. Obtain a set of dependency equations by expressing each column vector of R that does not contain a leading 1 as a linear combination of preceding column vectors that do contain leading 1's.

Step 6. Replace the column vectors of R that appear in the dependency equations by the corresponding column vectors of A.

This completes the second part of the problem.

Concept Review

- Row vectors
- Column vectors
- Row space
- Column space
- Null space

- General solution
- Particular solution
- Relationships among linear systems and row spaces, column spaces, and null spaces

- Relationships among the row space, column space, and null space of a matrix
- Dependency equations

Skills

- Determine whether a given vector is in the column space of a matrix; if it is, express it as a linear combination of the column vectors of the matrix.
- Find a basis for the null space of a matrix.

- Find a basis for the row space of a matrix.
- Find a basis for the column space of a matrix.
- Find a basis for the span of a set of vectors in R^n.

Exercise Set 4.7

1. List the row vectors and column vectors of the matrix

$$\begin{bmatrix} 2 & -1 & 0 & 1 \\ 3 & 5 & 7 & -1 \\ 1 & 4 & 2 & 7 \end{bmatrix}$$

2. Express the product Ax as a linear combination of the column vectors of A.

(a) $\begin{bmatrix} 2 & 3 \\ -1 & 4 \end{bmatrix}\begin{bmatrix} 1 \\ 2 \end{bmatrix}$

(b) $\begin{bmatrix} 4 & 0 & -1 \\ 3 & 6 & 2 \\ 0 & -1 & 4 \end{bmatrix}\begin{bmatrix} -2 \\ 3 \\ 5 \end{bmatrix}$

(c) $\begin{bmatrix} -3 & 6 & 2 \\ 5 & -4 & 0 \\ 2 & 3 & -1 \\ 1 & 8 & 3 \end{bmatrix}\begin{bmatrix} -1 \\ 2 \\ 5 \end{bmatrix}$

(d) $\begin{bmatrix} 2 & 1 & 5 \\ 6 & 3 & -8 \end{bmatrix}\begin{bmatrix} 3 \\ 0 \\ -5 \end{bmatrix}$

3. Determine whether **b** is in the column space of A, and if so, express **b** as a linear combination of the column vectors of A.

(a) $A = \begin{bmatrix} 1 & 3 \\ 4 & -6 \end{bmatrix}$; $b = \begin{bmatrix} -2 \\ 10 \end{bmatrix}$

(b) $A = \begin{bmatrix} 1 & 1 & 2 \\ 1 & 0 & 1 \\ 2 & 1 & 3 \end{bmatrix}$; $b = \begin{bmatrix} -1 \\ 0 \\ 2 \end{bmatrix}$

(c) $A = \begin{bmatrix} 1 & -1 & 1 \\ 9 & 3 & 1 \\ 1 & 1 & 1 \end{bmatrix}$; $b = \begin{bmatrix} 5 \\ 1 \\ -1 \end{bmatrix}$

(d) $A = \begin{bmatrix} 1 & -1 & 1 \\ 1 & 1 & -1 \\ -1 & -1 & 1 \end{bmatrix}$; $b = \begin{bmatrix} 2 \\ 0 \\ 0 \end{bmatrix}$

(e) $A = \begin{bmatrix} 1 & 2 & 0 & 1 \\ 0 & 1 & 2 & 1 \\ 1 & 2 & 1 & 3 \\ 0 & 1 & 2 & 2 \end{bmatrix}$; $b = \begin{bmatrix} 4 \\ 3 \\ 5 \\ 7 \end{bmatrix}$

4. Suppose that $x_1 = -1, x_2 = 2, x_3 = 4, x_4 = -3$ is a solution of a nonhomogeneous linear system $Ax = b$ and that the solution set of the homogeneous system $Ax = 0$ is given by the formulas

$$x_1 = -3r + 4s, \quad x_2 = r - s, \quad x_3 = r, \quad x_4 = s$$

(a) Find a vector form of the general solution of $Ax = 0$.

(b) Find a vector form of the general solution of $Ax = b$.

5. In parts (a)–(d), find the vector form of the general solution of the given linear system $Ax = b$; then use that result to find the vector form of the general solution of $Ax = 0$.

(a) $\begin{aligned} x_1 - 3x_2 &= 1 \\ 2x_1 - 6x_2 &= 2 \end{aligned}$

(b) $\begin{aligned} x_1 + x_2 + 2x_3 &= 5 \\ x_1 \quad\quad + x_3 &= -2 \\ 2x_1 + x_2 + 3x_3 &= 3 \end{aligned}$

(c) $\begin{aligned} x_1 - 2x_2 + x_3 + 2x_4 &= -1 \\ 2x_1 - 4x_2 + 2x_3 + 4x_4 &= -2 \\ -x_1 + 2x_2 - x_3 - 2x_4 &= 1 \\ 3x_1 - 6x_2 + 3x_3 + 6x_4 &= -3 \end{aligned}$

(d)
$$x_1 + 2x_2 - 3x_3 + x_4 = 4$$
$$-2x_1 + x_2 + 2x_3 + x_4 = -1$$
$$-x_1 + 3x_2 - x_3 + 2x_4 = 3$$
$$4x_1 - 7x_2 \qquad\quad - 5x_4 = -5$$

6. Find a basis for the null space of A.

(a) $A = \begin{bmatrix} 1 & -1 & 3 \\ 5 & -4 & -4 \\ 7 & -6 & 2 \end{bmatrix}$ (b) $A = \begin{bmatrix} 2 & 0 & -1 \\ 4 & 0 & -2 \\ 0 & 0 & 0 \end{bmatrix}$

(c) $A = \begin{bmatrix} 1 & 4 & 5 & 2 \\ 2 & 1 & 3 & 0 \\ -1 & 3 & 2 & 2 \end{bmatrix}$

(d) $A = \begin{bmatrix} 1 & 4 & 5 & 6 & 9 \\ 3 & -2 & 1 & 4 & -1 \\ -1 & 0 & -1 & -2 & -1 \\ 2 & 3 & 5 & 7 & 8 \end{bmatrix}$

(e) $A = \begin{bmatrix} 1 & -3 & 2 & 2 & 1 \\ 0 & 3 & 6 & 0 & -3 \\ 2 & -3 & -2 & 4 & 4 \\ 3 & -6 & 0 & 6 & 5 \\ -2 & 9 & 2 & -4 & -5 \end{bmatrix}$

7. In each part, a matrix in row echelon form is given. By inspection, find bases for the row and column spaces of A.

(a) $\begin{bmatrix} 1 & 0 & 2 \\ 0 & 0 & 1 \\ 0 & 0 & 0 \end{bmatrix}$

(b) $\begin{bmatrix} 1 & -3 & 0 & 0 \\ 0 & 1 & 0 & 0 \\ 0 & 0 & 0 & 0 \\ 0 & 0 & 0 & 0 \end{bmatrix}$

(c) $\begin{bmatrix} 1 & 2 & 4 & 5 \\ 0 & 1 & -3 & 0 \\ 0 & 0 & 1 & -3 \\ 0 & 0 & 0 & 1 \\ 0 & 0 & 0 & 0 \end{bmatrix}$

(d) $\begin{bmatrix} 1 & 2 & -1 & 5 \\ 0 & 1 & 4 & 3 \\ 0 & 0 & 1 & -7 \\ 0 & 0 & 0 & 1 \end{bmatrix}$

8. For the matrices in Exercise 6, find a basis for the row space of A by reducing the matrix to row echelon form.

9. By inspection, find a basis for the row space and a basis for the column space of each matrix.

(a) $\begin{bmatrix} 1 & 0 & 2 \\ 0 & 0 & 1 \\ 0 & 0 & 0 \end{bmatrix}$

(b) $\begin{bmatrix} 1 & -3 & 0 & 0 \\ 0 & 1 & 0 & 0 \\ 0 & 0 & 0 & 0 \\ 0 & 0 & 0 & 0 \end{bmatrix}$

(c) $\begin{bmatrix} 1 & 2 & 4 & 5 \\ 0 & 1 & -3 & 0 \\ 0 & 0 & 1 & -3 \\ 0 & 0 & 0 & 1 \\ 0 & 0 & 0 & 0 \end{bmatrix}$

(d) $\begin{bmatrix} 1 & 2 & -1 & 5 \\ 0 & 1 & 4 & 3 \\ 0 & 0 & 1 & -7 \\ 0 & 0 & 0 & 1 \end{bmatrix}$

10. For the matrices in Exercise 6, find a basis for the row space of A consisting entirely of row vectors of A.

11. Find a basis for the subspace of R^4 spanned by the given vectors.

(a) $(1, 1, -4, -3)$, $(2, 0, 2, -2)$, $(2, -1, 3, 2)$

(b) $(-1, 1, -2, 0)$, $(3, 3, 6, 0)$, $(9, 0, 0, 3)$

(c) $(1, 1, 0, 0)$, $(0, 0, 1, 1)$, $(-2, 0, 2, 2)$, $(0, -3, 0, 3)$

12. Find a subset of the vectors that forms a basis for the space spanned by the vectors; then express each vector that is not in the basis as a linear combination of the basis vectors.

(a) $\mathbf{v}_1 = (1, 0, 1, 1)$, $\mathbf{v}_2 = (-3, 3, 7, 1)$,
$\mathbf{v}_3 = (-1, 3, 9, 3)$, $\mathbf{v}_4 = (-5, 3, 5, -1)$

(b) $\mathbf{v}_1 = (1, -2, 0, 3)$, $\mathbf{v}_2 = (2, -4, 0, 6)$,
$\mathbf{v}_3 = (-1, 1, 2, 0)$, $\mathbf{v}_4 = (0, -1, 2, 3)$

(c) $\mathbf{v}_1 = (1, -1, 5, 2)$, $\mathbf{v}_2 = (-2, 3, 1, 0)$,
$\mathbf{v}_3 = (4, -5, 9, 4)$, $\mathbf{v}_4 = (0, 4, 2, -3)$,
$\mathbf{v}_5 = (-7, 18, 2, -8)$

13. Prove that the row vectors of an $n \times n$ invertible matrix A form a basis for R^n.

14. Construct a matrix whose null space consists of all linear combinations of the vectors

$$\mathbf{v}_1 = \begin{bmatrix} 1 \\ -1 \\ 3 \\ 2 \end{bmatrix} \quad \text{and} \quad \mathbf{v}_2 = \begin{bmatrix} 2 \\ 0 \\ -2 \\ 4 \end{bmatrix}$$

15. (a) Let

$$A = \begin{bmatrix} 0 & 1 & 0 \\ 1 & 0 & 0 \\ 0 & 0 & 0 \end{bmatrix}$$

Show that relative to an xyz-coordinate system in 3-space the null space of A consists of all points on the z-axis and that the column space consists of all points in the xy-plane (see the accompanying figure).

(b) Find a 3×3 matrix whose null space is the x-axis and whose column space is the yz-plane.

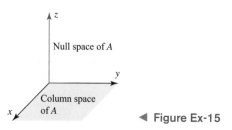

Null space of A

Column space of A

◀ Figure Ex-15

16. Find a 3×3 matrix whose null space is

(a) a point.   (b) a line.   (c) a plane.

17. (a) Find all 2×2 matrices whose null space is the line $3x - 5y = 0$.

(b) Sketch the null spaces of the following matrices:

$$A = \begin{bmatrix} 1 & 4 \\ 0 & 5 \end{bmatrix}, \quad B = \begin{bmatrix} 1 & 0 \\ 0 & 5 \end{bmatrix},$$

$$C = \begin{bmatrix} 6 & 2 \\ 3 & 1 \end{bmatrix}, \quad D = \begin{bmatrix} 0 & 0 \\ 0 & 0 \end{bmatrix}$$

18. The equation $x_1 + x_2 + x_3 = 1$ can be viewed as a linear system of one equation in three unknowns. Express its general solution as a particular solution plus the general solution of the corresponding homogeneous system. [*Suggestion:* Write the vectors in column form.]

19. Suppose that A and B are $n \times n$ matrices and A is invertible. Invent and prove a theorem that describes how the row spaces of AB and B are related.

True-False Exercises

In parts (a)–(j) determine whether the statement is true or false, and justify your answer.

(a) The span of $v_1, \ldots, v_n$ is the column space of the matrix whose column vectors are $v_1, \ldots, v_n$.

(b) The column space of a matrix A is the set of solutions of $Ax = b$.

(c) If R is the reduced row echelon form of A, then those column vectors of R that contain the leading 1's form a basis for the column space of A.

(d) The set of nonzero row vectors of a matrix A is a basis for the row space of A.

(e) If A and B are $n \times n$ matrices that have the same row space, then A and B have the same column space.

(f) If E is an $m \times m$ elementary matrix and A is an $m \times n$ matrix, then the null space of EA is the same as the null space of A.

(g) If E is an $m \times m$ elementary matrix and A is an $m \times n$ matrix, then the row space of EA is the same as the row space of A.

(h) If E is an $m \times m$ elementary matrix and A is an $m \times n$ matrix, then the column space of EA is the same as the column space of A.

(i) The system $Ax = b$ is inconsistent if and only if b is not in the column space of A.

(j) There is an invertible matrix A and a singular matrix B such that the row spaces of A and B are the same.

4.8 Rank, Nullity, and the Fundamental Matrix Spaces

In the last section we investigated relationships between a system of linear equations and the row space, column space, and null space of its coefficient matrix. In this section we will be concerned with the dimensions of those spaces. The results we obtain will provide a deeper insight into the relationship between a linear system and its coefficient matrix.

Row and Column Spaces Have Equal Dimensions   In Examples 6 and 7 of Section 4.7 we found that the row and column spaces of the matrix

$$A = \begin{bmatrix} 1 & -3 & 4 & -2 & 5 & 4 \\ 2 & -6 & 9 & -1 & 8 & 2 \\ 2 & -6 & 9 & -1 & 9 & 7 \\ -1 & 3 & -4 & 2 & -5 & -4 \end{bmatrix}$$

both have three basis vectors and hence are both three-dimensional. The fact that these spaces have the same dimension is not accidental, but rather a consequence of the following theorem.

THEOREM 4.8.1 *The row space and column space of a matrix A have the same dimension.*

Proof   Let R be any row echelon form of A. It follows from Theorems 4.7.4 and 4.7.6*b* that

$$\dim(\text{row space of } A) = \dim(\text{row space of } R)$$
$$\dim(\text{column space of } A) = \dim(\text{column space of } R)$$

so it suffices to show that the row and column spaces of R have the same dimension. But the dimension of the row space of R is the number of nonzero rows, and by Theorem 4.7.5 the dimension of the column space of R is the number of leading 1's. Since these two numbers are the same, the row and column space have the same dimension. ◄

Rank and Nullity The dimensions of the row space, column space, and null space of a matrix are such important numbers that there is some notation and terminology associated with them.

The proof of Theorem 4.8.1 shows that the rank of A can be interpreted as the number of leading 1's in any row echelon form of A.

DEFINITION 1 The common dimension of the row space and column space of a matrix A is called the *rank* of A and is denoted by rank(A); the dimension of the null space of A is called the *nullity* of A and is denoted by nullity(A).

▶ **EXAMPLE 1 Rank and Nullity of a 4 × 6 Matrix**

Find the rank and nullity of the matrix

$$A = \begin{bmatrix} -1 & 2 & 0 & 4 & 5 & -3 \\ 3 & -7 & 2 & 0 & 1 & 4 \\ 2 & -5 & 2 & 4 & 6 & 1 \\ 4 & -9 & 2 & -4 & -4 & 7 \end{bmatrix}$$

Solution The reduced row echelon form of A is

$$\begin{bmatrix} 1 & 0 & -4 & -28 & -37 & 13 \\ 0 & 1 & -2 & -12 & -16 & 5 \\ 0 & 0 & 0 & 0 & 0 & 0 \\ 0 & 0 & 0 & 0 & 0 & 0 \end{bmatrix} \tag{1}$$

(verify). Since this matrix has two leading 1's, its row and column spaces are two-dimensional and rank(A) = 2. To find the nullity of A, we must find the dimension of the solution space of the linear system $A\mathbf{x} = \mathbf{0}$. This system can be solved by reducing its augmented matrix to reduced row echelon form. The resulting matrix will be identical to (1), except that it will have an additional last column of zeros, and hence the corresponding system of equations will be

$$x_1 - 4x_3 - 28x_4 - 37x_5 + 13x_6 = 0$$
$$x_2 - 2x_3 - 12x_4 - 16x_5 + 5x_6 = 0$$

Solving these equations for the leading variables yields

$$x_1 = 4x_3 + 28x_4 + 37x_5 - 13x_6$$
$$x_2 = 2x_3 + 12x_4 + 16x_5 - 5x_6 \tag{2}$$

from which we obtain the general solution

$$x_1 = 4r + 28s + 37t - 13u$$
$$x_2 = 2r + 12s + 16t - 5u$$
$$x_3 = r$$
$$x_4 = s$$
$$x_5 = t$$
$$x_6 = u$$

or in column vector form

$$
\begin{bmatrix} x_1 \\ x_2 \\ x_3 \\ x_4 \\ x_5 \\ x_6 \end{bmatrix} = r \begin{bmatrix} 4 \\ 2 \\ 1 \\ 0 \\ 0 \\ 0 \end{bmatrix} + s \begin{bmatrix} 28 \\ 12 \\ 0 \\ 1 \\ 0 \\ 0 \end{bmatrix} + t \begin{bmatrix} 37 \\ 16 \\ 0 \\ 0 \\ 1 \\ 0 \end{bmatrix} + u \begin{bmatrix} -13 \\ -5 \\ 0 \\ 0 \\ 0 \\ 1 \end{bmatrix} \tag{3}
$$

Because the four vectors on the right side of (3) form a basis for the solution space, $\text{nullity}(A) = 4$.

▶ **EXAMPLE 2 Maximum Value for Rank**

What is the maximum possible rank of an $m \times n$ matrix A that is not square?

Solution Since the row vectors of A lie in R^n and the column vectors in R^m, the row space of A is at most n-dimensional and the column space is at most m-dimensional. Since the rank of A is the common dimension of its row and column space, it follows that the rank is at most the smaller of m and n. We denote this by writing

$$
\text{rank}(A) \leq \min(m, n)
$$

in which $\min(m, n)$ is the minimum of m and n. ◀

The following theorem establishes an important relationship between the rank and nullity of a matrix.

THEOREM 4.8.2 Dimension Theorem for Matrices

If A is a matrix with n columns, then

$$
rank(A) + nullity(A) = n \tag{4}
$$

Proof Since A has n columns, the homogeneous linear system $A\mathbf{x} = \mathbf{0}$ has n unknowns (variables). These fall into two distinct categories: the leading variables and the free variables. Thus,

$$
\begin{bmatrix} \text{number of leading} \\ \text{variables} \end{bmatrix} + \begin{bmatrix} \text{number of free} \\ \text{variables} \end{bmatrix} = n
$$

But the number of leading variables is the same as the number of leading 1's in the reduced row echelon form of A, which is the rank of A; and the number of free variables is the same as the number of parameters in the general solution of $A\mathbf{x} = \mathbf{0}$, which is the nullity of A. This yields Formula (4). ◀

▶ **EXAMPLE 3 The Sum of Rank and Nullity**

The matrix

$$
A = \begin{bmatrix} -1 & 2 & 0 & 4 & 5 & -3 \\ 3 & -7 & 2 & 0 & 1 & 4 \\ 2 & -5 & 2 & 4 & 6 & 1 \\ 4 & -9 & 2 & -4 & -4 & 7 \end{bmatrix}
$$

has 6 columns, so

$$\text{rank}(A) + \text{nullity}(A) = 6$$

This is consistent with Example 1, where we showed that

$$\text{rank}(A) = 2 \quad \text{and} \quad \text{nullity}(A) = 4 \ \blacktriangleleft$$

The following theorem, which summarizes results already obtained, interprets rank and nullity in the context of a homogeneous linear system.

THEOREM 4.8.3 *If A is an $m \times n$ matrix, then*

(a) *rank(A) = the number of leading variables in the general solution of $A\mathbf{x} = \mathbf{0}$.*

(b) *nullity(A) = the number of parameters in the general solution of $A\mathbf{x} = \mathbf{0}$.*

▶ **EXAMPLE 4 Number of Parameters in a General Solution**

Find the number of parameters in the general solution of $A\mathbf{x} = \mathbf{0}$ if A is a 5×7 matrix of rank 3.

Solution From (4),

$$\text{nullity}(A) = n - \text{rank}(A) = 7 - 3 = 4$$

Thus there are four parameters. ◀

Equivalence Theorem   In Theorem 2.3.8 we listed seven results that are equivalent to the invertibility of a square matrix A. We are now in a position to add eight more results to that list to produce a single theorem that summarizes most of the topics we have covered thus far.

THEOREM 4.8.4 Equivalent Statements

If A is an $n \times n$ matrix, then the following statements are equivalent.

(a)   *A is invertible.*

(b)   *$A\mathbf{x} = \mathbf{0}$ has only the trivial solution.*

(c)   *The reduced row echelon form of A is I_n.*

(d)   *A is expressible as a product of elementary matrices.*

(e)   *$A\mathbf{x} = \mathbf{b}$ is consistent for every $n \times 1$ matrix $\mathbf{b}$.*

(f)   *$A\mathbf{x} = \mathbf{b}$ has exactly one solution for every $n \times 1$ matrix $\mathbf{b}$.*

(g)   *$\det(A) \neq 0$.*

(h)   *The column vectors of A are linearly independent.*

(i)   *The row vectors of A are linearly independent.*

(j)   *The column vectors of A span R^n.*

(k)   *The row vectors of A span R^n.*

(l)   *The column vectors of A form a basis for R^n.*

(m)   *The row vectors of A form a basis for R^n.*

(n)   *A has rank n.*

(o)   *A has nullity 0.*

Proof The equivalence of (h) through (m) follows from Theorem 4.5.4 (we omit the details). To complete the proof we will show that (b), (n), and (o) are equivalent by proving the chain of implications (b) $\Rightarrow$ (o) $\Rightarrow$ (n) $\Rightarrow$ (b).

(b) $\Rightarrow$ (o) If $A\mathbf{x} = \mathbf{0}$ has only the trivial solution, then there are no parameters in that solution, so nullity(A) = 0 by Theorem 4.8.3b.

(o) $\Rightarrow$ (n) Theorem 4.8.2.

(n) $\Rightarrow$ (b) If A has rank n, then Theorem 4.8.3a implies that there are n leading variables (hence no free variables) in the general solution of $A\mathbf{x} = \mathbf{0}$. This leaves the trivial solution as the only possibility. ◄

Overdetermined and Underdetermined Systems

In engineering and other applications, the occurrence of an overdetermined or underdetermined linear system often signals that one or more variables were omitted in formulating the problem or that extraneous variables were included. This often leads to some kind of undesirable physical result.

In many applications the equations in a linear system correspond to physical constraints or conditions that must be satisfied. In general, the most desirable systems are those that have the same number of constraints as unknowns, since such systems often have a unique solution. Unfortunately, it is not always possible to match the number of constraints and unknowns, so researchers are often faced with linear systems that have more constraints than unknowns, called *overdetermined systems*, or with fewer constraints than unknowns, called *underdetermined systems*. The following two theorems will help us to analyze both overdetermined and underdetermined systems.

THEOREM 4.8.5 *If* $A\mathbf{x} = \mathbf{b}$ *is a consistent linear system of m equations in n unknowns, and if A has rank r, then the general solution of the system contains n − r parameters.*

Proof It follows from Theorem 4.7.2 that the number of parameters is equal to the nullity of A, which, by Theorem 4.8.2, is $n - r$. ◄

THEOREM 4.8.6 *Let A be an m × n matrix.*

(a) (*Overdetermined Case*). *If m > n, then the linear system* $A\mathbf{x} = \mathbf{b}$ *is inconsistent for at least one vector* **b** *in* R^n.

(b) (*Underdetermined Case*). *If m < n, then for each vector* **b** *in* R^m *the linear system* $A\mathbf{x} = \mathbf{b}$ *is either inconsistent or has infinitely many solutions.*

Proof (a) Assume that $m > n$, in which case the column vectors of A cannot span R^m (fewer vectors than the dimension of R^m). Thus, there is at least one vector **b** in R^m that is not in the column space of A, and for that **b** the system $A\mathbf{x} = \mathbf{b}$ is inconsistent by Theorem 4.7.1.

Proof (b) Assume that $m < n$. For each vector **b** in R^n there are two possibilities: either the system $A\mathbf{x} = \mathbf{b}$ is consistent or it is inconsistent. If it is inconsistent, then the proof is complete. If it is consistent, then Theorem 4.8.5 implies that the general solution has $n - r$ parameters, where $r = \text{rank}(A)$. But rank(A) is the smaller of m and n, so

$$n - r = n - m > 0$$

This means that the general solution has at least one parameter and hence there are infinitely many solutions. ◄

▶ **EXAMPLE 5 Overdetermined and Underdetermined Systems**

(a) What can you say about the solutions of an overdetermined system $A\mathbf{x} = \mathbf{b}$ of 7 equations in 5 unknowns in which A has rank $r = 4$?

(b) What can you say about the solutions of an underdetermined system $A\mathbf{x} = \mathbf{b}$ of 5 equations in 7 unknowns in which A has rank $r = 4$?

Solution (a) The system is consistent for some vector $\mathbf{b}$ in R^7, and for any such $\mathbf{b}$ the number of parameters in the general solution is $n - r = 5 - 4 = 1$.

Solution (b) The system may be consistent or inconsistent, but if it is consistent for the vector $\mathbf{b}$ in R^5, then the general solution has $n - r = 7 - 4 = 3$ parameters.

▶ **EXAMPLE 6 An Overdetermined System**

The linear system

$$
\begin{aligned}
x_1 - 2x_2 &= b_1 \\
x_1 - x_2 &= b_2 \\
x_1 + x_2 &= b_3 \\
x_1 + 2x_2 &= b_4 \\
x_1 + 3x_2 &= b_5
\end{aligned}
$$

is overdetermined, so it cannot be consistent for all possible values of b_1, b_2, b_3, b_4, and b_5. Exact conditions under which the system is consistent can be obtained by solving the linear system by Gauss–Jordan elimination. We leave it for you to show that the augmented matrix is row equivalent to

$$
\begin{bmatrix}
1 & 0 & 2b_2 - b_1 \\
0 & 1 & b_2 - b_1 \\
0 & 0 & b_3 - 3b_2 + 2b_1 \\
0 & 0 & b_4 - 4b_2 + 3b_1 \\
0 & 0 & b_5 - 5b_2 + 4b_1
\end{bmatrix}
\tag{5}
$$

Thus, the system is consistent if and only if b_1, b_2, b_3, b_4, and b_5 satisfy the conditions

$$
\begin{aligned}
2b_1 - 3b_2 + b_3 &= 0 \\
3b_1 - 4b_2 + b_4 &= 0 \\
4b_1 - 5b_2 + b_5 &= 0
\end{aligned}
$$

Solving this homogeneous linear system yields

$$
b_1 = 5r - 4s, \quad b_2 = 4r - 3s, \quad b_3 = 2r - s, \quad b_4 = r, \quad b_5 = s
$$

where r and s are arbitrary. ◀

Remark The coefficient matrix for the linear system in the last example has $n = 2$ columns, and it has rank $r = 2$ because there are two nonzero rows in its reduced row echelon form. This implies that when the system is consistent its general solution will contain $n - r = 0$ parameters; that is, the solution will be unique. With a moment's thought, you should be able to see that this is so from (5).

The Fundamental Spaces of a Matrix

There are six important vector spaces associated with a matrix A and its transpose A^T:

row space of A	row space of A^T
column space of A	column space of A^T
null space of A	null space of A^T

However, transposing a matrix converts row vectors into column vectors and conversely, so except for a difference in notation, the row space of A^T is the same as the column space of A, and the column space of A^T is the same as the row space of A. Thus, of the six spaces listed above, only the following four are distinct:

If A is an $m \times n$ matrix, then the row space and null space of A are subspaces of R^n, and the column space of A and the null space of A^T are subspaces of R^m.

$$\text{row space of } A \qquad \text{column space of } A$$
$$\text{null space of } A \qquad \text{null space of } A^T$$

These are called the ***fundamental spaces*** of a matrix A. We will conclude this section by discussing how these four subspaces are related.

Let us focus for a moment on the matrix A^T. Since the row space and column space of a matrix have the same dimension, and since transposing a matrix converts its columns to rows and its rows to columns, the following result should not be surprising.

> **THEOREM 4.8.7** *If A is any matrix, then $rank(A) = rank(A^T)$.*

Proof

$$\text{rank}(A) = \dim(\text{row space of } A) = \dim(\text{column space of } A^T) = \text{rank}(A^T). \;\blacktriangleleft$$

This result has some important implications. For example, if A is an $m \times n$ matrix, then applying Formula (4) to the matrix A^T and using the fact that this matrix has m columns yields

$$\text{rank}(A^T) + \text{nullity}(A^T) = m$$

which, by virtue of Theorem 4.8.7, can be rewritten as

$$\text{rank}(A) + \text{nullity}(A^T) = m \tag{6}$$

This alternative form of Formula (4) in Theorem 4.8.2 makes it possible to express the dimensions of all four fundamental spaces in terms of the size and rank of A. Specifically, if $\text{rank}(A) = r$, then

$$\dim[\text{row}(A)] = r \qquad \dim[\text{col}(A)] = r$$
$$\dim[\text{null}(A)] = n - r \qquad \dim[\text{null}(A^T)] = m - r \tag{7}$$

The four formulas in (7) provide an *algebraic* relationship between the size of a matrix and the dimensions of its fundamental spaces. Our next objective is to find a *geometric* relationship between the fundamental spaces themselves. For this purpose recall from Theorem 3.4.3 that if A is an $m \times n$ matrix, then the null space of A consists of those vectors that are orthogonal to each of the row vectors of A. To develop that idea in more detail, we make the following definition.

> **DEFINITION 2** If W is a subspace of R^n, then the set of all vectors in R^n that are orthogonal to every vector in W is called the ***orthogonal complement*** of W and is denoted by the symbol $W^{\perp}$.

The following theorem lists three basic properties of orthogonal complements. We will omit the formal proof because a more general version of this theorem will be given later in the text.

THEOREM 4.8.8 *If W is a subspace of R^n, then*:

(a) $W^\perp$ *is a subspace of R^n.*

(b) *The only vector common to W and $W^\perp$ is $\mathbf{0}$.*

(c) *The orthogonal complement of $W^\perp$ is W.*

▶ **EXAMPLE 7 Orthogonal Complements**

Explain why $\{\mathbf{0}\}$ and R^n are orthogonal complements.

In R^2 the orthogonal complement of a line W through the origin is the line through the origin that is perpendicular to W (Figure 4.8.1a); and in R^3 the orthogonal complement of a plane W through the origin is the line through the origin that is perpendicular to that plane (Figure 4.8.1b). ◀

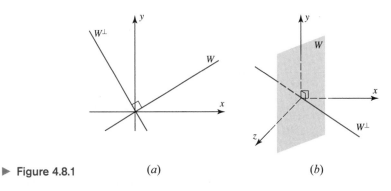

▶ **Figure 4.8.1** (a) (b)

A Geometric Link Between the Fundamental Spaces

The following theorem provides a geometric link between the fundamental spaces of a matrix. Part (a) is essentially a restatement of Theorem 3.4.3 in the language of orthogonal complements, and part (b), whose proof is left as an exercise, follows from part (a). The essential idea of the theorem is illustrated in Figure 4.8.2.

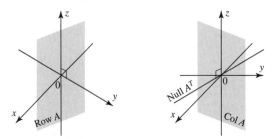

▶ **Figure 4.8.2**

THEOREM 4.8.9 *If A is an $m \times n$ matrix, then*:

(a) *The null space of A and the row space of A are orthogonal complements in R^n.*

(b) *The null space of A^T and the column space of A are orthogonal complements in R^m.*

More on the Equivalence
Theorem

As our final result in this section, we will add two more statements to Theorem 4.8.4. We leave the proof that those statements are equivalent to the rest as an exercise.

THEOREM 4.8.10 Equivalent Statements

If A is an $n \times n$ matrix, then the following statements are equivalent.

(*a*) *A is invertible.*

(*b*) *$A\mathbf{x} = \mathbf{0}$ has only the trivial solution.*

(*c*) *The reduced row echelon form of A is I_n.*

(*d*) *A is expressible as a product of elementary matrices.*

(*e*) *$A\mathbf{x} = \mathbf{b}$ is consistent for every $n \times 1$ matrix $\mathbf{b}$.*

(*f*) *$A\mathbf{x} = \mathbf{b}$ has exactly one solution for every $n \times 1$ matrix $\mathbf{b}$.*

(*g*) *$\det(A) \neq 0$.*

(*h*) *The column vectors of A are linearly independent.*

(*i*) *The row vectors of A are linearly independent.*

(*j*) *The column vectors of A span R^n.*

(*k*) *The row vectors of A span R^n.*

(*l*) *The column vectors of A form a basis for R^n.*

(*m*) *The row vectors of A form a basis for R^n.*

(*n*) *A has rank n.*

(*o*) *A has nullity 0.*

(*p*) *The orthogonal complement of the null space of A is R^n.*

(*q*) *The orthogonal complement of the row space of A is $\{\mathbf{0}\}$.*

Applications of Rank

The advent of the Internet has stimulated research on finding efficient methods for transmitting large amounts of digital data over communications lines with limited bandwidths. Digital data are commonly stored in matrix form, and many techniques for improving transmission speed use the rank of a matrix in some way. Rank plays a role because it measures the "redundancy" in a matrix in the sense that if A is an $m \times n$ matrix of rank k, then $n - k$ of the column vectors and $m - k$ of the row vectors can be expressed in terms of k linearly independent column or row vectors. The essential idea in many data compression schemes is to approximate the original data set by a data set with smaller rank that conveys nearly the same information, then eliminate redundant vectors in the approximating set to speed up the transmission time.

Concept Review

- Rank
- Nullity
- Dimension Theorem
- Overdetermined system

- Underdetermined system
- Fundamental spaces of a matrix
- Relationships among the fundamental spaces

- Orthogonal complement
- Equivalent characterizations of invertible matrices

Skills

- Find the rank and nullity of a matrix.

- Find the dimension of the row space of a matrix.

Exercise Set 4.8

1. Verify that $\text{rank}(A) = \text{rank}(A^T)$.

$$A = \begin{bmatrix} 1 & 2 & 4 & 0 \\ -3 & 1 & 5 & 2 \\ -2 & 3 & 9 & 2 \end{bmatrix}$$

2. Find the rank and nullity of the matrix; then verify that the values obtained satisfy Formula (4) in the Dimension Theorem.

(a) $A = \begin{bmatrix} 1 & -1 & 3 \\ 5 & -4 & -4 \\ 7 & -6 & 2 \end{bmatrix}$ (b) $A = \begin{bmatrix} 2 & 0 & -1 \\ 4 & 0 & -2 \\ 0 & 0 & 0 \end{bmatrix}$

(c) $A = \begin{bmatrix} 1 & 4 & 5 & 2 \\ 2 & 1 & 3 & 0 \\ -1 & 3 & 2 & 2 \end{bmatrix}$

(d) $A = \begin{bmatrix} 1 & 4 & 5 & 6 & 9 \\ 3 & -2 & 1 & 4 & -1 \\ -1 & 0 & -1 & -2 & -1 \\ 2 & 3 & 5 & 7 & 8 \end{bmatrix}$

(e) $A = \begin{bmatrix} 1 & -3 & 2 & 2 & 1 \\ 0 & 3 & 6 & 0 & -3 \\ 2 & -3 & -2 & 4 & 4 \\ 3 & -6 & 0 & 6 & 5 \\ -2 & 9 & 2 & -4 & -5 \end{bmatrix}$

3. In each part of Exercise 2, use the results obtained to find the number of leading variables and the number of parameters in the solution of $A\mathbf{x} = \mathbf{0}$ without solving the system.

4. In each part, use the information in the table to find the dimension of the row space of A, column space of A, null space of A, and null space of A^T.

	(a)	(b)	(c)	(d)	(e)	(f)	(g)
Size of A	3×3	3×3	3×3	5×9	9×5	4×4	6×2
Rank(A)	3	2	1	2	2	0	2

5. In each part, find the largest possible value for the rank of A and the smallest possible value for the nullity of A.

(a) A is 4×4 (b) A is 3×5 (c) A is 5×3

6. If A is an $m \times n$ matrix, what is the largest possible value for its rank and the smallest possible value for its nullity?

7. In each part, use the information in the table to determine whether the linear system $A\mathbf{x} = \mathbf{b}$ is consistent. If so, state the number of parameters in its general solution.

	(a)	(b)	(c)	(d)	(e)	(f)	(g)
Size of A	3×3	3×3	3×3	5×9	5×9	4×4	6×2
Rank(A)	3	2	1	2	2	0	2
Rank[A | b]	3	3	1	2	3	0	2

8. For each of the matrices in Exercise 7, find the nullity of A, and determine the number of parameters in the general solution of the homogeneous linear system $A\mathbf{x} = \mathbf{0}$.

9. What conditions must be satisfied by b_1, b_2, b_3, b_4, and b_5 for the overdetermined linear system

$$x_1 - 3x_2 = b_1$$
$$x_1 - 2x_2 = b_2$$
$$x_1 + x_2 = b_3$$
$$x_1 - 4x_2 = b_4$$
$$x_1 + 5x_2 = b_5$$

to be consistent?

10. Let

$$A = \begin{bmatrix} a_{11} & a_{12} & a_{13} \\ a_{21} & a_{22} & a_{23} \end{bmatrix}$$

Show that A has rank 2 if and only if one or more of the determinants

$$\begin{vmatrix} a_{11} & a_{12} \\ a_{21} & a_{22} \end{vmatrix}, \quad \begin{vmatrix} a_{11} & a_{13} \\ a_{21} & a_{23} \end{vmatrix}, \quad \begin{vmatrix} a_{12} & a_{13} \\ a_{22} & a_{23} \end{vmatrix}$$

is nonzero.

11. Suppose that A is a 3×3 matrix whose null space is a line through the origin in 3-space. Can the row or column space of A also be a line through the origin? Explain.

12. Discuss how the rank of A varies with t.

(a) $A = \begin{bmatrix} 1 & 1 & t \\ 1 & t & 1 \\ t & 1 & 1 \end{bmatrix}$ (b) $A = \begin{bmatrix} t & 3 & -1 \\ 3 & 6 & -2 \\ -1 & -3 & t \end{bmatrix}$

13. Are there values of r and s for which

$$\begin{bmatrix} 1 & 0 & 0 \\ 0 & r-2 & 2 \\ 0 & s-1 & r+2 \\ 0 & 0 & 3 \end{bmatrix}$$

has rank 1? Has rank 2? If so, find those values.

14. Use the result in Exercise 10 to show that the set of points (x, y, z) in R^3 for which the matrix

$$\begin{bmatrix} x & y & z \\ 1 & x & y \end{bmatrix}$$

has rank 1 is the curve with parametric equations $x = t$, $y = t^2$, $z = t^3$.

15. Prove: If $k \neq 0$, then A and kA have the same rank.

16. (a) Give an example of a 3×3 matrix whose column space is a plane through the origin in 3-space.

(b) What kind of geometric object is the null space of your matrix?

(c) What kind of geometric object is the row space of your matrix?

17. (a) If A is a 3×5 matrix, then the number of leading 1's in the reduced row echelon form of A is at most _____. Why?

   (b) If A is a 3×5 matrix, then the number of parameters in the general solution of $A\mathbf{x} = \mathbf{0}$ is at most _____. Why?

   (c) If A is a 5×3 matrix, then the number of leading 1's in the reduced row echelon form of A is at most _____. Why?

   (d) If A is a 5×3 matrix, then the number of parameters in the general solution of $A\mathbf{x} = \mathbf{0}$ is at most _____. Why?

18. (a) If A is a 3×5 matrix, then the rank of A is at most _____. Why?

   (b) If A is a 3×5 matrix, then the nullity of A is at most _____. Why?

   (c) If A is a 3×5 matrix, then the rank of A^T is at most _____. Why?

   (d) If A is a 3×5 matrix, then the nullity of A^T is at most _____. Why?

19. Find matrices A and B for which $\text{rank}(A) = \text{rank}(B)$, but $\text{rank}(A^2) \neq \text{rank}(B^2)$.

20. Prove: If a matrix A is not square, then either the row vectors or the column vectors of A are linearly dependent.

True-False Exercises

In parts (a)–(j) determine whether the statement is true or false, and justify your answer.

(a) Either the row vectors or the column vectors of a square matrix are linearly independent.

(b) A matrix with linearly independent row vectors and linearly independent column vectors is square.

(c) The nullity of a nonzero $m \times n$ matrix is at most m.

(d) Adding one additional column to a matrix increases its rank by one.

(e) The nullity of a square matrix with linearly dependent rows is at least one.

(f) If A is square and $A\mathbf{x} = \mathbf{b}$ is inconsistent for some vector $\mathbf{b}$, then the nullity of A is zero.

(g) If a matrix A has more rows than columns, then the dimension of the row space is greater than the dimension of the column space.

(h) If $\text{rank}(A^T) = \text{rank}(A)$, then A is square.

(i) There is no 3×3 matrix whose row space and null space are both lines in 3-space.

(j) If V is a subspace of R^n and W is a subspace of V, then $W^\perp$ is a subspace of $V^\perp$.

4.9 Matrix Transformations from R^n to R^m

In this section we will study functions of the form $\mathbf{w} = F(\mathbf{x})$, where the independent variable $\mathbf{x}$ is a vector in R^n and the dependent variable $\mathbf{w}$ is a vector in R^m. We will concentrate on a special class of such functions called "matrix transformations." Such transformations are fundamental in the study of linear algebra and have important applications in physics, engineering, social sciences, and various branches of mathematics.

Functions and Transformations

Recall that a ***function*** is a rule that associates with each element of a set A one and only one element in a set B. If f associates the element b with the element a, then we write

$$b = f(a)$$

and we say that b is the ***image*** of a under f or that $f(a)$ is the ***value*** of f at a. The set A is called the ***domain*** of f and the set B the ***codomain*** of f (Figure 4.9.1). The subset of the codomain that consists of all images of points in the domain is called the ***range*** of f.

For many common functions the domain and codomain are sets of real numbers, but in this text we will be concerned with functions for which the domain and codomain are vector spaces.

Domain
A

Codomain
B

▲ Figure 4.9.1

DEFINITION 1 If V and W are vector spaces, and if f is a function with domain V and codomain W, then we say that f is a ***transformation*** from V to W or that f ***maps*** V to W, which we denote by writing

$$f : V \rightarrow W$$

In the special case where $V = W$, the transformation is also called an ***operator*** on V.

In this section we will be concerned exclusively with transformations from R^n to R^m; transformations of general vector spaces will be considered in a later section. To illustrate one way in which such transformations can arise, suppose that $f_1, f_2, \ldots, f_m$ are real-valued functions of n variables, say

$$
\begin{aligned}
w_1 &= f_1(x_1, x_2, \ldots, x_n) \\
w_2 &= f_2(x_1, x_2, \ldots, x_n) \\
&\ \ \vdots \qquad\qquad \vdots \\
w_m &= f_m(x_1, x_2, \ldots, x_n)
\end{aligned}
\tag{1}
$$

These m equations assign a unique point $(w_1, w_2, \ldots, w_m)$ in R^m to each point $(x_1, x_2, \ldots, x_n)$ in R^n and thus define a transformation from R^n to R^m. If we denote this transformation by T, then $T : R^n \rightarrow R^m$ and

$$T(x_1, x_2, \ldots, x_n) = (w_1, w_2, \ldots, w_m)$$

Matrix Transformations In the special case where the equations in (1) are linear, they can be expressed in the form

$$
\begin{aligned}
w_1 &= a_{11}x_1 + a_{12}x_2 + \cdots + a_{1n}x_n \\
w_2 &= a_{21}x_1 + a_{22}x_2 + \cdots + a_{2n}x_n \\
&\ \ \vdots \qquad\ \ \vdots \qquad\ \ \vdots \qquad\qquad\ \ \vdots \\
w_m &= a_{m1}x_1 + a_{m2}x_2 + \cdots + a_{mn}x_n
\end{aligned}
\tag{2}
$$

which we can write in matrix notation as

$$
\begin{bmatrix} w_1 \\ w_2 \\ \vdots \\ w_m \end{bmatrix}
=
\begin{bmatrix}
a_{11} & a_{12} & \cdots & a_{1n} \\
a_{21} & a_{22} & \cdots & a_{2n} \\
\vdots & \vdots & & \vdots \\
a_{m1} & a_{m2} & \cdots & a_{mn}
\end{bmatrix}
\begin{bmatrix} x_1 \\ x_2 \\ \vdots \\ x_n \end{bmatrix}
\tag{3}
$$

or more briefly as

$$\mathbf{w} = A\mathbf{x} \tag{4}$$

Although we could view this as a linear system, we will view it instead as a transformation that maps the column vector $\mathbf{x}$ in R^n into the column vector $\mathbf{w}$ in R^m by multiplying $\mathbf{x}$ on the left by A. We call this a ***matrix transformation*** (or ***matrix operator*** if $m = n$), and we denote it by $T_A : R^n \rightarrow R^m$. With this notation, Equation (4) can be expressed as

$$\mathbf{w} = T_A(\mathbf{x}) \tag{5}$$

The matrix transformation T_A is called ***multiplication by*** A, and the matrix A is called the ***standard matrix*** for the transformation.

We will also find it convenient, on occasion, to express (5) in the schematic form

$$\mathbf{x} \xrightarrow{\ T_A\ } \mathbf{w} \tag{6}$$

which is read "T_A maps $\mathbf{x}$ into $\mathbf{w}$."

▶ **EXAMPLE 1 A Matrix Transformation from R^4 to R^3**

The matrix transformation $T: R^4 \rightarrow R^3$ defined by the equations

$$
\begin{aligned}
w_1 &= 2x_1 - 3x_2 + x_3 - 5x_4 \\
w_2 &= 4x_1 + x_2 - 2x_3 + x_4 \\
w_3 &= 5x_1 - x_2 + 4x_3
\end{aligned}
\tag{7}
$$

can be expressed in matrix form as

$$
\begin{bmatrix} w_1 \\ w_2 \\ w_3 \end{bmatrix} =
\begin{bmatrix} 2 & -3 & 1 & -5 \\ 4 & 1 & -2 & 1 \\ 5 & -1 & 4 & 0 \end{bmatrix}
\begin{bmatrix} x_1 \\ x_2 \\ x_3 \\ x_4 \end{bmatrix}
\tag{8}
$$

so the standard matrix for T is

$$
A = \begin{bmatrix} 2 & -3 & 1 & -5 \\ 4 & 1 & -2 & 1 \\ 5 & -1 & 4 & 0 \end{bmatrix}
$$

The image of a point (x_1, x_2, x_3, x_4) can be computed directly from the defining equations (7) or from (8) by matrix multiplication. For example, if

$$
(x_1, x_2, x_3, x_4) = (1, -3, 0, 2)
$$

then substituting in (7) yields $w_1 = 1$, $w_2 = 3$, $w_3 = 8$ (verify), or alternatively from (8),

$$
\begin{bmatrix} w_1 \\ w_2 \\ w_3 \end{bmatrix} =
\begin{bmatrix} 2 & -3 & 1 & -5 \\ 4 & 1 & -2 & 1 \\ 5 & -1 & 4 & 0 \end{bmatrix}
\begin{bmatrix} 1 \\ -3 \\ 0 \\ 2 \end{bmatrix} =
\begin{bmatrix} 1 \\ 3 \\ 8 \end{bmatrix} \quad ◀
$$

Some Notational Matters

Sometimes we will want to denote a matrix transformation without giving a name to the matrix itself. In such cases we will denote the standard matrix for $T: R^n \rightarrow R^m$ by the symbol $[T]$. Thus, the equation

$$
T(\mathbf{x}) = [T]\mathbf{x}
\tag{9}
$$

is simply the statement that T is a matrix transformation with standard matrix $[T]$, and the image of $\mathbf{x}$ under this transformation is the product of the matrix $[T]$ and the column vector $\mathbf{x}$.

Properties of Matrix Transformations

The following theorem lists four basic properties of matrix transformations that follow from properties of matrix multiplication.

THEOREM 4.9.1 *For every matrix A the matrix transformation $T_A: R^n \rightarrow R^m$ has the following properties for all vectors $\mathbf{u}$ and $\mathbf{v}$ in R^n and for every scalar k:*

(a) $T_A(\mathbf{0}) = \mathbf{0}$

(b) $T_A(k\mathbf{u}) = kT_A(\mathbf{u})$ **[Homogeneity property]**

(c) $T_A(\mathbf{u} + \mathbf{v}) = T_A(\mathbf{u}) + T_A(\mathbf{v})$ **[Additivity property]**

(d) $T_A(\mathbf{u} - \mathbf{v}) = T_A(\mathbf{u}) - T_A(\mathbf{v})$

Proof All four parts are restatements of familiar properties of matrix multiplication:

$$A\mathbf{0} = \mathbf{0}, \quad A(k\mathbf{u}) = k(A\mathbf{u}), \quad A(\mathbf{u} + \mathbf{v}) = A\mathbf{u} + A\mathbf{v}, \quad A(\mathbf{u} - \mathbf{v}) = A\mathbf{u} - A\mathbf{v} \quad \blacktriangleleft$$

It follows from Theorem 4.9.1 that a matrix transformation maps linear combinations of vectors in R^n into the corresponding linear combinations in R^m in the sense that

$$T_A(k_1\mathbf{u}_1 + k_2\mathbf{u}_2 + \cdots + k_r\mathbf{u}_r) = k_1 T_A(\mathbf{u}_1) + k_2 T_A(\mathbf{u}_2) + \cdots + k_r T_A(\mathbf{u}_r) \qquad (10)$$

Depending on whether n-tuples and m-tuples are regarded as vectors or points, the geometric effect of a matrix transformation $T_A: R^n \to R^m$ is to map each vector (point) in R^n into a vector (point) in R^m (Figure 4.9.2).

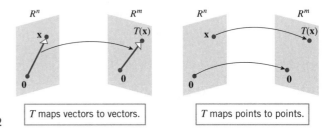

T maps vectors to vectors. T maps points to points.

▶ **Figure 4.9.2**

The following theorem states that if two matrix transformations from R^n to R^m have the same image at each point of R^n, then the matrices themselves must be the same.

THEOREM 4.9.2 *If $T_A: R^n \to R^m$ and $T_B: R^n \to R^m$ are matrix transformations, and if $T_A(\mathbf{x}) = T_B(\mathbf{x})$ for every vector $\mathbf{x}$ in R^n, then $A = B$.*

Proof To say that $T_A(\mathbf{x}) = T_B(\mathbf{x})$ for every vector in R^n is the same as saying that

$$A\mathbf{x} = B\mathbf{x}$$

for every vector $\mathbf{x}$ in R^n. This is true, in particular, if $\mathbf{x}$ is any of the standard basis vectors $\mathbf{e}_1, \mathbf{e}_2, \ldots, \mathbf{e}_n$ for R^n; that is,

$$A\mathbf{e}_j = B\mathbf{e}_j \quad (j = 1, 2, \ldots, n) \qquad (11)$$

Since every entry of $\mathbf{e}_j$ is 0 except for the jth, which is 1, it follows from Theorem 1.3.1 that $A\mathbf{e}_j$ is the jth column of A and $B\mathbf{e}_j$ is the jth column of B. Thus, it follows from (11) that corresponding columns of A and B are the same, and hence that $A = B$. $\blacktriangleleft$

▶ **EXAMPLE 2 Zero Transformations**

If 0 is the $m \times n$ zero matrix, then

$$T_0(\mathbf{x}) = 0\mathbf{x} = \mathbf{0}$$

so multiplication by zero maps every vector in R^n into the zero vector in R^m. We call T_0 the ***zero transformation*** from R^n to R^m.

▶ **EXAMPLE 3 Identity Operators**

If I is the $n \times n$ identity matrix, then

$$T_I(\mathbf{x}) = I\mathbf{x} = \mathbf{x}$$

so multiplication by I maps every vector in R^n into itself. We call T_I the *identity operator* on R^n. ◀

A Procedure for Finding Standard Matrices

There is a way of finding the standard matrix for a matrix transformation from R^n to R^m by considering the effect of that transformation on the standard basis vectors for R^n. To explain the idea, suppose that A is unknown and that

$$\mathbf{e}_1, \quad \mathbf{e}_2, \ldots, \quad \mathbf{e}_n$$

are the standard basis vectors for R^n. Suppose also that the images of these vectors under the transformation T_A are

$$T_A(\mathbf{e}_1) = A\mathbf{e}_1, \quad T_A(\mathbf{e}_2) = A\mathbf{e}_2, \ldots, \quad T_A(\mathbf{e}_n) = A\mathbf{e}_n$$

It follows from Theorem 1.3.1 that $A\mathbf{e}_j$ is a linear combination of the columns of A in which the successive coefficients are the entries of $\mathbf{e}_j$. But all entries of $\mathbf{e}_j$ are zero except the jth, so the product $A\mathbf{e}_j$ is just the jth column of the matrix A. Thus,

$$A = [T_A(\mathbf{e}_1) \mid T_A(\mathbf{e}_2) \mid \cdots \mid T_A(\mathbf{e}_n)] \tag{12}$$

In summary, we have the following procedure for finding the standard matrix for a matrix transformation:

Finding the Standard Matrix for a Matrix Transformation

Step 1. Find the images of the standard basis vectors $\mathbf{e}_1, \mathbf{e}_2, \ldots, \mathbf{e}_n$ for R^n in column form.

Step 2. Construct the matrix that has the images obtained in Step 1 as its successive columns. This matrix is the standard matrix for the transformation.

Reflection Operators

Some of the most basic matrix operators on R^2 and R^3 are those that map each point into its symmetric image about a fixed line or a fixed plane; these are called *reflection operators*. Table 1 shows the standard matrices for the reflections about the coordinate axes in R^2, and Table 2 shows the standard matrices for the reflections about the coordinate planes in R^3. In each case the standard matrix was obtained by finding the images of the standard basis vectors, converting those images to column vectors, and then using those column vectors as successive columns of the standard matrix.

Projection Operators

Matrix operators on R^2 and R^3 that map each point into its orthogonal projection on a fixed line or plane are called *projection operators* (or more precisely, *orthogonal projection* operators). Table 3 shows the standard matrices for the orthogonal projections on the coordinate axes in R^2, and Table 4 shows the standard matrices for the orthogonal projections on the coordinate planes in R^3.

Table 1

Operator	Illustration	Images of e_1 and e_2	Standard Matrix
Reflection about the y-axis $T(x, y) = (-x, y)$		$T(\mathbf{e}_1) = T(1, 0) = (-1, 0)$ $T(\mathbf{e}_2) = T(0, 1) = (0, 1)$	$\begin{bmatrix} -1 & 0 \\ 0 & 1 \end{bmatrix}$
Reflection about the x-axis $T(x, y) = (x, -y)$		$T(\mathbf{e}_1) = T(1, 0) = (1, 0)$ $T(\mathbf{e}_2) = T(0, 1) = (0, -1)$	$\begin{bmatrix} 1 & 0 \\ 0 & -1 \end{bmatrix}$
Reflection about the line $y = x$ $T(x, y) = (y, x)$		$T(\mathbf{e}_1) = T(1, 0) = (0, 1)$ $T(\mathbf{e}_2) = T(0, 1) = (1, 0)$	$\begin{bmatrix} 0 & 1 \\ 1 & 0 \end{bmatrix}$

Table 2

Operator	Illustration	e_1, e_2, e_3	Standard Matrix
Reflection about the xy-plane $T(x, y, z) = (x, y, -z)$		$T(\mathbf{e}_1) = T(1, 0, 0) = (1, 0, 0)$ $T(\mathbf{e}_2) = T(0, 1, 0) = (0, 1, 0)$ $T(\mathbf{e}_3) = T(0, 0, 1) = (0, 0, -1)$	$\begin{bmatrix} 1 & 0 & 0 \\ 0 & 1 & 0 \\ 0 & 0 & -1 \end{bmatrix}$
Reflection about the xz-plane $T(x, y, z) = (x, -y, z)$		$T(\mathbf{e}_1) = T(1, 0, 0) = (1, 0, 0)$ $T(\mathbf{e}_2) = T(0, 1, 0) = (0, -1, 0)$ $T(\mathbf{e}_3) = T(0, 0, 1) = (0, 0, 1)$	$\begin{bmatrix} 1 & 0 & 0 \\ 0 & -1 & 0 \\ 0 & 0 & 1 \end{bmatrix}$
Reflection about the yz-plane $T(x, y, z) = (-x, y, z)$		$T(\mathbf{e}_1) = T(1, 0, 0) = (-1, 0, 0)$ $T(\mathbf{e}_2) = T(0, 1, 0) = (0, 1, 0)$ $T(\mathbf{e}_3) = T(0, 0, 1) = (0, 0, 1)$	$\begin{bmatrix} -1 & 0 & 0 \\ 0 & 1 & 0 \\ 0 & 0 & 1 \end{bmatrix}$

Table 3

Operator	Illustration	Images of e_1 and e_2	Standard Matrix
Orthogonal projection on the x-axis $T(x, y) = (x, 0)$		$T(\mathbf{e}_1) = T(1, 0) = (1, 0)$ $T(\mathbf{e}_2) = T(0, 1) = (0, 0)$	$\begin{bmatrix} 1 & 0 \\ 0 & 0 \end{bmatrix}$
Orthogonal projection on the y-axis $T(x, y) = (0, y)$		$T(\mathbf{e}_1) = T(1, 0) = (0, 0)$ $T(\mathbf{e}_2) = T(0, 1) = (0, 1)$	$\begin{bmatrix} 0 & 0 \\ 0 & 1 \end{bmatrix}$

Table 4

Operator	Illustration	Images of e_1, e_2, e_3	Standard Matrix
Orthogonal projection on the xy-plane $T(x, y, z) = (x, y, 0)$		$T(\mathbf{e}_1) = T(1, 0, 0) = (1, 0, 0)$ $T(\mathbf{e}_2) = T(0, 1, 0) = (0, 1, 0)$ $T(\mathbf{e}_3) = T(0, 0, 1) = (0, 0, 0)$	$\begin{bmatrix} 1 & 0 & 0 \\ 0 & 1 & 0 \\ 0 & 0 & 0 \end{bmatrix}$
Orthogonal projection on the xz-plane $T(x, y, z) = (x, 0, z)$		$T(\mathbf{e}_1) = T(1, 0, 0) = (1, 0, 0)$ $T(\mathbf{e}_2) = T(0, 1, 0) = (0, 0, 0)$ $T(\mathbf{e}_3) = T(0, 0, 1) = (0, 0, 1)$	$\begin{bmatrix} 1 & 0 & 0 \\ 0 & 0 & 0 \\ 0 & 0 & 1 \end{bmatrix}$
Orthogonal projection on the yz-plane $T(x, y, z) = (0, y, z)$		$T(\mathbf{e}_1) = T(1, 0, 0) = (0, 0, 0)$ $T(\mathbf{e}_2) = T(0, 1, 0) = (0, 1, 0)$ $T(\mathbf{e}_3) = T(0, 0, 1) = (0, 0, 1)$	$\begin{bmatrix} 0 & 0 & 0 \\ 0 & 1 & 0 \\ 0 & 0 & 1 \end{bmatrix}$

Rotation Operators   Matrix operators on R^2 and R^3 that move points along circular arcs are called ***rotation operators***. Let us consider how to find the standard matrix for the rotation operator $T: R^2 \to R^2$ that moves points *counterclockwise* about the origin through an angle θ (Figure 4.9.3). As illustrated in Figure 4.9.3, the images of the standard basis vectors are

$$T(\mathbf{e}_1) = T(1, 0) = (\cos\theta, \sin\theta) \quad \text{and} \quad T(\mathbf{e}_2) = T(0, 1) = (-\sin\theta, \cos\theta)$$

so the standard matrix for T is

$$[T(\mathbf{e}_1) \mid T(\mathbf{e}_2)] = \begin{bmatrix} \cos\theta & -\sin\theta \\ \sin\theta & \cos\theta \end{bmatrix}$$

▶ Figure 4.9.3

In keeping with common usage we will denote this operator by R_θ and call

$$R_\theta = \begin{bmatrix} \cos\theta & -\sin\theta \\ \sin\theta & \cos\theta \end{bmatrix} \qquad (13)$$

the *rotation matrix* for R^2. If $\mathbf{x} = (x, y)$ is a vector in R^2, and if $\mathbf{w} = (w_1, w_2)$ is its image under the rotation, then the relationship $\mathbf{w} = R_\theta\mathbf{x}$ can be written in component form as

$$\begin{aligned} w_1 &= x\cos\theta - y\sin\theta \\ w_2 &= x\sin\theta + y\cos\theta \end{aligned} \qquad (14)$$

These are called the *rotation equations* for R^2. These ideas are summarized in Table 5.

Table 5

Operator	Illustration	Rotation Equations	Standard Matrix
Rotation through an angle θ		$\begin{aligned} w_1 &= x\cos\theta - y\sin\theta \\ w_2 &= x\sin\theta + y\cos\theta \end{aligned}$	$\begin{bmatrix} \cos\theta & -\sin\theta \\ \sin\theta & \cos\theta \end{bmatrix}$

In the plane, counterclockwise angles are positive and clockwise angles are negative. The rotation matrix for a *clockwise* rotation of $-\theta$ radians can be obtained by replacing θ by $-\theta$ in (12). After simplification this yields

$$R_{-\theta} = \begin{bmatrix} \cos\theta & \sin\theta \\ -\sin\theta & \cos\theta \end{bmatrix}$$

▶ **EXAMPLE 4 A Rotation Operator**

Find the image of $\mathbf{x} = (1, 1)$ under a rotation of $\pi/6$ radians ($= 30°$) about the origin.

Solution It follows from (13) with $\theta = \pi/6$ that

$$R_{\pi/6}\mathbf{x} = \begin{bmatrix} \frac{\sqrt{3}}{2} & -\frac{1}{2} \\ \frac{1}{2} & \frac{\sqrt{3}}{2} \end{bmatrix} \begin{bmatrix} 1 \\ 1 \end{bmatrix} = \begin{bmatrix} \frac{\sqrt{3}-1}{2} \\ \frac{1+\sqrt{3}}{2} \end{bmatrix} \approx \begin{bmatrix} 0.37 \\ 1.37 \end{bmatrix}$$

or in comma-delimited notation, $R_{\pi/6}(1, 1) = (0.37, 1.37)$. ◀

Rotations in R^3

A rotation of vectors in R^3 is usually described in relation to a ray emanating from the origin, called the *axis of rotation*. As a vector revolves around the axis of rotation, it sweeps out some portion of a cone (Figure 4.9.4*a*). The *angle of rotation*, which is measured in the base of the cone, is described as "clockwise" or "counterclockwise" in relation to a viewpoint that is along the axis of rotation *looking toward the origin*. For example, in Figure 4.9.4*a* the vector $\mathbf{w}$ results from rotating the vector $\mathbf{x}$ counterclockwise around the axis l through an angle θ. As in R^2, angles are *positive* if they are generated by counterclockwise rotations and *negative* if they are generated by clockwise rotations.

The most common way of describing a general axis of rotation is to specify a nonzero vector $\mathbf{u}$ that runs along the axis of rotation and has its initial point at the origin. The counterclockwise direction for a rotation about the axis can then be determined by a "right-hand rule" (Figure 4.9.4b): If the thumb of the right hand points in the direction of **u**, then the cupped fingers point in a counterclockwise direction.

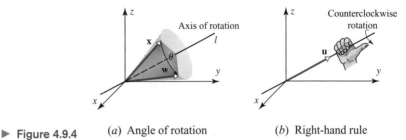

▶ **Figure 4.9.4**   (a) Angle of rotation   (b) Right-hand rule

A **rotation operator** on R^3 is a matrix operator that rotates each vector in R^3 about some rotation axis through a fixed angle θ. In Table 6 we have described the rotation operators on R^3 whose axes of rotation are the positive coordinate axes. For each of these rotations one of the components is unchanged, and the relationships between the other components can be derived by the same procedure used to derive (14). For example, in the rotation about the z-axis, the z-components of **x** and $\mathbf{w} = T(\mathbf{x})$ are the same, and the x- and y-components are related as in (14). This yields the rotation equation shown in the last row of Table 6.

Table 6

Operator	Illustration	Rotation Equations	Standard Matrix
Counterclockwise rotation about the positive x-axis through an angle θ		$\begin{aligned} w_1 &= x \\ w_2 &= y\cos\theta - z\sin\theta \\ w_3 &= y\sin\theta + z\cos\theta \end{aligned}$	$\begin{bmatrix} 1 & 0 & 0 \\ 0 & \cos\theta & -\sin\theta \\ 0 & \sin\theta & \cos\theta \end{bmatrix}$
Counterclockwise rotation about the positive y-axis through an angle θ		$\begin{aligned} w_1 &= x\cos\theta + z\sin\theta \\ w_2 &= y \\ w_3 &= -x\sin\theta + z\cos\theta \end{aligned}$	$\begin{bmatrix} \cos\theta & 0 & \sin\theta \\ 0 & 1 & 0 \\ -\sin\theta & 0 & \cos\theta \end{bmatrix}$
Counterclockwise rotation about the positive z-axis through an angle θ		$\begin{aligned} w_1 &= x\cos\theta - y\sin\theta \\ w_2 &= x\sin\theta + y\cos\theta \\ w_3 &= z \end{aligned}$	$\begin{bmatrix} \cos\theta & -\sin\theta & 0 \\ \sin\theta & \cos\theta & 0 \\ 0 & 0 & 1 \end{bmatrix}$

For completeness, we note that the standard matrix for a counterclockwise rotation through an angle θ about an axis in R^3, which is determined by an arbitrary *unit vector* $\mathbf{u} = (a, b, c)$ that has its initial point at the origin, is

$$\begin{bmatrix} a^2(1 - \cos\theta) + \cos\theta & ab(1 - \cos\theta) - c\sin\theta & ac(1 - \cos\theta) + b\sin\theta \\ ab(1 - \cos\theta) + c\sin\theta & b^2(1 - \cos\theta) + \cos\theta & bc(1 - \cos\theta) - a\sin\theta \\ ac(1 - \cos\theta) - b\sin\theta & bc(1 - \cos\theta) + a\sin\theta & c^2(1 - \cos\theta) + \cos\theta \end{bmatrix} \quad (15)$$

The derivation can be found in the book *Principles of Interactive Computer Graphics*, by W. M. Newman and R. F. Sproull (New York: McGraw-Hill, 1979). You may find it instructive to derive the results in Table 6 as special cases of this more general result.

Dilations and Contractions If k is a nonnegative scalar, then the operator $T(\mathbf{x}) = k\mathbf{x}$ on R^2 or R^3 has the effect of increasing or decreasing the length of each vector by a factor of k. If $0 \le k < 1$ the operator is called a ***contraction*** with factor k, and if $k > 1$ it is called a ***dilation*** with factor k (Figure 4.9.5). If $k = 1$, then T is the identity operator and can be regarded either as a contraction or a dilation. Tables 7 and 8 illustrate these operators.

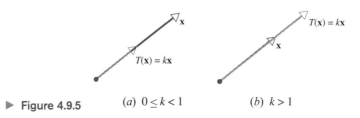

▶ **Figure 4.9.5**   (a) $0 \le k < 1$ (b) $k > 1$

Table 7

Operator	Illustration $T(x, y) = (kx, ky)$	Effect on the Standard Basis	Standard Matrix
Contraction with factor k on R^2 ($0 \le k < 1$)			$\begin{bmatrix} k & 0 \\ 0 & k \end{bmatrix}$
Dilation with factor k on R^2 ($k > 1$)			

Table 8

Operator	Illustration $T(x, y, z) = (kx, ky, kz)$	Standard Matrix
Contraction with factor k on R^3 $(0 \le k \le 1)$		$\begin{bmatrix} k & 0 & 0 \\ 0 & k & 0 \\ 0 & 0 & k \end{bmatrix}$
Dilation with factor k on R^3 $(k \ge 1)$		

Expansion and Compressions

In a dilation or contraction of R^2 or R^3, all coordinates are multiplied by a factor k. If only one of the coordinates is multiplied by k, then the resulting operator is called an ***expansion*** or ***compression*** with factor k. This is illustrated in Table 9 for R^2. You should have no trouble extending these results to R^3.

Table 9

Operator	Illustration $T(x, y) = (kx, y)$	Effect on the Standard Basis	Standard Matrix
Compression of R^2 in the x-direction with factor k $(0 \le k < 1)$			$\begin{bmatrix} k & 0 \\ 0 & 1 \end{bmatrix}$
Expansion of R^2 in the x-direction with factor k $(k > 1)$			

Operator	Illustration $T(x, y) = (x, ky)$	Effect on the Standard Basis	Standard Matrix
Compression of R^2 in the y-direction with factor k $(0 \le k < 1)$			$\begin{bmatrix} 1 & 0 \\ 0 & k \end{bmatrix}$
Expansion of R^2 in the y-direction with factor k $(k > 1)$			

Shears   A matrix operator of the form $T(x, y) = (x + ky, y)$ translates a point (x, y) in the xy-plane parallel to the x-axis by an amount ky that is proportional to the y-coordinate of the point. This operator leaves the points on the x-axis fixed (since $y = 0$), but as we progress away from the x-axis, the translation distance increases. We call this operator the ***shear in the x-direction with factor k***. Similarly, a matrix operator of the form $T(x, y) = (x, y + kx)$ is called the ***shear in the y-direction with factor k***. Table 10 illustrates the basic information about shears in R^2.

Table 10

Operator	Effect on the Standard Basis			Standard Matrix
Shear of R^2 in the x-direction with factor k $T(x, y) = (x + ky, y)$	(0, 1); (1, 0)	(k, 1); (1, 0) $(k > 0)$	(k, 1); (1, 0) $(k < 0)$	$\begin{bmatrix} 1 & k \\ 0 & 1 \end{bmatrix}$
Shear of R^2 in the y-direction with factor k $T(x, y) = (x, y + kx)$	(0, 1); (1, 0)	(0, 1); (1, k) $(k > 0)$	(0, 1); (1, k) $(k < 0)$	$\begin{bmatrix} 1 & 0 \\ k & 1 \end{bmatrix}$

▶ **EXAMPLE 5 Some Basic Matrix Operators on R^2**

In each part describe the matrix operator corresponding to A, and show its effect on the unit square.

(a) $A_1 = \begin{bmatrix} 1 & 2 \\ 0 & 1 \end{bmatrix}$ (b) $A_2 = \begin{bmatrix} 2 & 0 \\ 0 & 2 \end{bmatrix}$ (c) $A_3 = \begin{bmatrix} 2 & 0 \\ 0 & 1 \end{bmatrix}$

Solution   By comparing the forms of these matrices to those in Tables 7, 9, and 10, we see that the matrix A_1 corresponds to a shear in the x-direction with factor 2, the matrix A_2 corresponds to a dilation with factor 2, and A_3 corresponds to an expansion in the x-direction with factor 2. The effects of these operators on the unit square are shown in Figure 4.9.6. ◀

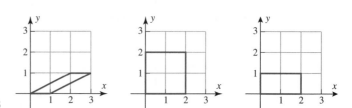

▶ Figure 4.9.6

OPTIONAL
Orthogonal Projections on Lines Through the Origin

In Table 3 we listed the standard matrices for the orthogonal projections on the coordinate axes in R^2. These are special cases of the more general operator $T: R^2 \to R^2$ that maps each point into its orthogonal projection on a line L through the origin that makes an

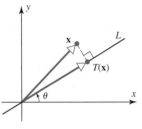

▲ Figure 4.9.7

angle θ with the positive x-axis (Figure 4.9.7). In Example 4 of Section 3.3 we used Formula (10) of that section to find the orthogonal projections of the standard basis vectors for R^2 on that line. Expressed in matrix form, we found those projections to be

$$T(\mathbf{e}_1) = \begin{bmatrix} \cos^2\theta \\ \sin\theta\cos\theta \end{bmatrix} \quad \text{and} \quad T(\mathbf{e}_2) = \begin{bmatrix} \sin\theta\cos\theta \\ \sin^2\theta \end{bmatrix}$$

Thus, the standard matrix for T is

$$[T] = [T(\mathbf{e}_1) \mid T(\mathbf{e}_2)] = \begin{bmatrix} \cos^2\theta & \sin\theta\cos\theta \\ \sin\theta\cos\theta & \sin^2\theta \end{bmatrix} = \begin{bmatrix} \cos^2\theta & \frac{1}{2}\sin 2\theta \\ \frac{1}{2}\sin 2\theta & \sin^2\theta \end{bmatrix}$$

In keeping with common usage, we will denote this operator by

$$P_\theta = \begin{bmatrix} \cos^2\theta & \sin\theta\cos\theta \\ \sin\theta\cos\theta & \sin^2\theta \end{bmatrix} = \begin{bmatrix} \cos^2\theta & \frac{1}{2}\sin 2\theta \\ \frac{1}{2}\sin 2\theta & \sin^2\theta \end{bmatrix} \tag{16}$$

> We have included two versions of Formula (16) because both are commonly used. Whereas the first version involves only the angle θ, the second involves both θ and 2θ.

▶ EXAMPLE 6 Orthogonal Projection on a Line Through the Origin

Use Formula (16) to find the orthogonal projection of the vector $\mathbf{x} = (1, 5)$ on the line through the origin that makes an angle of $\pi/6 (= 30°)$ with the x-axis.

Solution Since $\sin(\pi/6) = 1/2$ and $\cos(\pi/6) = \sqrt{3}/2$, it follows from (16) that the standard matrix for this projection is

$$P_{\pi/6} = \begin{bmatrix} \cos^2(\pi/6) & \sin(\pi/6)\cos(\pi/6) \\ \sin(\pi/6)\cos\theta & \sin^2(\pi/6) \end{bmatrix} = \begin{bmatrix} \frac{3}{4} & \frac{\sqrt{3}}{4} \\ \frac{\sqrt{3}}{4} & \frac{1}{4} \end{bmatrix}$$

Thus,

$$P_{\pi/6}\mathbf{x} = \begin{bmatrix} \frac{3}{4} & \frac{\sqrt{3}}{4} \\ \frac{\sqrt{3}}{4} & \frac{1}{4} \end{bmatrix}\begin{bmatrix} 1 \\ 5 \end{bmatrix} = \begin{bmatrix} \frac{3+5\sqrt{3}}{4} \\ \frac{\sqrt{3}+5}{4} \end{bmatrix} \approx \begin{bmatrix} 2.91 \\ 1.68 \end{bmatrix}$$

or in comma-delimited notation, $P_{\pi/6}(1, 5) \approx (2.91, 1.68)$ ◀

Reflections About Lines Through the Origin

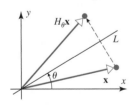

▲ Figure 4.9.8

In Table 1 we listed the reflections about the coordinate axes in R^2. These are special cases of the more general operator $H_\theta: R^2 \to R^2$ that maps each point into its reflection about a line L through the origin that makes an angle θ with the positive x-axis (Figure 4.9.8). We could find the standard matrix for H_θ by finding the images of the standard basis vectors, but instead we will take advantage of our work on orthogonal projections by using the Formula (16) for P_θ to find a formula for H_θ.

You should be able to see from Figure 4.9.9 that for every vector $\mathbf{x}$ in R^n

$$P_\theta\mathbf{x} - \mathbf{x} = \tfrac{1}{2}(H_\theta\mathbf{x} - \mathbf{x}) \quad \text{or equivalently} \quad H_\theta\mathbf{x} = (2P_\theta - I)\mathbf{x}$$

Thus, it follows from Theorem 4.9.2 that

$$H_\theta = 2P_\theta - I \tag{17}$$

and hence from (16) that

$$H_\theta = \begin{bmatrix} \cos 2\theta & \sin 2\theta \\ \sin 2\theta & -\cos 2\theta \end{bmatrix} \tag{18}$$

▲ Figure 4.9.9

▶ **EXAMPLE 7 Reflection About a Line Through the Origin**

Find the orthogonal projection of the vector $\mathbf{x} = (1, 5)$ on the line through the origin that makes an angle of $\pi/6(= 30°)$ with the x-axis.

Solution Since $\sin(\pi/3) = \sqrt{3}/2$ and $\cos(\pi/3) = 1/2$, it follows from (18) that the standard matrix for this projection is

$$H_{\pi/6} = \begin{bmatrix} \cos(\pi/3) & \sin(\pi/3) \\ \sin(\pi/3) & -\cos(\pi/3) \end{bmatrix} = \begin{bmatrix} \frac{1}{2} & \frac{\sqrt{3}}{2} \\ \frac{\sqrt{3}}{2} & -\frac{1}{2} \end{bmatrix}$$

Thus,

$$H_{\pi/6}\mathbf{x} = \begin{bmatrix} \frac{1}{2} & \frac{\sqrt{3}}{2} \\ \frac{\sqrt{3}}{2} & -\frac{1}{2} \end{bmatrix} \begin{bmatrix} 1 \\ 5 \end{bmatrix} = \begin{bmatrix} \frac{1+5\sqrt{3}}{2} \\ \frac{\sqrt{3}-5}{2} \end{bmatrix} \approx \begin{bmatrix} 4.83 \\ -1.63 \end{bmatrix}$$

| Show that the standard matri-
| ces in Tables 1 and 3 are special
| cases of (18) and (16).

or in comma-delimited notation, $H_{\pi/6}(1, 5) \approx (4.83, -1.63)$ ◀

Concept Review

- Function
- Image
- Value
- Domain
- Codomain
- Transformation
- Operator
- Matrix transformation
- Matrix operator

- Standard matrix
- Properties of matrix transformations
- Zero transformation
- Identity operator
- Reflection operator
- Projection operator
- Rotation operator
- Rotation matrix

- Rotation equations
- Axis of rotation in 3-space
- Angle of rotation in 3-space
- Expansion operator
- Compression operator
- Shear
- Dilation
- Contraction

Skills

- Find the domain and codomain of a transformation, and determine whether the transformation is linear.
- Find the standard matrix for a matrix transformation.

- Describe the effect of a matrix operator on the standard basis in R^n.

Exercise Set 4.9

▶ In Exercises 1–2, find the domain and codomain of the transformation $T_A(\mathbf{x}) = A\mathbf{x}$. ◀

1. (a) A has size 3×2.   (b) A has size 2×3.
   (c) A has size 3×3.   (d) A has size 1×6.

2. (a) A has size 4×5.   (b) A has size 5×4.
   (c) A has size 4×4.   (d) A has size 3×1.

3. If $T(x_1, x_2) = (x_1 + x_2, -x_2, 3x_1)$, then the domain of T is _____, the codomain of T is _____, and the image of $\mathbf{x} = (1, -2)$ under T is _____.

4. If $T(x_1, x_2, x_3) = (x_1 + 2x_2, x_1 - 2x_2)$, then the domain of T is _____, the codomain of T is _____, and the image of $\mathbf{x} = (0, -1, 4)$ under T is _____.

5. In each part, find the domain and codomain of the transformation defined by the equations, and determine whether the transformation is linear.

(a) $w_1 = 3x_1 - 2x_2 + 4x_3$
    $w_2 = 5x_1 - 8x_2 + x_3$

(b) $w_1 = 2x_1x_2 - x_2$
    $w_2 = x_1 + 3x_1x_2$
    $w_3 = x_1 + x_2$

(c) $w_1 = 5x_1 - x_2 + x_3$
    $w_2 = -x_1 + x_2 + 7x_3$
    $w_3 = 2x_1 - 4x_2 - x_3$

(d) $w_1 = x_1^2 - 3x_2 + x_3 - 2x_4$
    $w_2 = 3x_1 - 4x_2 - x_3^2 + x_4$

6. In each part, determine whether T is a matrix transformation.

(a) $T(x, y) = (2x, y)$ (b) $T(x, y) = (-y, x)$

(c) $T(x, y) = (2x + y, x - y)$

(d) $T(x, y) = (x^2, y)$ (e) $T(x, y) = (x, y + 1)$

7. In each part, determine whether T is a matrix transformation.

(a) $T(x, y, z) = (0, 0)$ (b) $T(x, y, z) = (1, 1)$

(c) $T(x, y, z) = (3x - 4y, 2x - 5z)$

(d) $T(x, y, z) = (y^2, z)$ (e) $T(x, y, z) = (y - 1, x)$

8. Find the standard matrix for the transformation defined by the equations.

(a) $w_1 = 2x_1 - 3x_2 + x_4$
    $w_2 = 3x_1 + 5x_2 - x_4$

(b) $w_1 = 7x_1 + 2x_2 - 8x_3$
    $w_2 = \quad\;\; - x_2 + 5x_3$
    $w_3 = 4x_1 + 7x_2 - \quad x_3$

(c) $w_1 = -x_1 + x_2$
    $w_2 = 3x_1 - 2x_2$
    $w_3 = 5x_1 - 7x_2$

(d) $w_1 = x_1$
    $w_2 = x_1 + x_2$
    $w_3 = x_1 + x_2 + x_3$
    $w_4 = x_1 + x_2 + x_3 + x_4$

9. Find the standard matrix for the operator $T: R^3 \to R^3$ defined by

$$w_1 = 3x_1 + 5x_2 - x_3$$
$$w_2 = 4x_1 - x_2 + x_3$$
$$w_3 = 3x_1 + 2x_2 - x_3$$

and then calculate $T(-1, 2, 4)$ by directly substituting in the equations and also by matrix multiplication.

10. Find the standard matrix for the operator T defined by the formula.

(a) $T(x_1, x_2) = (2x_1 - x_2, x_1 + x_2)$

(b) $T(x_1, x_2) = (x_1, x_2)$

(c) $T(x_1, x_2, x_3) = (x_1 + 2x_2 + x_3, x_1 + 5x_2, x_3)$

(d) $T(x_1, x_2, x_3) = (4x_1, 7x_2, -8x_3)$

11. Find the standard matrix for the transformation T defined by the formula.

(a) $T(x_1, x_2) = (x_2, -x_1, x_1 + 3x_2, x_1 - x_2)$

(b) $T(x_1, x_2, x_3, x_4) = (7x_1 + 2x_2 - x_3 + x_4, x_2 + x_3, -x_1)$

(c) $T(x_1, x_2, x_3) = (0, 0, 0, 0, 0)$

(d) $T(x_1, x_2, x_3, x_4) = (x_4, x_1, x_3, x_2, x_1 - x_3)$

12. In each part, find $T(x)$, and express the answer in matrix form.

(a) $[T] = \begin{bmatrix} 1 & 2 \\ 3 & 4 \end{bmatrix}$; $\mathbf{x} = \begin{bmatrix} 3 \\ -2 \end{bmatrix}$

(b) $[T] = \begin{bmatrix} -1 & 2 & 0 \\ 3 & 1 & 5 \end{bmatrix}$; $\mathbf{x} = \begin{bmatrix} -1 \\ 1 \\ 3 \end{bmatrix}$

(c) $[T] = \begin{bmatrix} -2 & 1 & 4 \\ 3 & 5 & 7 \\ 6 & 0 & -1 \end{bmatrix}$; $\mathbf{x} = \begin{bmatrix} x_1 \\ x_2 \\ x_3 \end{bmatrix}$

(d) $[T] = \begin{bmatrix} -1 & 1 \\ 2 & 4 \\ 7 & 8 \end{bmatrix}$; $\mathbf{x} = \begin{bmatrix} x_1 \\ x_2 \end{bmatrix}$

13. In each part, use the standard matrix for T to find $T(\mathbf{x})$; then check the result by calculating $T(\mathbf{x})$ directly.

(a) $T(x_1, x_2) = (-x_1 + x_2, x_2)$; $\mathbf{x} = (-1, 4)$

(b) $T(x_1, x_2, x_3) = (2x_1 - x_2 + x_3, x_2 + x_3, 0)$; $\mathbf{x} = (2, 1, -3)$

14. Use matrix multiplication to find the reflection of $(-1, 2)$ about

(a) the x-axis.   (b) the y-axis.   (c) the line $y = x$.

15. Use matrix multiplication to find the reflection of $(2, -5, 3)$ about

(a) the xy-plane.   (b) the xz-plane.   (c) the yz-plane.

16. Use matrix multiplication to find the orthogonal projection of $(2, -5)$ on

(a) the x-axis.   (b) the y-axis.

17. Use matrix multiplication to find the orthogonal projection of $(-2, 1, 3)$ on

(a) the xy-plane.   (b) the xz-plane.   (c) the yz-plane.

18. Use matrix multiplication to find the image of the vector $(3, -4)$ when it is rotated through an angle of

(a) $\theta = 30°$.   (b) $\theta = -60°$.

(c) $\theta = 45°$.   (d) $\theta = 90°$.

19. Use matrix multiplication to find the image of the vector $(-2, 1, 2)$ if it is rotated

(a) $30°$ about the x-axis.   (b) $45°$ about the y-axis.

(c) $90°$ about the z-axis.

20. Find the standard matrix for the operator that rotates a vector in R^3 through an angle of $-60°$ about

(a) the x-axis.   (b) the y-axis.   (c) the z-axis.

21. Use matrix multiplication to find the image of the vector $(-2, 1, 2)$ if it is rotated

(a) $-30°$ about the x-axis.   (b) $-45°$ about the y-axis.

(c) $-90°$ about the z-axis.

22. In R^3 the **orthogonal projections** on the x-axis, y-axis, and z-axis are defined by

$$T_1(x, y, z) = (x, 0, 0), \quad T_2(x, y, z) = (0, y, 0),$$
$$T_3(x, y, z) = (0, 0, z)$$

respectively.

(a) Show that the orthogonal projections on the coordinate axes are matrix operators, and find their standard matrices.

(b) Show that if $T: R^3 \rightarrow R^3$ is an orthogonal projection on one of the coordinate axes, then for every vector $\mathbf{x}$ in R^3, the vectors $T(\mathbf{x})$ and $\mathbf{x} - T(\mathbf{x})$ are orthogonal.

(c) Make a sketch showing $\mathbf{x}$ and $\mathbf{x} - T(\mathbf{x})$ in the case where T is the orthogonal projection on the x-axis.

23. Use Formula (14) to derive the standard matrices for the rotations about the x-axis, y-axis, and z-axis in R^3.

24. Use Formula (15) to find the standard matrix for a rotation of $\pi/2$ radians about the axis determined by the vector $\mathbf{v} = (1, 1, 1)$. [*Note:* Formula (15) requires that the vector defining the axis of rotation have length 1.]

25. Use Formula (15) to find the standard matrix for a rotation of $180°$ about the axis determined by the vector $\mathbf{v} = (2, 2, 1)$. [*Note:* Formula (15) requires that the vector defining the axis of rotation have length 1.]

26. It can be proved that if A is a 2×2 matrix with orthonormal column vectors and for which $\det(A) = 1$, then multiplication by A is a rotation through some angle θ. Verify that

$$A = \begin{bmatrix} -\frac{1}{\sqrt{2}} & -\frac{1}{\sqrt{2}} \\ \frac{1}{\sqrt{2}} & -\frac{1}{\sqrt{2}} \end{bmatrix}$$

satisfies the stated conditions and find the angle of rotation.

27. The result stated in Exercise 26 can be extended to R^3; that is, it can be proved that if A is a 3×3 matrix with orthonormal column vectors and for which $\det(A) = 1$, then multiplication by A is a rotation about some axis through some angle θ. Use Formula (14) to show that the angle of rotation satisfies the equation

$$\cos \theta = \frac{\text{tr}(A) - 1}{2}$$

28. Let A be a 3×3 matrix (other than the identity matrix) satisfying the conditions stated in Exercise 27. It can be shown that if $\mathbf{x}$ is any nonzero vector in R^3, then the vector $\mathbf{u} = A\mathbf{x} + A^T\mathbf{x} + [1 - \text{tr}(A)]\mathbf{x}$ determines an axis of rotation when $\mathbf{u}$ is positioned with its initial point at the origin. [See "The Axis of Rotation: Analysis, Algebra, Geometry," by Dan Kalman, *Mathematics Magazine*, Vol. 62, No. 4, October 1989.]

(a) Show that multiplication by

$$A = \begin{bmatrix} \frac{1}{9} & -\frac{4}{9} & \frac{8}{9} \\ \frac{8}{9} & \frac{4}{9} & \frac{1}{9} \\ -\frac{4}{9} & \frac{7}{9} & \frac{4}{9} \end{bmatrix}$$

is a rotation.

(b) Find a vector of length 1 that defines an axis for the rotation.

(c) Use the result in Exercise 27 to find the angle of rotation about the axis obtained in part (b).

29. In words, describe the geometric effect of multiplying a vector $\mathbf{x}$ by the matrix A.

(a) $A = \begin{bmatrix} 2 & 0 \\ 0 & 0 \end{bmatrix}$ (b) $A = \begin{bmatrix} 2 & 0 \\ 0 & -2 \end{bmatrix}$

30. In words, describe the geometric effect of multiplying a vector $\mathbf{x}$ by the matrix A.

(a) $A = \begin{bmatrix} 2 & 0 \\ 0 & 3 \end{bmatrix}$ (b) $A = \begin{bmatrix} \frac{\sqrt{3}}{2} & -\frac{1}{2} \\ \frac{1}{2} & \frac{\sqrt{3}}{2} \end{bmatrix}$

31. In words, describe the geometric effect of multiplying a vector $\mathbf{x}$ by the matrix

$$A = \begin{bmatrix} \cos^2 \theta - \sin^2 \theta & -2 \sin \theta \cos \theta \\ 2 \sin \theta \cos \theta & \cos^2 \theta - \sin^2 \theta \end{bmatrix}$$

32. If multiplication by A rotates a vector $\mathbf{x}$ in the xy-plane through an angle θ, what is the effect of multiplying $\mathbf{x}$ by A^T? Explain your reasoning.

33. Let $\mathbf{x}_0$ be a nonzero column vector in R^2, and suppose that $T: R^2 \rightarrow R^2$ is the transformation defined by the formula $T(\mathbf{x}) = \mathbf{x}_0 + R_\theta \mathbf{x}$, where R_θ is the standard matrix of the rotation of R^2 about the origin through the angle θ. Give a geometric description of this transformation. Is it a matrix transformation? Explain.

34. A function of the form $f(x) = mx + b$ is commonly called a "linear function" because the graph of $y = mx + b$ is a line. Is f a matrix transformation on R?

35. Let $\mathbf{x} = \mathbf{x}_0 + t\mathbf{v}$ be a line in R^n, and let $T: R^n \rightarrow R^n$ be a matrix operator on R^n. What kind of geometric object is the image of this line under the operator T? Explain your reasoning.

True-False Exercises

In parts (a)–(i) determine whether the statement is true or false, and justify your answer.

(a) If A is a 2×3 matrix, then the domain of the transformation T_A is R^2.

(b) If A is an $m \times n$ matrix, then the codomain of the transformation T_A is R^n.

(c) If $T: R^n \rightarrow R^m$ and $T(\mathbf{0}) = \mathbf{0}$, then T is a matrix transformation.

(d) If $T: R^n \rightarrow R^m$ and $T(c_1\mathbf{x} + c_2\mathbf{y}) = c_1 T(\mathbf{x}) + c_2 T(\mathbf{y})$ for all scalars c_1 and c_2 and all vectors $\mathbf{x}$ and $\mathbf{y}$ in R^n, then T is a matrix transformation.

(e) There is only one matrix transformation $T: R^n \rightarrow R^m$ such that $T(-\mathbf{x}) = -T(\mathbf{x})$ for every vector $\mathbf{x}$ in R^n.

(f) There is only one matrix transformation $T: R^n \rightarrow R^m$ such that $T(\mathbf{x} + \mathbf{y}) = T(\mathbf{x} - \mathbf{y})$ for all vectors $\mathbf{x}$ and $\mathbf{y}$ in R^n.

(g) If $\mathbf{b}$ is a nonzero vector in R^n, then $T(\mathbf{x}) = \mathbf{x} + \mathbf{b}$ is a matrix operator on R^n.

(h) The matrix $\begin{bmatrix} \frac{1}{2} & -\frac{1}{2} \\ \frac{1}{2} & \frac{1}{2} \end{bmatrix}$ is the standard matrix for a rotation.

(i) The standard matrices of the reflections about the coordinate axes in 2-space have the form $\begin{bmatrix} a & 0 \\ 0 & -a \end{bmatrix}$, where $a = \pm 1$.

4.10 Properties of Matrix Transformations

In this section we will discuss properties of matrix transformations. We will show, for example, that if several matrix transformations are performed in succession, then the same result can be obtained by a single matrix transformation that is chosen appropriately. We will also explore the relationship between the invertibility of a matrix and properties of the corresponding transformation.

Compositions of Matrix Transformations

Suppose that T_A is a matrix transformation from R^n to R^k and T_B is a matrix transformation from R^k to R^m. If $\mathbf{x}$ is a vector in R^n, then T_A maps this vector into a vector $T_A(\mathbf{x})$ in R^k, and T_B, in turn, maps that vector into the vector $T_B(T_A(\mathbf{x}))$ in R^m. This process creates a transformation from R^n to R^m that we call the ***composition of T_B with T_A*** and denote by the symbol

$$T_B \circ T_A$$

which is read "T_B circle T_A". As illustrated in Figure 4.10.1, the transformation T_A in the formula is performed first; that is,

$$(T_B \circ T_A)(\mathbf{x}) = T_B(T_A(\mathbf{x})) \tag{1}$$

This composition is itself a matrix transformation since

$$(T_B \circ T_A)(\mathbf{x}) = T_B(T_A(\mathbf{x})) = B(T_A(\mathbf{x})) = B(A\mathbf{x}) = (BA)\mathbf{x}$$

which shows that it is multiplication by BA. This is expressed by the formula

$$T_B \circ T_A = T_{BA} \tag{2}$$

> **WARNING** Just as it is *not* true, in general, that
> $$AB = BA$$
> so it is *not* true, in general, that
> $$T_B \circ T_A = T_A \circ T_B$$
> That is, *order matters when matrix transformations are composed.*

Compositions can be defined for any finite succession of matrix transformations whose domains and ranges have the appropriate dimensions. For example, to extend Formula (2) to three factors, consider the matrix transformations

$$T_A: R^n \rightarrow R^k, \quad T_B: R^k \rightarrow R^l, \quad T_C: R^l \rightarrow R^m$$

We define the composition $(T_C \circ T_B \circ T_A): R^n \rightarrow R^m$ by

$$(T_C \circ T_B \circ T_A)(\mathbf{x}) = T_C(T_B(T_A(\mathbf{x})))$$

As above, it can be shown that this is a matrix transformation whose standard matrix is CBA and that

$$T_C \circ T_B \circ T_A = T_{CBA} \tag{3}$$

▶ **Figure 4.10.1**

As in Formula (9) of Section 4.9, we can use square brackets to denote a matrix transformation without referencing a specific matrix. Thus, for example, the formula

$$[T_2 \circ T_1] = [T_2][T_1] \tag{4}$$

is a restatement of Formula (2) which states that the standard matrix for a composition is the product of the standard matrices in the appropriate order. Similarly,

$$[T_3 \circ T_2 \circ T_1] = [T_3][T_2][T_1] \tag{5}$$

is a restatement of Formula (3).

▶ **EXAMPLE 1** Composition of Two Rotations

Let $T_1: R^2 \to R^2$ and $T_2: R^2 \to R^2$ be the matrix operators that rotate vectors through the angles θ_1 and θ_2, respectively. Thus the operation

$$(T_2 \circ T_1)(\mathbf{x}) = T_2(T_1(\mathbf{x}))$$

first rotates $\mathbf{x}$ through the angle θ_1, then rotates $T_1(\mathbf{x})$ through the angle θ_2. It follows that the net effect of $T_2 \circ T_1$ is to rotate each vector in R^2 through the angle $\theta_1 + \theta_2$ (Figure 4.10.2). Thus, the standard matrices for these matrix operators are

$$[T_1] = \begin{bmatrix} \cos\theta_1 & -\sin\theta_1 \\ \sin\theta_1 & \cos\theta_1 \end{bmatrix}, \quad [T_2] = \begin{bmatrix} \cos\theta_2 & -\sin\theta_2 \\ \sin\theta_2 & \cos\theta_2 \end{bmatrix},$$

$$[T_2 \circ T_1] = \begin{bmatrix} \cos(\theta_1 + \theta_2) & -\sin(\theta_1 + \theta_2) \\ \sin(\theta_1 + \theta_2) & \cos(\theta_1 + \theta_2) \end{bmatrix}$$

These matrices should satisfy (4). With the help of some basic trigonometric identities, we can confirm that this is so as follows:

$$[T_2][T_1] = \begin{bmatrix} \cos\theta_2 & -\sin\theta_2 \\ \sin\theta_2 & \cos\theta_2 \end{bmatrix} \begin{bmatrix} \cos\theta_1 & -\sin\theta_1 \\ \sin\theta_1 & \cos\theta_1 \end{bmatrix}$$

$$= \begin{bmatrix} \cos\theta_2 \cos\theta_1 - \sin\theta_2 \sin\theta_1 & -(\cos\theta_2 \sin\theta_1 + \sin\theta_2 \cos\theta_1) \\ \sin\theta_2 \cos\theta_1 + \cos\theta_2 \sin\theta_1 & -\sin\theta_2 \sin\theta_1 + \cos\theta_2 \cos\theta_1 \end{bmatrix}$$

$$= \begin{bmatrix} \cos(\theta_1 + \theta_2) & -\sin(\theta_1 + \theta_2) \\ \sin(\theta_1 + \theta_2) & \cos(\theta_1 + \theta_2) \end{bmatrix}$$

$$= [T_2 \circ T_1]$$

▶ **EXAMPLE 2** Composition Is Not Commutative

Let $T_1: R^2 \to R^2$ be the reflection about the line $y = x$, and let $T_2: R^2 \to R^2$ be the orthogonal projection on the y-axis. Figure 4.10.3 illustrates graphically that $T_1 \circ T_2$ and $T_2 \circ T_1$ have different effects on a vector $\mathbf{x}$. This same conclusion can be reached by showing that the standard matrices for T_1 and T_2 do not commute:

$$[T_1 \circ T_2] = [T_1][T_2] = \begin{bmatrix} 0 & 1 \\ 1 & 0 \end{bmatrix} \begin{bmatrix} 0 & 0 \\ 0 & 1 \end{bmatrix} = \begin{bmatrix} 0 & 1 \\ 0 & 0 \end{bmatrix}$$

$$[T_2 \circ T_1] = [T_2][T_1] = \begin{bmatrix} 0 & 0 \\ 0 & 1 \end{bmatrix} \begin{bmatrix} 0 & 1 \\ 1 & 0 \end{bmatrix} = \begin{bmatrix} 0 & 0 \\ 1 & 0 \end{bmatrix}$$

so $[T_2 \circ T_1] \neq [T_1 \circ T_2]$.

▲ Figure 4.10.2

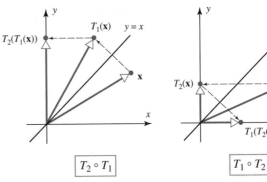

▲ Figure 4.10.3

▶ **EXAMPLE 3** **Composition of Two Reflections**

Let $T_1: R^2 \to R^2$ be the reflection about the y-axis, and let $T_2: R^2 \to R^2$ be the reflection about the x-axis. In this case $T_1 \circ T_2$ and $T_2 \circ T_1$ are the same; both map every vector $\mathbf{x} = (x, y)$ into its negative $-\mathbf{x} = (-x, -y)$ (Figure 4.10.4):

$$(T_1 \circ T_2)(x, y) = T_1(x, -y) = (-x, -y)$$
$$(T_2 \circ T_1)(x, y) = T_2(-x, y) = (-x, -y)$$

The equality of $T_1 \circ T_2$ and $T_2 \circ T_1$ can also be deduced by showing that the standard matrices for T_1 and T_2 commute:

$$[T_1 \circ T_2] = [T_1][T_2] = \begin{bmatrix} -1 & 0 \\ 0 & 1 \end{bmatrix} \begin{bmatrix} 1 & 0 \\ 0 & -1 \end{bmatrix} = \begin{bmatrix} -1 & 0 \\ 0 & -1 \end{bmatrix}$$

$$[T_2 \circ T_1] = [T_2][T_1] = \begin{bmatrix} 1 & 0 \\ 0 & -1 \end{bmatrix} \begin{bmatrix} -1 & 0 \\ 0 & 1 \end{bmatrix} = \begin{bmatrix} -1 & 0 \\ 0 & -1 \end{bmatrix}$$

The operator $T(\mathbf{x}) = -\mathbf{x}$ on R^2 or R^3 is called the ***reflection about the origin***. As the foregoing computations show, the standard matrix for this operator on R^2 is

$$[T] = \begin{bmatrix} -1 & 0 \\ 0 & -1 \end{bmatrix}$$

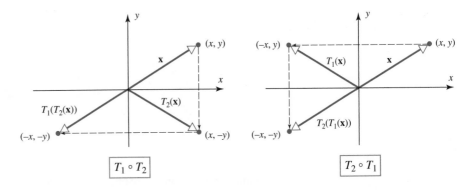

▶ Figure 4.10.4

▶ **EXAMPLE 4** **Composition of Three Transformations**

Find the standard matrix for the operator $T: R^3 \to R^3$ that first rotates a vector counterclockwise about the z-axis through an angle θ, then reflects the resulting vector about the yz-plane, and then projects that vector orthogonally onto the xy-plane.

Solution The operator T can be expressed as the composition

$$T = T_3 \circ T_2 \circ T_1$$

where T_1 is the rotation about the z-axis, T_2 is the reflection about the yz-plane, and T_3 is the orthogonal projection on the xy-plane. From Tables 6, 2, and 4 of Section 4.9, the standard matrices for these operators are

$$[T_1] = \begin{bmatrix} \cos\theta & -\sin\theta & 0 \\ \sin\theta & \cos\theta & 0 \\ 0 & 0 & 1 \end{bmatrix}, \quad [T_2] = \begin{bmatrix} -1 & 0 & 0 \\ 0 & 1 & 0 \\ 0 & 0 & 1 \end{bmatrix}, \quad [T_3] = \begin{bmatrix} 1 & 0 & 0 \\ 0 & 1 & 0 \\ 0 & 0 & 0 \end{bmatrix}$$

Thus, it follows from (5) that the standard matrix for T is

$$[T] = \begin{bmatrix} 1 & 0 & 0 \\ 0 & 1 & 0 \\ 0 & 0 & 0 \end{bmatrix} \begin{bmatrix} -1 & 0 & 0 \\ 0 & 1 & 0 \\ 0 & 0 & 1 \end{bmatrix} \begin{bmatrix} \cos\theta & -\sin\theta & 0 \\ \sin\theta & \cos\theta & 0 \\ 0 & 0 & 1 \end{bmatrix}$$

$$= \begin{bmatrix} -\cos\theta & \sin\theta & 0 \\ \sin\theta & \cos\theta & 0 \\ 0 & 0 & 0 \end{bmatrix} \blacktriangleleft$$

One-to-One Matrix Transformations

Our next objective is to establish a link between the invertibility of a matrix A and properties of the corresponding matrix transformation T_A.

DEFINITION 1 A matrix transformation $T_A : R^n \rightarrow R^m$ is said to be ***one-to-one*** if T_A maps distinct vectors (points) in R^n into distinct vectors (points) in R^m.

(See Figure 4.10.5). This idea can be expressed in various ways. For example, you should be able to see that the following are just restatements of Definition 1:

1. T_A is one-to-one if for each vector $\mathbf{b}$ in the range of A there is exactly one vector $\mathbf{x}$ in R^n such that $T_A\mathbf{x} = \mathbf{b}$.

2. T_A is one-to-one if the equality $T_A(\mathbf{u}) = T_A(\mathbf{v})$ implies that $\mathbf{u} = \mathbf{v}$.

▶ **Figure 4.10.5**   One-to-one   Not one-to-one

Rotation operators on R^2 are one-to-one since distinct vectors that are rotated through the same angle have distinct images (Figure 4.10.6). In contrast, the orthogonal projection of R^3 on the xy-plane is not one-to-one because it maps distinct points on the same vertical line into the same point (Figure 4.10.7).

The following theorem establishes a fundamental relationship between the invertibility of a matrix and properties of the corresponding matrix transformation.

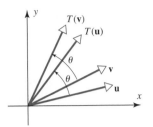

▲ **Figure 4.10.6** Distinct vectors **u** and **v** are rotated into distinct vectors $T(\mathbf{u})$ and $T(\mathbf{v})$.

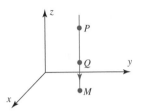

▲ **Figure 4.10.7** The distinct points P and Q are mapped into the same point M.

THEOREM 4.10.1 *If A is an $n \times n$ matrix and $T_A: R^n \to R^n$ is the corresponding matrix operator, then the following statements are equivalent.*

(*a*)   *A is invertible.*

(*b*)   *The range of T_A is R^n.*

(*c*)   *T_A is one-to-one.*

Proof   We will establish the chain of implications $(a) \Rightarrow (b) \Rightarrow (c) \Rightarrow (a)$.

(*a*) $\Rightarrow$ (*b*)   Assume that A is invertible. By parts (a) and (e) of Theorem 4.8.10, the system $A\mathbf{x} = \mathbf{b}$ is consistent for every $n \times 1$ matrix $\mathbf{b}$ in R^n. This implies that T_A maps $\mathbf{x}$ into the arbitrary vector $\mathbf{b}$ in R^n, which in turn implies that the range of T_A is all of R^n.

(*b*) $\Rightarrow$ (*c*)   Assume that the range of T_A is R^n. This implies that for every vector $\mathbf{b}$ in R^n there is some vector $\mathbf{x}$ in R^n for which $T_A(\mathbf{x}) = \mathbf{b}$ and hence that the linear system $A\mathbf{x} = \mathbf{b}$ is consistent for every vector $\mathbf{b}$ in R^n. But the equivalence of parts (e) and (f) of Theorem 4.8.10 implies that $A\mathbf{x} = \mathbf{b}$ has a unique solution for every vector $\mathbf{b}$ in R^n and hence that for every vector $\mathbf{b}$ in the range of T_A there is exactly one vector $\mathbf{x}$ in R^n such that $T_A \mathbf{x} = \mathbf{b}$.

(*c*) $\Rightarrow$ (*a*)   Assume that T_A is one-to-one. Thus, if $\mathbf{b}$ is a vector in the range of T_A, there is a unique vector $\mathbf{x}$ in R^n for which $T_A(\mathbf{x}) = \mathbf{b}$. We leave it for you to complete the proof using Exercise 30.   ◄

▶ **EXAMPLE 5 Properties of a Rotation Operator**

As indicated in Figure 4.10.6, the operator $T: R^n \to R^n$ that rotates vectors in R^2 through an angle θ is one-to-one. Confirm that $[T]$ is invertible in accordance with Theorem 4.10.1.

Solution   From Table 5 of Section 4.9 the standard matrix for T is

$$[T] = \begin{bmatrix} \cos\theta & -\sin\theta \\ \sin\theta & \cos\theta \end{bmatrix}$$

This matrix is invertible because

$$\det[T] = \begin{vmatrix} \cos\theta & -\sin\theta \\ \sin\theta & \cos\theta \end{vmatrix} = \cos^2\theta + \sin^2\theta = 1 \neq 0$$

▶ **EXAMPLE 6 Properties of a Projection Operator**

As indicated in Figure 4.10.7, the operator $T: R^n \to R^n$ that projects each vector in R^3 orthogonally on the xy-plane is not one-to-one. Confirm that $[T]$ is not invertible in accordance with Theorem 4.10.1.

Solution   From Table 4 of Section 4.9 the standard matrix for T is

$$[T] = \begin{bmatrix} 1 & 0 & 0 \\ 0 & 1 & 0 \\ 0 & 0 & 0 \end{bmatrix}$$

This matrix is not invertible since $\det[T] = 0$.   ◄

Inverse of a One-to-One Matrix Operator

If $T_A\colon R^n \to R^n$ is a one-to-one matrix operator, then it follows from Theorem 4.10.1 that A is invertible. The matrix operator

$$T_{A^{-1}}\colon R^n \to R^n$$

that corresponds to A^{-1} is called the ***inverse operator*** or (more simply) the ***inverse*** of T_A. This terminology is appropriate because T_A and $T_{A^{-1}}$ cancel the effect of each other in the sense that if $\mathbf{x}$ is any vector in R^n, then

$$T_A(T_{A^{-1}}(\mathbf{x})) = AA^{-1}\mathbf{x} = I\mathbf{x} = \mathbf{x}$$
$$T_{A^{-1}}(T_A(\mathbf{x})) = A^{-1}A\mathbf{x} = I\mathbf{x} = \mathbf{x}$$

or, equivalently,

$$T_A \circ T_{A^{-1}} = T_{AA^{-1}} = T_I$$
$$T_{A^{-1}} \circ T_A = T_{A^{-1}A} = T_I$$

From a more geometric viewpoint, if $\mathbf{w}$ is the image of $\mathbf{x}$ under T_A, then $T_{A^{-1}}$ maps $\mathbf{w}$ back into $\mathbf{x}$, since

$$T_{A^{-1}}(\mathbf{w}) = T_{A^{-1}}(T_A(\mathbf{x})) = \mathbf{x}$$

(Figure 4.10.8).

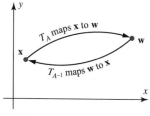

▲ Figure 4.10.8

Before considering examples, it will be helpful to touch on some notational matters. If $T_A\colon R^n \to R^n$ is a one-to-one matrix operator, and if $T_{A^{-1}}\colon R^n \to R^n$ is its inverse, then the standard matrices for these operators are related by the equation

$$T_{A^{-1}} = T_A^{-1} \tag{6}$$

In cases where it is preferable not to assign a name to the matrix, we will write this equation as

$$[T^{-1}] = [T]^{-1} \tag{7}$$

▶ **EXAMPLE 7 Standard Matrix for T^{-1}**

Let $T\colon R^2 \to R^2$ be the operator that rotates each vector in R^2 through the angle θ, so from Table 5 of Section 4.9,

$$[T] = \begin{bmatrix} \cos\theta & -\sin\theta \\ \sin\theta & \cos\theta \end{bmatrix} \tag{8}$$

It is evident geometrically that to undo the effect of T, one must rotate each vector in R^2 through the angle $-\theta$. But this is exactly what the operator T^{-1} does, since the standard matrix for T^{-1} is

$$[T^{-1}] = [T]^{-1} = \begin{bmatrix} \cos\theta & \sin\theta \\ -\sin\theta & \cos\theta \end{bmatrix} = \begin{bmatrix} \cos(-\theta) & -\sin(-\theta) \\ \sin(-\theta) & \cos(-\theta) \end{bmatrix}$$

(verify), which is the standard matrix for a rotation through the angle $-\theta$.

▶ **EXAMPLE 8 Finding T^{-1}**

Show that the operator $T\colon R^2 \to R^2$ defined by the equations

$$w_1 = 2x_1 + x_2$$
$$w_2 = 3x_1 + 4x_2$$

is one-to-one, and find $T^{-1}(w_1, w_2)$.

Solution The matrix form of these equations is

$$\begin{bmatrix} w_1 \\ w_2 \end{bmatrix} = \begin{bmatrix} 2 & 1 \\ 3 & 4 \end{bmatrix} \begin{bmatrix} x_1 \\ x_2 \end{bmatrix}$$

so the standard matrix for T is

$$[T] = \begin{bmatrix} 2 & 1 \\ 3 & 4 \end{bmatrix}$$

This matrix is invertible (so T is one-to-one) and the standard matrix for T^{-1} is

$$[T^{-1}] = [T]^{-1} = \begin{bmatrix} \frac{4}{5} & -\frac{1}{5} \\ -\frac{3}{5} & \frac{2}{5} \end{bmatrix}$$

Thus

$$[T^{-1}]\begin{bmatrix} w_1 \\ w_2 \end{bmatrix} = \begin{bmatrix} \frac{4}{5} & -\frac{1}{5} \\ -\frac{3}{5} & \frac{2}{5} \end{bmatrix} \begin{bmatrix} w_1 \\ w_2 \end{bmatrix} = \begin{bmatrix} \frac{4}{5}w_1 - \frac{1}{5}w_2 \\ -\frac{3}{5}w_1 + \frac{2}{5}w_2 \end{bmatrix}$$

from which we conclude that

$$T^{-1}(w_1, w_2) = \left(\tfrac{4}{5}w_1 - \tfrac{1}{5}w_2, -\tfrac{3}{5}w_1 + \tfrac{2}{5}w_2 \right) \quad \blacktriangleleft$$

Linearity Properties   Up to now we have focused exclusively on matrix transformations from R^n to R^m. However, these are not the only kinds of transformations from R^n to R^m. For example, if $f_1, f_2, \ldots, f_m$ are any functions of the n variables $x_1, x_2, \ldots, x_n$, then the equations

$$w_1 = f_1(x_1, x_2, \ldots, x_n)$$
$$w_2 = f_2(x_1, x_2, \ldots, x_n)$$
$$\vdots$$
$$w_m = f_m(x_1, x_2, \ldots, x_n)$$

define a transformation $T: R^n \to R^m$ that maps the vector $\mathbf{x} = (x_1, x_2, \ldots, x_n)$ into the vector $(w_1, w_2, \ldots, w_m)$. But it is only in the case where these equations are *linear* that T is a matrix transformation. The question that we will now consider is this:

> **Question** Are there algebraic properties of a transformation $T: R^n \to R^m$ that can be used to determine whether T is a matrix transformation?

The answer is provided by the following theorem.

> **THEOREM 4.10.2** $T: R^n \to R^m$ *is a matrix transformation if and only if the following relationships hold for all vectors* $\mathbf{u}$ *and* $\mathbf{v}$ *in* R^n *and for every scalar* k:
>
> (i) $T(\mathbf{u} + \mathbf{v}) = T(\mathbf{u}) + T(\mathbf{v})$ **[Additivity property]**
>
> (ii) $T(k\mathbf{u}) = kT(\mathbf{u})$ **[Homogeneity property]**

Proof   If T is a matrix transformation, then properties (i) and (ii) follow respectively from parts (c) and (b) of Theorem 4.9.1.

Conversely, assume that properties (i) and (ii) hold. We must show that there exists an $m \times n$ matrix A such that

$$T(\mathbf{x}) = A\mathbf{x}$$

for every vector $\mathbf{x}$ in R^n. As a first step, recall from Formula (10) of Section 4.9 that the additivity and homogeneity properties imply that

$$T(k_1\mathbf{u}_1 + k_2\mathbf{u}_2 + \cdots + k_r\mathbf{u}_r) = k_1T(\mathbf{u}_1) + k_2T(\mathbf{u}_2) + \cdots + k_rT(\mathbf{u}_r) \quad (9)$$

for all scalars $k_1, k_2, \ldots, k_r$ and all vectors $\mathbf{u}_1, \mathbf{u}_2, \ldots, \mathbf{u}_r$ in R^n. Let A be the matrix

$$A = [T(\mathbf{e}_1) \mid T(\mathbf{e}_2) \mid \cdots \mid T(\mathbf{e}_n)]$$

in which $\mathbf{e}_1, \mathbf{e}_2, \ldots, \mathbf{e}_n$ are the standard basis vectors for R^n. It follows from Theorem 1.3.1 that $A\mathbf{x}$ is a linear combination of the columns of A in which the successive coefficients are the entries $x_1, x_2, \ldots, x_n$ of $\mathbf{x}$. That is,

$$A\mathbf{x} = x_1 T(\mathbf{e}_1) + x_2 T(\mathbf{e}_2) + \cdots + x_n T(\mathbf{e}_n)$$

Using (9) we can rewrite this as

$$A\mathbf{x} = T(x_1\mathbf{e}_1 + x_2\mathbf{e}_2 + \cdots + x_n\mathbf{e}_n) = T(\mathbf{x})$$

which completes the proof. ◀

The additivity and homogeneity properties in Theorem 4.10.2 are called *linearity conditions*, and a transformation that satisfies these conditions is called a *linear transformation*. Using this terminology Theorem 4.10.2 can be restated as follows.

> **THEOREM 4.10.3** *Every linear transformation from R^n to R^m is a matrix transformation, and conversely, every matrix transformation from R^n to R^m is a linear transformation.*

More on the Equivalence Theorem

As our final result in this section, we will add parts (*b*) and (*c*) of Theorem 4.10.1 to Theorem 4.8.10.

> **THEOREM 4.10.4 Equivalent Statements**
> *If A is an $n \times n$ matrix, then the following statements are equivalent.*
> (*a*) A *is invertible.*
> (*b*) $A\mathbf{x} = \mathbf{0}$ *has only the trivial solution.*
> (*c*) *The reduced row echelon form of A is I_n.*
> (*d*) A *is expressible as a product of elementary matrices.*
> (*e*) $A\mathbf{x} = \mathbf{b}$ *is consistent for every $n \times 1$ matrix $\mathbf{b}$.*
> (*f*) $A\mathbf{x} = \mathbf{b}$ *has exactly one solution for every $n \times 1$ matrix $\mathbf{b}$.*
> (*g*) $\det(A) \neq 0$.
> (*h*) *The column vectors of A are linearly independent.*
> (*i*) *The row vectors of A are linearly independent.*
> (*j*) *The column vectors of A span R^n.*
> (*k*) *The row vectors of A span R^n.*
> (*l*) *The column vectors of A form a basis for R^n.*
> (*m*) *The row vectors of A form a basis for R^n.*
> (*n*) A *has rank n.*
> (*o*) A *has nullity 0.*
> (*p*) *The orthogonal complement of the null space of A is R^n.*
> (*q*) *The orthogonal complement of the row space of A is $\{\mathbf{0}\}$.*
> (*r*) *The range of T_A is R^n.*
> (*s*) T_A *is one-to-one.*

Concept Review

- Composition of matrix transformations
- Reflection about the origin
- One-to-one transformation
- Inverse of a matrix operator
- Linearity conditions
- Linear transformation
- Equivalent characterizations of invertible matrices

Skills

- Find the standard matrix for a composition of matrix transformations.
- Determine whether a matrix operator is one-to-one; if it is, then find the inverse operator.
- Determine whether a transformation is a linear transformation.

Exercise Set 4.10

▷ In Exercises 1–2, let T_A and T_B be the operators whose standard matrices are given. Find the standard matrices for $T_B \circ T_A$ and $T_A \circ T_B$. ◁

1. $A = \begin{bmatrix} 1 & -2 & 0 \\ 4 & 1 & -3 \\ 5 & 2 & 4 \end{bmatrix}$, $B = \begin{bmatrix} 2 & -3 & 3 \\ 5 & 0 & 1 \\ 6 & 1 & 7 \end{bmatrix}$

2. $A = \begin{bmatrix} 6 & 3 & -1 \\ 2 & 0 & 1 \\ 4 & -3 & 6 \end{bmatrix}$, $B = \begin{bmatrix} 4 & 0 & 4 \\ -1 & 5 & 2 \\ 2 & -3 & 8 \end{bmatrix}$

3. Let $T_1(x_1, x_2) = (x_1 + x_2, x_1 - x_2)$ and $T_2(x_1, x_2) = (3x_1, 2x_1 + 4x_2)$.

    (a) Find the standard matrices for T_1 and T_2.

    (b) Find the standard matrices for $T_2 \circ T_1$ and $T_1 \circ T_2$.

    (c) Use the matrices obtained in part (b) to find formulas for $T_1(T_2(x_1, x_2))$ and $T_2(T_1(x_1, x_2))$.

4. Let $T_1(x_1, x_2, x_3) = (4x_1, -2x_1 + x_2, -x_1 - 3x_2)$ and $T_2(x_1, x_2, x_3) = (x_1 + 2x_2, -x_3, 4x_1 - x_3)$.

    (a) Find the standard matrices for T_1 and T_2.

    (b) Find the standard matrices for $T_2 \circ T_1$ and $T_1 \circ T_2$.

    (c) Use the matrices obtained in part (b) to find formulas for $T_1(T_2(x_1, x_2, x_3))$ and $T_2(T_1(x_1, x_2, x_3))$.

5. Find the standard matrix for the stated composition in R^2.

    (a) A rotation of $90°$, followed by a reflection about the line $y = x$.

    (b) An orthogonal projection on the y-axis, followed by a contraction with factor $k = \frac{1}{2}$.

    (c) A reflection about the x-axis, followed by a dilation with factor $k = 3$.

6. Find the standard matrix for the stated composition in R^2.

    (a) A rotation of $60°$, followed by an orthogonal projection on the x-axis, followed by a reflection about the line $y = x$.

    (b) A dilation with factor $k = 2$, followed by a rotation of $45°$, followed by a reflection about the y-axis.

    (c) A rotation of $15°$, followed by a rotation of $105°$, followed by a rotation of $60°$.

7. Find the standard matrix for the stated composition in R^3.

    (a) A reflection about the yz-plane, followed by an orthogonal projection on the xz-plane.

    (b) A rotation of $45°$ about the y-axis, followed by a dilation with factor $k = \sqrt{2}$.

    (c) An orthogonal projection on the xy-plane, followed by a reflection about the yz-plane.

8. Find the standard matrix for the stated composition in R^3.

    (a) A rotation of $30°$ about the x-axis, followed by a rotation of $30°$ about the z-axis, followed by a contraction with factor $k = \frac{1}{4}$.

    (b) A reflection about the xy-plane, followed by a reflection about the xz-plane, followed by an orthogonal projection on the yz-plane.

    (c) A rotation of $270°$ about the x-axis, followed by a rotation of $90°$ about the y-axis, followed by a rotation of $180°$ about the z-axis.

9. Determine whether $T_1 \circ T_2 = T_2 \circ T_1$.

    (a) $T_1 : R^2 \rightarrow R^2$ is the orthogonal projection on the x-axis, and $T_2 : R^2 \rightarrow R^2$ is the orthogonal projection on the y-axis.

    (b) $T_1 : R^2 \rightarrow R^2$ is the rotation through an angle θ_1, and $T_2 : R^2 \rightarrow R^2$ is the rotation through an angle θ_2.

    (c) $T_1 : R^2 \rightarrow R^2$ is the orthogonal projection on the x-axis, and $T_2 : R^2 \rightarrow R^2$ is the rotation through an angle θ.

10. Determine whether $T_1 \circ T_2 = T_2 \circ T_1$.

    (a) $T_1 : R^3 \rightarrow R^3$ is a dilation by a factor k, and $T_2 : R^3 \rightarrow R^3$ is the rotation about the z-axis through an angle θ.

    (b) $T_1 : R^3 \rightarrow R^3$ is the rotation about the x-axis through an angle θ_1, and $T_2 : R^3 \rightarrow R^3$ is the rotation about the z-axis through an angle θ_2.

11. By inspection, determine whether the matrix operator is one-to-one.

   (a) the orthogonal projection on the x-axis in R^2

   (b) the reflection about the y-axis in R^2

   (c) the reflection about the line $y = x$ in R^2

   (d) a contraction with factor $k > 0$ in R^2

   (e) a rotation about the z-axis in R^3

   (f) a reflection about the xy-plane in R^3

   (g) a dilation with factor $k > 0$ in R^3

12. Find the standard matrix for the matrix operator defined by the equations, and use Theorem 4.10.4 to determine whether the operator is one-to-one.

   (a) $w_1 = 8x_1 + 4x_2$
       $w_2 = 2x_1 + \ x_2$

   (b) $w_1 = 2x_1 - 3x_2$
       $w_2 = 5x_1 + \ x_2$

   (c) $w_1 = -x_1 + 3x_2 + 2x_3$
       $w_2 = \ 2x_1 \quad\ + 4x_3$
       $w_3 = \ x_1 + 3x_2 + 6x_3$

   (d) $w_1 = \ x_1 + 2x_2 + 3x_3$
       $w_2 = 2x_1 + 5x_2 + 3x_3$
       $w_3 = \ x_1 \quad\quad\ + 8x_3$

13. Determine whether the matrix operator $T: R^2 \to R^2$ defined by the equations is one-to-one; if so, find the standard matrix for the inverse operator, and find $T^{-1}(w_1, w_2)$.

   (a) $w_1 = \ x_1 + 2x_2$
       $w_2 = -x_1 + \ x_2$

   (b) $w_1 = \ 4x_1 - 6x_2$
       $w_2 = -2x_1 + 3x_2$

   (c) $w_1 = -x_2$
       $w_2 = -x_1$

   (d) $w_1 = \ 3x_1$
       $w_2 = -5x_1$

14. Determine whether the matrix operator $T: R^3 \to R^3$ defined by the equations is one-to-one; if so, find the standard matrix for the inverse operator, and find $T^{-1}(w_1, w_2, w_3)$.

   (a) $w_1 = \ x_1 - 2x_2 + 2x_3$
       $w_2 = 2x_1 + \ x_2 + \ x_3$
       $w_3 = \ x_1 + \ x_2$

   (b) $w_1 = \ x_1 - 3x_2 + 4x_3$
       $w_2 = -x_1 + \ x_2 + \ x_3$
       $w_3 = \quad\quad\ - 2x_2 + 5x_3$

   (c) $w_1 = \ x_1 + 4x_2 - x_3$
       $w_2 = 2x_1 + 7x_2 + x_3$
       $w_3 = \ x_1 + 3x_2$

   (d) $w_1 = \quad\ x_1 + 2x_2 + \ x_3$
       $w_2 = -2x_1 + \ x_2 + 4x_3$
       $w_3 = \quad 7x_1 + 4x_2 - 5x_3$

15. By inspection, find the inverse of the given one-to-one matrix operator.

   (a) The reflection about the x-axis in R^2.

   (b) The rotation through an angle of $\pi/4$ in R^2.

   (c) The dilation by a factor of 3 in R^2.

   (d) The reflection about the yz-plane in R^3.

   (e) The contraction by a factor of $\frac{1}{5}$ in R^3.

▶ In Exercises 16–17, use Theorem 4.10.2 to determine whether $T: R^2 \to R^2$ is a matrix operator. ◀

16. (a) $T(x, y) = (2x, y)$
    (b) $T(x, y) = (x^2, y)$
    (c) $T(x, y) = (-y, x)$
    (d) $T(x, y) = (x, 0)$

17. (a) $T(x, y) = (2x + y, x - y)$
    (b) $T(x, y) = (x + 1, y)$ (c) $T(x, y) = (y, y)$
    (d) $T(x, y) = (\sqrt[3]{x}, \sqrt[3]{y})$

▶ In Exercises 18–19, use Theorem 4.10.2 to determine whether $T: R^3 \to R^2$ is a matrix transformation. ◀

18. (a) $T(x, y, z) = (x, x + y + z)$
    (b) $T(x, y, z) = (1, 1)$

19. (a) $T(x, y, z) = (0, 0)$
    (b) $T(x, y, z) = (3x - 4y, 2x - 5z)$

20. In each part, use Theorem 4.10.3 to find the standard matrix for the matrix operator from the images of the standard basis vectors.

   (a) The reflection operators on R^2 in Table 1 of Section 4.9.

   (b) The reflection operators on R^3 in Table 2 of Section 4.9.

   (c) The projection operators on R^2 in Table 3 of Section 4.9.

   (d) The projection operators on R^3 in Table 4 of Section 4.9.

   (e) The rotation operators on R^2 in Table 5 of Section 4.9.

   (f) The dilation and contraction operators on R^3 in Table 8 of Section 4.9.

21. Find the standard matrix for the given matrix operator.

   (a) $T: R^2 \to R^2$ projects a vector orthogonally onto the x-axis and then reflects that vector about the y-axis.

   (b) $T: R^2 \to R^2$ reflects a vector about the line $y = x$ and then reflects that vector about the x-axis.

   (c) $T: R^2 \to R^2$ dilates a vector by a factor of 3, then reflects that vector about the line $y = x$, and then projects that vector orthogonally onto the y-axis.

22. Find the standard matrix for the given matrix operator.

   (a) $T: R^3 \to R^3$ reflects a vector about the xz-plane and then contracts that vector by a factor of $\frac{1}{5}$.

   (b) $T: R^3 \to R^3$ projects a vector orthogonally onto the xz-plane and then projects that vector orthogonally onto the xy-plane.

   (c) $T: R^3 \to R^3$ reflects a vector about the xy-plane, then reflects that vector about the xz-plane, and then reflects that vector about the yz-plane.

23. Let $T_A: R^3 \to R^3$ be multiplication by

$$A = \begin{bmatrix} -1 & 3 & 0 \\ 2 & 1 & 2 \\ 4 & 5 & -3 \end{bmatrix}$$

and let e_1, e_2, and e_3 be the standard basis vectors for R^3. Find the following vectors by inspection.

   (a) $T_A(e_1)$, $T_A(e_2)$, and $T_A(e_3)$

   (b) $T_A(e_1 + e_2 + e_3)$ (c) $T_A(7e_3)$

24. Determine whether multiplication by A is a one-to-one matrix transformation.

(a) $A = \begin{bmatrix} 1 & -1 \\ 2 & 0 \\ 3 & -4 \end{bmatrix}$ 　　(b) $A = \begin{bmatrix} 1 & 2 & 3 \\ -1 & 0 & -4 \end{bmatrix}$

(c) $A = \begin{bmatrix} 1 & 2 & 1 \\ 0 & 1 & 1 \\ 1 & 1 & 0 \\ 1 & 0 & -1 \end{bmatrix}$

25. (a) Is a composition of one-to-one matrix transformations one-to-one? Justify your conclusion.

(b) Can the composition of a one-to-one matrix transformation and a matrix transformation that is not one-to-one be one-to-one? Account for both possible orders of composition and justify your conclusion.

26. Show that $T(x, y) = (0, 0)$ defines a matrix operator on R^2 but $T(x, y) = (1, 1)$ does not.

27. (a) Prove: If $T: R^n \to R^m$ is a matrix transformation, then $T(\mathbf{0}) = \mathbf{0}$; that is, T maps the zero vector in R^n into the zero vector in R^m.

(b) The converse of this is not true. Find an example of a function that satisfies $T(\mathbf{0}) = \mathbf{0}$ but is not a matrix transformation.

28. Prove: An $n \times n$ matrix A is invertible if and only if the linear system $A\mathbf{x} = \mathbf{w}$ has exactly one solution for every vector $\mathbf{w}$ in R^n for which the system is consistent.

29. Let A be an $n \times n$ matrix such that $\det(A) = 0$, and let $T: R^n \to R^n$ be multiplication by A.

(a) What can you say about the codomain of the matrix operator T? Give an example that illustrates your conclusion.

(b) What can you say about the number of vectors that T maps into $\mathbf{0}$?

30. Prove: If the matrix transformation $T_A: R^n \to R^n$ is one-to-one, then A is invertible.

True-False Exercises

In parts (a)–(f) determine whether the statement is true or false, and justify your answer.

(a) If $T: R^n \to R^m$ and $T(\mathbf{0}) = \mathbf{0}$, then T is a matrix transformation.

(b) If $T: R^n \to R^m$ and $T(c_1\mathbf{x} + c_2\mathbf{y}) = c_1 T(\mathbf{x}) + c_2 T(\mathbf{y})$ for all scalars c_1 and c_2 and all vectors $\mathbf{x}$ and $\mathbf{y}$ in R^n, then T is a matrix transformation.

(c) If $T: R^n \to R^m$ is a one-to-one matrix transformation, then there are no distinct vectors $\mathbf{x}$ and $\mathbf{y}$ for which $T(\mathbf{x} - \mathbf{y}) = \mathbf{0}$.

(d) If $T: R^n \to R^m$ is a matrix transformation and $m > n$, then T is one-to-one.

(e) If $T: R^n \to R^m$ is a matrix transformation and $m = n$, then T is one-to-one.

(f) If $T: R^n \to R^m$ is a matrix transformation and $m < n$, then T is one-to-one.

4.11　Geometry of Matrix Operators on R^2

In this optional section we will discuss matrix operators on R^2 in a little more depth. The ideas that we will develop here have important applications to computer graphics.

Transformations of Regions　In Section 4.9 we focused on the effect that a matrix operator has on individual vectors in R^2 and R^3. However, it is also important to understand how such operators affect the shapes of regions. For example, Figure 4.11.1 shows a famous picture of Albert Einstein and three computer-generated modifications of that image that result from matrix operators on R^2. The original picture was scanned and then digitized to decompose it into a rectangular array of pixels. The pixels were then transformed as follows:

- The program MATLAB was used to assign coordinates and a gray level to each pixel.
- The coordinates of the pixels were transformed by matrix multiplication.
- The pixels were then assigned their original gray levels to produce the transformed picture.

The overall effect of a matrix operator on R^2 can often be ascertained by graphing the images of the vertices $(0, 0)$, $(1, 0)$, $(0, 1)$, and $(1, 1)$ of the unit square (Figure 4.11.2).

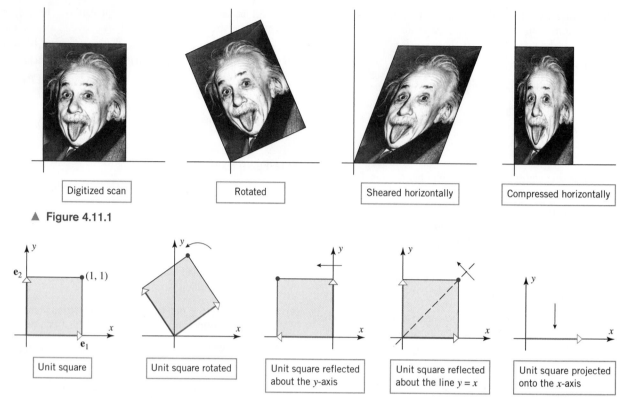

▲ Figure 4.11.1

▲ Figure 4.11.2

Table 1 shows the effect that some of the matrix operators studied in Section 4.9 have on the unit square. For clarity, we have shaded a portion of the original square and its corresponding image.

▶ **EXAMPLE 1** Transforming with Diagonal Matrices

Suppose that the xy-plane first is compressed or expanded by a factor of k_1 in the x-direction and then is compressed or expanded by a factor of k_2 in the y-direction. Find a single matrix operator that performs both operations.

Solution The standard matrices for the two operations are

$$\begin{bmatrix} k_1 & 0 \\ 0 & 1 \end{bmatrix} \qquad\qquad \begin{bmatrix} 1 & 0 \\ 0 & k_2 \end{bmatrix}$$

x-compression (expansion) y-compression (expansion)

Thus, the standard matrix for the composition of the x-operation followed by the y-operation is

$$A = \begin{bmatrix} 1 & 0 \\ 0 & k_2 \end{bmatrix}\begin{bmatrix} k_1 & 0 \\ 0 & 1 \end{bmatrix} = \begin{bmatrix} k_1 & 0 \\ 0 & k_2 \end{bmatrix} \tag{1}$$

This shows that multiplication by a diagonal 2×2 matrix compresses or expands the plane in the x-direction and also in the y-direction. In the special case where k_1 and k_2 are the same, say $k_1 = k_2 = k$, Formula (1) simplifies to

$$A = \begin{bmatrix} k & 0 \\ 0 & k \end{bmatrix}$$

which is a contraction or a dilation (Table 7 of Section 4.9). ◀

Table 1

Operator	Standard Matrix	Effect on the Unit Square
Reflection about the y-axis	$\begin{bmatrix} -1 & 0 \\ 0 & 1 \end{bmatrix}$	
Reflection about the x-axis	$\begin{bmatrix} 1 & 0 \\ 0 & -1 \end{bmatrix}$	
Reflection about the line $y = x$	$\begin{bmatrix} 0 & 1 \\ 1 & 0 \end{bmatrix}$	
Counterclockwise rotation through an angle θ	$\begin{bmatrix} \cos\theta & -\sin\theta \\ \sin\theta & \cos\theta \end{bmatrix}$	
Compression in the x-direction by a factor of k $(0 < k < 1)$	$\begin{bmatrix} k & 0 \\ 0 & 0 \end{bmatrix}$	
Expansion in the x-direction by a factor of k $(k > 1)$	$\begin{bmatrix} k & 0 \\ 0 & 0 \end{bmatrix}$	
Shear in the x-direction with factor $k > 0$	$\begin{bmatrix} 1 & k \\ 0 & 1 \end{bmatrix}$	
Shear in the x-direction with factor $k < 0$	$\begin{bmatrix} 1 & 0 \\ k & 1 \end{bmatrix}$	

▶ **EXAMPLE 2 Finding Matrix Operators**

(a) Find the standard matrix for the operator on R^2 that first shears by a factor of 2 in the x-direction and then reflects the result about the line $y = x$. Sketch the image of the unit square under this operator.

(b) Find the standard matrix for the operator on R^2 that first reflects about $y = x$ and then shears by a factor of 2 in the x-direction. Sketch the image of the unit square under this operator.

(c) Confirm that the shear and the reflection in parts (a) and (b) do not commute.

Solution (a) The standard matrix for the shear is

$$A_1 = \begin{bmatrix} 1 & 2 \\ 0 & 1 \end{bmatrix}$$

and for the reflection is

$$A_2 = \begin{bmatrix} 0 & 1 \\ 1 & 0 \end{bmatrix}$$

Thus, the standard matrix for the shear followed by the reflection is

$$A_2 A_1 = \begin{bmatrix} 0 & 1 \\ 1 & 0 \end{bmatrix}\begin{bmatrix} 1 & 2 \\ 0 & 1 \end{bmatrix} = \begin{bmatrix} 0 & 1 \\ 1 & 2 \end{bmatrix}$$

Solution (b) The standard matrix for the reflection followed by the shear is

$$A_1 A_2 = \begin{bmatrix} 1 & 2 \\ 0 & 1 \end{bmatrix}\begin{bmatrix} 0 & 1 \\ 1 & 0 \end{bmatrix} = \begin{bmatrix} 2 & 1 \\ 1 & 0 \end{bmatrix}$$

Solution (c) The computations in Solutions (a) and (b) show that $A_1 A_2 \neq A_2 A_1$, so the standard matrices, and hence the operators, do not commute. The same conclusion follows from Figures 4.11.3 and 4.11.4, since the two operators produce different images of the unit square. ◀

▶ **Figure 4.11.3**

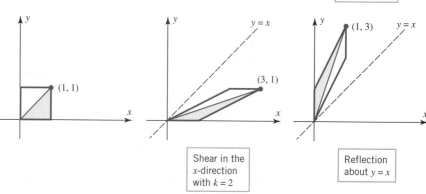

▶ **Figure 4.11.4**

Geometry of One-to-One Matrix Operators

We will now turn our attention to one-to-one matrix operators on R^2, which are important because they map distinct points into distinct points. Recall from Theorem 4.10.4 (the Equivalence Theorem) that a matrix transformation T_A is one-to-one if and only if A can be expressed as a product of elementary matrices. Thus, we can analyze the effect of any one-to-one transformation T_A by first factoring the matrix A into a product of elementary matrices, say

$$A = E_1 E_2 \cdots E_r$$

and then expressing T_A as the composition

$$T_A = T_{E_1 E_2 \cdots E_r} = T_{E_1} \circ T_{E_2} \circ \cdots \circ T_{E_r} \tag{2}$$

The following theorem explains the geometric effect of matrix operators corresponding to elementary matrices.

THEOREM 4.11.1 *If E is an elementary matrtix, then $T_E : R^2 \to R^2$ is one of the following:*

(a)   *A shear along a coordinate axis.*

(b)   *A reflection about $y = x$.*

(c)   *A compression along a coordinate axis.*

(d)   *An expansion along a coordinate axis.*

(e)   *A reflection about a coordinate axis.*

(f)   *A compression or expansion along a coordinate axis followed by a reflection about a coordinate axis.*

Proof   Because a 2×2 elementary matrix results from performing a single elementary row operation on the 2×2 identity matrix, such a matrix must have one of the following forms (verify):

$$\begin{bmatrix} 1 & 0 \\ k & 1 \end{bmatrix}, \quad \begin{bmatrix} 1 & k \\ 0 & 1 \end{bmatrix}, \quad \begin{bmatrix} 0 & 1 \\ 1 & 0 \end{bmatrix}, \quad \begin{bmatrix} k & 0 \\ 0 & 1 \end{bmatrix}, \quad \begin{bmatrix} 1 & 0 \\ 0 & k \end{bmatrix}$$

The first two matrices represent shears along coordinate axes, and the third represents a reflection about $y = x$. If $k > 0$, the last two matrices represent compressions or expansions along coordinate axes, depending on whether $0 \le k < 1$ or $k > 1$. If $k < 0$, and if we express k in the form $k = -k_1$, where $k_1 > 0$, then the last two matrices can be written as

$$\begin{bmatrix} k & 0 \\ 0 & 1 \end{bmatrix} = \begin{bmatrix} -k_1 & 0 \\ 0 & 1 \end{bmatrix} = \begin{bmatrix} -1 & 0 \\ 0 & 1 \end{bmatrix} \begin{bmatrix} k_1 & 0 \\ 0 & 1 \end{bmatrix} \tag{3}$$

$$\begin{bmatrix} 1 & 0 \\ 0 & k \end{bmatrix} = \begin{bmatrix} 1 & 0 \\ 0 & -k_1 \end{bmatrix} = \begin{bmatrix} 1 & 0 \\ 0 & -1 \end{bmatrix} \begin{bmatrix} 1 & 0 \\ 0 & k_1 \end{bmatrix} \tag{4}$$

Since $k_1 > 0$, the product in (3) represents a compression or expansion along the x-axis followed by a reflection about the y-axis, and (4) represents a compression or expansion along the y-axis followed by a reflection about the x-axis. In the case where $k = -1$, transformations (3) and (4) are simply reflections about the y-axis and x-axis, respectively. ◄

Since every invertible matrix is a product of elementary matrices, the following result follows from Theorem 4.11.1 and Formula (2).

> **THEOREM 4.11.2** *If $T_A: R^2 \to R^2$ is multiplication by an invertible matrix A, then the geometric effect of T_A is the same as an appropriate succession of shears, compressions, expansions, and reflections.*

▶ **EXAMPLE 3** Analyzing the Geometric Effect of a Matrix Operator

Assuming that k_1 and k_2 are positive, express the diagonal matrix

$$A = \begin{bmatrix} k_1 & 0 \\ 0 & k_2 \end{bmatrix}$$

as a product of elementary matrices, and describe the geometric effect of multiplication by A in terms of compressions and expansions.

Solution From Example 1 we have

$$A = \begin{bmatrix} k_1 & 0 \\ 0 & k_2 \end{bmatrix} = \begin{bmatrix} 1 & 0 \\ 0 & k_2 \end{bmatrix} \begin{bmatrix} k_1 & 0 \\ 0 & 1 \end{bmatrix}$$

which shows that multiplication by A has the geometric effect of compressing or expanding by a factor of k_1 in the x-direction and then compressing or expanding by a factor of k_2 in the y-direction. ◀

▶ **EXAMPLE 4** Analyzing the Geometric Effect of a Matrix Operator

Express

$$A = \begin{bmatrix} 1 & 2 \\ 3 & 4 \end{bmatrix}$$

as a product of elementary matrices, and then describe the geometric effect of multiplication by A in terms of shears, compressions, expansions, and reflections.

Solution A can be reduced to I as follows:

$$\begin{bmatrix} 1 & 2 \\ 3 & 4 \end{bmatrix} \longrightarrow \begin{bmatrix} 1 & 2 \\ 0 & -2 \end{bmatrix} \longrightarrow \begin{bmatrix} 1 & 2 \\ 0 & 1 \end{bmatrix} \longrightarrow \begin{bmatrix} 1 & 0 \\ 0 & 1 \end{bmatrix}$$

| Add -3 times the first row to the second. | Multiply the second row by $-\frac{1}{2}$. | Add -2 times the second row to the first. |

The three successive row operations can be performed by multiplying A on the left successively by

$$E_1 = \begin{bmatrix} 1 & 0 \\ -3 & 1 \end{bmatrix}, \quad E_2 = \begin{bmatrix} 1 & 0 \\ 0 & -\frac{1}{2} \end{bmatrix}, \quad E_3 = \begin{bmatrix} 1 & -2 \\ 0 & 1 \end{bmatrix}$$

Inverting these matrices and using Formula (4) of Section 1.5 yields

$$A = E_1^{-1} E_2^{-1} E_3^{-1} = \begin{bmatrix} 1 & 0 \\ 3 & 1 \end{bmatrix} \begin{bmatrix} 1 & 0 \\ 0 & -2 \end{bmatrix} \begin{bmatrix} 1 & 2 \\ 0 & 1 \end{bmatrix}$$

Reading from right to left and noting that

$$\begin{bmatrix} 1 & 0 \\ 0 & -2 \end{bmatrix} = \begin{bmatrix} 1 & 0 \\ 0 & -1 \end{bmatrix} \begin{bmatrix} 1 & 0 \\ 0 & 2 \end{bmatrix}$$

it follows that the effect of multiplying by A is equivalent to

1. shearing by a factor of 2 in the x-direction,
2. then expanding by a factor of 2 in the y-direction,
3. then reflecting about the x-axis,
4. then shearing by a factor of 3 in the y-direction. ◀

Images of Lines Under
Matrix Operators

Many images in computer graphics are constructed by connecting points with line segments. The following theorem, some of whose parts are proved in the exercises, is helpful for understanding how matrix operators transform such figures.

> **THEOREM 4.11.3** *If $T: R^2 \rightarrow R^2$ is multiplication by an invertible matrix, then:*
>
> *(a) The image of a straight line is a straight line.*
>
> *(b) The image of a straight line through the origin is a straight line through the origin.*
>
> *(c) The images of parallel straight lines are parallel straight lines.*
>
> *(d) The image of the line segment joining points P and Q is the line segment joining the images of P and Q.*
>
> *(e) The images of three points lie on a line if and only if the points themselves lie on a line.*

Note that it follows from Theorem 4.11.3 that if A is an invertible 2×2 matrix, then multiplication by A maps triangles into triangles and parallelograms into parallelograms.

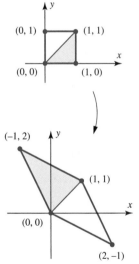

▲ Figure 4.11.5

▶ EXAMPLE 5 Image of a Square

Sketch the image of the square with vertices $(0, 0)$, $(1, 0)$, $(1, 1)$, and $(0, 1)$ under multiplication by

$$A = \begin{bmatrix} -1 & 2 \\ 2 & -1 \end{bmatrix}$$

Solution Since

$$\begin{bmatrix} -1 & 2 \\ 2 & -1 \end{bmatrix}\begin{bmatrix} 0 \\ 0 \end{bmatrix} = \begin{bmatrix} 0 \\ 0 \end{bmatrix}, \quad \begin{bmatrix} -1 & 2 \\ 2 & -1 \end{bmatrix}\begin{bmatrix} 1 \\ 0 \end{bmatrix} = \begin{bmatrix} -1 \\ 2 \end{bmatrix}$$

$$\begin{bmatrix} -1 & 2 \\ 2 & -1 \end{bmatrix}\begin{bmatrix} 0 \\ 1 \end{bmatrix} = \begin{bmatrix} 2 \\ -1 \end{bmatrix}, \quad \begin{bmatrix} -1 & 2 \\ 2 & -1 \end{bmatrix}\begin{bmatrix} 1 \\ 1 \end{bmatrix} = \begin{bmatrix} 1 \\ 1 \end{bmatrix}$$

the image of the square is a parallelogram with vertices $(0, 0)$, $(-1, 2)$, $(2, -1)$, and $(1, 1)$ (Figure 4.11.5).

▶ EXAMPLE 6 Image of a Line

According to Theorem 4.11.3, the invertible matrix

$$A = \begin{bmatrix} 3 & 1 \\ 2 & 1 \end{bmatrix}$$

maps the line $y = 2x + 1$ into another line. Find its equation.

Solution Let (x, y) be a point on the line $y = 2x + 1$, and let (x', y') be its image under multiplication by A. Then

$$\begin{bmatrix} x' \\ y' \end{bmatrix} = \begin{bmatrix} 3 & 1 \\ 2 & 1 \end{bmatrix}\begin{bmatrix} x \\ y \end{bmatrix} \quad \text{and} \quad \begin{bmatrix} x \\ y \end{bmatrix} = \begin{bmatrix} 3 & 1 \\ 2 & 1 \end{bmatrix}^{-1}\begin{bmatrix} x' \\ y' \end{bmatrix} = \begin{bmatrix} 1 & -1 \\ -2 & 3 \end{bmatrix}\begin{bmatrix} x' \\ y' \end{bmatrix}$$

so

$$x = \quad x' - \quad y'$$
$$y = -2x' + 3y'$$

Substituting in $y = 2x + 1$ yields

$$-2x' + 3y' = 2(x' - y') + 1 \quad \text{or equivalently} \quad y' = \tfrac{4}{5}x' + \tfrac{1}{5}$$

Thus (x', y') satisfies

$$y = \tfrac{4}{5}x + \tfrac{1}{5}$$

which is the equation we want. ◀

Concept Review

- Effect of a matrix operator on the unit square
- Geometry of one-to-one matrix operators
- Images of lines under matrix operators

Skills

- Find standard matrices for geometric transformations of R^2.
- Describe the geometric effect of an invertible matrix operator.
- Find the image of the unit square under a matrix operator.
- Find the image of a line under a matrix operator.

Exercise Set 4.11

1. Find the standard matrix for the operator $T: R^2 \to R^2$ that maps a point (x, y) into

   (a) its reflection about the line $y = -x$.

   (b) its reflection through the origin.

   (c) its orthogonal projection on the x-axis.

   (d) its orthogonal projection on the y-axis.

2. For each part of Exercise 1, use the matrix you have obtained to compute $T(2, 1)$. Check your answers geometrically by plotting the points $(2, 1)$ and $T(2, 1)$.

3. Find the standard matrix for the operator $T: R^3 \to R^3$ that maps a point (x, y, z) into

   (a) its reflection through the xy-plane.

   (b) its reflection through the xz-plane.

   (c) its reflection through the yz-plane.

4. For each part of Exercise 3, use the matrix you have obtained to compute $T(1, 1, 1)$. Check your answers geometrically by plotting the points $(1, 1, 1)$ and $T(1, 1, 1)$.

5. Find the standard matrix for the operator $T: R^3 \to R^3$ that

   (a) rotates each vector $90°$ counterclockwise about the z-axis (looking along the positive z-axis toward the origin).

   (b) rotates each vector $90°$ counterclockwise about the x-axis (looking along the positive x-axis toward the origin).

   (c) rotates each vector $90°$ counterclockwise about the y-axis (looking along the positive y-axis toward the origin).

6. Sketch the image of the rectangle with vertices $(0, 0)$, $(1, 0)$, $(1, 2)$, and $(0, 2)$ under

   (a) a reflection about the x-axis.

   (b) a reflection about the y-axis.

   (c) a compression of factor $k = \tfrac{1}{4}$ in the y-direction.

   (d) an expansion of factor $k = 2$ in the x-direction.

   (e) a shear of factor $k = 3$ in the x-direction.

   (f) a shear of factor $k = 2$ in the y-direction.

7. Sketch the image of the square with vertices $(0, 0)$, $(1, 0)$, $(0, 1)$, and $(1, 1)$ under multiplication by

$$A = \begin{bmatrix} -3 & 0 \\ 0 & 1 \end{bmatrix}$$

8. Find the matrix that rotates a point (x, y) about the origin through

   (a) $45°$ (b) $90°$ (c) $180°$ (d) $270°$ (e) $-30°$

9. Find the matrix that shears by

   (a) a factor of $k = 4$ in the y-direction.

   (b) a factor of $k = -2$ in the x-direction.

10. Find the matrix that compresses or expands by

   (a) a factor of $\tfrac{1}{3}$ in the y-direction.

   (b) a factor of 6 in the x-direction.

11. In each part, describe the geometric effect of multiplication by A.

(a) $A = \begin{bmatrix} 3 & 0 \\ 0 & 1 \end{bmatrix}$ (b) $A = \begin{bmatrix} 1 & 0 \\ 0 & -5 \end{bmatrix}$ (c) $A = \begin{bmatrix} 1 & 4 \\ 0 & 1 \end{bmatrix}$

12. In each part, express the matrix as a product of elementary matrices, and then describe the effect of multiplication by A in terms of compressions, expansions, reflections, and shears.

(a) $A = \begin{bmatrix} 2 & 0 \\ 0 & 3 \end{bmatrix}$ (b) $A = \begin{bmatrix} 1 & 4 \\ 2 & 9 \end{bmatrix}$

(c) $A = \begin{bmatrix} 0 & -2 \\ 4 & 0 \end{bmatrix}$ (d) $A = \begin{bmatrix} 1 & -3 \\ 4 & 6 \end{bmatrix}$

13. In each part, find a single matrix that performs the indicated succession of operations.

(a) Compresses by a factor of $\frac{1}{2}$ in the x-direction, then expands by a factor of 5 in the y-direction.

(b) Expands by a factor of 5 in the y-direction, then shears by a factor of 2 in the y-direction.

(c) Reflects about $y = x$, then rotates through an angle of $180°$ about the origin.

14. In each part, find a single matrix that performs the indicated succession of operations.

(a) Reflects about the y-axis, then expands by a factor of 5 in the x-direction, and then reflects about $y = x$.

(b) Rotates through $30°$ about the origin, then shears by a factor of -2 in the y-direction, and then expands by a factor of 3 in the y-direction.

15. Use matrix inversion to show the following.

(a) The inverse transformation for a reflection about $y = x$ is a reflection about $y = x$.

(b) The inverse transformation for a compression along an axis is an expansion along that axis.

(c) The inverse transformation for a reflection about a coordinate axis is a reflection about that axis.

(d) The inverse transformation for a shear along a coordinate axis is a shear along that axis.

16. Find an equation of the image of the line $y = -4x + 3$ under multiplication by
$$A = \begin{bmatrix} 4 & -3 \\ 3 & -2 \end{bmatrix}$$

17. In parts (a) through (e), find an equation of the image of the line $y = 2x$ under

(a) a shear of factor 3 in the x-direction.

(b) a compression of factor $\frac{1}{2}$ in the y-direction.

(c) a reflection about $y = x$.

(d) a reflection about the y-axis.

(e) a rotation of $60°$ about the origin.

18. Find the matrix for a shear in the x-direction that transforms the triangle with vertices $(0, 0)$, $(2, 1)$, and $(3, 0)$ into a right triangle with the right angle at the origin.

19. (a) Show that multiplication by
$$A = \begin{bmatrix} 3 & 1 \\ 6 & 2 \end{bmatrix}$$
maps each point in the plane onto the line $y = 2x$.

(b) It follows from part (a) that the noncollinear points $(1, 0)$, $(0, 1)$, $(-1, 0)$ are mapped onto a line. Does this violate part (e) of Theorem 4.11.3?

20. Prove part (a) of Theorem 4.11.3. [*Hint:* A line in the plane has an equation of the form $Ax + By + C = 0$, where A and B are not both zero. Use the method of Example 6 to show that the image of this line under multiplication by the invertible matrix
$$\begin{bmatrix} a & b \\ c & d \end{bmatrix}$$
has the equation $A'x + B'y + C = 0$, where
$$A' = (dA - cB)/(ad - bc)$$
and
$$B' = (-bA + aB)/(ad - bc)$$
Then show that A' and B' are not both zero to conclude that the image is a line.]

21. Use the hint in Exercise 20 to prove parts (b) and (c) of Theorem 4.11.3.

22. In each part of the accompanying figure, find the standard matrix for the operator described.

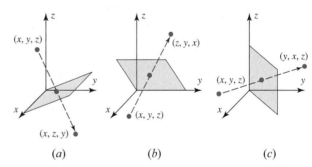

▲ **Figure Ex-22**

23. In R^3 the **shear in the xy-direction with factor k** is the matrix transformation that moves each point (x, y, z) parallel to the xy-plane to the new position $(x + kz, y + kz, z)$. (See the accompanying figure.)

(a) Find the standard matrix for the shear in the xy-direction with factor k.

(b) How would you define the shear in the xz-direction with factor k and the shear in the yz-direction with factor k?

Find the standard matrices for these matrix transformations.

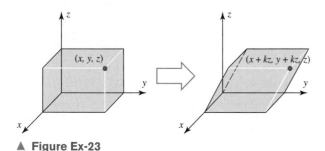

▲ **Figure Ex-23**

True-False Exercises

In parts (a)–(g) determine whether the statement is true or false, and justify your answer.

(a) The image of the unit square under a one-to-one matrix operator is a square.

(b) A 2×2 invertible matrix operator has the geometric effect of a succession of shears, compressions, expansions, and reflections.

(c) The image of a line under a one-to-one matrix operator is a line.

(d) Every reflection operator on R^2 is its own inverse.

(e) The matrix $\begin{bmatrix} 1 & 1 \\ 1 & -1 \end{bmatrix}$ represents reflection about a line.

(f) The matrix $\begin{bmatrix} 1 & -2 \\ 2 & 1 \end{bmatrix}$ represents a shear.

(g) The matrix $\begin{bmatrix} 1 & 0 \\ 0 & 3 \end{bmatrix}$ represents an expansion.

4.12 Dynamical Systems and Markov Chains

In this optional section we will show how matrix methods can be used to analyze the behavior of physical systems that evolve over time. The methods that we will study here have been applied to problems in business, ecology, demographics, sociology, and most of the physical sciences.

Dynamical Systems

A ***dynamical system*** is a finite set of variables whose values change with time. The value of a variable at a point in time is called the ***state of the variable*** at that time, and the vector formed from these states is called the ***state of the dynamical system*** at that time. Our primary objective in this section is to analyze how the state of a dynamical system changes with time. Let us begin with an example.

▶ **EXAMPLE 1 Market Share as a Dynamical System**

Suppose that two competing television channels, channel 1 and channel 2, each have 50% of the viewer market at some initial point in time. Assume that over each one-year period channel 1 captures 10% of channel 2's share, and channel 2 captures 20% of channel 1's share (see Figure 4.12.1). What is each channel's market share after one year?

Solution Let us begin by introducing the time-dependent variables

$$x_1(t) = \text{fraction of the market held by channel 1 at time } t$$
$$x_2(t) = \text{fraction of the market held by channel 2 at time } t$$

and the column vector

$$\mathbf{x}(t) = \begin{bmatrix} x_1(t) \\ x_2(t) \end{bmatrix} \quad \begin{matrix} \leftarrow \textbf{Channel 1's fraction of the market at time } t \textbf{ in years} \\ \leftarrow \textbf{Channel 2's fraction of the market at time } t \textbf{ in years} \end{matrix}$$

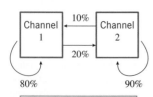

Channel 1 loses 20% and holds 80%.
Channel 2 loses 10% and holds 90%.

▲ **Figure 4.12.1**

The variables $x_1(t)$ and $x_2(t)$ form a dynamical system whose state at time t is the vector $\mathbf{x}(t)$. If we take $t = 0$ to be the starting point at which the two channels had 50% of the market, then the state of the system at that time is

$$\mathbf{x}(0) = \begin{bmatrix} x_1(0) \\ x_2(0) \end{bmatrix} = \begin{bmatrix} 0.5 \\ 0.5 \end{bmatrix} \quad \begin{array}{l} \leftarrow \textbf{Channel 1's fraction of the market at time } t = 0 \\ \leftarrow \textbf{Channel 2's fraction of the market at time } t = 0 \end{array} \tag{1}$$

Now let us try to find the state of the system at time $t = 1$ (one year later). Over the one-year period, channel 1 retains 80% of its initial 50%, and it gains 10% of channel 2's initial 50%. Thus,

$$\mathbf{x}_1(1) = 0.8(0.5) + 0.1(0.5) = 0.45 \tag{2}$$

Similarly, channel 2 gains 20% of channel 1's initial 50%, and retains 90% of its initial 50%. Thus,

$$x_2(1) = 0.2(0.5) + 0.9(0.5) = 0.55 \tag{3}$$

Therefore, the state of the system at time $t = 1$ is

$$\mathbf{x}(1) = \begin{bmatrix} x_1(1) \\ x_2(1) \end{bmatrix} = \begin{bmatrix} 0.45 \\ 0.55 \end{bmatrix} \quad \begin{array}{l} \leftarrow \textbf{Channel 1's fraction of the market at time } t = 1 \\ \leftarrow \textbf{Channel 2's fraction of the market at time } t = 1 \end{array} \tag{4}$$

▶ EXAMPLE 2 Evolution of Market Share over Five Years

Track the market shares of channels 1 and 2 in Example 1 over a five-year period.

Solution To solve this problem suppose that we have already computed the market share of each channel at time $t = k$ and we are interested in using the known values of $x_1(k)$ and $x_2(k)$ to compute the market shares $x_1(k + 1)$ and $x_2(k + 1)$ one year later. The analysis is exactly the same as that used to obtain Equations (2) and (3). Over the one-year period, channel 1 retains 80% of its starting fraction $x_1(k)$ and gains 10% of channel 2's starting fraction $x_2(k)$. Thus,

$$x_1(k + 1) = (0.8)x_1(k) + (0.1)x_2(k) \tag{5}$$

Similarly, channel 2 gains 20% of channel 1's starting fraction $x_1(k)$ and retains 90% of its own starting fraction $x_2(k)$. Thus,

$$x_2(k + 1) = (0.2)x_1(k) + (0.9)x_2(k) \tag{6}$$

Equations (5) and (6) can be expressed in matrix form as

$$\begin{bmatrix} x_1(k + 1) \\ x_2(k + 1) \end{bmatrix} = \begin{bmatrix} 0.8 & 0.1 \\ 0.2 & 0.9 \end{bmatrix} \begin{bmatrix} x_1(k) \\ x_2(k) \end{bmatrix} \tag{7}$$

which provides a way of using matrix multiplication to compute the state of the system at time $t = k + 1$ from the state at time $t = k$. For example, using (1) and (7) we obtain

$$\mathbf{x}(1) = \begin{bmatrix} 0.8 & 0.1 \\ 0.2 & 0.9 \end{bmatrix} \mathbf{x}(0) = \begin{bmatrix} 0.8 & 0.1 \\ 0.2 & 0.9 \end{bmatrix} \begin{bmatrix} 0.5 \\ 0.5 \end{bmatrix} = \begin{bmatrix} 0.45 \\ 0.55 \end{bmatrix}$$

which agrees with (4). Similarly,

$$\mathbf{x}(2) = \begin{bmatrix} 0.8 & 0.1 \\ 0.2 & 0.9 \end{bmatrix} \mathbf{x}(1) = \begin{bmatrix} 0.8 & 0.1 \\ 0.2 & 0.9 \end{bmatrix} \begin{bmatrix} 0.45 \\ 0.55 \end{bmatrix} = \begin{bmatrix} 0.415 \\ 0.585 \end{bmatrix}$$

We can now continue this process, using Formula (7) to compute $\mathbf{x}(3)$ from $\mathbf{x}(2)$, then $\mathbf{x}(4)$ from $\mathbf{x}(3)$, and so on. This yields (verify)

$$\mathbf{x}(3) = \begin{bmatrix} 0.3905 \\ 0.6095 \end{bmatrix}, \quad \mathbf{x}(4) = \begin{bmatrix} 0.37335 \\ 0.62665 \end{bmatrix}, \quad \mathbf{x}(5) = \begin{bmatrix} 0.361345 \\ 0.638655 \end{bmatrix} \tag{8}$$

Thus, after five years, channel 1 will hold about 36% of the market and channel 2 will hold about 64% of the market. ◀

If desired, we can continue the market analysis in the last example beyond the five-year period and explore what happens to the market share over the long term. We did so, using a computer, and obtained the following state vectors (rounded to six decimal places):

$$\mathbf{x}(10) \approx \begin{bmatrix} 0.338041 \\ 0.661959 \end{bmatrix}, \quad \mathbf{x}(20) \approx \begin{bmatrix} 0.333466 \\ 0.666534 \end{bmatrix}, \quad \mathbf{x}(40) \approx \begin{bmatrix} 0.333333 \\ 0.666667 \end{bmatrix} \qquad (9)$$

All subsequent state vectors, when rounded to six decimal places, are the same as $\mathbf{x}(40)$, so we see that the market shares eventually stabilize with channel 1 holding about one-third of the market and channel 2 holding about two-thirds. Later in this section, we will explain why this stabilization occurs.

Markov Chains   In many dynamical systems the states of the variables are not known with certainty but can be expressed as probabilities; such dynamical systems are called ***stochastic processes*** (from the Greek word *stokastikos*, meaning "proceeding by guesswork"). A detailed study of stochastic processes requires a precise definition of the term *probability*, which is outside the scope of this course. However, the following interpretation will suffice for our present purposes:

> Stated informally, the **probability** that an experiment or observation will have a certain outcome is approximately the fraction of the time that the outcome would occur if the experiment were to be repeated many times under constant conditions—the greater the number of repetitions, the more accurately the probability describes the fraction of occurrences.

For example, when we say that the probability of tossing heads with a fair coin is $\frac{1}{2}$, we mean that if the coin were tossed many times under constant conditions, then we would expect about half of the outcomes to be heads. Probabilities are often expressed as decimals or percentages. Thus, the probability of tossing heads with a fair coin can also be expressed as 0.5 or 50%.

If an experiment or observation has n possible outcomes, then the probabilities of those outcomes must be nonnegative fractions whose sum is 1. The probabilities are nonnegative because each describes the fraction of occurrences of an outcome over the long term, and the sum is 1 because they account for all possible outcomes. For example, if a box containing 10 balls has one red ball, three green balls, and six yellow balls, and if a ball is drawn at random from the box, then the probabilities of the various outcomes are

$$p_1 = \text{prob(red)} = 1/10 = 0.1$$
$$p_2 = \text{prob(green)} = 3/10 = 0.3$$
$$p_3 = \text{prob(yellow)} = 6/10 = 0.6$$

Each probability is a nonnegative fraction and

$$p_1 + p_2 + p_3 = 0.1 + 0.3 + 0.6 = 1$$

In a stochastic process with n possible states, the state vector at each time t has the form

$$\mathbf{x}(t) = \begin{bmatrix} x_1(t) \\ x_2(t) \\ \vdots \\ x_n(t) \end{bmatrix} \quad \begin{matrix} \text{Probability that the system is in state 1} \\ \text{Probability that the system is in state 2} \\ \vdots \\ \text{Probability that the system is in state } n \end{matrix}$$

The entries in this vector must add up to 1 since they account for all n possibilities. In general, a vector with nonnegative entries that add up to 1 is called a ***probability vector***.

▶ **EXAMPLE 3** **Example 1 Revisited from the Probability Viewpoint**

Observe that the state vectors in Examples 1 and 2 are all probability vectors. This is to be expected since the entries in each state vector are the fractional market shares of the channels, and together they account for the entire market. In practice, it is preferable to interpret the entries in the state vectors as probabilities rather than exact market fractions, since market information is usually obtained by statistical sampling procedures with intrinsic uncertainties. Thus, for example, the state vector

$$\mathbf{x}(1) = \begin{bmatrix} x_1(1) \\ x_2(1) \end{bmatrix} = \begin{bmatrix} 0.45 \\ 0.55 \end{bmatrix}$$

which we interpreted in Example 1 to mean that channel 1 has 45% of the market and channel 2 has 55%, can also be interpreted to mean that an individual picked at random from the market will be a channel 1 viewer with probability 0.45 and a channel 2 viewer with probability 0.55. ◀

A square matrix, each of whose columns is a probability vector, is called a **_stochastic matrix_**. Such matrices commonly occur in formulas that relate successive states of a stochastic process. For example, the state vectors $\mathbf{x}(k + 1)$ and $\mathbf{x}(k)$ in (7) are related by an equation of the form $\mathbf{x}(k + 1) = P\mathbf{x}(k)$ in which

$$P = \begin{bmatrix} 0.8 & 0.1 \\ 0.2 & 0.9 \end{bmatrix} \tag{10}$$

is a stochastic matrix. It should not be surprising that the column vectors of P are probability vectors, since the entries in each column provide a breakdown of what happens to each channel's market share over the year—the entries in column 1 convey that each year channel 1 retains 80% of its market share and loses 20%; and the entries in column 2 convey that each year channel 2 retains 90% of its market share and loses 10%. The entries in (10) can also be viewed as probabilities:

$p_{11} = 0.8 =$ probability that a channel 1 viewer remains a channel 1 viewer

$p_{21} = 0.2 =$ probability that a channel 1 viewer becomes a channel 2 viewer

$p_{12} = 0.1 =$ probability that a channel 2 viewer becomes a channel 1 viewer

$p_{22} = 0.9 =$ probability that a channel 2 viewer remains a channel 2 viewer

Example 1 is a special case of a large class of stochastic processes, called *Markov chains*.

Historical Note Markov chains are named in honor of the Russian mathematician A. A. Markov, a lover of poetry, who used them to analyze the alternation of vowels and consonants in the poem *Eugene Onegin* by Pushkin. Markov believed that the only applications of his chains were to the analysis of literary works, so he would be astonished to learn that his discovery is used today in the social sciences, quantum theory, and genetics!

[*Image: wikipedia*]

Andrei Andreyevich Markov (1856–1922)

State at time $t = k$

The entry p_{ij} is the probability that the system is in state i at time $t = k + 1$ if it is in state j at time $t = k$.

▲ **Figure 4.12.2**

DEFINITION 1 A *Markov chain* is a dynamical system whose state vectors at a succession of time intervals are probability vectors and for which the state vectors at successive time intervals are related by an equation of the form

$$\mathbf{x}(k + 1) = P\mathbf{x}(k)$$

in which $P = [p_{ij}]$ is a stochastic matrix and p_{ij} is the probability that the system will be in state i at time $t = k + 1$ if it is in state j at time $t = k$. The matrix P is called the *transition matrix* for the system.

Remark Note that in this definition the row index i corresponds to the later state and the column index j to the earlier state (Figure 4.12.2).

▶ **EXAMPLE 4 Wildlife Migration as a Markov Chain**

Suppose that a tagged lion can migrate over three adjacent game reserves in search of food, reserve 1, reserve 2, and reserve 3. Based on data about the food resources, researchers conclude that the monthly migration pattern of the lion can be modeled by a Markov chain with transition matrix

Reserve at time $t = k$

$$P = \begin{array}{c} \\ \\ \\ \end{array} \begin{bmatrix} 0.5 & 0.4 & 0.6 \\ 0.2 & 0.2 & 0.3 \\ 0.3 & 0.4 & 0.1 \end{bmatrix} \begin{array}{c} 1 \\ 2 \\ 3 \end{array} \quad \text{Reserve at time } t = k + 1$$

(see Figure 4.12.3). That is,

$p_{11} = 0.5 =$ probability that the lion will stay in reserve 1 when it is in reserve 1
$p_{12} = 0.4 =$ probability that the lion will move from reserve 2 to reserve 1
$p_{13} = 0.6 =$ probability that the lion will move from reserve 3 to reserve 1
$p_{21} = 0.2 =$ probability that the lion will move from reserve 1 to reserve 2
$p_{22} = 0.2 =$ probability that the lion will stay in reserve 2 when it is in reserve 2
$p_{23} = 0.3 =$ probability that the lion will move from reserve 3 to reserve 2
$p_{31} = 0.3 =$ probability that the lion will move from reserve 1 to reserve 3
$p_{32} = 0.4 =$ probability that the lion will move from reserve 2 to reserve 3
$p_{33} = 0.1 =$ probability that the lion will stay in reserve 3 when it is in reserve 3

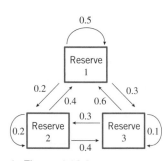

▲ **Figure 4.12.3**

Assuming that t is in months and the lion is released in reserve 2 at time $t = 0$, track its probable locations over a six-month period.

Solution Let $x_1(k)$, $x_2(k)$, and $x_3(k)$ be the probabilities that the lion is in reserve 1, 2, or 3, respectively, at time $t = k$, and let

$$\mathbf{x}(k) = \begin{bmatrix} x_1(k) \\ x_2(k) \\ x_3(k) \end{bmatrix}$$

be the state vector at that time. Since we know with certainty that the lion is in reserve 2 at time $t = 0$, the initial state vector is

$$\mathbf{x}(0) = \begin{bmatrix} 0 \\ 1 \\ 0 \end{bmatrix}$$

We leave it for you to show that the state vectors over a six-month period are

$$\mathbf{x}(1) = P\mathbf{x}(0) = \begin{bmatrix} 0.400 \\ 0.200 \\ 0.400 \end{bmatrix}, \quad \mathbf{x}(2) = P\mathbf{x}(1) = \begin{bmatrix} 0.520 \\ 0.240 \\ 0.240 \end{bmatrix}, \quad \mathbf{x}(3) = P\mathbf{x}(2) = \begin{bmatrix} 0.500 \\ 0.224 \\ 0.276 \end{bmatrix}$$

$$\mathbf{x}(4) = P\mathbf{x}(3) \approx \begin{bmatrix} 0.505 \\ 0.228 \\ 0.267 \end{bmatrix}, \quad \mathbf{x}(5) = P\mathbf{x}(4) \approx \begin{bmatrix} 0.504 \\ 0.227 \\ 0.269 \end{bmatrix}, \quad \mathbf{x}(6) = P\mathbf{x}(5) \approx \begin{bmatrix} 0.504 \\ 0.227 \\ 0.269 \end{bmatrix}$$

As in Example 2, the state vectors here seem to stabilize over time with a probability of approximately 0.504 that the lion is in reserve 1, a probability of approximately 0.227 that it is in reserve 2, and a probability of approximately 0.269 that it is in reserve 3. ◀

Markov Chains in Terms of Powers of the Transition Matrix

In a Markov chain with an initial state of $\mathbf{x}(0)$, the successive state vectors are

$$\mathbf{x}(1) = P\mathbf{x}(0), \quad \mathbf{x}(2) = P\mathbf{x}(1), \quad \mathbf{x}(3) = P\mathbf{x}(2), \quad \mathbf{x}(4) = P\mathbf{x}(3), \dots$$

For brevity, it is common to denote $\mathbf{x}(k)$ by $\mathbf{x}_k$, which allows us to write the successive state vectors more briefly as

$$\mathbf{x}_1 = P\mathbf{x}_0, \quad \mathbf{x}_2 = P\mathbf{x}_1, \quad \mathbf{x}_3 = P\mathbf{x}_2, \quad \mathbf{x}_4 = P\mathbf{x}_3, \dots \quad (11)$$

Alternatively, these state vectors can be expressed in terms of the initial state vector $\mathbf{x}_0$ as

$$\mathbf{x}_1 = P\mathbf{x}_0, \quad \mathbf{x}_2 = P(P\mathbf{x}_0) = P^2\mathbf{x}_0, \quad \mathbf{x}_3 = P(P^2\mathbf{x}_0) = P^3\mathbf{x}_0, \quad \mathbf{x}_4 = P(P^3\mathbf{x}_0) = P^4\mathbf{x}_0, \dots$$

from which it follows that

$$\mathbf{x}_k = P^k\mathbf{x}_0 \quad (12)$$

Note that Formula (12) makes it possible to compute the state vector $\mathbf{x}_k$ without first computing the earlier state vectors as required in Formula (11).

▶ **EXAMPLE 5 Finding a State Vector Directly from $\mathbf{x}_0$**

Use Formula (12) to find the state vector $\mathbf{x}(3)$ in Example 2.

Solution From (1) and (7), the initial state vector and transition matrix are

$$\mathbf{x}_0 = \mathbf{x}(0) = \begin{bmatrix} 0.5 \\ 0.5 \end{bmatrix} \quad \text{and} \quad P = \begin{bmatrix} 0.8 & 0.1 \\ 0.2 & 0.9 \end{bmatrix}$$

We leave it for you to calculate P^3 and show that

$$\mathbf{x}(3) = \mathbf{x}_3 = P^3\mathbf{x}_0 = \begin{bmatrix} 0.562 & 0.219 \\ 0.438 & 0.781 \end{bmatrix} \begin{bmatrix} 0.5 \\ 0.5 \end{bmatrix} = \begin{bmatrix} 0.3905 \\ 0.6095 \end{bmatrix}$$

which agrees with the result in (8). ◀

Long-Term Behavior of a Markov Chain

We have seen two examples of Markov chains in which the state vectors seem to stabilize after a period of time. Thus, it is reasonable to ask whether all Markov chains have this property. The following example shows that this is not the case.

▶ **EXAMPLE 6 A Markov Chain That Does Not Stabilize**

The matrix

$$P = \begin{bmatrix} 0 & 1 \\ 1 & 0 \end{bmatrix}$$

is stochastic and hence can be regarded as the transition matrix for a Markov chain. A simple calculation shows that $P^2 = I$, from which it follows that

$$I = P^2 = P^4 = P^6 = \cdots \quad \text{and} \quad P = P^3 = P^5 = P^7 = \cdots$$

Thus, the successive states in the Markov chain with initial vector $\mathbf{x}_0$ are

$$\mathbf{x}_0, \quad P\mathbf{x}_0, \quad \mathbf{x}_0, \quad P\mathbf{x}_0, \quad \mathbf{x}_0, \ldots$$

which oscillate between $\mathbf{x}_0$ and $P\mathbf{x}_0$. Thus, the Markov chain does not stabilize unless both components of $\mathbf{x}_0$ are $\frac{1}{2}$ (verify). ◀

A precise definition of what it means for a sequence of numbers or vectors to stabilize is given in calculus; however, that level of precision will not be needed here. Stated informally, we will say that a sequence of vectors

$$\mathbf{x}_1, \quad \mathbf{x}_2, \ldots, \quad \mathbf{x}_k, \ldots$$

approaches a ***limit*** $\mathbf{q}$ or that it ***converges*** to $\mathbf{q}$ if all entries in $\mathbf{x}_k$ can be made as close as we like to the corresponding entries in the vector $\mathbf{q}$ by taking k sufficiently large. We denote this by writing $\mathbf{x}_k \to \mathbf{q}$ as $k \to \infty$.

We saw in Example 6 that the state vectors of a Markov chain need not approach a limit in all cases. However, by imposing a mild condition on the transition matrix of a Markov chain, we can guarantee that the state vectors will approach a limit.

DEFINITION 2 A stochastic matrix P is said to be ***regular*** if P or some positive power of P has all positive entries, and a Markov chain whose transition matrix is regular is said to be a ***regular Markov chain***.

▶ **EXAMPLE 7** Regular Stochastic Matrices

The transition matrices in Examples 2 and 4 are regular because their entries are positive. The matrix

$$P = \begin{bmatrix} 0.5 & 1 \\ 0.5 & 0 \end{bmatrix}$$

is regular because

$$P^2 = \begin{bmatrix} 0.75 & 0.5 \\ 0.25 & 0.5 \end{bmatrix}$$

has positive entries. The matrix P in Example 6 is not regular because P and every positive power of P have some zero entries (verify). ◀

The following theorem, which we state without proof, is the fundamental result about the long-term behavior of Markov chains.

THEOREM 4.12.1 *If P is the transition matrix for a regular Markov chain, then*:

(a) *There is a unique probability vector $\mathbf{q}$ such that $P\mathbf{q} = \mathbf{q}$.*

(b) *For any initial probability vector $\mathbf{x}_0$, the sequence of state vectors*

$$\mathbf{x}_0, \quad P\mathbf{x}_0, \ldots, \quad P^k\mathbf{x}_0, \ldots$$

converges to $\mathbf{q}$.

The vector $\mathbf{q}$ in this theorem is called the ***steady-state*** vector of the Markov chain. It can be found by rewriting the equation in part (a) as

$$(I - P)\mathbf{q} = \mathbf{0}$$

and then solving this equation for $\mathbf{q}$ subject to the requirement that $\mathbf{q}$ be a probability vector. Here are some examples.

▶ **EXAMPLE 8 Examples 1 and 2 Revisited**

The transition matrix for the Markov chain in Example 2 is

$$P = \begin{bmatrix} 0.8 & 0.1 \\ 0.2 & 0.9 \end{bmatrix}$$

Since the entries of P are positive, the Markov chain is regular and hence has a unique steady-state vector $\mathbf{q}.$ To find $\mathbf{q}$ we will solve the system $(I - P)\mathbf{q} = \mathbf{0}$, which we can write as

$$\begin{bmatrix} 0.2 & -0.1 \\ -0.2 & 0.1 \end{bmatrix} \begin{bmatrix} q_1 \\ q_2 \end{bmatrix} = \begin{bmatrix} 0 \\ 0 \end{bmatrix}$$

The general solution of this system is

$$q_1 = 0.5s, \quad q_2 = s$$

(verify), which we can write in vector form as

$$\mathbf{q} = \begin{bmatrix} q_1 \\ q_2 \end{bmatrix} = \begin{bmatrix} 0.5s \\ s \end{bmatrix} = \begin{bmatrix} \frac{1}{2}s \\ s \end{bmatrix} \tag{13}$$

For $\mathbf{q}$ to be a probability vector, we must have

$$1 = q_1 + q_2 = \tfrac{3}{2}s$$

which implies that $s = \tfrac{2}{3}$. Substituting this value in (13) yields the steady-state vector

$$\mathbf{q} = \begin{bmatrix} \frac{1}{3} \\ \frac{2}{3} \end{bmatrix}$$

which is consistent with the numerical results obtained in (9).

▶ **EXAMPLE 9 Example 4 Revisited**

The transition matrix for the Markov chain in Example 4 is

$$P = \begin{bmatrix} 0.5 & 0.4 & 0.6 \\ 0.2 & 0.2 & 0.3 \\ 0.3 & 0.4 & 0.1 \end{bmatrix}$$

Since the entries of P are positive, the Markov chain is regular and hence has a unique steady-state vector $\mathbf{q}.$ To find $\mathbf{q}$ we will solve the system $(I - P)\mathbf{q} = \mathbf{0}$, which we can write (using fractions) as

$$\begin{bmatrix} \frac{1}{2} & -\frac{2}{5} & -\frac{3}{5} \\ -\frac{1}{5} & \frac{4}{5} & -\frac{3}{10} \\ -\frac{3}{10} & -\frac{2}{5} & \frac{9}{10} \end{bmatrix} \begin{bmatrix} q_1 \\ q_2 \\ q_3 \end{bmatrix} = \begin{bmatrix} 0 \\ 0 \\ 0 \end{bmatrix} \tag{14}$$

(We have converted to fractions to avoid roundoff error in this illustrative example.) We leave it for you to confirm that the reduced row echelon form of the coefficient matrix is

$$\begin{bmatrix} 1 & 0 & -\frac{15}{8} \\ 0 & 1 & -\frac{27}{32} \\ 0 & 0 & 0 \end{bmatrix}$$

and that the general solution of (14) is

$$q_1 = \tfrac{15}{8}s, \quad q_2 = \tfrac{27}{32}s, \quad q_3 = s \tag{15}$$

For $\mathbf{q}$ to be a probability vector we must have $q_1 + q_2 + q_3 = 1$, from which it follows that $s = \frac{32}{119}$ (verify). Substituting this value in (15) yields the steady-state vector

$$\mathbf{q} = \begin{bmatrix} \frac{60}{119} \\ \frac{27}{119} \\ \frac{32}{119} \end{bmatrix} \approx \begin{bmatrix} 0.5042 \\ 0.2269 \\ 0.2689 \end{bmatrix}$$

(verify), which is consistent with the results obtained in Example 4. ◄

Concept Review

- Dynamical system
- State of a variable
- State of a dynamical system
- Stochastic process

- Probability
- Probability vector
- Stochastic matrix
- Markov chain

- Transition matrix
- Regular stochastic matrix
- Regular Markov chain
- Steady-state vector

Skills

- Determine whether a matrix is stochastic.
- Compute the state vectors from a transition matrix and an initial state.
- Determine whether a stochastic matrix is regular.

- Determine whether a Markov chain is regular.
- Find the steady-state vector for a regular transition matrix.

Exercise Set 4.12

▶ In Exercises 1–2, determine whether A is a stochastic matrix. If A is not stochastic, then explain why not. ◄

1. (a) $A = \begin{bmatrix} 0.4 & 0.3 \\ 0.6 & 0.7 \end{bmatrix}$ (b) $A = \begin{bmatrix} 0.4 & 0.6 \\ 0.3 & 0.7 \end{bmatrix}$

(c) $A = \begin{bmatrix} 1 & \frac{1}{2} & \frac{1}{3} \\ 0 & 0 & \frac{1}{3} \\ 0 & \frac{1}{2} & \frac{1}{3} \end{bmatrix}$ (d) $A = \begin{bmatrix} \frac{1}{3} & \frac{1}{3} & \frac{1}{2} \\ \frac{1}{6} & \frac{1}{3} & -\frac{1}{2} \\ \frac{1}{2} & \frac{1}{3} & 1 \end{bmatrix}$

2. (a) $A = \begin{bmatrix} 0.2 & 0.9 \\ 0.8 & 0.1 \end{bmatrix}$ (b) $A = \begin{bmatrix} 0.2 & 0.8 \\ 0.9 & 0.1 \end{bmatrix}$

(c) $A = \begin{bmatrix} \frac{1}{12} & \frac{1}{9} & \frac{1}{6} \\ \frac{1}{2} & 0 & \frac{5}{6} \\ \frac{5}{12} & \frac{8}{9} & 0 \end{bmatrix}$ (d) $A = \begin{bmatrix} -1 & \frac{1}{3} & \frac{1}{2} \\ 0 & \frac{1}{3} & \frac{1}{2} \\ 2 & \frac{1}{3} & 0 \end{bmatrix}$

▶ In Exercises 3–4, use Formulas (11) and (12) to compute the state vector $\mathbf{x}_4$ in two different ways. ◄

3. $P = \begin{bmatrix} 0.5 & 0.6 \\ 0.5 & 0.4 \end{bmatrix}$; $\mathbf{x}_0 = \begin{bmatrix} 0.5 \\ 0.5 \end{bmatrix}$

4. $P = \begin{bmatrix} 0.8 & 0.5 \\ 0.2 & 0.5 \end{bmatrix}$; $\mathbf{x}_0 = \begin{bmatrix} 1 \\ 0 \end{bmatrix}$

▶ In Exercises 5–6, determine whether P is a regular stochastic matrix. ◄

5. (a) $P = \begin{bmatrix} \frac{1}{5} & \frac{1}{7} \\ \frac{4}{5} & \frac{6}{7} \end{bmatrix}$ (b) $P = \begin{bmatrix} \frac{1}{5} & 0 \\ \frac{4}{5} & 1 \end{bmatrix}$ (c) $P = \begin{bmatrix} \frac{1}{5} & 1 \\ \frac{4}{5} & 0 \end{bmatrix}$

6. (a) $P = \begin{bmatrix} \frac{1}{2} & 1 \\ \frac{1}{2} & 0 \end{bmatrix}$ (b) $P = \begin{bmatrix} 1 & \frac{2}{3} \\ 0 & \frac{1}{3} \end{bmatrix}$ (c) $P = \begin{bmatrix} \frac{3}{4} & \frac{1}{3} \\ \frac{1}{4} & \frac{2}{3} \end{bmatrix}$

In Exercises 7–10, verify that P is a regular stochastic matrix, and find the steady-state vector for the associated Markov chain.

7. $P = \begin{bmatrix} \frac{1}{4} & \frac{2}{3} \\ \frac{3}{4} & \frac{1}{3} \end{bmatrix}$

8. $P = \begin{bmatrix} 0.2 & 0.6 \\ 0.8 & 0.4 \end{bmatrix}$

9. $P = \begin{bmatrix} \frac{1}{2} & \frac{1}{2} & 0 \\ \frac{1}{4} & \frac{1}{2} & \frac{1}{3} \\ \frac{1}{4} & 0 & \frac{2}{3} \end{bmatrix}$

10. $P = \begin{bmatrix} \frac{1}{3} & \frac{1}{4} & \frac{2}{5} \\ 0 & \frac{3}{4} & \frac{2}{5} \\ \frac{2}{3} & 0 & \frac{1}{5} \end{bmatrix}$

11. Consider a Markov process with transition matrix

	State 1	State 2
State 1	0.2	0.1
State 2	0.8	0.9

(a) What does the entry 0.2 represent?

(b) What does the entry 0.1 represent?

(c) If the system is in state 1 initially, what is the probability that it will be in state 2 at the next observation?

(d) If the system has a 50% chance of being in state 1 initially, what is the probability that it will be in state 2 at the next observation?

12. Consider a Markov process with transition matrix

	State 1	State 2
State 1	0	$\frac{1}{7}$
State 2	1	$\frac{6}{7}$

(a) What does the entry $\frac{6}{7}$ represent?

(b) What does the entry 0 represent?

(c) If the system is in state 1 initially, what is the probability that it will be in state 1 at the next observation?

(d) If the system has a 50% chance of being in state 1 initially, what is the probability that it will be in state 2 at the next observation?

13. On a given day the air quality in a certain city is either good or bad. Records show that when the air quality is good on one day, then there is a 95% chance that it will be good the next day, and when the air quality is bad on one day, then there is a 45% chance that it will be bad the next day.

(a) Find a transition matrix for this phenomenon.

(b) If the air quality is good today, what is the probability that it will be good two days from now?

(c) If the air quality is bad today, what is the probability that it will be bad three days from now?

(d) If there is a 20% chance that the air quality will be good today, what is the probability that it will be good tomorrow?

14. In a laboratory experiment, a mouse can choose one of two food types each day, type I or type II. Records show that if the mouse chooses type I on a given day, then there is a 75%

chance that it will choose type I the next day, and if it chooses type II on one day, then there is a 50% chance that it will choose type II the next day.

(a) Find a transition matrix for this phenomenon.

(b) If the mouse chooses type I today, what is the probability that it will choose type I two days from now?

(c) If the mouse chooses type II today, what is the probability that it will choose type II three days from now?

(d) If there is a 10% chance that the mouse will choose type I today, what is the probability that it will choose type I tomorrow?

15. Suppose that at some initial point in time 100,000 people live in a certain city and 25,000 people live in its suburbs. The Regional Planning Commission determines that each year 5% of the city population moves to the suburbs and 3% of the suburban population moves to the city.

(a) Assuming that the total population remains constant, make a table that shows the populations of the city and its suburbs over a five-year period (round to the nearest integer).

(b) Over the long term, how will the population be distributed between the city and its suburbs?

16. Suppose that two competing television stations, station 1 and station 2, each have 50% of the viewer market at some initial point in time. Assume that over each one-year period station 1 captures 5% of station 2's market share and station 2 captures 10% of station 1's market share.

(a) Make a table that shows the market share of each station over a five-year period.

(b) Over the long term, how will the market share be distributed between the two stations?

17. Suppose that a car rental agency has three locations, numbered 1, 2, and 3. A customer may rent a car from any of the three locations and return it to any of the three locations. Records show that cars are rented and returned in accordance with the following probabilities:

		Rented from Location		
		1	2	3
Returned to Location	1	$\frac{1}{10}$	$\frac{1}{5}$	$\frac{3}{5}$
	2	$\frac{4}{5}$	$\frac{3}{10}$	$\frac{1}{5}$
	3	$\frac{1}{10}$	$\frac{1}{2}$	$\frac{1}{5}$

(a) Assuming that a car is rented from location 1, what is the probability that it will be at location 1 after two rentals?

(b) Assuming that this dynamical system can be modeled as a Markov chain, find the steady-state vector.

(c) If the rental agency owns 120 cars, how many parking spaces should it allocate at each location to be reasonably certain that it will have enough spaces for the cars over the long term? Explain your reasoning.

18. Physical traits are determined by the genes that an offspring receives from its parents. In the simplest case a trait in the offspring is determined by one pair of genes, one member of the pair inherited from the male parent and the other from the female parent. Typically, each gene in a pair can assume one of two forms, called *alleles*, denoted by A and a. This leads to three possible pairings:

$$AA, \quad Aa, \quad aa$$

called *genotypes* (the pairs Aa and aA determine the same trait and hence are not distinguished from one another). It is shown in the study of heredity that if a parent of known genotype is crossed with a random parent of unknown genotype, then the offspring will have the genotype probabilities given in the following table, which can be viewed as a transition matrix for a Markov process:

<table>
<tr><td></td><td></td><td colspan="3">**Genotype of Parent**</td></tr>
<tr><td></td><td></td><td>AA</td><td>Aa</td><td>aa</td></tr>
<tr><td rowspan="3">**Genotype of Offspring**</td><td>AA</td><td>$\frac{1}{2}$</td><td>$\frac{1}{4}$</td><td>0</td></tr>
<tr><td>Aa</td><td>$\frac{1}{2}$</td><td>$\frac{1}{2}$</td><td>$\frac{1}{2}$</td></tr>
<tr><td>aa</td><td>0</td><td>$\frac{1}{4}$</td><td>$\frac{1}{2}$</td></tr>
</table>

Thus, for example, the offspring of a parent of genotype AA that is crossed at random with a parent of unknown genotype will have a 50% chance of being AA, a 50% chance of being Aa, and no chance of being aa.

(a) Show that the transition matrix is regular.

(b) Find the steady-state vector, and discuss its physical interpretation.

19. Fill in the missing entries of the stochastic matrix

$$P = \begin{bmatrix} \frac{7}{10} & * & \frac{1}{5} \\ * & \frac{3}{10} & * \\ \frac{1}{10} & \frac{3}{5} & \frac{3}{10} \end{bmatrix}$$

and find its steady-state vector.

20. If P is an $n \times n$ stochastic matrix, and if M is a $1 \times n$ matrix whose entries are all 1's, then $MP = $ _____.

21. If P is a regular stochastic matrix with steady-state vector $\mathbf{q}$, what can you say about the sequence of products

$$P\mathbf{q}, \quad P^2\mathbf{q}, \quad P^3\mathbf{q}, \dots, \quad P^k\mathbf{q}, \dots$$

as $k \to \infty$?

22. (a) If P is a regular $n \times n$ stochastic matrix with steady-state vector $\mathbf{q}$, and if $\mathbf{e}_1, \mathbf{e}_2, \dots, \mathbf{e}_n$ are the standard unit vectors in column form, what can you say about the behavior of the sequence

$$P\mathbf{e}_i, \quad P^2\mathbf{e}_i, \quad P^3\mathbf{e}_i, \dots, \quad P^k\mathbf{e}_i, \dots$$

as $k \to \infty$ for each $i = 1, 2, \dots, n$?

(b) What does this tell you about the behavior of the column vectors of P^k as $k \to \infty$?

23. Prove that the product of two stochastic matrices is a stochastic matrix. [*Hint:* Write each column of the product as a linear combination of the columns of the first factor.]

24. Prove that if P is a stochastic matrix whose entries are all greater than or equal to ρ, then the entries of P^2 are greater than or equal to ρ.

True-False Exercises

In parts (a)–(e) determine whether the statement is true or false, and justify your answer.

(a) The vector $\begin{bmatrix} \frac{1}{3} \\ 0 \\ \frac{2}{3} \end{bmatrix}$ is a probability vector.

(b) The matrix $\begin{bmatrix} 0.2 & 1 \\ 0.8 & 0 \end{bmatrix}$ is a regular stochastic matrix.

(c) The column vectors of a transition matrix are probability vectors.

(d) A steady-state vector for a Markov chain with transition matrix P is any solution of the linear system $(I - P)\mathbf{q} = \mathbf{0}$.

(e) The square of every regular stochastic matrix is stochastic.

Chapter 4 Supplementary Exercises

1. Let V be the set of all ordered pairs of real numbers, and consider the following addition and scalar multiplication operations on $\mathbf{u} = (u_1, u_2, u_3)$ and $\mathbf{v} = (v_1, v_2, v_3)$:

$$\mathbf{u} + \mathbf{v} = (u_1 + v_1, u_2 + v_2, u_3 + v_3), \quad k\mathbf{u} = (ku_1, 0, 0)$$

(a) Compute $\mathbf{u} + \mathbf{v}$ and $k\mathbf{u}$ for $\mathbf{u} = (3, -2, 4)$, $\mathbf{v} = (1, 5, -2)$, and $k = -1$.

(b) In words, explain why V is closed under addition and scalar multiplication.

(c) Since the addition operation on V is the standard addition operation on R^3, certain vector space axioms hold for V because they are known to hold for R^3. Which axioms in Definition 1 of Section 4.1 are they?

(d) Show that Axioms 7, 8, and 9 hold.

(e) Show that Axiom 10 fails for the given operations.

2. In each part, the solution space of the system is a subspace of R^3 and so must be a line through the origin, a plane through the origin, all of R^3, or the origin only. For each system, determine which is the case. If the subspace is a plane, find an equation for it, and if it is a line, find parametric equations.

(a) $0x + 0y + 0z = 0$

(b) $\begin{aligned} 2x - 3y + z &= 0 \\ 6x - 9y + 3z &= 0 \\ -4x + 6y - 2z &= 0 \end{aligned}$

(c) $\begin{aligned} x - 2y + 7z &= 0 \\ -4x + 8y + 5z &= 0 \\ 2x - 4y + 3z &= 0 \end{aligned}$

(d) $\begin{aligned} x + 4y + 8z &= 0 \\ 2x + 5y + 6z &= 0 \\ 3x + y - 4z &= 0 \end{aligned}$

3. For what values of s is the solution space of

$$x_1 + x_2 + sx_3 = 0$$
$$x_1 + sx_2 + x_3 = 0$$
$$sx_1 + x_2 + x_3 = 0$$

the origin only, a line through the origin, a plane through the origin, or all of R^3?

4. (a) Express $(4a, a - b, a + 2b)$ as a linear combination of $(4, 1, 1)$ and $(0, -1, 2)$.

(b) Express $(3a + b + 3c, -a + 4b - c, 2a + b + 2c)$ as a linear combination of $(3, -1, 2)$ and $(1, 4, 1)$.

(c) Express $(2a - b + 4c, 3a - c, 4b + c)$ as a linear combination of three nonzero vectors.

5. Let W be the space spanned by $\mathbf{f} = \sin x$ and $\mathbf{g} = \cos x$.

(a) Show that for any value of θ, $\mathbf{f}_1 = \sin(x + \theta)$ and $\mathbf{g}_1 = \cos(x + \theta)$ are vectors in W.

(b) Show that $\mathbf{f}_1$ and $\mathbf{g}_1$ form a basis for W.

6. (a) Express $\mathbf{v} = (1, 1)$ as a linear combination of $\mathbf{v}_1 = (1, -1)$, $\mathbf{v}_2 = (3, 0)$, and $\mathbf{v}_3 = (2, 1)$ in two different ways.

(b) Explain why this does not violate Theorem 4.4.1.

7. Let A be an $n \times n$ matrix, and let $\mathbf{v}_1, \mathbf{v}_2, \ldots, \mathbf{v}_n$ be linearly independent vectors in R^n expressed as $n \times 1$ matrices. What must be true about A for $A\mathbf{v}_1, A\mathbf{v}_2, \ldots, A\mathbf{v}_n$ to be linearly independent?

8. Must a basis for P_n contain a polynomial of degree k for each $k = 0, 1, 2, \ldots, n$? Justify your answer.

9. For the purpose of this exercise, let us define a "checkerboard matrix" to be a square matrix $A = [a_{ij}]$ such that

$$a_{ij} = \begin{cases} 1 & \text{if } i + j \text{ is even} \\ 0 & \text{if } i + j \text{ is odd} \end{cases}$$

Find the rank and nullity of the following checkerboard matrices.

(a) The 3×3 checkerboard matrix.

(b) The 4×4 checkerboard matrix.

(c) The $n \times n$ checkerboard matrix.

10. For the purpose of this exercise, let us define an "X-matrix" to be a square matrix with an odd number of rows and columns that has 0's everywhere except on the two diagonals where it has 1's. Find the rank and nullity of the following X-matrices.

(a) $\begin{bmatrix} 1 & 0 & 1 \\ 0 & 1 & 0 \\ 1 & 0 & 1 \end{bmatrix}$

(b) $\begin{bmatrix} 1 & 0 & 0 & 0 & 1 \\ 0 & 1 & 0 & 1 & 0 \\ 0 & 0 & 1 & 0 & 0 \\ 0 & 1 & 0 & 1 & 0 \\ 1 & 0 & 0 & 0 & 1 \end{bmatrix}$

(c) the X-matrix of size $(2n + 1) \times (2n + 1)$

11. In each part, show that the stated set of polynomials is a subspace of P_n and find a basis for it.

(a) All polynomials in P_n such that $p(-x) = p(x)$.

(b) All polynomials in P_n such that $p(0) = 0$.

12. (*Calculus required*) Show that the set of all polynomials in P_n that have a horizontal tangent at $x = 0$ is a subspace of P_n. Find a basis for this subspace.

13. (a) Find a basis for the vector space of all 3×3 symmetric matrices.

(b) Find a basis for the vector space of all 3×3 skew-symmetric matrices.

14. Various advanced texts in linear algebra prove the following determinant criterion for rank: *The rank of a matrix A is r if and only if A has some $r \times r$ submatrix with a nonzero determinant, and all square submatrices of larger size have determinant zero.* [*Note:* A submatrix of A is any matrix obtained by deleting rows or columns of A. The matrix A itself is also considered to be a submatrix of A.] In each part, use this criterion to find the rank of the matrix.

(a) $\begin{bmatrix} 1 & 2 & 0 \\ 2 & 4 & -1 \end{bmatrix}$

(b) $\begin{bmatrix} 1 & 2 & 3 \\ 2 & 4 & 6 \end{bmatrix}$

(c) $\begin{bmatrix} 1 & 0 & 1 \\ 2 & -1 & 3 \\ 3 & -1 & 4 \end{bmatrix}$

(d) $\begin{bmatrix} 1 & -1 & 2 & 0 \\ 3 & 1 & 0 & 0 \\ -1 & 2 & 4 & 0 \end{bmatrix}$

15. Use the result in Exercise 14 above to find the possible ranks for matrices of the form

$$\begin{bmatrix} 0 & 0 & 0 & 0 & 0 & a_{16} \\ 0 & 0 & 0 & 0 & 0 & a_{26} \\ 0 & 0 & 0 & 0 & 0 & a_{36} \\ 0 & 0 & 0 & 0 & 0 & a_{46} \\ a_{51} & a_{52} & a_{53} & a_{54} & a_{55} & a_{56} \end{bmatrix}$$

16. Prove: If S is a basis for a vector space V, then for any vectors $\mathbf{u}$ and $\mathbf{v}$ in V and any scalar k, the following relationships hold.

(a) $(\mathbf{u} + \mathbf{v})_S = (\mathbf{u})_S + (\mathbf{v})_S$

(b) $(k\mathbf{u})_S = k(\mathbf{u})_S$

CHAPTER 5

Eigenvalues and Eigenvectors

CHAPTER CONTENTS   5.1 Eigenvalues and Eigenvectors 295
5.2 Diagonalization 305
5.3 Complex Vector Spaces 315
5.4 Differential Equations 327

INTRODUCTION   In this chapter we will focus on classes of scalars and vectors known as "eigenvalues" and "eigenvectors," terms derived from the German word *eigen*, meaning "own," "peculiar to," "characteristic," or "individual." The underlying idea first appeared in the study of rotational motion but was later used to classify various kinds of surfaces and to describe solutions of certain differential equations. In the early 1900s it was applied to matrices and matrix transformations, and today it has applications in such diverse fields as computer graphics, mechanical vibrations, heat flow, population dynamics, quantum mechanics, and economics to name just a few.

5.1 Eigenvalues and Eigenvectors

In this section we will define the notions of "eigenvalue" and "eigenvector" and discuss some of their basic properties.

Definition of Eigenvalue and Eigenvector

We begin with the main definition in this section.

> **DEFINITION 1** If A is an $n \times n$ matrix, then a nonzero vector $\mathbf{x}$ in R^n is called an *eigenvector* of A (or of the matrix operator T_A) if $A\mathbf{x}$ is a scalar multiple of $\mathbf{x}$; that is,
>
> $$A\mathbf{x} = \lambda\mathbf{x}$$
>
> for some scalar λ. The scalar λ is called an *eigenvalue* of A (or of T_A), and $\mathbf{x}$ is said to be an *eigenvector corresponding to* λ.

The requirement that an eigenvector be nonzero is imposed to avoid the unimportant case $A\mathbf{0} = \lambda\mathbf{0}$, which holds for every A and λ.

In general, the image of a vector $\mathbf{x}$ under multiplication by a square matrix A differs from $\mathbf{x}$ in both magnitude and direction. However, in the special case where $\mathbf{x}$ is an eigenvector of A, multiplication by A leaves the direction unchanged. For example, in R^2 or R^3 multiplication by A maps each eigenvector $\mathbf{x}$ of A (if any) along the same line through the origin as $\mathbf{x}$. Depending on the sign and magnitude of the eigenvalue λ corresponding to $\mathbf{x}$, the operation $A\mathbf{x} = \lambda\mathbf{x}$ compresses or stretches $\mathbf{x}$ by a factor of λ, with a reversal of direction in the case where λ is negative (Figure 5.1.1).

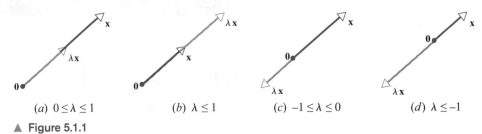

(a) $0 \le \lambda \le 1$ (b) $\lambda \le 1$ (c) $-1 \le \lambda \le 0$ (d) $\lambda \le -1$

▲ Figure 5.1.1

▶ **EXAMPLE 1 Eigenvector of a 2 × 2 Matrix**

The vector $\mathbf{x} = \begin{bmatrix} 1 \\ 2 \end{bmatrix}$ is an eigenvector of

$$A = \begin{bmatrix} 3 & 0 \\ 8 & -1 \end{bmatrix}$$

corresponding to the eigenvalue $\lambda = 3$, since

$$A\mathbf{x} = \begin{bmatrix} 3 & 0 \\ 8 & -1 \end{bmatrix} \begin{bmatrix} 1 \\ 2 \end{bmatrix} = \begin{bmatrix} 3 \\ 6 \end{bmatrix} = 3\mathbf{x}$$

▲ Figure 5.1.2

Geometrically, multiplication by A has stretched the vector $\mathbf{x}$ by a factor of 3 (Figure 5.1.2). ◀

Computing Eigenvalues and Eigenvectors

Our next objective is to obtain a general procedure for finding eigenvalues and eigenvectors of an $n \times n$ matrix A. We will begin with the problem of finding the eigenvalues of A. Note first that the equation $A\mathbf{x} = \lambda\mathbf{x}$ can be rewritten as $A\mathbf{x} = \lambda I \mathbf{x}$, or equivalently, as

$$(\lambda I - A)\mathbf{x} = 0$$

For λ to be an eigenvalue of A this equation must have a nonzero solution for $\mathbf{x}$. But it follows from parts (b) and (g) of Theorem 4.9.4 that this is so if and only if the coefficient matrix $\lambda I - A$ has a zero determinant. Thus, we have the following result.

THEOREM 5.1.1 *If A is an $n \times n$ matrix, then λ is an eigenvalue of A if and only if it satisfies the equation*

$$\det(\lambda I - A) = 0 \qquad (1)$$

*This is called the **characteristic equation** of A.*

▶ **EXAMPLE 2 Finding Eigenvalues**

In Example 1 we observed that $\lambda = 3$ is an eigenvalue of the matrix

$$A = \begin{bmatrix} 3 & 0 \\ 8 & -1 \end{bmatrix}$$

but we did not explain how we found it. Use the characteristic equation to find all eigenvalues of this matrix.

Solution It follows from Formula (1) that the eigenvalues of A are the solutions of the equation $\det(\lambda I - A) = 0$, which we can write as

$$\begin{vmatrix} \lambda - 3 & 0 \\ -8 & \lambda + 1 \end{vmatrix} = 0$$

from which we obtain

$$(\lambda - 3)(\lambda + 1) = 0 \tag{2}$$

This shows that the eigenvalues of A are $\lambda = 3$ and $\lambda = -1$. Thus, in addition to the eigenvalue $\lambda = 3$ noted in Example 1, we have discovered a second eigenvalue $\lambda = -1$.
◄

When the determinant $\det(\lambda I - A)$ that appears on the left side of (1) is expanded, the result is a polynomial $p(\lambda)$ of degree n that is called the ***characteristic polynomial*** of A. For example, it follows from (2) that the characteristic polynomial of the 2×2 matrix A in Example 2 is

$$p(\lambda) = (\lambda - 3)(\lambda + 1) = \lambda^2 - 2\lambda - 3$$

which is a polynomial of degree 2. In general, the characteristic polynomial of an $n \times n$ matrix has the form

$$p(\lambda) = \lambda^n + c_1\lambda^{n-1} + \cdots + c_n$$

in which the coefficient of λ^n is 1 (Exercise 17). Since a polynomial of degree n has at most n distinct roots, it follows that the equation

$$\lambda^n + c_1\lambda^{n-1} + \cdots + c_n = 0 \tag{3}$$

has at most n distinct solutions and consequently that an $n \times n$ matrix has at most n distinct eigenvalues. Since some of these solutions may be complex numbers, it is possible for a matrix to have complex eigenvalues, even if that matrix itself has real entries. We will discuss this issue in more detail later, but for now we will focus on examples in which the eigenvalues are real numbers.

► **EXAMPLE 3 Eigenvalues of a 3 × 3 Matrix**

Find the eigenvalues of

$$A = \begin{bmatrix} 0 & 1 & 0 \\ 0 & 0 & 1 \\ 4 & -17 & 8 \end{bmatrix}$$

Solution The characteristic polynomial of A is

$$\det(\lambda I - A) = \det \begin{bmatrix} \lambda & -1 & 0 \\ 0 & \lambda & -1 \\ -4 & 17 & \lambda - 8 \end{bmatrix} = \lambda^3 - 8\lambda^2 + 17\lambda - 4$$

The eigenvalues of A must therefore satisfy the cubic equation

$$\lambda^3 - 8\lambda^2 + 17\lambda - 4 = 0 \tag{4}$$

To solve this equation, we will begin by searching for integer solutions. This task can be simplified by exploiting the fact that all integer solutions (if there are any) of a polynomial equation with *integer coefficients*

$$\lambda^n + c_1\lambda^{n-1} + \cdots + c_n = 0$$

must be divisors of the constant term, c_n. Thus, the only possible integer solutions of (4) are the divisors of -4, that is, $\pm 1, \pm 2, \pm 4$. Successively substituting these values in (4) shows that $\lambda = 4$ is an integer solution. As a consequence, $\lambda - 4$ must be a factor of the left side of (4). Dividing $\lambda - 4$ into $\lambda^3 - 8\lambda^2 + 17\lambda - 4$ shows that (4) can be rewritten as

$$(\lambda - 4)(\lambda^2 - 4\lambda + 1) = 0$$

In applications involving large matrices it is often not feasible to compute the characteristic equation directly so other methods must be used to find eigenvalues. We will consider such methods in Chapter 9.

Thus, the remaining solutions of (4) satisfy the quadratic equation

$$\lambda^2 - 4\lambda + 1 = 0$$

which can be solved by the quadratic formula. Thus the eigenvalues of A are

$$\lambda = 4, \quad \lambda = 2 + \sqrt{3}, \quad \text{and} \quad \lambda = 2 - \sqrt{3}$$

▶ **EXAMPLE 4** **Eigenvalues of an Upper Triangular Matrix**

Find the eigenvalues of the upper triangular matrix

$$A = \begin{bmatrix} a_{11} & a_{12} & a_{13} & a_{14} \\ 0 & a_{22} & a_{23} & a_{24} \\ 0 & 0 & a_{33} & a_{34} \\ 0 & 0 & 0 & a_{44} \end{bmatrix}$$

Solution Recalling that the determinant of a triangular matrix is the product of the entries on the main diagonal (Theorem 2.1.2), we obtain

$$\det(\lambda I - A) = \det \begin{bmatrix} \lambda - a_{11} & -a_{12} & -a_{13} & -a_{14} \\ 0 & \lambda - a_{22} & -a_{23} & -a_{24} \\ 0 & 0 & \lambda - a_{33} & -a_{34} \\ 0 & 0 & 0 & \lambda - a_{44} \end{bmatrix}$$

$$= (\lambda - a_{11})(\lambda - a_{22})(\lambda - a_{33})(\lambda - a_{44})$$

Thus, the characteristic equation is

$$(\lambda - a_{11})(\lambda - a_{22})(\lambda - a_{33})(\lambda - a_{44}) = 0$$

and the eigenvalues are

$$\lambda = a_{11}, \quad \lambda = a_{22}, \quad \lambda = a_{33}, \quad \lambda = a_{44}$$

which are precisely the diagonal entries of A. ◀

The following general theorem should be evident from the computations in the preceding example.

THEOREM 5.1.2 *If A is an $n \times n$ triangular matrix (upper triangular, lower triangular, or diagonal), then the eigenvalues of A are the entries on the main diagonal of A.*

▶ **EXAMPLE 5** **Eigenvalues of a Lower Triangular Matrix**

By inspection, the eigenvalues of the lower triangular matrix

$$A = \begin{bmatrix} \frac{1}{2} & 0 & 0 \\ -1 & \frac{2}{3} & 0 \\ 5 & -8 & -\frac{1}{4} \end{bmatrix}$$

Had Theorem 5.1.2 been available earlier, we could have anticipated the result obtained in Example 2.

are $\lambda = \frac{1}{2}$, $\lambda = \frac{2}{3}$, and $\lambda = -\frac{1}{4}$. ◀

THEOREM 5.1.3 *If A is an n × n matrix, the following statements are equivalent.*

(a) *λ is an eigenvalue of A.*

(b) *The system of equations $(\lambda I - A)\mathbf{x} = \mathbf{0}$ has nontrivial solutions.*

(c) *There is a nonzero vector $\mathbf{x}$ such that $A\mathbf{x} = \lambda \mathbf{x}$.*

(d) *λ is a solution of the characteristic equation $\det(\lambda I - A) = 0$.*

Finding Eigenvectors and Bases for Eigenspaces

Now that we know how to find the eigenvalues of a matrix, we will consider the problem of finding the corresponding eigenvectors. Since the eigenvectors corresponding to an eigenvalue λ of a matrix A are the nonzero vectors that satisfy the equation

$$(\lambda I - A)\mathbf{x} = \mathbf{0}$$

Notice that $\mathbf{x} = \mathbf{0}$ is in every eigenspace even though it is not an eigenvector. Thus, it is the *nonzero* vectors in an eigenspace that are the eigenvectors.

these eigenvectors are the nonzero vectors in the null space of the matrix $\lambda I - A$. We call this null space the **eigenspace** of A corresponding to λ. Stated another way, *the eigenspace of A corresponding to the eigenvalue λ is the solution space of the homogeneous system $(\lambda I - A)\mathbf{x} = \mathbf{0}$.*

▶ **EXAMPLE 6 Bases for Eigenspaces**

Find bases for the eigenspaces of the matrix

$$A = \begin{bmatrix} 3 & 0 \\ 8 & -1 \end{bmatrix}$$

Solution In Example 1 we found the characteristic equation of A to be

$$(\lambda - 3)(\lambda + 1) = 0$$

from which we obtained the eigenvalues $\lambda = 3$ and $\lambda = -1$. Thus, there are two eigenspaces of A, one corresponding to each of these eigenvalues.

By definition,

$$\mathbf{x} = \begin{bmatrix} x_1 \\ x_2 \end{bmatrix}$$

is an eigenvector of A corresponding to an eigenvalue λ if and only if $\mathbf{x}$ is a nontrivial solution of $(\lambda I - A)\mathbf{x} = \mathbf{0}$, that is, of

$$\begin{bmatrix} \lambda - 3 & 0 \\ -8 & \lambda + 1 \end{bmatrix}\begin{bmatrix} x_1 \\ x_2 \end{bmatrix} = \begin{bmatrix} 0 \\ 0 \end{bmatrix}$$

If $\lambda = 3$, then this equation becomes

$$\begin{bmatrix} 0 & 0 \\ -8 & 4 \end{bmatrix}\begin{bmatrix} x_1 \\ x_2 \end{bmatrix} = \begin{bmatrix} 0 \\ 0 \end{bmatrix}$$

Historical Note Methods of linear algebra are used in the emerging field of computerized face recognition. Researchers are working with the idea that every human face in a racial group is a combination of a few dozen primary shapes. For example, by analyzing three-dimensional scans of many faces, researchers at Rockefeller University have produced both an average head shape in the Caucasian group–dubbed the **meanhead** (top row left in the figure to the left)–and a set of standardized variations from that shape, called **eigenheads** (15 of which are shown in the picture). These are so named because they are eigenvectors of a certain matrix that stores digitized facial information. Face shapes are represented mathematically as linear combinations of the eigenheads.

[Image: Courtesy Dr. Joseph Atick, Dr. Norman Redlich, and Dr. Paul Griffith]

whose general solution is

$$x_1 = \tfrac{1}{2}t, \quad x_2 = t$$

(verify) or in matrix form,

$$\begin{bmatrix} x_1 \\ x_2 \end{bmatrix} = \begin{bmatrix} \tfrac{1}{2}t \\ t \end{bmatrix} = t \begin{bmatrix} \tfrac{1}{2} \\ 1 \end{bmatrix}$$

Thus,

$$\begin{bmatrix} \tfrac{1}{2} \\ 1 \end{bmatrix}$$

is a basis for the eigenspace corresponding to $\lambda = 3$. We leave it as an exercise for you to follow the pattern of these computations and show that

$$\begin{bmatrix} 0 \\ 1 \end{bmatrix}$$

is a basis for the eigenspace corresponding to $\lambda = -1$.

▶ **EXAMPLE 7 Eigenvectors and Bases for Eigenspaces**

Find bases for the eigenspaces of

$$A = \begin{bmatrix} 0 & 0 & -2 \\ 1 & 2 & 1 \\ 1 & 0 & 3 \end{bmatrix}$$

Solution The characteristic equation of A is $\lambda^3 - 5\lambda^2 + 8\lambda - 4 = 0$, or in factored form, $(\lambda - 1)(\lambda - 2)^2 = 0$ (verify). Thus, the distinct eigenvalues of A are $\lambda = 1$ and $\lambda = 2$, so there are two eigenspaces of A.

By definition,

$$\mathbf{x} = \begin{bmatrix} x_1 \\ x_2 \\ x_3 \end{bmatrix}$$

is an eigenvector of A corresponding to λ if and only if $\mathbf{x}$ is a nontrivial solution of $(\lambda I - A)\mathbf{x} = \mathbf{0}$, or in matrix form,

$$\begin{bmatrix} \lambda & 0 & 2 \\ -1 & \lambda - 2 & -1 \\ -1 & 0 & \lambda - 3 \end{bmatrix} \begin{bmatrix} x_1 \\ x_2 \\ x_3 \end{bmatrix} = \begin{bmatrix} 0 \\ 0 \\ 0 \end{bmatrix} \tag{5}$$

In the case where $\lambda = 2$, Formula (5) becomes

$$\begin{bmatrix} 2 & 0 & 2 \\ -1 & 0 & -1 \\ -1 & 0 & -1 \end{bmatrix} \begin{bmatrix} x_1 \\ x_2 \\ x_3 \end{bmatrix} = \begin{bmatrix} 0 \\ 0 \\ 0 \end{bmatrix}$$

Solving this system using Gaussian elimination yields (verify)

$$x_1 = -s, \quad x_2 = t, \quad x_3 = s$$

Thus, the eigenvectors of A corresponding to $\lambda = 2$ are the nonzero vectors of the form

$$\mathbf{x} = \begin{bmatrix} -s \\ t \\ s \end{bmatrix} = \begin{bmatrix} -s \\ 0 \\ s \end{bmatrix} + \begin{bmatrix} 0 \\ t \\ 0 \end{bmatrix} = s \begin{bmatrix} -1 \\ 0 \\ 1 \end{bmatrix} + t \begin{bmatrix} 0 \\ 1 \\ 0 \end{bmatrix}$$

Since

$$\begin{bmatrix} -1 \\ 0 \\ 1 \end{bmatrix} \quad \text{and} \quad \begin{bmatrix} 0 \\ 1 \\ 0 \end{bmatrix}$$

are linearly independent (why?), these vectors form a basis for the eigenspace corresponding to $\lambda = 2$.

If $\lambda = 1$, then (5) becomes

$$\begin{bmatrix} 1 & 0 & 2 \\ -1 & -1 & -1 \\ -1 & 0 & -2 \end{bmatrix} \begin{bmatrix} x_1 \\ x_2 \\ x_3 \end{bmatrix} = \begin{bmatrix} 0 \\ 0 \\ 0 \end{bmatrix}$$

Solving this system yields (verify)

$$x_1 = -2s, \quad x_2 = s, \quad x_3 = s$$

Thus, the eigenvectors corresponding to $\lambda = 1$ are the nonzero vectors of the form

$$\begin{bmatrix} -2s \\ s \\ s \end{bmatrix} = s \begin{bmatrix} -2 \\ 1 \\ 1 \end{bmatrix} \quad \text{so that} \quad \begin{bmatrix} -2 \\ 1 \\ 1 \end{bmatrix}$$

is a basis for the eigenspace corresponding to $\lambda = 1$. ◀

Powers of a Matrix   Once the eigenvalues and eigenvectors of a matrix A are found, it is a simple matter to find the eigenvalues and eigenvectors of any positive integer power of A; for example, if λ is an eigenvalue of A and $\mathbf{x}$ is a corresponding eigenvector, then

$$A^2\mathbf{x} = A(A\mathbf{x}) = A(\lambda\mathbf{x}) = \lambda(A\mathbf{x}) = \lambda(\lambda\mathbf{x}) = \lambda^2\mathbf{x}$$

which shows that λ^2 is an eigenvalue of A^2 and that $\mathbf{x}$ is a corresponding eigenvector. In general, we have the following result.

THEOREM 5.1.4 *If k is a positive integer, λ is an eigenvalue of a matrix A, and $\mathbf{x}$ is a corresponding eigenvector, then λ^k is an eigenvalue of A^k and $\mathbf{x}$ is a corresponding eigenvector.*

▶ **EXAMPLE 8 Powers of a Matrix**

In Example 7 we showed that the eigenvalues of

$$A = \begin{bmatrix} 0 & 0 & -2 \\ 1 & 2 & 1 \\ 1 & 0 & 3 \end{bmatrix}$$

are $\lambda = 2$ and $\lambda = 1$, so from Theorem 5.1.4 both $\lambda = 2^7 = 128$ and $\lambda = 1^7 = 1$ are eigenvalues of A^7. We also showed that

$$\begin{bmatrix} -1 \\ 0 \\ 1 \end{bmatrix} \quad \text{and} \quad \begin{bmatrix} 0 \\ 1 \\ 0 \end{bmatrix}$$

are eigenvectors of A corresponding to the eigenvalue $\lambda = 2$, so from Theorem 5.1.4 they are also eigenvectors of A^7 corresponding to $\lambda = 2^7 = 128$. Similarly, the eigenvector

$$\begin{bmatrix} -2 \\ 1 \\ 1 \end{bmatrix}$$

of A corresponding to the eigenvalue $\lambda = 1$ is also an eigenvector of A^7 corresponding to $\lambda = 1^7 = 1$. ◀

Eigenvalues and Invertibility   The next theorem establishes a relationship between eigenvalues and the invertibility of a matrix.

THEOREM 5.1.5 *A square matrix A is invertible if and only if $\lambda = 0$ is not an eigenvalue of A.*

Proof Assume that A is an $n \times n$ matrix and observe first that $\lambda = 0$ is a solution of the characteristic equation

$$\lambda^n + c_1\lambda^{n-1} + \cdots + c_n = 0$$

if and only if the constant term c_n is zero. Thus, it suffices to prove that A is invertible if and only if $c_n \neq 0$. But

$$\det(\lambda I - A) = \lambda^n + c_1\lambda^{n-1} + \cdots + c_n$$

or, on setting $\lambda = 0$,

$$\det(-A) = c_n \quad \text{or} \quad (-1)^n \det(A) = c_n$$

It follows from the last equation that $\det(A) = 0$ if and only if $c_n = 0$, and this in turn implies that A is invertible if and only if $c_n \neq 0$. ◄

► EXAMPLE 9 Eigenvalues and Invertibility

The matrix A in Example 7 is invertible since it has eigenvalues $\lambda = 1$ and $\lambda = 2$, neither of which is zero. We leave it for you to check this conclusion by showing that $\det(A) \neq 0$. ◄

More on the Equivalence Theorem   As our final result in this section, we will use Theorem 5.1.5 to add one additional part to Theorem 4.10.4.

THEOREM 5.1.6 Equivalent Statements

If A is an $n \times n$ matrix, then the following statements are equivalent.

(a)   A is invertible.

(b) $A\mathbf{x} = \mathbf{0}$ has only the trivial solution.

(c)   The reduced row echelon form of A is I_n.

(d)   A is expressible as a product of elementary matrices.

(e) $A\mathbf{x} = \mathbf{b}$ is consistent for every $n \times 1$ matrix $\mathbf{b}$.

(f) $A\mathbf{x} = \mathbf{b}$ has exactly one solution for every $n \times 1$ matrix $\mathbf{b}$.

(g) $\det(A) \neq 0$.

(h)   The column vectors of A are linearly independent.

(i)   The row vectors of A are linearly independent.

(j)   The column vectors of A span R^n.

(k)   The row vectors of A span R^n.

(l)   The column vectors of A form a basis for R^n.

(m)   The row vectors of A form a basis for R^n.

(n)   A has rank n.

(o)   A has nullity 0.

(p)   The orthogonal complement of the null space of A is R^n.

(q)   The orthogonal complement of the row space of A is $\{\mathbf{0}\}$.

(r)   The range of T_A is R^n.

(s) T_A is one-to-one.

(t) $\lambda = 0$ is not an eigenvalue of A.

This theorem relates all of the major topics we have studied thus far.

Concept Review

- Eigenvector
- Eigenvalue
- Characteristic equation
- Characteristic polynomial
- Eigenspace
- Equivalence Theorem

Skills

- Find the eigenvalues of a matrix.
- Find bases for the eigenspaces of a matrix.

Exercise Set 5.1

▶ In Exercises 1–2, confirm by multiplication that $\mathbf{x}$ is an eigenvector of A, and find the corresponding eigenvalue. ◀

1. $A = \begin{bmatrix} 4 & 0 & 1 \\ 2 & 3 & 2 \\ 1 & 0 & 4 \end{bmatrix}; \; \mathbf{x} = \begin{bmatrix} 1 \\ 2 \\ 1 \end{bmatrix}$

2. $A = \begin{bmatrix} 2 & -1 & -1 \\ -1 & 2 & -1 \\ -1 & -1 & 2 \end{bmatrix}; \; \mathbf{x} = \begin{bmatrix} 1 \\ 1 \\ 1 \end{bmatrix}$

3. Find the characteristic equations of the following matrices:

(a) $\begin{bmatrix} 3 & 0 \\ 8 & -1 \end{bmatrix}$ (b) $\begin{bmatrix} 10 & -9 \\ 4 & -2 \end{bmatrix}$ (c) $\begin{bmatrix} 0 & 3 \\ 4 & 0 \end{bmatrix}$

(d) $\begin{bmatrix} -2 & -7 \\ 1 & 2 \end{bmatrix}$ (e) $\begin{bmatrix} 0 & 0 \\ 0 & 0 \end{bmatrix}$ (f) $\begin{bmatrix} 1 & 0 \\ 0 & 1 \end{bmatrix}$

4. Find the eigenvalues of the matrices in Exercise 3.

5. Find bases for the eigenspaces of the matrices in Exercise 3.

6. Find the characteristic equations of the following matrices:

(a) $\begin{bmatrix} 4 & 0 & 1 \\ -2 & 1 & 0 \\ -2 & 0 & 1 \end{bmatrix}$ (b) $\begin{bmatrix} 3 & 0 & -5 \\ \frac{1}{5} & -1 & 0 \\ 1 & 1 & -2 \end{bmatrix}$

(c) $\begin{bmatrix} -2 & 0 & 1 \\ -6 & -2 & 0 \\ 19 & 5 & -4 \end{bmatrix}$ (d) $\begin{bmatrix} -1 & 0 & 1 \\ -1 & 3 & 0 \\ -4 & 13 & -1 \end{bmatrix}$

(e) $\begin{bmatrix} 5 & 0 & 1 \\ 1 & 1 & 0 \\ -7 & 1 & 0 \end{bmatrix}$ (f) $\begin{bmatrix} 5 & 6 & 2 \\ 0 & -1 & -8 \\ 1 & 0 & -2 \end{bmatrix}$

7. Find the eigenvalues of the matrices in Exercise 6.

8. Find bases for the eigenspaces of the matrices in Exercise 6.

9. Find the characteristic equations of the following matrices:

(a) $\begin{bmatrix} 0 & 0 & 2 & 0 \\ 1 & 0 & 1 & 0 \\ 0 & 1 & -2 & 0 \\ 0 & 0 & 0 & 1 \end{bmatrix}$ (b) $\begin{bmatrix} 10 & -9 & 0 & 0 \\ 4 & -2 & 0 & 0 \\ 0 & 0 & -2 & -7 \\ 0 & 0 & 1 & 2 \end{bmatrix}$

10. Find the eigenvalues of the matrices in Exercise 9.

11. Find bases for the eigenspaces of the matrices in Exercise 9.

12. By inspection, find the eigenvalues of the following matrices:

(a) $\begin{bmatrix} -1 & 6 \\ 0 & 5 \end{bmatrix}$ (b) $\begin{bmatrix} 3 & 0 & 0 \\ -2 & 7 & 0 \\ 4 & 8 & 1 \end{bmatrix}$

(c) $\begin{bmatrix} -\frac{1}{3} & 0 & 0 & 0 \\ 0 & -\frac{1}{3} & 0 & 0 \\ 0 & 0 & 1 & 0 \\ 0 & 0 & 0 & \frac{1}{2} \end{bmatrix}$

13. Find the eigenvalues of A^9 for

$$A = \begin{bmatrix} 1 & 3 & 7 & 11 \\ 0 & \frac{1}{2} & 3 & 8 \\ 0 & 0 & 0 & 4 \\ 0 & 0 & 0 & 2 \end{bmatrix}$$

14. Find the eigenvalues and bases for the eigenspaces of A^{25} for

$$A = \begin{bmatrix} -1 & -2 & -2 \\ 1 & 2 & 1 \\ -1 & -1 & 0 \end{bmatrix}$$

15. Let A be a 2×2 matrix, and call a line through the origin of R^2 *invariant* under A if $A\mathbf{x}$ lies on the line when $\mathbf{x}$ does. Find equations for all lines in R^2, if any, that are invariant under the given matrix.

(a) $A = \begin{bmatrix} 4 & -1 \\ 2 & 1 \end{bmatrix}$ (b) $A = \begin{bmatrix} 0 & 1 \\ -1 & 0 \end{bmatrix}$ (c) $A = \begin{bmatrix} 2 & 3 \\ 0 & 2 \end{bmatrix}$

16. Find $\det(A)$ given that A has $p(\lambda)$ as its characteristic polynomial.

(a) $p(\lambda) = \lambda^3 - 2\lambda^2 + \lambda + 5$

(b) $p(\lambda) = \lambda^4 - \lambda^3 + 7$

[*Hint:* See the proof of Theorem 5.1.5.]

17. Let A be an $n \times n$ matrix.

(a) Prove that the characteristic polynomial of A has degree n.

(b) Prove that the coefficient of λ^n in the characteristic polynomial is 1.

18. Show that the characteristic equation of a 2×2 matrix A can be expressed as $\lambda^2 - \mathrm{tr}(A)\lambda + \det(A) = 0$, where $\mathrm{tr}(A)$ is the trace of A.

19. Use the result in Exercise 18 to show that if

$$A = \begin{bmatrix} a & b \\ c & d \end{bmatrix}$$

then the solutions of the characteristic equation of A are

$$\lambda = \tfrac{1}{2}\left[(a + d) \pm \sqrt{(a - d)^2 + 4bc} \right]$$

Use this result to show that A has

(a) two distinct real eigenvalues if $(a - d)^2 + 4bc > 0$.

(b) two repeated real eigenvalues if $(a - d)^2 + 4bc = 0$.

(c) complex conjugate eigenvalues if $(a - d)^2 + 4bc < 0$.

20. Let A be the matrix in Exercise 19. Show that if $b \neq 0$, then

$$\mathbf{x}_1 = \begin{bmatrix} -b \\ a - \lambda_1 \end{bmatrix} \quad \text{and} \quad \mathbf{x}_2 = \begin{bmatrix} -b \\ a - \lambda_2 \end{bmatrix}$$

are eigenvectors of A that correspond, respectively, to the eigenvalues

$$\lambda_1 = \tfrac{1}{2}\left[(a + d) + \sqrt{(a - d)^2 + 4bc} \right]$$

and

$$\lambda_2 = \tfrac{1}{2}\left[(a + d) - \sqrt{(a - d)^2 + 4bc} \right]$$

21. Use the result of Exercise 18 to prove that if $p(\lambda)$ is the characteristic polynomial of a 2×2 matrix A, then $p(A) = 0$.

22. Prove: If a, b, c, and d are integers such that $a + b = c + d$, then

$$A = \begin{bmatrix} a & b \\ c & d \end{bmatrix}$$

has integer eigenvalues—namely, $\lambda_1 = a + b$ and $\lambda_2 = a - c$.

23. Prove: If λ is an eigenvalue of an invertible matrix A, and $\mathbf{x}$ is a corresponding eigenvector, then $1/\lambda$ is an eigenvalue of A^{-1}, and $\mathbf{x}$ is a corresponding eigenvector.

24. Prove: If λ is an eigenvalue of A, $\mathbf{x}$ is a corresponding eigenvector, and s is a scalar, then $\lambda - s$ is an eigenvalue of $A - sI$, and $\mathbf{x}$ is a corresponding eigenvector.

25. Prove: If λ is an eigenvalue of A and $\mathbf{x}$ is a corresponding eigenvector, then $s\lambda$ is an eigenvalue of sA for every scalar s, and $\mathbf{x}$ is a corresponding eigenvector.

26. Find the eigenvalues and bases for the eigenspaces of

$$A = \begin{bmatrix} -2 & 2 & 3 \\ -2 & 3 & 2 \\ -4 & 2 & 5 \end{bmatrix}$$

and then use Exercises 23 and 24 to find the eigenvalues and bases for the eigenspaces of

(a) A^{-1} (b) $A - 3I$ (c) $A + 2I$

27. (a) Prove that if A is a square matrix, then A and A^T have the same eigenvalues. [*Hint:* Look at the characteristic equation $\det(\lambda I - A) = 0$.]

(b) Show that A and A^T need not have the same eigenspaces. [*Hint:* Use the result in Exercise 20 to find a 2×2 matrix for which A and A^T have different eigenspaces.]

28. Suppose that the characteristic polynomial of some matrix A is found to be $p(\lambda) = (\lambda - 1)(\lambda - 3)^2(\lambda - 4)^3$. In each part, answer the question and explain your reasoning.

(a) What is the size of A?

(b) Is A invertible?

(c) How many eigenspaces does A have?

29. The eigenvectors that we have been studying are sometimes called *right eigenvectors* to distinguish them from *left eigenvectors*, which are $n \times 1$ column matrices $\mathbf{x}$ that satisfy the equation $\mathbf{x}^T A = \mu \mathbf{x}^T$ for some scalar μ. What is the relationship, if any, between the right eigenvectors and corresponding eigenvalues λ of A and the left eigenvectors and corresponding eigenvalues μ of A?

True-False Exercises

In parts (a)–(g) determine whether the statement is true or false, and justify your answer.

(a) If A is a square matrix and $A\mathbf{x} = \lambda\mathbf{x}$ for some nonzero scalar λ, then $\mathbf{x}$ is an eigenvector of A.

(b) If λ is an eigenvalue of a matrix A, then the linear system $(\lambda I - A)\mathbf{x} = \mathbf{0}$ has only the trivial solution.

(c) If the characteristic polynomial of a matrix A is $p(\lambda) = \lambda^2 + 1$, then A is invertible.

(d) If λ is an eigenvalue of a matrix A, then the eigenspace of A corresponding to λ is the set of eigenvectors of A corresponding to λ.

(e) If 0 is an eigenvalue of a matrix A, then A^2 is singular.

(f) The eigenvalues of a matrix A are the same as the eigenvalues of the reduced row echelon form of A.

(g) If 0 is an eigenvalue of a matrix A, then the set of columns of A is linearly independent.

5.2 Diagonalization

In this section we will be concerned with the problem of finding a basis for R^n that consists of eigenvectors of an $n \times n$ matrix A. Such bases can be used to study geometric properties of A and to simplify various numerical computations. These bases are also of physical significance in a wide variety of applications, some of which will be considered later in this text.

The Matrix Diagonalization Problem

Our first objective in this section is to show that the following two seemingly different problems are equivalent.

> **Problem 1** Given an $n \times n$ matrix A, does there exist an invertible matrix P such that $P^{-1}AP$ is diagonal?
>
> **Problem 2** Given an $n \times n$ matrix A, does A have n linearly independent eigenvectors?

Similarity

The matrix product $P^{-1}AP$ that appears in Problem 1 is called a ***similarity transformation*** of the matrix A. Such products are important in the study of eigenvectors and eigenvalues, so we will begin with some terminology about them.

> **DEFINITION 1** If A and B are square matrices, then we say that **B is similar to A** if there is an invertible matrix P such that $B = P^{-1}AP$.

Note that if B is similar to A, then it is also true that A is similar to B, since we can express B as $B = Q^{-1}AQ$ by taking $Q = P^{-1}$. This being the case, we will usually say that A and B are **similar matrices** if either is similar to the other.

Similarity Invariants

Similar matrices have many properties in common. For example, if $B = P^{-1}AP$, then it follows that A and B have the same determinant, since

$$\det(B) = \det(P^{-1}AP) = \det(P^{-1})\det(A)\det(P)$$

$$= \frac{1}{\det(P)}\det(A)\det(P) = \det(A)$$

In general, any property that is shared by all similar matrices is called a **similarity invariant** or is said to be **invariant under similarity**. Table 1 lists the most important similarity invariants. The proofs of some of these results are given as exercises.

Expressed in the language of similarity, Problem 1 posed above is equivalent to asking whether the matrix A is similar to a diagonal matrix. If so, the diagonal matrix will have all of the similarity-invariant properties of A, but will have a simpler form, making it easier to analyze and work with. This important idea has some associated terminology.

> **DEFINITION 2** A square matrix A is said to be **diagonalizable** if it is similar to some diagonal matrix; that is, if there exists an invertible matrix P such that $P^{-1}AP$ is diagonal. In this case the matrix P is said to **diagonalize** A.

Table 1 Similarity Invariants

Property	Description
Determinant	A and $P^{-1}AP$ have the same determinant.
Invertibility	A is invertible if and only if $P^{-1}AP$ is invertible.
Rank	A and $P^{-1}AP$ have the same rank.
Nullity	A and $P^{-1}AP$ have the same nullity.
Trace	A and $P^{-1}AP$ have the same trace.
Characteristic polynomial	A and $P^{-1}AP$ have the same characteristic polynomial.
Eigenvalues	A and $P^{-1}AP$ have the same eigenvalues.
Eigenspace dimension	If λ is an eigenvalue of A and hence of $P^{-1}AP$, then the eigenspace of A corresponding to λ and the eigenspace of $P^{-1}AP$ corresponding to λ have the same dimension.

The following theorem shows that Problems 1 and 2 posed above are actually two different forms of the same mathematical problem.

> **THEOREM 5.2.1** *If A is an $n \times n$ matrix, the following statements are equivalent.*
>
> (a) *A is diagonalizable.*
>
> (b) *A has n linearly independent eigenvectors.*

Part (b) of Theorem 5.2.1 is equivalent to saying that there is a *basis* for R^n consisting of eigenvectors of A. Why?

Proof (a) $\Rightarrow$ (b) Since A is assumed to be diagonalizable, it follows that there exists an invertible matrix P and a diagonal matrix D such that $P^{-1}AP = D$ or, equivalently,

$$AP = PD \tag{1}$$

If we denote the column vectors of P by $\mathbf{p}_1, \mathbf{p}_2, \ldots, \mathbf{p}_n$, and if we assume that the diagonal entries of D are $\lambda_1, \lambda_2, \ldots, \lambda_n$, then by Formula (6) of Section 1.3 the left side of (1) can be expressed as

$$AP = A[\mathbf{p}_1 \quad \mathbf{p}_2 \quad \cdots \quad \mathbf{p}_n] = [A\mathbf{p}_1 \quad A\mathbf{p}_2 \quad \cdots \quad A\mathbf{p}_n]$$

and, as noted in the comment following Example 1 of Section 1.7, the right side of (1) can be expressed as

$$PD = [\lambda_1\mathbf{p}_1 \quad \lambda_2\mathbf{p}_2 \quad \cdots \quad \lambda_n\mathbf{p}_n]$$

Thus, it follows from (1) that

$$A\mathbf{p}_1 = \lambda_1\mathbf{p}_1, \quad A\mathbf{p}_2 = \lambda_2\mathbf{p}_2, \ldots, \quad A\mathbf{p}_n = \lambda_n\mathbf{p}_n \tag{2}$$

Since P is invertible, we know from Theorem 5.1.6 that its column vectors $\mathbf{p}_1, \mathbf{p}_2, \ldots, \mathbf{p}_n$ are linearly independent (and hence nonzero). Thus, it follows from (2) that these n column vectors are eigenvectors of A.

Proof (b) $\Rightarrow$ (a) Assume that A has n linearly independent eigenvectors, $\mathbf{p}_1, \mathbf{p}_2, \ldots, \mathbf{p}_n$, and that $\lambda_1, \lambda_2, \ldots, \lambda_n$ are the corresponding eigenvalues. If we let

$$P = [\mathbf{p}_1 \quad \mathbf{p}_2 \quad \cdots \quad \mathbf{p}_n]$$

and if we let D be the diagonal matrix that has $\lambda_1, \lambda_2, \ldots, \lambda_n$ as its successive diagonal entries, then

$$AP = A[\mathbf{p}_1 \quad \mathbf{p}_2 \quad \cdots \quad \mathbf{p}_n] = [A\mathbf{p}_1 \quad A\mathbf{p}_2 \quad \cdots \quad A\mathbf{p}_n]$$
$$= [\lambda_1\mathbf{p}_1 \quad \lambda_2\mathbf{p}_2 \quad \cdots \quad \lambda_n\mathbf{p}_n] = PD$$

Since the column vectors of P are linearly independent, it follows from Theorem 5.1.6 that P is invertible, so that this last equation can be rewritten as $P^{-1}AP = D$, which shows that A is diagonalizable. ◀

Procedure for Diagonalizing a Matrix   The preceding theorem guarantees that an $n \times n$ matrix A with n linearly independent eigenvectors is diagonalizable, and the proof suggests the following method for diagonalizing A.

> **Procedure for Diagonalizing a Matrix**
>
> **Step 1.**   Confirm that the matrix is actually diagonalizable by finding n linearly independent eigenvectors. One way to do this is by finding a basis for each eigenspace and merging these basis vectors into a single set S. If this set has fewer than n vectors, then the matrix is not diagonalizable.
>
> **Step 2.**   Form the matrix $P = [\mathbf{p}_1 \quad \mathbf{p}_2 \quad \cdots \quad \mathbf{p}_n]$ that has the vectors in S as its column vectors.
>
> **Step 3.**   The matrix $P^{-1}AP$ will be diagonal and have the eigenvalues $\lambda_1, \lambda_2, \ldots, \lambda_n$ corresponding to the eigenvectors $\mathbf{p}_1, \mathbf{p}_2, \ldots, \mathbf{p}_n$ as its successive diagonal entries.

▶ **EXAMPLE 1 Finding a Matrix P That Diagonalizes a Matrix A**

Find a matrix P that diagonalizes

$$A = \begin{bmatrix} 0 & 0 & -2 \\ 1 & 2 & 1 \\ 1 & 0 & 3 \end{bmatrix}$$

Solution In Example 7 of the preceding section we found the characteristic equation of A to be

$$(\lambda - 1)(\lambda - 2)^2 = 0$$

and we found the following bases for the eigenspaces:

$$\lambda = 2: \quad \mathbf{p}_1 = \begin{bmatrix} -1 \\ 0 \\ 1 \end{bmatrix}, \quad \mathbf{p}_2 = \begin{bmatrix} 0 \\ 1 \\ 0 \end{bmatrix}; \qquad \lambda = 1: \quad \mathbf{p}_3 = \begin{bmatrix} -2 \\ 1 \\ 1 \end{bmatrix}$$

There are three basis vectors in total, so the matrix

$$P = \begin{bmatrix} -1 & 0 & -2 \\ 0 & 1 & 1 \\ 1 & 0 & 1 \end{bmatrix}$$

diagonalizes A. As a check, you should verify that

$$P^{-1}AP = \begin{bmatrix} 1 & 0 & 2 \\ 1 & 1 & 1 \\ -1 & 0 & -1 \end{bmatrix} \begin{bmatrix} 0 & 0 & -2 \\ 1 & 2 & 1 \\ 21 & 0 & 3 \end{bmatrix} \begin{bmatrix} -1 & 0 & -2 \\ 0 & 1 & 1 \\ 1 & 0 & 1 \end{bmatrix} = \begin{bmatrix} 2 & 0 & 0 \\ 0 & 2 & 0 \\ 0 & 0 & 1 \end{bmatrix} ◀$$

In general, there is no preferred order for the columns of P. Since the ith diagonal entry of $P^{-1}AP$ is an eigenvalue for the ith column vector of P, changing the order of the columns of P just changes the order of the eigenvalues on the diagonal of $P^{-1}AP$. Thus, had we written

$$P = \begin{bmatrix} -1 & -2 & 0 \\ 0 & 1 & 1 \\ 1 & 1 & 0 \end{bmatrix}$$

in the preceding example, we would have obtained

$$P^{-1}AP = \begin{bmatrix} 2 & 0 & 0 \\ 0 & 1 & 0 \\ 0 & 0 & 2 \end{bmatrix}$$

▶ **EXAMPLE 2 A Matrix That Is Not Diagonalizable**

Find a matrix P that diagonalizes

$$A = \begin{bmatrix} 1 & 0 & 0 \\ 1 & 2 & 0 \\ -3 & 5 & 2 \end{bmatrix}$$

Solution The characteristic polynomial of A is

$$\det(\lambda I - A) = \begin{vmatrix} \lambda - 1 & 0 & 0 \\ -1 & \lambda - 2 & 0 \\ 3 & -5 & \lambda - 2 \end{vmatrix} = (\lambda - 1)(\lambda - 2)^2$$

so the characteristic equation is

$$(\lambda - 1)(\lambda - 2)^2 = 0$$

Thus, the distinct eigenvalues of A are $\lambda = 1$ and $\lambda = 2$. We leave it for you to show that bases for the eigenspaces are

$$\lambda = 1: \quad \mathbf{p}_1 = \begin{bmatrix} \frac{1}{8} \\ -\frac{1}{8} \\ 1 \end{bmatrix}; \qquad \lambda = 2: \quad \mathbf{p}_2 = \begin{bmatrix} 0 \\ 0 \\ 1 \end{bmatrix}$$

Since A is a 3×3 matrix and there are only two basis vectors in total, A is not diagonalizable.

Alternative Solution If you are concerned only in determining whether a matrix is diagonalizable and not with actually finding a diagonalizing matrix P, then it is not necessary to compute bases for the eigenspaces—it suffices to find the dimensions of the eigenspaces. For this example, the eigenspace corresponding to $\lambda = 1$ is the solution space of the system

$$\begin{bmatrix} 0 & 0 & 0 \\ -1 & -1 & 0 \\ 3 & -5 & -1 \end{bmatrix} \begin{bmatrix} x_1 \\ x_2 \\ x_3 \end{bmatrix} = \begin{bmatrix} 0 \\ 0 \\ 0 \end{bmatrix}$$

Since the coefficient matrix has rank 2 (verify), the nullity of this matrix is 1 by Theorem 4.8.2, and hence the eigenspace corresponding to $\lambda = 1$ is one-dimensional.

The eigenspace corresponding to $\lambda = 2$ is the solution space of the system

$$\begin{bmatrix} 1 & 0 & 0 \\ -1 & 0 & 0 \\ 3 & -5 & 0 \end{bmatrix} \begin{bmatrix} x_1 \\ x_2 \\ x_3 \end{bmatrix} = \begin{bmatrix} 0 \\ 0 \\ 0 \end{bmatrix}$$

This coefficient matrix also has rank 2 and nullity 1 (verify), so the eigenspace corresponding to $\lambda = 2$ is also one-dimensional. Since the eigenspaces produce a total of two basis vectors, and since three are needed, the matrix A is not diagonalizable. ◀

There is an assumption in Example 1 that the column vectors of P, which are made up of basis vectors from the various eigenspaces of A, are linearly independent. The following theorem, proved at the end of this section, shows that this is so.

> **THEOREM 5.2.2** *If* $\mathbf{v}_1, \mathbf{v}_2, \ldots, \mathbf{v}_k$ *are eigenvectors of a matrix A corresponding to distinct eigenvalues, then* $\{\mathbf{v}_1, \mathbf{v}_2, \ldots, \mathbf{v}_k\}$ *is a linearly independent set.*

Remark   Theorem 5.2.2 is a special case of a more general result: Suppose that $\lambda_1, \lambda_2, \ldots, \lambda_k$ are distinct eigenvalues and that we choose a linearly independent set in each of the corresponding eigenspaces. If we then merge all these vectors into a single set, the result will still be a linearly independent set. For example, if we choose three linearly independent vectors from one eigenspace and two linearly independent vectors from another eigenspace, then the five vectors together form a linearly independent set. We omit the proof.

As a consequence of Theorem 5.2.2, we obtain the following important result.

> **THEOREM 5.2.3** *If an $n \times n$ matrix A has n distinct eigenvalues, then A is diagonalizable.*

Proof   If $\mathbf{v}_1, \mathbf{v}_2, \ldots, \mathbf{v}_n$ are eigenvectors corresponding to the distinct eigenvalues $\lambda_1, \lambda_2, \ldots, \lambda_n$, then by Theorem 5.2.2, $\mathbf{v}_1, \mathbf{v}_2, \ldots, \mathbf{v}_n$ are linearly independent. Thus, A is diagonalizable by Theorem 5.2.1. ◄

► EXAMPLE 3 Using Theorem 5.2.3

We saw in Example 3 of the preceding section that

$$A = \begin{bmatrix} 0 & 1 & 0 \\ 0 & 0 & 1 \\ 4 & -17 & 8 \end{bmatrix}$$

has three distinct eigenvalues: $\lambda = 4$, $\lambda = 2 + \sqrt{3}$, and $\lambda = 2 - \sqrt{3}$. Therefore, A is diagonalizable and

$$P^{-1}AP = \begin{bmatrix} 4 & 0 & 0 \\ 0 & 2 + \sqrt{3} & 0 \\ 0 & 0 & 2 - \sqrt{3} \end{bmatrix}$$

for some invertible matrix P. If needed, the matrix P can be found using the method shown in Example 1 of this section.

► EXAMPLE 4 Diagonalizability of Triangular Matrices

From Theorem 5.1.2, the eigenvalues of a triangular matrix are the entries on its main diagonal. Thus, a triangular matrix with distinct entries on the main diagonal is diagonalizable. For example,

$$A = \begin{bmatrix} -1 & 2 & 4 & 0 \\ 0 & 3 & 1 & 7 \\ 0 & 0 & 5 & 8 \\ 0 & 0 & 0 & -2 \end{bmatrix}$$

is a diagonalizable matrix with eigenvalues $\lambda_1 = -1$, $\lambda_2 = 3$, $\lambda_3 = 5$, $\lambda_4 = -2$. ◄

Computing Powers of a Matrix

There are many applications in which it is necessary to compute high powers of a square matrix A. We will show next that if A happens to be diagonalizable, then the computations can be simplified by diagonalizing A.

To start, suppose that A is a diagonalizable $n \times n$ matrix, that P diagonalizes A, and that

$$P^{-1}AP = \begin{bmatrix} \lambda_1 & 0 & \cdots & 0 \\ 0 & \lambda_2 & \cdots & 0 \\ \vdots & \vdots & & \vdots \\ 0 & 0 & \cdots & \lambda_n \end{bmatrix} = D$$

Squaring both sides of this equation yields

$$(P^{-1}AP)^2 = \begin{bmatrix} \lambda_1^2 & 0 & \cdots & 0 \\ 0 & \lambda_2^2 & \cdots & 0 \\ \vdots & \vdots & & \vdots \\ 0 & 0 & \cdots & \lambda_n^2 \end{bmatrix} = D^2$$

We can rewrite the left side of this equation as

$$(P^{-1}AP)^2 = P^{-1}APP^{-1}AP = P^{-1}AIAP = P^{-1}A^2P$$

from which we obtain the relationship $P^{-1}A^2P = D^2$. More generally, if k is a positive integer, then a similar computation will show that

$$P^{-1}A^kP = D^k = \begin{bmatrix} \lambda_1^k & 0 & \cdots & 0 \\ 0 & \lambda_2^k & \cdots & 0 \\ \vdots & \vdots & & \vdots \\ 0 & 0 & \cdots & \lambda_n^k \end{bmatrix}$$

which we can rewrite as

$$A^k = PD^kP^{-1} = P \begin{bmatrix} \lambda_1^k & 0 & \cdots & 0 \\ 0 & \lambda_2^k & \cdots & 0 \\ \vdots & \vdots & & \vdots \\ 0 & 0 & \cdots & \lambda_n^k \end{bmatrix} P^{-1} \tag{3}$$

Formula (3) reveals that raising a diagonalizable matrix A to a positive integer power has the effect of raising its eigenvalues to that power.

Note that computing the right side of this formula involves only three matrix multiplications and the powers of the diagonal entries of D. For matrices of large size and high powers of λ, this involves substantially fewer operations than computing A^k directly.

▶ **EXAMPLE 5** Power of a Matrix

Use (3) to find A^{13}, where

$$A = \begin{bmatrix} 0 & 0 & -2 \\ 1 & 2 & 1 \\ 1 & 0 & 3 \end{bmatrix}$$

Solution We showed in Example 1 that the matrix A is diagonalized by

$$P = \begin{bmatrix} -1 & 0 & -2 \\ 0 & 1 & 1 \\ 1 & 0 & 1 \end{bmatrix}$$

and that

$$D = P^{-1}AP = \begin{bmatrix} 2 & 0 & 0 \\ 0 & 2 & 0 \\ 0 & 0 & 1 \end{bmatrix}$$

Thus, it follows from (3) that

$$A^{13} = PD^{13}P^{-1} = \begin{bmatrix} -1 & 0 & -2 \\ 0 & 1 & 1 \\ 1 & 0 & 1 \end{bmatrix} \begin{bmatrix} 2^{13} & 0 & 0 \\ 0 & 2^{13} & 0 \\ 0 & 0 & 1^{13} \end{bmatrix} \begin{bmatrix} 1 & 0 & 2 \\ 1 & 1 & 1 \\ -1 & 0 & -1 \end{bmatrix} \quad (4)$$

$$= \begin{bmatrix} -8190 & 0 & -16382 \\ 8191 & 8192 & 8191 \\ 8191 & 0 & 16383 \end{bmatrix} \blacktriangleleft$$

Remark With the method in the preceding example, most of the work is in diagonalizing A. Once that work is done, it can be used to compute any power of A. Thus, to compute A^{1000} we need only change the exponents from 13 to 1000 in (4).

Eigenvalues of Powers of a Matrix

Once the eigenvalues and eigenvectors of any square matrix A are found, it is a simple matter to find the eigenvalues and eigenvectors of any positive integer power of A. For example, if λ is an eigenvalue of A and $\mathbf{x}$ is a corresponding eigenvector, then

$$A^2\mathbf{x} = A(A\mathbf{x}) = A(\lambda\mathbf{x}) = \lambda(A\mathbf{x}) = \lambda(\lambda\mathbf{x}) = \lambda^2\mathbf{x}$$

which shows not only that λ^2 is an eigenvalue of A^2 but that $\mathbf{x}$ is a corresponding eigenvector. In general, we have the following result.

Note that diagonalizability is not a requirement in Theorem 5.2.4.

THEOREM 5.2.4 *If λ is an eigenvalue of a square matrix A and $\mathbf{x}$ is a corresponding eigenvector, and if k is any positive integer, then λ^k is an eigenvalue of A^k and $\mathbf{x}$ is a corresponding eigenvector.*

Some problems that use this theorem are given in the exercises.

Geometric and Algebraic Multiplicity

Theorem 5.2.3 does not completely settle the diagonalizability question since it only guarantees that a square matrix with n distinct eigenvalues is diagonalizable, but does not preclude the possibility that there may exist diagonalizable matrices with fewer than n distinct eigenvalues. The following example shows that this is indeed the case.

▶ **EXAMPLE 6** **The Converse of Theorem 5.2.3 Is False**

Consider the matrices

$$I = \begin{bmatrix} 1 & 0 & 0 \\ 0 & 1 & 0 \\ 0 & 0 & 1 \end{bmatrix} \quad \text{and} \quad J = \begin{bmatrix} 1 & 1 & 0 \\ 0 & 1 & 1 \\ 0 & 0 & 1 \end{bmatrix}$$

It follows from Theorem 5.1.2 that both of these matrices have only one distinct eigenvalue, namely $\lambda = 1$, and hence only one eigenspace. We leave it as an exercise for you to solve the characteristic equations

$$(\lambda I - I)\mathbf{x} = \mathbf{0} \quad \text{and} \quad (\lambda J - I)\mathbf{x} = \mathbf{0}$$

with $\lambda = 1$ and show that for I the eigenspace is three-dimensional (all of R^3) and for J it is one-dimensional, consisting of all scalar multiples of

$$\mathbf{x} = \begin{bmatrix} 1 \\ 0 \\ 0 \end{bmatrix}$$

This shows that the converse of Theorem 5.2.3 is false, since we have produced two 3×3 matrices with fewer than three distinct eigenvalues, one of which is diagonalizable and the other of which is not. ◄

A full excursion into the study of diagonalizability is left for more advanced courses, but we will touch on one theorem that is important to a fuller understanding of diagonalizability. It can be proved that if λ_0 is an eigenvalue of A, then the dimension of the eigenspace corresponding to λ_0 cannot exceed the number of times that $\lambda - \lambda_0$ appears as a factor of the characteristic polynomial of A. For example, in Examples 1 and 2 the characteristic polynomial is

$$(\lambda - 1)(\lambda - 2)^2$$

Thus, the eigenspace corresponding to $\lambda = 1$ is at most (hence exactly) one-dimensional, and the eigenspace corresponding to $\lambda = 2$ is at most two-dimensional. In Example 1 the eigenspace corresponding to $\lambda = 2$ actually had dimension 2, resulting in diagonalizability, but in Example 2 the eigenspace corresponding to $\lambda = 2$ had only dimension 1, resulting in nondiagonalizability.

There is some terminology that is related to these ideas. If λ_0 is an eigenvalue of an $n \times n$ matrix A, then the dimension of the eigenspace corresponding to λ_0 is called the *geometric multiplicity* of λ_0, and the number of times that $\lambda - \lambda_0$ appears as a factor in the characteristic polynomial of A is called the *algebraic multiplicity* of A. The following theorem, which we state without proof, summarizes the preceding discussion.

> **THEOREM 5.2.5 Geometric and Algebraic Multiplicity**
>
> *If A is a square matrix, then*:
>
> (a) *For every eigenvalue of A, the geometric multiplicity is less than or equal to the algebraic multiplicity.*
>
> (b) *A is diagonalizable if and only if the geometric multiplicity of every eigenvalue is equal to the algebraic multiplicity.*

OPTIONAL

We will complete this section with an optional proof of Theorem 5.2.2.

Proof of Theorem 5.2.2   Let $\mathbf{v}_1, \mathbf{v}_2, \ldots, \mathbf{v}_k$ be eigenvectors of A corresponding to distinct eigenvalues $\lambda_1, \lambda_2, \ldots, \lambda_k$. We will assume that $\mathbf{v}_1, \mathbf{v}_2, \ldots, \mathbf{v}_k$ are linearly dependent and obtain a contradiction. We can then conclude that $\mathbf{v}_1, \mathbf{v}_2, \ldots, \mathbf{v}_k$ are linearly independent.

Since an eigenvector is nonzero by definition, $\{\mathbf{v}_1\}$ is linearly independent. Let r be the largest integer such that $\{\mathbf{v}_1, \mathbf{v}_2, \ldots, \mathbf{v}_r\}$ is linearly independent. Since we are assuming that $\{\mathbf{v}_1, \mathbf{v}_2, \ldots, \mathbf{v}_k\}$ is linearly dependent, r satisfies $1 \leq r < k$. Moreover, by the definition of r, $\{\mathbf{v}_1, \mathbf{v}_2, \ldots, \mathbf{v}_{r+1}\}$ is linearly dependent. Thus, there are scalars $c_1, c_2, \ldots, c_{r+1}$, not all zero, such that

$$c_1\mathbf{v}_1 + c_2\mathbf{v}_2 + \cdots + c_{r+1}\mathbf{v}_{r+1} = \mathbf{0} \tag{5}$$

Multiplying both sides of (5) by A and using the fact that

$$A\mathbf{v}_1 = \lambda_1\mathbf{v}_1, \quad A\mathbf{v}_2 = \lambda_2\mathbf{v}_2, \ldots, \quad A\mathbf{v}_{r+1} = \lambda_{r+1}\mathbf{v}_{r+1}$$

we obtain

$$c_1\lambda_1\mathbf{v}_1 + c_2\lambda_2\mathbf{v}_2 + \cdots + c_{r+1}\lambda_{r+1}\mathbf{v}_{r+1} = \mathbf{0} \tag{6}$$

If we now multiply both sides of (5) by λ_{r+1} and subtract the resulting equation from (6) we obtain

$$c_1(\lambda_1 - \lambda_{r+1})\mathbf{v}_1 + c_2(\lambda_2 - \lambda_{r+1})\mathbf{v}_2 + \cdots + c_r(\lambda_r - \lambda_{r+1})\mathbf{v}_r = \mathbf{0}$$

Since $\{\mathbf{v}_1, \mathbf{v}_2, \ldots, \mathbf{v}_r\}$ is a linearly independent set, this equation implies that

$$c_1(\lambda_1 - \lambda_{r+1}) = c_2(\lambda_2 - \lambda_{r+1}) = \cdots = c_r(\lambda_r - \lambda_{r+1}) = 0$$

and since $\lambda_1, \lambda_2, \ldots, \lambda_{r+1}$ are assumed to be distinct, it follows that

$$c_1 = c_2 = \cdots = c_r = 0 \tag{7}$$

Substituting these values in (5) yields

$$c_{r+1}\mathbf{v}_{r+1} = \mathbf{0}$$

Since the eigenvector $\mathbf{v}_{r+1}$ is nonzero, it follows that

$$c_{r+1} = 0 \tag{8}$$

But equations (7) and (8) contradict the fact that $c_1, c_2, \ldots, c_{r+1}$ are not all zero so the proof is complete. ◄

Concept Review

- Similarity transformation
- Similarity invariant
- Similar matrices
- Diagonalizable matrix
- Geometric multiplicity
- Algebraic multiplicity

Skills

- Determine whether a square matrix A is diagonalizable.
- Diagonalize a square matrix A.
- Find powers of a matrix using similarity.
- Find the geometric multiplicity and the algebraic multiplicity of an eigenvalue.

Exercise Set 5.2

▶ In Exercises 1–4, show that A and B are not similar matrices.

1. $A = \begin{bmatrix} 1 & 1 \\ 3 & 2 \end{bmatrix}$, $B = \begin{bmatrix} 1 & 0 \\ 3 & -2 \end{bmatrix}$

2. $A = \begin{bmatrix} 4 & -1 \\ 2 & 4 \end{bmatrix}$, $B = \begin{bmatrix} 4 & 1 \\ 2 & 4 \end{bmatrix}$

3. $A = \begin{bmatrix} 1 & 2 & 3 \\ 0 & 1 & 2 \\ 0 & 0 & 1 \end{bmatrix}$, $B = \begin{bmatrix} 1 & 2 & 0 \\ \frac{1}{2} & 1 & 0 \\ 0 & 0 & 1 \end{bmatrix}$

4. $A = \begin{bmatrix} 1 & 0 & 1 \\ 2 & 0 & 2 \\ 3 & 0 & 3 \end{bmatrix}$, $B = \begin{bmatrix} 1 & 1 & 0 \\ 2 & 2 & 0 \\ 0 & 1 & 1 \end{bmatrix}$

5. Let A be a 6×6 matrix with characteristic equation $\lambda^2(\lambda - 1)(\lambda - 2)^3 = 0$. What are the possible dimensions for eigenspaces of A?

6. Let

$$A = \begin{bmatrix} 4 & 0 & 1 \\ 2 & 3 & 2 \\ 1 & 0 & 4 \end{bmatrix}$$

(a) Find the eigenvalues of A.

(b) For each eigenvalue λ, find the rank of the matrix $\lambda I - A$.

(c) Is A diagonalizable? Justify your conclusion.

▶ In Exercises 7–11, use the method of Exercise 6 to determine whether the matrix is diagonalizable. ◄

7. $\begin{bmatrix} 2 & 0 \\ 1 & 2 \end{bmatrix}$ **8.** $\begin{bmatrix} 2 & -3 \\ 1 & -1 \end{bmatrix}$ **9.** $\begin{bmatrix} 3 & 0 & 0 \\ 0 & 2 & 0 \\ 0 & 1 & 2 \end{bmatrix}$

10. $\begin{bmatrix} -1 & 0 & 1 \\ -1 & 3 & 0 \\ -4 & 13 & -1 \end{bmatrix}$ **11.** $\begin{bmatrix} 2 & -1 & 0 & 1 \\ 0 & 2 & 1 & -1 \\ 0 & 0 & 3 & 2 \\ 0 & 0 & 0 & 3 \end{bmatrix}$

▶ In Exercises 12–15, find a matrix P that diagonalizes A, and compute $P^{-1}AP$. ◄

12. $A = \begin{bmatrix} -14 & 12 \\ -20 & 17 \end{bmatrix}$ **13.** $A = \begin{bmatrix} 1 & 0 \\ 6 & -1 \end{bmatrix}$

14. $A = \begin{bmatrix} 1 & 0 & 0 \\ 0 & 1 & 1 \\ 0 & 1 & 1 \end{bmatrix}$ **15.** $A = \begin{bmatrix} 2 & 0 & -2 \\ 0 & 3 & 0 \\ 0 & 0 & 3 \end{bmatrix}$

▶ In Exercises 16–21, find the geometric and algebraic multiplicity of each eigenvalue of the matrix A, and determine whether A is diagonalizable. If A is diagonalizable, then find a matrix P that diagonalizes A, and find $P^{-1}AP$. ◀

16. $A = \begin{bmatrix} 19 & -9 & -6 \\ 25 & -11 & -9 \\ 17 & -9 & -4 \end{bmatrix}$ **17.** $A = \begin{bmatrix} -1 & 4 & -2 \\ -3 & 4 & 0 \\ -3 & 1 & 3 \end{bmatrix}$

18. $A = \begin{bmatrix} 5 & 0 & 0 \\ 1 & 5 & 0 \\ 0 & 1 & 5 \end{bmatrix}$ **19.** $A = \begin{bmatrix} 0 & 0 & 0 \\ 0 & 0 & 0 \\ 3 & 0 & 1 \end{bmatrix}$

20. $A = \begin{bmatrix} -2 & 0 & 0 & 0 \\ 0 & -2 & 0 & 0 \\ 0 & 0 & 3 & 0 \\ 0 & 0 & 1 & 3 \end{bmatrix}$

21. $A = \begin{bmatrix} -2 & 0 & 0 & 0 \\ 0 & -2 & 5 & -5 \\ 0 & 0 & 3 & 0 \\ 0 & 0 & 0 & 3 \end{bmatrix}$

22. Use the method of Example 5 to compute A^{10}, where

$$A = \begin{bmatrix} 1 & 0 \\ -1 & 2 \end{bmatrix}$$

23. Use the method of Example 5 to compute A^{11}, where

$$A = \begin{bmatrix} -1 & 7 & -1 \\ 0 & 1 & 0 \\ 0 & 15 & -2 \end{bmatrix}$$

24. In each part, compute the stated power of

$$A = \begin{bmatrix} 1 & -2 & 8 \\ 0 & -1 & 0 \\ 0 & 0 & -1 \end{bmatrix}$$

(a) A^{1000} (b) A^{-1000} (c) A^{2301} (d) A^{-2301}

25. Find A^n if n is a positive integer and

$$A = \begin{bmatrix} 3 & -1 & 0 \\ -1 & 2 & -1 \\ 0 & -1 & 3 \end{bmatrix}$$

26. Let

$$A = \begin{bmatrix} a & b \\ c & d \end{bmatrix}$$

Show that

(a) A is diagonalizable if $(a - d)^2 + 4bc > 0$.

(b) A is not diagonalizable if $(a - d)^2 + 4bc < 0$.

[*Hint:* See Exercise 19 of Section 5.1.]

27. In the case where the matrix A in Exercise 26 is diagonalizable, find a matrix P that diagonalizes A. [*Hint:* See Exercise 20 of Section 5.1.]

28. Prove that similar matrices have the same rank.

29. Prove that similar matrices have the same nullity.

30. Prove that similar matrices have the same trace.

31. Prove that if A is diagonalizable, then so is A^k for every positive integer k.

32. Prove that if A is a diagonalizable matrix, then the rank of A is the number of nonzero eigenvalues of A.

33. Suppose that the characteristic polynomial of some matrix A is found to be $p(\lambda) = (\lambda - 1)(\lambda - 3)^2(\lambda - 4)^3$. In each part, answer the question and explain your reasoning.

(a) What can you say about the dimensions of the eigenspaces of A?

(b) What can you say about the dimensions of the eigenspaces if you know that A is diagonalizable?

(c) If $\{v_1, v_2, v_3\}$ is a linearly independent set of eigenvectors of A all of which correspond to the same eigenvalue of A, what can you say about the eigenvalue?

34. This problem will lead you through a proof of the fact that the algebraic multiplicity of an eigenvalue of an $n \times n$ matrix A is greater than or equal to the geometric multiplicity. For this purpose, assume that λ_0 is an eigenvalue with geometric multiplicity k.

(a) Prove that there is a basis $B = \{u_1, u_2, \ldots, u_n\}$ for R^n in which the first k vectors of B form a basis for the eigenspace corresponding to λ_0.

(b) Let P be the matrix having the vectors in B as columns. Prove that the product AP can be expressed as

$$AP = P \begin{bmatrix} \lambda_0 I_k & X \\ 0 & Y \end{bmatrix}$$

[*Hint:* Compare the first k column vectors on both sides.]

(c) Use the result in part (b) to prove that A is similar to

$$C = \begin{bmatrix} \lambda_0 I_k & X \\ 0 & Y \end{bmatrix}$$

and hence that A and C have the same characteristic polynomial.

(d) By considering $\det(\lambda I - C)$, prove that the characteristic polynomial of C (and hence A) contains the factor $(\lambda - \lambda_0)$ at least k times, thereby proving that the algebraic multiplicity of λ_0 is greater than or equal to the geometric multiplicity k.

True-False Exercises

In parts (a)–(h) determine whether the statement is true or false, and justify your answer.

(a) Every square matrix is similar to itself.

(b) If A, B, and C are matrices for which A is similar to B and B is similar to C, then A is similar to C.

(c) If A and B are similar invertible matrices, then A^{-1} and B^{-1} are similar.

(d) If A is diagonalizable, then there is a unique matrix P such that $P^{-1}AP$ is diagonal.

(e) If A is diagonalizable and invertible, then A^{-1} is diagonalizable.

(f) If A is diagonalizable, then A^T is diagonalizable.

(g) If there is a basis for R^n consisting of eigenvectors of an $n \times n$ matrix A, then A is diagonalizable.

(h) If every eigenvalue of a matrix A has algebraic multiplicity 1, then A is diagonalizable.

5.3 Complex Vector Spaces

Because the characteristic equation of any square matrix can have complex solutions, the notions of complex eigenvalues and eigenvectors arise naturally, even within the context of matrices with real entries. In this section we will discuss this idea and apply our results to study symmetric matrices in more detail. A review of the essentials of complex numbers appears in the back of this text.

Review of Complex Numbers

Recall that if $z = a + bi$ is a complex number, then:

- $\text{Re}(z) = a$ and $\text{Im}(z) = b$ are called the **real part** of z and the **imaginary part** of z, respectively,
- $|z| = \sqrt{a^2 + b^2}$ is called the **modulus** (or **absolute value**) of z,
- $\bar{z} = a - bi$ is called the **complex conjugate** of z,
- $z\bar{z} = a^2 + b^2 = |z|^2$,
- the angle ϕ in Figure 5.3.1 is called an **argument** of z,
- $\text{Re}(z) = |z| \cos \phi$
- $\text{Im}(z) = |z| \sin \phi$
- $z = |z|(\cos \phi + i \sin \phi)$ is called the **polar form** of z.

▲ Figure 5.3.1

Complex Eigenvalues

In Formula (3) of Section 5.1 we observed that the characteristic equation of a general $n \times n$ matrix A has the form

$$\lambda^n + c_1\lambda^{n-1} + \cdots + c_n = 0 \tag{1}$$

in which the highest power of λ has a coefficient of 1. Up to now we have limited our discussion to matrices in which the solutions of (1) are real numbers. However, it is possible for the characteristic equation of a matrix A with real entries to have imaginary solutions; for example, the characteristic equation of the matrix

$$A = \begin{bmatrix} -2 & -1 \\ 5 & 2 \end{bmatrix}$$

is

$$\begin{vmatrix} \lambda + 2 & 1 \\ -5 & \lambda - 2 \end{vmatrix} = \lambda^2 + 1 = 0$$

which has the imaginary solutions $\lambda = i$ and $\lambda = -i$. To deal with this case we will need to explore the notion of a complex vector space and some related ideas.

Vectors in C^n   A vector space in which scalars are allowed to be complex numbers is called a ***complex vector space***. In this section we will be concerned only with the following complex generalization of the real vector space R^n.

> **DEFINITION 1** If n is a positive integer, then a ***complex n-tuple*** is a sequence of n complex numbers $(v_1, v_2, \ldots, v_n)$. The set of all complex n-tuples is called ***complex n-space*** and is denoted by C^n. Scalars are complex numbers, and the operations of addition, subtraction, and scalar multiplication are performed componentwise.

The terminology used for n-tuples of real numbers applies to complex n-tuples without change. Thus, if $v_1, v_2, \ldots, v_n$ are complex numbers, then we call $\mathbf{v} = (v_1, v_2, \ldots, v_n)$ a ***vector*** in C^n and $v_1, v_2, \ldots, v_n$ its ***components***. Some examples of vectors in C^3 are

$$\mathbf{u} = (1 + i, -4i, 3 + 2i), \quad \mathbf{v} = (0, i, 5), \quad \mathbf{w} = \left(6 - \sqrt{2}i, 9 + \tfrac{1}{2}i, \pi i\right)$$

Every vector

$$\mathbf{v} = (v_1, v_2, \ldots, v_n) = (a_1 + b_1 i, a_2 + b_2 i, \ldots, a_n + b_n i)$$

in C^n can be split into ***real*** and ***imaginary parts*** as

$$\mathbf{v} = (a_1, a_2, \ldots, a_n) + i(b_1, b_2, \ldots, b_n)$$

which we also denote as

$$\mathbf{v} = \operatorname{Re}(\mathbf{v}) + i \operatorname{Im}(\mathbf{v})$$

where

$$\operatorname{Re}(\mathbf{v}) = (a_1, a_2, \ldots, a_n) \quad \text{and} \quad \operatorname{Im}(\mathbf{v}) = (b_1, b_2, \ldots, b_n)$$

The vector

$$\overline{\mathbf{v}} = (\overline{v}_1, \overline{v}_2, \ldots, \overline{v}_n) = (a_1 - b_1 i, a_2 - b_2 i, \ldots, a_n - b_n i)$$

is called the ***complex conjugate*** of $\mathbf{v}$ and can be expressed in terms of $\operatorname{Re}(\mathbf{v})$ and $\operatorname{Im}(\mathbf{v})$ as

$$\overline{\mathbf{v}} = (a_1, a_2, \ldots, a_n) - i(b_1, b_2, \ldots, b_n) = \operatorname{Re}(\mathbf{v}) - i \operatorname{Im}(\mathbf{v}) \tag{2}$$

It follows that the vectors in R^n can be viewed as those vectors in C^n whose imaginary part is zero; or stated another way, a vector $\mathbf{v}$ in C^n is in R^n if and only if $\overline{\mathbf{v}} = \mathbf{v}$.

In this section we will also need to consider matrices with complex entries, so henceforth we will call a matrix A a ***real matrix*** if its entries are required to be real numbers and a ***complex matrix*** if its entries are allowed to be complex numbers. The standard operations on real matrices carry over to complex matrices without change, and all of the familiar properties of matrices continue to hold.

If A is a complex matrix, then $\operatorname{Re}(A)$ and $\operatorname{Im}(A)$ are the matrices formed from the real and imaginary parts of the entries of A, and $\overline{A}$ is the matrix formed by taking the complex conjugate of each entry in A.

▶ **EXAMPLE 1**   **Real and Imaginary Parts of Vectors and Matrices**

Let

$$\mathbf{v} = (3 + i, -2i, 5) \quad \text{and} \quad A = \begin{bmatrix} 1 + i & -i \\ 4 & 6 - 2i \end{bmatrix}$$

Then
$$\overline{\mathbf{v}} = (3 - i, 2i, 5), \quad \text{Re}(\mathbf{v}) = (3, 0, 5), \quad \text{Im}(\mathbf{v}) = (1, -2, 0)$$

$$\overline{A} = \begin{bmatrix} 1 - i & i \\ 4 & 6 + 2i \end{bmatrix}, \quad \text{Re}(A) = \begin{bmatrix} 1 & 0 \\ 4 & 6 \end{bmatrix}, \quad \text{Im}(A) = \begin{bmatrix} 1 & -1 \\ 0 & -2 \end{bmatrix}$$

$$\det(A) = \begin{vmatrix} 1 + i & -i \\ 4 & 6 - 2i \end{vmatrix} = (1 + i)(6 - 2i) - (-i)(4) = 8 + 8i \quad \blacktriangleleft$$

Algebraic Properties of the Complex Conjugate

The next two theorems list some properties of complex vectors and matrices that we will need in this section. Some of the proofs are given as exercises.

THEOREM 5.3.1 *If* $\mathbf{u}$ *and* $\mathbf{v}$ *are vectors in* C^n, *and if* k *is a scalar, then:*

(a) $\overline{\overline{\mathbf{u}}} = \mathbf{u}$

(b) $\overline{k\mathbf{u}} = \overline{k}\,\overline{\mathbf{u}}$

(c) $\overline{\mathbf{u} + \mathbf{v}} = \overline{\mathbf{u}} + \overline{\mathbf{v}}$

(d) $\overline{\mathbf{u} - \mathbf{v}} = \overline{\mathbf{u}} - \overline{\mathbf{v}}$

THEOREM 5.3.2 *If* A *is an* $m \times k$ *complex matrix and* B *is a* $k \times n$ *complex matrix, then:*

(a) $\overline{\overline{A}} = A$

(b) $\overline{(A^T)} = (\overline{A})^T$

(c) $\overline{AB} = \overline{A}\,\overline{B}$

The Complex Euclidean Inner Product

The following definition extends the notions of dot product and norm to C^n.

DEFINITION 2 If $\mathbf{u} = (u_1, u_2, \ldots, u_n)$ and $\mathbf{v} = (v_1, v_2, \ldots, v_n)$ are vectors in C^n, then the *complex Euclidean inner product* of $\mathbf{u}$ and $\mathbf{v}$ (also called the *complex dot product*) is denoted by $\mathbf{u} \cdot \mathbf{v}$ and is defined as

$$\mathbf{u} \cdot \mathbf{v} = u_1 \overline{v}_1 + u_2 \overline{v}_2 + \cdots + u_n \overline{v}_n \tag{3}$$

We also define the *Euclidean norm* on C^n to be

$$\|\mathbf{v}\| = \sqrt{\mathbf{v} \cdot \mathbf{v}} = \sqrt{|v_1|^2 + |v_2|^2 + \cdots + |v_n|^2} \tag{4}$$

The complex conjugates in (3) ensure that $\|\mathbf{v}\|$ is a real number, for without them the quantity $\mathbf{v} \cdot \mathbf{v}$ in (4) might be imaginary.

As in the real case, we call $\mathbf{v}$ a *unit vector* in C^n if $\|\mathbf{v}\| = 1$, and we say two vectors $\mathbf{u}$ and $\mathbf{v}$ are *orthogonal* if $\mathbf{u} \cdot \mathbf{v} = 0$.

▶ **EXAMPLE 2 Complex Euclidean Inner Product and Norm**

Find $\mathbf{u} \cdot \mathbf{v}$, $\mathbf{v} \cdot \mathbf{u}$, $\|\mathbf{u}\|$, and $\|\mathbf{v}\|$ for the vectors

$$\mathbf{u} = (1 + i, i, 3 - i) \quad \text{and} \quad \mathbf{v} = (1 + i, 2, 4i)$$

Solution

$$\mathbf{u} \cdot \mathbf{v} = (1+i)(\overline{1+i}) + i(\overline{2}) + (3-i)(\overline{4i}) = (1+i)(1-i) + 2i + (3-i)(-4i) = -2 - 10i$$
$$\mathbf{v} \cdot \mathbf{u} = (1+i)(\overline{1+i}) + 2(\overline{i}) + (4i)(\overline{3-i}) = (1+i)(1-i) - 2i + 4i(3+i) = -2 + 10i$$
$$\|\mathbf{u}\| = \sqrt{|1+i|^2 + |i|^2 + |3-i|^2} = \sqrt{2+1+10} = \sqrt{13}$$
$$\|\mathbf{v}\| = \sqrt{|1+i|^2 + |2|^2 + |4i|^2} = \sqrt{2+4+16} = \sqrt{22} \blacktriangleleft$$

Recall from Table 1 of Section 3.2 that if $\mathbf{u}$ and $\mathbf{v}$ are *column vectors* in R^n, then their dot product can be expressed as

$$\mathbf{u} \cdot \mathbf{v} = \mathbf{u}^T \mathbf{v} = \mathbf{v}^T \mathbf{u}$$

The analogous formulas in C^n are (verify)

$$\mathbf{u} \cdot \mathbf{v} = \mathbf{u}^T \overline{\mathbf{v}} = \overline{\mathbf{v}}^T \mathbf{u} \tag{5}$$

Example 2 reveals a major difference between the dot product on R^n and the complex dot product on C^n. For the dot product on R^n we always have $\mathbf{v} \cdot \mathbf{u} = \mathbf{u} \cdot \mathbf{v}$ (the *symmetry property*), but for the complex dot product the corresponding relationship is given by $\mathbf{u} \cdot \mathbf{v} = \overline{\mathbf{v} \cdot \mathbf{u}}$, which is called its *antisymmetry* property. The following theorem is an analog of Theorem 3.2.2.

THEOREM 5.3.3 *If* $\mathbf{u}, \mathbf{v},$ *and* $\mathbf{w}$ *are vectors in* C^n, *and if* k *is a scalar, then the complex Euclidean inner product has the following properties:*

(a) $\mathbf{u} \cdot \mathbf{v} = \overline{\mathbf{v} \cdot \mathbf{u}}$ [Antisymmetry property]

(b) $\mathbf{u} \cdot (\mathbf{v} + \mathbf{w}) = \mathbf{u} \cdot \mathbf{v} + \mathbf{u} \cdot \mathbf{w}$ [Distributive property]

(c) $k(\mathbf{u} \cdot \mathbf{v}) = (k\mathbf{u}) \cdot \mathbf{v}$ [Homogeneity property]

(d) $\mathbf{u} \cdot k\mathbf{v} = \overline{k}(\mathbf{u} \cdot \mathbf{v})$ [Antihomogeneity property]

(e) $\mathbf{v} \cdot \mathbf{v} \geq 0$ *and* $\mathbf{v} \cdot \mathbf{v} = 0$ *if and only if* $\mathbf{v} = \mathbf{0}$.   [Positivity property]

Parts (c) and (d) of this theorem state that a scalar multiplying a complex Euclidean inner product can be regrouped with the first vector, but to regroup it with the second vector you must first take its complex conjugate. We will prove part (d), and leave the others as exercises.

Proof (d)

$$k(\mathbf{u} \cdot \mathbf{v}) = k(\overline{\mathbf{v} \cdot \mathbf{u}}) = \overline{\overline{k}}\,(\overline{\mathbf{v} \cdot \mathbf{u}}) = \overline{\overline{k}\,(\mathbf{v} \cdot \mathbf{u})} = \overline{(\overline{k}\mathbf{v}) \cdot \mathbf{u}} = \mathbf{u} \cdot (\overline{k}\mathbf{v})$$

To complete the proof. substitute $\overline{k}$ for k and use the fact that $\overline{\overline{k}} = k$. $\blacktriangleleft$

Vector Concepts in C^n

Except for the use of complex scalars, the notions of linear combination, linear independence, subspace, spanning, basis, and dimension carry over without change to C^n.

Eigenvalues and eigenvectors are defined for complex matrices exactly as for real matrices. If A is an $n \times n$ matrix with complex entries, then the complex roots of the characteristic equation $\det(\lambda I - A) = 0$ are called *complex eigenvalues* of A. As in the real case, λ is a complex eigenvalue of A if and only if there exists a nonzero vector $\mathbf{x}$ in C^n such that $A\mathbf{x} = \lambda\mathbf{x}$. Each such $\mathbf{x}$ is called a *complex eigenvector* of A corresponding to λ. The complex eigenvectors of A corresponding to λ are the nonzero solutions of the linear system $(\lambda I - A)\mathbf{x} = \mathbf{0}$, and the set of all such solutions is a subspace of C^n, called the *eigenspace* of A corresponding to λ.

Is R^n a subspace of C^n? Explain.

The following theorem states that if a *real matrix* has complex eigenvalues, then those eigenvalues and their corresponding eigenvectors occur in conjugate pairs.

THEOREM 5.3.4 *If* λ *is an eigenvalue of a real* $n \times n$ *matrix* A, *and if* $\mathbf{x}$ *is a corresponding eigenvector, then* $\bar{\lambda}$ *is also an eigenvalue of* A, *and* $\bar{\mathbf{x}}$ *is a corresponding eigenvector.*

Proof   Since λ is an eigenvalue of A and $\mathbf{x}$ is a corresponding eigenvector, we have

$$\overline{A\mathbf{x}} = \overline{\lambda\mathbf{x}} = \bar{\lambda}\bar{\mathbf{x}} \tag{6}$$

However, $\bar{A} = A$, since A has real entries, so it follows from part (c) of Theorem 5.3.2 that

$$\overline{A\mathbf{x}} = \bar{A}\bar{\mathbf{x}} = A\bar{\mathbf{x}} \tag{7}$$

Equations (6) and (7) together imply that

$$A\bar{\mathbf{x}} = \overline{A\mathbf{x}} = \bar{\lambda}\bar{\mathbf{x}}$$

in which $\bar{\mathbf{x}} \neq \mathbf{0}$ (why?); this tells us that $\bar{\lambda}$ is an eigenvalue of A and $\bar{\mathbf{x}}$ is a corresponding eigenvector. ◄

▶ **EXAMPLE 3 Complex Eigenvalues and Eigenvectors**

Find the eigenvalues and bases for the eigenspaces of

$$A = \begin{bmatrix} -2 & -1 \\ 5 & 2 \end{bmatrix}$$

Solution   The characteristic polynomial of A is

$$\begin{vmatrix} \lambda + 2 & 1 \\ -5 & \lambda - 2 \end{vmatrix} = \lambda^2 + 1 = (\lambda - i)(\lambda + i)$$

so the eigenvalues of A are $\lambda = i$ and $\lambda = -i$. Note that these eigenvalues are complex conjugates, as guaranteed by Theorem 5.3.4. To find the eigenvectors we must solve the system

$$\begin{bmatrix} \lambda + 2 & 1 \\ -5 & \lambda - 2 \end{bmatrix} \begin{bmatrix} x_1 \\ x_2 \end{bmatrix} = \begin{bmatrix} 0 \\ 0 \end{bmatrix}$$

with $\lambda = i$ and then with $\lambda = -i$. With $\lambda = i$, this system becomes

$$\begin{bmatrix} i + 2 & 1 \\ -5 & i - 2 \end{bmatrix} \begin{bmatrix} x_1 \\ x_2 \end{bmatrix} = \begin{bmatrix} 0 \\ 0 \end{bmatrix} \tag{8}$$

We could solve this system by reducing the augmented matrix

$$\begin{bmatrix} i + 2 & 1 & 0 \\ -5 & i - 2 & 0 \end{bmatrix} \tag{9}$$

to reduced row echelon form by Gauss–Jordan elimination, though the complex arithmetic is somewhat tedious. A simpler procedure here is first to observe that the reduced row echelon form of (9) must have a row of zeros because (8) has nontrivial solutions. This being the case, each row of (9) must be a scalar multiple of the other, and hence the first row can be made into a row of zeros by adding a suitable multiple of the second row to it. Accordingly, we can simply set the entries in the first row to zero, then interchange the rows, and then multiply the new first row by $-\frac{1}{5}$ to obtain the reduced row echelon form

$$\begin{bmatrix} 1 & \frac{2}{5} - \frac{1}{5}i & 0 \\ 0 & 0 & 0 \end{bmatrix}$$

Thus, a general solution of the system is

$$x_1 = \left(-\tfrac{2}{5} + \tfrac{1}{5}i\right)t, \quad x_2 = t$$

This tells us that the eigenspace corresponding to $\lambda = i$ is one-dimensional and consists of all complex scalar multiples of the basis vector

$$\mathbf{x} = \begin{bmatrix} -\frac{2}{5} + \frac{1}{5}i \\ 1 \end{bmatrix} \tag{10}$$

As a check, let us confirm that $A\mathbf{x} = i\mathbf{x}$. We obtain

$$A\mathbf{x} = \begin{bmatrix} -2 & -1 \\ 5 & 2 \end{bmatrix} \begin{bmatrix} -\frac{2}{5} + \frac{1}{5}i \\ 1 \end{bmatrix} = \begin{bmatrix} -2\left(-\frac{2}{5} + \frac{1}{5}i\right) - 1 \\ 5\left(-\frac{2}{5} + \frac{1}{5}i\right) + 2 \end{bmatrix} = \begin{bmatrix} -\frac{1}{5} - \frac{2}{5}i \\ i \end{bmatrix} = i\mathbf{x}$$

We could find a basis for the eigenspace corresponding to $\lambda = -i$ in a similar way, but the work is unnecessary, since Theorem 5.3.4 implies that

$$\bar{\mathbf{x}} = \begin{bmatrix} -\frac{2}{5} - \frac{1}{5}i \\ 1 \end{bmatrix} \tag{11}$$

must be a basis for this eigenspace. The following computations confirm that $\bar{\mathbf{x}}$ is an eigenvector of A corresponding to $\lambda = -i$:

$$A\bar{\mathbf{x}} = \begin{bmatrix} -2 & -1 \\ 5 & 2 \end{bmatrix} \begin{bmatrix} -\frac{2}{5} - \frac{1}{5}i \\ 1 \end{bmatrix}$$

$$= \begin{bmatrix} -2\left(-\frac{2}{5} - \frac{1}{5}i\right) - 1 \\ 5\left(-\frac{2}{5} - \frac{1}{5}i\right) + 2 \end{bmatrix} = \begin{bmatrix} -\frac{1}{5} + \frac{2}{5}i \\ -i \end{bmatrix} = -i\bar{\mathbf{x}} \blacktriangleleft$$

Since a number of our subsequent examples will involve 2×2 matrices with real entries, it will be useful to discuss some general results about the eigenvalues of such matrices. Observe first that the characteristic polynomial of the matrix

$$A = \begin{bmatrix} a & b \\ c & d \end{bmatrix}$$

is

$$\det(\lambda I - A) = \begin{vmatrix} \lambda - a & -b \\ -c & \lambda - d \end{vmatrix} = (\lambda - a)(\lambda - d) - bc = \lambda^2 - (a+d)\lambda + (ad - bc)$$

We can express this in terms of the trace and determinant of A as

$$\det(\lambda I - A) = \lambda^2 - \operatorname{tr}(A)\lambda + \det(A) \qquad (12)$$

from which it follows that the characteristic equation of A is

$$\lambda^2 - \operatorname{tr}(A)\lambda + \det(A) = 0 \qquad (13)$$

Now recall from algebra that if $ax^2 + bx + c = 0$ is a quadratic equation with real coefficients, then the **discriminant** $b^2 - 4ac$ determines the nature of the roots:

$$b^2 - 4ac > 0 \qquad \textbf{[Two distinct real roots]}$$
$$b^2 - 4ac = 0 \qquad \textbf{[One repeated real root]}$$
$$b^2 - 4ac < 0 \qquad \textbf{[Two conjugate imaginary roots]}$$

Applying this to (13) with $a = 1$, $b = -\operatorname{tr}(A)$, and $c = \det(A)$ yields the following theorem.

> **THEOREM 5.3.5** *If A is a 2×2 matrix with real entries, then the characteristic equation of A is $\lambda^2 - \operatorname{tr}(A)\lambda + \det(A) = 0$ and*
>
> *(a) A has two distinct real eigenvalues if $\operatorname{tr}(A)^2 - 4\det(A) > 0$;*
>
> *(b) A has one repeated real eigenvalue if $\operatorname{tr}(A)^2 - 4\det(A) = 0$;*
>
> *(c) A has two complex conjugate eigenvalues if $\operatorname{tr}(A)^2 - 4\det(A) < 0$.*

▶ **EXAMPLE 4 Eigenvalues of a 2 × 2 Matrix**

In each part, use Formula (13) for the characteristic equation to find the eigenvalues of

(a) $A = \begin{bmatrix} 2 & 2 \\ -1 & 5 \end{bmatrix}$ (b) $A = \begin{bmatrix} 0 & -1 \\ 1 & 2 \end{bmatrix}$ (c) $A = \begin{bmatrix} 2 & 3 \\ -3 & 2 \end{bmatrix}$

Solution (a) We have $\operatorname{tr}(A) = 7$ and $\det(A) = 12$, so the characteristic equation of A is

$$\lambda^2 - 7\lambda + 12 = 0$$

Factoring yields $(\lambda - 4)(\lambda - 3) = 0$, so the eigenvalues of A are $\lambda = 4$ and $\lambda = 3$.

Solution (b) We have $\operatorname{tr}(A) = 2$ and $\det(A) = 1$, so the characteristic equation of A is

$$\lambda^2 - 2\lambda + 1 = 0$$

Factoring this equation yields $(\lambda - 1)^2 = 0$, so $\lambda = 1$ is the only eigenvalue of A; it has algebraic multiplicity 2.

Solution (c) We have $\operatorname{tr}(A) = 4$ and $\det(A) = 13$, so the characteristic equation of A is

$$\lambda^2 - 4\lambda + 13 = 0$$

Solving this equation by the quadratic formula yields

$$\lambda = \frac{4 \pm \sqrt{(-4)^2 - 4(13)}}{2} = \frac{4 \pm \sqrt{-36}}{2} = 2 \pm 3i$$

Thus, the eigenvalues of A are $\lambda = 2 + 3i$ and $\lambda = 2 - 3i$. ◀

Symmetric Matrices Have
Real Eigenvalues

Our next result, which is concerned with the eigenvalues of real symmetric matrices, is important in a wide variety of applications. The key to its proof is to think of a real symmetric matrix as a complex matrix whose entries have an imaginary part of zero.

THEOREM 5.3.6 *If A is a real symmetric matrix, then A has real eigenvalues.*

Proof Suppose that λ is an eigenvalue of A and $\mathbf{x}$ is a corresponding eigenvector, where we allow for the possibility that λ is complex and $\mathbf{x}$ is in C^n. Thus,

$$A\mathbf{x} = \lambda\mathbf{x}$$

where $\mathbf{x} \neq \mathbf{0}$. If we multiply both sides of this equation by $\bar{\mathbf{x}}^T$ and use the fact that

$$\bar{\mathbf{x}}^T A\mathbf{x} = \bar{\mathbf{x}}^T(\lambda\mathbf{x}) = \lambda(\bar{\mathbf{x}}^T\mathbf{x}) = \lambda(\mathbf{x} \cdot \mathbf{x}) = \lambda\|\mathbf{x}\|^2$$

then we obtain

$$\lambda = \frac{\bar{\mathbf{x}}^T A\mathbf{x}}{\|\mathbf{x}\|^2}$$

Since the denominator in this expression is real, we can prove that λ is real by showing that

$$\overline{\bar{\mathbf{x}}^T A\mathbf{x}} = \bar{\mathbf{x}}^T A\mathbf{x} \tag{14}$$

But, A is symmetric and has real entries, so it follows from the second equality in (14) and properties of the conjugate that

$$\overline{\bar{\mathbf{x}}^T A\mathbf{x}} = \bar{\bar{\mathbf{x}}}^T \overline{A\mathbf{x}} = \mathbf{x}^T \overline{A\mathbf{x}} = (\overline{A\mathbf{x}})^T \mathbf{x} = (\overline{A}\bar{\mathbf{x}})^T \mathbf{x} = (A\bar{\mathbf{x}})^T \mathbf{x} = \bar{\mathbf{x}}^T A^T \mathbf{x} = \bar{\mathbf{x}}^T A\mathbf{x} \blacktriangleleft$$

A Geometric Interpretation
of Complex Eigenvalues

The following theorem is the key to understanding the geometric significance of complex eigenvalues of real 2×2 matrices.

THEOREM 5.3.7 *The eigenvalues of the real matrix*

$$C = \begin{bmatrix} a & -b \\ b & a \end{bmatrix} \tag{15}$$

are $\lambda = a \pm bi$. If a and b are not both zero, then this matrix can be factored as

$$\begin{bmatrix} a & -b \\ b & a \end{bmatrix} = \begin{bmatrix} |\lambda| & 0 \\ 0 & |\lambda| \end{bmatrix} \begin{bmatrix} \cos\phi & -\sin\phi \\ \sin\phi & \cos\phi \end{bmatrix} \tag{16}$$

where ϕ is the angle from the positive x-axis to the ray that joins the origin to the point (a, b) (Figure 5.3.2).

▲ Figure 5.3.2

Geometrically, this theorem states that multiplication by a matrix of form (15) can be viewed as a rotation through the angle ϕ followed by a scaling with factor $|\lambda|$ (Figure 5.3.3).

Proof The characteristic equation of C is $(\lambda - a)^2 + b^2 = 0$ (verify), from which it follows that the eigenvalues of C are $\lambda = a \pm bi$. Assuming that a and b are not both zero, let ϕ be the angle from the positive x-axis to the ray that joins the origin to the point (a, b). The angle ϕ is an argument of the eigenvalue $\lambda = a + bi$, so we see from Figure 5.3.2 that

$$a = |\lambda|\cos\phi \quad \text{and} \quad b = |\lambda|\sin\phi$$

▲ Figure 5.3.3

It follows from this that the matrix in (15) can be written as

$$\begin{bmatrix} a & -b \\ b & a \end{bmatrix} = \begin{bmatrix} |\lambda| & 0 \\ 0 & |\lambda| \end{bmatrix} \begin{bmatrix} \dfrac{a}{|\lambda|} & -\dfrac{b}{|\lambda|} \\ \dfrac{b}{|\lambda|} & \dfrac{a}{|\lambda|} \end{bmatrix} = \begin{bmatrix} |\lambda| & 0 \\ 0 & |\lambda| \end{bmatrix} \begin{bmatrix} \cos\phi & -\sin\phi \\ \sin\phi & \cos\phi \end{bmatrix} \quad \blacktriangleleft$$

The following theorem, whose proof is considered in the exercises, shows that every real 2×2 matrix with complex eigenvalues is similar to a matrix of form (15).

> **THEOREM 5.3.8** *Let A be a real 2×2 matrix with complex eigenvalues $\lambda = a \pm bi$ (where $b \neq 0$). If $\mathbf{x}$ is an eigenvector of A corresponding to $\lambda = a - bi$, then the matrix $P = [\mathrm{Re}(\mathbf{x}) \quad \mathrm{Im}(\mathbf{x})]$ is invertible and*
>
> $$A = P \begin{bmatrix} a & -b \\ b & a \end{bmatrix} P^{-1} \tag{17}$$

▶ **EXAMPLE 5 A Matrix Factorization Using Complex Eigenvalues**

Factor the matrix in Example 3 into form (17) using the eigenvalue $\lambda = -i$ and the corresponding eigenvector that was given in (11).

Solution For consistency with the notation in Theorem 5.3.8, let us denote the eigenvector in (11) that corresponds to $\lambda = -i$ by $\mathbf{x}$ (rather than $\bar{\mathbf{x}}$ as before). For this λ and $\mathbf{x}$ we have

$$a = 0, \quad b = 1, \quad \mathrm{Re}(\mathbf{x}) = \begin{bmatrix} -\frac{2}{5} \\ 1 \end{bmatrix}, \quad \mathrm{Im}(\mathbf{x}) = \begin{bmatrix} -\frac{1}{5} \\ 0 \end{bmatrix}$$

Thus,

$$P = [\mathrm{Re}(\mathbf{x}) \quad \mathrm{Im}(\mathbf{x})] = \begin{bmatrix} -\frac{2}{5} & -\frac{1}{5} \\ 1 & 0 \end{bmatrix}$$

so A can be factored in form (17) as

$$\begin{bmatrix} -2 & -1 \\ 5 & 2 \end{bmatrix} = \begin{bmatrix} -\frac{2}{5} & -\frac{1}{5} \\ 1 & 0 \end{bmatrix} \begin{bmatrix} 0 & -1 \\ 1 & 0 \end{bmatrix} \begin{bmatrix} 0 & 1 \\ -5 & -2 \end{bmatrix}$$

You may want to confirm this by multiplying out the right side. ◀

A Geometric Interpretation of Theorem 5.3.8

To clarify what Theorem 5.3.8 says geometrically, let us denote the matrices on the right side of (16) by S and R_ϕ, respectively, and then use (16) to rewrite (17) as

$$A = PSR_\phi P^{-1} = P \begin{bmatrix} |\lambda| & 0 \\ 0 & |\lambda| \end{bmatrix} \begin{bmatrix} \cos\phi & -\sin\phi \\ \sin\phi & \cos\phi \end{bmatrix} P^{-1} \tag{18}$$

If we now view P as the transition matrix from the basis $B = \{\mathrm{Re}(\mathbf{x}), \mathrm{Im}(\mathbf{x})\}$ to the standard basis, then (18) tells us that computing a product $A\mathbf{x}_0$ can be broken down into a three-step process:

Step 1. Map $\mathbf{x}_0$ from standard coordinates into B-coordinates by forming the product $P^{-1}\mathbf{x}_0$.

Step 2. Rotate and scale the vector $P^{-1}\mathbf{x}_0$ by forming the product $SR_\phi P^{-1}\mathbf{x}_0$.

Step 3. Map the rotated and scaled vector back to standard coordinates to obtain $A\mathbf{x}_0 = PSR_\phi P^{-1}\mathbf{x}_0$.

Power Sequences   There are many problems in which one is interested in how successive applications of a matrix transformation affect a specific vector. For example, if A is the standard matrix for an operator on R^n and $\mathbf{x}_0$ is some fixed vector in R^n, then one might be interested in the behavior of the power sequence

$$\mathbf{x}_0, \quad A\mathbf{x}_0, \quad A^2\mathbf{x}_0, \ldots, \quad A^k\mathbf{x}_0, \ldots$$

For example, if

$$A = \begin{bmatrix} \frac{1}{2} & \frac{3}{4} \\ -\frac{3}{5} & \frac{11}{10} \end{bmatrix} \quad \text{and} \quad \mathbf{x}_0 = \begin{bmatrix} 1 \\ 1 \end{bmatrix}$$

then with the help of a computer or calculator one can show that the first four terms in the power sequence are

$$\mathbf{x}_0 = \begin{bmatrix} 1 \\ 1 \end{bmatrix}, \quad A\mathbf{x}_0 = \begin{bmatrix} 1.25 \\ 0.5 \end{bmatrix}, \quad A^2\mathbf{x}_0 = \begin{bmatrix} 1.0 \\ -0.2 \end{bmatrix}, \quad A^3\mathbf{x}_0 = \begin{bmatrix} 0.35 \\ -0.82 \end{bmatrix}$$

With the help of MATLAB or a computer algebra system one can show that if the first 100 terms are plotted as ordered pairs (x, y), then the points move along the elliptical path shown in Figure 5.3.4*a*.

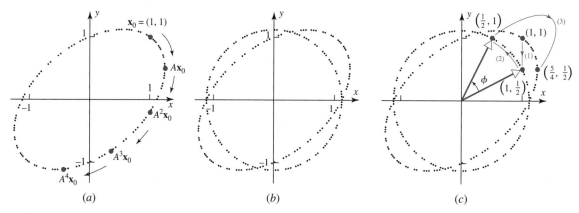

(a)   (b)   (c)

▲ Figure 5.3.4

To understand why the points move along an elliptical path, we will need to examine the eigenvalues and eigenvectors of A. We leave it for you to show that the eigenvalues of A are $\lambda = \frac{4}{5} \pm \frac{3}{5}i$ and that the corresponding eigenvectors are

$$\lambda_1 = \frac{4}{5} - \frac{3}{5}i: \quad \mathbf{v}_1 = \left(\frac{1}{2} + i, 1\right) \quad \text{and} \quad \lambda_2 = \frac{4}{5} + \frac{3}{5}i: \quad \mathbf{v}_2 = \left(\frac{1}{2} - i, 1\right)$$

If we take $\lambda = \lambda_1 = \frac{4}{5} - \frac{3}{5}i$ and $\mathbf{x} = \mathbf{v}_1 = \left(\frac{1}{2} + i, 1\right)$ in (17) and use the fact that $|\lambda| = 1$, then we obtain the factorization

$$\underbrace{\begin{bmatrix} \frac{1}{2} & \frac{3}{4} \\ -\frac{3}{5} & \frac{11}{10} \end{bmatrix}}_{A} = \underbrace{\begin{bmatrix} \frac{1}{2} & 1 \\ 1 & 0 \end{bmatrix}}_{P} \underbrace{\begin{bmatrix} \frac{4}{5} & -\frac{3}{5} \\ \frac{3}{5} & \frac{4}{5} \end{bmatrix}}_{R_\phi} \underbrace{\begin{bmatrix} 0 & 1 \\ 1 & -\frac{1}{2} \end{bmatrix}}_{P^{-1}} \tag{19}$$

where R_ϕ is a rotation about the origin through the angle ϕ whose tangent is

$$\tan\phi = \frac{\sin\phi}{\cos\phi} = \frac{3/5}{4/5} = \frac{3}{4} \quad \left(\phi = \tan^{-1}\frac{3}{4} \approx 36.9°\right)$$

The matrix P in (19) is the transition matrix from the basis

$$B = \{\text{Re}(\mathbf{x}), \text{Im}(\mathbf{x})\} = \left\{\left(\frac{1}{2}, 1\right), (1, 0)\right\}$$

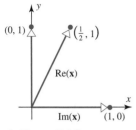

▲ Figure 5.3.5

to the standard basis, and P^{-1} is the transition matrix from the standard basis to the basis B (Figure 5.3.5). Next, observe that if n is a positive integer, then (19) implies that

$$A^n \mathbf{x}_0 = (PR_\phi P^{-1})^n \mathbf{x}_0 = PR_\phi^{\ n} P^{-1} \mathbf{x}_0$$

so the product $A^n \mathbf{x}_0$ can be computed by first mapping $\mathbf{x}_0$ into the point $P^{-1}\mathbf{x}_0$ in B-coordinates, then multiplying by $R_\phi^{\ n}$ to rotate this point about the origin through the angle $n\phi$, and then multiplying $R_\phi^{\ n} P^{-1}\mathbf{x}_0$ by P to map the resulting point back to standard coordinates. We can now see what is happening geometrically: In B-coordinates each successive multiplication by A causes the point $P^{-1}\mathbf{x}_0$ to advance through an angle ϕ, thereby tracing a circular orbit about the origin. However, the basis B is *skewed* (not orthogonal), so when the points on the circular orbit are transformed back to standard coordinates, the effect is to distort the circular orbit into the elliptical orbit traced by $A^n \mathbf{x}_0$ (Figure 5.3.4b). Here are the computations for the first step (successive steps are illustrated in Figure 5.3.4c):

$$\begin{bmatrix} \frac{1}{2} & \frac{3}{4} \\ -\frac{3}{5} & \frac{11}{10} \end{bmatrix} \begin{bmatrix} 1 \\ 1 \end{bmatrix} = \begin{bmatrix} \frac{1}{2} & 1 \\ 1 & 0 \end{bmatrix} \begin{bmatrix} \frac{4}{5} & -\frac{3}{5} \\ \frac{3}{5} & \frac{4}{5} \end{bmatrix} \begin{bmatrix} 0 & 1 \\ 1 & -\frac{1}{2} \end{bmatrix} \begin{bmatrix} 1 \\ 1 \end{bmatrix}$$

$$= \begin{bmatrix} \frac{1}{2} & 1 \\ 1 & 0 \end{bmatrix} \begin{bmatrix} \frac{4}{5} & -\frac{3}{5} \\ \frac{3}{5} & \frac{4}{5} \end{bmatrix} \begin{bmatrix} 1 \\ \frac{1}{2} \end{bmatrix} \qquad \text{[x_0 is mapped to B-coordinates.]}$$

$$= \begin{bmatrix} \frac{1}{2} & 1 \\ 1 & 0 \end{bmatrix} \begin{bmatrix} \frac{1}{2} \\ 1 \end{bmatrix} \qquad \text{[The point $\left(1, \frac{1}{2}\right)$ is rotated through the angle ϕ.]}$$

$$= \begin{bmatrix} \frac{5}{4} \\ \frac{1}{2} \end{bmatrix} \qquad \text{[The point $\left(\frac{1}{2}, 1\right)$ is mapped to standard coordinates.]}$$

Concept Review

- Real part of z
- Imaginary part of z
- Modulus of z
- Complex conjugate of z
- Argument of z
- Polar form of z

- Complex vector space
- Complex n-tuple
- Complex n-space
- Real matrix
- Complex matrix
- Complex Euclidean inner product

- Euclidean norm on C^n
- Antisymmetry property
- Complex eigenvalue
- Complex eigenvector
- Eigenspace in C^n
- Discriminant

Skills

- Find the real part, imaginary part, and complex conjugate of a complex matrix or vector.
- Find the determinant of a complex matrix.
- Find complex inner products and norms of complex vectors.

- Find the eigenvalues and bases for the eigenspaces of complex matrices.
- Factor a 2×2 real matrix with complex eigenvalues into a product of a scaling matrix and a rotation matrix.

Exercise Set 5.3

▶ In Exercises 1–2, find $\bar{\mathbf{u}}$, Re($\mathbf{u}$), Im($\mathbf{u}$), and $\|\mathbf{u}\|$. ◀

1. $\mathbf{u} = (2 - i, 4i, 1 + i)$ **2.** $\mathbf{u} = (6, 1 + 4i, 6 - 2i)$

▶ In Exercises 3–4, show that $\mathbf{u}$, $\mathbf{v}$, and k satisfy Theorem 5.3.1.

3. $\mathbf{u} = (3 - 4i, 2 + i, -6i)$, $\mathbf{v} = (1 + i, 2 - i, 4)$, $k = i$

4. $\mathbf{u} = (6, 1 + 4i, 6 - 2i)$, $\mathbf{v} = (4, 3 + 2i, i - 3)$, $k = -i$

5. Solve the equation $i\mathbf{x} - 3\mathbf{v} = \bar{\mathbf{u}}$ for $\mathbf{x}$, where $\mathbf{u}$ and $\mathbf{v}$ are the vectors in Exercise 3.

6. Solve the equation $(1 + i)\mathbf{x} + 2\mathbf{u} = \bar{\mathbf{v}}$ for $\mathbf{x}$, where $\mathbf{u}$ and $\mathbf{v}$ are the vectors in Exercise 4.

▶ In Exercises 7–8, find $\bar{A}$, Re(A), Im(A), det(A), and tr(A). ◀

7. $A = \begin{bmatrix} -5i & 4 \\ 2 - i & 1 + 5i \end{bmatrix}$ **8.** $A = \begin{bmatrix} 4i & 2 - 3i \\ 2 + 3i & 1 \end{bmatrix}$

9. Let A be the matrix given in Exercise 7, and let B be the matrix

$$B = \begin{bmatrix} 1 - i \\ 2i \end{bmatrix}$$

Confirm that these matrices have the properties stated in Theorem 5.3.2.

10. Let A be the matrix given in Exercise 8, and let B be the matrix

$$B = \begin{bmatrix} 5i \\ 1 - 4i \end{bmatrix}$$

Confirm that these matrices have the properties stated in Theorem 5.3.2.

▶ In Exercises 11–12, compute $\mathbf{u} \cdot \mathbf{v}$, $\mathbf{u} \cdot \mathbf{w}$, and $\mathbf{v} \cdot \mathbf{w}$, and show that the vectors satisfy Formula (5) and parts (a), (b), and (c) of Theorem 5.3.3. ◀

11. $\mathbf{u} = (i, 2i, 3)$, $\mathbf{v} = (4, -2i, 1 + i)$, $\mathbf{w} = (2 - i, 2i, 5 + 3i)$, $k = 2i$

12. $\mathbf{u} = (1 + i, 4, 3i)$, $\mathbf{v} = (3, -4i, 2 + 3i)$, $\mathbf{w} = (1 - i, 4i, 4 - 5i)$, $k = 1 + i$

13. Compute $\overline{(\mathbf{u} \cdot \mathbf{v})} - \overline{\mathbf{w} \cdot \mathbf{u}}$ for the vectors $\mathbf{u}$, $\mathbf{v}$, and $\mathbf{w}$ in Exercise 11.

14. Compute $\overline{(i\mathbf{u} \cdot \mathbf{w})} + (\|\mathbf{u}\|\mathbf{v}) \cdot \mathbf{u}$ for the vectors $\mathbf{u}$, $\mathbf{v}$, and $\mathbf{w}$ in Exercise 12.

▶ In Exercises 15–18, find the eigenvalues and bases for the eigenspaces of A. ◀

15. $A = \begin{bmatrix} 4 & -5 \\ 1 & 0 \end{bmatrix}$ **16.** $A = \begin{bmatrix} -1 & -5 \\ 4 & 7 \end{bmatrix}$

17. $A = \begin{bmatrix} 5 & -2 \\ 1 & 3 \end{bmatrix}$ **18.** $A = \begin{bmatrix} 8 & 6 \\ -3 & 2 \end{bmatrix}$

▶ In Exercises 19–22, each matrix C has form (15). Theorem 5.3.7 implies that C is the product of a scaling matrix with factor $|\lambda|$ and a rotation matrix with angle ϕ. Find $|\lambda|$ and ϕ for which $-\pi < \phi \le \pi$. ◀

19. $C = \begin{bmatrix} 1 & -1 \\ 1 & 1 \end{bmatrix}$ **20.** $C = \begin{bmatrix} 0 & 5 \\ -5 & 0 \end{bmatrix}$

21. $C = \begin{bmatrix} 1 & \sqrt{3} \\ -\sqrt{3} & 1 \end{bmatrix}$ **22.** $C = \begin{bmatrix} \sqrt{2} & \sqrt{2} \\ -\sqrt{2} & \sqrt{2} \end{bmatrix}$

▶ In Exercises 23–26, find an invertible matrix P and a matrix C of form (15) such that $A = PCP^{-1}$. ◀

23. $A = \begin{bmatrix} -1 & -5 \\ 4 & 7 \end{bmatrix}$ **24.** $A = \begin{bmatrix} 4 & -5 \\ 1 & 0 \end{bmatrix}$

25. $A = \begin{bmatrix} 8 & 6 \\ -3 & 2 \end{bmatrix}$ **26.** $A = \begin{bmatrix} 5 & -2 \\ 1 & 3 \end{bmatrix}$

27. Find all complex scalars k, if any, for which $\mathbf{u}$ and $\mathbf{v}$ are orthogonal in C^3.

(a) $\mathbf{u} = (2i, i, 3i)$, $\mathbf{v} = (i, 6i, k)$

(b) $\mathbf{u} = (k, k, 1 + i)$, $\mathbf{v} = (1, -1, 1 - i)$

28. Show that if A is a real $n \times n$ matrix and $\mathbf{x}$ is a column vector in C^n, then Re($A\mathbf{x}$) = A(Re($\mathbf{x}$)) and Im($A\mathbf{x}$) = A(Im($\mathbf{x}$)).

29. The matrices

$$\sigma_1 = \begin{bmatrix} 0 & 1 \\ 1 & 0 \end{bmatrix}, \quad \sigma_2 = \begin{bmatrix} 0 & -i \\ i & 0 \end{bmatrix}, \quad \sigma_3 = \begin{bmatrix} 1 & 0 \\ 0 & -1 \end{bmatrix}$$

called *Pauli spin matrices*, are used in quantum mechanics to study particle spin. The *Dirac matrices*, which are also used in quantum mechanics, are expressed in terms of the Pauli spin matrices and the 2×2 identity matrix I_2 as

$$\beta = \begin{bmatrix} I_2 & 0 \\ 0 & -I_2 \end{bmatrix}, \quad \alpha_x = \begin{bmatrix} 0 & \sigma_1 \\ \sigma_1 & 0 \end{bmatrix},$$

$$\alpha_y = \begin{bmatrix} 0 & \sigma_2 \\ \sigma_2 & 0 \end{bmatrix}, \quad \alpha_z = \begin{bmatrix} 0 & \sigma_3 \\ \sigma_3 & 0 \end{bmatrix}$$

(a) Show that $\beta^2 = \alpha_x^2 = \alpha_y^2 = \alpha_z^2$.

(b) Matrices A and B for which $AB = -BA$ are said to be *anticommutative*. Show that the Dirac matrices are anticommutative.

30. If k is a real scalar and $\mathbf{v}$ is a vector in R^n, then Theorem 3.2.1 states that $\|k\mathbf{v}\| = |k|\|\mathbf{v}\|$. Is this relationship also true if k is a complex scalar and $\mathbf{v}$ is a vector in C^n? Justify your answer.

31. Prove part (c) of Theorem 5.3.1.

32. Prove Theorem 5.3.2.

33. Prove that if $\mathbf{u}$ and $\mathbf{v}$ are vectors in C^n, then

$$\mathbf{u} \cdot \mathbf{v} = \frac{1}{4}\|\mathbf{u} + \mathbf{v}\|^2 - \frac{1}{4}\|\mathbf{u} - \mathbf{v}\|^2$$
$$+ \frac{i}{4}\|\mathbf{u} + i\mathbf{v}\|^2 - \frac{i}{4}\|\mathbf{u} - i\mathbf{v}\|^2$$

34. It follows from Theorem 5.3.7 that the eigenvalues of the rotation matrix

$$R_\phi = \begin{bmatrix} \cos\phi & -\sin\phi \\ \sin\phi & \cos\phi \end{bmatrix}$$

are $\lambda = \cos\phi \pm i\sin\phi$. Prove that if $\mathbf{x}$ is an eigenvector corresponding to either eigenvalue, then $\mathrm{Re}(\mathbf{x})$ and $\mathrm{Im}(\mathbf{x})$ are orthogonal and have the same length. [*Note:* This implies that $P = [\mathrm{Re}(\mathbf{x}) \mid \mathrm{Im}(\mathbf{x})]$ is a real scalar multiple of an orthogonal matrix.]

35. The two parts of this exercise lead you through a proof of Theorem 5.3.8.

(a) For notational simplicity, let

$$M = \begin{bmatrix} a & -b \\ b & a \end{bmatrix}$$

and let $\mathbf{u} = \mathrm{Re}(\mathbf{x})$ and $\mathbf{v} = \mathrm{Im}(\mathbf{x})$, so $P = [\mathbf{u} \mid \mathbf{v}]$. Show that the relationship $A\mathbf{x} = \lambda\mathbf{x}$ implies that

$$A\mathbf{x} = (a\mathbf{u} + b\mathbf{v}) + i(-b\mathbf{u} + a\mathbf{v})$$

and then equate real and imaginary parts in this equation to show that

$$AP = [A\mathbf{u} \mid A\mathbf{v}] = [a\mathbf{u} + b\mathbf{v} \mid -b\mathbf{u} + a\mathbf{v}] = PM$$

(b) Show that P is invertible, thereby completing the proof, since the result in part (a) implies that $A = PMP^{-1}$. [*Hint:* If P is not invertible, then one of its column vectors is a real scalar multiple of the other, say $\mathbf{v} = c\mathbf{u}$. Substitute this into the equations $A\mathbf{u} = a\mathbf{u} + b\mathbf{v}$

and $A\mathbf{v} = -b\mathbf{u} + a\mathbf{v}$ obtained in part (a), and show that $(1 + c^2)b\mathbf{u} = \mathbf{0}$. Finally, show that this leads to a contradiction, thereby proving that P is invertible.]

36. In this problem you will prove the complex analog of the Cauchy–Schwarz inequality.

(a) Prove: If k is a complex number, and $\mathbf{u}$ and $\mathbf{v}$ are vectors in C^n, then

$$(\mathbf{u} - k\mathbf{v}) \cdot (\mathbf{u} - k\mathbf{v}) = \mathbf{u} \cdot \mathbf{u} - \bar{k}(\mathbf{u} \cdot \mathbf{v}) - k\overline{(\mathbf{u} \cdot \mathbf{v})} + k\bar{k}(\mathbf{v} \cdot \mathbf{v})$$

(b) Use the result in part (a) to prove that

$$0 \le \mathbf{u} \cdot \mathbf{u} - \bar{k}(\mathbf{u} \cdot \mathbf{v}) - k\overline{(\mathbf{u} \cdot \mathbf{v})} + k\bar{k}(\mathbf{v} \cdot \mathbf{v})$$

(c) Take $k = (\mathbf{u} \cdot \mathbf{v})/(\mathbf{v} \cdot \mathbf{v})$ in part (b) to prove that

$$|\mathbf{u} \cdot \mathbf{v}| \le \|\mathbf{u}\|\,\|\mathbf{v}\|$$

True-False Exercises

In parts (a)–(f) determine whether the statement is true or false, and justify your answer.

(a) There is a real 5×5 matrix with no real eigenvalues.

(b) The eigenvalues of a 2×2 complex matrix are the solutions of the equation $\lambda^2 - \mathrm{tr}(A)\lambda + \det(A) = 0$.

(c) Matrices that have the same complex eigenvalues with the same algebraic multiplicities have the same trace.

(d) If λ is a complex eigenvalue of a real matrix A with a corresponding complex eigenvector $\mathbf{v}$, then $\bar{\lambda}$ is a complex eigenvalue of A and $\bar{\mathbf{v}}$ is a complex eigenvector of A corresponding to $\bar{\lambda}$.

(e) Every eigenvalue of a complex symmetric matrix is real.

(f) If a 2×2 real matrix A has complex eigenvalues and $\mathbf{x}_0$ is a vector in R^2, then the vectors $\mathbf{x}_0, A\mathbf{x}_0, A^2\mathbf{x}_0, \ldots, A^n\mathbf{x}_0, \ldots$ lie on an ellipse.

5.4 Differential Equations

Many laws of physics, chemistry, biology, engineering, and economics are described in terms of "differential equations"—that is, equations involving functions and their derivatives. In this section we will illustrate one way in which linear algebra, eigenvalues and eigenvectors can be applied to solving systems of differential equations. Calculus is a prerequisite for this section.

Terminology   Recall from calculus that a ***differential equation*** is an equation involving unknown functions and their derivatives. The ***order*** of a differential equation is the order of the highest derivative it contains. The simplest differential equations are the first-order equations of the form

$$y' = ay \qquad (1)$$

where $y = f(x)$ is an unknown differentiable function to be determined, $y' = dy/dx$ is its derivative, and a is a constant. As with most differential equations, this equation has infinitely many solutions; they are the functions of the form

$$y = ce^{ax} \tag{2}$$

where c is an arbitrary constant. That every function of this form is a solution of (1) follows from the computation

$$y' = cae^{ax} = ay$$

and that these are the only solution is shown in the exercises. Accordingly, we call (2) the *general solution* of (1). As an example, the general solution of the differential equation $y' = 5y$ is

$$y = ce^{5x} \tag{3}$$

Often, a physical problem that leads to a differential equation imposes some conditions that enable us to isolate one particular solution from the general solution. For example, if we require that solution (3) of the equation $y' = 5y$ satisfy the added condition

$$y(0) = 6 \tag{4}$$

(that is, $y = 6$ when $x = 0$), then on substituting these values in (3), we obtain $6 = ce^0 = c$, from which we conclude that

$$y = 6e^{5x}$$

is the only solution $y' = 5y$ that satisfies (4).

A condition such as (4), which specifies the value of the general solution at a point is called an *initial condition*, and the problem of solving a differential equation subject to an initial condition is called an *initial-value problem*.

First-Order Linear Systems

In this section we will be concerned with solving systems of differential equations of the form

$$
\begin{aligned}
y_1' &= a_{11}y_1 + a_{12}y_2 + \cdots + a_{1n}y_n \\
y_2' &= a_{21}y_1 + a_{22}y_2 + \cdots + a_{2n}y_n \\
&\ \ \vdots \qquad \vdots \qquad \vdots \qquad\qquad \vdots \\
y_n' &= a_{n1}y_1 + a_{n2}y_2 + \cdots + a_{nn}y_n
\end{aligned}
\tag{5}
$$

where $y_1 = f_1(x)$, $y_2 = f_2(x)$, ..., $y_n = f_n(x)$ are functions to be determined, and the a_{ij}'s are constants. In matrix notation, (5) can be written as

$$
\begin{bmatrix} y_1' \\ y_2' \\ \vdots \\ y_n' \end{bmatrix}
=
\begin{bmatrix}
a_{11} & a_{12} & \cdots & a_{1n} \\
a_{21} & a_{22} & \cdots & a_{2n} \\
\vdots & \vdots & & \vdots \\
a_{n1} & a_{n2} & \cdots & a_{nn}
\end{bmatrix}
\begin{bmatrix} y_1 \\ y_2 \\ \vdots \\ y_n \end{bmatrix}
$$

A system of differential equations of form (5) is called a *first-order linear system*.

or, more briefly as

$$\mathbf{y}' = A\mathbf{y} \tag{6}$$

where the notation $\mathbf{y}'$ denotes the vector obtained by differentiating each component of y.

▶ **EXAMPLE 1** **Solution of a Linear System with Initial Conditions**

(a)   Write the following system in matrix form:

$$
\begin{aligned}
y_1' &= 3y_1 \\
y_2' &= -2y_2 \\
y_3' &= 5y_3
\end{aligned}
\tag{7}
$$

(b)   Solve the system.

(c)   Find a solution of the system that satisfies the initial conditions $y_1(0) = 1$, $y_2(0) = 4$, and $y_3(0) = -2$.

Solution (a)

$$\begin{bmatrix} y_1' \\ y_2' \\ y_3' \end{bmatrix} = \begin{bmatrix} 3 & 0 & 0 \\ 0 & -2 & 0 \\ 0 & 0 & 5 \end{bmatrix} \begin{bmatrix} y_1 \\ y_2 \\ y_3 \end{bmatrix} \tag{8}$$

or

$$\mathbf{y}' = \begin{bmatrix} 3 & 0 & 0 \\ 0 & -2 & 0 \\ 0 & 0 & 5 \end{bmatrix} \mathbf{y} \tag{9}$$

Solution (b)   Because each equation in (7) involves only one unknown function, we can solve the equations individually. It follows from (2) that these solutions are

$$y_1 = c_1 e^{3x}$$
$$y_2 = c_2 e^{-2x}$$
$$y_3 = c_3 e^{5x}$$

or, in matrix notation,

$$\mathbf{y} = \begin{bmatrix} y_1 \\ y_2 \\ y_3 \end{bmatrix} = \begin{bmatrix} c_1 e^{3x} \\ c_2 e^{-2x} \\ c_3 e^{5x} \end{bmatrix} \tag{10}$$

Solution (c)   From the given initial conditions, we obtain

$$1 = y_1(0) = c_1 e^0 = c_1$$
$$4 = y_2(0) = c_2 e^0 = c_2$$
$$-2 = y_3(0) = c_3 e^0 = c_3$$

so the solution satisfying these conditions

$$y_1 = e^{3x}, \quad y_2 \quad e^{-2x}, \quad y_3 = -2e^{5x}$$

or, in matrix notation,

$$\mathbf{y} = \begin{bmatrix} y_1 \\ y_2 \\ y_3 \end{bmatrix} = \begin{bmatrix} e^{3x} \\ 4e^{-2x} \\ -2e^{5x} \end{bmatrix} \blacktriangleleft$$

Solution by Diagonalization

What made the system in Example 1 easy solve was the fact that each equation involved only one of the unknown functions, so matrix formulation, $\mathbf{y}' = A\mathbf{y}$, had a *diagonal* coefficient matrix A [Formula (9)]. A n complicated situation occurs when some or all of the equations in the system invol ore than one of the unknown functions, for in this case the coefficient matrix is no agonal. Let us now consider how we might solve such a system.

The basic idea for solving a system $\mathbf{y}'$ $A\mathbf{y}$ whose coefficient matrix A is not diagonal is to introduce a new unknown vector at is related to the unknown vector $\mathbf{y}$ by an equation of the form $\mathbf{y} = P\mathbf{u}$ in which an invertible matrix that diagonalizes A. Of course, such a matrix may or may not ex out if it does then we can rewrite the equation $\mathbf{y}' = A\mathbf{y}$ as

$$P\mathbf{u} \quad A(P\mathbf{u})$$

or alternatively as

$$\mathbf{u}' = (P^{-1}AP)\mathbf{u}$$

Since P is assumed to diagonalize A, this equation has the form

$$\mathbf{u}' = D\mathbf{u}$$

where D is diagonal. We can now solve this equation for $\mathbf{u}$ using the method of Example 1, and then obtain $\mathbf{y}$ by matrix multiplication using the relationship $\mathbf{y} = P\mathbf{u}$.

In summary, we have the following procedure for solving a system $\mathbf{y}' = A\mathbf{y}$ in the case were A is diagonalizable.

A Procedure for Solving $\mathbf{y}' - A\mathbf{y}$ if A is Diagonalizable

Step 1.   Find a matrix P that diagonalizes A.

Step 2.   Make the substitutions $\mathbf{y} = P\mathbf{u}$ and $\mathbf{y}' = P\mathbf{u}'$ to obtain a new "diagonal system" $\mathbf{u}' = D\mathbf{u}$, where $D = P^{-1}AP$.

Step 3.   Solve $\mathbf{u}' = D\mathbf{u}$.

Step 4.   Determine $\mathbf{y}$ from the equation $\mathbf{y} = P\mathbf{u}$.

▶ **EXAMPLE 2 Solution Using Diagonalization**

(a)   Solve the system

$$\begin{aligned} y_1' &= y_1 + y_2 \\ y_2' &= 4y_1 - 2y_2 \end{aligned}$$

(b)   Find the solution that satisfies the initial conditions $y_1(0) = 1$, $y_2(0) = 6$.

Solution (a) The coefficient matrix for the system is

$$A = \begin{bmatrix} 1 & 1 \\ 4 & -2 \end{bmatrix}$$

As discussed in Section 5.2, A will be diagonalized by any matrix P whose columns are linearly independent eigenvectors of A. Since

$$\det(\lambda I - A) = \begin{vmatrix} \lambda - 1 & -1 \\ -4 & \lambda + 2 \end{vmatrix} = \lambda^2 + \lambda - 6 = (\lambda + 3)(\lambda - 2)$$

the eigenvalues of A are $\lambda = 2$ and $\lambda = -3$. By definition,

$$\mathbf{x} = \begin{bmatrix} x_1 \\ x_2 \end{bmatrix}$$

is an eigenvector of A corresponding to λ if and only if $\mathbf{x}$ is a nontrivial solution of

$$\begin{bmatrix} \lambda - 1 & -1 \\ -4 & \lambda + 2 \end{bmatrix} \begin{bmatrix} x_1 \\ x_2 \end{bmatrix} = \begin{bmatrix} 0 \\ 0 \end{bmatrix}$$

If $\lambda = 2$, this system becomes

$$\begin{bmatrix} 1 & -1 \\ -4 & 4 \end{bmatrix} \begin{bmatrix} x_1 \\ x_2 \end{bmatrix} = \begin{bmatrix} 0 \\ 0 \end{bmatrix}$$

Solving this system yields $x_1 = t$, $x_2 = t$, so

$$\begin{bmatrix} x_1 \\ x_2 \end{bmatrix} = \begin{bmatrix} t \\ t \end{bmatrix} = t \begin{bmatrix} 1 \\ 1 \end{bmatrix}$$

Thus,

$$\mathbf{p}_1 = \begin{bmatrix} 1 \\ 1 \end{bmatrix}$$

is a basis for the eigenspace corresponding to $\lambda = 2$. Similarly, you can show that

$$\mathbf{p}_2 = \begin{bmatrix} -\frac{1}{4} \\ 1 \end{bmatrix}$$

is a basis for the eigenspace corresponding to $\lambda = -3$. Thus,

$$P = \begin{bmatrix} 1 & -\frac{1}{4} \\ 1 & 1 \end{bmatrix}$$

diagonalizes A, and

$$D = P^{-1}AP = \begin{bmatrix} 2 & 0 \\ 0 & -3 \end{bmatrix}$$

Thus, as noted in Step 2 of the procedure stated above, the substitution

$$\mathbf{y} = P\mathbf{u} \quad \text{and} \quad \mathbf{y}' = P\mathbf{u}'$$

yields the "diagonal system"

$$\mathbf{u}' = D\mathbf{u} = \begin{bmatrix} 2 & 0 \\ 0 & -3 \end{bmatrix}\mathbf{u} \quad \text{or} \quad \begin{matrix} u_1' = 2u_1 \\ u_2' = -3u_2 \end{matrix}$$

From (2) the solution of this system is

$$\begin{matrix} u_1 = c_1 e^{2x} \\ u_2 = c_2 e^{-3x} \end{matrix} \quad \text{or} \quad \mathbf{u} = \begin{bmatrix} c_1 e^{2x} \\ c_2 e^{-3x} \end{bmatrix}$$

so the equation $\mathbf{y} = P\mathbf{u}$ yields, as the solution for $\mathbf{y}$,

$$\mathbf{y} = \begin{bmatrix} y_1 \\ y_2 \end{bmatrix} = \begin{bmatrix} 1 & -\frac{1}{4} \\ 1 & 1 \end{bmatrix}\begin{bmatrix} c_1 e^{2x} \\ c_2 e^{-3x} \end{bmatrix} = \begin{bmatrix} c_1 e^{2x} - \frac{1}{4}c_2 e^{-3x} \\ c_1 e^{2x} + c_2 e^{-3x} \end{bmatrix}$$

or

$$\begin{matrix} y_1 = c_1 e^{2x} - \frac{1}{4}c_2 e^{-3x} \\ y_2 = c_1 e^{2x} + c_2 e^{-3x} \end{matrix} \tag{11}$$

Solution (b) If we substitute the given initial conditions in (11), we obtain

$$\begin{matrix} c_1 - \frac{1}{4}c_2 = 1 \\ c_1 + c_2 = 6 \end{matrix}$$

Solving this system, we obtain $c_1 = 2$, $c_2 = 4$, so it follows from (11) that the solution satisfying the initial conditions is

$$\begin{matrix} y_1 = 2e^{2x} - e^{-3x} \\ y_2 = 2e^{2x} + 4e^{-3x} \end{matrix} \blacktriangleleft$$

Remark Keep in mind that the method of Example 2 works because the coefficient matrix of the system can be diagonalized. In cases where this is not so, other methods are required. These are typically discussed in books devoted to differential equations.

Concept Review

- Differential equation
- Order of a differential equation
- General solution
- Particular solution
- Initial condition
- Initial-value problem
- First-order linear system

Skills

- Find the matrix form of a system of linear differential equations.
- Find the general solution of a system of linear differential equations by diagonalization.
- Find the particular solution of a system of linear differential equations satisfying an initial condition.

Exercise Set 5.4

1. (a) Solve the system
$$y'_1 = y_1 + 4y_2$$
$$y'_2 = 2y_1 + 3y_2$$

  (b) Find the solution that satisfies the initial conditions $y_1(0) = 0$, $y_2(0) = 0$.

2. (a) Solve the system
$$y'_1 = y_1 + 3y_2$$
$$y'_2 = 4y_1 + 5y_2$$

  (b) Find the solution that satisfies the conditions $y_1(0) = 2$, $y'_2(0) = 1$.

3. (a) Solve the system
$$y'_1 = 4y_1 \qquad + y_3$$
$$y'_2 = -2y_1 + y_2$$
$$y'_3 = -2y_1 \qquad + y_3$$

  (b) Find the solution that satisfies the initial conditions $y_1(0) = -1$, $y_2(0) = 1$, $y_3(0) = 0$.

4. Solve the system
$$y'_1 = 4y_1 + 2y_2 + 2y_3$$
$$y'_2 = 2y_1 + 4y_2 + 2y_3$$
$$y'_3 = 2y_1 + 2y_2 + 4y_3$$

5. Show that every solution of $y' = ay$ has the form $y = ce^{ax}$. [*Hint:* Let $y = f(x)$ be a solution of the equation, and show that $f(x)e^{-ax}$ is constant.]

6. Show that if A is diagonalizable and
$$\mathbf{y} = \begin{bmatrix} y_1 \\ y_2 \\ \vdots \\ y_n \end{bmatrix}$$
is a solution of the system $\mathbf{y}' = A\mathbf{y}$, then each y_i is a linear combination of $e^{\lambda_1 x}, e^{\lambda_2 x}, \ldots, e^{\lambda_n x}$, where $\lambda_1, \lambda_2, \ldots, \lambda_n$ are the eigenvalues of A.

7. Sometimes it is possible to solve a single higher-order linear differential equation with constant coefficients by expressing it as a system and applying the methods of this section. For the differential equation $y'' - y' - 6y = 0$, show that the substitutions $y_1 = y$ and $y_2 = y'$ lead to the system
$$y'_1 = y_2$$
$$y'_2 = 6y_1 + y_2$$

Solve this system, and use the result to solve the original differential equation.

8. Use the procedure in Exercise 7 to solve $y'' + y' - 12y = 0$.

9. Explain how you might use the procedure in Exercise 7 to solve $y''' - 6y'' + 11y' - 6y = 0$. Use your procedure to solve the equation.

10. (a) By rewriting (11) in matrix form, show that the solution of the system in Example 2 can be expressed as
$$\mathbf{y} = c_1 e^{2x} \begin{bmatrix} 1 \\ 1 \end{bmatrix} + c_2 e^{-3x} \begin{bmatrix} -\frac{1}{4} \\ 1 \end{bmatrix}$$

This is called the **general solution** of the system.

  (b) Note that in part (a), the vector in the first term is an eigenvector corresponding to the eigenvalue $\lambda_1 = 2$, and the vector in the second term is an eigenvector corresponding to the eigenvalue $\lambda_2 = -3$. This is a special case of the following general result:

> **Theorem.** *If the coefficient matrix A of the system $\mathbf{y}' = A\mathbf{y}$ is diagonalizable, then the general solution of the system can be expressed as*
> $$\mathbf{y} = c_1 e^{\lambda_1 x} \mathbf{x}_1 + c_2 e^{\lambda_2 x} \mathbf{x}_2 + \cdots + c_n e^{\lambda_n x} \mathbf{x}_n$$
> *where $\lambda_1, \lambda_2, \ldots, \lambda_n$ are the eigenvalues of A, and $\mathbf{x}_i$ is an eigenvector of A corresponding to λ_i.*

Prove this result by tracing through the four-step procedure preceding Example 2 with

$$D = \begin{bmatrix} \lambda_1 & 0 & \cdots & 0 \\ 0 & \lambda_2 & \cdots & 0 \\ \vdots & \vdots & & \vdots \\ 0 & 0 & \cdots & \lambda_n \end{bmatrix} \quad \text{and} \quad P = [\mathbf{x}_1 \mid \mathbf{x}_2 \mid \cdots \mid \mathbf{x}_n]$$

11. Consider the system of differential equations $\mathbf{y}' = A\mathbf{y}$, where A is a 2×2 matrix. For what values of $a_{11}, a_{12}, a_{21}, a_{22}$ do the component solutions $y_1(t)$, $y_2(t)$ tend to zero as $t \to \infty$? In particular, what must be true about the determinant and the trace of A for this to happen?

12. Solve the nondiagonalizable system

$$y_1' = y_1 + y_2$$
$$y_2' = y_2$$

True-False Exercises

In parts (a)–(e) determine whether the statement is true or false, and justify your answer.

(a) Every system of differential equations $\mathbf{y}' = A\mathbf{y}$ has a solution.

(b) If $\mathbf{x}' = A\mathbf{x}$ and $\mathbf{y}' = A\mathbf{y}$, then $\mathbf{x} = \mathbf{y}$.

(c) If $\mathbf{x}' = A\mathbf{x}$ and $\mathbf{y}' = A\mathbf{y}$, then $(c\mathbf{x} + d\mathbf{y})' = A(c\mathbf{x} + d\mathbf{y})$ for all scalars c and d.

(d) If A is a square matrix with distinct real eigenvalues, then it is possible to solve $\mathbf{x}' = A\mathbf{x}$ by diagonalization.

(e) If A and P are similar matrices, then $\mathbf{y}' = A\mathbf{y}$ and $\mathbf{u}' = P\mathbf{u}$ have the same solutions.

Chapter 5   Supplementary Exercises

1. (a) Show that if $0 < \theta < \pi$, then

$$A = \begin{bmatrix} \cos\theta & -\sin\theta \\ \sin\theta & \cos\theta \end{bmatrix}$$

   has no eigenvalues and consequently no eigenvectors.

   (b) Give a geometric explanation of the result in part (a).

2. Find the eigenvalues of

$$A = \begin{bmatrix} 0 & 1 & 0 \\ 0 & 0 & 1 \\ k^3 & -3k^2 & 3k \end{bmatrix}$$

3. (a) Show that if D is a diagonal matrix with nonnegative entries on the main diagonal, then there is a matrix S such that $S^2 = D$.

   (b) Show that if A is a diagonalizable matrix with nonnegative eigenvalues, then there is a matrix S such that $S^2 = A$.

   (c) Find a matrix S such that $S^2 = A$, given that

$$A = \begin{bmatrix} 1 & 3 & 1 \\ 0 & 4 & 5 \\ 0 & 0 & 9 \end{bmatrix}$$

4. Prove: If A is a square matrix, then A and A^T have the same characteristic polynomial.

5. Prove: If A is a square matrix and $p(\lambda) = \det(\lambda I - A)$ is the characteristic polynomial of A, then the coefficient of λ^{n-1} in $p(\lambda)$ is the negative of the trace of A.

6. Prove: If $b \neq 0$, then

$$A = \begin{bmatrix} a & b \\ 0 & a \end{bmatrix}$$

   is not diagonalizable.

7. In advanced linear algebra, one proves the **Cayley–Hamilton Theorem**, which states that a square matrix A satisfies its characteristic equation; that is, if

$$c_0 + c_1\lambda + c_2\lambda^2 + \cdots + c_{n-1}\lambda^{n-1} + \lambda^n = 0$$

   is the characteristic equation of A, then

$$c_0 I + c_1 A + c_2 A^2 + \cdots + c_{n-1}A^{n-1} + A^n = 0$$

   Verify this result for

   (a) $A = \begin{bmatrix} 3 & 6 \\ 1 & 2 \end{bmatrix}$ (b) $A = \begin{bmatrix} 0 & 1 & 0 \\ 0 & 0 & 1 \\ 1 & -3 & 3 \end{bmatrix}$

   ▶ In Exercises 8–10, use the Cayley–Hamilton Theorem, stated in Exercise 7. ◀

8. (a) Use Exercise 18 of Section 5.1 to prove the Cayley–Hamilton Theorem for 2×2 matrices.

   (b) Prove the Cayley–Hamilton Theorem for $n \times n$ diagonalizable matrices.

9. The Cayley–Hamilton Theorem provides a method for calculating powers of a matrix. For example, if A is a 2×2 matrix with characteristic equation

$$c_0 + c_1\lambda + \lambda^2 = 0$$

   then $c_0 I + c_1 A + A^2 = 0$, so

$$A^2 = -c_1 A - c_0 I$$

   Multiplying through by A yields $A^3 = -c_1 A^2 - c_0 A$, which expresses A^3 in terms of A^2 and A, and multiplying through by A^2 yields $A^4 = -c_1 A^3 - c_0 A^2$, which expresses A^4 in terms of A^3 and A^2. Continuing in this way, we can calculate successive powers of A by expressing them in terms of lower powers. Use this procedure to calculate A^2, A^3, A^4, and A^5 for

$$A = \begin{bmatrix} 3 & 6 \\ 1 & 2 \end{bmatrix}$$

10. Use the method of the preceding exercise to calculate A^3 and A^4 for

$$A = \begin{bmatrix} 0 & 1 & 0 \\ 0 & 0 & 1 \\ 1 & -3 & 3 \end{bmatrix}$$

11. Find the eigenvalues of the matrix

$$A = \begin{bmatrix} c_1 & c_2 & \cdots & c_n \\ c_1 & c_2 & \cdots & c_n \\ \vdots & \vdots & & \vdots \\ c_1 & c_2 & \cdots & c_n \end{bmatrix}$$

12. (a) It was shown in Exercise 17 of Section 5.1 that if A is an $n \times n$ matrix, then the coefficient of λ^n in the characteristic polynomial of A is 1. (A polynomial with this property is called **monic**.) Show that the matrix

$$\begin{bmatrix} 0 & 0 & 0 & \cdots & 0 & -c_0 \\ 1 & 0 & 0 & \cdots & 0 & -c_1 \\ 0 & 1 & 0 & \cdots & 0 & -c_2 \\ \vdots & \vdots & \vdots & & \vdots & \vdots \\ 0 & 0 & 0 & \cdots & 1 & -c_{n-1} \end{bmatrix}$$

has characteristic polynomial

$$p(\lambda) = c_0 + c_1\lambda + \cdots + c_{n-1}\lambda^{n-1} + \lambda^n$$

This shows that every monic polynomial is the characteristic polynomial of some matrix. The matrix in this example is called the **companion matrix** of $p(\lambda)$. [*Hint:* Evaluate all determinants in the problem by adding a multiple of the second row to the first to introduce a zero at the top of the first column, and then expanding by cofactors along the first column.]

(b) Find a matrix with characteristic polynomial

$$p(\lambda) = 1 - 2\lambda + \lambda^2 + 3\lambda^3 + \lambda^4$$

13. A square matrix A is called **nilpotent** if $A^n = 0$ for some positive integer n. What can you say about the eigenvalues of a nilpotent matrix?

14. Prove: If A is an $n \times n$ matrix and n is odd, then A has at least one real eigenvalue.

15. Find a 3×3 matrix A that has eigenvalues $\lambda = 0$, 1, and -1 with corresponding eigenvectors

$$\begin{bmatrix} 0 \\ 1 \\ -1 \end{bmatrix}, \quad \begin{bmatrix} 1 \\ -1 \\ 1 \end{bmatrix}, \quad \begin{bmatrix} 0 \\ 1 \\ 1 \end{bmatrix}$$

respectively.

16. Suppose that a 4×4 matrix A has eigenvalues $\lambda_1 = 1$, $\lambda_2 = -2$, $\lambda_3 = 3$, and $\lambda_4 = -3$.

(a) Use the method of Exercise 16 of Section 5.1 to find $\det(A)$.

(b) Use Exercise 5 above to find $\text{tr}(A)$.

17. Let A be a square matrix such that $A^3 = A$. What can you say about the eigenvalues of A?

18. (a) Solve the system

$$\begin{aligned} y_1' &= y_1 + 3y_2 \\ y_2' &= 2y_1 + 4y_2 \end{aligned}$$

(b) Find the solution satisfying the initial conditions $y_1(0) = 5$ and $y_2(0) = 6$.

CHAPTER 6

Inner Product Spaces

CHAPTER CONTENTS
6.1 Inner Products **335**
6.2 Angle and Orthogonality in Inner Product Spaces **345**
6.3 Gram–Schmidt Process; *QR*-Decomposition **352**
6.4 Best Approximation; Least Squares **366**
6.5 Least Squares Fitting to Data **376**
6.6 Function Approximation; Fourier Series **382**

INTRODUCTION   In Chapter 3 we defined the dot product of vectors in R^n, and we used that concept to define notions of length, angle, distance, and orthogonality. In this chapter we will generalize those ideas so they are applicable in any vector space, not just R^n. We will also discuss various applications of these ideas.

6.1 Inner Products

In this section we will use the most important properties of the dot product on R^n as axioms, which, if satisfied by the vectors in a vector space V, will enable us to extend the notions of length, distance, angle, and perpendicularity to general vector spaces.

General Inner Products   In Definition 4 of Section 3.2 we defined the dot product of two vectors in R^n, and in Theorem 3.2.2 we listed four fundamental properties of such products. Our first goal in this section is to extend the notion of a dot product to general real vector spaces by using those four properties as axioms. We make the following definition.

Note that Definition 1 applies only to *real* vector spaces. A definition of inner products on *complex* vector spaces is given in the exercises. Since we will have little need for complex vector spaces from this point on, you can assume that all vector spaces under discussion are real, even though some of the theorems are also valid in complex vector spaces.

DEFINITION 1 An *inner product* on a real vector space V is a function that associates a real number $\langle \mathbf{u}, \mathbf{v} \rangle$ with each pair of vectors in V in such a way that the following axioms are satisfied for all vectors $\mathbf{u}$, $\mathbf{v}$, and $\mathbf{w}$ in V and all scalars k.

1. $\langle \mathbf{u}, \mathbf{v} \rangle = \langle \mathbf{v}, \mathbf{u} \rangle$ [Symmetry axiom]

2. $\langle \mathbf{u} + \mathbf{v}, \mathbf{w} \rangle = \langle \mathbf{u}, \mathbf{w} \rangle + \langle \mathbf{v}, \mathbf{w} \rangle$ [Additivity axiom]

3. $\langle k\mathbf{u}, \mathbf{v} \rangle = k\langle \mathbf{u}, \mathbf{v} \rangle$ [Homogeneity axiom]

4. $\langle \mathbf{v}, \mathbf{v} \rangle \geq 0$ and $\langle \mathbf{v}, \mathbf{v} \rangle = 0$ if and only if $\mathbf{v} = \mathbf{0}$ [Positivity axiom]

A real vector space with an inner product is called a *real inner product space*.

Because the axioms for a real inner product space are based on properties of the dot product, these inner product space axioms will be satisfied automatically if we define the inner product of two vectors $\mathbf{u}$ and $\mathbf{v}$ in R^n to be

$$\langle \mathbf{u}, \mathbf{v} \rangle = \mathbf{u} \cdot \mathbf{v} = u_1 v_1 + u_2 v_2 + \cdots + u_n v_n$$

335

This inner product is commonly called the ***Euclidean inner product*** (or the ***standard inner product***) on R^n to distinguish it from other possible inner products that might be defined on R^n. We call R^n with the Euclidean inner product ***Euclidean n-space***.

Inner products can be used to define notions of norm and distance in a general inner product space just as we did with dot products in R^n. Recall from Formulas (11) and (19) of Section 3.2 that if $\mathbf{u}$ and $\mathbf{v}$ are vectors in Euclidean n-space, then norm and distance can be expressed in terms of the dot product as

$$\|\mathbf{v}\| = \sqrt{\mathbf{v} \cdot \mathbf{v}} \quad \text{and} \quad d(\mathbf{u}, \mathbf{v}) = \|\mathbf{u} - \mathbf{v}\| = \sqrt{(\mathbf{u} - \mathbf{v}) \cdot (\mathbf{u} - \mathbf{v})}$$

Motivated by these formulas we make the following definition.

DEFINITION 2 If V is a real inner product space, then the ***norm*** (or ***length***) of a vector $\mathbf{v}$ in V is denoted by $\|\mathbf{v}\|$ and is defined by

$$\|\mathbf{v}\| = \sqrt{\langle \mathbf{v}, \mathbf{v} \rangle}$$

and the ***distance*** between two vectors is denoted by $d(\mathbf{u}, \mathbf{v})$ and is defined by

$$d(\mathbf{u}, \mathbf{v}) = \|\mathbf{u} - \mathbf{v}\| = \sqrt{\langle \mathbf{u} - \mathbf{v}, \mathbf{u} - \mathbf{v} \rangle}$$

A vector of norm 1 is called a ***unit vector***.

The following theorem, which we state without proof, shows that norms and distances in real inner product spaces have many of the properties that you might expect.

THEOREM 6.1.1 *If $\mathbf{u}$ and $\mathbf{v}$ are vectors in a real inner product space V, and if k is a scalar, then:*
(a) $\|\mathbf{v}\| \geq 0$ *with equality if and only if* $\mathbf{v} = \mathbf{0}$.
(b) $\|k\mathbf{v}\| = |k|\|\mathbf{v}\|$.
(c) $d(\mathbf{u}, \mathbf{v}) = d(\mathbf{v}, \mathbf{u})$.
(d) $d(\mathbf{u}, \mathbf{v}) \geq 0$ *with equality if and only if* $\mathbf{u} = \mathbf{v}$.

Although the Euclidean inner product is the most important inner product on R^n, there are various applications in which it is desirable to modify it by *weighting* each term differently. More precisely, if

$$w_1, w_2, \ldots, w_n$$

are *positive* real numbers, which we will call ***weights***, and if $\mathbf{u} = (u_1, u_2, \ldots, u_n)$ and $\mathbf{v} = (v_1, v_2, \ldots, v_n)$ are vectors in R^n, then it can be shown that the formula

$$\langle \mathbf{u}, \mathbf{v} \rangle = w_1 u_1 v_1 + w_2 u_2 v_2 + \cdots + w_n u_n v_n \tag{1}$$

defines an inner product on R^n that we call the ***weighted Euclidean inner product with weights*** $w_1, w_2, \ldots, w_n$.

Note that the standard Euclidean inner product is the special case of the weighted Euclidean inner product in which all the weights are 1.

▶ **EXAMPLE 1 Weighted Euclidean Inner Product**

Let $\mathbf{u} = (u_1, u_2)$ and $\mathbf{v} = (v_1, v_2)$ be vectors in R^2. Verify that the weighted Euclidean inner product

$$\langle \mathbf{u}, \mathbf{v} \rangle = 3u_1 v_1 + 2u_2 v_2 \tag{2}$$

satisfies the four inner product axioms.

Solution

Axiom 1: Interchanging $\mathbf{u}$ and $\mathbf{v}$ in Formula (2) does not change the sum on the right side, so $\langle \mathbf{u}, \mathbf{v} \rangle = \langle \mathbf{v}, \mathbf{u} \rangle$.

Axiom 2: If $\mathbf{w} = (w_1, w_2)$, then

$$\langle \mathbf{u} + \mathbf{v}, \mathbf{w} \rangle = 3(u_1 + v_1)w_1 + 2(u_2 + v_2)w_2$$
$$= 3(u_1 w_1 + v_1 w_1) + 2(u_2 w_2 + v_2 w_2)$$
$$= (3u_1 w_1 + 2u_2 w_2) + (3v_1 w_1 + 2v_2 w_2)$$
$$= \langle \mathbf{u}, \mathbf{w} \rangle + \langle \mathbf{v}, \mathbf{w} \rangle$$

Axiom 3: $\langle k\mathbf{u}, \mathbf{v} \rangle = 3(ku_1)v_1 + 2(ku_2)v_2$
$$= k(3u_1 v_1 + 2u_2 v_2)$$
$$= k\langle \mathbf{u}, \mathbf{v} \rangle$$

Axiom 4: $\langle \mathbf{v}, \mathbf{v} \rangle = 3(v_1 v_1) + 2(v_2 v_2) = 3v_1^2 + 2v_2^2 \geq 0$ with equality if and only if $v_1 = v_2 = 0$; that is, if and only if $\mathbf{v} = \mathbf{0}$. ◀

> In Example 1, we are using subscripted w's to denote the components of the vector $\mathbf{w}$, not the weights. The weights are the numbers 3 and 2 in Formula (2).

An Application of Weighted Euclidean Inner Products

To illustrate one way in which a weighted Euclidean inner product can arise, suppose that some physical experiment has n possible numerical outcomes

$$x_1, x_2, \ldots, x_n$$

and that a series of m repetitions of the experiment yields these values with various frequencies. Specifically, suppose that x_1 occurs f_1 times, x_2 occurs f_2 times, and so forth. Since there are a total of m repetitions of the experiment, it follows that

$$f_1 + f_2 + \cdots + f_n = m$$

Thus, the **arithmetic average** of the observed numerical values (denoted by $\bar{x}$) is

$$\bar{x} = \frac{f_1 x_1 + f_2 x_2 + \cdots + f_n x_n}{f_1 + f_2 + \cdots + f_n} = \frac{1}{m}(f_1 x_1 + f_2 x_2 + \cdots + f_n x_n) \tag{3}$$

If we let

$$\mathbf{f} = (f_1, f_2, \ldots, f_n)$$
$$\mathbf{x} = (x_1, x_2, \ldots, x_n)$$
$$w_1 = w_2 = \cdots = w_n = 1/m$$

then (3) can be expressed as the weighted Euclidean inner product

$$\bar{x} = \langle \mathbf{f}, \mathbf{x} \rangle = w_1 f_1 x_1 + w_2 f_2 x_2 + \cdots + w_n f_n x_n$$

▶ EXAMPLE 2 Using a Weighted Euclidean Inner Product

It is important to keep in mind that norm and distance depend on the inner product being used. If the inner product is changed, then the norms and distances between vectors also change. For example, for the vectors $\mathbf{u} = (1, 0)$ and $\mathbf{v} = (0, 1)$ in R^2 with the Euclidean inner product we have

$$\|\mathbf{u}\| = \sqrt{1^2 + 0^2} = 1$$

and

$$d(\mathbf{u}, \mathbf{v}) = \|\mathbf{u} - \mathbf{v}\| = \|(1, -1)\| = \sqrt{1^2 + (-1)^2} = \sqrt{2}$$

but if we change to the weighted Euclidean inner product

$$\langle \mathbf{u}, \mathbf{v} \rangle = 3u_1 v_1 + 2u_2 v_2$$

we have

$$\|\mathbf{u}\| = \langle \mathbf{u}, \mathbf{u} \rangle^{1/2} = [3(1)(1) + 2(0)(0)]^{1/2} = \sqrt{3}$$

and

$$d(\mathbf{u}, \mathbf{v}) = \|\mathbf{u} - \mathbf{v}\| = \langle (1, -1), (1, -1) \rangle^{1/2}$$
$$= [3(1)(1) + 2(-1)(-1)]^{1/2} = \sqrt{5} \blacktriangleleft$$

Unit Circles and Spheres in Inner Product Spaces

If V is an inner product space, then the set of points in V that satisfy

$$\|\mathbf{u}\| = 1$$

is called the **unit sphere** or sometimes the **unit circle** in V.

(a) The unit circle using the standard Euclidean inner product.

▶ **EXAMPLE 3 Unusual Unit Circles in R^2**

(a) Sketch the unit circle in an xy-coordinate system in R^2 using the Euclidean inner product $\langle \mathbf{u}, \mathbf{v} \rangle = u_1 v_1 + u_2 v_2$.

(b) Sketch the unit circle in an xy-coordinate system in R^2 using the weighted Euclidean inner product $\langle \mathbf{u}, \mathbf{v} \rangle = \frac{1}{9} u_1 v_1 + \frac{1}{4} u_2 v_2$.

Solution (a) If $\mathbf{u} = (x, y)$, then $\|\mathbf{u}\| = \langle \mathbf{u}, \mathbf{u} \rangle^{1/2} = \sqrt{x^2 + y^2}$, so the equation of the unit circle is $\sqrt{x^2 + y^2} = 1$, or, on squaring both sides,

$$x^2 + y^2 = 1$$

As expected, the graph of this equation is a circle of radius 1 centered at the origin (Figure 6.1.1a).

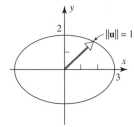

(b) The unit circle using a weighted Euclidean inner product.

▲ **Figure 6.1.1**

Solution (b) If $\mathbf{u} = (x, y)$, then $\|\mathbf{u}\| = \langle \mathbf{u}, \mathbf{u} \rangle^{1/2} = \sqrt{\frac{1}{9}x^2 + \frac{1}{4}y^2}$, so the equation of the unit circle is $\sqrt{\frac{1}{9}x^2 + \frac{1}{4}y^2} = 1$, or, on squaring both sides,

$$\frac{x^2}{9} + \frac{y^2}{4} = 1$$

The graph of this equation is the ellipse shown in Figure 6.1.1b. ◀

Remark It may seem odd that the "unit circle" in the second part of the last example turned out to have an elliptical shape. This will make more sense if you think of circles and spheres in general vector spaces *algebraically* ($\|\mathbf{u}\| = 1$) rather than geometrically. The change in geometry occurs because the norm, not being Euclidean, has the effect of distorting the space that we are used to seeing through "Euclidean eyes."

Inner Products Generated by Matrices

The Euclidean inner product and the weighted Euclidean inner products are special cases of a general class of inner products on R^n called **matrix inner products**. To define this class of inner products, let $\mathbf{u}$ and $\mathbf{v}$ be vectors in R^n that are expressed in *column form*, and let A be an *invertible* $n \times n$ matrix. It can be shown (Exercise 31) that if $\mathbf{u} \cdot \mathbf{v}$ is the Euclidean inner product on R^n, then the formula

$$\langle \mathbf{u}, \mathbf{v} \rangle = A\mathbf{u} \cdot A\mathbf{v} \tag{4}$$

also defines an inner product; it is called the **inner product on R^n generated by A**.

Recall from Table 1 of Section 3.2 that if $\mathbf{u}$ and $\mathbf{v}$ are in column form, then $\mathbf{u} \cdot \mathbf{v}$ can be written as $\mathbf{v}^T \mathbf{u}$ from which it follows that (4) can be expressed as

$$\langle \mathbf{u}, \mathbf{v} \rangle = (A\mathbf{v})^T A\mathbf{u}$$

or, equivalently as

$$\langle \mathbf{u}, \mathbf{v} \rangle = \mathbf{v}^T A^T A \mathbf{u} \tag{5}$$

▶ **EXAMPLE 4 Matrices Generating Weighted Euclidean Inner Products**

The standard Euclidean and weighted Euclidean inner products are examples of matrix inner products. The standard Euclidean inner product on R^n is generated by the $n \times n$ identity matrix, since setting $A = I$ in Formula (4) yields

$$\langle \mathbf{u}, \mathbf{v} \rangle = I\mathbf{u} \cdot I\mathbf{v} = \mathbf{u} \cdot \mathbf{v}$$

and the weighted Euclidean inner product

$$\langle \mathbf{u}, \mathbf{v} \rangle = w_1 u_1 v_1 + w_2 u_2 v_2 + \cdots + w_n u_n v_n \tag{6}$$

is generated by the matrix

$$A = \begin{bmatrix} \sqrt{w_1} & 0 & 0 & \cdots & 0 \\ 0 & \sqrt{w_2} & 0 & \cdots & 0 \\ \vdots & \vdots & \vdots & & \vdots \\ 0 & 0 & 0 & \cdots & \sqrt{w_n} \end{bmatrix} \tag{7}$$

This can be seen by first observing that $A^T A$ is the $n \times n$ diagonal matrix whose diagonal entries are the weights $w_1, w_2, \ldots, w_n$ and then observing that (5) simplifies to (6) when A is the matrix in Formula (7).

▶ **EXAMPLE 5 Example 1 Revisited**

Every diagonal matrix with positive diagonal entries generates a weighted inner product. Why?

The weighted Euclidean inner product $\langle \mathbf{u}, \mathbf{v} \rangle = 3u_1 v_1 + 2u_2 v_2$ discussed in Example 1 is the inner product on R^2 generated by

$$A = \begin{bmatrix} \sqrt{3} & 0 \\ 0 & \sqrt{2} \end{bmatrix} \quad ◀$$

Other Examples of Inner Products

So far, we have only considered examples of inner products on R^n. We will now consider examples of inner products on some of the other kinds of vector spaces that we discussed earlier.

▶ **EXAMPLE 6 An Inner Product on M_{nn}**

If U and V are $n \times n$ matrices, then the formula

$$\langle U, V \rangle = \text{tr}(U^T V) \tag{8}$$

defines an inner product on the vector space M_{nn} (see Definition 8 of Section 1.3 for a definition of trace). This can be proved by confirming that the four inner product space axioms are satisfied, but you can visualize why this is so by computing (8) for the 2×2 matrices

$$U = \begin{bmatrix} u_1 & u_2 \\ u_3 & u_4 \end{bmatrix} \quad \text{and} \quad V = \begin{bmatrix} v_1 & v_2 \\ v_3 & v_4 \end{bmatrix}$$

This yields

$$\langle U, V \rangle = \text{tr}(U^T V) = u_1 v_1 + u_2 v_2 + u_3 v_3 + u_v v_4$$

which is just the dot product of the corresponding entries in the two matrices. For example, if

$$U = \begin{bmatrix} 1 & 2 \\ 3 & 4 \end{bmatrix} \quad \text{and} \quad V = \begin{bmatrix} -1 & 0 \\ 3 & 2 \end{bmatrix}$$

then

$$\langle U, V \rangle = 1(-1) + 2(0) + 3(3) + 4(2) = 16$$

The norm of a matrix U relative to this inner product is

$$\|U\| = \langle U, U \rangle^{1/2} = \sqrt{u_1^2 + u_2^2 + u_3^2 + u_4^2}$$

▶ **EXAMPLE 7 The Standard Inner Product on P_n**

If

$$\mathbf{p} = a_0 + a_1 x + \cdots + a_n x^n \quad \text{and} \quad \mathbf{q} = b_0 + b_1 x + \cdots + b_n x^n$$

are polynomials in P_n, then the following formula defines an inner product on P_n (verify) that we will call the *standard inner product* on this space:

$$\langle \mathbf{p}, \mathbf{q} \rangle = a_0 b_0 + a_1 b_1 + \cdots + a_n b_n \tag{9}$$

The norm of a polynomial $\mathbf{p}$ relative to this inner product is

$$\|\mathbf{p}\| = \sqrt{\langle \mathbf{p}, \mathbf{p} \rangle} = \sqrt{a_0^2 + a_1^2 + \cdots + a_n^2}$$

▶ **EXAMPLE 8 The Evaluation Inner Product on P_n**

If

$$\mathbf{p} = p(x) = a_0 + a_1 x + \cdots + a_n x^n \quad \text{and} \quad \mathbf{q} = q(x) = b_0 + b_1 x + \cdots + b_n x^n$$

are polynomials in P_n, and if $x_0, x_1, \ldots, x_n$ are distinct real numbers (called *sample points*), then the formula

$$\langle \mathbf{p}, \mathbf{q} \rangle = p(x_0)q(x_0) + p(x_1)q(x_1) + \cdots + p(x_n)q(x_n) \tag{10}$$

defines an inner product on P_n called the *evaluation inner product* at $x_0, x_1, \ldots, x_n$. Algebraically, this can be viewed as the dot product in R^n of the n-tuples

$$\big(p(x_0), p(x_1), \ldots, p(x_n)\big) \quad \text{and} \quad \big(q(x_0), q(x_1), \ldots, q(x_n)\big)$$

and hence the first three inner product axioms follow from properties of the dot product. The fourth inner product axiom follows from the fact that

$$\langle \mathbf{p}, \mathbf{p} \rangle = [p(x_0)]^2 + [p(x_1)]^2 + \cdots + [p(x_n)]^2 \geq 0$$

with equality holding if and only if

$$p(x_0) = p(x_1) = \cdots = p(x_n) = 0$$

But a nonzero polynomial of degree n or less can have at most n distinct roots, so it must be that $\mathbf{p} = \mathbf{0}$, which proves that the fourth inner product axiom holds.

The norm of a polynomial $\mathbf{p}$ relative to the evaluation inner product is

$$\|\mathbf{p}\| = \sqrt{\langle \mathbf{p}, \mathbf{p} \rangle} = \sqrt{[p(x_0)]^2 + [p(x_1)]^2 + \cdots + [p(x_n)]^2} \tag{11}$$

▶ **EXAMPLE 9 Working with the Evaluation Inner Product**

Let P_2 have the evaluation inner product at the points

$$x_0 = -2, \quad x_1 = 0, \quad \text{and} \quad x_2 = 2$$

Compute $\langle \mathbf{p}, \mathbf{q} \rangle$ and $\|\mathbf{p}\|$ for the polynomials $\mathbf{p} = p(x) = x^2$ and $\mathbf{q} = q(x) = 1 + x$.

Solution It follows from (10) and (11) that

$$\langle \mathbf{p}, \mathbf{q} \rangle = p(-2)q(-2) + p(0)q(0) + p(2)q(2) = (4)(-1) + (0)(1) + (4)(3) = 8$$

$$\|\mathbf{p}\| = \sqrt{[p(x_0)]^2 + [p(x_1)]^2 + [p(x_2)]^2} = \sqrt{[p(-2)]^2 + [p(0)]^2 + [p(2)]^2}$$

$$= \sqrt{4^2 + 0^2 + 4^2} = \sqrt{32} = 4\sqrt{2}$$

CALCULUS REQUIRED

▶ **EXAMPLE 10 An Inner Product on C[a, b]**

Let $\mathbf{f} = f(x)$ and $\mathbf{g} = g(x)$ be two functions in $C[a, b]$ and define

$$\langle \mathbf{f}, \mathbf{g} \rangle = \int_a^b f(x)g(x)\,dx \tag{12}$$

We will show that this formula defines an inner product on $C[a, b]$ by verifying the four inner product axioms for functions $\mathbf{f} = f(x)$, $\mathbf{g} = g(x)$, and $\mathbf{h} = h(x)$ in $C[a, b]$:

1. $\langle \mathbf{f}, \mathbf{g} \rangle = \displaystyle\int_a^b f(x)g(x)\,dx = \int_a^b g(x)f(x)\,dx = \langle \mathbf{g}, \mathbf{f} \rangle$

   which proves that Axiom 1 holds.

2. $\langle \mathbf{f} + \mathbf{g}, \mathbf{h} \rangle = \displaystyle\int_a^b (f(x) + g(x))h(x)\,dx$

   $$= \int_a^b f(x)h(x)\,dx + \int_a^b g(x)h(x)\,dx$$

   $$= \langle \mathbf{f}, \mathbf{h} \rangle + \langle \mathbf{g}, \mathbf{h} \rangle$$

   which proves that Axiom 2 holds.

3. $\langle k\mathbf{f}, \mathbf{g} \rangle = \displaystyle\int_a^b kf(x)g(x)\,dx = k\int_a^b f(x)g(x)\,dx = k\langle \mathbf{f}, \mathbf{g} \rangle$

   which proves that Axiom 3 holds.

4. If $\mathbf{f} = f(x)$ is any function in $C[a, b]$, then

   $$\langle \mathbf{f}, \mathbf{f} \rangle = \int_a^b f^2(x)\,dx \geq 0 \tag{13}$$

   since $f^2(x) \geq 0$ for all x in the interval $[a, b]$. Moreover because f is continuous on $[a, b]$, the equality holds in Formula (13) if and only if the function f is identically zero on $[a, b]$, that is, if and only if $\mathbf{f} = \mathbf{0}$; and this proves that Axiom 4 holds.

CALCULUS REQUIRED

▶ **EXAMPLE 11 Norm of a Vector in C[a, b]**

If $C[a, b]$ has the inner product that was defined in Example 10, then the norm of a function $\mathbf{f} = f(x)$ relative to this inner product is

$$\|\mathbf{f}\| = \langle \mathbf{f}, \mathbf{f} \rangle^{1/2} = \sqrt{\int_a^b f^2(x)\,dx} \tag{14}$$

and the unit sphere in this space consists of all functions $\mathbf{f}$ in $C[a, b]$ that satisfy the equation

$$\int_a^b f^2(x)\,dx = 1 \quad ◀$$

Remark Note that the vector space P_n is a subspace of $C[a, b]$ because polynomials are continuous functions. Thus, Formula (12) defines an inner product on P_n.

Remark Recall from calculus that the arc length of a curve $y = f(x)$ over an interval $[a, b]$ is given by the formula

$$L = \int_a^b \sqrt{1 + [f'(x)]^2}\, dx \tag{15}$$

Do not confuse this concept of arc length with $\|\mathbf{f}\|$, which is the length (norm) of $\mathbf{f}$ when $\mathbf{f}$ is viewed as a vector in $C[a, b]$. Formulas (14) and (15) are quite different.

Algebraic Properties of Inner Products   The following theorem lists some of the algebraic properties of inner products that follow from the inner product axioms. This result is a generalization of Theorem 3.2.3, which applied only to the dot product on R^n.

THEOREM 6.1.2 *If* $\mathbf{u}$, $\mathbf{v}$, *and* $\mathbf{w}$ *are vectors in a real inner product space V, and if k is a scalar, then*:

(a) $\langle \mathbf{0}, \mathbf{v} \rangle = \langle \mathbf{v}, \mathbf{0} \rangle = 0$

(b) $\langle \mathbf{u}, \mathbf{v} + \mathbf{w} \rangle = \langle \mathbf{u}, \mathbf{v} \rangle + \langle \mathbf{u}, \mathbf{w} \rangle$

(c) $\langle \mathbf{u}, \mathbf{v} - \mathbf{w} \rangle = \langle \mathbf{u}, \mathbf{v} \rangle - \langle \mathbf{u}, \mathbf{w} \rangle$

(d) $\langle \mathbf{u} - \mathbf{v}, \mathbf{w} \rangle = \langle \mathbf{u}, \mathbf{w} \rangle - \langle \mathbf{v}, \mathbf{w} \rangle$

(e) $k\langle \mathbf{u}, \mathbf{v} \rangle = \langle \mathbf{u}, k\mathbf{v} \rangle$

Proof   We will prove part (b) and leave the proofs of the remaining parts as exercises.

$$
\begin{aligned}
\langle \mathbf{u}, \mathbf{v} + \mathbf{w} \rangle &= \langle \mathbf{v} + \mathbf{w}, \mathbf{u} \rangle && \text{[By symmetry]} \\
&= \langle \mathbf{v}, \mathbf{u} \rangle + \langle \mathbf{w}, \mathbf{u} \rangle && \text{[By additivity]} \\
&= \langle \mathbf{u}, \mathbf{v} \rangle + \langle \mathbf{u}, \mathbf{w} \rangle && \text{[By symmetry]} \quad \blacktriangleleft
\end{aligned}
$$

The following example illustrates how Theorem 6.1.2 and the defining properties of inner products can be used to perform algebraic computations with inner products. As you read through the example, you will find it instructive to justify the steps.

▶ **EXAMPLE 12 Calculating with Inner Products**

$$
\begin{aligned}
\langle \mathbf{u} - 2\mathbf{v}, 3\mathbf{u} + 4\mathbf{v} \rangle &= \langle \mathbf{u}, 3\mathbf{u} + 4\mathbf{v} \rangle - \langle 2\mathbf{v}, 3\mathbf{u} + 4\mathbf{v} \rangle \\
&= \langle \mathbf{u}, 3\mathbf{u} \rangle + \langle \mathbf{u}, 4\mathbf{v} \rangle - \langle 2\mathbf{v}, 3\mathbf{u} \rangle - \langle 2\mathbf{v}, 4\mathbf{v} \rangle \\
&= 3\langle \mathbf{u}, \mathbf{u} \rangle + 4\langle \mathbf{u}, \mathbf{v} \rangle - 6\langle \mathbf{v}, \mathbf{u} \rangle - 8\langle \mathbf{v}, \mathbf{v} \rangle \\
&= 3\|\mathbf{u}\|^2 + 4\langle \mathbf{u}, \mathbf{v} \rangle - 6\langle \mathbf{u}, \mathbf{v} \rangle - 8\|\mathbf{v}\|^2 \\
&= 3\|\mathbf{u}\|^2 - 2\langle \mathbf{u}, \mathbf{v} \rangle - 8\|\mathbf{v}\|^2 \quad \blacktriangleleft
\end{aligned}
$$

Concept Review

- Inner product axioms
- Euclidean inner product
- Euclidean n-space
- Weighted Euclidean inner product

- Unit circle (sphere)
- Matrix inner product
- Norm in an inner product space
- Distance between two vectors in an inner product space

- Examples of inner products
- Properties of inner products

Skills

- Compute the inner product of two vectors.
- Find the norm of a vector.
- Find the distance between two vectors.

- Show that a given formula defines an inner product.
- Show that a given formula does not define an inner product by demonstrating that at least one of the inner product space axioms fails.

Exercise Set 6.1

1. Let $\langle \mathbf{u}, \mathbf{v} \rangle$ be the Euclidean inner product on R^2, and let $\mathbf{u} = (1, 1)$, $\mathbf{v} = (3, 2)$, $\mathbf{w} = (0, -1)$, and $k = 3$. Compute the following.

(a) $\langle \mathbf{u}, \mathbf{v} \rangle$ (b) $\langle k\mathbf{v}, \mathbf{w} \rangle$ (c) $\langle \mathbf{u} + \mathbf{v}, \mathbf{w} \rangle$

(d) $\|\mathbf{v}\|$ (e) $d(\mathbf{u}, \mathbf{v})$ (f) $\|\mathbf{u} - k\mathbf{v}\|$

2. Repeat Exercise 1 for the weighted Euclidean inner product $\langle \mathbf{u}, \mathbf{v} \rangle = 2u_1v_1 + 3u_2v_2$.

3. Let $\langle \mathbf{u}, \mathbf{v} \rangle$ be the Euclidean inner product on R^2, and let $\mathbf{u} = (3, -2)$, $\mathbf{v} = (4, 5)$, $\mathbf{w} = (-1, 6)$, and $k = -4$. Verify the following.

(a) $\langle \mathbf{u}, \mathbf{v} \rangle = \langle \mathbf{v}, \mathbf{u} \rangle$

(b) $\langle \mathbf{u} + \mathbf{v}, \mathbf{w} \rangle = \langle \mathbf{u}, \mathbf{w} \rangle + \langle \mathbf{v}, \mathbf{w} \rangle$

(c) $\langle \mathbf{u}, \mathbf{v} + \mathbf{w} \rangle = \langle \mathbf{u}, \mathbf{v} \rangle + \langle \mathbf{u}, \mathbf{w} \rangle$

(d) $\langle k\mathbf{u}, \mathbf{v} \rangle = k\langle \mathbf{u}, \mathbf{v} \rangle = \langle \mathbf{u}, k\mathbf{v} \rangle$

(e) $\langle \mathbf{0}, \mathbf{v} \rangle = \langle \mathbf{v}, \mathbf{0} \rangle = 0$

4. Repeat Exercise 3 for the weighted Euclidean inner product $\langle \mathbf{u}, \mathbf{v} \rangle = 4u_1v_1 + 5u_2v_2$.

5. Let $\langle \mathbf{u}, \mathbf{v} \rangle$ be the inner product on R^2 generated by $\begin{bmatrix} 2 & 1 \\ 1 & 1 \end{bmatrix}$, and let $\mathbf{u} = (2, 1)$, $\mathbf{v} = (-1, 1)$, $\mathbf{w} = (0, -1)$. Compute the following.

(a) $\langle \mathbf{u}, \mathbf{v} \rangle$ (b) $\langle \mathbf{v}, \mathbf{w} \rangle$ (c) $\langle \mathbf{u} + \mathbf{v}, \mathbf{w} \rangle$

(d) $\|\mathbf{v}\|$ (e) $d(\mathbf{v}, \mathbf{w})$ (f) $\|\mathbf{v} - \mathbf{w}\|^2$

6. Repeat Exercise 5 for the inner product on R^2 generated by $\begin{bmatrix} 1 & 0 \\ 2 & -1 \end{bmatrix}$.

7. Compute $\langle \mathbf{u}, \mathbf{v} \rangle$ using the inner product in Example 6.

(a) $\mathbf{u} = \begin{bmatrix} 3 & -2 \\ 4 & 8 \end{bmatrix}$, $\mathbf{v} = \begin{bmatrix} -1 & 3 \\ 1 & 1 \end{bmatrix}$

(b) $\mathbf{u} = \begin{bmatrix} 1 & 2 \\ -3 & 5 \end{bmatrix}$, $\mathbf{v} = \begin{bmatrix} 4 & 6 \\ 0 & 8 \end{bmatrix}$

8. Compute $\langle \mathbf{p}, \mathbf{q} \rangle$ using the inner product in Example 7.

(a) $\mathbf{p} = -2 + x + 3x^2$, $\mathbf{q} = 4 - 7x^2$

(b) $\mathbf{p} = -5 + 2x + x^2$, $\mathbf{q} = 3 + 2x - 4x^2$

9. (a) Use Formula (4) to show that $\langle \mathbf{u}, \mathbf{v} \rangle = 9u_1v_1 + 4u_2v_2$ is the inner product on R^2 generated by

$$A = \begin{bmatrix} 3 & 0 \\ 0 & 2 \end{bmatrix}$$

(b) Use the inner product in part (a) to compute $\langle \mathbf{u}, \mathbf{v} \rangle$ if $\mathbf{u} = (-3, 2)$ and $\mathbf{v} = (1, 7)$.

10. (a) Use Formula (4) to show that

$$\langle \mathbf{u}, \mathbf{v} \rangle = 5u_1v_1 - u_1v_2 - u_2v_1 + 10u_2v_2$$

is the inner product on R^2 generated by

$$A = \begin{bmatrix} 2 & 1 \\ -1 & 3 \end{bmatrix}$$

(b) Use the inner product in part (a) to compute $\langle \mathbf{u}, \mathbf{v} \rangle$ if $\mathbf{u} = (0, -3)$ and $\mathbf{v} = (6, 2)$.

11. Let $\mathbf{u} = (u_1, u_2)$ and $\mathbf{v} = (v_1, v_2)$. In each part, the given expression is an inner product on R^2. Find a matrix that generates it.

(a) $\langle \mathbf{u}, \mathbf{v} \rangle = 3u_1v_1 + 5u_2v_2$ (b) $\langle \mathbf{u}, \mathbf{v} \rangle = 4u_1v_1 + 6u_2v_2$

12. Let P_2 have the inner product in Example 7. In each part, find $\|\mathbf{p}\|$.

(a) $\mathbf{p} = -2 + 3x + 2x^2$ (b) $\mathbf{p} = 4 - 3x^2$

13. Let M_{22} have the inner product in Example 6. In each part, find $\|A\|$.

(a) $A = \begin{bmatrix} -2 & 5 \\ 3 & 6 \end{bmatrix}$ (b) $A = \begin{bmatrix} 0 & 0 \\ 0 & 0 \end{bmatrix}$

14. Let P_2 have the inner product in Example 7. Find $d(\mathbf{p}, \mathbf{q})$.

$$\mathbf{p} = 3 - x + x^2, \quad \mathbf{q} = 2 + 5x^2$$

15. Let M_{22} have the inner product in Example 6. Find $d(A, B)$.

(a) $A = \begin{bmatrix} 2 & 6 \\ 9 & 4 \end{bmatrix}$, $B = \begin{bmatrix} -4 & 7 \\ 1 & 6 \end{bmatrix}$

(b) $A = \begin{bmatrix} -2 & 4 \\ 1 & 0 \end{bmatrix}$, $B = \begin{bmatrix} -5 & 1 \\ 6 & 2 \end{bmatrix}$

16. Let P_2 have the inner product of Example 9, and let $\mathbf{p} = 1 + x + x^2$ and $\mathbf{q} = 1 - 2x^2$. Compute the following.

(a) $\langle \mathbf{p}, \mathbf{q} \rangle$ (b) $\|\mathbf{p}\|$ (c) $d(\mathbf{p}, \mathbf{q})$

17. Let P_3 have the evaluation inner product at the sample points

$$x_0 = -1, \quad x_1 = 0, \quad x_2 = 1, \quad x_3 = 2$$

Find $\langle \mathbf{p}, \mathbf{q} \rangle$ and $\|\mathbf{p}\|$ for $\mathbf{p} = x + x^3$ and $\mathbf{q} = 1 + x^2$.

18. In each part, use the given inner product on R^2 to find $\|\mathbf{w}\|$, where $\mathbf{w} = (-1, 3)$.

(a) the Euclidean inner product

(b) the weighted Euclidean inner product $\langle \mathbf{u}, \mathbf{v} \rangle = 3u_1v_1 + 2u_2v_2$, where $\mathbf{u} = (u_1, u_2)$ and $\mathbf{v} = (v_1, v_2)$

(c) the inner product generated by the matrix

$$A = \begin{bmatrix} 1 & 2 \\ -1 & 3 \end{bmatrix}$$

19. Use the inner products in Exercise 18 to find $d(\mathbf{u}, \mathbf{v})$ for $\mathbf{u} = (-1, 2)$ and $\mathbf{v} = (2, 5)$.

20. Suppose that $\mathbf{u}$, $\mathbf{v}$, and $\mathbf{w}$ are vectors such that

$$\langle \mathbf{u}, \mathbf{v} \rangle = 2, \quad \langle \mathbf{v}, \mathbf{w} \rangle = -3, \quad \langle \mathbf{u}, \mathbf{w} \rangle = 5$$
$$\|\mathbf{u}\| = 1, \quad \|\mathbf{v}\| = 2, \quad \|\mathbf{w}\| = 7$$

Evaluate the given expression.

(a) $\langle \mathbf{u} + \mathbf{v}, \mathbf{v} + \mathbf{w} \rangle$ (b) $\langle 2\mathbf{v} - \mathbf{w}, 3\mathbf{u} + 2\mathbf{w} \rangle$

(c) $\langle \mathbf{u} - \mathbf{v} - 2\mathbf{w}, 4\mathbf{u} + \mathbf{v} \rangle$ (d) $\|\mathbf{u} + \mathbf{v}\|$

(e) $\|2\mathbf{w} - \mathbf{v}\|$ (f) $\|\mathbf{u} - 2\mathbf{v} + 4\mathbf{w}\|$

21. Sketch the unit circle in R^2 using the given inner product.

(a) $\langle \mathbf{u}, \mathbf{v} \rangle = \frac{1}{4}u_1v_1 + \frac{1}{16}u_2v_2$

(b) $\langle \mathbf{u}, \mathbf{v} \rangle = 2u_1v_1 + u_2v_2$

22. Find a weighted Euclidean inner product on R^2 for which the unit circle is the ellipse shown in the accompanying figure.

◀ **Figure Ex-22**

23. Let $\mathbf{u} = (u_1, u_2)$ and $\mathbf{v} = (v_1, v_2)$. Show that the following are inner products on R^2 by verifying that the inner product axioms hold.

(a) $\langle \mathbf{u}, \mathbf{v} \rangle = 3u_1v_1 + 5u_2v_2$

(b) $\langle \mathbf{u}, \mathbf{v} \rangle = 4u_1v_1 + u_2v_1 + u_1v_2 + 4u_2v_2$

24. Let $\mathbf{u} = (u_1, u_2, u_3)$ and $\mathbf{v} = (v_1, v_2, v_3)$. Determine which of the following are inner products on R^3. For those that are not, list the axioms that do not hold.

(a) $\langle \mathbf{u}, \mathbf{v} \rangle = u_1v_1 + u_3v_3$

(b) $\langle \mathbf{u}, \mathbf{v} \rangle = u_1^2v_1^2 + u_2^2v_2^2 + u_3^2v_3^2$

(c) $\langle \mathbf{u}, \mathbf{v} \rangle = 2u_1v_1 + u_2v_2 + 4u_3v_3$

(d) $\langle \mathbf{u}, \mathbf{v} \rangle = u_1v_1 - u_2v_2 + u_3v_3$

25. Show that the following identity holds for vectors in any inner product space.

$$\|\mathbf{u} + \mathbf{v}\|^2 + \|\mathbf{u} - \mathbf{v}\|^2 = 2\|\mathbf{u}\|^2 + 2\|\mathbf{v}\|^2$$

26. Show that the following identity holds for vectors in any inner product space.

$$\langle \mathbf{u}, \mathbf{v} \rangle = \frac{1}{4}\|\mathbf{u} + \mathbf{v}\|^2 - \frac{1}{4}\|\mathbf{u} - \mathbf{v}\|^2$$

27. Let $U = \begin{bmatrix} u_1 & u_2 \\ u_3 & u_4 \end{bmatrix}$ and $V = \begin{bmatrix} v_1 & v_2 \\ v_3 & v_4 \end{bmatrix}$.

Show that $\langle U, V \rangle = u_1v_1 + u_2v_3 + u_3v_2 + u_4v_4$ is *not* an inner product on M_{22}.

28. (*Calculus required*) Let the vector space P_2 have the inner product

$$\langle \mathbf{p}, \mathbf{q} \rangle = \int_{-1}^{1} p(x)q(x)\, dx$$

(a) Find $\|\mathbf{p}\|$ for $\mathbf{p} = 1$, $\mathbf{p} = x$, and $\mathbf{p} = x^2$.

(b) Find $d(\mathbf{p}, \mathbf{q})$ if $\mathbf{p} = 1$ and $\mathbf{q} = x$.

29. (*Calculus required*) Use the inner product

$$\langle \mathbf{p}, \mathbf{q} \rangle = \int_{-1}^{1} p(x)q(x)\, dx$$

on P_3, to compute $\langle \mathbf{p}, \mathbf{q} \rangle$.

(a) $\mathbf{p} = 1 - x + x^2 + 5x^3$, $\mathbf{q} = x - 3x^2$

(b) $\mathbf{p} = x - 5x^3$, $\mathbf{q} = 2 + 8x^2$

30. (*Calculus required*) In each part, use the inner product

$$\langle \mathbf{f}, \mathbf{g} \rangle = \int_{0}^{1} f(x)g(x)\, dx$$

on $C[0, 1]$ to compute the following.

(a) $\mathbf{f} = \cos 2\pi x$, $\mathbf{g} = \sin 2\pi x$

(b) $\mathbf{f} = x$, $\mathbf{g} = e^x$

(c) $\mathbf{f} = \tan \dfrac{\pi}{4}x$, $\mathbf{g} = 1$

31. Prove that Formula (4) defines an inner product on R^n.

32. The definition of a complex vector space was given in the first margin note in Section 4.1. The definition of a **complex inner product** on a complex vector space V is identical to Definition 1 except that scalars are allowed to be complex numbers, and Axiom 1 is replaced by $\langle \mathbf{u}, \mathbf{v} \rangle = \overline{\langle \mathbf{v}, \mathbf{u} \rangle}$. The remaining axioms are unchanged. A complex vector space with a complex inner product is called a **complex inner product space**. Prove that if V is a complex inner product space then $\langle \mathbf{u}, k\mathbf{v} \rangle = \bar{k}\langle \mathbf{u}, \mathbf{v} \rangle$.

True-False Exercises

In parts (a)–(g) determine whether the statement is true or false, and justify your answer.

(a) The dot product on R^2 is an example of a weighted inner product.

(b) The inner product of two vectors cannot be a negative real number.

(c) $\langle \mathbf{u}, \mathbf{v} + \mathbf{w} \rangle = \langle \mathbf{v}, \mathbf{u} \rangle + \langle \mathbf{w}, \mathbf{u} \rangle$.

(d) $\langle k\mathbf{u}, k\mathbf{v} \rangle = k^2 \langle \mathbf{u}, \mathbf{v} \rangle$.

(e) If $\langle \mathbf{u}, \mathbf{v} \rangle = 0$, then $\mathbf{u} = \mathbf{0}$ or $\mathbf{v} = \mathbf{0}$.

(f) If $\|\mathbf{v}\|^2 = 0$, then $\mathbf{v} = \mathbf{0}$.

(g) If A is an $n \times n$ matrix, then $\langle \mathbf{u}, \mathbf{v} \rangle = A\mathbf{u} \cdot A\mathbf{v}$ defines an inner product on R^n.

6.2 Angle and Orthogonality in Inner Product Spaces

In Section 3.2 we defined the notion of "angle" between vectors in R^n. In this section we will extend this idea to general vector spaces. This will enable us to extend the notion of orthogonality as well, thereby setting the groundwork for a variety of new applications.

Cauchy–Schwarz Inequality   Recall from Formula (20) of Section 3.2 that the angle θ between two vectors $\mathbf{u}$ and $\mathbf{v}$ in R^n is

$$\theta = \cos^{-1}\left(\frac{\mathbf{u} \cdot \mathbf{v}}{\|\mathbf{u}\|\|\mathbf{v}\|}\right) \tag{1}$$

We were assured that this formula was valid because it followed from the Cauchy–Schwarz inequality (Theorem 3.2.4) that

$$-1 \le \frac{\mathbf{u} \cdot \mathbf{v}}{\|\mathbf{u}\|\|\mathbf{v}\|} \le 1 \tag{2}$$

as required for the inverse cosine to be defined. The following generalization of Theorem 3.2.4 will enable us to define the angle between two vectors in *any* real inner product space.

> **THEOREM 6.2.1 Cauchy–Schwarz Inequality**
>
> *If $\mathbf{u}$ and $\mathbf{v}$ are vectors in a real inner product space V, then*
>
> $$|\langle \mathbf{u}, \mathbf{v}\rangle| \le \|\mathbf{u}\|\|\mathbf{v}\| \tag{3}$$

Proof   We warn you in advance that the proof presented here depends on a clever trick that is not easy to motivate.

In the case where $\mathbf{u} = \mathbf{0}$ the two sides of (3) are equal since $\langle \mathbf{u}, \mathbf{v}\rangle$ and $\|\mathbf{u}\|$ are both zero. Thus, we need only consider the case where $\mathbf{u} \ne \mathbf{0}$. Making this assumption, let

$$a = \langle \mathbf{u}, \mathbf{u}\rangle, \quad b = 2\langle \mathbf{u}, \mathbf{v}\rangle, \quad c = \langle \mathbf{v}, \mathbf{v}\rangle$$

and let t be any real number. Since the positivity axiom states that the inner product of any vector with itself is nonnegative, it follows that

$$0 \le \langle t\mathbf{u} + \mathbf{v}, t\mathbf{u} + \mathbf{v}\rangle = \langle \mathbf{u}, \mathbf{u}\rangle t^2 + 2\langle \mathbf{u}, \mathbf{v}\rangle t + \langle \mathbf{v}, \mathbf{v}\rangle$$
$$= at^2 + bt + c$$

This inequality implies that the quadratic polynomial $at^2 + bt + c$ has either no real roots or a repeated real root. Therefore, its discriminant must satisfy the inequality $b^2 - 4ac \le 0$. Expressing the coefficients a, b, and c in terms of the vectors $\mathbf{u}$ and $\mathbf{v}$ gives $4\langle \mathbf{u}, \mathbf{v}\rangle^2 - 4\langle \mathbf{u}, \mathbf{u}\rangle\langle \mathbf{v}, \mathbf{v}\rangle \le 0$ or, equivalently,

$$\langle \mathbf{u}, \mathbf{v}\rangle^2 \le \langle \mathbf{u}, \mathbf{u}\rangle\langle \mathbf{v}, \mathbf{v}\rangle$$

Taking square roots of both sides and using the fact that $\langle \mathbf{u}, \mathbf{u}\rangle$ and $\langle \mathbf{v}, \mathbf{v}\rangle$ are nonnegative yields

$$|\langle \mathbf{u}, \mathbf{v}\rangle| \le \langle \mathbf{u}, \mathbf{u}\rangle^{1/2}\langle \mathbf{v}, \mathbf{v}\rangle^{1/2} \quad \text{or equivalently} \quad |\langle \mathbf{u}, \mathbf{v}\rangle| \le \|\mathbf{u}\|\|\mathbf{v}\|$$

which completes the proof. ◀

The following two alternative forms of the Cauchy–Schwarz inequality are useful to know:

$$\langle \mathbf{u}, \mathbf{v} \rangle^2 \leq \langle \mathbf{u}, \mathbf{u} \rangle \langle \mathbf{v}, \mathbf{v} \rangle \tag{4}$$

$$\langle \mathbf{u}, \mathbf{v} \rangle^2 \leq \|\mathbf{u}\|^2 \|\mathbf{v}\|^2 \tag{5}$$

The first of these formulas was obtained in the proof of Theorem 6.2.1, and the second is a variation of the first.

Angle Between Vectors   Our next goal is to define what is meant by the "angle" between vectors in a real inner product space. As the first step, we leave it for you to use the Cauchy–Schwarz inequality to show that

$$-1 \leq \frac{\langle \mathbf{u}, \mathbf{v} \rangle}{\|\mathbf{u}\| \|\mathbf{v}\|} \leq 1 \tag{6}$$

This being the case, there is a unique angle θ in radian measure for which

$$\cos \theta = \frac{\langle \mathbf{u}, \mathbf{v} \rangle}{\|\mathbf{u}\| \|\mathbf{v}\|} \quad \text{and} \quad 0 \leq \theta \leq \pi \tag{7}$$

(Figure 6.2.1). This enables us to *define* the **angle θ between u and v** to be

$$\theta = \cos^{-1} \left(\frac{\langle \mathbf{u}, \mathbf{v} \rangle}{\|\mathbf{u}\| \|\mathbf{v}\|} \right) \tag{8}$$

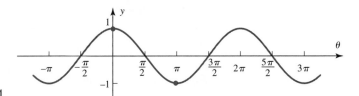

▶ Figure 6.2.1

▶ **EXAMPLE 1 Cosine of an Angle Between Two Vectors in R^4**

Let R^4 have the Euclidean inner product. Find the cosine of the angle θ between the vectors $\mathbf{u} = (4, 3, 1, -2)$ and $\mathbf{v} = (-2, 1, 2, 3)$.

Solution We leave it for you to verify that

$$\|\mathbf{u}\| = \sqrt{30}, \quad \|\mathbf{v}\| = \sqrt{18}, \quad \text{and} \quad \langle \mathbf{u}, \mathbf{v} \rangle = -9$$

from which it follows that

$$\cos \theta = \frac{\langle \mathbf{u}, \mathbf{v} \rangle}{\|\mathbf{u}\| \|\mathbf{v}\|} = -\frac{9}{\sqrt{30}\sqrt{18}} = -\frac{3}{2\sqrt{15}} \quad ◀$$

Properties of Length and Distance in General Inner Product Spaces   In Section 3.2 we used the dot product to extend the notions of length and distance to R^n, and we showed that various familiar theorems remained valid (see Theorems 3.2.5, 3.2.6, and 3.2.7). By making only minor adjustments to the proofs of those theorems, we can show that they remain valid in any real inner product space. For example, here is the generalization of Theorem 3.2.5 (the triangle inequalities).

THEOREM 6.2.2 *If* **u**, **v**, *and* **w** *are vectors in a real inner product space V, and if k is any scalar, then:*

(a) $\|\mathbf{u} + \mathbf{v}\| \leq \|\mathbf{u}\| + \|\mathbf{v}\|$ [Triangle inequality for vectors]

(b) $d(\mathbf{u}, \mathbf{v}) \leq d(\mathbf{u}, \mathbf{w}) + d(\mathbf{w}, \mathbf{v})$ [Triangle inequality for distances]

Proof (a)

$$\|\mathbf{u} + \mathbf{v}\|^2 = \langle \mathbf{u} + \mathbf{v}, \mathbf{u} + \mathbf{v} \rangle$$
$$= \langle \mathbf{u}, \mathbf{u} \rangle + 2\langle \mathbf{u}, \mathbf{v} \rangle + \langle \mathbf{v}, \mathbf{v} \rangle$$
$$\leq \langle \mathbf{u}, \mathbf{u} \rangle + 2|\langle \mathbf{u}, \mathbf{v} \rangle| + \langle \mathbf{v}, \mathbf{v} \rangle \quad \text{[Property of absolute value]}$$
$$\leq \langle \mathbf{u}, \mathbf{u} \rangle + 2\|\mathbf{u}\|\|\mathbf{v}\| + \langle \mathbf{v}, \mathbf{v} \rangle \quad \text{[By (3)]}$$
$$= \|\mathbf{u}\|^2 + 2\|\mathbf{u}\|\|\mathbf{v}\| + \|\mathbf{v}\|^2$$
$$= (\|\mathbf{u}\| + \|\mathbf{v}\|)^2$$

Taking square roots gives $\|\mathbf{u} + \mathbf{v}\| \leq \|\mathbf{u}\| + \|\mathbf{v}\|$.

Proof (b)   Identical to the proof of part (b) of Theorem 3.2.5. ◀

Orthogonality   Although Example 1 is a useful mathematical exercise, there is only an occasional need to compute angles in vector spaces other than R^2 and R^3. A problem of more interest in general vector spaces is ascertaining whether the angle between vectors is $\pi/2$. You should be able to see from Formula (8) that if $\mathbf{u}$ and $\mathbf{v}$ are *nonzero* vectors, then the angle between them is $\theta = \pi/2$ if and only if $\langle \mathbf{u}, \mathbf{v} \rangle = 0$. Accordingly, we make the following definition (which is applicable even if one or both of the vectors is zero).

DEFINITION 1   Two vectors $\mathbf{u}$ and $\mathbf{v}$ in an inner product space are called *orthogonal* if $\langle \mathbf{u}, \mathbf{v} \rangle = 0$.

As the following example shows, orthogonality depends on the inner product in the sense that for different inner products two vectors can be orthogonal with respect to one but not the other.

▶ **EXAMPLE 2 Orthogonality Depends on the Inner Product**

The vectors $\mathbf{u} = (1, 1)$ and $\mathbf{v} = (1, -1)$ are orthogonal with respect to the Euclidean inner product on R^2, since

$$\mathbf{u} \cdot \mathbf{v} = (1)(1) + (1)(-1) = 0$$

However, they are not orthogonal with respect to the weighted Euclidean inner product $\langle \mathbf{u}, \mathbf{v} \rangle = 3u_1v_1 + 2u_2v_2$, since

$$\langle \mathbf{u}, \mathbf{v} \rangle = 3(1)(1) + 2(1)(-1) = 1 \neq 0$$

▶ **EXAMPLE 3 Orthogonal Vectors in M_{22}**

If M_{22} has the inner product of Example 6 in the preceding section, then the matrices

$$U = \begin{bmatrix} 1 & 0 \\ 1 & 1 \end{bmatrix} \quad \text{and} \quad V = \begin{bmatrix} 0 & 2 \\ 0 & 0 \end{bmatrix}$$

are orthogonal, since

$$\langle U, V \rangle = 1(0) + 0(2) + 1(0) + 1(0) = 0$$

CALCULUS REQUIRED   ▶ **EXAMPLE 4 Orthogonal Vectors in P_2**

Let P_2 have the inner product

$$\langle \mathbf{p}, \mathbf{q} \rangle = \int_{-1}^{1} p(x)q(x)\,dx$$

and let $\mathbf{p} = x$ and $\mathbf{q} = x^2$. Then

$$\|\mathbf{p}\| = \langle \mathbf{p}, \mathbf{p}\rangle^{1/2} = \left[\int_{-1}^{1} xx\,dx\right]^{1/2} = \left[\int_{-1}^{1} x^2\,dx\right]^{1/2} = \sqrt{\frac{2}{3}}$$

$$\|\mathbf{q}\| = \langle \mathbf{q}, \mathbf{q}\rangle^{1/2} = \left[\int_{-1}^{1} x^2 x^2\,dx\right]^{1/2} = \left[\int_{-1}^{1} x^4\,dx\right]^{1/2} = \sqrt{\frac{2}{5}}$$

$$\langle \mathbf{p}, \mathbf{q}\rangle = \int_{-1}^{1} xx^2\,dx = \int_{-1}^{1} x^3\,dx = 0$$

Because $\langle \mathbf{p}, \mathbf{q}\rangle = 0$, the vectors $\mathbf{p} = x$ and $\mathbf{q} = x^2$ are orthogonal relative to the given inner product. ◄

In Section 3.3 we proved the Theorem of Pythagoras for vectors in Euclidean n-space. The following theorem extends this result to vectors in any real inner product space.

> **THEOREM 6.2.3** Generalized Theorem of Pythagoras
>
> *If $\mathbf{u}$ and $\mathbf{v}$ are orthogonal vectors in an inner product space, then*
>
> $$\|\mathbf{u} + \mathbf{v}\|^2 = \|\mathbf{u}\|^2 + \|\mathbf{v}\|^2$$

Proof The orthogonality of $\mathbf{u}$ and $\mathbf{v}$ implies that $\langle \mathbf{u}, \mathbf{v}\rangle = 0$, so

$$\|\mathbf{u} + \mathbf{v}\|^2 = \langle \mathbf{u} + \mathbf{v}, \mathbf{u} + \mathbf{v}\rangle = \|\mathbf{u}\|^2 + 2\langle \mathbf{u}, \mathbf{v}\rangle + \|\mathbf{v}\|^2$$
$$= \|\mathbf{u}\|^2 + \|\mathbf{v}\|^2 \quad ◄$$

CALCULUS REQUIRED

▶ **EXAMPLE 5** Theorem of Pythagoras in P_2

In Example 4 we showed that $\mathbf{p} = x$ and $\mathbf{q} = x^2$ are orthogonal with respect to the inner product

$$\langle \mathbf{p}, \mathbf{q}\rangle = \int_{-1}^{1} p(x)q(x)\,dx$$

on P_2. It follows from Theorem 6.2.3 that

$$\|\mathbf{p} + \mathbf{q}\|^2 = \|\mathbf{p}\|^2 + \|\mathbf{q}\|^2$$

Thus, from the computations in Example 4, we have

$$\|\mathbf{p} + \mathbf{q}\|^2 = \left(\sqrt{\frac{2}{3}}\right)^2 + \left(\sqrt{\frac{2}{5}}\right)^2 = \frac{2}{3} + \frac{2}{5} = \frac{16}{15}$$

We can check this result by direct integration:

$$\|\mathbf{p} + \mathbf{q}\|^2 = \langle \mathbf{p} + \mathbf{q}, \mathbf{p} + \mathbf{q}\rangle = \int_{-1}^{1} (x + x^2)(x + x^2)\,dx$$
$$= \int_{-1}^{1} x^2\,dx + 2\int_{-1}^{1} x^3\,dx + \int_{-1}^{1} x^4\,dx = \frac{2}{3} + 0 + \frac{2}{5} = \frac{16}{15} \quad ◄$$

Orthogonal Complements In Section 4.8 we defined the notion of an *orthogonal complement* for subspaces of R^n, and we used that definition to establish a geometric link between the fundamental spaces of a matrix. The following definition extends that idea to general inner product spaces.

> **DEFINITION 2** If W is a subspace of an inner product space V, then the set of all vectors in V that are orthogonal to every vector in W is called the ***orthogonal complement*** of W and is denoted by the symbol $W^{\perp}$.

In Theorem 4.8.8 we stated three properties of orthogonal complements in R^n. The following theorem generalizes parts (a) and (b) of that theorem to general inner product spaces.

> **THEOREM 6.2.4** *If W is a subspace of an inner product space V, then:*
> *(a) $W^{\perp}$ is a subspace of V.*
> *(b) $W \cap W^{\perp} = \{\mathbf{0}\}$.*

Proof (a) The set $W^{\perp}$ contains at least the zero vector, since $\langle \mathbf{0}, \mathbf{w} \rangle = 0$ for every vector $\mathbf{w}$ in W. Thus, it remains to show that $W^{\perp}$ is closed under addition and scalar multiplication. To do this, suppose that $\mathbf{u}$ and $\mathbf{v}$ are vectors in $W^{\perp}$, so that for every vector $\mathbf{w}$ in W we have $\langle \mathbf{u}, \mathbf{w} \rangle = 0$ and $\langle \mathbf{v}, \mathbf{w} \rangle = 0$. It follows from the additivity and homogeneity axioms of inner products that

$$\langle \mathbf{u} + \mathbf{v}, \mathbf{w} \rangle = \langle \mathbf{u}, \mathbf{w} \rangle + \langle \mathbf{v}, \mathbf{w} \rangle = 0 + 0 = 0$$
$$\langle k\mathbf{u}, \mathbf{w} \rangle = k\langle \mathbf{u}, \mathbf{w} \rangle = k(0) = 0$$

which proves that $\mathbf{u} + \mathbf{v}$ and $k\mathbf{u}$ are in $W^{\perp}$.

Proof (b) If $\mathbf{v}$ is any vector in both W and $W^{\perp}$, then $\mathbf{v}$ is orthogonal to itself; that is, $\langle \mathbf{v}, \mathbf{v} \rangle = 0$. It follows from the positivity axiom for inner products that $\mathbf{v} = \mathbf{0}$. ◀

The next theorem, which we state without proof, generalizes part (c) of Theorem 4.8.8. Note, however, that this theorem applies only to finite-dimensional inner product spaces, whereas Theorem 6.2.5 does not have this restriction.

Theorem 6.2.5 implies that in a finite-dimensional inner product space orthogonal complements occur in pairs, each being orthogonal to the other (Figure 6.2.2).

> **THEOREM 6.2.5** *If W is a subspace of a finite-dimensional inner product space V, then the orthogonal complement of $W^{\perp}$ is W; that is,*
>
> $$(W^{\perp})^{\perp} = W$$

In our study of the fundamental spaces of a matrix in Section 4.8 we showed that the row space and null space of a matrix are orthogonal complements with respect to the Euclidean inner product on R^n (Theorem 4.8.9). The following example takes advantage of that fact.

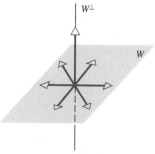

▲ **Figure 6.2.2** Each vector in W is orthogonal to each vector in $W^{\perp}$ and conversely.

▶ **EXAMPLE 6** **Basis for an Orthogonal Complement**

Let W be the subspace of R^6 spanned by the vectors

$$\mathbf{w}_1 = (1, 3, -2, 0, 2, 0), \qquad \mathbf{w}_2 = (2, 6, -5, -2, 4, -3),$$
$$\mathbf{w}_3 = (0, 0, 5, 10, 0, 15), \qquad \mathbf{w}_4 = (2, 6, 0, 8, 4, 18)$$

Find a basis for the orthogonal complement of W.

Solution The space W is the same as the row space of the matrix

$$A = \begin{bmatrix} 1 & 3 & -2 & 0 & 2 & 0 \\ 2 & 6 & -5 & -2 & 4 & -3 \\ 0 & 0 & 5 & 10 & 0 & 15 \\ 2 & 6 & 0 & 8 & 4 & 18 \end{bmatrix}$$

Since the row space and null space of A are orthogonal complements, our problem reduces to finding a basis for the null space of this matrix. In Example 4 of Section 4.7 we showed that

$$\mathbf{v}_1 = \begin{bmatrix} -3 \\ 1 \\ 0 \\ 0 \\ 0 \\ 0 \end{bmatrix}, \quad \mathbf{v}_2 = \begin{bmatrix} -4 \\ 0 \\ -2 \\ 1 \\ 0 \\ 0 \end{bmatrix}, \quad \mathbf{v}_3 = \begin{bmatrix} -2 \\ 0 \\ 0 \\ 0 \\ 1 \\ 0 \end{bmatrix}$$

form a basis for this null space. Expressing these vectors in comma-delimited form (to match that of $\mathbf{w}_1$, $\mathbf{w}_2$, $\mathbf{w}_3$, and $\mathbf{w}_4$), we obtain the basis vectors

$$\mathbf{v}_1 = (-3, 1, 0, 0, 0, 0), \quad \mathbf{v}_2 = (-4, 0, -2, 1, 0, 0), \quad \mathbf{v}_3 = (-2, 0, 0, 0, 1, 0)$$

You may want to check that these vectors are orthogonal to $\mathbf{w}_1$, $\mathbf{w}_2$, $\mathbf{w}_3$, and $\mathbf{w}_4$ by computing the necessary dot products. ◀

Concept Review

- Cauchy–Schwarz inequality
- Angle between vectors
- Orthogonal vectors
- Orthogonal complement

Skills

- Find the angle between two vectors in an inner product space.
- Determine whether two vectors in an inner product space are orthogonal.
- Find a basis for the orthogonal complement of a subspace of an inner product space.

Exercise Set 6.2

1. Let R^2, R^3, and R^4 have the Euclidean inner product. In each part, find the cosine of the angle between $\mathbf{u}$ and $\mathbf{v}$.

(a) $\mathbf{u} = (1, -3)$, $\mathbf{v} = (2, 4)$

(b) $\mathbf{u} = (-1, 0)$, $\mathbf{v} = (3, 8)$

(c) $\mathbf{u} = (-1, 5, 2)$, $\mathbf{v} = (2, 4, -9)$

(d) $\mathbf{u} = (4, 1, 8)$, $\mathbf{v} = (1, 0, -3)$

(e) $\mathbf{u} = (1, 0, 1, 0)$, $\mathbf{v} = (-3, -3, -3, -3)$

(f) $\mathbf{u} = (2, 1, 7, -1)$, $\mathbf{v} = (4, 0, 0, 0)$

2. Let P_2 have the inner product in Example 7 of Section 6.1. Find the cosine of the angle between $\mathbf{p}$ and $\mathbf{q}$.

(a) $\mathbf{p} = -1 + 5x + 2x^2$, $\mathbf{q} = 2 + 4x - 9x^2$

(b) $\mathbf{p} = x - x^2$, $\mathbf{q} = 7 + 3x + 3x^2$

3. Let M_{22} have the inner product in Example 6 of Section 6.1. Find the cosine of the angle between A and B.

(a) $A = \begin{bmatrix} 2 & 6 \\ 1 & -3 \end{bmatrix}$, $B = \begin{bmatrix} 3 & 2 \\ 1 & 0 \end{bmatrix}$

(b) $A = \begin{bmatrix} 2 & 4 \\ -1 & 3 \end{bmatrix}$, $B = \begin{bmatrix} -3 & 1 \\ 4 & 2 \end{bmatrix}$

4. In each part, determine whether the given vectors are orthogonal with respect to the Euclidean inner product.

(a) $\mathbf{u} = (-1, 3, 2)$, $\mathbf{v} = (4, 2, -1)$

(b) $\mathbf{u} = (-2, -2, -2)$, $\mathbf{v} = (1, 1, 1)$

(c) $\mathbf{u} = (u_1, u_2, u_3)$, $\mathbf{v} = (0, 0, 0)$

(d) $\mathbf{u} = (-4, 6, -10, 1)$, $\mathbf{v} = (2, 1, -2, 9)$

(e) $\mathbf{u} = (0, 3, -2, 1)$, $\mathbf{v} = (5, 2, -1, 0)$

(f) $\mathbf{u} = (a, b)$, $\mathbf{v} = (-b, a)$

5. Show that $\mathbf{p} = 1 - x + 2x^2$ and $\mathbf{q} = 2x + x^2$ are orthogonal with respect to the inner product in Exercise 2.

6. Let

$$A = \begin{bmatrix} 2 & 1 \\ -1 & 3 \end{bmatrix}$$

Which of the following matrices are orthogonal to A with respect to the inner product in Exercise 3?

(a) $\begin{bmatrix} -3 & 0 \\ 0 & 2 \end{bmatrix}$

(b) $\begin{bmatrix} 1 & 1 \\ 0 & -1 \end{bmatrix}$

(c) $\begin{bmatrix} 0 & 0 \\ 0 & 0 \end{bmatrix}$

(d) $\begin{bmatrix} 2 & 1 \\ 5 & 2 \end{bmatrix}$

7. Do there exist scalars k and l such that the vectors $\mathbf{u} = (2, k, 6)$, $\mathbf{v} = (l, 5, 3)$, and $\mathbf{w} = (1, 2, 3)$ are mutually orthogonal with respect to the Euclidean inner product?

8. Let R^3 have the Euclidean inner product, and suppose that $\mathbf{u} = (1, 1, -1)$ and $\mathbf{v} = (6, 7, -15)$. Find a value of k for which $\|k\mathbf{u} + \mathbf{v}\| = 13$.

9. Let R^3 have the Euclidean inner product. For which values of k are $\mathbf{u}$ and $\mathbf{v}$ orthogonal?

(a) $\mathbf{u} = (2, 1, 3)$, $\mathbf{v} = (1, 7, k)$

(b) $\mathbf{u} = (k, k, 1)$, $\mathbf{v} = (k, 5, 6)$

10. Let R^4 have the Euclidean inner product. Find two unit vectors that are orthogonal to all three of the vectors $\mathbf{u} = (2, 1, -4, 0)$, $\mathbf{v} = (-1, -1, 2, 2)$, and $\mathbf{w} = (3, 2, 5, 4)$.

11. In each part, verify that the Cauchy–Schwarz inequality holds for the given vectors using the Euclidean inner product.

(a) $\mathbf{u} = (3, 2)$, $\mathbf{v} = (4, -1)$

(b) $\mathbf{u} = (-3, 1, 0)$, $\mathbf{v} = (2, -1, 3)$

(c) $\mathbf{u} = (-4, 2, 1)$, $\mathbf{v} = (8, -4, -2)$

(d) $\mathbf{u} = (0, -2, 2, 1)$, $\mathbf{v} = (-1, -1, 1, 1)$

12. In each part, verify that the Cauchy–Schwarz inequality holds for the given vectors.

(a) $\mathbf{u} = (-2, 1)$ and $\mathbf{v} = (1, 0)$ using the inner product of Example 1 of Section 6.1.

(b) $U = \begin{bmatrix} -1 & 2 \\ 6 & 1 \end{bmatrix}$ and $V = \begin{bmatrix} 1 & 0 \\ 3 & 3 \end{bmatrix}$ using the inner product in Example 6 of Section 6.1.

(c) $\mathbf{p} = -1 + 2x + x^2$ and $\mathbf{q} = 2 - 4x^2$ using the inner product given in Example 7 of Section 6.1.

13. Let R^4 have the Euclidean inner product, and let $\mathbf{u} = (-1, 1, 0, 2)$. Determine whether the vector $\mathbf{u}$ is orthogonal to the subspace spanned by the vectors $\mathbf{w}_1 = (0, 0, 0, 0)$, $\mathbf{w}_2 = (1, -1, 3, 0)$, and $\mathbf{w}_3 = (4, 0, 9, 2)$.

▷ In Exercises 14–15, assume that R^n has the Euclidean inner product. ◁

14. Let W be the line in R^2 with equation $y = 2x$. Find an equation for $W^\perp$.

15. (a) Let W be the plane in R^3 with equation $x - 2y - 3z = 0$. Find parametric equations for $W^\perp$.

(b) Let W be the line in R^3 with parametric equations

$$x = 2t, \quad y = -5t, \quad z = 4t$$

Find an equation for $W^\perp$.

(c) Let W be the intersection of the two planes

$$x + y + z = 0 \quad \text{and} \quad x - y + z = 0$$

in R^3. Find an equation for $W^\perp$.

16. Find a basis for the orthogonal complement of the subspace of R^n spanned by the vectors.

(a) $\mathbf{v}_1 = (1, -1, 3)$, $\mathbf{v}_2 = (5, -4, -4)$, $\mathbf{v}_3 = (7, -6, 2)$

(b) $\mathbf{v}_1 = (2, 0, -1)$, $\mathbf{v}_2 = (4, 0, -2)$

(c) $\mathbf{v}_1 = (1, 4, 5, 2)$, $\mathbf{v}_2 = (2, 1, 3, 0)$, $\mathbf{v}_3 = (-1, 3, 2, 2)$

(d) $\mathbf{v}_1 = (1, 4, 5, 6, 9)$, $\mathbf{v}_2 = (3, -2, 1, 4, -1)$, $\mathbf{v}_3 = (-1, 0, -1, -2, -1)$, $\mathbf{v}_4 = (2, 3, 5, 7, 8)$

17. Let V be an inner product space. Show that if $\mathbf{u}$ and $\mathbf{v}$ are orthogonal unit vectors in V, then $\|\mathbf{u} - \mathbf{v}\| = \sqrt{2}$.

18. Let V be an inner product space. Show that if $\mathbf{w}$ is orthogonal to both $\mathbf{u}_1$ and $\mathbf{u}_2$, then it is orthogonal to $k_1\mathbf{u}_1 + k_2\mathbf{u}_2$ for all scalars k_1 and k_2. Interpret this result geometrically in the case where V is R^3 with the Euclidean inner product.

19. Let V be an inner product space. Show that if $\mathbf{w}$ is orthogonal to each of the vectors $\mathbf{u}_1, \mathbf{u}_2, \ldots, \mathbf{u}_r$, then it is orthogonal to every vector in span$\{\mathbf{u}_1, \mathbf{u}_2, \ldots, \mathbf{u}_r\}$.

20. Let $\{\mathbf{v}_1, \mathbf{v}_2, \ldots, \mathbf{v}_r\}$ be a basis for an inner product space V. Show that the zero vector is the only vector in V that is orthogonal to all of the basis vectors.

21. Let $\{\mathbf{w}_1, \mathbf{w}_2, \ldots, \mathbf{w}_k\}$ be a basis for a subspace W of V. Show that $W^\perp$ consists of all vectors in V that are orthogonal to every basis vector.

22. Prove the following generalization of Theorem 6.2.3: If $\mathbf{v}_1, \mathbf{v}_2, \ldots, \mathbf{v}_r$ are pairwise orthogonal vectors in an inner product space V, then

$$\|\mathbf{v}_1 + \mathbf{v}_2 + \cdots + \mathbf{v}_r\|^2 = \|\mathbf{v}_1\|^2 + \|\mathbf{v}_2\|^2 + \cdots + \|\mathbf{v}_r\|^2$$

23. Prove: If $\mathbf{u}$ and $\mathbf{v}$ are $n \times 1$ matrices and A is an $n \times n$ matrix, then

$$(\mathbf{v}^T A^T A \mathbf{u})^2 \leq (\mathbf{u}^T A^T A \mathbf{u})(\mathbf{v}^T A^T A \mathbf{v})$$

24. Use the Cauchy–Schwarz inequality to prove that for all real values of a, b, and θ,

$$(a \cos \theta + b \sin \theta)^2 \leq a^2 + b^2$$

25. Prove: If $w_1, w_2, \ldots, w_n$ are positive real numbers, and if $\mathbf{u} = (u_1, u_2, \ldots, u_n)$ and $\mathbf{v} = (v_1, v_2, \ldots, v_n)$ are any two vectors in R^n, then

$$|w_1 u_1 v_1 + w_2 u_2 v_2 + \cdots + w_n u_n v_n|$$
$$\leq (w_1 u_1^2 + w_2 u_2^2 + \cdots + w_n u_n^2)^{1/2} (w_1 v_1^2 + w_2 v_2^2 + \cdots + w_n v_n^2)^{1/2}$$

26. Show that equality holds in the Cauchy–Schwarz inequality if and only if $\mathbf{u}$ and $\mathbf{v}$ are linearly dependent.

27. Use vector methods to prove that a triangle that is inscribed in a circle so that it has a diameter for a side must be a right triangle. [*Hint:* Express the vectors $\overrightarrow{AB}$ and $\overrightarrow{BC}$ in the accompanying figure in terms of $\mathbf{u}$ and $\mathbf{v}$.]

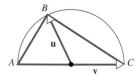

◀ Figure Ex-27

28. As illustrated in the accompanying figure, the vectors $\mathbf{u} = (1, \sqrt{3})$ and $\mathbf{v} = (-1, \sqrt{3})$ have norm 2 and an angle of 60° between them relative to the Euclidean inner product. Find a weighted Euclidean inner product with respect to which $\mathbf{u}$ and $\mathbf{v}$ are orthogonal unit vectors.

◀ Figure Ex-28

29. (*Calculus required*) Let $f(x)$ and $g(x)$ be continuous functions on $[0, 1]$. Prove:

(a) $\left[\int_0^1 f(x) g(x) \, dx \right]^2 \leq \left[\int_0^1 f^2(x) \, dx \right] \left[\int_0^1 g^2(x) \, dx \right]$

(b) $\left[\int_0^1 [f(x) + g(x)]^2 \, dx \right]^{1/2} \leq \left[\int_0^1 f^2(x) \, dx \right]^{1/2}$
$$+ \left[\int_0^1 g^2(x) \, dx \right]^{1/2}$$

[*Hint:* Use the Cauchy–Schwarz inequality.]

30. (*Calculus required*) Let $C[0, \pi]$ have the inner product

$$\langle \mathbf{f}, \mathbf{g} \rangle = \int_0^\pi f(x) g(x) \, dx$$

and let $\mathbf{f}_n = \cos nx$ $(n = 0, 1, 2, \ldots)$. Show that if $k \neq l$, then $\mathbf{f}_k$ and $\mathbf{f}_l$ are orthogonal vectors.

31. (a) Let W be the line $y = x$ in an xy-coordinate system in R^2. Describe the subspace $W^\perp$.

(b) Let W be the y-axis in an xyz-coordinate system in R^3. Describe the subspace $W^\perp$.

(c) Let W be the yz-plane of an xyz-coordinate system in R^3. Describe the subspace $W^\perp$.

32. Prove that Formula (4) holds for all nonzero vectors $\mathbf{u}$ and $\mathbf{v}$ in an inner product space V.

True-False Exercises

In parts (a)–(f) determine whether the statement is true or false, and justify your answer.

(a) If $\mathbf{u}$ is orthogonal to every vector of a subspace W, then $\mathbf{u} = \mathbf{0}$.

(b) If $\mathbf{u}$ is a vector in both W and $W^\perp$, then $\mathbf{u} = \mathbf{0}$.

(c) If $\mathbf{u}$ and $\mathbf{v}$ are vectors in $W^\perp$, then $\mathbf{u} + \mathbf{v}$ is in $W^\perp$.

(d) If $\mathbf{u}$ is a vector in $W^\perp$ and k is a real number, then $k\mathbf{u}$ is in $W^\perp$.

(e) If $\mathbf{u}$ and $\mathbf{v}$ are orthogonal, then $|\langle \mathbf{u}, \mathbf{v} \rangle| = \|\mathbf{u}\| \|\mathbf{v}\|$.

(f) If $\mathbf{u}$ and $\mathbf{v}$ are orthogonal, then $\|\mathbf{u} + \mathbf{v}\| = \|\mathbf{u}\| + \|\mathbf{v}\|$.

6.3 Gram–Schmidt Process; *QR*-Decomposition

In many problems involving vector spaces, the problem solver is free to choose any basis for the vector space that seems appropriate. In inner product spaces, the solution of a problem is often greatly simplified by choosing a basis in which the vectors are orthogonal to one another. In this section we will show how such bases can be obtained.

Orthogonal and Orthonormal Sets

Recall from Section 6.2 that two vectors in an inner product space are said to be *orthogonal* if their inner product is zero. The following definition extends the notion of orthogonality to sets of vectors in an inner product space.

DEFINITION 1 A set of two or more vectors in a real inner product space is said to be *orthogonal* if all pairs of distinct vectors in the set are orthogonal. An orthogonal set in which each vector has norm 1 is said to be *orthonormal*.

▶ **EXAMPLE 1 An Orthogonal Set in R^3**

Let

$$\mathbf{u}_1 = (0, 1, 0), \quad \mathbf{u}_2 = (1, 0, 1), \quad \mathbf{u}_3 = (1, 0, -1)$$

and assume that R^3 has the Euclidean inner product. It follows that the set of vectors $S = \{\mathbf{u}_1, \mathbf{u}_2, \mathbf{u}_3\}$ is orthogonal since $\langle \mathbf{u}_1, \mathbf{u}_2 \rangle = \langle \mathbf{u}_1, \mathbf{u}_3 \rangle = \langle \mathbf{u}_2, \mathbf{u}_3 \rangle = 0.$ ◀

If $\mathbf{v}$ is a nonzero vector in an inner product space, then it follows from Theorem 6.1.1*b* with $k = \|\mathbf{v}\|$ that

$$\left\| \frac{1}{\|\mathbf{v}\|} \mathbf{v} \right\| = \left| \frac{1}{\|\mathbf{v}\|} \right| \|\mathbf{v}\| = \frac{1}{\|\mathbf{v}\|} \|\mathbf{v}\| = 1$$

from which we see that multiplying a nonzero vector by the reciprocal of its norm produces a vector of norm 1. This process is called *normalizing* $\mathbf{v}$. It follows that *any orthogonal set of nonzero vectors can be converted to an orthonormal set by normalizing each of its vectors.*

▶ **EXAMPLE 2 Constructing an Orthonormal Set**

The Euclidean norms of the vectors in Example 1 are

$$\|\mathbf{u}_1\| = 1, \quad \|\mathbf{u}_2\| = \sqrt{2}, \quad \|\mathbf{u}_3\| = \sqrt{2}$$

Consequently, normalizing $\mathbf{u}_1$, $\mathbf{u}_2$, and $\mathbf{u}_3$ yields

$$\mathbf{v}_1 = \frac{\mathbf{u}_1}{\|\mathbf{u}_1\|} = (0, 1, 0), \quad \mathbf{v}_2 = \frac{\mathbf{u}_2}{\|\mathbf{u}_2\|} = \left(\frac{1}{\sqrt{2}}, 0, \frac{1}{\sqrt{2}} \right),$$

$$\mathbf{v}_3 = \frac{\mathbf{u}_3}{\|\mathbf{u}_3\|} = \left(\frac{1}{\sqrt{2}}, 0, -\frac{1}{\sqrt{2}} \right)$$

We leave it for you to verify that the set $S = \{\mathbf{v}_1, \mathbf{v}_2, \mathbf{v}_3\}$ is orthonormal by showing that

$$\langle \mathbf{v}_1, \mathbf{v}_2 \rangle = \langle \mathbf{v}_1, \mathbf{v}_3 \rangle = \langle \mathbf{v}_2, \mathbf{v}_3 \rangle = 0 \quad \text{and} \quad \|\mathbf{v}_1\| = \|\mathbf{v}_2\| = \|\mathbf{v}_3\| = 1 \quad ◀$$

In R^2 any two nonzero perpendicular vectors are linearly independent because neither is a scalar multiple of the other; and in R^3 any three nonzero mutually perpendicular vectors are linearly independent because no one lies in the plane of the other two (and hence is not expressible as a linear combination of the other two). The following theorem generalizes these observations.

THEOREM 6.3.1 *If $S = \{\mathbf{v}_1, \mathbf{v}_2, \ldots, \mathbf{v}_n\}$ is an orthogonal set of nonzero vectors in an inner product space, then S is linearly independent.*

Proof Assume that

$$k_1 \mathbf{v}_1 + k_2 \mathbf{v}_2 + \cdots + k_n \mathbf{v}_n = \mathbf{0} \tag{1}$$

To demonstrate that $S = \{\mathbf{v}_1, \mathbf{v}_2, \ldots, \mathbf{v}_n\}$ is linearly independent, we must prove that $k_1 = k_2 = \cdots = k_n = 0$.

For each $\mathbf{v}_i$ in S, it follows from (1) that

$$\langle k_1\mathbf{v}_1 + k_2\mathbf{v}_2 + \cdots + k_n\mathbf{v}_n, \mathbf{v}_i \rangle = \langle \mathbf{0}, \mathbf{v}_i \rangle = 0$$

or, equivalently,

$$k_1\langle \mathbf{v}_1, \mathbf{v}_i \rangle + k_2\langle \mathbf{v}_2, \mathbf{v}_i \rangle + \cdots + k_n\langle \mathbf{v}_n, \mathbf{v}_i \rangle = 0$$

From the orthogonality of S it follows that $\langle \mathbf{v}_j, \mathbf{v}_i \rangle = 0$ when $j \neq i$, so this equation reduces to

$$k_i\langle \mathbf{v}_i, \mathbf{v}_i \rangle = 0$$

Since the vectors in S are assumed to be nonzero, it follows from the positivity axiom for inner products that $\langle \mathbf{v}_i, \mathbf{v}_i \rangle \neq 0$. Thus, the preceding equation implies that each k_i in Equation (1) is zero, which is what we wanted to prove. ◄

Since an orthonormal set is orthogonal, and since its vectors are nonzero (norm 1), it follows from Theorem 6.3.1 that every *orthonormal* set is linearly independent.

In an inner product space, a basis consisting of orthonormal vectors is called an *orthonormal basis*, and a basis consisting of orthogonal vectors is called an *orthogonal basis*. A familiar example of an orthonormal basis is the standard basis for R^n with the Euclidean inner product:

$$\mathbf{e}_1 = (1, 0, 0, \ldots, 0), \quad \mathbf{e}_2 = (0, 1, 0, \ldots, 0), \ldots, \quad \mathbf{e}_n = (0, 0, 0, \ldots, 1)$$

▶ **EXAMPLE 3 An Orthonormal Basis**

In Example 2 we showed that the vectors

$$\mathbf{v}_1 = (0, 1, 0), \quad \mathbf{v}_2 = \left(\frac{1}{\sqrt{2}}, 0, \frac{1}{\sqrt{2}} \right), \quad \text{and} \quad \mathbf{v}_3 = \left(\frac{1}{\sqrt{2}}, 0, -\frac{1}{\sqrt{2}} \right)$$

form an orthonormal set with respect to the Euclidean inner product on R^3. By Theorem 6.3.1, these vectors form a linearly independent set, and since R^3 is three-dimensional, it follows from Theorem 4.5.4 that $S = \{\mathbf{v}_1, \mathbf{v}_2, \mathbf{v}_3\}$ is an orthonormal basis for R^3. ◄

Coordinates Relative to Orthonormal Bases

One way to express a vector $\mathbf{u}$ as a linear combination of basis vectors

$$S = \{\mathbf{v}_1, \mathbf{v}_2, \ldots, \mathbf{v}_n\}$$

is to convert the vector equation

$$\mathbf{u} = c_1\mathbf{v}_1 + c_2\mathbf{v}_2 + \cdots + c_n\mathbf{v}_n$$

to a linear system and solve for the coefficients $c_1, c_2, \ldots, c_n$. However, if the basis happens to be orthogonal or orthonormal, then the following theorem shows that the coefficients can be obtained more simply by computing appropriate inner products.

THEOREM 6.3.2

(a) *If $S = \{\mathbf{v}_1, \mathbf{v}_2, \ldots, \mathbf{v}_n\}$ is an orthogonal basis for an inner product space V, and if $\mathbf{u}$ is any vector in V, then*

$$\mathbf{u} = \frac{\langle \mathbf{u}, \mathbf{v}_1 \rangle}{\|\mathbf{v}_1\|^2}\mathbf{v}_1 + \frac{\langle \mathbf{u}, \mathbf{v}_2 \rangle}{\|\mathbf{v}_2\|^2}\mathbf{v}_2 + \cdots + \frac{\langle \mathbf{u}, \mathbf{v}_n \rangle}{\|\mathbf{v}_n\|^2}\mathbf{v}_n \tag{2}$$

(b) *If $S = \{\mathbf{v}_1, \mathbf{v}_2, \ldots, \mathbf{v}_n\}$ is an orthonormal basis for an inner product space V, and if $\mathbf{u}$ is any vector in V, then*

$$\mathbf{u} = \langle \mathbf{u}, \mathbf{v}_1 \rangle\mathbf{v}_1 + \langle \mathbf{u}, \mathbf{v}_2 \rangle\mathbf{v}_2 + \cdots + \langle \mathbf{u}, \mathbf{v}_n \rangle\mathbf{v}_n \tag{3}$$

Proof (a)   Since $S = \{\mathbf{v}_1, \mathbf{v}_2, \ldots, \mathbf{v}_n\}$ is a basis for V, every vector $\mathbf{u}$ in V can be expressed in the form

$$\mathbf{u} = c_1 \mathbf{v}_1 + c_2 \mathbf{v}_2 + \cdots + c_n \mathbf{v}_n$$

We will complete the proof by showing that

$$c_i = \frac{\langle \mathbf{u}, \mathbf{v}_i \rangle}{\|\mathbf{v}_i\|^2} \tag{4}$$

for $i = 1, 2, \ldots, n$. To do this, observe first that

$$\langle \mathbf{u}, \mathbf{v}_i \rangle = \langle c_1 \mathbf{v}_1 + c_2 \mathbf{v}_2 + \cdots + c_n \mathbf{v}_n, \mathbf{v}_i \rangle$$
$$= c_1 \langle \mathbf{v}_1, \mathbf{v}_i \rangle + c_2 \langle \mathbf{v}_2, \mathbf{v}_i \rangle + \cdots + c_n \langle \mathbf{v}_n, \mathbf{v}_i \rangle$$

Since S is an orthogonal set, all of the inner products in the last equality are zero except the ith, so we have

$$\langle \mathbf{u}, \mathbf{v}_i \rangle = c_i \langle \mathbf{v}_i, \mathbf{v}_i \rangle = c_i \|\mathbf{v}_i\|^2$$

Solving this equation for c_i yields (4), which completes the proof.

Proof (b)   In this case, $\|\mathbf{v}_1\| = \|\mathbf{v}_2\| = \cdots = \|\mathbf{v}_n\| = 1$, so Formula (2) simplifies to Formula (3). ◄

Using the terminology and notation from Definition 2 of Section 4.4, it follows from Theorem 6.3.2 that the coordinate vector of a vector $\mathbf{u}$ in V relative to an orthogonal basis $S = \{\mathbf{v}_1, \mathbf{v}_2, \ldots, \mathbf{v}_n\}$ is

$$(\mathbf{u})_S = \left(\frac{\langle \mathbf{u}, \mathbf{v}_1 \rangle}{\|\mathbf{v}_1\|^2}, \frac{\langle \mathbf{u}, \mathbf{v}_2 \rangle}{\|\mathbf{v}_2\|^2}, \ldots, \frac{\langle \mathbf{u}, \mathbf{v}_n \rangle}{\|\mathbf{v}_n\|^2} \right) \tag{5}$$

and relative to an orthonormal basis $S = \{\mathbf{v}_1, \mathbf{v}_2, \ldots, \mathbf{v}_n\}$ is

$$(\mathbf{u})_S = (\langle \mathbf{u}, \mathbf{v}_1 \rangle, \langle \mathbf{u}, \mathbf{v}_2 \rangle, \ldots, \langle \mathbf{u}, \mathbf{v}_n \rangle) \tag{6}$$

▶ **EXAMPLE 4 A Coordinate Vector Relative to an Orthonormal Basis**

Let

$$\mathbf{v}_1 = (0, 1, 0), \quad \mathbf{v}_2 = \left(-\tfrac{4}{5}, 0, \tfrac{3}{5} \right), \quad \mathbf{v}_3 = \left(\tfrac{3}{5}, 0, \tfrac{4}{5} \right)$$

It is easy to check that $S = \{\mathbf{v}_1, \mathbf{v}_2, \mathbf{v}_3\}$ is an orthonormal basis for R^3 with the Euclidean inner product. Express the vector $\mathbf{u} = (1, 1, 1)$ as a linear combination of the vectors in S, and find the coordinate vector $(\mathbf{u})_S$.

Solution   We leave it for you to verify that

$$\langle \mathbf{u}, \mathbf{v}_1 \rangle = 1, \quad \langle \mathbf{u}, \mathbf{v}_2 \rangle = -\tfrac{1}{5}, \quad \text{and} \quad \langle \mathbf{u}, \mathbf{v}_3 \rangle = \tfrac{7}{5}$$

Therefore, by Theorem 6.3.2 we have

$$\mathbf{u} = \mathbf{v}_1 - \tfrac{1}{5} \mathbf{v}_2 + \tfrac{7}{5} \mathbf{v}_3$$

that is,

$$(1, 1, 1) = (0, 1, 0) - \tfrac{1}{5}\left(-\tfrac{4}{5}, 0, \tfrac{3}{5} \right) + \tfrac{7}{5}\left(\tfrac{3}{5}, 0, \tfrac{4}{5} \right)$$

Thus, the coordinate vector of $\mathbf{u}$ relative to S is

$$(\mathbf{u})_S = (\langle \mathbf{u}, \mathbf{v}_1 \rangle, \langle \mathbf{u}, \mathbf{v}_2 \rangle, \langle \mathbf{u}, \mathbf{v}_3 \rangle) = \left(1, -\tfrac{1}{5}, \tfrac{7}{5} \right)$$

▶ **EXAMPLE 5** **An Orthonormal Basis from an Orthogonal Basis**

(a)   Show that the vectors

$$\mathbf{w}_1 = (0, 2, 0), \quad \mathbf{w}_2 = (3, 0, 3), \quad \mathbf{w}_3 = (-4, 0, 4)$$

form an orthogonal basis for R^3 with the Euclidean inner product, and use that basis to find an orthonormal basis by normalizing each vector.

(b)   Express the vector $\mathbf{u} = (1, 2, 4)$ as a linear combination of the orthonormal basis vectors obtained in part (a).

Solution (a) The given vectors form an orthogonal set since

$$\langle \mathbf{w}_1, \mathbf{w}_2 \rangle = 0, \quad \langle \mathbf{w}_1, \mathbf{w}_3 \rangle = 0, \quad \langle \mathbf{w}_2, \mathbf{w}_3 \rangle = 0$$

It follows from Theorem 6.3.1 that these vectors are linearly independent and hence form a basis for R^3 by Theorem 4.5.4. We leave it for you to calculate the norms of $\mathbf{w}_1$, $\mathbf{w}_2$, and $\mathbf{w}_3$ and then obtain the orthonormal basis

$$\mathbf{v}_1 = \frac{\mathbf{w}_1}{\|\mathbf{w}_1\|} = (0, 1, 0), \quad \mathbf{v}_2 = \frac{\mathbf{w}_2}{\|\mathbf{w}_2\|} = \left(\frac{1}{\sqrt{2}}, 0, \frac{1}{\sqrt{2}} \right),$$

$$\mathbf{v}_3 = \frac{\mathbf{w}_3}{\|\mathbf{w}_3\|} = \left(-\frac{1}{\sqrt{2}}, 0, \frac{1}{\sqrt{2}} \right)$$

Solution (b) It follows from Formula (3) that

$$\mathbf{u} = \langle \mathbf{u}, \mathbf{v}_1 \rangle \mathbf{v}_1 + \langle \mathbf{u}, \mathbf{v}_2 \rangle \mathbf{v}_2 + \langle \mathbf{u}, \mathbf{v}_3 \rangle \mathbf{v}_3$$

We leave it for you to confirm that

$$\langle \mathbf{u}, \mathbf{v}_1 \rangle = (1, 2, 4) \cdot (0, 1, 0) = 2$$

$$\langle \mathbf{u}, \mathbf{v}_2 \rangle = (1, 2, 4) \cdot \left(\frac{1}{\sqrt{2}}, 0, \frac{1}{\sqrt{2}} \right) = \frac{5}{\sqrt{2}}$$

$$\langle \mathbf{u}, \mathbf{v}_3 \rangle = (1, 2, 4) \cdot \left(-\frac{1}{\sqrt{2}}, 0, \frac{1}{\sqrt{2}} \right) = \frac{3}{\sqrt{2}}$$

and hence that

$$(1, 2, 4) = 2(0, 1, 0) + \frac{5}{\sqrt{2}} \left(\frac{1}{\sqrt{2}}, 0, \frac{1}{\sqrt{2}} \right) + \frac{3}{\sqrt{2}} \left(-\frac{1}{\sqrt{2}}, 0, \frac{1}{\sqrt{2}} \right) \quad ◀$$

Orthogonal Projections   Many applied problems are best solved by working with orthogonal or orthonormal basis vectors. Such bases are typically found by starting with some simple basis (say a standard basis) and then converting that basis into an orthogonal or orthonormal basis. To explain exactly how that is done will require some preliminary ideas about orthogonal projections.

In Section 3.3 we proved a result called the *Projection Theorem* (see Theorem 3.3.2) which dealt with the problem of decomposing a vector $\mathbf{u}$ in R^n into a sum of two terms, $\mathbf{w}_1$ and $\mathbf{w}_2$, in which $\mathbf{w}_1$ is the orthogonal projection of $\mathbf{u}$ on some nonzero vector $\mathbf{a}$ and $\mathbf{w}_2$ is orthogonal to $\mathbf{w}_1$ (Figure 3.3.2). That result is a special case of the following more general theorem.

THEOREM 6.3.3 **Projection Theorem**

If W is a finite-dimensional subspace of an inner product space V, then every vector $\mathbf{u}$ in V can be expressed in exactly one way as

$$\mathbf{u} = \mathbf{w}_1 + \mathbf{w}_2 \tag{7}$$

where $\mathbf{w}_1$ is in W and $\mathbf{w}_2$ is in $W^{\perp}$.

The vectors $\mathbf{w}_1$ and $\mathbf{w}_2$ in Formula (7) are commonly denoted by

$$\mathbf{w}_1 = \text{proj}_W \mathbf{u} \quad \text{and} \quad \mathbf{w}_2 = \text{proj}_{W^\perp} \mathbf{u} \tag{8}$$

They are called the ***orthogonal projection of* u *on W*** and the ***orthogonal projection of* u *on* $W^\perp$**, respectively. The vector $\mathbf{w}_2$ is also called the ***component of* u *orthogonal to*** *W*. Using the notation in (8), Formula (7) can be expressed as

$$\mathbf{u} = \text{proj}_W \mathbf{u} + \text{proj}_{W^\perp} \mathbf{u} \tag{9}$$

▲ **Figure 6.3.1**

(Figure 6.3.1). Moreover, since $\text{proj}_{W^\perp} \mathbf{u} = \mathbf{u} - \text{proj}_W \mathbf{u}$, we can also express Formula (9) as

$$\mathbf{u} = \text{proj}_W \mathbf{u} + (\mathbf{u} - \text{proj}_W \mathbf{u}) \tag{10}$$

The following theorem provides formulas for calculating orthogonal projections.

THEOREM 6.3.4 *Let W be a finite-dimensional subspace of an inner product space V.*

(a) *If $\{\mathbf{v}_1, \mathbf{v}_2, \ldots, \mathbf{v}_r\}$ is an orthogonal basis for W, and* **u** *is any vector in V, then*

$$\text{proj}_W \mathbf{u} = \frac{\langle \mathbf{u}, \mathbf{v}_1 \rangle}{\|\mathbf{v}_1\|^2} \mathbf{v}_1 + \frac{\langle \mathbf{u}, \mathbf{v}_2 \rangle}{\|\mathbf{v}_2\|^2} \mathbf{v}_2 + \cdots + \frac{\langle \mathbf{u}, \mathbf{v}_r \rangle}{\|\mathbf{v}_r\|^2} \mathbf{v}_r \tag{11}$$

(b) *If $\{\mathbf{v}_1, \mathbf{v}_2, \ldots, \mathbf{v}_r\}$ is an orthonormal basis for W, and* **u** *is any vector in V, then*

$$\text{proj}_W \mathbf{u} = \langle \mathbf{u}, \mathbf{v}_1 \rangle \mathbf{v}_1 + \langle \mathbf{u}, \mathbf{v}_2 \rangle \mathbf{v}_2 + \cdots + \langle \mathbf{u}, \mathbf{v}_r \rangle \mathbf{v}_r \tag{12}$$

Proof (a)   It follows from Theorem 6.3.3 that the vector **u** can be expressed in the form $\mathbf{u} = \mathbf{w}_1 + \mathbf{w}_2$, where $\mathbf{w}_1 = \text{proj}_W \mathbf{u}$ is in *W* and $\mathbf{w}_2$ is in $W^\perp$; and it follows from Theorem 6.3.2 that the component $\text{proj}_W \mathbf{u} = \mathbf{w}_1$ can be expressed in terms of the basis vectors for *W* as

$$\text{proj}_W \mathbf{u} = \mathbf{w}_1 = \frac{\langle \mathbf{w}_1, \mathbf{v}_1 \rangle}{\|\mathbf{v}_1\|^2} \mathbf{v}_1 + \frac{\langle \mathbf{w}_1, \mathbf{v}_2 \rangle}{\|\mathbf{v}_2\|^2} \mathbf{v}_2 + \cdots + \frac{\langle \mathbf{w}_1, \mathbf{v}_r \rangle}{\|\mathbf{v}_r\|^2} \mathbf{v}_r \tag{13}$$

Since $\mathbf{w}_2$ is orthogonal to *W*, it follows that

$$\langle \mathbf{w}_2, \mathbf{v}_1 \rangle = \langle \mathbf{w}_2, \mathbf{v}_2 \rangle = \cdots = \langle \mathbf{w}_2, \mathbf{v}_r \rangle = 0$$

so we can rewrite (13) as

$$\text{proj}_W \mathbf{u} = \mathbf{w}_1 = \frac{\langle \mathbf{w}_1 + \mathbf{w}_2, \mathbf{v}_1 \rangle}{\|\mathbf{v}_1\|^2} \mathbf{v}_1 + \frac{\langle \mathbf{w}_1 + \mathbf{w}_2, \mathbf{v}_2 \rangle}{\|\mathbf{v}_2\|^2} \mathbf{v}_2 + \cdots + \frac{\langle \mathbf{w}_1 + \mathbf{w}_2, \mathbf{v}_r \rangle}{\|\mathbf{v}_r\|^2} \mathbf{v}_r$$

or, equivalently, as

$$\text{proj}_W \mathbf{u} = \mathbf{w}_1 = \frac{\langle \mathbf{u}, \mathbf{v}_1 \rangle}{\|\mathbf{v}_1\|^2} \mathbf{v}_1 + \frac{\langle \mathbf{u}, \mathbf{v}_2 \rangle}{\|\mathbf{v}_2\|^2} \mathbf{v}_2 + \cdots + \frac{\langle \mathbf{u}, \mathbf{v}_r \rangle}{\|\mathbf{v}_r\|^2} \mathbf{v}_r$$

Proof (b)   In this case, $\|\mathbf{v}_1\| = \|\mathbf{v}_2\| = \cdots = \|\mathbf{v}_r\| = 1$, so Formula (13) simplifies to Formula (12). ◀

▶ **EXAMPLE 6 Calculating Projections**

Let R^3 have the Euclidean inner product, and let *W* be the subspace spanned by the orthonormal vectors $\mathbf{v}_1 = (0, 1, 0)$ and $\mathbf{v}_2 = \left(-\frac{4}{5}, 0, \frac{3}{5}\right)$. From Formula (12) the orthogonal projection of $\mathbf{u} = (1, 1, 1)$ on *W* is

$$\begin{aligned}
\text{proj}_W \mathbf{u} &= \langle \mathbf{u}, \mathbf{v}_1 \rangle \mathbf{v}_1 + \langle \mathbf{u}, \mathbf{v}_2 \rangle \mathbf{v}_2 \\
&= (1)(0, 1, 0) + \left(-\tfrac{1}{5}\right)\left(-\tfrac{4}{5}, 0, \tfrac{3}{5}\right) \\
&= \left(\tfrac{4}{25}, 1, -\tfrac{3}{25}\right)
\end{aligned}$$

The component of $\mathbf{u}$ orthogonal to W is

$$\text{proj}_{W^\perp}\,\mathbf{u} = \mathbf{u} - \text{proj}_W\,\mathbf{u} = (1, 1, 1) - \left(\tfrac{4}{25}, 1, -\tfrac{3}{25}\right) = \left(\tfrac{21}{25}, 0, \tfrac{28}{25}\right)$$

Observe that $\text{proj}_{W^\perp}\,\mathbf{u}$ is orthogonal to both $\mathbf{v}_1$ and $\mathbf{v}_2$, so this vector is orthogonal to each vector in the space W spanned by $\mathbf{v}_1$ and $\mathbf{v}_2$, as it should be. ◀

A Geometric Interpretation of Orthogonal Projections

If W is a one-dimensional subspace of an inner product space V, say span$\{\mathbf{a}\}$, then Formula (11) has only the one term

$$\text{proj}_W\,\mathbf{u} = \frac{\langle \mathbf{u}, \mathbf{a} \rangle}{\|\mathbf{a}\|^2}\mathbf{a}$$

In the special case where V is R^3 with the Euclidean inner product, this is exactly Formula (10) of Section 3.3 for the orthogonal projection of $\mathbf{u}$ along $\mathbf{a}$. This suggests that we can think of (11) as the sum of orthogonal projections on "axes" determined by the basis vectors for the subspace W (Figure 6.3.2).

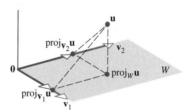

▶ **Figure 6.3.2**

The Gram–Schmidt Process

We have seen that orthonormal bases exhibit a variety of useful properties. Our next theorem, which is the main result in this section, shows that every nonzero finite-dimensional vector space has an orthonormal basis. The proof of this result is extremely important, since it provides an algorithm, or method, for converting an arbitrary basis into an orthonormal basis.

> **THEOREM 6.3.5** *Every nonzero finite-dimensional inner product space has an orthonormal basis.*

Proof Let W be any nonzero finite-dimensional subspace of an inner product space, and suppose that $\{\mathbf{u}_1, \mathbf{u}_2, \ldots, \mathbf{u}_r\}$ is any basis for W. It suffices to show that W has an orthogonal basis, since the vectors in that basis can be normalized to obtain an orthonormal basis. The following sequence of steps will produce an orthogonal basis $\{\mathbf{v}_1, \mathbf{v}_2, \ldots, \mathbf{v}_r\}$ for W: for EXSL

Step 1. Let $\mathbf{v}_1 = \mathbf{u}_1$.

Step 2. As illustrated in Figure 6.3.3, we can obtain a vector $\mathbf{v}_2$ that is orthogonal to $\mathbf{v}_1$ by computing the component of $\mathbf{u}_2$ that is orthogonal to the space W_1 spanned by $\mathbf{v}_1$. Using Formula (11) to perform this computation we obtain

$$\mathbf{v}_2 = \mathbf{u}_2 - \text{proj}_{W_1}\,\mathbf{u}_2 = \mathbf{u}_2 - \frac{\langle \mathbf{u}_2, \mathbf{v}_1 \rangle}{\|\mathbf{v}_1\|^2}\mathbf{v}_1$$

$\mathbf{v}_2 = \mathbf{u}_2 - \text{proj}_{W_1}\,\mathbf{u}_2$

▲ **Figure 6.3.3**

Of course, if $\mathbf{v}_2 = \mathbf{0}$, then $\mathbf{v}_2$ is not a basis vector. But this cannot happen, since it would then follow from the above formula for $\mathbf{v}_2$ that

$$\mathbf{u}_2 = \frac{\langle \mathbf{u}_2, \mathbf{v}_1 \rangle}{\|\mathbf{v}_1\|^2}\mathbf{v}_1 = \frac{\langle \mathbf{u}_2, \mathbf{v}_1 \rangle}{\|\mathbf{u}_1\|^2}\mathbf{u}_1$$

which implies that $\mathbf{u}_2$ is a multiple of $\mathbf{u}_1$, contradicting the linear independence of the basis $S = \{\mathbf{u}_1, \mathbf{u}_2, \ldots, \mathbf{u}_n\}$.

Step 3. To construct a vector $\mathbf{v}_3$ that is orthogonal to both $\mathbf{v}_1$ and $\mathbf{v}_2$, we compute the

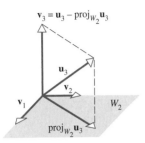

$\mathbf{v}_3 = \mathbf{u}_3 - \text{proj}_{W_2}\mathbf{u}_3$

▲ **Figure 6.3.4**

Step 3. To construct a vector $\mathbf{v}_3$ that is orthogonal to both $\mathbf{v}_1$ and $\mathbf{v}_2$, we compute the component of $\mathbf{u}_3$ orthogonal to the space W_2 spanned by $\mathbf{v}_1$ and $\mathbf{v}_2$ (Figure 6.3.4). Using Formula (11) to perform this computation we obtain

$$\mathbf{v}_3 = \mathbf{u}_3 - \text{proj}_{W_2}\mathbf{u}_3 = \mathbf{u}_3 - \frac{\langle \mathbf{u}_3, \mathbf{v}_1 \rangle}{\|\mathbf{v}_1\|^2}\mathbf{v}_1 - \frac{\langle \mathbf{u}_3, \mathbf{v}_2 \rangle}{\|\mathbf{v}_2\|^2}\mathbf{v}_2$$

As in Step 2, the linear independence of $\{\mathbf{u}_1, \mathbf{u}_2, \ldots, \mathbf{u}_n\}$ ensures that $\mathbf{v}_3 \neq \mathbf{0}$. We leave the details for you.

Step 4. To determine a vector $\mathbf{v}_4$ that is orthogonal to $\mathbf{v}_1$, $\mathbf{v}_2$, and $\mathbf{v}_3$, we compute the component of $\mathbf{u}_4$ orthogonal to the space W_3 spanned by $\mathbf{v}_1$, $\mathbf{v}_2$, and $\mathbf{v}_3$. From (11),

$$\mathbf{v}_4 = \mathbf{u}_4 - \text{proj}_{W_3}\mathbf{u}_4 = \mathbf{u}_4 - \frac{\langle \mathbf{u}_4, \mathbf{v}_1 \rangle}{\|\mathbf{v}_1\|^2}\mathbf{v}_1 - \frac{\langle \mathbf{u}_4, \mathbf{v}_2 \rangle}{\|\mathbf{v}_2\|^2}\mathbf{v}_2 - \frac{\langle \mathbf{u}_4, \mathbf{v}_3 \rangle}{\|\mathbf{v}_3\|^2}\mathbf{v}_3$$

Continuing in this way we will produce an orthogonal set of vectors $\{\mathbf{v}_1, \mathbf{v}_2, \ldots, \mathbf{v}_r\}$ after r steps. Since orthogonal sets are linearly independent, this set will be an orthogonal basis for the r-dimensional space W. By normalizing these basis vectors we can obtain an orthonormal basis. ◀

The step-by-step construction of an orthogonal (or orthonormal) basis given in the foregoing proof is called the ***Gram–Schmidt process***. For reference, we provide the following summary of the steps.

The Gram–Schmidt Process

To convert a basis $\{\mathbf{u}_1, \mathbf{u}_2, \ldots, \mathbf{u}_r\}$ into an orthogonal basis $\{\mathbf{v}_1, \mathbf{v}_2, \ldots, \mathbf{v}_r\}$, perform the following computations:

Step 1. $\mathbf{v}_1 = \mathbf{u}_1$

Step 2. $\mathbf{v}_2 = \mathbf{u}_2 - \dfrac{\langle \mathbf{u}_2, \mathbf{v}_1 \rangle}{\|\mathbf{v}_1\|^2}\mathbf{v}_1$

Step 3. $\mathbf{v}_3 = \mathbf{u}_3 - \dfrac{\langle \mathbf{u}_3, \mathbf{v}_1 \rangle}{\|\mathbf{v}_1\|^2}\mathbf{v}_1 - \dfrac{\langle \mathbf{u}_3, \mathbf{v}_2 \rangle}{\|\mathbf{v}_2\|^2}\mathbf{v}_2$

Step 4. $\mathbf{v}_4 = \mathbf{u}_4 - \dfrac{\langle \mathbf{u}_4, \mathbf{v}_1 \rangle}{\|\mathbf{v}_1\|^2}\mathbf{v}_1 - \dfrac{\langle \mathbf{u}_4, \mathbf{v}_2 \rangle}{\|\mathbf{v}_2\|^2}\mathbf{v}_2 - \dfrac{\langle \mathbf{u}_4, \mathbf{v}_3 \rangle}{\|\mathbf{v}_3\|^2}\mathbf{v}_3$
  ⋮

(continue for r steps)

Optional Step. To convert the orthogonal basis into an orthonormal basis $\{\mathbf{q}_1, \mathbf{q}_2, \ldots, \mathbf{q}_r\}$, normalize the orthogonal basis vectors.

▶ **EXAMPLE 7 Using the Gram–Schmidt Process**

Assume that the vector space R^3 has the Euclidean inner product. Apply the Gram–Schmidt process to transform the basis vectors

$$\mathbf{u}_1 = (1, 1, 1), \quad \mathbf{u}_2 = (0, 1, 1), \quad \mathbf{u}_3 = (0, 0, 1)$$

into an orthogonal basis $\{\mathbf{v}_1, \mathbf{v}_2, \mathbf{v}_3\}$, and then normalize the orthogonal basis vectors to obtain an orthonormal basis $\{\mathbf{q}_1, \mathbf{q}_2, \mathbf{q}_3\}$.

Solution

Step 1. $\mathbf{v}_1 = \mathbf{u}_1 = (1, 1, 1)$

Step 2. $\mathbf{v}_2 = \mathbf{u}_2 - \text{proj}_{W_1}\mathbf{u}_2 = \mathbf{u}_2 - \dfrac{\langle \mathbf{u}_2, \mathbf{v}_1 \rangle}{\|\mathbf{v}_1\|^2}\mathbf{v}_1$

$$= (0, 1, 1) - \frac{2}{3}(1, 1, 1) = \left(-\frac{2}{3}, \frac{1}{3}, \frac{1}{3}\right)$$

Step 3. $\mathbf{v}_3 = \mathbf{u}_3 - \text{proj}_{W_2} \mathbf{u}_3 = \mathbf{u}_3 - \dfrac{\langle \mathbf{u}_3, \mathbf{v}_1 \rangle}{\|\mathbf{v}_1\|^2}\mathbf{v}_1 - \dfrac{\langle \mathbf{u}_3, \mathbf{v}_2 \rangle}{\|\mathbf{v}_2\|^2}\mathbf{v}_2$

$$= (0, 0, 1) - \frac{1}{3}(1, 1, 1) - \frac{1/3}{2/3}\left(-\frac{2}{3}, \frac{1}{3}, \frac{1}{3}\right)$$

$$= \left(0, -\frac{1}{2}, \frac{1}{2}\right)$$

Thus,

$$\mathbf{v}_1 = (1, 1, 1), \quad \mathbf{v}_2 = \left(-\frac{2}{3}, \frac{1}{3}, \frac{1}{3}\right), \quad \mathbf{v}_3 = \left(0, -\frac{1}{2}, \frac{1}{2}\right)$$

form an orthogonal basis for R^3. The norms of these vectors are

$$\|\mathbf{v}_1\| = \sqrt{3}, \quad \|\mathbf{v}_2\| = \frac{\sqrt{6}}{3}, \quad \|\mathbf{v}_3\| = \frac{1}{\sqrt{2}}$$

so an orthonormal basis for R^3 is

$$\mathbf{q}_1 = \frac{\mathbf{v}_1}{\|\mathbf{v}_1\|} = \left(\frac{1}{\sqrt{3}}, \frac{1}{\sqrt{3}}, \frac{1}{\sqrt{3}}\right), \quad \mathbf{q}_2 = \frac{\mathbf{v}_2}{\|\mathbf{v}_2\|} = \left(-\frac{2}{\sqrt{6}}, \frac{1}{\sqrt{6}}, \frac{1}{\sqrt{6}}\right),$$

$$\mathbf{q}_3 = \frac{\mathbf{v}_3}{\|\mathbf{v}_3\|} = \left(0, -\frac{1}{\sqrt{2}}, \frac{1}{\sqrt{2}}\right) \blacktriangleleft$$

Remark   In the last example we normalized at the end to convert the orthogonal basis into an orthonormal basis. Alternatively, we could have normalized each orthogonal basis vector as soon as it was obtained, thereby producing an orthonormal basis step by step. However, that procedure generally has the disadvantage in hand calculation of producing more square roots to manipulate. A more useful variation is to "scale" the orthogonal basis vectors at each step to eliminate some of the fractions. For example, after Step 2 above, we could have multiplied by 3 to produce $(-2, 1, 1)$ as the second orthogonal basis vector, thereby simplifying the calculations in Step 3.

Erhardt Schmidt
(1875–1959)

Jorgen Pederson Gram
(1850–1916)

Historical Note   Schmidt was a German mathematician who studied for his doctoral degree at Göttingen University under David Hilbert, one of the giants of modern mathematics. For most of his life he taught at Berlin University where, in addition to making important contributions to many branches of mathematics, he fashioned some of Hilbert's ideas into a general concept, called a *Hilbert space*–a fundamental idea in the study of infinite-dimensional vector spaces. He first described the process that bears his name in a paper on integral equations that he published in 1907.

[*Image: Archives of the Mathematisches Forschungsinst*]

Historical Note   Gram was a Danish actuary whose early education was at village schools supplemented by private tutoring. He obtained a doctorate degree in mathematics while working for the Hafnia Life Insurance Company, where he specialized in the mathematics of accident insurance. It was in his dissertation that his contributions to the Gram–Schmidt process were formulated. He eventually became interested in abstract mathematics and received a gold medal from the Royal Danish Society of Sciences and Letters in recognition of his work. His lifelong interest in applied mathematics never wavered, however, and he produced a variety of treatises on Danish forest management.

[*Image: wikipedia*]

CALCULUS REQUIRED ▶ **EXAMPLE 8 Legendre Polynomials**

Let the vector space P_2 have the inner product

$$\langle \mathbf{p}, \mathbf{q} \rangle = \int_{-1}^{1} p(x)q(x)\,dx$$

Apply the Gram–Schmidt process to transform the standard basis $\{1, x, x^2\}$ for P_2 into an orthogonal basis $\{\phi_1(x), \phi_2(x), \phi_3(x)\}$.

Solution Take $\mathbf{u}_1 = 1$, $\mathbf{u}_2 = x$, and $\mathbf{u}_3 = x^2$.

Step 1. $\mathbf{v}_1 = \mathbf{u}_1 = 1$

Step 2. We have

$$\langle \mathbf{u}_2, \mathbf{v}_1 \rangle = \int_{-1}^{1} x\,dx = 0$$

so

$$\mathbf{v}_2 = \mathbf{u}_2 - \frac{\langle \mathbf{u}_2, \mathbf{v}_1 \rangle}{\|\mathbf{v}_1\|^2}\mathbf{v}_1 = \mathbf{u}_2 = x$$

Step 3. We have

$$\langle \mathbf{u}_3, \mathbf{v}_1 \rangle = \int_{-1}^{1} x^2\,dx = \frac{x^3}{3}\bigg]_{-1}^{1} = \frac{2}{3}$$

$$\langle \mathbf{u}_3, \mathbf{v}_2 \rangle = \int_{-1}^{1} x^3\,dx = \frac{x^4}{4}\bigg]_{-1}^{1} = 0$$

$$\|\mathbf{v}_1\|^2 = \langle \mathbf{v}_1, \mathbf{v}_1 \rangle = \int_{-1}^{1} 1\,dx = x\bigg]_{-1}^{1} = 2$$

so

$$\mathbf{v}_3 = \mathbf{u}_3 - \frac{\langle \mathbf{u}_3, \mathbf{v}_1 \rangle}{\|\mathbf{v}_1\|^2}\mathbf{v}_1 - \frac{\langle \mathbf{u}_3, \mathbf{v}_2 \rangle}{\|\mathbf{v}_2\|^2}\mathbf{v}_2 = x^2 - \frac{1}{3}$$

Thus, we have obtained the orthogonal basis $\{\phi_1(x), \phi_2(x), \phi_3(x)\}$ in which

$$\phi_1(x) = 1, \quad \phi_2(x) = x, \quad \phi_3(x) = x^2 - \frac{1}{3} \quad \blacktriangleleft$$

Remark The orthogonal basis vectors in the foregoing example are often scaled so all three functions have a value of 1 at $x = 1$. The resulting polynomials

$$1, \quad x, \quad \frac{1}{2}(3x^2 - 1)$$

which are known as the first three ***Legendre polynomials***, play an important role in a variety of applications. The scaling does not affect the orthogonality.

Extending Orthonormal Sets to Orthonormal Bases Recall from part (*b*) of Theorem 4.5.5 that a linearly independent set in a finite-dimensional vector space can be enlarged to a basis by adding appropriate vectors. The following theorem is an analog of that result for orthogonal and orthonormal sets in finite-dimensional inner product spaces.

> **THEOREM 6.3.6** *If W is a finite-dimensional inner product space, then:*
>
> (a) *Every orthogonal set of nonzero vectors in W can be enlarged to an orthogonal basis for W.*
>
> (b) *Every orthonormal set in W can be enlarged to an orthonormal basis for W.*

We will prove part (b) and leave part (a) as an exercise.

Proof (b) Suppose that $S = \{\mathbf{v}_1, \mathbf{v}_2, \ldots, \mathbf{v}_s\}$ is an orthonormal set of vectors in W. Part (b) of Theorem 4.5.5 tells us that we can enlarge S to some basis

$$S' = \{\mathbf{v}_1, \mathbf{v}_2, \ldots, \mathbf{v}_s, \mathbf{v}_{s+1}, \ldots, \mathbf{v}_k\}$$

for W. If we now apply the Gram–Schmidt process to the set S', then the vectors $\mathbf{v}_1, \mathbf{v}_2, \ldots, \mathbf{v}_s$, will not be affected since they are already orthonormal, and the resulting set

$$S'' = \{\mathbf{v}_1, \mathbf{v}_2, \ldots, \mathbf{v}_s, \mathbf{v}_{s+1}, \ldots, \mathbf{v}_k\}$$

will be an orthonormal basis for W. ◄

OPTIONAL

QR-Decomposition

In recent years a numerical algorithm based on the Gram–Schmidt process, and known as **QR-decomposition**, has assumed growing importance as the mathematical foundation for a wide variety of numerical algorithms, including those for computing eigenvalues of large matrices. The technical aspects of such algorithms are discussed in textbooks that specialize in the numerical aspects of linear algebra. However, we will discuss some of the underlying ideas here. We begin by posing the following problem.

> **Problem** If A is an $m \times n$ matrix with linearly independent column vectors, and if Q is the matrix that results by applying the Gram–Schmidt process to the column vectors of A, what relationship, if any, exists between A and Q?

To solve this problem, suppose that the column vectors of A are $\mathbf{u}_1, \mathbf{u}_2, \ldots, \mathbf{u}_n$ and the orthonormal column vectors of Q are $\mathbf{q}_1, \mathbf{q}_2, \ldots, \mathbf{q}_n$. Thus, A and Q can be written in partitioned form as

$$A = [\mathbf{u}_1 \mid \mathbf{u}_2 \mid \cdots \mid \mathbf{u}_n] \quad \text{and} \quad Q = [\mathbf{q}_1 \mid \mathbf{q}_2 \mid \cdots \mid \mathbf{q}_n]$$

It follows from Theorem 6.3.2b that $\mathbf{u}_1, \mathbf{u}_2, \ldots, \mathbf{u}_n$ are expressible in terms of the vectors $\mathbf{q}_1, \mathbf{q}_2, \ldots, \mathbf{q}_n$ as

$$\mathbf{u}_1 = \langle \mathbf{u}_1, \mathbf{q}_1 \rangle \mathbf{q}_1 + \langle \mathbf{u}_1, \mathbf{q}_2 \rangle \mathbf{q}_2 + \cdots + \langle \mathbf{u}_1, \mathbf{q}_n \rangle \mathbf{q}_n$$
$$\mathbf{u}_2 = \langle \mathbf{u}_2, \mathbf{q}_1 \rangle \mathbf{q}_1 + \langle \mathbf{u}_2, \mathbf{q}_2 \rangle \mathbf{q}_2 + \cdots + \langle \mathbf{u}_2, \mathbf{q}_n \rangle \mathbf{q}_n$$
$$\vdots \qquad \vdots \qquad \vdots \qquad \vdots$$
$$\mathbf{u}_n = \langle \mathbf{u}_n, \mathbf{q}_1 \rangle \mathbf{q}_1 + \langle \mathbf{u}_n, \mathbf{q}_2 \rangle \mathbf{q}_2 + \cdots + \langle \mathbf{u}_n, \mathbf{q}_n \rangle \mathbf{q}_n$$

Recalling from Section 1.3 (Example 9) that the jth column vector of a matrix product is a linear combination of the column vectors of the first factor with coefficients coming from the jth column of the second factor, it follows that these relationships can be expressed in matrix form as

$$[\mathbf{u}_1 \mid \mathbf{u}_2 \mid \cdots \mid \mathbf{u}_n] = [\mathbf{q}_1 \mid \mathbf{q}_2 \mid \cdots \mid \mathbf{q}_n] \begin{bmatrix} \langle \mathbf{u}_1, \mathbf{q}_1 \rangle & \langle \mathbf{u}_2, \mathbf{q}_1 \rangle & \cdots & \langle \mathbf{u}_n, \mathbf{q}_1 \rangle \\ \langle \mathbf{u}_1, \mathbf{q}_2 \rangle & \langle \mathbf{u}_2, \mathbf{q}_2 \rangle & \cdots & \langle \mathbf{u}_n, \mathbf{q}_2 \rangle \\ \vdots & \vdots & & \vdots \\ \langle \mathbf{u}_1, \mathbf{q}_n \rangle & \langle \mathbf{u}_2, \mathbf{q}_n \rangle & \cdots & \langle \mathbf{u}_n, \mathbf{q}_n \rangle \end{bmatrix}$$

or more briefly as

$$A = QR \tag{14}$$

where R is the second factor in the product. However, it is a property of the Gram–Schmidt process that for $j \geq 2$, the vector $\mathbf{q}_j$ is orthogonal to $\mathbf{u}_1, \mathbf{u}_2, \ldots, \mathbf{u}_{j-1}$. Thus, all entries below the main diagonal of R are zero, and R has the form

$$R = \begin{bmatrix} \langle \mathbf{u}_1, \mathbf{q}_1 \rangle & \langle \mathbf{u}_2, \mathbf{q}_1 \rangle & \cdots & \langle \mathbf{u}_n, \mathbf{q}_1 \rangle \\ 0 & \langle \mathbf{u}_2, \mathbf{q}_2 \rangle & \cdots & \langle \mathbf{u}_n, \mathbf{q}_2 \rangle \\ \vdots & \vdots & & \vdots \\ 0 & 0 & \cdots & \langle \mathbf{u}_n, \mathbf{q}_n \rangle \end{bmatrix} \tag{15}$$

We leave it for you to show that R is invertible by showing that its diagonal entries are nonzero. Thus, Equation (14) is a factorization of A into the product of a matrix Q with orthonormal column vectors and an invertible upper triangular matrix R. We call Equation (14) the ***QR-decomposition of A***. In summary, we have the following theorem.

THEOREM 6.3.7 *QR*-Decomposition

If A is an $m \times n$ matrix with linearly independent column vectors, then A can be factored as

$$A = QR$$

where Q is an $m \times n$ matrix with orthonormal column vectors, and R is an $n \times n$ invertible upper triangular matrix.

It is common in numerical linear algebra to say that a matrix with linearly independent columns has ***full column rank***.

Recall from Theorem 5.1.6 (the Equivalence Theorem) that a *square* matrix has linearly independent column vectors if and only if it is invertible. Thus, it follows from the foregoing theorem that *every invertible matrix has a QR-decomposition.*

▶ **EXAMPLE 9** *QR*-Decomposition of a 3 × 3 Matrix

Find the *QR*-decomposition of

$$A = \begin{bmatrix} 1 & 0 & 0 \\ 1 & 1 & 0 \\ 1 & 1 & 1 \end{bmatrix}$$

Solution The column vectors of A are

$$\mathbf{u}_1 = \begin{bmatrix} 1 \\ 1 \\ 1 \end{bmatrix}, \quad \mathbf{u}_2 = \begin{bmatrix} 0 \\ 1 \\ 1 \end{bmatrix}, \quad \mathbf{u}_3 = \begin{bmatrix} 0 \\ 0 \\ 1 \end{bmatrix}$$

Applying the Gram–Schmidt process with normalization to these column vectors yields the orthonormal vectors (see Example 7)

$$\mathbf{q}_1 = \begin{bmatrix} \frac{1}{\sqrt{3}} \\ \frac{1}{\sqrt{3}} \\ \frac{1}{\sqrt{3}} \end{bmatrix}, \quad \mathbf{q}_2 = \begin{bmatrix} -\frac{2}{\sqrt{6}} \\ \frac{1}{\sqrt{6}} \\ \frac{1}{\sqrt{6}} \end{bmatrix}, \quad \mathbf{q}_3 = \begin{bmatrix} 0 \\ -\frac{1}{\sqrt{2}} \\ \frac{1}{\sqrt{2}} \end{bmatrix}$$

Thus, it follows from Formula (15) that R is

$$R = \begin{bmatrix} \langle \mathbf{u}_1, \mathbf{q}_1 \rangle & \langle \mathbf{u}_2, \mathbf{q}_1 \rangle & \langle \mathbf{u}_3, \mathbf{q}_1 \rangle \\ 0 & \langle \mathbf{u}_2, \mathbf{q}_2 \rangle & \langle \mathbf{u}_3, \mathbf{q}_2 \rangle \\ 0 & 0 & \langle \mathbf{u}_3, \mathbf{q}_3 \rangle \end{bmatrix} = \begin{bmatrix} \frac{3}{\sqrt{3}} & \frac{2}{\sqrt{3}} & \frac{1}{\sqrt{3}} \\ 0 & \frac{2}{\sqrt{6}} & \frac{1}{\sqrt{6}} \\ 0 & 0 & \frac{1}{\sqrt{2}} \end{bmatrix}$$

from which it follows that the QR-decomposition of A is

> Show that the matrix Q in Example 9 has the property $QQ^T = I$, and show that *every* $m \times n$ matrix with orthonormal column vectors has this property.

$$\begin{bmatrix} 1 & 0 & 0 \\ 1 & 1 & 0 \\ 1 & 1 & 1 \end{bmatrix} = \begin{bmatrix} \frac{1}{\sqrt{3}} & -\frac{2}{\sqrt{6}} & 0 \\ \frac{1}{\sqrt{3}} & \frac{1}{\sqrt{6}} & -\frac{1}{\sqrt{2}} \\ \frac{1}{\sqrt{3}} & \frac{1}{\sqrt{6}} & \frac{1}{\sqrt{2}} \end{bmatrix} \begin{bmatrix} \frac{3}{\sqrt{3}} & \frac{2}{\sqrt{3}} & \frac{1}{\sqrt{3}} \\ 0 & \frac{2}{\sqrt{6}} & \frac{1}{\sqrt{6}} \\ 0 & 0 & \frac{1}{\sqrt{2}} \end{bmatrix} \blacktriangleleft$$

$$A \qquad = \qquad\qquad Q \qquad\qquad\qquad R$$

Concept Review

- Orthogonal and orthonormal sets
- Normalizing a vector
- Orthogonal projections
- Gram–Schmidt process
- QR-decomposition

Skills

- Determine whether a set of vectors is orthogonal (or orthonormal).
- Compute the coordinates of a vector with respect to an orthogonal (or orthonormal) basis.
- Find the orthogonal projection of a vector onto a subspace.
- Use the Gram–Schmidt process to construct an orthogonal (or orthonormal) basis for an inner product space.
- Find the QR-decomposition of an invertible matrix.

Exercise Set 6.3

1. Which of the following sets of vectors are orthogonal with respect to the Euclidean inner product on R^2?

(a) $(0, 1)$, $(2, 0)$

(b) $\left(-\frac{1}{\sqrt{2}}, \frac{1}{\sqrt{2}}\right)$, $\left(\frac{1}{\sqrt{2}}, \frac{1}{\sqrt{2}}\right)$

(c) $\left(-\frac{1}{\sqrt{2}}, -\frac{1}{\sqrt{2}}\right)$, $\left(\frac{1}{\sqrt{2}}, \frac{1}{\sqrt{2}}\right)$

(d) $(0, 0)$, $(0, 1)$

2. Which of the sets in Exercise 1 are orthonormal with respect to the Euclidean inner product on R^2?

3. Which of the following sets of vectors are orthogonal with respect to the Euclidean inner product on R^3?

(a) $\left(\frac{1}{\sqrt{2}}, 0, \frac{1}{\sqrt{2}}\right)$, $\left(\frac{1}{\sqrt{3}}, \frac{1}{\sqrt{3}}, -\frac{1}{\sqrt{3}}\right)$, $\left(-\frac{1}{\sqrt{2}}, 0, \frac{1}{\sqrt{2}}\right)$

(b) $\left(\frac{2}{3}, -\frac{2}{3}, \frac{1}{3}\right)$, $\left(\frac{2}{3}, \frac{1}{3}, -\frac{2}{3}\right)$, $\left(\frac{1}{3}, \frac{2}{3}, \frac{2}{3}\right)$

(c) $(1, 0, 0)$, $\left(0, \frac{1}{\sqrt{2}}, \frac{1}{\sqrt{2}}\right)$, $(0, 0, 1)$

(d) $\left(\frac{1}{\sqrt{6}}, \frac{1}{\sqrt{6}}, -\frac{2}{\sqrt{6}}\right)$, $\left(\frac{1}{\sqrt{2}}, -\frac{1}{\sqrt{2}}, 0\right)$

4. Which of the sets in Exercise 3 are orthonormal with respect to the Euclidean inner product on R^3?

5. Which of the following sets of polynomials are orthonormal with respect to the inner product on P_2 discussed in Example 7 of Section 6.1?

(a) $p_1(x) = \frac{2}{3} - \frac{2}{3}x + \frac{1}{3}x^2$, $p_2(x) = \frac{2}{3} + \frac{1}{3}x - \frac{2}{3}x^2$, $p_3(x) = \frac{1}{3} + \frac{2}{3}x + \frac{2}{3}x^2$

(b) $p_1(x) = 1$, $p_2(x) = \frac{1}{\sqrt{2}}x + \frac{1}{\sqrt{2}}x^2$, $p_3(x) = x^2$

6. Which of the following sets of matrices are orthonormal with respect to the inner product on M_{22} discussed in Example 6 of Section 6.1?

(a) $\begin{bmatrix} 1 & 0 \\ 0 & 0 \end{bmatrix}$, $\begin{bmatrix} 0 & \frac{2}{3} \\ \frac{1}{3} & -\frac{2}{3} \end{bmatrix}$, $\begin{bmatrix} 0 & \frac{2}{3} \\ -\frac{2}{3} & \frac{1}{3} \end{bmatrix}$, $\begin{bmatrix} 0 & \frac{1}{3} \\ \frac{2}{3} & \frac{2}{3} \end{bmatrix}$

(b) $\begin{bmatrix} 1 & 0 \\ 0 & 0 \end{bmatrix}$, $\begin{bmatrix} 0 & 1 \\ 0 & 0 \end{bmatrix}$, $\begin{bmatrix} 0 & 0 \\ 1 & 1 \end{bmatrix}$, $\begin{bmatrix} 0 & 0 \\ 1 & -1 \end{bmatrix}$

7. Verify that the given vectors form an orthogonal set with respect to the Euclidean inner product; then convert it to an orthonormal set by normalizing the vectors.

(a) $(-1, 2)$, $(6, 3)$

(b) $(1, 0, -1)$, $(2, 0, 2)$, $(0, 5, 0)$

(c) $\left(\frac{1}{5}, \frac{1}{5}, \frac{1}{5}\right)$, $\left(-\frac{1}{2}, \frac{1}{2}, 0\right)$, $\left(\frac{1}{3}, \frac{1}{3}, -\frac{2}{3}\right)$

8. Verify that the set of vectors $\{(1, 0), (0, 1)\}$ is orthogonal with respect to the inner product $\langle \mathbf{u}, \mathbf{v} \rangle = 4u_1v_1 + u_2v_2$ on R^2; then convert it to an orthonormal set by normalizing the vectors.

9. Verify that the vectors

$$\mathbf{v}_1 = \left(-\frac{3}{5}, \frac{4}{5}, 0\right), \quad \mathbf{v}_2 = \left(\frac{4}{5}, \frac{3}{5}, 0\right), \quad \mathbf{v}_3 = (0, 0, 1)$$

form an orthonormal basis for R^3 with the Euclidean inner product; then use Theorem 6.3.2b to express each of the following as linear combinations of $\mathbf{v}_1$, $\mathbf{v}_2$, and $\mathbf{v}_3$.

(a) $(1, -1, 2)$ (b) $(3, -7, 4)$ (c) $\left(\frac{1}{7}, -\frac{3}{7}, \frac{5}{7}\right)$

10. Verify that the vectors

$$\mathbf{v}_1 = (1, -1, 2, -1), \quad \mathbf{v}_2 = (-2, 2, 3, 2),$$
$$\mathbf{v}_3 = (1, 2, 0, -1), \quad \mathbf{v}_4 = (1, 0, 0, 1)$$

form an orthogonal basis for R^4 with the Euclidean inner product; then use Theorem 6.3.2a to express each of the following as linear combinations of $\mathbf{v}_1$, $\mathbf{v}_2$, $\mathbf{v}_3$, and $\mathbf{v}_4$.

(a) $(1, 1, 1, 1)$

(b) $(\sqrt{2}, -3\sqrt{2}, 5\sqrt{2}, -\sqrt{2})$

(c) $\left(-\frac{1}{3}, \frac{2}{3}, -\frac{1}{3}, \frac{4}{3}\right)$

11. (a) Show that the vectors

$$\mathbf{v}_1 = (1, -2, 3, -4), \quad \mathbf{v}_2 = (2, 1, -4, -3),$$
$$\mathbf{v}_3 = (-3, 4, 1, -2), \quad \mathbf{v}_4 = (4, 3, 2, 1)$$

form an orthogonal basis for R^4 with the Euclidean inner product.

(b) Use Theorem 6.3.2a to express $\mathbf{u} = (-1, 2, 3, 7)$ as a linear combination of the vectors in part (a).

▶ In Exercises 12–13, an orthonormal basis with respect to the Euclidean inner product is given. Use Theorem 6.3.2b to find the coordinate vector of $\mathbf{w}$ with respect to that basis. ◀

12. (a) $\mathbf{w} = (3, 7)$; $\mathbf{u}_1 = \left(\frac{1}{\sqrt{2}}, -\frac{1}{\sqrt{2}}\right)$, $\mathbf{u}_2 = \left(\frac{1}{\sqrt{2}}, \frac{1}{\sqrt{2}}\right)$

(b) $\mathbf{w} = (-1, 0, 2)$; $\mathbf{u}_1 = \left(\frac{2}{3}, -\frac{2}{3}, \frac{1}{3}\right)$,

$\mathbf{u}_2 = \left(\frac{2}{3}, \frac{1}{3}, -\frac{2}{3}\right)$, $\mathbf{u}_3 = \left(\frac{1}{3}, \frac{2}{3}, \frac{2}{3}\right)$

13. (a) $\mathbf{w} = (2, 0, 5)$; $\mathbf{u}_1 = \left(\frac{2}{3}, \frac{1}{3}, \frac{2}{3}\right)$, $\mathbf{u}_2 = \left(\frac{1}{3}, \frac{2}{3}, -\frac{2}{3}\right)$,

$\mathbf{u}_3 = \left(\frac{2}{3}, -\frac{2}{3}, -\frac{1}{3}\right)$

(b) $\mathbf{w} = (-1, 1, 2)$; $\mathbf{u}_1 = \left(\frac{3}{\sqrt{11}}, \frac{1}{\sqrt{11}}, \frac{1}{\sqrt{11}}\right)$,

$\mathbf{u}_2 = \left(-\frac{1}{\sqrt{6}}, \frac{2}{\sqrt{6}}, \frac{1}{\sqrt{6}}\right)$,

$\mathbf{u}_3 = \left(-\frac{1}{\sqrt{66}}, -\frac{4}{\sqrt{66}}, \frac{7}{\sqrt{66}}\right)$

▶ In Exercises 14–15, the given vectors are orthogonal with respect to the Euclidean inner product. Find $\text{proj}_W \mathbf{x}$, where $\mathbf{x} = (1, 2, 0, -2)$ and W is the subspace of R^4 spanned by the vectors. ◀

14. (a) $\mathbf{v}_1 = (1, 1, 1, 1)$, $\mathbf{v}_2 = (1, 1, -1, -1)$

(b) $\mathbf{v}_1 = (0, 1, -4, -1)$, $\mathbf{v}_2 = (3, 5, 1, 1)$

15. (a) $\mathbf{v}_1 = (1, 1, 1, 1)$, $\mathbf{v}_2 = (1, 1, -1, -1)$,
$\mathbf{v}_3 = (1, -1, 1, -1)$

(b) $\mathbf{v}_1 = (0, 1, -4, -1)$, $\mathbf{v}_2 = (3, 5, 1, 1)$,
$\mathbf{v}_3 = (1, 0, 1, -4)$

▶ In Exercises 16–17, the given vectors are orthonormal with respect to the Euclidean inner product. Use Theorem 6.3.4b to find $\text{proj}_W \mathbf{x}$, where $\mathbf{x} = (1, 2, 0, -1)$ and W is the subspace of R^4 spanned by the vectors. ◀

16. (a) $\mathbf{v}_1 = \left(0, \frac{1}{\sqrt{18}}, -\frac{4}{\sqrt{18}}, -\frac{1}{\sqrt{18}}\right)$, $\mathbf{v}_2 = \left(\frac{1}{2}, \frac{5}{6}, \frac{1}{6}, \frac{1}{6}\right)$

(b) $\mathbf{v}_1 = \left(\frac{1}{2}, \frac{1}{2}, \frac{1}{2}, \frac{1}{2}\right)$, $\mathbf{v}_2 = \left(\frac{1}{2}, \frac{1}{2}, -\frac{1}{2}, -\frac{1}{2}\right)$

17. (a) $\mathbf{v}_1 = \left(0, \frac{1}{\sqrt{18}}, -\frac{4}{\sqrt{18}}, -\frac{1}{\sqrt{18}}\right)$, $\mathbf{v}_2 = \left(\frac{1}{2}, \frac{5}{6}, \frac{1}{6}, \frac{1}{6}\right)$,

$\mathbf{v}_3 = \left(\frac{1}{\sqrt{18}}, 0, \frac{1}{\sqrt{18}}, -\frac{4}{\sqrt{18}}\right)$

(b) $\mathbf{v}_1 = \left(\frac{1}{2}, \frac{1}{2}, \frac{1}{2}, \frac{1}{2}\right)$, $\mathbf{v}_2 = \left(\frac{1}{2}, \frac{1}{2}, -\frac{1}{2}, -\frac{1}{2}\right)$,

$\mathbf{v}_3 = \left(\frac{1}{2}, -\frac{1}{2}, \frac{1}{2}, -\frac{1}{2}\right)$

18. In Example 6 of Section 4.9 we found the orthogonal projection of the vector $\mathbf{x} = (1, 5)$ onto the line through the origin making an angle of $\pi/6$ radians with the *x*-axis. Solve that same problem using Theorem 6.3.4.

19. Find the vectors $\mathbf{w}_1$ in W and $\mathbf{w}_2$ in $W^\perp$ such that $\mathbf{x} = \mathbf{w}_1 + \mathbf{w}_2$, where $\mathbf{x}$ and W are as given in

(a) Exercise 14(a).   (b) Exercise 15(a).

20. Find the vectors $\mathbf{w}_1$ in W and $\mathbf{w}_2$ in $W^\perp$ such that $\mathbf{x} = \mathbf{w}_1 + \mathbf{w}_2$, where $\mathbf{x}$ and W are as given in

(a) Exercise 16(a).   (b) Exercise 17(a).

21. Let R^2 have the Euclidean inner product. Use the Gram–Schmidt process to transform the basis $\{\mathbf{u}_1, \mathbf{u}_2\}$ into an orthonormal basis. Draw both sets of basis vectors in the *xy*-plane.

(a) $\mathbf{u}_1 = (1, -3)$, $\mathbf{u}_2 = (2, 2)$

(b) $\mathbf{u}_1 = (1, 0)$, $\mathbf{u}_2 = (3, -5)$

22. Let R^3 have the Euclidean inner product. Use the Gram–Schmidt process to transform the basis $\{\mathbf{u}_1, \mathbf{u}_2, \mathbf{u}_3\}$ into an orthonormal basis.

(a) $\mathbf{u}_1 = (1, 1, 1)$, $\mathbf{u}_2 = (-1, 1, 0)$, $\mathbf{u}_3 = (1, 2, 1)$

(b) $\mathbf{u}_1 = (1, 0, 0)$, $\mathbf{u}_2 = (3, 7, -2)$, $\mathbf{u}_3 = (0, 4, 1)$

23. Let R^4 have the Euclidean inner product. Use the Gram–Schmidt process to transform the basis $\{\mathbf{u}_1, \mathbf{u}_2, \mathbf{u}_3, \mathbf{u}_4\}$ into an orthonormal basis.

$$\mathbf{u}_1 = (0, 2, 1, 0), \qquad \mathbf{u}_2 = (1, -1, 0, 0),$$
$$\mathbf{u}_3 = (1, 2, 0, -1), \qquad \mathbf{u}_4 = (1, 0, 0, 1)$$

24. Let R^3 have the Euclidean inner product. Find an orthonormal basis for the subspace spanned by $(0, 1, 2)$, $(-1, 0, 1)$, $(-1, 1, 3)$.

25. Let R^3 have the inner product

$$\langle \mathbf{u}, \mathbf{v} \rangle = u_1 v_1 + 2u_2 v_2 + 3u_3 v_3$$

Use the Gram–Schmidt process to transform $\mathbf{u}_1 = (1, 1, 1)$, $\mathbf{u}_2 = (1, 1, 0)$, $\mathbf{u}_3 = (1, 0, 0)$ into an orthonormal basis.

26. The subspace of R^3 spanned by the vectors $\mathbf{u}_1 = \left(\frac{4}{5}, 0, -\frac{3}{5}\right)$ and $\mathbf{u}_2 = (0, 1, 0)$ is a plane passing through the origin. Express $\mathbf{w} = (1, 2, 3)$ in the form $\mathbf{w} = \mathbf{w}_1 + \mathbf{w}_2$, where $\mathbf{w}_1$ lies in the plane and $\mathbf{w}_2$ is perpendicular to the plane.

27. Repeat Exercise 26 with $\mathbf{u}_1 = (1, 1, 1)$ and $\mathbf{u}_2 = (2, 0, -1)$.

28. Let R^4 have the Euclidean inner product. Express the vector $\mathbf{w} = (-1, 2, 6, 0)$ in the form $\mathbf{w} = \mathbf{w}_1 + \mathbf{w}_2$, where $\mathbf{w}_1$ is in the space W spanned by $\mathbf{u}_1 = (-1, 0, 1, 2)$ and $\mathbf{u}_2 = (0, 1, 0, 1)$, and $\mathbf{w}_2$ is orthogonal to W.

29. Find the QR-decomposition of the matrix, where possible.

(a) $\begin{bmatrix} 1 & -1 \\ 2 & 3 \end{bmatrix}$ (b) $\begin{bmatrix} 1 & 2 \\ 0 & 1 \\ 1 & 4 \end{bmatrix}$ (c) $\begin{bmatrix} 1 & 1 \\ -2 & 1 \\ 2 & 1 \end{bmatrix}$

(d) $\begin{bmatrix} 1 & 0 & 2 \\ 0 & 1 & 1 \\ 1 & 2 & 0 \end{bmatrix}$ (e) $\begin{bmatrix} 1 & 2 & 1 \\ 1 & 1 & 1 \\ 0 & 3 & 1 \end{bmatrix}$ (f) $\begin{bmatrix} 1 & 0 & 1 \\ -1 & 1 & 1 \\ 1 & 0 & 1 \\ -1 & 1 & 1 \end{bmatrix}$

30. In Step 3 of the proof of Theorem 6.3.5, it was stated that "the linear independence of $\{\mathbf{u}_1, \mathbf{u}_2, \ldots, \mathbf{u}_n\}$ ensures that $\mathbf{v}_3 \neq \mathbf{0}$." Prove this statement.

31. Prove that the diagonal entries of R in Formula (15) are nonzero.

32. (*Calculus required*) Use Theorem 6.3.2a to express the following polynomials as linear combinations of the first three Legendre polynomials (see the Remark following Example 8).

(a) $1 + x + 4x^2$ (b) $2 - 7x^2$ (c) $4 + 3x$

33. (*Calculus required*) Let P_2 have the inner product

$$\langle \mathbf{p}, \mathbf{q} \rangle = \int_0^1 p(x)q(x)\, dx$$

Apply the Gram–Schmidt process to transform the standard basis $S = \{1, x, x^2\}$ into an orthonormal basis.

34. Find vectors $\mathbf{x}$ and $\mathbf{y}$ in R^2 that are orthonormal with respect to the inner product $\langle \mathbf{u}, \mathbf{v} \rangle = 3u_1 v_1 + 2u_2 v_2$ but are not orthonormal with respect to the Euclidean inner product.

True-False Exercises

In parts (a)–(f) determine whether the statement is true or false, and justify your answer.

(a) Every linearly independent set of vectors in an inner product space is orthogonal.

(b) Every orthogonal set of vectors in an inner product space is linearly independent.

(c) Every nontrivial subspace of R^3 has an orthonormal basis with respect to the Euclidean inner product.

(d) Every nonzero finite-dimensional inner product space has an orthonormal basis.

(e) $\text{proj}_W \mathbf{x}$ is orthogonal to every vector of W.

(f) If A is an $n \times n$ matrix with a nonzero determinant, then A has a QR-decomposition.

6.4 Best Approximation; Least Squares

In this section we will be concerned with linear systems that cannot be solved exactly and for which an approximate solution is needed. Such systems commonly occur in applications where measurement errors "perturb" the coefficients of a consistent system sufficiently to produce inconsistency.

Least Squares Solutions of Linear Systems

Suppose that $A\mathbf{x} = \mathbf{b}$ is an *inconsistent* linear system of m equations in n unknowns in which we suspect the inconsistency to be caused by measurement errors in the coefficients of A. Since no exact solution is possible, we will look for a vector $\mathbf{x}$ that comes as "close as possible" to being a solution in the sense that it minimizes $\|\mathbf{b} - A\mathbf{x}\|$ with respect to the Euclidean inner product on R^m. You can think of $A\mathbf{x}$ as an approximation to $\mathbf{b}$ and $\|\mathbf{b} - A\mathbf{x}\|$ as the *error* in that approximation—the smaller the error, the better the approximation. This leads to the following problem.

Least Squares Problem Given a linear system $A\mathbf{x} = \mathbf{b}$ of m equations in n unknowns, find a vector $\mathbf{x}$ that minimizes $\|\mathbf{b} - A\mathbf{x}\|$ with respect to the Euclidean inner product on R^m. We call such an $\mathbf{x}$ a *least squares solution* of the system, we call $\mathbf{b} - A\mathbf{x}$ the *least squares error vector*, and we call $\|\mathbf{b} - A\mathbf{x}\|$ the *least squares error*.

To clarify the above terminology, suppose that the matrix form of $\mathbf{b} - A\mathbf{x}$ is

$$\mathbf{b} - A\mathbf{x} = \begin{bmatrix} e_1 \\ e_2 \\ \vdots \\ e_m \end{bmatrix}$$

The term "least squares solution" results from the fact that minimizing $\|\mathbf{b} - A\mathbf{x}\|$ also minimizes $\|\mathbf{b} - A\mathbf{x}\|^2 = e_1^2 + e_2^2 + \cdots + e_m^2$.

Best Approximation   Suppose that $\mathbf{b}$ is a fixed vector in R^3 that we would like to approximate by a vector $\mathbf{w}$ that is required to lie in some subspace W of R^3. Unless $\mathbf{b}$ happens to be in W, then any such approximation will result in an "error vector" $\mathbf{b} - \mathbf{w}$ that cannot be made equal to $\mathbf{0}$ no matter how $\mathbf{w}$ is chosen (Figure 6.4.1a). However, by choosing

$$\mathbf{w} = \text{proj}_W \mathbf{b}$$

we can make the length of the error vector

$$\|\mathbf{b} - \mathbf{w}\| = \|\mathbf{b} - \text{proj}_W \mathbf{b}\|$$

as small as possible (Figure 6.4.1b).

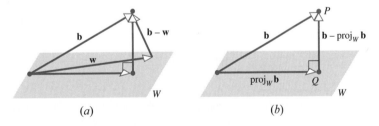

▶ Figure 6.4.1   (a)   (b)

These geometric ideas suggest the following general theorem.

THEOREM 6.4.1 Best Approximation Theorem

*If W is a finite-dimensional subspace of an inner product space V, and if $\mathbf{b}$ is a vector in V, then $\text{proj}_W \mathbf{b}$ is the **best approximation** to $\mathbf{b}$ from W in the sense that*

$$\|\mathbf{b} - \text{proj}_W \mathbf{b}\| < \|\mathbf{b} - \mathbf{w}\|$$

for every vector $\mathbf{w}$ in W that is different from $\text{proj}_W \mathbf{b}$.

Proof For every vector $\mathbf{w}$ in W, we can write

$$\mathbf{b} - \mathbf{w} = (\mathbf{b} - \text{proj}_W \mathbf{b}) + (\text{proj}_W \mathbf{b} - \mathbf{w}) \tag{1}$$

But $\text{proj}_W \mathbf{b} - \mathbf{w}$ being a difference of vectors in W is itself in W; and since $\mathbf{b} - \text{proj}_W \mathbf{b}$ is orthogonal to W, the two terms on the right side of (1) are orthogonal. Thus, it follows from the Theorem of Pythagoras (Theorem 6.2.3) that

$$\|\mathbf{b} - \mathbf{w}\|^2 = \|\mathbf{b} - \text{proj}_W \mathbf{b}\|^2 + \|\text{proj}_W \mathbf{b} - \mathbf{w}\|^2$$

Since $\mathbf{w} \neq \text{proj}_W \mathbf{b}$, it follows that the second term in this sum is positive, and hence that

$$\|\mathbf{b} - \text{proj}_W \mathbf{b}\|^2 < \|\mathbf{b} - \mathbf{w}\|^2$$

Since norms are nonnegative, it follows (from a property of inequalities) that

$$\|\mathbf{b} - \text{proj}_W \mathbf{b}\| < \|\mathbf{b} - \mathbf{w}\| \quad \blacktriangleleft$$

Least Squares Solutions of Linear Systems

One way to find a least squares solution of $A\mathbf{x} = \mathbf{b}$ is to calculate the orthogonal projection $\text{proj}_W \mathbf{b}$ on the column space W of the matrix A and then solve the equation

$$A\mathbf{x} = \text{proj}_W \mathbf{b} \tag{2}$$

However, we can avoid the need to calculate the projection by rewriting (2) as

$$\mathbf{b} - A\mathbf{x} = \mathbf{b} - \text{proj}_W \mathbf{b}$$

and then multiplying both sides of this equation by A^T to obtain

$$A^T(\mathbf{b} - A\mathbf{x}) = A^T(\mathbf{b} - \text{proj}_W \mathbf{b}) \tag{3}$$

Since $\mathbf{b} - \text{proj}_W \mathbf{b}$ is the component of $\mathbf{b}$ that is orthogonal to the column space of A, it follows from Theorem 4.8.9b that this vector lies in the null space of A^T, and hence that

$$A^T(\mathbf{b} - \text{proj}_W \mathbf{b}) = \mathbf{0}$$

Thus, (3) simplifies to

$$A^T(\mathbf{b} - A\mathbf{x}) = \mathbf{0}$$

which we can rewrite as

$$A^T A\mathbf{x} = A^T \mathbf{b} \tag{4}$$

This is called the ***normal equation*** or the ***normal system*** associated with $A\mathbf{x} = \mathbf{b}$. When viewed as a linear system, the individual equations are called the ***normal equations*** associated with $A\mathbf{x} = \mathbf{b}$.

In summary, we have established the following result.

THEOREM 6.4.2 *For every linear system $A\mathbf{x} = \mathbf{b}$, the associated normal system*

$$A^T A\mathbf{x} = A^T \mathbf{b} \tag{5}$$

is consistent, and all solutions of (5) are least squares solutions of $A\mathbf{x} = \mathbf{b}$. Moreover, if W is the column space of A, and $\mathbf{x}$ is any least squares solution of $A\mathbf{x} = \mathbf{b}$, then the orthogonal projection of $\mathbf{b}$ on W is

$$\text{proj}_W \mathbf{b} = A\mathbf{x} \tag{6}$$

If a linear system is consistent, then its exact solutions are the same as its least squares solutions, in which case the error is zero.

▶ **EXAMPLE 1 Least Squares Solution**

(a) Find all least squares solutions of the linear system

$$\begin{aligned} x_1 - x_2 &= 4 \\ 3x_1 + 2x_2 &= 1 \\ -2x_1 + 4x_2 &= 3 \end{aligned}$$

(b) Find the error vector and the error.

Solution (a) It will be convenient to express the system in the matrix form $A\mathbf{x} = \mathbf{b}$, where

$$A = \begin{bmatrix} 1 & -1 \\ 3 & 2 \\ -2 & 4 \end{bmatrix} \quad \text{and} \quad \mathbf{b} = \begin{bmatrix} 4 \\ 1 \\ 3 \end{bmatrix}$$

It follows that

$$A^T A = \begin{bmatrix} 1 & 3 & -2 \\ -1 & 2 & 4 \end{bmatrix} \begin{bmatrix} 1 & -1 \\ 3 & 2 \\ -2 & 4 \end{bmatrix} = \begin{bmatrix} 14 & -3 \\ -3 & 21 \end{bmatrix}$$

$$A^T \mathbf{b} = \begin{bmatrix} 1 & 3 & -2 \\ -1 & 2 & 4 \end{bmatrix} \begin{bmatrix} 4 \\ 1 \\ 3 \end{bmatrix} = \begin{bmatrix} 1 \\ 10 \end{bmatrix}$$

so the normal system $A^T A \mathbf{x} = A^T \mathbf{b}$ is

$$\begin{bmatrix} 14 & -3 \\ -3 & 21 \end{bmatrix} \begin{bmatrix} x_1 \\ x_2 \end{bmatrix} = \begin{bmatrix} 1 \\ 10 \end{bmatrix}$$

Solving this system yields a unique least squares solution, namely,

$$x_1 = \tfrac{17}{95}, \quad x_2 = \tfrac{143}{285}$$

Solution (b) The error vector is

$$\mathbf{b} - A\mathbf{x} = \begin{bmatrix} 4 \\ 1 \\ 3 \end{bmatrix} - \begin{bmatrix} 1 & -1 \\ 3 & 2 \\ -2 & 4 \end{bmatrix} \begin{bmatrix} \frac{17}{95} \\ \frac{143}{285} \end{bmatrix} = \begin{bmatrix} 4 \\ 1 \\ 3 \end{bmatrix} - \begin{bmatrix} -\frac{92}{285} \\ \frac{439}{285} \\ \frac{95}{57} \end{bmatrix} = \begin{bmatrix} \frac{1232}{285} \\ -\frac{154}{285} \\ \frac{4}{3} \end{bmatrix}$$

and the error is

$$\|\mathbf{b} - A\mathbf{x}\| \approx 4.556$$

▶ **EXAMPLE 2** **Orthogonal Projection on a Subspace**

Find the orthogonal projection of the vector $\mathbf{u} = (-3, -3, 8, 9)$ on the subspace of R^4 spanned by the vectors

$$\mathbf{u}_1 = (3, 1, 0, 1), \quad \mathbf{u}_2 = (1, 2, 1, 1), \quad \mathbf{u}_3 = (-1, 0, 2, -1)$$

Solution We could solve this problem by first using the Gram–Schmidt process to convert $\{\mathbf{u}_1, \mathbf{u}_2, \mathbf{u}_3\}$ into an orthonormal basis and then applying the method used in Example 6 of Section 6.3. However, the following method is more efficient.

The subspace W of R^4 spanned by $\mathbf{u}_1, \mathbf{u}_2$, and $\mathbf{u}_3$ is the column space of the matrix

$$A = \begin{bmatrix} 3 & 1 & -1 \\ 1 & 2 & 0 \\ 0 & 1 & 2 \\ 1 & 1 & -1 \end{bmatrix}$$

Thus, if $\mathbf{u}$ is expressed as a column vector, we can find the orthogonal projection of $\mathbf{u}$ on W by finding a least squares solution of the system $A\mathbf{x} = \mathbf{u}$ and then calculating

$\text{proj}_W \mathbf{u} = A\mathbf{x}$ from the least squares solution. The computations are as follows: The system $A\mathbf{x} = \mathbf{u}$ is

$$\begin{bmatrix} 3 & 1 & -1 \\ 1 & 2 & 0 \\ 0 & 1 & 2 \\ 1 & 1 & -1 \end{bmatrix} \begin{bmatrix} x_1 \\ x_2 \\ x_3 \end{bmatrix} = \begin{bmatrix} -3 \\ -3 \\ 8 \\ 9 \end{bmatrix}$$

so

$$A^T A = \begin{bmatrix} 3 & 1 & 0 & 1 \\ 1 & 2 & 1 & 1 \\ -1 & 0 & 2 & -1 \end{bmatrix} \begin{bmatrix} 3 & 1 & -1 \\ 1 & 2 & 0 \\ 0 & 1 & 2 \\ 1 & 1 & -1 \end{bmatrix} = \begin{bmatrix} 11 & 6 & -4 \\ 6 & 7 & 0 \\ -4 & 0 & 6 \end{bmatrix}$$

$$A^T \mathbf{u} = \begin{bmatrix} 3 & 1 & 0 & 1 \\ 1 & 2 & 1 & 1 \\ -1 & 0 & 2 & -1 \end{bmatrix} \begin{bmatrix} -3 \\ -3 \\ 8 \\ 9 \end{bmatrix} = \begin{bmatrix} -3 \\ 8 \\ 10 \end{bmatrix}$$

The normal system $A^T A \mathbf{x} = A^T \mathbf{u}$ in this case is

$$\begin{bmatrix} 11 & 6 & -4 \\ 6 & 7 & 0 \\ -4 & 0 & 6 \end{bmatrix} \begin{bmatrix} x_1 \\ x_2 \\ x_3 \end{bmatrix} = \begin{bmatrix} -3 \\ 8 \\ 10 \end{bmatrix}$$

Solving this system yields

$$\mathbf{x} = \begin{bmatrix} x_1 \\ x_2 \\ x_3 \end{bmatrix} = \begin{bmatrix} -1 \\ 2 \\ 1 \end{bmatrix}$$

as the least squares solution of $A\mathbf{x} = \mathbf{u}$ (verify), so

$$\text{proj}_W \mathbf{u} = A\mathbf{x} = \begin{bmatrix} 3 & 1 & -1 \\ 1 & 2 & 0 \\ 0 & 1 & 2 \\ 1 & 1 & -1 \end{bmatrix} \begin{bmatrix} -1 \\ 2 \\ 1 \end{bmatrix} = \begin{bmatrix} -2 \\ 3 \\ 4 \\ 0 \end{bmatrix}$$

or, in comma-delimited notation, $\text{proj}_W \mathbf{u} = (-2, 3, 4, 0)$. ◀

Uniqueness of Least Squares Solutions

In general, least squares solutions of linear systems are not unique. Although the linear system in Example 1 turned out to have a unique least squares solution, that occurred only because the coefficient matrix of the system happened to satisfy certain conditions that guarantee uniqueness. Our next theorem will show what those conditions are.

THEOREM 6.4.3 *If A is an $m \times n$ matrix, then the following are equivalent.*

(*a*) *A has linearly independent column vectors.*

(*b*) *$A^T A$ is invertible.*

Proof We will prove that (*a*) $\Rightarrow$ (*b*) and leave the proof that (*b*) $\Rightarrow$ (*a*) as an exercise.

(*a*) $\Rightarrow$ (*b*) Assume that A has linearly independent column vectors. The matrix $A^T A$ has size $n \times n$, so we can prove that this matrix is invertible by showing that the linear system $A^TA\mathbf{x} = \mathbf{0}$ has only the trivial solution. But if $\mathbf{x}$ is any solution of this system, then $A\mathbf{x}$ is in the null space of A^T and also in the column space of A. By Theorem 4.8.9b these spaces are orthogonal complements, so part (b) of Theorem 6.2.4 implies that $A\mathbf{x} = \mathbf{0}$. But A is assumed to have linearly independent column vectors, so $\mathbf{x} = \mathbf{0}$ by Theorem 1.3.1. ◀

The next theorem, which follows directly from Theorems 6.4.2 and 6.4.3, gives an explicit formula for the least squares solution of a linear system in which the coefficient matrix has linearly independent column vectors.

> As an exercise, try using Formula (7) to solve the problem in part (a) of Example 1.

THEOREM 6.4.4 *If A is an m × n matrix with linearly independent column vectors, then for every m × 1 matrix* $\mathbf{b}$*, the linear system $A\mathbf{x} = \mathbf{b}$ has a unique least squares solution. This solution is given by*

$$\mathbf{x} = (A^TA)^{-1}A^T\mathbf{b} \tag{7}$$

Moreover, if W is the column space of A, then the orthogonal projection of $\mathbf{b}$ *on W is*

$$\text{proj}_W \mathbf{b} = A\mathbf{x} = A(A^TA)^{-1}A^T\mathbf{b} \tag{8}$$

OPTIONAL

The Role of QR-Decomposition in Least Squares Problems

Formulas (7) and (8) have theoretical use but are not well suited for numerical computation. In practice, least squares solutions of $A\mathbf{x} = \mathbf{b}$ are typically found by using some variation of Gaussian elimination to solve the normal equations or by using QR-decomposition and the following theorem.

THEOREM 6.4.5 *If A is an m × n matrix with linearly independent column vectors, and if A = QR is a QR-decomposition of A (see Theorem 6.3.7), then for each* $\mathbf{b}$ *in R^m the system $A\mathbf{x} = \mathbf{b}$ has a unique least squares solution given by*

$$\mathbf{x} = R^{-1}Q^T\mathbf{b} \tag{9}$$

A proof of this theorem and a discussion of its use can be found in many books on numerical methods of linear algebra. However, you can obtain Formula (9) by making the substitution $A = QR$ in (7) and using the fact that $Q^TQ = I$ to obtain

$$\begin{aligned}
\mathbf{x} &= \left((QR)^T(QR)\right)^{-1}(QR)^T\mathbf{b} \\
&= (R^TQ^TQR)^{-1}(QR)^T\mathbf{b} \\
&= R^{-1}(R^T)^{-1}R^TQ^T\mathbf{b} \\
&= R^{-1}Q^T\mathbf{b}
\end{aligned}$$

Orthogonal Projections on Subspaces of R^m

In Section 4.8 we showed how to compute orthogonal projections on the coordinate axes of a rectangular coordinate system in R^3 and more generally on lines through the origin of R^3. We will now consider the problem of finding orthogonal projections on *subspaces* of R^m. We begin with the following definition.

DEFINITION 1 If W is a subspace of R^m, then the linear transformation $P: R^m \rightarrow W$ that maps each vector $\mathbf{x}$ in R^m into its orthogonal projection $\text{proj}_W \mathbf{x}$ in W is called the ***orthogonal projection of R^m on W***.

It follows from Formula (7) that the standard matrix for the transformation P is

$$[P] = A(A^T A)^{-1} A^T \tag{10}$$

where A is constructed using any basis for W as its column vectors.

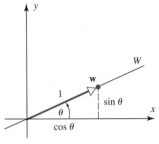

▲ Figure 6.4.2

▶ **EXAMPLE 3 The Standard Matrix for an Orthogonal Projection on a Line**

We showed in Formula (16) of Section 4.9 that

$$P_\theta = \begin{bmatrix} \cos^2\theta & \sin\theta\cos\theta \\ \sin\theta\cos\theta & \sin^2\theta \end{bmatrix}$$

is the standard matrix for the orthogonal projection on the line W through the origin of R^2 that makes an angle θ with the positive x-axis. Derive this result using Formula (10).

Solution The column vectors of A can be formed from any basis for W. Since W is one-dimensional, we can take $\mathbf{w} = (\cos\theta, \sin\theta)$ as the basis vector (Figure 6.4.2), so

$$A = \begin{bmatrix} \cos\theta \\ \sin\theta \end{bmatrix}$$

We leave it for you to show that $A^T A$ is the 1×1 identity matrix. Thus, Formula (10) simplifies to

$$[P] = A(A^T A)^{-1} A^T = AA^T = \begin{bmatrix} \cos\theta \\ \sin\theta \end{bmatrix} [\cos\theta \quad \sin\theta]$$

$$= \begin{bmatrix} \cos^2\theta & \sin\theta\cos\theta \\ \sin\theta\cos\theta & \sin^2\theta \end{bmatrix} = P_\theta \blacktriangleleft$$

Another View of Least Squares

Recall from Theorem 4.8.9 that the null space and row space of an $m \times n$ matrix A are orthogonal complements, as are the null space of A^T and the column space of A. Thus, given a linear system $A\mathbf{x} = \mathbf{b}$ in which A is an $m \times n$ matrix, the Projection Theorem (6.3.3) tells us that the vectors $\mathbf{x}$ and $\mathbf{b}$ can each be decomposed into sums of orthogonal terms as

$$\mathbf{x} = \mathbf{x}_{\text{row}(A)} + \mathbf{x}_{\text{null}(A)} \quad \text{and} \quad \mathbf{b} = \mathbf{b}_{\text{null}(A^T)} + \mathbf{b}_{\text{col}(A)}$$

where $\mathbf{x}_{\text{row}(A)}$ and $\mathbf{x}_{\text{null}(A)}$ are the orthogonal projections of $\mathbf{x}$ on the row space of A and the null space of A, and the vectors $\mathbf{b}_{\text{null}(A^T)}$ and $\mathbf{b}_{\text{col}(A)}$ are the orthogonal projections of $\mathbf{b}$ on the null space of A^T and the column space of A.

In Figure 6.4.3 we have represented the fundamental spaces of A by perpendicular lines in R^n and R^m on which we indicated the orthogonal projections of $\mathbf{x}$ and $\mathbf{b}$. (This, of course, is only pictorial since the fundamental spaces need not be one-dimensional.) The figure shows $A\mathbf{x}$ as a point in the column space of A and conveys that $\mathbf{b}_{\text{col}(A)}$ is the point in col(A) that is closest to $\mathbf{b}$. This illustrates that the least squares solutions of $A\mathbf{x} = \mathbf{b}$ are the exact solutions of the equation $A\mathbf{x} = \mathbf{b}_{\text{col}(A)}$.

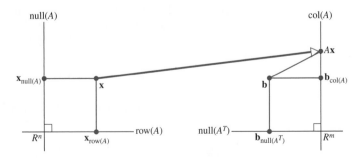

▶ Figure 6.4.3

More on the Equivalence
Theorem

As our final result in the main part of this section we will add one additional part to Theorem 5.1.6.

THEOREM 6.4.6 Equivalent Statements

If A is an n × n matrix, then the following statements are equivalent.

(a) *A is invertible.*

(b) $A\mathbf{x} = \mathbf{0}$ *has only the trivial solution.*

(c) *The reduced row echelon form of A is I_n.*

(d) *A is expressible as a product of elementary matrices.*

(e) $A\mathbf{x} = \mathbf{b}$ *is consistent for every n × 1 matrix* $\mathbf{b}$.

(f) $A\mathbf{x} = \mathbf{b}$ *has exactly one solution for every n × 1 matrix* $\mathbf{b}$.

(g) $\det(A) \neq 0$.

(h) *The column vectors of A are linearly independent.*

(i) *The row vectors of A are linearly independent.*

(j) *The column vectors of A span R^n.*

(k) *The row vectors of A span R^n.*

(l) *The column vectors of A form a basis for R^n.*

(m) *The row vectors of A form a basis for R^n.*

(n) *A has rank n.*

(o) *A has nullity 0.*

(p) *The orthogonal complement of the null space of A is R^n.*

(q) *The orthogonal complement of the row space of A is $\{\mathbf{0}\}$.*

(r) *The range of T_A is R^n.*

(s) T_A *is one-to-one.*

(t) $\lambda = 0$ *is not an eigenvalue of A.*

(u) A^TA *is invertible.*

The proof of part (u) follows from part (h) of this theorem and Theorem 6.4.3 applied to square matrices.

OPTIONAL

We now have all the ingredients needed to prove Theorem 6.3.3 in the special case where V is the vector space R^m.

Proof of Theorem 6.3.3   We will leave the case where $W = \{\mathbf{0}\}$ as an exercise, so assume that $W \neq \{\mathbf{0}\}$. Let $\{\mathbf{v}_1, \mathbf{v}_2, \ldots, \mathbf{v}_k\}$ be any basis for W, and form the $m \times k$ matrix M that has these basis vectors as successive columns. This makes W the column space of M and hence $W^\perp$ the null space of M^T. We will complete the proof by showing that every vector $\mathbf{u}$ in R^m can be written in exactly one way as

$$\mathbf{u} = \mathbf{w}_1 + \mathbf{w}_2$$

where $\mathbf{w}_1$ is in the column space of M and $M^T\mathbf{w}_2 = \mathbf{0}$. However, to say that $\mathbf{w}_1$ is in the column space of M is equivalent to saying $\mathbf{w}_1 = M\mathbf{x}$ for some vector $\mathbf{x}$ in R^m, and to say that $M^T\mathbf{w}_2 = \mathbf{0}$ is equivalent to saying that $M^T(\mathbf{u} - \mathbf{w}_1) = \mathbf{0}$. Thus, if we can show that the equation

$$M^T(\mathbf{u} - M\mathbf{x}) = \mathbf{0} \tag{11}$$

has a unique solution for $\mathbf{x}$, then $\mathbf{w}_1 = M\mathbf{x}$ and $\mathbf{w}_2 = \mathbf{x} - \mathbf{w}_1$ will be uniquely determined vectors with the required properties. To do this, let us rewrite (11) as

$$M^T M\mathbf{x} = M^T\mathbf{u}$$

Since the matrix M has linearly independent column vectors, the matrix $M^T M$ is invertible by Theorem 6.4.6 and hence the equation has a unique solution as required to complete the proof. ◄

Concept Review

- Least squares problem
- Least squares solution
- Least squares error vector
- Least squares error
- Best approximation
- Normal equation
- Orthogonal projection

Skills

- Find the least squares solution of a linear system.
- Find the error and error vector associated with a least squares solution to a linear system.
- Use the techniques developed in this section to compute orthogonal projections.
- Find the standard matrix of an orthogonal projection.

Exercise Set 6.4

1. Find the normal system associated with the given linear system.

(a) $\begin{bmatrix} 1 & -1 \\ 2 & 3 \\ 4 & 5 \end{bmatrix} \begin{bmatrix} x_1 \\ x_2 \end{bmatrix} = \begin{bmatrix} 2 \\ -1 \\ 5 \end{bmatrix}$

(b) $\begin{bmatrix} 2 & -1 & 0 \\ 3 & 1 & 2 \\ -1 & 4 & 5 \\ 1 & 2 & 4 \end{bmatrix} \begin{bmatrix} x_1 \\ x_2 \\ x_3 \end{bmatrix} = \begin{bmatrix} -1 \\ 0 \\ 1 \\ 2 \end{bmatrix}$

▶ In Exercises 2–4, find the least squares solution of the linear equation $A\mathbf{x} = \mathbf{b}$. ◄

2. (a) $A = \begin{bmatrix} 1 & -1 \\ 2 & 3 \\ 4 & 5 \end{bmatrix}$; $\mathbf{b} = \begin{bmatrix} 2 \\ -1 \\ 5 \end{bmatrix}$

(b) $A = \begin{bmatrix} 2 & -2 \\ 1 & 1 \\ 3 & 1 \end{bmatrix}$; $\mathbf{b} = \begin{bmatrix} 2 \\ -1 \\ 1 \end{bmatrix}$

3. (a) $A = \begin{bmatrix} 1 & 1 \\ -1 & 1 \\ -1 & 2 \end{bmatrix}$; $\mathbf{b} = \begin{bmatrix} 7 \\ 0 \\ -7 \end{bmatrix}$

(b) $A = \begin{bmatrix} 1 & 0 & -1 \\ 2 & 1 & -2 \\ 1 & 1 & 0 \\ 1 & 1 & -1 \end{bmatrix}$; $\mathbf{b} = \begin{bmatrix} 6 \\ 0 \\ 9 \\ 3 \end{bmatrix}$

4. (a) $A = \begin{bmatrix} 3 & 2 & -1 \\ 1 & -4 & 3 \\ 1 & 10 & -7 \end{bmatrix}$; $\mathbf{b} = \begin{bmatrix} 2 \\ -2 \\ 1 \end{bmatrix}$

(b) $A = \begin{bmatrix} 2 & 0 & -1 \\ 1 & -2 & 2 \\ 2 & -1 & 0 \\ 0 & 1 & -1 \end{bmatrix}$; $\mathbf{b} = \begin{bmatrix} 0 \\ 6 \\ 0 \\ 6 \end{bmatrix}$

▶ In Exercises 5–6, find the least squares error vector $\mathbf{e} = \mathbf{b} - A\mathbf{x}$ resulting from the least squares solution $\mathbf{x}$ and verify that it is orthogonal to the column space of A. ◄

5. (a) A and $\mathbf{b}$ are as in Exercise 3(a).

(b) A and $\mathbf{b}$ are as in Exercise 3(b).

6. (a) A and $\mathbf{b}$ are as in Exercise 4(a).

(b) A and $\mathbf{b}$ are as in Exercise 4(b).

7. Find all least squares solutions of $A\mathbf{x} = \mathbf{b}$ and confirm that all of the solutions have the same error vector. Compute the least squares error.

(a) $A = \begin{bmatrix} 2 & 1 \\ 4 & 2 \\ -2 & 1 \end{bmatrix}$; $\mathbf{b} = \begin{bmatrix} 3 \\ 2 \\ 1 \end{bmatrix}$

(b) $A = \begin{bmatrix} 1 & 3 \\ -2 & -6 \\ 3 & 9 \end{bmatrix}$; $\mathbf{b} = \begin{bmatrix} 1 \\ 0 \\ 1 \end{bmatrix}$

(c) $A = \begin{bmatrix} -1 & 3 & 2 \\ 2 & 1 & 3 \\ 0 & 1 & 1 \end{bmatrix}$; $\mathbf{b} = \begin{bmatrix} 7 \\ 0 \\ -7 \end{bmatrix}$

8. Find the orthogonal projection of $\mathbf{u}$ on the subspace of R^3 spanned by the vectors $\mathbf{v}_1$ and $\mathbf{v}_2$.

(a) $\mathbf{u} = (2, 1, 3);\quad \mathbf{v}_1 = (1, 1, 0),\quad \mathbf{v}_2 = (1, 2, 1)$

(b) $\mathbf{u} = (1, -6, 1);\quad \mathbf{v}_1 = (-1, 2, 1),\quad \mathbf{v}_2 = (2, 2, 4)$

9. Find the orthogonal projection of $\mathbf{u}$ on the subspace of R^4 spanned by the vectors $\mathbf{v}_1$, $\mathbf{v}_2$, and $\mathbf{v}_3$.

(a) $\mathbf{u} = (6, 3, 9, 6);\ \mathbf{v}_1 = (2, 1, 1, 1),\ \mathbf{v}_2 = (1, 0, 1, 1),$
$\mathbf{v}_3 = (-2, -1, 0, -1)$

(b) $\mathbf{u} = (-2, 0, 2, 4);\ \mathbf{v}_1 = (1, 1, 3, 0),$
$\mathbf{v}_2 = (-2, -1, -2, 1),\ \mathbf{v}_3 = (-3, -1, 1, 3)$

10. Find the orthogonal projection of $\mathbf{u} = (5, 6, 7, 2)$ on the solution space of the homogeneous linear system

$$x_1 + x_2 + x_3 \qquad = 0$$
$$2x_2 + x_3 + x_4 = 0$$

11. In each part, find $\det(A^TA)$, and apply Theorem 6.4.3 to determine whether A has linearly independent column vectors.

(a) $A = \begin{bmatrix} -1 & 3 & 2 \\ 2 & 1 & 3 \\ 0 & 1 & 1 \end{bmatrix}$ (b) $A = \begin{bmatrix} 2 & -1 & 3 \\ 0 & 1 & 1 \\ -1 & 0 & -2 \\ 4 & -5 & 3 \end{bmatrix}$

12. Use Formula (10) and the method of Example 3 to find the standard matrix for the orthogonal projection $P\colon R^2 \to R^2$ onto

(a) the x-axis. (b) the y-axis.

[*Note:* Compare your results to Table 3 of Section 4.9.]

13. Use Formula (10) and the method of Example 3 to find the standard matrix for the orthogonal projection $P\colon R^3 \to R^3$ onto

(a) the xz-plane. (b) the yz-plane.

[*Note:* Compare your results to Table 4 of Section 4.9.]

14. Show that if $\mathbf{w} = (a, b, c)$ is a nonzero vector, then the standard matrix for the orthogonal projection of R^3 on the line span$\{\mathbf{w}\}$ is

$$P = \frac{1}{a^2 + b^2 + c^2} \begin{bmatrix} a^2 & ab & ac \\ ab & b^2 & bc \\ ac & bc & c^2 \end{bmatrix}$$

15. Let W be the plane with equation $5x - 3y + z = 0$.

(a) Find a basis for W.

(b) Use Formula (10) to find the standard matrix for the orthogonal projection on W.

(c) Use the matrix obtained in part (b) to find the orthogonal projection of a point $P_0(x_0, y_0, z_0)$ on W.

(d) Find the distance between the point $P_0(1, -2, 4)$ and the plane W, and check your result using Theorem 3.3.3.

16. Let W be the line with parametric equations

$$x = 2t, \quad y = -t, \quad z = 4t$$

(a) Find a basis for W.

(b) Use Formula (10) to find the standard matrix for the orthogonal projection on W.

(c) Use the matrix obtained in part (b) to find the orthogonal projection of a point $P_0(x_0, y_0, z_0)$ on W.

(d) Find the distance between the point $P_0(2, 1, -3)$ and the line W.

17. In R^3, consider the line l given by the equations

$$x = t, \quad y = t, \quad z = t$$

and the line m given by the equations

$$x = s, \quad y = 2s - 1, \quad z = 1$$

Let P be a point on l, and let Q be a point on m. Find the values of t and s that minimize the distance between the lines by minimizing the squared distance $\|P - Q\|^2$.

18. Prove: If A has linearly independent column vectors, and if $A\mathbf{x} = \mathbf{b}$ is consistent, then the least squares solution of $A\mathbf{x} = \mathbf{b}$ and the exact solution of $A\mathbf{x} = \mathbf{b}$ are the same.

19. Prove: If A has linearly independent column vectors, and if $\mathbf{b}$ is orthogonal to the column space of A, then the least squares solution of $A\mathbf{x} = \mathbf{b}$ is $\mathbf{x} = \mathbf{0}$.

20. Let $P\colon R^m \to W$ be the orthogonal projection of R^m onto a subspace W.

(a) Prove that $[P]^2 = [P]$.

(b) What does the result in part (a) imply about the composition $P \circ P$?

(c) Show that $[P]$ is symmetric.

21. Let A be an $m \times n$ matrix with linearly independent row vectors. Find a standard matrix for the orthogonal projection of R^n onto the row space of A. [*Hint:* Start with Formula (10).]

22. Prove the implication $(b) \Rightarrow (a)$ of Theorem 6.4.3.

True-False Exercises

In parts (a)–(h) determine whether the statement is true or false, and justify your answer.

(a) If A is an $m \times n$ matrix, then A^TA is a square matrix.

(b) If A^TA is invertible, then A is invertible.

(c) If A is invertible, then A^TA is invertible.

(d) If $A\mathbf{x} = \mathbf{b}$ is a consistent linear system, then $A^TA\mathbf{x} = A^T\mathbf{b}$ is also consistent.

(e) If $A\mathbf{x} = \mathbf{b}$ is an inconsistent linear system, then $A^TA\mathbf{x} = A^T\mathbf{b}$ is also inconsistent.

(f) Every linear system has a least squares solution.

(g) Every linear system has a unique least squares solution.

(h) If A is an $m \times n$ matrix with linearly independent columns and $\mathbf{b}$ is in R^m, then $A\mathbf{x} = \mathbf{b}$ has a unique least squares solution.

6.5 Least Squares Fitting to Data

In this section we will use results about orthogonal projections in inner product spaces to obtain a technique for fitting a line or other polynomial curve to a set of experimentally determined points in the plane.

Fitting a Curve to Data   A common problem in experimental work is to obtain a mathematical relationship $y = f(x)$ between two variables x and y by "fitting" a curve to points in the plane corresponding to various experimentally determined values of x and y, say

$$(x_1, y_1), (x_2, y_2), \ldots, (x_n, y_n)$$

On the basis of theoretical considerations or simply by observing the pattern of the points, the experimenter decides on the general form of the curve $y = f(x)$ to be fitted. Some possibilities are (Figure 6.5.1)

(a)   A straight line: $y = a + bx$

(b)   A quadratic polynomial: $y = a + bx + cx^2$

(c)   A cubic polynomial: $y = a + bx + cx^2 + dx^3$

Because the points are obtained experimentally, there is often some measurement "error" in the data, making it impossible to find a curve of the desired form that passes through all the points. Thus, the idea is to choose the curve (by determining its coefficients) that "best" fits the data. We begin with the simplest and most common case: fitting a straight line to data points.

  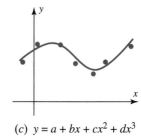

▶ Figure 6.5.1   (a) $y = a + bx$ (b) $y = a + bx + cx^2$ (c) $y = a + bx + cx^2 + dx^3$

Least Squares Fit of a   Suppose we want to fit a straight line $y = a + bx$ to the experimentally determined points
Straight Line

$$(x_1, y_1), (x_2, y_2), \ldots, (x_n, y_n)$$

If the data points were collinear, the line would pass through all n points, and the unknown coefficients a and b would satisfy the equations

$$y_1 = a + bx_1$$
$$y_2 = a + bx_2$$
$$\vdots$$
$$y_n = a + bx_n$$

We can write this system in matrix form as

$$\begin{bmatrix} 1 & x_1 \\ 1 & x_2 \\ \vdots & \vdots \\ 1 & x_n \end{bmatrix} \begin{bmatrix} a \\ b \end{bmatrix} = \begin{bmatrix} y_1 \\ y_2 \\ \vdots \\ y_n \end{bmatrix}$$

or more compactly as

$$Mv = y \tag{1}$$

where

$$\mathbf{y} = \begin{bmatrix} y_1 \\ y_2 \\ \vdots \\ y_n \end{bmatrix}, \quad M = \begin{bmatrix} 1 & x_1 \\ 1 & x_2 \\ \vdots & \vdots \\ 1 & x_n \end{bmatrix}, \quad \mathbf{v} = \begin{bmatrix} a \\ b \end{bmatrix} \tag{2}$$

If the data points are not collinear, then it is impossible to find coefficients a and b that satisfy system (1) exactly; that is, the system is inconsistent. In this case we will look for a least squares solution

$$\mathbf{v} = \mathbf{v}^* = \begin{bmatrix} a^* \\ b^* \end{bmatrix}$$

We call a line $y = a^* + b^*x$ whose coefficients come from a least squares solution a *regression line* or a *least squares straight line fit* to the data. To explain this terminology, recall that a least squares solution of (1) minimizes

$$\|\mathbf{y} - M\mathbf{v}\| \tag{3}$$

If we express the square of (3) in terms of components, we obtain

$$\|\mathbf{y} - M\mathbf{v}\|^2 = (y_1 - a - bx_1)^2 + (y_2 - a - bx_2)^2 + \cdots + (y_n - a - bx_n)^2 \tag{4}$$

If we now let

$$d_1 = |y_1 - a - bx_1|, \quad d_2 = |y_2 - a - bx_2|, \dots, \quad d_n = |y_n - a - bx_n|$$

then (4) can be written as

$$\|\mathbf{y} - M\mathbf{v}\|^2 = d_1^2 + d_2^2 + \cdots + d_n^2 \tag{5}$$

As illustrated in Figure 6.5.2, the number d_i can be interpreted as the vertical distance between the line $y = a + bx$ and the data point (x_i, y_i). This distance is a measure of the "error" at the point (x_i, y_i) resulting from the inexact fit of $y = a + bx$ to the data points, the assumption being that the x_i are known exactly and that all the error is in the measurement of the y_i. Since (3) and (5) are minimized by the same vector $\mathbf{v}^*$, the least squares straight line fit minimizes the sum of the squares of the estimated errors d_i, hence the name *least squares straight line fit*.

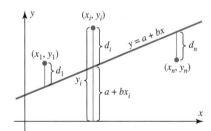

▶ **Figure 6.5.2** d_i measures the vertical error in the least squares straight line.

Normal Equations   Recall from Theorem 6.4.2 that the least squares solutions of (1) can be obtained by solving the associated normal system

$$M^T M \mathbf{v} = M^T \mathbf{y}$$

the equations of which are called the *normal equations*.

In the exercises it will be shown that the column vectors of M are linearly independent if and only if the n data points do not lie on a vertical line in the xy-plane. In this case it follows from Theorem 6.4.4 that the least squares solution is unique and is given by

$$\mathbf{v}^* = (M^T M)^{-1} M^T \mathbf{y}$$

In summary, we have the following theorem.

THEOREM 6.5.1 **Uniqueness of the Least Squares Solution**

Let $(x_1, y_1), (x_2, y_2), \ldots, (x_n, y_n)$ be a set of two or more data points, not all lying on a vertical line, and let

$$M = \begin{bmatrix} 1 & x_1 \\ 1 & x_2 \\ \vdots & \vdots \\ 1 & x_n \end{bmatrix} \quad and \quad \mathbf{y} = \begin{bmatrix} y_1 \\ y_2 \\ \vdots \\ y_n \end{bmatrix}$$

Then there is a unique least squares straight line fit

$$y = a^* + b^* x$$

to the data points. Moreover,

$$\mathbf{v}^* = \begin{bmatrix} a^* \\ b^* \end{bmatrix}$$

is given by the formula

$$\mathbf{v}^* = (M^T M)^{-1} M^T \mathbf{y} \tag{6}$$

which expresses the fact that $\mathbf{v} = \mathbf{v}^*$ is the unique solution of the normal equations

$$M^T M \mathbf{v} = M^T \mathbf{y} \tag{7}$$

▶ **EXAMPLE 1** **Least Squares Straight Line Fit**

Find the least squares straight line fit to the four points $(0, 1)$, $(1, 3)$, $(2, 4)$, and $(3, 4)$. (See Figure 6.5.3.)

Solution We have

$$M = \begin{bmatrix} 1 & 0 \\ 1 & 1 \\ 1 & 2 \\ 1 & 3 \end{bmatrix}, \quad M^T M = \begin{bmatrix} 4 & 6 \\ 6 & 14 \end{bmatrix}, \quad and \quad (M^T M)^{-1} = \frac{1}{10} \begin{bmatrix} 7 & -3 \\ -3 & 2 \end{bmatrix}$$

$$\mathbf{v}^* = (M^T M)^{-1} M^T \mathbf{y} = \frac{1}{10} \begin{bmatrix} 7 & -3 \\ -3 & 2 \end{bmatrix} \begin{bmatrix} 1 & 1 & 1 & 1 \\ 0 & 1 & 2 & 3 \end{bmatrix} \begin{bmatrix} 1 \\ 3 \\ 4 \\ 4 \end{bmatrix} = \begin{bmatrix} 1.5 \\ 1 \end{bmatrix}$$

so the desired line is $y = 1.5 + x$.

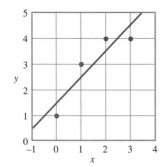

▲ Figure 6.5.3

▶ **EXAMPLE 2** **Spring Constant**

Hooke's law in physics states that the length x of a uniform spring is a linear function of the force y applied to it. If we express this relationship as $y = a + bx$, then the coefficient b is called the ***spring constant***. Suppose a particular unstretched spring has a measured length of 6.1 inches (i.e., $x = 6.1$ when $y = 0$). Forces of 2 pounds, 4 pounds,

Force y

▲ Figure 6.5.4

x_i	y_i
6.1	0
7.6	2
8.7	4
10.4	6

and 6 pounds are then applied to the spring, and the corresponding lengths are found to be 7.6 inches, 8.7 inches, and 10.4 inches (see Figure 6.5.4). Find the spring constant.

Solution We have

$$
M = \begin{bmatrix} 1 & 6.1 \\ 1 & 7.6 \\ 1 & 8.7 \\ 1 & 10.4 \end{bmatrix}, \quad \mathbf{y} = \begin{bmatrix} 0 \\ 2 \\ 4 \\ 6 \end{bmatrix},
$$

and

$$
\mathbf{v}^* = \begin{bmatrix} a^* \\ b^* \end{bmatrix} = (M^T M)^{-1} M^T \mathbf{y} \approx \begin{bmatrix} -8.6 \\ 1.4 \end{bmatrix}
$$

where the numerical values have been rounded to one decimal place. Thus, the estimated value of the spring constant is $b^* \approx 1.4$ pounds/inch. ◀

Least Squares Fit of a Polynomial

The technique described for fitting a straight line to data points can be generalized to fitting a polynomial of specified degree to data points. Let us attempt to fit a polynomial of fixed degree m

$$
y = a_0 + a_1 x + \cdots + a_m x^m \tag{8}
$$

to n points

$$
(x_1, y_1), (x_2, y_2), \ldots, (x_n, y_n)
$$

Substituting these n values of x and y into (8) yields the n equations

$$
\begin{aligned}
y_1 &= a_0 + a_1 x_1 + \cdots + a_m x_1^m \\
y_2 &= a_0 + a_1 x_2 + \cdots + a_m x_2^m \\
&\;\;\vdots \qquad \vdots \qquad \vdots \qquad\;\; \vdots \\
y_n &= a_0 + a_1 x_n + \cdots + a_m x_n^m
\end{aligned}
$$

or, in matrix form,

$$
\mathbf{y} = M\mathbf{v} \tag{9}
$$

where

$$
\mathbf{y} = \begin{bmatrix} y_1 \\ y_2 \\ \vdots \\ y_n \end{bmatrix}, \quad M = \begin{bmatrix} 1 & x_1 & x_1^2 & \cdots & x_1^m \\ 1 & x_2 & x_2^2 & \cdots & x_2^m \\ \vdots & \vdots & \vdots & & \vdots \\ 1 & x_n & x_n^2 & \cdots & x_n^m \end{bmatrix}, \quad \mathbf{v} = \begin{bmatrix} a_0 \\ a_1 \\ \vdots \\ a_m \end{bmatrix} \tag{10}
$$

Altitude h (km)

Source: NASA

Historical Note On October 5, 1991 the *Magellan* spacecraft entered the atmosphere of Venus and transmitted the temperature T in kelvins (K) versus the altitude h in kilometers (km) until its signal was lost at an altitude of about 34 km. Discounting the initial erratic signal, the data strongly suggested a linear relationship, so a least squares straight line fit was used on the linear part of the data to obtain the equation

$$
T = 737.5 - 8.125h
$$

By setting $h = 0$ in this equation, the surface temperature of Venus was estimated at $T \approx 737.5$ K.

As before, the solutions of the normal equations

$$M^T M \mathbf{v} = M^T \mathbf{y}$$

determine the coefficients of the polynomial, and the vector $\mathbf{v}$ minimizes

$$\|\mathbf{y} - M\mathbf{v}\|$$

Conditions that guarantee the invertibility of $M^T M$ are discussed in the exercises (Exercise 7). If $M^T M$ is invertible, then the normal equations have a unique solution $\mathbf{v} = \mathbf{v}^*$, which is given by

$$\mathbf{v}^* = (M^T M)^{-1} M^T \mathbf{y} \tag{11}$$

▶ **EXAMPLE 3** Fitting a Quadratic Curve to Data

According to Newton's second law of motion, a body near the Earth's surface falls vertically downward according to the equation

$$s = s_0 + v_0 t + \tfrac{1}{2} g t^2 \tag{12}$$

where

s = vertical displacement downward relative to some fixed point

s_0 = initial displacement at time $t = 0$

v_0 = initial velocity at time $t = 0$

g = acceleration of gravity at the Earth's surface

from Equation (12) by releasing a weight with unknown initial displacement and velocity and measuring the distance it has fallen at certain times relative to a fixed reference point. Suppose that a laboratory experiment is performed to evaluate g. Suppose it is found that at times $t = .1, .2, .3, .4,$ and $.5$ seconds the weight has fallen $s = -0.18, 0.31, 1.03, 2.48,$ and 3.73 feet, respectively, from the reference point. Find an approximate value of g using these data.

Solution The mathematical problem is to fit a quadratic curve

$$s = a_0 + a_1 t + a_2 t^2 \tag{13}$$

to the five data points:

$$(.1, -0.18), \quad (.2, 0.31), \quad (.3, 1.03), \quad (.4, 2.48), \quad (.5, 3.73)$$

With the appropriate adjustments in notation, the matrices M and $\mathbf{y}$ in (10) are

$$M = \begin{bmatrix} 1 & t_1 & t_1^2 \\ 1 & t_2 & t_2^2 \\ 1 & t_3 & t_3^2 \\ 1 & t_4 & t_4^2 \\ 1 & t_5 & t_5^2 \end{bmatrix} = \begin{bmatrix} 1 & .1 & .01 \\ 1 & .2 & .04 \\ 1 & .3 & .09 \\ 1 & .4 & .16 \\ 1 & .5 & .25 \end{bmatrix}, \quad \mathbf{y} = \begin{bmatrix} s_1 \\ s_2 \\ s_3 \\ s_4 \\ s_5 \end{bmatrix} = \begin{bmatrix} -0.18 \\ 0.31 \\ 1.03 \\ 2.48 \\ 3.73 \end{bmatrix}$$

Thus, from (11),

$$\mathbf{v}^* = \begin{bmatrix} a_0^* \\ a_1^* \\ a_2^* \end{bmatrix} = (M^T M)^{-1} M^T \mathbf{y} \doteq \begin{bmatrix} -0.40 \\ 0.35 \\ 16.1 \end{bmatrix}$$

From (12) and (13), we have $a_2 = \tfrac{1}{2} g$, so the estimated value of g is

$$g = 2a_2^* = 2(16.1) = 32.2 \text{ feet/second}^2$$

If desired, we can also estimate the initial displacement and initial velocity of the weight:

$$s_0 = a_0^* = -0.40 \text{ feet}$$
$$v_0 = a_1^* = 0.35 \text{ feet/second}$$

In Figure 6.5.5 we have plotted the five data points and the approximating polynomial.

◀

Concept Review

- Least squares straight line fit
- Regression line
- Least squares polynomial fit

Skills

- Find the least squares straight line fit to a set of data points.
- Find the least squares polynomial fit to a set of data points.
- Use the techniques of this section to solve applied problems.

Exercise Set 6.5

1. Find the least squares straight line fit to the three points $(0, 0)$, $(1, 2)$, and $(2, 7)$.

2. Find the least squares straight line fit to the four points $(0, 1)$, $(2, 0)$, $(3, 1)$, and $(3, 2)$.

3. Find the quadratic polynomial that best fits the four points $(2, 0)$, $(3, -10)$, $(5, -48)$, and $(6, -76)$.

4. Find the cubic polynomial that best fits the five points $(-1, -14)$, $(0, -5)$, $(1, -4)$, $(2, 1)$, and $(3, 22)$.

5. Show that the matrix M in Equation (2) has linearly independent columns if and only if at least two of the numbers $x_1, x_2, \ldots, x_n$ are distinct.

6. Show that the columns of the $n \times (m + 1)$ matrix M in Equation (10) are linearly independent if $n > m$ and at least $m + 1$ of the numbers $x_1, x_2, \ldots, x_n$ are distinct. [*Hint:* A nonzero polynomial of degree m has at most m distinct roots.]

7. Let M be the matrix in Equation (10). Using Exercise 6, show that a sufficient condition for the matrix $M^T M$ to be invertible is that $n > m$ and that at least $m + 1$ of the numbers $x_1, x_2, \ldots, x_n$ are distinct.

8. The owner of a rapidly expanding business finds that for the first five months of the year the sales (in thousands) are $4.0, $4.4, $5.2, $6.4, and $8.0. The owner plots these figures on a graph and conjectures that for the rest of the year, the sales curve can be approximated by a quadratic polynomial. Find the least squares quadratic polynomial fit to the sales curve, and use it to project the sales for the twelfth month of the year.

9. A corporation obtains the following data relating the number of sales representatives on its staff to annual sales:

Number of Sales Representatives	5	10	15	20	25	30
Annual Sales (millions)	3.4	4.3	5.2	6.1	7.2	8.3

Explain how you might use least squares methods to estimate the annual sales with 45 representatives, and discuss the assumptions that you are making. (You need not perform the actual computations.)

10. *Pathfinder* is an experimental, lightweight, remotely piloted, solar-powered aircraft that was used in a series of experiments by NASA to determine the feasibility of applying solar power for long-duration, high-altitude flight. In August 1997 *Pathfinder* recorded the data in the accompanying table relating altitude H and temperature T. Show that a linear model is reasonable by plotting the data, and then find the least squares line $H = H_0 + kT$ of best fit.

Table Ex-10

Altitude H (thousands of feet)	15	20	25	30	35	40	45
Temperature T (°C)	4.5	-5.9	-16.1	-27.6	-39.8	-50.2	-62.9

11. Find a curve of the form $y = a + (b/x)$ that best fits the data points $(1, 7)$, $(3, 3)$, $(6, 1)$ by making the substitution $X = 1/x$. Draw the curve and plot the data points in the same coordinate system.

True-False Exercises

In parts (a)–(d) determine whether the statement is true or false, and justify your answer.

(a) Every set of data points has a unique least squares straight line fit.

(b) If the data points (x_1, y_1), (x_2, y_2), $\ldots$, (x_n, y_n) are not collinear, then (1) is an inconsistent system.

(c) If $y = a + bx$ is the least squares line fit to the data points (x_1, y_1), (x_2, y_2), $\ldots$, (x_n, y_n), then $d_i = |y_i - (a + bx_i)|$ is minimal for every $1 \le i \le n$.

(d) If $y = a + bx$ is the least squares line fit to the data points (x_1, y_1), (x_2, y_2), $\ldots$, (x_n, y_n), then $\sum_{i=1}^{n} |y_i - (a + bx_i)|^2$ is minimal.

6.6 Function Approximation; Fourier Series

In this section we will show how orthogonal projections can be used to approximate certain types of functions by simpler functions that are easier to work with. The ideas explained here have important applications in engineering and science. Calculus is required.

Best Approximations All of the problems that we will study in this section will be special cases of the following general problem.

> **Approximation Problem** Given a function f that is continuous on an interval $[a, b]$, find the "best possible approximation" to f using only functions from a specified subspace W of $C[a, b]$.

Here are some examples of such problems:

(a) Find the best possible approximation to e^x over $[0, 1]$ by a polynomial of the form $a_0 + a_1 x + a_2 x^2$.

(b) Find the best possible approximation to $\sin \pi x$ over $[-1, 1]$ by a function of the form $a_0 + a_1 e^x + a_2 e^{2x} + a_3 e^{3x}$.

(c) Find the best possible approximation to x over $[0, 2\pi]$ by a function of the form $a_0 + a_1 \sin x + a_2 \sin 2x + b_1 \cos x + b_2 \cos 2x$.

In the first example W is the subspace of $C[0, 1]$ spanned by 1, x, and x^2; in the second example W is the subspace of $C[-1, 1]$ spanned by 1, e^x, e^{2x}, and e^{3x}; and in the third example W is the subspace of $C[0, 2\pi]$ spanned by 1, $\sin x$, $\sin 2x$, $\cos x$, and $\cos 2x$.

Measurements of Error To solve approximation problems of the preceding types, we first need to make the phrase "best approximation over $[a, b]$" mathematically precise. To do this we will need some way of quantifying the error that results when one continuous function is approximated by another over an interval $[a, b]$. If we were to approximate $f(x)$ by $g(x)$, and if we were concerned only with the error in that approximation at a *single point* x_0, then it would be natural to define the error to be

$$\text{error} = |f(x_0) - g(x_0)|$$

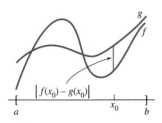

▲ **Figure 6.6.1** The deviation between f and g at x_0.

sometimes called the *deviation* between f and g at x_0 (Figure 6.6.1). However, we are not concerned simply with measuring the error at a single point but rather with measuring it over the *entire* interval $[a, b]$. The problem is that an approximation may have small deviations in one part of the interval and large deviations in another. One possible way of accounting for this is to integrate the deviation $|f(x) - g(x)|$ over the interval $[a, b]$ and define the error over the interval to be

$$\text{error} = \int_a^b |f(x) - g(x)| \, dx \tag{1}$$

Geometrically, (1) is the area between the graphs of $f(x)$ and $g(x)$ over the interval $[a, b]$ (Figure 6.6.2); the greater the area, the greater the overall error.

Although (1) is natural and appealing geometrically, most mathematicians and scientists generally favor the following alternative measure of error, called the ***mean square error***:

$$\text{mean square error} = \int_a^b [f(x) - g(x)]^2 \, dx$$

Mean square error emphasizes the effect of larger errors because of the squaring and has the added advantage that it allows us to bring to bear the theory of inner product spaces. To see how, suppose that $\mathbf{f}$ is a continuous function on $[a, b]$ that we want to approximate by a function $\mathbf{g}$ from a subspace W of $C[a, b]$, and suppose that $C[a, b]$ is given the inner product

$$\langle \mathbf{f}, \mathbf{g} \rangle = \int_a^b f(x)g(x) \, dx$$

It follows that

$$\|\mathbf{f} - \mathbf{g}\|^2 = \langle \mathbf{f} - \mathbf{g}, \mathbf{f} - \mathbf{g} \rangle = \int_a^b [f(x) - g(x)]^2 \, dx = \text{mean square error}$$

so minimizing the mean square error is the same as minimizing $\|\mathbf{f} - \mathbf{g}\|^2$. Thus the approximation problem posed informally at the beginning of this section can be restated more precisely as follows.

▲ **Figure 6.6.2** The area between the graphs of **f** and **g** over $[a, b]$ measures the error in approximating f by g over $[a, b]$.

Least Squares Approximation

Least Squares Approximation Problem Let **f** be a function that is continuous on an interval $[a, b]$, let $C[a, b]$ have the inner product

$$\langle \mathbf{f}, \mathbf{g} \rangle = \int_a^b f(x)g(x) \, dx$$

and let W be a finite-dimensional subspace of $C[a, b]$. Find a function **g** in W that minimizes

$$\|\mathbf{f} - \mathbf{g}\|^2 = \int_a^b [f(x) - g(x)]^2 \, dx$$

Since $\|\mathbf{f} - \mathbf{g}\|^2$ and $\|\mathbf{f} - \mathbf{g}\|$ are minimized by the same function **g**, this problem is equivalent to looking for a function **g** in W that is closest to **f**. But we know from Theorem 6.4.1 that $\mathbf{g} = \text{proj}_W \mathbf{f}$ is such a function (Figure 6.6.3).

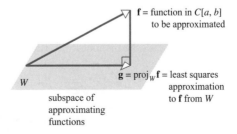

f = function in $C[a, b]$ to be approximated

g = $\text{proj}_W \mathbf{f}$ = least squares approximation to **f** from W

W

subspace of approximating functions

▶ **Figure 6.6.3**

Thus, we have the following result.

THEOREM 6.6.1 *If* **f** *is a continuous function on* $[a, b]$, *and* W *is a finite-dimensional subspace of* $C[a, b]$, *then the function* **g** *in* W *that minimizes the mean square error*

$$\int_a^b [f(x) - g(x)]^2 \, dx$$

is $\mathbf{g} = \mathrm{proj}_W \mathbf{f}$, *where the orthogonal projection is relative to the inner product*

$$\langle \mathbf{f}, \mathbf{g} \rangle = \int_a^b f(x)g(x) \, dx$$

The function $\mathbf{g} = \mathrm{proj}_W \mathbf{f}$ *is called the **least squares approximation** to* **f** *from* W.

Fourier Series   A function of the form

$$T(x) = c_0 + c_1 \cos x + c_2 \cos 2x + \cdots + c_n \cos nx$$
$$+ d_1 \sin x + d_2 \sin 2x + \cdots + d_n \sin nx \tag{2}$$

is called a ***trigonometric polynomial***; if c_n and d_n are not both zero, then $T(x)$ is said to have **order n**. For example,

$$T(x) = 2 + \cos x - 3 \cos 2x + 7 \sin 4x$$

is a trigonometric polynomial of order 4 with

$$c_0 = 2, \quad c_1 = 1, \quad c_2 = -3, \quad c_3 = 0, \quad c_4 = 0, \quad d_1 = 0, \quad d_2 = 0, \quad d_3 = 0, \quad d_4 = 7$$

It is evident from (2) that the trigonometric polynomials of order n or less are the various possible linear combinations of

$$1, \quad \cos x, \quad \cos 2x, \ldots, \quad \cos nx, \qquad \sin x, \quad \sin 2x, \ldots, \quad \sin nx \tag{3}$$

It can be shown that these $2n + 1$ functions are linearly independent and thus form a basis for a $(2n + 1)$-dimensional subspace of $C[a, b]$.

Let us now consider the problem of finding the least squares approximation of a continuous function $f(x)$ over the interval $[0, 2\pi]$ by a trigonometric polynomial of order n or less. As noted above, the least squares approximation to **f** from W is the orthogonal projection of **f** on W. To find this orthogonal projection, we must find an orthonormal basis $\mathbf{g}_0, \mathbf{g}_1, \ldots, \mathbf{g}_{2n}$ for W, after which we can compute the orthogonal projection on W from the formula

$$\mathrm{proj}_W \mathbf{f} = \langle \mathbf{f}, \mathbf{g}_0 \rangle \mathbf{g}_0 + \langle \mathbf{f}, \mathbf{g}_1 \rangle \mathbf{g}_1 + \cdots + \langle \mathbf{f}, \mathbf{g}_{2n} \rangle \mathbf{g}_{2n} \tag{4}$$

(see Theorem 6.3.4b). An orthonormal basis for W can be obtained by applying the Gram–Schmidt process to the basis vectors in (3) using the inner product

$$\langle \mathbf{f}, \mathbf{g} \rangle = \int_0^{2\pi} f(x)g(x) \, dx$$

This yields the orthonormal basis

$$\mathbf{g}_0 = \frac{1}{\sqrt{2\pi}}, \quad \mathbf{g}_1 = \frac{1}{\sqrt{\pi}} \cos x, \ldots, \quad \mathbf{g}_n = \frac{1}{\sqrt{\pi}} \cos nx,$$

$$\mathbf{g}_{n+1} = \frac{1}{\sqrt{\pi}} \sin x, \ldots, \quad \mathbf{g}_{2n} = \frac{1}{\sqrt{\pi}} \sin nx \tag{5}$$

(see Exercise 6). If we introduce the notation

$$a_0 = \frac{2}{\sqrt{2\pi}}\langle \mathbf{f}, \mathbf{g}_0 \rangle, \quad a_1 = \frac{1}{\sqrt{\pi}}\langle \mathbf{f}, \mathbf{g}_1 \rangle, \ldots, \quad a_n = \frac{1}{\sqrt{\pi}}\langle \mathbf{f}, \mathbf{g}_n \rangle$$

$$b_1 = \frac{1}{\sqrt{\pi}}\langle \mathbf{f}, \mathbf{g}_{n+1} \rangle, \ldots, \quad b_n = \frac{1}{\sqrt{\pi}}\langle \mathbf{f}, \mathbf{g}_{2n} \rangle \tag{6}$$

then on substituting (5) in (4), we obtain

$$\text{proj}_W\,\mathbf{f} = \frac{a_0}{2} + [a_1 \cos x + \cdots + a_n \cos nx] + [b_1 \sin x + \cdots + b_n \sin nx] \tag{7}$$

where

$$a_0 = \frac{2}{\sqrt{2\pi}}\langle \mathbf{f}, \mathbf{g}_0 \rangle = \frac{2}{\sqrt{2\pi}} \int_0^{2\pi} f(x) \frac{1}{\sqrt{2\pi}}\,dx = \frac{1}{\pi}\int_0^{2\pi} f(x)\,dx$$

$$a_1 = \frac{1}{\sqrt{\pi}}\langle \mathbf{f}, \mathbf{g}_1 \rangle = \frac{1}{\sqrt{\pi}} \int_0^{2\pi} f(x) \frac{1}{\sqrt{\pi}}\cos x\,dx = \frac{1}{\pi}\int_0^{2\pi} f(x) \cos x\,dx$$

$$\vdots$$

$$a_n = \frac{1}{\sqrt{\pi}}\langle \mathbf{f}, \mathbf{g}_n \rangle = \frac{1}{\sqrt{\pi}} \int_0^{2\pi} f(x) \frac{1}{\sqrt{\pi}}\cos nx\,dx = \frac{1}{\pi}\int_0^{2\pi} f(x) \cos nx\,dx$$

$$b_1 = \frac{1}{\sqrt{\pi}}\langle \mathbf{f}, \mathbf{g}_{n+1} \rangle = \frac{1}{\sqrt{\pi}} \int_0^{2\pi} f(x) \frac{1}{\sqrt{\pi}}\sin x\,dx = \frac{1}{\pi}\int_0^{2\pi} f(x) \sin x\,dx$$

$$\vdots$$

$$b_n = \frac{1}{\sqrt{\pi}}\langle \mathbf{f}, \mathbf{g}_{2n} \rangle = \frac{1}{\sqrt{\pi}} \int_0^{2\pi} f(x) \frac{1}{\sqrt{\pi}}\sin nx\,dx = \frac{1}{\pi}\int_0^{2\pi} f(x) \sin nx\,dx$$

In short,

$$a_k = \frac{1}{\pi}\int_0^{2\pi} f(x) \cos kx\,dx, \quad b_k = \frac{1}{\pi}\int_0^{2\pi} f(x) \sin kx\,dx \tag{8}$$

The numbers $a_0, a_1, \ldots, a_n, b_1, \ldots, b_n$ are called the ***Fourier coefficients*** of $\mathbf{f}$.

▶ **EXAMPLE 1** **Least Squares Approximations**

Find the least squares approximation of $f(x) = x$ on $[0, 2\pi]$ by

(a)   a trigonometric polynomial of order 2 or less;

(b)   a trigonometric polynomial of order n or less.

Solution (a)

$$a_0 = \frac{1}{\pi}\int_0^{2\pi} f(x)\,dx = \frac{1}{\pi}\int_0^{2\pi} x\,dx = 2\pi \tag{9a}$$

For $k = 1, 2, \ldots$, integration by parts yields (verify)

$$a_k = \frac{1}{\pi}\int_0^{2\pi} f(x) \cos kx\,dx = \frac{1}{\pi}\int_0^{2\pi} x \cos kx\,dx = 0 \tag{9b}$$

$$b_k = \frac{1}{\pi}\int_0^{2\pi} f(x) \sin kx\,dx = \frac{1}{\pi}\int_0^{2\pi} x \sin kx\,dx = -\frac{2}{k} \tag{9c}$$

Thus, the least squares approximation to x on $[0, 2\pi]$ by a trigonometric polynomial of order 2 or less is

$$x \approx \frac{a_0}{2} + a_1 \cos x + a_2 \cos 2x + b_1 \sin x + b_2 \sin 2x$$

or, from (9a), (9b), and (9c),

$$x \approx \pi - 2 \sin x - \sin 2x$$

Solution (b) The least squares approximation to x on $[0, 2\pi]$ by a trigonometric polynomial of order n or less is

$$x \approx \frac{a_0}{2} + [a_1 \cos x + \cdots + a_n \cos nx] + [b_1 \sin x + \cdots + b_n \sin nx]$$

or, from (9a), (9b), and (9c),

$$x \approx \pi - 2 \left(\sin x + \frac{\sin 2x}{2} + \frac{\sin 3x}{3} + \cdots + \frac{\sin nx}{n} \right)$$

The graphs of $y = x$ and some of these approximations are shown in Figure 6.6.4.

**Jean Baptiste
Fourier (1768–1830)**

Historical Note Fourier was a French mathematician and physicist who discovered the Fourier series and related ideas while working on problems of heat diffusion. This discovery was one of the most influential in the history of mathematics; it is the cornerstone of many fields of mathematical research and a basic tool in many branches of engineering. Fourier, a political activist during the French revolution, spent time in jail for his defense of many victims during the Terror. He later became a favorite of Napoleon and was named a baron.

[Image: The Granger Collection, New York]

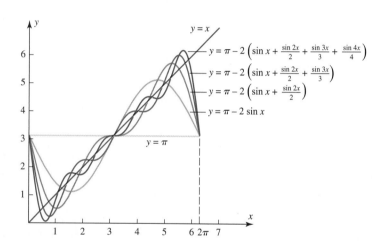

▶ **Figure 6.6.4**

It is natural to expect that the mean square error will diminish as the number of terms in the least squares approximation

$$f(x) \approx \frac{a_0}{2} + \sum_{k=1}^{n} (a_k \cos kx + b_k \sin kx)$$

increases. It can be proved that for functions f in $C[0, 2\pi]$, the mean square error approaches zero as $n \to +\infty$; this is denoted by writing

$$f(x) = \frac{a_0}{2} + \sum_{k=1}^{\infty} (a_k \cos kx + b_k \sin kx)$$

The right side of this equation is called the **Fourier series** for f over the interval $[0, 2\pi]$. Such series are of major importance in engineering, science, and mathematics. ◀

Concept Review

- Approximation of functions
- Mean square error

- Least squares approximation
- Trigonometric polynomial

- Fourier coefficients
- Fourier series

Skills

- Find the least squares approximation of a function.
- Find the mean square error of the least squares approximation of a function.
- Compute the Fourier series of a function.

Exercise Set 6.6

1. Find the least squares approximation of $f(x) = 1 + x$ over the interval $[0, 2\pi]$ by
   (a) a trigonometric polynomial of order 2 or less.
   (b) a trigonometric polynomial of order n or less.

2. Find the least squares approximation of $f(x) = x^2$ over the interval $[0, 2\pi]$ by
   (a) a trigonometric polynomial of order 3 or less.
   (b) a trigonometric polynomial of order n or less.

3. (a) Find the least squares approximation of x over the interval $[0, 1]$ by a function of the form $a + be^x$.
   (b) Find the mean square error of the approximation.

4. (a) Find the least squares approximation of e^x over the interval $[0, 1]$ by a polynomial of the form $a_0 + a_1 x$.
   (b) Find the mean square error of the approximation.

5. (a) Find the least squares approximation of $\sin \pi x$ over the interval $[-1, 1]$ by a polynomial of the form $a_0 + a_1 x + a_2 x^2$.
   (b) Find the mean square error of the approximation.

6. Use the Gram–Schmidt process to obtain the orthonormal basis (5) from the basis (3).

7. Carry out the integrations indicated in Formulas (9a), (9b), and (9c).

8. Find the Fourier series of $f(x) = \pi - x$ over the interval $[0, 2\pi]$.

9. Find the Fourier series of $f(x) = 1$, $0 < x < \pi$ and $f(x) = 0$, $\pi \le x \le 2\pi$ over the interval $[0, 2\pi]$.

10. What is the Fourier series of $\sin(3x)$?

True-False Exercises

In parts (a)–(e) determine whether the statement is true or false, and justify your answer.

(a) If a function $\mathbf{f}$ in $C[a, b]$ is approximated by the function $\mathbf{g}$, then the mean square error is the same as the area between the graphs of $f(x)$ and $g(x)$ over the interval $[a, b]$.

(b) Given a finite-dimensional subspace W of $C[a, b]$, the function $\mathbf{g} = \text{proj}_W \mathbf{f}$ minimizes the mean square error.

(c) $\{1, \cos x, \sin x, \cos 2x, \sin 2x\}$ is an orthogonal subset of the vector space $C[0, 2\pi]$ with respect to the inner product $\langle \mathbf{f}, \mathbf{g} \rangle = \int_0^{2\pi} f(x)g(x)\,dx$.

(d) $\{1, \cos x, \sin x, \cos 2x, \sin 2x\}$ is an orthonormal subset of the vector space $C[0, 2\pi]$ with respect to the inner product $\langle \mathbf{f}, \mathbf{g} \rangle = \int_0^{2\pi} f(x)g(x)\,dx$.

(e) $\{1, \cos x, \sin x, \cos 2x, \sin 2x\}$ is a linearly independent subset of $C[0, 2\pi]$.

Chapter 6 Supplementary Exercises

1. Let R^4 have the Euclidean inner product.
   (a) Find a vector in R^4 that is orthogonal to $\mathbf{u}_1 = (1, 0, 0, 0)$ and $\mathbf{u}_4 = (0, 0, 0, 1)$ and makes equal angles with $\mathbf{u}_2 = (0, 1, 0, 0)$ and $\mathbf{u}_3 = (0, 0, 1, 0)$.
   (b) Find a vector $\mathbf{x} = (x_1, x_2, x_3, x_4)$ of length 1 that is orthogonal to $\mathbf{u}_1$ and $\mathbf{u}_4$ above and such that the cosine of the angle between $\mathbf{x}$ and $\mathbf{u}_2$ is twice the cosine of the angle between $\mathbf{x}$ and $\mathbf{u}_3$.

2. Prove: If $\langle \mathbf{u}, \mathbf{v} \rangle$ is the Euclidean inner product on R^n, and if A is an $n \times n$ matrix, then

$$\langle \mathbf{u}, A\mathbf{v} \rangle = \langle A^T \mathbf{u}, \mathbf{v} \rangle$$

[*Hint:* Use the fact that $\langle \mathbf{u}, \mathbf{v} \rangle = \mathbf{u} \cdot \mathbf{v} = \mathbf{v}^T \mathbf{u}$.]

3. Let M_{22} have the inner product $\langle U, V \rangle = \text{tr}(U^T V) = \text{tr}(V^T U)$ that was defined in Example 6 of Section 6.1. Describe the orthogonal complement of
   (a) the subspace of all diagonal matrices.
   (b) the subspace of symmetric matrices.

4. Let $A\mathbf{x} = \mathbf{0}$ be a system of m equations in n unknowns. Show that

$$\mathbf{x} = \begin{bmatrix} x_1 \\ x_2 \\ \vdots \\ x_n \end{bmatrix}$$

is a solution of this system if and only if the vector $\mathbf{x} = (x_1, x_2, \ldots, x_n)$ is orthogonal to every row vector of A with respect to the Euclidean inner product on R^n.

5. Use the Cauchy–Schwarz inequality to show that if $a_1, a_2, \ldots, a_n$ are positive real numbers, then

$$(a_1 + a_2 + \cdots + a_n)\left(\frac{1}{a_1} + \frac{1}{a_2} + \cdots + \frac{1}{a_n}\right) \geq n^2$$

6. Show that if $\mathbf{x}$ and $\mathbf{y}$ are vectors in an inner product space and c is any scalar, then

$$\|c\mathbf{x} + \mathbf{y}\|^2 = c^2\|\mathbf{x}\|^2 + 2c\langle\mathbf{x}, \mathbf{y}\rangle + \|\mathbf{y}\|^2$$

7. Let R^3 have the Euclidean inner product. Find two vectors of length 1 that are orthogonal to all three of the vectors $\mathbf{u}_1 = (1, 1, -1)$, $\mathbf{u}_2 = (-2, -1, 2)$, and $\mathbf{u}_3 = (-1, 0, 1)$.

8. Find a weighted Euclidean inner product on R^n such that the vectors

$$\mathbf{v}_1 = (1, 0, 0, \ldots, 0)$$
$$\mathbf{v}_2 = (0, \sqrt{2}, 0, \ldots, 0)$$
$$\mathbf{v}_3 = (0, 0, \sqrt{3}, \ldots, 0)$$
$$\vdots$$
$$\mathbf{v}_n = (0, 0, 0, \ldots, \sqrt{n})$$

form an orthonormal set.

9. Is there a weighted Euclidean inner product on R^2 for which the vectors $(1, 2)$ and $(3, -1)$ form an orthonormal set? Justify your answer.

10. If $\mathbf{u}$ and $\mathbf{v}$ are vectors in an inner product space V, then $\mathbf{u}$, $\mathbf{v}$, and $\mathbf{u} - \mathbf{v}$ can be regarded as sides of a "triangle" in V (see the accompanying figure). Prove that the law of cosines holds for any such triangle; that is,

$$\|\mathbf{u} - \mathbf{v}\|^2 = \|\mathbf{u}\|^2 + \|\mathbf{v}\|^2 - 2\|\mathbf{u}\|\|\mathbf{v}\|\cos\theta$$

where θ is the angle between $\mathbf{u}$ and $\mathbf{v}$.

◀ Figure Ex-10

11. (a) As shown in Figure 3.2.6, the vectors $(k, 0, 0)$, $(0, k, 0)$, and $(0, 0, k)$ form the edges of a cube in R^3 with diagonal (k, k, k). Similarly, the vectors

$$(k, 0, 0, \ldots, 0), \quad (0, k, 0, \ldots, 0), \ldots, \quad (0, 0, 0, \ldots, k)$$

can be regarded as edges of a "cube" in R^n with diagonal $(k, k, k, \ldots, k)$. Show that each of the above edges makes an angle of θ with the diagonal, where $\cos\theta = 1/\sqrt{n}$.

(b) (*Calculus required*) What happens to the angle θ in part (a) as the dimension of R^n approaches $+\infty$?

12. Let $\mathbf{u}$ and $\mathbf{v}$ be vectors in an inner product space.

(a) Prove that $\|\mathbf{u}\| = \|\mathbf{v}\|$ if and only if $\mathbf{u} + \mathbf{v}$ and $\mathbf{u} - \mathbf{v}$ are orthogonal.

(b) Give a geometric interpretation of this result in R^2 with the Euclidean inner product.

13. Let $\mathbf{u}$ be a vector in an inner product space V, and let $\{\mathbf{v}_1, \mathbf{v}_2, \ldots, \mathbf{v}_n\}$ be an orthonormal basis for V. Show that if α_i is the angle between $\mathbf{u}$ and $\mathbf{v}_i$, then

$$\cos^2\alpha_1 + \cos^2\alpha_2 + \cdots + \cos^2\alpha_n = 1$$

14. Prove: If $\langle\mathbf{u}, \mathbf{v}\rangle_1$ and $\langle\mathbf{u}, \mathbf{v}\rangle_2$ are two inner products on a vector space V, then the quantity $\langle\mathbf{u}, \mathbf{v}\rangle = \langle\mathbf{u}, \mathbf{v}\rangle_1 + \langle\mathbf{u}, \mathbf{v}\rangle_2$ is also an inner product.

15. Prove Theorem 6.2.5.

16. Prove: If A has linearly independent column vectors, and if $\mathbf{b}$ is orthogonal to the column space of A, then the least squares solution of $A\mathbf{x} = \mathbf{b}$ is $\mathbf{x} = \mathbf{0}$.

17. Is there any value of s for which $x_1 = 1$ and $x_2 = 2$ is the least squares solution of the following linear system?

$$\begin{aligned} x_1 - x_2 &= 1 \\ 2x_1 + 3x_2 &= 1 \\ 4x_1 + 5x_2 &= s \end{aligned}$$

Explain your reasoning.

18. Show that if p and q are distinct positive integers, then the functions $f(x) = \sin px$ and $g(x) = \sin qx$ are orthogonal with respect to the inner product

$$\langle\mathbf{f}, \mathbf{g}\rangle = \int_0^{2\pi} f(x)g(x)\,dx$$

19. Show that if p and q are positive integers, then the functions $f(x) = \cos px$ and $g(x) = \sin qx$ are orthogonal with respect to the inner product

$$\langle\mathbf{f}, \mathbf{g}\rangle = \int_0^{2\pi} f(x)g(x)\,dx$$

CHAPTER 7

Diagonalization and Quadratic Forms

CHAPTER CONTENTS
7.1 Orthogonal Matrices **389**
7.2 Orthogonal Diagonalization **397**
7.3 Quadratic Forms **405**
7.4 Optimization Using Quadratic Forms **417**
7.5 Hermitian, Unitary, and Normal Matrices **424**

INTRODUCTION In Section 5.2 we found conditions that guaranteed the diagonalizability of an $n \times n$ matrix, but we did not consider what class or classes of matrices might actually satisfy those conditions. In this chapter we will show that every symmetric matrix is diagonalizable. This is an extremely important result because many applications utilize it in some essential way.

7.1 Orthogonal Matrices

In this section we will discuss the class of matrices whose inverses can be obtained by transposition. Such matrices occur in a variety of applications and arise as well as transition matrices when one orthonormal basis is changed to another.

Orthogonal Matrices We begin with the following definition.

Recall from Theorem 1.6.3 that if either product in (1) holds, then so does the other. Thus, A is orthogonal if *either* $AA^T = I$ or $A^TA = I$.

> **DEFINITION 1** A square matrix A is said to be ***orthogonal*** if its transpose is the same as its inverse, that is, if
> $$A^{-1} = A^T$$
> or, equivalently, if
> $$AA^T = A^TA = I \qquad (1)$$

▶ **EXAMPLE 1 A 3 × 3 Orthogonal Matrix**

The matrix
$$A = \begin{bmatrix} \frac{3}{7} & \frac{2}{7} & \frac{6}{7} \\ -\frac{6}{7} & \frac{3}{7} & \frac{2}{7} \\ \frac{2}{7} & \frac{6}{7} & -\frac{3}{7} \end{bmatrix}$$

is orthogonal since

$$A^TA = \begin{bmatrix} \frac{3}{7} & -\frac{6}{7} & \frac{2}{7} \\ \frac{2}{7} & \frac{3}{7} & \frac{6}{7} \\ \frac{6}{7} & \frac{2}{7} & -\frac{3}{7} \end{bmatrix} \begin{bmatrix} \frac{3}{7} & \frac{2}{7} & \frac{6}{7} \\ -\frac{6}{7} & \frac{3}{7} & \frac{2}{7} \\ \frac{2}{7} & \frac{6}{7} & -\frac{3}{7} \end{bmatrix} = \begin{bmatrix} 1 & 0 & 0 \\ 0 & 1 & 0 \\ 0 & 0 & 1 \end{bmatrix}$$

▶ **EXAMPLE 2 Rotation and Reflection Matrices are Orthogonal**

Recall from Table 5 of Section 4.9 that the standard matrix for the counterclockwise rotation of R^2 through an angle θ is

$$A = \begin{bmatrix} \cos\theta & -\sin\theta \\ \sin\theta & \cos\theta \end{bmatrix}$$

This matrix is orthogonal for all choices of θ since

$$A^T A = \begin{bmatrix} \cos\theta & \sin\theta \\ -\sin\theta & \cos\theta \end{bmatrix} \begin{bmatrix} \cos\theta & -\sin\theta \\ \sin\theta & \cos\theta \end{bmatrix} = \begin{bmatrix} 1 & 0 \\ 0 & 1 \end{bmatrix}$$

We leave it for you to verify that the reflection matrices in Tables 1 and 2 and the rotation matrices in Table 6 of Section 4.9 are all orthogonal. ◀

Observe that for the orthogonal matrices in Examples 1 and 2, both the row vectors and the column vectors form orthonormal sets with respect to the Euclidean inner product. This is a consequence of the following theorem.

THEOREM 7.1.1 *The following are equivalent for an $n \times n$ matrix A.*

(a) *A is orthogonal.*

(b) *The row vectors of A form an orthonormal set in R^n with the Euclidean inner product.*

(c) *The column vectors of A form an orthonormal set in R^n with the Euclidean inner product.*

Proof We will prove the equivalence of (a) and (b) and leave the equivalence of (a) and (c) as an exercise.

(a) ⇔ (b) The entry in the ith row and jth column of the matrix product AA^T is the dot product of the ith row vector of A and the jth column vector of A^T (see Formula (5) of Section 1.3). But except for a difference in form, the jth column vector of A^T is the jth row vector of A. Thus, if the row vectors of A are $\mathbf{r}_1, \mathbf{r}_2, \ldots, \mathbf{r}_n$, then the matrix product AA^T can be expressed as

$$AA^T = \begin{bmatrix} \mathbf{r}_1 \cdot \mathbf{r}_1 & \mathbf{r}_1 \cdot \mathbf{r}_2 & \cdots & \mathbf{r}_1 \cdot \mathbf{r}_n \\ \mathbf{r}_2 \cdot \mathbf{r}_1 & \mathbf{r}_2 \cdot \mathbf{r}_2 & \cdots & \mathbf{r}_2 \cdot \mathbf{r}_n \\ \vdots & \vdots & & \vdots \\ \mathbf{r}_n \cdot \mathbf{r}_1 & \mathbf{r}_n \cdot \mathbf{r}_2 & \cdots & \mathbf{r}_n \cdot \mathbf{r}_n \end{bmatrix}$$

[see Formula (28) of Section 3.2]. Thus, it follows that $AA^T = I$ if and only if

$$\mathbf{r}_1 \cdot \mathbf{r}_1 = \mathbf{r}_2 \cdot \mathbf{r}_2 = \cdots = \mathbf{r}_n \cdot \mathbf{r}_n = 1$$

and

$$\mathbf{r}_i \cdot \mathbf{r}_j = 0 \quad \text{when } i \neq j$$

which are true if and only if $\{\mathbf{r}_1, \mathbf{r}_2, \ldots, \mathbf{r}_n\}$ is an orthonormal set in R^n. ◀

WARNING Note that an orthogonal matrix is one with *orthonormal* rows and columns—not simply orthogonal rows and columns.

The following theorem lists three more fundamental properties of orthogonal matrices. The proofs are all straightforward and are left as exercises.

THEOREM 7.1.2

(a) *The inverse of an orthogonal matrix is orthogonal.*

(b) *A product of orthogonal matrices is orthogonal.*

(c) *If A is orthogonal, then $\det(A) = 1$ or $\det(A) = -1$.*

▶ **EXAMPLE 3** $\det(A) = \pm 1$ for an Orthogonal Matrix *A*

The matrix

$$A = \begin{bmatrix} \frac{1}{\sqrt{2}} & \frac{1}{\sqrt{2}} \\ -\frac{1}{\sqrt{2}} & \frac{1}{\sqrt{2}} \end{bmatrix}$$

is orthogonal since its row (and column) vectors form orthonormal sets in R^2 with the Euclidean inner product. We leave it for you to verify that $\det(A) = 1$ and that interchanging the rows produces an orthogonal matrix whose determinant is -1. ◀

Orthogonal Matrices as Linear Operators

We observed in Example 2 that the standard matrices for the basic reflection and rotation operators on R^2 and R^3 are orthogonal. The next theorem will explain why this is so.

THEOREM 7.1.3 *If A is an $n \times n$ matrix, then the following are equivalent.*

(a) *A is orthogonal.*

(b) $\|A\mathbf{x}\| = \|\mathbf{x}\|$ *for all $\mathbf{x}$ in R^n.*

(c) $A\mathbf{x} \cdot A\mathbf{y} = \mathbf{x} \cdot \mathbf{y}$ *for all $\mathbf{x}$ and $\mathbf{y}$ in R^n.*

Proof We will prove the sequence of implications $(a) \Rightarrow (b) \Rightarrow (c) \Rightarrow (a)$.

(a) ⇒ (b) Assume that A is orthogonal, so that $A^T A = I$. It follows from Formula (26) of Section 3.2 that

$$\|A\mathbf{x}\| = (A\mathbf{x} \cdot A\mathbf{x})^{1/2} = (\mathbf{x} \cdot A^T A\mathbf{x})^{1/2} = (\mathbf{x} \cdot \mathbf{x})^{1/2} = \|\mathbf{x}\|$$

(b) ⇒ (c) Assume that $\|A\mathbf{x}\| = \|\mathbf{x}\|$ for all $\mathbf{x}$ in R^n. From Theorem 3.2.7 we have

$$A\mathbf{x} \cdot A\mathbf{y} = \tfrac{1}{4}\|A\mathbf{x} + A\mathbf{y}\|^2 - \tfrac{1}{4}\|A\mathbf{x} - A\mathbf{y}\|^2 = \tfrac{1}{4}\|A(\mathbf{x} + \mathbf{y})\|^2 - \tfrac{1}{4}\|A(\mathbf{x} - \mathbf{y})\|^2$$
$$= \tfrac{1}{4}\|\mathbf{x} + \mathbf{y}\|^2 - \tfrac{1}{4}\|\mathbf{x} - \mathbf{y}\|^2 = \mathbf{x} \cdot \mathbf{y}$$

(c) ⇒ (a) Assume that $A\mathbf{x} \cdot A\mathbf{y} = \mathbf{x} \cdot \mathbf{y}$ for all $\mathbf{x}$ and $\mathbf{y}$ in R^n. It follows from Formula (26) of Section 3.2 that

$$\mathbf{x} \cdot \mathbf{y} = \mathbf{x} \cdot A^T A\mathbf{y}$$

which can be rewritten as $\mathbf{x} \cdot (A^T A\mathbf{y} - \mathbf{y}) = 0$ or as

$$\mathbf{x} \cdot (A^T A - I)\mathbf{y} = 0$$

Since this equation holds for all $\mathbf{x}$ in R^n, it holds in particular if $\mathbf{x} = (A^T A - I)\mathbf{y}$, so

$$(A^T A - I)\mathbf{y} \cdot (A^T A - I)\mathbf{y} = 0$$

Thus, it follows from the positivity axiom for inner products that

$$(A^T A - I)\mathbf{y} = \mathbf{0}$$

Since this equation is satisfied by every vector $\mathbf{y}$ in R^n, it must be that $A^T A - I$ is the zero matrix (why?) and hence that $A^T A = I$. Thus, A is orthogonal. ◀

Theorem 7.1.3 has a useful geometric interpretation when considered from the viewpoint of matrix transformations: If A is an orthogonal matrix and $T_A : R^n \to R^n$ is multiplication by A, then we will call T_A an ***orthogonal operator*** on R^n. It follows from parts (a) and (b) of Theorem 7.1.3 that the orthogonal operators on R^n are precisely those operators that leave the lengths of all vectors unchanged. This explains why, in Example 2, we found the standard matrices for the basic reflections and rotations of R^2 and R^3 to be orthogonal.

Parts (a) and (c) of Theorem 7.1.3 imply that orthogonal operators leave the *angle* between two vectors unchanged. Why?

Change of Orthonormal Basis

Orthonormal bases for inner product spaces are convenient because, as the following theorem shows, many familiar formulas hold for such bases. We leave the proof as an exercise.

> **THEOREM 7.1.4** *If S is an orthonormal basis for an n-dimensional inner product space V, and if*
>
> $$(\mathbf{u})_S = (u_1, u_2, \ldots, u_n) \quad and \quad (\mathbf{v})_S = (v_1, v_2, \ldots, v_n)$$
>
> *then:*
>
> (a) $\|\mathbf{u}\| = \sqrt{u_1^2 + u_2^2 + \cdots + u_n^2}$
>
> (b) $d(\mathbf{u}, \mathbf{v}) = \sqrt{(u_1 - v_1)^2 + (u_2 - v_2)^2 + \cdots + (u_n - v_n)^2}$
>
> (c) $\langle \mathbf{u}, \mathbf{v} \rangle = u_1 v_1 + u_2 v_2 + \cdots + u_n v_n$

Remark   Note that the three parts of Theorem 7.1.4 can be expressed as

$$\|\mathbf{u}\| = \|(\mathbf{u})_S\| \qquad d(\mathbf{u}, \mathbf{v}) = d\big((\mathbf{u})_S, (\mathbf{v})_S\big) \qquad \langle \mathbf{u}, \mathbf{v} \rangle = \langle (\mathbf{u})_S, (\mathbf{v})_S \rangle$$

where the norm, distance, and inner product on the left sides are relative to the inner product on V and on the right sides are relative to the Euclidean inner product on R^n.

Transitions between orthonormal bases for an inner product space are of special importance in geometry and various applications. The following theorem, whose proof is deferred to the end of this section, is concerned with transitions of this type.

> **THEOREM 7.1.5** *Let V be a finite-dimensional inner product space. If P is the transition matrix from one orthonormal basis for V to another orthonormal basis for V, then P is an orthogonal matrix.*

▶ **EXAMPLE 4   Rotation of Axes in 2-Space**

In many problems a rectangular xy-coordinate system is given, and a new $x'y'$-coordinate system is obtained by rotating the xy-system counterclockwise about the origin through an angle θ. When this is done, each point Q in the plane has two sets of coordinates—coordinates (x, y) relative to the xy-system and coordinates (x', y') relative to the $x'y'$-system (Figure 7.1.1a).

By introducing unit vectors $\mathbf{u}_1$ and $\mathbf{u}_2$ along the positive x- and y-axes and unit vectors $\mathbf{u}_1'$ and $\mathbf{u}_2'$ along the positive x'- and y'-axes, we can regard this rotation as a change from an old basis $B = \{\mathbf{u}_1, \mathbf{u}_2\}$ to a new basis $B' = \{\mathbf{u}_1', \mathbf{u}_2'\}$ (Figure 7.1.1b). Thus, the new coordinates (x', y') and the old coordinates (x, y) of a point Q will be related by

$$\begin{bmatrix} x' \\ y' \end{bmatrix} = P^{-1} \begin{bmatrix} x \\ y \end{bmatrix} \tag{2}$$

where P is the transition from B' to B. To find P we must determine the coordinate matrices of the new basis vectors $\mathbf{u}_1'$ and $\mathbf{u}_2'$ relative to the old basis. As indicated in Figure 7.1.1c, the components of $\mathbf{u}_1'$ in the old basis are $\cos\theta$ and $\sin\theta$, so

$$[\mathbf{u}_1']_B = \begin{bmatrix} \cos\theta \\ \sin\theta \end{bmatrix}$$

Similarly, from Figure 7.1.1d, we see that the components of $\mathbf{u}_2'$ in the old basis are $\cos(\theta + \pi/2) = -\sin\theta$ and $\sin(\theta + \pi/2) = \cos\theta$, so

$$[\mathbf{u}_2']_B = \begin{bmatrix} -\sin\theta \\ \cos\theta \end{bmatrix}$$

Thus the transition matrix from B' to B is

$$P = \begin{bmatrix} \cos\theta & -\sin\theta \\ \sin\theta & \cos\theta \end{bmatrix} \tag{3}$$

Observe that P is an orthogonal matrix, as expected, since B and B' are orthonormal bases. Thus

$$P^{-1} = P^T = \begin{bmatrix} \cos\theta & \sin\theta \\ -\sin\theta & \cos\theta \end{bmatrix}$$

so (2) yields

$$\begin{bmatrix} x' \\ y' \end{bmatrix} = \begin{bmatrix} \cos\theta & \sin\theta \\ -\sin\theta & \cos\theta \end{bmatrix} \begin{bmatrix} x \\ y \end{bmatrix} \tag{4}$$

or, equivalently,

$$\begin{aligned} x' &= x\cos\theta + y\sin\theta \\ y' &= -x\sin\theta + y\cos\theta \end{aligned} \tag{5}$$

These are sometimes called the ***rotation equations*** for R^2.

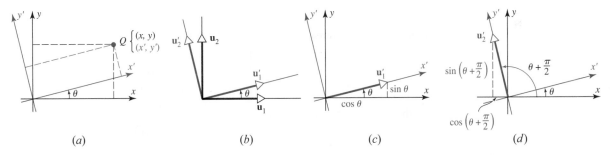

(a) (b) (c) (d)

▲ Figure 7.1.1

▶ **EXAMPLE 5 Rotation of Axes in 2-Space**

Use form (4) of the rotation equations for R^2 to find the new coordinates of the point $Q(2, 1)$ if the coordinate axes of a rectangular coordinate system are rotated through an angle of $\theta = \pi/4$.

Solution Since

$$\sin\frac{\pi}{4} = \cos\frac{\pi}{4} = \frac{1}{\sqrt{2}}$$

the equation in (4) becomes

$$\begin{bmatrix} x' \\ y' \end{bmatrix} = \begin{bmatrix} \frac{1}{\sqrt{2}} & \frac{1}{\sqrt{2}} \\ -\frac{1}{\sqrt{2}} & \frac{1}{\sqrt{2}} \end{bmatrix} \begin{bmatrix} x \\ y \end{bmatrix}$$

Thus, if the old coordinates of a point Q are $(x, y) = (2, -1)$, then

$$\begin{bmatrix} x' \\ y' \end{bmatrix} = \begin{bmatrix} \frac{1}{\sqrt{2}} & \frac{1}{\sqrt{2}} \\ -\frac{1}{\sqrt{2}} & \frac{1}{\sqrt{2}} \end{bmatrix} \begin{bmatrix} 2 \\ -1 \end{bmatrix} = \begin{bmatrix} \frac{1}{\sqrt{2}} \\ -\frac{3}{\sqrt{2}} \end{bmatrix}$$

so the new coordinates of Q are $(x', y') = \left(\frac{1}{\sqrt{2}}, -\frac{3}{\sqrt{2}} \right)$. ◄

Remark Observe that the coefficient matrix in (4) is the same as the standard matrix for the linear operator that rotates the vectors of R^2 through the angle $-\theta$ (see margin note for Table 5 of Section 4.9). This is to be expected since rotating the coordinate axes through the angle θ with the vectors of R^2 kept fixed has the same effect as rotating the vectors in R^2 through the angle $-\theta$ with the axes kept fixed.

▲ Figure 7.1.2

▶ **EXAMPLE 6** Application to Rotation of Axes in 3-Space

Suppose that a rectangular xyz-coordinate system is rotated around its z-axis counterclockwise (looking down the positive z-axis) through an angle θ (Figure 7.1.2). If we introduce unit vectors $\mathbf{u}_1$, $\mathbf{u}_2$, and $\mathbf{u}_3$ along the positive x-, y-, and z-axes and unit vectors $\mathbf{u}_1'$, $\mathbf{u}_2'$, and $\mathbf{u}_3'$ along the positive x'-, y'-, and z'-axes, we can regard the rotation as a change from the old basis $B = \{\mathbf{u}_1, \mathbf{u}_2, \mathbf{u}_3\}$ to the new basis $B' = \{\mathbf{u}_1', \mathbf{u}_2', \mathbf{u}_3'\}$. In light of Example 4, it should be evident that

$$[\mathbf{u}_1']_B = \begin{bmatrix} \cos \theta \\ \sin \theta \\ 0 \end{bmatrix} \quad \text{and} \quad [\mathbf{u}_2']_B = \begin{bmatrix} -\sin \theta \\ \cos \theta \\ 0 \end{bmatrix}$$

Moreover, since $\mathbf{u}_3'$ extends 1 unit up the positive z'-axis,

$$[\mathbf{u}_3']_B = \begin{bmatrix} 0 \\ 0 \\ 1 \end{bmatrix}$$

It follows that the transition matrix from B' to B is

$$P = \begin{bmatrix} \cos \theta & -\sin \theta & 0 \\ \sin \theta & \cos \theta & 0 \\ 0 & 0 & 1 \end{bmatrix}$$

and the transition matrix from B to B' is

$$P^{-1} = \begin{bmatrix} \cos \theta & \sin \theta & 0 \\ -\sin \theta & \cos \theta & 0 \\ 0 & 0 & 1 \end{bmatrix}$$

(verify). Thus, the new coordinates (x', y', z') of a point Q can be computed from its old coordinates (x, y, z) by

$$\begin{bmatrix} x' \\ y' \\ z' \end{bmatrix} = \begin{bmatrix} \cos \theta & \sin \theta & 0 \\ -\sin \theta & \cos \theta & 0 \\ 0 & 0 & 1 \end{bmatrix} \begin{bmatrix} x \\ y \\ z \end{bmatrix}$$ ◄

OPTIONAL

We conclude this section with an optional proof of Theorem 7.1.5.

Proof of Theorem 7.1.5 Assume that V is an n-dimensional inner product space and that P is the transition matrix from an orthonormal basis B' to an orthonormal basis B. We will denote the norm relative to the inner product on V by the symbol $\| \ \|_V$ to distinguish it from the norm relative to the Euclidean inner product on R^n, which we will denote by $\| \ \|$.

To prove that P is orthogonal, we will use Theorem 7.1.3 and show that $\|P\mathbf{x}\| = \|\mathbf{x}\|$ for every vector $\mathbf{x}$ in R^n. As a first step in this direction, recall from Theorem 7.1.4a that for any orthonormal basis for V the norm of any vector $\mathbf{u}$ in V is the same as the norm of its coordinate vector with respect to the Euclidean inner product, that is

$$\|\mathbf{u}\|_V = \|[\mathbf{u}]_{B'}\| = \|[\mathbf{u}]_B\|$$

or

$$\|\mathbf{u}\|_V = \|[\mathbf{u}]_{B'}\| = \|P[\mathbf{u}]_{B'}\| \tag{6}$$

Now let $\mathbf{x}$ be any vector in R^n, and let $\mathbf{u}$ be the vector in V whose coordinate vector with respect to the basis B' is $\mathbf{x}$; that is, $[\mathbf{u}]_{B'} = \mathbf{x}$. Thus, from (6),

$$\|\mathbf{u}\| = \|\mathbf{x}\| = \|P\mathbf{x}\|$$

which proves that P is orthogonal. ◀

> Recall that $(\mathbf{u})_S$ denotes a coordinate vector expressed in comma-delimited form whereas $[\mathbf{u}]_S$ denotes a coordinate vector expressed in column form.

Concept Review

- Orthogonal matrix
- Orthogonal operator
- Properties of orthogonal matrices.
- Geometric properties of an orthogonal operator
- Properties of transition matrices from one orthonormal basis to another.

Skills

- Be able to identify an orthogonal matrix.
- Know the possible values for the determinant of an orthogonal matrix.
- Find the new coordinates of a point resulting from a rotation of axes.

Exercise Set 7.1

1. (a) Show that the matrix

$$A = \begin{bmatrix} \frac{4}{5} & 0 & -\frac{3}{5} \\ -\frac{9}{25} & \frac{4}{5} & -\frac{12}{25} \\ \frac{12}{25} & \frac{3}{5} & \frac{16}{25} \end{bmatrix}$$

is orthogonal in three ways: by calculating A^TA, by using part (b) of Theorem 7.1.1, and by using part (c) of Theorem 7.1.1.

(b) Find the inverse of the matrix A in part (a).

2. (a) Show that the matrix

$$A = \begin{bmatrix} \frac{1}{3} & \frac{2}{3} & \frac{2}{3} \\ \frac{2}{3} & -\frac{2}{3} & \frac{1}{3} \\ -\frac{2}{3} & -\frac{1}{3} & \frac{2}{3} \end{bmatrix}$$

is orthogonal.

(b) Let $T : R^3 \to R^3$ be multiplication by the matrix A in part (a). Find $T(\mathbf{x})$ for the vector $\mathbf{x} = (-2, 3, 5)$. Using the Euclidean inner product on R^3, verify that $\|T(\mathbf{x})\| = \|\mathbf{x}\|$.

3. Determine which of the following matrices are orthogonal. For those that are orthogonal, find the inverse.

(a) $\begin{bmatrix} 1 & 0 \\ 0 & 1 \end{bmatrix}$

(b) $\begin{bmatrix} \frac{1}{\sqrt{2}} & -\frac{1}{\sqrt{2}} \\ \frac{1}{\sqrt{2}} & \frac{1}{\sqrt{2}} \end{bmatrix}$

(c) $\begin{bmatrix} 0 & 1 & \frac{1}{\sqrt{2}} \\ 1 & 0 & 0 \\ 0 & 0 & \frac{1}{\sqrt{2}} \end{bmatrix}$

(d) $\begin{bmatrix} -\frac{1}{\sqrt{2}} & \frac{1}{\sqrt{6}} & \frac{1}{\sqrt{3}} \\ 0 & -\frac{2}{\sqrt{6}} & \frac{1}{\sqrt{3}} \\ \frac{1}{\sqrt{2}} & \frac{1}{\sqrt{6}} & \frac{1}{\sqrt{3}} \end{bmatrix}$

(e) $\begin{bmatrix} \frac{1}{2} & \frac{1}{2} & \frac{1}{2} & \frac{1}{2} \\ \frac{1}{2} & -\frac{5}{6} & \frac{1}{6} & \frac{1}{6} \\ \frac{1}{2} & \frac{1}{6} & \frac{1}{6} & -\frac{5}{6} \\ \frac{1}{2} & \frac{1}{6} & -\frac{5}{6} & \frac{1}{6} \end{bmatrix}$

(f) $\begin{bmatrix} 1 & 0 & 0 & 0 \\ 0 & \frac{1}{\sqrt{3}} & -\frac{1}{2} & 0 \\ 0 & \frac{1}{\sqrt{3}} & 0 & 1 \\ 0 & \frac{1}{\sqrt{3}} & \frac{1}{2} & 0 \end{bmatrix}$

4. Prove that if A is orthogonal, then A^T is orthogonal.

5. Verify that the reflection matrices in Tables 1 and 2 of Section 4.9 are orthogonal.

6. Let a rectangular $x'y'$-coordinate system be obtained by rotating a rectangular xy-coordinate system counterclockwise through the angle $\theta = 3\pi/4$.

  (a) Find the $x'y'$-coordinates of the point whose xy-coordinates are $(-2, 6)$.

  (b) Find the xy-coordinates of the point whose $x'y'$-coordinates are $(5, 2)$.

7. Repeat Exercise 6 with $\theta = \pi/3$.

8. Let a rectangular $x'y'z'$-coordinate system be obtained by rotating a rectangular xyz-coordinate system counterclockwise about the z-axis (looking down the z-axis) through the angle $\theta = \pi/4$.

  (a) Find the $x'y'z'$-coordinates of the point whose xyz-coordinates are $(-1, 2, 5)$.

  (b) Find the xyz-coordinates of the point whose $x'y'z'$-coordinates are $(1, 6, -3)$.

9. Repeat Exercise 8 for a rotation of $\theta = \pi/3$ counterclockwise about the y-axis (looking along the positive y-axis toward the origin).

10. Repeat Exercise 8 for a rotation of $\theta = 3\pi/4$ counterclockwise about the x-axis (looking along the positive x-axis toward the origin).

11. (a) A rectangular $x'y'z'$-coordinate system is obtained by rotating an xyz-coordinate system counterclockwise about the y-axis through an angle θ (looking along the positive y-axis toward the origin). Find a matrix A such that

$$\begin{bmatrix} x' \\ y' \\ z' \end{bmatrix} = A \begin{bmatrix} x \\ y \\ z \end{bmatrix}$$

where (x, y, z) and (x', y', z') are the coordinates of the same point in the xyz- and $x'y'z'$-systems, respectively.

  (b) Repeat part (a) for a rotation about the x-axis.

12. A rectangular $x''y''z''$-coordinate system is obtained by first rotating a rectangular xyz-coordinate system 60° counterclockwise about the z-axis (looking down the positive z-axis) to obtain an $x'y'z'$-coordinate system, and then rotating the $x'y'z'$-coordinate system 45° counterclockwise about the y'-axis (looking along the positive y'-axis toward the origin). Find a matrix A such that

$$\begin{bmatrix} x'' \\ y'' \\ z'' \end{bmatrix} = A \begin{bmatrix} x \\ y \\ z \end{bmatrix}$$

where (x, y, z) and (x'', y'', z'') are the xyz- and $x''y''z''$-coordinates of the same point.

13. What conditions must a and b satisfy for the matrix

$$\begin{bmatrix} a+b & b-a \\ a-b & b+a \end{bmatrix}$$

to be orthogonal?

14. Prove that a 2×2 orthogonal matrix A has only one of two possible forms:

$$A = \begin{bmatrix} \cos\theta & -\sin\theta \\ \sin\theta & \cos\theta \end{bmatrix} \quad \text{or} \quad A = \begin{bmatrix} \cos\theta & \sin\theta \\ \sin\theta & -\cos\theta \end{bmatrix}$$

where $0 \le \theta < 2\pi$. [*Hint:* Start with a general 2×2 matrix $A = (a_{ij})$, and use the fact that the column vectors form an orthonormal set in R^2.]

15. (a) Use the result in Exercise 14 to prove that multiplication by a 2×2 orthogonal matrix is either a rotation or a rotation followed by a reflection about the x-axis.

  (b) Prove that multiplication by A is a rotation if $\det(A) = 1$ and that a rotation followed by a reflection if $\det(A) = -1$.

16. Use the result in Exercise 15 to determine whether multiplication by A is a rotation or a rotation followed by a reflection about the x-axis. Find the angle of rotation in either case.

  (a) $A = \begin{bmatrix} -\frac{1}{\sqrt{2}} & \frac{1}{\sqrt{2}} \\ -\frac{1}{\sqrt{2}} & -\frac{1}{\sqrt{2}} \end{bmatrix}$ (b) $A = \begin{bmatrix} -\frac{1}{2} & \frac{\sqrt{3}}{2} \\ \frac{\sqrt{3}}{2} & \frac{1}{2} \end{bmatrix}$

17. Find a, b, and c for which the matrix

$$\begin{bmatrix} a & \frac{1}{\sqrt{2}} & -\frac{1}{\sqrt{2}} \\ b & \frac{1}{\sqrt{6}} & \frac{1}{\sqrt{6}} \\ c & \frac{1}{\sqrt{3}} & \frac{1}{\sqrt{3}} \end{bmatrix}$$

is orthogonal. Are the values of a, b, and c unique? Explain.

18. The result in Exercise 15 has an analog for 3×3 orthogonal matrices: It can be proved that multiplication by a 3×3 orthogonal matrix A is a rotation about some axis if $\det(A) = 1$ and is a rotation about some axis followed by a reflection about some coordinate plane if $\det(A) = -1$. Determine whether multiplication by A is a rotation or a rotation followed by a reflection.

  (a) $A = \begin{bmatrix} \frac{3}{7} & \frac{2}{7} & \frac{6}{7} \\ -\frac{6}{7} & \frac{3}{7} & \frac{2}{7} \\ \frac{2}{7} & \frac{6}{7} & -\frac{3}{7} \end{bmatrix}$ (b) $A = \begin{bmatrix} \frac{2}{7} & \frac{3}{7} & \frac{6}{7} \\ \frac{3}{7} & -\frac{6}{7} & \frac{2}{7} \\ \frac{6}{7} & \frac{2}{7} & -\frac{3}{7} \end{bmatrix}$

19. Use the fact stated in Exercise 18 and part (*b*) of Theorem 7.1.2 to show that a composition of rotations can always be accomplished by a single rotation about some appropriate axis.

20. Prove the equivalence of statements (*a*) and (*c*) in Theorem 7.1.1.

21. A linear operator on R^2 is called **rigid** if it does not change the lengths of vectors, and it is called **angle preserving** if it does not change the angle between nonzero vectors.

  (a) Name two different types of linear operators that are rigid.

  (b) Name two different types of linear operators that are angle preserving.

(c) Are there any linear operators on R^2 that are rigid and not angle preserving? Angle preserving and not rigid? Justify your answer.

True-False Exercises

In parts (a)–(h) determine whether the statement is true or false, and justify your answer.

(a) The matrix $\begin{bmatrix} 1 & 0 \\ 0 & 1 \\ 0 & 0 \end{bmatrix}$ is orthogonal.

(b) The matrix $\begin{bmatrix} 1 & -2 \\ 2 & 1 \end{bmatrix}$ is orthogonal.

(c) An $m \times n$ matrix A is orthogonal if $A^TA = I$.

(d) A square matrix whose columns form an orthogonal set is orthogonal.

(e) Every orthogonal matrix is invertible.

(f) If A is an orthogonal matrix, then A^2 is orthogonal and $(\det A)^2 = 1$.

(g) Every eigenvalue of an orthogonal matrix has absolute value 1.

(h) If A is a square matrix and $\|A\mathbf{u}\| = 1$ for all unit vectors $\mathbf{u}$, then A is orthogonal.

7.2 Orthogonal Diagonalization

In this section we will be concerned with the problem of diagonalizing a symmetric matrix A. As we will see, this problem is closely related to that of finding an orthonormal basis for R^n that consists of eigenvectors of A. Problems of this type are important because many of the matrices that arise in applications are symmetric.

The Orthogonal Diagonalization Problem

In Definition 1 of Section 5.2 we defined two square matrices, A and B, to be *similar* if there is an *invertible* matrix P such that $P^{-1}AP = B$. In this section we will be concerned with the special case in which it is possible to find an *orthogonal* matrix P for which this relationship holds.

We begin with the following definition.

DEFINITION 1 If A and B are square matrices, then we say that A and B are **orthogonally similar** if there is an orthogonal matrix P such that $P^TAP = B$.

If A is orthogonally similar to some diagonal matrix, say

$$P^TAP = D$$

then we say that A is **orthogonally diagonalizable** and that P **orthogonally diagonalizes** A.

Our first goal in this section is to determine what conditions a matrix must satisfy to be orthogonally diagonalizable. As a first step, observe that there is no hope of orthogonally diagonalizing a matrix that is not symmetric. To see why this is so, suppose that

$$P^TAP = D \tag{1}$$

where P is an orthogonal matrix and D is a diagonal matrix. Multiplying the left side of (1) by P, the right side by P^T, and then using the fact that $PP^T = P^TP = I$, we can rewrite this equation as

$$A = PDP^T \tag{2}$$

Now transposing both sides of this equation and using the fact that a diagonal matrix is the same as its transpose we obtain

$$A^T = (PDP^T)^T = (P^T)^T D^T P^T = PDP^T = A$$

so A must be symmetric.

Conditions for Orthogonal Diagonalizability

The following theorem shows that every symmetric matrix is, in fact, orthogonally diagonalizable. In this theorem, and for the remainder of this section, *orthogonal* will mean orthogonal with respect to the Euclidean inner product on R^n.

> **THEOREM 7.2.1** *If A is an $n \times n$ matrix, then the following are equivalent.*
>
> (a) *A is orthogonally diagonalizable.*
>
> (b) *A has an orthonormal set of n eigenvectors.*
>
> (c) *A is symmetric.*

Proof (a) $\Rightarrow$ (b) Since A is orthogonally diagonalizable, there is an orthogonal matrix P such that $P^{-1}AP$ is diagonal. As shown in the proof of Theorem 5.2.1, the n column vectors of P are eigenvectors of A. Since P is orthogonal, these column vectors are orthonormal, so A has n orthonormal eigenvectors.

(b) $\Rightarrow$ (a) Assume that A has an orthonormal set of n eigenvectors $\{\mathbf{p}_1, \mathbf{p}_2, \ldots, \mathbf{p}_n\}$. As shown in the proof of Theorem 5.2.1, the matrix P with these eigenvectors as columns diagonalizes A. Since these eigenvectors are orthonormal, P is orthogonal and thus orthogonally diagonalizes A.

(a) $\Rightarrow$ (c) In the proof that $(a) \Rightarrow (b)$ we showed that an orthogonally diagonalizable $n \times n$ matrix A is orthogonally diagonalized by an $n \times n$ matrix P whose columns form an orthonormal set of eigenvectors of A. Let D be the diagonal matrix

$$D = P^T A P$$

from which it follows that

$$A = PDP^T$$

Thus,

$$A^T = (PDP^T)^T = PD^T P^T = PDP^T = A$$

which shows that A is symmetric.

(c) $\Rightarrow$ (a) The proof of this part is beyond the scope of this text and will be omitted.

◀

Properties of Symmetric Matrices

Our next goal is to devise a procedure for orthogonally diagonalizing a symmetric matrix, but before we can do so, we need the following critical theorem about eigenvalues and eigenvectors of symmetric matrices.

> **THEOREM 7.2.2** *If A is a symmetric matrix, then:*
>
> (a) *The eigenvalues of A are all real numbers.*
>
> (b) *Eigenvectors from different eigenspaces are orthogonal.*

Part (a), which requires results about complex vector spaces, will be discussed in Section 7.5.

Proof (b) Let $\mathbf{v}_1$ and $\mathbf{v}_2$ be eigenvectors corresponding to distinct eigenvalues λ_1 and λ_2 of the matrix A. We want to show that $\mathbf{v}_1 \cdot \mathbf{v}_2 = 0$. Our proof of this involves the trick of starting with the expression $A\mathbf{v}_1 \cdot \mathbf{v}_2$. It follows from Formula (26) of Section 3.2 and the symmetry of A that

$$A\mathbf{v}_1 \cdot \mathbf{v}_2 = \mathbf{v}_1 \cdot A^T \mathbf{v}_2 = \mathbf{v}_1 \cdot A\mathbf{v}_2 \tag{3}$$

But $\mathbf{v}_1$ is an eigenvector of A corresponding to λ_1, and $\mathbf{v}_2$ is an eigenvector of A corresponding to λ_2, so (3) yields the relationship

$$\lambda_1 \mathbf{v}_1 \cdot \mathbf{v}_2 = \mathbf{v}_1 \cdot \lambda_2 \mathbf{v}_2$$

which can be rewritten as

$$(\lambda_1 - \lambda_2)(\mathbf{v}_1 \cdot \mathbf{v}_2) = 0 \qquad (4)$$

But $\lambda_1 - \lambda_2 \neq 0$, since λ_1 and λ_2 were assumed distinct. Thus, it follows from (4) that $\mathbf{v}_1 \cdot \mathbf{v}_2 = 0$. ◄

Theorem 7.2.2 yields the following procedure for orthogonally diagonalizing a symmetric matrix.

Orthogonally Diagonalizing an $n \times n$ Symmetric Matrix

Step 1. Find a basis for each eigenspace of A.

Step 2. Apply the Gram–Schmidt process to each of these bases to obtain an orthonormal basis for each eigenspace.

Step 3. Form the matrix P whose columns are the vectors constructed in Step 2. This matrix will orthogonally diagonalize A, and the eigenvalues on the diagonal of $D = P^T A P$ will be in the same order as their corresponding eigenvectors in P.

Remark The justification of this procedure should be clear: Theorem 7.2.2 ensures that eigenvectors from *different* eigenspaces are orthogonal, and applying the Gram–Schmidt process ensures that the eigenvectors within the *same* eigenspace are orthonormal. It follows that the *entire* set of eigenvectors obtained by this procedure will be orthonormal.

▶ **EXAMPLE 1 Orthogonally Diagonalizing a Symmetric Matrix**

Find an orthogonal matrix P that diagonalizes

$$A = \begin{bmatrix} 4 & 2 & 2 \\ 2 & 4 & 2 \\ 2 & 2 & 4 \end{bmatrix}$$

Solution We leave it for you to verify that the characteristic equation of A is

$$\det(\lambda I - A) = \det \begin{bmatrix} \lambda - 4 & -2 & -2 \\ -2 & \lambda - 4 & -2 \\ -2 & -2 & \lambda - 4 \end{bmatrix} = (\lambda - 2)^2(\lambda - 8) = 0$$

Thus, the distinct eigenvalues of A are $\lambda = 2$ and $\lambda = 8$. By the method used in Example 7 of Section 5.1, it can be shown that

$$\mathbf{u}_1 = \begin{bmatrix} -1 \\ 1 \\ 0 \end{bmatrix} \quad \text{and} \quad \mathbf{u}_2 = \begin{bmatrix} -1 \\ 0 \\ 1 \end{bmatrix} \qquad (5)$$

form a basis for the eigenspace corresponding to $\lambda = 2$. Applying the Gram–Schmidt process to $\{\mathbf{u}_1, \mathbf{u}_2\}$ yields the following orthonormal eigenvectors (verify):

$$\mathbf{v}_1 = \begin{bmatrix} -\frac{1}{\sqrt{2}} \\ \frac{1}{\sqrt{2}} \\ 0 \end{bmatrix} \quad \text{and} \quad \mathbf{v}_2 = \begin{bmatrix} -\frac{1}{\sqrt{6}} \\ -\frac{1}{\sqrt{6}} \\ \frac{2}{\sqrt{6}} \end{bmatrix} \qquad (6)$$

The eigenspace corresponding to $\lambda = 8$ has

$$\mathbf{u}_3 = \begin{bmatrix} 1 \\ 1 \\ 1 \end{bmatrix}$$

as a basis. Applying the Gram–Schmidt process to $\{\mathbf{u}_3\}$ (i.e., normalizing $\mathbf{u}_3$) yields

$$\mathbf{v}_3 = \begin{bmatrix} \frac{1}{\sqrt{3}} \\ \frac{1}{\sqrt{3}} \\ \frac{1}{\sqrt{3}} \end{bmatrix}$$

Finally, using $\mathbf{v}_1$, $\mathbf{v}_2$, and $\mathbf{v}_3$ as column vectors, we obtain

$$P = \begin{bmatrix} -\frac{1}{\sqrt{2}} & -\frac{1}{\sqrt{6}} & \frac{1}{\sqrt{3}} \\ \frac{1}{\sqrt{2}} & -\frac{1}{\sqrt{6}} & \frac{1}{\sqrt{3}} \\ 0 & \frac{2}{\sqrt{6}} & \frac{1}{\sqrt{3}} \end{bmatrix}$$

which orthogonally diagonalizes A. As a check, we leave it for you to confirm that

$$P^T A P = \begin{bmatrix} -\frac{1}{\sqrt{2}} & \frac{1}{\sqrt{2}} & 0 \\ -\frac{1}{\sqrt{6}} & -\frac{1}{\sqrt{6}} & \frac{2}{\sqrt{6}} \\ \frac{1}{\sqrt{3}} & \frac{1}{\sqrt{3}} & \frac{1}{\sqrt{3}} \end{bmatrix} \begin{bmatrix} 4 & 2 & 2 \\ 2 & 4 & 2 \\ 2 & 2 & 4 \end{bmatrix} \begin{bmatrix} -\frac{1}{\sqrt{2}} & -\frac{1}{\sqrt{6}} & \frac{1}{\sqrt{3}} \\ \frac{1}{\sqrt{2}} & -\frac{1}{\sqrt{6}} & \frac{1}{\sqrt{3}} \\ 0 & \frac{2}{\sqrt{6}} & \frac{1}{\sqrt{3}} \end{bmatrix} = \begin{bmatrix} 2 & 0 & 0 \\ 0 & 2 & 0 \\ 0 & 0 & 8 \end{bmatrix} \blacktriangleleft$$

Spectral Decomposition   If A is a symmetric matrix that is orthogonally diagonalized by

$$P = [\mathbf{u}_1 \quad \mathbf{u}_2 \quad \cdots \quad \mathbf{u}_n]$$

and if $\lambda_1, \lambda_2, \ldots, \lambda_n$ are the eigenvalues of A corresponding to the unit eigenvectors $\mathbf{u}_1, \mathbf{u}_2, \ldots, \mathbf{u}_n$, then we know that $D = P^T A P$, where D is a diagonal matrix with the eigenvalues in the diagonal positions. It follows from this that the matrix A can be expressed as

$$A = PDP^T = [\mathbf{u}_1 \quad \mathbf{u}_2 \quad \cdots \quad \mathbf{u}_n] \begin{bmatrix} \lambda_1 & 0 & \cdots & 0 \\ 0 & \lambda_2 & \cdots & 0 \\ \vdots & \vdots & \ddots & \vdots \\ 0 & 0 & \cdots & \lambda_n \end{bmatrix} \begin{bmatrix} \mathbf{u}_1^T \\ \mathbf{u}_2^T \\ \vdots \\ \mathbf{u}_n^T \end{bmatrix}$$

$$= [\lambda_1 \mathbf{u}_1 \quad \lambda_2 \mathbf{u}_2 \quad \cdots \quad \lambda_n \mathbf{u}_n] \begin{bmatrix} \mathbf{u}_1^T \\ \mathbf{u}_2^T \\ \vdots \\ \mathbf{u}_n^T \end{bmatrix}$$

Multiplying out, we obtain the formula

$$A = \lambda_1 \mathbf{u}_1 \mathbf{u}_1^T + \lambda_2 \mathbf{u}_2 \mathbf{u}_2^T + \cdots + \lambda_n \mathbf{u}_n \mathbf{u}_n^T \tag{7}$$

which is called a ***spectral decomposition of*** A.[*]

---

[*] The terminology *spectral decomposition* is derived from the fact that the set of all eigenvalues of a matrix A is sometimes called the **spectrum** of A. The terminology *eigenvalue decomposition* is due to Professor Dan Kalman, who introduced it in an award-winning paper entitled "A Singularly Valuable Decomposition: The SVD of a Matrix," *The College Mathematics Journal*, Vol. 27, No. 1, January 1996.

Note that in each term of the spectral decomposition of A has the form $\lambda \mathbf{u}\mathbf{u}^T$, where $\mathbf{u}$ is a unit eigenvector of A in column form, and λ is an eigenvalue of A corresponding to $\mathbf{u}$. Since $\mathbf{u}$ has size $n \times 1$, it follows that the product $\mathbf{u}\mathbf{u}^T$ has size $n \times n$. It can be proved (though we will not do it) that $\mathbf{u}\mathbf{u}^T$ is the standard matrix for the orthogonal projection of R^n on the subspace spanned by the vector $\mathbf{u}$. Accepting this to be so, the spectral decomposition of A tells that the image of a vector $\mathbf{x}$ under multiplication by a symmetric matrix A can be obtained by projecting $\mathbf{x}$ orthogonally on the lines (one-dimensional subspaces) determined by the eigenvectors of A, then scaling those projections by the eigenvalues, and then adding the scaled projections. Here is an example.

▶ **EXAMPLE 2 A Geometric Interpretation of a Spectral Decomposition**

The matrix

$$A = \begin{bmatrix} 1 & 2 \\ 2 & -2 \end{bmatrix}$$

has eigenvalues $\lambda_1 = -3$ and $\lambda_2 = 2$ with corresponding eigenvectors

$$\mathbf{x}_1 = \begin{bmatrix} 1 \\ -2 \end{bmatrix} \quad \text{and} \quad \mathbf{x}_2 = \begin{bmatrix} 2 \\ 1 \end{bmatrix}$$

(verify). Normalizing these basis vectors yields

$$\mathbf{u}_1 = \frac{\mathbf{x}_1}{\|\mathbf{x}_1\|} = \begin{bmatrix} \frac{1}{\sqrt{5}} \\ -\frac{2}{\sqrt{5}} \end{bmatrix} \quad \text{and} \quad \mathbf{u}_2 = \frac{\mathbf{x}_2}{\|\mathbf{x}_2\|} = \begin{bmatrix} \frac{2}{\sqrt{5}} \\ \frac{1}{\sqrt{5}} \end{bmatrix}$$

so a spectral decomposition of A is

$$\begin{bmatrix} 1 & 2 \\ 2 & -2 \end{bmatrix} = \lambda_1 \mathbf{u}_1 \mathbf{u}_1^T + \lambda_2 \mathbf{u}_2 \mathbf{u}_2^T = (-3) \begin{bmatrix} \frac{1}{\sqrt{5}} \\ -\frac{2}{\sqrt{5}} \end{bmatrix} \begin{bmatrix} \frac{1}{\sqrt{5}} & -\frac{2}{\sqrt{5}} \end{bmatrix} + (2) \begin{bmatrix} \frac{2}{\sqrt{5}} \\ \frac{1}{\sqrt{5}} \end{bmatrix} \begin{bmatrix} \frac{2}{\sqrt{5}} & \frac{1}{\sqrt{5}} \end{bmatrix}$$

$$= (-3) \begin{bmatrix} \frac{1}{5} & -\frac{2}{5} \\ -\frac{2}{5} & \frac{4}{5} \end{bmatrix} + (2) \begin{bmatrix} \frac{4}{5} & \frac{2}{5} \\ \frac{2}{5} & \frac{1}{5} \end{bmatrix} \tag{8}$$

where, as noted above, the 2×2 matrices on the right side of (8) are the standard matrices for the orthogonal projections onto the eigenspaces corresponding to $\lambda_1 = -3$ and $\lambda_2 = 2$, respectively.

Now let us see what this spectral decomposition tells us about the image of the vector $\mathbf{x} = (1, 1)$ under multiplication by A. Writing $\mathbf{x}$ in column form, it follows that

$$A\mathbf{x} = \begin{bmatrix} 1 & 2 \\ 2 & -2 \end{bmatrix} \begin{bmatrix} 1 \\ 1 \end{bmatrix} = \begin{bmatrix} 3 \\ 0 \end{bmatrix} \tag{9}$$

and from (8) that

$$A\mathbf{x} = \begin{bmatrix} 1 & 2 \\ 2 & -2 \end{bmatrix} \begin{bmatrix} 1 \\ 1 \end{bmatrix} = (-3) \begin{bmatrix} \frac{1}{5} & -\frac{2}{5} \\ -\frac{2}{5} & \frac{4}{5} \end{bmatrix} \begin{bmatrix} 1 \\ 1 \end{bmatrix} + (2) \begin{bmatrix} \frac{4}{5} & \frac{2}{5} \\ \frac{2}{5} & \frac{1}{5} \end{bmatrix} \begin{bmatrix} 1 \\ 1 \end{bmatrix}$$

$$= (-3) \begin{bmatrix} -\frac{1}{5} \\ \frac{2}{5} \end{bmatrix} + (2) \begin{bmatrix} \frac{6}{5} \\ \frac{3}{5} \end{bmatrix}$$

$$= \begin{bmatrix} \frac{3}{5} \\ -\frac{6}{5} \end{bmatrix} + \begin{bmatrix} \frac{12}{5} \\ \frac{6}{5} \end{bmatrix} = \begin{bmatrix} 3 \\ 0 \end{bmatrix} \tag{10}$$

Formulas (9) and (10) provide two different ways of viewing the image of the vector $(1, 1)$ under multiplication by A: Formula (9) tells us directly that the image of this vector is $(3, 0)$, whereas Formula (10) tells us that this image can also be obtained by projecting $(1, 1)$ onto the eigenspaces corresponding to $\lambda_1 = -3$ and $\lambda_2 = 2$ to obtain the vectors $\left(-\frac{1}{5}, \frac{2}{5}\right)$ and $\left(\frac{6}{5}, \frac{3}{5}\right)$, then scaling by the eigenvalues to obtain $\left(\frac{3}{5}, -\frac{6}{5}\right)$ and $\left(\frac{12}{5}, \frac{6}{5}\right)$, and then adding these vectors (see Figure 7.2.1). ◄

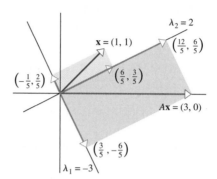

▶ **Figure 7.2.1**

The Nondiagonalizable Case If A is an $n \times n$ matrix that is not orthogonally diagonalizable, it may still be possible to achieve considerable simplification in the form of $P^T A P$ by choosing the orthogonal matrix P appropriately. We will consider two theorems (without proof) that illustrate this. The first, due to the German mathematician Isaai Schur, states that every square matrix A is orthogonally similar to an upper triangular matrix that has the eigenvalues of A on the main diagonal.

THEOREM 7.2.3 Schur's Theorem

If A is an $n \times n$ matrix with real entries and real eigenvalues, then there is an orthogonal matrix P such that $P^T A P$ is an upper triangular matrix of the form

$$P^T A P = \begin{bmatrix} \lambda_1 & \times & \times & \cdots & \times \\ 0 & \lambda_2 & \times & \cdots & \times \\ 0 & 0 & \lambda_3 & \cdots & \times \\ \vdots & \vdots & \vdots & \ddots & \vdots \\ 0 & 0 & 0 & \cdots & \lambda_n \end{bmatrix} \tag{11}$$

in which $\lambda_1, \lambda_2, \ldots, \lambda_n$ are the eigenvalues of the matrix A repeated according to multiplicity.

Historical Note   The life of the German mathematician Issai Schur is a sad reminder of the effect that Nazi policies had on Jewish intellectuals during the 1930s. Schur was a brilliant mathematician and a popular lecturer who attracted many students and researchers to the University of Berlin, where he worked and taught. His lectures sometimes attracted so many students that opera glasses were needed to see him from the back row. Schur's life became increasingly difficult under Nazi rule, and in April of 1933 he was forced to "retire" from the university under a law that prohibited non-Aryans from holding "civil service" positions. There was an outcry from many of his students and colleagues who respected and liked him, but it did not stave off his complete dismissal in 1935. Schur, who thought of himself as a loyal German never understood the persecution and humiliation he received at Nazi hands. He left Germany for Palestine in 1939, a broken man. Lacking in financial resources, he had to sell his beloved mathematics books and lived in poverty until his death in 1941.

Issai Schur (1875–1941)

[*Image: Courtesy Electronic Publishing Services, Inc., New York City*]

It is common to denote the upper triangular matrix in (11) by S (for Schur), in which case that equation can be rewritten as

$$A = PSP^T \tag{12}$$

which is called a **Schur decomposition** of A.

The next theorem, due to the German mathematician and engineer Karl Hessenberg (1904–1959), states that every square matrix with real entries is orthogonally similar to a matrix in which each entry below the first **subdiagonal** is zero (Figure 7.2.2). Such a matrix is said to be in **upper Hessenberg form**.

First subdiagonal

▲ Figure 7.2.2

Note that unlike those in (11), the diagonal entries in (13) are usually *not* the eigenvalues of A.

THEOREM 7.2.4 Hessenberg's Theorem

If A is an $n \times n$ matrix, then there is an orthogonal matrix P such that P^TAP is a matrix of the form

$$P^TAP = \begin{bmatrix} \times & \times & \cdots & \times & \times & \times \\ \times & \times & \cdots & \times & \times & \times \\ 0 & \times & \ddots & \times & \times & \times \\ \vdots & \vdots & \ddots & \vdots & \vdots & \vdots \\ 0 & 0 & \cdots & \times & \times & \times \\ 0 & 0 & \cdots & 0 & \times & \times \end{bmatrix} \tag{13}$$

It is common to denote the upper Hessenberg matrix in (13) by H (for Hessenberg), in which case that equation can be rewritten as

$$A = PHP^T \tag{14}$$

which is called an **upper Hessenberg decomposition** of A.

Remark In many numerical algorithms the initial matrix is first converted to upper Hessenberg form to reduce the amount of computation in subsequent parts of the algorithm. Many computer packages have built-in commands for finding Schur and Hessenberg decompositions.

Concept Review

- Orthogonally similar matrices
- Orthogonally diagonalizable matrix
- Spectral decomposition (or eigenvalue decomposition)
- Schur decomposition
- Subdiagonal
- Upper Hessenburg form
- Upper Hessenburg decomposition

Skills

- Be able to recognize an orthogonally diagonalizable matrix.
- Know that eigenvalues of symmetric matrices are real numbers.
- Know that for a symmetric matrix eigenvectors from different eigenspaces are orthogonal.
- Be able to orthogonally diagonalize a symmetric matrix.
- Be able to find the spectral decomposition of a symmetric matrix.
- Know the statement of Schur's Theorem.
- Know the statement of Hessenburg's Theorem.

Exercise Set 7.2

1. Find the characteristic equation of the given symmetric matrix, and then by inspection determine the dimensions of the eigenspaces.

(a) $\begin{bmatrix} 1 & 2 \\ 2 & 4 \end{bmatrix}$

(b) $\begin{bmatrix} 1 & -4 & 2 \\ -4 & 1 & -2 \\ 2 & -2 & -2 \end{bmatrix}$

(c) $\begin{bmatrix} 1 & 1 & 1 \\ 1 & 1 & 1 \\ 1 & 1 & 1 \end{bmatrix}$

(d) $\begin{bmatrix} 4 & 2 & 2 \\ 2 & 4 & 2 \\ 2 & 2 & 4 \end{bmatrix}$

(e) $\begin{bmatrix} 4 & 4 & 0 & 0 \\ 4 & 4 & 0 & 0 \\ 0 & 0 & 0 & 0 \\ 0 & 0 & 0 & 0 \end{bmatrix}$

(f) $\begin{bmatrix} 2 & -1 & 0 & 0 \\ -1 & 2 & 0 & 0 \\ 0 & 0 & 2 & -1 \\ 0 & 0 & -1 & 2 \end{bmatrix}$

▶ In Exercises 2–9, find a matrix P that orthogonally diagonalizes A, and determine $P^{-1}AP$. ◀

2. $A = \begin{bmatrix} 3 & 1 \\ 1 & 3 \end{bmatrix}$

3. $A = \begin{bmatrix} 6 & 2\sqrt{3} \\ 2\sqrt{3} & 7 \end{bmatrix}$

4. $A = \begin{bmatrix} 6 & -2 \\ -2 & 3 \end{bmatrix}$

5. $A = \begin{bmatrix} -2 & 0 & -36 \\ 0 & -3 & 0 \\ -36 & 0 & -23 \end{bmatrix}$

6. $A = \begin{bmatrix} 1 & 1 & 0 \\ 1 & 1 & 0 \\ 0 & 0 & 0 \end{bmatrix}$

7. $A = \begin{bmatrix} 2 & -1 & -1 \\ -1 & 2 & -1 \\ -1 & -1 & 2 \end{bmatrix}$

8. $PA = \begin{bmatrix} 3 & 1 & 0 & 0 \\ 1 & 3 & 0 & 0 \\ 0 & 0 & 0 & 0 \\ 0 & 0 & 0 & 0 \end{bmatrix}$

9. $A = \begin{bmatrix} -7 & 24 & 0 & 0 \\ 24 & 7 & 0 & 0 \\ 0 & 0 & -7 & 24 \\ 0 & 0 & 24 & 7 \end{bmatrix}$

10. Assuming that $b \neq 0$, find a matrix that orthogonally diagonalizes

$$\begin{bmatrix} a & b \\ b & a \end{bmatrix}$$

11. Prove that if A is any $m \times n$ matrix, then A^TA has an orthonormal set of n eigenvectors.

12. (a) Show that if v is any $n \times 1$ matrix and I is the $n \times n$ identity matrix, then $I - vv^T$ is orthogonally diagonalizable.

(b) Find a matrix P that orthogonally diagonalizes $I - vv^T$ if

$$v = \begin{bmatrix} 1 \\ 0 \\ 1 \end{bmatrix}$$

13. Use the result in Exercise 19 of Section 5.1 to prove Theorem 7.2.2a for 2×2 symmetric matrices.

14. Does there exist a 3×3 symmetric matrix with eigenvalues $\lambda_1 = -1, \lambda_2 = 3, \lambda_3 = 7$ and corresponding eigenvectors

$$\begin{bmatrix} 0 \\ 1 \\ -1 \end{bmatrix}, \begin{bmatrix} 1 \\ 0 \\ 0 \end{bmatrix}, \begin{bmatrix} 0 \\ 1 \\ 1 \end{bmatrix}?$$

If so, find such a matrix; if not, explain why not.

15. Is the converse of Theorem 7.2.2b true? Explain.

16. Find the spectral decomposition of each matrix.

(a) $\begin{bmatrix} 3 & 1 \\ 1 & 3 \end{bmatrix}$

(b) $\begin{bmatrix} 6 & -2 \\ -2 & 3 \end{bmatrix}$

(c) $\begin{bmatrix} -3 & 1 & 2 \\ 1 & -3 & 2 \\ 2 & 2 & 0 \end{bmatrix}$

(d) $\begin{bmatrix} -2 & 0 & -36 \\ 0 & -3 & 0 \\ -36 & 0 & -23 \end{bmatrix}$

17. Show that if A is a symmetric orthogonal matrix, then 1 and -1 are the only possible eigenvalues.

18. (a) Find a 3×3 symmetric matrix whose eigenvalues are $\lambda_1 = -1, \lambda_2 = 3, \lambda_3 = 7$ and for which the corresponding eigenvectors are $v_1 = (0, 1, -1), v_2 = (1, 0, 0),$ $v_3 = (0, 1, 1)$.

(b) Is there a 3×3 symmetric matrix with eigenvalues $\lambda_1 = -1, \lambda_2 = 3, \lambda_3 = 7$ and corresponding eigenvectors $v_1 = (0, 1, -1), v_2 = (1, 0, 0), v_3 = (1, 1, 1)$? Explain your reasoning.

19. Let A be a diagonalizable matrix with the property that eigenvectors from distinct eigenvalues are orthogonal. Must A be symmetric? Explain you reasoning.

20. Prove: If $\{u_1, u_2, \ldots, u_n\}$ is an orthonormal basis for R^n, and if A can be expressed as

$$A = c u_1 u_1^T + c_2 u_2 u_2^T + \cdots + c_n u_n u_n^T$$

then A is symmetric and has eigenvalues $c_1, c_2, \ldots, c_n$.

21. In this exercise we will establish that a matrix A is orthogonally diagonalizable if and only if it is symmetric. We have shown that an orthogonally diagonalizable matrix is symmetric. The harder part is to prove that a symmetric matrix A is orthogonally diagonalizable. We will proceed in two steps: first we will show that A is diagonalizable, and then we will build on that result to show that A is orthogonally diagonalizable.

(a) Assume that A is a symmetric $n \times n$ matrix. One way to prove that A is diagonalizable is to show that for each eigenvalue λ_0 the geometric multiplicity is equal to the algebraic multiplicity. For this purpose, assume that the geometric multiplicity of λ_0 is k, let $B_0 = \{u_1, u_2, \ldots, u_k\}$ be an orthonormal basis for the eigenspace corresponding to λ_0, extend this to an orthonormal basis

$B = \{\mathbf{u}_1, \mathbf{u}_2, \ldots, \mathbf{u}_n\}$ for R^n, and let P be the matrix having the vectors of B as columns. As shown in Exercise 34(b) of Section 5.2, the product AP can be written as

$$AP = P \begin{bmatrix} \lambda_0 I_k & X \\ 0 & Y \end{bmatrix}$$

Use the fact that B is an orthonormal basis to prove that $X = 0$ [a zero matrix of size $n \times (n - k)$].

(b) It follows from part (a) and Exercise 34(c) of Section 5.2 that A has the same characteristic polynomial as

$$C = \begin{bmatrix} \lambda_0 I_k & 0 \\ 0 & Y \end{bmatrix}$$

Use this fact and Exercise 34(d) of Section 5.2 to prove that the algebraic multiplicity of λ_0 is the same as the geometric multiplicity of λ_0. This establishes that A is diagonalizable.

(c) Use Theorem 7.2.2(b) and the fact that A is diagonalizable to prove that A is orthogonally diagonalizable.

True-False Exercises

In parts (a)–(g) determine whether the statement is true or false, and justify your answer.

(a) If A is a square matrix, then AA^T and A^TA are orthogonally diagonalizable.

(b) If $\mathbf{v}_1$ and $\mathbf{v}_2$ are eigenvectors from distinct eigenspaces of a symmetric matrix, then $\|\mathbf{v}_1 + \mathbf{v}_2\|^2 = \|\mathbf{v}_1\|^2 + \|\mathbf{v}_2\|^2$.

(c) Every orthogonal matrix is orthogonally diagonalizable.

(d) If A is both invertible and orthogonally diagonalizable, then A^{-1} is orthogonally diagonalizable.

(e) Every eigenvalue of an orthogonal matrix has absolute value 1.

(f) If A is an $n \times n$ orthogonally diagonalizable matrix, then there exists an orthonormal basis for R^n consisting of eigenvectors of A.

(g) If A is orthogonally diagonalizable, then A has real eigenvalues.

7.3 Quadratic Forms

In this section we will use matrix methods to study real-valued functions of several variables in which each term is either the square of a variable or the product of two variables. Such functions arise in a variety of applications, including geometry, vibrations of mechanical systems, statistics, and electrical engineering.

Definition of a Quadratic Form

Expressions of the form

$$a_1 x_1 + a_2 x_2 + \cdots + a_n x_n$$

occurred in our study of linear equations and linear systems. If $a_1, a_2, \ldots, a_n$ are treated as fixed constants, then this expression is a real-valued function of the n variables $x_1, x_2, \ldots, x_n$ and is called a *linear form* on R^n. All variables in a linear form occur to the first power and there are no products of variables. Here we will be concerned with *quadratic forms* on R^n, which are functions of the form

$$a_1 x_1^2 + a_2 x_2^2 + \cdots + a_n x_n^2 + \text{(all possible terms } a_k x_i x_j \text{ in which } x_i \neq x_j)$$

The terms of the form $a_k x_i x_j$ are called *cross product terms*. It is common to combine the cross product terms involving $x_i x_j$ with those involving $x_j x_i$ to avoid duplication. Thus, a general quadratic form on R^2 would typically be expressed as

$$a_1 x_1^2 + a_2 x_2^2 + 2a_3 x_1 x_2 \tag{1}$$

and a general quadratic form on R^3 as

$$a_1 x_1^2 + a_2 x_2^2 + a_3 x_3^2 + 2a_4 x_1 x_2 + 2a_5 x_1 x_3 + 2a_6 x_2 x_3 \tag{2}$$

If, as usual, we do not distinguish between the number a and the 1×1 matrix $[a]$, and if we let $\mathbf{x}$ be the column vector of variables, then (1) and (2) can be expressed in matrix form as

$$\begin{bmatrix} x_1 & x_2 \end{bmatrix} \begin{bmatrix} a_1 & a_3 \\ a_3 & a_2 \end{bmatrix} \begin{bmatrix} x_1 \\ x_2 \end{bmatrix} = \mathbf{x}^T A \mathbf{x}$$

$$\begin{bmatrix} x_1 & x_2 & x_3 \end{bmatrix} \begin{bmatrix} a_1 & a_4 & a_5 \\ a_4 & a_2 & a_6 \\ a_5 & a_6 & a_3 \end{bmatrix} \begin{bmatrix} x_1 \\ x_2 \\ x_3 \end{bmatrix} = \mathbf{x}^T A \mathbf{x}$$

(verify). Note that the matrix A in these formulas is symmetric, that its diagonal entries are the coefficients of the squared terms, and its off-diagonal entries are half the coefficients of the cross product terms. In general, if A is a symmetric $n \times n$ matrix and $\mathbf{x}$ is an $n \times 1$ column vector of variables, then we call the function

$$Q_A(\mathbf{x}) = \mathbf{x}^T A \mathbf{x} \tag{3}$$

the *quadratic form associated with A*. When convenient, (3) can be expressed in dot product notation as

$$\mathbf{x}^T A \mathbf{x} = \mathbf{x} \cdot A\mathbf{x} = A\mathbf{x} \cdot \mathbf{x} \tag{4}$$

In the case where A is a diagonal matrix, the quadratic form $\mathbf{x}^T A \mathbf{x}$ has no cross product terms; for example, if A has diagonal entries $\lambda_1, \lambda_2, \ldots, \lambda_n$, then

$$\mathbf{x}^T A \mathbf{x} = [x_1 \quad x_2 \quad \cdots \quad x_n] \begin{bmatrix} \lambda_1 & 0 & \cdots & 0 \\ 0 & \lambda_2 & \cdots & 0 \\ \vdots & \vdots & \ddots & \vdots \\ 0 & 0 & \cdots & \lambda_n \end{bmatrix} \begin{bmatrix} x_1 \\ x_2 \\ \vdots \\ x_n \end{bmatrix} = \lambda_1 x_1^2 + \lambda_2 x_2^2 + \cdots + \lambda_n x_n^2$$

▶ **EXAMPLE 1 Expressing Quadratic Forms in Matrix Notation**

In each part, express the quadratic form in the matrix notation $\mathbf{x}^T A \mathbf{x}$, where A is symmetric.

(a) $2x^2 + 6xy - 5y^2$ (b) $x_1^2 + 7x_2^2 - 3x_3^2 + 4x_1x_2 - 2x_1x_3 + 8x_2x_3$

Solution The diagonal entries of A are the coefficients of the squared terms, and the off-diagonal entries are half the coefficients of the cross product terms, so

$$2x^2 + 6xy - 5y^2 = [x \quad y] \begin{bmatrix} 2 & 3 \\ 3 & -5 \end{bmatrix} \begin{bmatrix} x \\ y \end{bmatrix}$$

$$x_1^2 + 7x_2^2 - 3x_3^2 + 4x_1x_2 - 2x_1x_3 + 8x_2x_3 = [x_1 \quad x_2 \quad x_3] \begin{bmatrix} 1 & 2 & -1 \\ 2 & 7 & 4 \\ -1 & 4 & -3 \end{bmatrix} \begin{bmatrix} x_1 \\ x_2 \\ x_3 \end{bmatrix} \ ◀$$

Change of Variable in a Quadratic Form

There are three important kinds of problems that occur in applications of quadratic forms:

> **Problem 1** If $\mathbf{x}^T A \mathbf{x}$ is a quadratic form on R^2 or R^3, what kind of curve or surface is represented by the equation $\mathbf{x}^T A \mathbf{x} = k$?
>
> **Problem 2** If $\mathbf{x}^T A \mathbf{x}$ is a quadratic form on R^n, what conditions must A satisfy for $\mathbf{x}^T A \mathbf{x}$ to have positive values for $\mathbf{x} \neq \mathbf{0}$?
>
> **Problem 3** If $\mathbf{x}^T A \mathbf{x}$ is a quadratic form on R^n, what are its maximum and minimum values if $\mathbf{x}$ is constrained to satisfy $\|\mathbf{x}\| = 1$?

We will consider the first two problems in this section and the third problem in the next section.

Many of the techniques for solving these problems are based on simplifying the quadratic form $\mathbf{x}^T A \mathbf{x}$ by making a substitution

$$\mathbf{x} = P\mathbf{y} \tag{5}$$

that expresses the variables $x_1, x_2, \ldots, x_n$ in terms of new variables $y_1, y_2, \ldots, y_n$. If P is invertible, then we call (5) a **change of variable**, and if P is orthogonal, then we call (5) an **orthogonal change of variable**.

If we make the change of variable $\mathbf{x} = P\mathbf{y}$ in the quadratic form $\mathbf{x}^T A\mathbf{x}$, then we obtain

$$\mathbf{x}^T A\mathbf{x} = (P\mathbf{y})^T A(P\mathbf{y}) = \mathbf{y}^T P^T A P\mathbf{y} = \mathbf{y}^T(P^T A P)\mathbf{y} \tag{6}$$

Since the matrix $B = P^T A P$ is symmetric (verify), the effect of the change of variable is to produce a new quadratic form $\mathbf{y}^T B\mathbf{y}$ in the variables $y_1, y_2, \ldots, y_n$. In particular, if we choose P to orthogonally diagonalize A, then the new quadratic form will be $\mathbf{y}^T D\mathbf{y}$, where D is a diagonal matrix with the eigenvalues of A on the main diagonal; that is,

$$\mathbf{x}^T A\mathbf{x} = \mathbf{y}^T D\mathbf{y} = [y_1 \quad y_2 \quad \cdots \quad y_n] \begin{bmatrix} \lambda_1 & 0 & \cdots & 0 \\ 0 & \lambda_2 & \cdots & 0 \\ \vdots & \vdots & \ddots & \vdots \\ 0 & 0 & \cdots & \lambda_n \end{bmatrix} \begin{bmatrix} y_1 \\ y_2 \\ \vdots \\ y_n \end{bmatrix}$$

$$= \lambda_1 y_1^2 + \lambda_2 y_2^2 + \cdots + \lambda_n y_n^2$$

Thus, we have the following result, called the **principal axes theorem**.

THEOREM 7.3.1 The Principal Axes Theorem

If A is a symmetric $n \times n$ matrix, then there is an orthogonal change of variable that transforms the quadratic form $\mathbf{x}^T A\mathbf{x}$ into a quadratic form $\mathbf{y}^T D\mathbf{y}$ with no cross product terms. Specifically, if P orthogonally diagonalizes A, then making the change of variable $\mathbf{x} = P\mathbf{y}$ in the quadratic form $\mathbf{x}^T A\mathbf{x}$ yields the quadratic form

$$\mathbf{x}^T A\mathbf{x} = \mathbf{y}^T D\mathbf{y} = \lambda_1 y_1^2 + \lambda_2 y_2^2 + \cdots + \lambda_n y_n^2$$

in which $\lambda_1, \lambda_2, \ldots, \lambda_n$ are the eigenvalues of A corresponding to the eigenvectors that form the successive columns of P.

▶ **EXAMPLE 2 An Illustration of the Principal Axes Theorem**

Find an orthogonal change of variable that eliminates the cross product terms in the quadratic form $Q = x_1^2 - x_3^2 - 4x_1 x_2 + 4x_2 x_3$, and express Q in terms of the new variables.

Solution The quadratic form can be expressed in matrix notation as

$$Q = \mathbf{x}^T A\mathbf{x} = [x_1 \quad x_2 \quad x_3] \begin{bmatrix} 1 & -2 & 0 \\ -2 & 0 & 2 \\ 0 & 2 & -1 \end{bmatrix} \begin{bmatrix} x_1 \\ x_2 \\ x_3 \end{bmatrix}$$

The characteristic equation of the matrix A is

$$\begin{vmatrix} \lambda - 1 & 2 & 0 \\ 2 & \lambda & -2 \\ 0 & -2 & \lambda + 1 \end{vmatrix} = \lambda^3 - 9\lambda = \lambda(\lambda + 3)(\lambda - 3) = 0$$

so the eigenvalues are $\lambda = 0, -3, 3$. We leave it for you to show that orthonormal bases for the three eigenspaces are

$$\lambda = 0: \begin{bmatrix} \frac{2}{3} \\ \frac{1}{3} \\ \frac{2}{3} \end{bmatrix}, \quad \lambda = -3: \begin{bmatrix} -\frac{1}{3} \\ -\frac{2}{3} \\ \frac{2}{3} \end{bmatrix}, \quad \lambda = 3: \begin{bmatrix} -\frac{2}{3} \\ \frac{2}{3} \\ \frac{1}{3} \end{bmatrix}$$

Thus, a substitution $\mathbf{x} = P\mathbf{y}$ that eliminates the cross product terms is

$$
\begin{bmatrix} x_1 \\ x_2 \\ x_3 \end{bmatrix} = \begin{bmatrix} \frac{2}{3} & -\frac{1}{3} & -\frac{2}{3} \\ \frac{1}{3} & -\frac{2}{3} & \frac{2}{3} \\ \frac{2}{3} & \frac{2}{3} & \frac{1}{3} \end{bmatrix} \begin{bmatrix} y_1 \\ y_2 \\ y_3 \end{bmatrix}
$$

This produces the new quadratic form

$$
Q = \mathbf{y}^T(P^TAP)\mathbf{y} = \begin{bmatrix} y_1 & y_2 & y_3 \end{bmatrix} \begin{bmatrix} 0 & 0 & 0 \\ 0 & -3 & 0 \\ 0 & 0 & 3 \end{bmatrix} \begin{bmatrix} y_1 \\ y_2 \\ y_3 \end{bmatrix} = -3y_2^2 + 3y_3^2
$$

in which there are no cross product terms. ◀

Remark If A is a symmetric $n \times n$ matrix, then the quadratic form $\mathbf{x}^TA\mathbf{x}$ is a real-valued function whose range is the set of all possible values for $\mathbf{x}^TA\mathbf{x}$ as $\mathbf{x}$ varies over R^n. It can be shown that an orthogonal change of variable $\mathbf{x} = P\mathbf{y}$ does not alter the range of a quadratic form; that is, the set of all values for $\mathbf{x}^TA\mathbf{x}$ as $\mathbf{x}$ varies over R^n is the same as the set of all values for $\mathbf{y}^T(P^TAP)\mathbf{y}$ as $\mathbf{y}$ varies over R^n.

Quadratic Forms in Geometry

Recall that a ***conic section*** or ***conic*** is a curve that results by cutting a double-napped cone with a plane (Figure 7.3.1). The most important conic sections are ellipses, hyperbolas, and parabolas, which result when the cutting plane does not pass through the vertex. Circles are special cases of ellipses that result when the cutting plane is perpendicular to the axis of symmetry of the cone. If the cutting plane passes through the vertex, then the resulting intersection is called a ***degenerate conic***. The possibilities are a point, a pair of intersecting lines, or a single line.

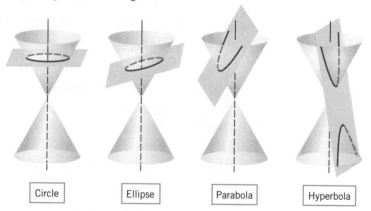

| Circle | Ellipse | Parabola | Hyperbola |

▶ **Figure 7.3.1**

Quadratic forms in R^2 arise naturally in the study of conic sections. For example, it is shown in analytic geometry that an equation of the form

$$
ax^2 + 2bxy + cy^2 + dx + ey + f = 0 \tag{7}
$$

in which a, b, and c are not all zero, represents a conic section.* If $d = e = 0$ in (7), then there are no linear terms, so the equation becomes

$$
ax^2 + 2bxy + cy^2 + f = 0 \tag{8}
$$

---

*We must also allow for the possibility that there are no real values of x and y that satisfy the equation, as with $x^2 + y^2 + 1 = 0$. In such cases we say that the equation has ***no graph*** or has ***an empty graph***.

and is said to represent a *central conic*. These include circles, ellipses, and hyperbolas, but not parabolas. Furthermore, if $b = 0$ in (8), then there is no cross product term (i.e., term involving xy), and the equation

$$ax^2 + cy^2 + f = 0 \qquad (9)$$

is said to represent a *central conic in standard position*. The most important conics of this type are shown in Table 1.

Table 1

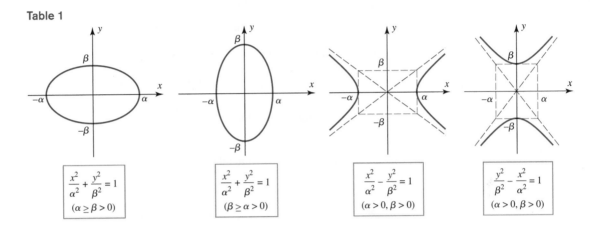

| $\dfrac{x^2}{\alpha^2} + \dfrac{y^2}{\beta^2} = 1$ $(\alpha \geq \beta > 0)$ | $\dfrac{x^2}{\alpha^2} + \dfrac{y^2}{\beta^2} = 1$ $(\beta \geq \alpha > 0)$ | $\dfrac{x^2}{\alpha^2} - \dfrac{y^2}{\beta^2} = 1$ $(\alpha > 0, \beta > 0)$ | $\dfrac{y^2}{\beta^2} - \dfrac{x^2}{\alpha^2} = 1$ $(\alpha > 0, \beta > 0)$ |

If we take the constant f in Equations (8) and (9) to the right side and let $k = -f$, then we can rewrite these equations in matrix form as

$$\begin{bmatrix} x & y \end{bmatrix} \begin{bmatrix} a & b \\ b & c \end{bmatrix} \begin{bmatrix} x \\ y \end{bmatrix} = k \quad \text{and} \quad \begin{bmatrix} x & y \end{bmatrix} \begin{bmatrix} a & 0 \\ 0 & c \end{bmatrix} \begin{bmatrix} x \\ y \end{bmatrix} = k \qquad (10)$$

The first of these corresponds to Equation (8) in which there is a cross product term $2bxy$, and the second corresponds to Equation (9) in which there is no cross product term. Geometrically, the existence of a cross product term signals that the graph of the quadratic form is rotated about the origin, as in Figure 7.3.2. The three-dimensional analogs of the equations in (10) are

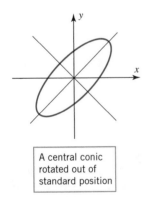

A central conic rotated out of standard position

▲ Figure 7.3.2

$$\begin{bmatrix} x & y & z \end{bmatrix} \begin{bmatrix} a & d & e \\ d & b & f \\ e & f & c \end{bmatrix} \begin{bmatrix} x \\ y \\ z \end{bmatrix} = k \quad \text{and} \quad \begin{bmatrix} x & y & z \end{bmatrix} \begin{bmatrix} a & 0 & 0 \\ 0 & b & 0 \\ 0 & 0 & c \end{bmatrix} \begin{bmatrix} x \\ y \\ z \end{bmatrix} = k \qquad (11)$$

If a, b, and c are not all zero, then the graphs of these equations in R^3 are called *central quadrics in standard position*.

Identifying Conic Sections   We are now ready to consider the first of the three problems posed earlier, identifying the curve or surface represented by an equation $\mathbf{x}^T A \mathbf{x} = k$ in two or three variables. We will focus on the two-variable case. We noted above that an equation of the form

$$ax^2 + 2bxy + cy^2 + f = 0 \qquad (12)$$

represents a central conic. If $b = 0$, then the conic is in standard position, and if $b \neq 0$, it is rotated. It is an easy matter to identify central conics in standard position by matching the equation with one of the standard forms. For example, the equation

$$9x^2 + 16y^2 - 144 = 0$$

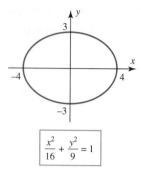

$$\frac{x^2}{16} + \frac{y^2}{9} = 1$$

▲ Figure 7.3.3

can be rewritten as

$$\frac{x^2}{16} + \frac{y^2}{9} = 1$$

which, by comparison with Table 1, is the ellipse shown in Figure 7.3.3.

If a central conic is rotated out of standard position, then it can be identified by first rotating the coordinate axes to put it in standard position and then matching the resulting equation with one of the standard forms in Table 1. To find a rotation that eliminates the cross product term in the equation

$$ax^2 + 2bxy + cy^2 = k \tag{13}$$

it will be convenient to express the equation in the matrix form

$$\mathbf{x}^T A \mathbf{x} = [x \quad y] \begin{bmatrix} a & b \\ b & c \end{bmatrix} \begin{bmatrix} x \\ y \end{bmatrix} = k \tag{14}$$

and look for a change of variable

$$\mathbf{x} = P\mathbf{x}'$$

that diagonalizes A and for which $\det(P) = 1$. Since we saw in Example 4 of Section 7.1 that the transition matrix

$$P = \begin{bmatrix} \cos\theta & -\sin\theta \\ \sin\theta & \cos\theta \end{bmatrix} \tag{15}$$

has the effect of rotating the xy-axes of a rectangular coordinate system through an angle θ, our problem reduces to finding θ that diagonalizes A, thereby eliminating the cross product term in (13). If we make this change of variable, then in the $x'y'$-coordinate system, Equation (14) will become

$$\mathbf{x}'^T D \mathbf{x}' = [x' \quad y'] \begin{bmatrix} \lambda_1 & 0 \\ 0 & \lambda_2 \end{bmatrix} \begin{bmatrix} x' \\ y' \end{bmatrix} = k \tag{16}$$

where λ_1 and λ_2 are the eigenvalues of A. The conic can now be identified by writing (16) in the form

$$\lambda_1 x'^2 + \lambda_2 y'^2 = k \tag{17}$$

and performing the necessary algebra to match it with one of the standard forms in Table 1. For example, if λ_1, λ_2, and k are positive, then (17) represents an ellipse with an axis of length $2\sqrt{k/\lambda_1}$ in the x'-direction and $2\sqrt{k/\lambda_2}$ in the y'-direction. The first column vector of P, which is a unit eigenvector corresponding to λ_1, is along the positive x'-axis; and the second column vector of P, which is a unit eigenvector corresponding to λ_2, is a unit vector along the y'-axis. These are called the **principal axes** of the ellipse, which explains why Theorem 7.3.1 is called "the principal axes theorem." (See Figure 7.3.4.)

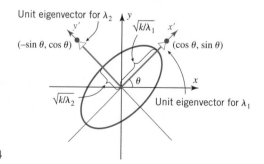

▶ Figure 7.3.4

▶ **EXAMPLE 3 Identifying a Conic by Eliminating the Cross Product Term**

(a) Identify the conic whose equation is $5x^2 - 4xy + 8y^2 - 36 = 0$ by rotating the xy-axes to put the conic in standard position.

(b) Find the angle θ through which you rotated the xy-axes in part (a).

Solution (a) The given equation can be written in the matrix form

$$\mathbf{x}^T A \mathbf{x} = 36$$

where

$$A = \begin{bmatrix} 5 & -2 \\ -2 & 8 \end{bmatrix}$$

The characteristic polynomial of A is

$$\begin{vmatrix} \lambda - 5 & 2 \\ 2 & \lambda - 8 \end{vmatrix} = (\lambda - 4)(\lambda - 9)$$

so the eigenvalues are $\lambda = 4$ and $\lambda = 9$. We leave it for you to show that orthonormal bases for the eigenspaces are

$$\lambda = 4: \begin{bmatrix} \frac{2}{\sqrt{5}} \\ \frac{1}{\sqrt{5}} \end{bmatrix}, \quad \lambda = 9: \begin{bmatrix} -\frac{1}{\sqrt{5}} \\ \frac{2}{\sqrt{5}} \end{bmatrix}$$

Thus, A is orthogonally diagonalized by

$$P = \begin{bmatrix} \frac{2}{\sqrt{5}} & -\frac{1}{\sqrt{5}} \\ \frac{1}{\sqrt{5}} & \frac{2}{\sqrt{5}} \end{bmatrix} \tag{18}$$

Had it turned out that $\det(P) = -1$, then we would have interchanged the columns to reverse the sign.

Moreover, it happens by chance that $\det(P) = 1$, so we are assured that the substitution $\mathbf{x} = P\mathbf{x}'$ performs a rotation of axes. It follows from (16) that the equation of the conic in the $x'y'$-coordinate system is

$$[x' \quad y'] \begin{bmatrix} 4 & 0 \\ 0 & 9 \end{bmatrix} \begin{bmatrix} x' \\ y' \end{bmatrix} = 36$$

which we can write as

$$4x'^2 + 9y'^2 = 36 \quad \text{or} \quad \frac{x'^2}{9} + \frac{y'^2}{4} = 1$$

We can now see from Table 1 that the conic is an ellipse whose axis has length $2\alpha = 6$ in the x'-direction and length $2\beta = 4$ in the y'-direction.

Solution (b) It follows from (15) that

$$P = \begin{bmatrix} \frac{2}{\sqrt{5}} & -\frac{1}{\sqrt{5}} \\ \frac{1}{\sqrt{5}} & \frac{2}{\sqrt{5}} \end{bmatrix} = \begin{bmatrix} \cos\theta & -\sin\theta \\ \sin\theta & \cos\theta \end{bmatrix}$$

which implies that

$$\cos\theta = \frac{2}{\sqrt{5}}, \quad \sin\theta = \frac{1}{\sqrt{5}}, \quad \tan\theta = \frac{\sin\theta}{\cos\theta} = \frac{1}{2}$$

Thus, $\theta = \tan^{-1}\frac{1}{2} \approx 26.6°$ (Figure 7.3.5). ◀

▲ Figure 7.3.5

Remark In the exercises we will ask you to show that if $b \neq 0$, then the cross product term in the equation

$$ax^2 + 2bxy + cy^2 = k$$

can be eliminated by a rotation through an angle θ that satisfies

$$\cot 2\theta = \frac{a - c}{2b} \qquad (19)$$

We leave it for you to confirm that this is consistent with part (b) of the last example.

Positive Definite Quadratic Forms

We will now consider the second of the two problems posed earlier, determining conditions under which $\mathbf{x}^T A \mathbf{x} > 0$ for all nonzero values of $\mathbf{x}$. We will explain why this is important shortly, but first we introduce some terminology.

The terminology in Definition 1 also applies to the matrix A; that is, A is positive definite, negative definite, or indefinite in accordance with whether the associated quadratic form has that property.

DEFINITION 1 A quadratic form $\mathbf{x}^T A \mathbf{x}$ is said to be

positive definite if $\mathbf{x}^T A \mathbf{x} > 0$ for $\mathbf{x} \neq \mathbf{0}$

negative definite if $\mathbf{x}^T A \mathbf{x} < 0$ for $\mathbf{x} \neq \mathbf{0}$

indefinite if $\mathbf{x}^T A \mathbf{x}$ has both positive and negative values

The following theorem, whose proof is deferred to the end of the section, provides a way of using eigenvalues to determine whether a matrix A and its associated quadratic form $\mathbf{x}^T A \mathbf{x}$ are positive definite, negative definite, or indefinite.

THEOREM 7.3.2 *If A is a symmetric matrix, then*:

(a) $\mathbf{x}^T A \mathbf{x}$ *is positive definite if and only if all eigenvalues of A are positive.*

(b) $\mathbf{x}^T A \mathbf{x}$ *is negative definite if and only if all eigenvalues of A are negative.*

(c) $\mathbf{x}^T A \mathbf{x}$ *is indefinite if and only if A has at least one positive eigenvalue and at least one negative eigenvalue.*

Remark The three classifications in Definition 1 do not exhaust all of the possibilities. For example, a quadratic form for which $\mathbf{x}^T A \mathbf{x} \geq 0$ if $\mathbf{x} \neq \mathbf{0}$ is called *positive semidefinite*, and one for which $\mathbf{x}^T A \mathbf{x} \leq 0$ if $\mathbf{x} \neq \mathbf{0}$ is called *negative semidefinite*. Every positive definite form is positive semidefinite, but not conversely, and every negative definite form is negative semidefinite, but not conversely (why?). By adjusting the proof of Theorem 7.3.2 appropriately, one can prove that $\mathbf{x}^T A \mathbf{x}$ is positive semidefinite if and only if all eigenvalues of A are nonnegative and is negative semidefinite if and only if all eigenvalues of A are nonpositive.

▶ **EXAMPLE 4** Positive Definite Quadratic Forms

It is not usually possible to tell from the signs of the entries in a symmetric matrix A whether that matrix is positive definite, negative definite, or indefinite. For example, the entries of the matrix

$$A = \begin{bmatrix} 3 & 1 & 1 \\ 1 & 0 & 2 \\ 1 & 2 & 0 \end{bmatrix}$$

are nonnegative, but the matrix is indefinite since its eigenvalues are $\lambda = 1,\ 4,\ -2$ (verify). To see this another way, let us write out the quadratic form as

Positive definite and negative definite matrices are invertible. Why?

$$\mathbf{x}^T A \mathbf{x} = [x_1 \quad x_2 \quad x_3] \begin{bmatrix} 3 & 1 & 1 \\ 1 & 0 & 2 \\ 1 & 2 & 0 \end{bmatrix} \begin{bmatrix} x_1 \\ x_2 \\ x_3 \end{bmatrix} = 3x_1^2 + 2x_1 x_2 + 2x_1 x_3 + 4x_2 x_3$$

We can now see, for example, that

$$\mathbf{x}^T A \mathbf{x} = 4 \quad \text{for} \quad x_1 = 0, \quad x_2 = 1, \quad x_3 = 1$$

and

$$\mathbf{x}^T A \mathbf{x} = -4 \quad \text{for} \quad x_1 = 0, \quad x_2 = 1, \quad x_3 = -1 \quad \blacktriangleleft$$

Classifying Conic Sections Using Eigenvalues

If $\mathbf{x}^T B \mathbf{x} = k$ is the equation of a conic, and if $k \neq 0$, then we can divide through by k and rewrite the equation in the form

$$\mathbf{x}^T A \mathbf{x} = 1 \tag{20}$$

where $A = (1/k)B$. If we now rotate the coordinate axes to eliminate the cross product term (if any) in this equation, then the equation of the conic in the new coordinate system will be of the form

$$\lambda_1 x'^2 + \lambda_2 y'^2 = 1 \tag{21}$$

in which λ_1 and λ_2 are the eigenvalues of A. The particular type of conic represented by this equation will depend on the signs of the eigenvalues λ_1 and λ_2. For example, you should be able to see from (21) that:

- $\mathbf{x}^T A \mathbf{x} = 1$ represents an ellipse if $\lambda_1 > 0$ and $\lambda_2 > 0$.
- $\mathbf{x}^T A \mathbf{x} = 1$ has no graph if $\lambda_1 < 0$ and $\lambda_2 < 0$.
- $\mathbf{x}^T A \mathbf{x} = 1$ represents a hyperbola if λ_1 and λ_2 have opposite signs.

In the case of the ellipse, Equation (21) can be rewritten as

$$\frac{x'^2}{(1/\sqrt{\lambda_1})^2} + \frac{y'^2}{(1/\sqrt{\lambda_2})^2} = 1 \tag{22}$$

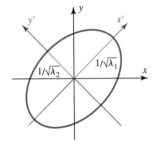

▲ Figure 7.3.6

so the axes of the ellipse have lengths $2/\sqrt{\lambda_1}$ and $2/\sqrt{\lambda_2}$ (Figure 7.3.6).

The following theorem is an immediate consequence of this discussion and Theorem 7.3.2.

THEOREM 7.3.3 *If A is a symmetric 2×2 matrix, then:*

(a) $\mathbf{x}^T A \mathbf{x} = 1$ *represents an ellipse if A is positive definite.*

(b) $\mathbf{x}^T A \mathbf{x} = 1$ *has no graph if A is negative definite.*

(c) $\mathbf{x}^T A \mathbf{x} = 1$ *represents a hyperbola if A is indefinite.*

In Example 3 we performed a rotation to show that the equation

$$5x^2 - 4xy + 8y^2 - 36 = 0$$

represents an ellipse with a major axis of length 6 and a minor axis of length 4. This conclusion can also be obtained by rewriting the equation in the form

$$\tfrac{5}{36}x^2 - \tfrac{1}{9}xy + \tfrac{2}{9}y^2 = 1$$

and showing that the associated matrix

$$A = \begin{bmatrix} \frac{5}{36} & -\frac{1}{18} \\ -\frac{1}{18} & \frac{2}{9} \end{bmatrix}$$

has eigenvalues $\lambda_1 = \frac{1}{9}$ and $\lambda_2 = \frac{1}{4}$. These eigenvalues are positive, so the matrix A is positive definite and the equation represents an ellipse. Moreover, it follows from (21) that the axes of the ellipse have lengths $2/\sqrt{\lambda_1} = 6$ and $2/\sqrt{\lambda_2} = 4$, which is consistent with Example 3.

Identifying Positive Definite Matrices

Positive definite matrices are the most important symmetric matrices in applications, so it will be useful to learn a little more about them. We already know that a symmetric matrix is positive definite if and only if its eigenvalues are all positive; now we will give a criterion that can be used to determine whether a symmetric matrix is positive definite *without* finding the eigenvalues. For this purpose we define the **kth *principal submatrix*** of an $n \times n$ matrix A to be the $k \times k$ submatrix consisting of the first k rows and columns of A. For example, here are the principal submatrices of a general 4×4 matrix:

$$
\begin{bmatrix} a_{11} & a_{12} & a_{13} & a_{14} \\ a_{21} & a_{22} & a_{23} & a_{24} \\ a_{31} & a_{32} & a_{33} & a_{34} \\ a_{41} & a_{42} & a_{43} & a_{44} \end{bmatrix}
\quad
\begin{bmatrix} a_{11} & a_{12} & a_{13} & a_{14} \\ a_{21} & a_{22} & a_{23} & a_{24} \\ a_{31} & a_{32} & a_{33} & a_{34} \\ a_{41} & a_{42} & a_{43} & a_{44} \end{bmatrix}
\quad
\begin{bmatrix} a_{11} & a_{12} & a_{13} & a_{14} \\ a_{21} & a_{22} & a_{23} & a_{24} \\ a_{31} & a_{32} & a_{33} & a_{34} \\ a_{41} & a_{42} & a_{43} & a_{44} \end{bmatrix}
\quad
\begin{bmatrix} a_{11} & a_{12} & a_{13} & a_{14} \\ a_{21} & a_{22} & a_{23} & a_{24} \\ a_{31} & a_{32} & a_{33} & a_{34} \\ a_{41} & a_{42} & a_{43} & a_{44} \end{bmatrix}
$$

| First principal submatrix | Second principal submatrix | Third principal submatrix | Fourth principal submatrix $= A$ |

The following theorem, which we state without proof, provides a determinant test for ascertaining whether a symmetric matrix is positive definite.

THEOREM 7.3.4 *A symmetric matrix A is positive definite if and only if the determinant of every principal submatrix is positive.*

▶ **EXAMPLE 5** **Working with Principal Submatrices**

The matrix

$$
A = \begin{bmatrix} 2 & -1 & -3 \\ -1 & 2 & 4 \\ -3 & 4 & 9 \end{bmatrix}
$$

is positive definite since the determinants

$$
|2| = 2, \quad \begin{vmatrix} 2 & -1 \\ -1 & 2 \end{vmatrix} = 3, \quad \begin{vmatrix} 2 & -1 & -3 \\ -1 & 2 & 4 \\ -3 & 4 & 9 \end{vmatrix} = 1
$$

are all positive. Thus, we are guaranteed that all eigenvalues of A are positive and $\mathbf{x}^T A \mathbf{x} > 0$ for $\mathbf{x} \neq \mathbf{0}$. ◀

OPTIONAL

We conclude this section with an optional proof of Theorem 7.3.2.

Proofs of Theorem 7.3.2(a) and (b) It follows from the principal axes theorem (Theorem 7.3.1) that there is an orthogonal change of variable $\mathbf{x} = P\mathbf{y}$ for which

$$
\mathbf{x}^T A \mathbf{x} = \mathbf{y}^T D \mathbf{y} = \lambda_1 y_1^2 + \lambda_2 y_2^2 + \cdots + \lambda_n y_n^2 \tag{23}
$$

where the λ's are the eigenvalues of A. Moreover, it follows from the invertibility of P that $\mathbf{y} \neq \mathbf{0}$ if and only if $\mathbf{x} \neq \mathbf{0}$, so the values of $\mathbf{x}^T A \mathbf{x}$ for $\mathbf{x} \neq \mathbf{0}$ are the same as the values of $\mathbf{y}^T D \mathbf{y}$ for $\mathbf{y} \neq \mathbf{0}$. Thus, it follows from (23) that $\mathbf{x}^T A \mathbf{x} > 0$ for $\mathbf{x} \neq \mathbf{0}$ if and only if all of the λ's in that equation are positive, and that $\mathbf{x}^T A \mathbf{x} < 0$ for $\mathbf{x} \neq \mathbf{0}$ if and only if all of the λ's are negative. This proves parts (*a*) and (*b*).

Proof (c) Assume that A has at least one positive eigenvalue and at least one negative eigenvalue, and to be specific, suppose that $\lambda_1 > 0$ and $\lambda_2 < 0$ in (23). Then

$$
\mathbf{x}^T A \mathbf{x} > 0 \quad \text{if} \quad y_1 = 1 \text{ and all other } y\text{'s are } 0
$$

and

$$
\mathbf{x}^T A \mathbf{x} < 0 \quad \text{if} \quad y_2 = 1 \text{ and all other } y\text{'s are } 0
$$

which proves that $\mathbf{x}^T A \mathbf{x}$ is indefinite. Conversely, if $\mathbf{x}^T A \mathbf{x} > 0$ for some $\mathbf{x}$, then $\mathbf{y}^T D \mathbf{y} > 0$ for some $\mathbf{y}$, so at least one of the λ's in (23) must be positive. Similarly, if $\mathbf{x}^T A \mathbf{x} < 0$ for some $\mathbf{x}$, then $\mathbf{y}^T D \mathbf{y} < 0$ for some $\mathbf{y}$, so at least one of the λ's in (23) must be negative, which completes the proof. ◀

Concept Review

- Linear form
- Quadratic form
- Cross product term
- Quadratic form associated with a matrix
- Change of variable
- Orthogonal change of variable
- Principal Axes Theorem

- Conic section
- Degenerate conic
- Central conic
- Standard position of a central conic
- Standard form of a central conic
- Central quadric
- Principal axes of an ellipse

- Positive definite quadratic form
- Negative definite quadratic form
- Indefinite quadratic form
- Positive semidefinite quadratic form
- Negative semidefinite quadratic form
- Principal submatrix

Skills

- Express a quadratic form in the matrix notation $\mathbf{x}^T A \mathbf{x}$, where A is a symmetric matrix.
- Find an orthogonal change of variable that eliminates the cross product terms in a quadratic form, and express the quadratic form in terms of the new variable.
- Identify a conic section from an equation by rotating axes to place the conic in standard position, and find the angle of rotation.

- Identify a conic section using eigenvalues.
- Classify matrices and quadratic forms as positive definite, negative definite, indefinite, positive semidefinite or negative semidefinite.

Exercise Set 7.3

▶ In Exercises 1–2, express the quadratic form in the matrix notation $\mathbf{x}^T A \mathbf{x}$, where A is a symmetric matrix. ◀

1. (a) $3x_1^2 + 7x_2^2$
(b) $4x_1^2 - 9x_2^2 - 6x_1x_2$
(c) $9x_1^2 - x_2^2 + 4x_3^2 + 6x_1x_2 - 8x_1x_3 + x_2x_3$

2. (a) $5x_1^2 + 5x_1x_2$
(b) $-7x_1x_2$
(c) $x_1^2 + x_2^2 - 3x_3^2 - 5x_1x_2 + 9x_1x_3$

▶ In Exercises 3–4, find a formula for the quadratic form that does not use matrices. ◀

3. $[x \quad y] \begin{bmatrix} 2 & -3 \\ -3 & 5 \end{bmatrix} \begin{bmatrix} x \\ y \end{bmatrix}$

4. $[x_1 \quad x_2 \quad x_3] \begin{bmatrix} -2 & \frac{7}{2} & 1 \\ \frac{7}{2} & 0 & 6 \\ 1 & 6 & 3 \end{bmatrix} \begin{bmatrix} x_1 \\ x_2 \\ x_3 \end{bmatrix}$

▶ In Exercises 5–8, find an orthogonal change of variables that eliminates the cross product terms in the quadratic form Q, and express Q in terms of the new variables. ◀

5. $Q = 2x_1^2 + 2x_2^2 - 2x_1x_2$

6. $Q = 5x_1^2 + 2x_2^2 + 4x_3^2 + 4x_1x_2$

7. $Q = 3x_1^2 + 4x_2^2 + 5x_3^2 + 4x_1x_2 - 4x_2x_3$

8. $Q = 2x_1^2 + 5x_2^2 + 5x_3^2 + 4x_1x_2 - 4x_1x_3 - 8x_2x_3$

▶ In Exercises 9–10, express the quadratic equation in the matrix form $\mathbf{x}^T A \mathbf{x} + K \mathbf{x} + f = 0$, where $\mathbf{x}^T A \mathbf{x}$ is the associated quadratic form and K is an appropriate matrix. ◀

9. (a) $2x^2 + xy + x - 6y + 2 = 0$
(b) $y^2 + 7x - 8y - 5 = 0$

10. (a) $x^2 - xy + 5x + 8y - 3 = 0$
(b) $5xy = 8$

▶ In Exercises 11–12, identify the conic section represented by the equation. ◀

11. (a) $2x^2 + 5y^2 = 20$
(b) $x^2 - y^2 - 8 = 0$
(c) $7y^2 - 2x = 0$
(d) $x^2 + y^2 - 25 = 0$

12. (a) $4x^2 + 9y^2 = 1$
(b) $4x^2 - 5y^2 = 20$
(c) $-x^2 = 2y$
(d) $x^2 - 3 = -y^2$

▶ In Exercises 13–16, identify the conic section represented by the equation by rotating axes to place the conic in standard position. Find an equation of the conic in the rotated coordinates, and find the angle of rotation. ◀

13. $2x^2 - 4xy - y^2 + 8 = 0$ **14.** $5x^2 + 4xy + 5y^2 = 9$

15. $11x^2 + 24xy + 4y^2 - 15 = 0$

16. $x^2 + xy + y^2 = \frac{1}{2}$

▶ In Exercises 17–18, determine by inspection whether the matrix is positive definite, negative definite, indefinite, positive semidefinite, or negative semidefinite. ◀

17. (a) $\begin{bmatrix} 1 & 0 \\ 0 & 2 \end{bmatrix}$ (b) $\begin{bmatrix} -1 & 0 \\ 0 & -2 \end{bmatrix}$ (c) $\begin{bmatrix} -1 & 0 \\ 0 & 2 \end{bmatrix}$

   (d) $\begin{bmatrix} 1 & 0 \\ 0 & 0 \end{bmatrix}$ (e) $\begin{bmatrix} 0 & 0 \\ 0 & -2 \end{bmatrix}$

18. (a) $\begin{bmatrix} 2 & 0 \\ 0 & -5 \end{bmatrix}$ (b) $\begin{bmatrix} -2 & 0 \\ 0 & -5 \end{bmatrix}$ (c) $\begin{bmatrix} 2 & 0 \\ 0 & 5 \end{bmatrix}$

   (d) $\begin{bmatrix} 0 & 0 \\ 0 & -5 \end{bmatrix}$ (e) $\begin{bmatrix} 2 & 0 \\ 0 & 0 \end{bmatrix}$

▶ In Exercises 19–24, classify the quadratic form as positive definite, negative definite, indefinite, positive semidefinite, or negative semidefinite. ◀

19. $x_1^2 + x_2^2$ **20.** $-x_1^2 - 3x_2^2$ **21.** $(x_1 - x_2)^2$

22. $-(x_1 - x_2)^2$ **23.** $x_1^2 - x_2^2$ **24.** $x_1 x_2$

▶ In Exercises 25–26, show that the matrix A is positive definite first by using Theorem 7.3.2 and second by using Theorem 7.3.4. ◀

25. (a) $A = \begin{bmatrix} 5 & -2 \\ -2 & 5 \end{bmatrix}$ (b) $A = \begin{bmatrix} 2 & -1 & 0 \\ -1 & 2 & 0 \\ 0 & 0 & 5 \end{bmatrix}$

26. (a) $A = \begin{bmatrix} 2 & 1 \\ 1 & 2 \end{bmatrix}$ (b) $A = \begin{bmatrix} 3 & -1 & 0 \\ -1 & 2 & -1 \\ 0 & -1 & 3 \end{bmatrix}$

▶ In Exercises 27–28, find all values of k for which the quadratic form is positive definite. ◀

27. $5x_1^2 + x_2^2 + kx_3^2 + 4x_1x_2 - 2x_1x_3 - 2x_2x_3$

28. $3x_1^2 + x_2^2 + 2x_3^2 - 2x_1x_3 + 2kx_2x_3$

29. Let $\mathbf{x}^T A\mathbf{x}$ be a quadratic form in the variables $x_1, x_2, \ldots, x_n$, and define $T : R^n \to R$ by $T(\mathbf{x}) = \mathbf{x}^T A\mathbf{x}$.

   (a) Show that $T(\mathbf{x} + \mathbf{y}) = T(\mathbf{x}) + 2\mathbf{x}^T A\mathbf{y} + T(\mathbf{y})$.

   (b) Show that $T(c\mathbf{x}) = c^2 T(\mathbf{x})$.

30. Express the quadratic form $(c_1x_1 + c_2x_2 + \cdots + c_nx_n)^2$ in the matrix notation $\mathbf{x}^T A\mathbf{x}$, where A is symmetric.

31. In statistics, the quantities

$$\overline{x} = \frac{1}{n}(x_1 + x_2 + \cdots + x_n)$$

and

$$s_x^2 = \frac{1}{n-1}\left[(x_1 - \overline{x})^2 + (x_2 - \overline{x})^2 + \cdots + (x_n - \overline{x})^2\right]$$

are called, respectively, the *sample mean* and *sample variance* of $\mathbf{x} = (x_1, x_2, \ldots, x_n)$.

   (a) Express the quadratic form s_x^2 in the matrix notation $\mathbf{x}^T A\mathbf{x}$, where A is symmetric.

   (b) Is s_x^2 a positive definite quadratic form? Explain.

32. The graph in an xyz-coordinate system of an equation of form $ax^2 + by^2 + cz^2 = 1$ in which a, b, and c are positive is a surface called a *central ellipsoid in standard position* (see the accompanying figure). This is the three-dimensional generalization of the ellipse $ax^2 + by^2 = 1$ in the xy-plane. The intersections of the ellipsoid $ax^2 + by^2 + cz^2 = 1$ with the coordinate axes determine three line segments called the *axes* of the ellipsoid. If a central ellipsoid is rotated about the origin so two or more of its axes do not coincide with any of the coordinate axes, then the resulting equation will have one or more cross product terms.

   (a) Show that the equation

$$\tfrac{4}{3}x^2 + \tfrac{4}{3}y^2 + \tfrac{4}{3}z^2 + \tfrac{4}{3}xy + \tfrac{4}{3}xz + \tfrac{4}{3}yz = 1$$

represents an ellipsoid, and find the lengths of its axes. [*Suggestion:* Write the equation in the form $\mathbf{x}^T A\mathbf{x} = 1$ and make an orthogonal change of variable to eliminate the cross product terms.]

   (b) What property must a symmetric 3×3 matrix have in order for the equation $\mathbf{x}^T A\mathbf{x} = 1$ to represent an ellipsoid?

◀ Figure Ex-32

33. What property must a symmetric 2×2 matrix A have for $\mathbf{x}^T A\mathbf{x} = 1$ to represent a circle?

34. Prove: If $b \neq 0$, then the cross product term can be eliminated from the quadratic form $ax^2 + 2bxy + cy^2$ by rotating the coordinate axes through an angle θ that satisfies the equation

$$\cot 2\theta = \frac{a - c}{2b}$$

35. Prove that if A is an $n \times n$ symmetric matrix all of whose eigenvalues are nonnegative, then $\mathbf{x}^T A\mathbf{x} \geq 0$ for all nonzero $\mathbf{x}$ in R^n.

True-False Exercises

In parts (a)–(l) determine whether the statement is true or false, and justify your answer.

(a) A symmetric matrix with positive definite eigenvalues is positive definite.

(b) $x_1^2 - x_2^2 + x_3^2 + 4x_1x_2x_3$ is a quadratic form.

(c) $(x_1 - 3x_2)^2$ is a quadratic form.

(d) A positive definite matrix is invertible.

(e) A symmetric matrix is either positive definite, negative definite, or indefinite.

(f) If A is positive definite, then $-A$ is negative definite.

(g) $\mathbf{x} \cdot \mathbf{x}$ is a quadratic form for all $\mathbf{x}$ in R^n.

(h) If $\mathbf{x}^T A \mathbf{x}$ is a positive definite quadratic form, then so is $\mathbf{x}^T A^{-1} \mathbf{x}$.

(i) If A is a matrix with only positive eigenvalues, then $\mathbf{x}^T A \mathbf{x}$ is a positive definite quadratic form.

(j) If A is a 2×2 symmetric matrix with positive entries and $\det(A) > 0$, then A is positive definite.

(k) If $\mathbf{x}^T A \mathbf{x}$ is a quadratic form with no cross product terms, then A is a diagonal matrix.

(l) If $\mathbf{x}^T A \mathbf{x}$ is a positive definite quadratic form in two variables and $c \neq 0$, then the graph of the equation $\mathbf{x}^T A \mathbf{x} = c$ is an ellipse.

7.4 Optimization Using Quadratic Forms

Quadratic forms arise in various problems in which the maximum or minimum value of some quantity is required. In this section we will discuss some problems of this type.

Constrained Extremum Problems

Our first goal in this section is to consider the problem of finding the maximum and minimum values of a quadratic form $\mathbf{x}^T A \mathbf{x}$ subject of the constraint $\|\mathbf{x}\| = 1$. Problems of this type arise in a wide variety of applications.

To visualize this problem geometrically in the case where $\mathbf{x}^T A \mathbf{x}$ is a quadratic form on R^2, view $z = \mathbf{x}^T A \mathbf{x}$ as the equation of some surface in a rectangular xyz-coordinate system and view $\|\mathbf{x}\| = 1$ as the unit circle centered at the origin of the xy-plane. Geometrically, the problem of finding the maximum and minimum values of $\mathbf{x}^T A \mathbf{x}$ subject to the requirement $\|\mathbf{x}\| = 1$ amounts to finding the highest and lowest points on the intersection of the surface with the right circular cylinder determined by the circle (Figure 7.4.1).

The following theorem, whose proof is deferred to the end of the section, is the key result for solving problems of this type.

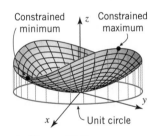

Constrained minimum   Constrained maximum   Unit circle

▲ Figure 7.4.1

THEOREM 7.4.1 Constrained Extremum Theorem

Let A be a symmetric $n \times n$ matrix whose eigenvalues in order of decreasing size are $\lambda_1 \geq \lambda_2 \geq \cdots \geq \lambda_n$. Then:

(a) the quadratic form $\mathbf{x}^T A \mathbf{x}$ attains a maximum value and a minimum value on the set of vectors for which $\|\mathbf{x}\| = 1$;

(b) the maximum value attained in part (a) occurs at a unit vector corresponding to the eigenvalue λ_1;

(c) the minimum value attained in part (a) occurs at a unit vector corresponding to the eigenvalue λ_n.

Remark The condition $\|\mathbf{x}\| = 1$ in this theorem is called a ***constraint***, and the maximum or minimum value of $\mathbf{x}^T A \mathbf{x}$ subject to the constraint is called a ***constrained extremum***. This constraint can also be expressed as $\mathbf{x}^T \mathbf{x} = 1$ or as $x_1^2 + x_2^2 + \cdots + x_n^2 = 1$, when convenient.

▶ **EXAMPLE 1** Finding Constrained Extrema

Find the maximum and minimum values of the quadratic form

$$z = 5x^2 + 5y^2 + 4xy$$

subject to the constraint $x^2 + y^2 = 1$.

Solution The quadratic form can be expressed in matrix notation as

$$z = 5x^2 + 5y^2 + 4xy = \mathbf{x}^T A \mathbf{x} = \begin{bmatrix} x & y \end{bmatrix} \begin{bmatrix} 5 & 2 \\ 2 & 5 \end{bmatrix} \begin{bmatrix} x \\ y \end{bmatrix}$$

We leave it for you to show that the eigenvalues of A are $\lambda_1 = 7$ and $\lambda_2 = 3$ and that corresponding eigenvectors are

$$\lambda_1 = 7: \begin{bmatrix} 1 \\ 1 \end{bmatrix}, \quad \lambda_2 = 3: \begin{bmatrix} -1 \\ 1 \end{bmatrix}$$

Normalizing these eigenvectors yields

$$\lambda_1 = 7: \begin{bmatrix} \frac{1}{\sqrt{2}} \\ \frac{1}{\sqrt{2}} \end{bmatrix}, \quad \lambda_2 = 3: \begin{bmatrix} -\frac{1}{\sqrt{2}} \\ \frac{1}{\sqrt{2}} \end{bmatrix} \tag{1}$$

Thus, the constrained extrema are

constrained maximum: $z = 7$ at $(x, y) = \left(\frac{1}{\sqrt{2}}, \frac{1}{\sqrt{2}}\right)$

constrained minimum: $z = 3$ at $(x, y) = \left(-\frac{1}{\sqrt{2}}, \frac{1}{\sqrt{2}}\right)$ ◀

Remark Since the negatives of the eigenvectors in (1) are also unit eigenvectors, they too produce the maximum and minimum values of z; that is, the constrained maximum $z = 7$ also occurs at the point $(x, y) = \left(-\frac{1}{\sqrt{2}}, -\frac{1}{\sqrt{2}}\right)$ and the constrained minimum $z = 3$ at $(x, y) = \left(\frac{1}{\sqrt{2}}, -\frac{1}{\sqrt{2}}\right)$.

▶ **EXAMPLE 2** A Constrained Extremum Problem

A rectangle is to be inscribed in the ellipse $4x^2 + 9y^2 = 36$, as shown in Figure 7.4.2. Use eigenvalue methods to find nonnegative values of x and y that produce the inscribed rectangle with maximum area.

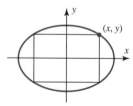

▲ **Figure 7.4.2** A rectangle inscribed in the ellipse $4x^2 + 9y^2 = 36$.

Solution The area z of the inscribed rectangle is given by $z = 4xy$, so the problem is to maximize the quadratic form $z = 4xy$ subject to the constraint $4x^2 + 9y^2 = 36$. In this problem, the graph of the constraint equation is an ellipse rather than the unit circle as required in Theorem 7.4.1, but we can remedy this problem by rewriting the constraint as

$$\left(\frac{x}{3}\right)^2 + \left(\frac{y}{2}\right)^2 = 1$$

and defining new variables, x_1 and y_1, by the equations

$$x = 3x_1 \quad \text{and} \quad y = 2y_1$$

This enables us to reformulate the problem as follows:

maximize $z = 4xy = 24x_1 y_1$

subject to the constraint

$$x_1^2 + y_1^2 = 1$$

To solve this problem, we will write the quadratic form $z = 24x_1 y_1$ as

$$z = \mathbf{x}^T A \mathbf{x} = \begin{bmatrix} x_1 & y_1 \end{bmatrix} \begin{bmatrix} 0 & 12 \\ 12 & 0 \end{bmatrix} \begin{bmatrix} x_1 \\ y_1 \end{bmatrix}$$

We now leave it for you to show that the largest eigenvalue of A is $\lambda = 12$ and that the only corresponding unit eigenvector with nonnegative entries is

$$\mathbf{x} = \begin{bmatrix} x_1 \\ y_1 \end{bmatrix} = \begin{bmatrix} \frac{1}{\sqrt{2}} \\ \frac{1}{\sqrt{2}} \end{bmatrix}$$

Thus, the maximum area is $z = 12$, and this occurs when

$$x = 3x_1 = \frac{3}{\sqrt{2}} \quad \text{and} \quad y = 2y_1 = \frac{2}{\sqrt{2}} \blacktriangleleft$$

Constrained Extrema and Level Curves

▲ Figure 7.4.3

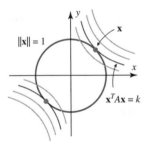

▲ Figure 7.4.4

A useful way of visualizing the behavior of a function $f(x, y)$ of two variables is to consider the curves in the xy-plane along which $f(x, y)$ is constant. These curves have equations of the form

$$f(x, y) = k$$

and are called the *level curves* of f (Figure 7.4.3). In particular, the level curves of a quadratic form $\mathbf{x}^T A \mathbf{x}$ on R^2 have equations of the form

$$\mathbf{x}^T A \mathbf{x} = k \tag{2}$$

so the maximum and minimum values of $\mathbf{x}^T A \mathbf{x}$ subject to the constraint $\|\mathbf{x}\| = 1$ are the largest and smallest values of k for which the graph of (2) intersects the unit circle. Typically, such values of k produce level curves that just touch the unit circle (Figure 7.4.4), and the coordinates of the points where the level curves just touch produce the vectors that maximize or minimize $\mathbf{x}^T A \mathbf{x}$ subject to the constraint $\|\mathbf{x}\| = 1$.

▶ **EXAMPLE 3 Example 1 Revisited Using Level Curves**

In Example 1 (and its following remark) we found the maximum and minimum values of the quadratic form

$$z = 5x^2 + 5y^2 + 4xy$$

subject to the constraint $x^2 + y^2 = 1$. We showed that the constrained maximum is $z = 7$, and this is attained at the points

$$(x, y) = \left(\frac{1}{\sqrt{2}}, \frac{1}{\sqrt{2}} \right) \quad \text{and} \quad (x, y) = \left(-\frac{1}{\sqrt{2}}, -\frac{1}{\sqrt{2}} \right) \tag{3}$$

and that the constrained minimum $z = 3$, and this is attained at the points

$$(x, y) = \left(-\frac{1}{\sqrt{2}}, \frac{1}{\sqrt{2}} \right) \quad \text{and} \quad (x, y) = \left(\frac{1}{\sqrt{2}}, -\frac{1}{\sqrt{2}} \right) \tag{4}$$

Geometrically, this means that the level curve $5x^2 + 5y^2 + 4xy = 7$ should just touch the unit circle at the points in (3), and the level curve $5x^2 + 5y^2 + 4xy = 3$ should just touch it at the points in (4). All of this is consistent with Figure 7.4.5. ◀

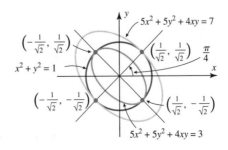

▶ Figure 7.4.5

We will conclude this section by showing how quadratic forms can be used to study characteristics of real-valued functions of two variables.

Recall that if a function $f(x, y)$ has first-order partial derivatives, then its relative maxima and minima, if any, occur at points where

$$f_x(x, y) = 0 \quad \text{and} \quad f_y(x, y) = 0$$

These are called *critical points* of f. The specific behavior of f at a critical point (x_0, y_0) is determined by the sign of

$$D(x, y) = f(x, y) - f(x_0, y_0) \tag{5}$$

at points (x, y) that are close to, but different from, (x_0, y_0):

Relative minimum at $(0, 0)$

(a)

- If $D(x, y) > 0$ at points (x, y) that are sufficiently close to, but different from, (x_0, y_0), then $f(x_0, y_0) < f(x, y)$ at such points and f is said to have a *relative minimum* at (x_0, y_0) (Figure 7.4.6a).

- If $D(x, y) < 0$ at points (x, y) that are sufficiently close to, but different from, (x_0, y_0), then $f(x_0, y_0) > f(x, y)$ at such points and f is said to have a *relative maximum* at (x_0, y_0) (Figure 7.4.6b).

- If $D(x, y)$ has both positive and negative values inside *every* circle centered at (x_0, y_0), then there are points (x, y) that are arbitrarily close to (x_0, y_0) at which $f(x_0, y_0) < f(x, y)$ and points (x, y) that are arbitrarily close to (x_0, y_0) at which $f(x_0, y_0) > f(x, y)$. In this case we say that f has a *saddle point* at (x_0, y_0) (Figure 7.4.6c).

Relative maximum at $(0, 0)$

(b)

In general, it can be difficult to determine the sign of (5) directly. However, the following theorem, which is proved in calculus, makes it possible to analyze critical points using derivatives.

Saddle point at $(0, 0)$

(c)

▲ Figure 7.4.6

THEOREM 7.4.2 Second Derivative Test

Suppose that (x_0, y_0) is a critical point of $f(x, y)$ and that f has continuous second-order partial derivatives in some circular region centered at (x_0, y_0). Then:

(a) f has a relative minimum at (x_0, y_0) if

$$f_{xx}(x_0, y_0) f_{yy}(x_0, y_0) - f_{xy}^2(x_0, y_0) > 0 \quad \text{and} \quad f_{xx}(x_0, y_0) > 0$$

(b) f has a relative maximum at (x_0, y_0) if

$$f_{xx}(x_0, y_0) f_{yy}(x_0, y_0) - f_{xy}^2(x_0, y_0) > 0 \quad \text{and} \quad f_{xx}(x_0, y_0) < 0$$

(c) f has a saddle point at (x_0, y_0) if

$$f_{xx}(x_0, y_0) f_{yy}(x_0, y_0) - f_{xy}^2(x_0, y_0) < 0$$

(d) The test is inconclusive if

$$f_{xx}(x_0, y_0) f_{yy}(x_0, y_0) - f_{xy}^2(x_0, y_0) = 0$$

Our interest here is in showing how to reformulate this theorem using properties of symmetric matrices. For this purpose we consider the symmetric matrix

$$H(x, y) = \begin{bmatrix} f_{xx}(x, y) & f_{xy}(x, y) \\ f_{xy}(x, y) & f_{yy}(x, y) \end{bmatrix}$$

which is called the *Hessian* or *Hessian matrix* of f in honor of the German mathematician and scientist Ludwig Otto Hesse (1811–1874). The notation $H(x, y)$ emphasizes that the entries in the matrix depend on x and y. The Hessian is of interest because

$$\det[H(x_0, y_0)] = \begin{vmatrix} f_{xx}(x_0, y_0) & f_{xy}(x_0, y_0) \\ f_{xy}(x_0, y_0) & f_{yy}(x_0, y_0) \end{vmatrix} = f_{xx}(x_0, y_0) f_{yy}(x_0, y_0) - f_{xy}^2(x_0, y_0)$$

is the expression that appears in Theorem 7.4.2. We can now reformulate the second derivative test as follows.

THEOREM 7.4.3 Hessian Form of the Second Derivative Test

Suppose that (x_0, y_0) is a critical point of $f(x, y)$ and that f has continuous second-order partial derivatives in some circular region centered at (x_0, y_0). If $H(x_0, y_0)$ is the Hessian of f at (x_0, y_0), then:

(a) f has a relative minimum at (x_0, y_0) if $H(x_0, y_0)$ is positive definite.

(b) f has a relative maximum at (x_0, y_0) if $H(x_0, y_0)$ is negative definite.

(c) f has a saddle point at (x_0, y_0) if $H(x_0, y_0)$ is indefinite.

(d) The test is inconclusive otherwise.

We will prove part (a). The proofs of the remaining parts will be left as exercises.

Proof (a) If $H(x_0, y_0)$ is positive definite, then Theorem 7.3.4 implies that the principal submatrices of $H(x_0, y_0)$ have positive determinants. Thus,

$$\det[H(x_0, y_0)] = \begin{vmatrix} f_{xx}(x_0, y_0) & f_{xy}(x_0, y_0) \\ f_{xy}(x_0, y_0) & f_{yy}(x_0, y_0) \end{vmatrix} = f_{xx}(x_0, y_0) f_{yy}(x_0, y_0) - f_{xy}^2(x_0, y_0) > 0$$

and

$$\det[f_{xx}(x_0, y_0)] = f_{xx}(x_0, y_0) > 0$$

so f has a relative minimum at (x_0, y_0) by part (a) of Theorem 7.4.2. ◄

▶ **EXAMPLE 4 Using the Hessian to Classify Relative Extrema**

Find the critical points of the function

$$f(x, y) = \tfrac{1}{3}x^3 + xy^2 - 8xy + 3$$

and use the eigenvalues of the Hessian matrix at those points to determine which of them, if any, are relative maxima, relative minima, or saddle points.

Solution To find both the critical points and the Hessian matrix we will need to calculate the first and second partial derivatives of f. These derivatives are

$$f_x(x, y) = x^2 + y^2 - 8y, \quad f_y(x, y) = 2xy - 8x, \quad f_{xy}(x, y) = 2y - 8$$
$$f_{xx}(x, y) = 2x, \qquad\qquad f_{yy}(x, y) = 2x$$

Thus, the Hessian matrix is

$$H(x, y) = \begin{bmatrix} f_{xx}(x, y) & f_{xy}(x, y) \\ f_{xy}(x, y) & f_{yy}(x, y) \end{bmatrix} = \begin{bmatrix} 2x & 2y - 8 \\ 2y - 8 & 2x \end{bmatrix}$$

To find the critical points we set f_x and f_y equal to zero. This yields the equations

$$f_x(x, y) = x^2 + y^2 - 8y = 0 \quad \text{and} \quad f_y(x, y) = 2xy - 8x = 2x(y - 4) = 0$$

Solving the second equation yields $x = 0$ or $y = 4$. Substituting $x = 0$ in the first equation and solving for y yields $y = 0$ or $y = 8$; and substituting $y = 4$ into the first equation and solving for x yields $x = 4$ or $x = -4$. Thus, we have four critical points:

$$(0, 0), \quad (0, 8), \quad (4, 4), \quad (-4, 4)$$

Evaluating the Hessian matrix at these points yields

$$H(0, 0) = \begin{bmatrix} 0 & -8 \\ -8 & 0 \end{bmatrix}, \quad H(0, 8) = \begin{bmatrix} 0 & 8 \\ 8 & 0 \end{bmatrix}$$

$$H(4, 4) = \begin{bmatrix} 8 & 0 \\ 0 & 8 \end{bmatrix}, \quad H(-4, 4) = \begin{bmatrix} -8 & 0 \\ 0 & -8 \end{bmatrix}$$

We leave it for you to find the eigenvalues of these matrices and deduce the following classifications of the stationary points:

Critical Point (x_0, y_0)	λ_1	λ_2	Classification
(0, 0)	8	−8	Saddle point
(0, 8)	8	−8	Saddle point
(4, 4)	8	8	Relative minimum
(−4, 4)	−8	−8	Relative maximum

◀

OPTIONAL

We conclude this section with an optional proof of Theorem 7.4.1.

Proof of Theorem 7.4.1 The first step in the proof is to show that $A\mathbf{x}$ has constrained maximum and minimum values for $\|\mathbf{x}\| = 1$. Since A is symmetric, the principal axes theorem (Theorem 7.3.1) implies that there is an orthogonal change of variable $\mathbf{x} = P\mathbf{y}$ such that

$$\mathbf{x}^T A \mathbf{x} = \lambda_1 y_1^2 + \lambda_2 y_2^2 + \cdots + \lambda_n y_n^2 \tag{6}$$

in which $\lambda_1, \lambda_2, \ldots, \lambda_n$ are the eigenvalues of A. Let us assume that $\|\mathbf{x}\| = 1$ and that the column vectors of P (which are unit eigenvectors of A) have been ordered so that

$$\lambda_1 \geq \lambda_2 \geq \cdots \geq \lambda_n \tag{7}$$

Since the matrix P is orthogonal, multiplication by P is length preserving, so that $\|\mathbf{y}\| = \|\mathbf{x}\| = 1$; that is,

$$y_1^2 + y_2^2 + \cdots + y_n^2 = 1$$

It follows from this equation and (7) that

$$\lambda_n = \lambda_n(y_1^2 + y_2^2 + \cdots + y_n^2) \leq \lambda_1 y_1^2 + \lambda_2 y_2^2 + \cdots + \lambda_n y_n^2$$
$$\leq \lambda_1(y_1^2 + y_2^2 + \cdots + y_n^2) = \lambda_1$$

and hence from (6) that

$$\lambda_n \leq \mathbf{x}^T A \mathbf{x} \leq \lambda_1$$

This shows that all values of $\mathbf{x}^T A\mathbf{x}$ for which $\|\mathbf{x}\| = 1$ lie between the largest and smallest eigenvalues of A. Now let $\mathbf{x}$ be a unit eigenvector corresponding to λ_1. Then

$$\mathbf{x}^T A \mathbf{x} = \mathbf{x}^T (\lambda_1 \mathbf{x}) = \lambda_1 \mathbf{x}^T \mathbf{x} = \lambda_1 \|\mathbf{x}\|^2 = \lambda_1$$

which shows that $\mathbf{x}^T A \mathbf{x}$ has λ_1 as a constrained maximum and that this maximum occurs if $\mathbf{x}$ is a unit eigenvector of A corresponding to λ_1. Similarly, if $\mathbf{x}$ is a unit eigenvector corresponding to λ_n, then

$$\mathbf{x}^T A \mathbf{x} = \mathbf{x}^T (\lambda_n \mathbf{x}) = \lambda_n \mathbf{x}^T \mathbf{x} = \lambda_n \|\mathbf{x}\|^2 = \lambda_n$$

so $\mathbf{x}^T A\mathbf{x}$ has λ_n as a constrained minimum and this minimum occurs if $\mathbf{x}$ is a unit eigenvector of A corresponding to λ_n. This completes the proof. ◀

Concept Review

- Constraint
- Constrained extremum
- Level curve

- Critical point
- Relative minimum
- Relative maximum

- Saddle point
- Second derivative test
- Hessian matrix

Skills

- Find the maximum and minimum values of a quadratic form subject to a constraint.

- Find the critical points of a real-valued function of two variables, and use the eigenvalues of the Hessian matrix at the critical points to classify them as relative maxima, relative minima, or saddle points.

Exercise Set 7.4

▶ In Exercises 1–4, find the maximum and minimum values of the given quadratic form subject to the constraint $x^2 + y^2 = 1$, and determine the values of x and y at which the maximum and minimum occur. ◀

1. $5x^2 - y^2$ **2.** xy **3.** $3x^2 + 7y^2$ **4.** $5x^2 + 5xy$

▶ In Exercises 5–6, find the maximum and minimum values of the given quadratic form subject to the constraint

$$x^2 + y^2 + z^2 = 1$$

and determine the values of x, y, and z at which the maximum and minimum occur. ◀

5. $9x^2 + 4y^2 + 3z^2$ **6.** $2x^2 + y^2 + z^2 + 2xy + 2xz$

7. Use the method of Example 2 to find the maximum and minimum values of xy subject to the constraint $4x^2 + 8y^2 = 16$.

8. Use the method of Example 2 to find the maximum and minimum values of $x^2 + xy + 2y^2$ subject to the constraint $x^2 + 3y^2 = 16$.

▶ In Exercises 9–10, draw the unit circle and the level curves corresponding to the given quadratic form. Show that the unit circle intersects each of these curves in exactly two places, label the intersection points, and verify that the constrained extrema occur at those points. ◀

9. $5x^2 - y^2$ **10.** xy

11. (a) Show that the function $f(x, y) = 4xy - x^4 - y^4$ has critical points at $(0, 0)$, $(1, 1)$, and $(-1, -1)$.

   (b) Use the Hessian form of the second derivative test to show f has relative maxima at $(1, 1)$ and $(-1, -1)$ and a saddle point at $(0, 0)$.

12. (a) Show that the function $f(x, y) = x^3 - 6xy - y^3$ has critical points at $(0, 0)$ and $(-2, 2)$.

   (b) Use the Hessian form of the second derivative test to show f has a relative maximum at $(-2, 2)$ and a saddle point at $(0, 0)$.

▶ In Exercises 13–16, find the critical points of f, if any, and classify them as relative maxima, relative minima, or saddle points.

13. $f(x, y) = x^3 - 3xy - y^3|$

14. $f(x, y) = x^3 - 3xy + y^3$

15. $f(x, y) = x^2 + 2y^2 - x^2y$

16. $f(x, y) = x^3 + y^3 - 3x - 3y$

17. A rectangle whose center is at the origin and whose sides are parallel to the coordinate axes is to be inscribed in the ellipse $x^2 + 25y^2 = 25$. Use the method of Example 2 to find nonnegative values of x and y that produce the inscribed rectangle with maximum area.

18. Suppose that the temperature at a point (x, y) on a metal plate is $T(x, y) = 4x^2 - 4xy + y^2$. An ant, walking on the plate, traverses a circle of radius 5 centered at the origin. What are the highest and lowest temperatures encountered by the ant?

19. (a) Show that the functions

$$f(x, y) = x^4 + y^4 \quad \text{and} \quad g(x, y) = x^4 - y^4$$

   have a critical point at $(0, 0)$ but the second derivative test is inconclusive at that point.

   (b) Give a reasonable argument to show that f has a relative minimum at $(0, 0)$ and g has a saddle point at $(0, 0)$.

20. Suppose that the Hessian matrix of a certain quadratic form $f(x, y)$ is

$$H = \begin{bmatrix} 2 & 4 \\ 4 & 2 \end{bmatrix}$$

   What can you say about the location and classification of the critical points of f?

21. Suppose that A is an $n \times n$ symmetric matrix and

$$q(\mathbf{x}) = \mathbf{x}^T A \mathbf{x}$$

   where $\mathbf{x}$ is a vector in R^n that is expressed in column form. What can you say about the value of q if $\mathbf{x}$ is a unit eigenvector corresponding to an eigenvalue λ of A?

22. Prove: If $\mathbf{x}^T A \mathbf{x}$ is a quadratic form whose minimum and maximum values subject to the constraint $\|\mathbf{x}\| = 1$ are m and M, respectively, then for each number c in the interval $m \leq c \leq M$, there is a unit vector $\mathbf{x}_c$ such that $\mathbf{x}_c^T A \mathbf{x}_c = c$. [*Hint:* In the case where $m < M$, let $\mathbf{u}_m$ and $\mathbf{u}_M$ be unit eigenvectors of A such that $\mathbf{u}_m^T A \mathbf{u}_m = m$ and $\mathbf{u}_M^T A \mathbf{u}_M = M$, and let

$$\mathbf{x}_c = \sqrt{\frac{M-c}{M-m}}\,\mathbf{u}_m + \sqrt{\frac{c-m}{M-m}}\,\mathbf{u}_M$$

Show that $\mathbf{x}_c^T A \mathbf{x}_c = c$.]

True-False Exercises

In parts (a)–(e) determine whether the statement is true or false, and justify your answer.

(a) A quadratic form must have either a maximum or minimum value.

(b) The maximum value of a quadratic form $\mathbf{x}^T A \mathbf{x}$ subject to the constraint $\|\mathbf{x}\| = 1$ occurs at a unit eigenvector corresponding to the largest eigenvalue of A.

(c) The Hessian matrix of a function f with continuous second-order partial derivatives is a symmetric matrix.

(d) If (x_0, y_0) is a critical point of a function f and the Hessian of f at (x_0, y_0) is 0, then f has neither a relative maximum nor a relative minimum at (x_0, y_0).

(e) If A is a symmetric matrix and $\det A < 0$, then the minimum of $\mathbf{x}^T A \mathbf{x}$ subject to the constraint $\|\mathbf{x}\| = 1$ is negative.

7.5 Hermitian, Unitary, and Normal Matrices

We know that every real symmetric matrix is orthogonally diagonalizable and that the real symmetric matrices are the only orthogonally diagonalizable matrices. In this section we will consider the diagonalization problem for complex matrices.

Hermitian and Unitary Matrices

The transpose operation is less important for complex matrices than for real matrices. A more useful operation for complex matrices is given in the following definition.

> **DEFINITION 1** If A is a complex matrix, then the ***conjugate transpose*** of A, denoted by A^*, is defined by
>
> $$A^* = \overline{A}^T \tag{1}$$

Remark Since part (*b*) of Theorem 5.3.2 states that $\overline{(A^T)} = (\overline{A})^T$, the order in which the transpose and conjugation operations are performed in computing $A^* = \overline{A}^T$ does not matter. Moreover, in the case where A has real entries we have $A^* = (\overline{A})^T = A^T$, so A^* is the same as A^T for real matrices.

▶ **EXAMPLE 1 Conjugate Transpose**
Find the conjugate transpose A^* of the matrix

$$A = \begin{bmatrix} 1+i & -i & 0 \\ 2 & 3-2i & i \end{bmatrix}$$

Solution We have

$$\overline{A} = \begin{bmatrix} 1-i & i & 0 \\ 2 & 3+2i & -i \end{bmatrix} \quad \text{and hence} \quad A^* = \overline{A}^T = \begin{bmatrix} 1-i & 2 \\ i & 3+2i \\ 0 & -i \end{bmatrix} \blacktriangleleft$$

The following theorem, parts of which are given as exercises, shows that the basic algebraic properties of the conjugate transpose operation are similar to those of the transpose (compare to Theorem 1.4.8).

THEOREM 7.5.1 *If k is a complex scalar, and if A, B, and C are complex matrices whose sizes are such that the stated operations can be performed, then:*

(a) $(A^*)^* = A$

(b) $(A + B)^* = A^* + B^*$

(c) $(A - B)^* = A^* - B^*$

(d) $(kA)^* = \bar{k}A^*$

(e) $(AB)^* = B^*A^*$

Remark Note that the relationship $\mathbf{u} \cdot \mathbf{v} = \bar{\mathbf{v}}^T\mathbf{u}$ in Formula (5) of Section 5.3 can be expressed in terms of the conjugate transpose as

$$\mathbf{u} \cdot \mathbf{v} = \mathbf{v}^*\mathbf{u} \qquad (2)$$

We are now ready to define two new classes of matrices that will be important in our study of diagonalization in C^n.

DEFINITION 2 A square complex matrix A is said to be **unitary** if

$$A^{-1} = A^* \qquad (3)$$

and is said to be **Hermitian**[*] if

$$A^* = A \qquad (4)$$

Note that a unitary matrix can also be defined as a square complex matrix A for which

$$AA^* = A^*A = I$$

If A is a real matrix, then $A^* = A^T$, in which case (3) becomes $A^{-1} = A^T$ and (4) becomes $A^T = A$. Thus, the unitary matrices are complex generalizations of the real orthogonal matrices and Hermitian matrices are complex generalizations of the real symmetric matrices.

▶ **EXAMPLE 2 Recognizing Hermitian Matrices**

Hermitian matrices are easy to recognize because their diagonal entries are real (why?), and the entries that are symmetrically positioned across the main diagonal are complex conjugates. Thus, for example, we can tell by inspection that

$$A = \begin{bmatrix} 1 & i & 1+i \\ -i & -5 & 2-i \\ 1-i & 2+i & 3 \end{bmatrix}$$

is Hermitian. ◀

The fact that real symmetric matrices have real eigenvalues is a special case of the following more general result about Hermitian matrices, the proof of which is left for the exercises.

THEOREM 7.5.2 *The eigenvalues of a Hermitian matrix are real numbers.*

The fact that eigenvectors from different eigenspaces of a real symmetric matrix are orthogonal is a special case of the following more general result about Hermitian matrices.

---

[*]In honor of the French mathematician Charles Hermite (1822–1901).

THEOREM 7.5.3 *If A is a Hermitian matrix, then eigenvectors from different eigen-spaces are orthogonal.*

Proof Let $\mathbf{v}_1$ and $\mathbf{v}_2$ be eigenvectors of A corresponding to distinct eigenvalues λ_1 and λ_2. Using Formula (2) and the facts that $\lambda_1 = \bar{\lambda}_1$, $\lambda_2 = \bar{\lambda}_2$, and $A = A^*$, we can write

$$\lambda_1(\mathbf{v}_2 \cdot \mathbf{v}_1) = (\lambda_1 \mathbf{v}_1)^* \mathbf{v}_2 = (A\mathbf{v}_1)^* \mathbf{v}_2 = (\mathbf{v}_1^* A^*)\mathbf{v}_2$$
$$= (\mathbf{v}_1^* A)\mathbf{v}_2 \;=\; \mathbf{v}_1^*(A\mathbf{v}_2)$$
$$= \mathbf{v}_1^*(\lambda_2 \mathbf{v}_2) \;=\; \lambda_2(\mathbf{v}_1^* \mathbf{v}_2) = \lambda_2(\mathbf{v}_2 \cdot \mathbf{v}_1)$$

This implies that $(\lambda_1 - \lambda_2)(\mathbf{v}_2 \cdot \mathbf{v}_1) = 0$ and hence that $\mathbf{v}_2 \cdot \mathbf{v}_1 = 0$ (since $\lambda_1 \neq \lambda_2$). ◀

▶ **EXAMPLE 3 Eigenvalues and Eigenvectors of a Hermitian Matrix**

Confirm that the Hermitian matrix

$$A = \begin{bmatrix} 2 & 1+i \\ 1-i & 3 \end{bmatrix}$$

has real eigenvalues and that eigenvectors from different eigenspaces are orthogonal.

Solution The characteristic polynomial of A is

$$\det(\lambda I - A) = \begin{vmatrix} \lambda - 2 & -1-i \\ -1+i & \lambda - 3 \end{vmatrix}$$
$$= (\lambda - 2)(\lambda - 3) - (-1-i)(-1+i)$$
$$= (\lambda^2 - 5\lambda + 6) - 2 = (\lambda - 1)(\lambda - 4)$$

so the eigenvalues of A are $\lambda = 1$ and $\lambda = 4$, which are real. Bases for the eigenspaces of A can be obtained by solving the linear system

$$\begin{bmatrix} \lambda - 2 & -1-i \\ -1+i & \lambda - 3 \end{bmatrix} \begin{bmatrix} x_1 \\ x_2 \end{bmatrix} = \begin{bmatrix} 0 \\ 0 \end{bmatrix}$$

with $\lambda = 1$ and with $\lambda = 4$. We leave it for you to do this and to show that the general solutions of these systems are

$$\lambda = 1: \; \begin{bmatrix} x_1 \\ x_2 \end{bmatrix} = t \begin{bmatrix} -1-i \\ 1 \end{bmatrix} \quad \text{and} \quad \lambda = 4: \; \begin{bmatrix} x_1 \\ x_2 \end{bmatrix} = t \begin{bmatrix} \frac{1}{2}(1+i) \\ 1 \end{bmatrix}$$

Thus, bases for these eigenspaces are

$$\lambda = 1: \; \mathbf{v}_1 = \begin{bmatrix} -1-i \\ 1 \end{bmatrix} \quad \text{and} \quad \lambda = 4: \; \mathbf{v}_2 = \begin{bmatrix} \frac{1}{2}(1+i) \\ 1 \end{bmatrix}$$

The vectors $\mathbf{v}_1$ and $\mathbf{v}_2$ are orthogonal since

$$\mathbf{v}_1 \cdot \mathbf{v}_2 = (-1-i)\left(\overline{\tfrac{1}{2}(1+i)}\right) + (1)(1) = \tfrac{1}{2}(-1-i)(1-i) + 1 = 0$$

and hence all scalar multiples of them are also orthogonal. ◀

Unitary matrices are not usually easy to recognize by inspection. However, the following analog of Theorems 7.1.1 and 7.1.3, part of which is proved in the exercises, provides a way of ascertaining whether a matrix is unitary without computing its inverse.

THEOREM 7.5.4 *If A is an n × n matrix with complex entries, then the following are equivalent.*

(a) *A is unitary.*

(b) $\|A\mathbf{x}\| = \|\mathbf{x}\|$ *for all* $\mathbf{x}$ *in* C^n.

(c) $A\mathbf{x} \cdot A\mathbf{y} = \mathbf{x} \cdot \mathbf{y}$ *for all* $\mathbf{x}$ *and* $\mathbf{y}$ *in* C^n.

(d) *The column vectors of A form an orthonormal set in* C^n *with respect to the complex Euclidean inner product.*

(e) *The row vectors of A form an orthonormal set in* C^n *with respect to the complex Euclidean inner product.*

▶ **EXAMPLE 4 A Unitary Matrix**

Use Theorem 7.5.4 to show that

$$A = \begin{bmatrix} \frac{1}{2}(1+i) & \frac{1}{2}(1+i) \\ \frac{1}{2}(1-i) & \frac{1}{2}(-1+i) \end{bmatrix}$$

is unitary, and then find A^{-1}.

Solution We will show that the row vectors

$$\mathbf{r}_1 = \begin{bmatrix} \frac{1}{2}(1+i) & \frac{1}{2}(1+i) \end{bmatrix} \quad \text{and} \quad \mathbf{r}_2 = \begin{bmatrix} \frac{1}{2}(1-i) & \frac{1}{2}(-1+i) \end{bmatrix}$$

are orthonormal. The relevant computations are

$$\|\mathbf{r}_1\| = \sqrt{\left|\tfrac{1}{2}(1+i)\right|^2 + \left|\tfrac{1}{2}(1+i)\right|^2} = \sqrt{\tfrac{1}{2} + \tfrac{1}{2}} = 1$$

$$\|\mathbf{r}_2\| = \sqrt{\left|\tfrac{1}{2}(1-i)\right|^2 + \left|\tfrac{1}{2}(-1+i)\right|^2} = \sqrt{\tfrac{1}{2} + \tfrac{1}{2}} = 1$$

$$\mathbf{r}_1 \cdot \mathbf{r}_2 = \left(\tfrac{1}{2}(1+i)\right)\left(\overline{\tfrac{1}{2}(1-i)}\right) + \left(\tfrac{1}{2}(1+i)\right)\left(\overline{\tfrac{1}{2}(-1+i)}\right)$$

$$= \left(\tfrac{1}{2}(1+i)\right)\left(\tfrac{1}{2}(1+i)\right) + \left(\tfrac{1}{2}(1+i)\right)\left(\tfrac{1}{2}(-1-i)\right) = \tfrac{1}{2}i - \tfrac{1}{2}i = 0$$

Since we now know that A is unitary, it follows that

$$A^{-1} = A^* = \begin{bmatrix} \frac{1}{2}(1-i) & \frac{1}{2}(1+i) \\ \frac{1}{2}(1-i) & \frac{1}{2}(-1-i) \end{bmatrix}$$

You can confirm the validity of this result by showing that $AA^* = A^*A = I$. ◀

Unitary Diagonalizability Since unitary matrices are the complex analogs of the real orthogonal matrices, the following definition is a natural generalization of orthogonal diagonalizability for real matrices.

DEFINITION 3 A square complex matrix is said to be ***unitarily diagonalizable*** if there is a unitary matrix P such that $P^*AP = D$ is a complex diagonal matrix. Any such matrix P is said to ***unitarily diagonalize*** A.

Recall that a real symmetric $n \times n$ matrix A has an orthonormal set of n eigenvectors and is orthogonally diagonalized by any $n \times n$ matrix whose column vectors are an orthonormal set of eigenvectors of A. Here is the complex analog of that result.

THEOREM 7.5.5 *Every $n \times n$ Hermitian matrix A has an orthonormal set of n eigenvectors and is unitarily diagonalized by any $n \times n$ matrix P whose column vectors form an orthonormal set of eigenvectors of A.*

The procedure for unitarily diagonalizing a Hermitian matrix A is exactly the same as that for orthogonally diagonalizing a symmetric matrix:

Unitarily Diagonalizing a Hermitian Matrix

Step 1. Find a basis for each eigenspace of A.

Step 2. Apply the Gram–Schmidt process to each of these bases to obtain orthonormal bases for the eigenspaces.

Step 3. Form the matrix P whose column vectors are the basis vectors obtained in Step 2. This will be a unitary matrix (Theorem 7.5.4) and will unitarily diagonalize A.

▶ **EXAMPLE 5 Unitary Diagonalization of a Hermitian Matrix**

Find a matrix P that unitarily diagonalizes the Hermitian matrix

$$A = \begin{bmatrix} 2 & 1+i \\ 1-i & 3 \end{bmatrix}$$

Solution We showed in Example 3 that the eigenvalues of A are $\lambda = 1$ and $\lambda = 4$ and that bases for the corresponding eigenspaces are

$$\lambda = 1: \ \mathbf{v}_1 = \begin{bmatrix} -1-i \\ 1 \end{bmatrix} \quad \text{and} \quad \lambda = 4: \ \mathbf{v}_2 = \begin{bmatrix} \frac{1}{2}(1+i) \\ 1 \end{bmatrix}$$

Since each eigenspace has only one basis vector, the Gram–Schmidt process is simply a matter of normalizing these basis vectors. We leave it for you to show that

$$\mathbf{p}_1 = \frac{\mathbf{v}_1}{\|\mathbf{v}_1\|} = \begin{bmatrix} \frac{-1-i}{\sqrt{3}} \\ \frac{1}{\sqrt{3}} \end{bmatrix} \quad \text{and} \quad \mathbf{p}_2 = \frac{\mathbf{v}_2}{\|\mathbf{v}_2\|} = \begin{bmatrix} \frac{1+i}{\sqrt{6}} \\ \frac{2}{\sqrt{6}} \end{bmatrix}$$

Thus, A is unitarily diagonalized by the matrix

$$P = [\mathbf{p}_1 \quad \mathbf{p}_2] = \begin{bmatrix} \frac{-1-i}{\sqrt{3}} & \frac{1+i}{\sqrt{6}} \\ \frac{1}{\sqrt{3}} & \frac{2}{\sqrt{6}} \end{bmatrix}$$

Although it is a little tedious, you may want to check this result by showing that

$$P^*AP = \begin{bmatrix} \frac{-1+i}{\sqrt{3}} & \frac{1}{\sqrt{3}} \\ \frac{1-i}{\sqrt{6}} & \frac{2}{\sqrt{6}} \end{bmatrix} \begin{bmatrix} 2 & 1+i \\ 1-i & 3 \end{bmatrix} \begin{bmatrix} \frac{-1-i}{\sqrt{3}} & \frac{1+i}{\sqrt{6}} \\ \frac{1}{\sqrt{3}} & \frac{2}{\sqrt{6}} \end{bmatrix} = \begin{bmatrix} 1 & 0 \\ 0 & 4 \end{bmatrix} \ ◀$$

Skew-Symmetric and Skew-Hermitian Matrices In Exercise 37 of Section 1.7 we defined a square matrix with real entries to be **skew-symmetric** if $A^T = -A$. A skew-symmetric matrix must have zeros on the main diagonal (why?), and each entry off the main diagonal must be the negative of its mirror image about the main diagonal. Here is an example.

$$A = \begin{bmatrix} 0 & 1 & -2 \\ -1 & 0 & 4 \\ 2 & -4 & 0 \end{bmatrix} \quad \text{[skew-symmetric]}$$

We leave it for you to confirm that $A^T = -A$.

The complex analogs of the skew-symmetric matrices are the matrices for which $A^* = -A$. Such matrices are said to be **skew-Hermitian**.

Since a skew-Hermitian matrix A has the property

$$A^* = \overline{A}^T = -A$$

it must be that A has zeros or pure imaginary numbers on the main diagonal (why?), and that the complex conjugate of each entry off the main diagonal is the negative of its mirror image about the main diagonal. Here is an example.

$$A = \begin{bmatrix} i & 1-i & 5 \\ -1-i & 2i & i \\ -5 & i & 0 \end{bmatrix} \quad \text{[skew-Hermitian]}$$

Normal Matrices   Hermitian matrices enjoy many, but not all, of the properties of real symmetric matrices. For example, we know that real symmetric matrices are orthogonally diagonalizable and Hermitian matrices are unitarily diagonalizable. However, whereas the real symmetric matrices are the only orthogonally diagonalizable matrices, the Hermitian matrices do not constitute the entire class of unitarily diagonalizable complex matrices; that is, there exist unitarily diagonalizable matrices that are not Hermitian. Specifically, it can be proved that a square complex matrix A is unitarily diagonalizable if and only if

$$AA^* = A^*A \tag{5}$$

Matrices with this property are said to be **normal**. Normal matrices include the Hermitian, skew-Hermitian, and unitary matrices in the complex case and the symmetric, skew-symmetric, and orthogonal matrices in the real case. The nonzero skew-symmetric matrices are particularly interesting because they are examples of real matrices that are not orthogonally diagonalizable but are unitarily diagonalizable.

A Comparison of Eigenvalues   We have seen that Hermitian matrices have real eigenvalues. In the exercises we will ask you to show that the eigenvalues of a skew-Hermitian matrix are either zero or purely imaginary (have real part of zero) and that the eigenvalues of unitary matrices have modulus 1. These ideas are illustrated schematically in Figure 7.5.1.

▶ **Figure 7.5.1**

Concept Review

- Conjugate transpose
- Unitary matrix
- Hermitian matrix
- Unitarily diagonalizable matrix
- Skew-symmetric matrix
- Skew-Hermitian matrix
- Normal matrix

Skills

- Find the conjugate transpose of a matrix.
- Be able to identify Hermitian matrices.
- Find the inverse of a unitary matrix.
- Find a unitary matrix that diagonalizes a Hermitian matrix.

Exercise Set 7.5

▶ In Exercises 1–2, find A^*. ◀

1. $A = \begin{bmatrix} 2i & 1-i \\ 4 & 3+i \\ 5+i & 0 \end{bmatrix}$

2. $A = \begin{bmatrix} 2i & 1-i & -1+i \\ 4 & 5-7i & -i \end{bmatrix}$

▶ In Exercises 3–4, substitute numbers for the ×'s so that A is Hermitian. ◀

3. $A = \begin{bmatrix} 1 & i & 2-3i \\ \times & -3 & 1 \\ \times & \times & 2 \end{bmatrix}$

4. $A = \begin{bmatrix} 2 & 0 & 3+5i \\ \times & -4 & -i \\ \times & \times & 6 \end{bmatrix}$

▶ In Exercises 5–6, show that A is not Hermitian for any choice of the ×'s. ◀

5. (a) $A = \begin{bmatrix} 1 & i & 2-3i \\ -i & -3 & \times \\ 2-3i & \times & \times \end{bmatrix}$

(b) $A = \begin{bmatrix} \times & \times & 3+5i \\ 0 & i & -i \\ 3-5i & i & \times \end{bmatrix}$

6. (a) $A = \begin{bmatrix} 1 & 1+i & \times \\ 1+i & 7 & \times \\ 6-2i & \times & 0 \end{bmatrix}$

(b) $A = \begin{bmatrix} 1 & \times & 3+5i \\ \times & 3 & 1-i \\ 3-5i & \times & 2+i \end{bmatrix}$

▶ In Exercises 7–8, verify that the eigenvalues of the Hermitian matrix A are real and that eigenvectors from different eigenspaces are orthogonal (see Theorem 7.5.3). ◀

7. $A = \begin{bmatrix} 3 & 2-3i \\ 2+3i & -1 \end{bmatrix}$

8. $A = \begin{bmatrix} 0 & 2i \\ -2i & 2 \end{bmatrix}$

▶ In Exercises 9–12, show that A is unitary, and find A^{-1}. ◀

9. $A = \begin{bmatrix} \frac{3}{5} & \frac{4}{5}i \\ -\frac{4}{5} & \frac{3}{5}i \end{bmatrix}$

10. $A = \begin{bmatrix} \frac{1}{\sqrt{2}} & \frac{1}{\sqrt{2}} \\ -\frac{1}{2}(1+i) & \frac{1}{2}(1+i) \end{bmatrix}$

11. $A = \begin{bmatrix} \frac{1}{2\sqrt{2}}\left(\sqrt{3}+i\right) & \frac{1}{2\sqrt{2}}\left(1-i\sqrt{3}\right) \\ \frac{1}{2\sqrt{2}}\left(1+i\sqrt{3}\right) & \frac{1}{2\sqrt{2}}\left(i-\sqrt{3}\right) \end{bmatrix}$

12. $A = \begin{bmatrix} \frac{1}{\sqrt{3}}(-1+i) & \frac{1}{\sqrt{6}}(1-i) \\ \frac{1}{\sqrt{3}} & \frac{2}{\sqrt{6}} \end{bmatrix}$

▶ In Exercises 13–18, find a unitary matrix P that diagonalizes the Hermitian matrix A, and determine $P^{-1}AP$. ◀

13. $A = \begin{bmatrix} 4 & 1-i \\ 1+i & 5 \end{bmatrix}$

14. $A = \begin{bmatrix} 3 & -i \\ i & 3 \end{bmatrix}$

15. $A = \begin{bmatrix} 6 & 2+2i \\ 2-2i & 4 \end{bmatrix}$

16. $A = \begin{bmatrix} 0 & 3+i \\ 3-i & -3 \end{bmatrix}$

17. $A = \begin{bmatrix} 5 & 0 & 0 \\ 0 & -1 & -1+i \\ 0 & -1-i & 0 \end{bmatrix}$

18. $A = \begin{bmatrix} 2 & \frac{1}{\sqrt{2}}i & -\frac{1}{\sqrt{2}}i \\ -\frac{1}{\sqrt{2}}i & 2 & 0 \\ \frac{1}{\sqrt{2}}i & 0 & 2 \end{bmatrix}$

▶ In Exercises 19–20, substitute numbers for the ×'s so that A is skew-Hermitian. ◀

19. $A = \begin{bmatrix} 0 & i & 2-3i \\ \times & 0 & 1 \\ \times & \times & 4i \end{bmatrix}$

20. $A = \begin{bmatrix} 0 & 0 & 3-5i \\ \times & 0 & -i \\ \times & \times & 0 \end{bmatrix}$

▶ In Exercises 21–22, show that A is not skew-Hermitian for any choice of the ×'s. ◀

21. (a) $A = \begin{bmatrix} 0 & i & 2-3i \\ -i & 0 & \times \\ 2+3i & \times & \times \end{bmatrix}$

(b) $A = \begin{bmatrix} 1 & \times & 3-5i \\ \times & 2i & -i \\ -3+5i & i & 3i \end{bmatrix}$

22. (a) $A = \begin{bmatrix} i & \times & 2-3i \\ \times & 0 & 1+i \\ 2+3i & -1-i & \times \end{bmatrix}$

(b) $A = \begin{bmatrix} 0 & -i & 4+7i \\ \times & 0 & \times \\ -4-7i & \times & 1 \end{bmatrix}$

▶ In Exercises **23–24**, verify that the eigenvalues of the skew-Hermitian matrix A are pure imaginary numbers. ◀

23. $A = \begin{bmatrix} 0 & -1+i \\ 1+i & i \end{bmatrix}$ **24.** $A = \begin{bmatrix} 0 & 3i \\ 3i & 0 \end{bmatrix}$

▶ In Exercises **25–26**, show that A is normal. ◀

25. $A = \begin{bmatrix} 1+2i & 2+i & -2-i \\ 2+i & 1+i & -i \\ -2-i & -i & 1+i \end{bmatrix}$

26. $A = \begin{bmatrix} 2+2i & i & 1-i \\ i & -2i & 1-3i \\ 1-i & 1-3i & -3+8i \end{bmatrix}$

27. Show that the matrix

$$A = \frac{1}{\sqrt{2}} \begin{bmatrix} e^{i\theta} & e^{-i\theta} \\ ie^{i\theta} & -ie^{-i\theta} \end{bmatrix}$$

is unitary for all real values of θ. [*Note:* See Formula (17) in Appendix B for the definition of $e^{i\theta}$.]

28. Prove that each entry on the main diagonal of a skew-Hermitian matrix is either zero or a pure imaginary number.

29. Let A be any $n \times n$ matrix with complex entries, and define the matrices B and C to be

$$B = \frac{1}{2}(A + A^*) \quad \text{and} \quad C = \frac{1}{2i}(A - A^*)$$

(a) Show that B and C are Hermitian.

(b) Show that $A = B + iC$ and $A^* = B - iC$.

(c) What condition must B and C satisfy for A to be normal?

30. Show that if A is an $n \times n$ matrix with complex entries, and if $\mathbf{u}$ and $\mathbf{v}$ are vectors in C^n that are expressed in column form, then

$$A\mathbf{u} \cdot \mathbf{v} = \mathbf{u} \cdot A^*\mathbf{v} \quad \text{and} \quad \mathbf{u} \cdot A\mathbf{v} = A^*\mathbf{u} \cdot \mathbf{v}$$

31. Show that if A is a unitary matrix, then so is A^*.

32. Show that the eigenvalues of a skew-Hermitian matrix are either zero or purely imaginary.

33. Show that the eigenvalues of a unitary matrix have modulus 1.

34. Show that if $\mathbf{u}$ is a nonzero vector in C^n that is expressed in column form, then $P = \mathbf{u}\mathbf{u}^*$ is Hermitian.

35. Show that if $\mathbf{u}$ is a unit vector in C^n that is expressed in column form, then $H = I - 2\mathbf{u}\mathbf{u}^*$ is Hermitian and unitary.

36. What can you say about the inverse of a matrix A that is both Hermitian and unitary?

37. Find a 2×2 matrix that is both Hermitian and unitary and whose entries are not all real numbers.

38. Under what conditions is the following matrix normal?

$$A = \begin{bmatrix} a & 0 & 0 \\ 0 & 0 & c \\ 0 & b & 0 \end{bmatrix}$$

39. What geometric interpretations might you reasonably give to multiplication by the matrices $P = \mathbf{u}\mathbf{u}^*$ and $H = I - 2\mathbf{u}\mathbf{u}^*$ in Exercises 34 and 35?

40. Prove that if A is an invertible matrix, then A^* is invertible, and $(A^*)^{-1} = (A^{-1})^*$.

41. (a) Prove that $\det(\overline{A}) = \overline{\det(A)}$.

(b) Use the result in part (a) and the fact that a square matrix and its transpose have the same determinant to prove that $\det(A^*) = \overline{\det(A)}$.

42. Use part (b) of Exercise 41 to prove:

(a) If A is Hermitian, then $\det(A)$ is real.

(b) If A is unitary, then $|\det(A)| = 1$.

43. Use properties of the transpose and complex conjugate to prove parts (*a*) and (*e*) of Theorem 7.5.1.

44. Use properties of the transpose and complex conjugate to prove parts (*b*) and (*d*) of Theorem 7.5.1.

45. Prove that an $n \times n$ matrix with complex entries is unitary if and only if the columns of A form an orthonormal set in C^n.

46. Prove that the eigenvalues of a Hermitian matrix are real.

True-False Exercises

In parts (a)–(e) determine whether the statement is true or false, and justify your answer.

(a) The matrix $\begin{bmatrix} 0 & i \\ i & 2 \end{bmatrix}$ is Hermitian.

(b) The matrix $\begin{bmatrix} -\frac{i}{\sqrt{2}} & \frac{i}{\sqrt{6}} & \frac{i}{\sqrt{3}} \\ 0 & -\frac{i}{\sqrt{6}} & \frac{i}{\sqrt{3}} \\ \frac{i}{\sqrt{2}} & \frac{i}{\sqrt{6}} & \frac{i}{\sqrt{3}} \end{bmatrix}$ is unitary.

(c) The conjugate transpose of a unitary matrix is unitary.

(d) Every unitarily diagonalizable matrix is Hermitian.

(e) A positive integer power of a skew-Hermitian matrix is skew-Hermitian.

Chapter 7   Supplementary Exercises

1. Verify that each matrix is orthogonal, and find its inverse.

(a) $\begin{bmatrix} \frac{3}{5} & -\frac{4}{5} \\ \frac{4}{5} & \frac{3}{5} \end{bmatrix}$

(b) $\begin{bmatrix} \frac{4}{5} & 0 & -\frac{3}{5} \\ -\frac{9}{25} & \frac{4}{5} & -\frac{12}{25} \\ \frac{12}{25} & \frac{3}{5} & \frac{16}{25} \end{bmatrix}$

2. Prove: If Q is an orthogonal matrix, then each entry of Q is the same as its cofactor if $\det(Q) = 1$ and is the negative of its cofactor if $\det(Q) = -1$.

3. Prove that if A is a positive definite symmetric matrix, and if $\mathbf{u}$ and $\mathbf{v}$ vectors in R^n in column form, then

$$\langle \mathbf{u}, \mathbf{v} \rangle = \mathbf{u}^T A \mathbf{v}$$

is an inner product on R^n.

4. Find the characteristic polynomial and the dimensions of the eigenspaces of the symmetric matrix

$$\begin{bmatrix} 3 & 2 & 2 \\ 2 & 3 & 2 \\ 2 & 2 & 3 \end{bmatrix}$$

5. Find a matrix P that orthogonally diagonalizes

$$A = \begin{bmatrix} 1 & 0 & 1 \\ 0 & 1 & 0 \\ 1 & 0 & 1 \end{bmatrix}$$

and determine the diagonal matrix $D = P^T A P$.

6. Express each quadratic form in the matrix notation $\mathbf{x}^T A \mathbf{x}$.

(a) $-4x_1^2 + 16x_2^2 - 15x_1 x_2$

(b) $9x_1^2 - x_2^2 + 4x_3^2 + 6x_1 x_2 - 8x_1 x_3 + x_2 x_3$

7. Classify the quadradic form

$$x_1^2 - 3x_1 x_2 + 4x_2^2$$

as positive definite, negative definite, indefinite, positive semi-definite, or negative semidefinite.

8. Find an orthogonal change of variable that eliminates the cross product terms in each quadratic form, and express the quadratic form in terms of the new variables.

(a) $-3x_1^2 + 5x_2^2 + 2x_1 x_2$

(b) $-5x_1^2 + x_2^2 - x_3^2 + 6x_1 x_3 + 4x_1 x_2$

9. Identify the type of conic section represented by each equation.

(a) $y - x^2 = 0$

(b) $3x - 11y^2 = 0$

10. Find a unitary matrix U that diagonalizes

$$A = \begin{bmatrix} 1 & 1 & 0 \\ 0 & 1 & 1 \\ 1 & 0 & 1 \end{bmatrix}$$

and determine the diagonal matrix $D = U^{-1} A U$.

11. Show that if U is an $n \times n$ unitary matrix and

$$|z_1| = |z_2| = \cdots = |z_n| = 1$$

then the product

$$U \begin{bmatrix} z_1 & 0 & 0 & \cdots & 0 \\ 0 & z_2 & 0 & \cdots & 0 \\ \vdots & \vdots & \vdots & & \vdots \\ 0 & 0 & 0 & \cdots & z_n \end{bmatrix}$$

is also unitary.

12. Suppose that $A^* = -A$.

(a) Show that iA is Hermitian.

(b) Show that A is unitarily diagonalizable and has pure imaginary eigenvalues.

CHAPTER 8

Linear Transformations

CHAPTER CONTENTS

8.1 General Linear Transformations **433**

8.2 Isomorphism **445**

8.3 Compositions and Inverse Transformations **452**

8.4 Matrices for General Linear Transformations **458**

8.5 Similarity **468**

INTRODUCTION   In Sections 4.9 and 4.10 we studied linear transformations from R^n to R^m. In this chapter we will define and study linear transformations from a general vector space V to a general vector space W. The results we obtain here have important applications in physics, engineering, and various branches of mathematics.

8.1 General Linear Transformations

Up to now our study of linear transformations has focused on transformations from R^n to R^m. In this section we will turn our attention to linear transformations involving general vector spaces. We will illustrate ways in which such transformations arise, and we will establish a fundamental relationship between general n-dimensional vector spaces and R^n.

Definitions and Terminology   In Section 4.9 we defined a *matrix transformation* $T_A: R^n \rightarrow R^m$ to be a mapping of the form

$$T_A(\mathbf{x}) = A\mathbf{x}$$

in which A is an $m \times n$ matrix. We subsequently established in Theorems 4.10.2 and 4.10.3 that the matrix transformations are precisely the *linear transformations* from R^n to R^m, that is, the transformations with the linearity properties

$$T(\mathbf{u} + \mathbf{v}) = T(\mathbf{u}) + T(\mathbf{v}) \quad \text{and} \quad T(k\mathbf{u}) = kT(\mathbf{u})$$

We will use these two properties as the starting point for defining more general linear transformations.

DEFINITION 1   If $T : V \rightarrow W$ is a function from a vector space V to a vector space W, then T is called a ***linear transformation*** from V to W if the following two properties hold for all vectors $\mathbf{u}$ and $\mathbf{v}$ in V and for all scalars k:

(i) $T(k\mathbf{u}) = kT(\mathbf{u})$ **[Homogeneity property]**

(ii) $T(\mathbf{u} + \mathbf{v}) = T(\mathbf{u}) + T(\mathbf{v})$ **[Additivity property]**

In the special case where $V = W$, the linear transformation T is called a ***linear operator*** on the vector space V.

The homogeneity and additivity properties of a linear transformation $T : V \to W$ can be used in combination to show that if $\mathbf{v}_1$ and $\mathbf{v}_2$ are vectors in V and k_1 and k_2 are any scalars, then

$$T(k_1\mathbf{v}_1 + k_2\mathbf{v}_2) = k_1 T(\mathbf{v}_1) + k_2 T(\mathbf{v}_2)$$

More generally, if $\mathbf{v}_1, \mathbf{v}_2, \ldots, \mathbf{v}_r$ are vectors in V and $k_1, k_2, \ldots, k_r$ are any scalars, then

$$T(k_1\mathbf{v}_1 + k_2\mathbf{v}_2 + \cdots + k_r\mathbf{v}_r) = k_1 T(\mathbf{v}_1) + k_2 T(\mathbf{v}_2) + \cdots + k_r T(\mathbf{v}_r) \qquad (1)$$

The following theorem is an analog of parts (a) and (d) of Theorem 4.9.1.

THEOREM 8.1.1 *If* $T : V \to W$ *is a linear transformation, then*:

(a) $T(\mathbf{0}) = \mathbf{0}$.

(b) $T(\mathbf{u} - \mathbf{v}) = T(\mathbf{u}) - T(\mathbf{v})$ *for all* $\mathbf{u}$ *and* $\mathbf{v}$ *in* V.

Use the two parts of Theorem 8.1.1 to prove that

$$T(-\mathbf{v}) = -\mathbf{v}$$

for all $\mathbf{v}$ in V.

Proof Let $\mathbf{u}$ be any vector in V. Since $0\mathbf{u} = \mathbf{0}$, it follows from the homogeneity property in Definition 1 that

$$T(\mathbf{0}) = T(0\mathbf{u}) = 0T(\mathbf{u}) = \mathbf{0}$$

which proves (a).

We can prove part (b) by rewriting $T(\mathbf{u} - \mathbf{v})$ as

$$
\begin{aligned}
T(\mathbf{u} - \mathbf{v}) &= T\big(\mathbf{u} + (-1)\mathbf{v}\big) \\
&= T(\mathbf{u}) + (-1)T(\mathbf{v}) \\
&= T(\mathbf{u}) - T(\mathbf{v})
\end{aligned}
$$

We leave it for you to justify each step. ◄

► **EXAMPLE 1 Matrix Transformations**

Because we have based the definition of a general linear transformation on the homogeneity and additivity properties of *matrix transformations*, it follows that a matrix transformation $T_A : R^n \to R^m$ is also a linear transformation in this more general sense with $V = R^n$ and $W = R^m$.

► **EXAMPLE 2 The Zero Transformation**

Let V and W be any two vector spaces. The mapping $T : V \to W$ such that $T(\mathbf{v}) = \mathbf{0}$ for every $\mathbf{v}$ in V is a linear transformation called the **zero transformation**. To see that T is linear, observe that

$$T(\mathbf{u} + \mathbf{v}) = \mathbf{0}, \quad T(\mathbf{u}) = \mathbf{0}, \quad T(\mathbf{v}) = \mathbf{0}, \quad \text{and} \quad T(k\mathbf{u}) = \mathbf{0}$$

Therefore,

$$T(\mathbf{u} + \mathbf{v}) = T(\mathbf{u}) + T(\mathbf{v}) \quad \text{and} \quad T(k\mathbf{u}) = kT(\mathbf{u})$$

► **EXAMPLE 3 The Identity Operator**

Let V be any vector space. The mapping $I : V \to V$ defined by $I(\mathbf{v}) = \mathbf{v}$ is called the **identity operator** on V. We will leave it for you to verify that I is linear.

▶ **EXAMPLE 4 Dilation and Contraction Operators**

If V is a vector space and k is any scalar, then the mapping $T : V \to V$ given by $T(\mathbf{x}) = k\mathbf{x}$ is a linear operator on V, for if c is any scalar and if $\mathbf{u}$ and $\mathbf{v}$ are any vectors in V, then

$$T(c\mathbf{u}) = k(c\mathbf{u}) = c(k\mathbf{u}) = cT(\mathbf{u})$$
$$T(\mathbf{u} + \mathbf{v}) = k(\mathbf{u} + \mathbf{v}) = k\mathbf{u} + k\mathbf{v} = T(\mathbf{u}) + T(\mathbf{v})$$

If $0 < k < 1$, then T is called the **contraction** of V with factor k, and if $k > 1$, it is called the **dilation** of V with factor k (Figure 8.1.1).

▶ **Figure 8.1.1**

Dilation of V Contraction of V

▶ **EXAMPLE 5 A Linear Transformation from P_n to P_{n+1}**

Let $\mathbf{p} = p(x) = c_0 + c_1 x + \cdots + c_n x^n$ be a polynomial in P_n, and define the transformation $T : P_n \to P_{n+1}$ by

$$T(\mathbf{p}) = T(p(x)) = xp(x) = c_0 x + c_1 x^2 + \cdots + c_n x^{n+1}$$

This transformation is linear because for any scalar k and any polynomials $\mathbf{p}_1$ and $\mathbf{p}_2$ in P_n we have

$$T(k\mathbf{p}) = T(kp(x)) = x(kp(x)) = k(xp(x)) = kT(\mathbf{p})$$

and

$$T(\mathbf{p}_1 + \mathbf{p}_2) = T(p_1(x) + p_2(x)) = x(p_1(x) + p_2(x))$$
$$= xp_1(x) + xp_2(x) = T(\mathbf{p}_1) + T(\mathbf{p}_2)$$

▶ **EXAMPLE 6 A Linear Transformation Using an Inner Product**

Let V be an inner product space, let $\mathbf{v}_0$ be any fixed vector in V, and let $T : V \to R$ be the transformation

$$T(\mathbf{x}) = \langle \mathbf{x}, \mathbf{v}_0 \rangle$$

that maps a vector $\mathbf{x}$ into its inner product with $\mathbf{v}_0$. This transformation is linear, for if k is any scalar, and if $\mathbf{u}$ and $\mathbf{v}$ are any vectors in V, then it follows from properties of inner products that

$$T(k\mathbf{u}) = \langle k\mathbf{u}, \mathbf{v}_0 \rangle = k\langle \mathbf{u}, \mathbf{v}_0 \rangle = kT(\mathbf{u})$$
$$T(\mathbf{u} + \mathbf{v}) = \langle \mathbf{u} + \mathbf{v}, \mathbf{v}_0 \rangle = \langle \mathbf{u}, \mathbf{v}_0 \rangle + \langle \mathbf{v}, \mathbf{v}_0 \rangle = T(\mathbf{u}) + T(\mathbf{v})$$

▶ **EXAMPLE 7 Transformations on Matrix Spaces**

Let M_{nn} be the vector space of $n \times n$ matrices. In each part determine whether the transformation is linear.

(a) $T_1(A) = A^T$ (b) $T_2(A) = \det(A)$

Solution (a) It follows from parts (b) and (d) of Theorem 1.4.8 that

$$T_1(kA) = (kA)^T = kA^T = kT_1(A)$$
$$T_1(A + B) = (A + B)^T = A^T + B^T = T_1(A) + T_1(B)$$

so T_1 is linear.

Solution (b) It follows from Formula (1) of Section 2.3 that

$$T_2(kA) = \det(kA) = k^n \det(A) = k^n T_2(A)$$

Thus, T_2 is not homogeneous and hence not linear if $n > 1$. Note that additivity also fails because we showed in Example 1 of Section 2.3 that $\det(A + B)$ and $\det(A) + \det(B)$ are not generally equal.

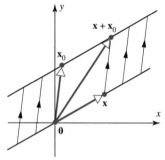

▲ **Figure 8.1.2** $T(\mathbf{x}) = \mathbf{x} + \mathbf{x}_0$ translates each point $\mathbf{x}$ along a line parallel to $\mathbf{x}_0$ through a distance $\|\mathbf{x}_0\|$.

▶ **EXAMPLE 8 Translation Is Not Linear**

Part (*a*) of Theorem 8.1.1 states that a linear transformation maps $\mathbf{0}$ to $\mathbf{0}$. This property is useful for identifying transformations that are *not* linear. For example, if $\mathbf{x}_0$ is a fixed nonzero vector in R^2, then the transformation

$$T(\mathbf{x}) = \mathbf{x} + \mathbf{x}_0$$

has the geometric effect of translating each point $\mathbf{x}$ in a direction parallel to $\mathbf{x}_0$ through a distance of $\|\mathbf{x}_0\|$ (Figure 8.1.2). This cannot be a linear transformation since $T(\mathbf{0}) = \mathbf{x}_0$, so T does not map $\mathbf{0}$ to $\mathbf{0}$.

▶ **EXAMPLE 9 The Evaluation Transformation**

Let V be a subspace of $F(-\infty, \infty)$, let

$$x_1, x_2, \ldots, x_n$$

be distinct real numbers, and let $T : V \to R^n$ be the transformation

$$T(f) = \big(f(x_1), f(x_2), \ldots, f(x_n)\big) \tag{2}$$

that associates with f the n-tuple of function values at $x_1, x_2, \ldots, x_n$. We call this the ***evaluation transformation*** on V at $x_1, x_2, \ldots, x_n$. Thus, for example, if

$$x_1 = -1, \quad x_2 = 2, \quad x_3 = 4$$

and if $f(x) = x^2 - 1$, then

$$T(f) = \big(f(x_1), f(x_2), f(x_3)\big) = (0, 3, 15)$$

The evaluation transformation in (2) is linear, for if k is any scalar, and if f and g are any functions in V, then

$$\begin{aligned} T(kf) &= \big((kf)(x_1), (kf)(x_2), \ldots, (kf)(x_n)\big) \\ &= \big(kf(x_1), kf(x_2), \ldots, kf(x_n)\big) \\ &= k\big(f(x_1), f(x_2), \ldots, f(x_n)\big) = kT(f) \end{aligned}$$

and

$$\begin{aligned} T(f + g) &= \big((f + g)(x_1), (f + g)(x_2), \ldots, (f + g)(x_n)\big) \\ &= \big(f(x_1) + g(x_1), f(x_2) + g(x_2), \ldots, f(x_n) + g(x_n)\big) \\ &= \big(f(x_1), f(x_2), \ldots, f(x_n)\big) + \big(g(x_1), g(x_2), \ldots, g(x_n)\big) \\ &= T(f) + T(g) \blacktriangleleft \end{aligned}$$

Finding Linear Transformations from Images of Basis Vectors

We saw in Formula (11) of Section 4.10 that if $T : R^n \to R^m$ is a matrix transformation, say multiplication by A, and if $\mathbf{e}_1, \mathbf{e}_2, \ldots, \mathbf{e}_n$ are the standard basis vectors for R^n, then A can be expressed as

$$A = [T(\mathbf{e}_1) \mid T(\mathbf{e}_2) \mid \ldots \mid T(\mathbf{e}_n)]$$

It follows from this that the image of any vector $\mathbf{v} = (c_1, c_2, \ldots, c_n)$ in R^n under multiplication by A can be expressed as

$$T(\mathbf{v}) = c_1 T(\mathbf{e}_1) + c_2 T(\mathbf{e}_2) + \cdots + c_n T(\mathbf{e}_n)$$

This formula tells us that for a matrix transformation the image of any vector is expressible as a linear combination of the images of the standard basis vectors. This is a special case of the following more general result.

THEOREM 8.1.2 *Let $T : V \to W$ be a linear transformation, where V is finite dimensional. If $S = \{\mathbf{v}_1, \mathbf{v}_2, \ldots, \mathbf{v}_n\}$ is a basis for V, then the image of any vector $\mathbf{v}$ in V can be expressed as*

$$T(\mathbf{v}) = c_1 T(\mathbf{v}_1) + c_2 T(\mathbf{v}_2) + \cdots + c_n T(\mathbf{v}_n) \tag{3}$$

where $c_1, c_2, \ldots, c_n$ are the coefficients required to express $\mathbf{v}$ as a linear combination of the vectors in S.

Proof Express $\mathbf{v}$ as $\mathbf{v} = c_1 \mathbf{v}_1 + c_2 \mathbf{v}_2 + \cdots + c_n \mathbf{v}_n$ and use the linearity of T. ◀

▶ **EXAMPLE 10** Computing with Images of Basis Vectors

Consider the basis $S = \{\mathbf{v}_1, \mathbf{v}_2, \mathbf{v}_3\}$ for R^3, where

$$\mathbf{v}_1 = (1, 1, 1), \quad \mathbf{v}_2 = (1, 1, 0), \quad \mathbf{v}_3 = (1, 0, 0)$$

Let $T : R^3 \to R^2$ be the linear transformation for which

$$T(\mathbf{v}_1) = (1, 0), \quad T(\mathbf{v}_2) = (2, -1), \quad T(\mathbf{v}_3) = (4, 3)$$

Find a formula for $T(x_1, x_2, x_3)$, and then use that formula to compute $T(2, -3, 5)$.

Solution We first need to express $\mathbf{x} = (x_1, x_2, x_3)$ as a linear combination of $\mathbf{v}_1, \mathbf{v}_2$, and $\mathbf{v}_3$. If we write

$$(x_1, x_2, x_3) = c_1(1, 1, 1) + c_2(1, 1, 0) + c_3(1, 0, 0)$$

then on equating corresponding components, we obtain

$$c_1 + c_2 + c_3 = x_1$$
$$c_1 + c_2 \quad\quad = x_2$$
$$c_1 \quad\quad\quad = x_3$$

which yields $c_1 = x_3$, $c_2 = x_2 - x_3$, $c_3 = x_1 - x_2$, so

$$(x_1, x_2, x_3) = x_3(1, 1, 1) + (x_2 - x_3)(1, 1, 0) + (x_1 - x_2)(1, 0, 0)$$
$$= x_3 \mathbf{v}_1 + (x_2 - x_3)\mathbf{v}_2 + (x_1 - x_2)\mathbf{v}_3$$

Thus

$$T(x_1, x_2, x_3) = x_3 T(\mathbf{v}_1) + (x_2 - x_3)T(\mathbf{v}_2) + (x_1 - x_2)T(\mathbf{v}_3)$$
$$= x_3(1, 0) + (x_2 - x_3)(2, -1) + (x_1 - x_2)(4, 3)$$
$$= (4x_1 - 2x_2 - x_3, 3x_1 - 4x_2 + x_3)$$

From this formula, we obtain

$$T(2, -3, 5) = (9, 23)$$

► **EXAMPLE 11 A Linear Transformation from $C^1(-\infty, \infty)$ to $F(-\infty, \infty)$**

Let $V = C^1(-\infty, \infty)$ be the vector space of functions with continuous first derivatives on $(-\infty, \infty)$, and let $W = F(-\infty, \infty)$ be the vector space of all real-valued functions defined on $(-\infty, \infty)$. Let $D: V \to W$ be the transformation that maps a function $\mathbf{f} = f(x)$ into its derivative—that is,

$$D(\mathbf{f}) = f'(x)$$

From the properties of differentiation, we have

$$D(\mathbf{f} + \mathbf{g}) = D(k\mathbf{f}) = kD(\mathbf{f}) \quad \text{and} \quad D(\mathbf{f}) + D(\mathbf{g})$$

Thus, D is a linear transformation.

► **EXAMPLE 12 An Integral Transformation**

Let $V = C(-\infty, \infty)$ be the vector space of continuous functions on the interval $(-\infty, \infty)$, let $W = C^1(-\infty, \infty)$ be the vector space of functions with continuous first derivatives on $(-\infty, \infty)$, and let $J: V \to W$ be the transformation that maps a function f in V into

$$J(f) = \int_0^x f(t)\, dt$$

For example, if $f(x) = x^2$, then

$$J(f) = \int_0^x t^2\, dt = \frac{t^3}{3}\Bigg]_0^x = \frac{x^3}{3}$$

The transformation $J: V \to W$ is linear, for if k is any constant, and if f and g are any functions in V, then properties of the integral imply that

$$J(kf) = \int_0^x kf(t)\, dt = k\int_0^x f(t)\, dt = kJ(f)$$

$$J(f + g) = \int_0^x (f(t) + g(t))\, dt = \int_0^x f(t)\, dt + \int_0^x g(t)\, dt = J(f) + J(g) \;\blacktriangleleft$$

Kernel and Range

Recall that if A is an $m \times n$ matrix, then the null space of A consists of all vectors $\mathbf{x}$ in R^n such that $A\mathbf{x} = \mathbf{0}$, and by Theorem 4.7.1 the column space of A consists of all vectors $\mathbf{b}$ in R^m for which there is at least one vector $\mathbf{x}$ in R^n such that $A\mathbf{x} = \mathbf{b}$. From the viewpoint of matrix transformations, the null space of A consists of all vectors in R^n that multiplication by A maps into $\mathbf{0}$, and the column space of A consists of all vectors in R^m that are images of at least one vector in R^n under multiplication by A. The following definition extends these ideas to general linear transformations.

DEFINITION 2 If $T: V \to W$ is a linear transformation, then the set of vectors in V that T maps into $\mathbf{0}$ is called the **kernel** of T and is denoted by $\ker(T)$. The set of all vectors in W that are images under T of at least one vector in V is called the **range** of T and is denoted by $R(T)$.

► **EXAMPLE 13 Kernel and Range of a Matrix Transformation**

If $T_A: R^n \to R^m$ is multiplication by the $m \times n$ matrix A, then, as discussed above, the kernel of T_A is the null space of A, and the range of T_A is the column space of A.

▶ **EXAMPLE 14** **Kernel and Range of the Zero Transformation**

Let $T : V \rightarrow W$ be the zero transformation. Since T maps *every* vector in V into $\mathbf{0}$, it follows that $\ker(T) = V$. Moreover, since $\mathbf{0}$ is the *only* image under T of vectors in V, it follows that $R(T) = \{\mathbf{0}\}$.

▶ **EXAMPLE 15** **Kernel and Range of the Identity Operator**

Let $I : V \rightarrow V$ be the identity operator. Since $I(\mathbf{v}) = \mathbf{v}$ for all vectors in V, *every* vector in V is the image of some vector (namely, itself); thus $R(I) = V$. Since the *only* vector that I maps into $\mathbf{0}$ is $\mathbf{0}$, it follows that $\ker(I) = \{\mathbf{0}\}$.

▶ **EXAMPLE 16** **Kernel and Range of an Orthogonal Projection**

As illustrated in Figure 8.1.3a, the points that T maps into $\mathbf{0} = (0, 0, 0)$ are precisely those on the z-axis, so $\ker(T)$ is the set of points of the form $(0, 0, z)$. As illustrated in Figure 8.1.3b, T maps the points in R^3 to the xy-plane, where each point in that plane is the image of each point on the vertical line above it. Thus, $R(T)$ is the set of points of the form $(x, y, 0)$.

▶ **EXAMPLE 17** **Kernel and Range of a Rotation**

Let $T : R^2 \rightarrow R^2$ be the linear operator that rotates each vector in the xy-plane through the angle θ (Figure 8.1.4). Since *every* vector in the xy-plane can be obtained by rotating some vector through the angle θ, it follows that $R(T) = R^2$. Moreover, the only vector that rotates into $\mathbf{0}$ is $\mathbf{0}$, so $\ker(T) = \{\mathbf{0}\}$.

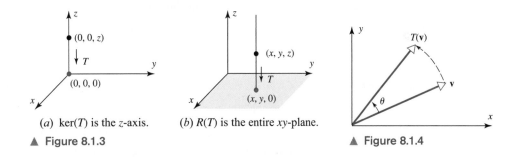

(a) $\ker(T)$ is the z-axis.   (b) $R(T)$ is the entire xy-plane.

▲ **Figure 8.1.3**   ▲ **Figure 8.1.4**

CALCULUS REQUIRED

▶ **EXAMPLE 18** **Kernel of a Differentiation Transformation**

Let $V = C^1(-\infty, \infty)$ be the vector space of functions with continuous first derivatives on $(-\infty, \infty)$, let $W = F(-\infty, \infty)$ be the vector space of all real-valued functions defined on $(-\infty, \infty)$, and let $D : V \rightarrow W$ be the differentiation transformation $D(\mathbf{f}) = f'(x)$. The kernel of D is the set of functions in V with derivative zero. From calculus, this is the set of constant functions on $(-\infty, \infty)$. ◀

Properties of Kernel and Range

In all of the preceding examples, $\ker(T)$ and $R(T)$ turned out to be *subspaces*. In Examples 14, 15, and 17 they were either the zero subspace or the entire vector space. In Example 16 the kernel was a line through the origin, and the range was a plane through the origin, both of which are subspaces of R^3. All of this is a consequence of the following general theorem.

> **THEOREM 8.1.3** *If $T: V \to W$ is a linear transformation, then*:
>
> (a) *The kernel of T is a subspace of V.*
> (b) *The range of T is a subspace of W.*

Proof (a)   To show that $\ker(T)$ is a subspace, we must show that it contains at least one vector and is closed under addition and scalar multiplication. By part (a) of Theorem 8.1.1, the vector $\mathbf{0}$ is in $\ker(T)$, so the kernel contains at least one vector. Let $\mathbf{v}_1$ and $\mathbf{v}_2$ be vectors in $\ker(T)$, and let k be any scalar. Then

$$T(\mathbf{v}_1 + \mathbf{v}_2) = T(\mathbf{v}_1) + T(\mathbf{v}_2) = \mathbf{0} + \mathbf{0} = \mathbf{0}$$

so $\mathbf{v}_1 + \mathbf{v}_2$ is in $\ker(T)$. Also,

$$T(k\mathbf{v}_1) = kT(\mathbf{v}_1) = k\mathbf{0} = \mathbf{0}$$

so $k\mathbf{v}_1$ is in $\ker(T)$.

Proof (b)   To show that $R(T)$ is a subspace of W, we must show that it contains at least one vector and is closed under addition and scalar multiplication. However, it contains at least the zero vector of W since $T(\mathbf{0}) = (\mathbf{0})$ by part (a) of Theorem 8.1.1. To prove that it is closed under addition and scalar multiplication, we must show that if $\mathbf{w}_1$ and $\mathbf{w}_2$ are vectors in $R(T)$, and if k is any scalar, then there exist vectors $\mathbf{a}$ and $\mathbf{b}$ in V for which

$$T(\mathbf{a}) = \mathbf{w}_1 + \mathbf{w}_2 \quad \text{and} \quad T(\mathbf{b}) = k\mathbf{w}_1 \tag{4}$$

But the fact $\mathbf{w}_1$ and $\mathbf{w}_2$ are in $R(T)$ tells us that there exist vectors $\mathbf{v}_1$ and $\mathbf{v}_2$ in V such that

$$T(\mathbf{v}_1) = \mathbf{w}_1 \quad \text{and} \quad T(\mathbf{v}_2) = \mathbf{w}_2$$

The following computations complete the proof by showing that the vectors $\mathbf{a} = \mathbf{v}_1 + \mathbf{v}_2$ and $\mathbf{b} = k\mathbf{v}_1$ satisfy the equations in (4):

$$T(\mathbf{a}) = T(\mathbf{v}_1 + \mathbf{v}_2) = T(\mathbf{v}_1) + T(\mathbf{v}_2) = \mathbf{w}_1 + \mathbf{w}_2$$
$$T(\mathbf{b}) = T(k\mathbf{v}_1) = kT(\mathbf{v}_1) = k\mathbf{w}_1 \ \blacktriangleleft$$

CALCULUS REQUIRED   ▶ **EXAMPLE 19 Application to Differential Equations**

Differential equations of the form

$$y'' + \omega^2 y = 0 \qquad (\omega \text{ a positive constant}) \tag{5}$$

arise in the study of vibrations. The set of all solutions of this equation on the interval $(-\infty, \infty)$ is the kernel of the linear transformation $D: C^2(-\infty, \infty) \to C(-\infty, \infty)$, given by

$$D(y) = y'' + \omega^2 y$$

It is proved in standard textbooks on differential equations that the kernel is a two-dimensional subspace of $C^2(-\infty, \infty)$, so that if we can find two linearly independent solutions of (5), then all other solutions can be expressed as linear combinations of those two. We leave it for you to confirm by differentiating that

$$y_1 = \cos \omega x \quad \text{and} \quad y_2 = \sin \omega x$$

are solutions of (5). These functions are linearly independent since neither is a scalar multiple of the other, and thus

$$y = c_1 \cos \omega x + c_2 \sin \omega x \tag{6}$$

is a "general solution" of (5) in the sense that every choice of c_1 and c_2 produces a solution, and every solution is of this form. ◀

Rank and Nullity of Linear Transformations

In Definition 1 of Section 4.8 we defined the notions of *rank* and *nullity* for an $m \times n$ matrix, and in Theorem 4.8.2, which we called the *Dimension Theorem*, we proved that the sum of the rank and nullity is n. We will show next that this result is a special case of a more general result about linear transformations. We start with the following definition.

DEFINITION 3 Let $T : V \rightarrow W$ be a linear transformation. If the range of T is finite-dimensional, then its dimension is called the **rank of T**; and if the kernel of T is finite-dimensional, then its dimension is called the **nullity of T**. The rank of T is denoted by $\mathrm{rank}(T)$ and the nullity of T by $\mathrm{nullity}(T)$.

The following theorem, whose proof is optional, generalizes Theorem 4.8.2.

THEOREM 8.1.4 Dimension Theorem for Linear Transformations

If $T : V \rightarrow W$ is a linear transformation from an n-dimensional vector space V to a vector space W, then

$$\mathrm{rank}(T) + \mathrm{nullity}(T) = n \tag{7}$$

In the special case where A is an $m \times n$ matrix and $T_A : R^n \rightarrow R^m$ is multiplication by A, the kernel of T_A is the null space of A, and the range of T_A is the column space of A. Thus, it follows from Theorem 8.1.4 that

$$\mathrm{rank}(T_A) + \mathrm{nullity}(T_A) = n$$

OPTIONAL

Proof of Theorem 8.1.4   We must show that

$$\dim(R(T)) + \dim(\ker(T)) = n$$

We will give the proof for the case where $1 \leq \dim(\ker(T)) < n$. The cases where $\dim(\ker(T)) = 0$ and $\dim(\ker(T)) = n$ are left as exercises. Assume $\dim(\ker(T)) = r$, and let $\mathbf{v}_1, \ldots, \mathbf{v}_r$ be a basis for the kernel. Since $\{\mathbf{v}_1, \ldots, \mathbf{v}_r\}$ is linearly independent, Theorem 4.5.5b states that there are $n - r$ vectors, $\mathbf{v}_{r+1}, \ldots, \mathbf{v}_n$, such that the extended set $\{\mathbf{v}_1, \ldots, \mathbf{v}_r, \mathbf{v}_{r+1}, \ldots, \mathbf{v}_n\}$ is a basis for V. To complete the proof, we will show that the $n - r$ vectors in the set $S = \{T(\mathbf{v}_{r+1}), \ldots, T(\mathbf{v}_n)\}$ form a basis for the range of T. It will then follow that

$$\dim(R(T)) + \dim(\ker(T)) = (n - r) + r = n$$

First we show that S spans the range of T. If $\mathbf{b}$ is any vector in the range of T, then $\mathbf{b} = T(\mathbf{v})$ for some vector $\mathbf{v}$ in V. Since $\{\mathbf{v}_1, \ldots, \mathbf{v}_r, \mathbf{v}_{r+1}, \ldots, \mathbf{v}_n\}$ is a basis for V, the vector $\mathbf{v}$ can be written in the form

$$\mathbf{v} = c_1 \mathbf{v}_1 + \cdots + c_r \mathbf{v}_r + c_{r+1} \mathbf{v}_{r+1} + \cdots + c_n \mathbf{v}_n$$

Since $\mathbf{v}_1, \ldots, \mathbf{v}_r$ lie in the kernel of T, we have $T(\mathbf{v}_1) = \cdots = T(\mathbf{v}_r) = \mathbf{0}$, so

$$\mathbf{b} = T(\mathbf{v}) = c_{r+1} T(\mathbf{v}_{r+1}) + \cdots + c_n T(\mathbf{v}_n)$$

Thus S spans the range of T.

Finally, we show that S is a linearly independent set and consequently forms a basis for the range of T. Suppose that some linear combination of the vectors in S is zero; that is,

$$k_{r+1} T(\mathbf{v}_{r+1}) + \cdots + k_n T(\mathbf{v}_n) = \mathbf{0} \tag{8}$$

We must show that $k_{r+1} = \cdots = k_n = 0$. Since T is linear, (8) can be rewritten as

$$T(k_{r+1} \mathbf{v}_{r+1} + \cdots + k_n \mathbf{v}_n) = \mathbf{0}$$

which says that $k_{r+1}\mathbf{v}_{r+1} + \cdots + k_n\mathbf{v}_n$ is in the kernel of T. This vector can therefore be written as a linear combination of the basis vectors $\{\mathbf{v}_1, \ldots, \mathbf{v}_r\}$, say

$$k_{r+1}\mathbf{v}_{r+1} + \cdots + k_n\mathbf{v}_n = k_1\mathbf{v}_1 + \cdots + k_r\mathbf{v}_r$$

Thus,

$$k_1\mathbf{v}_1 + \cdots + k_r\mathbf{v}_r - k_{r+1}\mathbf{v}_{r+1} - \cdots - k_n\mathbf{v}_n = \mathbf{0}$$

Since $\{\mathbf{v}_1, \ldots, \mathbf{v}_n\}$ is linearly independent, all of the k's are zero; in particular, $k_{r+1} = \cdots = k_n = 0$, which completes the proof. ◀

Concept Review

- Linear transformation
- Linear operator
- Zero transformation
- Identity operator

- Contraction
- Dilation
- Evaluation transformation
- Kernel

- Range
- Rank
- Nullity

Skills

- Determine whether a function is a linear transformation.
- Find a formula for a linear transformation $T : V \to W$ given the values of T on a basis for V.
- Find a basis for the kernel of a linear transformation.

- Find a basis for the range of a linear transformation.
- Find the rank of a linear transformation.
- Find the nullity of a linear transformation.

Exercise Set 8.1

▶ In Exercises 1–8, determine whether the function is a linear transformation. Justify your answer. ◀

1. $T : V \to R$, where V is an inner product space, and $T(\mathbf{u}) = \|\mathbf{u}\|$.

2. $T : R^3 \to R^3$, where $\mathbf{v}_0$ is a fixed vector in R^3 and $T(\mathbf{u}) = \mathbf{u} \times \mathbf{v}_0$.

3. $T : M_{22} \to M_{23}$, where B is a fixed 2×3 matrix and $T(A) = AB$.

4. $T : M_{nn} \to R$, where $T(A) = \text{tr}(A)$.

5. $F : M_{mn} \to M_{nm}$, where $F(A) = A^T$.

6. $T : M_{22} \to R$, where

(a) $T\left(\begin{bmatrix} a & b \\ c & d \end{bmatrix}\right) = 3a - 4b + c - d$

(b) $T\left(\begin{bmatrix} a & b \\ c & d \end{bmatrix}\right) = a^2 + b^2$

7. $T : P_2 \to P_2$, where

(a) $T(a_0 + a_1x + a_2x^2) = a_0 + a_1(x + 1) + a_2(x + 1)^2$

(b) $T(a_0 + a_1x + a_2x^2)$
    $= (a_0 + 1) + (a_1 + 1)x + (a_2 + 1)x^2$

8. $T : F(-\infty, \infty) \to F(-\infty, \infty)$, where

(a) $T(f(x)) = 1 + f(x)$ (b) $T(f(x)) = f(x + 1)$

9. Consider the basis $S = \{\mathbf{v}_1, \mathbf{v}_2\}$ for R^2, where $\mathbf{v}_1 = (1, 1)$ and $\mathbf{v}_2 = (1, 0)$, and let $T : R^2 \to R^2$ be the linear operator for which

$$T(\mathbf{v}_1) = (1, -2) \quad \text{and} \quad T(\mathbf{v}_2) = (-4, 1)$$

Find a formula for $T(x_1, x_2)$, and use that formula to find $T(5, -3)$.

10. Consider the basis $S = \{\mathbf{v}_1, \mathbf{v}_2\}$ for R^2, where $\mathbf{v}_1 = (-2, 1)$ and $\mathbf{v}_2 = (1, 3)$, and let $T : R^2 \to R^3$ be the linear transformation such that

$$T(\mathbf{v}_1) = (-1, 2, 0) \quad \text{and} \quad T(\mathbf{v}_2) = (0, -3, 5)$$

Find a formula for $T(x_1, x_2)$, and use that formula to find $T(2, -3)$.

11. Consider the basis $S = \{\mathbf{v}_1, \mathbf{v}_2, \mathbf{v}_3\}$ for R^3, where $\mathbf{v}_1 = (1, 1, 1)$, $\mathbf{v}_2 = (1, 1, 0)$, and $\mathbf{v}_3 = (1, 0, 0)$, and let $T : R^3 \to R^3$ be the linear operator for which

$$T(\mathbf{v}_1) = (2, -1, 4), \quad T(\mathbf{v}_2) = (3, 0, 1),$$
$$T(\mathbf{v}_3) = (-1, 5, 1)$$

Find a formula for $T(x_1, x_2, x_3)$, and use that formula to find $T(2, 4, -1)$.

12. Consider the basis $S = \{\mathbf{v}_1, \mathbf{v}_2, \mathbf{v}_3\}$ for R^3, where
$\mathbf{v}_1 = (1, 2, 1)$, $\mathbf{v}_2 = (2, 9, 0)$, and $\mathbf{v}_3 = (3, 3, 4)$, and let
$T : R^3 \rightarrow R^2$ be the linear transformation for which

$$T(\mathbf{v}_1) = (1, 0), \quad T(\mathbf{v}_2) = (-1, 1), \quad T(\mathbf{v}_3) = (0, 1)$$

Find a formula for $T(x_1, x_2, x_3)$, and use that formula to find $T(7, 13, 7)$.

13. Let $\mathbf{v}_1$, $\mathbf{v}_2$, and $\mathbf{v}_3$ be vectors in a vector space V, and let $T : V \rightarrow R^3$ be a linear transformation for which

$$T(\mathbf{v}_1) = (1, -1, 2), \quad T(\mathbf{v}_2) = (0, 3, 2),$$
$$T(\mathbf{v}_3) = (-3, 1, 2)$$

Find $T(2\mathbf{v}_1 - 3\mathbf{v}_2 + 4\mathbf{v}_3)$.

14. Let $T : R^2 \rightarrow R^2$ be the linear operator given by the formula

$$T(x, y) = (2x - y, -8x + 4y)$$

Which of the following vectors are in $R(T)$?

(a) $(1, -4)$ (b) $(5, 0)$ (c) $(-3, 12)$

15. Let $T : R^2 \rightarrow R^2$ be the linear operator in Exercise 14. Which of the following vectors are in $\ker(T)$?

(a) $(5, 10)$ (b) $(3, 2)$ (c) $(1, 1)$

16. Let $T : R^4 \rightarrow R^3$ be the linear transformation given by the formula

$$T(x_1, x_2, x_3, x_4) = (4x_1 + x_2 - 2x_3 - 3x_4,$$
$$2x_1 + x_2 + x_3 - 4x_4, 6x_1 - 9x_3 + 9x_4)$$

Which of the following are in $R(T)$?

(a) $(0, 0, 6)$ (b) $(1, 3, 0)$ (c) $(2, 4, 1)$

17. Let $T : R^4 \rightarrow R^3$ be the linear transformation in Exercise 16. Which of the following are in $\ker(T)$?

(a) $(3, -8, 2, 0)$ (b) $(0, 0, 0, 1)$ (c) $(0, -4, 1, 0)$

18. Let $T : P_2 \rightarrow P_3$ be the linear transformation defined by $T(p(x)) = xp(x)$. Which of the following are in $\ker(T)$?

(a) x^2 (b) 0 (c) $1 + x$

19. Let $T : P_2 \rightarrow P_3$ be the linear transformation in Exercise 18. Which of the following are in $R(T)$?

(a) $x + x^2$ (b) $1 + x$ (c) $3 - x^2$

20. Find a basis for the kernel of

(a) the linear operator in Exercise 14.

(b) the linear transformation in Exercise 16.

(c) the linear transformation in Exercise 18.

21. Find a basis for the range of

(a) the linear operator in Exercise 14.

(b) the linear transformation in Exercise 16.

(c) the linear transformation in Exercise 18.

22. Verify Formula (7) of the dimension theorem for

(a) the linear operator in Exercise 14.

(b) the linear transformation in Exercise 16.

(c) the linear transformation in Exercise 18.

▶ In Exercises 23–26, let T be multiplication by the matrix A. Find

(a) a basis for the range of T.

(b) a basis for the kernel of T.

(c) the rank and nullity of T.

(d) the rank and nullity of A. ◀

23. $A = \begin{bmatrix} 1 & -1 & 3 \\ 5 & 6 & -4 \\ 7 & 4 & 2 \end{bmatrix}$ 24. $A = \begin{bmatrix} 2 & 0 & -1 \\ 4 & 0 & -2 \\ 20 & 0 & 0 \end{bmatrix}$

25. $A = \begin{bmatrix} 4 & 1 & 5 & 2 \\ 1 & 2 & 3 & 0 \end{bmatrix}$

26. $A = \begin{bmatrix} 1 & 4 & 5 & 0 & 9 \\ 3 & -2 & 1 & 0 & -1 \\ -1 & 0 & -1 & 0 & -1 \\ 2 & 3 & 5 & 1 & 8 \end{bmatrix}$

27. Describe the kernel and range of

(a) the orthogonal projection on the xz-plane.

(b) the orthogonal projection on the yz-plane.

(c) the orthogonal projection on the plane defined by the equation $y = x$.

28. Let V be any vector space, and let $T : V \rightarrow V$ be defined by $T(\mathbf{v}) = 3\mathbf{v}$.

(a) What is the kernel of T?

(b) What is the range of T?

29. In each part, use the given information to find the nullity of the linear transformation T.

(a) $T : R^5 \rightarrow R^7$ has rank 3.

(b) $T : P_4 \rightarrow P_3$ has rank 1.

(c) The range of $T : R^6 \rightarrow R^3$ is R^3.

(d) $T : M_{22} \rightarrow M_{22}$ has rank 3.

30. Let A be a 7×6 matrix such that $A\mathbf{x} = \mathbf{0}$ has only the trivial solution, and let $T : R^6 \rightarrow R^7$ be multiplication by A. Find the rank and nullity of T.

31. Let A be a 5×7 matrix with rank 4.

(a) What is the dimension of the solution space of $A\mathbf{x} = \mathbf{0}$?

(b) Is $A\mathbf{x} = \mathbf{b}$ consistent for all vectors $\mathbf{b}$ in R^5? Explain.

32. Let $T : R^3 \rightarrow W$ be a linear transformation from R^3 to any vector space. Give a geometric description of $\ker(T)$.

33. Let $T:V \to R^3$ be a linear transformation from any vector space to R^3. Give a geometric description of $R(T)$.

34. Let $T:R^3 \to R^3$ be multiplication by

$$\begin{bmatrix} 1 & 3 & 4 \\ 3 & 4 & 7 \\ -2 & 2 & 0 \end{bmatrix}$$

(a) Show that the kernel of T is a line through the origin, and find parametric equations for it.

(b) Show that the range of T is a plane through the origin, and find an equation for it.

35. (a) Show that if a_1, a_2, b_1, and b_2 are any scalars, then the formula

$$F(x, y) = (a_1 x + b_1 y, a_2 x + b_2 y)$$

defines a linear operator on R^2.

(b) Does the formula $F(x, y) = (a_1 x^2 + b_1 y^2, a_2 x^2 + b_2 y^2)$ define a linear operator on R^2? Explain.

36. Let $\{v_1, v_2, \ldots, v_n\}$ be a basis for a vector space V, and let $T:V \to W$ be a linear transformation. Show that if

$$T(v_1) = T(v_2) = \cdots = T(v_n) = 0$$

then T is the zero transformation.

37. Let $\{v_1, v_2, \ldots, v_n\}$ be a basis for a vector space V, and let $T:V \to V$ be a linear operator. Show that if

$$T(v_1) = v_1, \quad T(v_2) = v_2, \ldots, \quad T(v_n) = v_n$$

then T is the identity transformation on V.

38. For a positive integer $n > 1$, let $T:M_{nn} \to R$ be the linear transformation defined by $T(A) = \text{tr}(A)$, where A is an $n \times n$ matrix with real entries. Determine the dimension of $\ker(T)$.

39. Prove: If $\{v_1, v_2, \ldots, v_n\}$ is a basis for V and $w_1, w_2, \ldots, w_n$ are vectors in W, not necessarily distinct, then there exists a linear transformation $T:V \to W$ such that

$$T(v_1) = w_1, \quad T(v_2) = w_2, \ldots, \quad T(v_n) = w_n$$

40. (*Calculus required*) Let $V = C[a, b]$ be the vector space of functions continuous on $[a, b]$, and let $T:V \to V$ be the transformation defined by

$$T(\mathbf{f}) = 5f(x) + 3 \int_a^x f(t)\, dt$$

Is T a linear operator?

41. (*Calculus required*) Let $D:P_3 \to P_2$ be the differentiation transformation $D(\mathbf{p}) = p'(x)$. What is the kernel of D?

42. (*Calculus required*) Let $J:P_1 \to R$ be the integration transformation $J(\mathbf{p}) = \int_{-1}^1 p(x)\, dx$. What is the kernel of J?

43. (*Calculus required*) Let V be the vector space of real-valued functions with continuous derivatives of all orders on the interval $(-\infty, \infty)$, and let $W = F(-\infty, \infty)$ be the vector space of real-valued functions defined on $(-\infty, \infty)$.

(a) Find a linear transformation $T:V \to W$ whose kernel is P_3.

(b) Find a linear transformation $T:V \to W$ whose kernel is P_n.

44. If A is an $m \times n$ matrix, and if the linear system $A\mathbf{x} = \mathbf{b}$ is consistent for every vector $\mathbf{b}$ in R^m, what can you say about the range of $T_A:R^n \to R^m$?

True-False Exercises

In parts (a)–(i) determine whether the statement is true or false, and justify your answer.

(a) If $T(c_1 v_1 + c_2 v_2) = c_1 T(v_1) + c_2 T(v_2)$ for all vectors v_1 and v_2 in V and all scalars c_1 and c_2, then T is a linear transformation.

(b) If v is a nonzero vector in V, then there is exactly one linear transformation $T:V \to W$ such that $T(-v) = -T(v)$.

(c) There is exactly one linear transformation $T:V \to W$ for which $T(u + v) = T(u - v)$ for all vectors u and v in V.

(d) If v_0 is a nonzero vector in V, then the formula $T(v) = v_0 + v$ defines a linear operator on V.

(e) The kernel of a linear transformation is a vector space.

(f) The range of a linear transformation is a vector space.

(g) If $T:P_6 \to M_{22}$ is a linear transformation, then the nullity of T is 3.

(h) The function $T:M_{22} \to R$ defined by $T(A) = \det A$ is a linear transformation.

(i) The linear transformation $T:M_{22} \to M_{22}$ defined by

$$T(A) = \begin{bmatrix} 1 & 3 \\ 2 & 6 \end{bmatrix} A$$

has rank 1.

8.2 Isomorphism

In this section we will establish a fundamental connection between real finite-dimensional vector spaces and the Euclidean space R^n. This connection is not only important theoretically, but it has practical applications in that it allows us to perform vector computations in general vector spaces by working with the vectors in R^n.

One-to-One and Onto

Although many of the theorems in this text have been concerned exclusively with the vector space R^n, this is not as limiting as it might seem. As we will show, the vector space R^n is the "mother" of all real n-dimensional vector spaces in the sense that any such space might differ from R^n in the notation used to represent vectors, but not in its algebraic structure. To explain what we mean by this, we will need two definitions, the first of which is a generalization of Definition 1 in Section 4.10. (See Figure 8.2.1).

DEFINITION 1 If $T : V \to W$ is a linear transformation from a vector space V to a vector space W, then T is said to be **one-to-one** if T maps distinct vectors in V into distinct vectors in W.

DEFINITION 2 If $T : V \to W$ is a linear transformation from a vector space V to a vector space W, then T is said to be **onto** (or **onto W**) if every vector in W is the image of at least one vector in V.

One-to-one. Distinct vectors in V have distinct images in W.

Not one-to-one. There exist distinct vectors in V with the same image.

Onto W. Every vector in W is the image of some vector in V.

Not onto W. Not every vector in W is the image of some vector in V.

▲ Figure 8.2.1

The following theorem provides a useful way of telling whether a linear transformation is one-to-one by examining its kernel.

THEOREM 8.2.1 *If $T : V \to W$ is a linear transformation, then the following statements are equivalent.*

(a) T is one-to-one.

(b) $\ker(T) = \{\mathbf{0}\}$.

Proof (a) $\Rightarrow$ (b) Since T is linear, we know that $T(\mathbf{0}) = \mathbf{0}$ by Theorem 8.1.1a. Since T is one-to-one, there can be no other vectors in V that map into $\mathbf{0}$, so $\ker(T) = \{\mathbf{0}\}$.

(b) $\Rightarrow$ **(a)** Assume that $\ker(T) = \{0\}$. If $\mathbf{u}$ and $\mathbf{v}$ are distinct vectors in V, then $\mathbf{u} - \mathbf{v} \neq \mathbf{0}$. This implies that $T(\mathbf{u} - \mathbf{v}) \neq \mathbf{0}$, for otherwise $\ker(T)$ would contain a nonzero vector. Since T is linear, it follows that

$$T(\mathbf{u}) - T(\mathbf{v}) = T(\mathbf{u} - \mathbf{v}) \neq \mathbf{0}$$

so T maps distinct vectors in V into distinct vectors in W and hence is one-to-one. ◄

In the special case where V is finite-dimensional and T is a linear operator on V, then we can add a third statement to those in Theorem 8.2.1.

THEOREM 8.2.2 *If V is a finite-dimensional vector space, and if $T : V \to V$ is a linear operator, then the following statements are equivalent.*

(a) *T is one-to-one.*

(b) *$\ker(T) = \{0\}$.*

(c) *T is onto [i.e., $R(T) = V$].*

Proof We already know that (a) and (b) are equivalent by Theorem 8.2.1, so it suffices to show that (b) and (c) are equivalent. We leave it for you to do this by assuming that $\dim(V) = n$ and applying Theorem 8.1.4. ◄

▶ EXAMPLE 1 Dilations and Contractions Are One-to-One and Onto

Show that if V is a finite-dimensional vector space and c is any *nonzero* scalar, then the linear operator $T : V \to V$ defined by $T(\mathbf{v}) = c\mathbf{v}$ is one-to-one and onto.

Solution The operator T is onto (and hence one-to-one) for if $\mathbf{v}$ is any vector in V then that vector is the image of the vector $(1/c)\mathbf{v}$.

▶ EXAMPLE 2 Matrix Operators

If $T_A : R^n \to R^n$ is the matrix operator $T_A(\mathbf{x}) = A\mathbf{x}$, then it follows from parts (r) and (s) of Theorem 5.1.6 that T_A is one-to-one and onto if and only if A is invertible.

▶ EXAMPLE 3 Shifting Operators

Let $V = R^\infty$ be the sequence space discussed in Example 3 of Section 4.1, and consider the linear "shifting operators" on V defined by

$$T_1(u_1, u_2, \ldots, u_n, \ldots) = (0, u_1, u_2, \ldots, u_n, \ldots)$$
$$T_2(u_1, u_2, \ldots, u_n, \ldots) = (u_2, u_3, \ldots, u_n, \ldots)$$

(a) Show that T_1 is one-to-one but not onto.

(b) Show that T_2 is onto but not one to one.

Solution (a) The operator T_1 is one-to-one because distinct sequences in R^∞ obviously have distinct images. This operator is not onto because no vector in R^∞ maps into the sequence $(1, 0, 0, \ldots, 0, \ldots)$, for example.

Solution (b) The operator T_2 is not one-to-one because, for example, the vectors $(1, 0, 0, \ldots, 0, \ldots)$ and $(2, 0, 0, \ldots, 0, \ldots)$ both map into $(0, 0, 0, \ldots, 0, \ldots)$. This operator is onto because every possible sequence of real numbers can be obtained with an appropriate choice of the numbers $u_2, u_3, \ldots, u_n, \ldots$.

Why does Example 3 not violate Theorem 8.2.2?

▶ **EXAMPLE 4 Basic Transformations That Are One-to-One and Onto**

The linear transformations $T_1: P_3 \to R^4$ and $T_2: M_{22} \to R^4$ defined by

$$T_1(a + bx + cx^2 + dx^3) = (a, b, c, d)$$

$$T_2\left(\begin{bmatrix} a & b \\ c & d \end{bmatrix}\right) = (a, b, c, d)$$

are both one-to-one and onto (verify by showing that their kernels contain only the zero vector).

▶ **EXAMPLE 5 A One-to-One Linear Transformation**

Let $T: P_n \to P_{n+1}$ be the linear transformation

$$T(\mathbf{p}) = T(p(x)) = xp(x)$$

discussed in Example 5 of Section 8.1. If

$$\mathbf{p} = p(x) = c_0 + c_1 x + \cdots + c_n x^n \quad \text{and} \quad \mathbf{q} = q(x) = d_0 + d_1 x + \cdots + d_n x^n$$

are distinct polynomials, then they differ in at least one coefficient. Thus,

$$T(\mathbf{p}) = c_0 x + c_1 x^2 + \cdots + c_n x^{n+1} \quad \text{and} \quad T(\mathbf{q}) = d_0 x + d_1 x^2 + \cdots + d_n x^{n+1}$$

also differ in at least one coefficient. It follows that T is one-to-one since it maps distinct polynomials $\mathbf{p}$ and $\mathbf{q}$ into distinct polynomials $T(\mathbf{p})$ and $T(\mathbf{q})$.

CALCULUS REQUIRED

▶ **EXAMPLE 6 A Transformation That Is Not One-to-One**

Let

$$D: C^1(-\infty, \infty) \to F(-\infty, \infty)$$

be the differentiation transformation discussed in Example 11 of Section 8.1. This linear transformation is *not* one-to-one because it maps functions that differ by a constant into the same function. For example,

$$D(x^2) = D(x^2 + 1) = 2x \quad ◀$$

Dimension and Linear Transformations

In the exercises we will ask you to prove the following two important facts about a linear transformation $T: V \to W$ in the case where V and W are finite-dimensional:

1. If $\dim(W) < \dim(V)$, then T cannot be one-to-one.
2. If $\dim(V) < \dim(W)$, then T cannot be onto.

Stated informally, if a linear transformation maps a "bigger" space to a "smaller" space, then some points in the "bigger" space must have the same image; and if a linear transformation maps a "smaller" space to a "bigger" space, then there must be points in the "bigger" space that are not images of any points in the "smaller" space.

Remark These observations tell us, for example, that any linear transformation from R^3 to R^2 must map some distinct points of R^3 into the *same* point in R^2, and it also tells us that there is no linear transformation that maps R^2 onto *all* of R^3.

Isomorphism

Our next definition paves the way for the main result in this section.

DEFINITION 3 If a linear transformation $T: V \to W$ is both one-to-one and onto, then T is said to be an *isomorphism*, and the vector spaces V and W are said to be *isomorphic*.

The word *isomorphic* is derived from the Greek words *iso*, meaning "identical," and *morphe*, meaning "form." This terminology is appropriate because, as we will now explain, isomorphic vector spaces have the same "algebraic form," even though they may consist of different kinds of objects. To illustrate this idea, examine Table 1 in which we have shown how the isomorphism

$$a_0 + a_1 x + a_2 x^2 \xrightarrow{T} (a_0, a_1, a_2)$$

matches up vector operations in P_2 and R^3.

Table 1

Operation in P_2	Operation in R^3
$3(1 - 2x + 3x^2) = 3 - 6x + 9x^2$	$3(1, -2, 3) = (3, -6, 9)$
$(2 + x - x^2) + (1 - x + 5x^2) = 3 + 4x^2$	$(2, 1, -1) + (1, -1, 5) = (3, 0, 4)$
$(4 + 2x + 3x^2) - (2 - 4x + 3x^2) = 2 + 6x$	$(4, 2, 3) - (2, -4, 3) = (2, 6, 0)$

The following theorem, which is one of the most important results in linear algebra, reveals the fundamental importance of the vector space R^n.

THEOREM 8.2.3 *Every real n-dimensional vector space is isomorphic to R^n.*

Theorem 8.2.3 tells us that a real *n*-dimensional vector space may differ from R^n in notation, but its algebraic structure will be the same.

Proof Let V be a real *n*-dimensional vector space. To prove that V is isomorphic to R^n we must find a linear transformation $T : V \rightarrow R^n$ that is one-to-one and onto. For this purpose, let

$$\mathbf{v}_1, \mathbf{v}_2, \ldots, \mathbf{v}_n$$

be any basis for V, let

$$\mathbf{u} = k_1 \mathbf{v}_1 + k_2 \mathbf{v}_2 + \cdots + k_n \mathbf{v}_n \tag{1}$$

be the representation of a vector $\mathbf{u}$ in V as a linear combination of the basis vectors, and define the transformation $T : V \rightarrow R^n$ by

$$T(\mathbf{u}) = (k_1, k_2, \ldots, k_n) \tag{2}$$

We will show that T is an isomorphism (linear, one-to-one, and onto). To prove the linearity, let $\mathbf{u}$ and $\mathbf{v}$ be vectors in V, let c be a scalar, and let

$$\mathbf{u} = k_1 \mathbf{v}_1 + k_2 \mathbf{v}_2 + \cdots + k_n \mathbf{v}_n \quad \text{and} \quad \mathbf{v} = d_1 \mathbf{v}_1 + d_2 \mathbf{v}_2 + \cdots + d_n \mathbf{v}_n \tag{3}$$

be the representations of $\mathbf{u}$ and $\mathbf{v}$ as linear combinations of the basis vectors. Then it follows from (1) that

$$\begin{aligned} T(c\mathbf{u}) &= T(ck_1 \mathbf{v}_1 + ck_2 \mathbf{v}_2 + \cdots + ck_n \mathbf{v}_n) \\ &= (ck_1, ck_2, \ldots, ck_n) \\ &= c(k_1, k_2, \ldots, k_n) = cT(\mathbf{u}) \end{aligned}$$

and it follows from (2) that

$$\begin{aligned} T(\mathbf{u} + \mathbf{v}) &= T\big((k_1 + d_1)\mathbf{v}_1 + (k_2 + d_2)\mathbf{v}_2 + \cdots + (k_n + d_n)\mathbf{v}_n\big) \\ &= (k_1 + d_1, k_2 + d_2, \ldots, k_n + d_n) \\ &= (k_1, k_2, \ldots, k_n) + (d_1, d_2, \ldots, d_n) \\ &= T(\mathbf{u}) + T(\mathbf{v}) \end{aligned}$$

which shows that T is linear. To show that T is one-to-one, we must show that if $\mathbf{u}$ and $\mathbf{v}$ are distinct vectors in V, then so are their images in R^n. But if $\mathbf{u} \neq \mathbf{v}$, and if the representations of these vectors in terms of the basis vectors are as in (3), then we must have $k_i \neq d_i$ for at least one i. Thus,

$$T(\mathbf{u}) = (k_1, k_2, \ldots, k_n) \neq (d_1, d_2, \ldots, d_n) = T(\mathbf{v})$$

which shows that $\mathbf{u}$ and $\mathbf{v}$ have distinct images under T. Finally, the transformation T is onto, for if

$$\mathbf{w} = (k_1, k_2, \ldots, k_n)$$

is any vector in R^n, then it follows from (2) that $\mathbf{w}$ is the image under T of the vector

$$\mathbf{u} = k_1 \mathbf{v}_1 + k_2 \mathbf{v}_2 + \cdots + k_n \mathbf{v}_n \quad \blacktriangleleft$$

Remark Note that the isomorphism T in Formula (2) of the foregoing proof is the coordinate map

$$\mathbf{u} \xrightarrow{T} (k_1, k_2, \ldots, k_n) = (\mathbf{u})_S$$

that maps $\mathbf{u}$ into its coordinate vector with respect to the basis $S = \{\mathbf{v}_1, \mathbf{v}_2, \ldots, \mathbf{v}_n\}$. Since there are generally many possible bases for a given vector space V, there are generally many possible isomorphisms between V and R^n, one for each different basis.

▶ **EXAMPLE 7 The Natural Isomorphism from P_{n-1} to R^n**

We leave it for you to verify that the mapping

$$a_0 + a_1 x + \cdots + a_{n-1} x^{n-1} \xrightarrow{T} (a_0, a_1, \ldots, a_{n-1})$$

from P_{n-1} to R^n is one-to-one, onto, and linear. This is called the **natural isomorphism** from P_{n-1} to R^n because, as the following computations show, it maps the natural basis $\{1, x, x^2, \ldots, x^{n-1}\}$ for P_{n-1} into the standard basis for R^n:

$$1 = 1 + 0x + 0x^2 + \cdots + 0x^{n-1} \xrightarrow{T} (1, 0, 0, \ldots, 0)$$
$$x = 0 + x + 0x^2 + \cdots + 0x^{n-1} \xrightarrow{T} (0, 1, 0, \ldots, 0)$$
$$\vdots \qquad\qquad \vdots \qquad\qquad \vdots \qquad\qquad \vdots$$
$$x^{n-1} = 0 + 0x + 0x^2 + \cdots + x^{n-1} \xrightarrow{T} (0, 0, 0, \ldots, 1)$$

▶ **EXAMPLE 8 The Natural Isomorphism from M_{22} to R^4**

The matrices

$$E_1 = \begin{bmatrix} 1 & 0 \\ 0 & 0 \end{bmatrix}, \quad E_2 = \begin{bmatrix} 0 & 1 \\ 0 & 0 \end{bmatrix}, \quad E_3 = \begin{bmatrix} 0 & 0 \\ 1 & 0 \end{bmatrix}, \quad E_4 = \begin{bmatrix} 0 & 0 \\ 0 & 1 \end{bmatrix}$$

form a basis for the vector space M_{22} of 2×2 matrices. An isomorphism $T : M_{22} \to R^4$ can be constructed by first writing a matrix A in M_{22} in terms of the basis vectors as

$$A = \begin{bmatrix} a_1 & a_2 \\ a_3 & a_4 \end{bmatrix} = a_1 \begin{bmatrix} 1 & 0 \\ 0 & 0 \end{bmatrix} + a_2 \begin{bmatrix} 0 & 1 \\ 0 & 0 \end{bmatrix} + a_3 \begin{bmatrix} 0 & 0 \\ 1 & 0 \end{bmatrix} + a_4 \begin{bmatrix} 0 & 0 \\ 0 & 1 \end{bmatrix}$$

and then defining T as

$$T(A) = (a_1, a_2, a_3, a_4)$$

Thus, for example,

$$\begin{bmatrix} 1 & -3 \\ 4 & 6 \end{bmatrix} \xrightarrow{T} (1, -3, 4, 6)$$

More generally, this idea can be used to show that the vector space M_{mn} of $m \times n$ matrices with real entries is isomorphic to R^{mn}.

▶ **EXAMPLE 9 Differentiation by Matrix Multiplication**

Consider the differentiation transformation $D: P_3 \to P_2$ on the vector space of polynomials of degree three or less. If we map P_3 and P_2 into R^4 and R^3, respectively, by the natural isomorphisms, then the transformation D produces a corresponding matrix transformation from R^4 to R^3. Specifically, the derivative transformation

$$a_0 + a_1 x + a_2 x^2 + a_3 x^3 \xrightarrow{D} a_1 + 2a_2 x + 3a_3 x^2$$

produces the matrix transformation

$$\begin{bmatrix} 0 & 1 & 0 & 0 \\ 0 & 0 & 2 & 0 \\ 0 & 0 & 0 & 3 \end{bmatrix} \begin{bmatrix} a_0 \\ a_1 \\ a_2 \\ a_3 \end{bmatrix} = \begin{bmatrix} a_1 \\ 2a_2 \\ 3a_3 \end{bmatrix}$$

Thus, for example, the derivative

$$\frac{d}{dx}(2 + x + 4x^2 - x^3) = 1 + 8x - 3x^2$$

can be calculated as the matrix product

$$\begin{bmatrix} 0 & 1 & 0 & 0 \\ 0 & 0 & 2 & 0 \\ 0 & 0 & 0 & 3 \end{bmatrix} \begin{bmatrix} 2 \\ 1 \\ 4 \\ -1 \end{bmatrix} = \begin{bmatrix} 1 \\ 8 \\ -3 \end{bmatrix}$$

This idea is useful for constructing numerical algorithms to perform derivative calculations. ◀

Inner Product Space
Isomorphisms

In the case where V is a real n-dimensional inner product space, both V and R^n have, in addition to their algebraic structure, a geometric structure arising from their respective inner products. Thus, it is reasonable to inquire if there exists an isomorphism from V to R^n that preserves the geometric structure as well as the algebraic structure. For example, we would want orthogonal vectors in V to have orthogonal counterparts in R^n, and we would want orthonormal sets in V to correspond to orthonormal sets in R^n.

In order for an isomorphism to preserve geometric structure, it obviously has to preserve inner products, since notions of length, angle, and orthogonality are all based on the inner product. Thus, if V and W are inner product spaces, then we call an isomorphism $T: V \to W$ an ***inner product space isomorphism*** if

$$\langle T(\mathbf{u}), T(\mathbf{v}) \rangle = \langle \mathbf{u}, \mathbf{v} \rangle$$

It can be proved that if V is any real n-dimensional inner product space and R^n has the Euclidean inner product (the dot product), then there exists an inner product space isomorphism from V to R^n. Under such an isomorphism, the inner product space V has the same algebraic and geometric structure as R^n. In this sense, every n-dimensional inner product space is a "carbon copy" of R^n with the Euclidean inner product that differs only in the notation used to represent vectors.

▶ **EXAMPLE 10 An Inner Product Space Isomorphism**

Let R^n be the vector space of real n-tuples in comma-delimited form, let M_n be the vector space of real $n \times 1$ matrices, let R^n have the Euclidean inner product $\langle \mathbf{u}, \mathbf{v} \rangle = \mathbf{u} \cdot \mathbf{v}$, and let M_n have the inner product $\langle \mathbf{u}, \mathbf{v} \rangle = \mathbf{u}^T \mathbf{v}$ in which $\mathbf{u}$ and $\mathbf{v}$ are expressed in column form. The mapping $T: R^n \to M_n$ defined by

$$(v_1, v_2, \ldots, v_n) \xrightarrow{T} \begin{bmatrix} v_1 \\ v_2 \\ \vdots \\ v_n \end{bmatrix}$$

is an inner product space isomorphism, so the distinction between the inner product space R^n and the inner product space M_n is essentially notational, a fact that we have used many times in this text. ◀

Concept Review

- One-to-one
- Onto
- Isomorphism
- Isomorphic vector spaces
- Natural isomorphism
- Inner product space isomorphism

Skills

- Determine whether a linear transformation is one-to-one.
- Determine whether a linear transformation is onto.
- Determine whether a linear transformation is an isomorphism.

Exercise Set 8.2

1. In each part, find $\ker(T)$, and determine whether the linear transformation T is one-to-one.

   (a) $T: R^2 \to R^2$, where $T(x, y) = (y, x)$

   (b) $T: R^2 \to R^2$, where $T(x, y) = (0, 2x + 3y)$

   (c) $T: R^2 \to R^2$, where $T(x, y) = (x + y, x - y)$

   (d) $T: R^2 \to R^3$, where $T(x, y) = (x, y, x + y)$

   (e) $T: R^2 \to R^3$, where $T(x, y) = (x - y, y - x, 2x - 2y)$

   (f) $T: R^3 \to R^2$, where $T(x, y, z) = (x + y + z, x - y - z)$

2. Which of the transformations in Exercise 1 are onto?

3. In each part, determine whether multiplication by A is a one-to-one linear transformation.

   (a) $A = \begin{bmatrix} 1 & -2 \\ 2 & -4 \\ -3 & 6 \end{bmatrix}$

   (b) $A = \begin{bmatrix} 1 & 3 & 5 & 7 \\ 2 & -1 & 2 & 4 \\ -1 & 3 & 0 & 0 \end{bmatrix}$

   (c) $A = \begin{bmatrix} 4 & -2 \\ 1 & 5 \\ 5 & 3 \end{bmatrix}$

4. Which of the transformations in Exercise 3 are onto?

5. As indicated in the accompanying figure, let $T: R^2 \to R^2$ be the orthogonal projection on the line $y = x$.

   (a) Find the kernel of T.

   (b) Is T one-to-one? Justify your conclusion.

◀ Figure Ex-5

6. As indicated in the accompanying figure, let $T: R^2 \to R^2$ be the linear operator that reflects each point about the y-axis.

   (a) Find the kernel of T.

   (b) Is T one-to-one? Justify your conclusion.

◀ Figure Ex-6

7. In each part, use the given information to determine whether the linear transformation T is one-to-one.

   (a) $T: R^m \to R^m$; nullity$(T) = 0$

   (b) $T: R^n \to R^n$; rank$(T) = n - 1$

   (c) $T: R^m \to R^n$; $n < m$

   (d) $T: R^n \to R^n$; $R(T) = R^n$

8. In each part, determine whether the linear transformation T is one-to-one.

(a) $T: P_2 \rightarrow P_3$, where $T(a_0 + a_1 x + a_2 x^2) =$
$$x(a_0 + a_1 x + a_2 x^2)$$

(b) $T: P_2 \rightarrow P_2$, where $T(p(x)) = p(x+1)$

9. Prove: If V and W are finite-dimensional vector spaces such that $\dim(W) < \dim(V)$, then there is no one-to-one linear transformation $T: V \rightarrow W$.

10. Prove: There can be an onto linear transformation from V to W only if $\dim(V) \geq \dim(W)$.

11. (a) Find an isomorphism between the vector space of all 3×3 symmetric matrices and R^6.

(b) Find two different isomorphisms between the vector space of all 2×2 matrices and R^4.

(c) Find an isomorphism between the vector space of all polynomials of degree at most 3 such that $p(0) = 0$ and R^3.

(d) Find an isomorphism between the vector spaces $\text{span}\{1, \sin(x), \cos(x)\}$ and R^3.

12. (*Calculus required*) Let $J: P_1 \rightarrow R$ be the integration transformation $J(\mathbf{p}) = \int_{-1}^{1} p(x)\, dx$. Determine whether J is one-to-one. Justify your conclusion.

13. (*Calculus required*) Let V be the vector space $C^1[0, 1]$ and let $T: V \rightarrow R$ be defined by

$$T(\mathbf{f}) = f(0) + 2f'(0) + 3f'(1)$$

Verify that T is a linear transformation. Determine whether T is one-to-one, and justify your conclusion.

14. (*Calculus required*) Devise a method for using matrix multiplication to differentiate functions in the vector space $\text{span}\{1, \sin(x), \cos(x), \sin(2x), \cos(2x)\}$. Use your method to find the derivative of $3 - 4\sin(x) + \sin(2x) + 5\cos(2x)$.

15. Does the formula $T(a, b, c) = ax^2 + bx + c$ define a one-to-one linear transformation from R^3 to P_2? Explain your reasoning.

16. Let E be a fixed 2×2 elementary matrix. Does the formula $T(A) = EA$ define a one-to-one linear operator on M_{22}? Explain your reasoning.

17. Let $\mathbf{a}$ be a fixed vector in R^3. Does the formula $T(\mathbf{v}) = \mathbf{a} \times \mathbf{v}$ define a one-to-one linear operator on R^3? Explain your reasoning.

18. Prove that an inner product space isomorphism preserves angles and distances—that is, the angle between $\mathbf{u}$ and $\mathbf{v}$ in V is equal to the angle between $T(\mathbf{u})$ and $T(\mathbf{v})$ in W, and $\|\mathbf{u} - \mathbf{v}\|_V = \|T(\mathbf{u}) - T(\mathbf{v})\|_W$.

19. Does an inner product space isomorphism map orthonormal sets to orthonormal sets? Justify your answer.

20. Find an inner product space isomorphism between P_5 and M_{23}.

True-False Exercises

In parts (a)–(f) determine whether the statement is true or false, and justify your answer.

(a) The vector spaces R^2 and P_2 are isomorphic.

(b) If the kernel of a linear transformation $T: P_3 \rightarrow P_3$ is $\{0\}$, then T is an isomorphism.

(c) Every linear transformation from M_{33} to P_9 is an isomorphism.

(d) There is a subspace of M_{23} that is isomorphic to R^4.

(e) There is a 2×2 matrix P such that $T: M_{22} \rightarrow M_{22}$ defined by $T(A) = AP - PA$ is an isomorphism.

(f) There is a linear transformation $T: P_4 \rightarrow P_4$ such that the kernel of T is isomorphic to the range of T.

8.3 Compositions and Inverse Transformations

In Section 4.10 we discussed compositions and inverses of matrix transformations. In this section we will extend some of those ideas to general linear transformations.

Composition of Linear Transformations

The following definition extends Formula (1) of Section 4.10 to general linear transformations.

Note that the word "with" establishes the order of the operations in a composition. The composition of T_2 *with* T_1 is

$$(T_2 \circ T_1)(\mathbf{u}) = T_2(T_1(\mathbf{u}))$$

whereas the composition of T_1 *with* T_2 is

$$(T_1 \circ T_2)(\mathbf{u}) = T_1(T_2(\mathbf{u}))$$

DEFINITION 1 If $T_1: U \rightarrow V$ and $T_2: V \rightarrow W$ are linear transformations, then the *composition of T_2 with T_1*, denoted by $T_2 \circ T_1$ (which is read "T_2 circle T_1"), is the function defined by the formula

$$(T_2 \circ T_1)(\mathbf{u}) = T_2(T_1(\mathbf{u})) \tag{1}$$

where $\mathbf{u}$ is a vector in U.

Remark Observe that this definition requires that the domain of T_2 (which is V) contain the range of T_1. This is essential for the formula $T_2(T_1(\mathbf{u}))$ to make sense (Figure 8.3.1).

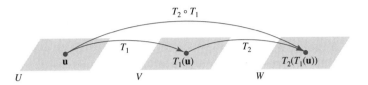

► **Figure 8.3.1** The composition of T_2 with T_1.

Our first theorem shows that the composition of two linear transformations is itself a linear transformation.

THEOREM 8.3.1 *If* $T_1: U \to V$ *and* $T_2: V \to W$ *are linear transformations, then* $(T_2 \circ T_1): U \to W$ *is also a linear transformation.*

Proof If $\mathbf{u}$ and $\mathbf{v}$ are vectors in U and c is a scalar, then it follows from (1) and the linearity of T_1 and T_2 that

$$(T_2 \circ T_1)(\mathbf{u} + \mathbf{v}) = T_2(T_1(\mathbf{u} + \mathbf{v})) = T_2(T_1(\mathbf{u}) + T_1(\mathbf{v}))$$
$$= T_2(T_1(\mathbf{u})) + T_2(T_1(\mathbf{v}))$$
$$= (T_2 \circ T_1)(\mathbf{u}) + (T_2 \circ T_1)(\mathbf{v})$$

and

$$(T_2 \circ T_1)(c\mathbf{u}) = T_2(T_1(c\mathbf{u})) = T_2(cT_1(\mathbf{u}))$$
$$= cT_2(T_1(\mathbf{u})) = c(T_2 \circ T_1)(\mathbf{u})$$

Thus, $T_2 \circ T_1$ satisfies the two requirements of a linear transformation. ◄

► **EXAMPLE 1 Composition of Linear Transformations**

Let $T_1: P_1 \to P_2$ and $T_2: P_2 \to P_2$ be the linear transformations given by the formulas

$$T_1(p(x)) = xp(x) \quad \text{and} \quad T_2(p(x)) = p(2x + 4)$$

Then the composition $(T_2 \circ T_1): P_1 \to P_2$ is given by the formula

$$(T_2 \circ T_1)(p(x)) = T_2(T_1(p(x))) = T_2(xp(x)) = (2x + 4)p(2x + 4)$$

In particular, if $p(x) = c_0 + c_1 x$, then

$$(T_2 \circ T_1)(p(x)) = (T_2 \circ T_1)(c_0 + c_1 x) = (2x + 4)(c_0 + c_1(2x + 4))$$
$$= c_0(2x + 4) + c_1(2x + 4)^2$$

► **EXAMPLE 2 Composition with the Identity Operator**

If $T: V \to V$ is any linear operator, and if $I: V \to V$ is the identity operator (Example 1 of Section 8.1), then for all vectors $\mathbf{v}$ in V, we have

$$(T \circ I)(\mathbf{v}) = T(I(\mathbf{v})) = T(\mathbf{v})$$
$$(I \circ T)(\mathbf{v}) = I(T(\mathbf{v})) = T(\mathbf{v})$$

It follows that $T \circ I$ and $I \circ T$ are the same as T; that is,

$$T \circ I = T \quad \text{and} \quad I \circ T = T \quad ◄ \qquad (2)$$

As illustrated in Figure 8.3.2, compositions can be defined for more than two linear transformations. For example, if

$$T_1: U \to V, \quad T_2: V \to W, \quad \text{and} \quad T_3: W \to Y$$

are linear transformations, then the composition $T_3 \circ T_2 \circ T_1$ is defined by

$$(T_3 \circ T_2 \circ T_1)(\mathbf{u}) = T_3(T_2(T_1(\mathbf{u}))) \tag{3}$$

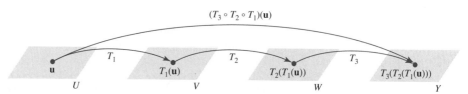

▲ **Figure 8.3.2** The composition of three linear transformations.

Inverse Linear Transformations

In Theorem 4.10.1 we showed that a matrix operator $T_A: R^n \to R^n$ is one-to-one if and only if the matrix A is invertible, in which case the inverse operator is $T_{A^{-1}}$. We then showed that if $\mathbf{w}$ is the image of a vector $\mathbf{x}$ under the operator T_A, then $\mathbf{x}$ is the image under $T_{A^{-1}}$ of the vector $\mathbf{w}$ (see Figure 4.10.8). Our next objective is to extend the notion of invertibility to general linear transformations.

Recall that if $T: V \to W$ is a linear transformation, then the range of T, denoted by $R(T)$, is the subspace of W consisting of all images under T of vectors in V. If T is one-to-one, then each vector $\mathbf{w}$ in $R(T)$ is the image of a unique vector $\mathbf{v}$ in V. This uniqueness allows us to define a new function, called the ***inverse of*** T and denoted by T^{-1}, that maps $\mathbf{w}$ back into $\mathbf{v}$ (Figure 8.3.3).

▶ **Figure 8.3.3** The inverse of T maps $T(\mathbf{v})$ back into $\mathbf{v}$.

It can be proved (Exercise 19) that $T^{-1}: R(T) \to V$ is a linear transformation. More-over, it follows from the definition of T^{-1} that

$$T^{-1}(T(\mathbf{v})) = T^{-1}(\mathbf{w}) = \mathbf{v} \tag{4}$$

$$T(T^{-1}(\mathbf{w})) = T(\mathbf{v}) = \mathbf{w} \tag{5}$$

so that T and T^{-1}, when applied in succession in either order, cancel the effect of each other.

Remark It is important to note that if $T: V \to W$ is a one-to-one linear transformation, then the domain of T^{-1} is the *range* of T, where the range may or may not be all of W. However, in the special case where $T: V \to V$ is a one-to-one *linear operator* and V is n-dimensional, then it follows from Theorem 8.2.2 that T must also be onto, so the domain of T^{-1} is *all* of V.

▶ **EXAMPLE 3 An Inverse Transformation**

In Example 5 of Section 8.2 we showed that the linear transformation $T : P_n \to P_{n+1}$ given by

$$T(\mathbf{p}) = T(p(x)) = xp(x)$$

is one-to-one; thus, T has an inverse. In this case the range of T is *not* all of P_{n+1} but rather the subspace of P_{n+1} consisting of polynomials with a zero constant term. This is evident from the formula for T:

$$T(c_0 + c_1x + \cdots + c_nx^n) = c_0x + c_1x^2 + \cdots + c_nx^{n+1}$$

It follows that $T^{-1} : R(T) \to P_n$ is given by the formula

$$T^{-1}(c_0x + c_1x^2 + \cdots + c_nx^{n+1}) = c_0 + c_1x + \cdots + c_nx^n$$

For example, in the case where $n \geq 3$,

$$T^{-1}(2x - x^2 + 5x^3 + 3x^4) = 2 - x + 5x^2 + 3x^3$$

▶ **EXAMPLE 4 An Inverse Transformation**

Let $T : R^3 \to R^3$ be the linear operator defined by the formula

$$T(x_1, x_2, x_3) = (3x_1 + x_2, -2x_1 - 4x_2 + 3x_3, 5x_1 + 4x_2 - 2x_3)$$

Determine whether T is one-to-one; if so, find $T^{-1}(x_1, x_2, x_3)$.

Solution It follows from Formula (12) of Section 4.9 that the standard matrix for T is

$$[T] = \begin{bmatrix} 3 & 1 & 0 \\ -2 & -4 & 3 \\ 5 & 4 & -2 \end{bmatrix}$$

(verify). This matrix is invertible, and from Formula (8) of Section 4.10 the standard matrix for T^{-1} is

$$[T^{-1}] = [T]^{-1} = \begin{bmatrix} 4 & -2 & -3 \\ -11 & 6 & 9 \\ -12 & 7 & 10 \end{bmatrix}$$

It follows that

$$T^{-1}\left(\begin{bmatrix} x_1 \\ x_2 \\ x_3 \end{bmatrix} \right) = [T^{-1}] \begin{bmatrix} x_1 \\ x_2 \\ x_3 \end{bmatrix} = \begin{bmatrix} 4 & -2 & -3 \\ -11 & 6 & 9 \\ -12 & 7 & 10 \end{bmatrix} \begin{bmatrix} x_1 \\ x_2 \\ x_3 \end{bmatrix} = \begin{bmatrix} 4x_1 - 2x_2 - 3x_3 \\ -11x_1 + 6x_2 + 9x_3 \\ -12x_1 + 7x_2 + 10x_3 \end{bmatrix}$$

Expressing this result in horizontal notation yields

$$T^{-1}(x_1, x_2, x_3) = (4x_1 - 2x_2 - 3x_3, -11x_1 + 6x_2 + 9x_3, -12x_1 + 7x_2 + 10x_3) \blacktriangleleft$$

Composition of One-To-One Linear Transformations

The following theorem shows that a composition of one-to-one linear transformations is one-to-one, and it relates the inverse of a composition to the inverses of its individual linear transformations.

THEOREM 8.3.2 *If $T_1 : U \to V$ and $T_2 : V \to W$ are one-to-one linear transformations, then*

(a) $T_2 \circ T_1$ *is one-to-one.*

(b) $(T_2 \circ T_1)^{-1} = T_1^{-1} \circ T_2^{-1}$.

Proof (a) We want to show that $T_2 \circ T_1$ maps distinct vectors in U into distinct vectors in W. But if $\mathbf{u}$ and $\mathbf{v}$ are distinct vectors in U, then $T_1(\mathbf{u})$ and $T_1(\mathbf{v})$ are distinct vectors in V since T_1 is one-to-one. This and the fact that T_2 is one-to-one imply that

$$T_2(T_1(\mathbf{u})) \quad \text{and} \quad T_2(T_1(\mathbf{v}))$$

are also distinct vectors. But these expressions can also be written as

$$(T_2 \circ T_1)(\mathbf{u}) \quad \text{and} \quad (T_2 \circ T_1)(\mathbf{v})$$

so $T_2 \circ T_1$ maps $\mathbf{u}$ and $\mathbf{v}$ into distinct vectors in W.

Proof (b) We want to show that

$$(T_2 \circ T_1)^{-1}(\mathbf{w}) = (T_1^{-1} \circ T_2^{-1})(\mathbf{w})$$

for every vector $\mathbf{w}$ in the range of $T_2 \circ T_1$. For this purpose, let

$$\mathbf{u} = (T_2 \circ T_1)^{-1}(\mathbf{w}) \tag{6}$$

so our goal is to show that

$$\mathbf{u} = (T_1^{-1} \circ T_2^{-1})(\mathbf{w})$$

But it follows from (6) that

$$(T_2 \circ T_1)(\mathbf{u}) = \mathbf{w}$$

or, equivalently,

$$T_2(T_1(\mathbf{u})) = \mathbf{w}$$

Now, taking T_2^{-1} of each side of this equation, then taking T_1^{-1} of each side of the result, and then using (4) yields (verify)

$$\mathbf{u} = T_1^{-1}(T_2^{-1}(\mathbf{w}))$$

or, equivalently,

$$\mathbf{u} = (T_1^{-1} \circ T_2^{-1})(\mathbf{w}) \blacktriangleleft$$

In words, part (b) of Theorem 8.3.2 states that *the inverse of a composition is the composition of the inverses in the reverse order*. This result can be extended to compositions of three or more linear transformations; for example,

$$(T_3 \circ T_2 \circ T_1)^{-1} = T_1^{-1} \circ T_2^{-1} \circ T_3^{-1} \tag{7}$$

In the case where T_A, T_B, and T_C are matrix operators on R^n, Formula (7) can be written as

$$(T_C \circ T_B \circ T_A)^{-1} = T_A^{-1} \circ T_B^{-1} \circ T_C^{-1}$$

Note the order of the subscripts on the two sides of Formula (8).

or alternatively as

$$(T_{CBA})^{-1} = T_{A^{-1}B^{-1}C^{-1}} \tag{8}$$

Concept Review
- Composition of linear transformations
- Inverse of a linear transformation

Skills
- Find the domain and range of the composition of two linear transformations.
- Find the composition of two linear transformations.
- Determine whether a linear transformation has an inverse.
- Find the inverse of a linear transformation.

Exercise Set 8.3

1. Find $(T_2 \circ T_1)(x, y)$.

(a) $T_1(x, y) = (2x, 3y),\ T_2(x, y) = (x - y, x + y)$

(b) $T_1(x, y) = (x - 3y, 0),\ T_2(x, y) = (4x - 5y, 3x - 6y)$

(c) $T_1(x, y) = (2x, -3y, x + y),$
$T_2(x, y, z) = (x - y, y + z)$

(d) $T_1(x, y) = (x - y, y + z, x - z),$
$T_2(x, y, z) = (0, x + y + z)$

2. Find $(T_3 \circ T_2 \circ T_1)(x, y)$.

(a) $T_1(x, y) = (-2y, 3x, x - 2y),\ T_2(x, y, z) = (y, z, x),$
$T_3(x, y, z) = (x + z, y - z)$

(b) $T_1(x, y) = (x + y, y, -x),$
$T_2(x, y, z) = (0, x + y + z, 3y),$
$T_3(x, y, z) = (3x + 2y, 4z - x - 3y)$

3. Let $T_1: M_{22} \to R$ and $T_2: M_{22} \to M_{22}$ be the linear transformations given by $T_1(A) = \operatorname{tr}(A)$ and $T_2(A) = A^T$.

(a) Find $(T_1 \circ T_2)(A)$, where $A = \begin{bmatrix} a & b \\ c & d \end{bmatrix}$.

(b) Can you find $(T_2 \circ T_1)(A)$? Explain.

4. Let $T_1: P_n \to P_n$ and $T_2: P_n \to P_n$ be the linear operators given by $T_1(p(x)) = p(x - 1)$ and $T_2(p(x)) = p(x + 1)$. Find $(T_1 \circ T_2)(p(x))$ and $(T_2 \circ T_1)(p(x))$.

5. Let $T_1: V \to V$ be the dilation $T_1(\mathbf{v}) = 4\mathbf{v}$. Find a linear operator $T_2: V \to V$ such that $T_1 \circ T_2 = I$ and $T_2 \circ T_1 = I$.

6. Suppose that the linear transformations $T_1: P_2 \to P_2$ and $T_2: P_2 \to P_3$ are given by the formulas $T_1(p(x)) = p(x + 1)$ and $T_2(p(x)) = xp(x)$. Find $(T_2 \circ T_1)(a_0 + a_1 x + a_2 x^2)$.

7. Let $q_0(x)$ be a fixed polynomial of degree m, and define a function T with domain P_n by the formula $T(p(x)) = p(q_0(x))$. Show that T is a linear transformation.

8. Use the definition of $T_3 \circ T_2 \circ T_1$ given by Formula (3) to prove that

(a) $T_3 \circ T_2 \circ T_1$ is a linear transformation.

(b) $T_3 \circ T_2 \circ T_1 = (T_3 \circ T_2) \circ T_1$.

(c) $T_3 \circ T_2 \circ T_1 = T_3 \circ (T_2 \circ T_1)$.

9. Let $T: R^3 \to R^3$ be the orthogonal projection of R^3 onto the xy-plane. Show that $T \circ T = T$.

10. In each part, let $T: R^2 \to R^2$ be multiplication by A. Determine whether T has an inverse; if so, find

$$T^{-1}\left(\begin{bmatrix} x_1 \\ x_2 \end{bmatrix}\right)$$

(a) $A = \begin{bmatrix} 5 & 2 \\ 2 & 1 \end{bmatrix}$
(b) $A = \begin{bmatrix} 6 & -3 \\ 4 & -2 \end{bmatrix}$
(c) $A = \begin{bmatrix} 4 & 7 \\ -1 & 3 \end{bmatrix}$

11. In each part, let $T: R^3 \to R^3$ be multiplication by A. Determine whether T has an inverse; if so, find

$$T^{-1}\left(\begin{bmatrix} x_1 \\ x_2 \\ x_3 \end{bmatrix}\right)$$

(a) $A = \begin{bmatrix} 1 & 5 & 2 \\ 1 & 2 & 1 \\ -1 & 1 & 0 \end{bmatrix}$
(b) $A = \begin{bmatrix} 1 & 4 & -1 \\ 1 & 2 & 1 \\ -1 & 1 & 0 \end{bmatrix}$

(c) $A = \begin{bmatrix} 1 & 0 & 1 \\ 0 & 1 & 1 \\ 1 & 1 & 0 \end{bmatrix}$
(d) $A = \begin{bmatrix} 1 & -1 & 1 \\ 0 & 2 & -1 \\ 2 & 3 & 0 \end{bmatrix}$

12. In each part, determine whether the linear operator $T: R^n \to R^n$ is one-to-one; if so, find $T^{-1}(x_1, x_2, \ldots, x_n)$.

(a) $T(x_1, x_2, \ldots, x_n) = (0, x_1, x_2, \ldots, x_{n-1})$

(b) $T(x_1, x_2, \ldots, x_n) = (x_n, x_{n-1}, \ldots, x_2, x_1)$

(c) $T(x_1, x_2, \ldots, x_n) = (x_2, x_3, \ldots, x_n, x_1)$

13. Let $T: R^n \to R^n$ be the linear operator defined by the formula

$$T(x_1, x_2, \ldots, x_n) = (a_1 x_1, a_2 x_2, \ldots, a_n x_n)$$

where $a_1, \ldots, a_n$ are constants.

(a) Under what conditions will T have an inverse?

(b) Assuming that the conditions determined in part (a) are satisfied, find a formula for $T^{-1}(x_1, x_2, \ldots, x_n)$.

14. Let $T_1: R^2 \to R^2$ and $T_2: R^2 \to R^2$ be the linear operators given by the formulas

$$T_1(x, y) = (x + y, x - y) \quad \text{and} \quad T_2(x, y) = (2x + y, x - 2y)$$

(a) Show that T_1 and T_2 are one-to-one.

(b) Find formulas for

$$T_1^{-1}(x, y), \quad T_2^{-1}(x, y), \quad (T_2 \circ T_1)^{-1}(x, y)$$

(c) Verify that $(T_2 \circ T_1)^{-1} = T_1^{-1} \circ T_2^{-1}$.

15. Let $T_1: P_2 \to P_3$ and $T_2: P_3 \to P_3$ be the linear transformations given by the formulas

$$T_1(p(x)) = xp(x) \quad \text{and} \quad T_2(p(x)) = p(x + 1)$$

(a) Find formulas for $T_1^{-1}(p(x)), T_2^{-1}(p(x))$, and $(T_2 \circ T_1)^{-1}(p(x))$.

(b) Verify that $(T_2 \circ T_1)^{-1} = T_1^{-1} \circ T_2^{-1}$.

16. Let $T_A: R^3 \to R^3$, $T_B: R^3 \to R^3$, and $T_C: R^3 \to R^3$ be the reflections about the xy-plane, the xz-plane, and the yz-plane, respectively. Verify Formula (8) for these linear operators.

17. Let $T : P_1 \to R^2$ be the function defined by the formula

$$T(p(x)) = (p(0), p(1)).$$

  (a) Find $T(1 - 2x)$.

  (b) Show that T is a linear transformation.

  (c) Show that T is one-to-one.

  (d) Find $T^{-1}(2, 3)$, and sketch its graph.

18. Let $T : R^2 \to R^2$ be the linear operator given by the formula $T(x, y) = (x + ky, -y)$. Show that T is one-to-one and that $T^{-1} = T$ for every real value of k.

19. Prove: If $T : V \to W$ is a one-to-one linear transformation, then $T^{-1} : R(T) \to V$ is a one-to-one linear transformation.

▶ In Exercises 20–21, determine whether $T_1 \circ T_2 = T_2 \circ T_1$. ◀

20. (a) $T_1 : R^2 \to R^2$ is the orthogonal projection on the x-axis, and $T_2 : R^2 \to R^2$ is the orthogonal projection on the y-axis.

  (b) $T_1 : R^2 \to R^2$ is the rotation about the origin through an angle θ_1, and $T_2 : R^2 \to R^2$ is the rotation about the origin through an angle θ_2.

  (c) $T_1 : R^3 \to R^3$ is the rotation about the x-axis through an angle θ_1, and $T_2 : R^3 \to R^3$ is the rotation about the z-axis through an angle θ_2.

21. (a) $T_1 : R^2 \to R^2$ is the reflection about the x-axis, and $T_2 : R^2 \to R^2$ is the reflection about the y-axis.

  (b) $T_1 : R^2 \to R^2$ is the orthogonal projection on the x-axis, and $T_2 : R^2 \to R^2$ is the counterclockwise rotation through an angle θ.

  (c) $T_1 : R^3 \to R^3$ is a dilation by a factor k, and $T_2 : R^3 \to R^3$ is the counterclockwise rotation about the z-axis through an angle θ.

22. (*Calculus required*) Let

$$D(\mathbf{f}) = f'(x) \quad \text{and} \quad J(\mathbf{f}) = \int_0^x f(t)\, dt$$

be the linear transformations in Examples 9 and 10 of Sec-

tion 8.1. Find $(J \circ D)(\mathbf{f})$ for

  (a) $\mathbf{f}(x) = x^2 + 3x + 2$ (b) $\mathbf{f}(x) = \sin x$

  (c) $\mathbf{f}(x) = e^x + 3$

23. (*Calculus required*) The Fundamental Theorem of Calculus implies that integration and differentiation reverse the actions of each other. Define a transformation $D : P_n \to P_{n-1}$ by $D(p(x)) = p'(x)$, and define $J : P_{n-1} \to P_n$ by

$$J(p(x)) = \int_0^x p(t)\, dt$$

  (a) Show that D and J are linear transformations.

  (b) Explain why J is not the inverse transformation of D.

  (c) Can the domains and/or codomains of D and J be restricted so they are inverse linear transformations?

True-False Exercises

In parts (a)–(f) determine whether the statement is true or false, and justify your answer.

  (a) The composition of two linear transformations is also a linear transformation.

  (b) If $T_1 : V \to V$ and $T_2 : V \to V$ are any two linear operators, then $T_1 \circ T_2 = T_2 \circ T_1$.

  (c) The inverse of a linear transformation is a linear transformation.

  (d) If a linear transformation T has an inverse, then the kernel of T is the zero subspace.

  (e) If $T : R^2 \to R^2$ is the orthogonal projection onto the x-axis, then $T^{-1} : R^2 \to R^2$ maps each point on the x-axis onto a line that is perpendicular to the x-axis.

  (f) If $T_1 : U \to V$ and $T_2 : V \to W$ are linear transformations, and if T_1 is not one-to-one, then neither is $T_2 \circ T_1$.

8.4 Matrices for General Linear Transformations

In this section we will show that a general linear transformation from any n-dimensional vector space V to any m-dimensional vector space W can be performed using an appropriate matrix transformation from R^n to R^m. This idea is used in computer computations since computers are well suited for performing matrix computations.

Matrices of Linear Transformations

Suppose that V is an n-dimensional vector space, W is an m-dimensional vector space, and that $T : V \to W$ is a linear transformation. Suppose further that B is a basis for V, that B' is a basis for W, and that for each vector $\mathbf{x}$ in V, the coordinate matrices for $\mathbf{x}$ and $T(\mathbf{x})$ are $[\mathbf{x}]_B$ and $[T(\mathbf{x})]_{B'}$, respectively (Figure 8.4.1).

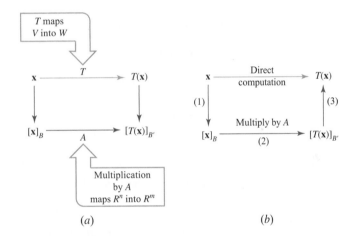

◄ **Figure 8.4.1**

It will be our goal to find an $m \times n$ matrix A such that multiplication by A maps the vector $[\mathbf{x}]_B$ into the vector $[T(\mathbf{x})]_{B'}$ for each $\mathbf{x}$ in V (Figure 8.4.2a). If we can do so, then, as illustrated in Figure 8.4.2b, we will be able to execute the linear transformation T by using matrix multiplication and the following *indirect* procedure:

Finding $T(\mathbf{x})$ Indirectly

Step 1. Compute the coordinate vector $[\mathbf{x}]_B$.

Step 2. Multiply $[\mathbf{x}]_B$ on the left by A to produce $[T(\mathbf{x})]_{B'}$.

Step 3. Reconstruct $T(\mathbf{x})$ from its coordinate vector $[T(\mathbf{x})]_{B'}$.

◄ **Figure 8.4.2**

(a)   *(b)*

The key to executing this plan is to find an $m \times n$ matrix A with the property that

$$A[\mathbf{x}]_B = [T(\mathbf{x})]_{B'} \qquad (1)$$

For this purpose, let $B = \{\mathbf{u}_1, \mathbf{u}_2, \ldots, \mathbf{u}_n\}$ be a basis for the n-dimensional space V and $B' = \{\mathbf{v}_1, \mathbf{v}_2, \ldots, \mathbf{v}_m\}$ a basis for the m-dimensional space W. Since Equation (1) must hold for all vectors in V, it must hold, in particular, for the basis vectors in B; that is,

$$A[\mathbf{u}_1]_B = [T(\mathbf{u}_1)]_{B'}, \quad A[\mathbf{u}_2]_B = [T(\mathbf{u}_2)]_{B'}, \ldots, \quad A[\mathbf{u}_n]_B = [T(\mathbf{u}_n)]_{B'} \qquad (2)$$

But

$$[\mathbf{u}_1]_B = \begin{bmatrix} 1 \\ 0 \\ 0 \\ \vdots \\ 0 \end{bmatrix}, \quad [\mathbf{u}_2]_B = \begin{bmatrix} 0 \\ 1 \\ 0 \\ \vdots \\ 0 \end{bmatrix}, \ldots, \quad [\mathbf{u}_n]_B = \begin{bmatrix} 0 \\ 0 \\ 0 \\ \vdots \\ 1 \end{bmatrix}$$

so

$$A[\mathbf{u}_1]_B = \begin{bmatrix} a_{11} & a_{12} & \cdots & a_{1n} \\ a_{21} & a_{22} & \cdots & a_{2n} \\ \vdots & \vdots & & \vdots \\ a_{m1} & a_{m2} & \cdots & a_{mn} \end{bmatrix} \begin{bmatrix} 1 \\ 0 \\ 0 \\ \vdots \\ 0 \end{bmatrix} = \begin{bmatrix} a_{11} \\ a_{21} \\ \vdots \\ a_{m1} \end{bmatrix}$$

$$A[\mathbf{u}_2]_B = \begin{bmatrix} a_{11} & a_{12} & \cdots & a_{1n} \\ a_{21} & a_{22} & \cdots & a_{2n} \\ \vdots & \vdots & & \vdots \\ a_{m1} & a_{m2} & \cdots & a_{mn} \end{bmatrix} \begin{bmatrix} 0 \\ 1 \\ 0 \\ \vdots \\ 0 \end{bmatrix} = \begin{bmatrix} a_{12} \\ a_{22} \\ \vdots \\ a_{m2} \end{bmatrix}$$

$$\vdots \qquad\qquad \vdots \qquad\qquad \vdots$$

$$A[\mathbf{u}_n]_B = \begin{bmatrix} a_{11} & a_{12} & \cdots & a_{1n} \\ a_{21} & a_{22} & \cdots & a_{2n} \\ \vdots & \vdots & & \vdots \\ a_{m1} & a_{m2} & \cdots & a_{mn} \end{bmatrix} \begin{bmatrix} 0 \\ 0 \\ 0 \\ \vdots \\ 1 \end{bmatrix} = \begin{bmatrix} a_{1n} \\ a_{2n} \\ \vdots \\ a_{mn} \end{bmatrix}$$

Substituting these results into (2) yields

$$\begin{bmatrix} a_{11} \\ a_{21} \\ \vdots \\ a_{m1} \end{bmatrix} = [T(\mathbf{u}_1)]_{B'}, \quad \begin{bmatrix} a_{12} \\ a_{22} \\ \vdots \\ a_{m2} \end{bmatrix} = [T(\mathbf{u}_2)]_{B'}, \dots, \quad \begin{bmatrix} a_{1n} \\ a_{2n} \\ \vdots \\ a_{mn} \end{bmatrix} = [T(\mathbf{u}_n)]_{B'}$$

which shows that the successive columns of A are the coordinate vectors of

$$T(\mathbf{u}_1), T(\mathbf{u}_2), \dots, T(\mathbf{u}_n)$$

with respect to the basis B'. Thus, the matrix A that completes the link in Figure 8.4.2*a* is

$$A = \left[[T(\mathbf{u}_1)]_{B'} \mid [T(\mathbf{u}_2)]_{B'} \mid \cdots \mid [T(\mathbf{u}_n)]_{B'} \right] \tag{3}$$

We will call this *the matrix for T relative to the bases B and B'* and will denote it by the symbol $[T]_{B',B}$. Using this notation, Formula (3) can be written as

$$[T]_{B',B} = \left[[T(\mathbf{u}_1)]_{B'} \mid [T(\mathbf{u}_2)]_{B'} \mid \cdots \mid [T(\mathbf{u}_n)]_{B'} \right] \tag{4}$$

and from (1), this matrix has the property

$$[T]_{B',B}[\mathbf{x}]_B = [T(\mathbf{x})]_{B'} \tag{5}$$

We leave it as an exercise to show that in the special case where $T_A\colon R^n \to R^m$ is multiplication by A, and where B and B' are the *standard bases* for R^n and R^m, respectively, then

$$[T]_{B',B} = A \tag{6}$$

Remark Observe that in the notation $[T]_{B',B}$ the right subscript is a basis for the domain of T, and the left subscript is a basis for the image space of T (Figure 8.4.3). Moreover, observe how the subscript B seems to "cancel out" in Formula (5) (Figure 8.4.4).

$[T]_{B',B}$

| Basis for the image space | Basis for the domain |

▲ Figure 8.4.3

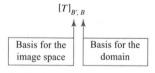

$[T]_{B',B}[\mathbf{x}]_B = [T(\mathbf{x})]_B$

Cancellation

▲ Figure 8.4.4

▶ **EXAMPLE 1 Matrix for a Linear Transformation**

Let $T : P_1 \to P_2$ be the linear transformation defined by

$$T(p(x)) = xp(x)$$

Find the matrix for T with respect to the standard bases

$$B = \{\mathbf{u}_1, \mathbf{u}_2\} \quad \text{and} \quad B' = \{\mathbf{v}_1, \mathbf{v}_2, \mathbf{v}_3\}$$

where

$$\mathbf{u}_1 = 1, \quad \mathbf{u}_2 = x; \qquad \mathbf{v}_1 = 1, \quad \mathbf{v}_2 = x, \quad \mathbf{v}_3 = x^2$$

Solution From the given formula for T we obtain

$$T(\mathbf{u}_1) = T(1) = (x)(1) = x$$
$$T(\mathbf{u}_2) = T(x) = (x)(x) = x^2$$

By inspection, the coordinate vectors for $T(\mathbf{u}_1)$ and $T(\mathbf{u}_2)$ relative to B' are

$$[T(\mathbf{u}_1)]_{B'} = \begin{bmatrix} 0 \\ 1 \\ 0 \end{bmatrix}, \quad [T(\mathbf{u}_2)]_{B'} = \begin{bmatrix} 0 \\ 0 \\ 1 \end{bmatrix}$$

Thus, the matrix for T with respect to B and B' is

$$[T]_{B',B} = \left[[T(\mathbf{u}_1)]_{B'} \mid [T(\mathbf{u}_2)]_{B'} \right] = \begin{bmatrix} 0 & 0 \\ 1 & 0 \\ 0 & 1 \end{bmatrix}$$

▶ **EXAMPLE 2 The Three-Step Procedure**

Let $T : P_1 \to P_2$ be the linear transformation in Example 1, and use the three-step procedure described in the following figure to perform the computation

$$T(a + bx) = x(a + bx) = ax + bx^2$$

Solution

Step 1. The coordinate matrix for $\mathbf{x} = a + bx$ relative to the basis $B = \{1, x\}$ is

$$[\mathbf{x}]_B = \begin{bmatrix} a \\ b \end{bmatrix}$$

Step 2. Multiplying $[\mathbf{x}]_B$ by the matrix $[T]_{B',B}$ found in Example 1 we obtain

$$[T]_{B',B}[\mathbf{x}]_B = \begin{bmatrix} 0 & 0 \\ 1 & 0 \\ 0 & 1 \end{bmatrix} \begin{bmatrix} a \\ b \end{bmatrix} = \begin{bmatrix} 0 \\ a \\ b \end{bmatrix} = [T(\mathbf{x})]_{B'}$$

Although Example 2 is simple, the procedure that it illustrates is applicable to problems of great complexity.

Step 3. Reconstructing $T(\mathbf{x}) = T(a + bx)$ from $[T(\mathbf{x})]_{B'}$ we obtain

$$T(a + bx) = 0 + ax + bx^2 = ax + bx^2$$

▶ **EXAMPLE 3 Matrix for a Linear Transformation**

Let $T: R^2 \rightarrow R^3$ be the linear transformation defined by

$$T\left(\begin{bmatrix} x_1 \\ x_2 \end{bmatrix}\right) = \begin{bmatrix} x_2 \\ -5x_1 + 13x_2 \\ -7x_1 + 16x_2 \end{bmatrix} = \begin{bmatrix} 0 & 1 \\ -5 & 13 \\ -7 & 16 \end{bmatrix} \begin{bmatrix} x_1 \\ x_2 \end{bmatrix}$$

Find the matrix for the transformation T with respect to the bases $B = \{\mathbf{u}_1, \mathbf{u}_2\}$ for R^2 and $B' = \{\mathbf{v}_1, \mathbf{v}_2, \mathbf{v}_3\}$ for R^3, where

$$\mathbf{u}_1 = \begin{bmatrix} 3 \\ 1 \end{bmatrix}, \quad \mathbf{u}_2 = \begin{bmatrix} 5 \\ 2 \end{bmatrix}; \quad \mathbf{v}_1 = \begin{bmatrix} 1 \\ 0 \\ -1 \end{bmatrix}, \quad \mathbf{v}_2 = \begin{bmatrix} -1 \\ 2 \\ 2 \end{bmatrix}, \quad \mathbf{v}_3 = \begin{bmatrix} 0 \\ 1 \\ 2 \end{bmatrix}$$

Solution From the formula for T,

$$T(\mathbf{u}_1) = \begin{bmatrix} 1 \\ -2 \\ -5 \end{bmatrix}, \quad T(\mathbf{u}_2) = \begin{bmatrix} 2 \\ 1 \\ -3 \end{bmatrix}$$

Expressing these vectors as linear combinations of $\mathbf{v}_1$, $\mathbf{v}_2$, and $\mathbf{v}_3$, we obtain (verify)

$$T(\mathbf{u}_1) = \mathbf{v}_1 - 2\mathbf{v}_3, \quad T(\mathbf{u}_2) = 3\mathbf{v}_1 + \mathbf{v}_2 - \mathbf{v}_3$$

Thus,

$$[T(\mathbf{u}_1)]_{B'} = \begin{bmatrix} 1 \\ 0 \\ -2 \end{bmatrix}, \quad [T(\mathbf{u}_2)]_{B'} = \begin{bmatrix} 3 \\ 1 \\ -1 \end{bmatrix}$$

so

$$[T]_{B',B} = \left[[T(\mathbf{u}_1)]_{B'} \mid [T(\mathbf{u}_2)]_{B'} \right] = \begin{bmatrix} 1 & 3 \\ 0 & 1 \\ -2 & -1 \end{bmatrix} \blacktriangleleft$$

Remark Example 3 illustrates that a fixed linear transformation generally has multiple representations, each depending on the bases chosen. In this case the matrices

$$[T] = \begin{bmatrix} 0 & 1 \\ -5 & 13 \\ -7 & 16 \end{bmatrix} \quad \text{and} \quad [T]_{B',B} = \begin{bmatrix} 1 & 3 \\ 0 & 1 \\ -2 & -1 \end{bmatrix}$$

both represent the transformation T, the first relative to the standard bases for R^2 and R^3, the second relative to the bases B and B' stated in the example.

Matrices of Linear Operators

In the special case where $V = W$ (so that $T: V \rightarrow V$ is a linear operator), it is usual to take $B = B'$ when constructing a matrix for T. In this case the resulting matrix is called the **matrix for T relative to the basis B** and is usually denoted by $[T]_B$ rather than $[T]_{B,B}$. If $B = \{\mathbf{u}_1, \mathbf{u}_2, \ldots, \mathbf{u}_n\}$, then Formulas (4) and (5) become

Phrased informally, Formulas (7) and (8) state that *the matrix for T, when multiplied by the coordinate vector for* $\mathbf{x}$, *produces the coordinate vector for T($\mathbf{x}$).*

$$[T]_B = \left[[T(\mathbf{u}_1)]_B \mid [T(\mathbf{u}_2)]_B \mid \cdots \mid [T(\mathbf{u}_n)]_B \right] \tag{7}$$

$$[T]_B[\mathbf{x}]_B = [T(\mathbf{x})]_B \tag{8}$$

In the special case where $T: R^n \rightarrow R^n$ is a matrix operator, say multiplication by A, and B is the standard basis for R^n, then Formula (7) simplifies to

$$[T]_B = A \tag{9}$$

Matrices of Identity Operators

Recall that the identity operator $I: V \to V$ maps every vector in V into itself, that is, $I(\mathbf{x}) = \mathbf{x}$ for every vector $\mathbf{x}$ in V. The following example shows that if V is n-dimensional, then the matrix for I relative to *any* basis B for V is the $n \times n$ identity matrix.

▶ **EXAMPLE 4 Matrices of Identity Operators**

If $B = \{\mathbf{u}_1, \mathbf{u}_2, \ldots, \mathbf{u}_n\}$ is a basis for a finite-dimensional vector space V, and if $I: V \to V$ is the identity operator on V, then

$$I(\mathbf{u}_1) = \mathbf{u}_1, \quad I(\mathbf{u}_2) = \mathbf{u}_2, \ldots, \quad I(\mathbf{u}_n) = \mathbf{u}_n$$

Therefore,

$$[I]_B = \begin{bmatrix} 1 & 0 & \cdots & 0 \\ 0 & 1 & \cdots & 0 \\ 0 & 0 & \cdots & 0 \\ \vdots & \vdots & & \vdots \\ 0 & 0 & \cdots & 1 \end{bmatrix} = I$$

$$\uparrow \qquad \uparrow \qquad \uparrow$$
$$[I(\mathbf{u}_1)]_B \quad [I(\mathbf{u}_2)]_B \quad [I(\mathbf{u}_n)]_B$$

▶ **EXAMPLE 5 Linear Operator on P_2**

Let $T: P_2 \to P_2$ be the linear operator defined by

$$T(p(x)) = p(3x - 5)$$

that is, $T(c_0 + c_1 x + c_2 x^2) = c_0 + c_1(3x - 5) + c_2(3x - 5)^2$.

(a) Find $[T]_B$ relative to the basis $B = \{1, x, x^2\}$.

(b) Use the indirect procedure to compute $T(1 + 2x + 3x^2)$.

(c) Check the result in (b) by computing $T(1 + 2x + 3x^2)$ directly.

Solution (a) From the formula for T,

$$T(1) = 1, \quad T(x) = 3x - 5, \quad T(x^2) = (3x - 5)^2 = 9x^2 - 30x + 25$$

so

$$[T(1)]_B = \begin{bmatrix} 1 \\ 0 \\ 0 \end{bmatrix}, \quad [T(x)]_B = \begin{bmatrix} -5 \\ 3 \\ 0 \end{bmatrix}, \quad [T(x^2)]_B = \begin{bmatrix} 25 \\ -30 \\ 9 \end{bmatrix}$$

Thus,

$$[T]_B = \begin{bmatrix} 1 & -5 & 25 \\ 0 & 3 & -30 \\ 0 & 0 & 9 \end{bmatrix}$$

Solution (b)

Step 1. The coordinate matrix for $\mathbf{p} = 1 + 2x + 3x^2$ relative to the basis $B = \{1, x, x^2\}$ is

$$[\mathbf{p}]_B = \begin{bmatrix} 1 \\ 2 \\ 3 \end{bmatrix}$$

Step 2. Multiplying $[\mathbf{p}]_B$ by the matrix $[T]_B$ found in part (a) we obtain

$$[T]_B[\mathbf{p}]_B = \begin{bmatrix} 1 & -5 & 25 \\ 0 & 3 & -30 \\ 0 & 0 & 9 \end{bmatrix} \begin{bmatrix} 1 \\ 2 \\ 3 \end{bmatrix} = \begin{bmatrix} 66 \\ -84 \\ 27 \end{bmatrix} = [T(\mathbf{p})]_B$$

Step 3. Reconstructing $T(\mathbf{p}) = T(1 + 2x + 3x^2)$ from $[T(\mathbf{p})]_B$ we obtain

$$T(1 + 2x + 3x^2) = 66 - 84x + 27x^2$$

Solution (c) By direct computation,

$$\begin{aligned} T(1 + 2x + 3x^2) &= 1 + 2(3x - 5) + 3(3x - 5)^2 \\ &= 1 + 6x - 10 + 27x^2 - 90x + 75 \\ &= 66 - 84x + 27x^2 \end{aligned}$$

which agrees with the result in (b). ◀

Matrices of Compositions and Inverse Transformations

We will conclude this section by mentioning two theorems without proof that are generalizations of Formulas (4) and (7) of Section 4.10.

THEOREM 8.4.1 *If $T_1: U \to V$ and $T_2: V \to W$ are linear transformations, and if B, B'', and B' are bases for U, V, and W, respectively, then*

$$[T_2 \circ T_1]_{B', B} = [T_2]_{B', B''}[T_1]_{B'', B} \tag{10}$$

THEOREM 8.4.2 *If $T: V \to V$ is a linear operator, and if B is a basis for V, then the following are equivalent.*

(a) T is one-to-one.

(b) $[T]_B$ is invertible.

Moreover, when these equivalent conditions hold,

$$[T^{-1}]_B = [T]_B^{-1} \tag{11}$$

$[T_2 \circ T_1]_{B', B} = [T_2]_{B', B''} [T_1]_{B'', B}$

| Cancellation | | Cancellation |

▲ Figure 8.4.5

Remark In (10), observe how the interior subscript B'' (the basis for the intermediate space V) seems to "cancel out," leaving only the bases for the domain and image space of the composition as subscripts (Figure 8.4.5). This cancellation of interior subscripts suggests the following extension of Formula (10) to compositions of three linear transformations (Figure 8.4.6):

$$[T_3 \circ T_2 \circ T_1]_{B', B} = [T_3]_{B', B'''}[T_2]_{B''', B''}[T_1]_{B'', B} \tag{12}$$

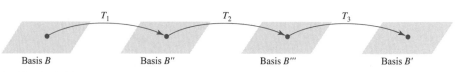

Basis B T_1 Basis B'' T_2 Basis B''' T_3 Basis B'

▲ Figure 8.4.6

The following example illustrates Theorem 8.4.1.

▶ **EXAMPLE 6 Composition**

Let $T_1: P_1 \to P_2$ be the linear transformation defined by

$$T_1(p(x)) = xp(x)$$

and let $T_2: P_2 \to P_2$ be the linear operator defined by

$$T_2(p(x)) = p(3x - 5)$$

Then the composition $(T_2 \circ T_1): P_1 \to P_2$ is given by

$$(T_2 \circ T_1)(p(x)) = T_2(T_1(p(x))) = T_2(xp(x)) = (3x - 5)p(3x - 5)$$

Thus, if $p(x) = c_0 + c_1 x$, then

$$(T_2 \circ T_1)(c_0 + c_1 x) = (3x - 5)(c_0 + c_1(3x - 5))$$
$$= c_0(3x - 5) + c_1(3x - 5)^2 \quad (13)$$

In this example, P_1 plays the role of U in Theorem 8.4.1, and P_2 plays the roles of both V and W; thus we can take $B' = B''$ in (10) so that the formula simplifies to

$$[T_2 \circ T_1]_{B',B} = [T_2]_{B'}[T_1]_{B',B} \quad (14)$$

Let us choose $B = \{1, x\}$ to be the basis for P_1 and choose $B' = \{1, x, x^2\}$ to be the basis for P_2. We showed in Examples 1 and 5 that

$$[T_1]_{B',B} = \begin{bmatrix} 0 & 0 \\ 1 & 0 \\ 0 & 1 \end{bmatrix} \quad \text{and} \quad [T_2]_{B'} = \begin{bmatrix} 1 & -5 & 25 \\ 0 & 3 & -30 \\ 0 & 0 & 9 \end{bmatrix}$$

Thus, it follows from (14) that

$$[T_2 \circ T_1]_{B',B} = \begin{bmatrix} 1 & -5 & 25 \\ 0 & 3 & -30 \\ 0 & 0 & 9 \end{bmatrix} \begin{bmatrix} 0 & 0 \\ 1 & 0 \\ 0 & 1 \end{bmatrix} = \begin{bmatrix} -5 & 25 \\ 3 & -30 \\ 0 & 9 \end{bmatrix} \quad (15)$$

As a check, we will calculate $[T_2 \circ T_1]_{B',B}$ directly from Formula (4). Since $B = \{1, x\}$, it follows from Formula (4) with $\mathbf{u}_1 = 1$ and $\mathbf{u}_2 = x$ that

$$[T_2 \circ T_1]_{B',B} = \left[[(T_2 \circ T_1)(1)]_{B'} \mid [(T_2 \circ T_1)(x)]_{B'} \right] \quad (16)$$

Using (13) yields

$$(T_2 \circ T_1)(1) = 3x - 5 \quad \text{and} \quad (T_2 \circ T_1)(x) = (3x - 5)^2 = 9x^2 - 30x + 25$$

From this and the fact that $B' = \{1, x, x^2\}$, it follows that

$$[(T_2 \circ T_1)(1)]_{B'} = \begin{bmatrix} -5 \\ 3 \\ 0 \end{bmatrix} \quad \text{and} \quad [(T_2 \circ T_1)(x)]_{B'} = \begin{bmatrix} 25 \\ -30 \\ 9 \end{bmatrix}$$

Substituting in (16) yields

$$[T_2 \circ T_1]_{B',B} = \begin{bmatrix} -5 & 25 \\ 3 & -30 \\ 0 & 9 \end{bmatrix}$$

which agrees with (15). ◀

Concept Review

- Matrix for a linear transformation relative to bases
- Matrix for a linear operator relative to a basis
- The three-step procedure for finding $T(\mathbf{x})$

Skills

- Find the matrix for a linear transformation $T : V \rightarrow W$ relative to bases of V and W.
- For a linear transformation $T : V \rightarrow W$ find $T(\mathbf{x})$ using the matrix for T relative to bases of V and W.

Exercise Set 8.4

1. Let $T : P_2 \rightarrow P_3$ be the linear transformation defined by $T(p(x)) = xp(x)$.

   (a) Find the matrix for T relative to the standard bases
   $$B = \{\mathbf{u}_1, \mathbf{u}_2, \mathbf{u}_3\} \quad \text{and} \quad B' = \{\mathbf{v}_1, \mathbf{v}_2, \mathbf{v}_3, \mathbf{v}_4\}$$
   where
   $$\mathbf{u}_1 = 1, \quad \mathbf{u}_2 = x, \quad \mathbf{u}_3 = x^2$$
   $$\mathbf{v}_1 = 1, \quad \mathbf{v}_2 = x, \quad \mathbf{v}_3 = x^2, \quad \mathbf{v}_4 = x^3$$

   (b) Verify that the matrix $[T]_{B',B}$ obtained in part (a) satisfies Formula (5) for every vector $\mathbf{x} = c_0 + c_1 x + c_2 x^2$ in P_2.

2. Let $T : P_2 \rightarrow P_1$ be the linear transformation defined by
   $$T(a_0 + a_1 x + a_2 x^2) = (a_0 + a_1) - (2a_1 + 3a_2)x$$

   (a) Find the matrix for T relative to the standard bases $B = \{1, x, x^2\}$ and $B' = \{1, x\}$ for P_2 and P_1.

   (b) Verify that the matrix $[T]_{B',B}$ obtained in part (a) satisfies Formula (5) for every vector $\mathbf{x} = c_0 + c_1 x + c_2 x^2$ in P_2.

3. Let $T : P_2 \rightarrow P_2$ be the linear operator defined by
   $$T(a_0 + a_1 x + a_2 x^2) = a_0 + a_1(x - 1) + a_2(x - 1)^2$$

   (a) Find the matrix for T relative to the standard basis $B = \{1, x, x^2\}$ for P_2.

   (b) Verify that the matrix $[T]_B$ obtained in part (a) satisfies Formula (8) for every vector $\mathbf{x} = a_0 + a_1 x + a_2 x^2$ in P_2.

4. Let $T : R^2 \rightarrow R^2$ be the linear operator defined by
   $$T\left(\begin{bmatrix} x_1 \\ x_2 \end{bmatrix}\right) = \begin{bmatrix} x_1 - x_2 \\ x_1 + x_2 \end{bmatrix}$$
   and let $B = \{\mathbf{u}_1, \mathbf{u}_2\}$ be the basis for which
   $$\mathbf{u}_1 = \begin{bmatrix} 1 \\ 1 \end{bmatrix} \quad \text{and} \quad \mathbf{u}_2 = \begin{bmatrix} -1 \\ 0 \end{bmatrix}$$

   (a) Find $[T]_B$.

   (b) Verify that Formula (8) holds for every vector $\mathbf{x}$ in R^2.

5. Let $T : R^2 \rightarrow R^3$ be defined by
   $$T\left(\begin{bmatrix} x_1 \\ x_2 \end{bmatrix}\right) = \begin{bmatrix} x_1 + 2x_2 \\ -x_1 \\ 0 \end{bmatrix}$$

   (a) Find the matrix $[T]_{B',B}$ relative to the bases $B = \{\mathbf{u}_1, \mathbf{u}_2\}$ and $B' = \{\mathbf{v}_1, \mathbf{v}_2, \mathbf{v}_3\}$, where
   $$\mathbf{u}_1 = \begin{bmatrix} 1 \\ 3 \end{bmatrix}, \quad \mathbf{u}_2 = \begin{bmatrix} -2 \\ 4 \end{bmatrix}$$
   $$\mathbf{v}_1 = \begin{bmatrix} 1 \\ 1 \\ 1 \end{bmatrix}, \quad \mathbf{v}_2 = \begin{bmatrix} 2 \\ 2 \\ 0 \end{bmatrix}, \quad \mathbf{v}_3 = \begin{bmatrix} 3 \\ 0 \\ 0 \end{bmatrix}$$

   (b) Verify that Formula (5) holds for every vector in R^2.

6. Let $T : R^3 \rightarrow R^3$ be the linear operator defined by
   $$T(x_1, x_2, x_3) = (x_1 - x_2, x_2 - x_1, x_1 - x_3)$$

   (a) Find the matrix for T with respect to the basis $B = \{\mathbf{v}_1, \mathbf{v}_2, \mathbf{v}_3\}$, where
   $$\mathbf{v}_1 = (1, 0, 1), \quad \mathbf{v}_2 = (0, 1, 1), \quad \mathbf{v}_3 = (1, 1, 0)$$

   (b) Verify that Formula (8) holds for every vector $\mathbf{x} = (x_1, x_2, x_3)$ in R^3.

   (c) Is T one-to-one? If so, find the matrix of T^{-1}.

7. Let $T : P_2 \rightarrow P_2$ be the linear operator defined by $T(p(x)) = p(2x + 1)$, that is,
   $$T(c_0 + c_1 x + c_2 x^2) = c_0 + c_1(2x + 1) + c_2(2x + 1)^2$$

   (a) Find $[T]_B$ with respect to the basis $B = \{1, x, x^2\}$.

   (b) Use the three-step procedure illustrated in Example 2 to compute $T(2 - 3x + 4x^2)$.

   (c) Check the result obtained in part (b) by computing $T(2 - 3x + 4x^2)$ directly.

8. Let $T : P_2 \rightarrow P_3$ be the linear transformation defined by $T(p(x)) = xp(x - 3)$, that is,
   $$T(c_0 + c_1 x + c_2 x^2) = x(c_0 + c_1(x - 3) + c_2(x - 3)^2)$$

   (a) Find $[T]_{B',B}$ relative to the bases $B = \{1, x, x^2\}$ and $B' = \{1, x, x^2, x^3\}$.

   (b) Use the three-step procedure illustrated in Example 2 to compute $T(1 + x - x^2)$.

   (c) Check the result obtained in part (b) by computing $T(1 + x - x^2)$ directly.

9. Let $\mathbf{v}_1 = \begin{bmatrix} 1 \\ 3 \end{bmatrix}$ and $\mathbf{v}_2 = \begin{bmatrix} -1 \\ 4 \end{bmatrix}$, and let

$$A = \begin{bmatrix} 1 & 3 \\ -2 & 5 \end{bmatrix}$$

be the matrix for $T: R^2 \to R^2$ relative to the basis $B = \{\mathbf{v}_1, \mathbf{v}_2\}$.

(a) Find $[T(\mathbf{v}_1)]_B$ and $[T(\mathbf{v}_2)]_B$.

(b) Find $T(\mathbf{v}_1)$ and $T(\mathbf{v}_2)$.

(c) Find a formula for $T\left(\begin{bmatrix} x_1 \\ x_2 \end{bmatrix}\right)$.

(d) Use the formula obtained in (c) to compute $T\left(\begin{bmatrix} 1 \\ 1 \end{bmatrix}\right)$.

10. Let $A = \begin{bmatrix} 3 & -2 & 1 & 0 \\ 1 & 6 & 2 & 1 \\ -3 & 0 & 7 & 1 \end{bmatrix}$ be the matrix for

$T: R^4 \to R^3$ relative to the bases $B = \{\mathbf{v}_1, \mathbf{v}_2, \mathbf{v}_3, \mathbf{v}_4\}$ and $B' = \{\mathbf{w}_1, \mathbf{w}_2, \mathbf{w}_3\}$, where

$$\mathbf{v}_1 = \begin{bmatrix} 0 \\ 1 \\ 1 \\ 1 \end{bmatrix}, \quad \mathbf{v}_2 = \begin{bmatrix} 2 \\ 1 \\ -1 \\ -1 \end{bmatrix}, \quad \mathbf{v}_3 = \begin{bmatrix} 1 \\ 4 \\ -1 \\ 2 \end{bmatrix}, \quad \mathbf{v}_4 = \begin{bmatrix} 6 \\ 9 \\ 4 \\ 2 \end{bmatrix}$$

$$\mathbf{w}_1 = \begin{bmatrix} 0 \\ 8 \\ 8 \end{bmatrix}, \quad \mathbf{w}_2 = \begin{bmatrix} -7 \\ 8 \\ 1 \end{bmatrix}, \quad \mathbf{w}_3 = \begin{bmatrix} -6 \\ 9 \\ 1 \end{bmatrix}$$

(a) Find $[T(\mathbf{v}_1)]_{B'}$, $[T(\mathbf{v}_2)]_{B'}$, $[T(\mathbf{v}_3)]_{B'}$, and $[T(\mathbf{v}_4)]_{B'}$.

(b) Find $T(\mathbf{v}_1)$, $T(\mathbf{v}_2)$, $T(\mathbf{v}_3)$, and $T(\mathbf{v}_4)$.

(c) Find a formula for $T\left(\begin{bmatrix} x_1 \\ x_2 \\ x_3 \\ x_4 \end{bmatrix}\right)$.

(d) Use the formula obtained in (c) to compute $T\left(\begin{bmatrix} 2 \\ 2 \\ 0 \\ 0 \end{bmatrix}\right)$.

11. Let $A = \begin{bmatrix} 1 & 3 & -1 \\ 2 & 0 & 5 \\ 6 & -2 & 4 \end{bmatrix}$ be the matrix for $T: P_2 \to P_2$ with

respect to the basis $B = \{\mathbf{v}_1, \mathbf{v}_2, \mathbf{v}_3\}$, where $\mathbf{v}_1 = 3x + 3x^2$, $\mathbf{v}_2 = -1 + 3x + 2x^2$, $\mathbf{v}_3 = 3 + 7x + 2x^2$.

(a) Find $[T(\mathbf{v}_1)]_B$, $[T(\mathbf{v}_2)]_B$, and $[T(\mathbf{v}_3)]_B$.

(b) Find $T(\mathbf{v}_1)$, $T(\mathbf{v}_2)$, and $T(\mathbf{v}_3)$.

(c) Find a formula for $T(a_0 + a_1x + a_2x^2)$.

(d) Use the formula obtained in (c) to compute $T(1 + x^2)$.

12. Let $T_1: P_1 \to P_2$ be the linear transformation defined by

$$T_1(p(x)) = xp(x)$$

and let $T_2: P_2 \to P_2$ be the linear operator defined by

$$T_2(p(x)) = p(2x + 1)$$

Let $B = \{1, x\}$ and $B' = \{1, x, x^2\}$ be the standard bases for P_1 and P_2.

(a) Find $[T_2 \circ T_1]_{B', B}$, $[T_2]_{B'}$, and $[T_1]_{B', B}$.

(b) State a formula relating the matrices in part (a).

(c) Verify that the matrices in part (a) satisfy the formula you stated in part (b).

13. Let $T_1: P_1 \to P_2$ be the linear transformation defined by

$$T_1(c_0 + c_1x) = 2c_0 - 3c_1x$$

and let $T_2: P_2 \to P_3$ be the linear transformation defined by

$$T_2(c_0 + c_1x + c_2x^2) = 3c_0x + 3c_1x^2 + 3c_2x^3$$

Let $B = \{1, x\}$, $B'' = \{1, x, x^2\}$, and $B' = \{1, x, x^2, x^3\}$.

(a) Find $[T_2 \circ T_1]_{B', B}$, $[T_2]_{B', B''}$, and $[T_1]_{B'', B}$.

(b) State a formula relating the matrices in part (a).

(c) Verify that the matrices in part (a) satisfy the formula you stated in part (b).

14. Show that if $T: V \to W$ is the zero transformation, then the matrix for T with respect to any bases for V and W is a zero matrix.

15. Show that if $T: V \to V$ is a contraction or a dilation of V (Example 4 of Section 8.1), then the matrix for T relative to any basis for V is a positive scalar multiple of the identity matrix.

16. Let $B = \{\mathbf{v}_1, \mathbf{v}_2, \mathbf{v}_3, \mathbf{v}_4\}$ be a basis for a vector space V. Find the matrix with respect to B of the linear operator $T: V \to V$ defined by $T(\mathbf{v}_1) = \mathbf{v}_2$, $T(\mathbf{v}_2) = \mathbf{v}_3$, $T(\mathbf{v}_3) = \mathbf{v}_4$, $T(\mathbf{v}_4) = \mathbf{v}_1$.

17. Prove that if B and B' are the standard bases for R^n and R^m, respectively, then the matrix for a linear transformation $T: R^n \to R^m$ relative to the bases B and B' is the standard matrix for T.

18. (*Calculus required*) Let $D: P_2 \to P_2$ be the differentiation operator $D(\mathbf{p}) = p'(x)$. In parts (a) and (b), find the matrix of D relative to the basis $B = \{\mathbf{p}_1, \mathbf{p}_2, \mathbf{p}_3\}$.

(a) $\mathbf{p}_1 = 1$, $\mathbf{p}_2 = x$, $\mathbf{p}_3 = x^2$

(b) $\mathbf{p}_1 = 2$, $\mathbf{p}_2 = 2 - 3x$, $\mathbf{p}_3 = 2 - 3x + 8x^2$

(c) Use the matrix in part (a) to compute $D(6 - 6x + 24x^2)$.

(d) Repeat the directions for part (c) for the matrix in part (b).

19. (*Calculus required*) In each part, suppose that $B = \{\mathbf{f}_1, \mathbf{f}_2, \mathbf{f}_3\}$ is a basis for a subspace V of the vector space of real-valued functions defined on the real line. Find the matrix with respect to B for differentiation operator $D: V \to V$.

(a) $\mathbf{f}_1 = 1$, $\mathbf{f}_2 = \sin x$, $\mathbf{f}_3 = \cos x$

(b) $\mathbf{f}_1 = 1$, $\mathbf{f}_2 = e^x$, $\mathbf{f}_3 = e^{2x}$

(c) $\mathbf{f}_1 = e^{2x}$, $\mathbf{f}_2 = xe^{2x}$, $\mathbf{f}_3 = x^2e^{2x}$

(d) Use the matrix in part (c) to compute $D(4e^{2x} + 6xe^{2x} - 10x^2e^{2x})$.

20. Let V be a four-dimensional vector space with basis B, let W be a seven-dimensional vector space with basis B', and let $T : V \rightarrow W$ be a linear transformation. Identify the four vector spaces that contain the vectors at the corners of the accompanying diagram.

◀ **Figure Ex-20**

21. In each part, fill in the missing part of the equation.

(a) $[T_2 \circ T_1]_{B',B} = [T_2]\underline{\quad?\quad}[T_1]_{B'',B}$

(b) $[T_3 \circ T_2 \circ T_1]_{B',B} = [T_3]\underline{\quad?\quad}[T_2]_{B''',B''}[T_1]_{B'',B}$

True-False Exercises

In parts (a)–(e) determine whether the statement is true or false, and justify your answer.

(a) If the matrix of a linear transformation $T : V \rightarrow W$ relative to some bases of V and W is $\begin{bmatrix} 2 & 4 \\ 0 & 3 \end{bmatrix}$, then there is a nonzero vector $\mathbf{x}$ in V such that $T(\mathbf{x}) = 2\mathbf{x}$.

(b) If the matrix of a linear transformation $T : V \rightarrow W$ relative to bases for V and W is $\begin{bmatrix} 2 & 4 \\ 0 & 3 \end{bmatrix}$, then there is a nonzero vector $\mathbf{x}$ in V such that $T(\mathbf{x}) = 4\mathbf{x}$.

(c) If the matrix of a linear transformation $T : V \rightarrow W$ relative to certain bases for V and W is $\begin{bmatrix} 1 & 4 \\ 2 & 3 \end{bmatrix}$, then T is one-to-one.

(d) If $S : V \rightarrow V$ and $T : V \rightarrow V$ are linear operators and B is a basis for V, then the matrix of $S \circ T$ relative to B is $[T]_B[S]_B$.

(e) If $T : V \rightarrow V$ is an invertible linear operator and B is a basis for V, then the matrix for T^{-1} relative to B is $[T]_B^{-1}$.

8.5 Similarity

The matrix for a linear operator $T : V \rightarrow V$ depends on the basis selected for V. One of the fundamental problems of linear algebra is to choose a basis for V that makes the matrix for T as simple as possible—a diagonal or a triangular matrix, for example. In this section we will study this problem.

Simple Matrices for Linear Operators

Standard bases do not necessarily produce the simplest matrices for linear operators. For example, consider the matrix operator $T : R^2 \rightarrow R^2$ whose standard matrix is

$$[T] = \begin{bmatrix} 1 & 1 \\ -2 & 4 \end{bmatrix} \tag{1}$$

and view $[T]$ as the matrix for T relative to the standard basis $B = \{\mathbf{e}_1, \mathbf{e}_2\}$ for R^2. Let us compare this to the matrix for T relative to the basis $B' = \{\mathbf{u}_1', \mathbf{u}_2'\}$ for R^2 in which

$$\mathbf{u}_1' = \begin{bmatrix} 1 \\ 1 \end{bmatrix}, \quad \mathbf{u}_2' = \begin{bmatrix} 1 \\ 2 \end{bmatrix} \tag{2}$$

Since

$$T(\mathbf{u}_1') = \begin{bmatrix} 1 & 1 \\ -2 & 4 \end{bmatrix}\begin{bmatrix} 1 \\ 1 \end{bmatrix} = \begin{bmatrix} 2 \\ 2 \end{bmatrix} = 2\mathbf{u}_1' \quad \text{and} \quad T(\mathbf{u}_2') = \begin{bmatrix} 1 & 1 \\ -2 & 4 \end{bmatrix}\begin{bmatrix} 1 \\ 2 \end{bmatrix} = \begin{bmatrix} 3 \\ 6 \end{bmatrix} = 3\mathbf{u}_2'$$

it follows that

$$[T(\mathbf{u}_1')]_{B'} = \begin{bmatrix} 2 \\ 0 \end{bmatrix} \quad \text{and} \quad [T(\mathbf{u}_2')]_{B'} = \begin{bmatrix} 0 \\ 3 \end{bmatrix}$$

so the matrix for T relative to the basis B' is

$$[T]_{B'} = \begin{bmatrix} T(\mathbf{u}_1')_{B'} \mid T(\mathbf{u}_2')_{B'} \end{bmatrix} = \begin{bmatrix} 2 & 0 \\ 0 & 3 \end{bmatrix}$$

This matrix, being diagonal, has a simpler form than $[T]$ and conveys clearly that the operator T scales $\mathbf{u}_1'$ by a factor of 2 and $\mathbf{u}_2'$ by a factor of 3, information that is not immediately evident from $[T]$.

One of the major themes in more advanced linear algebra courses is to determine the "simplest possible form" that can be obtained for the matrix of a linear operator by choosing the basis appropriately. Sometimes it is possible to obtain a diagonal matrix (as above, for example), whereas other times one must settle for a triangular matrix or some other form. We will only be able to touch on this important topic in this text.

The problem of finding a basis that produces the simplest possible matrix for a linear operator $T: V \to V$ can be attacked by first finding a matrix for T relative to *any* basis, typically a standard basis, where applicable, and then changing the basis in a way that simplifies the matrix. Before pursuing this idea, it will be helpful to revisit some concepts about changing bases.

A New View of Transition Matrices

Recall from Formulas (7) and (8) of Section 4.6 that if $B = \{\mathbf{u}_1, \mathbf{u}_2, \ldots, \mathbf{u}_n\}$ and $B' = \{\mathbf{u}_1', \mathbf{u}_2', \ldots, \mathbf{u}_n'\}$ are bases for a vector space V, then the transition matrices from B to B' and from B' to B are

$$P_{B \to B'} = \left[[\mathbf{u}_1]_{B'} \mid [\mathbf{u}_2]_{B'} \mid \cdots \mid [\mathbf{u}_n]_{B'} \right] \tag{3}$$

$$P_{B' \to B} = \left[[\mathbf{u}_1']_B \mid [\mathbf{u}_2']_B \mid \cdots \mid [\mathbf{u}_n']_B \right] \tag{4}$$

where the matrices $P_{B \to B'}$ and $P_{B' \to B}$ are inverses of each other. We also showed in Formulas (9) and (10) of that section that if $\mathbf{v}$ is any vector in V, then

$$P_{B \to B'}[\mathbf{v}]_B = [\mathbf{v}]_{B'} \tag{5}$$

$$P_{B' \to B}[\mathbf{v}]_{B'} = [\mathbf{v}]_B \tag{6}$$

The following theorem shows that transition matrices in Formulas (3) and (4) can be viewed as matrices for identity operators.

THEOREM 8.5.1 *If B and B' are bases for a finite-dimensional vector space V, and if $I: V \to V$ is the identity operator on V, then*

$$P_{B \to B'} = [I]_{B',B} \quad and \quad P_{B' \to B} = [I]_{B,B'}$$

Proof Suppose that $B = \{\mathbf{u}_1, \mathbf{u}_2, \ldots, \mathbf{u}_n\}$ and $B' = \{\mathbf{u}_1', \mathbf{u}_2', \ldots, \mathbf{u}_n'\}$ are bases for V. Using the fact that $I(\mathbf{v}) = \mathbf{v}$ for all $\mathbf{v}$ in V, it follows from Formula (4) of Section 8.4 that

$$[I]_{B',B} = \left[[I(\mathbf{u}_1)]_{B'} \mid [I(\mathbf{u}_2)]_{B'} \mid \cdots \mid [I(\mathbf{u}_n)]_{B'} \right]$$
$$= \left[[\mathbf{u}_1]_{B'} \mid [\mathbf{u}_2]_{B'} \mid \cdots \mid [\mathbf{u}_n]_{B'} \right]$$
$$= P_{B \to B'} \quad \text{[Formula (3) above]}$$

The proof that $[I]_{B,B'} = P_{B' \to B}$ is similar. ◄

Effect of Changing Bases on Matrices of Linear Operators

We are now ready to consider the main problem in this section.

Problem If B and B' are two bases for a finite-dimensional vector space V, and if $T: V \to V$ is a linear operator, what relationship, if any, exists between the matrices $[T]_B$ and $[T]_{B'}$?

The answer to this question can be obtained by considering the composition of the three linear operators on V pictured in Figure 8.5.1.

$$I \qquad T \qquad I$$

V
Basis = B'
V
Basis = B
V
Basis = B
V
Basis = B'

v   v   T(v)   T(v)

► **Figure 8.5.1**

In this figure, **v** is first mapped into itself by the identity operator, then **v** is mapped into $T(\mathbf{v})$ by T, and then $T(\mathbf{v})$ is mapped into itself by the identity operator. All four vector spaces involved in the composition are the same (namely, V), but the bases for the spaces vary. Since the starting vector is **v** and the final vector is $T(\mathbf{v})$, the composition produces the same result as applying T directly; that is,

$$T = I \circ T \circ I \tag{7}$$

If, as illustrated in Figure 8.5.1, if the first and last vector spaces are assigned the basis B' and the middle two spaces are assigned the basis B, then it follows from (7) and Formula (12) of Section 8.4 (with an appropriate adjustment to the names of the bases) that

$$[T]_{B',B'} = [I \circ T \circ I]_{B',B'} = [I]_{B',B}[T]_{B,B}[I]_{B,B'} \tag{8}$$

or, in simpler notation,

$$[T]_{B'} = [I]_{B',B}[T]_B[I]_{B,B'} \tag{9}$$

We can simplify this formula even further by using Theorem 8.5.1 to rewrite it as

$$[T]_{B'} = P_{B \to B'}[T]_B P_{B' \to B} \tag{10}$$

In summary, we have the following theorem.

THEOREM 8.5.2 *Let $T: V \to V$ be a linear operator on a finite-dimensional vector space V, and let B and B' be bases for V. Then*

$$[T]_{B'} = P^{-1}[T]_B P \tag{11}$$

where $P = P_{B' \to B}$ and $P^{-1} = P_{B \to B'}$.

$$[T]_{B'} = P_{B \to B'}[T]_B P_{B' \to B}$$

Exterior subscripts

▲ **Figure 8.5.2**

Warning   When applying Theorem 8.5.2, it is easy to forget whether $P = P_{B' \to B}$ (correct) or $P = P_{B \to B'}$ (incorrect). It may help to use the diagram in Figure 8.5.2 and observe that the *exterior* subscripts of the transition matrices match the subscript of the matrix they enclose.

In the terminology of Definition 1 of Section 5.2, Theorem 8.5.2 tells us that matrices representing the same linear operator relative to different bases must be similar. The following theorem is a rephrasing of Theorem 8.5.2 in the language of similarity.

THEOREM 8.5.3 *Two matrices, A and B, are similar if and only if they represent the same linear operator. Moreover, if $B = P^{-1}AP$, then P is the transition matrix from the basis relative to matrix B to the basis relative to matrix A.*

► **EXAMPLE 1 Similar Matrices Represent the Same Linear Operator**

We showed at the beginning of this section that the matrices

$$C = \begin{bmatrix} 1 & 1 \\ -2 & 4 \end{bmatrix} \quad \text{and} \quad D = \begin{bmatrix} 2 & 0 \\ 0 & 3 \end{bmatrix}$$

represent the same linear operator $T: R^2 \to R^2$. Verify that these matrices are similar by finding a matrix P for which $D = P^{-1}CP$.

Solution We need to find the transition matrix

$$P = P_{B' \to B} = \left[[\mathbf{u}_1']_B \mid [\mathbf{u}_2']_B \right]$$

where $B' = \{\mathbf{u}_1', \mathbf{u}_2'\}$ is the basis for R^2 given by (2) and $B = \{\mathbf{e}_1, \mathbf{e}_2\}$ is the standard basis for R^2. We see by inspection that

$$\mathbf{u}_1' = \mathbf{e}_1 + \mathbf{e}_2$$
$$\mathbf{u}_2' = \mathbf{e}_1 + 2\mathbf{e}_2$$

from which it follows that

$$[\mathbf{u}_1']_B = \begin{bmatrix} 1 \\ 1 \end{bmatrix} \quad \text{and} \quad [\mathbf{u}_2']_B = \begin{bmatrix} 1 \\ 2 \end{bmatrix}$$

Thus,

$$P = P_{B' \to B} = \left[[\mathbf{u}_1']_B \mid [\mathbf{u}_2']_B \right] = \begin{bmatrix} 1 & 1 \\ 1 & 2 \end{bmatrix}$$

We leave it for you to verify that

$$P^{-1} = \begin{bmatrix} 2 & -1 \\ -1 & 1 \end{bmatrix}$$

and hence that

$$\underbrace{\begin{bmatrix} 2 & 0 \\ 0 & 3 \end{bmatrix}}_{D} = \underbrace{\begin{bmatrix} 2 & -1 \\ -1 & 1 \end{bmatrix}}_{P^{-1}} \underbrace{\begin{bmatrix} 1 & 1 \\ -2 & 4 \end{bmatrix}}_{C} \underbrace{\begin{bmatrix} 1 & 1 \\ 1 & 2 \end{bmatrix}}_{P} \blacktriangleleft$$

Similarity Invariants Recall from Section 5.2 that a property of a square matrix is called a *similarity invariant* if that property is shared by all similar matrices. In Table 1 of that section (table reproduced below), we listed the most important similarity invariants. Since we know from Theorem 8.5.3 that two matrices are similar if and only if they represent the same linear operator $T: V \to V$, it follows that if B and B' are bases for V, then every similarity invariant

Table 1 Similarity Invariants

Property	Description
Determinant	A and $P^{-1}AP$ have the same determinant.
Invertibility	A is invertible if and only if $P^{-1}AP$ is invertible.
Rank	A and $P^{-1}AP$ have the same rank.
Nullity	A and $P^{-1}AP$ have the same nullity.
Trace	A and $P^{-1}AP$ have the same trace.
Characteristic polynomial	A and $P^{-1}AP$ have the same characteristic polynomial.
Eigenvalues	A and $P^{-1}AP$ have the same eigenvalues.
Eigenspace dimension	If λ is an eigenvalue of A and $P^{-1}AP$, then the eigenspace of A corresponding to λ and the eigenspace of $P^{-1}AP$ corresponding to λ have the same dimension.

property of $[T]_B$ is also a similarity invariant property of $[T]_{B'}$ for any other basis B' for V. For example, for any two bases B and B' we must have

$$\det([T]_B) = \det([T]_{B'})$$

It follows from this equation that the value of the determinant depends on T, but not on the particular basis that is used to obtain the matrix for T. Thus, the determinant can be regarded as a property of the linear operator T; indeed, if V is a finite-dimensional vector space, then we can *define* the **determinant of the linear operator** T to be

$$\det(T) = \det([T]_B) \tag{12}$$

where B is *any* basis for V.

▶ **EXAMPLE 2 Determinant of a Linear Operator**

At the beginning of this section we showed that the matrices

$$[T] = \begin{bmatrix} 1 & 1 \\ -2 & 4 \end{bmatrix} \quad \text{and} \quad [T]_{B'} = \begin{bmatrix} 2 & 0 \\ 0 & 3 \end{bmatrix}$$

represent the same linear operator relative to different bases, the first relative to the standard basis $B = \{\mathbf{e}_1, \mathbf{e}_2\}$ for R^2 and the second relative to the basis $B' = \{\mathbf{u}_1', \mathbf{u}_2'\}$ for which

$$\mathbf{u}_1' = \begin{bmatrix} 1 \\ 1 \end{bmatrix}, \quad \mathbf{u}_2' = \begin{bmatrix} 1 \\ 2 \end{bmatrix}$$

This means that $[T]$ and $[T]_{B'}$ must be similar matrices and hence must have the same similarity invariant properties. In particular, they must have the same determinant. We leave it for you to verify that

$$\det[T] = \begin{vmatrix} 1 & 1 \\ -2 & 4 \end{vmatrix} = 6 \quad \text{and} \quad \det[T]_{B'} = \begin{vmatrix} 2 & 0 \\ 0 & 3 \end{vmatrix} = 6$$

▶ **EXAMPLE 3 Eigenvalues and Bases for Eigenspaces**

Find the eigenvalues and bases for the eigenspaces of the linear operator $T : P_2 \rightarrow P_2$ defined by

$$T(a + bx + cx^2) = -2c + (a + 2b + c)x + (a + 3c)x^2$$

Solution We leave it for you to show that the matrix for T with respect to the standard basis $B = \{1, x, x^2\}$ is

$$[T]_B = \begin{bmatrix} 0 & 0 & -2 \\ 1 & 2 & 1 \\ 1 & 0 & 3 \end{bmatrix}$$

The eigenvalues of T are $\lambda = 1$ and $\lambda = 2$ (Example 7 of Section 5.1). Also from that example, the eigenspace of $[T]_B$ corresponding to $\lambda = 2$ has the basis $\{\mathbf{u}_1, \mathbf{u}_2\}$, where

$$\mathbf{u}_1 = \begin{bmatrix} -1 \\ 0 \\ 1 \end{bmatrix}, \quad \mathbf{u}_2 = \begin{bmatrix} 0 \\ 1 \\ 0 \end{bmatrix}$$

and the eigenspace of $[T]_B$ corresponding to $\lambda = 1$ has the basis $\{\mathbf{u}_3\}$, where

$$\mathbf{u}_3 = \begin{bmatrix} -2 \\ 1 \\ 1 \end{bmatrix}$$

The matrices $\mathbf{u}_1$, $\mathbf{u}_2$, and $\mathbf{u}_3$ are the coordinate matrices relative to B of

$$\mathbf{p}_1 = -1 + x^2, \quad \mathbf{p}_2 = x, \quad \mathbf{p}_3 = -2 + x + x^2$$

Thus, the eigenspace of T corresponding to $\lambda = 2$ has the basis

$$\{\mathbf{p}_1, \mathbf{p}_2\} = \{-1 + x^2, x\}$$

and that corresponding to $\lambda = 1$ has the basis

$$\{\mathbf{p}_3\} = \{-2 + x + x^2\}$$

As a check, you can use the given formula for T to verify that

$$T(\mathbf{p}_1) = 2\mathbf{p}_1, \quad T(\mathbf{p}_2) = 2\mathbf{p}_2, \quad \text{and} \quad T(\mathbf{p}_3) = \mathbf{p}_3 \blacktriangleleft$$

Concept Review

• Similarity of matrices representing a linear operator

• Similarity invariant

• Determinant of a linear operator

Skills

• Show that two matrices A and B represent the same linear operator, and find a transition matrix P so that $B = P^{-1}AP$.

• Find the eigenvalues and bases for the eigenspaces of a linear operator on a finite-dimensional vector space.

Exercise Set 8.5

▶ In Exercises 1–7, find the matrix for T relative to the basis B, and use Theorem 8.5.2 to compute the matrix for T relative to the basis B'. ◀

1. $T: R^2 \to R^2$ is defined by

$$T\left(\begin{bmatrix} x_1 \\ x_2 \end{bmatrix}\right) = \begin{bmatrix} x_1 - 2x_2 \\ -x_2 \end{bmatrix}$$

and $B = \{\mathbf{u}_1, \mathbf{u}_2\}$ and $B' = \{\mathbf{v}_1, \mathbf{v}_2\}$, where

$$\mathbf{u}_1 = \begin{bmatrix} 1 \\ 0 \end{bmatrix}, \quad \mathbf{u}_2 = \begin{bmatrix} 0 \\ 1 \end{bmatrix}; \quad \mathbf{v}_1 = \begin{bmatrix} 2 \\ 1 \end{bmatrix}, \quad \mathbf{v}_2 = \begin{bmatrix} -3 \\ 4 \end{bmatrix}$$

2. $T: R^2 \to R^2$ is defined by

$$T\left(\begin{bmatrix} x_1 \\ x_2 \end{bmatrix}\right) = \begin{bmatrix} x_1 + 7x_2 \\ 3x_1 - 4x_2 \end{bmatrix}$$

and $B = \{\mathbf{u}_1, \mathbf{u}_2\}$ and $B' = \{\mathbf{v}_1, \mathbf{v}_2\}$, where

$$\mathbf{u}_1 = \begin{bmatrix} 2 \\ 2 \end{bmatrix}, \quad \mathbf{u}_2 = \begin{bmatrix} 4 \\ -1 \end{bmatrix}; \quad \mathbf{v}_1 = \begin{bmatrix} 1 \\ 3 \end{bmatrix}, \quad \mathbf{v}_2 = \begin{bmatrix} -1 \\ -1 \end{bmatrix}$$

3. $T: R^2 \to R^2$ is the rotation about the origin through an angle of $45°$; B and B' are the bases in Exercise 1.

4. $T: R^3 \to R^3$ is defined by

$$T\left(\begin{bmatrix} x_1 \\ x_2 \\ x_3 \end{bmatrix}\right) = \begin{bmatrix} x_1 + 2x_2 - x_3 \\ -x_2 \\ x_1 + 7x_3 \end{bmatrix}$$

and B is the standard basis for R^3 and $B' = \{\mathbf{v}_1, \mathbf{v}_2, \mathbf{v}_3\}$, where

$$\mathbf{v}_1 = \begin{bmatrix} 1 \\ 0 \\ 0 \end{bmatrix}, \quad \mathbf{v}_2 = \begin{bmatrix} 1 \\ 1 \\ 0 \end{bmatrix}, \quad \mathbf{v}_3 = \begin{bmatrix} 1 \\ 1 \\ 1 \end{bmatrix}$$

5. $T: R^3 \to R^3$ is the orthogonal projection on the xy-plane, and B and B' are as in Exercise 4.

6. $T: R^2 \to R^2$ is defined by $T(\mathbf{x}) = 5\mathbf{x}$, and B and B' are the bases in Exercise 2.

7. $T: P_1 \to P_1$ is defined by $T(a_0 + a_1 x) = a_0 + a_1(x + 1)$, and $B = \{\mathbf{p}_1, \mathbf{p}_2\}$ and $B' = \{\mathbf{q}_1, \mathbf{q}_2\}$, where $\mathbf{p}_1 = 6 + 3x$, $\mathbf{p}_2 = 10 + 2x$, $\mathbf{q}_1 = 2$, $\mathbf{q}_2 = 3 + 2x$.

8. Find $\det(T)$.

(a) $T: R^2 \to R^2$, where
$$T(x_1, x_2) = (3x_1 - 4x_2, -x_1 + 7x_2)$$

(b) $T: R^3 \to R^3$, where
$$T(x_1, x_2, x_3) = (x_1 - x_2, x_2 - x_3, x_3 - x_1)$$

(c) $T: P_2 \to P_2$, where
$$T(p(x)) = p(x - 1)$$

9. Prove that the following are similarity invariants:

(a) rank   (b) nullity   (c) invertibility

10. Let $T: P_4 \to P_4$ be the linear operator given by the formula $T(p(x)) = p(2x + 1)$.

(a) Find a matrix for T relative to some convenient basis, and then use it to find the rank and nullity of T.

(b) Use the result in part (a) to determine whether T is one-to-one.

11. In each part, find a basis for R^2 relative to which the matrix for T is diagonal.

(a) $T\left(\begin{bmatrix} x_1 \\ x_2 \end{bmatrix}\right) = \begin{bmatrix} x_1 - x_2 \\ 2x_1 + 4x_2 \end{bmatrix}$

(b) $T\left(\begin{bmatrix} x_1 \\ x_2 \end{bmatrix}\right) = \begin{bmatrix} 4x_1 - x_2 \\ -3x_1 + x_2 \end{bmatrix}$

12. In each part, find a basis for R^3 relative to which the matrix for T is diagonal.

(a) $T\left(\begin{bmatrix} x_1 \\ x_2 \\ x_3 \end{bmatrix}\right) = \begin{bmatrix} -2x_1 + x_2 - x_3 \\ x_1 - 2x_2 - x_3 \\ -x_1 - x_2 - 2x_3 \end{bmatrix}$

(b) $T\left(\begin{bmatrix} x_1 \\ x_2 \\ x_3 \end{bmatrix}\right) = \begin{bmatrix} -x_2 + x_3 \\ -x_1 + x_3 \\ x_1 + x_2 \end{bmatrix}$

(c) $T\left(\begin{bmatrix} x_1 \\ x_2 \\ x_3 \end{bmatrix}\right) = \begin{bmatrix} 4x_1 + x_3 \\ 2x_1 + 3x_2 + 2x_3 \\ x_1 + 4x_3 \end{bmatrix}$

13. Let $T: P_2 \to P_2$ be defined by

$$T(a_0 + a_1x + a_2x^2) = (5a_0 + 6a_1 + 2a_2)$$
$$- (a_1 + 8a_2)x + (a_0 - 2a_2)x^2$$

(a) Find the eigenvalues of T.

(b) Find bases for the eigenspaces of T.

14. Let $T: M_{22} \to M_{22}$ be defined by

$$T\left(\begin{bmatrix} a & b \\ c & d \end{bmatrix}\right) = \begin{bmatrix} 2c & a+c \\ b-2c & d \end{bmatrix}$$

(a) Find the eigenvalues of T.

(b) Find bases for the eigenspaces of T.

15. Let λ be an eigenvalue of a linear operator $T: V \to V$. Prove that the eigenvectors of T corresponding to λ are the nonzero vectors in the kernel of $\lambda I - T$.

16. (a) Prove that if A and B are similar matrices, then A^2 and B^2 are also similar. More generally, prove that A^k and B^k are similar if k is any positive integer.

(b) If A^2 and B^2 are similar, must A and B be similar? Explain.

17. Let C and D be $m \times n$ matrices, and let $B = \{\mathbf{v}_1, \mathbf{v}_2, \dots, \mathbf{v}_n\}$ be a basis for a vector space V. Show that if $C[\mathbf{x}]_B = D[\mathbf{x}]_B$ for all $\mathbf{x}$ in V, then $C = D$.

18. Find two nonzero 2×2 matrices that are not similar, and explain why they are not.

19. Complete the proof below by justifying each step.

Hypothesis: A and B are similar matrices.

Conclusion: A and B have the same characteristic polynomial.

Proof: (1) $\det(\lambda I - B) = \det(\lambda I - P^{-1}AP)$

(2) $\qquad\qquad = \det(\lambda P^{-1}P - P^{-1}AP)$

(3) $\qquad\qquad = \det(P^{-1}(\lambda I - A)P)$

(4) $\qquad\qquad = \det(P^{-1})\det(\lambda I - A)\det(P)$

(5) $\qquad\qquad = \det(P^{-1})\det(P)\det(\lambda I - A)$

(6) $\qquad\qquad = \det(\lambda I - A)$

20. If A and B are similar matrices, say $B = P^{-1}AP$, then it follows from Exercise 19 that A and B have the same eigenvalues. Suppose that λ is one of the common eigenvalues and $\mathbf{x}$ is a corresponding eigenvector of A. See if you can find an eigenvector of B corresponding to λ (expressed in terms of λ, $\mathbf{x}$, and P).

21. Since the standard basis for R^n is so simple, why would one want to represent a linear operator on R^n in another basis?

22. Prove that trace is a similarity invariant.

True-False Exercises

In parts (a)–(h) determine whether the statement is true or false, and justify your answer.

(a) A matrix cannot be similar to itself.

(b) If A is similar to B, and B is similar to C, then A is similar to C.

(c) If A and B are similar and B is singular, then A is singular.

(d) If A and B are invertible and similar, then A^{-1} and B^{-1} are similar.

(e) If $T_1: R^n \to R^n$ and $T_2: R^n \to R^n$ are linear operators, and if $[T_1]_{B',B} = [T_2]_{B',B}$ with respect to two bases B and B' for R^n, then $T_1(\mathbf{x}) = T_2(\mathbf{x})$ for every vector $\mathbf{x}$ in R^n.

(f) If $T_1: R^n \to R^n$ is a linear operator, and if $[T_1]_B = [T_1]_{B'}$ with respect to two bases B and B' for R^n, then $B = B'$.

(g) If $T: R^n \to R^n$ is a linear operator, and if $[T]_B = I_n$ with respect to some basis B for R^n, then T is the identity operator on R^n.

(h) If $T: R^n \to R^n$ is a linear operator, and if $[T]_{B',B} = I_n$ with respect to two bases B and B' for R^n, then T is the identity operator on R^n.

Chapter 8 Supplementary Exercises

1. Let A be an $n \times n$ matrix, B a nonzero $n \times 1$ matrix, and $\mathbf{x}$ a vector in R^n expressed in matrix notation. Is $T(\mathbf{x}) = A\mathbf{x} + B$ a linear operator on R^n? Justify your answer.

2. Let

$$A = \begin{bmatrix} \cos\theta & -\sin\theta \\ \sin\theta & \cos\theta \end{bmatrix}$$

(a) Show that

$$A^2 = \begin{bmatrix} \cos 2\theta & -\sin 2\theta \\ \sin 2\theta & \cos 2\theta \end{bmatrix} \quad \text{and} \quad A^3 = \begin{bmatrix} \cos 3\theta & -\sin 3\theta \\ \sin 3\theta & \cos 3\theta \end{bmatrix}$$

(b) Based on your answer to part (a), make a guess at the form of the matrix A^n for any positive integer n.

(c) By considering the geometric effect of multiplication by A, obtain the result in part (b) geometrically.

3. Let $T: V \to V$ be defined by $T(\mathbf{v}) = \|\mathbf{v}\|\mathbf{v}$. Show that T is not a linear operator on V.

4. Let $\mathbf{v}_1, \mathbf{v}_2, \ldots, \mathbf{v}_m$ be fixed vectors in R^n, and let $T: R^n \to R^m$ be the function defined by $T(\mathbf{x}) = (\mathbf{x} \cdot \mathbf{v}_1, \mathbf{x} \cdot \mathbf{v}_2, \ldots, \mathbf{x} \cdot \mathbf{v}_m)$, where $\mathbf{x} \cdot \mathbf{v}_i$ is the Euclidean inner product on R^n.

(a) Show that T is a linear transformation.

(b) Show that the matrix with row vectors $\mathbf{v}_1, \mathbf{v}_2, \ldots, \mathbf{v}_m$ is the standard matrix for T.

5. Let $\{\mathbf{e}_1, \mathbf{e}_2, \mathbf{e}_3, \mathbf{e}_4\}$ be the standard basis for R^4, and let $T: R^4 \to R^3$ be the linear transformation for which

$$T(\mathbf{e}_1) = (1, 2, 1), \quad T(\mathbf{e}_2) = (0, 1, 0),$$
$$T(\mathbf{e}_3) = (1, 3, 0), \quad T(\mathbf{e}_4) = (1, 1, 1)$$

(a) Find bases for the range and kernel of T.

(b) Find the rank and nullity of T.

6. Suppose that vectors in R^3 are denoted by 1×3 matrices, and define $T: R^3 \to R^3$ by

$$T([x_1 \quad x_2 \quad x_3]) = [x_1 \quad x_2 \quad x_3] \begin{bmatrix} -1 & 2 & 4 \\ 3 & 0 & 1 \\ 2 & 2 & 5 \end{bmatrix}$$

(a) Find a basis for the kernel of T.

(b) Find a basis for the range of T.

7. Let $B = \{\mathbf{v}_1, \mathbf{v}_2, \mathbf{v}_3, \mathbf{v}_4\}$ be a basis for a vector space V, and let $T: V \to V$ be the linear operator for which

$$T(\mathbf{v}_1) = \mathbf{v}_1 + \mathbf{v}_2 + \mathbf{v}_3 + 3\mathbf{v}_4$$
$$T(\mathbf{v}_2) = \mathbf{v}_1 - \mathbf{v}_2 + 2\mathbf{v}_3 + 2\mathbf{v}_4$$
$$T(\mathbf{v}_3) = 2\mathbf{v}_1 - 4\mathbf{v}_2 + 5\mathbf{v}_3 + 3\mathbf{v}_4$$
$$T(\mathbf{v}_4) = -2\mathbf{v}_1 + 6\mathbf{v}_2 - 6\mathbf{v}_3 - 2\mathbf{v}_4$$

(a) Find the rank and nullity of T.

(b) Determine whether T is one-to-one.

8. Let V and W be vector spaces, let T, T_1, and T_2 be linear transformations from V to W, and let k be a scalar. Define new transformations, $T_1 + T_2$ and kT, by the formulas

$$(T_1 + T_2)(\mathbf{x}) = T_1(\mathbf{x}) + T_2(\mathbf{x})$$
$$(kT)(\mathbf{x}) = k(T(\mathbf{x}))$$

(a) Show that $(T_1 + T_2): V \to W$ and $kT: V \to W$ are both linear transformations.

(b) Show that the set of all linear transformations from V to W with the operations in part (a) is a vector space.

9. Let A and B be similar matrices. Prove:

(a) A^T and B^T are similar.

(b) If A and B are invertible, then A^{-1} and B^{-1} are similar.

10. (**Fredholm Alternative Theorem**) Let $T: V \to V$ be a linear operator on an n-dimensional vector space. Prove that *exactly one* of the following statements holds:

(i) The equation $T(\mathbf{x}) = \mathbf{b}$ has a solution for all vectors $\mathbf{b}$ in V.

(ii) Nullity of $T > 0$.

11. Let $T: M_{22} \to M_{22}$ be the linear operator defined by

$$T(X) = \begin{bmatrix} 1 & 1 \\ 0 & 0 \end{bmatrix} X + X \begin{bmatrix} 0 & 0 \\ 1 & 1 \end{bmatrix}$$

Find the rank and nullity of T.

12. Prove: If A and B are similar matrices, and if B and C are also similar matrices, then A and C are similar matrices.

13. Let $L: M_{22} \to M_{22}$ be the linear operator that is defined by $L(M) = M^T$. Find the matrix for L with respect to the standard basis for M_{22}.

14. Let $B = \{\mathbf{u}_1, \mathbf{u}_2, \mathbf{u}_3\}$ and $B' = \{\mathbf{v}_1, \mathbf{v}_2, \mathbf{v}_3\}$ be bases for a vector space V, and let

$$P = \begin{bmatrix} 2 & -1 & 3 \\ 1 & 1 & 4 \\ 0 & 1 & 2 \end{bmatrix}$$

be the transition matrix from B' to B.

(a) Express $\mathbf{v}_1, \mathbf{v}_2, \mathbf{v}_3$ as linear combinations of $\mathbf{u}_1, \mathbf{u}_2, \mathbf{u}_3$.

(b) Express $\mathbf{u}_1, \mathbf{u}_2, \mathbf{u}_3$ as linear combinations of $\mathbf{v}_1, \mathbf{v}_2, \mathbf{v}_3$.

15. Let $B = \{\mathbf{u}_1, \mathbf{u}_2, \mathbf{u}_3\}$ be a basis for a vector space V, and let $T: V \to V$ be a linear operator for which

$$[T]_B = \begin{bmatrix} -3 & 4 & 7 \\ 1 & 0 & -2 \\ 0 & 1 & 0 \end{bmatrix}$$

Find $[T]_{B'}$, where $B' = \{\mathbf{v}_1, \mathbf{v}_2, \mathbf{v}_3\}$ is the basis for V defined by

$$\mathbf{v}_1 = \mathbf{u}_1, \quad \mathbf{v}_2 = \mathbf{u}_1 + \mathbf{u}_2, \quad \mathbf{v}_3 = \mathbf{u}_1 + \mathbf{u}_2 + \mathbf{u}_3$$

16. Show that the matrices

$$\begin{bmatrix} 1 & 1 \\ -1 & 4 \end{bmatrix} \quad \text{and} \quad \begin{bmatrix} 2 & 1 \\ 1 & 3 \end{bmatrix}$$

are similar but that

$$\begin{bmatrix} 3 & 1 \\ -6 & -2 \end{bmatrix} \quad \text{and} \quad \begin{bmatrix} -1 & 2 \\ 1 & 0 \end{bmatrix}$$

are not.

17. Suppose that $T: V \to V$ is a linear operator, and B is a basis for V for which

$$[T(\mathbf{x})]_B = \begin{bmatrix} x_1 - x_2 + x_3 \\ x_2 \\ x_1 - x_3 \end{bmatrix} \quad \text{if} \quad [\mathbf{x}]_B = \begin{bmatrix} x_1 \\ x_2 \\ x_3 \end{bmatrix}$$

Find $[T]_B$.

18. Let $T: V \to V$ be a linear operator. Prove that T is one-to-one if and only if $\det(T) \neq 0$.

19. (*Calculus required*)

(a) Show that if $\mathbf{f} = f(x)$ is twice differentiable, then the function $D: C^2(-\infty, \infty) \to F(-\infty, \infty)$ defined by $D(\mathbf{f}) = f''(x)$ is a linear transformation.

(b) Find a basis for the kernel of D.

(c) Show that the set of functions satisfying the equation $D(\mathbf{f}) = f(x)$ is a two-dimensional subspace of $C^2(-\infty, \infty)$, and find a basis for this subspace.

20. Let $T: P_2 \to R^3$ be the function defined by the formula

$$T(p(x)) = \begin{bmatrix} p(-1) \\ p(0) \\ p(1) \end{bmatrix}$$

(a) Find $T(x^2 + 5x + 6)$.

(b) Show that T is a linear transformation.

(c) Show that T is one-to-one.

(d) Find $T^{-1}(0, 3, 0)$.

(e) Sketch the graph of the polynomial in part (d).

21. Let x_1, x_2, and x_3 be distinct real numbers such that

$$x_1 < x_2 < x_3$$

and let $T: P_2 \to R^3$ be the function defined by the formula

$$T(p(x)) = \begin{bmatrix} p(x_1) \\ p(x_2) \\ p(x_3) \end{bmatrix}$$

(a) Show that T is a linear transformation.

(b) Show that T is one-to-one.

(c) Verify that if a_1, a_2, and a_3 are any real numbers, then

$$T^{-1}\left(\begin{bmatrix} a_1 \\ a_2 \\ a_3 \end{bmatrix} \right) = a_1 P_1(x) + a_2 P_2(x) + a_3 P_3(x)$$

where

$$P_1(x) = \frac{(x - x_2)(x - x_3)}{(x_1 - x_2)(x_1 - x_3)}$$

$$P_2(x) = \frac{(x - x_1)(x - x_3)}{(x_2 - x_1)(x_2 - x_3)}$$

$$P_3(x) = \frac{(x - x_1)(x - x_2)}{(x_3 - x_1)(x_3 - x_2)}$$

(d) What relationship exists between the graph of the function

$$a_1 P_1(x) + a_2 P_2(x) + a_3 P_3(x)$$

and the points (x_1, a_1), (x_2, a_2), and (x_3, a_3)?

22. (*Calculus required*) Let $p(x)$ and $q(x)$ be continuous functions, and let V be the subspace of $C(-\infty, +\infty)$ consisting of all twice differentiable functions. Define $L: V \to V$ by

$$L(y(x)) = y''(x) + p(x)y'(x) + q(x)y(x)$$

(a) Show that L is a linear transformation.

(b) Consider the special case where $p(x) = 0$ and $q(x) = 1$. Show that the function

$$\phi(x) = c_1 \sin x + c_2 \cos x$$

is in the null space of L for all real values of c_1 and c_2.

23. (*Calculus required*) Let $D: P_n \to P_n$ be the differentiation operator $D(\mathbf{p}) = \mathbf{p}'$. Show that the matrix for D relative to the basis $B = \{1, x, x^2, \ldots, x^n\}$ is

$$\begin{bmatrix} 0 & 1 & 0 & 0 & \cdots & 0 \\ 0 & 0 & 2 & 0 & \cdots & 0 \\ 0 & 0 & 0 & 3 & \cdots & 0 \\ \vdots & \vdots & \vdots & \vdots & & \vdots \\ 0 & 0 & 0 & 0 & \cdots & n \\ 0 & 0 & 0 & 0 & \cdots & 0 \end{bmatrix}$$

24. (*Calculus required*) It can be shown that for any real number c, the vectors

$$1, \quad x - c, \quad \frac{(x - c)^2}{2!}, \ldots, \frac{(x - c)^n}{n!}$$

form a basis for P_n. Find the matrix for the differentiation operator of Exercise 23 with respect to this basis.

25. (*Calculus required*) Let $J: P_n \to P_{n+1}$ be the integration transformation defined by

$$J(\mathbf{p}) = \int_0^x (a_0 + a_1 t + \cdots + a_n t^n)\, dt$$

$$= a_0 x + \frac{a_1}{2} x^2 + \cdots + \frac{a_n}{n+1} x^{n+1}$$

where $\mathbf{p} = a_0 + a_1 x + \cdots + a_n x^n$. Find the matrix for J with respect to the standard bases for P_n and P_{n+1}.

CHAPTER 9

Numerical Methods

CHAPTER CONTENTS
9.1 *LU*-Decompositions 477
9.2 The Power Method 487
9.3 Internet Search Engines 496
9.4 Comparison of Procedures for Solving Linear Systems 501
9.5 Singular Value Decomposition 506
9.6 Data Compression Using Singular Value Decomposition 514

INTRODUCTION This chapter is concerned with "numerical methods" of linear algebra, an area of study that encompasses techniques for solving large-scale linear systems and for finding numerical approximations of various kinds. It is not our objective to discuss algorithms and technical issues in fine detail, since there are many excellent books on the subject. Rather, we will be concerned with introducing some of the basic ideas and exploring important contemporary applications that rely heavily on numerical ideas—singular value decomposition and data compression. A computing utility such as MATLAB, *Mathematica*, or Maple is recommended for Sections 9.2 to 9.6.

9.1 *LU*-Decompositions

Up to now, we have focused on two methods for solving linear systems, Gaussian elimination (reduction to row echelon form) and Gauss–Jordan elimination (reduction to reduced row echelon form). While these methods are fine for the small-scale problems in this text, they are not suitable for large-scale problems in which computer roundoff error, memory usage, and speed are concerns. In this section we will discuss a method for solving a linear system of n equations in n unknowns that is based on factoring its coefficient matrix into a product of lower and upper triangular matrices. This method, called "*LU*-decomposition," is the basis for many computer algorithms in common use.

Solving Linear Systems by Factoring
Our first goal in this section is to show how to solve a linear system $A\mathbf{x} = \mathbf{b}$ of n equations in n unknowns by factoring the coefficient matrix A into a product

$$A = LU \tag{1}$$

where L is lower triangular and U is upper triangular. Once we understand how to do this, we will discuss how to obtain the factorization itself.

Assuming that we have somehow obtained the factorization in (1), the linear system $A\mathbf{x} = \mathbf{b}$ can be solved by the following procedure, called ***LU-decomposition***.

The Method of LU-Decomposition

Step 1. Rewrite the system $A\mathbf{x} = \mathbf{b}$ as

$$LU\mathbf{x} = \mathbf{b} \qquad (2)$$

Step 2. Define a new $n \times 1$ matrix $\mathbf{y}$ by

$$U\mathbf{x} = \mathbf{y} \qquad (3)$$

Step 3. Use (3) to rewrite (2) as $L\mathbf{y} = \mathbf{b}$ and solve this system for $\mathbf{y}$.

Step 4. Substitute $\mathbf{y}$ in (3) and solve for $\mathbf{x}$.

This procedure, which is illustrated in Figure 9.1.1, replaces the single linear system $A\mathbf{x} = \mathbf{b}$ by a pair of linear systems

$$U\mathbf{x} = \mathbf{y}$$
$$L\mathbf{y} = \mathbf{b}$$

that must be solved in succession. However, since each of these systems has a triangular coefficient matrix, it generally turns out to involve no more computation to solve the two systems than to solve the original system directly.

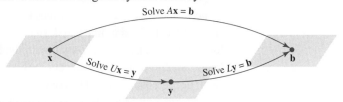

▶ **Figure 9.1.1**

▶ **EXAMPLE 1 Solving $A\mathbf{x} = \mathbf{b}$ by LU-Decomposition**

Later in this section we will derive the factorization

$$\underbrace{\begin{bmatrix} 2 & 6 & 2 \\ -3 & -8 & 0 \\ 4 & 9 & 2 \end{bmatrix}}_{A} = \underbrace{\begin{bmatrix} 2 & 0 & 0 \\ -3 & 1 & 0 \\ 4 & -3 & 7 \end{bmatrix}}_{L} \underbrace{\begin{bmatrix} 1 & 3 & 1 \\ 0 & 1 & 3 \\ 0 & 0 & 1 \end{bmatrix}}_{U} \qquad (4)$$

Use this result to solve the linear system

$$\underbrace{\begin{bmatrix} 2 & 6 & 2 \\ -3 & -8 & 0 \\ 4 & 9 & 2 \end{bmatrix}}_{A} \underbrace{\begin{bmatrix} x_1 \\ x_2 \\ x_3 \end{bmatrix}}_{\mathbf{x}} = \underbrace{\begin{bmatrix} 2 \\ 2 \\ 3 \end{bmatrix}}_{\mathbf{b}}$$

From (4) we can rewrite this system as

$$\underbrace{\begin{bmatrix} 2 & 0 & 0 \\ -3 & 1 & 0 \\ 4 & -3 & 7 \end{bmatrix}}_{L} \underbrace{\begin{bmatrix} 1 & 3 & 1 \\ 0 & 1 & 3 \\ 0 & 0 & 1 \end{bmatrix}}_{U} \underbrace{\begin{bmatrix} x_1 \\ x_2 \\ x_3 \end{bmatrix}}_{\mathbf{x}} = \underbrace{\begin{bmatrix} 2 \\ 2 \\ 3 \end{bmatrix}}_{\mathbf{b}} \qquad (5)$$

Historical Note In 1979 an important library of machine-independent linear algebra programs called LINPACK was developed at Argonne National Laboratories. Many of the programs in that library use the decomposition methods that we will study in this section. Variations of the LINPACK routines are used in many computer programs, including MATLAB, *Mathematica*, and Maple.

As specified in Step 2 above, let us define y_1, y_2, and y_3 by the equation

$$\underbrace{\begin{bmatrix} 1 & 3 & 1 \\ 0 & 1 & 3 \\ 0 & 0 & 1 \end{bmatrix}}_{U} \underbrace{\begin{bmatrix} x_1 \\ x_2 \\ x_3 \end{bmatrix}}_{\mathbf{x}} = \underbrace{\begin{bmatrix} y_1 \\ y_2 \\ y_3 \end{bmatrix}}_{\mathbf{y}} \tag{6}$$

which allows us to rewrite (5) as

$$\underbrace{\begin{bmatrix} 2 & 0 & 0 \\ -3 & 1 & 0 \\ 4 & -3 & 7 \end{bmatrix}}_{L} \underbrace{\begin{bmatrix} y_1 \\ y_2 \\ y_3 \end{bmatrix}}_{\mathbf{y}} = \underbrace{\begin{bmatrix} 2 \\ 2 \\ 3 \end{bmatrix}}_{\mathbf{b}} \tag{7}$$

or equivalently as

$$\begin{aligned} 2y_1 & & & = 2 \\ -3y_1 + \; & y_2 & & = 2 \\ 4y_1 - \; & 3y_2 + 7y_3 & & = 3 \end{aligned}$$

This system can be solved by a procedure that is similar to back substitution, except that we solve the equations from the top down instead of from the bottom up. This procedure, called *forward substitution*, yields

$$y_1 = 1, \quad y_2 = 5, \quad y_3 = 2$$

(verify). As indicated in Step 4 above, we substitute these values into (6), which yields the linear system

$$\begin{bmatrix} 1 & 3 & 1 \\ 0 & 1 & 3 \\ 0 & 0 & 1 \end{bmatrix} \begin{bmatrix} x_1 \\ x_2 \\ x_3 \end{bmatrix} = \begin{bmatrix} 1 \\ 5 \\ 2 \end{bmatrix}$$

or, equivalently,

$$\begin{aligned} x_1 + 3x_2 + \; x_3 &= 1 \\ x_2 + 3x_3 &= 5 \\ x_3 &= 2 \end{aligned}$$

Solving this system by back substitution yields

$$x_1 = 2, \quad x_2 = -1, \quad x_3 = 2$$

(verify). ◄

Alan Mathison Turing (1912–1954)

Historical Note Although the ideas were known earlier, credit for popularizing the matrix formulation of the *LU*-decomposition is often given to the British mathematician Alan Turing for his work on the subject in 1948. Turing, one of the great geniuses of the twentieth century, is the founder of the field of artificial intelligence. Among his many accomplishments in that field, he developed the concept of an internally programmed computer before the practical technology had reached the point where the construction of such a machine was possible. During World War II Turing was secretly recruited by the British government's Code and Cypher School at Bletchley Park to help break the Nazi Enigma codes; it was Turing's statistical approach that provided the breakthrough. In addition to being a brilliant mathematician, Turing was a world-class runner who competed successfully with Olympic-level competition. Sadly, Turing, a homosexual, was tried and convicted of "gross indecency" in 1952, in violation of the then-existing British statutes. Depressed, he committed suicide at age 41 by eating an apple laced with cyanide.

[*Image: Time & Life Pictures/Getty Images, Inc.*]

Finding LU-Decompositions

Example 1 makes it clear that after A is factored into lower and upper triangular matrices, the system $A\mathbf{x} = \mathbf{b}$ can be solved by one forward substitution and one back substitution. We will now show how to obtain such factorizations. We begin with some terminology.

DEFINITION 1 A factorization of a square matrix A as $A = LU$, where L is lower triangular and U is upper triangular is called an *LU-decomposition* (or *LU-factorization*) of A.

Not every square matrix has an LU-decomposition. However, we will see that if it is possible to reduce a square matrix A to row echelon form by Gaussian elimination *without performing any row interchanges*, then A will have an LU-decomposition, though it may not be unique. To see why this is so, assume that A has been reduced to a row echelon form U using a sequence of row operations that does not include row interchanges. We know from Theorem 1.5.1 that these operations can be accomplished by multiplying A on the left by an appropriate sequence of elementary matrices; that is, there exist elementary matrices $E_1, E_2, \ldots, E_k$ such that

$$E_k \cdots E_2 E_1 A = U \tag{8}$$

Since elementary matrices are invertible, we can solve (8) for A as

$$A = E_1^{-1} E_2^{-1} \cdots E_k^{-1} U$$

or more briefly as

$$A = LU \tag{9}$$

where

$$L = E_1^{-1} E_2^{-1} \cdots E_k^{-1} \tag{10}$$

We now have all of the ingredients to prove the following result.

THEOREM 9.1.1 *If A is a square matrix that can be reduced to a row echelon form U by Gaussian elimination without row interchanges, then A can be factored as $A = LU$, where L is a lower triangular matrix.*

Proof Let L and U be the matrices in Formulas (10) and (8), respectively. The matrix U is upper triangular because it is a row echelon form of a square matrix (so all entries below its main diagonal are zero). To prove that L is lower triangular, it suffices to prove that each factor on the right side of (10) is lower triangular, since Theorem 1.7.1*b* will then imply that L itself is lower triangular. Since row interchanges are excluded, each E_j results either by adding a scalar multiple of one row of an identity matrix to a row below or by multiplying one row of an identity matrix by a nonzero scalar. In either case, the resulting matrix E_j is lower triangular and hence so is E_j^{-1} by Theorem 1.7.1*d*. This completes the proof. ◄

▶ **EXAMPLE 2** An *LU*-Decomposition

Find an LU-decomposition of

$$A = \begin{bmatrix} 2 & 6 & 2 \\ -3 & -8 & 0 \\ 4 & 9 & 2 \end{bmatrix}$$

Solution To obtain an *LU*-decomposition, $A = LU$, we will reduce A to a row echelon form U using Gaussian elimination and then calculate L from (10). The steps are as follows:

	Reduction to Row Echelon Form	Row Operation	Elementary Matrix Corresponding to the Row Operation	Inverse of the Elementary Matrix

$$\begin{bmatrix} 2 & 6 & 2 \\ -3 & -8 & 0 \\ 4 & 9 & 2 \end{bmatrix}$$

Step 1 $\tfrac{1}{2} \times$ row 1 $E_1 = \begin{bmatrix} \tfrac{1}{2} & 0 & 0 \\ 0 & 1 & 0 \\ 0 & 0 & 1 \end{bmatrix}$ $E_1^{-1} = \begin{bmatrix} 2 & 0 & 0 \\ 0 & 1 & 0 \\ 0 & 0 & 1 \end{bmatrix}$

$$\begin{bmatrix} 1 & 3 & 1 \\ -3 & -8 & 0 \\ 4 & 9 & 2 \end{bmatrix}$$

Step 2 $(3 \times$ row 1$) +$ row 2 $E_2 = \begin{bmatrix} 1 & 0 & 0 \\ 3 & 1 & 0 \\ 0 & 0 & 1 \end{bmatrix}$ $E_2^{-1} = \begin{bmatrix} 1 & 0 & 0 \\ -3 & 1 & 0 \\ 0 & 0 & 1 \end{bmatrix}$

$$\begin{bmatrix} 1 & 3 & 1 \\ 0 & 1 & 3 \\ 4 & 9 & 2 \end{bmatrix}$$

Step 3 $(-4 \times$ row 1$) +$ row 3 $E_3 = \begin{bmatrix} 1 & 0 & 0 \\ 0 & 1 & 0 \\ -4 & 0 & 1 \end{bmatrix}$ $E_3^{-1} = \begin{bmatrix} 1 & 0 & 0 \\ 0 & 1 & 0 \\ 4 & 0 & 1 \end{bmatrix}$

$$\begin{bmatrix} 1 & 3 & 1 \\ 0 & 1 & 3 \\ 0 & -3 & -2 \end{bmatrix}$$

Step 4 $(3 \times$ row 2$) +$ row 3 $E_4 = \begin{bmatrix} 1 & 0 & 0 \\ 0 & 1 & 0 \\ 0 & 3 & 1 \end{bmatrix}$ $E_4^{-1} = \begin{bmatrix} 1 & 0 & 0 \\ 0 & 1 & 0 \\ 0 & -3 & 1 \end{bmatrix}$

$$\begin{bmatrix} 1 & 3 & 1 \\ 0 & 1 & 3 \\ 0 & 0 & 7 \end{bmatrix}$$

Step 5 $\tfrac{1}{7} \times$ row 3 $E_5 = \begin{bmatrix} 1 & 0 & 0 \\ 0 & 1 & 0 \\ 0 & 0 & \tfrac{1}{7} \end{bmatrix}$ $E_5^{-1} = \begin{bmatrix} 1 & 0 & 0 \\ 0 & 1 & 0 \\ 0 & 0 & 7 \end{bmatrix}$

$$\begin{bmatrix} 1 & 3 & 1 \\ 0 & 1 & 3 \\ 0 & 0 & 1 \end{bmatrix} = U$$

and, from (10),

$$
L = \begin{bmatrix} 2 & 0 & 0 \\ 0 & 1 & 0 \\ 0 & 0 & 1 \end{bmatrix} \begin{bmatrix} 1 & 0 & 0 \\ -3 & 1 & 0 \\ 0 & 0 & 1 \end{bmatrix} \begin{bmatrix} 1 & 0 & 0 \\ 0 & 1 & 0 \\ 4 & 0 & 1 \end{bmatrix} \begin{bmatrix} 1 & 0 & 0 \\ 0 & 1 & 0 \\ 0 & -3 & 1 \end{bmatrix} \begin{bmatrix} 1 & 0 & 0 \\ 0 & 1 & 0 \\ 0 & 0 & 7 \end{bmatrix}
$$

$$
= \begin{bmatrix} 2 & 0 & 0 \\ -3 & 1 & 0 \\ 4 & -3 & 7 \end{bmatrix}
$$

so

$$
\begin{bmatrix} 2 & 6 & 2 \\ -3 & -8 & 0 \\ 4 & 9 & 2 \end{bmatrix} = \begin{bmatrix} 2 & 0 & 0 \\ -3 & 1 & 0 \\ 4 & -3 & 7 \end{bmatrix} \begin{bmatrix} 1 & 3 & 1 \\ 0 & 1 & 3 \\ 0 & 0 & 1 \end{bmatrix}
$$

is an LU-decomposition of A. ◀

Bookkeeping   As Example 2 shows, most of the work in constructing an LU-decomposition is expended in calculating L. However, *all* this work can be eliminated by some careful bookkeeping of the operations used to reduce A to U.

Because we are assuming that no row interchanges are required to reduce A to U, there are only two types of operations involved—multiplying a row by a nonzero constant, and adding a scalar multiple of one row to another. The first operation is used to introduce the leading 1's and the second to introduce zeros below the leading 1's.

In Example 2, a multiplier of $\frac{1}{2}$ was needed in Step 1 to introduce a leading 1 in the first row, and a multiplier of $\frac{1}{7}$ was needed in Step 5 to introduce a leading 1 in the third row. No actual multiplier was required to introduce a leading 1 in the second row because it was already a 1 at the end of Step 2, but for convenience let us say that the multiplier was 1. Comparing these multipliers with the successive diagonal entries of L, we see that these diagonal entries are precisely the reciprocals of the multipliers used to construct U:

$$
L = \begin{bmatrix} ② & 0 & 0 \\ -3 & ① & 0 \\ 4 & -3 & ⑦ \end{bmatrix} \tag{11}
$$

Also observe in Example 2 that to introduce zeros below the leading 1 in the first row, we used the operations

add 3 times the first row to the second

add −4 times the first row to the third

and to introduce the zero below the leading 1 in the second row, we used the operation

add 3 times the second row to the third

Now note in (12) that in each position below the main diagonal of L, the entry is the *negative* of the multiplier in the operation that introduced the zero in that position in U:

$$
L = \begin{bmatrix} 2 & 0 & 0 \\ -3 & 1 & 0 \\ 4 & -3 & 7 \end{bmatrix} \tag{12}
$$

This suggests the following procedure for constructing an LU-decomposition of a square matrix A, assuming that this matrix can be reduced to row echelon form without row interchanges.

Procedure for Constructing an *LU*-Decomposition

Step 1. Reduce A to a row echelon form U by Gaussian elimination without row interchanges, keeping track of the multipliers used to introduce the leading 1's and the multipliers used to introduce the zeros below the leading 1's.

Step 2. In each position along the main diagonal of L, place the reciprocal of the multiplier that introduced the leading 1 in that position in U.

Step 3. In each position below the main diagonal of L, place the negative of the multiplier used to introduce the zero in that position in U.

Step 4. Form the decomposition $A = LU$.

▶ **EXAMPLE 3** Constructing an *LU*-Decomposition

Find an LU-decomposition of

$$A = \begin{bmatrix} 6 & -2 & 0 \\ 9 & -1 & 1 \\ 3 & 7 & 5 \end{bmatrix}$$

Solution We will reduce A to a row echelon form U and at each step we will fill in an entry of L in accordance with the four-step procedure above.

$$A = \begin{bmatrix} 6 & -2 & 0 \\ 9 & -1 & 1 \\ 3 & 7 & 5 \end{bmatrix} \qquad \begin{bmatrix} \bullet & 0 & 0 \\ \bullet & \bullet & 0 \\ \bullet & \bullet & \bullet \end{bmatrix}$$ $\bullet$ denotes an unknown entry of L.

$$\begin{bmatrix} ① & -\tfrac{1}{3} & 0 \\ 9 & -1 & 1 \\ 3 & 7 & 5 \end{bmatrix} \longleftarrow \text{multiplier} = \tfrac{1}{6} \qquad \begin{bmatrix} 6 & 0 & 0 \\ \bullet & \bullet & 0 \\ \bullet & \bullet & \bullet \end{bmatrix}$$

$$\begin{bmatrix} 1 & -\tfrac{1}{3} & 0 \\ ⓪ & 2 & 1 \\ ⓪ & 8 & 5 \end{bmatrix} \begin{matrix} \\ \longleftarrow \text{multiplier} = -9 \\ \longleftarrow \text{multiplier} = -3 \end{matrix} \qquad \begin{bmatrix} 6 & 0 & 0 \\ 9 & \bullet & 0 \\ 3 & \bullet & \bullet \end{bmatrix}$$

$$\begin{bmatrix} 1 & -\tfrac{1}{3} & 0 \\ 0 & ① & \tfrac{1}{2} \\ 0 & 8 & 5 \end{bmatrix} \longleftarrow \text{multiplier} = \tfrac{1}{2} \qquad \begin{bmatrix} 6 & 0 & 0 \\ 9 & 2 & 0 \\ 3 & \bullet & \bullet \end{bmatrix}$$

$$\begin{bmatrix} 1 & -\tfrac{1}{3} & 0 \\ 0 & 1 & \tfrac{1}{2} \\ 0 & ⓪ & 1 \end{bmatrix} \longleftarrow \text{multiplier} = -8 \qquad \begin{bmatrix} 6 & 0 & 0 \\ 9 & 2 & 0 \\ 3 & 8 & \bullet \end{bmatrix}$$

$$U = \begin{bmatrix} 1 & -\tfrac{1}{3} & 0 \\ 0 & 1 & \tfrac{1}{2} \\ 0 & 0 & ① \end{bmatrix} \longleftarrow \text{multiplier} = 1 \qquad L = \begin{bmatrix} 6 & 0 & 0 \\ 9 & 2 & 0 \\ 3 & 8 & 1 \end{bmatrix}$$ No actual operation is performed here since there is already a leading 1 in the third row.

Thus, we have constructed the LU-decomposition

$$A = LU = \begin{bmatrix} 6 & 0 & 0 \\ 9 & 2 & 0 \\ 3 & 8 & 1 \end{bmatrix} \begin{bmatrix} 1 & -\frac{1}{3} & 0 \\ 0 & 1 & \frac{1}{2} \\ 0 & 0 & 1 \end{bmatrix}$$

We leave it for you to confirm this end result by multiplying the factors. ◄

LU-Decompositions Are Not Unique

In the absence of restrictions, LU-decompositions are not unique. For example, if

$$A = LU = \begin{bmatrix} l_{11} & 0 & 0 \\ l_{21} & l_{22} & 0 \\ l_{31} & l_{32} & l_{33} \end{bmatrix} \begin{bmatrix} 1 & u_{12} & u_{13} \\ 0 & 1 & u_{23} \\ 0 & 0 & 1 \end{bmatrix}$$

and L has nonzero diagonal entries, then we can shift the diagonal entries from the left factor to the right factor by writing

$$A = \begin{bmatrix} 1 & 0 & 0 \\ l_{21}/l_{11} & 1 & 0 \\ l_{31}/l_{11} & l_{32}/l_{22} & 1 \end{bmatrix} \begin{bmatrix} l_{11} & 0 & 0 \\ 0 & l_{22} & 0 \\ 0 & 0 & l_{33} \end{bmatrix} \begin{bmatrix} 1 & u_{12} & u_{13} \\ 0 & 1 & u_{23} \\ 0 & 0 & 1 \end{bmatrix}$$

$$= \begin{bmatrix} 1 & 0 & 0 \\ l_{21}/l_{11} & 1 & 0 \\ l_{31}/l_{11} & l_{32}/l_{22} & 1 \end{bmatrix} \begin{bmatrix} l_{11} & l_{11}u_{12} & l_{11}u_{13} \\ 0 & l_{22} & l_{22}u_{23} \\ 0 & 0 & l_{33} \end{bmatrix}$$

which is another LU-decomposition of A.

LDU-Decompositions

The method we have described for computing LU-decompositions may result in an "asymmetry" in that the matrix U has 1's on the main diagonal but L need not. However, if it is preferred to have 1's on the main diagonal of the lower triangular factor, then we can "shift" the diagonal entries of L to a diagonal matrix D and write L as

$$L = L'D$$

where L' is a lower triangular matrix with 1's on the main diagonal. For example, a general 3×3 lower triangular matrix with nonzero entries on the main diagonal can be factored as

$$\underbrace{\begin{bmatrix} a_{11} & 0 & 0 \\ a_{21} & a_{22} & 0 \\ a_{31} & a_{32} & a_{33} \end{bmatrix}}_{L} = \underbrace{\begin{bmatrix} 1 & 0 & 0 \\ a_{21}/a_{11} & 1 & 0 \\ a_{31}/a_{11} & a_{32}/a_{22} & 1 \end{bmatrix}}_{L'} \underbrace{\begin{bmatrix} a_{11} & 0 & 0 \\ 0 & a_{22} & 0 \\ 0 & 0 & a_{33} \end{bmatrix}}_{D}$$

Note that the columns of L' are obtained by dividing each entry in the corresponding column of L by the diagonal entry in the column. Thus, for example, we can rewrite (4) as

$$\begin{bmatrix} 2 & 6 & 2 \\ -3 & -8 & 0 \\ 4 & 9 & 2 \end{bmatrix} = \begin{bmatrix} 2 & 0 & 0 \\ -3 & 1 & 0 \\ 4 & -3 & 7 \end{bmatrix} \begin{bmatrix} 1 & 3 & 1 \\ 0 & 1 & 3 \\ 0 & 0 & 1 \end{bmatrix}$$

$$= \begin{bmatrix} 1 & 0 & 0 \\ -\frac{3}{2} & 1 & 0 \\ 2 & -3 & 1 \end{bmatrix} \begin{bmatrix} 2 & 0 & 0 \\ 0 & 1 & 0 \\ 0 & 0 & 7 \end{bmatrix} \begin{bmatrix} 1 & 3 & 1 \\ 0 & 1 & 3 \\ 0 & 0 & 1 \end{bmatrix}$$

One can prove that if A is a square matrix that can be reduced to row echelon form without row interchanges, then A can be factored *uniquely* as

$$A = LDU$$

where L is a lower triangular matrix with 1's on the main diagonal, D is a diagonal matrix, and U is an upper triangular matrix with 1's on the main diagonal. This is called the ***LDU-decomposition*** (or ***LDU-factorization***) of A.

PLU-Decompositions Many computer algorithms for solving linear systems perform row interchanges to reduce roundoff error, in which case the existence of an LU-decomposition is not guaranteed. However, it is possible to work around this problem by "preprocessing" the coefficient matrix A so that the row interchanges are performed *prior* to computing the LU-decomposition itself. More specifically, the idea is to create a matrix Q (called a ***permutation matrix***) by multiplying, in sequence, those elementary matrices that produce the row interchanges and then execute them by computing the product QA. This product can then be reduced to row echelon form *without* row interchanges, so it is assured to have an LU-decomposition

$$QA = LU \tag{13}$$

Because the matrix Q is invertible (being a product of elementary matrices), the systems $A\mathbf{x} = \mathbf{b}$ and $QA\mathbf{x} = Q\mathbf{b}$ will have the same solutions. But it follows from (13) that the latter system can be rewritten as $LU\mathbf{x} = Q\mathbf{b}$ and hence can be solved using LU-decomposition.

It is common to see Equation (13) expressed as

$$A = PLU \tag{14}$$

in which $P = Q^{-1}$. This is called a ***PLU-decomposition*** or (***PLU-factorization***) of A.

Concept Review

- LU-decomposition
- LDU-decomposition
- PLU-decomposition

Skills

- Determine whether a square matrix has an LU-decomposition.
- Find an LU-decomposition of a square matrix.
- Use the method of LU-decomposition to solve linear systems.

- Find the LDU-decomposition of a square matrix.
- Find a PLU-decomposition of a square matrix.

Exercise Set 9.1

1. Use the method of Example 1 and the LU-decomposition

$$\begin{bmatrix} 3 & -6 \\ -2 & 5 \end{bmatrix} = \begin{bmatrix} 3 & 0 \\ -2 & 1 \end{bmatrix} \begin{bmatrix} 1 & -2 \\ 0 & 1 \end{bmatrix}$$

to solve the system

$$3x_1 - 6x_2 = 0$$
$$-2x_1 + 5x_2 = 1$$

2. Use the method of Example 1 and the LU-decomposition

$$\begin{bmatrix} 3 & -6 & -3 \\ 2 & 0 & 6 \\ -4 & 7 & 4 \end{bmatrix} = \begin{bmatrix} 3 & 0 & 0 \\ 2 & 4 & 0 \\ -4 & -1 & 2 \end{bmatrix} \begin{bmatrix} 1 & -2 & -1 \\ 0 & 1 & 2 \\ 0 & 0 & 1 \end{bmatrix}$$

to solve the system

$$3x_1 - 6x_2 - 3x_3 = -3$$
$$2x_1 \qquad\quad + 6x_3 = -22$$
$$-4x_1 + 7x_2 + 4x_3 = \quad 3$$

▶ In Exercises 3–10, find an LU-decomposition of the coefficient matrix, and then use the method of Example 1 to solve the system. ◀

3. $\begin{bmatrix} 2 & 8 \\ -1 & -1 \end{bmatrix} \begin{bmatrix} x_1 \\ x_2 \end{bmatrix} = \begin{bmatrix} -2 \\ -2 \end{bmatrix}$

4. $\begin{bmatrix} -5 & -10 \\ 6 & 5 \end{bmatrix} \begin{bmatrix} x_1 \\ x_2 \end{bmatrix} = \begin{bmatrix} -10 \\ 19 \end{bmatrix}$

5. $\begin{bmatrix} 2 & -2 & -2 \\ 0 & -2 & 2 \\ -1 & 5 & 2 \end{bmatrix} \begin{bmatrix} x_1 \\ x_2 \\ x_3 \end{bmatrix} = \begin{bmatrix} -4 \\ -2 \\ 6 \end{bmatrix}$

6. $\begin{bmatrix} -3 & 12 & -6 \\ 1 & -2 & 2 \\ 0 & 1 & 1 \end{bmatrix} \begin{bmatrix} x_1 \\ x_2 \\ x_3 \end{bmatrix} = \begin{bmatrix} -33 \\ 7 \\ -1 \end{bmatrix}$

7. $\begin{bmatrix} 5 & 5 & 10 \\ -8 & -7 & -9 \\ 0 & 4 & 26 \end{bmatrix} \begin{bmatrix} x_1 \\ x_2 \\ x_3 \end{bmatrix} = \begin{bmatrix} 0 \\ 1 \\ 4 \end{bmatrix}$

8. $\begin{bmatrix} -1 & -3 & -4 \\ 3 & 10 & -10 \\ -2 & -4 & 11 \end{bmatrix} \begin{bmatrix} x_1 \\ x_2 \\ x_3 \end{bmatrix} = \begin{bmatrix} -6 \\ -3 \\ 9 \end{bmatrix}$

9. $\begin{bmatrix} -1 & 0 & 1 & 0 \\ 2 & 3 & -2 & 6 \\ 0 & -1 & 2 & 0 \\ 0 & 0 & 1 & 5 \end{bmatrix} \begin{bmatrix} x_1 \\ x_2 \\ x_3 \\ x_4 \end{bmatrix} = \begin{bmatrix} 5 \\ -1 \\ 3 \\ 7 \end{bmatrix}$

10. $\begin{bmatrix} 2 & -4 & 0 & 0 \\ 1 & 2 & 12 & 0 \\ 0 & -1 & -4 & -5 \\ 0 & 0 & 2 & 11 \end{bmatrix} \begin{bmatrix} x_1 \\ x_2 \\ x_3 \\ x_4 \end{bmatrix} = \begin{bmatrix} 8 \\ 0 \\ 1 \\ 0 \end{bmatrix}$

11. Let

$$A = \begin{bmatrix} 2 & 1 & -1 \\ -2 & -1 & 2 \\ 2 & 1 & 0 \end{bmatrix}$$

(a) Find an LU-decomposition of A.

(b) Express A in the form $A = L_1DU_1$, where L_1 is lower triangular with 1's along the main diagonal, U_1 is upper triangular, and D is a diagonal matrix.

(c) Express A in the form $A = L_2U_2$, where L_2 is lower triangular with 1's along the main diagonal and U_2 is upper triangular.

▶ In Exercises 12–13, find an LDU-decomposition of A ◀

12. $A = \begin{bmatrix} 2 & 2 \\ 4 & 1 \end{bmatrix}$

13. $A = \begin{bmatrix} 3 & -12 & 6 \\ 0 & 2 & 0 \\ 6 & -28 & 13 \end{bmatrix}$

14. (a) Show that the matrix

$$\begin{bmatrix} 0 & 1 \\ 1 & 0 \end{bmatrix}$$

has no LU-decomposition.

(b) Find a PLU-decomposition of this matrix.

▶ In Exercises 15–16, use the given PLU-decomposition of A to solve the linear system $Ax = b$ by rewriting it as $P^{-1}Ax = P^{-1}b$ and solving this system by LU-decomposition. ◀

15. $b = \begin{bmatrix} 2 \\ 1 \\ 5 \end{bmatrix}$; $A = \begin{bmatrix} 0 & 1 & 4 \\ 1 & 2 & 2 \\ 3 & 1 & 3 \end{bmatrix}$;

$A = \begin{bmatrix} 0 & 1 & 0 \\ 1 & 0 & 0 \\ 0 & 0 & 1 \end{bmatrix} \begin{bmatrix} 1 & 0 & 0 \\ 0 & 1 & 0 \\ 3 & -5 & 1 \end{bmatrix} \begin{bmatrix} 1 & 2 & 2 \\ 0 & 1 & 4 \\ 0 & 0 & 17 \end{bmatrix} = PLU$

16. $b = \begin{bmatrix} 3 \\ 0 \\ 6 \end{bmatrix}$; $A = \begin{bmatrix} 4 & 1 & 2 \\ 0 & 2 & 1 \\ 8 & 1 & 8 \end{bmatrix}$;

$A = \begin{bmatrix} 1 & 0 & 0 \\ 0 & 0 & 1 \\ 0 & 1 & 0 \end{bmatrix} \begin{bmatrix} 1 & 0 & 0 \\ 2 & 1 & 0 \\ 0 & -2 & 1 \end{bmatrix} \begin{bmatrix} 4 & 1 & 2 \\ 0 & -1 & 4 \\ 0 & 0 & 9 \end{bmatrix} = PLU$

▶ In Exercises 17–18, find a PLU-decomposition of A, and use it to solve the linear system $Ax = b$ by the method of Exercises 15 and 16. ◀

17. $A = \begin{bmatrix} 3 & -1 & 0 \\ 3 & -1 & 1 \\ 0 & 2 & 1 \end{bmatrix}$; $b = \begin{bmatrix} -2 \\ 1 \\ 4 \end{bmatrix}$

18. $A = \begin{bmatrix} 0 & 3 & -2 \\ 1 & 1 & 4 \\ 2 & 2 & 5 \end{bmatrix}$; $b = \begin{bmatrix} 7 \\ 5 \\ -2 \end{bmatrix}$

19. Let

$$A = \begin{bmatrix} a & b \\ c & d \end{bmatrix}$$

(a) Prove: If $a \neq 0$, then the matrix A has a unique LU-decomposition with 1's along the main diagonal of L.

(b) Find the LU-decomposition described in part (a).

20. Let $Ax = b$ be a linear system of n equations in n unknowns, and assume that A is an invertible matrix that can be reduced to row-echelon form without row interchanges. How many additions and multiplications are required to solve the system by the method of Example 1?

21. Prove: If A is any $n \times n$ matrix, then A can be factored as $A = PLU$, where L is lower triangular, U is upper triangular, and P can be obtained by interchanging the rows of I_n appropriately. [*Hint:* Let U be a row echelon form of A, and let all row interchanges required in the reduction of A to U be performed first.]

True-False Exercises

In parts (a)–(e) determine whether the statement is true or false, and justify your answer.

(a) Every square matrix has an LU-decomposition.

(b) If a square matrix A is row equivalent to an upper triangular matrix U, then A has an LU-decomposition.

(c) If $L_1, L_2, \ldots, L_k$ are $n \times n$ lower triangular matrices, then the product $L_1 L_2 \cdots L_k$ is lower triangular.

(d) If a square matrix A has an LU-decomposition, then A has a unique LDU-decomposition.

(e) Every square matrix has a PLU-decomposition.

9.2 The Power Method

The eigenvalues of a square matrix can, in theory, be found by solving the characteristic equation. However, this procedure has so many computational difficulties that it is almost never used in applications. In this section we will discuss an algorithm that can be used to approximate the eigenvalue with greatest absolute value and a corresponding eigenvector. This particular eigenvalue and its corresponding eigenvectors are important because they arise naturally in many iterative processes. The methods we will study in this section have recently been used to create Internet search engines such as Google. We will discuss this application in the next section.

The Power Method There are many applications in which some vector $\mathbf{x}_0$ in R^n is multiplied repeatedly by an $n \times n$ matrix A to produce a sequence

$$\mathbf{x}_0, \quad A\mathbf{x}_0, \quad A^2\mathbf{x}_0, \ldots, \quad A^k\mathbf{x}_0, \ldots$$

We call a sequence of this form a ***power sequence generated by A***. In this section we will be concerned with the convergence of power sequences and how such sequences can be used to approximate eigenvalues and eigenvectors. For this purpose, we make the following definition.

> **DEFINITION 1** If the *distinct* eigenvalues of a matrix A are $\lambda_1, \lambda_2, \ldots, \lambda_k$, and if $|\lambda_1|$ is larger than $|\lambda_2|, \ldots, |\lambda_k|$, then λ_1 is called a ***dominant eigenvalue*** of A. Any eigenvector corresponding to a dominant eigenvalue is called a ***dominant eigenvector*** of A.

▶ **EXAMPLE 1 Dominant Eigenvalues**

Some matrices have dominant eigenvalues and some do not. For example, if the distinct eigenvalues of a matrix are

$$\lambda_1 = -4, \quad \lambda_2 = -2, \quad \lambda_3 = 1, \quad \lambda_4 = 3$$

then $\lambda_1 = -4$ is dominant since $|\lambda_1| = 4$ is greater than the absolute values of all the other eigenvalues; but if the distinct eigenvalues of a matrix are

$$\lambda_1 = 7, \quad \lambda_2 = -7, \quad \lambda_3 = -2, \quad \lambda_4 = 5$$

then $|\lambda_1| = |\lambda_2| = 7$, so there is no eigenvalue whose absolute value is greater than the absolute value of all the other eigenvalues. ◀

The most important theorems about convergence of power sequences apply to $n \times n$ matrices with n linearly independent eigenvectors (symmetric matrices, for example), so we will limit our discussion to this case in this section.

> **THEOREM 9.2.1** *Let A be a symmetric $n \times n$ matrix with a positive[*] dominant eigenvalue λ. If $\mathbf{x}_0$ is a unit vector in R^n that is not orthogonal to the eigenspace corresponding to λ, then the normalized power sequence*
>
> $$\mathbf{x}_0, \quad \mathbf{x}_1 = \frac{A\mathbf{x}_0}{\|A\mathbf{x}_0\|}, \quad \mathbf{x}_2 = \frac{A\mathbf{x}_1}{\|A\mathbf{x}_1\|}, \dots, \quad \mathbf{x}_k = \frac{A\mathbf{x}_{k-1}}{\|A\mathbf{x}_{k-1}\|}, \dots \quad (1)$$
>
> *converges to a unit dominant eigenvector, and the sequence*
>
> $$A\mathbf{x}_1 \cdot \mathbf{x}_1, \quad A\mathbf{x}_2 \cdot \mathbf{x}_2, \quad A\mathbf{x}_3 \cdot \mathbf{x}_3, \dots, \quad A\mathbf{x}_k \cdot \mathbf{x}_k, \dots \quad (2)$$
>
> *converges to the dominant eigenvalue λ.*

Remark In the exercises we will ask you to show that (1) can also be expressed as

$$\mathbf{x}_0, \quad \mathbf{x}_1 = \frac{A\mathbf{x}_0}{\|A\mathbf{x}_0\|}, \quad \mathbf{x}_2 = \frac{A^2\mathbf{x}_0}{\|A^2\mathbf{x}_0\|}, \dots, \quad \mathbf{x}_k = \frac{A^k\mathbf{x}_0}{\|A^k\mathbf{x}_0\|}, \dots \quad (3)$$

This form of the power sequence expresses each iterate in terms of the starting vector $\mathbf{x}_0$, rather than in terms of its predecessor.

We will not prove Theorem 9.2.1, but we can make it plausible geometrically in the 2×2 case where A is a symmetric matrix with distinct positive eigenvalues, λ_1 and λ_2, one of which is dominant. To be specific, assume that λ_1 is dominant and

$$\lambda_1 > \lambda_2 > 0$$

Since we are assuming that A is symmetric and has distinct eigenvalues, it follows from Theorem 7.2.2 that the eigenspaces corresponding to λ_1 and λ_2 are perpendicular lines through the origin. Thus, the assumption that $\mathbf{x}_0$ is a unit vector that is not orthogonal to the eigenspace corresponding to λ_1 implies that $\mathbf{x}_0$ does not lie in the eigenspace corresponding to λ_2. To see the geometric effect of multiplying $\mathbf{x}_0$ by A, it will be useful to split $\mathbf{x}_0$ into the sum

$$\mathbf{x}_0 = \mathbf{v}_0 + \mathbf{w}_0 \quad (4)$$

where $\mathbf{v}_0$ and $\mathbf{w}_0$ are the orthogonal projections of $\mathbf{x}_0$ on the eigenspaces of λ_1 and λ_2, respectively (Figure 9.2.1*a*).

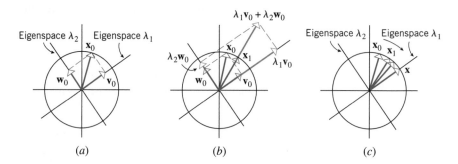

▶ Figure 9.2.1

(a) (b) (c)

This enables us to express $A\mathbf{x}_0$ as

$$A\mathbf{x}_0 = A\mathbf{v}_0 + A\mathbf{w}_0 = \lambda_1\mathbf{v}_0 + \lambda_2\mathbf{w}_0 \quad (5)$$

---

[*] If the dominant eigenvalue is not positive, sequence (2) will still converge to the dominant eigenvalue, but sequence (1) may not converge to a *specific* dominant eigenvector because of *alternation* (see Exercise 11). Nevertheless, each term of (1) will closely approximate *some* dominant eigenvector for sufficiently large values of k.

which tells us that multiplying $\mathbf{x}_0$ by A "scales" the terms $\mathbf{v}_0$ and $\mathbf{w}_0$ in (4) by λ_1 and λ_2, respectively. However, λ_1 is larger than λ_2, so the scaling is greater in the direction of $\mathbf{v}_0$ than in the direction of $\mathbf{w}_0$. Thus, multiplying $\mathbf{x}_0$ by A "pulls" $\mathbf{x}_0$ toward the eigenspace of λ_1, and normalizing produces a vector $\mathbf{x}_1 = A\mathbf{x}_0/\|A\mathbf{x}_0\|$, which is on the unit circle and is closer to the eigenspace of λ_1 than $\mathbf{x}_0$ (Figure 9.2.1b). Similarly, multiplying $\mathbf{x}_1$ by A and normalizing produces a unit vector $\mathbf{x}_2$ that is closer to the eigenspace of λ_1 than $\mathbf{x}_1$. Thus, it seems reasonable that by repeatedly multiplying by A and normalizing we will produce a sequence of vectors $\mathbf{x}_k$ that lie on the unit circle and converge to a unit vector $\mathbf{x}$ in the eigenspace of λ_1 (Figure 9.2.1c). Moreover, if $\mathbf{x}_k$ converges to $\mathbf{x}$, then it also seems reasonable that $A\mathbf{x}_k \cdot \mathbf{x}_k$ will converge to

$$A\mathbf{x} \cdot \mathbf{x} = \lambda_1 \mathbf{x} \cdot \mathbf{x} = \lambda_1 \|\mathbf{x}\|^2 = \lambda_1$$

which is the dominant eigenvalue of A.

The Power Method with Euclidean Scaling

Theorem 9.2.1 provides us with an algorithm for approximating the dominant eigenvalue and a corresponding unit eigenvector of a symmetric matrix A, provided the dominant eigenvalue is positive. This algorithm, called the ***power method with Euclidean scaling***, is as follows:

The Power Method with Euclidean Scaling

Step 1. Choose an arbitrary nonzero vector and normalize it, if need be, to obtain a unit vector $\mathbf{x}_0$.

Step 2. Compute $A\mathbf{x}_0$ and normalize it to obtain the first approximation $\mathbf{x}_1$ to a dominant unit eigenvector. Compute $A\mathbf{x}_1 \cdot \mathbf{x}_1$ to obtain the first approximation to the dominant eigenvalue.

Step 3. Compute $A\mathbf{x}_1$ and normalize it to obtain the second approximation $\mathbf{x}_2$ to a dominant unit eigenvector. Compute $A\mathbf{x}_2 \cdot \mathbf{x}_2$ to obtain the second approximation to the dominant eigenvalue.

Step 4. Compute $A\mathbf{x}_2$ and normalize it to obtain the third approximation $\mathbf{x}_3$ to a dominant unit eigenvector. Compute $A\mathbf{x}_3 \cdot \mathbf{x}_3$ to obtain the third approximation to the dominant eigenvalue.

Continuing in this way will usually generate a sequence of better and better approximations to the dominant eigenvalue and a corresponding unit eigenvector.[*]

▶ **EXAMPLE 2 The Power Method with Euclidean Scaling**

Apply the power method with Euclidean scaling to

$$A = \begin{bmatrix} 3 & 2 \\ 2 & 3 \end{bmatrix} \quad \text{with} \quad \mathbf{x}_0 = \begin{bmatrix} 1 \\ 0 \end{bmatrix}$$

Stop at $\mathbf{x}_5$ and compare the resulting approximations to the exact values of the dominant eigenvalue and eigenvector.

---

[*]If the vector $\mathbf{x}_0$ happens to be orthogonal to the eigenspace of the dominant eigenvalue, then the hypotheses of Theorem 9.2.1 will be violated and the method may fail. However, the reality is that computer roundoff errors usually perturb $\mathbf{x}_0$ enough to destroy any orthogonality and make the algorithm work. This is one instance in which errors help to obtain correct results!

Solution We will leave it for you to show that the eigenvalues of A are $\lambda = 1$ and $\lambda = 5$ and that the eigenspace corresponding to the dominant eigenvalue $\lambda = 5$ is the line represented by the parametric equations $x_1 = t$, $x_2 = t$, which we can write in vector form as

$$\mathbf{x} = t \begin{bmatrix} 1 \\ 1 \end{bmatrix} \tag{6}$$

Setting $t = 1/\sqrt{2}$ yields the normalized dominant eigenvector

$$\mathbf{v}_1 = \begin{bmatrix} \frac{1}{\sqrt{2}} \\ \frac{1}{\sqrt{2}} \end{bmatrix} \approx \begin{bmatrix} 0.707106781187\ldots \\ 0.707106781187\ldots \end{bmatrix} \tag{7}$$

Now let us see what happens when we use the power method, starting with the unit vector $\mathbf{x}_0$.

$$A\mathbf{x}_0 = \begin{bmatrix} 3 & 2 \\ 2 & 3 \end{bmatrix}\begin{bmatrix} 1 \\ 0 \end{bmatrix} = \begin{bmatrix} 3 \\ 2 \end{bmatrix} \qquad \mathbf{x}_1 = \frac{A\mathbf{x}_0}{\|A\mathbf{x}_0\|} = \frac{1}{\sqrt{13}}\begin{bmatrix} 3 \\ 2 \end{bmatrix} \approx \frac{1}{3.60555}\begin{bmatrix} 3 \\ 2 \end{bmatrix} \approx \begin{bmatrix} 0.83205 \\ 0.55470 \end{bmatrix}$$

$$A\mathbf{x}_1 \approx \begin{bmatrix} 3 & 2 \\ 2 & 3 \end{bmatrix}\begin{bmatrix} 0.83205 \\ 0.55470 \end{bmatrix} \approx \begin{bmatrix} 3.60555 \\ 3.32820 \end{bmatrix} \qquad \mathbf{x}_2 = \frac{A\mathbf{x}_1}{\|A\mathbf{x}_1\|} \approx \frac{1}{4.90682}\begin{bmatrix} 3.60555 \\ 3.32820 \end{bmatrix} \approx \begin{bmatrix} 0.73480 \\ 0.67828 \end{bmatrix}$$

$$A\mathbf{x}_2 \approx \begin{bmatrix} 3 & 2 \\ 2 & 3 \end{bmatrix}\begin{bmatrix} 0.73480 \\ 0.67828 \end{bmatrix} \approx \begin{bmatrix} 3.56097 \\ 3.50445 \end{bmatrix} \qquad \mathbf{x}_3 = \frac{A\mathbf{x}_2}{\|A\mathbf{x}_2\|} \approx \frac{1}{4.99616}\begin{bmatrix} 3.56097 \\ 3.50445 \end{bmatrix} \approx \begin{bmatrix} 0.71274 \\ 0.70143 \end{bmatrix}$$

$$A\mathbf{x}_3 \approx \begin{bmatrix} 3 & 2 \\ 2 & 3 \end{bmatrix}\begin{bmatrix} 0.71274 \\ 0.70143 \end{bmatrix} \approx \begin{bmatrix} 3.54108 \\ 3.52976 \end{bmatrix} \qquad \mathbf{x}_4 = \frac{A\mathbf{x}_3}{\|A\mathbf{x}_3\|} \approx \frac{1}{4.99985}\begin{bmatrix} 3.54108 \\ 3.52976 \end{bmatrix} \approx \begin{bmatrix} 0.70824 \\ 0.70597 \end{bmatrix}$$

$$A\mathbf{x}_4 \approx \begin{bmatrix} 3 & 2 \\ 2 & 3 \end{bmatrix}\begin{bmatrix} 0.70824 \\ 0.70597 \end{bmatrix} \approx \begin{bmatrix} 3.53666 \\ 3.53440 \end{bmatrix} \qquad \mathbf{x}_5 = \frac{A\mathbf{x}_4}{\|A\mathbf{x}_4\|} \approx \frac{1}{4.99999}\begin{bmatrix} 3.53666 \\ 3.53440 \end{bmatrix} \approx \begin{bmatrix} 0.70733 \\ 0.70688 \end{bmatrix}$$

$$\lambda^{(1)} = (A\mathbf{x}_1) \cdot \mathbf{x}_1 = (A\mathbf{x}_1)^T \mathbf{x}_1 \approx \begin{bmatrix} 3.60555 & 3.32820 \end{bmatrix}\begin{bmatrix} 0.83205 \\ 0.55470 \end{bmatrix} \approx 4.84615$$

$$\lambda^{(2)} = (A\mathbf{x}_2) \cdot \mathbf{x}_2 = (A\mathbf{x}_2)^T \mathbf{x}_2 \approx \begin{bmatrix} 3.56097 & 3.50445 \end{bmatrix}\begin{bmatrix} 0.73480 \\ 0.67828 \end{bmatrix} \approx 4.99361$$

$$\lambda^{(3)} = (A\mathbf{x}_3) \cdot \mathbf{x}_3 = (A\mathbf{x}_3)^T \mathbf{x}_3 \approx \begin{bmatrix} 3.54108 & 3.52976 \end{bmatrix}\begin{bmatrix} 0.71274 \\ 0.70143 \end{bmatrix} \approx 4.99974$$

$$\lambda^{(4)} = (A\mathbf{x}_4) \cdot \mathbf{x}_4 = (A\mathbf{x}_4)^T \mathbf{x}_4 \approx \begin{bmatrix} 3.53666 & 3.53440 \end{bmatrix}\begin{bmatrix} 0.70824 \\ 0.70597 \end{bmatrix} \approx 4.99999$$

$$\lambda^{(5)} = (A\mathbf{x}_5) \cdot \mathbf{x}_5 = (A\mathbf{x}_5)^T \mathbf{x}_5 \approx \begin{bmatrix} 3.53576 & 3.53531 \end{bmatrix}\begin{bmatrix} 0.70733 \\ 0.70688 \end{bmatrix} \approx 5.00000$$

It is accidental that $\lambda^{(5)}$ (the fifth approximation) produced five decimal place accuracy. In general, n iterations need not produce n decimal place accuracy.

Thus, $\lambda^{(5)}$ approximates the dominant eigenvalue to five decimal place accuracy and $\mathbf{x}_5$ approximates the dominant eigenvector in (7) correctly to three decimal place accuracy. ◄

The Power Method with Maximum Entry Scaling

There is a variation of the power method in which the iterates, rather than being normalized at each stage, are scaled to make the maximum entry 1. To describe this method, it will be convenient to denote the maximum *absolute value* of the entries in a vector $\mathbf{x}$ by

$\max(\mathbf{x})$. Thus, for example, if

$$\mathbf{x} = \begin{bmatrix} 5 \\ 3 \\ -7 \\ 2 \end{bmatrix}$$

then $\max(\mathbf{x}) = 7$. We will need the following variation of Theorem 9.2.1.

THEOREM 9.2.2 *Let A be a symmetric $n \times n$ matrix with a positive dominant[*] eigenvalue λ. If $\mathbf{x}_0$ is a nonzero vector in R^n that is not orthogonal to the eigenspace corresponding to λ, then the sequence*

$$\mathbf{x}_0, \quad \mathbf{x}_1 = \frac{A\mathbf{x}_0}{\max(A\mathbf{x}_0)}, \quad \mathbf{x}_2 = \frac{A\mathbf{x}_1}{\max(A\mathbf{x}_1)}, \dots, \quad \mathbf{x}_k = \frac{A\mathbf{x}_{k-1}}{\max(A\mathbf{x}_{k-1})}, \dots \quad (8)$$

converges to an eigenvector corresponding to λ, and the sequence

$$\frac{A\mathbf{x}_1 \cdot \mathbf{x}_1}{\mathbf{x}_1 \cdot \mathbf{x}_1}, \quad \frac{A\mathbf{x}_2 \cdot \mathbf{x}_2}{\mathbf{x}_2 \cdot \mathbf{x}_2}, \quad \frac{A\mathbf{x}_3 \cdot \mathbf{x}_3}{\mathbf{x}_3 \cdot \mathbf{x}_3}, \dots, \quad \frac{A\mathbf{x}_k \cdot \mathbf{x}_k}{\mathbf{x}_k \cdot \mathbf{x}_k}, \dots \quad (9)$$

converges to λ.

Remark   In the exercises we will ask you to show that (8) can be written in the alternative form

$$\mathbf{x}_0, \quad \mathbf{x}_1 = \frac{A\mathbf{x}_0}{\max(A\mathbf{x}_0)}, \quad \mathbf{x}_2 = \frac{A^2\mathbf{x}_0}{\max(A^2\mathbf{x}_0)}, \dots, \quad \mathbf{x}_k = \frac{A^k\mathbf{x}_0}{\max(A^k\mathbf{x}_0)}, \dots \quad (10)$$

which expresses the iterates in terms of the initial vector $\mathbf{x}_0$.

We will omit the proof of this theorem, but if we accept that (8) converges to an eigenvector of A, then it is not hard to see why (9) converges to the dominant eigenvalue. For this purpose we note that each term in (9) is of the form

$$\frac{A\mathbf{x} \cdot \mathbf{x}}{\mathbf{x} \cdot \mathbf{x}} \quad (11)$$

which is called a ***Rayleigh quotient*** of A. In the case where λ is an eigenvalue of A and $\mathbf{x}$ is a corresponding eigenvector, the Rayleigh quotient is

$$\frac{A\mathbf{x} \cdot \mathbf{x}}{\mathbf{x} \cdot \mathbf{x}} = \frac{\lambda \mathbf{x} \cdot \mathbf{x}}{\mathbf{x} \cdot \mathbf{x}} = \frac{\lambda(\mathbf{x} \cdot \mathbf{x})}{\mathbf{x} \cdot \mathbf{x}} = \lambda$$

Thus, if $\mathbf{x}_k$ converges to a dominant eigenvector $\mathbf{x}$, then it seems reasonable that

$$\frac{A\mathbf{x}_k \cdot \mathbf{x}_k}{\mathbf{x}_k \cdot \mathbf{x}_k} \quad \text{converges to} \quad \frac{A\mathbf{x} \cdot \mathbf{x}}{\mathbf{x} \cdot \mathbf{x}} = \lambda$$

which is the dominant eigenvalue.

Theorem 9.2.2 produces the following algorithm, called the ***power method with maximum entry scaling***.

---

[*]As in Theorem 9.2.1, if the dominant eigenvalue is not positive, sequence (9) will still converge to the dominant eigenvalue, but sequence (8) may not converge to a *specific* dominant eigenvector. Nevertheless, each term of (8) will closely approximate *some* dominant eigenvector for sufficiently large values of k.

John William Strutt Rayleigh (1842–1919)

The Power Method with Maximum Entry Scaling

Step 1.   Choose an arbitrary nonzero vector $\mathbf{x}_0$.

Step 2.   Compute $A\mathbf{x}_0$ and multiply it by the factor $1/\max(A\mathbf{x}_0)$ to obtain the first approximation $\mathbf{x}_1$ to a dominant eigenvector. Compute the Rayleigh quotient of $\mathbf{x}_1$ to obtain the first approximation to the dominant eigenvalue.

Step 3.   Compute $A\mathbf{x}_1$ and scale it by the factor $1/\max(A\mathbf{x}_1)$ to obtain the second approximation $\mathbf{x}_2$ to a dominant eigenvector. Compute the Rayleigh quotient of $\mathbf{x}_2$ to obtain the second approximation to the dominant eigenvalue.

Step 4.   Compute $A\mathbf{x}_2$ and scale it by the factor $1/\max(A\mathbf{x}_2)$ to obtain the third approximation $\mathbf{x}_3$ to a dominant eigenvector. Compute the Rayleigh quotient of $\mathbf{x}_3$ to obtain the third approximation to the dominant eigenvalue.

Continuing in this way will generate a sequence of better and better approximations to the dominant eigenvalue and a corresponding eigenvector.

▶ **EXAMPLE 3 Example 2 Revisited Using Maximum Entry Scaling**

Apply the power method with maximum entry scaling to

$$A = \begin{bmatrix} 3 & 2 \\ 2 & 3 \end{bmatrix} \quad \text{with} \quad \mathbf{x}_0 = \begin{bmatrix} 1 \\ 0 \end{bmatrix}$$

Stop at $\mathbf{x}_5$ and compare the resulting approximations to the exact values and to the approximations obtained in Example 2.

Solution We leave it for you to confirm that

$$A\mathbf{x}_0 = \begin{bmatrix} 3 & 2 \\ 2 & 3 \end{bmatrix}\begin{bmatrix} 1 \\ 0 \end{bmatrix} = \begin{bmatrix} 3 \\ 2 \end{bmatrix} \qquad \mathbf{x}_1 = \frac{A\mathbf{x}_0}{\max(A\mathbf{x}_0)} = \frac{1}{3}\begin{bmatrix} 3 \\ 2 \end{bmatrix} \approx \begin{bmatrix} 1.00000 \\ 0.66667 \end{bmatrix}$$

$$A\mathbf{x}_1 \approx \begin{bmatrix} 3 & 2 \\ 2 & 3 \end{bmatrix}\begin{bmatrix} 1.00000 \\ 0.66667 \end{bmatrix} \approx \begin{bmatrix} 4.33333 \\ 4.00000 \end{bmatrix} \qquad \mathbf{x}_2 = \frac{A\mathbf{x}_1}{\max(A\mathbf{x}_1)} \approx \frac{1}{4.33333}\begin{bmatrix} 4.33333 \\ 4.00000 \end{bmatrix} \approx \begin{bmatrix} 1.00000 \\ 0.92308 \end{bmatrix}$$

$$A\mathbf{x}_2 \approx \begin{bmatrix} 3 & 2 \\ 2 & 3 \end{bmatrix}\begin{bmatrix} 1.00000 \\ 0.92308 \end{bmatrix} \approx \begin{bmatrix} 4.84615 \\ 4.76923 \end{bmatrix} \qquad \mathbf{x}_3 = \frac{A\mathbf{x}_2}{\max(A\mathbf{x}_2)} \approx \frac{1}{4.84615}\begin{bmatrix} 4.84615 \\ 4.76923 \end{bmatrix} \approx \begin{bmatrix} 1.00000 \\ 0.98413 \end{bmatrix}$$

$$A\mathbf{x}_3 \approx \begin{bmatrix} 3 & 2 \\ 2 & 3 \end{bmatrix}\begin{bmatrix} 1.00000 \\ 0.98413 \end{bmatrix} \approx \begin{bmatrix} 4.96825 \\ 4.95238 \end{bmatrix} \qquad \mathbf{x}_4 = \frac{A\mathbf{x}_3}{\max(A\mathbf{x}_3)} \approx \frac{1}{4.96825}\begin{bmatrix} 4.96825 \\ 4.95238 \end{bmatrix} \approx \begin{bmatrix} 1.00000 \\ 0.99681 \end{bmatrix}$$

$$A\mathbf{x}_4 \approx \begin{bmatrix} 3 & 2 \\ 2 & 3 \end{bmatrix}\begin{bmatrix} 1.00000 \\ 0.99681 \end{bmatrix} \approx \begin{bmatrix} 4.99361 \\ 4.99042 \end{bmatrix} \qquad \mathbf{x}_5 = \frac{A\mathbf{x}_4}{\max(A\mathbf{x}_4)} \approx \frac{1}{4.99361}\begin{bmatrix} 4.99361 \\ 4.99042 \end{bmatrix} \approx \begin{bmatrix} 1.00000 \\ 0.99936 \end{bmatrix}$$

$$\lambda^{(1)} = \frac{A\mathbf{x}_1 \cdot \mathbf{x}_1}{\mathbf{x}_1 \cdot \mathbf{x}_1} = \frac{(A\mathbf{x}_1)^T \mathbf{x}_1}{\mathbf{x}_1^T \mathbf{x}_1} \approx \frac{7.00000}{1.44444} \approx 4.84615$$

$$\lambda^{(2)} = \frac{A\mathbf{x}_2 \cdot \mathbf{x}_2}{\mathbf{x}_2 \cdot \mathbf{x}_2} = \frac{(A\mathbf{x}_2)^T \mathbf{x}_2}{\mathbf{x}_2^T \mathbf{x}_2} \approx \frac{9.24852}{1.85207} \approx 4.99361$$

$$\lambda^{(3)} = \frac{A\mathbf{x}_3 \cdot \mathbf{x}_3}{\mathbf{x}_3 \cdot \mathbf{x}_3} = \frac{(A\mathbf{x}_3)^T \mathbf{x}_3}{\mathbf{x}_3^T \mathbf{x}_3} \approx \frac{9.84203}{1.96851} \approx 4.99974$$

$$\lambda^{(4)} = \frac{A\mathbf{x}_4 \cdot \mathbf{x}_4}{\mathbf{x}_4 \cdot \mathbf{x}_4} = \frac{(A\mathbf{x}_4)^T \mathbf{x}_4}{\mathbf{x}_4^T \mathbf{x}_4} \approx \frac{9.96808}{1.99362} \approx 4.99999$$

$$\lambda^{(5)} = \frac{A\mathbf{x}_5 \cdot \mathbf{x}_5}{\mathbf{x}_5 \cdot \mathbf{x}_5} = \frac{(A\mathbf{x}_5)^T \mathbf{x}_5}{\mathbf{x}_5^T \mathbf{x}_5} \approx \frac{9.99360}{1.99872} \approx 5.00000$$

Whereas the power method with Euclidean scaling produces a sequence that approaches a *unit* dominant eigenvector, maximum entry scaling produces a sequence that approaches an eigenvector whose largest component is 1.

Thus, $\lambda^{(5)}$ approximates the dominant eigenvalue correctly to five decimal places and $\mathbf{x}_5$ closely approximates the dominant eigenvector

$$\mathbf{x} = \begin{bmatrix} 1 \\ 1 \end{bmatrix}$$

that results by taking $t = 1$ in (6). ◀

Rate of Convergence   If A is a symmetric matrix whose distinct eigenvalues can be arranged so that

$$|\lambda_1| > |\lambda_2| \geq |\lambda_3| \geq \cdots \geq |\lambda_k|$$

then the "rate" at which the Rayleigh quotients converge to the dominant eigenvalue λ_1 depends on the ratio $|\lambda_1|/|\lambda_2|$; that is, the convergence is slow when this ratio is near 1 and rapid when it is large—the greater the ratio, the more rapid the convergence. For example, if A is a 2×2 symmetric matrix, then the greater the ratio $|\lambda_1|/|\lambda_2|$, the greater the disparity between the scaling effects of λ_1 and λ_2 in Figure 9.2.1, and hence the greater the effect that multiplication by A has on pulling the iterates toward the eigenspace of λ_1. Indeed, the rapid convergence in Example 3 is due to the fact that $|\lambda_1|/|\lambda_2| = 5/1 = 5$, which is considered to be a large ratio. In cases where the ratio is close to 1, the convergence of the power method may be so slow that other methods must be used.

Stopping Procedures   If λ is the exact value of the dominant eigenvalue, and if a power method produces the approximation $\lambda^{(k)}$ at the kth iteration, then we call

$$\left| \frac{\lambda - \lambda^{(k)}}{\lambda} \right| \tag{12}$$

the *relative error* in $\lambda^{(k)}$. If this is expressed as a percentage, then it is called the *percentage error* in $\lambda^{(k)}$. For example, if $\lambda = 5$ and the approximation after three iterations is $\lambda^{(3)} = 5.1$, then

$$\text{relative error in } \lambda^{(3)} = \left| \frac{\lambda - \lambda^{(3)}}{\lambda} \right| = \left| \frac{5 - 5.1}{5} \right| = |-0.02| = 0.02$$

$$\text{percentage error in } \lambda^{(3)} = 0.02 \times 100\% = 2\%$$

In applications one usually knows the relative error E that can be tolerated in the dominant eigenvalue, so the goal is to stop computing iterates once the relative error in the approximation to that eigenvalue is less than E. However, there is a problem in computing the relative error from (12) in that the eigenvalue λ is unknown. To circumvent this problem, it is usual to estimate λ by $\lambda^{(k)}$ and stop the computations when

$$\left| \frac{\lambda^{(k)} - \lambda^{(k-1)}}{\lambda^{(k)}} \right| < E \tag{13}$$

The quantity on the left side of (13) is called the *estimated relative error* in $\lambda^{(k)}$ and its percentage form is called the *estimated percentage error* in $\lambda^{(k)}$.

▶ **EXAMPLE 4 Estimated Relative Error**

For the computations in Example 3, find the smallest value of k for which the estimated percentage error in $\lambda^{(k)}$ is less than 0.1%.

Solution The estimated percentage errors in the approximations in Example 3 are as follows:

		APPROXIMATION	RELATIVE ERROR	PERCENTAGE ERROR
$\lambda^{(2)}$:	$\left\| \dfrac{\lambda^{(2)} - \lambda^{(1)}}{\lambda^{(2)}} \right\| \approx$	$\left\| \dfrac{4.99361 - 4.84615}{4.99361} \right\| \approx$	$0.02953 =$	2.953%
$\lambda^{(3)}$:	$\left\| \dfrac{\lambda^{(3)} - \lambda^{(2)}}{\lambda^{(3)}} \right\| \approx$	$\left\| \dfrac{4.99974 - 4.99361}{4.99974} \right\| \approx$	$0.00123 =$	0.123%
$\lambda^{(4)}$:	$\left\| \dfrac{\lambda^{(4)} - \lambda^{(3)}}{\lambda^{(4)}} \right\| \approx$	$\left\| \dfrac{4.99999 - 4.99974}{4.99999} \right\| \approx$	$0.00005 =$	0.005%
$\lambda^{(5)}$:	$\left\| \dfrac{\lambda^{(5)} - \lambda^{(4)}}{\lambda^{(5)}} \right\| \approx$	$\left\| \dfrac{5.00000 - 4.99999}{5.00000} \right\| \approx$	$0.00000 =$	0%

Thus, $\lambda^{(4)} = 4.99999$ is the first approximation whose estimated percentage error is less than 0.1%. ◀

Remark A rule for deciding when to stop an iterative process is called a *stopping procedure*. In the exercises, we will discuss stopping procedures for the power method that are based on the dominant eigenvector rather than the dominant eigenvalue.

Concept Review

- Power sequence
- Dominant eigenvalue
- Dominant eigenvector
- Power method with Euclidean scaling
- Rayleigh quotient
- Power method with maximum entry scaling
- Relative error
- Percentage error
- Estimated relative error
- Estimated percentage error
- Stopping procedure

Skills

- Identify the dominant eigenvalue of a matrix.
- Use the power methods described in this section to approximate a dominant eigenvector.
- Find the estimated relative and percentage errors associated with the power methods.

Exercise Set 9.2

▶ In Exercises 1–2, the distinct eigenvalues of a matrix are given. Determine whether A has a dominant eigenvalue, and if so, find it.

1. (a) $\lambda_1 = 7, \ \lambda_2 = 3, \ \lambda_3 = -8, \ \lambda_4 = 1$

  (b) $\lambda_1 = -5, \ \lambda_2 = 3, \ \lambda_3 = 2, \ \lambda_4 = 5$

2. (a) $\lambda_1 = 1, \ \lambda_2 = 0, \ \lambda_3 = -3, \ \lambda_4 = 2$

  (b) $\lambda_1 = -3, \ \lambda_2 = -2, \ \lambda_3 = -1, \ \lambda_4 = 3$

▶ In Exercises 3–4, apply the power method with Euclidean scaling to the matrix A, starting with x_0 and stopping at x_4. Compare the resulting approximations to the exact values of the dominant eigenvalue and the corresponding unit eigenvector. ◀

3. $A = \begin{bmatrix} 5 & -1 \\ -1 & -1 \end{bmatrix}$; $\ x_0 = \begin{bmatrix} 1 \\ 0 \end{bmatrix}$

4. $A = \begin{bmatrix} 7 & -2 & 0 \\ -2 & 6 & -2 \\ 0 & -2 & 5 \end{bmatrix}$; $\ x_0 = \begin{bmatrix} 1 \\ 0 \\ 0 \end{bmatrix}$

▶ In Exercises 5–6, apply the power method with maximum entry scaling to the matrix A, starting with x_0 and stopping at x_4. Compare the resulting approximations to the exact values of the dominant eigenvalue and the corresponding scaled eigenvector.

5. $A = \begin{bmatrix} 1 & -3 \\ -3 & 5 \end{bmatrix}$; $\ x_0 = \begin{bmatrix} 1 \\ 1 \end{bmatrix}$

6. $A = \begin{bmatrix} 3 & 2 & 2 \\ 2 & 2 & 0 \\ 2 & 0 & 4 \end{bmatrix}$; $\mathbf{x}_0 = \begin{bmatrix} 1 \\ 1 \\ 1 \end{bmatrix}$

7. Let

$$A = \begin{bmatrix} 2 & -1 \\ -1 & 2 \end{bmatrix}; \quad \mathbf{x}_0 = \begin{bmatrix} 1 \\ 0 \end{bmatrix}$$

(a) Use the power method with maximum entry scaling to approximate a dominant eigenvector of A. Start with $\mathbf{x}_0$, round off all computations to three decimal places, and stop after three iterations.

(b) Use the result in part (a) and the Rayleigh quotient to approximate the dominant eigenvalue of A.

(c) Find the exact values of the eigenvector and eigenvalue approximated in parts (a) and (b).

(d) Find the percentage error in the approximation of the dominant eigenvalue.

8. Repeat the directions of Exercise 7 with

$$A = \begin{bmatrix} 2 & 1 & 0 \\ 1 & 2 & 0 \\ 0 & 0 & 10 \end{bmatrix}; \quad \mathbf{x}_0 = \begin{bmatrix} 1 \\ 1 \\ 1 \end{bmatrix}$$

▶ In Exercises 9–10, a matrix A with a dominant eigenvalue and a sequence $\mathbf{x}_0, A\mathbf{x}_0, \ldots, A^5\mathbf{x}_0$ are given. Use Formulas (8) and (9) to approximate the dominant eigenvalue and a corresponding eigenvector. ◀

9. $A = \begin{bmatrix} 1 & 2 \\ 2 & 1 \end{bmatrix}$; $\mathbf{x}_0 = \begin{bmatrix} 1 \\ 0 \end{bmatrix}$, $A\mathbf{x}_0 = \begin{bmatrix} 1 \\ 2 \end{bmatrix}$, $A^2\mathbf{x}_0 = \begin{bmatrix} 5 \\ 4 \end{bmatrix}$,

$A^3\mathbf{x}_0 = \begin{bmatrix} 13 \\ 14 \end{bmatrix}$, $A^4\mathbf{x}_0 = \begin{bmatrix} 41 \\ 40 \end{bmatrix}$, $A^5\mathbf{x}_0 = \begin{bmatrix} 121 \\ 122 \end{bmatrix}$

10. $A = \begin{bmatrix} 1 & 2 \\ 2 & 1 \end{bmatrix}$; $\mathbf{x}_0 = \begin{bmatrix} 0 \\ 1 \end{bmatrix}$, $A\mathbf{x}_0 = \begin{bmatrix} 2 \\ 1 \end{bmatrix}$, $A^2\mathbf{x}_0 = \begin{bmatrix} 4 \\ 5 \end{bmatrix}$,

$A^3\mathbf{x}_0 = \begin{bmatrix} 14 \\ 13 \end{bmatrix}$, $A^4\mathbf{x}_0 = \begin{bmatrix} 40 \\ 41 \end{bmatrix}$, $A^5\mathbf{x}_0 = \begin{bmatrix} 122 \\ 121 \end{bmatrix}$

11. Consider matrices

$$A = \begin{bmatrix} -1 & 0 \\ 0 & 0 \end{bmatrix} \quad \text{and} \quad \mathbf{x}_0 = \begin{bmatrix} a \\ b \end{bmatrix}$$

where $\mathbf{x}_0$ is a unit vector and $a \neq 0$. Show that even though the matrix A is symmetric and has a dominant eigenvalue, the power sequence (1) in Theorem 9.2.1 does not converge. This shows that the requirement in that theorem that the dominant eigenvalue be positive is essential.

12. Use the power method with Euclidean scaling to approximate the dominant eigenvalue and a corresponding eigenvector of A. Choose your own starting vector, and stop when the estimated percentage error in the eigenvalue approximation is less than 0.1%.

(a) $\begin{bmatrix} 1 & 3 & 3 \\ 3 & 4 & -1 \\ 3 & -1 & 10 \end{bmatrix}$ (b) $\begin{bmatrix} 1 & 0 & 1 & 1 \\ 0 & 2 & -1 & 1 \\ 1 & -1 & 4 & 1 \\ 1 & 1 & 1 & 8 \end{bmatrix}$

13. Repeat Exercise 12, but this time stop when all corresponding entries in two successive eigenvector approximations differ by less than 0.01 in absolute value.

14. Repeat Exercise 12 using maximum entry scaling.

15. Prove: If A is a nonzero $n \times n$ matrix, then $A^T A$ and $A A^T$ have positive dominant eigenvalues.

16. (***For readers familiar with proof by induction***) Let A be an $n \times n$ matrix, let $\mathbf{x}_0$ be a unit vector in R^n, and define the sequence $\mathbf{x}_1, \mathbf{x}_2, \ldots, \mathbf{x}_k, \ldots$ by

$$\mathbf{x}_1 = \frac{A\mathbf{x}_0}{\|A\mathbf{x}_0\|}, \quad \mathbf{x}_2 = \frac{A\mathbf{x}_1}{\|A\mathbf{x}_1\|}, \ldots, \quad \mathbf{x}_k = \frac{A\mathbf{x}_{k-1}}{\|A\mathbf{x}_{k-1}\|}, \ldots$$

Prove by induction that $\mathbf{x}_k = A^k\mathbf{x}_0/\|A^k\mathbf{x}_0\|$.

17. (***For readers familiar with proof by induction***) Let A be an $n \times n$ matrix, let $\mathbf{x}_0$ be a nonzero vector in R^n, and define the sequence $\mathbf{x}_1, \mathbf{x}_2, \ldots, \mathbf{x}_k, \ldots$ by

$$\mathbf{x}_1 = \frac{A\mathbf{x}_0}{\max(A\mathbf{x}_0)}, \quad \mathbf{x}_2 = \frac{A\mathbf{x}_1}{\max(A\mathbf{x}_1)}, \ldots,$$

$$\mathbf{x}_k = \frac{A\mathbf{x}_{k-1}}{\max(A\mathbf{x}_{k-1})}, \ldots$$

Prove by induction that

$$\mathbf{x}_k = \frac{A^k\mathbf{x}_0}{\max(A^k\mathbf{x}_0)}$$

9.3 Internet Search Engines

Early search engines on the Internet worked by examining key words and phrases in pages and titles of posted documents. Today's most popular search engines use algorithms based on the power method to analyze hyperlinks (references) between documents. In this section we will discuss one of the ways in which this is done.

Google, the most widely used engine for searching the Internet, was developed in 1996 by Larry Page and Sergey Brin while both were graduate students at Stanford University. Google uses a procedure known as the *PageRank algorithm* to analyze how documents at relevant sites reference one another. It then assigns to each site a *PageRank score*, stores those scores as a matrix, and uses the components of the dominant eigenvector of that matrix to establish the relative importance of the sites to the search.

Google starts by using a standard text-based search engine to find an initial set S_0 of sites containing relevant pages. Since words can have multiple meanings, the set S_0 will typically contain irrelevant sites and miss others of relevance. To compensate for this, the set S_0 is expanded to a larger set S by adjoining all sites referenced by the pages in the sites of S_0. The underlying assumption is that S will contain the most important sites relevant to the search. This process is then repeated a number of times to refine the search information still further.

To be more specific, suppose that the search set S contains n sites, and define the *adjacency matrix* for S to be the $n \times n$ matrix $A = [a_{ij}]$ in which

$$a_{ij} = 1 \text{ if site } i \text{ references site } j$$
$$a_{ij} = 0 \text{ if site } i \text{ does not reference site } j$$

We will assume that no site references itself, so the diagonal entries of A will all be zero.

▶ **EXAMPLE 1 Adjacency Matrices**

Here is a typical adjacency matrix for a search set with four sites:

$$A = \begin{array}{c} \\ \\ \text{Referencing Site} \end{array} \overset{\displaystyle \begin{array}{cccc} \text{Referenced Site} \\ 1 \quad 2 \quad 3 \quad 4 \end{array}}{\begin{bmatrix} 0 & 0 & 1 & 1 \\ 1 & 0 & 0 & 0 \\ 1 & 0 & 0 & 1 \\ 1 & 1 & 1 & 0 \end{bmatrix}} \begin{array}{c} 1 \\ 2 \\ 3 \\ 4 \end{array} \tag{1}$$

Thus, Site 1 references Sites 3 and 4, Site 2 references Site 1, and so forth. ◀

There are two basic roles that a site can play in the search process—the site may be a *hub*, meaning that it *references* many other sites, or it may be an *authority*, meaning that it is *referenced by* many other sites. A given site will typically have both hub and authority properties in that it will both reference and be referenced.

Historical Note The term *google* is a variation of the word *googol*, which stands for the number 10^{100} (1 followed by 100 zeros). This term was invented by the American mathematician Edward Kasner (1878–1955) in 1938, and the story goes that it came about when Kasner asked his eight-year-old nephew to give a name to a really big number—he responded with "googol." Kasner then went on to define a *googolplex* to be 10^{googol} (1 followed by googol zeros).

In general, if A is an adjacency matrix for n sites, then the column sums of A measure the authority aspect of the sites and the row sums of A measure their hub aspect. For example, the column sums of the matrix in (1) are 3, 1, 2, and 2, which means that Site 1 is referenced by three other sites, Site 2 is referenced by one other site, and so forth. Similarly, the row sums of the matrix in (1) are 2, 1, 2, and 3, so Site 1 references two other sites, Site 2 references one other site, and so forth.

Accordingly, if A is an adjacency matrix, then we call the vector $\mathbf{h}_0$ of row sums of A the **initial hub vector** of A, and we call the vector $\mathbf{a}_0$ of column sums of A the **initial authority vector** of A. Alternatively, we can think of $\mathbf{a}_0$ as the vector of *row* sums of A^T, which turns out to be more convenient for computations. The entries in the hub vector are called **hub weights** and those in the authority vector **authority weights**.

▶ **EXAMPLE 2 Initial Hub and Authority Vectors of an Adjacency Matrix**

Find the initial hub and authority vectors for the adjacency matrix A in Example 1.

Solution The row sums of A yield the initial hub vector

$$\mathbf{h}_0 = \begin{bmatrix} 2 \\ 1 \\ 2 \\ 3 \end{bmatrix} \begin{array}{l} \text{Site 1} \\ \text{Site 2} \\ \text{Site 3} \\ \text{Site 4} \end{array} \tag{2}$$

and the row sums of A^T (the column sums of A) yield the initial authority vector

$$\mathbf{a}_0 = \begin{bmatrix} 3 \\ 1 \\ 2 \\ 2 \end{bmatrix} \begin{array}{l} \text{Site 1} \\ \text{Site 2} \\ \text{Site 3} \\ \text{Site 4} \end{array} \tag{3}$$ ◀

The link counting in Example 2 suggests that Site 4 is the major hub and Site 1 is the greatest authority. However, counting links does not tell the whole story; for example, it seems reasonable that if Site 1 is to be considered the greatest authority, then more weight should be given to hubs that link to that site, and if Site 4 is to be considered a major hub, then more weight should be given to sites to which it links. Thus, there is an interaction between hubs and authorities that needs to be accounted for in the search process. Accordingly, once the search engine has calculated the initial authority vector $\mathbf{a}_0$, it then uses the information in that vector to create new hub and authority vectors $\mathbf{h}_1$ and $\mathbf{a}_1$ using the formulas

$$\mathbf{h}_1 = \frac{A\mathbf{a}_0}{\|A\mathbf{a}_0\|} \quad \text{and} \quad \mathbf{a}_1 = \frac{A^T\mathbf{h}_1}{\|A^T\mathbf{h}_1\|} \tag{4}$$

The numerators in these formulas do the weighting, and the normalization serves to control the size of the entries. To understand how the numerators accomplish the weighting, view the product $A\mathbf{a}_0$ as a linear combination of the column vectors of A with coefficients from $\mathbf{a}_0$. For example, with the adjacency matrix in Example 1 and the authority vector calculated in Example 2 we have

$$A\mathbf{a}_0 = \begin{bmatrix} 0 & 0 & 1 & 1 \\ 1 & 0 & 0 & 0 \\ 1 & 0 & 0 & 1 \\ 1 & 1 & 1 & 0 \end{bmatrix} \begin{bmatrix} 3 \\ 1 \\ 2 \\ 2 \end{bmatrix} = 3 \begin{bmatrix} 0 \\ 1 \\ 1 \\ 1 \end{bmatrix} + 1 \begin{bmatrix} 0 \\ 0 \\ 0 \\ 1 \end{bmatrix} + 2 \begin{bmatrix} 1 \\ 0 \\ 0 \\ 1 \end{bmatrix} + 2 \begin{bmatrix} 1 \\ 0 \\ 1 \\ 0 \end{bmatrix} = \begin{bmatrix} 4 \\ 3 \\ 5 \\ 6 \end{bmatrix} \begin{array}{l} \text{Site 1} \\ \text{Site 2} \\ \text{Site 3} \\ \text{Site 4} \end{array}$$

Referenced Site
1 2 3 4

Thus, we see that the links to each referenced site are weighted by the authority values in $\mathbf{a}_0$. To control the size of the entries, the search engine normalizes $A\mathbf{a}_0$ to produce the updated hub vector

$$\mathbf{h}_1 = \frac{A\mathbf{a}_0}{\|A\mathbf{a}_0\|} = \frac{1}{\sqrt{86}} \begin{bmatrix} 4 \\ 3 \\ 5 \\ 6 \end{bmatrix} \approx \begin{bmatrix} 0.43133 \\ 0.32350 \\ 0.53916 \\ 0.64700 \end{bmatrix} \begin{matrix} \text{Site 1} \\ \text{Site 2} \\ \text{Site 3} \\ \text{Site 4} \end{matrix} \quad \textbf{New Hub Weights}$$

The new hub vector $\mathbf{h}_1$ can now be used to update the authority vector using Formula (4). The product $A^T\mathbf{h}_1$ performs the weighting, and the normalization controls the size:

$$A^T\mathbf{h}_1 \approx \begin{matrix} & \text{1} & \text{2} & \text{3} & \text{4} \end{matrix}$$

Referencing Site

$$A^T\mathbf{h}_1 \approx \begin{bmatrix} 0 & 1 & 1 & 1 \\ 0 & 0 & 0 & 1 \\ 1 & 0 & 0 & 1 \\ 1 & 0 & 1 & 0 \end{bmatrix} \begin{bmatrix} 0.43133 \\ 0.32350 \\ 0.53916 \\ 0.64700 \end{bmatrix} \approx 0.43133 \begin{bmatrix} 0 \\ 0 \\ 1 \\ 1 \end{bmatrix} + 0.32350 \begin{bmatrix} 1 \\ 0 \\ 0 \\ 0 \end{bmatrix} + 0.53916 \begin{bmatrix} 1 \\ 0 \\ 0 \\ 1 \end{bmatrix} + 0.64700 \begin{bmatrix} 1 \\ 1 \\ 1 \\ 0 \end{bmatrix} \approx \begin{bmatrix} 1.50966 \\ 0.64700 \\ 1.07833 \\ 0.97049 \end{bmatrix} \begin{matrix} \text{Site 1} \\ \text{Site 2} \\ \text{Site 3} \\ \text{Site 4} \end{matrix}$$

$$\mathbf{a}_1 = \frac{A^T\mathbf{h}_1}{\|A^T\mathbf{h}_1\|} \approx \frac{1}{2.19142} \begin{bmatrix} 1.50966 \\ 0.64700 \\ 1.07833 \\ 0.97049 \end{bmatrix} \approx \begin{bmatrix} 0.68889 \\ 0.29524 \\ 0.49207 \\ 0.44286 \end{bmatrix} \begin{matrix} \text{Site 1} \\ \text{Site 2} \\ \text{Site 3} \\ \text{Site 4} \end{matrix} \quad \textbf{New Authority Weights}$$

Once the updated hub and authority vectors, $\mathbf{h}_1$ and $\mathbf{a}_1$, are obtained, the search engine repeats the process and computes a succession of hub and authority vectors, thereby generating the interrelated sequences

$$\mathbf{h}_1 = \frac{A\mathbf{a}_0}{\|A\mathbf{a}_0\|}, \quad \mathbf{h}_2 = \frac{A\mathbf{a}_1}{\|A\mathbf{a}_1\|}, \quad \mathbf{h}_3 = \frac{A\mathbf{a}_2}{\|A\mathbf{a}_2\|}, \ldots, \quad \mathbf{h}_k = \frac{A\mathbf{a}_{k-1}}{\|A\mathbf{a}_{k-1}\|}, \ldots \quad (5)$$

$$\mathbf{a}_0, \quad \mathbf{a}_1 = \frac{A^T\mathbf{h}_1}{\|A^T\mathbf{h}_1\|}, \quad \mathbf{a}_2 = \frac{A^T\mathbf{h}_2}{\|A^T\mathbf{h}_2\|}, \quad \mathbf{a}_3 = \frac{A^T\mathbf{h}_3}{\|A^T\mathbf{h}_3\|}, \ldots, \quad \mathbf{a}_k = \frac{A^T\mathbf{h}_k}{\|A^T\mathbf{h}_k\|}, \ldots \quad (6)$$

However, each of these is a power sequence in disguise. For example, if we substitute the expression for $\mathbf{h}_k$ into the expression for $\mathbf{a}_k$, then we obtain

$$\mathbf{a}_k = \frac{A^T\mathbf{h}_k}{\|A^T\mathbf{h}_k\|} = \frac{A^T\left(\dfrac{A\mathbf{a}_{k-1}}{\|A\mathbf{a}_{k-1}\|}\right)}{\left\|A^T\left(\dfrac{A\mathbf{a}_{k-1}}{\|A\mathbf{a}_{k-1}\|}\right)\right\|} = \frac{(A^TA)\mathbf{a}_{k-1}}{\|(A^TA)\mathbf{a}_{k-1}\|}$$

which means that we can rewrite (6) as

$$\mathbf{a}_0, \quad \mathbf{a}_1 = \frac{(A^TA)\mathbf{a}_0}{\|(A^TA)\mathbf{a}_0\|}, \quad \mathbf{a}_2 = \frac{(A^TA)\mathbf{a}_1}{\|(A^TA)\mathbf{a}_1\|}, \ldots, \quad \mathbf{a}_k = \frac{(A^TA)\mathbf{a}_{k-1}}{\|(A^TA)\mathbf{a}_{k-1}\|}, \ldots \quad (7)$$

Similarly, we can rewrite (5) as

$$\mathbf{h}_1 = \frac{A\mathbf{a}_0}{\|A\mathbf{a}_0\|}, \quad \mathbf{h}_2 = \frac{(AA^T)\mathbf{h}_1}{\|(AA^T)\mathbf{h}_1\|}, \ldots, \quad \mathbf{h}_k = \frac{(AA^T)\mathbf{h}_{k-1}}{\|(AA^T)\mathbf{h}_{k-1}\|}, \ldots \quad (8)$$

Remark In Exercise 15 of Section 9.2 you were asked to show that A^TA and AA^T both have positive dominant eigenvalues. That being the case, Theorem 9.2.1 ensures that (7) and (8) converge to the dominant eigenvectors of A^TA and AA^T, respectively. The entries in those eigenvectors are the authority and hub weights that Google uses to rank the search sites in order of importance as hubs and authorities.

▶ **EXAMPLE 3 A Ranking Procedure**

Suppose that a search engine produces 10 Internet sites in its search set and that the adjacency matrix for those sites is

Referenced Site

$$
A = \begin{array}{c} \\ \\ \\ \\ \\ \\ \\ \\ \\ \\ \end{array}
\begin{array}{cccccccccc}
1 & 2 & 3 & 4 & 5 & 6 & 7 & 8 & 9 & 10 \\
\end{array}
$$

$$
A = \left[\begin{array}{cccccccccc}
0 & 1 & 0 & 0 & 1 & 0 & 0 & 1 & 0 & 0 \\
0 & 0 & 0 & 0 & 1 & 0 & 0 & 0 & 0 & 0 \\
0 & 0 & 0 & 0 & 1 & 0 & 0 & 0 & 0 & 0 \\
0 & 0 & 0 & 0 & 0 & 1 & 1 & 0 & 0 & 0 \\
0 & 0 & 0 & 0 & 0 & 0 & 0 & 1 & 0 & 0 \\
0 & 1 & 1 & 1 & 1 & 0 & 0 & 1 & 0 & 1 \\
0 & 0 & 0 & 0 & 0 & 0 & 0 & 0 & 0 & 1 \\
0 & 0 & 0 & 0 & 1 & 0 & 0 & 0 & 0 & 0 \\
0 & 0 & 0 & 0 & 0 & 1 & 0 & 0 & 0 & 0 \\
0 & 0 & 0 & 0 & 0 & 1 & 0 & 0 & 0 & 0 \\
\end{array}\right]
\begin{array}{c}
1 \\ 2 \\ 3 \\ 4 \\ 5 \\ 6 \\ 7 \\ 8 \\ 9 \\ 10
\end{array}
\quad \text{Referencing Site}
$$

Use Formula (7) to rank the sites in decreasing order of authority.

Solution We will take $\mathbf{a}_0$ to be the normalized vector of column sums of A, and then we will compute the iterates in (7) until the authority vectors seem to stabilize. We leave it for you to show that

$$
\mathbf{a}_0 = \frac{1}{\sqrt{54}}
\begin{bmatrix}
0 \\ 2 \\ 1 \\ 1 \\ 5 \\ 3 \\ 1 \\ 3 \\ 0 \\ 2
\end{bmatrix}
\approx
\begin{bmatrix}
0 \\ 0.27217 \\ 0.13608 \\ 0.13608 \\ 0.68041 \\ 0.40825 \\ 0.13608 \\ 0.40825 \\ 0 \\ 0.27217
\end{bmatrix}
$$

and that

$$
(A^T A)\mathbf{a}_0 \approx
\left[\begin{array}{cccccccccc}
0 & 0 & 0 & 0 & 0 & 0 & 0 & 0 & 0 & 0 \\
0 & 2 & 1 & 1 & 2 & 0 & 0 & 2 & 0 & 1 \\
0 & 1 & 1 & 1 & 1 & 0 & 0 & 1 & 0 & 1 \\
0 & 1 & 1 & 1 & 1 & 0 & 0 & 1 & 0 & 1 \\
0 & 2 & 1 & 1 & 5 & 0 & 0 & 2 & 0 & 1 \\
0 & 0 & 0 & 0 & 0 & 3 & 1 & 0 & 0 & 0 \\
0 & 0 & 0 & 0 & 0 & 1 & 1 & 0 & 0 & 0 \\
0 & 2 & 1 & 1 & 2 & 0 & 0 & 3 & 0 & 1 \\
0 & 0 & 0 & 0 & 0 & 0 & 0 & 0 & 0 & 0 \\
0 & 1 & 1 & 1 & 1 & 0 & 0 & 1 & 0 & 2 \\
\end{array}\right]
\begin{bmatrix}
0 \\ 0.27217 \\ 0.13608 \\ 0.13608 \\ 0.68041 \\ 0.40825 \\ 0.13608 \\ 0.40825 \\ 0 \\ 0.27217
\end{bmatrix}
\approx
\begin{bmatrix}
0 \\ 3.26599 \\ 1.90516 \\ 1.90516 \\ 5.30723 \\ 1.36083 \\ 0.54433 \\ 3.67423 \\ 0 \\ 2.17732
\end{bmatrix}
$$

Thus,

$$\mathbf{a}_1 = \frac{(A^T A)\mathbf{a}_0}{\|(A^T A)\mathbf{a}_0\|} \approx \frac{1}{8.15362} \begin{bmatrix} 0 \\ 3.26599 \\ 1.90516 \\ 1.90516 \\ 5.30723 \\ 1.36083 \\ 0.54433 \\ 3.67423 \\ 0 \\ 2.17732 \end{bmatrix} \approx \begin{bmatrix} 0 \\ 0.40056 \\ 0.23366 \\ 0.23366 \\ 0.65090 \\ 0.16690 \\ 0.06676 \\ 0.45063 \\ 0 \\ 0.26704 \end{bmatrix}$$

Continuing in this way yields the following authority iterates:

$$\mathbf{a}_0 \quad \mathbf{a}_1 = \frac{(A^T A)\mathbf{a}_0}{\|(A^T A)\mathbf{a}_0\|} \quad \mathbf{a}_2 = \frac{(A^T A)\mathbf{a}_1}{\|(A^T A)\mathbf{a}_1\|} \quad \mathbf{a}_3 = \frac{(A^T A)\mathbf{a}_2}{\|(A^T A)\mathbf{a}_2\|} \quad \mathbf{a}_4 = \frac{(A^T A)\mathbf{a}_3}{\|(A^T A)\mathbf{a}_3\|} \quad \cdots \quad \mathbf{a}_9 = \frac{(A^T A)\mathbf{a}_8}{\|(A^T A)\mathbf{a}_8\|} \quad \mathbf{a}_{10} = \frac{(A^T A)\mathbf{a}_9}{\|(A^T A)\mathbf{a}_9\|}$$

	$\mathbf{a}_0$	$\mathbf{a}_1$	$\mathbf{a}_2$	$\mathbf{a}_3$	$\mathbf{a}_4$	$\mathbf{a}_9$	$\mathbf{a}_{10}$	
	0	0	0	0	0	0	0	Site 1
	0.27217	0.40056	0.41652	0.41918	0.41973	0.41990	0.41990	Site 2
	0.13608	0.23366	0.24917	0.25233	0.25309	0.25337	0.25337	Site 3
	0.13608	0.23366	0.24917	0.25233	0.25309	0.25337	0.25337	Site 4
	0.68041	0.65090	0.63407	0.62836	0.62665	0.62597	0.62597	Site 5
	0.40825	0.16690	0.06322	0.02372	0.00889	0.00007	0.00002	Site 6
	0.13608	0.06676	0.02603	0.00981	0.00368	0.00003	0.00001	Site 7
	0.40825	0.45063	0.46672	0.47050	0.47137	0.47165	0.47165	Site 8
	0	0	0	0	0	0	0	Site 9
	0.27217	0.26704	0.27892	0.28300	0.28416	0.28460	0.28460	Site 10

The small changes between $\mathbf{a}_9$ and $\mathbf{a}_{10}$ suggest that the iterates have stabilized near a dominant eigenvector of $A^T A$. From the entries in $\mathbf{a}_{10}$ we conclude that Sites 1, 6, 7, and 9 are probably irrelevant to the search and that the remaining sites should be searched in order of decreasing importance as

$$\text{Site 5,} \quad \text{Site 8,} \quad \text{Site 2,} \quad \text{Site 10,} \quad \text{Sites 3 and 4 (a tie)} \quad \blacktriangleleft$$

Concept Review

- Adjacency matrix
- Hub vector
- Authority vector
- Hub weights
- Authority weights

Skills

- Find the initial hub and authority vectors of an adjacency matrix.
- Use the method of Example 3 to rank sites.

Exercise Set 9.3

▶ In Exercises 1–2, find the initial hub and authority vectors for the given adjacency matrix A. ◀

1.   **Referenced Site**

$$A = \begin{array}{c} \\ \\ \\ \end{array} \begin{array}{ccc} 1 & 2 & 3 \\ \end{array}$$
$$A = \begin{bmatrix} 0 & 0 & 1 \\ 1 & 0 & 1 \\ 1 & 0 & 1 \end{bmatrix} \begin{array}{c} 1 \\ 2 \\ 3 \end{array} \text{ Referencing Site}$$

2.   **Referenced Site**

$$\begin{array}{cccc} 1 & 2 & 3 & 4 \\ \end{array}$$
$$A = \begin{bmatrix} 0 & 1 & 0 & 1 \\ 1 & 0 & 0 & 1 \\ 1 & 0 & 0 & 1 \\ 1 & 1 & 1 & 0 \end{bmatrix} \begin{array}{c} 1 \\ 2 \\ 3 \\ 4 \end{array} \text{ Referencing Site}$$

In Exercises 3–4, find the updated hub and authority vectors in $\mathbf{h}_1$ and $\mathbf{a}_1$ for the adjacency matrix A. ◄

3. The matrix in Exercise 1.   **4.** The matrix in Exercise 2.

In Exercises 5–8, the adjacency matrix A of an Internet search engine is given. Use the method of Example 3 to rank the sites in decreasing order of authority. ◄

5.

Referenced Site

$$A = \begin{bmatrix} 0 & 0 & 1 & 0 \\ 1 & 0 & 0 & 0 \\ 1 & 1 & 0 & 0 \\ 0 & 1 & 0 & 0 \end{bmatrix} \begin{matrix} 1 \\ 2 \\ 3 \\ 4 \end{matrix}$$ Referencing Site

(columns: 1 2 3 4)

6.

Referenced Site

$$A = \begin{bmatrix} 0 & 1 & 1 & 0 \\ 0 & 0 & 1 & 0 \\ 1 & 0 & 0 & 1 \\ 1 & 0 & 0 & 0 \end{bmatrix} \begin{matrix} 1 \\ 2 \\ 3 \\ 4 \end{matrix}$$ Referencing Site

(columns: 1 2 3 4)

7.

Referenced Site

$$A = \begin{bmatrix} 0 & 1 & 1 & 1 & 0 \\ 1 & 0 & 0 & 0 & 1 \\ 0 & 0 & 0 & 0 & 1 \\ 0 & 1 & 0 & 0 & 0 \\ 0 & 1 & 1 & 0 & 0 \end{bmatrix} \begin{matrix} 1 \\ 2 \\ 3 \\ 4 \\ 5 \end{matrix}$$ Referencing Site

(columns: 1 2 3 4 5)

8.

Referenced Site

$$A = \begin{bmatrix} 0 & 1 & 1 & 0 & 1 & 1 & 0 & 0 & 0 & 1 \\ 0 & 0 & 1 & 0 & 0 & 0 & 0 & 0 & 0 & 0 \\ 0 & 0 & 0 & 0 & 0 & 0 & 0 & 0 & 0 & 1 \\ 0 & 1 & 1 & 0 & 0 & 1 & 1 & 0 & 0 & 1 \\ 0 & 0 & 0 & 1 & 0 & 0 & 0 & 0 & 0 & 0 \\ 0 & 1 & 0 & 0 & 0 & 0 & 0 & 0 & 0 & 0 \\ 0 & 0 & 0 & 0 & 0 & 0 & 0 & 0 & 1 & 0 \\ 0 & 0 & 0 & 0 & 0 & 1 & 0 & 0 & 0 & 0 \\ 0 & 1 & 1 & 0 & 0 & 1 & 0 & 1 & 0 & 1 \\ 0 & 0 & 0 & 0 & 0 & 1 & 0 & 0 & 0 & 0 \end{bmatrix} \begin{matrix} 1 \\ 2 \\ 3 \\ 4 \\ 5 \\ 6 \\ 7 \\ 8 \\ 9 \\ 10 \end{matrix}$$ Referencing Site

(columns: 1 2 3 4 5 6 7 8 9 10)

9.4 Comparison of Procedures for Solving Linear Systems

There is an old saying that "time is money." This is especially true in industry where the cost of solving a linear system is generally determined by the time it takes for a computer to perform the required computations. This typically depends both on the speed of the computer processor and on the number of operations required by the algorithm. Thus, choosing the right algorithm has important financial implication in an industrial or research setting. In this section we will discuss some of the factors that affect the choice of algorithms for solving large-scale linear systems.

Flops and the Cost of Solving a Linear System

In computer jargon, an arithmetic operation $(+, -, *, \div)$ on two real numbers is called a **flop**, which is an acronym for "floating-point operation."[*] The total number of flops required to solve a problem, which is called the **cost** of the solution, provides a convenient way of choosing between various algorithms for solving the problem. When needed, the cost in flops can be converted to units of time or money if the speed of the computer processor and the financial aspects of its operation are known. For example, many of today's personal computers are capable of performing in excess of 10 gigaflops per second (1 gigaflop $= 10^9$ flops). Thus, an algorithm that costs 1,000,000 flops would be executed in 0.0001 seconds.

---

[*]Real numbers are stored in computers as numerical approximations called **floating-point numbers**. In base 10, a floating-point number has the form $\pm .d_1 d_2 \cdots d_n \times 10^m$, where m is an integer, called the **mantissa**, and n is the number of digits to the right of the decimal point. The value of n varies with the computer. In some literature the term *flop* is used as a measure of processing speed and stands for "floating-point operations *per second*." In our usage it is interpreted as a counting unit.

To illustrate how costs (in flops) can be computed, let us count the number of flops required to solve a linear system of n equations in n unknowns by Gauss–Jordan elimination. For this purpose we will need the following formulas for the sum of the first n positive integers and the sum of the squares of the first n positive integers:

$$1 + 2 + 3 + \cdots + n = \frac{n(n+1)}{2} \tag{1}$$

$$1^2 + 2^2 + 3^2 + \cdots + n^2 = \frac{n(n+1)(2n+1)}{6} \tag{2}$$

Let $A\mathbf{x} = \mathbf{b}$ be a linear system of n equations in n unknowns to be solved by Gauss–Jordan elimination (or, equivalently, by Gaussian elimination with back substitution). For simplicity, let us assume that A is invertible and that no row interchanges are required to reduce the augmented matrix $[A \mid \mathbf{b}]$ to row echelon form. The diagrams that accompany the following analysis provide a convenient way of counting the operations required to introduce a leading 1 in the first row and then zeros below it. In our operation counts, we will lump divisions and multiplications together as "multiplications," and we will lump additions and subtractions together as "additions."

Step 1. It requires n flops (multiplications) to introduce the leading 1 in the first row.

$$\begin{bmatrix}
1 & \times & \times & \cdots & \times & \times & \times \\
\bullet & \bullet & \bullet & \cdots & \bullet & \bullet & \bullet \\
\bullet & \bullet & \bullet & \cdots & \bullet & \bullet & \bullet \\
\vdots & \vdots & \vdots & & \vdots & \vdots & \vdots \\
\bullet & \bullet & \bullet & \cdots & \bullet & \bullet & \bullet \\
\bullet & \bullet & \bullet & \cdots & \bullet & \bullet & \bullet
\end{bmatrix}$$

$$\begin{bmatrix} \times \text{ denotes a quantity that is being computed.} \\ \bullet \text{ denotes a quantity that is not being computed.} \\ \text{The augmented matrix size is } n \times (n+1). \end{bmatrix}$$

Step 2. It requires n multiplications and n additions to introduce a zero below the leading 1, and there are $n-1$ rows below the leading 1, so the number of flops required to introduce zeros below the leading 1 is $2n(n-1)$.

$$\begin{bmatrix}
1 & \bullet & \bullet & \cdots & \bullet & \bullet & \bullet \\
0 & \times & \times & \cdots & \times & \times & \times \\
0 & \times & \times & \cdots & \times & \times & \times \\
\vdots & \vdots & \vdots & & \vdots & \vdots & \vdots \\
0 & \times & \times & \cdots & \times & \times & \times \\
0 & \times & \times & \cdots & \times & \times & \times
\end{bmatrix}$$

Column 1. Combining Steps 1 and 2, the number of flops required for column 1 is

$$n + 2n(n-1) = 2n^2 - n$$

Column 2. The procedure for column 2 is the same as for column 1, except that now we are dealing with one less row and one less column. Thus, the number of flops required to introduce the leading 1 in row 2 and the zeros below it can be obtained by replacing n by $n - 1$ in the flop count for the first column. Thus, the number of flops required for column 2 is

$$2(n - 1)^2 - (n - 1)$$

Column 3. By the argument for column 2, the number of flops required for column 3 is

$$2(n - 2)^2 - (n - 2)$$

Total for all columns. The pattern should now be clear. The total number of flops required to create the n leading 1's and the associated zeros is

$$(2n^2 - n) + [2(n - 1)^2 - (n - 1)] + [2(n - 2)^2 - (n - 2)] + \cdots + (2 - 1)$$

which we can rewrite as

$$2[n^2 + (n - 1)^2 + \cdots + 1] - [n + (n - 1) + \cdots + 1]$$

or on applying Formulas (1) and (2) as

$$2\frac{n(n + 1)(2n + 1)}{6} - \frac{n(n + 1)}{2} = \frac{2}{3}n^3 + \frac{1}{2}n^2 - \frac{1}{6}n$$

Next, let us count the number of operations required to complete the backward phase (the back substitution).

Column n. It requires $n - 1$ multiplications and $n - 1$ additions to introduce zeros above the leading 1 in the nth column, so the total number of flops required for the column is $2(n - 1)$.

$$\begin{bmatrix} 1 & \bullet & \bullet & \cdots & \bullet & 0 & | & \times \\ 0 & 1 & \bullet & \cdots & \bullet & 0 & | & \times \\ 0 & 0 & 1 & \cdots & \bullet & 0 & | & \times \\ \vdots & \vdots & \vdots & & \vdots & \vdots & | & \vdots \\ 0 & 0 & 0 & \cdots & 1 & 0 & | & \times \\ 0 & 0 & 0 & \cdots & 0 & 1 & | & \bullet \end{bmatrix}$$

Column $(n - 1)$. The procedure is the same as for Step 1, except that now we are dealing with one less row. Thus, the number of flops required for the $(n - 1)$st column is $2(n - 2)$.

$$\begin{bmatrix} 1 & \bullet & \bullet & \cdots & 0 & 0 & | & \times \\ 0 & 1 & \bullet & \cdots & 0 & 0 & | & \times \\ 0 & 0 & 1 & \cdots & 0 & 0 & | & \times \\ \vdots & \vdots & \vdots & & \vdots & \vdots & | & \vdots \\ 0 & 0 & 0 & \cdots & 1 & 0 & | & \bullet \\ 0 & 0 & 0 & \cdots & 0 & 1 & | & \bullet \end{bmatrix}$$

Column $(n - 2)$. By the argument for column $(n - 1)$, the number of flops required for column $(n - 2)$ is $2(n - 3)$.

Total. The pattern should now be clear. The total number of flops to complete the backward phase is

$$2(n - 1) + 2(n - 2) + 2(n - 3) + \cdots + 2(n - n) = 2[n^2 - (1 + 2 + \cdots + n)]$$

which we can rewrite using Formula (1) as

$$2\left(n^2 - \frac{n(n+1)}{2}\right) = n^2 - n$$

In summary, we have shown that for Gauss–Jordan elimination the number of flops required for the forward and backward phases is

$$\text{flops for forward phase} = \tfrac{2}{3}n^3 + \tfrac{1}{2}n^2 - \tfrac{1}{6}n \qquad (3)$$

$$\text{flops for backward phase} = n^2 - n \qquad (4)$$

Thus, the total cost of solving a linear system by Gauss–Jordan elimination is

$$\text{flops for both phases} = \tfrac{2}{3}n^3 + \tfrac{3}{2}n^2 - \tfrac{7}{6}n \qquad (5)$$

Cost Estimates for Solving Large Linear Systems   It is a property of polynomials that for large values of the independent variable the term of highest power makes the major contribution to the value of the polynomial. Thus, for *large* linear systems we can use (3) and (4) to approximate the number of flops in the forward and backward phases as

$$\text{flops for forward phase} \approx \tfrac{2}{3}n^3 \qquad (6)$$

$$\text{flops for backward phase} \approx n^2 \qquad (7)$$

This shows that it is more costly to execute the forward phase than the backward phase for large linear systems. Indeed, the cost difference between the forward and backward phases can be enormous, as the next example shows.

▶ **EXAMPLE 1 Cost of Solving a Large Linear System**

Approximate the time required to execute the forward and backward phases of Gauss–Jordan elimination for a system of 10,000 ($= 10^4$) equations in 10,000 unknowns using a computer that can execute 10 gigaflops per second.

Solution We have $n = 10^4$ for the given system, so from (6) and (7) the number of gigaflops required for the forward and backward phases is

$$\text{gigaflops for forward phase} \approx \tfrac{2}{3}n^3 \times 10^{-9} = \tfrac{2}{3}(10^4)^3 \times 10^{-9} = \tfrac{2}{3} \times 10^3$$

$$\text{gigaflops for backward phase} \approx n^2 \times 10^{-9} = (10^4)^2 \times 10^{-9} = 10^{-1}$$

Thus, at 10 gigaflops/s the execution times for the forward and backward phases are

$$\text{time for forward phase} \approx \left(\tfrac{2}{3} \times 10^3\right) \times 10^{-1} \text{ s} \approx 66.67 \text{ s}$$

$$\text{time for backward phase} \approx (10^{-1}) \times 10^{-1} \text{ s} \approx 0.01 \text{ s} \quad ◀$$

We leave it as an exercise for you to confirm the results in Table 1.

Considerations in Choosing an Algorithm for Solving a Linear System   For a *single* linear system $A\mathbf{x} = \mathbf{b}$ of n equations in n unknowns, the methods of LU-decomposition and Gauss–Jordan elimination differ in bookkeeping but otherwise involve the same number of flops. Thus, neither method has a cost advantage over the other. However, LU-decomposition has other advantages that make it the method of choice:

- Gauss–Jordan elimination and Gaussian elimination both use the augmented matrix $[A \mid \mathbf{b}]$, so $\mathbf{b}$ must be known. In contrast, LU-decomposition uses only the matrix A, so once that decomposition is known it can be used with as many right-hand sides as are required, one at a time.

- The LU-decomposition that is computed to solve $A\mathbf{x} = \mathbf{b}$ can be used to compute A^{-1}, if needed, with little additional work.

- For large linear systems in which computer memory is at a premium, one can dispense with the storage of the 1's and zeros that appear on or below the main diagonal of U, since those entries are known from the form of U. The space that this opens up can then be used to store the entries of L, thereby reducing the amount of memory required to solve the system.

- If A is a large matrix consisting mostly of zeros, and if the nonzero entries are concentrated in a "band" around the main diagonal, then there are techniques that can be used to reduce the cost of LU-decomposition, giving it an advantage over Gauss–Jordan elimination.

Table 1

The cost in flops for Gaussian elimination is the same as that for the forward phase of Gauss–Jordan elimination.

Approximate Cost for an $n \times n$ Matrix A with Large n	
Algorithm	**Cost in Flops**
Gauss–Jordan elimination (forward phase)	$\approx \frac{2}{3}n^3$
Gauss–Jordan elimination (backward phase)	$\approx n^2$
LU-decomposition of A	$\approx \frac{2}{3}n^3$
Forward substitution to solve $L\mathbf{y} = \mathbf{b}$	$\approx n^2$
Backward substitution to solve $U\mathbf{x} = \mathbf{y}$	$\approx n^2$
A^{-1} by reducing $[A \mid I]$ to $[I \mid A^{-1}]$	$\approx 2n^3$
Compute $A^{-1}\mathbf{b}$	$\approx 2n^3$

Concept Review

- Flop
- Formula for the sum of the first n positive integers
- Formula for the sum of the squares of the first n positive integers
- Cost in flops for solving large linear systems by various methods
- Cost in flops for inverting a matrix by row reduction
- Issues to consider when choosing an algorithm to solve a large linear system

Skills

- Compute the cost of solving a linear system by Gauss–Jordan elimination.
- Approximate the time required to execute the forward and backward phases of Gauss–Jordan elimination.
- Approximate the time required to find an LU-decomposition of a matrix.
- Approximate the time required to find the inverse of an invertible matrix.

Exercise Set 9.4

1. A certain computer can execute 10 gigaflops per second. Use Formula (5) to find the time required to solve the system using Gauss–Jordan elimination.

   (a) A system of 1000 equations in 1000 unknowns.

   (b) A system of 10,000 equations in 10,000 unknowns.

   (c) A system of 100,000 equations in 100,000 unknowns.

2. A certain computer can execute 100 gigaflops per second. Use Formula (5) to find the time required to solve the system using Gauss–Jordan elimination.

   (a) A system of 10,000 equations in 10,000 unknowns.

   (b) A system of 100,000 equations in 100,000 unknowns.

   (c) A system of 1,000,000 equations in 1,000,000 unknowns.

3. Today's personal computers can execute 70 gigaflops per second. Use Table 1 to estimate the time required to perform the following operations on the invertible $10,000 \times 10,000$ matrix A.

   (a) Execute the forward phase of Gauss–Jordan elimination.

   (b) Execute the backward phase of Gauss–Jordan elimination.

   (c) LU-decomposition of A.

   (d) Find A^{-1} by reducing $[A \mid I]$ to $[I \mid A^{-1}]$.

4. The IBM Roadrunner computer can operate at speeds in excess of 1 petaflop per second (1 petaflop $= 10^{15}$ flops). Use Table 1 to estimate the time required to perform the following operations of the invertible $100,000 \times 100,000$ matrix A.

   (a) Execute the forward phase of Gauss–Jordan elimination.

   (b) Execute the backward phase of Gauss–Jordan elimination.

   (c) LU-decomposition of A.

   (d) Find A^{-1} by reducing $[A \mid I]$ to $[I \mid A^{-1}]$.

5. (a) Approximate the time required to execute the forward phase of Gauss–Jordan elimination for a system of 100,000 equations in 100,000 unknowns using a computer that can execute 1 gigaflop per second. Do the same for the backward phase. (See Table 1.)

   (b) How many gigaflops per second must a computer be able to execute to find the LU-decomposition of a matrix of size $10,000 \times 10,000$ in less than 0.5 s? (See Table 1.)

6. About how many teraflops per second must a computer be able to execute to find the inverse of a matrix of size $100,000 \times 100,000$ in less than 0.5 s? (1 teraflop $= 10^{12}$ flops.)

▷ In Exercises 7–10, A and B are $n \times n$ matrices and c is a real number. ◁

7. How many flops are required to compute cA?

8. How many flops are required to compute $A + B$?

9. How many flops are required to compute AB?

10. If A is a diagonal matrix and k is a positive integer, how many flops are required to compute A^k?

9.5 Singular Value Decomposition

In this section we will discuss an extension of the diagonalization theory for $n \times n$ symmetric matrices to general $m \times n$ matrices. The results that we will develop in this section have applications to compression, storage, and transmission of digitized information and form the basis for many of the best computational algorithms that are currently available for solving linear systems.

Decompositions of Square Matrices

We saw in Formula (2) of Section 7.2 that every symmetric matrix A can be expressed as

$$A = PDP^T \tag{1}$$

where P is an $n \times n$ orthogonal matrix of eigenvectors of A, and D is the diagonal matrix whose diagonal entries are the eigenvalues corresponding to the column vectors of P. In this section we will call (1) an *eigenvalue decomposition* of A (abbreviated EVD of A).

If an $n \times n$ matrix A is not symmetric, then it does not have an eigenvalue decomposition, but it does have a *Hessenberg decomposition*

$$A = PHP^T$$

in which P is an orthogonal matrix and H is in upper Hessenberg form (Theorem 7.2.4). Moreover, if A has real eigenvalues, then it has a **Schur decomposition**

$$A = PSP^T$$

in which P is an orthogonal matrix and S is upper triangular (Theorem 7.2.3).

The eigenvalue, Hessenberg, and Schur decompositions are important in numerical algorithms not only because the matrices D, H, and S have simpler forms than A, but also because the orthogonal matrices that appear in these factorizations do not magnify roundoff error. To see why this is so, suppose that $\hat{\mathbf{x}}$ is a column vector whose entries are known exactly and that

$$\mathbf{x} = \hat{\mathbf{x}} + \mathbf{e}$$

is the vector that results when roundoff error is present in the entries of $\hat{\mathbf{x}}$. If P is an orthogonal matrix, then the length-preserving property of orthogonal transformations implies that

$$\|P\mathbf{x} - P\hat{\mathbf{x}}\| = \|\mathbf{x} - \hat{\mathbf{x}}\| = \|\mathbf{e}\|$$

which tells us that the error in approximating $P\hat{\mathbf{x}}$ by $P\mathbf{x}$ has the same magnitude as the error in approximating $\hat{\mathbf{x}}$ by $\mathbf{x}$.

There are two main paths that one might follow in looking for other kinds of decompositions of a general square matrix A: One might look for decompositions of the form

$$A = PJP^{-1}$$

in which P is invertible but not necessarily orthogonal, or one might look for decompositions of the form

$$A = U\Sigma V^T$$

in which U and V are orthogonal but not necessarily the same. The first path leads to decompositions in which J is either diagonal or a certain kind of block diagonal matrix, called a **Jordan canonical form** in honor of the French mathematician Camille Jordan (see p. 510). Jordan canonical forms, which we will not consider in this text, are important theoretically and in certain applications, but they are of lesser importance numerically because of the roundoff difficulties that result from the lack of orthogonality in P. In this section we will focus on the second path.

Singular Values   Since matrix products of the form A^TA will play an important role in our work, we will begin with two basic theorems about them.

THEOREM 9.5.1 *If A is an $m \times n$ matrix, then*:

(a) A and A^TA have the same null space.

(b) A and A^TA have the same row space.

(c) A^T and A^TA have the same column space.

(d) A and A^TA have the same rank.

We will prove part (a) and leave the remaining proofs for the exercises.

Proof (a)   We must show that every solution of $A\mathbf{x} = \mathbf{0}$ is a solution of $A^TA\mathbf{x} = \mathbf{0}$, and conversely. If $\mathbf{x}_0$ is any solution of $A\mathbf{x} = \mathbf{0}$, then $\mathbf{x}_0$ is also a solution of $A^TA\mathbf{x} = \mathbf{0}$ since

$$A^TA\mathbf{x}_0 = A^T(A\mathbf{x}_0) = A^T\mathbf{0} = \mathbf{0}$$

Conversely, if $\mathbf{x}_0$ is any solution of $A^TA\mathbf{x} = \mathbf{0}$, then $\mathbf{x}_0$ is in the null space of A^TA and hence is orthogonal to all vectors in the row space of A^TA by part (q) of Theorem 4.8.10.

However, A^TA is symmetric, so $\mathbf{x}_0$ is also orthogonal to every vector in the column space of A^TA. In particular, $\mathbf{x}_0$ must be orthogonal to the vector $(A^TA)\mathbf{x}_0$; that is,

$$\mathbf{x}_0 \cdot (A^TA)\mathbf{x}_0 = 0$$

Using the first formula in Table 1 of Section 3.2 and properties of the transpose operation we can rewrite this as

$$\mathbf{x}_0^T(A^TA)\mathbf{x}_0 = (A\mathbf{x}_0)^T(A\mathbf{x}_0) = (A\mathbf{x}_0) \cdot (A\mathbf{x}_0) = \|A\mathbf{x}_0\|^2 = 0$$

which implies that $A\mathbf{x}_0 = 0$, thereby proving that $\mathbf{x}_0$ is a solution of $A\mathbf{x}_0 = 0$. ◄

THEOREM 9.5.2 *If A is an $m \times n$ matrix, then:*

(a) A^TA *is orthogonally diagonalizable.*

(b) *The eigenvalues of A^TA are nonnegative.*

Proof (a) The matrix A^TA, being symmetric, is orthogonally diagonalizable by Theorem 7.2.1.

Proof (b) Since A^TA is orthogonally diagonalizable, there is an orthonormal basis for R^n consisting of eigenvectors of A^TA, say $\{\mathbf{v}_1, \mathbf{v}_2, \ldots, \mathbf{v}_n\}$. If we let $\lambda_1, \lambda_2, \ldots, \lambda_n$ be the corresponding eigenvalues, then for $1 \leq i \leq n$ we have

$$\|A\mathbf{v}_i\|^2 = A\mathbf{v}_i \cdot A\mathbf{v}_i = \mathbf{v}_i \cdot A^TA\mathbf{v}_i \qquad \text{[Formula (26) of Section 3.2]}$$
$$= \mathbf{v}_i \cdot \lambda_i\mathbf{v}_i = \lambda_i(\mathbf{v}_i \cdot \mathbf{v}_i) = \lambda_i\|\mathbf{v}_i\|^2 = \lambda_i$$

It follows from this relationship that $\lambda_i \geq 0$. ◄

We will assume throughout this section that the eigenvalues of A^TA are named so that

$$\lambda_1 \geq \lambda_2 \geq \cdots \geq \lambda_n \geq 0$$

and hence that

$$\sigma_1 \geq \sigma_2 \geq \cdots \geq \sigma_n \geq 0$$

DEFINITION 1 If A is an $m \times n$ matrix, and if $\lambda_1, \lambda_2, \ldots, \lambda_n$ are the eigenvalues of A^TA, then the numbers

$$\sigma_1 = \sqrt{\lambda_1}, \quad \sigma_2 = \sqrt{\lambda_2}, \ldots, \quad \sigma_n = \sqrt{\lambda_n}$$

are called the *singular values* of A.

► **EXAMPLE 1 Singular Values**

Find the singular values of the matrix

$$\begin{bmatrix} 1 & 1 \\ 0 & 1 \\ 1 & 0 \end{bmatrix}$$

Solution The first step is to find the eigenvalues of the matrix

$$A^TA = \begin{bmatrix} 1 & 0 & 1 \\ 1 & 1 & 0 \end{bmatrix} \begin{bmatrix} 1 & 1 \\ 0 & 1 \\ 1 & 0 \end{bmatrix} = \begin{bmatrix} 2 & 1 \\ 1 & 2 \end{bmatrix}$$

The characteristic polynomial of A^TA is

$$\lambda^2 - 4\lambda + 3 = (\lambda - 3)(\lambda - 1)$$

so the eigenvalues of A^TA are $\lambda_1 = 3$ and $\lambda_2 = 1$ and the singular values of A in order of decreasing size are

$$\sigma_1 = \sqrt{\lambda_1} = \sqrt{3}, \quad \sigma_2 = \sqrt{\lambda_2} = 1 \quad ◄$$

Singular Value Decomposition

Main diagonal

▲ **Figure 9.5.1**

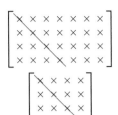

Harry Bateman
(1882–1946)

Historical Note The term *singular value* is apparently due to the British-born mathematician Harry Bateman, who used it in a research paper published in 1908. Bateman emigrated to the United States in 1910, teaching at Bryn Mawr College, Johns Hopkins University, and finally at the California Institute of Technology. Interestingly, he was awarded his Ph.D. in 1913 by Johns Hopkins at which point in time he was already an eminent mathematician with 60 publications to his name.
[*Image: Courtesy of the Archives, California Institute of Technology*]

Before turning to the main result in this section, we will find it useful to extend the notion of a "main diagonal" to matrices that are not square. We define the ***main diagonal*** of an $m \times n$ matrix to be the line of entries shown in Figure 9.5.1—it starts at the upper left corner and extends diagonally as far as it can go. We will refer to the entries on the main diagonal as the ***diagonal entries***.

We are now ready to consider the main result in this section, which is concerned with a specific way of factoring a general $m \times n$ matrix A. This factorization, called ***singular value decomposition*** (abbreviated SVD) will be given in two forms, a brief form that captures the main idea, and an expanded form that spells out the details. The proof is given at the end of this section.

THEOREM 9.5.3 Singular Value Decomposition

If A is an $m \times n$ matrix, then A can be expressed in the form

$$A = U\Sigma V^T$$

where U and V are orthogonal matrices and Σ is an $m \times n$ matrix whose diagonal entries are the singular values of A and whose other entries are zero.

THEOREM 9.5.4 Singular Value Decomposition (Expanded Form)

If A is an $m \times n$ matrix of rank k, then A can be factored as

$$A = U\Sigma V^T = [\mathbf{u}_1 \quad \mathbf{u}_2 \quad \cdots \quad \mathbf{u}_k \mid \mathbf{u}_{k+1} \quad \cdots \quad \mathbf{u}_m]
\begin{bmatrix}
\sigma_1 & 0 & \cdots & 0 & \\
0 & \sigma_2 & \cdots & 0 & 0_{k\times(n-k)} \\
\vdots & \vdots & \ddots & \vdots & \\
0 & 0 & \cdots & \sigma_k & \\
\hline
& 0_{(m-k)\times k} & & & 0_{(m-k)\times(n-k)}
\end{bmatrix}
\begin{bmatrix}
\mathbf{v}_1^T \\
\mathbf{v}_2^T \\
\vdots \\
\mathbf{v}_k^T \\
\mathbf{v}_{k+1}^T \\
\vdots \\
\mathbf{v}_n^T
\end{bmatrix}$$

in which U, Σ, and V have sizes $m \times m$, $m \times n$, and $n \times n$, respectively, and in which

(*a*) $V = [\mathbf{v}_1 \quad \mathbf{v}_2 \quad \cdots \quad \mathbf{v}_n]$ *orthogonally diagonalizes A^TA.*

(*b*) *The nonzero diagonal entries of Σ are $\sigma_1 = \sqrt{\lambda_1}$, $\sigma_2 = \sqrt{\lambda_2}, \ldots, \sigma_k = \sqrt{\lambda_k}$, where $\lambda_1, \lambda_2, \ldots, \lambda_k$ are the nonzero eigenvalues of A^TA corresponding to the column vectors of V.*

(*c*) *The column vectors of V are ordered so that $\sigma_1 \geq \sigma_2 \geq \cdots \geq \sigma_k > 0$.*

(*d*) $\mathbf{u}_i = \dfrac{A\mathbf{v}_i}{\|A\mathbf{v}_i\|} = \dfrac{1}{\sigma_i}A\mathbf{v}_i \quad (i = 1, 2, \ldots, k)$

(*e*) $\{\mathbf{u}_1, \mathbf{u}_2, \ldots, \mathbf{u}_k\}$ *is an orthonormal basis for* col(A).

(*f*) $\{\mathbf{u}_1, \mathbf{u}_2, \ldots, \mathbf{u}_k, \mathbf{u}_{k+1}, \ldots, \mathbf{u}_m\}$ *is an extension of $\{\mathbf{u}_1, \mathbf{u}_2, \ldots, \mathbf{u}_k\}$ to an orthonormal basis for R^m.*

The vectors $\mathbf{u}_1, \mathbf{u}_2, \ldots, \mathbf{u}_k$ are called the ***left singular vectors*** of A, and the vectors $\mathbf{v}_1, \mathbf{v}_2, \ldots, \mathbf{v}_k$ are called the ***right singular vectors*** of A.

▶ **EXAMPLE 2 Singular Value Decomposition if A Is Not Square**

Find a singular value decomposition of the matrix

$$A = \begin{bmatrix} 1 & 1 \\ 0 & 1 \\ 1 & 0 \end{bmatrix}$$

Solution We showed in Example 1 that the eigenvalues of $A^T A$ are $\lambda_1 = 3$ and $\lambda_2 = 1$ and that the corresponding singular values of A are $\sigma_1 = \sqrt{3}$ and $\sigma_2 = 1$. We leave it for you to verify that

$$\mathbf{v}_1 = \begin{bmatrix} \frac{\sqrt{2}}{2} \\ \frac{\sqrt{2}}{2} \end{bmatrix} \quad \text{and} \quad \mathbf{v}_2 = \begin{bmatrix} \frac{\sqrt{2}}{2} \\ -\frac{\sqrt{2}}{2} \end{bmatrix}$$

are eigenvectors corresponding to λ_1 and λ_2, respectively, and that $V = [\mathbf{v}_1 \mid \mathbf{v}_2]$ orthogonally diagonalizes $A^T A$. From part (d) of Theorem 9.5.4, the vectors

$$\mathbf{u}_1 = \frac{1}{\sigma_1} A \mathbf{v}_1 = \frac{\sqrt{3}}{3} \begin{bmatrix} 1 & 1 \\ 0 & 1 \\ 1 & 0 \end{bmatrix} \begin{bmatrix} \frac{\sqrt{2}}{2} \\ \frac{\sqrt{2}}{2} \end{bmatrix} = \begin{bmatrix} \frac{\sqrt{6}}{3} \\ \frac{\sqrt{6}}{6} \\ \frac{\sqrt{6}}{6} \end{bmatrix}$$

$$\mathbf{u}_2 = \frac{1}{\sigma_2} A \mathbf{v}_2 = (1) \begin{bmatrix} 1 & 1 \\ 0 & 1 \\ 1 & 0 \end{bmatrix} \begin{bmatrix} \frac{\sqrt{2}}{2} \\ -\frac{\sqrt{2}}{2} \end{bmatrix} = \begin{bmatrix} 0 \\ -\frac{\sqrt{2}}{2} \\ \frac{\sqrt{2}}{2} \end{bmatrix}$$

Eugenio Beltrami (1835–1900)

Camille Jordan (1838–1922)

Herman Klaus Weyl (1885–1955)

Gene H. Golub (1932–   )

Historical Note The theory of singular value decompositions can be traced back to the work of five people: the Italian mathematician Eugenio Beltrami, the French mathematician Camille Jordan, the English mathematician James Sylvester (see p. 34), and the German mathematicians Erhard Schmidt (see p. 360) and the mathematician Herman Weyl. More recently, the pioneering efforts of the American mathematician Gene Golub produced a stable and efficient algorithm for computing it. Beltrami and Jordan were the progenitors of the decomposition—Beltrami gave a proof of the result for real, invertible matrices with distinct singular values in 1873. Subsequently, Jordan refined the theory and eliminated the unnecessary restrictions imposed by Beltrami. Sylvester, apparently unfamiliar with the work of Beltrami and Jordan, rediscovered the result in 1889 and suggested its importance. Schmidt was the first person to show that the singular value decomposition could be used to approximate a matrix by another matrix with lower rank, and, in so doing, he transformed it from a mathematical curiosity to an important practical tool. Weyl showed how to find the lower rank approximations in the presence of error.
[*Images: wikipedia (Beltrami); The Granger Collection, New York (Jordan); Courtesy Electronic Publishing Services, Inc., New York City (Weyl); wikipedia (Golub)*]

are two of the three column vectors of U. Note that $\mathbf{u}_1$ and $\mathbf{u}_2$ are orthonormal, as expected. We could extend the set $\{\mathbf{u}_1, \mathbf{u}_2\}$ to an orthonormal basis for R^3. However, the computations will be easier if we first remove the messy radicals by multiplying $\mathbf{u}_1$ and $\mathbf{u}_2$ by appropriate scalars. Thus, we will look for a unit vector $\mathbf{u}_3$ that is orthogonal to

$$\sqrt{6}\,\mathbf{u}_1 = \begin{bmatrix} 2 \\ 1 \\ 1 \end{bmatrix} \quad \text{and} \quad \sqrt{2}\,\mathbf{u}_2 = \begin{bmatrix} 0 \\ -1 \\ 1 \end{bmatrix}$$

To satisfy these two orthogonality conditions, the vector $\mathbf{u}_3$ must be a solution of the homogeneous linear system

$$\begin{bmatrix} 2 & 1 & 1 \\ 0 & -1 & 1 \end{bmatrix} \begin{bmatrix} x_1 \\ x_2 \\ x_3 \end{bmatrix} = \begin{bmatrix} 0 \\ 0 \\ 0 \end{bmatrix}$$

We leave it for you to show that a general solution of this system is

$$\begin{bmatrix} x_1 \\ x_2 \\ x_3 \end{bmatrix} = t \begin{bmatrix} -1 \\ 1 \\ 1 \end{bmatrix}$$

Normalizing the vector on the right yields

$$\mathbf{u}_3 = \begin{bmatrix} -\frac{1}{\sqrt{3}} \\ \frac{1}{\sqrt{3}} \\ \frac{1}{\sqrt{3}} \end{bmatrix}$$

Thus, the singular value decomposition of A is

$$\underbrace{\begin{bmatrix} 1 & 1 \\ 0 & 1 \\ 1 & 0 \end{bmatrix}}_{A} = \underbrace{\begin{bmatrix} \frac{\sqrt{6}}{3} & 0 & -\frac{1}{\sqrt{3}} \\ \frac{\sqrt{6}}{6} & -\frac{\sqrt{2}}{2} & \frac{1}{\sqrt{3}} \\ \frac{\sqrt{6}}{6} & \frac{\sqrt{2}}{2} & \frac{1}{\sqrt{3}} \end{bmatrix}}_{U} \underbrace{\begin{bmatrix} \sqrt{3} & 0 \\ 0 & 1 \\ 0 & 0 \end{bmatrix}}_{\Sigma} \underbrace{\begin{bmatrix} \frac{\sqrt{2}}{2} & \frac{\sqrt{2}}{2} \\ \frac{\sqrt{2}}{2} & -\frac{\sqrt{2}}{2} \end{bmatrix}}_{V^T}$$

You may want to confirm the validity of this equation by multiplying out the matrices on the right side. ◀

OPTIONAL

We conclude this section with an optional proof of Theorem 9.5.4.

Proof of Theorem 9.5.4   For notational simplicity we will prove this theorem in the case where A is an $n \times n$ matrix. To modify the argument for an $m \times n$ matrix you need only make the notational adjustments required to account for the possibility that $m > n$ or $n > m$.

The matrix $A^T A$ is symmetric, so it has an eigenvalue decomposition

$$A^T A = VDV^T$$

in which the column vectors of

$$V = [\mathbf{v}_1 \mid \mathbf{v}_2 \mid \cdots \mid \mathbf{v}_n]$$

are unit eigenvectors of $A^T A$, and D is a diagonal matrix whose successive diagonal entries $\lambda_1, \lambda_2, \ldots, \lambda_n$ are the eigenvalues of $A^T A$ corresponding in succession to the column vectors of V. Since A is assumed to have rank k, it follows from Theorem 9.5.1

that A^TA also has rank k. It follows as well that D has rank k, since it is similar to A^TA and rank is a similarity invariant. Thus, D can be expressed in the form

$$
D = \begin{bmatrix} \lambda_1 & & & & & & & 0 \\ & \lambda_2 & & & & & & \\ & & \ddots & & & & & \\ & & & \lambda_k & & & & \\ & & & & 0 & & & \\ & & & & & \ddots & & \\ 0 & & & & & & & 0 \end{bmatrix} \tag{2}
$$

where $\lambda_1 \geq \lambda_2 \geq \cdots \geq \lambda_k > 0$. Now let us consider the set of image vectors

$$\{A\mathbf{v}_1, A\mathbf{v}_2, \ldots, A\mathbf{v}_n\} \tag{3}$$

This is an orthogonal set, for if $i \neq j$, then the orthogonality of $\mathbf{v}_i$ and $\mathbf{v}_j$ implies that

$$A\mathbf{v}_i \cdot A\mathbf{v}_j = \mathbf{v}_i \cdot A^TA\mathbf{v}_j = \mathbf{v}_i \cdot \lambda_j\mathbf{v}_j = \lambda_j(\mathbf{v}_i \cdot \mathbf{v}_j) = 0$$

Moreover, the first k vectors in (3) are nonzero since we showed in the proof of Theorem 9.5.2b that $\|A\mathbf{v}_i\|^2 = \lambda_i$ for $i = 1, 2, \ldots, n$, and we have assumed that the first k diagonal entries in (2) are positive. Thus,

$$S = \{A\mathbf{v}_1, A\mathbf{v}_2, \ldots, A\mathbf{v}_k\}$$

is an orthogonal set of *nonzero* vectors in the column space of A. But the column space of A has dimension k since

$$\text{rank}(A) = \text{rank}(A^TA) = k$$

and hence S, being a linearly independent set of k vectors, must be an orthogonal basis for col(A). If we now normalize the vectors in S, we will obtain an orthonormal basis $\{\mathbf{u}_1, \mathbf{u}_2, \ldots, \mathbf{u}_k\}$ for col(A) in which

$$\mathbf{u}_i = \frac{A\mathbf{v}_i}{\|A\mathbf{v}_i\|} = \frac{1}{\sqrt{\lambda_i}}A\mathbf{v}_i \qquad (1 \leq i \leq k)$$

or, equivalently, in which

$$A\mathbf{v}_1 = \sqrt{\lambda_1}\mathbf{u}_1 = \sigma_1\mathbf{u}_1, \quad A\mathbf{v}_2 = \sqrt{\lambda_2}\mathbf{u}_2 = \sigma_2\mathbf{u}_2, \ldots, \quad A\mathbf{v}_k = \sqrt{\lambda_k}\mathbf{u}_k = \sigma_k\mathbf{u}_k \tag{4}$$

It follows from Theorem 6.3.6 that we can extend this to an orthonormal basis

$$\{\mathbf{u}_1, \mathbf{u}_2, \ldots, \mathbf{u}_k, \mathbf{u}_{k+1}, \ldots, \mathbf{u}_n\}$$

for R^n. Now let U be the orthogonal matrix

$$U = \begin{bmatrix} \mathbf{u}_1 & \mathbf{u}_2 & \cdots & \mathbf{u}_k & \mathbf{u}_{k+1} & \cdots & \mathbf{u}_n \end{bmatrix}$$

and let Σ be the diagonal matrix

$$
\Sigma = \begin{bmatrix} \sigma_1 & & & & & & & 0 \\ & \sigma_2 & & & & & & \\ & & \ddots & & & & & \\ & & & \sigma_k & & & & \\ & & & & 0 & & & \\ & & & & & \ddots & & \\ 0 & & & & & & & 0 \end{bmatrix}
$$

It follows from (4), and the fact that $A\mathbf{v}_i = 0$ for $i > k$, that

$$U\Sigma = [\sigma_1\mathbf{u}_1 \quad \sigma_2\mathbf{u}_2 \quad \cdots \quad \sigma_k\mathbf{u}_k \quad 0 \quad \cdots \quad 0]$$
$$= [A\mathbf{v}_1 \quad A\mathbf{v}_2 \quad \cdots \quad A\mathbf{v}_k \quad A\mathbf{v}_{k+1} \quad \cdots \quad A\mathbf{v}_n]$$
$$= AV$$

which we can rewrite using the orthogonality of V as $A = U\Sigma V^T$. ◀

Concept Review

- Eigenvalue decomposition
- Hessenberg decomposition
- Schur decomposition
- Magnification of roundoff error

- Properties that A and $A^T A$ have in common
- $A^T A$ is orthogonally diagonalizable
- Eigenvalues of $A^T A$ are nonnegative

- Singular values
- Diagonal entries of a matrix that is not square
- Singular value decomposition

Skills

- Find the singular values of an $m \times n$ matrix.

- Find a singular value decomposition of an $m \times n$ matrix.

Exercise Set 9.5

▶ In Exercises 1–4, find the distinct singular values of A. ◀

1. $A = \begin{bmatrix} 1 & 2 & 0 \end{bmatrix}$

2. $A = \begin{bmatrix} 3 & 0 \\ 0 & 4 \end{bmatrix}$

3. $A = \begin{bmatrix} 1 & -2 \\ 2 & 1 \end{bmatrix}$

4. $A = \begin{bmatrix} \sqrt{2} & 0 \\ 1 & \sqrt{2} \end{bmatrix}$

▶ In Exercises 5–12, find a singular value decomposition of A.

5. $A = \begin{bmatrix} 1 & -1 \\ 1 & 1 \end{bmatrix}$

6. $A = \begin{bmatrix} -3 & 0 \\ 0 & -4 \end{bmatrix}$

7. $A = \begin{bmatrix} 4 & 6 \\ 0 & 4 \end{bmatrix}$

8. $A = \begin{bmatrix} 3 & 3 \\ 3 & 3 \end{bmatrix}$

9. $A = \begin{bmatrix} -2 & 2 \\ -1 & 1 \\ 2 & -2 \end{bmatrix}$

10. $A = \begin{bmatrix} -2 & -1 & 2 \\ 2 & 1 & -2 \end{bmatrix}$

11. $A = \begin{bmatrix} 1 & 0 \\ 1 & 1 \\ -1 & 1 \end{bmatrix}$

12. $A = \begin{bmatrix} 6 & 4 \\ 0 & 0 \\ 4 & 0 \end{bmatrix}$

13. Prove: If A is an $m \times n$ matrix, then $A^T A$ and $A A^T$ have the same rank.

14. Prove part (d) of Theorem 9.5.1 by using part (a) of the theorem and the fact that A and $A^T A$ have n columns.

15. (a) Prove part (b) of Theorem 9.5.1 by first showing that row$(A^T A)$ is a subspace of row(A).

(b) Prove part (c) of Theorem 9.5.1 by using part (b).

16. Let $T : R^n \rightarrow R^m$ be a linear transformation whose standard matrix A has the singular value decomposition $A = U\Sigma V^T$, and let $B = \{\mathbf{v}_1, \mathbf{v}_2, \ldots, \mathbf{v}_n\}$ and $B' = \{\mathbf{u}_1, \mathbf{u}_2, \ldots, \mathbf{u}_m\}$ be the column vectors of V and U, respectively. Show that $\Sigma = [T]_{B',B}$.

17. Show that the singular values of $A^T A$ are the squares of the singular values of A.

18. Show that if $A = U\Sigma V^T$ is a singular value decomposition of A, then U orthogonally diagonalizes AA^T.

True-False Exercises

In parts (a)–(g) determine whether the statement is true or false, and justify your answer.

(a) If A is an $m \times n$ matrix, then $A^T A$ is an $m \times m$ matrix.

(b) If A is an $m \times n$ matrix, then $A^T A$ is a symmetric matrix.

(c) If A is an $m \times n$ matrix, then the eigenvalues of $A^T A$ are positive real numbers.

(d) If A is an $n \times n$ matrix, then A is orthogonally diagonalizable.

(e) If A is an $m \times n$ matrix, then $A^T A$ is orthogonally diagonalizable.

(f) The eigenvalues of $A^T A$ are the singular values of A.

(g) Every $m \times n$ matrix has a singular value decomposition.

9.6 Data Compression Using Singular Value Decomposition

Efficient transmission and storage of large quantities of digital data has become a major problem in our technological world. In this section we will discuss the role that singular value decomposition plays in compressing digital data so that it can be transmitted more rapidly and stored in less space. We assume here that you have read Section 9.5.

Reduced Singular Value Decomposition

Algebraically, the zero rows and columns of the matrix Σ in Theorem 9.5.4 are superfluous and can be eliminated by multiplying out the expression $U\Sigma V^T$ using block multiplication and the partitioning shown in that formula. The products that involve zero blocks as factors drop out, leaving

$$A = [\mathbf{u}_1 \quad \mathbf{u}_2 \quad \cdots \quad \mathbf{u}_k] \begin{bmatrix} \sigma_1 & 0 & \cdots & 0 \\ 0 & \sigma_2 & \cdots & 0 \\ \vdots & \vdots & \ddots & \vdots \\ 0 & 0 & \cdots & \sigma_k \end{bmatrix} \begin{bmatrix} \mathbf{v}_1^T \\ \mathbf{v}_2^T \\ \vdots \\ \mathbf{v}_k^T \end{bmatrix} \tag{1}$$

which is called a ***reduced singular value decomposition*** of A. In this text we will denote the matrices on the right side of (1) by U_1, Σ_1, and V_1^T, respectively, and we will write this equation as

$$A = U_1 \Sigma_1 V_1^T \tag{2}$$

Note that the sizes of U_1, Σ_1, and V_1^T are $m \times k$, $k \times k$, and $k \times n$, respectively, and that the matrix Σ_1 is invertible, since its diagonal entries are positive.

If we multiply out on the right side of (1) using the column-row rule, then we obtain

$$A = \sigma_1 \mathbf{u}_1 \mathbf{v}_1^T + \sigma_2 \mathbf{u}_2 \mathbf{v}_2^T + \cdots + \sigma_k \mathbf{u}_k \mathbf{v}_k^T \tag{3}$$

which is called a ***reduced singular value expansion*** of A. This result applies to *all* matrices, whereas the spectral decomposition [Formula (7) of Section 7.2] applies only to symmetric matrices.

Remark It can be proved that an $m \times n$ matrix M has rank 1 if and only if it can be factored as $M = \mathbf{u}\mathbf{v}^T$, where $\mathbf{u}$ is a column vector in R^m and V is a column vector in R^n. Thus, a reduced singular value decomposition expresses a matrix A of rank k as a linear combination of k rank 1 matrices.

▶ **EXAMPLE 1 Reduced Singular Value Decomposition**

Find a reduced singular value decomposition and a reduced singular value expansion of the matrix

$$A = \begin{bmatrix} 1 & 1 \\ 0 & 1 \\ 1 & 0 \end{bmatrix}$$

Solution In Example 2 of Section 9.5 we found the singular value decomposition

$$
\begin{bmatrix} 1 & 1 \\ 0 & 1 \\ 1 & 0 \end{bmatrix} = \begin{bmatrix} \frac{\sqrt{6}}{3} & 0 & -\frac{1}{\sqrt{3}} \\ \frac{\sqrt{6}}{6} & -\frac{\sqrt{2}}{2} & \frac{1}{\sqrt{3}} \\ \frac{\sqrt{6}}{6} & \frac{\sqrt{2}}{2} & \frac{1}{\sqrt{3}} \end{bmatrix} \begin{bmatrix} \sqrt{3} & 0 \\ 0 & 1 \\ 0 & 0 \end{bmatrix} \begin{bmatrix} \frac{\sqrt{2}}{2} & \frac{\sqrt{2}}{2} \\ \frac{\sqrt{2}}{2} & -\frac{\sqrt{2}}{2} \end{bmatrix} \qquad (4)
$$

$$
A \quad = \qquad\qquad U \qquad\qquad\qquad \Sigma \qquad\qquad V^T
$$

Since A has rank 2 (verify), it follows from (1) with $k = 2$ that the reduced singular value decomposition of A corresponding to (4) is

$$
\begin{bmatrix} 1 & 1 \\ 0 & 1 \\ 1 & 0 \end{bmatrix} = \begin{bmatrix} \frac{\sqrt{6}}{3} & 0 \\ \frac{\sqrt{6}}{6} & -\frac{\sqrt{2}}{2} \\ \frac{\sqrt{6}}{6} & \frac{\sqrt{2}}{2} \end{bmatrix} \begin{bmatrix} \sqrt{3} & 0 \\ 0 & 1 \end{bmatrix} \begin{bmatrix} \frac{\sqrt{2}}{2} & \frac{\sqrt{2}}{2} \\ \frac{\sqrt{2}}{2} & -\frac{\sqrt{2}}{2} \end{bmatrix}
$$

This yields the reduced singular value expansion

$$
\begin{bmatrix} 1 & 1 \\ 0 & 1 \\ 1 & 0 \end{bmatrix} = \sigma_1 \mathbf{u}_1 \mathbf{v}_1^T + \sigma_2 \mathbf{u}_2 \mathbf{v}_2^T = \sqrt{3} \begin{bmatrix} \frac{\sqrt{6}}{3} \\ \frac{\sqrt{6}}{6} \\ \frac{\sqrt{6}}{6} \end{bmatrix} \begin{bmatrix} \frac{\sqrt{2}}{2} & \frac{\sqrt{2}}{2} \end{bmatrix} + (1) \begin{bmatrix} 0 \\ -\frac{\sqrt{2}}{2} \\ \frac{\sqrt{2}}{2} \end{bmatrix} \begin{bmatrix} \frac{\sqrt{2}}{2} & -\frac{\sqrt{2}}{2} \end{bmatrix}
$$

$$
= \sqrt{3} \begin{bmatrix} \frac{\sqrt{3}}{3} & \frac{\sqrt{3}}{3} \\ \frac{\sqrt{3}}{6} & \frac{\sqrt{3}}{6} \\ \frac{\sqrt{3}}{6} & \frac{\sqrt{3}}{6} \end{bmatrix} + (1) \begin{bmatrix} 0 & 0 \\ -\frac{1}{2} & \frac{1}{2} \\ \frac{1}{2} & -\frac{1}{2} \end{bmatrix}
$$

Note that the matrices in the expansion have rank 1, as expected. ◄

Data Compression and Image Processing

Singular value decompositions can be used to "compress" visual information for the purpose of reducing its required storage space and speeding up its electronic transmission. The first step in compressing a visual image is to represent it as a numerical matrix from which the visual image can be recovered when needed.

For example, a black and white photograph might be scanned as a rectangular array of pixels (points) and then stored as a matrix A by assigning each pixel a numerical value in accordance with its gray level. If 256 different gray levels are used ($0 =$ white to $255 =$ black), then the entries in the matrix would be integers between 0 and 255. The image can be recovered from the matrix A by printing or displaying the pixels with their assigned gray levels.

Original   Reconstruction

Historical Note In 1924 the U.S. Federal Bureau of Investigation (FBI) began collecting fingerprints and handprints and now has more than 30 million such prints in its files. To reduce the storage cost, the FBI began working with the Los Alamos National Laboratory, the National Bureau of Standards, and other groups in 1993 to devise rank based compression methods for storing prints in digital form. The following figure shows an original fingerprint and a reconstruction from digital data that was compressed at a ratio of 26:1.

If the matrix A has size $m \times n$, then one might store each of its mn entries individually. An alternative procedure is to compute the reduced singular value decomposition

$$A = \sigma_1 \mathbf{u}_1 \mathbf{v}_1^T + \sigma_2 \mathbf{u}_2 \mathbf{v}_2^T + \cdots + \sigma_k \mathbf{u}_k \mathbf{v}_k^T \tag{5}$$

in which $\sigma_1 \geq \sigma_2 \geq \cdots \geq \sigma_k$, and store the σ's, the $\mathbf{u}$'s, and the $\mathbf{v}$'s. When needed, the matrix A (and hence the image it represents) can be reconstructed from (5). Since each $\mathbf{u}_j$ has m entries and each $\mathbf{v}_j$ has n entries, this method requires storage space for

$$km + kn + k = k(m + n + 1)$$

numbers. Suppose, however, that the singular values $\sigma_{r+1}, \ldots, \sigma_k$ are sufficiently small that dropping the corresponding terms in (5) produces an acceptable approximation

$$A_r = \sigma_1 \mathbf{u}_1 \mathbf{v}_1^T + \sigma_2 \mathbf{u}_2 \mathbf{v}_2^T + \cdots + \sigma_r \mathbf{u}_r \mathbf{v}_r^T \tag{6}$$

to A and the image that it represents. We call (6) the ***rank r approximation of A***. This matrix requires storage space for only

$$rm + rn + r = r(m + n + 1)$$

numbers, compared to mn numbers required for entry-by-entry storage of A. For example, the rank 100 approximation of a 1000×1000 matrix A requires storage for only

$$100(1000 + 1000 + 1) = 200{,}100$$

numbers, compared to the 1,000,000 numbers required for entry-by-entry storage of A—a compression of almost 80%.

Figure 9.6.1 shows some approximations of a digitized mandrill image obtained using (6).

Rank 4

Rank 10

Rank 20

Rank 50

Rank 128

▲ Figure 9.6.1

Concept Review

- Reduced singular value decomposition
- Reduced singular value expansion
- Rank of an approximation

Skills

- Find the reduced singular value decomposition of an $m \times n$ matrix.
- Find the reduced singular value expansion of an $m \times n$ matrix.

Exercise Set 9.6

▶ In Exercises 1–4, find a reduced singular value decomposition of A. [*Note:* Each matrix appears in Exercise Set 9.5, where you were asked to find its (unreduced) singular value decomposition.]

1. $A = \begin{bmatrix} -2 & 2 \\ -1 & 1 \\ 2 & -2 \end{bmatrix}$ **2.** $A = \begin{bmatrix} -2 & -1 & 2 \\ 2 & 1 & -2 \end{bmatrix}$

3. $A = \begin{bmatrix} 1 & 0 \\ 1 & 1 \\ -1 & 1 \end{bmatrix}$ **4.** $A = \begin{bmatrix} 6 & 4 \\ 0 & 0 \\ 4 & 0 \end{bmatrix}$

▶ In Exercises 5–8, find a reduced singular value expansion of A.

5. The matrix A in Exercise 1.

6. The matrix A in Exercise 2.

7. The matrix A in Exercise 3.

8. The matrix A in Exercise 4.

9. Suppose A is a 200×500 matrix. How many numbers must be stored in the rank 100 approximation of A? Compare this with the number of entries of A.

True-False Exercises

In parts (a)–(c) determine whether the statement is true or false, and justify your answer. Assume that $U_1 \Sigma_1 V_1^T$ is a reduced singular value decomposition of an $m \times n$ matrix of rank k.

(a) U_1 has size $m \times k$.

(b) Σ_1 has size $k \times k$.

(c) V_1 has size $k \times n$.

Chapter 9 Supplementary Exercises

1. Find an LU-decomposition of $A = \begin{bmatrix} -6 & 2 \\ 6 & 0 \end{bmatrix}$.

2. Find the LDU-decomposition of the matrix A in Exercise 1.

3. Find an LU-decomposition of $A = \begin{bmatrix} 2 & 4 & 6 \\ 1 & 4 & 7 \\ 1 & 3 & 7 \end{bmatrix}$.

4. Find the LDU-decomposition of the matrix A in Exercise 3.

5. Let $A = \begin{bmatrix} 2 & 1 \\ 1 & 2 \end{bmatrix}$ and $\mathbf{x}_0 = \begin{bmatrix} 1 \\ 0 \end{bmatrix}$.

(a) Identify the dominant eigenvalue of A and then find the corresponding dominant unit eigenvector $\mathbf{v}$ with *positive* entries.

(b) Apply the power method with Euclidean scaling to A and $\mathbf{x}_0$, stopping at $\mathbf{x}_5$. Compare your value of $\mathbf{x}_5$ to the eigenvector $\mathbf{v}$ found in part (a).

(c) Apply the power method with maximum entry scaling to A and $\mathbf{x}_0$, stopping at $\mathbf{x}_5$. Compare your result with the eigenvector $\begin{bmatrix} 1 \\ 1 \end{bmatrix}$.

6. Consider the symmetric matrix

$$A = \begin{bmatrix} 0 & 1 \\ 1 & 0 \end{bmatrix}$$

Discuss the behavior of the power sequence

$$\mathbf{x}_0, \quad \mathbf{x}_1, \ldots, \quad \mathbf{x}_k, \ldots$$

with Euclidean scaling for a general *nonzero* vector $\mathbf{x}_0$. What is it about the matrix that causes the observed behavior?

7. Suppose that a symmetric matrix A has distinct eigenvalues $\lambda_1 = 8$, $\lambda_2 = 1.4$, $\lambda_3 = 2.3$, and $\lambda_4 = -8.1$. What can you say about the convergence of the Rayleigh quotients?

8. Find a singular value decomposition of $A = \begin{bmatrix} 1 & 1 \\ 1 & 1 \end{bmatrix}$.

9. Find a singular value decomposition of $A = \begin{bmatrix} 1 & 1 \\ 0 & 0 \\ 1 & 1 \end{bmatrix}$.

10. Find a reduced singular value decomposition and a reduced singular value expansion of the matrix A in Exercise 9.

11. Find the reduced singular value decomposition of the matrix whose singular value decomposition is

$$A = \begin{bmatrix} \frac{1}{2} & \frac{1}{2} & \frac{1}{2} & \frac{1}{2} \\ \frac{1}{2} & -\frac{1}{2} & -\frac{1}{2} & \frac{1}{2} \\ \frac{1}{2} & -\frac{1}{2} & \frac{1}{2} & -\frac{1}{2} \\ \frac{1}{2} & \frac{1}{2} & -\frac{1}{2} & -\frac{1}{2} \end{bmatrix} \begin{bmatrix} 24 & 0 & 0 \\ 0 & 12 & 0 \\ 0 & 0 & 0 \\ 0 & 0 & 0 \end{bmatrix} \begin{bmatrix} \frac{2}{3} & -\frac{1}{3} & \frac{2}{3} \\ \frac{2}{3} & \frac{2}{3} & -\frac{1}{3} \\ -\frac{1}{3} & \frac{2}{3} & \frac{2}{3} \end{bmatrix}$$

12. Do orthogonally similar matrices have the same singular values? Justify your answer.

13. If P is the standard matrix for the orthogonal projection of R^n onto a subspace W, what can you say about the singular values of P?

Since many of the most important concepts in linear algebra occur as theorem statements, it is important to be familiar with the various ways in which theorems can be structured. This appendix will help you to do that.

Contrapositive Form of a Theorem

The simplest theorems are of the form

$$\textit{If } H \textit{ is true, then } C \textit{ is true.} \tag{1}$$

where H is a statement, called the **hypothesis**, and C is a statement, called the **conclusion**. *The theorem is true if the conclusion is true whenever the hypothesis is true, and the theorem is false if there is some case where the hypothesis is true but the conclusion is false.* It is common to denote a theorem of form (1) as

$$H \Rightarrow C \tag{2}$$

(read, "H implies C"). As an example, the theorem

$$\textit{If } a \textit{ and } b \textit{ are both positive numbers, then } ab \textit{ is a positive number.} \tag{3}$$

is of form (2), where

$$H = a \textit{ and } b \textit{ are both positive numbers} \tag{4}$$

$$C = ab \textit{ is a positive number} \tag{5}$$

Sometimes it is desirable to phrase theorems in a *negative* way. For example, the theorem in (3) can be rephrased equivalently as

$$\textit{If } ab \textit{ is not a positive number, then } a \textit{ and } b \textit{ are not both positive numbers.} \tag{6}$$

If we write $\sim H$ to mean that (4) is false and $\sim C$ to mean that (5) is false, then the structure of the theorem in (6) is

$$\sim C \Rightarrow \sim H \tag{7}$$

In general, any theorem of form (2) can be rephrased in form (7), which is called the **contrapositive** of (2). If a theorem is true, then so is its contrapositive, and vice versa.

Converse of a Theorem

The **converse** of a theorem is the statement that results when the hypothesis and conclusion are interchanged. Thus, the converse of the theorem $H \Rightarrow C$ is the statement $C \Rightarrow H$. Whereas the contrapositive of a true theorem must itself be a true theorem, the converse of a true theorem may or may not be true. For example, the converse of (3) is the *false* statement

$$\textit{If } ab \textit{ is a positive number, then } a \textit{ and } b \textit{ are both positive numbers.}$$

but the converse of the true theorem

$$\textit{If } a > b, \textit{ then } 2a > 2b. \tag{8}$$

is the *true* theorem

$$\textit{If } 2a > 2b, \textit{ then } a > b. \tag{9}$$

Equivalent Statements

If a theorem $H \Rightarrow C$ and its converse $C \Rightarrow H$ are both true, then we say that H and C are **equivalent** statements, which we denote by writing

$$H \Leftrightarrow C \tag{10}$$

(read, "*H* and *C* are equivalent"). There are various ways of phrasing equivalent statements as a single theorem. Here are three ways in which (8) and (9) can be combined into a single theorem.

Form 1 *If a > b, then 2a > 2b, and conversely, if 2a > 2b, then a > b.*

Form 2 *a > b if and only if 2a > 2b.*

Form 3 *The following statements are equivalent.*

(i) $a > b$

(ii) $2a > 2b$

Theorems Involving Three or More Statements

Sometimes two true theorems will give you a third true theorem for free. Specifically, if $H \Rightarrow C$ is a true theorem, and $C \Rightarrow D$ is a true theorem, then $H \Rightarrow D$ must also be a true theorem. For example, the theorems

If opposite sides of a quadrilateral are parallel, then the quadrilateral is a parallelogram.

and

Opposite sides of a parallelogram have equal lengths.

imply the third theorem

If opposite sides of a quadrilateral are parallel, then they have equal lengths.

Sometimes three theorems yield equivalent statements for free. For example, if

$$H \Rightarrow C, \quad C \Rightarrow D, \quad D \Rightarrow H \tag{11}$$

then we have the *implication loop* in Figure A.1 from which we can conclude that

$$C \Rightarrow H, \quad D \Rightarrow C, \quad H \Rightarrow D \tag{12}$$

Combining this with (11) we obtain

$$H \Leftrightarrow C, \quad C \Leftrightarrow D, \quad D \Leftrightarrow H \tag{13}$$

In summary, if you want to prove the three equivalences in (13), you need only prove the three implications in (11).

▲ Figure 1.0.1

Complex numbers arise naturally in the course of solving polynomial equations. For example, the solutions of the quadratic equation $ax^2 + bx + c = 0$, which are given by the quadratic formula

$$x = \frac{-b \pm \sqrt{b^2 - 4ac}}{2a}$$

are complex numbers if the expression inside the radical is negative. In this appendix we will review some of the basic ideas about complex numbers that are used in this text.

Complex Numbers   To deal with the problem that the equation $x^2 = -1$ has no real solutions, mathematicians of the eighteenth century invented the "imaginary" number

$$i = \sqrt{-1}$$

which is assumed to have the property

$$i^2 = (\sqrt{-1})^2 = -1$$

but which otherwise has the algebraic properties of a real number. An expression of the form

$$a + bi \quad \text{or} \quad a + ib$$

in which a and b are *real* numbers is called a ***complex number***. Sometimes it will be convenient to use a single letter, typically z, to denote a complex number, in which case we write

$$z = a + bi \quad \text{or} \quad z = a + ib$$

The number a is called the ***real part*** of z and is denoted by $\operatorname{Re}(z)$, and the number b is called the ***imaginary part*** of z and is denoted by $\operatorname{Im}(z)$. Thus,

$$\operatorname{Re}(3 + 2i) = 3, \qquad \operatorname{Im}(3 + 2i) = 2$$
$$\operatorname{Re}(1 - 5i) = 1, \qquad \operatorname{Im}(1 - 5i) = \operatorname{Im}(1 + (-5)i) = -5$$
$$\operatorname{Re}(7i) = \operatorname{Re}(0 + 7i) = 0, \quad \operatorname{Im}(7i) = 7$$
$$\operatorname{Re}(4) = 4, \qquad \operatorname{Im}(4) = \operatorname{Im}(4 + 0i) = 0$$

Two complex numbers are considered ***equal*** if and only if their real parts are equal and their imaginary parts are equal; that is,

$$a + bi = c + di \quad \text{if and only if} \quad a = c \text{ and } b = d$$

A complex number $z = bi$ whose real part is zero is said to be ***pure imaginary***. A complex number $z = a$ whose imaginary part is zero is a real number, so the real numbers can be viewed as a subset of the complex numbers.

     Complex numbers are added, subtracted, and multiplied in accordance with the standard rules of algebra but with $i^2 = -1$:

$$(a + bi) + (c + di) = (a + c) + (b + d)i \tag{1}$$
$$(a + bi) - (c + di) = (a - c) + (b - d)i \tag{2}$$
$$(a + bi)(c + di) = (ac - bd) + (ad + bc)i \tag{3}$$

The multiplication formula is obtained by expanding the left side and using the fact that $i^2 = -1$. Also note that if $b = 0$, then the multiplication formula simplifies to

$$a(c + di) = ac + adi \tag{4}$$

The set of complex numbers with these operations is commonly denoted by the symbol C and is called the ***complex number system***.

▶ EXAMPLE 1 Multiplying Complex Numbers

As a practical matter, it is usually more convenient to compute products of complex numbers by expansion, rather than substituting in (3). For example,

$$(3 - 2i)(4 + 5i) = 12 + 15i - 8i - 10i^2 = (12 + 10) + 7i = 22 + 7i \quad ◀$$

The Complex Plane   A complex number $z = a + bi$ can be associated with the ordered pair (a, b) of real numbers and represented geometrically by a point or a vector in the xy-plane (Figure B.1). We call this the ***complex plane***. Points on the x-axis have an imaginary part of zero and hence correspond to real numbers, whereas points on the y-axis have a real part of zero and correspond to pure imaginary numbers. Accordingly, we call the x-axis the ***real axis*** and the y-axis the ***imaginary axis*** (Figure B.2).

▲ Figure B.1   ▲ Figure B.2

Complex numbers can be added, subtracted, or multiplied by real numbers geometrically by performing these operations on their associated vectors (Figure B.3, for example). In this sense the complex number system C is closely related to R^2, the main difference being that complex numbers can be multiplied to produce other complex numbers, whereas there is no multiplication operation on R^2 that produces other vectors in R^2 (the dot product produces a scalar, not a vector in R^2).

▶ Figure B.3

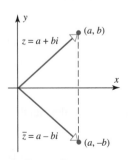

▲ Figure B.4

If $z = a + bi$ is a complex number, then the ***complex conjugate*** of z, or more simply, the ***conjugate*** of z, is denoted by $\bar{z}$ (read, "z bar") and is defined by

$$\bar{z} = a - bi \tag{5}$$

Numerically, $\bar{z}$ is obtained from z by reversing the sign of the imaginary part, and geometrically it is obtained by reflecting the vector for z about the real axis (Figure B.4).

▶ **EXAMPLE 2 Some Complex Conjugates**

$$z = 3 + 4i \qquad \bar{z} = 3 - 4i$$
$$z = -2 - 5i \qquad \bar{z} = -2 + 5i$$
$$z = i \qquad \bar{z} = -i$$
$$z = 7 \qquad \bar{z} = 7 \blacktriangleleft$$

Remark   The last computation in this example illustrates the fact that a real number is equal to its complex conjugate. More generally, $z = \bar{z}$ if and only if z is a real number.

The following computation shows that the product of a complex number $z = a + bi$ and its conjugate $z = a - bi$ is a nonnegative real number:

$$z\bar{z} = (a + bi)(a - bi) = a^2 - abi + bai - b^2 i^2 = a^2 + b^2 \tag{6}$$

You will recognize that

$$\sqrt{z\bar{z}} = \sqrt{a^2 + b^2}$$

is the length of the vector corresponding to z (Figure B.5); we call this length the **modulus** (or **absolute value** of z) and denote it by $|z|$. Thus,

$$|z| = \sqrt{z\bar{z}} = \sqrt{a^2 + b^2} \tag{7}$$

$|z| = \sqrt{a^2 + b^2}$

▲ **Figure B.5**

Note that if $b = 0$, then $z = a$ is a real number and $|z| = \sqrt{a^2} = |a|$, which tells us that the modulus of a real number is the same as its absolute value as defined in beginning algebra.

▶ **EXAMPLE 3 Some Modulus Computations**

$$z = 3 + 4i \qquad |z| = \sqrt{3^2 + 4^2} = 5$$
$$z = -4 - 5i \qquad |z| = \sqrt{(-4)^2 + (-5)^2} = \sqrt{41}$$
$$z = i \qquad |z| = \sqrt{0^2 + 1^2} = 1 \blacktriangleleft$$

Reciprocals and Division   If $z \neq 0$, then the **reciprocal** (or **multiplicative inverse**) of z is denoted by $1/z$ (or z^{-1}) and is defined by the property

$$\left(\frac{1}{z}\right) z = 1$$

This equation has a unique solution for $1/z$, which we can obtain by multiplying both sides by $\bar{z}$ and using the fact that $z\bar{z} = |z|^2$ [see (7)]. This yields

$$\frac{1}{z} = \frac{\bar{z}}{|z|^2} \tag{8}$$

If $z_2 \neq 0$, then the **quotient** z_1/z_2 is defined to be the product of z_1 and $1/z_2$. This yields the formula

$$\frac{z_1}{z_2} = \frac{\bar{z}_2}{|z_2|^2} z_1 = \frac{z_1 \bar{z}_2}{|z_2|^2} \tag{9}$$

Observe that the expression on the right side of (9) results if the numerator and denominator of z_1/z_2 are multiplied by $\bar{z}_2$. As a practical matter, this is often the best way to perform divisions of complex numbers.

▶ **EXAMPLE 4** **Division of Complex Numbers**

Let $z_1 = 3 + 4i$ and $z_2 = 1 - 2i$. Express z_1/z_2 in the form $a + bi$.

Solution We will multiply the numerator and denominator of z_1/z_2 by $\bar{z}_2$. This yields

$$\frac{z_1}{z_2} = \frac{z_1\bar{z}_2}{z_2\bar{z}_2} = \frac{3 + 4i}{1 - 2i} \cdot \frac{1 + 2i}{1 + 2i}$$

$$= \frac{3 + 6i + 4i + 8i^2}{1 - 4i^2}$$

$$= \frac{-5 + 10i}{5}$$

$$= -1 + 2i \blacktriangleleft$$

The following theorems list some useful properties of the modulus and conjugate operations.

THEOREM 2.0.1 *The following results hold for any complex numbers* z, z_1, *and* z_2.

(a) $\overline{z_1 + z_2} = \bar{z}_1 + \bar{z}_2$

(b) $\overline{z_1 - z_2} = \bar{z}_1 - \bar{z}_2$

(c) $\overline{z_1 z_2} = \bar{z}_1 \bar{z}_2$

(d) $\overline{z_1/z_2} = \bar{z}_1/\bar{z}_2$

(e) $\bar{\bar{z}} = z$

THEOREM 2.0.2 *The following results hold for any complex numbers* z, z_1, *and* z_2.

(a) $|\bar{z}| = |z|$

(b) $|z_1 z_2| = |z_1||z_2|$

(c) $|z_1/z_2| = |z_1|/|z_2|$

(d) $|z_1 + z_2| \le |z_1| + |z_2|$

Polar Form of a Complex Number

If $z = a + bi$ is a nonzero complex number, and if ϕ is an angle from the real axis to the vector z, then, as suggested in Figure B.6, the real and imaginary parts of z can be expressed as

$$a = |z| \cos \phi \quad \text{and} \quad b = |z| \sin \phi \tag{10}$$

Thus, the complex number $z = a + bi$ can be expressed as

$$z = |z|(\cos \phi + i \sin \phi) \tag{11}$$

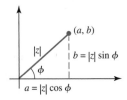

▲ Figure B.6

which is called a **polar form** of z. The angle ϕ in this formula is called an **argument** of z. The argument of z is not unique because we can add or subtract any multiple of 2π to it to obtain a different argument of z. However, there is only one argument whose radian measure satisfies

$$-\pi < \phi \le \pi \tag{12}$$

This is called the **principal argument** of z.

▶ **EXAMPLE 5 Polar Form of a Complex Number**

Express $z = 1 - \sqrt{3}i$ in polar form using the principal argument.

Solution The modulus of z is

$$|z| = \sqrt{1^2 + (-\sqrt{3})^2} = \sqrt{4} = 2$$

Thus, it follows from (10) with $a = 1$ and $b = -\sqrt{3}$ that

$$1 = 2\cos\phi \quad \text{and} \quad -\sqrt{3} = 2\sin\phi$$

and this implies that

$$\cos\phi = \frac{1}{2} \quad \text{and} \quad \sin\phi = -\frac{\sqrt{3}}{2}$$

The unique angle ϕ that satisfies these equations and whose radian measure satisfies (12) is $\phi = -\pi/3$ (Figure B.7). Thus, a polar form of z is

$$z = 2\left(\cos\left(-\frac{\pi}{3}\right) + i\sin\left(-\frac{\pi}{3}\right)\right) = 2\left(\cos\frac{\pi}{3} - i\sin\frac{\pi}{3}\right) \quad ◀$$

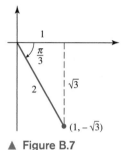

▲ Figure B.7

Geometric Interpretation of Multiplication and Division of Complex Numbers

We now show how polar forms of complex numbers provide geometric interpretations of multiplication and division. Let

$$z_1 = |z_1|(\cos\phi_1 + i\sin\phi_1) \quad \text{and} \quad z_2 = |z_2|(\cos\phi_2 + i\sin\phi_2)$$

be polar forms of the nonzero complex numbers z_1 and z_2. Multiplying, we obtain

$$z_1 z_2 = |z_1||z_2|[(\cos\phi_1\cos\phi_2 - \sin\phi_1\sin\phi_2) + i(\sin\phi_1\cos\phi_2 + \cos\phi_1\sin\phi_2)]$$

Now applying the trigonometric identities

$$\cos(\phi_1 + \phi_2) = \cos\phi_1\cos\phi_2 - \sin\phi_1\sin\phi_2$$
$$\sin(\phi_1 + \phi_2) = \sin\phi_1\cos\phi_2 + \cos\phi_1\sin\phi_2$$

▲ Figure B.8

yields

$$z_1 z_2 = |z_1||z_2|[\cos(\phi_1 + \phi_2) + i\sin(\phi_1 + \phi_2)] \tag{13}$$

which is a polar form of the complex number with modulus $|z_1||z_2|$ and argument $\phi_1 + \phi_2$. Thus, we have shown that *multiplying two complex numbers has the geometric effect of multiplying their moduli and adding their arguments* (Figure B.8).

Similar kinds of computations show that

$$\frac{z_1}{z_2} = \frac{|z_1|}{|z_2|}[\cos(\phi_1 - \phi_2) + i\sin(\phi_1 - \phi_2)] \tag{14}$$

which tells us that *dividing complex numbers has the geometric effect of dividing their moduli and subtracting their arguments* (both in the appropriate order).

▶ **EXAMPLE 6 Multiplying and Dividing in Polar Form**

Use polar forms of the complex numbers $z_1 = 1 + \sqrt{3}i$ and $z_2 = \sqrt{3} + i$ to compute $z_1 z_2$ and z_1/z_2.

Solution Polar forms of these complex numbers are

$$z_1 = 2\left(\cos\frac{\pi}{3} + i\sin\frac{\pi}{3}\right) \quad \text{and} \quad z_2 = 2\left(\cos\frac{\pi}{6} + i\sin\frac{\pi}{6}\right)$$

(verify). Thus, it follows from (13) that

$$z_1 z_2 = 4\left[\cos\left(\frac{\pi}{3} + \frac{\pi}{6}\right) + i\sin\left(\frac{\pi}{3} + \frac{\pi}{6}\right)\right] = 4\left[\cos\left(\frac{\pi}{2}\right) + i\sin\left(\frac{\pi}{2}\right)\right] = 4i$$

and from (14) that

$$\frac{z_1}{z_2} = 1 \cdot \left[\cos\left(\frac{\pi}{3} - \frac{\pi}{6}\right) + i\sin\left(\frac{\pi}{3} - \frac{\pi}{6}\right)\right] = \cos\left(\frac{\pi}{6}\right) + i\sin\left(\frac{\pi}{6}\right) = \frac{\sqrt{3}}{2} + \frac{1}{2}i$$

As a check, let us calculate $z_1 z_2$ and z_1/z_2 directly:

$$z_1 z_2 = (1 + \sqrt{3}i)(\sqrt{3} + i) = \sqrt{3} + i + 3i + \sqrt{3}i^2 = 4i$$

$$\frac{z_1}{z_2} = \frac{1 + \sqrt{3}i}{\sqrt{3} + i} = \frac{1 + \sqrt{3}i}{\sqrt{3} + i} \cdot \frac{\sqrt{3} - i}{\sqrt{3} - i} = \frac{\sqrt{3} - i + 3i - \sqrt{3}i^2}{3 - i^2} = \frac{2\sqrt{3} + 2i}{4} = \frac{\sqrt{3}}{2} + \frac{1}{2}i$$

which agrees with the results obtained using polar forms. ◀

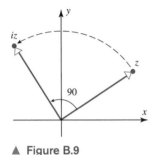

▲ Figure B.9

Remark The complex number i has a modulus of 1 and a principal argument of $\pi/2$. Thus, if z is a complex number, then iz has the same modulus as z but its argument is greater by $\pi/2$ ($= 90°$); that is, multiplication by i has the geometric effect of rotating the vector z counterclockwise by $90°$ (Figure B.9).

DeMoivre's Formula If n is a positive integer, and if z is a nonzero complex number with polar form

$$z = |z|(\cos\phi + i\sin\phi)$$

then raising z to the nth power yields

$$z^n = \underbrace{z \cdot z \cdots z}_{n \text{ factors}} = |z|^n[\cos(\underbrace{\phi + \phi + \cdots + \phi}_{n \text{ terms}})] + i[\sin(\underbrace{\phi + \phi + \cdots + \phi}_{n \text{ terms}})]$$

which we can write more succinctly as

$$z^n = |z|^n(\cos n\phi + i\sin n\phi) \tag{15}$$

In the special case where $|z| = 1$ this formula simplifies to

$$z^n = \cos n\phi + i\sin n\phi$$

which, using the polar form for z, becomes

$$(\cos\phi + i\sin\phi)^n = \cos n\phi + i\sin n\phi \tag{16}$$

This result is called **DeMoivre's formula**.

Euler's Formula If θ is a real number, say the radian measure of some angle, then the **complex exponential** function $e^{i\theta}$ is defined to be

$$e^{i\theta} = \cos\theta + i\sin\theta \tag{17}$$

which is sometimes called **Euler's formula**. One motivation for this formula comes from the Maclaurin series in calculus. Readers who have studied infinite series in calculus can deduce (17) by formally substituting $i\theta$ for x in the Maclaurin series for e^x and writing

$$e^{i\theta} = 1 + i\theta + \frac{(i\theta)^2}{2!} + \frac{(i\theta)^3}{3!} + \frac{(i\theta)^4}{4!} + \frac{(i\theta)^5}{5!} + \frac{(i\theta)^6}{6!} + \cdots$$

$$= 1 + i\theta - \frac{\theta^2}{2!} - i\frac{\theta^3}{3!} + \frac{\theta^4}{4!} + i\frac{\theta^5}{5!} - \frac{\theta^6}{6!} + \cdots$$

$$= \left(1 - \frac{\theta^2}{2!} + \frac{\theta^4}{4!} - \frac{\theta^6}{6!} + \cdots\right) + i\left(\theta - \frac{\theta^3}{3!} + \frac{\theta^5}{5!} - \cdots\right)$$

$$= \cos\theta + i\sin\theta$$

where the last step follows from the Maclaurin series for $\cos\theta$ and $\sin\theta$.

If $z = a + bi$ is any complex number, then the ***complex exponential*** e^z is defined to be

$$e^z = e^{a+bi} = e^a e^{ib} = e^a(\cos b + i\sin b) \tag{18}$$

It can be proved that complex exponentials satisfy the standard laws of exponents. Thus, for example,

$$e^{z_1}e^{z_2} = e^{z_1+z_2}, \qquad \frac{e^{z_1}}{e^{z_2}} = e^{z_1-z_2}, \qquad \frac{1}{e^z} = e^{-z}$$

ANSWERS TO EXERCISES

Exercise Set 1.1 (page 9)

1. (a), (c), and (f) are linear equations; (b), (d) and (e) are not linear equations

3. (a) and (d) are linear systems; (b) and (c) are not linear systems   **5.** (a) and (d) are both consistent

7. (a), (d), and (e) are solutions; (b) and (c) are not solutions   **9.** (a) $x = \frac{5}{7}t + \frac{3}{7}$ (b) $x_1 = \frac{1}{4}r - \frac{5}{8}s + \frac{3}{4}t - \frac{1}{8}$
$y = t$ $x_2 = r$
   $x_3 = s$
   $x_4 = t$

11. (a) $\begin{aligned} 2x_1 &= 0 \\ 3x_1 - 4x_2 &= 0 \\ x_2 &= 1 \end{aligned}$ (b) $\begin{aligned} 3x_1 \quad\quad - 2x_3 &= 5 \\ 7x_1 + x_2 + 4x_3 &= -3 \\ -2x_2 + x_3 &= 7 \end{aligned}$ (c) $\begin{aligned} 7x_1 + 2x_2 + x_3 - 3x_4 &= 5 \\ x_1 + 2x_2 + 4x_3 &= 1 \end{aligned}$

(d) $\begin{aligned} x_1 \quad\quad\quad &= 7 \\ x_2 \quad\quad &= -2 \\ x_3 \quad &= 3 \\ x_4 &= 4 \end{aligned}$

13. (a) $\begin{bmatrix} -2 & 6 \\ 3 & 8 \\ 9 & -3 \end{bmatrix}$ (b) $\begin{bmatrix} 6 & -1 & 3 & 4 \\ 0 & 5 & -1 & 1 \end{bmatrix}$ (c) $\begin{bmatrix} 0 & 2 & 0 & -3 & 1 & 0 \\ -3 & -1 & 1 & 0 & 0 & -1 \\ 6 & 2 & -1 & 2 & -3 & 6 \end{bmatrix}$

(d) $\begin{bmatrix} 1 & 0 & 0 & 0 & -1 & 7 \end{bmatrix}$

True/False 1.1

(a) True   (b) False   (c) True   (d) True   (e) False   (f) False   (g) True   (h) False

Exercise Set 1.2 (page 22)

1. (a) Both   (b) Both   (c) Both   (d) Both   (e) Both   (f) Both   (g) Row echelon

3. (a) $x_1 = -37$, $x_2 = -8$, $x_3 = 5$ (b) $x_1 = 13t - 10$, $x_2 = 13t - 5$, $x_3 = -t + 2$, $x_4 = t$
(c) $x_1 = -7s + 2t - 11$, $x_2 = s$, $x_3 = -3t - 4$, $x_4 = -3t + 9$, $x_5 = t$ (d) Inconsistent

5. $x_1 = 3$, $x_2 = 1$, $x_3 = 2$ **7.** $x = t - 1, y = 2s, z = s, w = t$ **9.** $x_1 = 3$, $x_2 = 1$, $x_3 = 2$

11. $x = t - 1, y = 2s, z = s, w = t$ **13.** Has nontrivial solutions   **15.** Has nontrivial solutions   **17.** $x_1 = 0$, $x_2 = 0$, $x_3 = 0$

19. $x_1 = -s$, $x_2 = -t - s$, $x_3 = 4s$, $x_4 = t$ **21.** $w = t, x = -t, y = t, z = 0$ **23.** $I_1 = -1$, $I_2 = 0$, $I_3 = 1$, $I_4 = 2$

25. If $a = 4$, there are infinitely many solutions; if $a = -4$, there are no solutions; if $a \neq \pm 4$, there is exactly one solution.

27. If $a = 3$, there are infinitely many solutions; if $a = -3$, there are no solutions; if $a \neq \pm 3$, there is exactly one solution.

29. $x = \frac{2a}{3} - \frac{b}{9}$, $y = -\frac{a}{3} + \frac{2b}{9}$ **31.** $\begin{bmatrix} 1 & 3 \\ 0 & 1 \end{bmatrix}$ and $\begin{bmatrix} 1 & 0 \\ 0 & 1 \end{bmatrix}$ are possible answers.   **35.** $x = \pm 1$, $y = \pm\sqrt{3}$, $z = \pm\sqrt{2}$

37. $a = 1, b = -6, c = 2, d = 10$ **39.** The nonhomogeneous system will either be inconsistent or have exactly one solution.

True/False 1.2

(a) True   (b) False   (c) False   (d) True   (e) True   (f) False   (g) True   (h) False   (i) False

Exercise Set 1.3 (page 35)

1. (a) Undefined   (b) 4×2 (c) Undefined   (d) Undefined   (e) 5×5 (f) 5×2 (g) Undefined   (h) 5×2

3. (a) $\begin{bmatrix} 7 & 6 & 5 \\ -2 & 1 & 3 \\ 7 & 3 & 7 \end{bmatrix}$ (b) $\begin{bmatrix} -5 & 4 & -1 \\ 0 & -1 & -1 \\ -1 & 1 & 1 \end{bmatrix}$ (c) $\begin{bmatrix} 15 & 0 \\ -5 & 10 \\ 5 & 5 \end{bmatrix}$ (d) $\begin{bmatrix} -7 & -28 & -14 \\ -21 & -7 & -35 \end{bmatrix}$ (e) Undefined

(f) $\begin{bmatrix} 22 & -6 & 8 \\ -2 & 4 & 6 \\ 10 & 0 & 4 \end{bmatrix}$ (g) $\begin{bmatrix} -39 & -21 & -24 \\ 9 & -6 & -15 \\ -33 & -12 & -30 \end{bmatrix}$ (h) $\begin{bmatrix} 0 & 0 \\ 0 & 0 \\ 0 & 0 \end{bmatrix}$ (i) 5 (j) -25 (k) 168 (l) Undefined

5. (a) $\begin{bmatrix} 12 & -3 \\ -4 & 5 \\ 4 & 1 \end{bmatrix}$ (b) Undefined (c) $\begin{bmatrix} 42 & 108 & 75 \\ 12 & -3 & 21 \\ 36 & 78 & 63 \end{bmatrix}$ (d) $\begin{bmatrix} 3 & 45 & 9 \\ 11 & -11 & 17 \\ 7 & 17 & 13 \end{bmatrix}$ (e) $\begin{bmatrix} 3 & 45 & 9 \\ 11 & -11 & 17 \\ 7 & 17 & 13 \end{bmatrix}$

(f) $\begin{bmatrix} 21 & 17 \\ 17 & 35 \end{bmatrix}$ (g) $\begin{bmatrix} 0 & -2 & 11 \\ 12 & 1 & 8 \end{bmatrix}$ (h) $\begin{bmatrix} 12 & 6 & 9 \\ 48 & -20 & 14 \\ 24 & 8 & 16 \end{bmatrix}$ (i) 61 (j) 35 (k) 28 (l) 99

7. (a) $[67\ 41\ 41]$ (b) $[63\ 67\ 57]$ (c) $\begin{bmatrix} 41 \\ 21 \\ 67 \end{bmatrix}$ (d) $\begin{bmatrix} 6 \\ 6 \\ 63 \end{bmatrix}$ (e) $[24\ 56\ 97]$ (f) $\begin{bmatrix} 76 \\ 98 \\ 97 \end{bmatrix}$

9. (a) $\begin{bmatrix} -3 \\ 48 \\ 24 \end{bmatrix} = 3\begin{bmatrix} 3 \\ 6 \\ 0 \end{bmatrix} + 6\begin{bmatrix} -2 \\ 5 \\ 4 \end{bmatrix}$; $\begin{bmatrix} 12 \\ 29 \\ 56 \end{bmatrix} = -2\begin{bmatrix} 3 \\ 6 \\ 0 \end{bmatrix} + 5\begin{bmatrix} -2 \\ 5 \\ 4 \end{bmatrix} + 4\begin{bmatrix} 7 \\ 4 \\ 9 \end{bmatrix}$;

$\begin{bmatrix} 76 \\ 98 \\ 97 \end{bmatrix} = 7\begin{bmatrix} 3 \\ 6 \\ 0 \end{bmatrix} + 4\begin{bmatrix} -2 \\ 5 \\ 4 \end{bmatrix} + 9\begin{bmatrix} 7 \\ 4 \\ 9 \end{bmatrix}$

(b) $\begin{bmatrix} 64 \\ 21 \\ 77 \end{bmatrix} = 6\begin{bmatrix} 6 \\ 0 \\ 7 \end{bmatrix} + 7\begin{bmatrix} 4 \\ 3 \\ 5 \end{bmatrix}$; $\begin{bmatrix} 14 \\ 22 \\ 28 \end{bmatrix} = -2\begin{bmatrix} 6 \\ 0 \\ 7 \end{bmatrix} + \begin{bmatrix} -2 \\ 1 \\ 7 \end{bmatrix} + 7\begin{bmatrix} 4 \\ 3 \\ 5 \end{bmatrix}$;

$\begin{bmatrix} 38 \\ 18 \\ 74 \end{bmatrix} = 4\begin{bmatrix} 6 \\ 0 \\ 7 \end{bmatrix} + 3\begin{bmatrix} -2 \\ 1 \\ 7 \end{bmatrix} + 5\begin{bmatrix} 4 \\ 3 \\ 5 \end{bmatrix}$

11. (a) $\begin{bmatrix} 2 & -3 & 5 \\ 9 & -1 & 1 \\ 1 & 5 & 4 \end{bmatrix}\begin{bmatrix} x_1 \\ x_2 \\ x_3 \end{bmatrix} = \begin{bmatrix} 7 \\ -1 \\ 0 \end{bmatrix}$ (b) $\begin{bmatrix} 4 & 0 & -3 & 1 \\ 5 & 1 & 0 & -8 \\ 2 & -5 & 9 & -1 \\ 0 & 3 & -1 & 7 \end{bmatrix}\begin{bmatrix} x_1 \\ x_2 \\ x_3 \\ x_4 \end{bmatrix} = \begin{bmatrix} 1 \\ 3 \\ 0 \\ 2 \end{bmatrix}$

13. (a)
$$\begin{aligned} 5x_1 &+ 6x_2 &- 7x_3 &= 2 \\ -x_1 &- 2x_2 &+ 3x_3 &= 0 \\ &\ 4x_2 &- x_3 &= 3 \end{aligned}$$
(b)
$$\begin{aligned} x_1 &+ x_2 &+ x_3 &= 2 \\ 2x_1 &+ 3x_2 & &= 2 \\ 5x_1 &- 3x_2 &- 6x_3 &= -9 \end{aligned}$$

15. -1 17. $a = 4, b = -6, c = -1, d = 1$

23. (a) $\begin{bmatrix} a_{11} & 0 & 0 & 0 & 0 & 0 \\ 0 & a_{22} & 0 & 0 & 0 & 0 \\ 0 & 0 & a_{33} & 0 & 0 & 0 \\ 0 & 0 & 0 & a_{44} & 0 & 0 \\ 0 & 0 & 0 & 0 & a_{55} & 0 \\ 0 & 0 & 0 & 0 & 0 & a_{66} \end{bmatrix}$ (b) $\begin{bmatrix} a_{11} & a_{12} & a_{13} & a_{14} & a_{15} & a_{16} \\ 0 & a_{22} & a_{23} & a_{24} & a_{25} & a_{26} \\ 0 & 0 & a_{33} & a_{34} & a_{35} & a_{36} \\ 0 & 0 & 0 & a_{44} & a_{45} & a_{46} \\ 0 & 0 & 0 & 0 & a_{55} & a_{56} \\ 0 & 0 & 0 & 0 & 0 & a_{66} \end{bmatrix}$

(c) $\begin{bmatrix} a_{11} & 0 & 0 & 0 & 0 & 0 \\ a_{21} & a_{22} & 0 & 0 & 0 & 0 \\ a_{31} & a_{32} & a_{33} & 0 & 0 & 0 \\ a_{41} & a_{42} & a_{43} & a_{44} & 0 & 0 \\ a_{51} & a_{52} & a_{53} & a_{54} & a_{55} & 0 \\ a_{61} & a_{62} & a_{63} & a_{64} & a_{65} & a_{66} \end{bmatrix}$ (d) $\begin{bmatrix} a_{11} & a_{12} & 0 & 0 & 0 & 0 \\ a_{21} & a_{22} & a_{23} & 0 & 0 & 0 \\ 0 & a_{32} & a_{33} & a_{34} & 0 & 0 \\ 0 & 0 & a_{43} & a_{44} & a_{45} & 0 \\ 0 & 0 & 0 & a_{54} & a_{55} & a_{56} \\ 0 & 0 & 0 & 0 & a_{65} & a_{66} \end{bmatrix}$

25. $f\begin{pmatrix} x_1 \\ x_2 \end{pmatrix} = \begin{pmatrix} x_1 + x_2 \\ x_2 \end{pmatrix}$

(a) $f\begin{pmatrix} 1 \\ 1 \end{pmatrix} = \begin{pmatrix} 2 \\ 1 \end{pmatrix}$ (b) $f\begin{pmatrix} 2 \\ 0 \end{pmatrix} = \begin{pmatrix} 2 \\ 0 \end{pmatrix}$ (c) $f\begin{pmatrix} 4 \\ 3 \end{pmatrix} = \begin{pmatrix} 7 \\ 3 \end{pmatrix}$ (d) $f\begin{pmatrix} 2 \\ -2 \end{pmatrix} = \begin{pmatrix} 0 \\ -2 \end{pmatrix}$

27. One; namely, $A = \begin{bmatrix} 1 & 1 & 0 \\ 1 & -1 & 0 \\ 0 & 0 & 0 \end{bmatrix}$

29. (a) $\begin{bmatrix} 1 & 1 \\ 1 & 1 \end{bmatrix}$ and $\begin{bmatrix} -1 & -1 \\ -1 & -1 \end{bmatrix}$ (b) Four; $\begin{bmatrix} \sqrt{5} & 0 \\ 0 & 3 \end{bmatrix}, \begin{bmatrix} -\sqrt{5} & 0 \\ 0 & 3 \end{bmatrix}, \begin{bmatrix} \sqrt{5} & 0 \\ 0 & -3 \end{bmatrix}, \begin{bmatrix} -\sqrt{5} & 0 \\ 0 & -3 \end{bmatrix}$

True/False 1.3

(a) True   (b) False   (c) False   (d) False   (e) True   (f) False   (g) False   (h) True   (i) True   (j) True

(k) True   (l) False   (m) True   (n) True   (o) False

Exercise Set 1.4   (page 49)

5. $B^{-1} = \begin{bmatrix} \frac{1}{5} & \frac{3}{20} \\ -\frac{1}{5} & \frac{1}{10} \end{bmatrix}$ **7.** $D^{-1} = \begin{bmatrix} \frac{1}{2} & 0 \\ 0 & \frac{1}{3} \end{bmatrix}$ **9.** $\begin{bmatrix} \frac{1}{2}(e^x + e^{-x}) & -\frac{1}{2}(e^x - e^{-x}) \\ -\frac{1}{2}(e^x - e^{-x}) & \frac{1}{2}(e^x + e^{-x}) \end{bmatrix}$

15. $A = \begin{bmatrix} \frac{2}{7} & 1 \\ \frac{1}{7} & \frac{3}{7} \end{bmatrix}$ **17.** $\begin{bmatrix} -\frac{9}{13} & \frac{1}{13} \\ \frac{2}{13} & -\frac{6}{13} \end{bmatrix}$

19. (a) $\begin{bmatrix} 41 & 15 \\ 30 & 11 \end{bmatrix}$ (b) $\begin{bmatrix} 11 & -15 \\ -30 & 41 \end{bmatrix}$ (c) $\begin{bmatrix} 6 & 2 \\ 4 & 2 \end{bmatrix}$ (d) $\begin{bmatrix} 1 & 1 \\ 2 & -1 \end{bmatrix}$ (e) $\begin{bmatrix} 20 & 7 \\ 14 & 6 \end{bmatrix}$ (f) $\begin{bmatrix} 39 & 13 \\ 26 & 13 \end{bmatrix}$

21. (a) $\begin{bmatrix} 27 & 0 & 0 \\ 0 & 26 & -18 \\ 0 & 18 & 26 \end{bmatrix}$ (b) $\begin{bmatrix} \frac{1}{27} & 0 & 0 \\ 0 & 0.026 & 0.018 \\ 0 & -0.018 & 0.026 \end{bmatrix}$ (c) $\begin{bmatrix} 4 & 0 & 0 \\ 0 & -5 & -12 \\ 0 & 12 & -5 \end{bmatrix}$ (d) $\begin{bmatrix} 1 & 0 & 0 \\ 0 & -3 & 3 \\ 0 & -3 & -3 \end{bmatrix}$

(e) $\begin{bmatrix} 16 & 0 & 0 \\ 0 & -14 & -15 \\ 0 & 15 & -14 \end{bmatrix}$ (f) $\begin{bmatrix} 25 & 0 & 0 \\ 0 & 32 & -24 \\ 0 & 24 & 32 \end{bmatrix}$

27. $\begin{bmatrix} \frac{1}{a_{11}} & 0 & \cdots & 0 \\ 0 & \frac{1}{a_{22}} & \cdots & 0 \\ \vdots & \vdots & & \vdots \\ 0 & 0 & \cdots & \frac{1}{a_{nn}} \end{bmatrix}$ **31.** $D = CA^{-1}B^{-1}A^{-2}BC^2(B^T)^{-1}A^2$ **33.** B^{-1} **35.** $A^{-1} = \begin{bmatrix} \frac{1}{2} & \frac{1}{2} & -\frac{1}{2} \\ -\frac{1}{2} & \frac{1}{2} & \frac{1}{2} \\ \frac{1}{2} & -\frac{1}{2} & \frac{1}{2} \end{bmatrix}$

37. $A^{-1} = \begin{bmatrix} \frac{1}{2} & \frac{1}{2} & -\frac{1}{2} \\ -\frac{1}{2} & \frac{1}{2} & \frac{1}{2} \\ 1 & 0 & 0 \end{bmatrix}$ **39.** $x_1 = \frac{1}{23}, x_2 = \frac{13}{23}$ **41.** $x_1 = -\frac{1}{11}, x_2 = \frac{6}{11}$

True/False 1.4

(a) False   (b) False   (c) False   (d) False   (e) False   (f) True   (g) True   (h) True   (i) False   (j) True

(k) False

Exercise Set 1.5   (page 58)

1. (a) Elementary   **(b)** Not elementary   **(c)** Not elementary   **(d)** Not elementary

3. (a) Add 3 times row 2 to row 1: $\begin{bmatrix} 1 & 3 \\ 0 & 1 \end{bmatrix}$ **(b)** Multiply row 1 by $-\frac{1}{7}$: $\begin{bmatrix} -\frac{1}{7} & 0 & 0 \\ 0 & 1 & 0 \\ 0 & 0 & 1 \end{bmatrix}$

(c) Add 5 times row 1 to row 3: $\begin{bmatrix} 1 & 0 & 0 \\ 0 & 1 & 0 \\ 5 & 0 & 1 \end{bmatrix}$ **(d)** Swap rows 1 and 3: $\begin{bmatrix} 0 & 0 & 1 & 0 \\ 0 & 1 & 0 & 0 \\ 1 & 0 & 0 & 0 \\ 0 & 0 & 0 & 1 \end{bmatrix}$

5. (a) Swap rows 1 and 2: $EA = \begin{bmatrix} 3 & -6 & -6 & -6 \\ -1 & -2 & 5 & -1 \end{bmatrix}$

(b) Add -3 times row 2 to row 3: $EA = \begin{bmatrix} 2 & -1 & 0 & -4 & -4 \\ 1 & -3 & -1 & 5 & 3 \\ -1 & 9 & 4 & -12 & -10 \end{bmatrix}$

(c) Add 4 times row 3 to row 1: $EA = \begin{bmatrix} 13 & 28 \\ 2 & 5 \\ 3 & 6 \end{bmatrix}$

7. (a) $\begin{bmatrix} 0 & 0 & 1 \\ 0 & 1 & 0 \\ 1 & 0 & 0 \end{bmatrix}$ **(b)** $\begin{bmatrix} 0 & 0 & 1 \\ 0 & 1 & 0 \\ 1 & 0 & 0 \end{bmatrix}$ **(c)** $\begin{bmatrix} 1 & 0 & 0 \\ 0 & 1 & 0 \\ -2 & 0 & 1 \end{bmatrix}$ **(d)** $\begin{bmatrix} 1 & 0 & 0 \\ 0 & 1 & 0 \\ 2 & 0 & 1 \end{bmatrix}$

9. $\begin{bmatrix} -7 & 4 \\ 2 & -1 \end{bmatrix}$ **11.** $\begin{bmatrix} \frac{2}{7} & \frac{3}{7} \\ \frac{3}{7} & \frac{1}{7} \end{bmatrix}$ **13.** $\begin{bmatrix} \frac{3}{2} & -\frac{11}{10} & -\frac{6}{5} \\ -1 & 1 & 1 \\ -\frac{1}{2} & \frac{7}{10} & \frac{2}{5} \end{bmatrix}$ **15.** No inverse   **17.** $\begin{bmatrix} \frac{1}{2} & -\frac{1}{2} & \frac{1}{2} \\ -\frac{1}{2} & \frac{1}{2} & \frac{1}{2} \\ \frac{1}{2} & \frac{1}{2} & -\frac{1}{2} \end{bmatrix}$

19. $\begin{bmatrix} \frac{7}{2} & 0 & -3 \\ -1 & 1 & 0 \\ 0 & -1 & 1 \end{bmatrix}$ **21.** $\begin{bmatrix} \frac{1}{4} & \frac{1}{2} & -3 & 0 \\ -\frac{1}{8} & \frac{1}{4} & -\frac{3}{2} & 0 \\ 0 & 0 & \frac{1}{2} & 0 \\ \frac{1}{40} & -\frac{1}{20} & -\frac{1}{10} & -\frac{1}{5} \end{bmatrix}$ **23.** $\begin{bmatrix} -\frac{7}{12} & \frac{5}{24} & \frac{5}{8} & -\frac{1}{4} \\ \frac{5}{6} & \frac{5}{12} & \frac{1}{4} & -\frac{1}{2} \\ \frac{5}{12} & \frac{5}{24} & \frac{5}{8} & -\frac{1}{4} \\ -\frac{1}{12} & -\frac{1}{24} & -\frac{1}{8} & \frac{1}{4} \end{bmatrix}$

25. (a) $\begin{bmatrix} \frac{1}{k_1} & 0 & 0 & 0 \\ 0 & \frac{1}{k_2} & 0 & 0 \\ 0 & 0 & \frac{1}{k_3} & 0 \\ 0 & 0 & 0 & \frac{1}{k_4} \end{bmatrix}$ **(b)** $\begin{bmatrix} \frac{1}{k} & -\frac{1}{k} & 0 & 0 \\ 0 & 1 & 0 & 0 \\ 0 & 0 & \frac{1}{k} & -\frac{1}{k} \\ 0 & 0 & 0 & 1 \end{bmatrix}$ **27.** $c \neq 0, 1$

29. $\begin{bmatrix} -3 & 1 \\ 2 & 2 \end{bmatrix} = \begin{bmatrix} 1 & 0 \\ 0 & 2 \end{bmatrix}\begin{bmatrix} 1 & 1 \\ 0 & 1 \end{bmatrix}\begin{bmatrix} -4 & 0 \\ 0 & 1 \end{bmatrix}\begin{bmatrix} 1 & 0 \\ 1 & 1 \end{bmatrix}$

31. $\begin{bmatrix} 1 & 0 & -2 \\ 0 & 4 & 3 \\ 0 & 0 & 1 \end{bmatrix} = \begin{bmatrix} 1 & 0 & -2 \\ 0 & 1 & 0 \\ 0 & 0 & 1 \end{bmatrix}\begin{bmatrix} 1 & 0 & 0 \\ 0 & 1 & 3 \\ 0 & 0 & 1 \end{bmatrix}\begin{bmatrix} 1 & 0 & 0 \\ 0 & 4 & 0 \\ 0 & 0 & 1 \end{bmatrix}$

33. $\begin{bmatrix} -\frac{1}{4} & \frac{1}{8} \\ \frac{1}{4} & \frac{3}{8} \end{bmatrix} = \begin{bmatrix} 1 & 0 \\ -1 & 1 \end{bmatrix}\begin{bmatrix} -\frac{1}{4} & 0 \\ 0 & 1 \end{bmatrix}\begin{bmatrix} 1 & -1 \\ 0 & 1 \end{bmatrix}\begin{bmatrix} 1 & 0 \\ 0 & \frac{1}{2} \end{bmatrix}$

35. $\begin{bmatrix} 1 & 0 & 2 \\ 0 & \frac{1}{4} & -\frac{3}{4} \\ 0 & 0 & 1 \end{bmatrix} = \begin{bmatrix} 1 & 0 & 0 \\ 0 & \frac{1}{4} & 0 \\ 0 & 0 & 1 \end{bmatrix}\begin{bmatrix} 1 & 0 & 0 \\ 0 & 1 & -3 \\ 0 & 0 & 1 \end{bmatrix}\begin{bmatrix} 1 & 0 & 2 \\ 0 & 1 & 0 \\ 0 & 0 & 1 \end{bmatrix}$

37. Add -1 times the first row to the second row. Add -1 times the first row to the third row. Add -1 times the second row to the first row. Add the second row to the third row.

True/False 1.5

(a) False **(b)** True **(c)** True **(d)** True **(e)** True **(f)** True **(g)** False

Exercise Set 1.6 (page 65)

1. $x_1 = 3$, $x_2 = -1$ **3.** $x_1 = -1$, $x_2 = 4$, $x_3 = -7$ **5.** $x_1 = 1$, $x_2 = 5$, $x_3 = -1$ **7.** $x_1 = 2b_1 - 5b_2$, $x_2 = -b_1 + 3b_2$

9. **(i)** $x_1 = \frac{22}{17}$, $x_2 = \frac{1}{17}$ **(ii)** $x_1 = \frac{21}{17}$, $x_2 = \frac{11}{17}$

11. **(i)** $x_1 = \frac{7}{15}$, $x_2 = \frac{4}{15}$ **(ii)** $x_1 = \frac{34}{15}$, $x_2 = \frac{28}{15}$ **(iii)** $x_1 = \frac{19}{15}$, $x_2 = \frac{13}{15}$ **(iv)** $x_1 = -\frac{1}{5}$, $x_2 = \frac{3}{5}$

13. No conditions on b_1 and b_2 **15.** $b_3 = b_1 - b_2$ **17.** $b_1 = b_3 + b_4$, $b_2 = 2b_3 + b_4$

19. $X = \begin{bmatrix} 11 & 12 & -3 & 27 & 26 \\ -6 & -8 & 1 & -18 & -17 \\ -15 & -21 & 9 & -38 & -35 \end{bmatrix}$

True/False 1.6

(a) True **(b)** True **(c)** True **(d)** True **(e)** True **(f)** True **(g)** True

Exercise Set 1.7 (page 71)

1. $\begin{bmatrix} \frac{1}{2} & 0 \\ 0 & -\frac{1}{5} \end{bmatrix}$ **3.** $\begin{bmatrix} -1 & 0 & 0 \\ 0 & \frac{1}{2} & 0 \\ 0 & 0 & 3 \end{bmatrix}$ **5.** $\begin{bmatrix} 6 & 3 \\ 4 & -1 \\ 4 & 10 \end{bmatrix}$ **7.** $\begin{bmatrix} -15 & 10 & 0 & 20 & -20 \\ 2 & -10 & 6 & 0 & 6 \\ 18 & -6 & -6 & -6 & -6 \end{bmatrix}$

9. $A^2 = \begin{bmatrix} 1 & 0 \\ 0 & 4 \end{bmatrix}$, $A^{-2} = \begin{bmatrix} 1 & 0 \\ 0 & \frac{1}{4} \end{bmatrix}$, $A^{-k} = \begin{bmatrix} 1 & 0 \\ 0 & 1/(-2)^k \end{bmatrix}$

11. $A^2 = \begin{bmatrix} \frac{1}{4} & 0 & 0 \\ 0 & \frac{1}{9} & 0 \\ 0 & 0 & \frac{1}{16} \end{bmatrix}$, $A^{-2} = \begin{bmatrix} 4 & 0 & 0 \\ 0 & 9 & 0 \\ 0 & 0 & 16 \end{bmatrix}$, $A^{-k} = \begin{bmatrix} 2^k & 0 & 0 \\ 0 & 3^k & 0 \\ 0 & 0 & 4^k \end{bmatrix}$

13. Not symmetric **15.** Symmetric **17.** Not symmetric **19.** Not symmetric **21.** Not invertible **23.** $a = -8$

25. $x \neq 1$, -2, 4 **27.** $\begin{bmatrix} 1 & 0 & 0 \\ 0 & -1 & 0 \\ 0 & 0 & -1 \end{bmatrix}$ **35.** **(a)** Yes **(b)** No (unless $n = 1$) **(c)** Yes **(d)** No (unless $n = 1$)

39. $\begin{bmatrix} 0 & 0 & -8 \\ 0 & 0 & -4 \\ 8 & 4 & 0 \end{bmatrix}$ **43.** $A = \begin{bmatrix} 1 & 10 \\ 0 & -2 \end{bmatrix}$

True/False 1.7

(a) True **(b)** False **(c)** False **(d)** True **(e)** True **(f)** False **(g)** False **(h)** True **(i)** True **(j)** False
(k) False **(l)** False **(m)** True

Exercise Set 1.8 (page 84)

1.

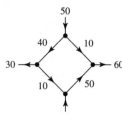

3. **(a)** $x_3 - x_4 = -500$, $-x_1 + x_4 = 100$, $x_1 - x_2 = 300$, $x_2 - x_3 = 100$

(b) $x_1 = -100 + t$, $x_2 = -400 + t$, $x_3 = -500 + t$, $x_4 = t$

(c) For all rates to be nonnegative, we need $t = 500$ cars per hour, so $x_1 = 400$, $x_2 = 100$, $x_3 = 0$, $x_4 = 500$

5. $I_1 = \frac{13}{5}$ A, $I_2 = -\frac{2}{5}$ A, $I_3 = \frac{11}{5}$ A   7. $I_1 = I_4 = I_5 = I_6 = \frac{1}{2}$ A, $I_2 = I_3 = 0$ A

9. $x_1 = 1$, $x_2 = 5$, $x_3 = 3$, and $x_4 = 4$; the balanced equation is $C_3H_8 + 5O_2 \rightarrow 3CO_2 + 4H_2O$

11. $x_1 = x_2 = x_3 = x_4 = t$; the balanced equation is $CH_3COF + H_2O \rightarrow CH_3COOH + HF$

13. $p(x) = x^2 - 2x + 2$ 15. $p(x) = 1 + \frac{13}{6}x - \frac{1}{6}x^3$

17. **(a)** Using $a_1 = k$ as a parameter, $p(x) = 1 + kx + (1 - k)x^2$ where $-\infty < k < \infty$.

(b) The graphs for k = 0, 1, 2, and 3 are shown.

True/False 1.8

(a) True   **(b)** False   **(c)** True   **(d)** False   **(e)** False

Exercise Set 1.9 (page 90)

1. **(a)** $\begin{bmatrix} 0.50 & 0.25 \\ 0.25 & 0.10 \end{bmatrix}$ **(b)** $\begin{bmatrix} \$25{,}290 \\ \$22{,}581 \end{bmatrix}$ 3. **(a)** $\begin{bmatrix} 0.1 & 0.6 & 0.4 \\ 0.3 & 0.2 & 0.3 \\ 0.4 & 0.1 & 0.2 \end{bmatrix}$ **(b)** $\begin{bmatrix} \$31{,}500 \\ \$26{,}500 \\ \$26{,}300 \end{bmatrix}$ 5. $\begin{bmatrix} 123.08 \\ 202.56 \end{bmatrix}$

True/False 1.9

(a) False   **(b)** True   **(c)** False   **(d)** True   **(e)** True

Chapter 1 Supplementary Exercises (page 91)

1. $\begin{aligned} 3x_1 - x_2 \qquad\quad + x_4 &= 1 \\ 2x_1 \qquad + 3x_3 + 3x_4 &= -1 \end{aligned}$

$x_1 = -\frac{3}{2}s - \frac{3}{2}t - \frac{1}{2}$, $x_2 = -\frac{9}{2}s - \frac{1}{2}t - \frac{5}{2}$, $x_3 = s$, $x_4 = t$

3. $\begin{aligned} 2x_1 - 4x_2 + x_3 &= 6 \\ -4x_1 \qquad\quad + 3x_3 &= -1 \\ x_2 - x_3 &= 3 \end{aligned}$

$x_1 = -\frac{17}{2}$, $x_2 = -\frac{26}{3}$, $x_3 = -\frac{35}{3}$

5. $x' = \frac{3}{5}x + \frac{4}{5}y$, $y' = -\frac{4}{5}x + \frac{3}{5}y$ 7. $x = 4$, $y = 2$, $z = 3$

9. **(a)** $a \neq 0$, $b \neq 2$ **(b)** $a \neq 0$, $b = 2$ **(c)** $a = 0$, $b = 2$ **(d)** $a = 0$, $b \neq 2$

11. $K = \begin{bmatrix} 0 & 2 \\ 1 & 1 \end{bmatrix}$ 13. **(a)** $X = \begin{bmatrix} -1 & 3 & -1 \\ 6 & 0 & 1 \end{bmatrix}$ **(b)** $X = \begin{bmatrix} 1 & -2 \\ 3 & 1 \end{bmatrix}$ **(c)** $X = \begin{bmatrix} -\frac{113}{37} & -\frac{160}{37} \\ -\frac{20}{37} & -\frac{46}{37} \end{bmatrix}$

15. $a = 1$, $b = -2$, $c = 3$

Exercise Set 2.1 (page 98)

1. $M_{11} = 29$, $C_{11} = 29$
$M_{12} = 21$, $C_{12} = -21$
$M_{13} = 27$, $C_{13} = 27$
$M_{21} = -11$, $C_{21} = 11$
$M_{22} = 13$, $C_{22} = 13$
$M_{23} = -5$, $C_{23} = 5$
$M_{31} = -19$, $C_{31} = -19$
$M_{32} = -19$, $C_{32} = 19$
$M_{33} = 19$, $C_{33} = 19$

3. **(a)** $M_{13} = 0$, $C_{13} = 0$
(b) $M_{23} = -96$, $C_{23} = 96$
(c) $M_{22} = -48$, $C_{22} = -48$
(d) $M_{21} = 72$, $C_{21} = -72$

5. 22; $\begin{bmatrix} \frac{2}{11} & -\frac{5}{22} \\ \frac{1}{11} & \frac{3}{22} \end{bmatrix}$
7. 59; $\begin{bmatrix} -\frac{2}{59} & -\frac{7}{59} \\ \frac{7}{59} & -\frac{5}{59} \end{bmatrix}$

9. $a^2 - 5a + 21$ **11.** -65 **13.** -123 **15.** $\lambda = 1$ or -3 **17.** $\lambda = 1$ or -1 **19.** (all parts) -123 **21.** -40
23. 0 **25.** -240 **27.** -1 **29.** 0 **31.** 6 **33.** The determinant is $\sin^2\theta + \cos^2\theta = 1$. **35.** $d_2 = d_1 + \lambda$

True/False 2.1

(a) False **(b)** False **(c)** True **(d)** True **(e)** True **(f)** True **(g)** False **(h)** False **(i)** False **(j)** True

Exercise Set 2.2 (page 105)

5. -5 **7.** -1 **9.** 1 **11.** 5 **13.** 33 **15.** 6 **17.** -2
19. Exercise 14: 39; Exercise 15: 6; Exercise 16: $-\frac{1}{6}$; Exercise 17: -2 **21.** -6 **23.** 72 **25.** -6 **27.** 18

True/False 2.2

(a) True **(b)** True **(c)** False **(d)** False **(e)** True **(f)** True

Exercise Set 2.3 (page 115)

7. Invertible **9.** Invertible **11.** Not invertible **13.** Invertible **15.** $k \neq \frac{5 \pm \sqrt{17}}{2}$ **17.** $k \neq -1$

19. $A^{-1} = \begin{bmatrix} 3 & -5 & -5 \\ -3 & 4 & 5 \\ 2 & -2 & -3 \end{bmatrix}$ **21.** $A^{-1} = \begin{bmatrix} \frac{1}{2} & \frac{3}{2} & 1 \\ 0 & 1 & \frac{3}{2} \\ 0 & 0 & \frac{1}{2} \end{bmatrix}$ **23.** $A^{-1} = \begin{bmatrix} -4 & 3 & 0 & -1 \\ 2 & -1 & 0 & 0 \\ -7 & 0 & -1 & 8 \\ 6 & 0 & 1 & -7 \end{bmatrix}$

25. $x = \frac{3}{11}$, $y = \frac{2}{11}$, $z = -\frac{1}{11}$ **27.** $x_1 = -\frac{30}{11}$, $x_2 = -\frac{38}{11}$, $x_3 = -\frac{40}{11}$ **29.** Cramer's rule does not apply. **31.** $y = 0$
35. **(a)** -189 **(b)** $-\frac{1}{7}$ **(c)** $-\frac{8}{7}$ **(d)** $-\frac{1}{56}$ **(e)** 7 **37.** **(a)** 189 **(b)** $\frac{1}{7}$ **(c)** $\frac{8}{7}$ **(d)** $\frac{1}{56}$

True/False 2.3

(a) False **(b)** False **(c)** True **(d)** False **(e)** True **(f)** True **(g)** True **(h)** True **(i)** True **(j)** True
(k) True **(l)** True

Chapter 2 Supplementary Exercises (page 117)

1. -18 **3.** 24 **5.** -10 **7.** 329 **9.** Exercise 3: 24; Exercise 4: 0; Exercise 5: -10; Exercise 6: -48
11. The matrices in Exercises 1–3 are invertible, the matrix in Exercise 4 is not.

13. $b^2 - 5b + 21$ **15.** -120 **17.** $\begin{bmatrix} -\frac{1}{6} & \frac{1}{9} \\ \frac{1}{6} & \frac{2}{9} \end{bmatrix}$ **19.** $\begin{bmatrix} \frac{1}{8} & -\frac{1}{8} & -\frac{3}{8} \\ \frac{1}{8} & \frac{5}{24} & -\frac{1}{24} \\ \frac{1}{4} & -\frac{7}{12} & -\frac{1}{12} \end{bmatrix}$ **21.** $\begin{bmatrix} \frac{1}{5} & \frac{2}{5} & -\frac{1}{10} \\ \frac{1}{5} & -\frac{3}{5} & \frac{2}{5} \\ -\frac{2}{5} & \frac{6}{5} & -\frac{3}{10} \end{bmatrix}$

23. $\begin{bmatrix} \frac{10}{329} & -\frac{2}{329} & \frac{52}{329} & -\frac{27}{329} \\ \frac{55}{329} & -\frac{11}{329} & -\frac{43}{329} & \frac{16}{329} \\ -\frac{3}{47} & \frac{10}{47} & -\frac{25}{47} & -\frac{6}{47} \\ -\frac{31}{329} & \frac{72}{329} & \frac{102}{329} & -\frac{15}{329} \end{bmatrix}$

25. $x' = \frac{3}{5}x + \frac{4}{5}y, \ y' = -\frac{4}{5}x + \frac{3}{5}y$ **29. (b)** $\cos\beta = \frac{c^2 + a^2 - b^2}{2ac}, \ \cos\gamma = \frac{a^2 + b^2 - c^2}{2ab}$

Exercise Set 3.1 (page 128)

1. (a)   **(b)**   **(c)**   **(d)**   **(e)**   **(f)**

3. (a)   **(b)**   **(c)**   **(d)**   **(e)**   **(f)**

5. (a)   **(b)**   **(c)**

7. (a) $\overrightarrow{P_1 P_2} = (-1, \ 3)$ **(b)** $\overrightarrow{P_1 P_2} = (-3, \ 6, \ 1)$

9. (a) The terminal point is $B(2, 3)$.   **(b)** The initial point is $A(-2, -2, -1)$.

11. (a) $\mathbf{u} = (-1, 2, -4)$ is one possible answer.   **(b)** $\mathbf{u} = (7, -2, -6)$ is one possible answer.

13. (a) $\mathbf{u} + \mathbf{w} = (1, -4)$ **(b)** $\mathbf{v} - 3\mathbf{u} = (-12, 8)$ **(c)** $2(\mathbf{u} - 5\mathbf{w}) = (38, 28)$ **(d)** $3\mathbf{v} - 2(\mathbf{u} + 2\mathbf{w}) = (4, 29)$
(e) $-3(\mathbf{w} - 2\mathbf{u} + \mathbf{v}) = (33, -12)$ **(f)** $(-2\mathbf{u} - \mathbf{v}) - 5(\mathbf{v} + 3\mathbf{w}) = (37, 17)$

15. (a) $(-1, 9, -11, 1)$ **(b)** $(22, 53, -19, 14)$ **(c)** $(-13, 13, -36, -2)$ **(d)** $(-90, -114, 60, -36)$ **(e)** $(-9, -5, -5, -3)$
(f) $(27, 29, -27, 9)$

17. (a) $\mathbf{w} - \mathbf{u} = (-9, 3, -3, -8, 5)$ **(b)** $2\mathbf{v} + 3\mathbf{u} = (13, -5, 14, 13, -9)$ **(c)** $-\mathbf{w} + 3(\mathbf{v} - \mathbf{u}) = (-14, -2, 24, 2, 7)$
(d) $5(-\mathbf{v} + 4\mathbf{u} - \mathbf{w}) = (125, -25, -20, 75, -70)$ **(e)** $-2(3\mathbf{w} + \mathbf{v}) + (2\mathbf{u} + \mathbf{w}) = (32, -10, 1, 27, -16)$
(f) $\frac{1}{2}(\mathbf{w} - 5\mathbf{v} + 2\mathbf{u}) + \mathbf{v} = (\frac{9}{2}, \frac{3}{2}, -12, -\frac{5}{2}, -2)$

19. (a) $\mathbf{v} - \mathbf{w} = (-2, 1, -4, -2, 7)$ **(b)** $6\mathbf{u} + 2\mathbf{v} = (-10, 6, -4, 26, 28)$ **(c)** $(2\mathbf{u} - 7\mathbf{w}) - (8\mathbf{v} + \mathbf{u}) = (-77, 8, 94, -25, 23)$

21. $\mathbf{x} = \left(-\frac{8}{3}, \frac{1}{2}, \frac{8}{3}, \frac{2}{3}, \frac{11}{6}\right)$ **23. (a)** Not parallel   **(b)** Parallel   **(c)** Parallel   **25.** $a = 3, b = -1$

27. $c_1 = 2, \ c_2 = -1, \ c_3 = 5$ **29.** $c_1 = 1, \ c_2 = 1, \ c_3 = -1, \ c_4 = 1$ **33. (a)** $\left(\frac{9}{2}, -\frac{1}{2}, -\frac{1}{2}\right)$ **(b)** $\left(\frac{23}{4}, -\frac{9}{4}, \frac{1}{4}\right)$

True/False 3.1

(a) False   **(b)** False   **(c)** False   **(d)** True   **(e)** True   **(f)** False   **(g)** False   **(h)** True   **(i)** False
(j) True   **(k)** False

Exercise Set 3.2 (page 141)

1. (a) $\|\mathbf{v}\| = 5, \ \frac{\mathbf{v}}{\|\mathbf{v}\|} = \left(\frac{4}{5}, -\frac{3}{5}\right), \ -\frac{\mathbf{v}}{\|\mathbf{v}\|} = \left(-\frac{4}{5}, \frac{3}{5}\right)$ **(b)** $\|\mathbf{v}\| = 2\sqrt{3}, \ \frac{\mathbf{v}}{\|\mathbf{v}\|} = \left(\frac{1}{\sqrt{3}}, \frac{1}{\sqrt{3}}, \frac{1}{\sqrt{3}}\right), \ -\frac{\mathbf{v}}{\|\mathbf{v}\|} = \left(-\frac{1}{\sqrt{3}}, -\frac{1}{\sqrt{3}}, -\frac{1}{\sqrt{3}}\right)$
(c) $\|\mathbf{v}\| = \sqrt{15}, \ \frac{\mathbf{v}}{\|\mathbf{v}\|} = \frac{1}{\sqrt{15}}(1, 0, 2, 1, 3), \ -\frac{\mathbf{v}}{\|\mathbf{v}\|} = -\frac{1}{\sqrt{15}}(1, 0, 2, 1, 3)$

3. (a) $\|\mathbf{u} + \mathbf{v}\| = \sqrt{83}$ **(b)** $\|\mathbf{u}\| + \|\mathbf{v}\| = \sqrt{17} + \sqrt{26}$ **(c)** $\|-2\mathbf{u} + 2\mathbf{v}\| = 2\sqrt{3}$ **(d)** $\|3\mathbf{u} - 5\mathbf{v} + \mathbf{w}\| = \sqrt{466}$

5. (a) $\|3\mathbf{u} - 5\mathbf{v} + \mathbf{w}\| = \sqrt{2570}$ **(b)** $\|3\mathbf{u}\| - 5\|\mathbf{v}\| + \|\mathbf{w}\| = 3\sqrt{46} - 10\sqrt{21} + \sqrt{42}$ **(c)** $\|-\|\mathbf{u}\|\mathbf{v}\| = 2\sqrt{966}$

7. $k = \frac{5}{7}$, $k = -\frac{5}{7}$ **9. (a)** $\mathbf{u} \cdot \mathbf{v} = -8$, $\mathbf{u} \cdot \mathbf{u} = 26$, $\mathbf{v} \cdot \mathbf{v} = 24$ **(b)** $\mathbf{u} \cdot \mathbf{v} = 0$, $\mathbf{u} \cdot \mathbf{u} = 54$, $\mathbf{v} \cdot \mathbf{v} = 21$

11. (a) $\|\mathbf{u} - \mathbf{v}\| = \sqrt{14}$ **(b)** $\|\mathbf{u} - \mathbf{v}\| = \sqrt{59}$ **(c)** $\|\mathbf{u} - \mathbf{v}\| = \sqrt{677}$

13. (a) $\cos\theta = \frac{15}{\sqrt{27}\sqrt{17}}$; θ is acute   **(b)** $\cos\theta = -\frac{4}{\sqrt{6}\sqrt{45}}$; θ is obtuse   **(c)** $\cos\theta = -\frac{136}{\sqrt{225}\sqrt{180}}$; θ is obtuse

15. $\mathbf{a} \cdot \mathbf{b} = 45\frac{\sqrt{3}}{2}$

17. (a) $\mathbf{u} \cdot (\mathbf{v} \cdot \mathbf{w})$ does not make sense because $\mathbf{v} \cdot \mathbf{w}$ is a scalar.   **(b)** $\mathbf{u} \cdot (\mathbf{v} + \mathbf{w})$ makes sense.
   (c) $\|\mathbf{u} \cdot \mathbf{v}\|$ does not make sense because the quantity inside the norm is a scalar.
   (d) $(\mathbf{u} \cdot \mathbf{v}) - \|\mathbf{u}\|$ makes sense since the terms are both scalars.

19. (a) $\left(-\frac{4}{5}, -\frac{3}{5}\right)$ **(b)** $\left(\frac{1}{5\sqrt{2}}, \frac{7}{5\sqrt{2}}\right)$ **(c)** $\left(-\frac{3}{4}, \frac{1}{2}, \frac{\sqrt{3}}{4}\right)$ **(d)** $\left(\frac{1}{\sqrt{55}}, \frac{2}{\sqrt{55}}, \frac{3}{\sqrt{55}}, \frac{4}{\sqrt{55}}, \frac{5}{\sqrt{55}}\right)$

23. (a) $\cos\theta = -\frac{11}{\sqrt{962}}$ **(b)** $\cos\theta = -\frac{3}{\sqrt{10}}$ **(c)** $\cos\theta = 0$ **(d)** $\cos\theta = 0$

25. (a) $|\mathbf{u} \cdot \mathbf{v}| = 10$, $\|\mathbf{u}\| \, \|\mathbf{v}\| = \sqrt{13}\sqrt{17} \approx 14.866$ **(b)** $|\mathbf{u} \cdot \mathbf{v}| = 7$, $\|\mathbf{u}\| \, \|\mathbf{v}\| = \sqrt{10}\sqrt{14} \approx 11.832$
   (c) $|\mathbf{u} \cdot \mathbf{v}| = 5$, $\|\mathbf{u}\| \, \|\mathbf{v}\| = (3)(2) = 6$

27. A sphere of radius 1 centered at $(x_0, \ y_0, \ z_0)$.

True/False 3.2

(a) True   **(b)** True   **(c)** False   **(d)** True   **(e)** True   **(f)** False   **(g)** False   **(h)** False   **(i)** True   **(j)** True

---

Exercise Set 3.3 (page 150)

1. (a) Orthogonal   **(b)** Not orthogonal   **(c)** Not orthogonal   **(d)** Not orthogonal

3. (a) Not an orthogonal set   **(b)** Orthogonal set   **(c)** Orthogonal set   **(d)** Not an orthogonal set

5. $\pm\left(\frac{1}{\sqrt{3}}, \frac{1}{\sqrt{3}}, -\frac{1}{\sqrt{3}}\right)$ **7.** No   **9.** $-2(x+1) + (y-3) - (z+2) = 0$ **11.** $2z = 0$ **13.** Not parallel   **15.** Parallel

17. Not perpendicular   **19. (a)** $\frac{2}{5}$ **(b)** $\frac{18}{\sqrt{22}}$ **21.** $(0, 0)$, $(6, 2)$ **23.** $\left(-\frac{16}{13}, 0, -\frac{80}{13}\right)$, $\left(\frac{55}{13}, 1, -\frac{11}{13}\right)$

25. $\left(0, \frac{2}{5}, -\frac{1}{5}\right)$, $\left(1, \frac{3}{5}, \frac{6}{5}\right)$ **27.** $\left(\frac{1}{5}, -\frac{1}{5}, \frac{1}{10}, -\frac{1}{10}\right)$, $\left(\frac{9}{5}, \frac{6}{5}, \frac{9}{10}, \frac{21}{10}\right)$ **29.** 1   **31.** $\frac{1}{\sqrt{17}}$ **33.** $\frac{5}{3}$ **35.** $\frac{1}{\sqrt{29}}$

37. $\frac{11}{\sqrt{6}}$ **39.** 0 (The planes coincide.)   **41. (b)** $\cos\beta = \frac{b}{\|\mathbf{v}\|}$, $\cos\gamma = \frac{c}{\|\mathbf{v}\|}$

True/False 3.3

(a) True   **(b)** True   **(c)** True   **(d)** True   **(e)** True   **(f)** False   **(g)** False

---

Exercise Set 3.4 (page 159)

1. Vector equation: $(x, y) = (-4, 1) + t(0, -8)$;
   parametric equations: $x = -4$, $y = 1 - 8t$

3. Vector equation: $(x, y, z) = t(-3, 0, 1)$;
   parametric equations: $x = -3t$, $y = 0$, $z = t$

5. Point: $(3, -6)$; parallel vector: $(-5, -1)$ **7.** Point: $(4, 6)$; parallel vector: $(-6, -6)$

9. Vector equation: $(x, y, z) = (-3, 1, 0) + t_1(0, -3, 6) + t_2(-5, 1, 2)$;
   parametric equations: $x = -3 - 5t_2$, $y = 1 - 3t_1 + t_2$, $z = 6t_1 + 2t_2$

11. Vector equation: $(x, y, z) = (-1, 1, 4) + t_1(6, -1, 0) + t_2(-1, 3, 1)$;
   parametric equations: $x = -1 + 6t_1 - t_2$, $y = 1 - t_1 + 3t_2$, $z = 4 + t_2$

13. A possible answer is vector equation: $(x, y) = t(3, 2)$;
   parametric equations: $x = 3t$, $y = 2t$

15. A possible answer is vector equation: $(x, y, z) = t_1(0, 1, 0) + t_2(5, 0, 4)$;
parametric equations: $x + 5t_2$, $y = t_1$, $z = 4t_2$

17. $x_1 = -s - t$, $x_2 = s$, $x_3 = t$

19. $x_1 = \frac{3}{7}r - \frac{19}{4}s - \frac{8}{7}t$, $x_2 = -\frac{2}{7}r + \frac{1}{7}s + \frac{3}{7}t$, $x_3 = r$, $x_4 = s$, $x_5 = t$

21. (a) $(1, 0, 0) + s(-1, 1, 0) + t(-1, 0, 1)$ **(b)** a plane in R^3 passing through $P(1, 0, 0)$ and parallel to $(-1, 1, 0)$ and $(-1, 0, 1)$

23. (a)
$$\begin{aligned} x + y + z &= 0 \\ -2x + 3y &= 0 \end{aligned}$$
(b) a line through the origin in R^3 **(c)** $x = -\frac{3}{5}t$, $y = -\frac{2}{5}t$, $z = t$

25. (a) $x_1 = -\frac{2}{3}s + \frac{1}{3}t$, $x_2 = s$, $x_3 = t$ **(c)** $x_1 = 1 - \frac{2}{3}s + \frac{1}{3}t$, $x_2 = s$, $x_3 = 1 + t$

27. $x_1 = \frac{1}{3} - \frac{4}{3}s - \frac{1}{3}t$, $x_2 = s$, $x_3 = t$, $x_4 = 1$; The general solution of the associated homogeneous system is $x_1 = -\frac{4}{3}s - \frac{1}{3}t$, $x_2 = s$, $x_3 = t$, $x_4 = 0$. A particular solution of the given system is $x_1 = \frac{1}{3}$, $x_2 = 0$, $x_3 = 0$, $x_4 = 1$.

True/False 3.4

(a) True   **(b)** False   **(c)** True   **(d)** True   **(e)** False   **(f)** True

Exercise Set 3.5 (page 168)

1. (a) $(32, -6, -4)$ **(b)** $(-14, -20, -82)$ **(c)** $(27, 40, -42)$ **3.** $(18, 36, -18)$ **5.** $(-3, 9, -3)$ **7.** $\sqrt{59}$ **9.** $\sqrt{101}$

11. 3   **13.** 7   **15.** $\frac{\sqrt{374}}{2}$ **17.** 16   **19.** The vectors do not lie in the same plane.   **21.** -92 **23.** abc

25. (a) -3 **(b)** 3   **(c)** 3   **27. (a)** $\frac{\sqrt{26}}{2}$ **(b)** $\frac{\sqrt{26}}{3}$ **29.** $2(\mathbf{v} \times \mathbf{u})$ **37. (a)** $\frac{17}{6}$ **(b)** $\frac{1}{2}$

True/False 3.5

(a) True   **(b)** True   **(c)** False   **(d)** True   **(e)** False   **(f)** False

Chapter 3 Supplementary Exercises (page 170)

1. (a) $3\mathbf{v} - 2\mathbf{u} = (13, -3, 10)$ **(b)** $\|\mathbf{u} + \mathbf{v} + \mathbf{w}\| = \sqrt{70}$ **(c)** $\sqrt{774}$ **(d)** $\text{proj}_{\mathbf{w}}\mathbf{u} = -\frac{12}{27}(2, -5, -5)$
(e) $\mathbf{u} \cdot (\mathbf{v} \times \mathbf{w}) = -122$ **(f)** $(-5\mathbf{v} + \mathbf{w}) \times ((\mathbf{u} \cdot \mathbf{v})\mathbf{w}) = (-3150, -2430, 1170)$

3. (a) $3\mathbf{v} - 2\mathbf{u} = (-5, -12, 20, -2)$ **(b)** $\|\mathbf{u} + \mathbf{v} + \mathbf{w}\| = \sqrt{106}$ **(c)** $\sqrt{2810}$ **(d)** $\text{proj}_{\mathbf{w}}\mathbf{u} = -\frac{15}{77}(9, 1, -6, -6)$

5. Not an orthogonal set

7. (a) A line through the origin, perpendicular to the given vector.
(b) A plane through the origin, perpendicular to the given vector.   **(c)** $\{0\}$ (the origin)
(d) A line through the origin, perpendicular to the plane containing the two noncollinear vectors.

9. True   **11.** $S(-1, -1, 5)$ **13.** $\sqrt{\frac{14}{17}}$ **15.** $\frac{11}{\sqrt{35}}$

17. Vector equation: $(x, y, z) = (-2, 1, 3) + t_1(1, -2, -2) + t_2(5, -1, -5)$;
parametric equations: $x = -2 + t_1 + 5t_2$, $y = 1 - 2t_1 - t_2$, $z = 3 - 2t_1 - 5t_2$

19. Vector equation: $(x, y) = (0, -3) + t(8, -1)$;
parametric equations: $x = 8t$, $y = -3 - t$

21. A possible answer is vector equation: $(x, y) = (0, -5) + t(1, 3)$; parametric equations: $x = t$, $y = -5 + 3t$

23. $3(x + 1) + 6(y - 5) + 2(z - 6) = 0$ **25.** $-18(x - 9) - 51y - 24(z - 4) = 0$ **29.** A plane

Exercise Set 4.1 (page 178)

1. (a) $\mathbf{u} + \mathbf{v} = (2, 6)$, $3\mathbf{u} = (0, 6)$ **(c)** Axioms 1–5   **3.** The set is a vector space with the given operations.

5. Not a vector space, Axioms 5 and 6 fail.   **7.** Not a vector space. Axiom 8 fails.

9. The set is a vector space with the given operations.   **11.** The set is a vector space with the given operations.

True/False 4.1

(a) False   **(b)** False   **(c)** True   **(d)** False   **(e)** False

---

Exercise Set 4.2   (page 188)

1. (a), (c), (e)   **3.** (a), (b), (d)   **5.** (a), (c), (d)   **7.** (a), (b), (d)   **9.** (a), (b), (c)

11. (a) The vectors span   **(b)** The vectors do not span   **(c)** The vectors do not span   **(d)** The vectors span

13. The polynomials do not span

15. (a) Line; $x = -\frac{1}{2}t$, $y = -\frac{3}{2}t$, $z = t$ **(b)** Line; $x = 2t$, $y = t$, $z = 0$ **(c)** Origin   **(d)** Origin
(e) Line; $x = -3t$, $y = -2t$, $z = t$ **(f)** Plane; $x - 3y + z = 0$

True/False 4.2

(a) True   **(b)** True   **(c)** False   **(d)** False   **(e)** False   **(f)** True   **(g)** True   **(h)** False   **(i)** False
(j) True   **(k)** False

Exercise Set 4.3   (page 199)

1. (a) $\mathbf{u}_2$ is a scalar multiple of $\mathbf{u}_1$.   **(b)** The vectors are linearly dependent by Theorem 4.3.3.
(c) $\mathbf{p}_2$ is a scalar multiple of $\mathbf{p}_1$.   **(d)** B is a scalar multiple of A.

3. None

5. (a) They do not lie in a plane.   **(b)** The do lie in a plane.   **7. (b)** $\mathbf{v}_1 = \frac{2}{7}\mathbf{v}_2 - \frac{3}{7}\mathbf{v}_3$, $\mathbf{v}_2 = \frac{7}{2}\mathbf{v}_1 + \frac{3}{2}\mathbf{v}_3$, $\mathbf{v}_3 = -\frac{7}{3}\mathbf{v}_1 + \frac{2}{3}\mathbf{v}_2$

9. $\lambda = -\frac{1}{2}$, $\lambda = 1$

19. (a) They are linearly independent since $\mathbf{v}_1$, $\mathbf{v}_2$, and $\mathbf{v}_3$ do not lie in the same plane when they are placed with their initial points at the origin.
(b) They are not linearly independent since $\mathbf{v}_1$, $\mathbf{v}_2$, and $\mathbf{v}_3$ line in the same plane when they are placed with their initial points at the origin.

21. $W(x) = -x \sin x - \cos x \neq 0$ for some x.   **23. (a)** $W(x) = e^x \neq 0$ **(b)** $W(x) = 2 \neq 0$
25. $W(x) = 2 \sin x \neq 0$ for some x.

True/False 4.3

(a) True   **(b)** True   **(c)** False   **(d)** True   **(e)** True   **(f)** False   **(g)** True   **(h)** False

---

Exercise Set 4.4   (page 207)

1. (a) A basis for R^2 has two linearly independent vectors.   **(b)** A basis for R^3 has three linearly independent vectors.
(c) A basis for P_2 has three linearly independent vectors.   **(d)** A basis for M_{22} has four linearly independent vectors.

3. (a), (b)   **7. (a)** $(\mathbf{w})_S = (3, -7)$ **(b)** $(\mathbf{w})_S = \left(\frac{5}{28}, \frac{3}{14}\right)$ **(c)** $(\mathbf{w})_S = \left(a, \frac{b-a}{2}\right)$

9. (a) $(\mathbf{v})_S = (3, -2, 1)$ **(b)** $(\mathbf{v})_S = (-2, 0, 1)$ **11.** $(A)_S = (-1, 1, -1, 3)$ **13.** $A = A_1 - A_2 + A_3 - A_4$

15. $\mathbf{p} = 7\mathbf{p}_1 - 8\mathbf{p}_2 + 3\mathbf{p}_3$ **17. (a)** $(2, 0)$ **(b)** $\left(\frac{2}{\sqrt{3}}, -\frac{1}{\sqrt{3}}\right)$ **(c)** $(0, 1)$ **(d)** $\left(\frac{2}{\sqrt{3}}a, b - \frac{a}{\sqrt{3}}\right)$

True/False 4.4

(a) False   **(b)** False   **(c)** True   **(d)** True   **(e)** False

Exercise Set 4.5 (page 216)

1. Basis: $(1, 0, 1)$; dimension $= 1$ **3.** Basis: $(4, 1, 0, 0)$, $(-3, 0, 1, 0)$, $(1, 0, 0, 1)$; dimension $= 3$ **5.** No basis; dimension $= 0$

7. (a) $\left(\frac{2}{3}, 1, 0\right)$, $\left(-\frac{5}{3}, 0, 1\right)$ **(b)** $(1, 1, 0)$, $(0, 0, 1)$ **(c)** $(2, -1, 4)$ **(d)** $(1, 1, 0)$, $(0, 1, 1)$

9. (a) n **(b)** $\frac{n(n+1)}{2}$ **(c)** $\frac{n(n+1)}{2}$ **13.** Any two of $(0, 1, 0, 0)$, $(0, 0, 1, 0)$, and $(0, 0, 0, 1)$ can be used.

15. $\mathbf{v}_3 = (a, b, c)$ with $9a - 3b - 5c \neq 0$

True/False 4.5

(a) True **(b)** True **(c)** False **(d)** True **(e)** True **(f)** True **(g)** True **(h)** True **(i)** False **(j)** False

---

Exercise Set 4.6 (page 222)

1. (a) $[\mathbf{w}]_S = \begin{bmatrix} 3 \\ -7 \end{bmatrix}$ **(b)** $[\mathbf{w}]_S = \begin{bmatrix} \frac{5}{28} \\ \frac{3}{14} \end{bmatrix}$ **(c)** $[\mathbf{w}]_S = \begin{bmatrix} a \\ \frac{b-a}{2} \end{bmatrix}$

3. (a) $(\mathbf{p})_S = (4, -3, 1)$, $[\mathbf{p}]_S = \begin{bmatrix} 4 \\ -3 \\ 1 \end{bmatrix}$ **(b)** $(\mathbf{p})_S = (0, 2, -1)$, $[\mathbf{p}]_S = \begin{bmatrix} 0 \\ 2 \\ -1 \end{bmatrix}$

5. (a) $\mathbf{w} = (16, 10, 12)$ **(b)** $\mathbf{q} = 3 + 4x^2$ **(c)** $B = \begin{bmatrix} 15 & -1 \\ 6 & 3 \end{bmatrix}$

7. (a) $\begin{bmatrix} \frac{13}{10} & -\frac{1}{2} \\ -\frac{2}{5} & 0 \end{bmatrix}$ **(b)** $\begin{bmatrix} 0 & -\frac{5}{2} \\ -2 & -\frac{13}{2} \end{bmatrix}$ **(c)** $[\mathbf{w}]_B = \begin{bmatrix} -\frac{17}{10} \\ \frac{8}{5} \end{bmatrix}$, $[\mathbf{w}]_{B'} = \begin{bmatrix} -4 \\ -7 \end{bmatrix}$

9. (a) $\begin{bmatrix} 3 & 2 & \frac{5}{2} \\ -2 & -3 & -\frac{1}{2} \\ 5 & 1 & 6 \end{bmatrix}$ **(b)** $[\mathbf{w}]_B = \begin{bmatrix} 9 \\ -9 \\ 5 \end{bmatrix}$, $[\mathbf{w}]_{B'} = \begin{bmatrix} -\frac{7}{2} \\ \frac{23}{2} \\ 6 \end{bmatrix}$

11. (b) $\begin{bmatrix} 2 & 0 \\ 1 & 3 \end{bmatrix}$ **(c)** $\begin{bmatrix} \frac{1}{2} & 0 \\ -\frac{1}{6} & \frac{1}{3} \end{bmatrix}$ **(d)** $[\mathbf{h}]_B = \begin{bmatrix} 2 \\ -5 \end{bmatrix}$, $[\mathbf{h}]_{B'} = \begin{bmatrix} 1 \\ -2 \end{bmatrix}$

13. (a) $\begin{bmatrix} 1 & 2 & 3 \\ 2 & 5 & 3 \\ 1 & 0 & 8 \end{bmatrix}$ **(b)** $\begin{bmatrix} -40 & 16 & 9 \\ 13 & -5 & -3 \\ 5 & -2 & -1 \end{bmatrix}$ **(d)** $[\mathbf{w}]_B = \begin{bmatrix} -239 \\ 77 \\ 30 \end{bmatrix}$, $[\mathbf{w}]_S = \begin{bmatrix} 5 \\ -3 \\ 1 \end{bmatrix}$

(e) $[\mathbf{w}]_S = \begin{bmatrix} 3 \\ -5 \\ 0 \end{bmatrix}$, $[\mathbf{w}]_B = \begin{bmatrix} -200 \\ 64 \\ 25 \end{bmatrix}$

15. (a) $\begin{bmatrix} 3 & 5 \\ -1 & -2 \end{bmatrix}$ **(b)** $\begin{bmatrix} 2 & 5 \\ -1 & -3 \end{bmatrix}$ **(d)** $[\mathbf{w}]_{B_1} = \begin{bmatrix} 2 \\ -1 \end{bmatrix}$, $[\mathbf{w}]_{B_2} = \begin{bmatrix} -1 \\ 1 \end{bmatrix}$ **(e)** $[\mathbf{w}]_{B_2} = \begin{bmatrix} 3 \\ -1 \end{bmatrix}$, $[\mathbf{w}]_{B_1} = \begin{bmatrix} 4 \\ -1 \end{bmatrix}$

17. (a) $\begin{bmatrix} 3 & 2 & \frac{5}{2} \\ -2 & -3 & -\frac{1}{2} \\ 5 & 1 & 6 \end{bmatrix}$ **(b)** $[\mathbf{w}]_{B_1} = \begin{bmatrix} 9 \\ -9 \\ -5 \end{bmatrix}$, $[\mathbf{w}]_{B_2} = \begin{bmatrix} -\frac{7}{2} \\ \frac{23}{2} \\ 6 \end{bmatrix}$ **19. (a)** $\begin{bmatrix} \cos 2\theta & \sin 2\theta \\ \sin 2\theta & -\cos 2\theta \end{bmatrix}$

23. (a) $B = \{(1, 1, 0), (1, 0, 2), (0, 2, 1)\}$ **(b)** $B = \left\{\left(\frac{4}{5}, \frac{1}{5}, -\frac{2}{5}\right), \left(\frac{1}{5}, -\frac{1}{5}, \frac{2}{5}\right), \left(-\frac{2}{5}, \frac{2}{5}, \frac{1}{5}\right)\right\}$

True/False 4.6

(a) True **(b)** True **(c)** True **(d)** True **(e)** False **(f)** False

Exercise Set 4.7 (page 235)

1. $\mathbf{r}_1 = (2, -1, 0, 1)$, $\mathbf{r}_2 = (3, 5, 7, -1)$, $\mathbf{r}_3 = (1, 4, 2, 7)$;

$$\mathbf{c}_1 = \begin{bmatrix} 2 \\ 3 \\ 1 \end{bmatrix}, \ \mathbf{c}_2 = \begin{bmatrix} -1 \\ 5 \\ 4 \end{bmatrix}, \ \mathbf{c}_3 = \begin{bmatrix} 0 \\ 7 \\ 2 \end{bmatrix}, \ \mathbf{c}_4 = \begin{bmatrix} 1 \\ -1 \\ 7 \end{bmatrix}$$

3. **(a)** $\begin{bmatrix} -2 \\ 10 \end{bmatrix} = \begin{bmatrix} 1 \\ 4 \end{bmatrix} - \begin{bmatrix} 3 \\ -6 \end{bmatrix}$ **(b)** **b** is not in the column space of A. **(c)** $\begin{bmatrix} 1 \\ 9 \\ 1 \end{bmatrix} - 3\begin{bmatrix} -1 \\ 3 \\ 1 \end{bmatrix} + \begin{bmatrix} 1 \\ 1 \\ 1 \end{bmatrix} = \begin{bmatrix} 5 \\ 1 \\ -1 \end{bmatrix}$

(d) $\begin{bmatrix} 2 \\ 0 \\ 0 \end{bmatrix} = \begin{bmatrix} 1 \\ 1 \\ -1 \end{bmatrix} + (t-1)\begin{bmatrix} -1 \\ 1 \\ -1 \end{bmatrix} + t\begin{bmatrix} 1 \\ -1 \\ 1 \end{bmatrix}$ **(e)** $\begin{bmatrix} 4 \\ 3 \\ 5 \\ 7 \end{bmatrix} = -26\begin{bmatrix} 1 \\ 0 \\ 1 \\ 0 \end{bmatrix} + 13\begin{bmatrix} 2 \\ 1 \\ 2 \\ 1 \end{bmatrix} - 7\begin{bmatrix} 0 \\ 2 \\ 1 \\ 2 \end{bmatrix} + 4\begin{bmatrix} 1 \\ 1 \\ 3 \\ 2 \end{bmatrix}$

5. **(a)** $\begin{bmatrix} 1 \\ 0 \end{bmatrix} + t\begin{bmatrix} 3 \\ 1 \end{bmatrix}$; $t\begin{bmatrix} 3 \\ 1 \end{bmatrix}$ **(b)** $\begin{bmatrix} -2 \\ 7 \\ 0 \end{bmatrix} + t\begin{bmatrix} -1 \\ -1 \\ 1 \end{bmatrix}$; $t\begin{bmatrix} -1 \\ -1 \\ 1 \end{bmatrix}$

(c) $\begin{bmatrix} -1 \\ 0 \\ 0 \\ 0 \end{bmatrix} + r\begin{bmatrix} 2 \\ 1 \\ 0 \\ 0 \end{bmatrix} + s\begin{bmatrix} -1 \\ 0 \\ 1 \\ 0 \end{bmatrix} + t\begin{bmatrix} -2 \\ 0 \\ 0 \\ 1 \end{bmatrix}$; $r\begin{bmatrix} 2 \\ 1 \\ 0 \\ 0 \end{bmatrix} + s\begin{bmatrix} -1 \\ 0 \\ 1 \\ 0 \end{bmatrix} + t\begin{bmatrix} -2 \\ 0 \\ 0 \\ 1 \end{bmatrix}$

(d) $\begin{bmatrix} \frac{6}{5} \\ \frac{7}{5} \\ 0 \\ 0 \end{bmatrix} + s\begin{bmatrix} \frac{7}{5} \\ \frac{4}{5} \\ 1 \\ 0 \end{bmatrix} + t\begin{bmatrix} \frac{1}{5} \\ -\frac{3}{5} \\ 0 \\ 1 \end{bmatrix}$; $s\begin{bmatrix} \frac{7}{5} \\ \frac{4}{5} \\ 1 \\ 0 \end{bmatrix} + t\begin{bmatrix} \frac{1}{5} \\ -\frac{3}{5} \\ 0 \\ 1 \end{bmatrix}$

7. **(a)** $\mathbf{r}_1 = [1\ 0\ 2]$, $\mathbf{r}_2 = [0\ 0\ 1]$, $\mathbf{c}_1 = \begin{bmatrix} 1 \\ 0 \\ 0 \end{bmatrix}$, $\mathbf{c}_2 = \begin{bmatrix} 2 \\ 1 \\ 0 \end{bmatrix}$

(b) $\mathbf{r}_1 = [1\ -3\ 0\ 0]$, $\mathbf{r}_2 = [0\ 1\ 0\ 0]$, $\mathbf{c}_1 = \begin{bmatrix} 1 \\ 0 \\ 0 \\ 0 \end{bmatrix}$, $\mathbf{c}_2 = \begin{bmatrix} -3 \\ 1 \\ 0 \\ 0 \end{bmatrix}$

(c) $\mathbf{r}_1 = [1\ 2\ 4\ 5]$, $\mathbf{r}_2 = [0\ 1\ -3\ 0]$, $\mathbf{r}_3 = [0\ 0\ 1\ -3]$, $\mathbf{r}_4 = [0\ 0\ 0\ 1]$,

$$\mathbf{c}_1 = \begin{bmatrix} 1 \\ 0 \\ 0 \\ 0 \\ 0 \end{bmatrix}, \ \mathbf{c}_2 = \begin{bmatrix} 2 \\ 1 \\ 0 \\ 0 \\ 0 \end{bmatrix}, \ \mathbf{c}_3 = \begin{bmatrix} 4 \\ -3 \\ 1 \\ 0 \\ 0 \end{bmatrix}, \ \mathbf{c}_4 = \begin{bmatrix} 5 \\ 0 \\ -3 \\ 1 \\ 0 \end{bmatrix}$$

(d) $\mathbf{r}_1 = [1\ 2\ -1\ 5]$, $\mathbf{r}_2 = [0\ 1\ 4\ 3]$, $\mathbf{r}_3 = [0\ 0\ 1\ -7]$, $\mathbf{r}_4 = [0\ 0\ 0\ 1]$

$$\mathbf{c}_1 = \begin{bmatrix} 1 \\ 0 \\ 0 \\ 0 \end{bmatrix}, \ \mathbf{c}_2 = \begin{bmatrix} 2 \\ 1 \\ 0 \\ 0 \end{bmatrix}, \ \mathbf{c}_3 = \begin{bmatrix} -1 \\ 4 \\ 1 \\ 0 \end{bmatrix}, \ \mathbf{c}_4 = \begin{bmatrix} 5 \\ 3 \\ -7 \\ 1 \end{bmatrix}$$

9. **(a)** $\mathbf{r}_1 = [1\ 0\ 2]$; $\mathbf{r}_2 = [0\ 0\ 1]$; $\mathbf{c}_1 = \begin{bmatrix} 1 \\ 0 \\ 0 \end{bmatrix}$; $\mathbf{c}_2 = \begin{bmatrix} 2 \\ 1 \\ 0 \end{bmatrix}$

(b) $\mathbf{r}_1 = [1\ -3\ 0\ 0]$; $\mathbf{r}_2 = [0\ 1\ 0\ 0]$; $\mathbf{c}_1 = \begin{bmatrix} 1 \\ 0 \\ 0 \\ 0 \end{bmatrix}$; $\mathbf{c}_2 = \begin{bmatrix} -3 \\ 1 \\ 0 \\ 0 \end{bmatrix}$

(c) $\mathbf{r}_1 = [\begin{array}{cccc} 1 & 2 & 4 & 5 \end{array}]$; $\mathbf{r}_2 = [\begin{array}{cccc} 0 & 1 & -3 & 0 \end{array}]$; $\mathbf{r}_3 = [\begin{array}{cccc} 0 & 0 & 1 & -3 \end{array}]$;

$$\mathbf{r}_4 = [\begin{array}{cccc} 0 & 0 & 0 & 1 \end{array}]; \quad \mathbf{c}_1 = \begin{bmatrix} 1 \\ 0 \\ 0 \\ 0 \\ 0 \end{bmatrix}; \quad \mathbf{c}_2 = \begin{bmatrix} 2 \\ 1 \\ 0 \\ 0 \\ 0 \end{bmatrix}; \quad \mathbf{c}_3 = \begin{bmatrix} 4 \\ -3 \\ 1 \\ 0 \\ 0 \end{bmatrix}; \quad \mathbf{c}_4 = \begin{bmatrix} 5 \\ 0 \\ -3 \\ 1 \\ 0 \end{bmatrix}$$

(d) $\mathbf{r}_1 = [\begin{array}{cccc} 1 & 2 & -1 & 5 \end{array}]$; $\mathbf{r}_2 = [\begin{array}{cccc} 0 & 1 & 4 & 3 \end{array}]$; $\mathbf{r}_3 = [\begin{array}{cccc} 0 & 0 & 1 & -7 \end{array}]$;

$$\mathbf{r}_4 = [\begin{array}{cccc} 0 & 0 & 0 & 1 \end{array}]; \quad \mathbf{c}_1 = \begin{bmatrix} 1 \\ 0 \\ 0 \\ 0 \end{bmatrix}; \quad \mathbf{c}_2 = \begin{bmatrix} 2 \\ 1 \\ 0 \\ 0 \end{bmatrix}; \quad \mathbf{c}_3 = \begin{bmatrix} -1 \\ 4 \\ 1 \\ 0 \end{bmatrix}; \quad \mathbf{c}_4 = \begin{bmatrix} 5 \\ 3 \\ -7 \\ 1 \end{bmatrix}$$

11. (a) $(1, 1, -4 - 3)$, $(0, 1, -5, -2)$, $(0, 0, 1, -\frac{1}{2})$ (b) $(1, -1, 2, 0)$, $(0, 1, 0, 0)$, $(0, 0, 1, -\frac{1}{6})$
   (c) $(1, 1, 0, 0)$, $(0, 1, 1, 1)$, $(0, 0, 1, 1)$, $(0, 0, 0, 1)$

15. (b) $\begin{bmatrix} 0 & 0 & 0 \\ 0 & 1 & 0 \\ 0 & 0 & 1 \end{bmatrix}$

17. (a) $\begin{bmatrix} 3a & -5a \\ 3b & -5b \end{bmatrix}$ for all real numbers a, b not both 0.

(b) Since A and B are invertible, their null spaces are the origin. The null space of C is the line $3x + y = 0$. The null space of D is the entire xy-plane.

True/False 4.7

(a) True   (b) False   (c) False   (d) False   (e) False   (f) True   (g) True   (h) False   (i) True   (j) False

Exercise Set 4.8 (page 246)

1. $\text{Rank}(A) = \text{Rank}(A^T) = 2$ 3. (a) 2; 1   (b) 1; 2   (c) 2; 2   (d) 2; 3   (e) 3; 2
5. (a) Rank $= 4$, nullity $= 0$ (b) Rank $= 3$, nullity $= 2$ (c) Rank $= 3$, nullity $= 0$
7. (a) Yes, 0   (b) No   (c) Yes, 2   (d) Yes, 7   (e) No   (f) Yes, 4   (g) Yes, 0
9. $b_1 = r$, $b_2 = s$, $b_3 = 4s - 3r$, $b_4 = 2r - s$, $b_5 = 8s - 7r$ 11. No   13. Rank is 2 if $r = 2$ and $s = 1$; the rank is never 1.
17. (a) 3   (b) 5   (c) 3   (d) 3   19. $A = \begin{bmatrix} 0 & 1 \\ 0 & 0 \end{bmatrix}$; $B = \begin{bmatrix} 1 & 2 \\ 2 & 4 \end{bmatrix}$

True/False 4.8

(a) False   (b) True   (c) False   (d) False   (e) True   (f) False   (g) False   (h) False   (i) True   (j) False

Exercise Set 4.9 (page 260)

1. (a) Domain: R^2; codomain: R^3 (b) Domain: R^3; codomain: R^2 (c) Domain: R^3; codomain: R^3
   (d) Domain: R^6; codomain: R^1

3. R^2, R^3, $(-1, 2, 3)$

5. (a) Linear; $R^3 \rightarrow R^2$ (b) Nonlinear; $R^2 \rightarrow R^3$ (c) Linear; $R^3 \rightarrow R^3$ (d) Nonlinear; $R^4 \rightarrow R^2$

7. (a) and (c) are matrix transformations; (b), (d), and (e) are not matrix transformations.

9. $\begin{bmatrix} 3 & 5 & -1 \\ 4 & -1 & 1 \\ 3 & 2 & -1 \end{bmatrix}$; $T(-1, 2, 4) = (3, -2, -3)$

11. (a) $\begin{bmatrix} 0 & 1 \\ -1 & 0 \\ 1 & 3 \\ 1 & -1 \end{bmatrix}$ **(b)** $\begin{bmatrix} 7 & 2 & -1 & 1 \\ 0 & 1 & 1 & 0 \\ -1 & 0 & 0 & 0 \end{bmatrix}$ **(c)** $\begin{bmatrix} 0 & 0 & 0 \\ 0 & 0 & 0 \\ 0 & 0 & 0 \\ 0 & 0 & 0 \\ 0 & 0 & 0 \end{bmatrix}$ **(d)** $\begin{bmatrix} 0 & 0 & 0 & 1 \\ 1 & 0 & 0 & 0 \\ 0 & 0 & 1 & 0 \\ 0 & 1 & 0 & 0 \\ 1 & 0 & -1 & 0 \end{bmatrix}$

13. (a) $T(-1, 4) = (5, 4)$ **(b)** $T(2, 1, -3) = (0, -2, 0)$ **15. (a)** $(2, -5, -3)$ **(b)** $(2, 5, 3)$ **(c)** $(-2, -5, 3)$

17. (a) $(-2, 1, 0)$ **(b)** $(-2, 0, 3)$ **(c)** $(0, 1, 3)$ **19. (a)** $\left(-2, \frac{\sqrt{3}-2}{2}, \frac{1+2\sqrt{3}}{2}\right)$ **(b)** $\left(0, 1, 2\sqrt{2}\right)$ **(c)** $(-1, -2, 2)$

21. (a) $\left(-2, \frac{\sqrt{3}+2}{2}, \frac{-1+2\sqrt{3}}{2}\right)$ **(b)** $\left(-2\sqrt{2}, 1, 0\right)$ **(c)** $(1, 2, 2)$

25. $\begin{bmatrix} -\frac{1}{9} & \frac{8}{9} & \frac{4}{9} \\ \frac{8}{9} & -\frac{1}{9} & \frac{4}{9} \\ \frac{4}{9} & \frac{4}{9} & -\frac{7}{9} \end{bmatrix}$ **29. (a)** Twice the orthogonal projection on the x-axis.   **(b)** Twice the reflection about the x-axis.

31. Rotation through the angle 2θ.

33. Rotation through the angle θ and translation by x_0; not a matrix transformation since x_0 is nonzero.   **35.** A line in R^n.

True/False 4.9

(a) False   **(b)** False   **(c)** False   **(d)** True   **(e)** False   **(f)** True   **(g)** False   **(h)** False   **(i)** True

---

Exercise Set 4.10   (page 271)

1. $T_B \circ T_A = \begin{bmatrix} 5 & -1 & 21 \\ 10 & -8 & 4 \\ 45 & 3 & 25 \end{bmatrix}$, $T_A \circ T_B = \begin{bmatrix} -8 & -3 & 1 \\ -5 & -15 & -8 \\ 44 & -11 & 45 \end{bmatrix}$

3. (a) $T_1 = \begin{bmatrix} 1 & 1 \\ 1 & -1 \end{bmatrix}$, $T_2 = \begin{bmatrix} 3 & 0 \\ 2 & 4 \end{bmatrix}$ **(b)** $T_2 \circ T_1 = \begin{bmatrix} 3 & 3 \\ 6 & -2 \end{bmatrix}$, $T_1 \circ T_2 = \begin{bmatrix} 5 & 4 \\ 1 & -4 \end{bmatrix}$

(c) $T_2(T_1(x_1, x_2)) = (3x_1 + 3x_2, 6x_1 - 2x_2)$,
$T_1(T_2(x_1, x_2)) = (5x_1 + 4x_2, x_1 - 4x_2)$

5. (a) $\begin{bmatrix} 1 & 0 \\ 0 & -1 \end{bmatrix}$ **(b)** $\begin{bmatrix} 0 & 0 \\ 0 & \frac{1}{2} \end{bmatrix}$ **(c)** $\begin{bmatrix} 3 & 0 \\ 0 & -3 \end{bmatrix}$

7. (a) $\begin{bmatrix} -1 & 0 & 0 \\ 0 & 0 & 0 \\ 0 & 0 & 1 \end{bmatrix}$ **(b)** $\begin{bmatrix} 1 & 0 & 1 \\ 0 & \sqrt{2} & 0 \\ -1 & 0 & 1 \end{bmatrix}$ **(c)** $\begin{bmatrix} -1 & 0 & 0 \\ 0 & 1 & 0 \\ 0 & 0 & 0 \end{bmatrix}$

9. (a) $T_1 \circ T_2 = T_2 \circ T_1$ **(b)** $T_1 \circ T_2 = T_2 \circ T_1$ **(c)** $T_1 \circ T_2 \neq T_2 \circ T_1$

11. (a) Not one-to-one **(b)** One-to-one **(c)** One-to-one **(d)** One-to-one **(e)** One-to-one **(f)** One-to-one
(g) One-to-one

13. (a) One-to-one; $\begin{bmatrix} \frac{1}{3} & -\frac{2}{3} \\ \frac{1}{3} & \frac{1}{3} \end{bmatrix}$; $T^{-1}(w_1, w_2) = \left(\frac{1}{3}w_1 - \frac{2}{3}w_2, \frac{1}{3}w_1 + \frac{1}{3}w_2\right)$

(b) Not one-to-one **(c)** One-to-one; $\begin{bmatrix} 0 & -1 \\ -1 & 0 \end{bmatrix}$; $T^{-1}(w_1, w_2) = (-w_2, -w_1)$

(d) Not one-to-one

15. (a) Reflection about the x-axis **(b)** Rotation through the angle $-\frac{\pi}{4}$ **(c)** Contraction by a factor of $\frac{1}{3}$
(d) Reflection about the yz-plane **(e)** Dilation by a factor of 5

17. (a) Matrix operator **(b)** Not a matrix operator **(c)** Matrix operator **(d)** Not a matrix operator

19. (a) Matrix transformation **(b)** Matrix transformation **21. (a)** $\begin{bmatrix} -1 & 0 \\ 0 & 0 \end{bmatrix}$ **(b)** $\begin{bmatrix} 0 & 1 \\ -1 & 0 \end{bmatrix}$ **(c)** $\begin{bmatrix} 0 & 0 \\ 3 & 0 \end{bmatrix}$

23. (a) $T_A(e_1) = (-1, 2, 4)$, $T_A(e_2) = (3, 1, 5)$, $T_A(e_3) = (0, 2, -3)$ **(b)** $T_A(e_1 + e_2 + e_3) = (2, 5, 6)$
(c) $T_A(7e_3) = (0, 14, -21)$

25. (a) Yes   **(b)** Yes   **27. (b)** $T(x_1, x_2) = \left(x_1^2 + x_2^2, x_1 x_2\right)$

29. (a) The range of T is a proper subset of R^n.   **(b)** T must map infinitely many vectors to 0.

True/False 4.10

(a) False   **(b)** True   **(c)** True   **(d)** False   **(e)** False   **(f)** False

Exercise Set 4.11 (page 280)

1. (a) $\begin{bmatrix} 0 & -1 \\ -1 & 0 \end{bmatrix}$ **(b)** $\begin{bmatrix} -1 & 0 \\ 0 & -1 \end{bmatrix}$ **(c)** $\begin{bmatrix} 1 & 0 \\ 0 & 0 \end{bmatrix}$ **(d)** $\begin{bmatrix} 0 & 0 \\ 0 & 1 \end{bmatrix}$

3. (a) $\begin{bmatrix} 1 & 0 & 0 \\ 0 & 1 & 0 \\ 0 & 0 & -1 \end{bmatrix}$ **(b)** $\begin{bmatrix} 1 & 0 & 0 \\ 0 & -1 & 0 \\ 0 & 0 & 1 \end{bmatrix}$ **(c)** $\begin{bmatrix} -1 & 0 & 0 \\ 0 & 1 & 0 \\ 0 & 0 & 1 \end{bmatrix}$

5. (a) $\begin{bmatrix} 0 & -1 & 0 \\ 1 & 0 & 0 \\ 0 & 0 & 1 \end{bmatrix}$ **(b)** $\begin{bmatrix} 1 & 0 & 0 \\ 0 & 0 & -1 \\ 0 & 1 & 0 \end{bmatrix}$ **(c)** $\begin{bmatrix} 0 & 0 & 1 \\ 0 & 1 & 0 \\ -1 & 0 & 0 \end{bmatrix}$

7. Rectangle with vertices at $(0, 0)$, $(-3, 0)$, $(0, 1)$, $(-3, 1)$ **9. (a)** $\begin{bmatrix} 1 & 0 \\ 4 & 1 \end{bmatrix}$ **(b)** $\begin{bmatrix} 1 & -2 \\ 0 & 1 \end{bmatrix}$

11. (a) Expansion by a factor of 3 in the x-direction
　　 (b) Expansion by a factor of 5 in the y-direction and reflection about the x-axis
　　 (c) Shearing by a factor of 4 in the x-direction

13. (a) $\begin{bmatrix} \frac{1}{2} & 0 \\ 0 & 5 \end{bmatrix}$ **(b)** $\begin{bmatrix} 1 & 0 \\ 2 & 5 \end{bmatrix}$ **(c)** $\begin{bmatrix} 0 & -1 \\ -1 & 0 \end{bmatrix}$

17. (a) $y = \frac{2}{7}x$ **(b)** $y = x$ **(c)** $y = \frac{1}{2}x$ **(d)** $y = -2x$ **(e)** $y = \left(-\frac{8 + 5\sqrt{3}}{11}\right)x$ **19. (b)** No

23. (a) $\begin{bmatrix} 1 & 0 & k \\ 0 & 1 & k \\ 0 & 0 & 1 \end{bmatrix}$

　　 (b) Shear in the xz-direction with

　　　 factor k maps (x, y, z) to $(x + ky, y, z + ky)$: $\begin{bmatrix} 1 & k & 0 \\ 0 & 1 & 0 \\ 0 & k & 1 \end{bmatrix}$.

　　　 Shear in the yz-direction with factor k maps (x, y, z) to $(x, y + kx, z + kx)$: $\begin{bmatrix} 1 & 0 & 0 \\ k & 1 & 0 \\ k & 0 & 1 \end{bmatrix}$.

True/False 4.11

(a) False   **(b)** True   **(c)** True   **(d)** True   **(e)** False   **(f)** False   **(g)** True

Exercise Set 4.12 (page 290)

1. (a) Stochastic   **(b)** Not stochastic   **(c)** Stochastic   **(d)** Not stochastic   **3.** $\begin{bmatrix} 0.54545 \\ 0.45455 \end{bmatrix}$

5. (a) Regular   **(b)** Not regular   **(c)** Regular   **7.** $\begin{bmatrix} \frac{8}{17} \\ \frac{9}{17} \end{bmatrix}$ **9.** $\begin{bmatrix} \frac{4}{11} \\ \frac{4}{11} \\ \frac{3}{11} \end{bmatrix}$

11. (a) Probability that something in state 1 stays in state 1   **(b)** Probability that something in state 2 moves to state 1
　　 (c) 0.8   **(d)** 0.85

13. (a) $\begin{bmatrix} 0.95 & 0.55 \\ 0.05 & 0.45 \end{bmatrix}$ **(b)** 0.93   **(c)** 0.142   **(d)** 0.63

15. (a)

Year	1	2	3	4	5
City	95,750	91,840	88,243	84,933	81,889
Suburbs	29,250	33,160	36,757	40,067	43,111

(b)

City	46,875
Suburbs	78,125

17. (a) $\frac{23}{100}$ **(b)** $\begin{bmatrix} \frac{46}{159} \\ \frac{22}{53} \\ \frac{47}{159} \end{bmatrix}$ **(c)** 35, 50, 35   **19.** $P = \begin{bmatrix} \frac{7}{10} & \frac{1}{10} & \frac{1}{5} \\ \frac{1}{5} & \frac{3}{10} & \frac{1}{2} \\ \frac{1}{10} & \frac{3}{5} & \frac{3}{10} \end{bmatrix}$; $\mathbf{q} = \begin{bmatrix} \frac{1}{3} \\ \frac{1}{3} \\ \frac{1}{3} \end{bmatrix}$

21. $P^k \mathbf{q} = \mathbf{q}$ for every positive integer k

True/False 4.12

(a) True   **(b)** True   **(c)** True   **(d)** False   **(e)** True

Chapter 4 Supplementary Exercises (page 292)

1. (a) $\mathbf{u} + \mathbf{v} = (4, 3, 2)$, $-\mathbf{u} = (-3, 0, 0)$ **(c)** Axioms 1–5

3. If $s \neq 1, -2$, the solution space is the origin. If $s = 1$, the solution space is a plane through the origin. If $s = -2$, the solution space is a line through the origin.

7. A must be invertible

9. (a) Rank $= 2$, nullity $= 1$ **(b)** Rank $= 2$, nullity $= 2$ **(c)** Rank $= 2$, nullity $= n - 2$

11. (a) $\{1, x^2, x^4, \ldots, x^{2m}\}$ where $2m = n$ if n is even and $2m = n - 1$ if n is odd.   **(b)** $\{x, x^2, x^3, \ldots, x^n\}$

13. (a) $\left\{ \begin{bmatrix} 1 & 0 & 0 \\ 0 & 0 & 0 \\ 0 & 0 & 0 \end{bmatrix}, \begin{bmatrix} 0 & 1 & 0 \\ 1 & 0 & 0 \\ 0 & 0 & 0 \end{bmatrix}, \begin{bmatrix} 0 & 0 & 1 \\ 0 & 0 & 0 \\ 1 & 0 & 0 \end{bmatrix}, \begin{bmatrix} 0 & 0 & 0 \\ 0 & 1 & 0 \\ 0 & 0 & 0 \end{bmatrix}, \begin{bmatrix} 0 & 0 & 0 \\ 0 & 0 & 1 \\ 0 & 1 & 0 \end{bmatrix}, \begin{bmatrix} 0 & 0 & 0 \\ 0 & 0 & 0 \\ 0 & 0 & 1 \end{bmatrix} \right\}$

(b) $\left\{ \begin{bmatrix} 0 & 1 & 0 \\ -1 & 0 & 0 \\ 0 & 0 & 0 \end{bmatrix}, \begin{bmatrix} 0 & 0 & 1 \\ 0 & 0 & 0 \\ -1 & 0 & 0 \end{bmatrix}, \begin{bmatrix} 0 & 0 & 0 \\ 0 & 0 & 1 \\ 0 & -1 & 0 \end{bmatrix} \right\}$

15. Possible ranks are 2, 1, and 0.

Exercise Set 5.1 (page 303)

1. 5

3. (a) $\lambda^2 - 2\lambda - 3 = 0$ **(b)** $\lambda^2 - 8\lambda + 16 = 0$ **(c)** $\lambda^2 - 12 = 0$ **(d)** $\lambda^2 + 3 = 0$ **(e)** $\lambda^2 = 0$ **(f)** $\lambda^2 - 2\lambda + 1 = 0$

5. (a) Basis for eigenspace corresponding to $\lambda = 3$: $\begin{bmatrix} \frac{1}{2} \\ 1 \end{bmatrix}$; basis for eigenspace corresponding to $\lambda = -1$: $\begin{bmatrix} 0 \\ 1 \end{bmatrix}$

(b) Basis for eigenspace corresponding to $\lambda = 4$: $\begin{bmatrix} \frac{3}{2} \\ 1 \end{bmatrix}$

(c) Basis for eigenspace corresponding to $\lambda = \sqrt{12}$: $\begin{bmatrix} \frac{3}{\sqrt{12}} \\ 1 \end{bmatrix}$; basis for eigenspace corresponding to $\lambda = -\sqrt{12}$: $\begin{bmatrix} -\frac{3}{\sqrt{12}} \\ 1 \end{bmatrix}$

(d) There are no eigenspaces.   **(e)** Basis for eigenspace corresponding to $\lambda = 0$: $\begin{bmatrix} 1 \\ 0 \end{bmatrix}$, $\begin{bmatrix} 0 \\ 1 \end{bmatrix}$

(f) Basis for eigenspace corresponding to $\lambda = 1$: $\begin{bmatrix} 1 \\ 0 \end{bmatrix}$, $\begin{bmatrix} 0 \\ 1 \end{bmatrix}$

7. (a) $1, 2, 3$ **(b)** $-\sqrt{2}, 0, \sqrt{2}$ **(c)** -8 **(d)** 2   **(e)** 2   **(f)** $-4, 3$

9. (a) $\lambda^4 + \lambda^3 - 3\lambda^2 - \lambda + 2 = 0$ **(b)** $\lambda^4 - 8\lambda^3 + 19\lambda^2 - 24\lambda + 48 = 0$

11. (a) $\lambda = 1$: basis $\begin{bmatrix} 2 \\ 3 \\ 1 \\ 0 \end{bmatrix}$, $\begin{bmatrix} 0 \\ 0 \\ 0 \\ 1 \end{bmatrix}$; $\lambda = -2$: basis $\begin{bmatrix} -1 \\ 0 \\ 1 \\ 0 \end{bmatrix}$; $\lambda = -1$: basis $\begin{bmatrix} -2 \\ 1 \\ 1 \\ 0 \end{bmatrix}$ **(b)** $\lambda = 4$: basis $\begin{bmatrix} \frac{3}{2} \\ 1 \\ 0 \\ 0 \end{bmatrix}$

13. 1, $\left(\frac{1}{2}\right)^9 = \frac{1}{512}$, $2^9 = 512$ **15. (a)** $y = x$ and $y = 2x$ **(b)** No lines **(c)** $y = 0$

True/False 5.1

(a) True **(b)** False **(c)** True **(d)** False **(e)** True **(f)** False **(g)** False

Exercise Set 5.2 (page 313)

1. Possible reason: Determinants are different. **3.** Possible reason: Ranks are different.

5. $\lambda = 0$: 1 or 2; $\lambda = 1$: 1; $\lambda = 2$: 1, 2, or 3 **7.** Not diagonalizable **9.** Not diagonalizable **11.** Not diagonalizable

13. $P = \begin{bmatrix} \frac{1}{3} & 0 \\ 1 & 1 \end{bmatrix}$; $P^{-1}AP = \begin{bmatrix} 1 & 0 \\ 0 & -1 \end{bmatrix}$ **15.** $P = \begin{bmatrix} -2 & 0 & 1 \\ 0 & 1 & 0 \\ 1 & 0 & 0 \end{bmatrix}$; $P^{-1}AP = \begin{bmatrix} 3 & 0 & 0 \\ 0 & 3 & 0 \\ 0 & 0 & 2 \end{bmatrix}$

17. $P = \begin{bmatrix} 1 & 2 & 1 \\ 1 & 3 & 3 \\ 1 & 3 & 4 \end{bmatrix}$; $P^{-1}AP = \begin{bmatrix} 1 & 0 & 0 \\ 0 & 2 & 0 \\ 0 & 0 & 3 \end{bmatrix}$ **19.** $P = \begin{bmatrix} 1 & 0 & 0 \\ 0 & 1 & 0 \\ -3 & 0 & 1 \end{bmatrix}$; $P^{-1}AP = \begin{bmatrix} 0 & 0 & 0 \\ 0 & 0 & 0 \\ 0 & 0 & 1 \end{bmatrix}$

21. $P = \begin{bmatrix} 1 & 1 & 0 & 0 \\ 0 & 1 & 1 & 0 \\ 0 & 0 & 1 & 1 \\ 0 & 0 & 0 & 1 \end{bmatrix}$; $P^{-1}AP = \begin{bmatrix} -2 & 0 & 0 & 0 \\ 0 & -2 & 0 & 0 \\ 0 & 0 & 3 & 0 \\ 0 & 0 & 0 & 3 \end{bmatrix}$ **23.** $\begin{bmatrix} -1 & 10237 & -2047 \\ 0 & 1 & 0 \\ 0 & 10245 & -2048 \end{bmatrix}$

25. $A^n = PD^nP^{-1} = \begin{bmatrix} 1 & 1 & 1 \\ 2 & 0 & -1 \\ 1 & -1 & 1 \end{bmatrix} \begin{bmatrix} 1^n & 0 & 0 \\ 0 & 3^n & 0 \\ 0 & 0 & 4^n \end{bmatrix} \begin{bmatrix} \frac{1}{6} & \frac{1}{3} & \frac{1}{6} \\ \frac{1}{2} & 0 & -\frac{1}{2} \\ \frac{1}{3} & -\frac{1}{3} & \frac{1}{3} \end{bmatrix}$

27. On possibility is $P = \begin{bmatrix} -b & -b \\ a - \lambda_1 & a - \lambda_2 \end{bmatrix}$ where λ_1 and λ_2 are as in Exercise 20 of Section 5.1.

33. (a) $\lambda = 1$: dimension $= 1$; $\lambda = 3$: dimension ≤ 2; $\lambda = 4$: dimension ≤ 3 **(b)** Dimensions will be exactly 1, 2, and 3.
(c) $\lambda = 4$

True/False 5.2

(a) True **(b)** True **(c)** True **(d)** False **(e)** True **(f)** True **(g)** True **(h)** True

Exercise Set 5.3 (page 326)

1. $\bar{\mathbf{u}} = (2 + i, -4i, 1 - i)$, $\text{Re}(\mathbf{u}) = (2, 0, 1)$, $\text{Im}(\mathbf{u}) = (-1, 4, 1)$, $\|\mathbf{u}\| = \sqrt{23}$ **5.** $x = (7 - 6i, -4 - 8i, 6 - 12i)$

7. $\bar{A} = \begin{bmatrix} 5i & 4 \\ 2 + i & 1 - 5i \end{bmatrix}$, $\text{Re}(A) = \begin{bmatrix} 0 & 4 \\ 2 & 1 \end{bmatrix}$, $\text{Im}(A) = \begin{bmatrix} -5 & 0 \\ -1 & 5 \end{bmatrix}$, $\det(A) = 17 - i$, $\text{tr}(A) = 1$

11. $\mathbf{u} \cdot \mathbf{v} = -1 + i$, $\mathbf{u} \cdot \mathbf{w} = 18 - 7i$, $\mathbf{v} \cdot \mathbf{w} = 12 + 6i$ **13.** $-11 - 14i$

15. $\lambda_1 = 2 - i$, $\mathbf{x}_1 = \begin{bmatrix} 2 - i \\ 1 \end{bmatrix}$; $\lambda_2 = 2 + i$, $\mathbf{x}_1 = \begin{bmatrix} 2 + i \\ 1 \end{bmatrix}$

17. $\lambda_1 = 4 - i$, $\mathbf{x}_1 = \begin{bmatrix} 1 - i \\ 1 \end{bmatrix}$; $\lambda_2 = 4 + i$, $\mathbf{x}_1 = \begin{bmatrix} 1 + i \\ 1 \end{bmatrix}$ **19.** $|\lambda| = \sqrt{2}$, $\phi = \frac{\pi}{4}$ **21.** $|\lambda| = 2$, $\phi = -\frac{\pi}{3}$

23. $P = \begin{bmatrix} 5 & 0 \\ -4 & 2 \end{bmatrix}$, $C = \begin{bmatrix} 3 & -2 \\ 2 & 3 \end{bmatrix}$ **25.** $P = \begin{bmatrix} -1 & -1 \\ 1 & 0 \end{bmatrix}$, $C = \begin{bmatrix} 5 & -3 \\ 3 & 5 \end{bmatrix}$ **27. (a)** $k = -\frac{8}{3}i$ **(b)** None

True/False 5.3

(a) False   **(b)** True   **(c)** False   **(d)** True   **(e)** False   **(f)** False

Exercise Set 5.4 (page 332)

1. (a) $y_1 = c_1 e^{5x} - 2c_2 e^{-x}$
$y_2 = c_1 e^{5x} + c_2 e^{-x}$
(b) $y_1 = 0$
$y_2 = 0$
3. (a) $y_1 = -c_2 e^{2x} + c_3 e^{3x}$
$y_2 = c_1 e^x + 2c_2 e^{2x} - c_3 e^{3x}$
$y_3 = 2c_2 e^{2x} - c_3 e^{3x}$
(b) $y_1 = e^{2x} - 2e^{3x}$
$y_2 = e^x - 2e^{2x} + 2e^{3x}$
$y_3 = -2e^{2x} + 2e^{3x}$

7. $y = c_1 e^{3x} + c_2 e^{-2x}$ **9.** $y = c_1 e^x + c_2 e^{2x} + c_3 e^{3x}$

True/False 5.4

(a) False   **(b)** False   **(c)** True   **(d)** True   **(e)** False

Chapter 5 Supplementary Exercises (page 333)

1. (b) The transformation rotates vectors through the angle θ; therefore, if $0 < \theta < \pi$, then no nonzero vector is transformed into a vector in the same or opposite direction.

3. (c) $\begin{bmatrix} 1 & 1 & 0 \\ 0 & 2 & 1 \\ 0 & 0 & 3 \end{bmatrix}$
9. $A^2 = \begin{bmatrix} 15 & 30 \\ 5 & 10 \end{bmatrix}$, $A^3 = \begin{bmatrix} 75 & 150 \\ 25 & 50 \end{bmatrix}$, $A^4 = \begin{bmatrix} 375 & 750 \\ 125 & 250 \end{bmatrix}$, $A^5 = \begin{bmatrix} 1875 & 3750 \\ 625 & 1250 \end{bmatrix}$

11. $0, \operatorname{tr}(A)$ **13.** They are all 0.   **15.** $\begin{bmatrix} 1 & 0 & 0 \\ -1 & -\frac{1}{2} & -\frac{1}{2} \\ 1 & -\frac{1}{2} & -\frac{1}{2} \end{bmatrix}$ **17.** They are all 0, 1, or -1.

Exercise Set 6.1 (page 343)

1. (a) 5   **(b)** -6 **(c)** -3 **(d)** $\sqrt{13}$ **(e)** $\sqrt{5}$ **(f)** $\sqrt{89}$ **3. (a)** 2   **(b)** 11   **(c)** -13 **(d)** -8 **(e)** 0
5. (a) -5 **(b)** 1   **(c)** -7 **(d)** 1   **(e)** 1   **(f)** 1   **7. (a)** 3   **(b)** 56   **9. (b)** 29
11. (a) $\begin{bmatrix} \sqrt{3} & 0 \\ 0 & \sqrt{5} \end{bmatrix}$ **(b)** $\begin{bmatrix} 2 & 0 \\ 0 & \sqrt{6} \end{bmatrix}$ **13. (a)** $\sqrt{74}$ **(b)** 0   **15. (a)** $\sqrt{105}$ **(b)** $\sqrt{47}$
17. $\langle \mathbf{p}, \mathbf{q} \rangle = 50$, $\|\mathbf{p}\| = 6\sqrt{3}$ **19. (a)** $3\sqrt{2}$ **(b)** $3\sqrt{5}$ **(c)** $3\sqrt{13}$
21. (a)   **(b)**

27. For $V = \begin{bmatrix} 0 & 1 \\ -1 & 0 \end{bmatrix}$, then $\langle V, V \rangle = -2 < 0$, so Axiom 4 fails.   **29. (a)** $-\frac{28}{15}$ **(b)** 0

True/False 6.1

(a) True   **(b)** False   **(c)** True   **(d)** True   **(e)** False   **(f)** True   **(g)** False

Exercise Set 6.2 (page 350)

1. **(a)** $-\frac{1}{\sqrt{2}}$ **(b)** $-\frac{3}{\sqrt{73}}$ **(c)** 0 **(d)** $-\frac{20}{9\sqrt{10}}$ **(e)** $-\frac{1}{\sqrt{2}}$ **(f)** $\frac{2}{\sqrt{55}}$ **3.** **(a)** $\frac{19}{10\sqrt{7}}$ **(b)** 0 **7.** No

9. **(a)** $k = -3$ **(b)** $k = -2, -3$ **13.** No **15.** **(a)** $x = t, y = -2t, z = -3t$ **(b)** $2x - 5y + 4z = 0$ **(c)** $x - z = 0$

31. **(a)** The line $y = -x$ **(b)** The xz-plane **(c)** The x-axis

True/False 6.2

(a) False **(b)** True **(c)** True **(d)** True **(e)** False **(f)** False

Exercise Set 6.3 (page 364)

1. (a), (b), (d) **3.** (b), (d) **5.** (a)

7. **(a)** $\left(-\frac{1}{\sqrt{5}}, \frac{2}{\sqrt{5}}\right)$, $\left(\frac{2}{\sqrt{5}}, \frac{1}{\sqrt{5}}\right)$ **(b)** $\left(\frac{1}{\sqrt{2}}, 0, -\frac{1}{\sqrt{2}}\right)$, $\left(\frac{1}{\sqrt{2}}, 0, \frac{1}{\sqrt{2}}\right)$, $(0, 1, 0)$

   (c) $\left(\frac{1}{\sqrt{3}}, \frac{1}{\sqrt{3}}, \frac{1}{\sqrt{3}}\right)$, $\left(-\frac{1}{\sqrt{2}}, \frac{1}{\sqrt{2}}, 0\right)$, $\left(\frac{1}{\sqrt{6}}, \frac{1}{\sqrt{6}}, -\frac{2}{\sqrt{6}}\right)$

9. **(a)** $-\frac{7}{5}\mathbf{v}_1 + \frac{1}{5}\mathbf{v}_2 + 2\mathbf{v}_3$ **(b)** $-\frac{37}{5}\mathbf{v}_1 - \frac{9}{5}\mathbf{v}_2 + 4\mathbf{v}_3$ **(c)** $-\frac{3}{7}\mathbf{v}_1 - \frac{1}{7}\mathbf{v}_2 + \frac{5}{7}\mathbf{v}_3$ **11.** **(b)** $\mathbf{u} = -\frac{4}{5}\mathbf{v}_1 - \frac{11}{10}\mathbf{v}_2 + 0\mathbf{v}_3 + \frac{1}{2}\mathbf{v}_4$

13. **(a)** $\mathbf{w} = \frac{14}{3}\mathbf{u}_1 - \frac{8}{3}\mathbf{u}_2 - \frac{1}{3}\mathbf{u}_3$ **(b)** $\mathbf{w} = \frac{5}{\sqrt{6}}\mathbf{u}_2 + \frac{11}{\sqrt{66}}\mathbf{u}_3$ **15.** **(a)** $\left(\frac{7}{4}, \frac{5}{4}, -\frac{3}{4}, -\frac{5}{4}\right)$ **(b)** $\left(\frac{17}{12}, \frac{7}{4}, -\frac{1}{12}, -\frac{23}{12}\right)$

17. **(a)** $\left(\frac{23}{18}, \frac{11}{6}, -\frac{1}{18}, -\frac{17}{18}\right)$ **(b)** $\left(\frac{3}{2}, \frac{3}{2}, -\frac{1}{2}, -\frac{1}{2}\right)$

19. **(a)** $\mathbf{w}_1 = \left(\frac{3}{2}, \frac{3}{2}, -1, -1\right)$, $\mathbf{w}_2 = \left(-\frac{1}{2}, \frac{1}{2}, 1, -1\right)$ **(b)** $\mathbf{w}_1 = \left(\frac{7}{4}, \frac{5}{4}, -\frac{3}{4}, -\frac{5}{4}\right)$, $\mathbf{w}_2 = \left(-\frac{3}{4}, \frac{3}{4}, \frac{3}{4}, -\frac{3}{4}\right)$

21. **(a)** $\mathbf{v}_1 = \left(\frac{1}{\sqrt{10}}, -\frac{3}{\sqrt{10}}\right)$, $\mathbf{v}_2 = \left(\frac{3}{\sqrt{10}}, \frac{1}{\sqrt{10}}\right)$ **(b)** $\mathbf{v}_1 = (1, 0)$, $\mathbf{v}_2 = (0, -1)$

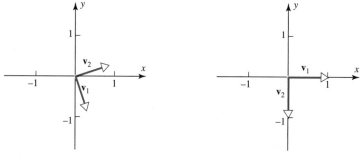

23. $\mathbf{v}_1 = \left(0, \frac{2}{\sqrt{5}}, \frac{1}{\sqrt{5}}, 0\right)$, $\mathbf{v}_2 = \left(\frac{5}{\sqrt{30}}, -\frac{1}{\sqrt{30}}, \frac{2}{\sqrt{30}}, 0\right)$,

   $\mathbf{v}_3 = \left(\frac{1}{\sqrt{10}}, \frac{1}{\sqrt{10}}, -\frac{2}{\sqrt{10}}, -\frac{2}{\sqrt{10}}\right)$, $\mathbf{v}_4 = \left(\frac{1}{\sqrt{15}}, \frac{1}{\sqrt{15}}, -\frac{2}{\sqrt{15}}, \frac{3}{\sqrt{15}}\right)$

25. $\mathbf{v}_1 = \left(\frac{1}{\sqrt{6}}, \frac{1}{\sqrt{6}}, \frac{1}{\sqrt{6}}\right)$, $\mathbf{v}_2 = \left(\frac{1}{\sqrt{6}}, \frac{1}{\sqrt{6}}, -\frac{1}{\sqrt{6}}\right)$, $\mathbf{v}_3 = \left(\frac{2}{\sqrt{6}}, -\frac{1}{\sqrt{6}}, 0\right)$ **27.** $\mathbf{w}_1 = \left(\frac{13}{14}, \frac{31}{14}, \frac{40}{14}\right)$, $\mathbf{w}_2 = \left(\frac{1}{14}, -\frac{3}{14}, \frac{2}{14}\right)$

29. **(a)** $\begin{bmatrix} \frac{1}{\sqrt{5}} & -\frac{2}{\sqrt{5}} \\ \frac{2}{\sqrt{5}} & \frac{1}{\sqrt{5}} \end{bmatrix} \begin{bmatrix} \sqrt{5} & \sqrt{5} \\ 0 & \sqrt{5} \end{bmatrix}$ **(b)** $\begin{bmatrix} \frac{1}{\sqrt{2}} & -\frac{1}{\sqrt{3}} \\ 0 & \frac{1}{\sqrt{3}} \\ \frac{1}{\sqrt{2}} & \frac{1}{\sqrt{3}} \end{bmatrix} \begin{bmatrix} \sqrt{2} & 3\sqrt{2} \\ 0 & \sqrt{3} \end{bmatrix}$ **(c)** $\begin{bmatrix} \frac{1}{3} & \frac{8}{\sqrt{234}} \\ -\frac{2}{3} & \frac{11}{\sqrt{234}} \\ \frac{2}{3} & \frac{7}{\sqrt{234}} \end{bmatrix} \begin{bmatrix} 3 & \frac{1}{3} \\ 0 & \frac{\sqrt{26}}{3} \end{bmatrix}$

   (d) $\begin{bmatrix} \frac{1}{\sqrt{2}} & -\frac{1}{\sqrt{3}} & \frac{1}{\sqrt{6}} \\ 0 & \frac{1}{\sqrt{3}} & \frac{2}{\sqrt{6}} \\ \frac{1}{\sqrt{2}} & \frac{1}{\sqrt{3}} & -\frac{1}{\sqrt{6}} \end{bmatrix} \begin{bmatrix} \sqrt{2} & \sqrt{2} & \sqrt{2} \\ 0 & \sqrt{3} & -\frac{1}{\sqrt{3}} \\ 0 & 0 & \frac{4}{\sqrt{6}} \end{bmatrix}$ **(e)** $\begin{bmatrix} \frac{1}{\sqrt{2}} & \frac{\sqrt{2}}{2\sqrt{19}} & -\frac{3}{\sqrt{19}} \\ \frac{1}{\sqrt{2}} & -\frac{\sqrt{2}}{2\sqrt{19}} & \frac{3}{\sqrt{19}} \\ 0 & \frac{3\sqrt{2}}{\sqrt{19}} & \frac{1}{\sqrt{19}} \end{bmatrix} \begin{bmatrix} \sqrt{2} & \frac{3}{\sqrt{2}} & \sqrt{2} \\ 0 & \frac{\sqrt{19}}{\sqrt{2}} & \frac{3\sqrt{2}}{\sqrt{19}} \\ 0 & 0 & \frac{1}{\sqrt{19}} \end{bmatrix}$

   (f) Columns not linearly independent

33. $\mathbf{v}_1 = 1$, $\mathbf{v}_2 = \sqrt{3}(2x - 1)$, $\mathbf{v}_3 = \sqrt{5}(6x^2 - 6x + 1)$

True/False 6.3

(a) False (b) True (c) True (d) True (e) False (f) True

Exercise Set 6.4 (page 374)

1. (a) $\begin{bmatrix} 21 & 25 \\ 25 & 35 \end{bmatrix} \begin{bmatrix} x_1 \\ x_2 \end{bmatrix} = \begin{bmatrix} 20 \\ 20 \end{bmatrix}$ (b) $\begin{bmatrix} 15 & -1 & 5 \\ -1 & 22 & 30 \\ 5 & 30 & 45 \end{bmatrix} \begin{bmatrix} x_1 \\ x_2 \\ x_3 \end{bmatrix} = \begin{bmatrix} -1 \\ 9 \\ 13 \end{bmatrix}$

3. (a) $x_1 = 5$, $x_2 = \frac{1}{2}$ (b) $x_1 = 12$, $x_2 = -3$, $x_3 = 9$ 5. (a) $\mathbf{e} = \begin{bmatrix} \frac{3}{2} \\ \frac{9}{2} \\ -3 \end{bmatrix}$ (b) $\mathbf{e} = \begin{bmatrix} 3 \\ -3 \\ 0 \\ 3 \end{bmatrix}$

7. (a) Solution: $\hat{\mathbf{x}} = \left(\frac{1}{10}, \frac{6}{5}\right)$; least squares error: $\frac{4}{5}\sqrt{5}$
   (b) Solution: $\hat{\mathbf{x}} = \left(\frac{2}{7}, 0\right) + t(-3, 1)$ (t a real number); least squares error: $\frac{1}{7}\sqrt{42}$
   (c) Solution: $\hat{\mathbf{x}} = \left(-\frac{7}{6}, \frac{7}{6}, 0\right) + t(-1, -1, 1)$ (t a real number); least squares error: $\frac{1}{2}\sqrt{294}$

9. (a) $(7, 2, 9, 5)$ (b) $\left(-\frac{12}{5}, -\frac{4}{5}, \frac{12}{5}, \frac{16}{5}\right)$

11. (a) $\det(A^T A) = 0$; A does not have linearly independent column vectors.
    (b) $\det(A^T A) = 0$; A does not have linearly independent column vectors.

13. (a) $[P] = \begin{bmatrix} 1 & 0 & 0 \\ 0 & 0 & 0 \\ 0 & 0 & 1 \end{bmatrix}$ (b) $[P] = \begin{bmatrix} 0 & 0 & 0 \\ 0 & 1 & 0 \\ 0 & 0 & 1 \end{bmatrix}$

15. (a) $(1, 0, -5)$, $(0, 1, 3)$ (b) $[P] = \frac{1}{35} \begin{bmatrix} 10 & 15 & -5 \\ 15 & 26 & 3 \\ -5 & 3 & 34 \end{bmatrix}$ (c) $\left(\frac{2x_0+3y_0-z_0}{7}, \frac{15x_0+26y_0+3z_0}{35}, \frac{-5x_0+3y_0+34z_0}{35}\right)$ (d) $\frac{3\sqrt{35}}{7}$

17. $s = t = 1$ 21. $[P] = A^T (AA^T)^{-1} A$

True/False 6.4

(a) True (b) False (c) True (d) True (e) False (f) True (g) False (h) True

Exercise Set 6.5 (page 381)

1. $y = -\frac{1}{2} + \frac{7}{2}x$ 3. $y = 2 + 5x - 3x^2$ 11. $y = \frac{5}{21} + \frac{48}{7x}$

True/False 6.5

(a) False (b) True (c) False (d) True

Exercise Set 6.6 (page 387)

1. (a) $(1 + \pi) - 2\sin x - \sin 2x$ (b) $(1 + \pi) - 2\left[\sin x + \frac{\sin 2x}{2} + \frac{\sin 3x}{3} + \cdots + \frac{\sin nx}{n}\right]$

3. (a) $-\frac{1}{2} + \frac{1}{e-1}e^x$ (b) $\frac{13}{12} + \frac{1+e}{2(1-e)}$ 5. (a) $\frac{3}{\pi}x$ (b) $1 - \frac{6}{\pi^2}$ 9. $\frac{1}{2} + \sum_{k=1}^{\infty} \frac{1}{k\pi}\left[1 - (-1)^k\right]\sin kx$

True/False 6.6

(a) False (b) True (c) True (d) False (e) True

Chapter 6 Supplementary Exercises (page 387)

1. (a) $(0, a, a, 0)$ with $a \neq 0$ **(b)** $\pm\left(0, \frac{2}{\sqrt{5}}, \frac{1}{\sqrt{5}}, 0\right)$

3. (a) The subspace of all matrices in M_{22} with only zeros on the diagonal.
  (b) The subspace of all skew-symmetric matrices in M_{22}.

7. $\pm\left(\frac{1}{\sqrt{2}}, 0, \frac{1}{\sqrt{2}}\right)$ **9.** No   **11. (b)** θ approaches $\frac{\pi}{2}$ **17.** No

Exercise Set 7.1 (page 395)

1. (b) $\begin{bmatrix} \frac{4}{5} & -\frac{9}{25} & \frac{12}{25} \\ 0 & \frac{4}{5} & \frac{3}{5} \\ -\frac{3}{5} & -\frac{12}{25} & \frac{16}{25} \end{bmatrix}$

3. (a) $\begin{bmatrix} 1 & 0 \\ 0 & 1 \end{bmatrix}$ **(b)** $\begin{bmatrix} \frac{1}{\sqrt{2}} & \frac{1}{\sqrt{2}} \\ -\frac{1}{\sqrt{2}} & \frac{1}{\sqrt{2}} \end{bmatrix}$ **(d)** $\begin{bmatrix} -\frac{1}{\sqrt{2}} & 0 & \frac{1}{\sqrt{2}} \\ \frac{1}{\sqrt{6}} & -\frac{2}{\sqrt{6}} & \frac{1}{\sqrt{6}} \\ \frac{1}{\sqrt{3}} & \frac{1}{\sqrt{3}} & \frac{1}{\sqrt{3}} \end{bmatrix}$ **(e)** $\begin{bmatrix} \frac{1}{2} & \frac{1}{2} & \frac{1}{2} & \frac{1}{2} \\ \frac{1}{2} & -\frac{5}{6} & \frac{1}{6} & \frac{1}{6} \\ \frac{1}{2} & \frac{1}{6} & \frac{1}{6} & -\frac{5}{6} \\ \frac{1}{2} & \frac{1}{6} & -\frac{5}{6} & \frac{1}{6} \end{bmatrix}$

7. (a) $\left(-1+3\sqrt{3}, \ 3+\sqrt{3}\right)$ **(b)** $\left(\frac{5}{2}-\sqrt{3}, \ \frac{5}{2}\sqrt{3}+1\right)$

9. (a) $\left(-\frac{1}{2}-\frac{5}{2}\sqrt{3}, \ 2, \ \frac{5}{2}-\frac{1}{2}\sqrt{3}\right)$ **(b)** $\left(\frac{1}{2}-\frac{3}{2}\sqrt{3}, \ 6, \ -\frac{3}{2}-\frac{1}{2}\sqrt{3}\right)$

11. (a) $A = \begin{bmatrix} \cos\theta & 0 & -\sin\theta \\ 0 & 1 & 0 \\ \sin\theta & 0 & \cos\theta \end{bmatrix}$ **(b)** $A = \begin{bmatrix} 1 & 0 & 0 \\ 0 & \cos\theta & \sin\theta \\ 0 & -\sin\theta & \cos\theta \end{bmatrix}$ **13.** $a^2+b^2 = \frac{1}{2}$

17. The only possibilities are $a = 0$, $b = -\frac{2}{\sqrt{6}}$, $c = \frac{1}{\sqrt{3}}$ or $a = 0$, $b = \frac{2}{\sqrt{6}}$, $c = -\frac{1}{\sqrt{3}}$.

21. (a) Rotations about the origin, reflections about any line through the origin, and any combination of these
  (b) Rotation about the origin, dilations, contractions, reflections about lines through the origin, and combinations of these
  (c) No; dilations and contractions

True/False 7.1

(a) False   **(b)** False   **(c)** False   **(d)** False   **(e)** True   **(f)** True   **(g)** True   **(h)** True

Exercise Set 7.2 (page 404)

1. (a) $\lambda^2 - 5\lambda = 0$; $\lambda = 0$: one-dimensional; $\lambda = 5$: one-dimensional
  (b) $\lambda^3 - 27\lambda - 54 = 0$; $\lambda = 6$: one-dimensional; $\lambda = -3$: two-dimensional
  (c) $\lambda^3 - 3\lambda^2 = 0$; $\lambda = 3$: one-dimensional; $\lambda = 0$: two-dimensional
  (d) $\lambda^3 - 12\lambda^2 + 36\lambda - 32 = 0$; $\lambda = 2$: two-dimensional; $\lambda = 8$: one-dimensional
  (e) $\lambda^4 - 8\lambda^3 = 0$; $\lambda = 0$: three-dimensional; $\lambda = 8$: one-dimensional
  (f) $\lambda^4 - 8\lambda^3 + 22\lambda^2 - 24\lambda + 9 = 0$; $\lambda = 1$: two-dimensional; $\lambda = 3$: two-dimensional

3. $P = \begin{bmatrix} -\frac{2}{\sqrt{7}} & \frac{\sqrt{3}}{\sqrt{7}} \\ \frac{\sqrt{3}}{\sqrt{7}} & \frac{2}{\sqrt{7}} \end{bmatrix}$; $P^{-1}AP = \begin{bmatrix} 3 & 0 \\ 0 & 10 \end{bmatrix}$ **5.** $P = \begin{bmatrix} -\frac{4}{5} & 0 & \frac{3}{5} \\ 0 & 1 & 0 \\ \frac{3}{5} & 0 & \frac{4}{5} \end{bmatrix}$; $P^{-1}AP = \begin{bmatrix} 25 & 0 & 0 \\ 0 & -3 & 0 \\ 0 & 0 & -50 \end{bmatrix}$

7. $P = \begin{bmatrix} \frac{1}{\sqrt{3}} & -\frac{1}{\sqrt{2}} & -\frac{1}{\sqrt{6}} \\ \frac{1}{\sqrt{3}} & \frac{1}{\sqrt{2}} & -\frac{1}{\sqrt{6}} \\ \frac{1}{\sqrt{3}} & 0 & \frac{2}{\sqrt{6}} \end{bmatrix}$; $P^{-1}AP = \begin{bmatrix} 0 & 0 & 0 \\ 0 & 3 & 0 \\ 0 & 0 & 3 \end{bmatrix}$

9. $P = \begin{bmatrix} -\frac{4}{5} & \frac{3}{5} & 0 & 0 \\ \frac{3}{5} & \frac{4}{5} & 0 & 0 \\ 0 & 0 & -\frac{4}{5} & \frac{3}{5} \\ 0 & 0 & \frac{3}{5} & \frac{4}{5} \end{bmatrix}$; $P^{-1}AP = \begin{bmatrix} -25 & 0 & 0 & 0 \\ 0 & 25 & 0 & 0 \\ 0 & 0 & -25 & 0 \\ 0 & 0 & 0 & 25 \end{bmatrix}$ **15.** No   **19.** Yes

True/False 7.2

(a) True   **(b)** True   **(c)** False   **(d)** True   **(e)** True   **(f)** False   **(g)** True

Exercise Set 7.3 (page 415)

1. (a) $[x_1 \; x_2] \begin{bmatrix} 3 & 0 \\ 0 & 7 \end{bmatrix} \begin{bmatrix} x_1 \\ x_2 \end{bmatrix}$ **(b)** $[x_1 \; x_2] \begin{bmatrix} 4 & -3 \\ -3 & -9 \end{bmatrix} \begin{bmatrix} x_1 \\ x_2 \end{bmatrix}$ **(c)** $[x_1 \; x_2 \; x_3] \begin{bmatrix} 9 & 3 & -4 \\ 3 & -1 & \frac{1}{2} \\ -4 & \frac{1}{2} & 4 \end{bmatrix} \begin{bmatrix} x_1 \\ x_2 \\ x_3 \end{bmatrix}$

3. $2x^2 + 5y^2 - 6xy$ **5.** $\begin{bmatrix} x_1 \\ x_2 \end{bmatrix} = \begin{bmatrix} \frac{1}{\sqrt{2}} & \frac{1}{\sqrt{2}} \\ -\frac{1}{\sqrt{2}} & \frac{1}{\sqrt{2}} \end{bmatrix} \begin{bmatrix} y_1 \\ y_2 \end{bmatrix}$; $Q = 3y_1^2 + y_2^2$

7. $\begin{bmatrix} x_1 \\ x_2 \\ x_3 \end{bmatrix} = \begin{bmatrix} -\frac{2}{3} & \frac{2}{3} & \frac{1}{3} \\ \frac{2}{3} & \frac{1}{3} & \frac{2}{3} \\ \frac{1}{3} & \frac{2}{3} & -\frac{2}{3} \end{bmatrix} \begin{bmatrix} y_1 \\ y_2 \\ y_3 \end{bmatrix}$; $Q = y_1^2 + 4y_2^2 + 7y_3^2$

9. (a) $[x \; y] \begin{bmatrix} 2 & \frac{1}{2} \\ \frac{1}{2} & 0 \end{bmatrix} \begin{bmatrix} x \\ y \end{bmatrix} + [-1 \; 6] \begin{bmatrix} x \\ y \end{bmatrix} + 2 = 0$ **(b)** $[x \; y] \begin{bmatrix} 0 & 0 \\ 0 & 1 \end{bmatrix} \begin{bmatrix} x \\ y \end{bmatrix} + [7 \; -8] \begin{bmatrix} x \\ y \end{bmatrix} - 5 = 0$

11. (a) ellipse   **(b)** hyperbola   **(c)** parabola   **(d)** circle   **13.** Hyperbola: $2(y')^2 - 3(x')^2 = 8$; $\theta \approx -26.6°$

15. Hyperbola: $4(x')^2 - (y')^2 = 3$; $\theta = 36.9°$

17. (a) Positive definite   **(b)** Negative definite   **(c)** Indefinite   **(d)** Positive semidefinite   **(e)** Negative semidefinite

19. Positive definite   **21.** Positive semidefinite   **23.** Indefinite   **27.** $k > 2$

31. (a) $A = \begin{bmatrix} \frac{1}{n} & -\frac{1}{n(n-1)} & \cdots & -\frac{1}{n(n-1)} \\ -\frac{1}{n(n-1)} & \frac{1}{n} & \cdots & -\frac{1}{n(n-1)} \\ \vdots & \vdots & \ddots & \vdots \\ -\frac{1}{n(n-1)} & -\frac{1}{n(n-1)} & \cdots & \frac{1}{n} \end{bmatrix}$ **(b)** Yes   **33.** A must have a positive eigenvalue of multiplicity 2.

True/False 7.3

(a) True   **(b)** False   **(c)** True   **(d)** True   **(e)** False   **(f)** True   **(g)** True   **(h)** True   **(i)** False   **(j)** True
(k) False   **(l)** False

Exercise Set 7.4 (page 423)

1. Maximum: 5 at $(1, 0)$ and $(-1, 0)$; minimum: -1 at $(0, 1)$ and $(0, -1)$

3. Maximum: 7 at $(0, 1)$ and $(0, -1)$; minimum: 3 at $(1, 0)$ and $(-1, 0)$

5. Maximum: 9 at $(1, 0, 0)$ and $(-1, 0, 0)$; minimum: 3 at $(0, 0, 1)$ and $(0, 0, -1)$

7. Maximum: $z = 4\sqrt{2}$ at $(x, y) = \left(2\sqrt{2}, 2\right)$ and $\left(-2\sqrt{2}, -2\right)$;
minimum: $z = -4\sqrt{2}$ at $(x, y) = \left(-2\sqrt{2}, 2\right)$ and $\left(2\sqrt{2}, -2\right)$

9.

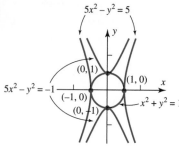

13. Critical points: $(-1, 1)$, relative maximum; $(0, 0)$, saddle point

15. Critical points: $(0, 0)$, relative minimum; $(2, 1)$ and $(-2, 1)$, saddle points

17. Corner points: $x = \frac{5}{\sqrt{2}}, \quad y = \frac{1}{\sqrt{2}}$ 21. $q(\mathbf{x}) = \lambda$

True/False 7.4

(a) False   (b) True   (c) True   (d) False   (e) True

Exercise Set 7.5   (page 430)

1. $A^* = \begin{bmatrix} -2i & 4 & 5-i \\ 1+i & 3-i & 0 \end{bmatrix}$ 3. $A = \begin{bmatrix} 1 & i & 2-3i \\ -i & -3 & 1 \\ 2+3i & 1 & 2 \end{bmatrix}$ 5. (a) $a_{13} \neq \overline{a}_{31}$ (b) $a_{22} \neq \overline{a}_{22}$

9. $A^* = A^{-1} = \begin{bmatrix} \frac{3}{5} & -\frac{4}{5} \\ -\frac{4}{5}i & -\frac{3}{5}i \end{bmatrix}$ 11. $A^* = A^{-1} = \begin{bmatrix} \frac{-i+\sqrt{3}}{2\sqrt{2}} & \frac{1-i\sqrt{3}}{2\sqrt{2}} \\ \frac{1+i\sqrt{3}}{2\sqrt{2}} & \frac{-i-\sqrt{3}}{2\sqrt{2}} \end{bmatrix}$ 13. $P = \begin{bmatrix} \frac{-1+i}{\sqrt{3}} & \frac{1-i}{\sqrt{6}} \\ \frac{1}{\sqrt{3}} & \frac{2}{\sqrt{6}} \end{bmatrix}$, $D = \begin{bmatrix} 3 & 0 \\ 0 & 6 \end{bmatrix}$

15. $P = \begin{bmatrix} \frac{-1-i}{\sqrt{6}} & \frac{1+i}{\sqrt{3}} \\ \frac{2}{\sqrt{6}} & \frac{1}{\sqrt{3}} \end{bmatrix}$, $D = \begin{bmatrix} 2 & 0 \\ 0 & 8 \end{bmatrix}$ 17. $P = \begin{bmatrix} 0 & 0 & 1 \\ \frac{2}{\sqrt{6}} & \frac{-1+i}{\sqrt{6}} & 0 \\ \frac{1+i}{\sqrt{6}} & \frac{2}{\sqrt{6}} & 0 \end{bmatrix}$, $D = \begin{bmatrix} -2 & 0 & 0 \\ 0 & 1 & 0 \\ 0 & 0 & 5 \end{bmatrix}$

19. $A = \begin{bmatrix} 0 & i & 2-3i \\ i & 0 & 1 \\ -2-3i & -1 & 4i \end{bmatrix}$ 21. (a) $a_{13} \neq -\overline{a}_{31}$ (b) $a_{11} \neq -\overline{a}_{11}$ 29. (c) B and C must commute.

37. $\begin{bmatrix} \frac{1}{\sqrt{2}} & -\frac{i}{\sqrt{2}} \\ \frac{i}{\sqrt{2}} & -\frac{1}{\sqrt{2}} \end{bmatrix}$

39. Multiplication of $\mathbf{x}$ by P corresponds to $\|\mathbf{u}\|^2$ times the orthogonal projection of $\mathbf{x}$ onto $W = \text{span}\{\mathbf{u}\}$. If $\|\mathbf{u}\| = 1$, then multiplications of $\mathbf{x}$ by $H = I - 2\mathbf{u}\mathbf{u}^*$ corresponds to reflection of $\mathbf{x}$ about the hyperplane $\mathbf{u}^\perp$.

True/False 7.5

(a) False   (b) False   (c) True   (d) False   (e) False

Chapter 7 Supplementary Exercises   (page 432)

1. (a) $\begin{bmatrix} \frac{3}{5} & -\frac{4}{5} \\ \frac{4}{5} & \frac{3}{5} \end{bmatrix}^{-1} = \begin{bmatrix} \frac{3}{5} & \frac{4}{5} \\ -\frac{4}{5} & \frac{3}{5} \end{bmatrix}$ (b) $\begin{bmatrix} \frac{4}{5} & 0 & -\frac{3}{5} \\ -\frac{9}{25} & \frac{4}{5} & -\frac{12}{25} \\ \frac{12}{25} & \frac{3}{5} & \frac{16}{25} \end{bmatrix}^{-1} = \begin{bmatrix} \frac{4}{5} & -\frac{9}{25} & \frac{12}{25} \\ 0 & \frac{4}{5} & \frac{3}{5} \\ -\frac{3}{5} & -\frac{12}{25} & \frac{16}{25} \end{bmatrix}$

5. $P = \begin{bmatrix} -\frac{1}{\sqrt{2}} & \frac{1}{\sqrt{2}} & 0 \\ 0 & 0 & 1 \\ \frac{1}{\sqrt{2}} & \frac{1}{\sqrt{2}} & 0 \end{bmatrix}$; $P^T A P = \begin{bmatrix} 0 & 0 & 0 \\ 0 & 2 & 0 \\ 0 & 0 & 1 \end{bmatrix}$

7. positive definite   9. (a) parabola   (b) parabola

Exercise Set 8.1 (page 442)

1. Nonlinear   **3.** Linear   **5.** Linear   **7. (a)** Linear   **(b)** Nonlinear

9. $T(x_1, x_2) = (-4x_1 + 5x_2, \ x_1 - 3x_2); \ T(2, -3) = (-23, 11)$

11. $T(x_1, x_2, x_3) = (-x_1 + 4x_2 - x_3, \ 5x_1 - 5x_2 - x_3, \ x_1 + 3x_3); \ T(2, 4, -1) = (15, -9, -1)$

13. $T(2\mathbf{v}_1 - 3\mathbf{v}_2 + 4\mathbf{v}_3) = (-10, -7, 6)$ **15. (a)**   **17. (a)**

19. (a)   **21. (a)** $(1, -4)$ **(b)** $(1, 0, 0), \ (0, 1, 0), \ \left(\frac{3}{2}, -4, 1\right)$ **(c)** x, x^2, x^3

23. (a) $\begin{bmatrix} 1 \\ 5 \\ 7 \end{bmatrix}, \begin{bmatrix} -1 \\ 6 \\ 4 \end{bmatrix}$ **(b)** $\begin{bmatrix} -14 \\ 19 \\ 11 \end{bmatrix}$ **(c)** $\text{Rank}(T) = 2, \text{nullity}(T) = 1$ **(d)** $\text{Rank}(A) = 2, \text{nullity}(A) = 1$

25. (a) $\begin{bmatrix} 1 \\ 0 \end{bmatrix}, \begin{bmatrix} 0 \\ 1 \end{bmatrix}$ **(b)** $\begin{bmatrix} -1 \\ -1 \\ 1 \\ 0 \end{bmatrix}, \begin{bmatrix} -4 \\ 2 \\ 0 \\ 7 \end{bmatrix}$ **(c)** $\text{Rank}(T) = \text{nullity}(T) = 2$ **(d)** $\text{Rank}(A) = \text{nullity}(A) = 2$

27. (a) Kernel: y-axis; range: xz-plane   **(b)** Kernel: x-axis; range: yz-plane
(c) Kernel: the line through the origin perpendicular to the plane $y = x$; range: plane $y = x$

29. (a) $\text{Nullity}(T) = 2$ **(b)** $\text{Nullity}(T) = 4$ **(c)** $\text{Nullity}(T) = 3$ **(d)** $\text{Nullity}(T) = 1$

31. (a) 3   **(b)** No   **33.** A line through the origin, a plane through the origin, the origin only, or all of R^3

35. (b) No   **41.** $\ker(D)$ consists of all constant polynomials.   **43. (a)** $T(f(x)) = f^{(4)}(x)$ **(b)** $T(f(x)) = f^{(n+1)}(x)$

True/False 8.1

(a) True   **(b)** False   **(c)** True   **(d)** False   **(e)** True   **(f)** True   **(g)** False   **(h)** False   **(i)** True

Exercise Set 8.2 (page 451)

1. (a) $\ker(T) = \{\mathbf{0}\}$; T is one-to-one   **(b)** $\ker(T) = \left\{k\left(-\frac{3}{2}, 1\right)\right\}$; T is not one-to-one   **(c)** $\ker(T) = \{\mathbf{0}\}$; T is one-to-one
(d) $\ker(T) = \{\mathbf{0}\}$; T is one-to-one   **(e)** $\ker(T) = \{k(1, 1)\}$; T is not one-to-one
(f) $\ker(T) = \{k(0, 1, -1)\}$; T is not one-to-one

3. (a) Not one-to-one   **(b)** Not one-to-one   **(c)** One-to-one

5. (a) $\ker(T) = \{k(-1, 1)\}$ **(b)** T is not one-to-one since $\ker(T) \neq \{\mathbf{0}\}$.

7. (a) T is one-to-one   **(b)** T is not one-to-one   **(c)** T is not one-to-one   **(d)** T is one-to-one

11. (a) $T\left(\begin{bmatrix} a & b & c \\ b & d & e \\ c & e & f \end{bmatrix}\right) = \begin{bmatrix} a \\ b \\ c \\ d \\ e \\ f \end{bmatrix}$ **(b)** $T\left(\begin{bmatrix} a & b \\ c & d \end{bmatrix}\right) = \begin{bmatrix} a \\ b \\ c \\ d \end{bmatrix}; \ T\left(\begin{bmatrix} a & b \\ c & d \end{bmatrix}\right) = \begin{bmatrix} a \\ c \\ b \\ d \end{bmatrix}$

(c) $T(ax^3 + bx^2 + cx) = \begin{bmatrix} a \\ b \\ c \end{bmatrix}$ **(d)** $T(a + b\sin(x) + c\cos(x)) = \begin{bmatrix} a \\ b \\ c \end{bmatrix}$

13. T is not one-to-one since, for example, $f(x) = x^2(x - 1)^2$ is in its kernel.   **15.** Yes; it is one-to-one

17. T is not one-to-one since, for example $\mathbf{a}$ is in its kernel.   **19.** No

True/False 8.2

(a) False   **(b)** True   **(c)** False   **(d)** True   **(e)** False   **(f)** False

Exercise Set 8.3 (page 457)

1. **(a)** $(T_2 \circ T_1)(x, y) = (2x - 3y, \ 2x + 3y)$ **(b)** $(T_2 \circ T_1)(x, y) = (4x - 12y, \ 3x - 9y)$
   (c) $(T_2 \circ T_1)(x, y) = (2x + 3y, \ x - 2y)$ **(d)** $(T_2 \circ T_1)(x, y) = (0, 2x)$

3. **(a)** $a + d$ **(b)** $(T_2 \circ T_1)(A)$ does not exist since $T_1(A)$ is not a 2×2 matrix.   5. $T_2(\mathbf{v}) = \tfrac{1}{4}\mathbf{v}$

11. **(a)** T has no inverse.   **(b)** $T^{-1}\begin{bmatrix} x_1 \\ x_2 \\ x_3 \end{bmatrix} = \begin{bmatrix} \tfrac{1}{8}x_1 + \tfrac{1}{8}x_2 - \tfrac{3}{4}x_3 \\ \tfrac{1}{8}x_1 + \tfrac{1}{8}x_2 + \tfrac{1}{4}x_3 \\ -\tfrac{3}{8}x_1 + \tfrac{5}{8}x_2 + \tfrac{1}{4}x_3 \end{bmatrix}$ **(c)** $T^{-1}\begin{bmatrix} x_1 \\ x_2 \\ x_3 \end{bmatrix} = \begin{bmatrix} \tfrac{1}{2}x_1 - \tfrac{1}{2}x_2 + \tfrac{1}{2}x_3 \\ -\tfrac{1}{2}x_1 + \tfrac{1}{2}x_2 + \tfrac{1}{2}x_3 \\ \tfrac{1}{2}x_1 + \tfrac{1}{2}x_2 - \tfrac{1}{2}x_3 \end{bmatrix}$

   (d) $T^{-1}\begin{bmatrix} x_1 \\ x_2 \\ x_3 \end{bmatrix} = \begin{bmatrix} 3x_1 + 3x_2 - x_3 \\ -2x_1 - 2x_2 + x_3 \\ -4x_1 - 5x_2 + 2x_3 \end{bmatrix}$

13. **(a)** $a_i \neq 0$ for $i = 1, 2, 3, \ldots, n$ **(b)** $T^{-1}(x_1, x_2, x_3, \ldots, x_n) = \left(\tfrac{1}{a_1}x_1, \tfrac{1}{a_2}x_2, \tfrac{1}{a_3}x_3, \ldots, \tfrac{1}{a_n}x_n \right)$

15. **(a)** $T_1^{-1}(p(x)) = \tfrac{p(x)}{x}$; $T_2^{-1}(p(x)) = p(x - 1)$; $(T_2 \circ T_1)^{-1}(p(x)) = \tfrac{1}{x}p(x - 1)$ 17. **(a)** $(1, -1)$ **(d)** $T^{-1}(2, 3) = 2 + x$

21. **(a)** $T_1 \circ T_2 = T_2 \circ T_1$ **(b)** $T_1 \circ T_2 \neq T_2 \circ T_1$ **(c)** $T_1 \circ T_2 = T_2 \circ T_1$

True/False 8.3

(a) True   **(b)** False   **(c)** False   **(d)** True   **(e)** False   **(f)** True

Exercise Set 8.4 (page 466)

1. **(a)** $\begin{bmatrix} 0 & 0 & 0 \\ 1 & 0 & 0 \\ 0 & 1 & 0 \\ 0 & 0 & 1 \end{bmatrix}$ 3. **(a)** $\begin{bmatrix} 1 & -1 & 1 \\ 0 & 1 & -2 \\ 0 & 0 & 1 \end{bmatrix}$ 5. **(a)** $\begin{bmatrix} 0 & 0 \\ -\tfrac{1}{2} & 1 \\ \tfrac{8}{3} & \tfrac{4}{3} \end{bmatrix}$ 7. **(a)** $\begin{bmatrix} 1 & 1 & 1 \\ 0 & 2 & 4 \\ 0 & 0 & 4 \end{bmatrix}$ **(b)** $3 + 10x + 16x^2$

9. **(a)** $[T(\mathbf{v}_1)]_B = \begin{bmatrix} 1 \\ -2 \end{bmatrix}$, $[T(\mathbf{v}_2)]_B = \begin{bmatrix} 3 \\ 5 \end{bmatrix}$ **(b)** $T(\mathbf{v}_1) = \begin{bmatrix} 3 \\ -5 \end{bmatrix}$, $T(\mathbf{v}_2) = \begin{bmatrix} -2 \\ 29 \end{bmatrix}$

   (c) $T\left(\begin{bmatrix} x_1 \\ x_2 \end{bmatrix} \right) = \begin{bmatrix} \tfrac{18}{7} & \tfrac{1}{7} \\ -\tfrac{107}{7} & \tfrac{24}{7} \end{bmatrix} \begin{bmatrix} x_1 \\ x_2 \end{bmatrix}$ **(d)** $\begin{bmatrix} \tfrac{19}{7} \\ -\tfrac{83}{7} \end{bmatrix}$

11. **(a)** $[T(\mathbf{v}_1)]_B = \begin{bmatrix} 1 \\ 2 \\ 6 \end{bmatrix}$, $[T(\mathbf{v}_2)]_B = \begin{bmatrix} 3 \\ 0 \\ -2 \end{bmatrix}$, $[T(\mathbf{v}_3)]_B = \begin{bmatrix} -1 \\ 5 \\ 4 \end{bmatrix}$

   (b) $T(\mathbf{v}_1) = 16 + 51x + 19x^2$, $T(\mathbf{v}_2) = -6 - 5x + 5x^2$, $T(\mathbf{v}_3) = 7 + 40x + 15x^2$

   (c) $T\left(a_0 + a_1 x + a_2 x^2 \right) = \tfrac{239a_0 - 161a_1 + 289a_2}{24} + \tfrac{201a_0 - 111a_1 + 247a_2}{8}x + \tfrac{61a_0 - 31a_1 + 107a_2}{12}x^2$

   (d) $T\left(1 + x^2 \right) = 22 + 56x + 14x^2$

13. **(a)** $[T_2 \circ T_1]_{B', B} = \begin{bmatrix} 0 & 0 \\ 6 & 0 \\ 0 & -9 \\ 0 & 0 \end{bmatrix}$, $[T_2]_{B', B''} = \begin{bmatrix} 0 & 0 & 0 \\ 3 & 0 & 0 \\ 0 & 3 & 0 \\ 0 & 0 & 3 \end{bmatrix}$, $[T_1]_{B'', B} = \begin{bmatrix} 2 & 0 \\ 0 & -3 \\ 0 & 0 \end{bmatrix}$ **(b)** $[T_2 \circ T_1]_{B', B} = [T_2]_{B', B''}[T_1]_{B'', B}$

19. **(a)** $\begin{bmatrix} 0 & 0 & 0 \\ 0 & 0 & -1 \\ 0 & 1 & 0 \end{bmatrix}$ **(b)** $\begin{bmatrix} 0 & 0 & 0 \\ 0 & 1 & 0 \\ 0 & 0 & 2 \end{bmatrix}$ **(c)** $\begin{bmatrix} 2 & 1 & 0 \\ 0 & 2 & 2 \\ 0 & 0 & 2 \end{bmatrix}$

   (d) $14e^{2x} - 8xe^{2x} - 20x^2 e^{2x}$ since $\begin{bmatrix} 2 & 1 & 0 \\ 0 & 2 & 2 \\ 0 & 0 & 2 \end{bmatrix} \begin{bmatrix} 4 \\ 6 \\ -10 \end{bmatrix} = \begin{bmatrix} 14 \\ -8 \\ -20 \end{bmatrix}$

21. **(a)** B', B'' **(b)** B', B'''

True/False 8.4

(a) False   **(b)** False   **(c)** True   **(d)** False   **(e)** True

---

Exercise Set 8.5 (page 473)

1. $[T]_B = \begin{bmatrix} 1 & -2 \\ 0 & -1 \end{bmatrix}$, $[T]_{B'} = \begin{bmatrix} -\frac{3}{11} & -\frac{56}{11} \\ -\frac{2}{11} & \frac{3}{11} \end{bmatrix}$ **3.** $[T]_B = \begin{bmatrix} \frac{1}{\sqrt{2}} & -\frac{1}{\sqrt{2}} \\ \frac{1}{\sqrt{2}} & \frac{1}{\sqrt{2}} \end{bmatrix}$, $[T]_{B'} = \begin{bmatrix} \frac{13}{11\sqrt{2}} & -\frac{25}{11\sqrt{2}} \\ \frac{5}{11\sqrt{2}} & \frac{9}{11\sqrt{2}} \end{bmatrix}$

5. $[T]_B = \begin{bmatrix} 1 & 0 & 0 \\ 0 & 1 & 0 \\ 0 & 0 & 0 \end{bmatrix}$, $[T]_{B'} = \begin{bmatrix} 1 & 0 & 0 \\ 0 & 1 & 1 \\ 0 & 0 & 0 \end{bmatrix}$ **7.** $[T]_B = \begin{bmatrix} \frac{2}{3} & -\frac{2}{9} \\ \frac{1}{2} & \frac{4}{3} \end{bmatrix}$, $[T]_{B'} = \begin{bmatrix} 1 & 1 \\ 0 & 1 \end{bmatrix}$

11. (a) $B' = \left\{ \begin{bmatrix} 1 \\ -1 \end{bmatrix}, \begin{bmatrix} 1 \\ -2 \end{bmatrix} \right\}$ **(b)** $B' = \left\{ \begin{bmatrix} \frac{-3-\sqrt{21}}{6} \\ 1 \end{bmatrix}, \begin{bmatrix} \frac{-3+\sqrt{21}}{6} \\ 1 \end{bmatrix} \right\}$

13. (a) $\lambda = -4$, $\lambda = 3$
  (b) Basis for eigenspace corresponding to $\lambda = -4$: $-2 + \frac{8}{3}x + x^2$; basis for eigenspace corresponding to $\lambda = 3$: $5 - 2x + x^2$

21. The choice of an appropriate basis can yield a better understanding of the linear operator.

True/False 8.5

(a) False   **(b)** True   **(c)** True   **(d)** True   **(e)** True   **(f)** False   **(g)** True   **(h)** False

---

Chapter 8 Supplementary Exercises (page 475)

1. No. $T(\mathbf{x}_1 + \mathbf{x}_2) = A(\mathbf{x}_1 + \mathbf{x}_2) + B \neq (A\mathbf{x}_1 + B) + (A\mathbf{x}_2 + B) = T(\mathbf{x}_1) + T(\mathbf{x}_2)$, and if $c \neq 1$, then
  $T(c\mathbf{x}) = cA\mathbf{x} + B \neq c(A\mathbf{x} + B) = cT(\mathbf{x})$.

5. (a) $T(\mathbf{e}_3)$ and any two of $T(\mathbf{e}_1)$, $T(\mathbf{e}_2)$, and $T(\mathbf{e}_4)$ form bases for the range; $(-1, 1, 0, 1)$ is a basis for the kernel.
  (b) Rank = 3, nullity = 1

7. (a) Rank$(T) = 2$ and nullity$(T) = 2$ **(b)** T is not one-to-one.   **11.** Rank = 3, nullity = 1   **13.** $\begin{bmatrix} 1 & 0 & 0 & 0 \\ 0 & 0 & 1 & 0 \\ 0 & 1 & 0 & 0 \\ 0 & 0 & 0 & 1 \end{bmatrix}$

15. $[T]_{B'} = \begin{bmatrix} -4 & 0 & 9 \\ 1 & 0 & -2 \\ 0 & 1 & 1 \end{bmatrix}$ **17.** $[T]_B = \begin{bmatrix} 1 & -1 & 1 \\ 0 & 1 & 0 \\ 1 & 0 & -1 \end{bmatrix}$ **19. (b)** $f(x) = x$, $g(x) = 1$

21. (d) The points are on the graph.   **25.** $\begin{bmatrix} 0 & 0 & 0 & \cdots & 0 \\ 1 & 0 & 0 & \cdots & 0 \\ 0 & \frac{1}{2} & 0 & \cdots & 0 \\ 0 & 0 & \frac{1}{3} & \cdots & 0 \\ \vdots & \vdots & \vdots & & \vdots \\ 0 & 0 & 0 & \cdots & \frac{1}{n+1} \end{bmatrix}$

Exercise Set 9.1 (page 485)

1. $x_1 = 2$, $x_2 = 1$ **3.** $x_1 = 3$, $x_2 = -1$ **5.** $x_1 = -1$, $x_2 = 1$, $x_3 = 0$
7. $x_1 = -1$, $x_2 = 1$, $x_3 = 0$ **9.** $x_1 = -3$, $x_2 = 1$, $x_3 = 2$, $x_4 = 1$

11. (a) $A = LU = \begin{bmatrix} 2 & 0 & 0 \\ -2 & 1 & 0 \\ 2 & 1 & 1 \end{bmatrix} \begin{bmatrix} 1 & \frac{1}{2} & -\frac{1}{2} \\ 0 & 0 & 1 \\ 0 & 0 & 0 \end{bmatrix}$ **(b)** $A = L_1 D U_1 = \begin{bmatrix} 1 & 0 & 0 \\ -1 & 1 & 0 \\ 1 & 1 & 1 \end{bmatrix} \begin{bmatrix} 2 & 0 & 0 \\ 0 & 1 & 0 \\ 0 & 0 & 1 \end{bmatrix} \begin{bmatrix} 1 & \frac{1}{2} & -\frac{1}{2} \\ 0 & 0 & 1 \\ 0 & 0 & 0 \end{bmatrix}$

(c) $A = L_2 U_2 = \begin{bmatrix} 1 & 0 & 0 \\ -1 & 1 & 0 \\ 1 & 1 & 1 \end{bmatrix} \begin{bmatrix} 2 & 1 & -1 \\ 0 & 0 & 1 \\ 0 & 0 & 0 \end{bmatrix}$

13. $A = \begin{bmatrix} 1 & 0 & 0 \\ 0 & 1 & 0 \\ 2 & -2 & 1 \end{bmatrix} \begin{bmatrix} 3 & 0 & 0 \\ 0 & 2 & 0 \\ 0 & 0 & 1 \end{bmatrix} \begin{bmatrix} 1 & -4 & 2 \\ 0 & 1 & 0 \\ 0 & 0 & 1 \end{bmatrix}$ **15.** $x_1 = \frac{21}{17}, \; x_2 = -\frac{14}{17}, \; x_3 = \frac{12}{17}$

17. $A = \begin{bmatrix} 1 & 0 & 0 \\ 0 & 0 & 1 \\ 0 & 1 & 0 \end{bmatrix} \begin{bmatrix} 3 & 0 & 0 \\ 0 & 2 & 0 \\ 3 & 0 & 1 \end{bmatrix} \begin{bmatrix} 1 & -\frac{1}{3} & 0 \\ 0 & 1 & \frac{1}{2} \\ 0 & 0 & 1 \end{bmatrix}$; $x_1 = -\frac{1}{2}, \; x_2 = \frac{1}{2}, \; x_3 = 3$

19. (b) $\begin{bmatrix} a & b \\ c & d \end{bmatrix} = \begin{bmatrix} 1 & 0 \\ \frac{c}{a} & 1 \end{bmatrix} \begin{bmatrix} a & b \\ 0 & \frac{ad-bc}{a} \end{bmatrix}$

True/False 9.1

(a) False **(b)** False **(c)** True **(d)** True **(e)** True

Exercise Set 9.2 (page 494)

1. (a) λ_3 dominant **(b)** No dominant eigenvalue

3. $x_1 \approx \begin{bmatrix} 0.98058 \\ -0.19612 \end{bmatrix}$; $x_2 \approx \begin{bmatrix} 0.98837 \\ -0.15206 \end{bmatrix}$; $x_3 \approx \begin{bmatrix} 0.98679 \\ -0.16201 \end{bmatrix}$; $x_4 \approx \begin{bmatrix} 0.98715 \\ -0.15977 \end{bmatrix}$;

dominant eigenvalue: $\lambda = 2 + \sqrt{10} \approx 5.16228$;

dominant eigenvector: $\begin{bmatrix} 1 \\ 3 - \sqrt{10} \end{bmatrix} \approx \begin{bmatrix} 1 \\ -0.16228 \end{bmatrix}$

5. $x_1 = \begin{bmatrix} -1 \\ 1 \end{bmatrix}$, $\lambda^{(1)} = 6$; $x_2 = \begin{bmatrix} -0.5 \\ 1 \end{bmatrix}$, $\lambda^{(2)} = 6.6$; $x_3 \approx \begin{bmatrix} -0.53846 \\ 1 \end{bmatrix}$, $\lambda^{(3)} \approx 6.60550$;

$x_4 \approx \begin{bmatrix} -0.53488 \\ 1 \end{bmatrix}$, $\lambda^{(4)} \approx 6.60555$;

dominant eigenvalue: $\lambda = 3 + \sqrt{13} \approx 6.60555$;

dominant eigenvector: $\begin{bmatrix} -\frac{3}{\sqrt{26+4\sqrt{13}}} \\ \frac{2+\sqrt{13}}{\sqrt{26+4\sqrt{13}}} \end{bmatrix} \approx \begin{bmatrix} -0.47186 \\ 0.88167 \end{bmatrix}$

7. (a) $x_1 = \begin{bmatrix} 11 \\ -0.5 \end{bmatrix}$, $x_2 = \begin{bmatrix} 1 \\ -0.8 \end{bmatrix}$, $x_3 \approx \begin{bmatrix} 1 \\ -0.929 \end{bmatrix}$ **(b)** $\lambda^{(1)} = 2.8, \; \lambda^{(2)} \approx 2.976, \; \lambda^{(3)} \approx 2.997$

(c) Dominant eigenvalue: $\lambda = 3$; dominant eigenvector: $\begin{bmatrix} 1 \\ -1 \end{bmatrix}$ **(d)** 0.1%

9. 2.99993; $\begin{bmatrix} 0.99180 \\ 1.00000 \end{bmatrix}$

13. (a) Starting with $\begin{bmatrix} 1 \\ 0 \\ 0 \end{bmatrix}$, it takes 8 iterations. **(b)** Starting with $\begin{bmatrix} 1 \\ 0 \\ 0 \\ 0 \end{bmatrix}$, it takes 8 iterations.

Exercise Set 9.3 (page 500)

1. $\mathbf{h}_0 = \begin{bmatrix} 1 \\ 2 \\ 2 \end{bmatrix}$, $\mathbf{a}_0 = \begin{bmatrix} 2 \\ 0 \\ 3 \end{bmatrix}$ 3. $\mathbf{h}_1 \approx \begin{bmatrix} 0.39057 \\ 0.65094 \\ 0.65094 \end{bmatrix}$, $\mathbf{a}_1 \approx \begin{bmatrix} 0.60971 \\ 0 \\ 0.79262 \end{bmatrix}$

5. Sites 1 and 2 (tie); sites 3 and 4 are irrelevant   7. Site 2, site 3, site 4; sites 1 and 5 are irrelevant

Exercise Set 9.4 (page 506)

1. (a) ≈ 0.067 second   (b) ≈ 66.68 seconds   (c) $\approx 66{,}668$ seconds, or about 18.5 hours

3. (a) ≈ 9.52 seconds   (b) ≈ 0.0014 second   (c) ≈ 9.52 seconds   (d) ≈ 28.6 seconds

5. (a) 6.67×10^5 s for forward phase, 10 s for backward phase   (b) 1334   7. n^2 flops   9. $2n^3 - n^2$ flops

Exercise Set 9.5 (page 513)

1. $0, \sqrt{5}$ 3. $\sqrt{5}$ 5. $A = \begin{bmatrix} \frac{1}{\sqrt{2}} & -\frac{1}{\sqrt{2}} \\ \frac{1}{\sqrt{2}} & \frac{1}{\sqrt{2}} \end{bmatrix} \begin{bmatrix} \sqrt{2} & 0 \\ 0 & \sqrt{2} \end{bmatrix} \begin{bmatrix} 1 & 0 \\ 0 & 1 \end{bmatrix}$ 7. $A = \begin{bmatrix} \frac{2}{\sqrt{5}} & -\frac{1}{\sqrt{5}} \\ \frac{1}{\sqrt{5}} & \frac{2}{\sqrt{5}} \end{bmatrix} \begin{bmatrix} 8 & 0 \\ 0 & 2 \end{bmatrix} \begin{bmatrix} \frac{1}{\sqrt{5}} & \frac{2}{\sqrt{5}} \\ -\frac{2}{\sqrt{5}} & \frac{1}{\sqrt{5}} \end{bmatrix}$

9. $A = \begin{bmatrix} \frac{2}{3} & \frac{1}{\sqrt{2}} & \frac{\sqrt{2}}{6} \\ \frac{1}{3} & 0 & -\frac{2\sqrt{2}}{3} \\ -\frac{2}{3} & \frac{1}{\sqrt{2}} & -\frac{\sqrt{2}}{6} \end{bmatrix} \begin{bmatrix} 3\sqrt{2} & 0 \\ 0 & 0 \\ 0 & 0 \end{bmatrix} \begin{bmatrix} -\frac{1}{\sqrt{2}} & \frac{1}{\sqrt{2}} \\ \frac{1}{\sqrt{2}} & \frac{1}{\sqrt{2}} \end{bmatrix}$ 11. $A = \begin{bmatrix} \frac{1}{\sqrt{3}} & 0 & \frac{2}{\sqrt{6}} \\ \frac{1}{\sqrt{3}} & \frac{1}{\sqrt{2}} & -\frac{1}{\sqrt{6}} \\ -\frac{1}{\sqrt{3}} & \frac{1}{\sqrt{2}} & \frac{1}{\sqrt{6}} \end{bmatrix} \begin{bmatrix} \sqrt{3} & 0 \\ 0 & \sqrt{2} \\ 0 & 0 \end{bmatrix} \begin{bmatrix} 1 & 0 \\ 0 & 1 \end{bmatrix}$

True/False 9.5

(a) False   (b) True   (c) False   (d) False   (e) True   (f) False   (g) True

Exercise Set 9.6 (page 517)

1. $\begin{bmatrix} \frac{2}{3} \\ \frac{1}{3} \\ -\frac{2}{3} \end{bmatrix} \begin{bmatrix} 3\sqrt{2} \end{bmatrix} \begin{bmatrix} -\frac{1}{\sqrt{2}} & \frac{1}{\sqrt{2}} \end{bmatrix}$ 3. $\begin{bmatrix} \frac{1}{\sqrt{3}} & 0 \\ \frac{1}{\sqrt{3}} & \frac{1}{\sqrt{2}} \\ -\frac{1}{\sqrt{3}} & \frac{1}{\sqrt{2}} \end{bmatrix} \begin{bmatrix} \sqrt{3} & 0 \\ 0 & \sqrt{2} \end{bmatrix} \begin{bmatrix} 1 & 0 \\ 0 & 1 \end{bmatrix}$ 5. $3\sqrt{2} \begin{bmatrix} \frac{2}{3} \\ \frac{1}{3} \\ -\frac{2}{3} \end{bmatrix} \begin{bmatrix} -\frac{1}{\sqrt{2}} & \frac{1}{\sqrt{2}} \end{bmatrix}$

7. $\sqrt{3} \begin{bmatrix} \frac{1}{\sqrt{3}} \\ \frac{1}{\sqrt{3}} \\ -\frac{1}{\sqrt{3}} \end{bmatrix} [1 \ \ 0] + \sqrt{2} \begin{bmatrix} 0 \\ \frac{1}{\sqrt{2}} \\ \frac{1}{\sqrt{2}} \end{bmatrix} [0 \ \ 1]$ 9. 70,100 numbers must be stored; A has 100,000 entries

True/False 9.6

(a) True   (b) True   (c) False

Chapter 9 Supplementary Exercises (page 517)

1. $\begin{bmatrix} 2 & 0 \\ -2 & 1 \end{bmatrix} \begin{bmatrix} -3 & 1 \\ 0 & 2 \end{bmatrix}$ 3. $\begin{bmatrix} 2 & 0 & 0 \\ 1 & 2 & 0 \\ 1 & 1 & 2 \end{bmatrix} \begin{bmatrix} 1 & 2 & 3 \\ 0 & 1 & 2 \\ 0 & 0 & 1 \end{bmatrix}$

5. (a) $\lambda = 3$, $\mathbf{v} = \begin{bmatrix} \frac{1}{\sqrt{2}} \\ \frac{1}{\sqrt{2}} \end{bmatrix}$ (b) $\mathbf{x}_5 \approx \begin{bmatrix} 0.7100 \\ 0.7041 \end{bmatrix}$, $\mathbf{v} \approx \begin{bmatrix} 0.7071 \\ 0.7071 \end{bmatrix}$ (c) $\mathbf{x}_5 \approx \begin{bmatrix} 1 \\ 0.9918 \end{bmatrix}$

9. $\begin{bmatrix} -\frac{1}{\sqrt{2}} & 0 & \frac{1}{\sqrt{2}} \\ 0 & 1 & 0 \\ -\frac{1}{\sqrt{2}} & 0 & -\frac{1}{\sqrt{2}} \end{bmatrix} \begin{bmatrix} 2 & 0 \\ 0 & 0 \\ 0 & 0 \end{bmatrix} \begin{bmatrix} -\frac{1}{\sqrt{2}} & -\frac{1}{\sqrt{2}} \\ -\frac{1}{\sqrt{2}} & \frac{1}{\sqrt{2}} \end{bmatrix}$

11. $\begin{bmatrix} 12 & 0 & 6 \\ 4 & -8 & 10 \\ 4 & -8 & 10 \\ 12 & 0 & 6 \end{bmatrix} = \begin{bmatrix} \frac{1}{2} & \frac{1}{2} \\ \frac{1}{2} & -\frac{1}{2} \\ \frac{1}{2} & -\frac{1}{2} \\ \frac{1}{2} & \frac{1}{2} \end{bmatrix} \begin{bmatrix} 24 & 0 \\ 0 & 12 \end{bmatrix} \begin{bmatrix} \frac{2}{3} & -\frac{1}{3} & \frac{2}{3} \\ \frac{2}{3} & \frac{2}{3} & -\frac{1}{3} \end{bmatrix}$

INDEX

A

Absolute value, 523
  of complex number, 315
  of determinant, 166
Addition
  associative law for, 38, 122
  of complex numbers, 522
  by scalars, 172
  of vectors in R^2 and R^3, 120, 122
  of vectors in R^n, 126
Additivity property
  of linear transformation, 434
  of matrix transformation, 249, 269, 270
Adjacency matrices, 496–497
Adjoint, of a matrix, 110–111
Aeronautics, yaw, pitch, and roll, 256
Algebraic multiplicity, 312
Algebraic operations, using vector
  components, 126–127
Algebraic properties of matrices, 38–48
Algebraic properties of vectors, dot
  product, 135–136
Alleles, 292
Amps (unit), 76
Angle
  in R^n, 137, 143
  between vectors, 134–135, 137, 346
Angle of rotation, 254
Anticommutativity, 326
Antihomogeneity property, of complex
  Euclidean inner product, 318
Antisymmetry property
  of complex Euclidean inner product,
    318
  of dot product, 318
Approximate integration, 83
Approximation problems, 382–384
Approximations, best approximations, 367
Area
  of parallelogram, 165
  of triangle, 154
Argument, of complex number, 315, 525
Arithmetic average, 337
Arithmetic operations
  complex numbers, 522–525
  matrices, 27–34, 38–42
  vectors in R^2 and R^3, 120–122
  vectors in R^n, 125
Artificial intelligence, 479
Associative law for addition, 38, 122
Associative law for matrix multiplication,
  38, 39–40
Astronautics, yaw, pitch, and roll, 256

Augmented matrices, 6–7, 11, 12, 18, 25,
  33
Authority, 496
Authority vectors, 497
Authority weights, 497
Axes
  rotation of in 2–space, 392–394
  rotation of in 3–space, 394–395
Axis of rotation, 254

B

Back-substitution, 19–20
Backward phase, 15
Balancing (of chemical equation), 79
Basis, 209–211
  change of, 217–222, 469–470
  coordinate system for vector space,
    201–202
  for eigenspaces, 299–301, 472
  finite basis, 201
  by inspection, 212
  linear combinations and, 233–234
  number of vectors in, 209, 211
  ordered basis, 205
  orthogonal basis, 354
  for orthogonal complement, 349–350
  orthonormal basis, 354–355
  for row and column spaces, 229–230,
    231
  by row reduction, 230, 231
  for row space of a matrix, 232–233
  standard basis, 202–203, 205–206
  transition matrix, 219
  uniqueness of basis representation, 204
  for vector space using row operations,
    232
Basis vectors, 201, 436–437
Bateman, Harry, 509
Battery, 76
Beltrami, Eugenio, 510
Best approximation theorem, 367
Bôcher, Maxime, 7, 184
Books, ISBN number of, 141
Branches (network), 73
Brightness, graphical images, 124
Brin, Sergey, 496
Bunyakovsky, Viktor Yakovlevich, 137

C

C^n, 318–321
Calculus of variations, 162
Cancellation law, 41
Carroll, Lewis, 96
Cauchy, Augustin, 109, 137, 172

Cauchy-Schwarz inequality, 137, 345–346
Cayley, Arthur, 29, 34, 43
Central conic, 409
Central conic in standard position, 409
Central ellipsoid in standard position, 416
Central quadrics in standard position, 409
Change-of-basis problem, 217–218, 469
Change of variable, 406–407
Characteristic equation, 296, 308
Characteristic polynomial, 297, 308
Chemical equations, balancing with linear
  systems, 78–80
Chemical formulas, 78
Clockwise closed-loop convention, 76
Closed economies, 86
Closure under addition, 172
Closure under scalar, 172
Codomain, 247
Coefficient matrices, 33, 308, 477
Coefficients
  of linear combination of matrices, 31
  of linear combination of vectors, 127,
    183
  literal, 44
Cofactor, 94
Cofactor expansion
  determinants by, 93–98
  elementary row operations and, 104
  of 2×2 matrices, 95
Collinear vectors, 121–122
Column matrices, 26
Column-matrix form of vectors, 128, 225
Column space, 225, 226, 228, 243
  basis for, 229–230, 231
  equal dimensions of row and column
    space, 237
Column vectors, 26, 27, 39
Columns, cofactor expansion and choice
  of column, 97
Combustion, linear systems to analyze
  combustion equation for methane,
  78–79
Comma-delimited form of vectors, 128,
  225
Common initial point, 122
Commutative law for addition, 38
Commutative law for multiplication, 40,
  46
Complete reaction (chemical), 78
Complex conjugates
  of complex numbers, 315, 522–523
  of vectors, 316, 317–318
Complex dot product, 317
Complex eigenvalues, 318, 319, 322–324

Complex eigenvectors, 318, 319

Complex Euclidean inner product, 317–318

Complex exponential, complex numbers, 527

Complex exponential function, 526

Complex inner product space, 344

Complex inner products, 344

Complex matrices, 316

Complex n-space, 316

Complex n-tuples, 316

Complex number system, 522

Complex numbers, 315–316, 521–527
  complex conjugates, 522–523
  defined, 521
  DeMoivre's formula, 526
  division of, 524, 525–526
  multiplication of, 521, 522, 525–526
  polar form of, 315, 524–525
  reciprocals, 523

Complex plane, 522

Complex vector spaces, 172, 315–325

Components (of a vector)
  algebraic operations using, 126–127
  calculating dot products using, 35
  complex n-tuples, 316
  finding, 123–124
  in R^2 and R^3, 122–123
  vector components of **u** along **a**, 147–148

Composition
  with identity operator, 453–454
  of linear transformations, 452–453, 455–456
  matrices of, 464, 465
  of matrix transformations, 263–267
  non-commutative nature of, 264
  of one-to-one linear transformations, 455–456
  of reflections, 265, 277
  of rotations, 264, 275
  of three transformations, 265–267

Compression operator, 257, 277

Computer programs, LU-decomposition and, 478

Computers
  Internet search engines, 496–500
  LINPACK, 478

Condensation, 96

Conic sections (conics), 408–412

Conjugate transpose, 424

Conjugates, of complex numbers, 522

Consistency, determining by elimination, 64–65

Consistent linear system, 3–4, 227

Constrained extremum, 417, 418

Constrained extremum problems, 417–419

Constrained extremum theorem, 417

Constraint, 417

Consumption matrix, 86

Consumption vectors, 87, 88

Continuous derivatives, functions with, 182

Contraction, 256, 257, 435, 446

Contraction operators, 435

Contrapositive, 519

Contrapositive form of theorem, 519

Convergence
  of power sequences, 487
  rate of, 493

Converse, 519

Coordinate map, 217

Coordinate matrices, 205

Coordinate systems, 200
  "basis vectors" for, 201
  units of measurement, 201

Coordinate vectors, 205
  computing, 220
  relative to orthonormal basis, 355
  relative to standard bases, 206

Coordinates, 205
  of generalized point, 124
  in R^3, 206–207
  relative to standard basis for R^n, 205–206

Corresponding linear systems, 158

Cramer, Gabriel, 113

Cramer's rule, 112–113

Critical points, 420

Cross product, 161–163
  calculating, 161
  determinant form of, 164
  geometric interpretation of, 164–165
  notation, 161
  properties of, 163

Cross product terms, 405, 411

Current (electrical), 76

D

Data compression, singular value decomposition, 514–516

Decomposition
  eigenvalue decomposition, 506, 507
  Hessenberg decomposition, 506, 507
  LDU-decomposition, 484–485
  LU-decompositions, 477–484, 504–505
  PLU-decomposition, 485
  Schur decomposition, 507
  singular value decomposition, 509–511, 514–516

of square matrices, 506–507

Degenerate conic, 408

Degrees of freedom, 209

DeMoivre, Abraham, 526

DeMoivre's formula, 526

Dependency equations, 234

Determinants, 43, 93–110
  by cofactor expansion, 93–98
  defined, 93
  of elementary matrices, 102
  equivalence theorem, 114
  evaluating by row reduction, 100–104
  general determinant, 95
  geometric interpretation of, 166–168
  of linear operator, 472
  of matrix product, 107–108
  properties of, 106–112
  sums of, 107
  of 3×3 matrices, 97–98
  of 2×2 matrices, 97–98
  of upper triangular matrix, 97

Deviation, 382

Diagonal coefficient matrices, 329

Diagonal entries, 509

Diagonal matrices, 66–68, 274–275

Diagonalizability
  nondiagonalizability of $n \times n$ matrix, 402
  orthogonal diagonalizability, 427–428
  of triangular matrices, 309–310

Diagonalization
  matrices, 305–313
  orthogonal diagonalization, 397
  solution of linear system by, 329–331

Dickson, Leonard Eugene, 111

Difference
  of complex numbers, 522
  matrices, 27
  vectors, 121, 126

Differential equations, 327–331, 440

Differentiation, by matrix multiplication, 450

Differentiation transformation, 439

Digital communications, matrix form and, 245

Dilation, 256, 257, 435, 446

Dilation operators, 435

Dimension theorem, for linear transformations, 441–442

Dimensions
  of spans, 209–210
  and transformations, 447
  of vector spaces, 209

Dirac matrices, 326

Direct product, 134

Discriminant, 321

Distance, 336
  between a point and a plane, 149–150
  general inner product spaces, 346–347
  orthogonal projections for, 148–149
  between parallel planes, 150
  real inner product spaces, 336
  in R^n, 132–133
  triangle inequality for, 138
Distinct eigenvalues, 487
Distributive property
  of complex Euclidean inner product, 318
  of dot product, 136
Division, of complex numbers, 524, 525–526
Dodgson, Charles Lutwidge, 96
Domain, 247
Dominant eigenvalue, 487–489
Dot product, 133–136
  algebraic properties of, 135–136
  antisymmetry property of, 318
  application of, 141
  calculating with, 136–137
  complex dot product, 317
  cross product and, 162
  dot product form of linear systems, 156–157
  as matrix multiplication, 139–140
  relationships involving, 162
  symmetry property of, 136, 318
  of vectors, 139
Dynamical system, 282–284

E

Echelon forms, 11–12, 21
Economic modeling, Leontief economic analysis with, 85–89
Economic sectors, 85
Economics, n-tuples and, 124
Eigenspaces, 299, 308, 318
  bases for, 299–301
  of real symmetric matrix, 425, 426
Eigenvalue decomposition (EVD), 506, 507
Eigenvalues, 295–296, 308, 319
  complex eigenvalues, 318, 319
  conic sections classified by using, 413
  dominant eigenvalues, 487–489
  of Hermitian, 425, 426
  of Hermitian matrices, 429
  invertibility and, 301–302
  of linear operators, 472
  of square matrix, 311
  of symmetric matrices, 398
  of 3×3 matrix, 297
  of triangular matrices, 298–299

of 2×2 matrix, 321
Eigenvectors, 295–296, 299
  of 2×2 vector, 296
  bases for eigenspaces and, 300–301
  complex eigenvectors, 318, 319
  left/right eigenvectors, 304
  of real symmetric matrix, 425, 426
  of square matrix, 311
  of symmetric matrices, 398
Einstein, Albert, 123, 124
Eisenstein, Gotthold, 29
Electrical circuits
  n-tuples and, 124
  network analysis with linear systems, 76–78
Electrical current, 76
Electrical potential, 76
Electrical resistance, 76
Elementary matrices, 51
  determinants, 102
  and homogeneous linear systems, 57
  invertibility, 53
  matrix operators corresponding to, 277
Elementary row operations, 7–8, 51, 228
  cofactor expansion and, 104
  determinants and, 100–104
  and inverse operations, 52–55
  and inverse row operations, 52–55
  for inverting matrices, 55–56
  matrix multiplication, 52
  row reduction and determinants, 100–104
Elimination methods, 14–16, 64–65
Ellipse, principal axes of, 410
Empty set, 209
Entries, 26
Equal complex numbers, 521
Equal matrices, 27–28, 39
Equal vectors, 120, 125
Equivalence theorem, 373
  determinants, 114
  invertibility, 53–54, 302
  $n \times n$ matrix, 240–241
Equivalent statements, 519–520
Equivalent vectors, 120, 125
Error vector, 369
Errors
  approximation problems, 384–385
  least squares error, 367
  mean square error, 383
  measurements of, 382–383
  percentage error, 493
  relative error, 493
  roundoff errors, 21–22
Estimated percentage error, 493
Estimated relative error, 493–494

Euclidean inner product, 336–339
  complex Euclidean inner product, 317–318
  of vectors in R^2 or R^3, 133
Euclidean n-space, 336
Euclidean norm, 317
Euclidean scaling, power method with, 489–490
Euler's formula, 527
Evaluation inner product, 340
Evaluation transformation, 436–437
EVD (eigenvalue decomposition), 506, 507
Expansion operator, 257, 277
Exponents, matrix laws, 46

F

Factoring, 447
Factorization, 480
Fingerprint storage, 515
Finite basis, 201
Finite-dimensional inner product space, 349, 362
Finite-dimensional vector space, 204, 212–213, 217
First-order linear system, 328
Floating-point numbers, 501
Floating-point operation, 501
Flops, 501–503
Flow conservation, in networks, 74
Forward phase, 15
Forward substitution, 479
4×6 matrix, rank and nullity of, 238
Fourier, Jean Baptiste, 386
Fourier coefficients, 385
Fourier series, 384–385, 386
Free variable theorem for homogeneous systems, 19
Free variables, 13, 239
Full column rank, 363
Function spaces, 182–183
Functions
  with continuous derivatives, 182
  defined, 247–248
  linear dependence of, 196–197
Fundamental spaces, 243, 244

G

Gauss, Carl Friedrich, 15, 29, 94
Gauss-Jordan elimination
  of augmented matrix, 319, 505
  described, 15
  for homogeneous system, 18
  polynomial interpolation by, 82
  roundoff errors, 21–22
  using, 44, 504, 505

Gaussian elimination, 11–16, 505
  defined, 16
  roundoff errors, 21–22
General determinant, 95
General solution, 13, 227, 240, 328
Generalized Theorem of Pythagoras, 348
Genotypes, 292
Geometric multiplicity, 312
Geometric vectors, 119
Geometry
  of linear systems, 152–159
  quadratic forms in, 408–409
  in R^n, 138
Gibbs, Josiah Willard, 134, 161
Golub, Gene H., 510
Google
  algorithms used by, 496
  origin of term, 496
Googol, 496
Gram, Jorgen Pederson, 360
Gram-Schmidt process, 356, 358–360,
  361, 362, 363, 384
Graphic images
  images of lines under matrix operators,
    279–280
  n-tuples and, 124
  RGB color model, 127
Grassmann, H.G., 172

H
Hermite, Charles, 426
Hermitian matrices, 425–429
Hesse, Ludwig Otto, 420
Hessenberg decomposition, 506, 507
Hessenberg's theorem, 403
Hessian matrices, 420, 421
Hilbert, David, 360
Hilbert space, 360
Hill, George William, 184
Homogeneity property
  of complex Euclidean inner product,
    318
  of dot product, 136
  of linear transformation, 434
  of matrix transformation, 249, 269, 270
Homogeneous equations, 145, 156
Homogeneous linear equations, 2
Homogeneous linear systems, 17–19, 227
  dimensions of solution space, 210
  and elementary matrices, 57
  free variable theorem for, 19
  solutions of, 186–187
Homogeneous systems, solutions spaces
  of, 187–188
Hooke's law, 378
Hub, 496

Hub vectors, 497
Hub weights, 497
Hue, graphical images, 124

I
Idempotency, 50
Identity matrices, 41–42
Identity operators
  about, 251, 434
  composition with, 453–454
  kernel and range of, 439
  matrices of, 463–464
Image processing, data compression and,
  515
Images, 247. *See also* Graphic images
  of a line, 279–280
  of a square, 279
Images of basis vectors, 436–437
Imaginary axis, 522
Imaginary numbers, 521. *See also*
  Complex numbers
Imaginary part
  of complex numbers, 315, 521
  of vectors and matrices, 316
Implication loop, 520
Inconsistent linear system, 3
Indefinite quadratic forms, 412
Infinite-dimensional vector space, 204
Initial authority vector, 497
Initial condition, 328
Initial hub vector, 497
Initial point, 119
Initial-value problem, 328
Inner product
  algebraic properties of, 342
  calculating, 342
  complex inner products, 344
  Euclidean inner product, 133, 317–318,
    336–339
  evaluation inner product, 340
  examples of, 336–341
  linear transformation using, 435
  matrix inner products, 338
  on M_{nn}, 339–340
  on real vector space, 335
  on R^n, 336–339
  standard inner products, 336, 340
Inner product space, 435
  complex inner product space, 344
  unit circle, 338
  unit sphere, 338
Inner product space isomorphism,
  450–451
Input-output analysis, 85
Inputs, in economics, 85
Instability, 22

Integer coefficients, 297
Integral transformation, 438
Integration, approximate, 83
Intermediate demand vector, 87
Internet search engines, 496–500
Interpolating polynomial, 80
Invariant under similarity, 305, 471–472
Inverse
  of a product, 45–46
  of diagonal matrices, 67
  of matrix using its adjoint, 111–112
  of 2×2 matrices, 44
Inverse linear transformations, 454–455
Inverse matrices, 41–45
Inverse operations, 52–55
Inverse row operations, 52–55
Inverse transformations, 464
Inversion, solving linear systems by,
  44–45, 60–62
Inversion algorithm, 55
Invertibility
  determinant test for, 108–110
  eigenvalues and, 301–302
  of elementary matrices, 53
  equivalence theorem, 53–54
  matrix transformation and, 266–267
  test for determinant, 108–110
  of transition matrices, 220
  of triangular matrices, 68
Invertibility test for determinant, 108–110
Invertible matrices, 41–45, 62–65
ISBN (books), 141
Isomorphism, 447–451

J
Jordan, Camille, 507, 510
Jordan, Wilhelm, 15
Jordan canonical form, 407
Junctions (network), 73, 76

K
Kalman, Dan, 400
Kasner, Edward, 496
Kernel, 438, 439–440, 445
Kirchhoff, Gustav, 77
Kirchhoff's current law, 76
Kirchhoff's voltage law, 76
*k*th principal submatrix, 414

L
Lagrange, Joseph Louis, 162
LDU-decomposition, 484–485
LDU-factorization, 485
Leading 1!, 11
Leading variables, 13, 239
Least squares, curve fitting, 376

Least squares approximation, 382–385
Least squares error, 367
Least squares error vector, 367
Least squares fit
  of polynomial, 379–380
  of quadratic curve to data, 380–381
  straight line fit, 376–377, 378
Least squares polynomial fit, 379–380
Least squares solutions, 378
  of linear systems, 366–371, 372
  QR-decomposition and, 371
  straight line fit, 376–377, 378
Least squares straight line fit, 377
Left distributive law, 38
Left eigenvectors, 304
Legendre polynomials, 361
Length, 130, 336, 346
Leontief, Wassily, 85, 86
Leontief equation, 87
Leontief input-output models, 85–89
Leontief matrices, 87
Level curves, 419
Line. See Lines
Line segment, from one point to another
  in R^2, 156
Linear algebra, 1. *See also* Linear
  equations; Linear systems
  coordinate systems, 200
Linear combinations
  basis and, 233–234
  history of term, 184
  of matrices, 31, 32–33
  of vectors, 127, 132–133, 183, 185–186
Linear dependence, 184, 193
Linear equations, 2–3, 156. *See also*
  Linear systems
Linear form, 405
Linear independence, 184, 190–198, 214
  examples of, 194
  of polynomials, 193
  of sets, 191–194
  of standard unit vectors in R^3, 191–192
  of standard unit vectors in R^4, 192
  of standard unit vectors in R^n, 191
  of two functions, 195
  use of terms, 193
  using the Wronskian, 197–198
Linear operators
  determinants of, 472
  matrices of, 462, 468–469
  orthogonal matrices as, 391–392
  on P^2, 463
Linear systems, 2–3. *See also*
  Homogeneous linear systems
  with a common coefficient matrix, 61
  applications, 73–83

augmented matrices, 6–7, 11, 12, 18,
  25, 33
  for balancing chemical equations,
    78–80
  coefficient matrix, 33
  comparison of procedures for solving,
    501–505
  computer solution, 1
  corresponding linear systems, 158
  cost estimate for solving, 501–504
  dot product form of, 156–157
  first-order linear system, 328
  general solution, 13
  geometry of, 152–159
  with infinitely many solutions, 5–7
  least squares solutions of, 366–371, 372
  network analysis with, 73–78
  with no solutions, 5
  non-homogeneous, 19
  number of solutions, 60
  overdetermined/underdetermined, 241
  polynomial interpolation, 80–82
  solution methods, 3, 4–7
  solutions, 3, 11
  solving by elimination row operations,
    7–8
  solving by Gaussian elimination,
    11–16, 21–22, 505
  solving by matrix inversion, 44–45,
    60–62
  solving with Cramer's rule, 112–113
  in three unknowns, 12–13
Linear transformations, 270
  composition of, 452–453, 455–456
  defined, 433
  dimension theorem for, 441–442
  examples of, 435, 438
  inverse linear transformations, 454–455
  matrices of, 458–462
  one-to-one, 445
  onto, 445
  from P_n to P_{n+1}, 435
  rank and nullity in, 441
  using inner product, 435
Linearity conditions, 270
Linearly dependent set, 191
Linearly independent set, 191, 192–193
Lines
  image of, 279–280
  line segment from one point to another
    in R^2, 156
  orthogonal projection on, 147
  orthogonal projection on lines through
    the origin, 258–259
  point-normal equations, 144–145
  through origin as subspaces, 180–181

through two points in R^3, 155–156
vector and parametric equations in R^2
  and R^3, 152–153, 154
vector and parametric equations of in
  R^4, 155
vector form of, 145, 154
vectors orthogonal to, 145
LINPACK, 478
Literal coefficients, 44
Lower triangular matrices, 68, 298
LU-decompositions, 477–484, 504–505
  constructing, 483–484
  examples of, 480–483
  finding, 480
  method, 478
LU-factorization, 480

M
M_{mn}. *See* $m \times n$ matrices
M_{nn}
  inner products on, 339–340
  standard basis, 203
  subspaces of, 181
$m \times n$ matrices, real vector spaces,
  174–175
Maclaurin series, 527
Magnitude (norm), 130
Main diagonal, 27, 509
Mantissa, 501
Mapping, 248
Markov, Andrei Andreyevich, 285
Markov chain, 285, 286–288
  steady-state vector of, 289
  transition matrix for, 289–290
MATLAB, 478
Matrices. *See also* Matrices of specific
  size, such as 2×2 matrices
  adjacency matrices, 496–497
  adjoint of, 110
  algebraic properties of, 38–48
  arithmetic operations with, 27–34
  coefficient matrices, 33, 308, 477
  column matrices, 26
  complex matrices, 316
  compositions of, 464, 465
  coordinate matrices, 205
  defined, 1, 6, 26
  determinants, 93–110
  diagonal coefficient matrices, 329
  diagonal matrices, 66–68, 274–275
  diagonalization, 305–313
  dimension theorem for matrices, 239
  elementary matrices, 51, 53, 57, 102,
    277
  entries, 26
  equality of, 27–28, 39

examples of, 26–27
fundamental spaces, 243, 244
Hermitian matrices, 425–427, 429
Hessian matrices, 420, 421
identity matrices, 41–42
of identity operators, 463–464
inner products generated by, 338–339
inverse matrices, 41–45
of inverse transformations, 464
invertibility, 53–54, 68, 108–110, 220
invertible matrices, 41–45, 62–65
inverting, 55–57
Leontief economic analysis with, 85–89
linear combination, 31, 32–33
of linear operators, 462, 468–469
of linear transformations, 458–462
normal matrices, 429
notation and terminology, 25–27, 33
orthogonal matrices, 389–395
orthogonally diagonalizable matrices, 397
partitioned, 30–31
permutation matrices, 485
positive definite matrices, 414
powers of, 45–46, 301–302, 310–311
with proportional rows or columns, 102–103
rank of, 239
real and imaginary parts of, 316–317
real matrices, 316, 322
redundancy in, 245
reflection matrices, 390
rotation matrices, 254, 390
row equivalents, 51
row matrices, 26
scalar multiples, 28
similar matrices, 305
singular/nonsingular matrices, 42
size of, 26, 39
skew-Hermitian matrices, 429
skew-symmetric matrices, 428–429
square matrices, 27, 34, 35, 42, 66, 68, 100–104, 311, 389, 506–507
standard matrices, 248, 251, 268–269, 274, 372
stochastic matrices, 288–289
submatrices, 30, 414
symmetric matrices, 69–70, 299, 322, 398, 420
trace, 35
transition matrices, 219–221, 469
transpose, 33–34
triangular matrices, 68–69, 298–299, 309–310
unitary matrices, 425, 426, 427
upper triangular matrices, 68, 97, 298

zero matrices, 40
Matrix factorization, 323
Matrix inner products, 338
Matrix multiplication. *See* Multiplication (matrices)
Matrix notation, 25–27, 33, 406
Matrix operators, 248, 446
  geometric effect of, 278
  graphics images of lines under matrix operators, 279–280
  on R^2, 273–280
Matrix polynomials, 46–47
Matrix spaces, transformations on, 435
Matrix transformations, 248, 434
  composition of, 263–267
  defined, 433
  kernel and range of, 438–439
  notation, 249
  properties of, 249–250
  from R^4 to R^3, 249
  standard matrix for, 251
  zero transformations, 250, 434, 439
Maximum entry scaling, power method with, 490–493
Mean square error, 383
Mechanical systems, n-tuples and, 125
Methane, linear systems to analyze combustion equation, 78–79
Minor, 94
Modulus, of complex numbers, 315, 523
Multiplication (complex numbers), 521, 522, 525–526
Multiplication (matrices), 28–31, 249–250. *See also* Product (of matrices)
  associative law for, 38, 39–40
  by columns and by rows, 30–31
  differentiation by, 450
  dot products as, 139–140
  elementary row operations, 52
  by invertible matrix, 278
  order and, 40
Multiplication (vectors). *See also* Cross product; Euclidean inner product; Inner product; Product (of vectors)
  in R^2 and R^3, 121
  by scalars, 172
Multiplication by A, 248
Multiplicative inverse, 523

N
N-dimensional vector space, 212
n-space, 123, 124. *See also* R^n
  $n \times n$ matrices
    equivalent statements, 240–241
    Hessenberg's theorem, 403

nondiagonalizability of, 402
Natural isomorphism, 449
Negative, of vector, 121
Negative definite quadratic forms, 412
Negative pole, 76
Negative semidefinite quadratic forms, 412
Network analysis, with linear systems, 73–78
Networks, defined, 73
Nodes (network), 73, 76
Non-homogeneous linear systems, 19
Nonsingular matrices, 42
Nontrivial solution, 17
Nonzero vectors, 188
Norm (length), 130, 148, 336
  calculating, 131
  complex Euclidean inner product and, 317–318
  Euclidean norm, 317
  real inner product spaces, 336
  of vector in $C[a, b]$, 341
Normal, 144
Normal equations, 368, 377–378
Normal matrices, 429
Normal system, 368
Normalization, 132, 364
Null space, 225, 228–229
Nullity, 441
  of 4×6 matrix, 238
  sum of, 239–240
Numerical analysis, 11
Numerical coefficients, 44

O
Ohm's law, 76
Ohms (unit), 76
One-to-one linear transformations, 445, 455–456
Onto linear transformations, 445
Open economies, Leontief analysis of, 86–89
Open sectors, 86
Operators, 248, 435, 446. *See also* Linear operators
Optimization, using quadratic forms, 417–422
Order
  of differential equation, 327
  matrix multiplication and, 40
Order n, 384
Ordered basis, 205
Ordered n-tuple, 124
Ordered n-tuple, 30
Ordered pair, 3
Ordered triple, 3

Orthogonal basis, 354, 356, 361
Orthogonal change of variable, 407
Orthogonal complement, 243–244, 348–350
Orthogonal diagonalizability, 427–428
Orthogonal diagonalization, 397
Orthogonal matrices, 389–395
Orthogonal operators, 392
Orthogonal projection operators, 251
Orthogonal projections, 146–147, 357
  on a subspace, 369–370
  geometric interpretation of, 358
  kernel and range of, 439
  on lines through the origin, 258–259
  for standard matrix, 372
  on subspaces of R^m, 371–372
Orthogonal sets, 143, 353
Orthogonal vectors, 143–145, 317
  in M_{22}, 347
  in P_2, 347–348
Orthogonality
  defined, 352–353
  inner product and, 347
  of row vectors and solution vectors, 157–158
Orthogonally diagonalizable matrices, 397
Orthonormal basis, 354–355, 358, 384
  change of, 392
  coordinate vectors relative to, 355
  from orthogonal basis, 356
  orthonormal sets extended to, 361–362
Orthonormal sets, 143, 354
  constructing, 353–354
  extended to orthonormal bases, 361–362
Orthonormality, 353
Outputs, in economics, 85
Outside demand vector, 87, 88
Overdetermined linear system, 241

P

P_n. See Polynomials
P_2
  linear operators on, 463
  orthogonal vectors in, 347–348
  Theorem of Pythagoras in, 348
Page, Larry, 496
PageRank algorithm, 496
PageRank score, 496
Parallel planes, distance between, 150
Parallel vectors, 121–122
Parallelogram, area of, 164
Parallelogram equation for vectors, 138
Parallelogram rule for vector addition, 120

Parameters, 5, 13, 152, 153
Parametric equations, 6
  of lines and planes in R^4, 155
  of lines in R^2 and R^3, 152–153, 154
  of planes in R^3, 153, 154–155
Particular solution, 227
Partitioned matrices, 30–31
Pauli spin matrices, 326
Percentage error, 493
Permutation matrices, 485
Perpendicular vectors, 143
Photographs, data compression and image processing, 515
Piazzi, Giuseppe, 15
Pitch (aircraft), 256
Pivot column, 21
Pivot position, 21
Pixels, data compression and image processing, 515
Planes
  distance between a point and a plane, 149–150
  distance between parallel planes, 150
  point-normal equations, 144–145
  through origin as subspaces, 181
  vector and parametric equations in R^3, 153, 154–155
  vector and parametric equations of in R^4, 155
  vector form of, 145, 154
  vectors orthogonal to, 145
PLU-decomposition, 485
PLU-factorization, 485
Plus-minus theorem, 211–212
Point-normal equations, 144–145
Points, distance between a point and a plane, 149–150
Polar form, of complex numbers, 315, 524–525
Poles (battery), 76
Polynomial interpolation, 80–82
Polynomials (P_n), 46–47
  characteristic polynomial, 297, 308
  least squares fit of, 379–380
  Legendre polynomials, 361
  linear independence of, 193
  linear transformation, 435
  linearly independent set in, 192–193
  spanning set for, 185
  standard basis for, 202
  standard inner product on, 340
  subspaces of, 182
  trigonometric polynomial, 384–385
Positive definite matrices, 414
Positive definite quadratic forms, 412–414
Positive pole, 76

Positive semidefinite quadratic forms, 412
Positivity property
  of complex Euclidean inner product, 318
  of dot product, 136
Power method, 487–494
  with Euclidean scaling, 489–490
  with maximum entry scaling, 490–493
  for search engine algorithms, 496–500
  stopping procedures, 494
Power sequence generated by A, 487
Powers of a matrix, 45–46, 67, 301–302, 310–311
Principal argument, of complex number, 525
Principal axes, 410
Principal axes theorem, 407–408, 410
Principal submatrices, 414
Probability, 284
Probability vector, 284
Product (of matrices), 28
  determinants of, 107–108
  inverse of, 45–46
  as linear combination, 31
  of lower triangular matrices, 68
  of symmetric matrices, 70
  transpose of, 47
Product (of vectors)
  cross product, 161–163
  scalar multiple in R^2 and R^3, 121
Production vector, 87, 88
Productive open economies, 88–89
Products (in chemical equation), 78
Profitable sectors, 89
Projection operators, 251, 253, 267–268
Projection theorem, 146–147, 356–357
Pure imaginary complex numbers, 521

Q

QR-decomposition, 362–364, 371
Quadratic curve, of least squares fit, 380–381
Quadratic form associated with A, 406
Quadratic forms, 405–409
  applications of, 406–409
  change of variable, 406–407
  conic sections, 408–409
  expressing in matrix notation, 406
  indefinite quadratic forms, 412
  negative definite quadratic forms, 412
  negative semidefinite quadratic forms, 412
  optimization using, 417–422
  positive definite quadratic forms, 412–414

positive semidefinite quadratic forms, 412

principal axes theorem, 407–408

Quotient, complex number division, 524

R

R_m, orthogonal projections on subspaces of, 371–372

R_n

  coordinates relative to standard basis for, 205–206

  distance in, 132–133

  Euclidean inner product, 336–339

  geometry in, 138

  linear independence of standard unit vectors in, 191

  norm of a vector, 131

  span in standard unit vector, 184

  spanning in, 184

  standard basis for, 202

  standard unit vectors in, 132

  Theorem of Pythagoras in, 148

  transition matrices for, 220–222

  two-point vector equations in, 155

  vector forms of lines and planes in, 154

  as vector space, 173

  vectors in, 123–125

R^2

  basic matrix operators on, 258

  dot product of vectors in, 133

  line segment from one point to another in, 156

  lines through origin are subspaces of, 180–181

  matrix operators on, 273–280

  norm of a vector, 131

  parametric equations, of lines in, 152–153, 154

  shears in, 258

  spanning in, 184–185

  unit circles in, 338

  vector addition in, 120, 122

  vectors in, 119–129

R^3

  coordinates in, 206–207

  dot product of vectors in, 133

  linear independence of standard unit vectors in, 191–192

  lines through origin are subspaces of, 180–181

  lines through two points in, 155–156

  matrix transformation from R^4 to R^3, 249

  norm of a vector, 131

  orthogonal set in, 353

  rotations in, 254–256

spanning in, 184–185

standard basis for, 202–203

vector addition in, 120, 122

vector and parametric equations of lines in, 152–153, 154

vector and parametric equations of planes in, 153, 154–155

vectors in, 119–129

R^4

  cosine of angle between two vectors in, 346

  linear independence of standard unit vectors in, 192

  matrix transformation from R^4 to R^3, 249

  Theorem of Pythagoras in, 148

  vector and parametric equations of lines and planes in, 155

Range, 247, 438, 439–440

Rank, 441

  of 4×6 matrix, 238

  dimension theorem for matrices, 239

  maximum value for, 239

  redundancy in a matrix and, 245

  sum of, 239–240

Rank of an approximation, 516

Rate of convergence, 493

Rayleigh, John William Strutt, 492

Rayleigh quotient, 491

Reactants (in chemical equation), 78

Real inner product space, 335, 345–346

Real line, 123

Real matrices, 316, 322

Real part

  of complex numbers, 315, 521

  of vectors and matrices, 316–317

Real-valued functions, vector space of, 175

Real vector space, 171, 172, 335

Reciprocals, complex numbers, 523

Rectangular coordinate systems, 200

Reduced row echelon forms, 11–12, 21, 319

Reduced singular value decomposition, 514–515

Reduced singular value expansion, 514

Redundancy. in matrices, 245

Reflection matrices, 390

Reflection operators, 251, 252, 259–260

Reflections, composition of, 265, 277

Regression line, 377

Regular Markov chain, 288

Regular stochastic matrices, 288–289

Relative error, 493

Relative maximum, 420, 421

Relative minimum, 420, 421

Resistance (electrical), 76

Resistor, 76

RGB color cube, 127

RGB color model, 127

RGB space, 127

Right distributive law, 38

Right eigenvectors, 304

Right-hand rule, 164

Roll (aircraft), 256

Rotation equations, 254, 393

Rotation matrices, 254, 390

Rotation of axes

  in 2–space, 392

  in 3–space, 394–395

Rotation operator, 253–255

  properties of, 267

  on R^3, 255

Rotations

  composition of, 264, 275

  kernel and range of, 439

  in R^3, 254–256

Roundoff errors, 21–22

Row-column method, 30–31

Row echelon form, 11–12, 14–15, 21, 229

Row equivalents, 51

Row matrices, 26

Row-matrix form of vectors, 128, 225

Row operations. *See* Elementary row operations

Row reduction

  basis by, 230, 231

  evaluating determinants by, 100–104

Row space, 225, 228–229, 243

  basis by row reduction, 230

  basis for, 229–230, 232–233

  equal dimensions of row and column space, 237

Row vectors, 26, 27, 39, 157, 225

Rows, cofactor expansion and choice of row, 97

S

Saddle points, 420, 421

Sample points, 340

Saturation, graphical images, 124

Scalar multiples, 28, 172

Scalar multiplication, 121, 172

Scalar triple product, 165–166

Scalars, 26, 119, 121

  from vector multiple, 161

  vector space scalars, 172

Scaling

  Euclidean scaling, 489–490

  maximum entry scaling, 490–493

Schmidt, Erhardt, 360, 510

Schur, Issai, 402

Schur decomposition, 403, 507
Schur's theorem, 402
Schwarz, Hermann Amandus, 137
Search engines, 496–500
Second derivative test, 420, 421
Sectors (economic), 85
Sets, linear independence, 191–194
Shear operators, 258, 277
Shifting operators, 446
Similar matrices, 305
Similarity invariants, 305, 306, 471–472
Similarity transformation, 305
Singular matrices, 42, 43
Singular value decomposition (SVD), 509–511, 514–516
Singular values, 507–508
Skew-Hermitian matrices, 429
Skew product, 161
Skew-symmetric matrices, 428–429
Solution vectors, 157
Solutions
    best approximations, 367
    comparison of procedures for solving linear systems, 501–505
    cost of, 501–504
    factoring, 477
    flops and, 501–504
    Gauss-Jordan elimination, 15, 18, 21–22, 44, 82, 319, 504, 505
    Gaussian elimination, 11–16, 21–22, 505
    general solution, 13, 227, 240, 328
    of homogeneous linear systems, 186–187
    least squares solutions, 366–371, 372
    of linear system by diagonalization, 329–331
    of linear systems, 3, 11
    of linear systems by factoring, 477
    of linear systems with initial conditions, 328–329
    particular solution, 227
    power method, 487–494
    trivial/nontrivial solutions, 17, 191
Solutions spaces, of homogeneous systems, 187–188
Spacecraft, yaw, pitch, and roll, 256
Spanning
    in R^2 and R^3, 184–185
    in R^n, 184
    testing for, 186
Spanning sets, 185, 188, 204
Spans, 184, 209–210
Spectral decomposition of A, 400, 401
Springs, constant, 378–379
Square, image of, 279

Square matrices, 42, 66, 68, 389
    decompositions of, 506–507
    determinants of, 100–104
    eigenvalues of, 311
    of order n, 27
    trace, 35
    transpose, 34
Standard basis
    coordinate vectors relative to, 206
    coordinates relative to standard basis
        for R^n, 205–206
        for M_{nn}, 203
        for polynomials, 202
        for R^3, 202–203
        for R^n, 202
Standard inner product, 336
    on polynomials, 340
Standard matrices, 248, 274
    for matrix transformation, 251
    for orthogonal projection, 372
    for T^{-1}, 268–269
Standard unit vectors, 132, 163
    linear independence in R^3, 191–192
    linear independence in R^4, 192
    linear independence in R^n, 191
    in span R^n, 184
State of a particle system, 125
State of the dynamical system, 272
State of the variable, 282
State vector, 287
Steady-state vector, of Markov chain, 289
Stochastic matrices, 288–289
Stochastic processes, 284
Stopping procedures, 494
String theory, 123, 124
Subdiagonal, 403
Submatrices, 30, 414
Subspaces, 179–188, 439
    creating, 183
    defined, 179
    examples of, 180–188
    of M_{nn}, 181
    orthogonal projections on, 369–370
    orthogonal projections on subspaces of R^m, 371–372
    of polynomials, 182
    of polynomials (P_n), 182
    of R^2 and R^3, 180–181
    zero subspace, 180
Subtraction
    of complex numbers, 522
    of vectors in R^2 and R^3, 121
    of vectors in R^n, 126
Sum
    of complex numbers, 522
    matrices, 27, 46

of rank and nullity, 239–240
of vectors in R^2 and R^3, 120, 122
of vectors in R^n, 126
SVD (singular value decomposition), 509–511, 514–516
Sylvester, James, 34, 94, 510
Symmetric matrices, 69–70, 322
    eigenvalues of, 398
    Hessian matrices, 420
    orthogonally diagonalizing, 299
Symmetry property, of dot product, 136, 318

T
T^{-1}, standard matrix for, 268–269
Taussky-Todd, Olga, 320
Technology Matrix, 86
Television, market share as dynamical system, 282–283
Terminal point, 119
Theorem
    contrapositive form of, 519
    converse of, 519
Theorem of Pythagoras
    generalized Theorem of Pythagoras, 348
    in R^4, 148
    in R^n, 148
Three-Step Procedure, 461
3×3 matrices
    adjoint, 111
    orthogonal matrix, 389–390
    QR-decomposition of, 363–364
    determinants, 97
    eigenvalues, 297
    3–space, 119
    cross product, 161–163
    scalar triple product, 165–166
    3–tuples, 123
Time, as fourth dimension, 123
Trace, square matrices, 35
Traffic flow, network analysis with linear systems, 74–75
Transformations, 248. See also Linear transformation; Matrix transformations
    differentiation transformation, 439
    dimensions and, 447
    evaluation transformation, 436–437
    integral transformation, 438
    inverse transformations, 464
    on matrix spaces, 435
    one-to-one linear transformation, 447
Transition matrices, 219–221, 469
    invertibility of, 220
    for R^n, 220–222
Translation, 120, 436

Transpose, 33–34
  determinant of, 101
  invertibility, 48
  of lower triangular matrix, 68
  properties, 47–48
  vector spaces, 242
Triangle, area of, 165
Triangle inequalities
  for distances, 138, 346
  for vectors, 138, 346
Triangle rule for vector addition, 120
Triangular matrices, 68–69
  diagonalizability of, 309–310
  eigenvalues of, 298–299
Trigonometric polynomial, 384
Trivial solution, 17, 191
Turing, Alan Mathison, 479
Two-point vector equations, in R^n, 155
2×2 matrices
  cofactor expansions of, 95
  determinants, 97–98
  eigenvalues of, 321, 322
  inverse of, 44
  vector space, 174
2×2 vector, eigenvectors, 296
2–space, 119
2–tuples, 123

U
Underdetermined linear system, 241
Unified field theory, 124
Unit circle, 338
Unit sphere, 338
Unit vectors, 131–132, 317, 336
Unitary diagonalization, of Hermitian
  matrices, 427–428
Unitary matrices, 425, 426, 427
Units of measurement, 201
Unknowns, 2
Unstable algorithms, 22
Upper Hessenberg decomposition, 403
Upper Hessenberg form, 403
Upper triangular matrices, 68, 97, 298

V
Value, 247
Vector addition
  parallelogram rule for, 120

in R^2 and R^3, 120, 122
  triangle rule for, 120
Vector equations
  of lines and planes in R^4, 155
  of lines in R^2 and R^3, 152–153, 154
  of planes in R^3, 153, 154–155
  two-point vector equations in R^n, 155
Vector forms, 154
Vector space, 171
  axioms, 171–172
  basis for using row operations, 232
  complex vector spaces, 172, 315–325
  dimensions of, 209
  examples of, 173–177, 204
  finite-dimensional vector space, 204,
    212–213
  infinite-dimensional/finite-dimensional,
    204
  of infinite sequences of real numbers,
    173
  isomorphic, 447
  of $m \times n$ matrices, 174–175
  n-dimensional vector space, 212
  of real-valued functions, 175
  real vector space, 171, 172
  subspaces, 179–188, 439
  transposes, 242–243
  of 2×2 matrices, 174
  zero vector space, 173, 209
Vector space scalars, 172
Vector subtraction, in R^2 and R^3, 121
Vectors, 119
  angle between, 134–135, 137, 346
  arithmetic operations, 120–122, 125
  "basis vectors," 201
  collinear vectors, 121–122
  column-matrix form of, 128, 225
  comma-delimited form of, 128, 225
  components of, 122–123
  in coordinate systems, 122–123
  coordinate vector, 205
  coordinate vectors, 206
  dot product, 133–137, 139–140
  equality, 120, 125
  equivalence, 120, 125
  geometric vectors, 119
  linear combination of, 127, 132–133,
    183, 185–186

linear independence, 184, 190–198
  nonzero vectors, 188
  norm of, 148
  normalizing, 132
  notation, 119, 128
  orthogonal vectors, 143–145, 317
  parallel vectors, 121–122
  parallelogram equation for, 138
  perpendicular vectors, 143
  probability vector, 284
  in R^2 and R^3, 119–129
  real and imaginary parts of, 316–317
  in R^n, 123–125
  row-matrix form of, 128, 225
  row vectors, 26, 27, 39, 157, 225
  solution vectors, 157
  standard unit vectors, 132, 163, 184,
    191–192
  state vector, 287
  triangle inequality for, 138
  unit vectors, 131–132, 317, 336
  zero vector, 120, 125
Voltage rises/drops, 76, 77
Volts (units), 76

W
Weight, 336
Weighted Euclidean inner products,
  336–338, 339
Weyl, Herman Klaus, 510
Wildlife migration, as Markov chain,
  286–287
Wilson, Edwin, 161
Wronski, Jozef Hoene de, 194
Wronskian, 197–198

Y
Yaw, 256

Z
Zero matrices, 40
Zero subspace, 180
Zero transformations, 250, 434, 439
Zero vector space, 173, 209
Zero vectors, 120, 125